T0376595

PROGRESS IN MARITIME TECHNOLOGY AND ENGINEERING

PROCEEDINGS OF THE 4TH INTERNATIONAL CONFERENCE ON MARITIME TECHNOLOGY AND ENGINEERING (MARTECH 2018), 7–9 MAY 2018, LISBON, PORTUGAL

Progress in Maritime Technology and Engineering

Editors

C. Guedes Soares
Centre for Marine Technology and Ocean Engineering (CENTEC), Instituto Superior Técnico, Universidade de Lisboa, Portugal

T.A. Santos
Ordem dos Engenheiros, Portugal

CRC Press
Taylor & Francis Group
Boca Raton London New York Leiden

CRC Press is an imprint of the
Taylor & Francis Group, an **informa** business
A BALKEMA BOOK

CRC Press/Balkema is an imprint of the Taylor & Francis Group, an informa business

Typeset by V Publishing Solutions Pvt Ltd., Chennai, India
Printed and bound in Great Britain by CPI Group (UK) Ltd, Croydon, CR0 4YY

Published by: CRC Press/Balkema
Schipholweg 107C, 2316 XC Leiden, The Netherlands
e-mail: Pub.NL@taylorandfrancis.com
www.crcpress.com – www.taylorandfrancis.com

ISBN: 978-1-138-58539-3 (Hbk + USB)
ISBN: 978-0-429-50529-4 (eBook)

Progress in Maritime Technology and Engineering – Guedes Soares & Santos (Eds)
© 2018 Taylor & Francis Group, London, ISBN 978-1-138-58539-3

Table of contents

Progress in Maritime Technology and Engineering – Guedes Soares & Santos (Eds)
© 2018 Taylor & Francis Group, London, ISBN 978-1-138-58539-3

Preface

Since 1987, the Naval Architecture and Marine Engineering branch of the Portuguese Association of Engineers (Ordem dos Engenheiros) and the Centre for Marine Technology and Ocean Engineering (CENTEC) of the Instituto Superior Técnico (IST), University of Lisbon (formerly Technical University of Lisbon) have been organizing national conferences on Naval Architecture and Marine Engineering. Initially, they were organised annually and later became biannual events.

These meetings had the objective of bringing together Portuguese professionals giving them an opportunity to present and discuss the ongoing technical activities. The meetings have been typically attended by 150 to 200 participants.

At the same time as the conferences have become more mature, the international contacts have also increased and the industry became more international in such a way that the fact that the conference was in Portuguese started to hinder its further development with wider participation. Therefore, a decision was made to experiment with having also papers in English, mixed with the usual papers in Portuguese. This was first implemented in the First International Conference of Maritime Technology and Engineering (MARTECH 2011), which was organized in the year that Instituto Superior Técnico completed 100 years. Subsequently, two more MARTECH conferences have been organized, namely in 2014 and 2016, always with a broadening of scope.

In this Fourth International Conference of Maritime Technology and Engineering (MARTECH 2018), a total of around 130 abstracts have been received and 80 papers were finally accepted.

The Scientific Committee had a major role in the review process of the papers although several other anonymous reviewers have also contributed and deserve our thanks for the detailed comments provided to the authors allowing them to improve their papers. The participation is coming from research and industry from almost every continent, which is also a demonstration of the wide geographical reach of the conference.

The contents of the present books are organized in the main subject areas corresponding to the sessions in the Conference and within each group the papers are listed by the alphabetic order of the authors.

We want to thank all contributors for their efforts and we hope that this Conference will be continued and improved in the future.

C. Guedes Soares & T.A. Santos

Progress in Maritime Technology and Engineering – Guedes Soares & Santos (Eds)
© 2018 Taylor & Francis Group, London, ISBN 978-1-138-58539-3

Organisation

CONFERENCE CHAIRMEN

Carlos Guedes Soares, *IST, Universidade de Lisboa, Portugal*
Pedro Ponte, *Ordem dos Engenheiros, Portugal*

ORGANIZING COMMITTEE

Yordan Garbatov, *IST, Universidade de Lisboa, Portugal*
Dina Dimas, *Ordem dos Engenheiros, Portugal*
Ângelo Teixeira, *IST, Universidade de Lisboa, Portugal*
Tiago A. Santos, *Ordem dos Engenheiros, Portugal*
Manuel Ventura, *IST, Universidade de Lisboa, Portugal*
Paulo Viana, *Ordem dos Engenheiros, Portugal*
Abel Simões, *ENIDH, Portugal*
António Oliveira, *Transinsular, Portugal*

TECHNICAL PROGRAMME COMMITTEE

Ermina Begovic, *Italy*
Kostas Belibassakis, *Greece*
Rui Carlos Botter, *Brazil*
Evangelos Boulougouris, *UK*
Dario Bruzzone, *Italy*
Nian Zhong Chen, *UK*
Matthew Collette, *USA*
Giorgio Contento, *Italy*
Vicente Diáz Casás, *Spain*
Leonard Domnisoru, *Romania*
Soren Ehlers, *Germany*
Selma Ergin, *Turkey*
Segen F. Estefen, *Brasil*
Pierre Ferrant, *France*
Juana Fortes, *Portugal*
Yordan Garbatov, *Portugal*
Sérgio Garcia, *Spain*
Peter Georgiev, *Bulgaria*
Hercules Haralambides, *The Netherlands*
Chunyan Ji, *China*
Xiaoli Jiang, *The Netherlands*
Jean-Marc Laurens, *France*
Luis Ramon Nuñez, *Spain*
Marcelo Ramos Martins, *Brazil*
Ould El Moctar, *Germany*
Thanos Pallis, *Greece*
Apostolos Papanikolau, *Greece*
Josko Parunov, *Croatia*
Preben T Pedersen, *Denmark*
Jasna Prpić-Oršić, *Croatia*
Harilaos Psaraftis, *Denmark*
Jonas Ringsberg, *Sweden*
Germán R. Rodríguez, *Spain*
Jani Romanoff, *Finland*
Xin Shi, *China*
Asgeir Johan Sørensen, *Norway*
Maciej Taczala, *Poland*
Michele Viviani, *Italy*
Alex Vredeveldt, *The Netherlands*
Decheng Wan, *China*
Duan Wenyang, *China*
Xinping Yan, *China*
Peilin Zhou, *UK*
Ling Zhu, *China*

TECHNICAL PROGRAMME & CONFERENCE SECRETARIAT

Sandra Ponce, *IST, Universidade de Lisboa, Portugal*
Maria de Fátima Pina, *IST, Universidade de Lisboa, Portugal*
Bruna Covelas, *IST, Universidade de Lisboa, Portugal*

Port performance I

Progress in Maritime Technology and Engineering – Guedes Soares & Santos (Eds)
 ISBN 978-1-138-58539-3

Comparative analysis of port performances between Italy and Brazil

A.N. Nascimento & A.M. Wahrhaftig
Federal University of Bahia, Salvador, Brazil

H.J.C. Ribeiro
Federal Institute of Education Science and Technology of Bahia, Salvador, Brazil

ABSTRACT: The State of Bahia has one of the largest port complexes in Brazil, that consists of public ports and private use terminals. One of which is specialized with a cargo carrying capacity of about 530 thousand TEUs (Twenty-foot Equivalent Unit) per year. On the other hand, the container terminal at the port of Genoa, Italy, has a similar capacity but has been performing better than the Brazilian one. The present study evaluates the differences and similarities between these ports, in the context of engineering, environmental sustainability and some topics of the port regulatory framework that can influence productivity. Based on the arrival and service rates of the ships and their respective probability distributions, a mathematical model of queuing theory was developed that indicates the port occupation rate, the time and the average number of ships in the queue, in a process with which one can assess the environmental impact of these terminals.

1 INTRODUCTION

The Container Terminal of the Port of Salvador (TCS), located in the northeast of Brazil, is a medium-sized structure port, with two berths (pier 1 and 2). Pier 1 is the most modern, with a greater depth and extension, is equipped with cranes which are able to serve ships of the super-postpanamax class, serving almost all the demands of the terminal. On the other hand, pier 2 has smaller extension and depth, operates with cranes which can serve ships of the panamax class only, meeting the needs of a small part of the terminal demand.

With similar characteristics to TCS, although with greater capacity, the Container Terminal of Genoa (TCG) is a benchmark in this work. It is located in the region of Liguria, Italy, having a modern structure, with berth length, seaport depth and cranes able to serve ships of the super-postpanamax class.

It should be noted that, not always, the entire cargo of the ship is destined for the port at which the vessel is arriving. Therefore, both loading and unloading may occur. Cargo may also be only loaded or unloaded, in whole or in part.

The Port of Salvador has a single container terminal, while in Genoa there are two important terminals. The results of the studies published here relate only to one of these Genoa's Terminals. Referring to the designation adopted—TCS and TCG—do not represent the commercial name of the companies involved herein.

Based on the study of these two port structures, this paper evaluates the differences and similarities of both maritime terminals, both in the engineering (naval-port infrastructure and queuing theory) and in environmental sustainability, as well as in some points of the regulatory framework and logistical infrastructure of both countries that can influence and reveal the competitiveness of these ports. Based on vessel arrival and service rates and their consequent probability distributions, in addition to other constraints, the queuing model reveals the time and average number of ships in process, the port occupation rate, and the expected probability of this finding variable, may provide a possibility of fine payments for delay in services, thus serving as the efficiency indicators of port operation.

The results obtained here, and their interpretations, are limited to the time of their respective data collection, as well as to the reliability of information that was possible to obtain at the time of the technical visits and professional meetings held. They are also limited to the consultations made to the electronic pages of the terminals and institutions that control local port operations.

The present work is in the context of other researches already carried out. Camelo et al. (2010), used queuing theory to simulate the behavior of the row of iron ore vessels in the port of Ponta da Madeira, Brazil, with the aid of the Arena® software and found high berth occupation rates, recommending investments in the expansion of its capacity to meet the expectations of growing demand for ores in the world market. In this same direction, Schoreder (2014) simulated the operational behavior of the container terminal at the port of Durban, South Africa, based on the operation of the queue system of the container terminal at the port of Rotterdam from the logic of model construction simulated with Simio® software. It is concluded, after validation, that the model represents appropriately the operational reality of the terminal. Navarro et al. (2015), on the other hand, applied a queuing network model in the container terminal of the port of Manila involving both the queue of container haulers to the port and the queue of ships awaiting loading with the aid of the software Promodel®. By demonstrating the usefulness of the model used, concludes by the adoption of vehicle reserves to support variability in the ship loading rhythm.

2 METHODOLOGY

The primary data for the construction of the queuing model, based on vessel movement in the TCS, was obtained in the statistics sector of the Docks Company of the State of Bahia—CODEBA, in the format of Excel® spreadsheets, comprising the years 2012 to 2016, except for the year 2015 that was excluded from the statistical database because of inconsistencies, resulting in 1,597 events exclusively for pier 1, the most important of this port. The pier 2, because of its smaller depth (only 12 m) as compared to pier 1 (14 m) and the available infra-structure, including the length, does not allow the docking of the ships planned to dock at pier 1. Hence, it configures a single server queue system (S = 1) since only the vessel movement data on pier 1 was considered for the purpose of comparison with the TCG in this article, highlighting that this specific terminal of Genoa has only 1 berth (S = 1). The Table 1 presents a brief of CODEBA data.

From this data, statistical tests were performed to evaluate adherence to certain probability distributions, according to Hillier, FS & Lieberman G. J. 1995, as a requirement to select the correct mathematical form for modeling the ship queue. Consequently, tests were applied for Poisson, exponential, Erlang and range, among others, for arrival and service time of the ships. The computational assistance for these tests was software R, version 3.4.1, and Quantitative Systems Business Plus (QSB+), as well as Excel® spreadsheets (Microsoft Corporation) were used for queuing system behavior calculations.

Based on this data with an aim to implement the mathematic model of queues, the average ship arrival rate (λ) and the average service rate (μ) were calculated. After calculating these input variables and evaluating their respective statistical behaviors (arrivals and services), the parameters of this queuing model were calculated using the QSB+, which are:

- L: Number of ships in the system (waiting and in service);
- Lq: Number of ships in queue (waiting to be attended);
- W: Waiting time in the system (waiting andin service);
- Wq: Waiting time in queue (waiting to be attended;
- ℓ: Terminal occupancy rate;
- Po: Probability of the terminal being idle;
- Pw: Probability of waiting to be attended.

The parameters implicit in Kendall's notation (A/B/s/N/m/Z), which together with (λ) and (μ), complete the information required for modeling, are selected by the QSB$^+$ software and are adopted for this work. They can be defined as:

- A describes the statistical distribution of the number of arrivals;
- B describes the statistical distribution of the time of service;
- s is the number of servers (Berth)
- N is the maximum capacity (maximum number of vessels allowed in the system);
- m is the size of the population that provides customers (ships);
- Z is the row discipline (how they are selected to be attended).

Considering an unlimited capacity of a given system (N), the population of clients that demand a single service of this system is also infinite (m), and its service is in order of arrival (Z): the notation of Kendall can be summarized as A/B/1. Thus, the notation of the type M/M/s, Markovian, is denoted as Poisson input and exponential service time with s servers. If it is M/G/1, it implies Poisson input and a general service distribution with 1 server (S = 1). However, it should be noted that the queuing models that are closest to reality, register values for $\ell < 1$.

During the survey, visits were made to the port of Salvador, Brazil, both to observe, in-situ operations and to discuss with them the details of the studies being carried out. Further to analyze the technical validation by experienced professionals, as well as

Table 1. Summary of data collected at CODEBA.

Vessel's Name	Pier	Hour and day: Input H:M/D	Hour and day: Exit H:M/D	Td H	C Σ	P C/H
ALIANÇA MANAUS	611	13:30/01	21:10/02	31.67	982	31.01
MONTE VERDE	611	22:40/02	6:45/03	8.08	255	31.56
LUTETIA VIAGEM	611	2:30/01	10:10/01	6.67	128	19.19
ER LONDON	611	9:30/03	4:35/04	19.08	723	37.89

H = Hour; M = Minute; D = day; Td = Time docked; C = Container; P = Productivity.

for data collection. As for the TCG, a general visit was made by the sea, accompanied by a representative of the Port Authority for an annotated observation of the infrastructure of the Genoa terminals. Data was also collected from queries to the public electronic pages of these terminals.

In addition to modeling the current state of the queuing system, two new simulations were carried out, one with a 50% increase in container demand in Brazil, based on the perspective of international trade growth in the next 10 years according to the Ministry of Mines and Energy of Brazil, 2016. The other simulation considered the possibility of unavailability of one of the cranes for both terminals due to machine failure.

The statistical treatment of this article was structured and started in the Laboratory of Integrated Production Systems—LABSIP, linked to the Department of Mechanical Engineering / Polytechnic School of the Federal University of Bahia (UFBA), and finalized in the Department of Civil and Environmental Engineering/Polytechnic of Milan (POLIMI).

As a consequence of the queuing model, the present study used scientific references and technical coefficients to calculate the mass of gaseous emissions generated by ships in the queuing system. Further it compared quantitative data in the regulatory frameworks and the World Bank in both countries (Italy and Brazil), to compare their performance.

3 OBTAINED RESULTS

3.1 *Results for the infrastructure*

The operational capacity of these terminals can be summarized as shown in Table 2.

According to Table 2, and due to the combination of factors such as water density, suction that the hulls of vessels are subjected to in squat, wave effects, background irregularities and sedimentation, the PIANC standard (PIANC, 1997) recommends a slack due to the squat that can be determined by the expression:

$$S_b = (C_z + 1/2C_\theta)\frac{V}{L_{pp}^2}(F_{nh}^2)\sqrt{(1-F_{nh}^2)}, \quad (1)$$

where:

Table 2. Infrastructure of the container terminals.

	Berth (m)	Depth (m)	Portainer[1] (unity)	Area (m^2)
TCS	377	14	3	120,000
TCR	526	15	5	174,000

1. Super-*post-panamax*.

S_b = maximum vertical displacement due to squat;

C_z and C_θ = coefficients recommended by the standard, with: C_z = 1.46 (due to heave) and C_θ = 1 (due to pitch).

V = buoyance volume in m^3;

L_{pp} = length between perpendiculars in m;

F_{nh} = Froude's number relative to local depth, given by:

$$F_{nh} = V_{ms}\sqrt{(gh)}, \quad (2)$$

where:

V_{ms} = speed of evolution in m / s;

g = acceleration due to gravity, 9.81 m / s^2; and

h = local depth in m;

To measure the clearances, vessels of the post-panamax type were adopted (PIANC,1997). Due to the similarity in terminal infrastructure, their gaps are equal (Sb = 1.15 m) and so the recommended maximum draft (T) for ships is: T = 12.85 m for TCS and T = 13.85 m for Genoa.

The maximum range for the booms of the cranes at both ports is 55 m, with 50 m being the maximum permissible molded breadth for ships.

Considering these dimensions, the berth lengths (Table 2), operating clearances and maximum length-to-beam ratio of 8, the maximum lengths for ships for these terminals are as follows: L = 350 m (TCS) and L = 400 m (TCG). Adopting the average block coefficient of 0.65 (PIANC, 1997), one can reach the full load displacements of:

$$\Delta_{TCS} \cong 1{,}298{,}881\ kN;\ \Delta_{TCG} \cong 1{,}601{,}360\ kN. \quad (3)$$

Assuming now the average recommended ratio (PIANC) for the relationship between the deadweight and the full displacement of 0.70, we arrive at the gross deadweights:

$$DWT_{TCS} \cong 907{,}437\ kN; \quad DWT_{TCG} \cong 1{,}120{,}952\ kN. \quad (4)$$

Considering that the variability of the load causes the actual net weight of the containers to vary, hence the average value of the gross weight of 133.k *kN* was adopted, thus estimating the following maximum ship loads at these terminals (measured in TEUs):

$$DWT_{TCS} \cong 6{,}800\ TEUs; \quad DWT_{TCG} \cong 8{,}400\ TEUs. \quad (5)$$

In addition to this dock infrastructure, the TCS and TCG have the following equipment for movement in the back area, as shown in Table 3.

With this infrastructure, the TCS informs the capacity of 530,000 TEUs/year, while the information of the TCG is 550,000 TEUs/year. In 2016 TCS moved 200,000 TEUs while TCG moved 300,000 TEUs.

3.2 *Results for statistical tests*

The application of the adhesion test to the number of ship arrivals proved to be validated for the Poisson distribution. The raw data was systematized for the application of the test and summaries of the results are available in Table 4.

The non-parametric *Kolmogorov-Smirnov* test was used for the time of service of the ships for different statistical distributions, including the gamma distribution. In this case, the shape parameter (α) and the rate parameter (β) are modeled by the probability density function given by Eq. (6)

$$f(x)=\begin{Bmatrix} \dfrac{\beta^{\alpha}x^{\alpha-1}e^{-\beta x}}{\gamma(\alpha)} \text{if } x \geq 0 \\ 0 \text{ if } x < 0 \end{Bmatrix} \tag{6}$$

where 'x' is the assumed value of the independent variable of the problem and the gamma function $\gamma(\alpha)$, on the other hand, is given by Eq. (7):

$$\gamma(\alpha)=\int_{0}^{\infty} x^{\alpha-1}.e^{-x}.dx \tag{7}$$

Based on this model, the adhesion test for the service time was applied, which was validated for the gamma distribution. It is reported that the treated data was imported from the Excel® worksheet into software R, which did not reject the null hypothesis of adherence between theoretical and observed behaviors. The results obtained can be seen in Figure 1.

As shown in Figure 1, the observed and theoretical (gamma) behavior results for significance of 5% and

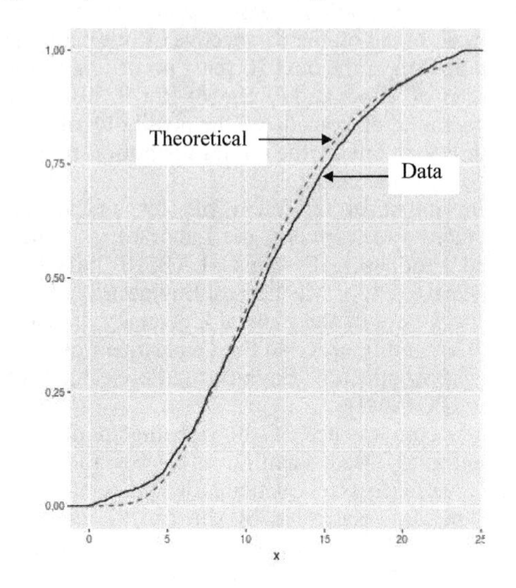

Figure 1. *Kolmogorov-Smir nov* test (working time)[1]. 1. Pier 611 for years 2012, 2013, 2014, 2016. p-value (0,983); test statistic (0.0577); observations (1597); shorter time (0.1667); maximum time (23.92); average time (11.60); Parameter shape-gamma (4.95); parameter rate-gamma (0.4264).

Table 3. Devices of moving (retro area).

	RTG	RMG	Reach Stackers	Access way
TCS	8	1	8	Road
TCR	8	6	17	Rodo-rail

Table 4. *Chi-squared* test for Poisson[1].

H_0: Is Poisson		<== adhered to Poisson distribution		significance levels (α)	
H_1: Is not Poisson				5%	degrees of freedom
					18
min	max	mean	standard deviation	sum χ^2	p-value
24	43	33.27083	4.76856	0.33581	0.9999999999997
		ships per month			
Ships	Events	Relative Observed Frequency	Poisson Relative Frequency	χ^2	Theoretical Expected Frequency
24	2	0.04167	0.01939	0.02561	0.64498
25	0	0.00000	0.02580	0.02580	0.85836
26	3	0.06250	0.03301	0.02634	1.09840
27	2	0.04167	0.04068	0.00002	1.35351
28	3	0.06250	0.04834	0.00415	1.60830
29	0	0.00000	0.05546	0.05546	1.84515
30	5	0.10417	0.06151	0.02959	2.04633
31	2	0.04167	0.06601	0.00898	2.19622
32	5	0.10417	0.06863	0.01840	2.28344
33	2	0.04167	0.06920	0.01095	2.30218
34	3	0.06250	0.06771	0.00040	2.25281
35	2	0.04167	0.06437	0.00801	2.14151
36	6	0.12500	0.05949	0.07215	1.97916
37	4	0.08333	0.05349	0.01665	1.77969
38	2	0.04167	0.04683	0.00057	1.55820
39	3	0.06250	0.03995	0.01272	1.32930
40	1	0.02083	0.03323	0.00463	1.10567
41	1	0.02083	0.02697	0.00140	0.89724
42	1	0.02083	0.02136	0.00001	0.71076
43	1	0.02083	0.01653	0.00112	0.54994

test statistics of 0.057, the p-value resulted in 0.983, indicating that there is no evidence to reject the null hypothesis. This test was applied to a gamma distribution with a shape parameter of approximately 4.95 and a scale parameter of approximately 0.43

3.3 *Results for queue output data*

Complementing the statistical results with the characteristics and infrastructure of the terminals, and considering unlimited vessel mooring capacities, it is possible to assume, for the purpose of the queuing system, and in the context of Kendall's notation, the M / G / 1 for the TCS, which is:

- A (distribution of number of arrivals): Poisson;
- B (distribution of time of service): gamma;
- s (number of berths): 1 (restricted to types of vessels);
- N (system capacity): unlimited (bay);
- m (size of ship population): infinite;
- Z (row discipline): FIFO / order of arrival

To compare the performance of the queue between the terminals, it was assumed that the statistical behavior for the TCG is equivalent to the TCS, i.e. M / G / 1. With the TCS data treatment the average arrival (λ) and service (μ) rates were calculated. For the TCG, inferences and calculations were made from the infrastructure and arrivals of ships in this terminal, available on the website in the global computer network of the Port of Genoa, 2017. The results, in ships/day, are: $\lambda_{TCS} = 1.10$; $\mu_{TCS} = 2.10$; $\lambda_{TCG} = 1.44$; $\mu_{TCG} = 2.88$.

With this data and through the QSB+ software, the behavior of the queue for both terminals was calculated, as shown in Figure 2, which represents the screen of the results of the application for the TCS.

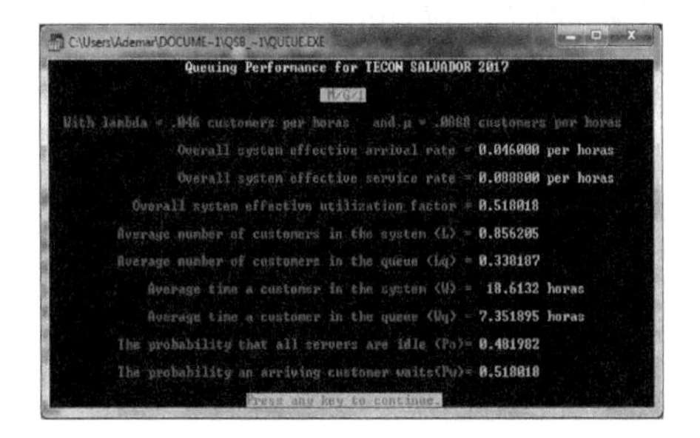

Figure 2. Results for the TCS1 Queue System[1].
1. Arrival rates and service period 2012, 2013, 2014, 2016. Standard deviation for service time equal to 5.22.

Table 5. Results of queue output parameters[1].

	ℓ (%)	L	Lq	W (h)	Wq (h)	P_0 (%)
TCS	52	1	0.5	19	7.5	48
TCG	50	1	0.5	14	6.0	50

1. It is observed that with S = 1, we have $P_w = \ell = 52\%$.

The summary of the approximated results, by QSB+, is in Table 5.

In order to forecast a 50% increase in demand, new calculations were made for the queuing system, maintaining service rates (μ), but assuming the following arrival rates (ships/day): $\lambda_{TCS} = 1.65$ e $\lambda_{TCG} = 2.16$. Table 6 summarizes the results obtained by QSB+ for this new scenario.

Considering the possible unavailability of one of the cranes at the terminals, it results in a proportional reduction of 1/3 of the service capacity for the TCS and 1/5 for the TCG, and new service rates of 1.4 ships/day and 2.30 ships/day, respectively for TCS and TCG. With this new input data, we get the results of Table 7.

With this data, other results are generated, such as the comparison between the two scenarios for the waiting time in the system (W) and the occupancy rate of the terminals (ℓ), as can be seen in Figures 3, 4 and 5.

The reduction in capacity over the service time of these terminals can be seen in Figure 5.

3.4 *Results of the impact of queuing on the environment*

While waiting in the queuing system, the ship's engines generate emissions that impact the environ-

Table 6. Result of queue with new demand[1].

	ℓ (%)	L	Lq	W (h)	Wq (h)	P_0 (%)
TCS	79	3	2	37	24	21
TCG	75	3	2	26	18	25

1. Considering the arrival rate 50% higher.

Table 7. Impact on queue with reduced capacity[1].

	ℓ (%)	L	Lq	W (h)	Wq (h)	P_0 (%)
TCS	77.5	3	2	50	32.5	22.5
TCG	62.5	2	1	21	11	37.5

1. Considering unavailability of 1 crane.

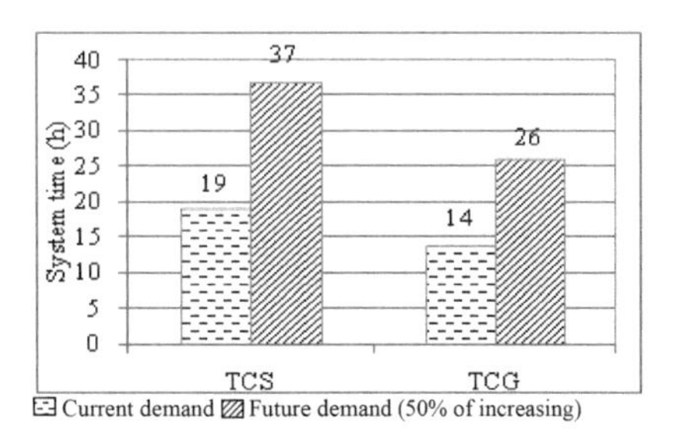

Figure 3. Impact of demand on waiting time.

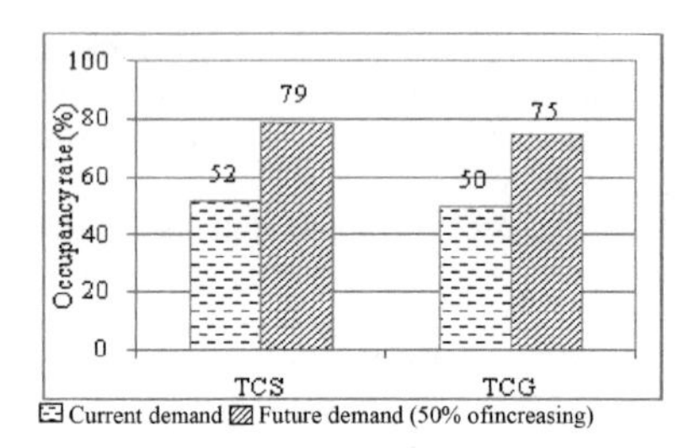

Figure 4. Impact of demand on the occupancy rate.

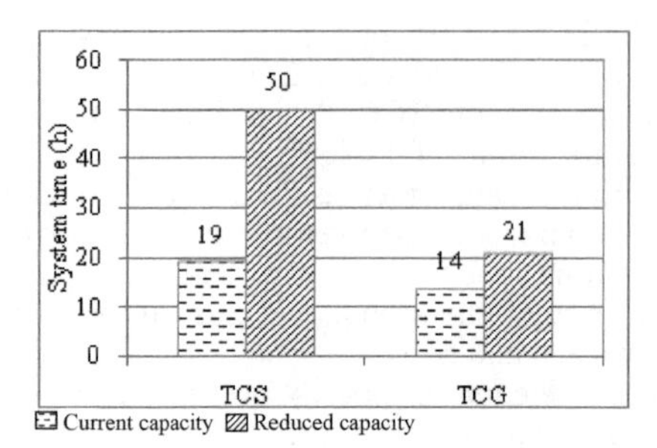

Figure 5. Effect of capacity reduction of ships in service.

Table 8. NOx emissions caused by the demand[1].

	Current demand (tNOx)	Future demand (tNOx)
TCS	0.033	0.197
TCG	0.024	0.138

1. EN = 320 kW·h; FP = 0.4; FE = 13.9 g/kWh.

ment, as reported by the International convention for the prevention of pollution from ships—MARPOL, 2005, highlighting nitrogen oxides among the pollutants. The mass (t) of combustion gases emitted by ships can be calculated by means of Eq. (8), as provided by the Environmental Protection Agency—EPA, 2009:

$$EG(t) = L \cdot [EN \cdot FP \cdot W \cdot FE], \tag{8}$$

in which: EG = gaseous emissions (t); L = number of ships in the queue system; EN = energy consumed (kW·h); FP = power factor (%); W = ship waiting time (h); FE = emission factor (g/kWh).

Based on Eq. (8) and considering the results of the queues for the TCS and the TCG, along with the NOx gas emission factor of 13.9 g/kWh, we get the following synthesized results in Table 8, with an aim to meet current and future demands.

The impact of this data can be best seen in the graphical comparison of Figure 6.

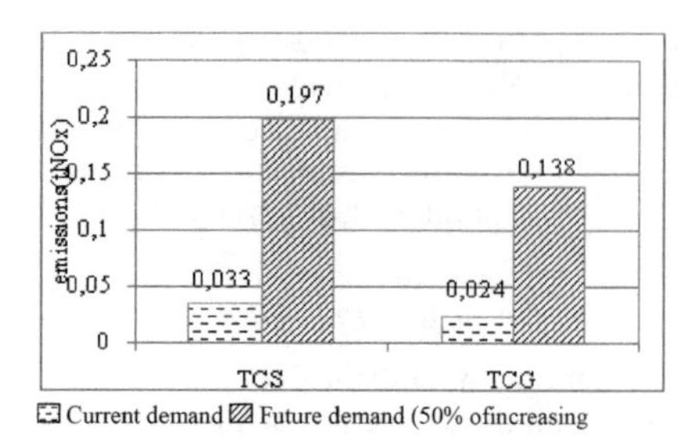

Figure 6. Gaseous NOx emissions caused by the demand.

Table 9. Accessibility and regulatory indicators.

	Railroad[1] (Km)	Highway[1] (Km)	Contract[2] (years)	Oversight[2] (agents)
TCS	0	4.3	25	7
TCG	1[3]	0.5	30	6

1. Available on the terminals web pages; 2. Provided in the Ports Act 12.815/2013 (Brazil) and Act 84/1994, updated by DL 169/2016 (Italia). 3. Ways: 3 × 370 m, and access by Italian railroad.

3.5 *Results of regulatory and logistical aspects*

Regarding the regulatory aspects and port accessibility, the data are shown in Table 9.

Another important result is available as a general logistics performance indicator, where Italy ranks 21st, and Brazil ranks 55th in the world ranking, according to the World Bank, 2016.

4 DISCUSSION OF RESULTS

From the results obtained, it can be observed that both terminals have similar characteristics, are housed inside a bay and have relatively close infrastructures, but with advantages for the TCG, which has more resources, and can serve larger vessels (8,400 TEUs) against 6,800 TEUs in the TCS, almost 25% higher, with better logistical accessibility.

However, it should be noted, that in this last resort, the TCS only has road access for container movement, an important exclusive logistics via (4.3 km). The TCG, in turn, has both a highway, just 500 m from the freeway, and trails (3 trails of 370 m) with access to the Italian railway network, which is an important and outstanding logistical advantage. With this infrastructure, the reported capacity of the TCG (550,000 TEUs) is approximately 4% higher than the capacity of the TCS (530,000 TEUs).

Regarding the demands for the terminal and the service capacity, it can be observed that while the TCS serves 2 vessels/day, the TCG can serve up to 3 vessels/day. Even though its arrival rate is approximately 1/3rd higher than the rate of TCS arrivals. This happens because the infrastructure of the Genoa terminal is more imposing than the terminal in Salvador. The availability of 2 more cranes than the TCS enables the TCG to handle a ship at around 8 h, while the TCS needs an average of 11.5 hours to process it. Thus, while in Genoa a ship waits at the terminal for an average of 14 hours between its arrival and departure, the same time in Salvador would be 19 hours, i.e. 5 hours more.

It can also be observed that due to their current arrival and service rates, at both terminals, two vessels will be present on average, one in service and the other coming to the queuing system. Of course, the service capacity makes the difference both in the waiting time at the terminal and in the load factor of the terminal. So, waiting in line at the TCG (6.0 h) would be 1.5 h less than at the TCS (7.5 h). This is linked to the infrastructure of these terminals and affects the results of their respective occupancy rates, making Salvador with 52%, a little higher than that of Genoa (50%).

Now, analyzing the results of the simulation with 50% increase in the arrival rate, a similar behavior was observed for both vessels in their respective systems, that is, 3 units, 1 ship in service and 2 ships in the queue. However, once again, the TCG presents advantages in attending these units. While the impact of this simulation causes the waiting time in Salvador to increase approximately by 95%, going from 19 h to 37 h, the impact in Genoa would be an increase of 86%. This would be a smaller increase in waiting time, jumping from 14 h to 26 h. In the same context, the occupancy rate in Genoa would also have less impact, jumping from 50% to 75%, thus a difference of 25%, while in Salvador the impact would be of 27%, jumping from 52% to 79%.

On the other hand, analyzing the results of the simulation with the reduction in service capacity, there is an even greater advantage for TCG. While at TCG the increase in the waiting time of a ship is 7 h, jumping from 14 h to 21 h, the impact on the TCS would be more significant, an increase of 31 h from 19 h to 50 h, i.e. double the current time. Also in this simulation, the impact on the queue is better absorbed by the TCG, which would have only 2 ships in the system, 1 more than in the current queuing state, while TCS would retain 3 ships, 2 more than its current operation.

Regarding the impact of the queue on the environment (NOx emissions), the results calculated for the TCG would once again have advantages on the TCS. At TCS, in the current state of operation 33 Kg is emitted, while in the TCG it is 27% smaller i.e. 24 Kg. However, this proportional difference would jump to 30% in favor of the TCG, with the increase of the demand in the terminals.

Regarding the normative regulatory aspects, it is observed that in Italy a revision in the port regulations occurred recently with the addition of Legislative Decree 169/2016, although Law 84/1994 is still in force. In Brazil, the complete legal framework is more recent and is based on Law 12,815 / 2013. The Italian regulatory revision implemented a service called "sportelo unico" which anticipates the services for the fulfillment of the documentary demands of the ship cargoes destined to Italian ports, 24 hours before arriving in the country. This already reduces the estimated time by 30% and 40%, indicating that in the port of Genoa the wait would have already been reduced from 4 to 5 days. Obviously, the queuing model presented here considers only vessels capable of being served in the system, disregarding the time of document processing. In this context, Brazil instituted in 2011 the procedure entitled "Paperless Harbor" to group together the necessary documentation for the processing of cargo ships in the so-called Virtual Single Document. Also in this legal aspect other operational similarities appear, so that in Brazil the concession contracts are of 25 years while in Italy the term is a little longer, of 30 years; the number of agents involved in ship liberations is also very close, with 7 in Brazil and only 6 in Italy.

As per the World Bank's logistic performance indicator (2016), evidenced in the ranking, Brazil is in a much less competitive position (55th) as compared to Italy (21st), i.e. an equivalent of 34 disadvantage positions. This indicator is part of a study conducted every 2 years and reveals that as compared to the previous edition (2014) Brazil improved 10 positions, while Italy lost only 1 position, and Germany occupies the first place, revealing itself as the country with the best logistics infrastructure. Among the criteria that make up this indicator are reliability of operations, cargo tracking, handling and port infrastructure.

5 CONCLUSIONS

A comparative study was carried out between the container terminal of the Port of Salvador-Brazil (TCS) and the container terminal of the Port of Genoa-Italy (TCG) by the modeling of discrete systems through queuing theory, complemented by aspects—the regulatory and logistical infrastructure of these countries. The queuing theory provides important results for the management of ongoing operations and for the planning of new guidelines that favors the improvement in the functioning of these productive systems.

Both the TCS and the TCG are important ports in both countries with infrastructure capable of ensuring competitiveness in their areas of influence. While the TCG has strong penetration in the markets of northern Italy and southern Europe, the TCS stands out in Brazil acting across the coast of the country and towards the North Atlantic.

The analysis of the data and the results indicate that the Genoa terminal, which has similarities with the Salvador terminal, can be a good reference for the latter. It can be seen that investments in dock infrastructure, such as the size and capacity of container handling, are very sensitive to the operational results. Thus, only one meter more depth in the cradle and the presence of two additional cranes, besides other important logistic complements, can make a significant difference in the operational results of the ports, as verified in this study, showing better yields for the TCG. Significant increases in demand would cause the TCS to operate close to the limits of its capacity with mechanical fatigue risks on the handling equipment, which could lead to interruptions with consequent payments of contractual fines for delay in the service of the ships.

Another aspect that deserves attention is the location of the TCS in a densely populated area of Salvador, an important tourist spot, which casts doubts on the security of the investments needed in the infrastructure to increase the capacity. The possibility of moving the Port to another area is still under evaluation and also its permanence to the present place with the extension of the berth is scrutinized. In Genoa, although the TCG is also located in the outskirts of the city, the railway infrastructure and the easy access to the Italian highways do not seem to present the same problems as for the TCS.

In the environmental context, as TCG presents a more efficient queuing system than the TCS, it releases less pollution in the region, even though these emissions should be below the limits recommended by international organizations.

Concerning the regulatory aspects, both in Italy and in Brazil there have been similar updates of its legal frameworks in order to reduce bureaucracy in the port system, although the effect seems to be faster in Italy than in Brazil (documentation of processing of ships). Also the concession period is similar, although in Brazil it is five years shorter than in Italy, with the idea of imposing a faster return on private investment. In general, Brazil has limited logistics infrastructure in the service of ports, being only served by the road, unlike Italy where the road and rail system show more availability for the transport of containers.

It should be noted that the methodology presented here, although supported by a consistent set of data regarding the port of Salvador, has its limitations due to the uncertainties of the data obtained in reference to the Italian port. Although the results obtained compose a representative model of port operations, very useful for the planning of such facilities.

REFERENCES

Brazil. Law number 12.815, on June 05, 2013. Regulates the exploration of ports by the Union. Official gazette of the Federative Republic of Brazil, Executive Branch, Brasília, DF. <Available in: http://www.planalto.gov.br/ccivil_03/_ato2011–2014/2013/lei/l12815.htm>, Access in: October 12, 2017.

Brazil. Ministry of Mines and Energy. 2016. Series "Economic Studies. Technical note. Characterization of the macroeconomic scenario for the next 10 years (2016–2015). Available in: < http://www.epe.gov.br>. Access in: November 09, 2017.

Camelo, G. R. et al. 2010. Queuing and simulation theory applied to the shipment of iron ore in the maritime terminal of the wood. Available in: www.e-publica-coes.uerj.br/index.php/cadest>. Access in: October24, 2016.

Chang, I.L. & Sullivan, R.S. 1994. Quantitative Systems for Business plus Version 2.1. Ed. Prentice Hall, New Jersey, USA.

Costa, F. Docks Company of the state of Bahia - CODEBA. Data base in spreadsheets. Electronic publishing [personal message]. Received by [annas@ufba.br] in March 07, 2017.

Hillier, F. S. & Lieberman, G.J. 1995. Introduction to operation research. New York, McGraw-Hill, 1995.

Itály. Legislative Decree 169/2016, August 31, 2016. Reorganizes, rationalizes and simplifies the discipline related to the Port Authority. Official diary of the Italian Republic. Rome. August 04, 2016. <Available in: http://www.gazzettaufficiale.it/eli/id/2016/08/31/16G00182/sg > Accessed in: October 12, 2017.

MARPOL-International Convention for the prevention of pollution from ships. 2005 (Anexo 6). Available in: <http://www.imo.org/en/about/conventions/listofconventions/pages/international-convention-for-the-prevention-of-pollution-from-ships-(marpol).aspx>. Accessin: November 04, 2017.

Navarro et al. 2015. Queuing Theory Application using Model Simulation: Solution to address Manila Port Congestion. Conference: Asia Pacific Industrial Engineering and Management Systems Conference, At Ho Chi Minh. Vietnam.

PIANC – World Association for Waterborne Transport Infrastructure. 1997– Appendix B. Available in Portuguese in: http://proamanaus.com.br/ohs/data/docs/3/Norma_Pianc_para_canais_de_acesso.pdf> Accessed in November 17, 2017.

Ports of Bahia. Displays information about container terminal. Available in: <http://www.codeba.com.br/eficiente/sites/portalcodeba/pt-br/home.php>. Access in: October, November and December. 2017.

Ports of Genoa. Displays information about container terminal. Available in: < https://www.portsofgenoa.com/it/terminal-merci/containers.html>. Access in: October, November and December. 2017.

R Development Core Team (2008) R Foundation for Statistical Computing: software R electronic version 3.4.1.

Schroeder, L. 2014. Applying queuing theory to the Port of Durban container terminal. Dissertation. B. Eng. Industrial and Systems Engineering, University of Pretoria.

USA Environmental Protection Agency. Current Methodologies in Preparing Mobile Source Port-Related Emission In-ventures. Final Report, April 2009. Available in: < https://archive.epa.gov/sectors/web/pdf/ports-emission-inv-april09.pdf>. Access in: October 13, 2017.

World Bank. International LPI Global Ranking. World Bank, 2016. Available in <https://lpi.worldbank.org/international/global> Access in: November10, 2017.

Progress in Maritime Technology and Engineering – Guedes Soares & Santos (Eds)
 ISBN 978-1-138-58539-3

Port of Santos, Brazil: Essential factors to implement a green port system

D.A. Moura
Federal University of ABC, Santo André, Brazil

R.C. Botter
University of SãoPaulo, São Paulo, Brazil

ABSTRACT: Having a port integrated with its surroundings, interacting harmoniously with the city in the surroundings, generating wealth and not denigrating the environment is a considerable challenge. Some ports in Europe and the United States can serve as a model for other ports in the world. These ports use renewable energy in their transport, storage and handling operations. They use renewable energies to load and unload ships, minimize or eliminate the use of fossil fuels in their day-to-day operations, and manage every chain that involves the movement of vessels at their terminals, as well as other modals integrated into the ports. This research analyze the current logistics operations of the port of Santos, the most significant port in Latin America, as well as its physical characteristics and legal aspects and public policies that govern Brazilian port activities. It aims to address what possible operations, technologies, and innovations are suitable in the medium and long-term to make it a port capable and integrated to other international ports with sustainable logistics operations.

1 INTRODUCTION

Some ports around the world are operating focused on activities that use clean and renewable energy. Increasingly, operations are being encouraged to use technologies and innovations that contribute to eliminating or reducing emissions of pollutants and chemicals into the environment and reducing noise pollution in the ports and their surroundings (Pavlic et al., 2014; Sciberras et al., 2017; Tang et al., 2017).

Promoting an environmental management system for each port is a necessary condition to accomplish the implementation of effective measures that contribute to green ports. Public policies focused on that area, involving companies that use private and public port terminals and state-owned ones that manage port operations, is a differential that can positively contribute to the search for sustainability and use of renewable and clean energy in port operations (Sislian et al., 2016; Pavlic et al., 2014; Puig et al., 2015).

Having a port integrated with its surroundings, interacting harmoniously with the city in the surroundings, generating wealth and not denigrating the environment is a considerable challenge.

Some ports in Europe and the United States can serve as a model for other ports in the world. These ports use renewable energy in their transport, storage and handling operations. They use renewable energies to load and unload ships, minimize or eliminate the use of fossil fuels in their day-to-day operations, and manage every chain that involves the movement of vessels at their terminals, as well as other modals integrated into the ports (Woodburn, 2017; Iannone, 2012).

The transition from infrastructure to a green port, however, is not straightforward. For example, replacing fossil fuels in their day-to-day operations with renewable and sustainable, clean energy is a task that requires time and dedication. Having public policies and involving all the actors that operate in the segment of port activities is an essential condition for success in this system. Understanding the current reality and being willing to face the new technological challenges is a level to be achieved that requires a lot of work, planning, and involvement in various areas of management. Evaluating possible market alternatives for renewable energy use and how these factors can impact port business is an enormous challenge (Cavallo et al, 2015; Zhang et al., 2017; Pavlic et al., 2014; Kuznetsov et al., 2015; Roh et al, 2016).

Having all the information of the procedures of the current state of a port and the practices of the ongoing environmental management, are essential conditions for the beginning of the work of implantation of a green port concept (Pavlic et al., 2014; Puig et al., 2015). Knowing in detail all the port's current energy use, having key performance information that portrays the energy consumption of operations and having indicators of environmental management is a prime condition and will allow to develop a sustainable environment using clean energies and generate a system with world-class ecological management practices (Kılkıs, 2015; Santos et al., 2016; Laxe et al., 2016; Lopes et al., 2013; Pavlic et al., 2014; Kuznetsov et al., 2015; Roh et al, 2016; Puig et al., 2015).

The high port administration must be committed to a system of sustainable logistic operations to implement a green port successfully. Otherwise, it will undoubtedly lead to failure (Pavlic et al., 2014; Roh et al., 2016).

This research analyze the current logistics operations of the port of Santos, the most significant port in Latin America, as well as its physical characteristics and legal aspects and public policies that govern Brazilian port activities. It aims to address what possible operations, technologies, and innovations are suitable in the medium and long-term to make it a port capable and integrated to other international ports with sustainable logistics operations.

When it comes to sustainable development logistically, it is essential to focus on three elements relevant to the system: economy, society, and the environment. A port must excel to balance these aspects (Kuznetsov et al., 2015; Roh et al., 2016).

2 SUSTAINABLE PORT ACTIVITIES

In Germany, the port of Bremerhaven and the ports in the city of Bremen are examples of focus on sustainable logistics. Its ports are models for the use of renewable energy and harmony between port operations and the local community. They are also examples concerning sound, air and water pollution control, as well as the use of technology and innovations to meet the prerequisites for a sustainable port (Peris-Mora et al., 2005; Cahoon et al., 2013; Kuznetsov et al., 2015; Roh et al., 2016; Tang et al., 2017).

The ports of Bremen and Bremerhaven offer clean fuel, the LNG, to vessels that dock in their ports, failing to use diesel, which is a fossil fuel. Also, the ports also provide electric power to the main engines of the vessels docked at the port terminals, avoiding once again the consumption of fossil fuel as well. This operation is called Onshore Power Supply—OPS (Hall et al., 2013; Tseng & Pilcher, 2015).

The use of solar energy, wind power and the use of improved engine technology with switching to diesel-electric units are operations that contribute to sustainable port logistics operations. Automatic guided battery vehicles are other technologies that provide to clean operations in ports (Peris-Mora et al., 2005; Lirn et al., 2013; Roh et al., 2016).

Since 2006, the port of Trelleborg in southern Sweden has been working with suppliers of reach stackers and terminal tractors to reduce the noise, noise pollution, of this equipment.

The port of Trelleborg has invested in the Intelligent Transport System, integrating information technology to improve processes and thereby eliminate operations that generate waste of energy. It integrated TOS (Terminal Operating System), WMS (Warehouse Management System) and (ERP—Enterprise Resource Planning).

The tractors terminals in the port of Trelleborg in Sweden work with Ad Blue catalytic converters and particulate filter which reduces emissions of particulate matter (PM) by almost 100%.

There are other examples of ports with the implementation of sustainable operations system. Among them is the port of Koper, Slovenia, which has implemented a system to replace diesel with renewable energy sources such as biomass and solar energy for water heating. The port of Koper commits to encourage and support initiatives in the introduction of new and modern technologies exclusively focused on clean, sustainable systems with a focus on environmental protection (Burskyte, et al., 2011; Roh et al., 2016; Puig et al., 2015).

This type of support has opened the door to a pilot project in the development of effective initiatives for the development of sustainable and clean projects with an active call for the reduction or elimination of emission gases that harm the environment. The port's board of directors endorsed the initial plan, whose the commitment served as an example for all employees (Davarzani et al., 2016; Roh et al., 2016).

The electrification of RTG (Rubber Tire Gantry—a kind of crane on tires) at the port of Koper was a process that helped immensely reduce the use of diesel and thereby reduce pollutants into the atmosphere. There was a reduction of noise pollution in port operations as well (Pavlic et al., 2014; Tang et al., 2017).

The port authority has an essential role in the implementation of a system of sustainable operations. The commitment of the port authority to handling, storing, loading and unloading processes is part of the success of a port considered internationally green (Bergmans et al., 2014; Hou & Geerlings, 2016; Kuznetsov et al., 2015; Roh et al., 2016).

A major civil engineering project in the Netherlands is the Maasvlakte II port project in Rotterdam. It involves the construction of a new port and support infrastructure on a brownfield site next to the Maasvlakte terminal. The size of the site is approximately 167 hectares, with 2,800 meters of wharf and an additional 500 meters for the operation of barges, with a capacity of 2.7 million TEUs (Frantzeskaki et al., 2014). The Maasvlakte

II terminal has the first Cargotec remote-control operated porteiners, as well as cranes for the barge terminal, Kuenz RMG cranes for the rail tracks on the quay line, Gottwald Automated Guided Vehicles (AGVs) with Gottwald Lifting Battery, Gottwald Robotic Battery Changing Stations and Management of Terminal Operating Systems (TOS), and Equipment Control System (ECS) by Navis (Cargotec) and TBA (Gottwald) simultaneously (Pavlic et al., 2014).

The terminal construction is being carried out according to the environmental performance certification rules of the Building Research Establishment Environmental Assessment Method (BREEAM) adopted by the Dutch Green Building Council. Another sustainable item of the terminal is related to the use of electrification of all equipment, allowing the terminal to be free of CO_2 emission, NOx, as well as other particles. In 2009, the Maasvlakte I terminal became the world's first container terminal powered by wind power (Iannone, 2012; Acciaro et al., 2014; Roh et al., 2016; Puig et al., 2015).

3 PRACTICAL STEPS FOR THE APPLICATION OF PERS (PORT ENVIRONMENTAL REVIEW SYSTEM)

The Self-Diagnostic Method (SDM) was first developed in the European research project ECOInformation and finalized by the ECOPORTS Foundation (EPF). Through this method, it is possible to evaluate the management of the port in question through a strategic questionnaire designated for environmental managers (Pavlic et al., 2014). The purposes of SDM are to define the different areas that need to be improved to reach an initial level of awareness. It helps port managers conduct a periodic review of the performance of their environmental management program compared to baseline data, and the European benchmark is to identify SWOT—based action priorities (strengths, weaknesses, Opportunity or Threat), and GAP analyzes, comparing current port provisions compared to other formal EMS – Environmental Management System (Laxe et al., 2017; Schippera et al„ 2017; Burskyte et al, 2011; Puig et al., 2015).

The SDM is a type of checklist of components that should be present in an environmental management program. It was designed as a resume, a snapshot of reality, and can be analysed independently and confidentially through the EPF (Pavlic et al., 2014; Puig et al., 2015). The checklist asks for a response on the following items:

- Environmental policy
- Management organization and staff
- Environmental training
- Operational management
- Emergency planning
- Monitoring and records
- Audit and environmental review

The final report of the SDM provides a concise overview of the current situation and focuses management efforts on the key areas that require attention in preparation for PERS—Port Environmental Review System (Pavlic et al., 2014; Puig et al., 2015).

The methodology used by the ports concerning the formulation and development of the PERS is summarized as follows:

- Creation of a team with academic skills and specialists of Port Authority.
- A series of consulting activities, workshops and training courses with port staff.
- Series of meetings and interviews with all stakeholders involved in the process.
- The conclusion of SDM, jointly by port executives and research team.
- The electronic format of requirements PERS.

Below are some of the main practical steps for a port to obtain the PERS (Environmental Review System of Port).

- Formulation of an Environmental Policy, that is, the public declaration by the port authority regarding the strategic environmental management of the port, establishing a framework for action and the environmental objectives to be achieved (Kuznetsov et al., 2015; Puig et al., 2015).
- Compilation and analysis of the port's environmental situation and identification of all environmental impacts of its operation. Significant environmental aspects should be identified for each port activity, and all relevant legislation should be made explicit. Thus, each executive or collaborator that acts in the port can acquire knowledge about the environment and the pertinent legislation. Also, appropriate indicators of environmental performance should be selected, allowing the monitoring and progress of port responses to new environmental challenges (Pavlic et al., 2014; Puig et al., 2015).
- Designation of specific responsibilities for port staff.
- Regular review of compliance with current environmental policy, so that the port authority's system/process about environmental protection is in line with environmental policy and current legislation (Puig et al., 2015).
- Definition of procedures for the determination of new objectives, actions, and initiatives that can also be developed.

- Production of Environmental Report, whose target is to provide information on the general management of environmental issues that concern the port, for all port personnel, the local community, and other interested groups.
- The most important stage of preparation for the application of PERS was to record and analyse the environmental situation in each port, as well as its ecological impacts that its operation may have locally and in the country (Puig et al., 2015).

This system acts as a useful, feasible tool for port environmental management and contributes substantially to the continuous improvement of the port environment. The voluntary self-regulation system of the PERS makes it flexible, familiar and well understood by the port authority and port staff. In addition, PERS gives the port authority the opportunity to identify environmental challenges and priorities, to respond in a timely, practical and cost-effective way to legislative and stakeholder pressures—city, population, business, etc. (Daamen & Vries, 2013; Pavlic et al., 2014; Puig et al., 2015).

Thus, based on the practical experience of some ports, the main benefits and lessons with the PERS system (in conjunction with the SDM questionnaire) are summarized as follows:

- Identification of the legal scenario related to environmental problems, setting up a mechanism to keep up with developments and better compliance with legislation.
- Recognition of the actual situation and identification of environmental priorities within the port (e.g., dust, waste, water runoff and noise).
- Identification of appropriate action plans to respond to legislative obligations and responsibilities. Also, identification of business risks and prevention of environmental accidents.
- Potential savings through energy efficiency, water consumption and waste treatment.
- Efficiency through large competitiveness.
- Improvement of the public image of the port.
- Recognition of the port's environmental credentials.
- Awareness of environmental issues and responsibilities throughout the Port Authority.
- Opportunities to improve contact with key port stakeholders, who are stakeholders, local administrators, the general public and environmental lobby groups, to promote a more transparent relationship (Puig et al., 2015).

Based on the PERS experience, the Port Authority may be encouraged to consider applying ISO 14001 in selected port areas. The PERS process is regarded as a positive intermediate step to an Integrated Environmental Management System, and the experience of its implementation creates internal capacity supported by the collaborative partnership (Pavlic et al., 2014; Puig et al., 2015).

Port authorities, particularly in the European Union, the United States and Canada, became familiar with the components of Environmental Management Systems (EMS) and the industry adopted well-established methodologies for port environmental management, such as Self Diagnosis Methodology (SDM) or the Port Environmental Review System—PERS (Puig et al., 2015).

4 METHODOLOGY

The methodology applied in this work was field research, with questionnaire application in some marine terminals in the Port of Santos. The purpose was to raise the current situation and analyse the positioning of each terminal about environmental aspects such as (Pavlic et al., 2014; Puig et al., 2015):

- Policy of treatment for solid or liquid waste and effluents.
- Policy to manage air quality, hazardous cargoes, noise pollution, water quality, waste from loading and unloading of ships, etc.
- The use of clean and renewable energy in terminal operations, use of technologies and innovations in sustainable operations (Hanssen et al, 2014; Roh et al, 2016).

Employees who hold management positions and operations directors, with a large knowledge of the environmental area of the port terminals were interviewed.

The questionnaire is composed of structured or open questions to obtain the maximum information of port operations.

5 PORT OF SANTOS, BRAZIL

The port of Santos, in Brazil, located in the state of São Paulo, the southeastern region of the country, is the most significant port in Latin America, as shown in Figures 1 and 2. It has 55 terminals and 17 container terminals. It has an extension of 13 km and is served by all the regular sea lines, which makes it possible to transport to any part of the world, starting from Santos.

Road access to this port is mostly by the Imigrantes and Anchieta highways, the Cônego Domênico Rangoni highway, the Rio-Santos highway and the Manoel da Nóbrega highway.

The concessionaires of the railways in the port of Santos are Rumo, MRS, and FCA. They

Figure 1. Port of Santos, São Paulo, Brazil.

Figure 2. Map of Brazil – Location of Port of Santos.

are responsible for the access to products/goods, through the rail mode. The FCA does not have direct access to the port of Santos. It serves the maritime complex by operating in partnership with Rumo, in Campinas, a city about 180 kilometers away from the port of Santos by road, or 280 kilometres by rail.

The integration of the Port of Santos with waterways is carried out using a railway. The Tietê-Paraná Waterway is responsible for most of the goods transported to the Port of Santos. The main products are soy, corn, and sugar. These goods usually originate in the Centre-West region of Brazil and are carried by barges to the city of Pederneiras, in São Paulo. From there, they travel to the Port of Santos, by railroad.

The left bank of the Port of Santos, in the city of Guarujá, has 3 km of wharf, while the right bank, in the city of Santos, has 12 km. In the left margin, there are ten warehouses. The right margin has sixty, which subdivide into 29 internal warehouses and 31 external warehouses. The central warehouses are situated on the estuary channel (first strip). The external warehouses are along the Saboó neighborhood to Ponta da Praia. The storage system also has 212 tanks and 12 patios to serve the Organized Port of Santos.

The main cargoes handled in the Port of Santos are:

- Liquid bulk: citrus juices, oils, alcohol, gasoline, diesel, LPG, vegetable oil, among others.
- Solid granules: sugar, corn, soy, coal, fertilizer and fertilizers, sulphur, pelleted citrus pulp, salt, wheat, among others.
- General cargo: sugar, cellulose, paper, aluminum.
- Special and containerized loads.
- Vehicles (roll-on-roll-off).
- Frozen beef.

The cargo handling at the Port of Santos scheduled for 2017 is 123 million tons of cargo, according to the Docks Company of the State of

Table 1. Characteristics of the Port of Santos.

AREA		
Area of port	Total	7,765,100 m^2
	Right bank	3,665,800 m^2
	Left bank	4,099,300 m^2
PIER		
Number of berths	CODESP (Public company)	53
	Private terminals	11
Local	Extension	Depth
Total	13.013 m	Between 5.0 and 13.5 m
CODESP	11.600 m	Between 6.6 and 13.5 m
Private terminals	1.413 m	Between 5.0 and 13.0 m

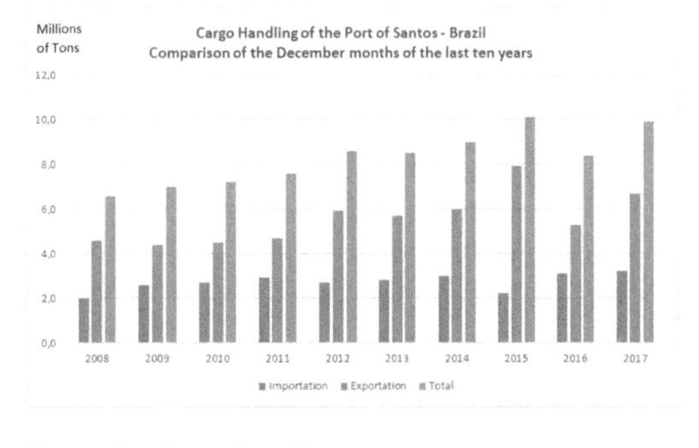

Figure 3. Cargo handling.

São Paulo (CODESP), the port authority. In the year 2016, the movement was 113.8 million tons.

The container handling projection for 2017 is to reach 3,66 million TEUs. In the year 2016, the Port of Santos handled 3,56 million TEUs.

Figure 3 shows the cargo handling of the Port of Santos, the comparison of the December months of the last ten years.

6 ENVIRONMENTAL CHARACTERISTICS OF THE PORT OF SANTOS

The port of Santos started a program of monitoring and noise control in twenty areas along the right and left bank of the port. This study is an inherent part of the Basic Environmental Plan of the port.

The pollution monitoring and control program is based on the following:

- Mapping of sources emitting noise;
- mapping of sensitive receptors, such as schools, hospitals, libraries and nursing homes;
- The monitoring of sound pressure levels in the vicinity of the Port of Santos;
- Actions to control the emission of noise at the source.

CODESP is promoting improvements in the pavements of the access roads and reducing the number of level crossings (road crossings). They aim at minimizing the impacts of congestion and reducing damage to the vehicles that travel there.

Logistic solutions such as the scheduling system for the access of cargo vehicles to port terminals also contribute to the reduction of congestion and queues, with consequent reduction of noise due to road traffic.

The Road Interference Mitigation Program aims to prevent port operations from harming the traffic in the cities of Santos and Guarujá through improvements in accesses, logistics solutions and other means.

In this program, which is still under development, new works and solutions are planned so that port traffic does not harm the well-being of the population of the cities of Baixada Santista.

The Risk Management Program aims at surveying the risks arising from the operations of storage, transportation, loading and unloading of dangerous chemicals, solid bulk and other goods. Proposing and taking actions to manage these risks and preparing for rapid and adequate response in the event of an accident resulted in three final products:

- Risk Analysis Study (RAS);
- Risk Management Plan (RMP)
- Emergency Action Plan (EAP) or Emergency Control Plan (ECP)

The Risk Analysis Study aims at analyzing the probability of occurrence of accidents, the magnitude of their consequences and the vulnerability of the surrounding populations. The Risk Management Program aims at the constant taking of actions, such as training to carry out handling activities based on appropriate procedures. The third, Emergency Action Plan or Emergency Control Plan is intended to plan and prepare for a prompt and adequate response to the occurrence of an accident involving such products.

7 CONCLUSIONS

The first and essential step for the Port of Santos is to analyze its current state concerning environmental aspects in all its terminals. Then, apply the Self Diagnosis Methodology (SDM). In this way, it is possible to initiate a process of implementation of specific public policies to obtain an ISO 14000 certification and, consequently, an international certification of a sustainable port.

The challenges are enormous. When it comes to the implementation of environmental systems in Brazilian companies, in general, it is not something so trivial, regardless of its segment. In the case of the Port of Santos, which is a government agency, this factor is even more crucial.

The barrier has a cultural aspect. Brazil is not a country that cultivates in its day-to-day cultural inherent factors related to the care of the environment: to think sustainably about its laws, daily routine of the people, the quality of life of the people, the concern with the fauna and flora, etc.

In addition to breaking the cultural barrier of the population, for the implementation of a sustainable port, the political boundary must be broken. Also, the focus on long-term planning and the involvement of all stakeholders in the same goal is critical. Therefore, it is clear that the implementation of innovations in the national port operations, starting with the Port of Santos, would be an essential step for Brazil's industrial policy, since it is the most significant port in Latin America.

ACKNOWLEDGEMENTS

Project 2015/00277-8, sponsored by FAPESP, a Foundation for Research Support of the State of São Paulo.

REFERENCES

Acciaro, M; Ghiara, H.; Cusano, M.I. 2014. Energy management in seaports: A new role for port authorities. Energy Policy (71): 4–12.

Bergmans, A.; Vandermoere, F.; Loots, I. 2014. Co-producing sustainability indicators for the port of Antwerp: How sustainability reporting creates new discursive spaces for concern and mobilization. ESSACHESS. Journal for Communication Studies, vol. 7, no. 1 (13): 107–123.
Burskyte, V.; Belous, O.; Stasiskiene, Z. 2011. Sustainable development of deep-water seaport: the case of Lithuania. Environmental Science and Pollution Research (18): 716–726.
Cahoon, S.; Pateman, H.; Chen, S.-L. 2013. Regional port authorities: leading players in innovation networks? Journal of Transport Geography (27): 66–75.
Cavallo, B.; D'apuzzo, L.; Squillante, M. 2015. A multi-criteria decision making method for sustainable development of Naples port city-area. Quality & Quantity—International Journal of Methodology (49): 1647–1659.
Daamen, T.A.; Vries, I. 2013. Governing the European port–city interface: institutional impacts on spatial projects between city and port. Journal of Transport Geography (27): 4–13.
Davarzani, H.; Fahimnia, B.; Bell, M.; Sarkis, J. 2016. Greening ports and maritime logistics: A review. Transportation Research Part D (48): 473–487.
Frantzeskaki, N.; Wittmayer, J.; Loorbach, D. 2014. The role of partnerships in 'realising' urban sustainability in Rotterdam's City Ports Area, The Netherlands. Journal of Cleaner Production (65): 406–417.
Hall, P.V.; O'Brien, T.; Woudsma, C. 2013. Environmental innovation and the role of stakeholder collaboration in West Coast port gateways. Research in Transportation Economics (42): 87–96.
Hanssen, L.; Vriend, H.; Gremmen, B. 2014. The role of biosolar technologies in future energy supply making scenarios for the Netherlands: Energy port and energy farm. Futures (63): 112–122.
Hou, L; Geerlings, H. 2016. Dynamics in sustainable port and hinterland operations: A conceptual framework and simulation of sustainability measures and their effectiveness, based on an application to the Port of Shanghai. Journal of Cleaner Production (135): 449–456.
Iannone, F. 2012. The private and social cost efficiency of port hinterland container distribution through a regional logistics system. Transportation Research Part A (46): 1424–1448.
Kılkıs, S. 2015. Composite index for benchmarking local energy systems of Mediterranean port cities. Energy (92): 622–638.
Kuznetsov, A.; John Dinwoodie, J.; Gibbs, D.; Sansom, M.; Knowles, H. 2015. Towards a sustainability management system for smaller ports. Marine Policy (54): 59–68.
Laxe, F.G.; Bermúdez, F.M.; Palmeroa, F.M.; Novo-Corti, I. 2016. Sustainability and the Spanish port system. Analysis of the relationship between economic and environmental indicators. Application to the Spanish case. Marine Pollution Bulletin (113): 232–239.
Laxe, F.G.; Bermúdez, F.M.; Palmeroa, F.M.; Novo-Corti, I. 2017. Assessment of port sustainability through synthetic indexes. Application to the Spanish case. Marine Pollution Bulletin (119): 220–225.
Lirn, T.-C.; Wu, Y.-C.; Chen, Y.J. 2013. Green performance criteria for sustainable ports in Asia. International Journal of Physical Distribution & Logistics Management, Vol. 43, No. 5/6: 427–451.
Lopes, C.; Antelo, L.T.; Franco-Uría A.; Botana, C.; Alonso, A.A. 2013. Sustainability of port activities within the framework of the fisheries sector: Port of Vigo (NW Spain). Ecological Indicators (30): 45–51.
Pavlic, B.; Cepak, F.; Sucic, B.; Peckaj, M.; Kandus, B. 2014. Sustainable port infrastructure, practical implementation of the green port concept. Thermal Science, Vol. 18, No. 3: 935–948.
Peris-Mora, E.; Orejas, J.M.D.; Subirats, A.; Ibánez, S.; Alvares, P. 2005. Development of a system of indicators for sustainable port management. Marine Pollution Bulletin (50): 1649–1660.
Puig, M.; Wooldridge, C.; Casal, J.; Darbra, R.M. 2015. Tool for the identification and assessment of Environmental Aspects in Ports (TEAP). Ocean & Coastal Management (113): 8–17.
Roh, S.; Thai, V.V.; Wong, Y.D. 2016. Towards Sustainable ASEAN Port Development: Challenges and Opportunities for Vietnamese Ports. The Asian Journal of Shipping and Logistics 32(2): 107–118.
Santos, S.; Rodrigues, L.L.; Branco, M.C. 2016. Online sustainability communication practices of European seaports. Journal of Cleaner Production (112): 2935–2942.
Schippera, C.C; Vreugdenhila, H.; Jong, M.P.C. 2017. A sustainability assessment of ports and port-city plans: Comparing ambitions with achievements. Transportation Research Part D (57): 84–111.
Sciberras, E.A.; Zahawi, B.; Atkinson, D.J. 2017. Reducing shipboard emissions—assessment of the role of electrical technologies. Transportation Research Part D (51): 227–239.
Sislian, L.; Jaegler, A.; Cariou, P. 2016. A literature review on port sustainability and ocean's carrier network problem. Research in Transportation Business & Management (19): 19–26.
Tang, J.; McNabola, A.; Misstear, B.; Caulfield, B. 2017. An evaluation of the impact of the Dublin port tunnel and HGV management strategy on air pollution emissions. Transportation Research Part D (52): 1–14.
Tseng, P.-H.; Pilcher, N. 2015. A study of the potential of shore power for the port of Kaohsiung, Taiwan: To introduce or not to introduce? Research in Transportation Business & Management (17): 83–91.
Woodburn, A. 2017. An analysis of rail freight operational efficiency and mode share in the British port-hinterland container market. Transportation Research Part D (51): 190–202.
Zhang, Y.; Kim, C.-W.; Tee, K.F.; Lam, J.S.L. 2017. Optimal sustainable life cycle maintenance strategies for port infrastructures. Journal of Cleaner Production (142): 1693–1709.

Progress in Maritime Technology and Engineering – Guedes Soares & Santos (Eds)
 ISBN 978-1-138-58539-3

Evaluation of port performance: Research opportunities from the systemic analysis of international literature

G.C. Fermino, A. Dutra & L. Ensslin
Universidade do Sul de Santa Catarina, Florianópolis, Santa Catarina, Brazil

S.R. Ensslin
Universidade Federal de Santa Catarina, Florianópolis, Santa Catarina, Brazil

ABSTRACT: This article aims to analyze the research opportunities on the topic of port industry performance evaluation, through the investigation of a Bibliographic Portfolio (BP) of international literature articles with relevance and scientific recognition. To achieve this goal, ProKnow-C (*Knowledge Development Process—Constructivist*) intervention instrument was used. The systemic analysis consists of a structured process of critical analysis of BP articles, based on a set of assumptions defined by the researcher informed by the theoretical affiliation adopted, in order to highlight gaps and research opportunities. It is an exploratory study, with a qualitative approach, involving the collection of primary and secondary data. Thus, it is possible to conclude that port performance evaluation studies, for the most part, apply standardized models, methods, tools and techniques, without taking into account the context specific needs, aiming particularly to measure the productivity and operational efficiency of the ports.

1 INTRODUCTION

The port sector is made up of large areas and among these are those of management and operations that relate to other organizations and actors with very distinct characteristics, forming a complex system of governance.

This complexity requires a process of performance evaluation to monitor the achievement of strategic objectives of areas and organizations, aiming to subsidize the decision-making process.

Dutra et al. (2015), report that performance evaluation is a process used to construct knowledge in the decision-maker (manager) on a specific context that he proposes to manage, through activities that: identify; organize; measure ordinally and cardinally; integrate the relevant criteria and permits monitoring and taking systematic actions for improvements.

When you can measure something about what we are talking about and express it in numbers, you know something about it [...] (otherwise) our knowledge is scarce and unsatisfactory; it is the beginning of knowledge, but, in thought, we would have finished entering the field of science (Neely et al., 1995).

A port performance can be assessed on the basis of efficiency and effectiveness dimensions. Efficiency is defined as the performance in the perspective of the port authority, while effectiveness involves the prospect of customers and all actors involved in the port environment (Brooks, et al., 2014). These are complementary dimensions, with a focus on internal performance and the other contemplates the perceived performance by external actors.

Chou & Liang (2001) emphasize that managers should adopt multicriteria evaluation models, contemplating performance objective and subjective dimensions, in a systemic and holistic way, by allowing better management of their organizations and also serve as a reference for potential investors, partners and customers.

A systemized and continuous process of a port performance evaluation does not only help to understand and improve its marketing and competitive position, but also provides a clear and solid basis for policy makers to local and regional development (Wu, et al., 2009).

Turner (2000) e Brooks & Pallis (2008) reported that a port must be seen as a system, rather than a set of terminals and independent operators, taking as its focus the system global performance, always recognizing the contribution and the interdependence of actors involved in the port environment.

In this context, from a theoretical affiliation adopted by the authors on the topic of performance evaluation, it is sought to answer the question: What are the research opportunities in maritime port performance evaluation, based on a systemic analysis of international literature? In order to answer this research question, the main objective is to identify and analyze opportunities for research on maritime port performance, through the research of bibliographical portfolio of articles with relevance and scientific recognition.

In order to achieve this objective, ProKnow-C (Knowledge Development Process—Constructivist) intervention instrument was used to make a systematic analysis of articles in the bibliographic portfolio, by comparing the theoretical affiliation on the subject of performance evaluation with the content of each scientific article.

The study is justified by the retrieval in the scientific literature of the paths taken by the researchers in the characterization of port sector performance evaluation, in terms of methodologies, types of indicators, form of measurement and evaluation management, making it possible to identify trends and gaps in the studies performed.

The present article is structured in the following way: Section 1 considers the introduction, section 2 the Theoretical Reference; Section 3 the Research Methodology; section 4 Presentation and Discussion of Results, containing opportunities for future research; Section 5 Conclusions and, finally, the References used in the course of the article.

2 SYSTEMATIC REVIEW OF SCIENTIFIC LITERATURE

In an integrated and systemic perspective, the reflection of ownership and management structures, different ports have different objectives and strategies, which can be affected not only by external factors such as global economic development and competition from other ports, but also by internal policy forces (Brooks & Pallis, 2008). Therefore, it is extremely important that researchers should take into account the specificities of objectives between different ports that require different evaluation processes, under penalty of obtaining biased conclusions. (Panayides & Song, 2008). According to Brooks & Schellinck (2013), depending on its corporate competitive strategy, the port can act in different ways.

Meeting expectations of improving port competitiveness, stakeholders interacting with customers and users have historically selected the most efficient ports and services, focusing on a limited number of ports capable of providing economies of scale. For this reason, the port competitiveness has concerned stakeholders. Port competitiveness measures can provide consistent information to all interested parties, characterizing them as measures of performance (Yeo et al., 2011).

A port or group of ports performance evaluation in a country is always significantly different from a group of ports in another country. This demonstrates the heterogeneity of a port performance (Wu et al., 2009). Therefore, the context delimitation by the manager is important to better characterize the port unit efficiency.

Casado-Martínez et al. (2009), states that designing simple models of performance evaluation as a starting point can be an excellent strategy, provided that improvement and expansion procedures are gradually incorporated. It stands out in this context the process of learning and generating results that stimulate specific interventions in any distortions.

According to Saengsupavanich et al. (2009), in a context facilitator of performance evaluation five aspects of management should be considered: (1) success; (ii) awareness; (iii) determination; (iv) promptly; and (vi) policy coverage.

However, Gómez et al. (2015) understand that ports are subject to changes in management policies that are more oriented towards the use of models in which economic and environmental factors can be considered as development variables.

In another composition form of studies in this area, a port performance can be integrated by three subgroups: service, operations and logistics (Woo et al., 2011). Lami & Beccuti (2010) argue that aspects related to transportation, environmental, logistical, financial, socioeconomic and urban planning issues must also be considered.

In measurement perspective, Panayide & Song (2008) use port efficiency as a performance measure, provided that port operations are integrated into the overall planning.

Also in performance evaluation context, Lam, & Song (2013), propose a hierarchical structure that classifies the performance indicators in three layers: (i) Evaluation: determinants of quality, punctuality, cost; (ii) Evaluation dimensions: functional, information and communication, relationship; (iii) Evaluation elements: shipping companies, other seaports, land transport corridors, freight, logistics service providers, consignees and the city where the port is located.

Wu et al. (2010) e Tetteh et al. (2016) report that in a comparative analysis between the efficiency of port operations in emerging markets and developed economies, conclude that the ports of advanced economies may not be models for those in developing markets, because many factors influence the port performance and these factors hinder the determination of performance analysis.

In this way, factors that influence the port competitiveness are difficult to measure. These measurement factors reflect the competitive situation dynamics (Yeo et al., 2011).

Finally, for Cerreta & Toro (2012), the evaluation process is becoming increasingly important in decision-making processes for planning and design of sustainable port plans. To Woo, Pettit & Beresford (2011), measuring instruments, or scales are developed when the phenomena that are believed to exist, from the theoretical understanding of the world, cannot be directly evaluated, but they must first be measured.

3 RESEARCH METHODOLOGIES

3.1 *Research framework*

The present study is based on constructivism, informed by the Roy's (1993) perspective, which seeks to construct, with and for the participants, a set of keys (Creswell 2014, p.194) based on their objectives and value systems in a way that helps them to understand the context in which they intend to intervene, through a qualitative approach (Creswell, 2014), based on researchers' choices and preferences, whether in the selection of bibliographic portfolio or in the systemic analysis of BP articles. It is an exploratory research (Richardson, 2008). As the nature it is theoretical-illustrative, since the BP (Articles) formation is done according to the researchers' perception of value and a structured analysis is performed on the researched topic (Flick & Netz, 2004). Regarding logic, the research was characterized as inductive so that the final considerations can not be applied to a universe (Da Silva & Menezes, 2005). The collection involved secondary data, from the databases researched (scientific articles). Finally, regarding the technique, bibliographic research and action research were used (Collis & Hussey, 2005).

3.2 *ProKnow-C intervention tool*

ProKnow-C (Knowledge Development Process—Constructivist) (Ensslin et al., 2015) intervention tool was used, whose objective is to build knowledge in the researcher about a specific topic that he/she is studying. In recent years, several studies have been published in different areas of knowledge, informed by ProKnow-C, for: (i) identify a fragment of relevant literature on the topic of interest to the researcher (Bibliographic Portfolio); (ii) understand the peculiarities of the study area; (iii) conduct BP critical analysis based on theoretical framework selected by the researcher; and (iv) suggest gaps in the literature that subsidize the formulation of future work (Dutra et al., 2015, Ensslin et al., 2012, Ensslin et al., 2013, Lacerda et al., 2012, Tasca et al., 2010).

3.2.1 *ProKnow-C steps*

Proknow-C comprises the steps provided in Figure 1.

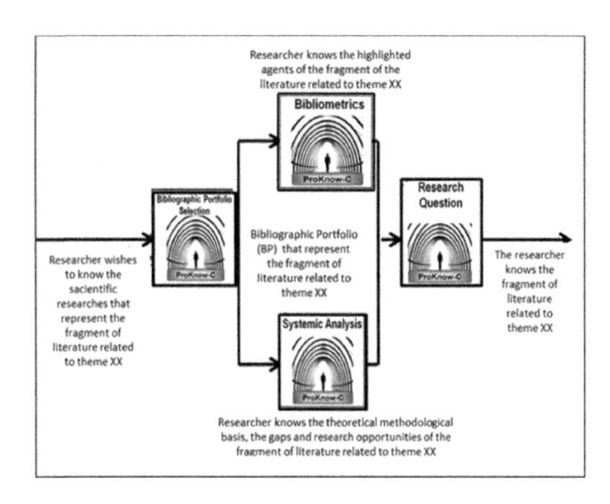

Figure 1. Proknow-C Steps. Source: Adapted from Ensslin, Ensslin & Pinto (2013), Lacerda, Ensslin & Ensslin (2012), Tasca et al. (2010).

The present study included the following steps: Selection of Bibliographic Portfolio—BP and Systemic Analysis.

The first step, Selection of BP, consists in defining the research theme, the research axes, with the respective keywords and after the definition of databases (Tasca et al., 2010, Ensslin et al., 2010). From this choice the researcher begins the knowledge expansion process about the theme, once that he acquires the scientific nomenclature understanding on his theme, (Ensslin et al., 2015). Time limitation, publications from 2008 to 2017.

BP selection resulted in the research topic: Evaluation of maritime port performance; Research Axes: Axis 1 - Performance Evaluation; key words: Measurement, Evaluation, Assessment, Appraisal, Indicator; Axis 2 - Ports; keywords: Port, Seaport, Harbor; Boolean equation (for research in databases). ("Performance" OR "Measurement" OR "Management" OR "Evaluation" OR "Assessment" OR "Appraisal" OR "Indicator*") AND ("Seaport*" OR "Port" OR "Harbor*"); Data Bases and quantities of articles searched: Scopus (1,870), Willey (1,621), ProQuest ASSIA (7), Web of Science (1,466) and Materials Science & Engineering Database (562), resulting in a database of 5,526 raw articles.

With the applied filters, we have the quantitative data: elimination of duplicate articles resulting in 5,492; articles with aligned titles 114; articles with confirmed scientific recognition and relevance, aligned abstracts and full text 52; Bibliographic Portfolio—BP 52 articles.

3.2.2 *Systemic analysis process*

The systemic analysis consists of a structured process of critical analysis of BP articles from a set of assumptions defined by the researcher informed by the adopted theoretical affiliation, with the objective of identifying research gaps.

Such assumptions can be characterized: (i) as lenses, according to Brunswik model adapted to the human judgment (Brunswik et al., 2001), (ii) as a vision of the world (Lacerda et al., 2014), to show how the researcher perceives the characteristics present in the environment, and (iii) as filters that result from different frames of interpretation shaped by beliefs, values, expectations and previous experience (Melão & Pidd, 2000).

Table 1. Theoretical affiliation of the lenses. Source: *Adapted from Dutra et al. 2015.*

Theoretical affiliation	Lenses	Definition of the lenses
Process of construction of knowledge in the decider...	1. Approach Context of the PE	Aims at the analysis of the approach used in the construction of the model PE, which can be normativist, descriptivist, prescriptivist or constructivist. Aims at the analysis of the context in which PE is applied, that is, the focus/emphasis of the evaluation.
... within a specific context...	2. Singularity	Aims at the analysis of whether the context of the PE is unique, i.e., specific of a reality and if it recognizes the decision-makers involved in the context.
... through the identification, organization, ...	3. Process of identification of PE criteria	Aims at the identification of the existence of a structured process for the identification of the evaluation criteria of the reality investigated and the participation of the decision-makers.
... measurement and ...	4. Criteria Measurement	Aims at the identification of the kind of scale used for the measurement of the criteria, if it is ordinal or cardinal, and also observe the properties of the measurement theory.
... integration of the aspects considered relevant to evidence performance ...	5. Criteria Integration	The purpose is to check if the criteria are integrated by dimension and/or area of performance, allowing the evidence of the global profile (status quo).
... contemplating the generation of actions of improvement from the profile of performance obtained.	6. Process of performance management	The purpose is to identify if it contemplates a structured process of performance management, focusing on the identification of actions of improvement.

Based on the theoretical affiliation of performance evaluation adopted, we have the lenses, object of the systemic analysis performed, as can be observed in Table 1. The definition of lenses requires a theoretical affiliation that supports the researchers' choices represented by the lenses that derive from the perceptions and values against the context that is intended to be analyzed, as shown in Table 1.

4 PRESENTATION AND DISCUSSION OF RESULTS

The results are presented below considering the systemic analysis of BP articles in relation to the six lenses resulting from the theoretical affiliation of performance evaluation.

Lens 1: Harmony between the Approach and its Use/Application. It is noted that 49 published articles (94.23%) adopted a realistic approach and 3 articles (5.77%) a prescriptivist one. Of these, in terms of use, 8 (15.38%) are generic and 44 articles (84.62%) are developed for specific use. In relation to harmony, 7 articles (13.46%) presented harmony between the approach and its use and 45 articles (86.54%) had no harmony between the approach and its use. According to Roy (1993), the constructivist approach is more suitable to support the decision by incorporating the decision maker values and perceptions. However, none of the articles made use of this approach.

Lens 2: Singularity. Regarding the singularity, it presents the following results: On the identification of the actors 29 articles (55.77%) identified the actors (decision maker) for whom the model was intended and 23 articles (44.23%) did not identify the actors (decision maker) for whom the model is intended. Regarding the identification of the context, 46 articles (88.46%) identified the physical context where the model will be used and 6 articles (11.54) did not identify the physical context where the model will be used. Now regarding the identification of actors/context, 34 articles (65.38%) identified the actors and the context and 18 articles (34.62%) did not identify the actors and the context.

Lens 3: Process to identify the criteria. In relation to the criteria identification, of the total of 52 articles, 15 (28.85%) recognize the need to expand the decision maker's knowledge and 37 articles (71.15%) did not recognize this need. In the continuous analysis, it was found that only 5 articles (9.62%) had the decision makers' participation in some way for the model construction and 47 articles (90.38%) had no decision makers participating in the process. Thus, standardized and generic models predominate.

Lens 4: Measurement. The analysis of the articles which carry out measurement in their performance evaluation models, it has been observed that 49 articles (94.23%) perform measurement and 3 (5.77%) of articles do not perform it. As for ordinal properties (nonambiguity, homogeneity, intelligibility and the possibility of distinguishing between better and worse performances), 20 articles (38.46%) met ordinal properties (qualitative dimension) and 32 (61.54%) articles do not meet ordinal properties (quantitative dimension). In relation to the articles regarding cardinal properties (the operations performed with the scales are contained in the statistical properties that the scale allows according to the Theory of Measurement) 46 (88.46%) of the articles meet cardinal properties (of a total of 16 that measure) and 6 (11.54%) of the articles do not meet them.

Lens 5: Integration. For the Integration lens, the analysis of the articles as the way they perform the integration, 20 (38.46%) of the articles perform numerical integration, whereas no article performs the integration in a descriptive way. 25 (48.07%) of the articles perform integration of two or more of the above forms (numerically/graphically/descriptively), 2 (3.85%) of articles performed the integration graphically and 5 (9.62%) of articles do not perform integration. For the analysis of the articles that determine reference levels for the determination of integration constants, 27 (51.92%) perform the integration with reference levels and 25 articles (48.08%) did not perform the integration from reference level.

Lens 6: Management (Performance Management Process). The generated knowledge can allow knowing the current profile, its monitoring and improvement of the researched studies. Of these 52, 47 (90.38%) indicate its profile, i.e., the current action situation, monitoring and improvement and 5 (9.62%) do not present.

4.1 *Opportunities for future research*

With the analysis of the lenses, it is seen that one must advance further studies, particularly noting that 94.23% of published articles have adopted a realistic approach and as Roy (1993), the constructivist approach is more suitable to support the decision by incorporating the decision maker's values and perceptions. However, none of the articles made use of this approach. Regarding the study singularity, a process to identify the criteria, measurement, integration, management—the process of performance management should be more studied in the face of expressive results.

Therefore, it is possible to conclude that studies on port performance evaluation, for the most part, apply standardized models, methods, tools and techniques, without taking into account the specific needs of the context, which aim in particular to measure the productivity and operational efficiency of the ports and the used instruments and evaluation criteria do not present evidence where and how to carry out the intervention to correct, develop, adapt or improve the evaluated performance.

5 CONCLUSIONS

The systemic analysis consisted of a structured process of critical evaluation of article BP, after the application of a compound of assumptions established by the researchers and filtered on the theoretical basis. In relation to the results, the main objectives of the study were to detect the main characteristics of the articles on the subject of port performance evaluation and to propose future research opportunities for researchers and managers in the port sector and impacted or influenced from the studied theoretical base.

Thus, it is possible to conclude that the studies about the port performance evaluation, in the most part, apply models, methods, tools and techniques, without taking into account the specific needs of the context, aiming especially to measure productivity and operational efficiency of ports and the used instruments and evaluation criteria do not present evidence where and how to perform the intervention to correct, develop, adapt or improve the competences and the evaluated performance.

For future work it is recommended (i) the proposition of port performance evaluation models focused on the constructivist approach; (ii) greater emphasis on performance evaluation, as instruments of support for port management at a strategic level.

REFERENCES

Brooks, M.R., Pallis, A.A. 2008. Assessing port governance models: process and performance components. *Maritime Policy & Management* 35(4): 411–432.

Brooks, M.R., Pallis, A.A. 2013. Assessing port governance models: process and performance components. *Maritime Policy & Management* 35(4): 411–432.

Brooks, M.R., Schellinck, T., Pallis, A.A. 2011. A systematic approach for evaluating port effectiveness. *Maritime Policy & Management* 38(3): 315–334.

Brunswik, E., Hammond, K., Stewart, T. 2001. *The essential Brunswik: beginnings, explications, applications*: Oxford University Press.

Casado-Martínez, M.C., Forja, J.M., DelValls, T.A. 2009. A multivariate assessment of sediment contamination in dredged materials from Spanish ports. *Journal of Hazardous Materials* 163(2): 1353–1359.

Cerreta, M., Toro, P. 2012. Strategic environmental assessment of port plans in Italy: Experiences, approaches, tools. *Sustainability* 4(11): 2888–2921.

Chou, T.Y., Liang, G.S. 2001. Application of a fuzzy multicriteria decision-making model for shipping company performance evaluation. *Maritime Policy & Management* 28(4): 375–392.

Collis, J.; Hussey, R. 2005. *Pesquisa em Administração: um guia prático para alunos de graduação e pós-graduação.* 2. ed. Porto Alegre: Bookman.

Creswell, J.W. 2014. *Research Design: Qualitative, Quantitative, and Mixed Methods Approaches.* 4th ed. SAGE Publications, Inc.

Da Silva, E.L., Menezes, E.M. 2005. *Metodologia da pesquisa e elaboração de dissertação.* UFSC, Florianópolis, 4a. edição, 123.

Dutra, A.; Ripoll-Feliu, V.M.; Fillol, A.G.; Ensslin, S.R.; Ensslin, L. 2015. The construction of knowledge from the scientific literature about the theme seaport performance evaluation. *The International Journal of Productivity and Performance Management* 64: 243–269.

Ensslin, L., Ensslin, S.R., Pacheco, G.C. 2012. *Um estudo sobre segurança em estádios de futebol baseado na análise bibliométrica da literatura internacional, Perspectivas em Perspectivas em Ciência da Informação* 17(2): 71–91.

Ensslin, L., Ensslin, S.R., Pinto, H.M. 2013. Processo de investigação e Análise bibliométrica: Avaliação da Qualidade dos Serviços Bancários. *RAC – Revista de Administração Contemporânea* 17(3): 325–349.

Flick, U., Netz, S. 2004. *Uma introdução à pesquisa qualitativa.* Bookman, Porto Alegre, Gadhia.

Gómez, A.G. et al. (2015). Environmental risk assessment of water quality in harbor areas: A new methodology applied to European ports. *Journal of environmental management* 155: 77–88.

Lacerda, R.T.O., Ensslin L., Ensslin, S.R. 2014. "Research opportunities in strategic management field: a performance measurement approach." *International Journal of Business Performance Management* 15(2): 158–174.

Lacerda, R.T.O., Ensslin, L., Ensslin, S.R. 2012. Uma análise bibliométrica da literatura sobre estratégia e avaliação de desempenho, *Gestão & Produção* 19(1): 59–78.

Lam, J.S.L., Song, D.W. 2013. Seaport network performance measurement in the context of global freight supply chains. *Polish Maritime Research* 20: 47–54.

Lami, I.M., Beccuti, B. 2010. Evaluation of a project for the radical transformation of the Port of Genoa-Italy: According to community impact evaluation (CIE). *Management of Environmental Quality: An International Journal* 21(1): 58–77.

Melão, N., Pidd, M.A. 2000. Conceptual Framework for Understanding Business Processes and Business Process Modelling. *Information Systems Journal* 10(2): 105–129.

Neely, A., Gregory, M., Platts, K. 1995. Performance measurement system design: a literature review and research agenda. *International journal of operations & production management* 15(4): 80–116.

Panayides, P.M., Song, D.-W. 2008. "Evaluating the integration of seaport container terminals in supply chains", *International Journal of Physical Distribution & Logistics Management* 38(7): 562–584.

Richardson, R. 2008. *Pesquisa social: métodos e técnicas.* 4 ed. São Paulo: Atlas.

Roy, B. 1993. Decision science or decision-aid science? *European Journal of Operational Research* 66(2): 184–203.

Saengsupavanich, C., Comanitwong, N., Gallardo, W.G., Lertsuchatavanich, C. 2009. Environmental performance evaluation of an industrial port and estate: ISO14001, port state control-derived indicators. *Journal of Cleaner Production* 17(2): 154–161.

Tasca, J.E., Ensslin, L., Ensslin, S.R., Alves, M.B.M. 2010. An approach for selecting a theoretical framework for the evaluation of training programs. *Journal of European Industrial Training* 34(7): 631–655.

Tetteh et al. 2016. Container Ports Throughput Analysis: A Comparative Evaluation of China and Five West African Countries' Seaports Efficiencies, *International Journal of Engineering Research in Africa* 22: 162–173.

Turner, H.S. 2000. Evaluating seaport policy alternatives: a simulation study of terminal leasing policy and system performance. *Maritime Policy & Management* 27(3): 283–301.

Woo, S.H., Pettit, S., Beresford, A.K. 2011. Port evolution and performance in changing logistics environments. *Maritime Economics & Logistics* 13(3): 250–277.

Wu, J., Yan, H., Liu, J. 2009. Groups in DEA based cross-evaluation: An application to Asian container ports. *Maritime Policy & Management* 36(6): 545–558.

Wu, J., Yan, H., Liu, J. 2010. DEA models for identifying sensitive performance measures in container port evaluation. *Maritime Economics & Logistics* 12(3): 215–236.

Yeo, G.T., Roe, M., Dinwoodie, J. 2011. Measuring the competitiveness of container ports: logisticians' perspectives. *European Journal of Marketing* 45(3): 455–470.

Progress in Maritime Technology and Engineering – Guedes Soares & Santos (Eds)
© 2018 Taylor & Francis Group, London, ISBN 978-1-138-58539-3

Performance evaluation of the infrastructure of ports from Santa Catarina State

J.P. Meirelles, S.R. Ensslin & E.M. Luz
Universidade Federal de Santa Catarina, Florianópolis, Santa Catarina, Brazil

A. Dutra & L. Ensslin
Universidade do Sul de Santa Catarina, Florianópolis, Santa Catarina, Brazil

ABSTRACT: The objective is to build a Constructivist Multi-Criteria Model of Performance Evaluation to support the decisions about Port Logistics, based on the Ports Infrastructure in the state of Santa Catarina. To do it so, the Multi-Criteria Decision Aiding-Constructivist (MCDA-C) methodology was used. With the results, it was possible to construct twelve performance indicators regarding the physical infrastructure and eight indicators regarding the infrastructure equipment. Among these indicators, eleven were in a compromising performance level. Thus, improvement actions were proposed in order to boost the unsatisfactory performance. Finally, the authors believe that the customized model for the assessment of the ports infrastructure of Santa Catarina will bring gains for the scientific community and professional benefits to the decider, assisting him in making his decisions regarding the choice of the most appropriate port, fulfilling the demand for each type of customer that seeks his advice.

1 INTRODUCTION

The total number of exported goods in the world, the volume transported by means of ports has surpassed the mark of 10 billion tons in 2015. Of this total, 71% is dry cargo, 17% crude oil and 11% oil and gas (UNCTAD 2015). Whereas regarding the containerized cargo in the world, more than 701 million TEUS's (Unit equivalent to a container of 20 feet) were moved in 2016, a relevant growth in relation to the bustling in the year of 2010 (approximately 549 million TEUS's) (UNCTAD 2016).

Among those responsible for this significant growth of maritime transport are the globalization and the entry of more nations in the World Trade Organization (Feng et al. 2012). Because of this, the increase of the demand for international trade, together with the intensive use of containers, resulted in the requirement of ports with better infrastructure and diversification in logistics areas (Petering 2009). To remain competitive and meet the new demands, the ports had to become more efficient, by applying lean and agile processes for their operations (Marlow & Casaca 2003).

The ports are essentially service providers, in particular for loads, ships and overland transport (Cullinane et al. 2004), being regarded as a complex center of activities, of difficult coordination due to the heterogeneity of services and services providers involved, characterizing the port logistics as a strategic link in trade among countries (Batista 2012, Wanke 2013). The port terminals work, therefore, as the bonds of land and sea transport, serving as transfer stations of multimodal transport. These terminals generally consist of berth areas (where the ships dock), storage areas (where the loads are stored) and areas of terrestrial interface (where trucks and trains are met) (Buhrkal et al. 2011, Stahlbock & Voß 2008).

In this context, making use of the information generated by the Performance Evaluation (PE) of ports can be useful to support decisions (Nam et al. 2002, Wanke 2013). As the majority of operating systems and management, the PE in ports starts with the selection of metrics at strategic, tactical and operational levels. A performance indicator is used to evaluate (ordinally and/or cardinally) one or several attributes of an object (product, process or any other relevant factor) and should allow the comparison via benchmarking and/or via historical data (Bichou 2006).

The PE of port infrastructure is not only a powerful management tool for port operators, but also constitutes an important contribution to the planning of regional and national port operations (Cullinane et al. 2004).

In this context, the following research question emerges, which indicators should be considered in PE of ports infrastructure of containerized cargoes to support the decisions of choosing the most appropriate port, according to the customer's demands? Thus, the objective of this study is to build a constructivist multi-criteria model of PE to support decisions on port infrastructure for import and export of containerized loads of ports in the state of Santa Catarina (Imbituba, Itajaí, Portonave and Itapoá, situated in the South of Brazil). For such purpose, the methodology Multi-Criteria Decision Aiding-Constructivist (MCDA-C) will be used.

2 METHODOLOGY

2.1 *Procedures for the selection of the theoretical reference*

The instrument used for selection of the Bibliographic Portfolio (BP), which informed the construction of the Theoretical Reference, was the Knowledge Development Process-Constructivist (ProKnow-C) (Thiel et al. 2017). According to Dutra et al. (2015, p. 250) "the main objective of ProKnow-C is to build knowledge to a specific researcher from his interests and delimitations, according to the constructivist view".

In this study, only the first step of the ProKnow-C was operationalized, which consists of the BP selection. This step seeks to identify and select relevant scientific publications that are aligned to the theme proposed by the author (Dutra et al. 2015). It is subdivided into three sub steps: Selecting the gross articles database; filtration of Database Items; and representativity test (Dutra et al. 2015). After the completion of the steps of the ProKnow-C, a BP of 26 articles was selected.

2.2 *Tool for construction of the MCDA-C model: Procedure for the data collection and treatment*

The methodology MCDA-C emerges as an offshoot of the traditional MCDA, aiming to support decision-making in complex, conflicting and uncertain contexts (Ensslin et al. 2001). The main vocation of the MCDA-C is the process of knowledge developing for the decider about the context in which he is inserted and needs to intervene (Longaray et al. 2018, Ensslin et al., 2010).

In the structuring stage the context of the problem in question is structured and organized, allowing to identify, organize and measure, ordinally, the concerns that the decision-maker considers relevant for the assessment in the context (Ensslin et al. 2010). For this, semi-structured interviews were carried out. Through these interviews, the Primary Evaluation Elements (PEEs) were identified, dealing with the characteristics and properties of the context in which the decision-maker considers important in the assessment process. So that it is possible a better understanding of the PEE, these are transformed into concepts geared to action, identifying its two poles: present polo, that is the direction of preference that the decider aims to achieve this dimension; and the opposite psychological pole, which is the minimum acceptable result and that the decision makers want to minimize (Ensslin et al. 2013).

The concepts constructed for each PEE are grouped in accordance with the strategic concern which they represent, forming the cognitive maps (Longaray et al. 2018, Ensslin et al. 2010). From the maps, a hierarchical structure of value is built that will highlight the Tree of Fundamental Points of View (Value Tree), which due to being very comprehensive, are decomposed until it is possible to obtain the Elementary Points of View (EPV). The EPV represent properties (objectives) of the context that can be measured objectively and not ambiguously (Ensslin et al. 2010). Thus, it is possible to construct the descriptors (ordinal scales) to measure qualitatively the objectives (Ensslin et al. 2013).

For each ordinal scale, the decider identifies the Reference Levels: Good level (from where the performance is regarded as excellent) and the Neutral level (which below the same the performance is regarded as Compromising). Between these two levels the performance is considered competitive, i.e., within the decider's expectations (Ensslin et al. 2001, Ensslin et al. 2000). On the basis of these indicators (goals + descriptors), it is possible to trace the performance profile, i.e., the *status quo* of each port performance (Ensslin et al. 2013). Based on qualitative knowledge promoted by structuring phase, it is moved to the Recommendations stage, where improvement actions are developed and suggested so that the port improves its performance in those indicators whose performance is compromising. Thus, both phases promote information that assist in the management (Ensslin et al. 2010).

3 THEORETICAL FRAMEWORK: PORT INFRASTRUCTURE

The port efficiency can be decomposed in some dimensions such as: cost efficiency (lower costs of production); efficiency of capital investment (related to the optimization of investments); allocative efficiency (seeks to allocate the inputs optimally); and the technical efficiency (indicates the ability to produce a maximum level of output from a minimum level of input) (Bichou 2006).

The characteristic factors of each port are differentiated by the quality of the existing infrastructure facilities and services. Some authors mention the location of the port and its infrastructure as critical variables to explain port efficiency and performance (Caldeirinha & Felício 2014, Tongzon & Heng 2005). Thus, the ports cannot be evaluated based on a single performance indicator (Feng et al. 2012, Petering 2009). Various infrastructure characteristics influence the choice of port for its customers and, many times, constitute restrictions for certain types of maritime traffic, such as, for example, the quay depth by restricting the size of vessels able to berth to the port (Russo & Rindone 2011). The efficiency of handling and stacking of containers is another characteristic of the infrastructure, being fundamental to any terminal of containers, because the better the infrastructure is employed related to such equipment, the greater the productivity of the port will be (Stahlbock & Voß 2008). Typically, there are three types of systems for handling and stacking of containers: Chassis, straddle-carrier and transtainers, being the later the most popular in the main terminals, due to the need for high capacity storage of containers. For the transtainer system, there are three types of equipment handling, the quayside cranes, yard cranes, and yard trucks. The quay crane is usually the most expensive handling equipment and also the obstacle in loading and unloading of ships in container terminals, requiring continuous evaluation of its performance (Dragović et al. 2006, Lee et al. 2009, Petering 2009).

Because the longer duration of operation, due to the increase in the vessels size, high expenses of staff, lack of qualified labor and efficient use of the available area, the ports have been investing in automated terminals, with intelligent vehicles used in the movement of containers, increasing the speed and, thus, the productivity of their operations (Yang et al. 2004).

Other authors make a relationship of inputs and outputs to assess the port infrastructure. In this respect, the following are listed as inputs: the berth length, the terminal area, the quantity of quayside and yard cranes, the quantity of straddle carriers, the quantity of reach stackers, the quantity of terminal tractors, number of access gates, and the navigation channel depth. As outputs we have the container throughput, occupation and accessibility of the berth, and productivity of quayside cranes (Cullinane et al. 2004, Cullinane et al. 2006, Cullinane & Wang 2006, Lu et al. 2015, Nam & Ha 2001, Russo & Rindone 2011, Tongzon & Heng 2005). In addition to these, other inputs are cited for the outputs container throughput, movements per hour and number of arrivals of vessels for a period of time, such as: number of berths; storage capacity of containers; location of the terminal; number of shipping lines and the number of port workers (Chen et al. 2016, Song & Han 2004, Tongzon 2001, Wanke 2013, Wu & Liang 2009, Wu et al. 2010, Wu et al. 2010).

Some characteristics that relate time of operations with the port infrastructure were also identified in the literature. Among the measures related to the time of operations are: average time of the vessel in the port and on the berth, time of the vessel in the line of waiting to dock, service time, time for berthing and unberthing, dwell time, turnaround time, and customs time (Buhrkal et al. 2011, Dragović et al. 2006, Madeira et al. 2012, Nam & Ha 2001, Nam et al. 2002, Sánchez et al. 2003).

Sánchez et al. (2003) examined the determinants of the maritime transport costs. In this study, the authors cite the ability of movement of containers, the rate of loading and unloading per hour and the average number of containers handled per vessel as the main determinants of shipping costs. These indicators are directly related with the port infrastructure, i.e., the better the performance of port infrastructure, the lower will be their overall costs.

4 RESULTS AND DISCUSSIONS

4.1 *Structuring stage*

Handling loads via ports in the state of Santa Catarina have a significant role in the current Brazilian logistics scenario. According to the National Agency of Waterway Transports—Antaq (2017), Santa Catarina occupied the eighth national position in handling loads, corresponding to approximately 41.5 million tons, for a total of 1 billion tons of cargo handled in 2016 in Brazil. The Ports of Santa Catarina Imbituba, Itajaí, Portonave and Itapoá moved about 20% of the total of TEU's in Brazil in 2016. In the modality of cargo moved in container, two ports from Santa Catarina occupy the second and sixth placing among the main Brazilian ports (Antaq 2017). The operational efficiency influences the choice of the port for export and/or import of goods. There is, then, the need to choose which option is the most appropriate for the

Table 1. PEE's and their constructed concepts.

PEE	Present Pole	...	Opposite Psychological Pole
Number of Berths	Having the greatest possible number of berths	...	Harming the customers' demand service
Cargo Handling Equipment	Having the greatest possible number of equipment for handling loads on the terminal.	...	Reducing the ability of cargo moving.

port logistics of Santa Catarina state. Given this scenario, the decider for whom the model will be built is a consultant in the area of port logistics. The Table 1 shows some PEE's of the total of 105 obtained during the interviews, with their concepts, where the ellipses (...) should be read as "instead of" and correspond to the psychological opposite.

With the knowledge generated with the identification of PEEs, it was possible to group them into areas of concern: port efficiency, port hinterland and costs of ports logistics chain. In the study herein, only the large area of Ports Efficiency was approached. In this area, six Fundamental Point of View (FPV) were identified, namely: Physical Infrastructure, Equipment Infrastructure, Information Technology, Operational Efficiency, Services Provision Availability, and Efficiency Management. Figures 1 and 2 show the cognitive maps, respectively, for the FPV's approached in this study 'Infrastructure Equipment' and 'Physical Infrastructure'.

The grouping of concepts enabled the construction of the Value Tree with their respective EPVs. With the construction of the Value Tree, it was possible identified 20 aspects considered as necessary by the decider and sufficient to evaluate the infrastructure of Santa Catarina state ports. For these aspects, the descriptors were built and established their respective reference levels—Good and Neutral—enabling the ordinal measurement of the performance of ports (Imbituba, Itajaí, Portonave and Itapoá) and the identification of the current performance profile, illustrated, respectively, in Figures 3 and 4.

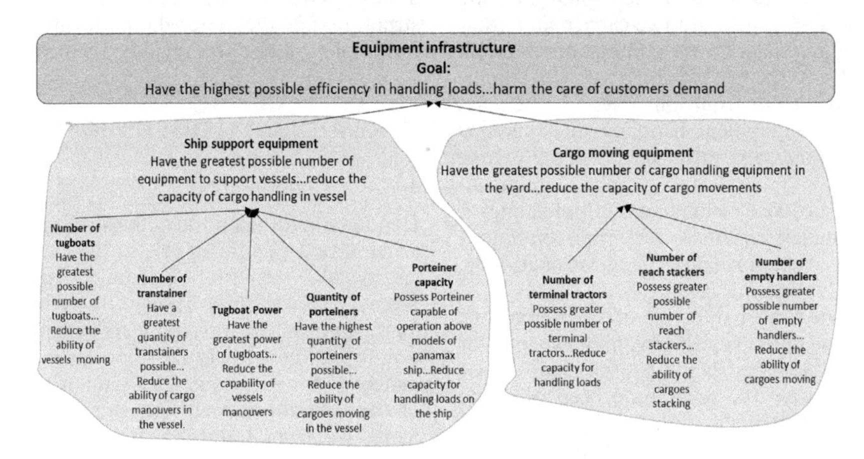

Figure 1. Clusters and sub clusters of cognitive map of FPV equipment infrastructure.

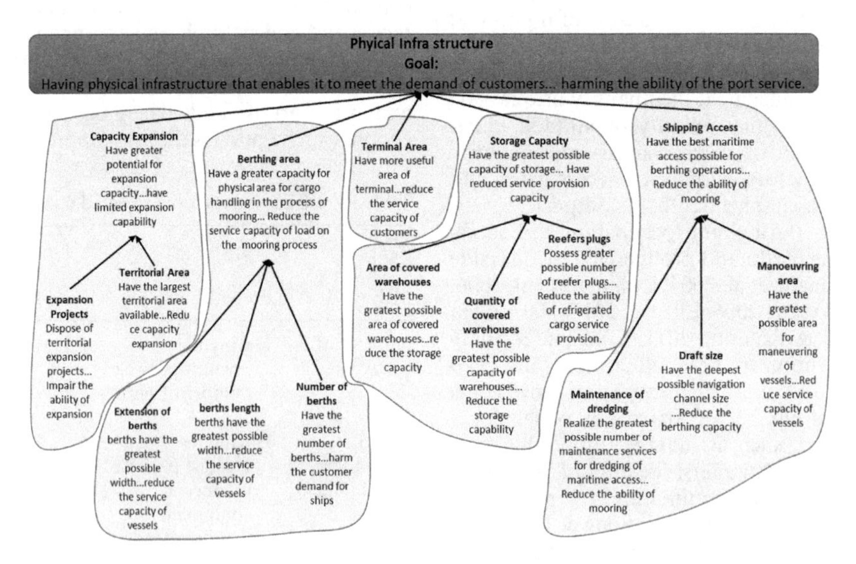

Figure 2. Clusters and sub clusters of cognitive map of FPV physical infrastructure.

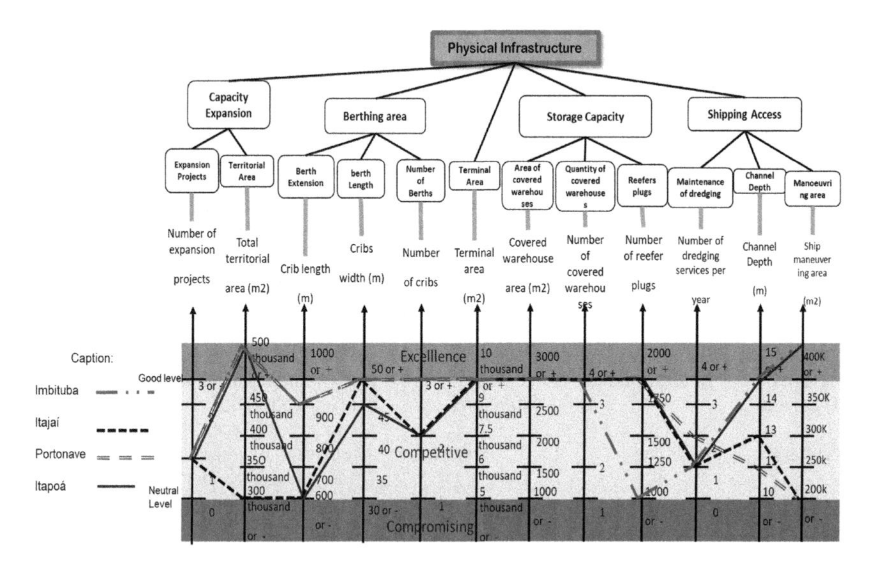

Figure 3. Descriptors and performance profile of the physical infrastructure for the ports of Imbituba, Itajai, Itapoá and Portonave.

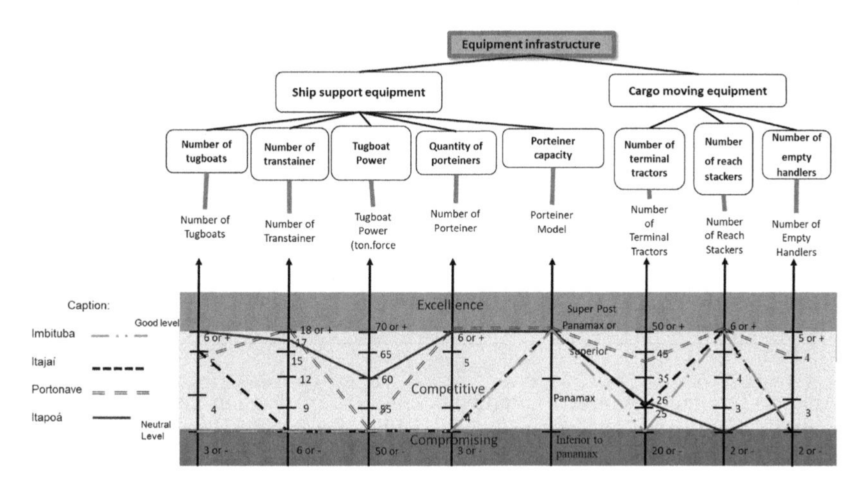

Figure 4. Descriptors and performance profile of the equipment infrastructure for the ports of Imbituba, Itajai, Itapoá and Portonave.

4.2 *Stage of recommendations*

Based on the performance profile of the ports of Santa Catarina, it is possible to identify that some aspects in which the performance of the port is in Compromising Level. In FPV Physical Infrastructure, the following indicators in Compromising Level are mentioned: Territorial Area (Itajaí), Length of Berth (Itajaí and Itapoá), Area of Maneuvering of Vessels (Itajaí and Portonave) and Number of Reefers Plugs (Imbituba). Whereas in FPV Equipment Infrastructure the following are mentioned as Compromising Level: Number of Tugs (Imbituba), Number of Transtainers (Imbituba and Itajaí), Tug Power (Imbituba, Itajaí, and Portonave), Number of Porteiners (Imbituba and Itajaí), Number of Terminal Tractors (Imbituba), Number of Reach Stackers (Itapoá) and Number of Empty Handlers (Imbituba and Itajaí).

On the basis of the indicators in which the ports presented compromising performance, it was possible to formulate proposals for actions for improvements to boost the performance of the ports on these indicators (Table 2).

Complementing the knowledge generated by *ad hoc* construction of the model, it is possible to relate the indicators constructed in this study with those found in the literature. It is noted that the

Table 2. Actions for improvement to performance improvement of the ports of Santa Catarina state in those indicators in Demanding level.

Indicators	Current status	Desired level: Competitive	Actions to reach the competitive level
Total Territorial Area (Itajaí)	300 thousand m²	350 thousand m² to 500 thousand m²	Design and implement extension of the total territorial area to increase the capacity of the general operation of the port.
Berth Length (Itajaí and Itapoá)	600 m	From 700 m to 1000 m	Design and execute the increase of the berth length of mooring to allow the receipt of larger vessels.
Number of Reefers plugs (Imbituba)	1000 plugs	1250 to 2000 plugs	Expand the number of reefers plugs to provide an increase in capacity of receiving/storage of loads that require refrigeration.
Area of Maneuvering of vessels (Itajaí and Portonave)	200 thousand m²	250 mil m² to 400 mil m²	Design and implement the expansion of the area of maneuvers of ships to provide a greater capacity for receiving larger ships.
Number of tugboats (Imbituba)	3 tugboats	4 to 6 tugboats	Acquisition of new tugs to increase the speed and ease of displacement of vessels for berthing and/or unberthing.
Number of Transtainers (Itajaí and Imbituba)	6 transtainers	9 to 18 transtainers	Acquisition of new transtainers to increase the speed and ease of loading and unloading of ships to the port.
Tugs power (Imbituba, Itajaí, and Portonave)	50 ton/power	55 ton/power to 70 ton/power	Acquisition of new tugs with a power greater than 50 ton/power to increase the capacity and speed of vessels for berthing and/or unberthing.
Number of porteiner (Imbituba and Itajaí)	3 Porteiners	4 to 6 Porteiners	Acquisition of new porteiners to increase the speed and ease of loading and unloading of ships to the port.
Number of terminal tractors (Imbituba)	20 Terminal Tractors	25 to 50 Terminal Tractors	Acquisition of new terminals tractors to increase the speed and ease of handling of cargo in the port area.
Number of Reach Stackers (Itapoá)	2 Reach Stackers	3 to 6 Reach Stackers	Acquisition of new Reach Stackers to increase the speed and ease of handling of cargo in the port area.
Number of Empty Handlers (Imbituba and Itajaí)	2 Empty Handlers	3 to 5 Empty Handlers	Acquisition of new Empty Handlers to increase the speed and ease of handling of cargo in the port area.

indicators: Area of Terminal, Number of Berths, Berth Width, Berth Length and Navigation Channel Depth, members of Physical Infrastructure are also present in the literature. The good performance of these indicators enables and demonstrates the capacity of the port to meet the demand for its services, as well as enable future expansion projects, and these are the decider's concerns.

With respect to the FPV Equipment infrastructure, the indicators: Number of Transtainers, Number of Terminal Tractors and Number of Reach Stackers were also found in the literature. These indicators are related to equipment used in loading and unloading, handling and stacking of containers in the port terminal. The good performance of these indicators allows and evidence a greater efficiency in the handling of containers, thereby increasing the overall efficiency of the ports.

On the other hand, some indicators constructed in this study were not identified, directly, in the consulted literature. In this specific context, these are indicators of great relevance, since besides being concerns of decision makers, at least one port is within the performance in Compromising Level. As is the case with the performance of the port(s) in the indicators: Number of Expansion Projects (Itajaí), Total Territorial Area (Itajaí), Area of Covered Warehouses (Imbituba), Number of Reefers Plugs (Imbituba), Maintenance of Dredging, Area for Maneuvering of Vessels (Itajaí and Portonave), Number of Tugs (Imbituba), Power of Tugs (Imbituba, Itajaí, and Portonave), Quantity of Porteiners (Imbituba and Itajaí), Capacity of Porteiners, and Number of Empty Handlers (Itapoá).

These performances in Compromising Level may be causing significant losses to these ports. This demonstrates the importance of the use of methodologies that take into account the specificities and needs in the context that is being assessed, because even being important aspects in this specific context, they are indicators without great relevance in the generic contexts consulted (literature). As quoted by Barros et al. (2012), the main shortages and concerns of Brazilian ports are related to technical improvements and infrastructure, which may differ considerably from the concerns of port managers in other parts of the world, because they are immersed in different contexts, with different concerns.

Such discrepancy between what is discussed in the literature as being the ideal about PE of ports and what has been built on this model of evaluation results from different models used in the PE process. Whereas, in the consulted literature, the PE models are generic, aiming at an optimal and pre-existing solution, often just adapted to different contexts, in this study a constructivist methodology was used, in which the decider has direct participation in the construction of the model, pointing out his main concerns and needs. Thus, a specific model was built that represents the *ad hoc* characteristics of the port context of Santa Catarina state. In addition to the supporting decision, the MCDA-C methodology used in this study also lists recommendations that highlight opportunities offered to Santa Catarina state ports' managers to improve their performance. Thus, the indicators constructed in this work are an important factor to be taken into consideration in the choice of port by the decider.

5 CONCLUSIONS

Due the need to choose the most appropriate port option to perform logistical operations, it is essential the use of methodologies that support the port managers in their decision making process. In contexts such as the Santa Catarina state, where there are several options of ports, each one presenting different levels of performance and service to their customers, it becomes relevant the choice of the most appropriate port for certain types of cargo, destinations and clients. Therefore, bearing in mind which indicators to use for PE is essential for a good decision-making. In this perspective, the objective of this study was to build a multi-criteria constructivist model to support the decisions of port logistics for importation and exportation of containerized cargo from the ports of Santa Catarina state. To ensure that this goal was reached, it was decided to build a model following the MCDA-C methodology, capable of representing the specificities of ports as to its physical and equipment infrastructure, as demanded by the decider.

During the preparation of the model, 20 indicators were constructed (objective + ordinal scale) that allowed the important aspects relating to port infrastructure to be measured and the profile of performance for each of the four ports of Santa Catarina state was evidenced. Whence it was possible to identify in which indicators the ports had excellent, competitive, or compromising performance. For the indicators whose performance is compromising actions for improvement were suggested.

It is highlighted for the decider's participation in the constructive process of this type of PE was of great importance, because it allowed the construction of knowledge and provided an effective gain for his professional activities, as well as for the scientific community as a means of expanding the studies on ports PE. In the authors' perception of this research, the construction of customized model for the evaluation of the operational efficiency of the ports from Santa Catarina state will assist the decision maker in making his decisions regarding the choice of which port is the most suitable for each type of client that looks for his advice.

This model of PE contains some limitations, among them: the construction of the model and identification of the performance of the ports of Santa Catarina state regarding only to FPVs 'physical infrastructure and equipment'; the construction of the model for the ports that operate only containerized loads. Therefore, some opportunities for future research were identified. The development of the Evaluation Stage, both for this model as for models containing the remaining FPVs identified, would be of great importance for the evaluation of the overall area of concern Efficiency Port, in addition to the recommendations of improvements could be based both on qualitative and quantitative data. Another possibility for future research would be the construction of the multi-criteria constructivist model for the other FPVs: Operational Efficiency, Information Technology, Service Provision Availability, and Managerial Efficiency.

REFERENCES

Antaq, Anuários estatísticos e sistema de informações gerenciais 2017. Disponível em: http://web.antaq.gov.br/Anuario/. Acessado em 1 Dez. 2017.

Barros, C.P., Felício, J.A., and Fernandes, R.L. 2012. Productivity analysis of Brazilian seaports. *Maritime Policy & Management* 39(5): 503–523.

Batista, L. 2012. Translating trade and transport facilitation into strategic operations performance objectives. *Supply Chain Management: An International Journal 17*(2): 124–137.

Bichou, K. 2006. Review of port performance approaches and a supply chain framework to port performance benchmarking. *Research in Transportation Economics 17*: 567–598.

Buhrkal, K., Zuglian, S., Ropke, S., Larsen, J., & Lusby, R. 2011. Models for the discrete berth allocation problem: A computational comparison. *Transportation Research Part E: Logistics and Transportation Review 47*(4): 461–473.

Caldeirinha, V.R., & Felício, J.A. 2014. The relationship between 'position-port', 'hard-port' and 'soft-port'characteristics and port performance: conceptual models. *Maritime Policy & Management 41*(6): 528–559.

Chen, L., Zhang, D., Ma, X., Wang, L., Li, S., Wu, Z., & Pan, G. 2016. Container port performance

measurement and comparison leveraging ship GPS traces and maritime open data. *IEEE Transactions on Intelligent Transportation Systems 17*(5): 1227–1242.
Cullinane, K., Song, D.-W., Ji, P., & Wang, T.-F. 2004. An application of DEA windows analysis to container port production efficiency. *Review of network Economics 3*(2): 184–206.
Cullinane, K., Wang, T.-F., Song, D.-W., & Ji, P. 2006. The technical efficiency of container ports: comparing data envelopment analysis and stochastic frontier analysis. *Transportation Research Part A: Policy and Practice 40*(4): 354–374.
Cullinane, K. & Wang, T.-F. 2006. The efficiency of European container ports: A cross-sectional data envelopment analysis. *International Journal of Logistics: Research and Applications 9*(1): 19–31.
Dragović, B., Park, N.K., & Radmilović, Z. 2006. Ship-berth link performance evaluation: simulation and analytical approaches. *Maritime Policy & Management 33*(3): 281–299.
Dutra, A., Ripoll-Feliu, V.M., Fillol, A.G., Ensslin, S.R., & Ensslin, L. 2015. The construction of knowledge from the scientific literature about the theme seaport performance evaluation. *International Journal of Productivity and Performance Management 64*(2): 243–269.
Ensslin, S.R., Ensslin, L., Back, F., & Lacerda, R.T.O 2013. Improved decision aiding in human resource management: a case using constructivist multi-criteria decision aiding. *International Journal of Productivity and Performance Management* 62(7): 735–757.
Ensslin, L., Giffhorn, E., Ensslin, S.R., Petri, S.M., & Vianna, W.B. 2010. Avaliação do desempenho de empresas terceirizadas com o uso da metodologia multicritério de apoio à decisão-construtivista. *Pesquisa Operacional 30*(1): 125–152.
Ensslin, L., Neto, G.M., & Noronha, S.M. 2001. *Apoio à decisão: metodologias para estruturação de problemas e avaliação multicritério de alternativas*: Insular.
Ensslin, L., Dutra, A. & Ensslin, S.R. 2000, MCDA: a constructivist approach to the management of human resources at a governmental agency. *International Transactions in Operational Research* 7: 79–100.
Feng, M., Mangan, J., & Lalwani, C. 2012. Comparing port performance: Western European versus Eastern Asian ports. *International Journal of Physical Distribution & Logistics Management 42*(5): 490–512.
Lee, D.-H., Cao, J.X., Shi, Q., & Chen, J.H. 2009. A heuristic algorithm for yard truck scheduling and storage allocation problems. *Transportation Research Part E: Logistics and Transportation Review 45*(5): 810–820.
Longaray, A., Ensslin, L., Ensslin, S., Alves, G., Dutra, A. & Munhoz, P. 2018. Using MCDA to evaluate the performance of the logistics process in public hospitals: the case of a Brazilian teaching hospital. *Intl. Trans. in Op. Res*. 25(1): 133–156.
Lu, B., Park, N.K., & Huo, Y. 2015. The evaluation of operational efficiency of the world's leading container seaports. *Journal of Coastal Research 73*(sp1): 248–254.
Madeira, A.G., Cardoso, M.M., Belderrain, M.C.N., Correia, A.R., & Schwanz, S.H. 2012. Multicriteria and multivariate analysis for port performance evaluation. *International Journal of Production Economics 140*(1): 450–456.
Marlow, P.B., & Casaca, A.C.P. 2003. Measuring lean ports performance. *International journal of transport management 1*(4): 189–202.
Nam, K.-C., & Ha, W.-I. 2001. Evaluation of handling systems for container terminals. *Journal of Waterway, Port, Coastal, and Ocean Engineering 127*(3): 171–175.
Nam, K.-C., Kwak, K.-S., & Yu, M.-S. 2002. Simulation study of container terminal performance. *Journal of Waterway, Port, Coastal, and Ocean Engineering 128*(3): 126–132.
Petering, M.E.H. 2009. Effect of block width and storage yard layout on marine container terminal performance. *Transportation Research Part E: Logistics and Transportation Review 45*(4): 591–610.
Russo, F., & Rindone, C. 2011. Container maritime transport on an international scale: Data envelopment analysis for transhipment port. *WIT Transactions on Ecology and the Environment 150*: 831–843.
Sánchez, R.J., Hoffmann, J., Micco, A., Pizzolitto, G.V., Sgut, M., & Wilmsmeier, G. 2003. Port efficiency and international trade: port efficiency as a determinant of maritime transport costs. *Maritime Economics & Logistics 5*(2): 199–218.
Song, D.-W., & Han, C.-H. 2004. An econometric approach to performance determinants of Asian container terminals. *International Journal of Transport Economics/Rivista internazionale di economia dei trasporti*. 39–53.
Stahlbock, R., & Voß, S. 2008. Operations research at container terminals: a literature update. *OR spectrum 30*(1): 1–52.
Thiel, G.G., Ensslin, S.R., & Ensslin, L. 2017. Street Lighting Management and Performance Evaluation: Opportunities and Challenges. *Lex Localis 15*(2): 303.
Tongzon, J. 2001. Efficiency measurement of selected Australian and other international ports using data envelopment analysis. *Transportation Research Part A: Policy and Practice 35*(2): 107–122.
Tongzon, J., & Heng, W. 2005. Port privatization, efficiency and competitiveness: Some empirical evidence from container ports (terminals). *Transportation Research Part A: Policy and Practice 39*(5): 405–424.
UNCTAD, Data Center 2015. Available in: http://unctadstat.unctad.org/wds/TableViewer/tableView.aspx?ReportId = 32363. Acessed in Sep 1. 2017.
UNCTAD, Data Center 2016. Available in: http://unctadstat.unctad.org/wds/TableViewer/tableView.aspx?ReportId = 13321. Acessed in Sep 1. 2017.
Wanke, P.F. 2013. Physical infrastructure and shipment consolidation efficiency drivers in Brazilian ports: A two-stage network-DEA approach. *Transport Policy 29:* 145–153.
Wu, J., & Liang, L. 2009. Performances and benchmarks of container ports using data envelopment analysis. *International Journal of Shipping and Transport Logistics 1*(3): 295–310.
Wu, J., Liang, L., & Song, M. 2010. Performance based clustering for benchmarking of container ports: an application of DEA and cluster analysis technique. *International Journal of Computational Intelligence Systems 3*(6) 709–722.
Wu, J., Yan, H., & Liu, J. 2010. DEA models for identifying sensitive performance measures in container port evaluation. *Maritime Economics & Logistics 12*(3): 215–236.
Yang, C., Choi, Y., & Ha, T. 2004. Simulation-based performance evaluation of transport vehicles at automated container terminals. *OR spectrum 26*(2): 149–170.

Port performance II

Progress in Maritime Technology and Engineering – Guedes Soares & Santos (Eds)
 ISBN 978-1-138-58539-3

Improving capacity of port shunting yard

A. Rusca, F. Rusca, E. Rosca, V. Dragu & M. Rosca
Faculty of Transports, University POLITEHNICA of Bucharest, Bucharest, Romania

ABSTRACT: The maritime ports connect transport land networks with maritime transport. When traffic volume is large it is necessary to develop a port shunting yard inside of maritime port area. These have some specific problems, like regulated and limited access, large volume in a short time for cargo flow that goes from territory to maritime ship and opposite, a rigorous timetable for the departure/arrival moments for maritime ships and for freight train, as well as limited land availability to develop new infrastructures to deserve flows between maritime and land networks. All these limitations conduce to a necessity for improving the operational capacity solutions which can be implemented in maritime ports with minimal material and human resources. In the paper is analyzed topological structure of port shunting yard and evaluated transit capacity through its various compartments. A discrete simulation model is developed with ARENA computer simulation software for wagons shunting process and various technologies for shunting process are tested to identify optimal solutions for increasing the transit capacity of the port shunting yard. The results and conclusion obtained through computer simulation allow a proper evaluation of transit capacity and help port administration to take the best solution in improving port services.

1 INTRODUCTION

The link between the rail and maritime transport is made within maritime ports. The specific features related to maritime transport (high shipping capacity, short stationary times in the port area, etc.) lead to the appearance of a large flow of goods in a limited time to/from shipping berths. For the railway station located in the port area, there is a need for an increased processing capacity for the trains coming from land territory, respectively for the wagons coming from maritime berths. The use of the humps in the railway station for shunting the wagons in relation to their destination leads to the increase of processing and transit capacities of the station. This is usually limited by the processing capacity of the shunting hump that allows a single shunting process in the same time. Unfortunately, the usually existing space limitations in the maritime ports area do not allow an optimal spatial development for maritime railway stations (needed to build several shunting humps). An alternative solution accordingly with the existing space, to grow the transit and the processing capacities is the use of new technological processes inside the railway station. Thus, wagons simple shunting process could be replaced by simultaneous shunting process that allows to be processed two train sets at the same time. The disadvantage of using this technology is the increase of the number of manoeuvres in the station, respectively the number of shunting operations. However, the station's transit capacity increases even under these conditions.

To verify the utility of these measures, it is necessary to test them by simulation. For this purpose, it is necessary to develop a logical model that present the activity of the maritime railway stations. Implemented in a simulation software, this model will allow to obtain results on qualitative parameters regarding the transit of freight wagons through the railway stations located inside maritime ports.

2 LITERATURE REVIEW

The scientific papers about shunting yards cover all levels, from global rail networks to marshaling operations inside the rail station. So, at the global level, the problems of train routing in shunting yards, number of shunting stages, single or in succeeding yards are resolved using dedicated algorithms (Allbrecht et al., 2015a; Jaehn et al., 2015; Jaehn & Michaelis, 2016). It is not part of our research, but must be mentioned also the problem of shunting process in case of passenger trains. The differences are not so great and the solutions identified can be useful in solving the incertitude of the shunting process from railway stations located in seaports (Haahr et al., 2017).

For operations inside the shunting stations an interesting review is made by Boysen et al. (2012). The applicability of theoretical aspects with respect to different yard characteristics are presented and are evaluated through the influence of practical requirements on the structure of decision problems. The authors identify three yard types:

hump yards, flat yards and gravity yards. For every type, the standard technology for shunting is presented like a mathematical decision problem. Strategic, tactical and operational tasks are developed concentrated on the hump yard. Completing the previous studies, Aldbrech et al. (2015b) presents a model for optimization of a shunting engine for the case of shunting without hump yard. Another survey, (Caballini et al., 2012) has the task to model and optimize the rail cycle in seaport terminals. The model proposed in this paper represents the transit of containers from maritime terminals to railway station, before the exit by train from seaport area to hinterland.

The optimization of operation process inside of seaports and dedicated railway stations is important, but the main objective for the administrator or owner is to dimension the capacity correctly, according to specifically aspects of maritime ports area. For sizing the capacity of shunting railway station, in terms of maximum number of wagons or trains that can be transited over a certain time horizon discrete-time model can be found in literature (Fioribello et al., 2016). Inside of container terminal the shunting process of containers are optimized according to capacity of handling equipment, number of slots inside of storage area, and enter/exit flows of container in/from seaports terminals (Vis & Koster, 2003; Kim & Park, 2004)

The discrete time model is developed considering the random nature of the distribution of train arrivals in the station (Marinov & Viegas, 2009). The activity quality factor inside of railway stations are evaluated using a simulation model. For seaports terminals, discrete simulation models are developed to evaluate the activity parameters under the influence of special conditions: the reliability of handling equipment, transit of dangerous goods, capacity of storage area, etc. (Guedes Soares & Teixeira 2001;da Silva & Botter, 2009; Angeloudis & Bell, 2011; Netto et al., 2015; Rusca et al., 2015; Rong et al., 2015; Rusca et al., 2016a; Dinu et al., 2017).

3 PORT SHUNTING YARD TECHNOLOGY

To reach the destination terminals from seaport, in port shunting yard, the inbound trains are decoupled and disassembled into individual freight cars. Then train is humped over the hump, whose inclination accelerates the freight cars by gravity. For the countries with large variances of temperature between seasons the number of lines over humps are two, one for the summer and one for the winter. Via a system of tracks and switching points the freight wagons are humped to given shunting tracks, where are waiting to be moved to area of freight maritime terminals.

The freight wagons from this area are also humped over the hump and shunting in correlation with their rail station of destination. Finally, a freight train is pulled from the shunting tracks to the departure tracks, where the cars are coupled, a break inspection for each car occurs, and where a locomotive is attached (Jaehn et al., 2014).

The typical shunting yard has the structure from Figure 1. The inbound trains are received in Receiving Tracks (group of lines dedicated for train arriving in rail station). Wagons are separated over the hump in Classification Tracks. From these, the wagons with destination in freight maritime

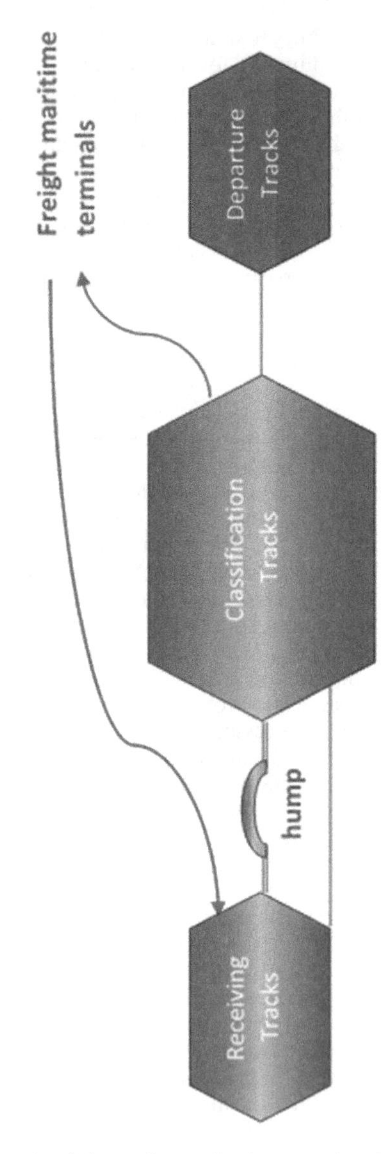

Figure 1. Scheme for a simple port shunting yard (red is output flow and blue is input flow to maritime terminals).

terminals are moved to gates of these terminals. After industrial process (wagons load/unload or both) is finish the freight wagons are moved to Receiving Tracks. After, over the hump are separated taking in consideration the destination. In Departure Tracks the outbound trains are created from wagons moved from Classification Tracks.

In some cases, in maritime ports are used a double port shunting yard, formatted from two symmetrical shunting yards (Fig. 2). First is for the input flow from hinterland to freight maritime terminals and second is for flow from maritime port to land destinations. Unfortunately, some problematic aspects are particular to port shunting yards, like limited access, large volume for cargo flow from territory to maritime ship and opposite, as well as limited land availability to develop new infrastructure to deserve flows between maritime and land networks. The last problems have a large impact over shunting yards development so as, the double port shunting yard is rarely found in maritime ports.

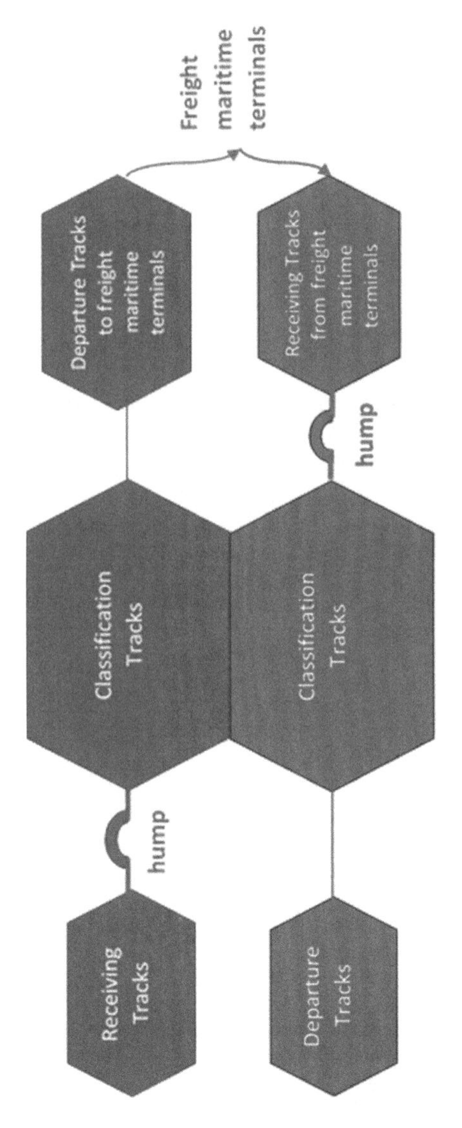

Figure 2. Scheme for a double port shunting yard (red is output flow and blue is input flow to maritime terminals).

In case of the simple shunting yard the transit capacity taking in consideration the shunting process is:

$$n_{wagons} = m\frac{1440-\sum T_{perm}^{dt}}{t_d + t_{pres} + t_{tint\,r}} \tag{1}$$

where

- 1440 is the number of minutes from a day;
- $\sum T_{perm}^{dt}$ is the time interval when hump is occupied with permanent operations (like shunting process for wagons from repair shop or from station storage area, to reload the engine, etc.);
- t_d is the time interval for a shunting process;
- t_{pres} is the time interval for a pressing process of wagons in Classification Tracks;
- $t_{int\,r}$ is the duration of interruptions of shunting process due to incompatible movements;
- m is the mean number of wagons from a train sets.

In case of simultaneous shunting the capacity is:

$$n_{wagons}^{simultaneous} = \left(\frac{T''}{t''_c} + \frac{T_r}{t_r} + \frac{1440-\sum T_{perm}^{dt} - T'' - T_r}{t'_c}\right) m \tag{2}$$

where:

- T'' is the occupied time of hump yard with simultaneous shunting of two train sets
- T_r is the occupied time of hump yard with repeated shunting process for railways cars from selection line;
- t''_c is the technologic interval for shunting installation in the case of simultaneous shunting;
- t_r is the time for a repeated shunting process.
- t'_c. is the technologic interval for shunting installation in the case of normal shunting:

$$t'_c = t_d + t_{pres} + t_{tint\,r} \tag{3}$$

Taking in consideration p ratio of trains with wagons separated by simultaneous shunting from total number of wagons, the value of T'' is calculated with relation:

$$T'' = \frac{p n_{wagons}^{simultaneous}}{2} \cdot \frac{1}{m} t''_c. \tag{4}$$

Also, if α is ratio of wagons repeated shunted from number of wagons simultaneous shunted, the occupied time of hump yard with repeated shunting process for railways cars from selection line is:

$$T_r = \frac{\alpha p n_{wagons}^{simultaneous}}{m_r} t_r \tag{5}$$

where m_r is the number of wagons from train set which are moved back to Receiving Tracks to be repeated shunted.

If value of α is too high, the value of transit capacity when using simultaneous shunting process decrease under the value obtained using simple shunting process. If we consider the inequality:

$$n_{wagons} \le n_{wagons}^{simultaneous} \tag{6}$$

and the expressions (1) to (5) result:

$$\frac{1440 - \sum T_{perm}^{dt}}{t_c'} m \le \left(\frac{p n_{wagons}^{simultaneous}}{2m} + \frac{\alpha p n_{wagons}^{simultaneous}}{m_r} + \frac{(1440 - \sum T_{perm}^{dt}) - T'' - T_r}{t_c'} \right) m. \tag{7}$$

If we introduce two new parameters γ and k with expression (Rusca et al. 2016b):

$$\gamma = \frac{t_c''}{t_c'} \tag{8}$$

and

$$k = \frac{m_r}{m} \tag{9}$$

result:

$$\frac{1440 - \sum T_{perm}^{dt}}{t_c'} m \le \frac{\left(1440 - \sum T_{perm}^{dt}\right) mk}{t_c' \left[k\left(1 - p + \frac{p\gamma}{2}\right) + \alpha p(\gamma - 1) \right]}. \tag{10}$$

In this case the limit value of $\propto$ for which the process of repeated shunting eliminates the increase obtained by simultaneous shunting of two train sets is:

$$\alpha_{max} = 0.5 \frac{(2-\gamma)k}{\gamma - 1}. \tag{11}$$

The best results for the transit capacity taking in consideration the simultaneous shunting process are obtained when $t_c' = t_c'' = t_r$:

$$n_{wagons}^{simultaneous} = \frac{\left(1440 - \sum T_{perm}^{dt}\right) m}{(t_d + t_{pres} + t_{tint\,r})\left(1 - \frac{p}{2}\right)} \tag{12}$$

However, simultaneous shunting process suppose the existence in the same time, in Receiving Tracks lines group of two train sets which can be simultaneous shunted. In case of trains arriving time according to a Poisson type, the probability that in the Receiving Tracks to be found a train set is:

$$p_1 = \left(1 - e^{-\frac{N_1 t_1}{24}}\right) \tag{13a}$$

where N_1 is the total train sets arriving from hinterland which are stationing in Receiving Tracks lines group for a time interval t_1, necessary in technological process.

For the case of the opposite direction, from maritime port to hinterland with also of Poisson type arrivals, the probability is:

$$p_2 = \left(1 - e^{-\frac{N_2 t_2}{24}}\right) \tag{13b}$$

And the probability of finding both simultaneously train sets in Receiving Tracks, to and from maritime port is:

$$p_{1,2} = \left(1 - e^{-\frac{N_1 t_1}{24}}\right)\left(1 - e^{-\frac{N_2 t_2}{24}}\right) \tag{14}$$

In case of arrivals type which don't correspond to Poisson it is difficult to find a mathematical model. The solution can be find using a discrete simulation model, adapted to case of a maritime port shunting yard.

4 DISCRETE SIMULATION MODEL

The main technologic process in a shunting yard is made in Receiving Tracks for the inbound train, in Classification Tracks for shunted wagons and in Departure Tracks for outbound trains. In the case of maritime ports, it is necessary to have the specifically technological process on the set of freight wagons from/to maritime terminals. In the case of a simulation model used to evaluate the transit through a maritime shunting yard it is necessary to take in consideration the duration of these technological processes. Also, the functional relations between main departments of the rail stations must be part of the logical model used in the simulation.

In our case, using Arena 12 simulation software we develop a simulation model taking in consideration inbound trains flow from hinterland and entering flow of wagons from maritime terminals. The model is built for the structure of typical shunting yard. Two situations for the model are envisaged, one with simple shunting process (Fig. 3) and one with simultaneous process (Fig. 4).

In case of simultaneous shunting process, the destination points of freight wagons from port or from hinterland are split in two sets and are allocated to shunting humps. If a wagon for second set is shunted on first hump is routed on a collector line and then re-shunted on second hump. To improve the process, the destination points from the port are grouped together. The same principle is used also for the destination points from hinterland.

The main input data for simulation are:

- Arrival time interval: Exponential ($\lambda = 0.4$ hours, 0.5 hours, 0.6 hours), Normal ($\lambda = 0.4$ hours and $\sigma = 0.1$ hours, $\lambda = 0.5$ hours and $\sigma = 0.1$ hours)
- Number of lines in Receiving Tracks (6 lines, 3 lines dedicated to inbound line and 3 lines dedicated to wagons sets from maritime terminals);

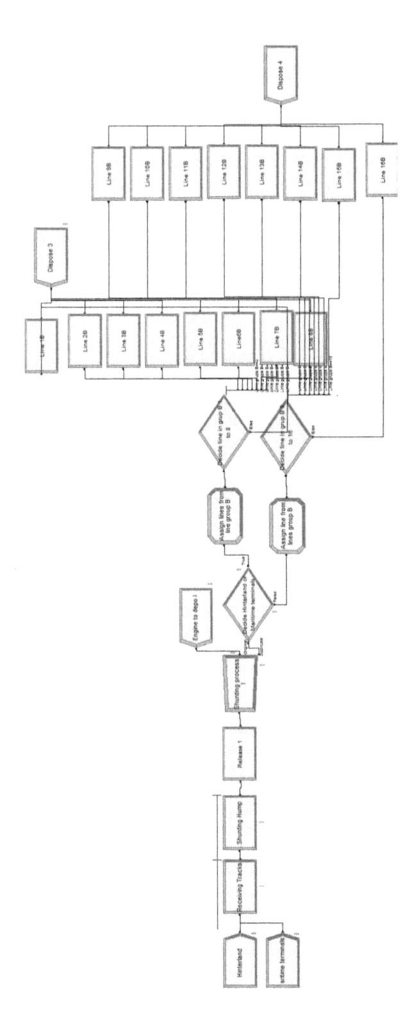

Figure 3. Simulation model for simple shunting process.

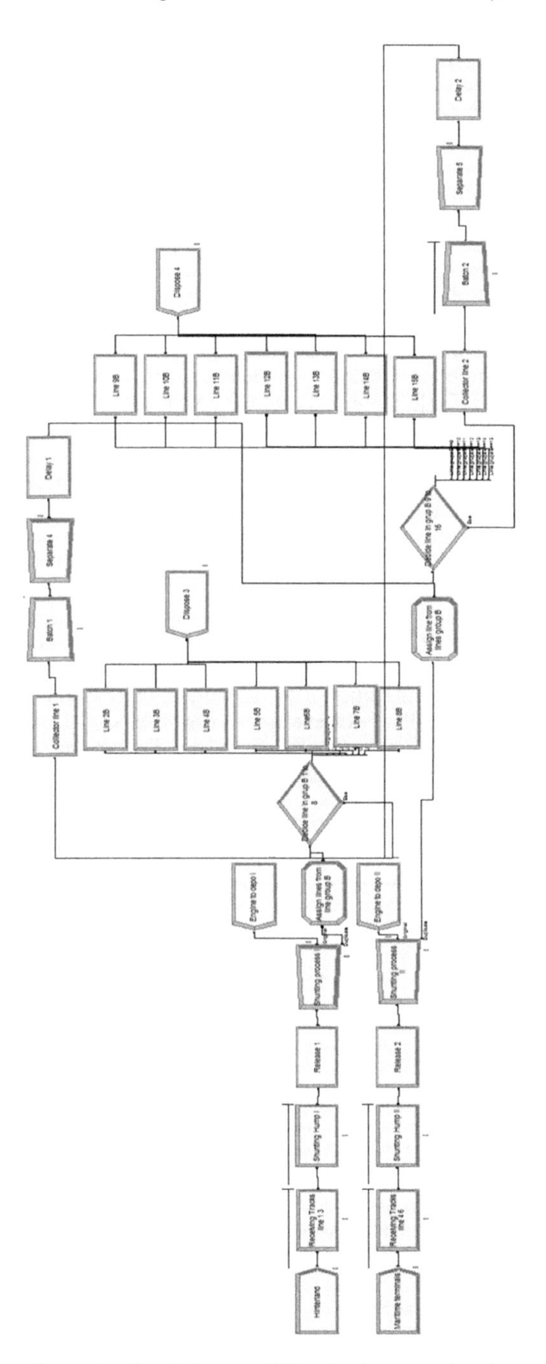

Figure 4. Simulation model for simultaneous shunting process.

- Number of lines in Classification Tracks (16 lines);
- Number of humps (2 hump lines);
- Number of wagons from a train sets $m = m_r = 60$ wagons; Duration of shunting process (25 minutes);
- Duration of transfer process from collector line to shunting hump (10 minutes).

Duration of technological process in Receiving Tracks is 50 minutes for inbound trains and 20 minutes for wagons groups from maritime terminals (the values are common for the most shunting rail station). The number of lines in Classification Tracks is 16 lines. Normally the number in this line group it is obtained by multiplying the 8th (8, 16, 24 or 32) for constructive reasons. The allocation of destination line it is made using a uniform distribution for every wagon. The Departure Tracks it is not limitative for transit capacity and it is not included in simulation model.

In Table 1 are presented the simulation results when is used the simple shunting process. The value in this case for hump utilization ratio is 1. This is hard to believe, because, in reality, it is not possible to have a full utilization for a resource like shunting hump. Also, the queue before Receiving Tracks it is very big with a waiting time over 20 hours (relatively, the value is common for Romanian port shunting yards). In comparison, when it is used the simultaneous shunting process, the results are improved (Table 2). The transit capacity through shunting rail station increase with 40-96% depending on attributes of entering flows (Table 3).

In the case when the entering flow value is not too big (Exponential with $\lambda = 1$ hour), from simulation we obtain for the situation when it is used simple shunting process a higher transit capacity with 4-6% than for the situation it is used simultaneous shunting process (Table 4). But the quality attributes like waiting time for a free line, or waiting time before shunting are significantly better in the second situation.

Table 1. The results for simulation-simple shunting process.

	E ($\lambda = 0.4$)	E ($\lambda = 0.5$)	E ($\lambda = 0.6$)	N ($\lambda = 0.4$, $\sigma = 0.1$)	N ($\lambda = 0.5$, $\sigma = 0.1$)
Nit	119	119	113	119	118
Nm	119	118	122	119	119
Wit	7125	7119	6792	7129	7061
Wm	7146	7058	7335	7151	7122
Qh	1	1	1	1	1
Twh	1.25	1.25	1.25	1.25	3
Tw	37.36	28.55	20.13	37.01	113

*Nit is mean of number of inbound train from all ten replicants, Nm is mean of number of wagons group from maritime terminals, Wit is mean of number of wagons from inbound trains, Wm is mean of number of wagons from wagons group from maritime terminals, Qh is hump utilization ratio, Twh is waiting time before shunting process, Tw is waiting time for a free line in Receiving Tracks.

Table 2. The result for simulation-simultaneous shunting process.

	E ($\lambda = 0.4$)	E ($\lambda = 0.5$)	E ($\lambda = 0.6$)	N ($\lambda = 0.4$, $\sigma = 0.1$)	N ($\lambda = 0.5$, $\sigma = 0.1$)
Nit	232	187	159	234	193
Nm	229	192	157	234	196
Wit	13925	11249	9552	14057	11602
Wm	13713	11528	9446	14063	11736
Qh1	0.974	0.799	0.679	0.999	0.825
Qh2	0.989	0.819	0.629	0.999	0.835
Twh1	0.005	0.04	0.06	0	0.01
Twh2	0.48	0.33	0.25	0.49	0.01
Twit	4.78	0.74	0.19	2.79	0.006
Twm	5.68	0.41	0.38	2.61	0

*h1 is first shunting hump and h2 is second shunting hump, Twit is waiting time of inbound train for a free line in Receiving Tracks, Twm is waiting time of wagons group from maritime terminals for a free line in Receiving Tracks.

Table 3. Variation of transit capacity simple shunting process versus simultaneous shunting process (%).

	E ($\lambda = 0.4$)	E ($\lambda = 0.5$)	E ($\lambda = 0.6$)	N ($\lambda = 0.4$, $\sigma = 0.1$)	N ($\lambda = 0.5$, $\sigma = 0.1$)
Nit	95	57	41	97	64
Nm	92	63	29	97	65

Table 4. The results for simulation-simultaneous shunting process.

Simultaneous shunting process	E($\lambda = 1$)	E($\lambda = 1$)	Simple shunting process
Nit	95	101	Nit
Nm	95	99	Nm
Wit	5715	6070	Wit
Wm	5697	5929	Wm
Qh1	0.40	0.853	Qh
Qh2	0.40		Twh
Twh1	0.06	0.70	Tw
Twh2	0.1		Nit
Twit	0.06	0.77	Nm
Twm	0.01		

5 CONCLUSIONS

The shunting yards located in the port area have a great importance in connecting terrestrial with maritime transport networks.

The specific particularities of the maritime ports induce the need to develop shunting yards so that these ones could manage variable transit flows, with volume's high fluctuations and indeterminate transit moments.

Thus, this requires the development of stations with high transit capacity. In the paper is presented the scheme of such a double shunting yard. (Figure 2). Unfortunately, in the case of maritime ports, the development of a double shunting yard is limited by the lack of availability in the port area of the land on which should be developed such a shunting yard.

The solution in this case is one of a technological nature. By changing the technological process used in a simple shunting yard (currently applied in most shunting yards) with a simultaneous shunting process, an increase of station's transit capacity can be obtained.

The decision makers of a port shunting yard should analyze the utility of implementing such a solution.

In the paper is presented a mathematical model by which can be determined the raise of capacity which can be obtained.

The use of simultaneous shunting process leads to additional repeated shunting operations, which under certain conditions may lead to a value for transit capacity under the value, obtained using simple shunting process.

Special importance's have also the characteristics of the flow of inbound trains and also the characteristics of the group of wagons coming from sea terminals.

The importance of this influence is analyzed using a discrete type simulation model. Five scenarios for the input flow are proposed and the values for transit process through the port shunting yard are compared.

The use of simultaneous shunting process leads in all five scenarios to 40–96% higher values for transit process through the station.

A sixth scenario is analyzed for the situation of lower input flows.

In this scenario, the transit capacity is higher by 4-6% when is used the simple shunting process, but that implies much higher waiting times.

In conclusion, getting a raise of transit capacity through maritime port shunting yards can be achieved both by constructive improvement (long term measure and difficult to be implemented) as well as by technological improvement. In the case of technological improvement, the cost of implementation is low and an important success is obtained in the case of existence of an input flow which is approaching to the maximum capacity of the station. Thus, it is up to decision makers to implement this solution.

ACKNOWLEDGEMENTS

This work has been funded by University Politehnica of Bucharest, through the "Excellence Research Grants" Program, UPB – GEX 2017. Identifier: UPB- GEX2017, Ctr. No. 66/25.09.2017 (GEX2017)".

REFERENCES

Adlbrecht, J.A. & Hüttler, B. & Ilo, N. & Gronalt, M. 2015. Train routing in shunting yards using Answer Set Programming, *Expert Systems with Applications* 42(21): 7292-7302.

Adlbrecht, J.A. & Hüttler, B. & Zazgornik, J. & Gronalt, M. 2015. The train marshalling by a single shunting engine problem, *Transportation Research Part C: Emerging Technologies* 58(A): 56-72.

Angeloudis, P. & Bell M.G.H. 2011. A review of container terminal simulation models, *Maritime Policy & Management* 38(5).

Boysen, N. & Fliedner, M. & Jaehn, F. & Pesch, E. 2012. Shunting yard operations: Theoretical aspects and applications, *European Journal of Operational Research* 220(1): 1-14.

Caballini, C. & Pasquale, C. & Sacone, S. & Siri, S. 2012. A discrete-time model for optimizing the rail port cycle, *IFAC Proceedings Volumes*, 45(24): 83-88.

Dinu, O. & Roşca, E. & Popa, M. & Roşca, M. A. & Rusca, A. 2017. Assessing materials handling and storage capacities in port terminals. In *IOP Conference Series: Materials Science and Engineering* 227(1): 012039. IOP Publishing.

Fioribello, S. & Caballini, C. & Sacone, S. & Siri, S. 2016. A planning approach for sizing the capacity of a port rail system: scenario analysis applied to La Spezia port network, *IFAC-PapersOnLine* 49(3): 371-376.

Haahr, J.T. & Lusby, R.M. & Wagenaar, J.C. 2017. Optimization methods for the Train Unit Shunting Problem, *European Journal of Operational Research*, 262(3): 981-995.

Jaehn, F. & Michaelis, S. 2016. Shunting of trains in succeeding yards, *Computers & Industrial Engineering*, 102: 1-9.

Jaehn, F. & Rieder, J. & Wiehl, A. 2015. Single-stage shunting minimizing weighted departure times, *Omega*, 52: 133-141.

Kim, K.H. & Park, Y.-M. 2004. A crane scheduling method for port container terminals. *European Journal of Operational Research,* 156: 752–768.

Marinov, M. & Viegas, J. 2009. A simulation modelling methodology for evaluating flat-shunted yard operations, *Simulation Modelling Practice and Theory*, 17(6): 1106-1129.

Netto, J. F. & Botter, R. C. & Medina, A. C. 2015. Analysis of capacity associated to levels of service at port terminals using systemic approach and simulation of discrete events. In *Winter Simulation Conference (WSC)*:3426-3437.

Rong, H. & Teixeira, A. & Soares, C. G. (2015). Evaluation of near-collisions in the Tagus River Estuary using a marine traffic simulation model. *Zeszyty Naukowe/Akademia Morska w Szczecinie*, 43 (115): 68-78.

Rusca, F. & Raicu, S. & Rosca, E. & Rosca, M. & Burciu, Ş. 2015. Risk assessment for dangerous goods in maritime transport. In Soares, C. G., Dejhalla, R., & Pavletic, D. (Eds.) *Towards Green Marine Technology and Transport-Proceedings of the 16th International Congress of the International Maritime Association of the Mediterranean, IMAM 2015*: 669-674. Pula, CRC Press.

Rusca, F. & Popa, M. & Rosca, E. & Rosca, M. A. & Rusca, A. 2016. Capacity analysis of storage area in a maritime container terminal. In *Maritime Technology and Engineering III: Proceedings of the 3rd International Conference on Maritime Technology and Engineering (MARTECH 2016, Lisbon, Portugal, 4-6 July 2016)*: 92-99. CRC Press.

Rusca, A. & Popa, M. & Rosca, E. & Rosca, M. & Dragu, V. & Rusca, F. 2016. Simulation model for port shunting yards. In *IOP Conference Series: Materials Science and Engineering*, 145(8):082003. IOP Publishing.

da Silva, A. & Botter, R. 2009. Method for assessing and selecting discrete event simulation software applied to the analysis of logistic systems In *Journal of Simulation* 3: 95.

Guedes Soares, C. & Teixeira, A.P. 2001. Risk assessment in maritime transportation, In *Reliability Engineering & System Safety*, 74(3):299-309.

Vis, I.F.A. & Koster, R. 2003. Transshipment of containers at a container terminal: An overview, *European Journal of Operational Research*, 147(1): 1-16.

Progress in Maritime Technology and Engineering – Guedes Soares & Santos (Eds)
© 2018 Taylor & Francis Group, London, ISBN 978-1-138-58539-3

Analysis of a new container terminal using a simulation approach

N.A.S. Mathias, T.A. Santos & C. Guedes Soares
Centre for Marine Technology and Ocean Engineering (CENTEC), Instituto Superior Técnico, Universidade de Lisboa, Lisbon, Portugal

ABSTRACT: The aim of this paper is to carry out the pre-dimensioning of a new container terminal with an empiric method and, subsequently, develop a simulation model of the same terminal using a discrete event simulation software and perform the simulation of its operations. The comprehensive model includes the appropriate accesses by road and rail as well as a container freight station. The model is used to study the flows of cargo and equipment along the container terminal, identifying bottlenecks in specific areas and is applied to the study of a new container terminal in the Port of Leixões. This terminal has a design capacity for 600.000 TEU and the land area available is scarce, so the aim of the study is to validate the method used for the pre-dimensioning, as well as determining by simulation any possible bottlenecks in layout and equipment. The application of simulation has indicated that this terminal, with some conditions, may be able to move the required number of TEUs while keeping within reasonable levels of utilization.

1 INTRODUCTION

1.1 *Historical background*

The port of Leixões is the major port infrastructure in the Northern Region of Portugal and of fundamental importance to the entire local industry. With the increased pressure on land vehicles, due to being responsible for major pollution, but also having an infrastructure with a high degree of maintenance and little transport capacity, the EU found it worthwhile to invest in rail networks. These have the main advantage that significant amounts of cargo can be transported at once, creating a scale economy, as well as reducing greenhouse gases production (electric propulsion). This rail network project (European Corridors, see Figure 1) implies that cargo will arrive at distant ports by moving it by rail. An example of this is the connection between Munich and Genova. For the Portuguese case, the PETI 3+ Project, see Ministério da Economia (2014), was created, integrating Portugal in Rail Freight Corridor n° 4 with two main connections (Sines, Lisbon and Setubal to the South, Aveiro and Leixões to the North).

However, at present, the container terminals in the port of Leixões have reached its maximum design capacity, so to meet the expectations created by the EU, a new container terminal is required, creating a need for a feasibility study. The terminal is to have a handling capacity of 600.000 TEU and is to be fitted with all necessary accesses (road, rail and maritime) and a container freight station (CFS). In order to better understand the maximum

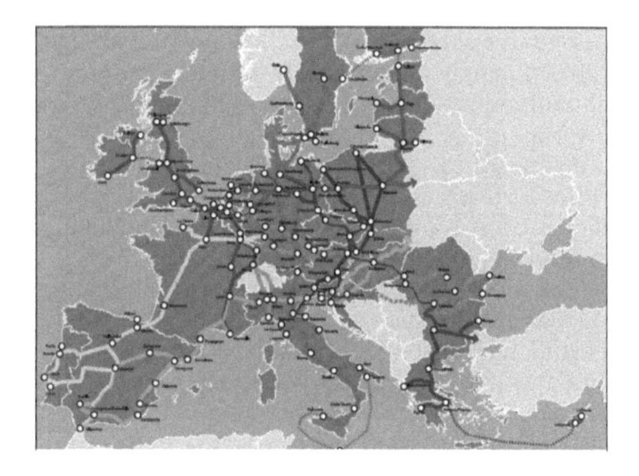

Figure 1. European Corridors Project (Source: ec.Europa.eu).

handling capacity of the new terminal, a discrete event simulation study will be carried out using software ARENA.

1.2 *Objectives and structure*

The objectives of this work are to properly design an entirely new container terminal (CT), based on information and models already publicly available (Brógueira Dias *et al.* (2011), Silva and Guedes Soares (2008)), making a preliminary design as a starting point followed by an optimization of several variables. The ultimate objective is to study the optimum number of equipment for the new terminal, in order to be able to handle the desired throughput. Also, interactions with other existing terminals will be considered in a succinct way. This paper will be divided in 8 sections. In the second section, the state-of-the-art will be presented, as well as the adopted theories from the information gathered. Thirdly, the need for a new terminal will be demonstrated and the available preliminary design, on which this work is based, will be presented. In section four, the modelling of the problem will be made with the necessary variables, formulations and the corresponding implications. In section five, the pre-processing data for the simulation will be shown. Section six will present the results obtained as well as the methods used to verify and validate the implemented model. The last section will show the conclusions obtained from the work produced, along with recommendations and suggestions for future work to be developed.

2 STATE OF THE ART

With the increase in freight transportation and an ever-growing competition among ports, the importance of having an optimal performing port has led to a substantial increase in research on this subject, mainly in methods to increase productivity and improvement of management policies, for every aspect of the terminals. Up until now, there is no integrated solution, rather optimization models for local issues. Nevertheless, one could classify the research into two major categories: simulation models and optimization models.

Few models have focused on the use of a generic simulation tool such as ARENA. For the Pusan East Container Terminal, a model was composed based on the ARENA software, where the terminal has been proposed in accordance with several distributions for ship inter-arrival (Erlang), ship time at berth (Beta), crane container handling (Gamma), queues and other processes for the interactions among the multiple components of the terminal, such as the berth-yard truck interaction. Moreover, the simulation was made from the point of view of supply chains, which consists of several interactions, as seen in Figure 2. The interactions and the impact of the supply chain characteristics are also discussed in Lee *et al.* (2003). The work of

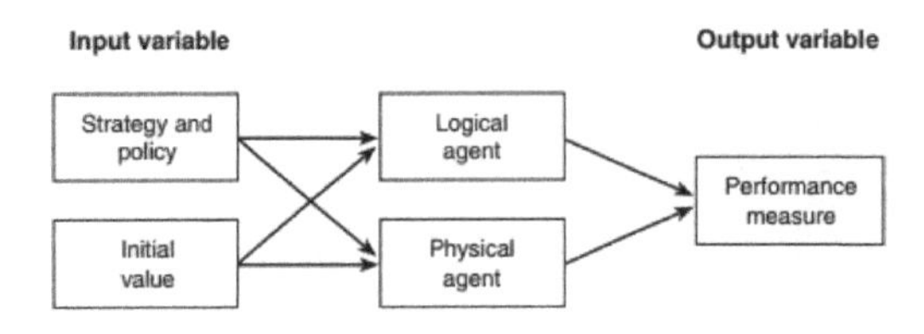

Figure 2. Supply chain management view of simulation (Source: Lee *et al.* (2003)).

Merkuryev (1998) is deeply focused on the optimization of the Baltic Container Terminal, by a progressive development of the basic model, having improved with several works over the years.

Another study using ARENA is presented by Cortés *et al.* (2007) where the main target was to relate the terminal with the traffic flow inside the river port of Seville. By considering timetables, number of vessels in the system, the vessel time in the dock, the maximum capacity of terminals, queues and, most importantly, the existence of other products, such as bulk, a highly-detailed model was obtained.

Silva and Guedes Soares (2008) describe an ARENA based simulation model for the optimization of an intermodal terminal layout. These authors aim to create the optimum terminal layout by varying parametrically several factors, such as the impact of the number of reach stackers (RS) and forklifts (FL) existing in the terminal, among others. The model is validated by real-time data, provided by Terminal XXI at the Port of Sines (Portugal). Silva and Guedes Soares (2006) studies have taken an approach to study the throughput, using a macroscopic model for the Port of Leixões, validated with data from 2003 to June 2005. Silva, Guedes Soares and Signoret (2015) have applied a Petri Net model with predicates to study a generic model of an intermodal cargo terminal, using as case study the port of Leixões.

In Portugal, the integration of terminal pre-dimensioning and simulation was never attempted and even simulation studies are scarce, and as both terminal pre-dimensioning and simulation are going to be carried out in this paper, a clear literature gap arises, validating the importance of this study.

3 PRE-DIMENSIONING OF THE TERMINAL

As a basis for this paper, the design sketches publically available were used (Brógueira *et al.* (2011), Silva *et al.* (2008)). The new terminal is going to occupy the existing Multi-Purpose Terminal and some reclaimed adjacent area, in order to achieve a terminal area suitable for receiving Post-Panamax vessels (2nd generation) and an annual throughput between 400.000 and 600.000 TEU. Also, dredging might be needed to support the Post-Panamax vessels that are supposed to arrive at this terminal.

3.1 *Arrival of ships*

The lack of space is the primary issue in the port of Leixões, determining which type of ships can be serviced, but also the area available. The solution found was, as a first step, to create a database of the existing ships arriving at the port, based on the AIS system. This database analysed the last three

Table 1. Container ships arriving at the new terminal.

Vessel type	Monthly average arrival	Weekly average arrival
Feeder	0	0
Feedermax	13.5	3
Handy-size	13.5	3
Sub-Panamax	13.5	3
Panamax	0	0
Post-Panamax	4.5	1

Table 2. Areas required for the terminal.

Characteristic	Value
Total Area [Ha]	15
CFS Area [Ha]	0.6
Number of Ground Slots [–]	2762
Number of Blocks [–]	12
Empty Containers [%]	11
Reefer Containers [%]	4
Hazard Containers [%]	5
Length of Quay [m]	553

months of the year 2015. The basic data provided showed all the ships in port.

The division of container ships in classes was then made according with the classification of Stopford (2009). Several regular lines were found and as an average, one can state that about 27 container ships arrive per week, about 61% of the ship are of Feedermax class, 4% are Feeder vessels, 23% are Handysize, 11% are Sub-Panamax and less than 1% are Panamax vessels. So, due to in-existence of references as a basis, the following ships were assumed to arrive at the port once the new terminal is in operation, based on the logic of serving mainly larger ships, as shown in Table 1.

Based on the ships indicated in Table 1 and the values of quay length (shown in Section 3.2), a schedule was made, following the rule of fitting the maximum number of ships to the quay.

3.2 *Areas required*

The areas required for the terminal were calculated using equations, based on empirical values, arriving at the results shown in Table 2. The details of these methods may be found in Rademaker (2007), Sharif (2011) and Thoresen (2003).

3.3 *Accessing the terminal*

Starting with the rail access, at present, there is a railway connecting the north side of the Fishing Docks with the existing South Terminal and to the Leixões Railway Track. Despite this access, it is impossible to deposit containers in the surrounding area of this railway track, due to the existence of a Bulk Terminal. Consequently, a new access, direct to the terminal, should be made, by using the existing tunnel that runs under the Fishing Docks and enlarging it, so that two trucks and a train (single-track) can pass through at the same exact instant, without damage to cargo and equipment. Only one track was chosen, due to the little current use of railway in Portugal. In fact, the INE (2016) states that only 2% of the total container movement is made by railway, which is obviously very little. In the specific case of the port of Leixões, only 7.000 TEU/year are moved by rail. It is assumed that a favourable development would be accomplished in upcoming years and, as such, the movement by rail should be of about 12%, in line with other European ports (iContainer(2016)). In Portugal, each train carries about 48 TEU and to achieve the 12% mark (72.000 TEU/year), 1500 connections should be made per year. This means two trains per day if we consider seven working days per week for rail workers. The internal roads of the port have access to the eastern highway, which runs in the North-South direction. At the terminal itself, the trucks enter the terminal by the East Entry and have a parking lot and CFS there. At the quay, there is also a 10 m wide road, in addition to the 7 m distance between blocks in the parking area, which allow the passage of two trucks at the same time.

The general layout of the new container terminal is shown in Figure 3. The terminal includes two berths at the quay (7), blocks for empty containers (6), a container freight station (5), a rail terminal (8) and blocks for containers waiting for the trains (4).

3.4 *Equipment*

A ship-to-shore (STS) crane has a gross productivity of 70.000–120.000 TEU/year according with Rademaker (2007). Based on the current gross productivity of the South Terminal, 85000 TEU/year with four cranes, and applying the same productivity for the new terminal, one can obtain the same number of cranes (four). For each STS crane, 5 yard trucks (YT) are needed, making a total of 20 YT. Adding another four for the rail terminal, which are necessary, this makes a total of 24 YT.

Two to three RMGs are needed per quay crane, which makes a total of nine in the yard. Besides this equipment, some reach stackers (RS) are also needed for the rail terminal, which represents 12% of the total movements; instead of 5.5 per crane, as stated by Rademaker (2007), only four will be needed for the rail terminal. The estimated value is an averaged one. For the park area, three RMGs

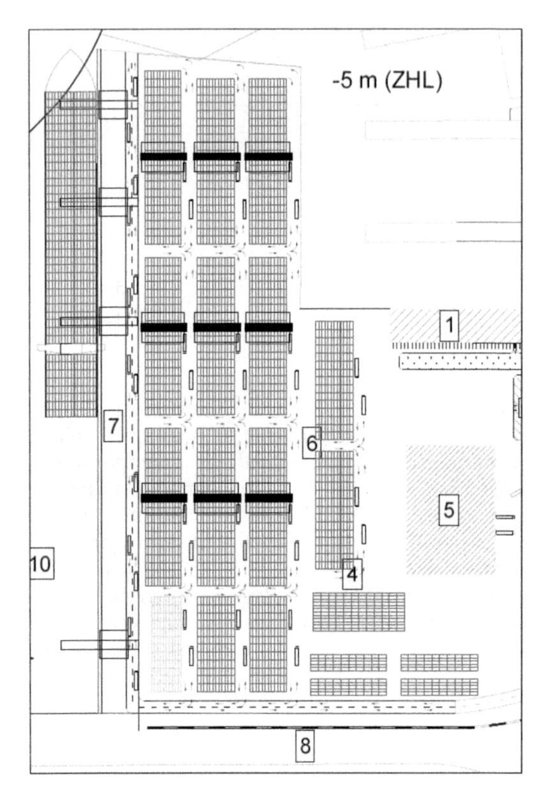

Figure 3. Preliminary plan of the container terminal.

Table 3. Rules of thumb for other terminals.

Data	TCN	TCS	Liscont	Sotagus	Vigo Guixar (Reefer)	New terminal
TEU/Area	4.1	2.2	2.9	2.7	1.4	4.0
TEU/Metre of Quay	694	648	556	606	328	1085
TEU/Gantry	125k	117k	117k	112.5k	62.5k	150k

were considered per row, due to their flexibility. The RMG is the most flexible solution available in the market for limited storage spaces, see Thoresen (2003).

The distance amongst blocks imposed is due to the need for RMGs to move freely across the rows of blocks. Considering that twelve containers are side-by-side (which gives a total of 25 m) and that the RMG still has to pass across these blocks, by checking the Liebherr catalogues (Liebherr (2016a,b)), one can easily conclude that this solution is valid: RMG: 35 to 50 m × 27 m(L × B).

In relation to the RS, only 3 were deemed necessary, since one reach stacker has an average 2 minutes time to (un-)load a train, see Kalmar (2016).

Having designed a preliminary outline of the terminal, a comparison with existing container terminals in the Iberian Peninsula is presented in Table 3.

4 GENERIC MODELLING OF A TERMINAL

When using simulation software, the formulation of the problem must be carefully planned, and variables, assumptions and other necessary data must be compilated, to guarantee the quality of the results. The adaptability of the created model itself depends on how many assumptions are made, the reliability of the data available and how realistically the variables represent the true functioning of the CT. In this context, flowcharts were chosen for the modelling of the new CT, due to their flexibility and adaptability to properly represent the functioning of a CT.

4.1 *External actors, entities and resources*

The determination of who the different actors, entities and resources are and in what way do they influence the diverse components of the entire system, is of considerable importance (several inputs can give different results). For clarity and conciseness, any (in-/out-) bound entity will be referred as an external actor, whereas the other objects that flow inside the terminal will be denominated entities (e.g. Reefers, Dry-TEU, etc.). The resources are the components that interconnect the different locations of the system. In a CT, the resources considered are the STS cranes, the YT and the RMGs, among others, and they interconnect the ship with the quay (STS Crane) and so on. To evaluate the performance of each of these components, different parameters were measured for the actors, resources and entities: Queuing Time; Waiting Time; Service Time; Saturation Levels; Number of entities in queues; Quay enlargement (quay only). The initial data for this terminal is difficult to determine. Nonetheless, from the existing terminals and using the AIS system, data for typical waiting time, queuing time and service time for ships can be obtained and, therefore, the typical performance values of some of the resources, such as the STS Crane, can be evaluated.

4.2 *Generic model*

A generic model is first delineated, as presented in Figure 4. The model is divided into three sections: seaborne operations, terminal and port operations

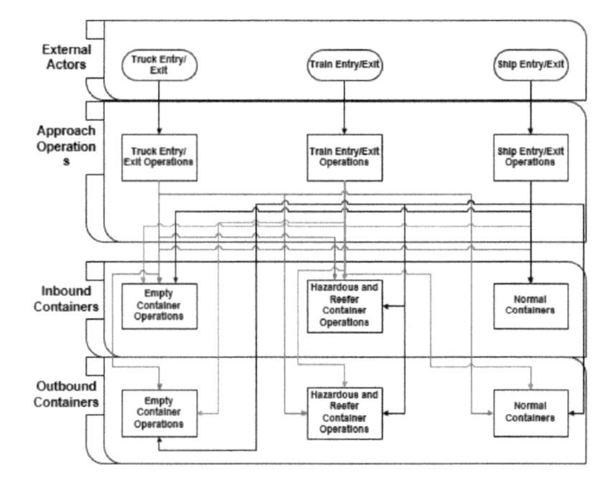

Figure 4. Generic flowchart of the container terminal operations.

and hinterland operations. These sections are submodels for the existing and interacting resources and actors in the different areas of the terminal. Each resource has a specific area of operation, in addition to a specific speed, transport capacity, performance, and so on, as well as a mean time to failure (MTTF) and a mean time to repair (MTTR).

5 PRE-PROCESSING

At a preliminary stage, to represent the actions and interactions of typical terminal operations, some considerations were made at the three major areas of interest at a CT: berth and seaside operations; yard and land operations and in- and outbound operations. These will imply assumptions and preliminary calculations, serving as input for the ARENA model.

5.1 *Pilot usage*

At present, the port of Leixões has four pilot boats and five pilots for every shift (8 hours), for a total number of 300 ships per month (avg. one pilot boat for every seventy-five ships). Adding another fifty-five monthly ships, the logical approach is to assign one more pilot boat. For pilot boats, the average speed (12.25 knots) of existing ones was considered. Based on the values available from the AIS system, it will be considered that the pilots will serve every type of ship, and that the arrival of other vessels will follow an exponential distribution, always considering the FIFO strategy.

5.2 *Other terminal ship arrivals and service times*

To model the pilot boat services, it was considered that ships would arrive and be served, according with a mean of the histogram distributions (AIS data). Both arrival rates and service times, follow an exponential distribution, with a mean time between arrivals of 2 hours and an average service time of 19 hours, at 95% confidence level, having applied Chi-square and K-S tests.

5.3 *Atmospheric conditions*

The tides and closure of the port, due to bad weather were considered, based on data provided by the Portuguese Hydrographic Institute, for tides (variation of 3 metres and by the Portuguese Ministry of Defence for port closure (once in the last 10 years).

5.4 *STS cranes performance and workability*

Due to little knowledge on the cycle times of STS cranes, the problem was tackled using speeds and distances of every component provided by the manufacturers Liebherr (2016b), culminating in the calculation of average values, using empirical equations of Rademaker (2007). Considering that a typical container is removed from the mid area of the ship and that the speeds are variable, for one cycle a 2.1 minutes average, with a given standard deviation was achieved. When comparing these values with the ones provided by other authors, similarity was achieved, as shown in Table 4.

5.5 *Horizontal transportation vehicles*

Horizontal Vehicles are divided into three categories: YT, RMGs and RS. For YT, typical values of speed are 7 km/h when loaded and 14 km/h when empty and constant speeds, Saanen (2004). For the RMGs the same logic was applied as for the STS cranes, having a mean value of 3.6 minutes.

5.6 *Failure, repair and maintenance*

Maintenance, repair and failure was considered for the following equipment: STS cranes, RMGs, YT

Table 4. Cycle times for STS gantry cranes.

Performance	Saanen (2004)	Silva and Guedes Soares (2008)	Rademaker (2007)	Cortés *et al.* (2007)	Sharif (2011)	This study
Cycle Time [min]	1.5	1.5	1.8	2.7	2.2	2.1
Moves/h	40	40	33	22	28	29

and RS. For each of the items above, three main parameters were considered: lifetime, MTTF and MTTR. For this study the failure rates and repair rates were considered constant and typical values from the literature were used, as presented by Martínez (2013), Rademaker (2007), Silva & Guedes Soares (2008) and World Cargo News (2006).

5.7 *Container freight station*

The Container Freight Station will have an average dwell time of 7 days, meaning that as the containers flow into and out of the CFS, 7 days will pass. The port of Leixões is not a major port in the East-West cargo flow, meaning that the seasonality of containers is probably very low. Consequently, by making an average of the CT throughput of various terminals with the CFS throughput, one can achieve that the average percentage is 1.53%, meaning that a good indicator for the CFS is as shown in Equation 1:

$$0.0153 \times 600{,}000 = 9180\ TEU/year \quad (1)$$

5.8 *Gate operations*

Currently, the gate of the port of Leixões has the following opening hours, divided into three shifts: 8h–12h; 13h–20h and 21h–24h. According to the Port Authority, the average service time of trucks at the gate is of 11 minutes. In addition, if the reservation is not previously made, the truck will have to wait an additional 5 minutes, according with Cardoso (2012). For the modelling of the gate, queuing theory will be used with an M/M/S model. Concerning the inter-arrival of trucks, typical variance goes from 1 to 10 days, transporting 1 FEU or 2 TEU, see Cardoso (2012). Consequently, the optimum number of inter-arrivals for this case are 66–87 trucks/hour, see Guan (2009), meaning that over a year, 528.000 trucks had to go through the gate. This would imply that 91 trucks had to pass each hour, making a total of 6 lanes in each direction necessary.

5.9 *Inspection*

For import containers, inspections must be carried out, as the containers are stacked, in accordance with the fact of having been inspected or not, presently at the port of Leixões. Inspection time follows a triangular distribution, as shown by Hyun *et al.* (2006):

$$T_{Inspection} = TRIA(4.54, 5.9, 28.2) \quad (2)$$

6 ANALYSIS OF RESULTS

The analysis of results focuses on the different components of the new terminal (Entry and quay operations; yard operations and in- and outbound operations), where queues, times and utilization rates, were studied. Furthermore, as a parametric study, three different scenarios for arrival rates were considered, to check the flexibility of the terminal. This parametric variation will appear last in this section. Each simulation was carried out for 10 cycles of 1 year.

6.1 *Entry and quay operations*

Starting with the maritime connection, two main components were evaluated: the queueing and waiting time. Queueing time is related to the time that a ship must wait for a terminal to be available, whereas a waiting time is the time a ship must wait for the tide to go up or a pilot boat to come. It is important to state that no or minimal dredging was considered. The actual draught for the port of Leixões is 11 metres, which with high-tide becomes 14 metres, enough for the ships to enter, with minimal dredging, if the ships are full. The maximum queueing time obtained for the terminal was of 0 hours, which is the ideal value, when comparing to values from the literature, presented by Bose (2011), Saanen (2004) and Sharif (2011). Nonetheless, as will be seen further ahead, when making parametric variations, these values tend to increase considerably.

When analysing waiting times for pilot boats, the results observed throughout the simulation show that the maximum waiting time for a pilot boat was of 0 hours and, consequently, the average also 0, meaning that a pilot boat is always available for the terminal to perform well with an average utilization rate of 0.4. When analysing the waiting times for tides to go up, the maximum value obtained was of 4.4 hours, whereas the average is 1.4 hours. The queue length, due to lack of draught, had a maximum value of 1 ship, being able to create a substantial queue, if the entrance of the port is not properly dredged. The results for waiting and queueing times are very satisfactory, when comparing with the present results of the port of Leixões (8.31 hours).

Analysing the quay, the focus will be on the quay utilization and the number of quay metres seized. So, for this simulation, the time average is at about 100 metres, meaning that the values of quay metres utilization level would be averaged along the time, in opposition, to instantaneous utilization values. But the main interest relies in the berth occupancy ratio, which implies a ratio of 35%, which is in line with the number of berths existing (Berths = 2), according to Thoresen (2003), as can be seen Table 5. When comparing these values from the literature and the initial values, one can state that an optimum was obtained.

Table 5. Desirable berth occupancy ratio (BOR) values.

Desirable BOR	Saanen (2004)	Thoresen (2003)	Thoresen (2003) 2 Berths
BOR	60–65%>	30–75%	35–50%

Table 6. Average service time per container.

Average service time per container	Scenario 1
Feedermax [hours/container]	0.053
Handy-size [hours/container]	0.050
Sub-Panamax [hours/container]	0.051
Post-Panamax [hours/container]	0.049

When considering the busy times of STS cranes, the crane utilization amounts to 50%, which is in line with what was expected, since the BOR has a moderate value. Concerning the maximum waiting times, the values are inside the expectable parameters, with a value of 0.29 hours. When comparing waiting times for cranes with values from the literature, where the waiting time for STS can reduce productivity up to 15%, meaning that 15% of the total service time is due to waiting time for STS cranes, see Guan (2009). Assuming 15% is 0.29 hours, the service time would be 1.9 hours/ship and service times are over this value, hence validating that the waiting time obtained in simulation has an impact of less than 15% on the ship service time.

The average service time per container is shown in Table 6 for the basic scenario that has been considered so far. Converting the values from Table 6 from hours to minutes and making an average out of it results in 2.02 minutes, implying that the values are perfectly aligned with the ones stated in Section 5.4.

6.2 *Yard operations*

6.2.1 *Yard occupation*

Beginning with the import yard, the occupation level is very high (92%), which was expected, when considering that the maximum throughput of the terminal is aimed. As a rule of thumb often used in the literature, the terminal capacity should not be larger than 80%, which happens in this case. The high average occupation ratio for import containers is due to the dwell time containers (5.3 days), which ideally should be of 3 to 5 days, see Voss *et al.* (2004), being clear that this is one of the bottlenecks of this terminal, which was expected, considering the low area for construction available.

When analysing the export yard, the export terminal utilization level is around 55%, which is more in line with the previously stated rule of thumb. This gives a global occupation level of the yard of about 73.5%, which is within the acceptable values. Typical values of truck arrivals were imposed to be of 0 to 6 days before the day of departure, implying that per week the number of blocks will never pass much more than 50%, due to the reservation of lines for different companies and lesser dwell time. When analysing the export dwell time, which was of 3.96 days, it is evident that the values are lower than the ones of the import containers.

Finally, for empty, hazard and reefer yard occupation levels, the occupation ratios are lower than the other blocks existing at the terminal, which is an optimal value, with exception made for the empty block, since it was considered that empty containers would not leave the terminal, except for CFS purpose, not being loaded to ships either, creating a considerable bottleneck and consequent problem.

The train terminal yard occupation levels, show similar patterns of occupation levels as the import and export yards, having an average ratio below 25%. The greater level of occupation of the import is due to the instant movement of a container after being cleared by inspection to the train terminal yard, whereas for export, they are first unloaded and then directly moved to the export yard, where they will wait before being loaded onto a ship, creating a low time-averaged occupation of the export train yard.

6.2.2 *Transporters*

Concerning RMGs, when a ship arrives, a higher productivity is required from the crane, when comparing with no ship situations. The results show that no RMG has a constant use and their average utilization, with a few exceptions tend to be around 50%, validating the number of RMGs present in the terminal, where one is present per block.

When evaluating the other transporter units' performance, the same conclusions shown beforehand were achieved. It became clear that the number of YT is sufficient to serve the terminal, after being increased to 30 units, due to long queueing times verified throughout the simulation process. The remaining number of transporter units was sufficient, considering the number of trains arriving at the terminal, served by the RS, and the number of FL present at the CFS.

6.2.3 *Inspection utilization*

Concerning the degree of utilization of the inspection service present in the new terminal, the values were considered sufficient (occupation levels are approximately 60%).

6.3 *In- and outbound operations*

In this section, the analysis will focus on the gate performance (In-/Out-perspective), as well as the train terminal performance, CFS performance and the service times of trains, CFS trucks and (I/O) trucks.

6.3.1 *Gate performance*

Starting with the entry gate, the average utilization rate was of 51%, meaning that no major queues were present, and the average waiting time close to zero (0.01 hours), which is in line with what was expected, as presently the port of Leixões does not suffer from a major bottleneck at the gate area, and with what was expected, Guan *et al.* (2009). For the exit gate, similar values were obtained. In a word, no major differences were verified due to high and not evenly spaced arrives.

6.3.2 *Train terminal performance*

The train terminal performance (occupancy plus time) have small values. Due to the low number of trains arriving per week and the schedule-based arrivals of trains, with little to no delays, an occupancy ratio of 0.03% was obtained with no waiting or queuing time. The speeds of access to the terminal, mentioned in section 3 do not have major influence in the performance of the train terminal.

6.3.3 *CFS performance*

The CFS working station has several stations to be worked on (10 in total), but only 5 are necessary for the CFS to work. Utilization ratios are at about 50% and the waiting time is almost non-existent.

6.3.4 *Turnaround times of trains, CFS trucks and trucks*

It is fundamental that turnaround times of trains and trucks stay as low as possible, to get a preferably high transporter utilization of the equipment of the terminal. Starting with the turnaround time of trucks it is recommended that values for the turnaround time are around 40 to 60 minutes, as mentioned by Giuliano *et al.* (2007). The values obtained by simulation are 0.57 hours meaning that they are perfectly in line with the expected ones.

In what concerns the CFS truck and train turnaround times, the CFS trucks have an average service time of 3 hours. Trains have a higher turnaround time, depending on the number of containers available in the yard and the number of slots available on the train.

6.4 *Parametric variation*

Two additional scenarios were considered: Scenario 2 with a smaller call-size (20.65% of TEU/ship instead of 20.65% containers/ship), implying an increment in the number of ships, so that the same number of containers is moved, and Scenario 3 with the same call-size as in Scenario 2, but only with Sub-Panamax and Panamax vessels. As a general perspective, quay parameters for the additional scenarios had a major increase for both average and maximum queueing times, which is due to the high values of BOR. The time-averaged STS cranes utilization also suffered an increase. Pilot boat utilization, which are call-size independent, have similar values, as expected.

There are significant queues present in both additional scenarios, leading to higher export block utilization rates and higher dwell times. In opposition, by the same logic, the values of the import yard are lower than expected, when comparing to the initial scenario, and the same applies to hazardous and reefer blocks. Also, since there are lesser ships and less containers, the degree of inspection use is lower, making the containers more quickly available. Hence the dwell time is significantly reduced.

The empty container yard is a major problem, requiring further insight and probably a hub outside the port for further storage. Concerning the use of the RMGs, their usage will reduce with less movement at the quay. In what concerns the other transporters, RS, FL and YT, a clear reduction of yard truck utilization is verified, in line with the other parameters directly related to ship arrival, whereas RS and FL have similar patterns in all three scenarios.

Train turnaround times decreased in Scenarios 2 and 3, due to the lower number of containers in the yard, whereas turnaround times of trucks have slightly decreased in value, when compared to the other values of Scenario 1, due to the lesser number in containers at the terminal and consequent lesser waiting time for an RMG. The train terminal occupation levels decrease, once again, in line with the number of ships which enter the terminal. Finally, the gate performance increases as numbers of trucks also increase, as expected. Still, as no bottleneck is created, no major issue from increasing the loads is verified.

6.5 *Throughput capacity of the terminal*

Results for the three scenarios are presented in Table 7. When analysing the throughput, it becomes clear that the first scenario is the one that presents the best values, due to having the greatest number of containers moved.

Table 7. Total throughput for the three scenarios.

Total throughput	Scenario 1	Scenario 2	Scenario 3
Throughput [TEU]	647.301	398.690	463.109

7 CONCLUSIONS

The port of Leixões is the key port in the Northern Region of Portugal. The present facilities are already at their maximum capacity and with the growth in traffic, due to the European Corridors, the need arises for a new container terminal. With the 2020 Horizon program, the financial resources will be available to assist the development of a new terminal at Leixões.

The present work analysed the feasibility of a new container terminal at the port of Leixões, located in the current location of the existing Multi-Purpose Terminal, with a handling capacity of 600.000 TEU. From the start it was clear that space was an issue, implying that nearly ideal working conditions need to be achieved. Up until now, no integrated study of container terminal pre-dimensioning and simulation had been published in the literature in Portugal and this study fills that literature gap.

As a basis, a pre-dimensioning was made, which revealed to be quite accurate, as a general perspective. In what concerns the results of discrete event simulation, it was revealed that, as a general perspective, the most appealing scenario is the first one, being the one that moves the biggest number of containers, as well as having modest values for utilization. One of the biggest bottlenecks is related to the quay which is limited to 553 metres, which, although never being passed, has times where it is almost fully occupied (550 metres), for scenarios 2 and 3.

The second major problem present at the terminal is the import yard which has a large occupation level. Due to lack of space at the yard, the new terminal should adopt active policies, such as to oblige companies to pick-up their containers early on, so that the dwell time is reduced. In addition, the empty containers yard is clearly under-dimensioned, being this the major bottleneck of the new terminal. A clear solution, such as a satellite terminal, or simply an increase in size and re-arrangement of the terminal is needed, to solve this issue, being the recommendation of the author to use a satellite terminal (empty container depot).

Improvements such as the insertion of detail to the level of the pile of containers, the modelling of all the processes at the CFS, as well as the inclusion of a special gate where trucks, who do not have clearance to enter can request their documents, should be studied. The ideal model would be the one that studies the whole port, including the modelling of the different depths of the port and the occupancy of the manoeuvring basin.

The present model can be used to test different additional strategies (stacking and others), as well as additional or other equipment, the addition of a satellite terminal and consequent further analysis of throughput. As a final remark, the current study demonstrated that, in optimal operation conditions, the new container terminal at the port of Leixões is able to move 600.000 TEU.

ACKOWLEDGEMENTS

This work was performed within the scope of the Strategic Research Plan of the Centre for Marine Technology and Ocean Engineering (CENTEC), which is financed by the Portuguese Foundation for Science and Technology (Fundação para a Ciência e Tecnologia—FCT).

REFERENCES

Böse, J.W. 2011. Handbook of Terminal Planning. Springer Science & Business Media.

Brógueira Dias, E., Estrada, J.L., Mealha, R.P. 2011. O Novo Terminal de Contentores de Leixões e a Remodelação do Porto de Pesca.

Cardoso, R.J.S. 2012. Gestão do Parque de Contentores do Porto de Leixões no Terminal de Contentores de Leixões. Universidade do Porto.

Cortés, P., Muñuzuri, J., Ibáñez, N., Guadix, J. 2007. Simulation of Freight Traffic in the Seville Inland Port. Simulation Modelling Practice and Theory 15:256–71.

Giuliano, G., O'Brien, T. 2007. Reducing Port-Related Truck Emissions: The Terminal Gate Appointment System at the Ports of Los Angeles and Long Beach. Transportation Research Part D: Transport and Environment 12(7):460–73.

Guan, C.Q. 2009. Analysis of Marine Container Terminal Gate Congestion, Truck Waiting Cost, and System Optimization. New Jersey Institute of Technology.

Hyun, J., Kap, Y., Kim, H. 2006. A Grouped Storage Method for Minimizing Relocations in Block Stacking Systems. Journal of Intelligent Manufacturing 17(4):453–63.

iContainer 2016. "Rotterdam Port Taps Central Europe Potential via Rail." Retrieved (http://www.icontainers.com/us/2016/10/25/rotterdam-port-taps-central-europe-potential-via-rail/).

INE 2016. Estatísticas dos Transportes.

Kalmar 2016. Technical Data Reach Stackers 42 to 45 tonnes.

Lee, T.-W., Park, N.-Y., Lee, D.-W. 2003. A Simulation Study for the Logistics Planning of a Container Terminal in View of SCM. Maritime Policy & Management 30(3), pp. 243–54.

Liebherr 2016a. Technical Description Rail Mounted Gantry Cranes.

Liebherr 2016b. Technical Description Ship-to-Shore Gantry Cranes.

Martínez, C.M. 2013. Metodologia para maximizar la rentabilidad de una terminal maritima de contenedores a traves de la optimizacion de su grado de automatizacion.

Ministério da Economia. 2014. Plano Estratégico dos Transportes e Infraestruturas Horizonte 2014–2020.

Rademaker, W.C.A. 2007. Container Terminal Automation: Feasibility of Terminal Automation for Mid-Sized Terminals. TU Delft, Delft University of Technology.

Saanen, Y.A. 2004. An Approach for Designing Robotized Marine Container Terminals. TU Delft, Delft University of Technology.

Sharif, M.N. 2011. Developing a Tool for Designing a Container Terminal Yard. TU Delft, Delft University of Technology.

Silva, C., Guedes Soares, C. 2006. Simulação da carga movimentada num terminal intermodal do Porto de Leixões, in *Inovação e Desenvolvimento nas Actividades Marítimas*, pp 189–206, Edições Salamandra, Lisboa.

Silva, C., Guedes Soares, C. 2008. Simulação e Validação da Carga Movimenta no Terminal Intermodal da PSA Sines. in *O Sector Marítimo Português*, pp. 367–384, Edições Salamandra, Lisbon, 2010.

Silva, C., Guedes Soares, C., Signoret, J.P. 2015. Intermodal terminal cargo handling simulation using Petri nets with predicates, Engineering for the Maritime Environment Vol. 229(4) pp. 323–339.

Silva, L.P.F., Veloso Gomes, F., Pinto, F.T., Santos, P.R., Lopes, H.G. 2008. Leixões Cruise Terminal: Architecture and Port Engineering. 3as Jornadas de Hidraulica, Recursos Hídricos e Ambiente.

Stopford, M. 2009. Maritime Economics, 3rd edition. Routledge.

Thoresen, C.A. 2003. Port Designer's Handbook: Recommendations and Guidelines. Thomas Telford.

Voβ, S., Stahlbock, R., Steenken, D. 2004. Container Terminal Operation and Operations Research—a Classification and Literature Review. OR Spectrum, 26(1):3–49.

World Cargo News 2006. Getting the best out of crane spreader.

Yuri Merkuryev, Y., Tolujew, J., Blümel, E., Novitsky, L., Ginters, E., Viktorova, E., Merkuryeva, G., Pronins, J. 1998. A Modelling and Simulation Methodology for Managing the Riga Harbour Container Terminal. Simulation 71(2):84–95.

Progress in Maritime Technology and Engineering – Guedes Soares & Santos (Eds)
 ISBN 978-1-138-58539-3

Operational and cost based analysis of ship to ship—transshipment in Brazil: An application to the iron ore in the port of Santos

P.C.M. Oliveira & R.C. Botter
University of Sao Paulo, Sao Paulo, Brazil

ABSTRACT: The Shipbuilding Industry has invested in larger vessels for economies of scale. These new ships require a revision of the port infrastructure and still require a greater depth in the ports. Furthermore, the investments result in high costs. A solution that is being used around the world is the transshipment. Brazil needs to be prepared for this challenge and the port of Santos, the main port of the country, has limitations for receiving larger ships. This work assessed operationally and from the point of view of logistics costs, the transshipment in Brazilian ports. To do so, it carried out an application study to the iron ore in the port of Santos, using simulation and the comparison of total logistical costs of the alternatives. The results obtained demonstrated the capacity of the transshipment to attend the demand and reduce cost in relation to the conventional operation.

1 INTRODUCTION

In recent decades, ships have been increasing in capacity, especially container or bulk vessels, a trend that has intensified over the last decade. Some of the benefits of using larger ships include more competitive ocean freight rates, lower energy consumption and CO2 emissions – (International Transport Forum, 2015).

Emerging economies have significantly influenced the increase in the demand for cargo in the global ocean trade, whose increase has resulted in greater competition among companies, servicing longer routes at lower costs (Port Technology, 2012). One of the solutions presented to accommodate increased demands has been to increase channel draught through dredging. However, its realization and maintenance incur high investment costs.

Larger ships also require terminal modernization, which depends on large investments, licenses, and the availability of areas for expansion. Furthermore, terminals concentrate cargo, impacting productivity (Liang et al., 2011).

In Brazil, ports usually cause bottlenecks and require investment to expand and recover existing infrastructure and to improve land access and dredging (IPEA, 2009).The maximum depth and width of the country's important port, the Port of Santos, is 13.2 meters and 200 meters, respectively (CODESP, 2015). This is not enough to allow the entry of these larger vessels that have been emerging.

Considering the revolution that these ships have caused in the naval sector, the transshipment can prove to be an alternative, solving problems related to depth, lack of capacity, and high investments.

Larger ships have been used to transport iron ore due to the high demand for this product, and simulation models present different possible transshipment configurations.

2 LITERATURE

2.1 *Ship to ship—transshipment*

Transshipment is defined as the direct transfer of merchandise from one vehicle to another.

According to Wang, (2015), it can be conducted to load exports or unload imports. Regarding types of ships, the most commonly used are those converted into platforms with on-board cranes that move cargo onto larger ships. Transshipment can be direct, i.e., transshipment units (TUs) transfers cargo from one ship to another, or indirect, which involves the same process but with temporary storage on board. Operations can occur with ships while stationary or underway, and also on open sea or sheltered waters.

2.2 *Literature on the theme*

Initially, few studies were found on the topic, leading to the conclusion that the theme is still underexplored. A more in-depth data collection process was required to find more complete studies on the theme. One of the first studies on the theme was conducted by Buckley et al., (1986), who created a mixed integer mathematical programming location-allocation model to compare dredging with transshipment operations to reduce the waterborne coal transportation costs from the United States and consequently, the final cost for European markets.

Silva et al. (2009) reported on a transshipment of coal from Moatize (Mozambique). A discrete event simulation technique was used to analyze the problem, and the developed simulation model presented several different transshipment configurations.

Cigolini et al., (2011) used simulation to size a transshipment system to supply coal to a power plant that services part of Italy's electricity needs, obtaining good results.

In another study, Liang et al., (2011) presented a mathematical model whose function was to minimize handling time, time spent at berth, departure and waiting time in transshipment operations.

Baird & Rother (2012) presented a preliminary technical and economic study of a floating container storage and transshipment terminal.

Brown (2013) presented an operation to export iron powder from New Zealand to the Asian market: after drilling and capturing the powder, it was allocated to a storage and unloading ship. The product was then transferred to a larger Capesize type ship and transported to its final destination, a Chinese port.

Cost analysis was conducted by Kurt et al., (2015) of a floating container terminal functioning as a conventional terminal.

Last, Wang (2015) analyzed a bulk transshipment operation based on a benchmark study of transshipment companies. Through several floating bulk transshipment operation (FBTO) configurations the most representative scenario was chosen: the use of a Capesize type ship and a transshipment unit with an onboard crane and a feeder ship.

In Brazil, Teixeira (2011) presented a computer simulation model of lay time for transshipment operations with oil exported from the Rio de Janeiro.

Souza (2012) detailed that oil tankers with dynamic positioning systems, i.e., which can carry out underway ship-to-ship transfers, are very expensive vessels. Thus, it is desirable to optimize their use by eliminating trips between offshore platforms and landside storage terminals. The author proposed two transshipment configurations: side-to-side and in convoy formation. Dynamic models were developed using simulations to compare the best results.

3 APPLYING THE SIMULATION TO THE PORT OF SANTOS-SP CASE

3.1 *Port operation*

Based on the premise that iron ore arrives at the Port of Santos for exportation, several analyses were conducted from the point of view of those loading the merchandise, i.e., exporters, the most common scenario involving iron ore in Santos. Thus, the aim of the study is to compare a conventional exportation operation with different transshipment configurations close to the port, using a larger ship, examining the viability of the transshipment scenario.

To obtain a large volume of data, annual export demand was set at 10 million tons/year, based on the average demand of a mine located in Corumbá, (MS), which could export its product through Santos-SP. After analyzing the literature, a simulation model was created to analyze the operations. Cost analysis took into account data relative to ocean freight, ship affreightment, fuel, port, and demurrage costs.

4 OPERATIONAL ANALYSIS OF A TRANSSHIPMENT SYSTEM

The simulation was developed based on the steps proposed by Botter (2002), and Chwif & Medina (2007), namely: conception, implementation, and analysis.

4.1 *Conception*

In this step, the system, objectives, data collection, and conceptual modeling are defined.

The current maximum operational draught of the Port of Santos is 13.2 meters. Thus, in a conventional port operational, the largest ship that can berth at a Santos terminal is the Panamax, as shown in Table 1, which presents the main classes of bulk ships. Partially loaded ships were not considered in the analysis, as the aim was to analyze maximum capacity.

In accordance with Banks et al., (2005), a conceptual model was constructed to present the system and prepare it for experiments. This model consists of a diagram that shows the relationships among entities, their attributes and specializations. Through the diagram, the system can be presented and prepared for experiments. Figure 1 presents the conventional operation model. A Panamax ship arrives at the port, waits for permission to enter the bay and, as soon as it is cleared to berth, heads towards the terminal, berths, unloads the iron ore and waits. If the channel is available, ship unmoors and continues on its way.

Figure 2 presents the conceptual model for the transshipment operation. The Port of Santos bay is deep enough to accommodate Capesize type ships;

Table 1. Types/classes of bulk ships and their dimensions.

Class	DWT	Medium Length (m)	Max medium draft (m)
Handymax	30 ~ 55,000	175.0	10.5
Panamax	60 ~ 80,000	227.0	12.3
Capesize	80 ~ 250,000	252.0	14.6
VLBC	250 ~ 400,000	309.5	18.6
Valemax	Over 400,000	365.0	23.0

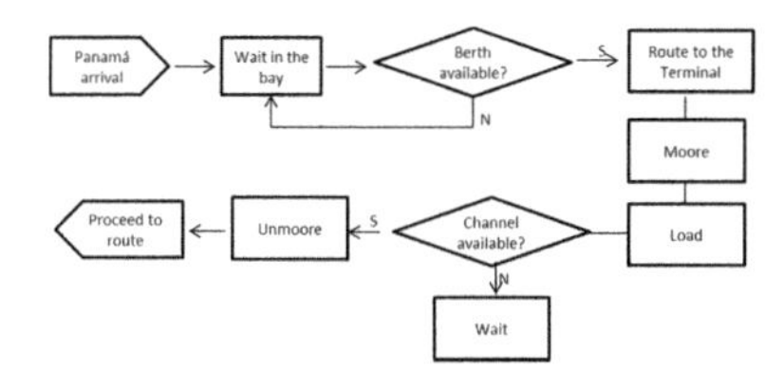

Figure 1. Conventional—Panamax operation model.

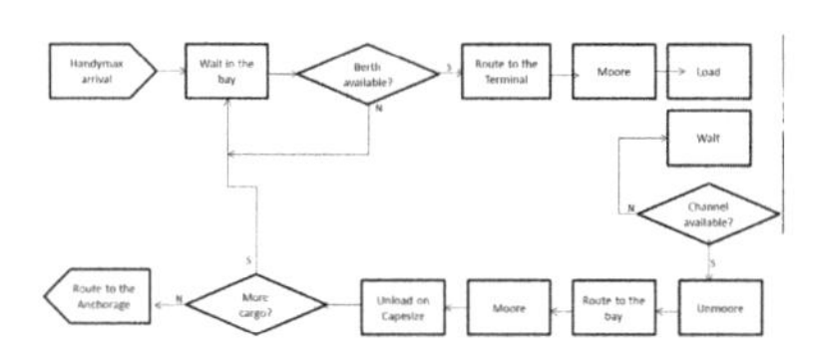

Figure 2. Conceptual model—transhipment.

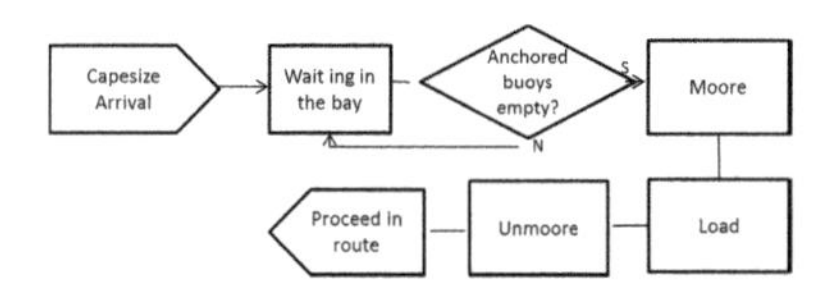

Figure 3. Conceptual model—Capesize operation.

thus, the simulation was set up using a Handymax TU to load the cargo at the terminal and take it to a larger ship moored at the bay.

In theory, this TU reaches the port, waits at the bay to berth, and as soon as possible, heads to the terminal, berths, and is loaded with iron ore. As soon as the channel is available, it unberths and heads to the bay, where it berths and unloads its cargo onto the Capesize. If there is more cargo to load, it repeats the operation, and if not, it anchors.

Figure 3 presents the operation model for the Capesize ocean-going vessel (OGV), which exports the product. It is anchored at the bay waiting for the TU to load and bring the cargo to it, making way to its final destination when it is fully loaded. Last, the simulation was based on the premise that both the systems were capable of meeting a minimum demand of 10 million t/year, based on the demand of a mine located in Corumbá-MS, which could export its iron ore through Santos-SP (Vetria, 2015).

Next, the system implementation scenarios were defined. To this end, the simulations were carried out at first with two configuration scenarios:

- Scenario A: Conventional direct operation at port terminal, using Panamax ships.
- Scenario B: Transshipment using 1 Handymax feeder and 1 Capesize as OGV moored at open sea off the shore of Santos.

The scenarios were set up based on the certain pre-defined processes, in order to:

- Represent an iron ore transshipment operation through the Port of Santos, in which a Handymax feeder ship is loaded at the port terminal and transports the cargo to the main vessel, which is secured by a closed-cycle mooring system approximately 30 km from the port.
- Determine the size of the fleet of ships required to move the iron ore demand in both scenarios.
- To conduct a sensitivity analysis for adverse operational and weather conditions that can delay or suspend the transshipment operation.
- Compare the transshipment operation with the conventional operation.

The following input parameters were incorporated into the model, based on Silva et al. (2009):

- Productivity rates of loading at the terminal and unloading at transshipment.
- Time of berth and unberth.
- Time of travel between terminal and transshipment location.
- Time of wait due to adverse operational and weather conditions.
- Time of feeding vessel (or TU) hold cleaning.

4.2 *Implementation*

In this study, a computer model was created, verified and validated. Considering that transshipment systems are dynamic and complex, computer simulation was chosen as the best tool for this analysis.

The simulations were executed using Arena software, which creates visual models based on modules suited to the simulation. These are then transformed into logic controls which can be tested numerous times (Chwif & Medina, 2007).

The following input parameters were defined for simulation scenarios (A) and (B). These were classified into one of two steps, either "from ship" or "from port/transshipment". Table 2 presents the input items defined for the ship.

Regarding the port and transshipment, the simulation was set up using one berth and one carrier vessel at the port, while terminal loading productivity and TU unloading was defined at 4,000 tph.

In terms of the carrier ship, one TU was made available, and closed-cycle time was used. The values set for other variables are presented in Table 3:

Table 2. Arena—Input items for the ships.

SHIPS	
Total demand (year)	
Total demand – year (t)	10 millions
Ships capacities	
Handymax T.U.	45,000 t
Ships capacities	
Panamax	70,000 t
Capesize	160,000 t
Contract – productivity	20,000 t/day
No of T.U feeders.	1

4.3 *Analysis*

In this step, the desired simulation was carried out, as well as sensitivity and results analysis, presented in Table 4. In scenario (A), the system met the demand with 143 Panamax. In (B), or the transshipment operation, the system partially met the demand (9MPTA), using 62 Capesize ships and with a mean OGV queuing time of 174 days.

Operationally, both scenarios were able to meet the demand.

However, the following observations must be made: even though scenario (A) met the demand, it generated a much higher number of OGV than scenario (B). On the other hand, in (B), the mean queuing time of moored ships reached 174 days, due to the waiting line generated while TUs completed their cycles and fully loaded the Capesize vessels. Thus, the demand was not met adequately. Approximately 60% of the TU cycle was due to terminal loading and transshipment operations conducted at the bay.

4.3.1 *Proposal for additional scenarios*

With the goal of using new parameters in the simulations to reflect in more realistic scenarios, an interview was conducted with a ship broker expert. Ship brokers serve as intermediaries between shipowners and charterers (Mendonça & Keedi, 1997).

The broker explained that in this type of operation, two TUs are usually used to increase productivity and reduce OGV queuing time. Around the world, this type of operation is carried out mainly on sheltered waters. The Santos bay is an area of open waters, presenting swell, which makes it difficult.

Furthermore, the expert explained that bad weather conditions can impact the operation in 25% of total time. He also suggested carrying out the simulation through the port of São Sebastião-SP, as it is closer to the terminal on sheltered sea and has a supply of Capesize ships arriving with imported coal.

With these observations, new scenarios were simulated, as presented below.

Table 3. Arena—Input items for port and transhipment.

PORT AND TRANSSHIPMENT	
Port – Terminal	
Number of berths	1
Loaders – ship	1
Load productivity – terminal	4,000 tph
Transshipment productivity (T.U.)	4,000 tph
T.U.	
T.U operating	1
Cleaning intervals	8 weeks
Cleaning time	6 hours
Additional time	
Time – terminal × anchorage	1.9 h
Time – anchorage × terminal	1.9 h
Mooring time – T.U – bay	1.1 h
Unmooring time T.U. – bay	1.1 h
Mooring time T.U. – terminal	0.9 h
Unmooring time T.U. – terminal	0.9 h

Table 4. Arena results for (A) and (B) Scenarios.

	(A) Panamax	(B) Capesize + 1 TU
Demand	Satisfied	Partial Satisfied
OGV generated	143	62
Berth occupation tax	37%	40%
OGV wait on the bay (days)	0	174
Medium Cycle T.U. (days)	–	1.59

4.3.2 *Simulating additional scenarios*

Three additional scenarios we simulated, namely:

- Scenario (C), using 2 T.U.s in the transshipment
- Scenario (D), using 2 T.U.s in the transshipment, a time impact of 25% under bad weather conditions.
- Scenario (E) using 2 TUs on the sheltered waters of São Sebastião-SP.

Table 5 presents the results of these simulations.

In the new scenarios, the demand was met, with the main difference found in the queuing time of the exporter ship at the bay. Furthermore, the results showed that the use of 2 TUs was necessary, and that at open sea, weather conditions greatly impacted the operation. Considering findings, scenario (D) was considered the most feasible. However, if a more conservative approach, scenario (E) is the best option.

4.3.3 *Sensitivity analysis*

Next, sensitivity analysis was conducted to measure how changes in certain parameters impacted the system. The two most important parameters were

Table 5. Comparative of transshipment scenarios simulated.

	(B) Capesize + 1 TU	(C) Capesize + 2 TU	(D) Cape + 2 TU + Climatic conditions	(E) Capesize + 2 TU S. Sebastião
Demand	Partial Satisfied	Satisfied	Satisfied	Satisfied
Ships year	62	62	62	63
Berth ocup%	40%	44%	41%	39%
OGV	174	2.5	3.4	15.5
TUcycle days	1.59	1.67	1.85	2.58

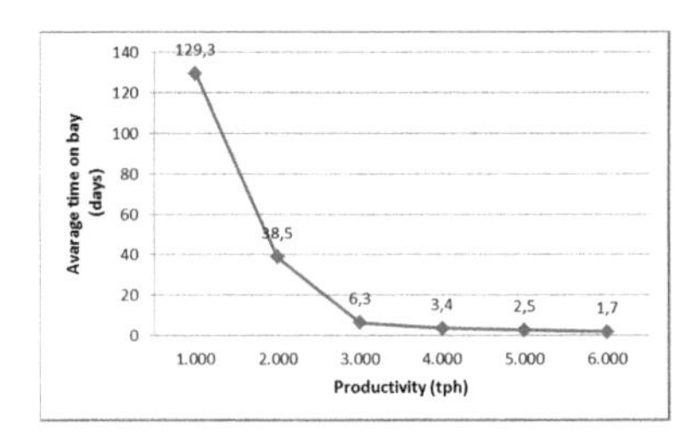

Figure 4. Productivity versus time on bay.

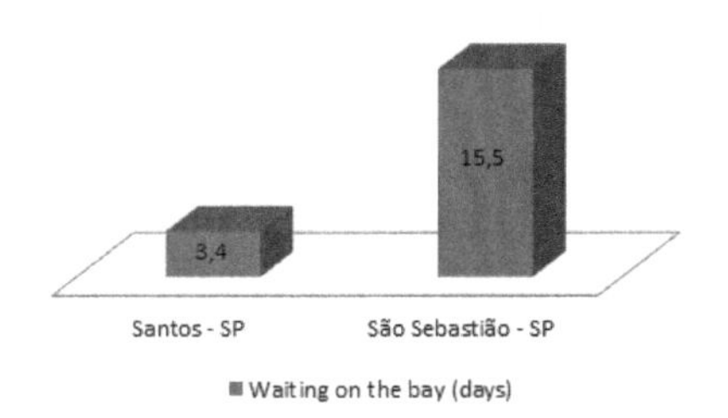

Figure 5. Transshipment place versus OGV queuing time.

chosen for this analysis: productivity rate and transshipment point. To determine the impact of productivity rates on the system, sensitivity analysis was carried out using capacities between 1,000 and 6,000 tph in scenario (D). Figure 4 presents these results.

Loading rates between 1,000 and 2,000 tph resulted in an unfeasible operation due to the OGV queuing time between loadings. At 3,000 tph or higher, queuing time was significantly reduced to under 10 days.

This analysis showed how the distance between the transshipment point and the port impacts mean queuing time of ships at the bay. Figure 5 presents these results. For Santos, mean queuing time was 3.4 days. For São Sebastião, jumped to 15.5 days.

4.4 *Conclusions*

The scenario with 2 TUs was more adequate. However, because the bay of the Santos is on open waters, the system can suffer interruptions. Therefore, scenario (D) was considered. The analysis of scenario (E), in which the operation is conducted in the sheltered waters of São Sebastião-SP, showed that the system was able to meet the demand; however, the TU cycle time increased in over 40%.

Next, we present the results of the total logistics cost analysis. The goal was to identify which scenarios were feasible both from the operational and cost point of view.

5 TOTAL LOGISTICS COST ANALYSIS

To analyze the costs of the transshipment, the concept of annual total logistics cost was used, as the intention was to assess the values of this type of operation from the point of view of shippers during a year's worth of operation.

In this analysis, the optimal situation (greatest client satisfaction at the lowest cost) is based on the total cost: the focus lies on reducing total logistic costs even if some partial costs are not the lowest possible (Bowersox et al., 2011).

5.1 *Logistics costs for maritime operations*

Ocean freight costs consist of the following components: time charter, bunker, and port charges (Bertoloto, 2010). Figure 6 shows the details.

5.1.1 *Freight*

Freight is paid in dollars per ton. Thus, in this analysis, the freight charge was considered for one Panamax and one Capesize vessel leaving from Santos-SP to Qingdao, China, in a voyage charter. Freight can vary significantly (Bertoloto, 2010).

These values were based on Platts and subtracted the port costs, that were negotiated apart. Qingdao was chosen because it is one of the greatest receivers of iron ore in China (Platts, 2017).

After obtaining the freight charges, they were multiplied by the total load capacity of the ships, in tons, to reach total freight. total.

5.1.2 *Time charter*

Time charter is calculated by its cost per day multiplied by the total days of travel, including those spent loading and unloading (Bertoloto, 2010).

In this analysis, was considered a quote from a broker for the affreightmentof a Handymax TU unit. On the other hand, the Capesize and Panamax vessels are under voyage charter, as mentioned on 5.1.1 section.

5.1.3 *Bunker*

The TU bunker was calculated using the average consumption weighted between transshipment and shipping. It was obtained based on a similar ship using simulation, through Oldendorff (2016).

5.1.4 *Port costs*

Conventional Panamax port operation costs were based on the proforma disbursement account (PDA) framework, a summary of port fees for maritime operations. The PDA presents the costs of an iron ore operation carried out at the port of Santos-SP for one Panamax ship. Table 6 presents these costs.

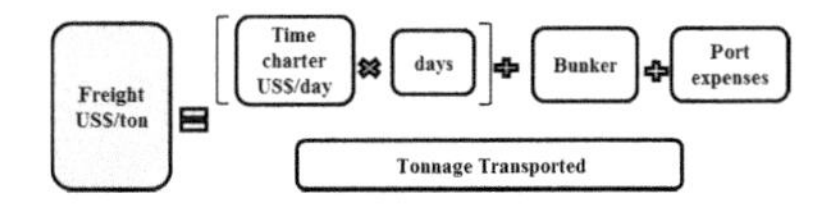

Figure 6. Ocean freight components.

To calculate transshipment port costs, we requested a quote from a broker. Table 7 presents the port costs considered for TU ships, and also the costs for an exporting Capesize ship moored at the bay.

5.1.5 *Additional costs*

In some of the simulation scenarios, ship queuing time was high; thus, our analysis also included demurrage as well as any possible despatch.

Table 8 presents the premises taken into account.

5.2 *Total annual costs of the scenarios*

Below is the analysis for the following systems:

- Scenario (A): Conventional operation directly at the port terminal, using a Panamax ship.
- Scenario (D): Transshipment operation using 2 Handymax as feeders and 1 Capesize ship as the main vessel moored at the bay of Santos, considering a 25% impact on operation time.
- Scenario (E): Transshipment operation using 2 Handymax as feeders and 1 Capesize as the main ship on the sheltered waters of São Sebastião-SP.

Table 6. Port expenses for a Panamax.

Port expenses	Panamax
Light dues	2,250.00
Funapol – Federal Police	342.29
Clearance	1,050.00
Pilotage in/out	18,564.08
Towage	40,733.00
Car hire	250.00
Communication/postage	250.00
Agency fee	3,000.00
Municipal tax	136.50
Certificates expenses	342.17
Ship owners tax – Sindamar	130.00

Table 7. Port expenses for a transshipment.

Port expense	T.U.	Bunker T.U	Capesize OGV
Light dues	2,250.00		2,250.00
Federal Police	313.00	313.00	350.00
Clearance	1,200.00		
Pilotage in/out	10,373.00	10,373.00	20,127.00
Pilotage for anchoring	8,360.00		
Towage	24,625.00		
Watchmen	740.00	740.00	
Channel dues	–	1,235.00	
Launch hire	750.00	975.00	
Car hire	200.00		
Postage	250.00		
Agency fee	2,500.00	1,500.00	3,500.00
Municipal tax	120.00		
IOF	10.53	5.75	
Port State Control Survey	529.42		
Certificates	1,588.24		
Ship owners tax	110.00	110.00	150.00

5.2.1 *Logistics costs for conventional operation (A) – Panamax*

Initially, considered costs for a conventional export operation Panamax ($Ft(p)$), namely: freight ($a(p)$) multiplied by ship ($x(p)$), as given by the formula:

$$Ft_{(p)} = a_{(p)} \cdot x_{(p)} \quad (1)$$

The total cost of the operation with Panamax ($Ct(p)$) was equal to the sum of port costs (b), as shown:

$$Ct_{(p)} = Ft_{(p)} + b \quad (2)$$

Next, the total annual cost was calculated (Cta(p)), considering the number of ships required (y(p)) to meet the total annual demand, as follows:

$$Cta_{(p)} = Ct_{(p)} + y_{(p)} \quad (3)$$

No results for demurrage result as output of simulation (d), but dispatch result of simulation did appear (e), obtaining the total annual logistics cost (A) ($Clt(A)$), for a total of US$ 221,848,662.80. Table 9 presents a summary of the analysis of scenario (A).

$$Clt(A) = Cta_{(p)} + d - e \quad (4)$$

5.2.2 *Logistics costs for scenario (D): transshipment considering bad weather conditions*

This analysis considered the costs of the TU carrier ship to load 1 Capesize ship:

- Time charter (tc)
- T.U. cycle time to load Capesize (z)
- Bunker needed to service 1 Capesize (bu)
- Eventual bunker at terminal when(bu) not available (bc)
- Port costs to service 1 Capesize (b)

Table 8. Premises.

Premise	Value
Turn time/free time (h)	12
Lay time (tpd)	20,000
Demurrage Panamax (US$/day)	15,000.00
Demurrage Capesize (US$/day)	25,000.00
Despatch Panamax (US$/day)	7,500.00
Despatch Capesize (US$/day)	12,500.00

Table 9. Total logistics costs—year—Panamax operation.

Description	Value US$
Panamax capacity $x_{(p)}$	70,000
Panamax – freight $x_{(p)}$	21.50
Total freight – 1 ship $x_{(p)}$	1,505,000.00
Port costs b	67,048.04
Total $Ct_{(p)}$	1,572,048.04
Panamax needed year $Ct_{(p)}$	143
Total cost – year $Ct_{(p)}$	224,802,869.72
Demurrage – year – d	0.00
Despatch – year – e	–2,954,206.92
Total logistics costs – year	**221.848.662,80** ***Clt(A)***

Despite of being moored, there were few compulsory Santos port costs. All above were multiplied by two T.U. to service one Capesize ($Ct(tu)$). Next, the number of Capesize required to meet the total annual demand ($y(c)$) was multiplied by the total cost of one operation ($Ct(tu)$). Estimated annual port costs for bunker was added (*bc*), for a total of US$ 33,602,103.22:

$$Clt(A) = Cta_{(p)} + d - e \tag{5}$$

Capesize freight ($Ft(c)$) was calculated using the freight costs in US$/t ($a(c)$) and multiplied by the number of ships ($x(c)$):

$$Ft_{(c)} = a_{(c)} . x_{(c)} \tag{6}$$

This was added to the port costs ($b(c)$), to obtain the total cost of one Capesize ship ($Ct(c)$). Next, the annual total cost was calculated, based on the ships necessary ($y(c)$) to meet the total annual demand:

$$Cta_{(c)} = Ct_{(c)} . y_{(c)} \tag{7}$$

In the simulation, demurrage output of simulation (d) was set at de US$ 1,686,963.11 and the dispatch result of simulation (e), at US$ 2,213,535.40. Table 10 presents the result.

$$Clt(C) = Cta_{(tu)} + (Cta_{(c)} + +d - e) \tag{8}$$

In this scenario, the cost of TUs increases due to longer cycle time, influenced by adverse conditions. Furthermore, costs relative to the Capesize were impacted by demurrage, as the exporting ships had to wait longer at bay for the carriers.

Table 10. Total logistics costs—year—Option D.

T.U.		Capesize	
Description	Value Us$	Description	Value Us$
T.U charter *tc*	25,000.00	Cape capcity $x(c)$	160,000
T.U. qtity *tu*	2	Cape freight $a(c)$	13.10
T.U cycle *z*	1.85	Total freight $Ct(c)$	2,096,000.00
Bunker *bu*	14,201.75	Port costs $b(c)$	26,377.00
Port costs *b*	215,676.76	Total $Ct(c)$	2,122,377.00
Total Ct (tu)	552,257.01	Cape year $y(c)$	62
Port bunkerbc	91,510.50	Total cost $Cta(c)$	131,587,374.00
Total cost – year *Cta (tu)*	34,331,445	Demurrage year *d*	1,686,963.11
		Despatch year *e*	–2,213,535.40
		Total year $Cta(c)$	131,060,801.71
Total logistics cost – year	**165,392,247.08**	***Clt(C)***	

Table 11. Total logistics costs—year—Option E.

T.U.		Capesize	
Description	Value Us$	Description	Value Us$
T.U charter *tc*	25,000.00	Cape capcity $x(c)$	160,000
T.U. qtity *tu*	2	Cape freight $a(c)$	13.10
T.U cycle *z*	2.58	Total freight $Ct(c)$	2,096,000.00
Bunker *bu*	19,805.68	Port costs $b(c)$	26,377.00
Port costs *b*	215,676.76	Total $Ct(c)$	2,122,377.00
Total Ct (tu)	599,964.88	Cape year $y(c)$	63
Port bunkerbc	91,510.50	Total cost $Cta(c)$	133,709,751.00
Total cost – year *Cta (tu)*	37,889,298	Demurrage year *d*	18,409,851.33
		Despatch year *e*	–703,863.48
		Total year $Cta(c)$	151,415,738.85
Total logistics cost – year	**189,305,036.74**	***Clt(C)***	

5.2.3 *Logistics costs for Scenario (E): transshipment in São Sebastião-SP*

This scenario considered the costs of transshipment with 2 TUs and 1 Capesize, in which cargo was loaded in Santos-SP and transshipped on the sheltered Waters of São Sebastião-SP. Table 11 presents a summary of these costs. The cost of the TUs was impacted by the longer distance between the Santos-SP terminal and São Sebastião-SP port.

For the Capesize, the highest impact was in demurrage by the longer queuing time of the OGV to receive cargo from TUs in Santos to Sebastião-SP.

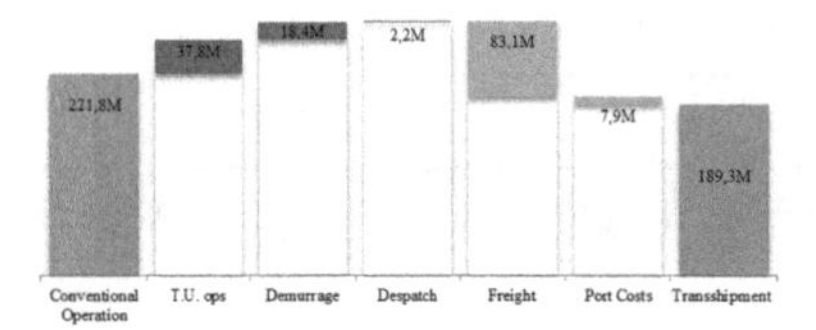

Figure 7. Comparison of costs—conventional and transhipment.

5.3 *Summary of results*

All the proposed transshipment scenarios incurred in lower annual total logistics costs when compared with the conventional scenario (A). However, scenario (D) presented the lowest costs. Considering that transshipment is usually conducted on sheltered waters, the option of using São Sebastião-SP (E) incurred in the second lowest logistic costs.

Figure 7 presents a summary of the additional costs (in red) and generated savings (in green) relative to one year's worth of operation. In conclusion, among all the impacting factors, the most relevant was annual savings with freight, which by itself as greater than the sum of all additional costs.

6 CONCLUSIONS

The operational and logistics cost analysis revealed that transshipment was the best option when compared to the conventional operation. The author suggests the use of scenario (D) and, when a lower-risk operation is preferred, scenario (E), as any associated additional costs were still lower than the costs of the conventional operation.

It is worth emphasizing that among the main cost impacts of this operation, the most relevant was the annual savings in maritime freights, which by itself was greater the sum of all additional costs. This is explained by the fact that the freight for Capesize ships is lower than Panamax ships, and less vessels are necessary in a year's time due to the Capesize's superior loading capacity.

In sum, future studies and projects should investigate relevant legislation, crewing, and the necessary licenses to carry out these operations in Brazil.

In conclusion, the present study showed that simulation scenarios can be used not only in the case of mineral ores in shallow draught ports, but also with other types of cargos and ports, especially in Brazil which has great potential to explore it.

We recommend future studies applied to other commodities, such as soy or sugar. Furthermore, with cargo subject to greater seasonality, the option of partially loaded ships could be explored, i.e., using vessels only in the peak shipping season, or partially-loaded or shared ships with other types of cargo. This could cheapen the operation and better meet the demand. Finally, the model can be used to explore the use of container operations.

REFERENCES

Baird, A.J.; Rother, D. (2012). Technical and economic evaluation of the floating container storage and transshipment terminal (FCSTT). Transportation Research part C. Elsevier.

Banks, J.; Carson II, J.S.; Nelson, B.L.; Nicol, D.M. (2005). Discrete-event system simulation. p. 9 to 11, Pearson

Bertoloto, L.P (2010). Modelo de previsão de frete marítimo de minério de ferro utilizando redes neurais artificiais. Master's Dissertation, UFF.

Botter, R.C. (2002). Tratamento de dados em simulação discreta. Thesis (Ph.D.) Poli—USP, Depto. De Engenharia Naval e Oceânica, São Paulo, Brazil.

Bowersox, D.J.; Closs, D.J.; Cooper, M.B. (2011). Supply Chain Logistics Management, 4th edition, McGraw-Hill, USA.

Brown, M. (2013). Exploration and Resource Definition of Offshore Titan-Magnetite Iron Sands, on the West Coast of New Zealand. Offshore Technology Conference, Houston.

Buckeley, P.; Lee, K.; Kuby, M. (1986). Evaluating dredging and offshore loading locations for U.S. coal exports using the local logistics system. Annals of Operations Research 6, p. 163–180.

Chwif, L., Medina, A.C. (2007). Modelagem e Simulação de Eventos Discretos 2nd edition. São Paulo, Brazil.

Cigolini, R.; Pero M.; Rossi, T. (2011). Sizing off-shore transshipment systems: a case study in maritime dry bulk transportation. Production Planning and Control, Taylor & Francis.

CODESP (2015). Panorama do Porto de Santos. Santos.

INTERNATIONAL TRANSPORT FORUM. (2015). The impact of Mega Ships. Paris.

IPEA (2009). Gargalos e demandas da infraestrutura portuária e os investimentos do PAC: mapeamento IPEA de obras portuárias. Brasília.

Kurt, I.; Boulougouris, E.; Turan, O. (2015). Cost based analysis of the offshore port system. International Conference on Ocean, Offshore and Arctic Engineering. Canada.

Liang, C.; Hwang, H.; Gen, M. (2011). A berth allocation planning problem with direct transshipment consideration. Journal of Intelligent Manufacturing, Springer US.

Mendonça, Paulo C.C.; Keedi, S. (1997). Transportes e Seguros no Comércio Exterior. Aduaneiras, Sao Paulo.

OLDENDORFF (2016). Transbordo de minério brasileiro em Trinidad. Trinidad.

PLATTS (2017). Índice de fretes marítimos.United States.

PORT TECHNOLOGY (2012). Berth productivity will have to keep up with shipping's supersized revolution. Maersk Line, 50 edition, p. 18–20, Denmark.

Silva, R.C.S.; Botter, R.C; Trevisan, E.F.C; Medina, A.C.; Pereira, N.; Netto, J.F. (2009). Planejamento de um Sistema de Transshipment para a exportação de carvão utilizando Simulação de eventos discretos. USP, São Paulo.

Souza, C.E.S. (2012). Modelagem e análise de duas alternativas para operações de transferência de petróleo entre dois navios em altomar. Master's Dissertation, USP, São Paulo.

Teixeira, V.B. (2011). Operações de transbordo de petróleo nacional na baía de Ilha Grande. Master's Dissertation UFRJ.

VETRIA (2015). Projeto Integrado. Brazil.

Wang, Y. (2015). Operability study of floating bulk transshipment operation. Delft University of Technology.

Maritime transportation and economics

Progress in Maritime Technology and Engineering – Guedes Soares & Santos (Eds)
 ISBN 978-1-138-58539-3

Evaluation of the Portuguese ocean economy using the Satellite Account for the Sea

A.S. Simões
Escola Superior Náutica Infante D. Henrique (ENIDH), Paço de Arcos, Portugal
Centre for Marine Technology and Ocean Engineering (CENTEC), Instituto Superior Técnico, Universidade de Lisboa, Lisbon, Portugal

M.R. Salvador & C. Guedes Soares
Centre for Marine Technology and Ocean Engineering (CENTEC), Instituto Superior Técnico, Universidade de Lisboa, Lisbon, Portugal

ABSTRACT: The drawing up of a Satellite Account for the Sea is the most appropriate instrument to estimate the dimension of the ocean economy, as well as to provide information on the production structure of the sea-related economic activities. Portugal became the first EU Member-State to prepare a Satellite Account for the Sea. This paper uses the published data to analyse the relative weight of the sectors of the Portuguese maritime economy. The results show the importance of the sectors "Fisheries, aquaculture, processing, whole-sale and retail of its products" and "Recreation, sports, culture and tourism"; together they represent 91% of all economic units, 61% of GVA and 67% of employment. International comparisons with other EU MS indicate that in Portugal the ocean economy weighs more heavily (in relative terms) than in its European partners. This is particularly true in terms of employment due to the dominance of labour-intensive sectors.

1 INTRODUCTION

The maritime industries are an important contributor to national economies. They are perceived as a significant and growing component of economic development and yet little reliable and consistent economic data exist to support this perception. But with the lack of steady concepts and definitions, as well as data, it is impossible to measure the ocean economy accurately. Maritime industries are difficult to quantify and the continued emergence of new activities prevent their inclusion in the national accounting systems.

Various attempts have been made to look at the maritime industries as forming a Maritime Cluster in various countries and in Portugal also (Ferreira et al. 2015; Salvador et al. 2016). However, there has been a clear difficulty in quantifying the relative importance of the sectors of the Cluster, as the economics of each sector of the cluster must be quantified on the basis of the official statistics of the country and it happens that the logic in which the economic sectors are organised in the national statistics does not allow direct extraction of data for the specific sectors of a Maritime Cluster.

As the OECD (2016:164) puts it: "*There are many reasons for wishing to put a value on ocean-based industries (…) It raises public awareness of the importance of the industries, offering them higher visibility; it raises awareness among policy makers, rendering the industries more amenable to policy action; it enables progress in their development to be tracked over time; it also enables their contribution to the overall economy to be tracked in monetary and employment terms; and finally, it lands weight to the perception of ocean-based industries as a set or clusters of activities whose defining common denominator is the ocean, its use and its resources*".

Several countries have been attempting to measure their national ocean economies with results astonishingly dissimilar. It is the case of the works by Pugh & Skinner (2002) and Pugh (2008) for the UK; Allen Consulting Group (2004) for Australia; Kalaydijan et al. (2009, 2011, 2014) for France. Also the "US National Ocean Economics Program" started to publish, since 2009, on a regular basis, the "State of the US Ocean and Coastal Economy".

The literature survey performed for this paper has identified a small number of countries that have already estimated their Satellite Account for the Sea (SAS). They are Australia, Canada, Japan and New Zealand. Also Belgium, China, Ireland, Portugal and South Korea have been developing active research in the attempt to measure their national ocean economies.

According to the OECD (1993), "*satellite accounts provide a framework linked to the central accounts and which enables attention to be focused on a certain field or aspect of economic and social life in the context of national accounts; common examples are satellite accounts for the environment, or tourism, or unpaid household work*".

The Satellite Accounts are an extension of the National Accounts System (NAS), allowing the integration of statistical data, classifications and alternative methods that do not correspond to the general model of the NAS.

The drawing up of a SAS is the most appropriate instrument to estimate the dimension and the importance of the ocean economy in the whole country, as well as to provide information on the production structure of the sea-related activities.

The Directorate-General for Maritime Policy (DGPM) and "Statistics Portugal" (INE) signed a Protocol, in June 2013, for the elaboration of a "Satellite Account for the Sea". For this purpose, a working group composed of representatives of different public institutions, universities and research centres was set up. The first data set was published in June 2016 relative to the period 2010–13. Portugal thus became the first EU Member-State to prepare a SAS.

According to DGPM, the objectives of this initiative were the following: (i) to measure the relevance of the ocean economy in Portugal; (ii) to support decision making regarding the coordination of public policies for the ocean; (iii) to monitor the National Ocean Strategy (NOS) 2013–2020; (iv) to provide reliable and adequate information for Portugal in the context of the EU Integrated Maritime Policy (IMP) and other processes where data for the ocean economy is decisive.

For this "account" data was gathered in the following groups, according to a value chain logic: 1) Fisheries, aquaculture, processing, whole-sale and retail of its products; 2) Non-living marine resources; 3) Ports, transports and logistics; 4) Recreation, sports, culture and tourism; 5) Shipbuilding, maintenance and repair; 6) Maritime equipment; 7) Infrastructures and maritime works; 8) Maritime services; and 9) New uses and resources of the ocean.

The variables that were estimated are the following: Output; Intermediate Consumption; Gross Value Added; Gross Operating Surplus; Employment; Compensation of Employees; Other Subsidies on Production; Other Taxes on Production; Final Consumption; Gross Fixed Capital Formation; and Exports and Imports of Goods and Services.

This paper uses the published SAS data of 2016 to show how the SAS can be used as an analytical tool to measure the ocean economy. Its objective is to provide an understanding of the global ocean economy measurement, with particular attention to its importance.

2 SATELLITE ACCOUNTS METHODOLOGICAL ISSUES

Satellite Accounts have emerged in response to the need to expand the capacity of the National Accounts to analyse certain areas of economic and/or social interest in a flexible manner without overloading or disorganising the integrated structure of the National Accounts System.

The EU Commission, together with the OECD, the United Nations, the IMF, and the World Bank jointly published the "1993 System of National Accounts (SNA)" to update issues of measurement of market economies. Since then, these organisations have improved Satellite Accounts' methodology (UN Statistical Division et al., 2008; EU Commission et al., 2013).

Specific sectors of economic activity (Tourism, Health, Culture, Social Economy, to refer the most common), in different countries, have elaborated Satellite Accounts, in order to detail relevant information aspects from each sector's own perspective.

In addition to the documents above, the Portuguese SAS also used as methodological references the "EU National and Regional Accounts System Manual" (2010), the Eurostat proposal for the IMP database (Ifremer et al., 2009) and several DG MARE reports, carried out within Blue Growth.

According to these references and taking into account some Portuguese specificity, the following levels of observation of the "Ocean" were defined (Fig. 1):

i. Nuclear activities (typical maritime industries where a major part of the operations takes place at sea or whose products come from or are intended to be used at sea or at the coastline);
ii. Cross-activities (industries which are not typical at sea but which have a strong impact on the development of nuclear activities. They are supporting the industries considered in the SAS. All activities related to the supply of goods and services classified as "seafarers" (fishing and

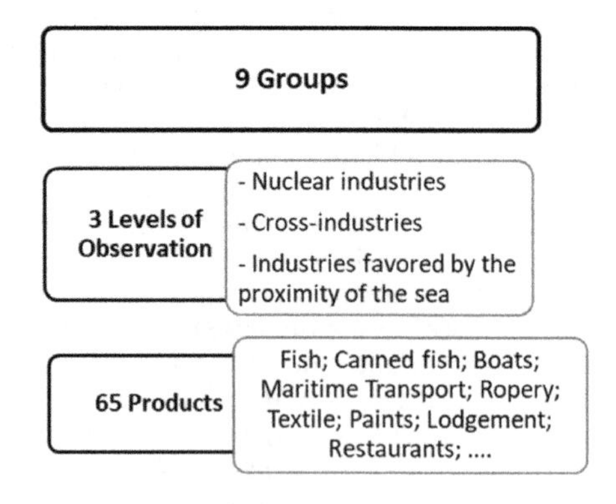

Figure 1. Outline of the SAS information.

aquaculture, fish processing, ports and maritime transport, shipbuilding and repair, etc.), as well as the whole structure of education, R & D, regulation and supervision of the maritime sector. Machinery and equipment for the maritime sectors, as well as engineering, insurance, classification, etc., should be measured here.

iii. Activities favoured by the proximity of the sea (coastal tourism).

The Portuguese SAS includes information from about 60,000 companies. The variables that could be calculated are: Output; Intermediate Consumption; Gross Value Added (GVA); Gross Operating Surplus (Gross Profit); Employment; "Compensation of Employees" (Wages); Other Subsidies on Production; Other Taxes on Production; Final Consumption; Gross Fixed Capital Formation (Investment); Exports and Imports of Goods and Services.

As referred by DGPM & Statistical Portugal (2014: 5), "*the SAS joins the set of economic activities carried out in the sea and others, dependent upon the sea, including non-tradable marine ecosystems*".

In addition, nine industry groups were identified, following a value chains' logic (Figs. 1 and 2): (i) Fisheries, aquaculture, processing, wholesale and retail of its products; (ii) Non-living marine resources; (iii) Ports, transports and logistics; (iv) Recreation, sports, culture and tourism; (v) Shipbuilding, maintenance and repair; (vi) Maritime equipment; (vii) Infrastructures and maritime works; (viii) Maritime services; (ix) New uses and resources of the ocean.

The chain of value logic was adapted to the greatest extent possible, taking into account, among other aspects, the level of disaggregation allowed by the National Statistical Office. Thus, it was a methodological option to consider the "Maritime Services" and the "Maritime Equipment" as autonomous groupings, containing economic activities transversal to the other groupings.

As expected, the group with the largest number of economic activity units was that of "Recreation, Sports, Culture and Tourism" (73.8% of the

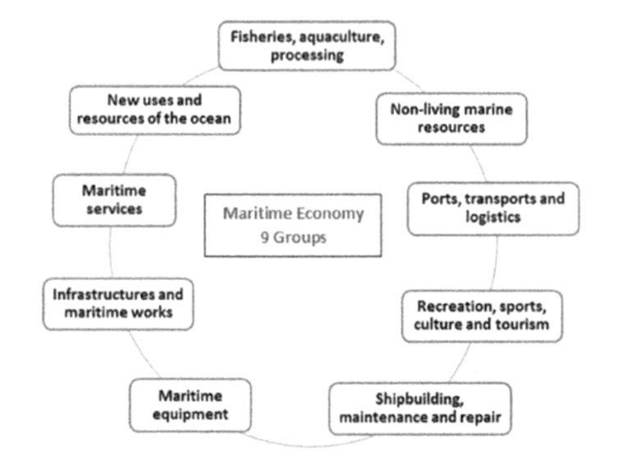

Figure 2. The 9 groups of the Portuguese ocean economy (according to SAS).

approximately 60,000 economic units considered in the Portuguese SAS), with emphasis in this group on hotels and restaurants (only for tourism purposes, in coastal areas).

The Eurostat and DG MARE traditionally defined as "coastal regions" (Classification of Territorial Units for Statistics—NUT's 3) those which have coastal areas or where more than half the population lives less than 50 km from the sea. Initially, the team that prepared the Portuguese SAS adopted this criterion. However, given the geography of the country, it soon became clear that this methodology overestimated the value of the Portuguese ocean economy.

Luckily, at that very moment, Eurostat (in the context of "Tourism Statistics") stated that its own criterion was very broad and proposed another one, where the parishes within each NUT 3, are classified as coastal or non-coastal, according to distance to sea: (i) if the parish is by the sea, it is an integral part of the coastal region; (ii) if the parish is not by the sea, but has more than 50% of its surface at a distance of 10 km from the sea, it is also considered a coastal parish.

As such, the Portuguese SAS team adopted this most recent criterion of Eurostat.

3 MAIN RESULTS

The Portuguese SAS includes industries located in the maritime space, in the coastal zones and also in areas far from the coast, as long as they are explicitly related to the sea.

According to the SAS final calculations, in the period 2010–13, the "Sea" represented 3.1% of Portuguese GVA (4.7 billion euros) and 3.6% of total employment (161 thousand Full-Time Equivalent—FTE), with the following composition:

- Nuclear activities—such as fishing and aquaculture, salting, shipbuilding, ports, maritime transports, coastal works, etc. – accounted for 1.7% of GVA and 2% of the employment.
- Cross-activities, i.e., maritime equipment and services, accounted for 0.6% of GVA and 0.7% of the employment.
- The activities favoured by the proximity of the sea, i.e., activities associated with coastal tourism, represented 0.8% of GVA and 0.9% of the employment.

In cumulative terms in 2010–13, "Ocean" GVA increased by 2.1% (largely driven by tourism and ports), while national GVA decreased by 5.4%. The "Ocean" employment decreased by 3.4% in the same period, which compares with the strong decrease of national economy as a whole (–10%).

With regard to productivity (GVA/Employment), SAS posted an accumulated growth of 6%

in the period under review (+ 5% for the national economy). However, the economy of the Ocean has levels of productivity below the national average (about 84% of that average, in 2013).

The period under consideration (2010–13) corresponded to a general contraction phase of the economic activity, with significant decreases in GDP and the employment. However, sea-related activities performed more favourably than the country average.

4 RESULTS BY GROUPS

4.1 *Number of economic activities*

The group with the highest number of units of economic activity is no. 4 – "Leisure, sports, culture and tourism" (73.8% of the 60,000 units included in the SAS). Of note in this group are hotel and catering industries (only for tourist purposes, in the coastal areas). This preponderance is also very pronounced in the GVA (35.5%) and in the employment (28.6%) of total SAS.

In second place is Cluster no. 1 – "Fishing, aquaculture, processing and marketing of its products", with 17.8% of the no. of economic units, 25.7% of GVA and 38.8% of employment. It should be noted that the processing and marketing activities play a decisive role in this group.

Table 1. Main indicators (by groups of industries) average values in the period 2010–13.

Groups of industries	No. of Econ. Units	GVA (10^6 euros)	Employment
Fisheries, aquaculture, processing, wholesale & retail of its products	10,296	1,203	62,414
Non-living marine resources	83	49	2,333
Ports, transports and logistics	1,092	676	15,086
Recreation, Sports, Culture, Tourism	43,370	1,660	45,950
Shipbuilding, maintenance, repair	373	119	4,404
Maritime equipment	495	159	9,028
Infrastructures and maritime works	772	65	2,850
Maritime services	2,235	741	18,615
New uses and resources ocean	22	7	88
Satellite Account for the Sea (SAS)	58,738	4,680	160,766
Portuguese Economy	–	152,425	4,409,186
SAS/National Economy	–	3.1%	3.6%

Source: Statistical Portugal (2016).

4.2 *Gross value added*

Group 4 (Tourism) generated 35.5% of total SAS GVA, followed by Group 1 (Fisheries and Processed Products) with 25.7% and Group 8 (Maritime services) with 15.8%.

Between 2010 and 2013, SAS GVA increased by 2.1%. This evolution was due mainly to Groups no. 1 (+ 4%), 3 (+ 30%) and 4 (+ 5.4%).

4.3 *Employment and wages*

Curiously, as regards Employment, the ranking of the main groups is exactly the reverse from that of GVA.

Group no. 1 (Fishing) represents 38.8% of the total, followed by Group no. 4 (Tourism) with 28.6%. Together, these two groups are responsible for almost 70% of the total Portuguese SAS employment. Groups 8 (Maritime Services) and 3 (Ports, Transports and Logistic) represent, respectively, 11.6% and 9.4% of the total.

In terms of average remuneration, Groups 8 ("Maritime Services") and 9 ("New uses and resources of the sea") pay the highest wages (+88.8% and +57.4% compared to national average). At the opposite extreme, Groups 1 and 6 ("Marine equipment") have the lowest remunerations in SAS. This dispersion reveals the heterogeneity of the qualifications of the human resources in the different groups.

Respectively, Groups no. 4 and 1 are responsible for the payment of 32.8% and 25.6% of the SAS global payroll. There was a general decline in all sectors (between 2010 and 2013, wages fell 1.2%), with the exception of Groups 1 (Fishing) and 5 (Shipbuilding and Repair). Wages in "Maritime Services" (Group 8) fell by 5.4%.

In terms of average remuneration, Groups 8 ("Maritime Services") and 9 ("New uses and resources of the sea") pay the highest wages (+88.8% and +57.4% compared to national aver-

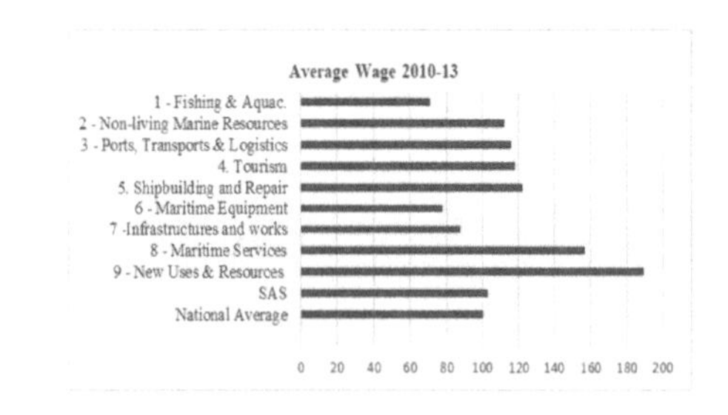

Figure 3. Average wages. Source: Statistical Portugal (2016).

age). At the opposite extreme, Groups 1 and 6 ("Marine equipment") have the lowest remunerations in SAS. This dispersion reveals the heterogeneity of the qualifications of the human resources in the different groups.

4.4 *Imports and exports*

Imports of maritime products and services fell by 35% between 2010 and 2013, from 4.3% national total to 2.8%. However, it should be noted that the year 2010 was exceptional due to the delivery of submarines to the Portuguese Navy. Considering only the period 2011–13, imports of the "Sea" fell by 1.5%.

Meanwhile, SAS exports increased by 12% between 2010 and 2013 (the national average was 25.2%). Thus, the weight of maritime exports declined in the national economy (3.3% in 2010 and 2.9% in 2013). With the exception of 2010, which had a negative trade balance, the remaining years present positive external balances.

The most important products in the structure of imports are "manufactured food products" (62.7% of the total) and "fishery and aquaculture products" (15%). Together these two products account for about 78% of the SAS imports.

In exports, Group 1 is also dominant, representing 41.4% of the total (32% manufactured food products—namely canned fish—and 9.4% fish and aquaculture products). This is followed by Groups 4 (24.7%), especially exports of accommodation services and 3 (12.4%), especially maritime transport.

4.5 *Gross fixed capital formation (Investment)*

Gross Fixed Capital Formation (GFCF) decreased by 74.3% in the period under review. However, not considering the year 2010, which is affected by the purchase of the submarines, the investment in SAS decreased 9.5% (–22.6% in the total national economy). Considering only the period 2011–13, the largest investments were made in "construction" and "civil engineering works" (38.7% of the total) and in "research and development services" (21.9%).

Table 2. Total GVA with and without Group 4 (Tourism) (M€).

2010	2011	2012	2013
4,616	4,699	4,689	4,715
SAS weight in national economy			
2.9%	3.1%	3.2%	3.2%
SAS without Tourism			
2010	2011	2012	2013
2,998	3,037	3,034	3,010
SAS weight in national economy			
2.5%	2.6%	2.8%	2.8%

Source: Statistical Portugal (2016).

5 EVALUATION OF THE PORTUGUESE OCEAN ECONOMY EXCLUDING TOURISM

Taking into account the weight that the "Recreation, Sports, Culture and Tourism" Group has in the total of the Portuguese SAS (73.8% of the number of economic units, 35.5% of GVA, and 28.6% of employment), it was decided to analyse the ocean economy by withdrawing this group. Tourism, in Portugal, has its own strong dynamics, clearly differentiated from the rest of the maritime sectors.

5.1 *SAS gross value added*

Even with the new more stringent Eurostat methodology, it was decided to measure the weight of the "Sea" in the national economy, removing Group 4. One concludes that "Recreation, sports, culture and tourism" – measured according to the new minimalist methodology proposed by Eurostat—represents, in average, mere 0.5 percentage points of national GVA. In 2013, without Tourism, the Groups with the highest GVA were Fisheries (no. 1), Ports (no. 3) and Maritime Services (No. 8) – (Fig. 4).

As can be seen in the Table 3, in relative terms, the weight of Group 4 in the Employment is higher

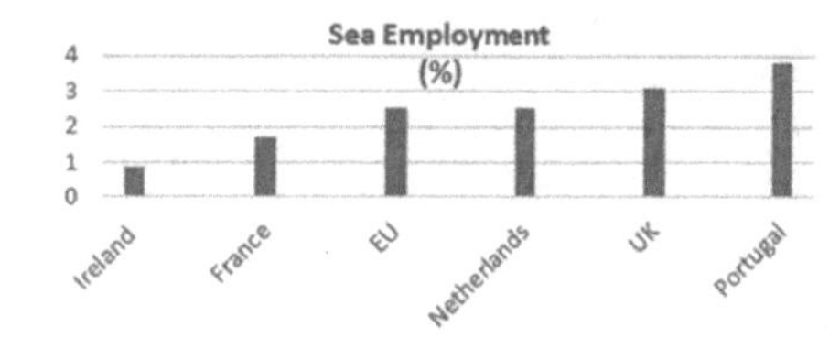

Figure 4. Ocean employment, in 2013. Source: OECD (2016).

Table 3. Total employment with and without Group 4 (Tourism).

			(FTE)
2010	2011	2012	2013
3,147.8	3,119.1	3.103.7	3,110.1
Weight in National Employment			
3.7%	3.8%	4.1%	4.1%
Total SAS without Tourism			
2010	2011	2012	2013
2.119,1	2.089,1	2.070,7	2.105,8
Weight in National Employment			
2.5%	2.6%	2.8%	2.8%

Source: Statistical Portugal (2016).

than its weight in the GVA. Tourism employment accounts for an average of 1.2% of national employment and 32–33% of total maritime employment.

From the above, it can be concluded that Eurostat's new minimalist methodology for quantifying Group 4 (Tourism) seems to be able to restrict it to activities directly linked to the sea.

6 INTERNATIONAL COMPARISONS

The OECD (2016) is also developing estimates about the value of the economy of the Ocean, although caution should be exercised with comparisons given the lack of harmonisation of classification of activities and methodologies between countries.

According to this source, the global ocean economy, measured in terms of the ocean-based industries' contribution to economic output and employment, is significant. Preliminary calculations value the ocean economy's output in 2010 at USD 1.5 trillion, or approximately 2.5% of world GVA. Offshore oil and gas accounted for one-third of total value added of the ocean based industries, followed by maritime and coastal tourism, maritime equipment and ports. Direct full-time employment in the ocean economy amounted to around 31 million jobs in 2010. The largest employers were industrial capture fisheries with over one-third of the total, and maritime and coastal tourism with almost one-quarter.

Many ocean-based industries have the potential to outperform the growth of the global economy as a whole, both in terms of value added and employment. The projections suggest that between 2010 and 2030 on a "business-as-usual" scenario basis, the ocean economy could more than double its contribution to global value added, reaching over USD 3 trillion

In what maritime employment is concerned Portugal ranks first among the EU countries (Fig. 4) largely due to the weight of the labour-intensive activities Groups of 1 and 4 (Fishing and Tourism) in its economic structure.

7 CONCLUSIONS

Economic activity in the ocean is expanding rapidly. The Portuguese Government has decided to proceed with the calculation of the EU's first Satellite Account for the Sea. It has thus placed itself at the forefront of international research in the area of maritime economy assessment.

However, the period studied in this paper corresponded to a deep economic crisis (Simões et al 2016). New calculations are needed for the post-crisis years. As such, much remains to be done with regard to the statistical and methodological base at national and international level for measuring the scale and performance of ocean-based industries and their contribution to the overall economy.

REFERENCES

Allen Consulting Group (2004). "The Economic Contribution of Australia's Marine Industries 1995–96 to 2002-03", The Allen Consulting Group Pty Ltd.

DGPM (2012). "The Maritime Economy in Portugal", Ed. DGPM, Lisbon [in Portuguese].

DGPM (2014). "The National Ocean Strategy 2013–2020", Lisbon [in Portuguese].

DGPM (2015), Annex A of the "National Ocean Strategy 2013–2020", Working Paper [in Portuguese].

DGPM/Statistical Portugal (2014). "Satellite Account for the Sea. Conceptual Definition of Economy of the Sea", Working Paper [in Portuguese].

EUROPEAN COMMISSION (2010). "The EU National and Regional Accounts System Manual", Luxembourg Publications Office of the European Union.

EUROPEAN COMMISSION (2013). "Tourism Satellite Accounts in Europe", Luxembourg Publications Office of the European Union.

Ferreira, A.; Guedes Soares, C., & Salvador, R. 2015; Features of the maritime clusters of the Atlantic Arc, in: Guedes Soares, C. & Santos T.A. (Eds.) Maritime Technology and Engineering, London, UK: Taylor & Francis Group; pp. 141–148.

Hara, T. (2008), "Quantitative Tourism Industry Analysis. Introduction to Input-Output, Social Accounting, Matrix Modelling and Tourism Satellite Accounts", Elsevier.

Ifremer et al. (2009). "Study in the field of Maritime Policy Approach towards an Integrated Maritime Policy Database", Eurostat, Luxembourg.

Kalaydijan et al. (2009). "French Maritime Economic Data 2009", Marine Economics Department, Ifremer, Paris.

Kalaydijan et al. (2011). "French Maritime Economic Data 2011", Marine Economics Department, Ifremer, Paris.

Kalaydijan et al. (2014). "French Maritime Economic Data 2014", Marine Economics Department, Ifremer, Paris.

OECD (2013). http://esa.un.org/unsd/sna1993/introduction.asp.

OECD (2016). "The Ocean Economy in 2030", Paris.

OECD, United Nations Statistical Division, IMF, World Bank & Commission of the European Communities (1993). "System of National Accounts US National Ocean Economics Program 1993", Brussels/Luxembourg, New York, Paris, Washington, D.C..

Pugh, D. & Skinner, L. (2002). "A New Analysis of Marine-Related Activities in the UK Economy", The Crown Estate, London.

Pugh, D. (2008). "Socio-economic Indicators of Marine-related Industries in the UK economy with Supporting Science and Technology", IACMST Information Document no.10, Inter-Agency Committee on Marine Science and Technology, Southampton.

Salvador, R.; Simões, A., & Guedes Soares, C. 2016. The economic features, internal structure and strategy of the emerging Portuguese mari-time cluster. Ocean and Coastal Management. 129, 25–35.

Simões, A.; Salvador. R., & Guedes Soares, C. 2016. The impact of the 2008 financial crisis on the Portuguese maritime cluster. In: Guedes Soares, C. & Santos T.A., (Eds.) Maritime Technology and Engineering 3. London, UK: Taylor & Francis Group; pp. 1197–1203.

United Nations Statistical Division, Eurostat, OECD & UNWTO (2008). "2008 Tourism Satellite Account: Recommended Methodological Framework", https://unstats.un.org/unsd/statcom/doc08/BG-TSA.pdf.

Progress in Maritime Technology and Engineering – Guedes Soares & Santos (Eds)
 ISBN 978-1-138-58539-3

Motorways of the sea

J.-M. Laurens & P.-M. Guilcher
ENSTA Bretagne, IRDL, Brest, France

ABSTRACT: The goal of the study is to trace the recent history of actions of the different governments to encourage Motorways of the Sea. The European Union has launched several funding programs to support MoS. An overview of the situation is given focusing on the examples of the *Saint-Nazaire/Gijón* and *Saint-Nazaire/Vigo* links. A solution is then proposed using six newly designed Ro-Ro vessels capable of transporting 110 semi-trailers each. The proposed logistics system implies that the tractor units will not be transported on board. Specialized Ro-Ro tractors will handle the installation of the semi-trailers on board whilst the tractor units are either coupled to newly arrived semi-trailers or returned to their base to fetch other semi-trailers. The six vessels are necessary to ensure three departures per day from each port. A complete ship design loop has been performed to prove the feasibility of the project. The vessels satisfy the rules and regulations of the IMO. The proposed twin screw motorization would run on LNG and allow the vessels to travel at 19 knots in sea states up to 5 in order to meet winter conditions in the Bay of Biscay.

1 INTRODUCTION

The number of motorized vehicles in the world has exploded in the last 10 years to around 1.3 billion vehicles today (source: OICA). This very important number raises several questions from an environmental and economic point of view as well as the logistical point of view. Estimates in this area are not always reliable, but the environmental footprint left by road transport on the planet is estimated at 13% of the global $CO2$ emissions, i.e. 6.4 Gt of $CO2$ in 2010 (IPCC, 2014), which corresponds to the consumption of 4.2 billion European households (SDES, 2017). These hallucinatory figures are directly linked to the most important problem that road transport is facing today: *traffic jams*. Motorists spent "*9% of their driving time stuck in traffic jams in 2016*", which is both a problem for our health and our economy. Indeed, a study by INRIX (a leading company in traffic data), aimed to raise "*the alarm bell to sensitize public awareness about the growing impact of traffic jams*" (INRIX, 2017), shows that by 2030, traffic problems will cost Western economies 221 billion Euros a year, hence 70 billion more than in 2013.All these figures are intended to sensitize the various governments to the growing danger of the saturation of roads and to encourage them to facilitate traffic flows. Diverting part of the traffic flow by means of motorways of the sea or short sea shipping is one of the various solutions offered.

2 DEFINITION OF MOTORWAYS OF THE SEA (MOS)

Short sea shipping has been examined for decades as an alternative to road freight. Since Europe and the USA were the first concerned by massive traffic jams, they produced the largest share of these studies, however Asia is also examining the possibility of increasing short sea shipping in its coastal regions. The many studies conducted in the USA underline the benefits of short sea shipping but they will not really materialize until the government massively invests in such a solution (Perakis, 2008). Europe is now coming to the same conclusion.

Motorways of the Sea officially appear for the first time in the European Commission's White Paper on Transport Policy for 2010 (EUROPEAN COMMISSION, 2001). This report was undoubtedly written as a consequence of the devastating fire of the Mont Blanc Tunnel in 1999, which was caused by a freight carrier. This report examines a program ending the same year, the Pilot Actions for Combine Transport (PACT) program (EUROPEAN COMMISSION, 1998). This program was a real start for the concept of MoS, in the sense that it "*encourages traffic to move from the road to other modes of transport in order to achieve a more balanced intermodal transport system*" (EUROPEAN COMMISSION, 1997). Indeed, thanks to the PACT program, the European Union has finally become aware of the existence of a common program in terms of freight transport in the member countries "*to reduce external costs of freight transport, such as those imposed by environmental damage, accident and congestion*" (EUROPEAN COMMISSION, 1997). Thus, the main idea of the PACT program, which was to financially support intermodal transport projects in order to make them more eco-friendly, is reflected in the conclusions of the 2001 report, evidence that the MoS have their own place in this program. The idea of MoS, as advanced in 2001, seems relatively simple: create motorways of the sea that will be as close as possible to the image of a highway, an easy and quick way to get from point A to point B. The MoS will have to propose "*a real competitive alternative to land transport*", while taking into account the PACT program's objectives, and in particular the nodal transfer for the transport of freight and not of passengers. Indeed, the European Union has estimated that "*the effects on traffic congestion of a transfer of passengers from the road to the seaway were minimal compared to what a transfer of freight traffic could accomplish*" (EUROPEAN COMMISSION, 1997). The European Union, by means of this White Paper and the introduction of MoS, seeks to raise the awareness among the various European players, while reiterating the desire to develop a real strategy for the use of transport means in Europe, in which the MoS could take part. However, no definition is clearly given to define the MoS: it was not until after 2004 that a real explanation of the terms "*motorways of the sea*" was provided (EUROPEAN COMMISSION, 1997).

3 EVOLUTION OF MOS WITHIN THE EUROPEAN UNION

After various consultations—which lasted two years—carried out in the different countries of the European Union with regard to the concept of MoS, the European Commission completed the Guide of Customs Procedures for Short Sea Shipping in April 2003 (EUROPEAN COMMISSION, 1997). This Guide will serve to draw up a working document explaining the customs procedures to adopt in order to simplify the formalities for the transport of freight between two States of the European Union as much as possible. It is therefore within the framework of an improvement of maritime transit that a new program was presented in 2003 encouraging the different Member States to develop MoS: the Marco Polo program. With a budget of 115 million Euros for the period 2003–2006, this program provided financial support to European governments which, lacking confidence, did not believe in the economic viability of the MoS, and hence did not commit themselves. This was planned by the European Commission: "*these lines will not develop spontaneously. It will be necessary, on the basis of the proposals of the Member States, to "label them", in particular through the granting of European funds to encourage their take-off and ensure an attractive commercial dimension*". Thus, thanks to the Marco Polo aid program, many MoS projects were carried out, notably in four priority regions that we will discuss later. In April 2004, a decision of the European Parliament and the Council clarified the expectations of the MoS, while reaffirming their strong desire to promote their development: "*support for the development of motorways of the sea should be considered complementary to the provision of an aid to promote the development of short sea transport activities under the Marco Polo program*". They provide a lengthy definition (extending to 6 points in total), with more details for MoS in Article 12-bis of this decision:

1. *The trans-European network of motorways of the sea within maritime flows on maritime routes for logistics, so as to improve maritime links by establishing new ones, reduce congestion and / or improve access to States and peripheral and island regions. Motorways of the sea should not exclude the combined transport of persons and freight, provided that cargo is predominant.*
2. *The trans-European network of Motorways of the Sea consists of equipment and infrastructure relating to at least two ports in two different Member States. Such equipment and infrastructure shall include elements, at least in one Member State, such as port equipment, electronic logistics management systems, safety and security procedures and administrative and customs procedures (...)*

In addition, this 2004 declaration lists specific projects, "*thus making definitive the concept of Motorways of the Sea*". These different projects take place in four different regions, namely:

- South-East Europe (motorway connecting the Adriatic Sea to the Eastern Mediterranean);
- South West Europe (motorway connecting the western Mediterranean to France, Spain, and Italy);
- Western Europe (motorway connecting the western Mediterranean to the North Sea);
- Baltic Sea (motorway connecting the Baltic States to central/western Europe).

All four have different issues: a MoS in the Eastern Mediterranean has to develop maritime transport, while in the Western Mediterranean, a MoS has to respond to congestion in the Alpine and Pyrenean regions. In view of these new directives, the Marco Polo II program has succeeded the Marco Polo program, and has a larger budget (450 million Euros for the period 2007–2013). It offers a wider geographical coverage, i.e. the required presence of only one of the two ports in the European Union and the other port in a neighboring country (EUROPEAN COMMISSION, 2006). The document also contains a new, clearer and more concise definition of MoS:

> "*Any innovative action to directly transfer freight from the road to short sea shipping or a combination of short sea shipping with other modes of transport where road journeys are as short as possible; actions of this type may include the modification or creation of ancillary infrastructures necessary for the implementation of a very high volume and high frequency intermodal maritime transport service, preferably including the use of the modes of transport the most environmentally friendly, such as inland navigation and rail transport, for the transport of freight in the hinterland and integrated door-to-door services. If possible, the resources of the outermost regions should also be integrated.*"

The Marco Polo II program ended in 2013 and we can only underline the very positive assessment made by the UE of this program through the scheme proposed in Figure 1.

This program was submitted for study to the European Commission and at present the last European plan promoting the MoS, the *Trans-European transport network* (TEN-T), has started. TEN-T is a program to improve the infrastructure of the European Union to transform all railways, roads, airports and shipping routes into a perfectly unified network. Its main goal will be to eliminate the fact that every day, "*7500 km of motorways are paralyzed by traffic jams*" (EUROPEAN COMMISSION, 2003) in Europe. It is therefore in May 2013 that the European Union agreed on 30 priority projects for 2020, representing an investment of 225 billion Euros by 2020. The MoS are part of these 30 projects (the 21st) and are included in the section "Maritime transport infrastructure and motorways of the sea" and can be qualified as "*Maritime leg of the Trans-European transport network*". A new definition, much shorter than the previous ones, is given and reinforces the idea of nodes and connections within the European Union: "*They (i.e. MoS) shall consist of short sea routes, ports, associated maritime infrastructures and equipment, and facilities as well as simplified administrative formalities enabling short sea shipping or sea-river services to operate between at least two spots, including a hinterland connection.*"

Thus since the 2000s, the European Union has implemented a policy to encourage interconnections between different types of traffic (road, maritime, rail ...) and in particular in the case of freight transport. The MoS are part of a true European policy and are intended to accelerate until 2020 with the TEN-T plan. But what about countries like France, which are directly concerned by this European policy?

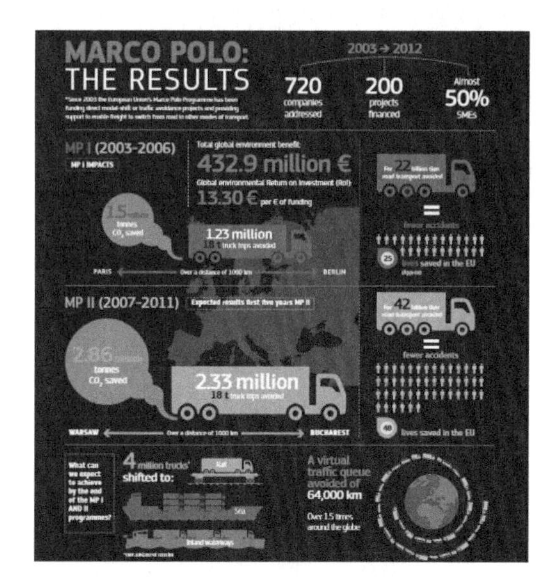

Figure 1. Results of the Marco Polo program *(according the UE, http://ec.europa.eu/transport/marcopolo/index_en.htm)*.

4 THE FRENCH CASE

France, like the rest of the European Union, has seen an explosion in the presence of vehicles on motorways, (Figure 2).

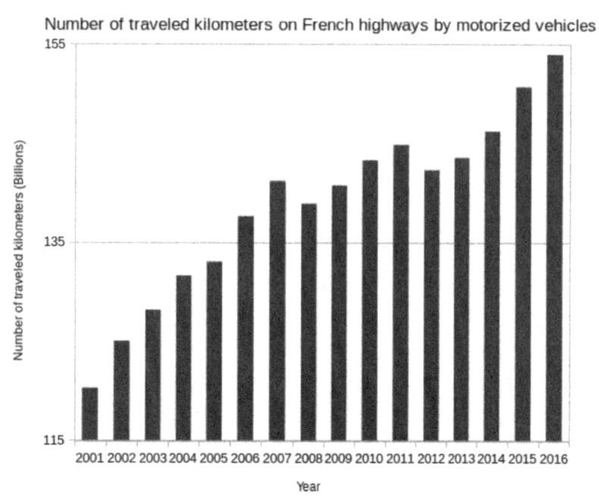

Figure 2. French road traffic.

Indeed, various axes are saturated like the Alps and the Pyrenees (natural obstacles), but also the Bordelaise region which ensures the transition between the North and South-West of France. France is therefore obviously at the heart of the various regions mentioned above, so it must participate in the different initiatives for the development of motorways of the sea. France aligns itself with the European policy in order to reduce the congestion of its motorways. Therefore, as expected, the desire to use motorways of the sea in order to smooth road traffic appeared in the objectives of the Grenelle of the sea (equivalent of the Grenelle of the environment for the maritime domain) in 2009, (Dang, 2014). As a reminder, the aim of the Grenelle of the sea is to propose new actions to be taken in the maritime sector, in order to ensure its economic and sustainable development. The Grenelle commits France, in the sense that it sets "*the objective to reroute from 5 to 10%*" the road traffic of the Alps and the Pyrenees towards a Mediterranean and Atlantic network of motorways of the sea. To this end, France joined forces with Spain, Italy and Portugal in 2015 "*to study the feasibility of an incentive for shippers using motorways of the sea*" (*Med Atlantic Ecobonus project*).

The Med Atlantic Ecobonus project has an estimated cost of approximately 1.5 million euros, half of which will be borne by the different countries of cooperation; the rest will be financed by the European Union. France is in charge of "*technical, environmental, economic and financial, regulatory, operational and technological specifications*". France is therefore a player in the development of motorways of the sea, shown by the Nantes-St-Nazaire/Gijón link.

5 THE MOTORWAY OF THE SEA NANTES-SAINT-NAZAIRE/GIJÓN

As part of the Marco Polo II project, the French and Spanish governments decided in 2009 to open a maritime link between the Montoir-de-Bretagne (Saint-Nazaire port) and Gijón terminals. This line aimed to unclog the motorway of Aquitaine as well as those crossing the Pyrenees, given the extreme traffic of trucks on these express ways: indeed, not less than 5.97 million trucks crossed the Pyrenees in 2010, according to the statistics of the Office of the General Commissioner General for Sustainable Development. It was not until September 2010, however, that this project could bear the "*MoS*" label, previously refused by the European Commission (EC), to allow the allocation of corresponding funding, and thus ensure the viability of the project. The refusal of the EC was based on the fact that the guidelines posted for this shipping line were directly linked to strict national requirements, in spite of the requirements of regional integration and innovation of the required transport model. To be as competitive as possible, the MoS Nantes-Saint-Nazaire/Gijón (see Figure 3) has to offer an attractive financial, economic, and ecological offer, otherwise all-road transport remains the preferred type of transport. To start the line, the French and Spanish States provided a cumulative aid of 30 million Euros (15 million for each country). The ship owner in charge of this line with a grant of more than 4 million Euros, provided by the European Union (EU) under the Marco program Polo II, commits itself to embark 150 trailers and 500 passengers with a frequency of 2–3 return trips/week to improve the Pyrenean and Atlantic road traffic.

To establish its economic model, the ship owner opted for the technique of the mixed ro-ro goods and passenger vessels. Indeed, ro-ro vessels (abbreviation of Roll-On, Roll-Off), are suitable for the transport of trucks, semi-trailers, tractors, agricultural machinery, construction equipment... They may also be suitable for containers or swap bodies. In particular, this type of vessel can ship heavy weights, the ship owner's core target for this MoS; this has allowed it to be able to offer two services for road carriers of all types: an escorted and an unaccompanied service. Detailed numbers are given in Dang's thesis (Dang, 2014). He came to the conclusion that: *a company that uses the MoS instead of traditional road transport would see its cost drop by **42.17%**! The saving of time is also substantial: **0.83 days**.*

Dang also underlines that MoS offer road carriers an efficient way to reduce their CO2 emissions. This judicious alternative makes it possible to significantly reduce the road distances traveled and, what's more, get rid of CO2 emissions, especially when transport vehicles are in traffic congestions. The studies have in fact been carried out by taking a constant CO2 emission over the entire

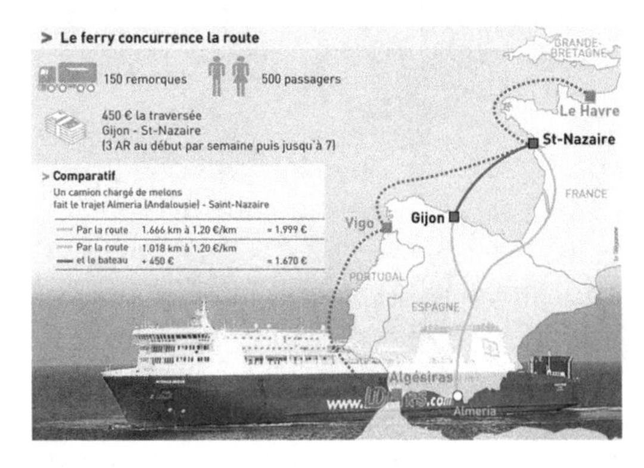

Figure 3. Ferry versus road.

duration of the road trip, which, in view of the previous elements, is certainly not the case. As a result, established rebates of 20 to 100% are very likely underestimated. The use of MoS Nantes-Saint-Nazaire / Gijón is a success from the ecological point of view: it amounts to reducing the CO2 emissions released into the atmosphere compared to the terrestrial way by **64.01%**. The MoS could therefore stress the environmental aspect to attract new partners.

Despite these promising prospects and the transport of more than 14,000 trucks in 2010–2011 alone, the MoS Nantes-Saint-Nazaire/Gijón was only able to maintain itself until September 2014, following the cutting off of European public subsidies.

Reasons for failure: The increased use of the line, due in part to the freight of Renault vehicles between the Spanish plant and France, was not enough to earn the six million Euros needed for the sustainability of the project. On the one hand, the filling rate was only 70%, which did not allow to reach the economic equilibrium and to face the very strong competitiveness of all-road transport. On the other hand, the line suffered from legal and political considerations intrinsic to the concept of MoS: a report of the European Parliament of December 2014 made "*the threefold finding that the concept is insufficiently known, including interested actors, that it is not viable without public funding and that cooperation between stakeholders is not necessarily present*" (Loyer, 2015).

Does that mean that we should give up and let road traffic continue to increase until total saturation? The main reason for failure is probably the timidity of the program.

To be permanently successful and to set an example to be followed, the Motorway of the Sea must show a significant reduction of the traffic. The statistics of the Marco-Polo programs bear witness to its efficiency but the public does not significantly perceive its effects and statistically it has been proved that the fluctuations of the economy are far more perceptible. To impact the traffic jams, it is estimated that the MoS must absorb at least 5% of truck traffic. In 2017, the ring road of Bordeaux is used by 12.000 transit trucks per day. The MoS must therefore be capable of transporting 600 trailers per day in each direction.

6 TWO SOLUTIONS

The most compact solution is to transport freight containers. The other solution is to transport the trailers using Ro-Ro vessels. To optimize these solutions, both types of vessels fitting the requirements are designed.

6.1 *Container ship design*

To transport the freight of 600 trailers per day in each direction, four container ships with a transport capacity between 300 and 360 TEU are needed. Larger container ships are not appropriate since they would require important harbor infrastructure modifications and with two departures per day in each direction we have less chance to run half empty ships. For the motorization, despite the fact that it uses more room, LNG is chosen because it generates far less pollution than Diesel. The main dimensions of the container vessel are given in Table 1.

The GA and capacity plans are given in Figure 4. The plans show where the containers are located; they also show the longitudinal and transverse bulkheads, the ballast tanks and the LNG propulsion system.

Stability calculations have been performed for two load conditions to verify that the vessel satisfies the IMO2008 and the SOLAS2009 rules. The scatter diagram of the Bay of Biscay weather conditions indicates that a sea state 4 should be considered for the design condition. In order to reach

Table 1. Main dimensions of container vessels[1].

Type	Container vessel 300–360 TEU
Crew	12
LOA	102 meters
Bext	18.20 meters
T	5.95 meters
L_{wl}	100.60 meters
B_{wl}	18.20 meters
C_b	0.74
C_p	0.80
Δ	8341 metric tons
S_w	3416square meters
V_{design}	14.5 knots

[1]Symbols are consistent with the ITTC Symbols and Terminology committee.

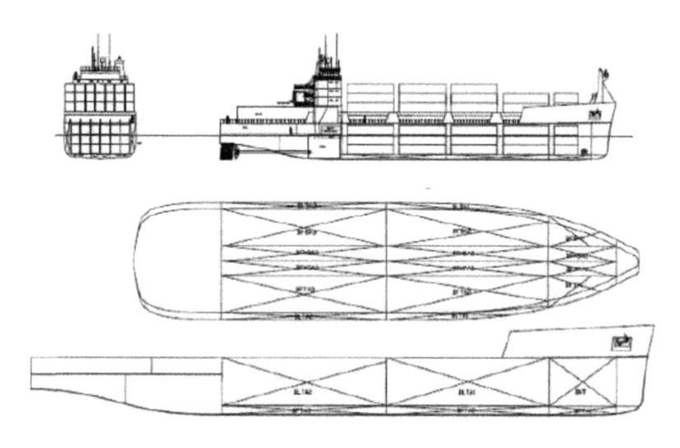

Figure 4. Container vessel's plans.

Table 2. Container vessel propulsive characteristics.

Engine	WARTSILA 6 L46F
Fuel	LNG
η_P	55%
1–w	0.72
1–t	0.83
Dprop	3.67 meters

the 14.5 knots design velocity in sea state 4, the chosen engine must have a power of more than 6 MW. The characteristics of the propulsive system are summarized in Table 2.

6.2 *Ro-Ro vessels design*

The ship is designed almost exclusively for the transportation of trailers without tractor. There is a gain of 20% in storage capacity compared to trucks. Because the vessel only transports trailers without drivers, it is far less complicated as there is no need for passenger accommodations. The unloading plus loading times will totalize no more than 6 hours using Ro-Ro tractors (Huub H.E. and Vrielink, 2016). The boarding of cars should remain exceptional and the number of passengers should not exceed 12 to avoid regulatory constraints of passenger ships amended by the SOLAS.

The target of the postponement of 600 trailers of the road to the sea (in both directions) will be made after a period of 3 years. Ultimately there will be six Ro-Ro vessels that will carry out the shuttle with three daily departures from each port. The main dimensions of the Ro-Ro vessels are given in Table 3.

The GA and capacity plans are given in Figure 5. The ship has three main decks. The lowest deck is located above the ballasts, 1.40 meters above the baseline. Furthermore, the gained space between the lowest deck and the next one offers enough room for the engine room, LNG tanks, MDO bunkers and other capacities. All ballasts are placed in the double bottom and double hulls. The two upper decks are reserved for the trailers. A retractable ramp is linking these two decks.

As previously, a complete design spiral procedure has been performed including stability, strength, maneuvering and seakeeping. The design speed in sea state 4 is taken to be 19 knots to achieve the same goal. It can also reach 18.3 and 17.5 knots for sea states 5 and 6 respectively. The ship is propelled by 2 dual fuel LNG engines with two fixed pitch propellers which offer better maneuvering capabilities. The propulsive system characteristics are summarized in Table 4.

Table 3. Main dimensions of ro-ro vessels.

Type	Ro-Ro 110–120 trailers
Crew	15
LOA	150 meters
Bext	25 meters
T	6.15 meters
L_{wl}	149 meters
B_{wl}	25 meters
C_b	0.57
C_p	0.60
Δ	13923 metric tons
S_w	4149 square meters
V_{design}	19 knots

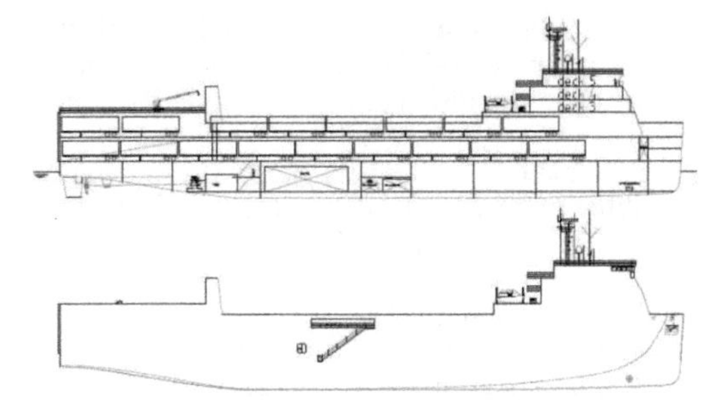

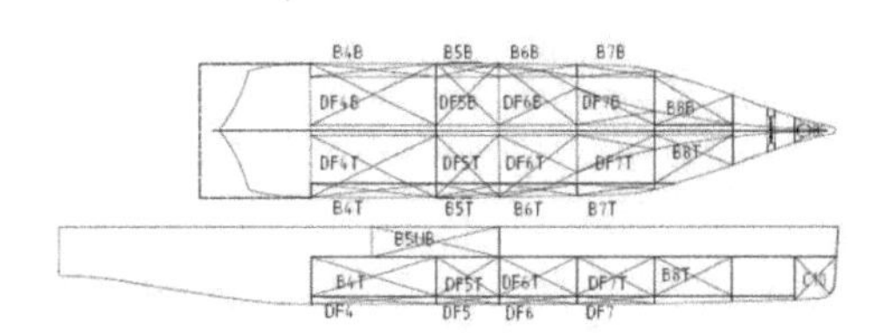

Figure 5. Ro-ro vessel plans.

Table 4. Ro-ro vessel propulsive characteristics.

Engine	WARTSILA 8 L46DF
Fuel	LNG
η_P	60%
1-w	0.88
1-t	0.87
Dprop.	3.67 meters

7 CONCLUSIONS

The previous Bay of Biscay Motorway of the Sea did not succeed in achieving a modal transfer of road to sea traffic to a significant extent. They failed in their target of decongesting the Bordeaux ring road and did not manage to establish a dominant position in the highly competitive freight transport market.

The primary suspect cause of failure is the lack of ambition of the project. Two solutions have been pre-designed and pre-dimensioned to reduce road traffic between the South-West of France and the North of Spain by at least 5%.

The most relevant solution seems to be transport by ro-ro vessels. The operating prices for this line are reduced by the attractive LNG fuel prices and reduced port charges due to regular and intense port traffic. By combining these regular departures (3 per day from each port) and attractive LNG prices, road transport companies should use this solution and gradually prefer it to classical road transport.

REFERENCES

Dang, K.L. (2014). *Les autoroutes de la mer. Géographie. Thèse de Doctorat.* Le Havre: Université du Havre.

EUROPEAN COMMISSION. (1997). *PACT—A user's guide.* Luxemburg: Office for Official Publications of the European Communities.

EUROPEAN COMMISSION. (1998). *Council Regulation (EC) No 2196/98 of 1 October 1998 on the granting of Community financial support for innovative measures in support of combined transport.* Luxemburg: OJ L277.

EUROPEAN COMMISSION. (2001). *Livre blanc—La politique européenne des transports à l'horizon 2010: l'heure des choix.* Luxembourg: Office des publications officielles des Communautés européennes.

EUROPEAN COMMISSION. (2003). *L'Europe à la croisée des chemins. Le transport durable: une nécessité. Série: L'Européen mouvement.* Luxemburg: Office des publications officielles des Communautés européennes.

EUROPEAN COMMISSION. (2006). *Annex 1 of Regulation, No 1692/2006 of the European Parliament and of the Council.* Luxemburg: Office for Official Publications of the European Communities.

Huub H.E. and Vrielink, O. (2016). *Exposure to whole-body vibrations of drivers of a roll-on/roll-of (RORO) tractor.* Ergolab research report.

INRIX. (2017). *Congestion is growing: so how do we tackle it ?* Online.

IPCC. (2014). *Fifth Assessment Report of the Intergovernmental Panel on Climate Change IPCC.* Geneva, Switzerland.

Loyer, E. (2015). *Les autoroutes de la mer en Méditerranée: une stratégie juridique pour un transport durable et une régulation compétitive du transport maritime.* Nice: Thèse de Doctorat en Droit.

Perakis, A.N. (2008). A survey of short sea shipping and its prospects in the USA. *Maritime Policy & Management*, 591–614.

SDES. (2017). *Service de la donnée et des études statistiques (SDES) du ministère de la Transition écologique et solidaire.* Online.

Progress in Maritime Technology and Engineering – Guedes Soares & Santos (Eds)
© 2018 Taylor & Francis Group, London, ISBN 978-1-138-58539-3

Characterizing the operation of a roll-on roll-off short sea shipping service

T.A. Santos & C. Guedes Soares
Centre for Marine Technology and Ocean Engineering (CENTEC), Instituto Superior Técnico, Universidade de Lisboa, Lisbon, Portugal

R.C. Botter
Universidade de São Paulo, São Paulo, Brazil

ABSTRACT: This paper presents a study on a roll-on roll-off (ro-ro) liner service connecting Leixões with Rotterdam. This line is currently unique in Portugal, but is considered highly important in the promotion of modal shift from road transportation to intermodal transportation, in line with European Union transport policies. The liner service is characterized in terms of frequency, size and characteristics of ships, navigation speeds, round voyage time and arrival and departure times. Uncertainties in some of these variables are assessed. The dedicated port terminals are characterized in terms of type of cargo units handled, technical characteristics of infrastructure, terminal equipment and operational procedures. This information will support a discrete event simulation of intermodal transportation between Portugal and Northern Europe. Conclusions regarding major features of the operation of this service are drawn.

1 INTRODUCTION

The European Union (EU) has been promoting, for many years now, the competitiveness and sustainability of supply chains in the union as a way to enhance the development of a common market and reduce the external costs related with freight transportation. Considering this, it is well known that transport modes such as short sea shipping (SSS), inland waterways and freight railways can provide significant advantages. A significant number of intermodal transport solutions have been promoted throughout the years under different financing programs, starting from 1992. Programs such as PACT, Marco Polo I and II, Motorways of the Seas and TEN-T have provided financing for such innovative solutions. Intermodality has also attracted significant interest from academia, resulting in a substantial body of literature, which has reviewed and analyzed the reasons for the successes and failures of intermodality and SSS in the EU, namely Baird (2007), Styhre (2009), Douet & Cappuccilli (2011), Baindur & Viegas (2011), Aperte & Baird (2013) and Ng *et al.* (2013).

On the practical side, it is well known that countries located in the periphery of the EU, such as Portugal and Spain, have to use the road networks of neighboring countries for conveying their exports and imports to central and northern Europe. Such countries are increasingly regulating road haulage (resting time, minimum wages and enforcement of cabotage protection), applying tolls and restricting the passage of freight vehicles in specific parts of the road network, thus attempting to reduce road congestion, road wear (maintenance costs), noise and pollution. Finally, bad weather or social unrest (strikes, blockades) frequently hamper the free passage of freight vehicles in EU roads and fuel prices are increasing volatile. All these circumstances add substantially to cost and uncertainty in road transportation.

It is therefore important to find competitive and sustainable alternative transport solutions for supporting the development of trade between peripheral countries and central and northern Europe. Intermodal transport solutions which include a maritime link could play a more significant role in alleviating the mounting pressures on fully road based freight transportation, particularly if the maritime link resorts to roll-on/roll-off (ro-ro) ships, characterized by faster cargo handling and low dwell times in port.

Several studies have been devoted to the evaluation of the technical and economic feasibility of intermodal solutions across specific corridors between this peripheral countries and central and northern Europe (Trant & Riordan, 2009). Another example, by Santos & Guedes Soares (2017a,b), presented two studies dedicated to the Portuguese case, considering the utilization of roll-on/roll-off (ro-ro) ships. In the first study, the transportation demand for an intermodal solution, comprising ro-ro ships and pre and post road haulage, was estimated for various combinations of ship speeds and freight rates, taking the perspective of the shipper. In a second study, a methodology was developed to identify the ro-ro ship and the required fleet size necessary for the estimated transport demand, allowing the identification of the optimal point of operation (ship speed and freight rate), taking the point of view of the shipping company rather than the shipper.

These studies considered mainly the point of view of the shippers and shipping companies, but it is also important to recognize that efficient roll-on/roll-off port terminals also play a fundamental role in providing cost and time efficient cargo handling operations in these fundamental intermodal nodes of supply chains. Therefore, it is under development a simulation study of integrated intermodal transportation using pre-haulage (road based), short sea shipping (ro-ro based) and post-haulage (road based). Such studies, see Keceli *et al.* (2013), typically require statistical data describing the operations in general and particularly in the port terminal, in addition to the fundamental technical characteristics of the ro-ro terminal. The main aim of this paper is to report research carried out to obtain such information and the ensuing data analysis.

The paper is organized in the following manner. Section 2 presents the methodology adopted in the research. Section 3 presents the basic information on the route under consideration, the characteristics of the port terminals, the ships employed in the line and the pricing policy. Section 4 presents data relative to the port schedule and operations. Finally, section 5 reports the main conclusions of the study.

2 METHODOLOGY

The characterization of intermodal transportation which includes a short sea link involves a significant number of deterministic and random variables, required in order to be able to set up a discrete event simulation model. This study covers mainly the variables directly related to this short sea link, thus neglecting the variables associated with pre-haulage and post-haulage. In the short sea link there are variables connected with the ship characteristics and operational details and variables related to the ro-ro port terminals. Finally, weather conditions at sea should also be taken in consideration. Table 1 synthetizes the main variables involved.

Table 1. Variables required for modeling short sea shipping operations.

Ship	Main dimensions
	Cargo capacity
	Operation speed
	Maneuvering time
Ro-ro terminal	Main dimensions
	Berth characteristics
	Cargo handling equipment
	Cargo handling productivity
Marine environment	Wind and wave conditions

The methodology to obtain information that enables characterizing these variables has been the following. Firstly, the website of the company running this regular liner service has been accessed and the information obtained there has been analyzed, enabling a first approximation to the characteristics of the service. Secondly, a visit was carried out to the ro-ro terminal in Leixões (Terminal Multiusos do Porto de Leixões) on the 6th December 2017, including also a visit to a cargo ro-ro ship, which was undertaking loading/unloading operations in that day. Interviews with shipping company representatives and crew allowed additional information to be obtained. Thirdly, information was gathered from port websites relating to the characteristics of some other port terminals called by ships in this line (Rotterdam and Santander). Fourthly, ships currently deployed in this line where tracked using the marine traffic website to evaluate typical speeds used in different sections of the route and its main characteristics were obtained from classification societies databases and equasis database. Fifth, an extensive analysis of the information provided by the Port Authority of Leixões (APDL) was carried out, covering the arrival and departure dates and times of ships deployed in this route over the period 2013–2017.

3 CASE STUDY: A RO-RO SHORT SEA SHIPPING SERVICE

3.1 *The short sea shipping route*

Given the importance of ro-ro services for conveying Portuguese cargos to Northern Europe, a case study is undertaken to examine the existing ro-ro service from the port of Leixões, in northern Portugal, to the port of Rotterdam in the Netherlands. Currently, this service is operated by a Belgian/Luxemburg shipowner. The company operates 25 ships, fully owned or under affreightment. It also operates ro-ro terminals and conducts logistics operations through a dedicated logistics company. The main terminal is located in Rotterdam, but other terminals in Gothenburg, Purfleet, Zeebrugge, Esbjerg, Killingholm, are also connected to Rotterdam, as shown in Figure 1. The company will receive shortly 3 new ro-ro ships built in Croatia and 3 ro-ro ships built in South Korea.

This regular line uses cargo ro-ro ships similar to the ones shown in Figure 2. The port of Leixões had already received a similar service, bound for

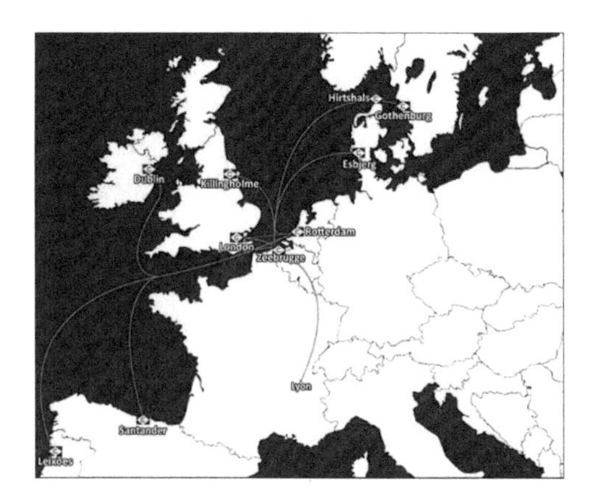

Figure 1. Ro-ro services network including the Leixões to Rotterdam route. (Source: Shipping company website).

Figure 2. Ships deployed in the current and in previous ro-ro services in Leixões.

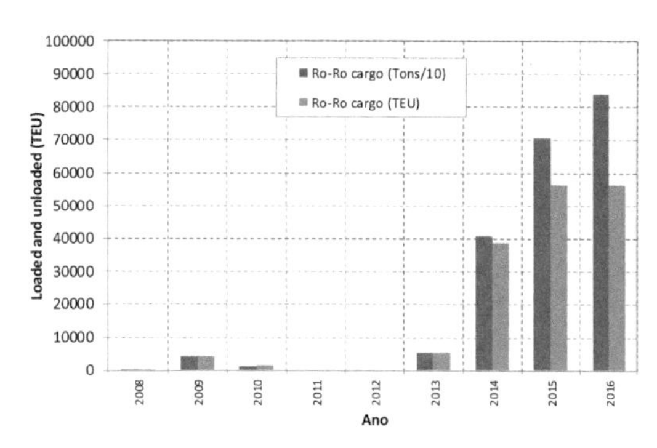

Figure 3. Roll-on/Roll-off cargo in the port of Leixões. (Source: APDL (2008–2016)).

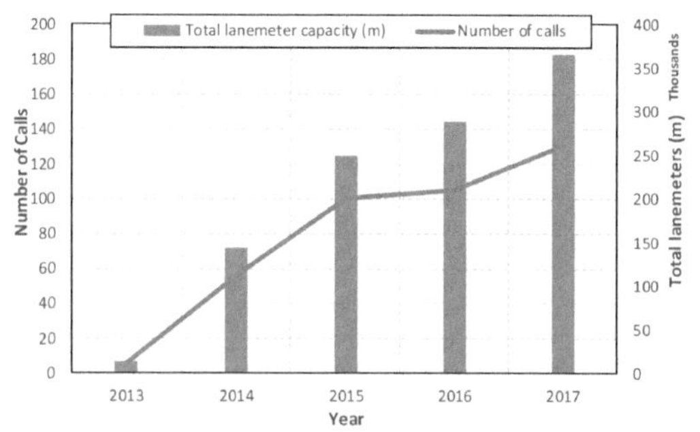

Figure 4. Number of calls and total lanemeter capacity of ships 2013–2017. (Source: APDL (2013–2017)).

Southampton (UK), in the past. The second ship in Figure 2 was used in such past services, but at the time the ro-ro terminal was on the north side of the port, as shown in Figure 9.

Currently (2017) the shipping company is carrying out 3 weekly departures from Leixões. The route between Leixões and Rotterdam is 980 miles long and takes about 2.5 days (each way). The round voyage is completed in one week, enabling each ship to ensure a weekly frequency. Crew of visited ship reported that the ship consumes 140 tons of MGO per leg and bunkering is carried out every three weeks in a northern European port.

It is worth pointing out that ro-ro services of this type have had varying degrees of success. The service in the past for Southampton, for example, was short-lived. The current one has however performed very well, being the main responsible for the growth in roll-on roll-off cargo in the port of Leixões, as shown in Figure 3.

As mentioned before, a survey has been carried out on the arrivals and departures of ships in this line covering the years between 2013 and 2017, using information from APDL. Figure 4 shows the number of calls and the total lanemeter capacity of ships deployed in the line, per year. It is possible to see a significant increase in number of calls and total capacity of ships. The number of calls is roughly equivalent to one weekly call offered in 2014, growing to two calls per week in 2015 and three calls per week in 2017.

Figure 5 shows the total lanemeter capacity per month throughout the period 2013 to 2017. It is possible to see significant fluctuations throughout the year but no seasonality in the capacity offered to the market. The average number of calls per month increases year after year, as is evident in the generalized upward trend shown in the graph. The current line started operations in late 2013, so calls shown in the figure correspond only to the ones that existed in the last two months of that year.

The analysis of data provided by APDL regarding ship origins and destinations shows that this line is not as simple as weekly round voyages to Rotterdam and back to Leixões. Figure 6 shows the origins and destinations of ships trading in this line during 2017. It may be seen that other ports of origin exist, such as Santander and Zeebrugge, and other ports of destination also exists,

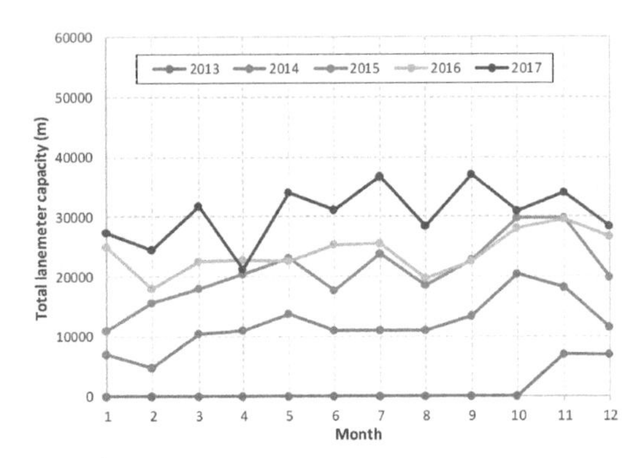

Figure 5. Total lanemeter capacity per month 2013–2017.

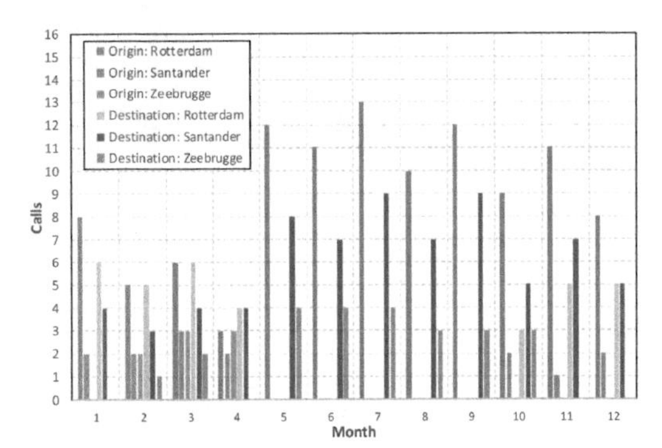

Figure 6. Origins and destinations of ships during 2017.

such as Santander and Zeebrugge. It is also possible to see that the pattern of operation changed during Spring-Summer. During this season (May to September) the ships came from Rotterdam directly and in the return voyage called frequently in Santander or went directly to Zeebrugge. During winter the ships came mainly from Rotterdam, but also in some cases from Santander and Zeebrugge. Destination during the Winter season was primarily Rotterdam, but in many cases, it was also Santander. Zeebrugge was rarely a direct destination in Winter.

3.2 *Characteristics of ro-ro ships deployed in this service*

The data from APDL shows that several different ships have been used in this ro-ro service over the period of 2013–2017. On the day of the visit to the terminal in Leixões, the ro-ro cargo ship *M/V Bore Sea* was moored alongside the quay. Table 2 shows the main characteristics of this ship and, for purpose of comparison, the characteristics of other ships deployed in this route. It is worth mentioning that most ships have overall lengths between 150 and 200 m, lanemeter capacities between 2000 and 3000 m and are capable of speeds above 20 knots.

M/V Bore Sea has a nominal capacity of 2863 lanemeters, to be used for containers on platforms and trailers. The ship has three roll-on roll-off decks. The hold is used for trailers only (no double stacking of containers is possible), while the main deck and upper deck may be used for trailers or containers (single or double stacked).

The ship may take, in practice, up to 31 trailers or containers in the hold, 40 trailers or mafis in the main deck and another 40 in the upper deck. In the upper deck, in bad weather, it is not possible to double stack in the first four forwardmost positions in each lane. Communication between decks is through ramps, which may take cars or small trucks. Lashing is carried out by the crew, which normally comprises 19 to 22 men. The ship can accommodate up to 12 passengers (used for truck drivers) as restricted by SOLAS regulations for cargo ships. However, in this line, trailers are overwhelmingly taken as unaccompanied cargo.

The ship is time chartered with a maximum speed of 21,5 knots, although the most economic speed is 17 knots, as reported by Bore Ltd (2013). During the visit, the crew informed that the ship is, not surprisingly, more economic at 15.5 knots.

This speed is used quite often in this route according with AIS data freely available through the marine traffic website. This ship and its sister (*M/V Bore Song*) reportedly (www.ship-techonology.com/proj-ects/mv-bore-song.-ro-ro-vessel/) costed over 100 million euros in 2011. Given the exchange rate at the time, the newbuilding price of each unit was approximately 65 million USD.

This value may be compared with some data obtained from shipbroking companies, shown in Figure 7. It may be concluded that a cargo ro-ro ship with capacity for 720 TEUs, as is the case of these ships, should cost about 40 million USD in the period 2012–2016, which is much less than the reported newbuilding price mentioned above. This is probably the result of the ship being built in a reputed German shipyard.

As may be seen, these ships are significantly more expensive than container ships of similar size, with a 720 TEUs ship costing only approximately 16 million USD. Since part of the ship deployed in this route are taken on time charter contracts, it is relevant to check the rates for such ships. Figure 8 shows the one-year time charter rates for cargo ro-ro ships from 1993 to 2015. *M/V Bore Sea* has 2863 lm of cargo capacity, so its time charter rate should not be much smaller than 14.000 €/day.

Table 2. Characteristics of ro-ro ships deployed in the line throughout 2013–2017.

	Catherine	Adeline Wilhelmine	Palatine Mazarine	Peregrine	Bore Sea	Louise Russ	Elizabeth Russ
IMO number	9209453	9539092	9376701	9376725	9443554	9226360	9186429
Length over all	182,2 m	152,0 m	195,4 m	195,4 m	195,4 m	172,9 m	153,4 m
Breadth	25,8 m	24,4 m	26,2 m	26,2 m	26,7 m	25,5 m	20,6 m
Draught	7,80 m	5,63 m	7,40 m	7,40 m	7,05 m	6.84 m	6,99 m
Depth	9,50 m	16,20 m	16,70 m	16,70 m	18,15 m	16,80 m	14,14 m
Lanemeters	2.750 m	2.342 m	2.907 m	3.468 m	2.863 m	2.500 m	1.624 m
Number of ro-ro decks	2 continuous decks + lower hold	3 continuous decks + lower hold	2 continuous decks + lower hold	3 continuous decks + lower hold	2 continuous decks + lower hold	2 continuous decks + lower hold	2 continuous decks + lower hold
Deadweight Tonnage	13.320 t	6.374 t	14.696 t	13.375 t	13.375 t	9.090 t	7.296 t
Gross Tonnage	21.287 t	21.020 t	25.325 t	25.235 t	25.586 t	18.256 t	10.471 t
Maximum continuous rating	12.600 kW	7.000 kW	10.800 kW	16.200 kW	12.800 kW	16.800 kW	12.600 kW
Speed	20,0 kn	17,0 kn	18,5 kn	22,3 kn	21,5 kn	22,5 kn	20,0 kn
Crew	19	22	19	20	22		14

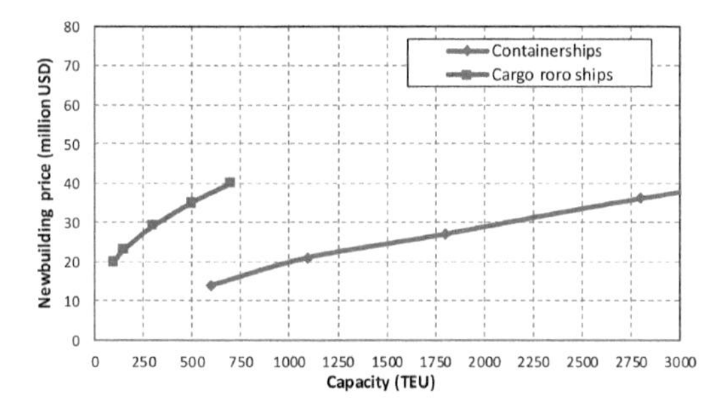

Figure 7. Newbuilding market prices of cargo ro-ro ships and container ships (Source: Maersk Brokers (2016), Bartlett (2012)).

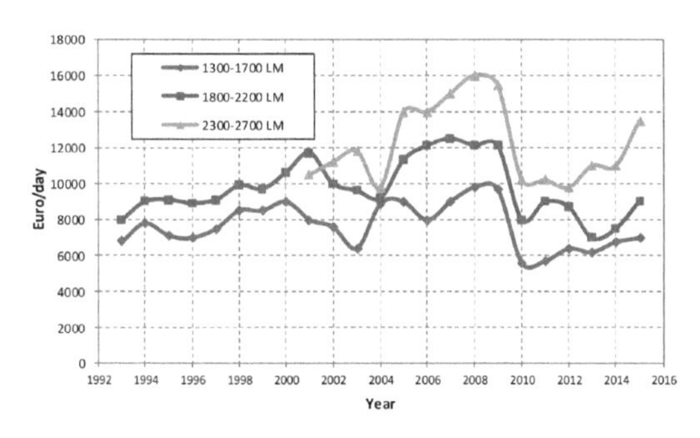

Figure 8. Time charter rates of cargo ro-ro ships (1993–2015).

3.3 *Characteristics of the ro-ro terminals*

3.3.1 *The Leixões terminal*

The port of Leixões includes a ro-ro terminal, called Terminal Multiusos do Porto de Leixões,

Figure 9. Location of Terminal Multiusos (port of Leixões).

covering an area of approximately 6 ha next to the south breakwater of the port, as shown in Figure 9. However, only about 3 ha are used for ro-ro operations, the remaining part being used by TCL for container storage. Using the empirical formulae given by PIANC (2014) it is possible to estimate the terminal area required for handling the total cargo amount indicated in Figure 3 as approximately 3.2 ha, which already more than the area currently available. However, as the terminal handles many containers that are stored using the traditional method (blocks) the pressure related to area shortage is not yet critical.

Previously to the construction of this terminal, another terminal in the north side of the port was used for ro-ro operations. This older terminal is still used for ro-ro cargo of other lines coming to Leixões (north-south lines going to West Africa). The area of Terminal Multiusos is licensed for this use to the Leixões container terminal operator, TCL. This company operates the cargo handling equipment used to load and unload the ro-ro ships, hires the stevedores as required for the operations and conducts the terminal planning as required. The terminal possesses a corner berth with a quay length of 310 m at a draft of 10 m. The ro-ro ramp can take ship ramps up to 26 m in breadth, representing a load of 260 tons. An additional quay, perpendicular to the first one has a length if 155 m at the same 10 m draft.

Figure 10 shows the general layout of the terminal, with three storage areas clearly visible. The first area comprises container storage in conventional blocks parallel to the quay; the second area comprises mafi trailers loaded with containers, parked at an angle to the quay; the third area, located along the south breakwater of the port, takes trailers parked also at angle to the quay.

The operation manager in the terminal reported that, predominantly, 45 ft containers are handled in this terminal. Container storage is done in the first 3 blocks of 45 ft containers, measuring 9 units in length, 3 units in breadth and 3 units in height. Total capacity was reported as being 243 containers (45 ft). Close inspection of the layout of the terminal, as shown in Figure 11, indicates, however, that 3 blocks of containers are available, with capacity for 20 × 3 × 3 TEUs (2 blocks) and 22 × 3 × 3 TEUs (1 block), indicating a total capacity of 558 TEUs (279 FEUs or approximately 250 45 ft).

There is also space, next to the quay, for mafi trailers already loaded and ready for embarkation, in 10 rows of 4 containers (45 ft). Each mafi can take two containers stacked. Therefore, the total capacity of this area is 80 containers (45 ft). Close inspection of Figure 11 shows that the area for mafi trailers appears to have 25 diagonal rows of 4 mafis (double stacked), indicating a total capacity for 200 units (45 ft containers).

However, this area is also used for stacking the mafi trailers themselves, so it is conceded that often it will not be available for containers. The overall number of slots in the terminal for 45 ft containers is therefore comprised somewhere between 323 and 450 containers of 45 ft.

Figure 10. General view of Terminal Multiusos de Leixões.

Figure 11. Plan of Terminal Multiusos of Leixões.

Figure 12. Trailer park of Terminal Multiusos de Leixões.

In the south breakwater, as shown in Figure 12, trailers are parked at an angle, to facilitate manoeuvring, with a capacity for approximately 48 trailers on the inside of the breakwater and 34 trailers on the outside. Operators reported that, at the moment, there is no cargo segregation (import/export) and there is a lack of storage space.

The operational procedures in the terminal consist basically in receiving the containers by road, brought by road-haulage companies, and storing them in the container blocks. The handling of the containers is carried out using reach stackers. In anticipation of the arrival of ships, some containers are doubled stacked on top of mafi trailers, in order to facilitate loading. When the ship actually arrives, mafi tractor heads are used to unload the containers (on top of mafi trailers), as shown in Figure 13. Typically, 6 tractor heads are used at the same time. Each one is able to take out 5–6 mafi

Figure 13. Mafi tractor head and trailer at work.

trailers per hour. Therefore, 36 mafi trailers may be taken out (or in) per working hour, implying 72 containers (45 ft). Trailers are also coming to the terminal by road and tractor heads are used to take the trailers to the ship or the other way around.

Operational staff reported during the visit that it would be possible to operate one such ship in 6–8 hours, enabling up to 3 departures per day. It was also reported that currently, no cargo-handling is done at night or during weekends for costs reasons. However, it will be shown in section 4 of this paper that the schedule has included ship calls on Saturdays. According with port regulations, TCL (2017), normal working days are only Monday to Saturday (8:00 until 24:00), but cargo handling is possible on Sundays and nights with a 90% increase in costs.

This explains the reluctance of the shipping company in using Sundays and night time for cargo handling operations. Possible enhancements to operations could include more storage space, segregation of imported and exported containers, improvements in the tractor heads and in the yard management.

3.3.2 *The rotterdam and santander terminals*

Rotterdam is the main port directly connected with Leixões. This ro-ro terminal serves several other lines of the same company, thus explaining the significantly larger size of the facility, with 27ha, as shown in Figure 14.

The quay in Rotterdam measures 1060 m and shows 4 berths (or 2 twin berths). This type of layout ensures a high degree of flexibility for the future, since it easily accommodates larger vessels. A portion of the quay length is lost (usually up to 60 metres), enlarging the total quay length necessary and representing a costly investment. In the yard, a significant portion of the space is taken by the trailer parking. Three container blocks are located further away from the quay. Access to the rail network is also available. On the contrary, in Santander, the quay length is only 400 m and the yard area is very small, about one hectare. This terminal is quite limited.

Figure 14. View of ro-ro terminal in Rotterdam.

3.4 *Pricing policies*

Cost and transit time are crucial parameters for the competitiveness of these services. The operation manager reported that an average freight of 800 euros was charged per trailer or container. Also, subsequently to the 2011 economic crisis in Portugal, export cargos grew significantly, surpassing imports. Today, imports are again recovering. It is anticipated that from 2018 price discrimination will be applied, with lower freights charged on the northbound cargos, to stimulate exports. Currently, the average utilization rate of the ships is in the region of 90%.

It is known that in the past ro-ro services there was price discrimination depending on whether the trailer belonged to the company or to the client and separate prices for containers. Currently, no price discrimination is applied in this service. Back in 2000, for the ro-ro service bound for Southampton, the freight rates were 1000£ when the trailer belonged to the company, 750£ when using a client trailer and 700£ for a 45 ft palletwide container.

4 OPERATIONAL ASPECTS

4.1 *Arrival and departure from port*

Pilots are mandatory in the port of Leixões due to the narrow entrance and considerable fishing vessel traffic. About half an hour is required to enter the port and moor alongside the quay. The same at departure. The ships currently being used in this route are fitted with bow and stern thrusters, so the assistance of tugs is generally not required.

Figure 15 shows the approach of one of these ships to Leixões, while Figure 16 shows the departure of the ship at the end of the day. The route heading northwest intends to promote a gradual approach to the traffic lanes of the traffic separation scheme in Finisterre, off Northwestern Spain.

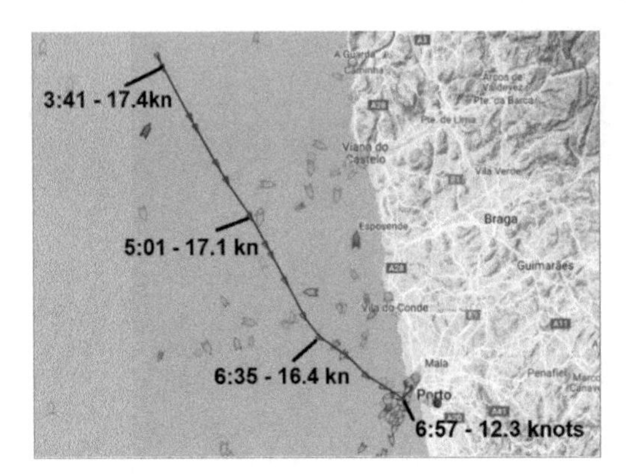

Figure 15. Arrival route to Leixões on 2017/12/06 (time and speed shown for selected points).

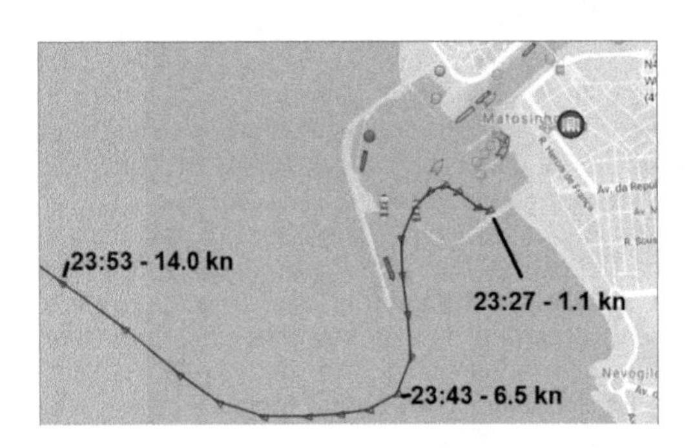

Figure 16. Departure route from Leixões on 2017/12/06 (time and speed shown for selected points).

Speed in high sea is seen to be in the range between 14 knots and 17.4 knots. Observations of marine traffic data further away from the coast and in the English Channel indicate that these ships typically use speeds in the region of 15 knots. ~

The visited ship had arrived slightly after 7:00 and initiated cargo handling operations at 8:00. Operations were anticipated to last until 22:30 and the ship to depart at 23:00. The shipping company is able to receive containers until 17:00 and trailers until 22:00, that is only half an hour before closing loading operations. This schedule corresponds with the schedule advertised in the company's website.

Information provided by APDL was used to produce Figure 17, which shows the arrival time of ship's calling in Leixões during 2017. It may be seen that most ships arrive around 7:30 in the morning (100 ship calls), while a few arrive around midday and a few others in the late afternoon (28 ship calls after 9:30). A maximum of 14 calls per month has been observed in 2017.

Figure 18 shows the arrival time of ship's calling in Leixões during 2017. It may be seen that most ships depart between 18:00 and 24:00, as might be expected since the majority arrives in the early morning. In fact, only 31 ships departed in 2017 before 18:00.

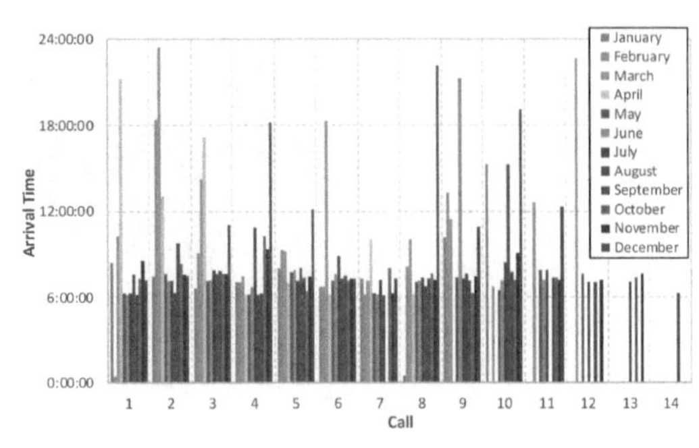

Figure 17. Arrival time in Leixões during 2017.

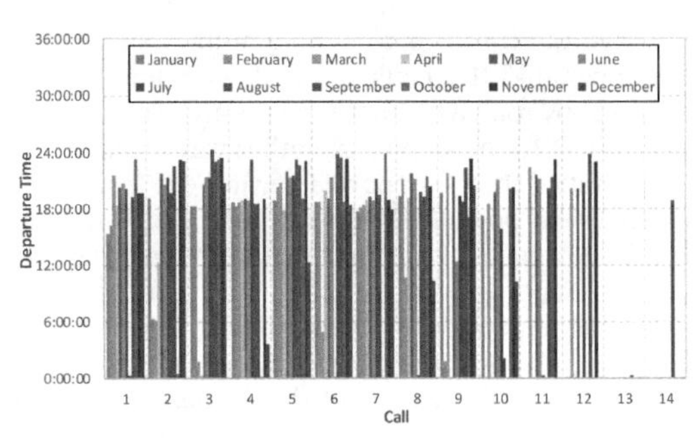

Figure 18. Departure time from Leixões during 2017.

4.2 *Port service and berthing times*

The information provided by APDL also allows a clearer and more comprehensive image of the company's pattern of operation. Information on ship arrival to port, berthing, unberthing and departure from port was collected for the years of 2013 to 2017 and allowed the calculation of several statistical parameters regarding this service, including for example the average service time. Figure 19 shows this time (measured from berthing to unberthing) for the period 2013–2017, discriminated per month. It may be seen that for months May to November the service time is around 12 hours, while between December and April the average service time often is 2 or 3 times higher. This is attributed to delayed departures due to bad weather conditions during winter time (probably along the route in the Bay of Biscay). It is worth mentioning that the port of Leixões itself very seldom closes due to bad weather, so this is not the cause of these high service times.

In addition, the information collected from APDL also allows the calculation of time from port entrance to berth and from berth to port entrance.

Both are generally considered as components of waiting time. Figure 20 shows the berthing times for the period 2013–2017. It may be seen that from March until October the average berthing time is around 18 minutes. For the months from November until February some more irregularity exists, which might be attributed to variable weather conditions in winter time. It is also important to note that an extremely high value (outlier) exists in November 2014, but this is clearly caused by some deficiency in the data obtained from website.

Figure 21 shows the unberthing times for the period 2013–2017. It may be seen that throughout the entire year of 2015 the unberthing time averaged 18 minutes. For the other 4 years, and excluding outliers in April 2017, November 2016 and December 2015, the average unberthing time was significantly and consistently lower, at an average of 14 minutes.

This leads to the conclusion that the unberthing process and port departure is a faster process than the berthing process. However, this difference is not too significant taking in consideration the total round voyage time.

Figure 22 shows the berth occupancy ratio for the 12 months of 2017. The values indicated as total are calculated over the 24 hours of the day

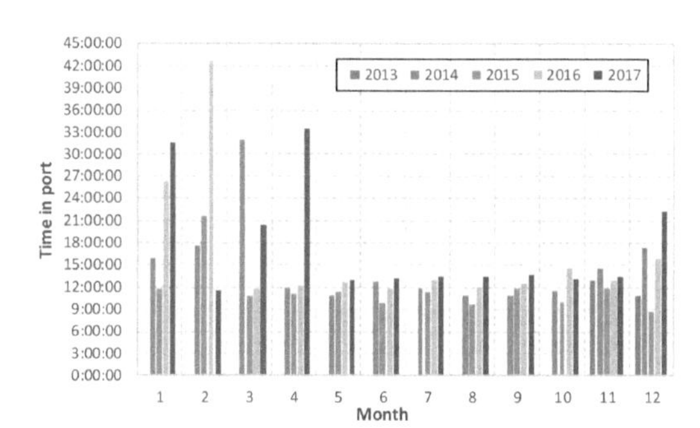

Figure 19. Average service time per month for period 2013–2017.

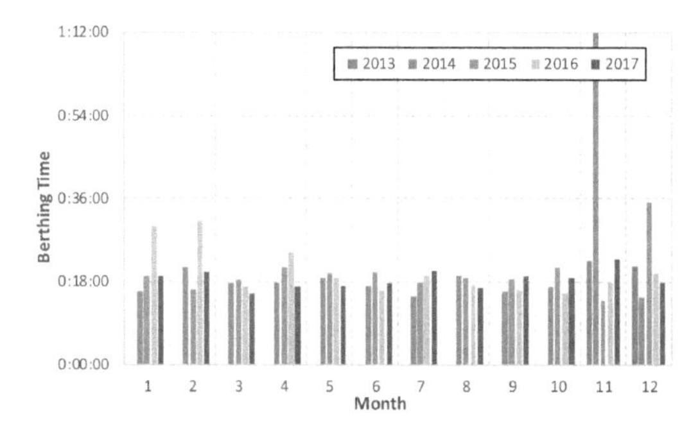

Figure 20. Average berthing time per month for period 2013–2017.

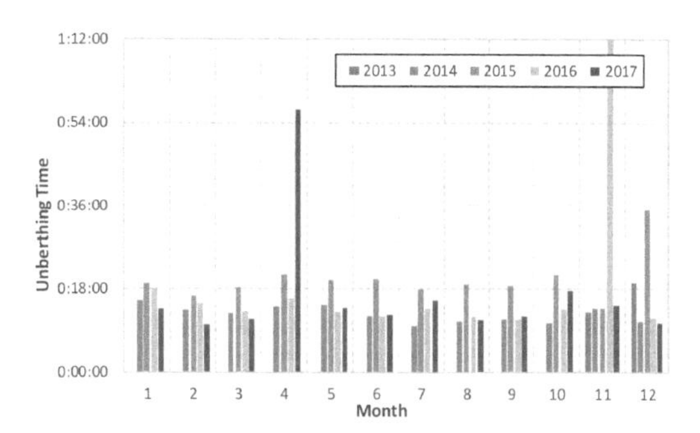

Figure 21. Average unberthing time per month for period 2013–2017.

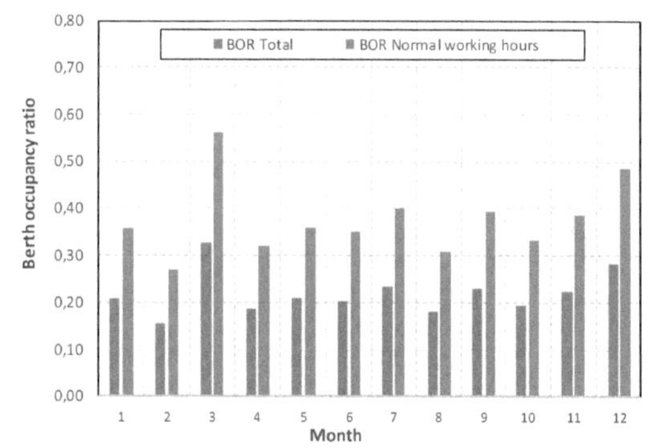

Figure 22. Berth occupancy ratio for 2017.

and seven days per week. It may be concluded from the figure that the berth occupancy ratio has an average value of around 20%, more or less constant. The values indicated as "normal working hours" are calculated over the 16 hours considered as normal working hours and excluding completely Sundays. This berth occupancy ratio thus represents the fraction of the available normal working time (with lower cargo handling costs) which is currently taken by ships of this line in the only available berth. It may be seen that in most months this berth occupancy ratio is between 35% and 40%. In two months it was slightly higher, while in two other months it was slightly lower.

5 CONCLUSIONS

This paper has presented the results of the characterization of an existing roll-on/roll-off short sea shipping service. It is an unaccompanied traffic service used by trucking companies and container freight forwarders and has enjoyed considerable success in recent years, handling over 800,000 tons of cargo in 2017. It resorts to modern cargo ro-ro ships able to take in some decks mafi trailers with doubled stacked 45 ft containers.

The ro-ro terminal in Leixões is somewhat limited in cargo capacity since it can hold only 200 containers with 45 ft. Mafi tractors are used to handle mafi trailers with containers and road trailers. Only one corner berth is available.

Currently, the service, quite often, is not direct to Rotterdam, but rather making an intermediate call in Santander. During summer it appears to be focused on less origins and destinations. The shipping company typically schedules the ships for arrival in early morning and departure in late afternoon/evening, avoiding night operations and Sundays, due to increased cargo handling costs.

Average service time is around 12 hours, with some much higher times in Winter. This is believed to be related to poor weather conditions at sea rather than slower cargo handling. Berthing time generally takes 18 minutes and unberthing 14 minutes. These low times are due to good maneuverability (bow thrusters) of the ships deployed in the route.

Berth occupancy ratio is situated between 20% and 35%, depending on how it is calculated, whether over 24 hours and 7 days per week or over normal working hours. In general, it appears feasible to accommodate 3 more calls per week, at least as far as the berth is concerned, but the container yard most certainly would constrain operations.

ACKNOWLEDGEMENTS

This work was performed within the scope of the Strategic Research Plan of the Centre for Marine Technology and Ocean Engineering (CENTEC), which is financed by the Portuguese Foundation for Science and Technology (Fundação para a Ciência e Tecnologia—FCT). The second author would like to acknowledge the financial support of the Brazilian Navy. The authors would like to acknowledge the valuable contribution of Pedro Martins in the compilation of the statistical data required to support this study.

REFERENCES

Aperte, X. G. and Baird, A. J. 2013. Motorways of the sea policy in Europe. Maritime Policy & Management, Volume 40, Issue 1, pp. 10–26.

Baindur, D. and Viegas, J. 2011. Challenges to implementing motorways of the sea concept—lessons from the past. Maritime Policy & Management, Volume 38, Issue 7, 673–690.

Baird, A. 2007. The economics of Motorways of the Seas, Maritime Policy and Management, Volume 34, Issue 4, pp. 287–310.

Baird, A. J. 2005. Maritime policy in Scotland. Maritime olicy & Management, Volume 32, Number 4, pp. 383–401.

Bartlett, R. 2012. The Valuation of Ships—Art and Science, Marine Money's 6th Annual Korean Ship Finance Forum, Busan.

Bore Ltd 2013. M/V Bore Sea technical specification.

Douet, M. and Cappuccilli, J. F. 2011. A review of Short Sea Shipping in the European Union, Journal of Transport Geography, Volume 19, pp. 968–976.

Keceli, Y., Aksoy, S., Aydogdu, V. 2013. A simulation model for decision support in Ro-Ro terminal operations. International Journal of Logistics Systems and Management, Vol. 15, No. 4, pp. 338–358.

Maersk Brokers 2016. Container Market—Weekly Report.

Ng, A.K.Y, Sauri, S. and Turró, M. 2013. Short Sea Shipping in Europe: Issues, Policies and Challenges. In Finger, M. and Holvad, T. (Eds.): Regulating Transport in Europe, Edward Elgar, Cheltenham, pp. 196–217.

PIANC 2014. Masterplans for the development of existing ports. Report n° 158 - Maritime Navigation Commission, Brussels, Belgium.

Santos, T.A., Guedes Soares, C. 2017a. Modeling of Transportation Demand in Short Sea Shipping, Maritime Economics and Logistics, Volume 19, Issue 4, pp 695–722.

Santos, T.A., Guedes Soares, C. 2017b. Ship and fleet sizing in short sea shipping, Maritime Policy and Management, Vol. 47, Issue 7, pp. 859–881.

Styhre, L. 2009. Strategies for capacity utilisation in short sea shipping. Maritime Economics & Logistics, Volume 11, Issue 4, pp. 418–437.

TCL 2017. Regulations for the operations in Terminal de Contentores de Leixões (in Portuguese).

Trant, G. and Riordan, K. 2009. Feasibility of New RoRo/RoPax Services between Ireland and Continental Europe.

Big data in shipping

Progress in Maritime Technology and Engineering – Guedes Soares & Santos (Eds)
 ISBN 978-1-138-58539-3

Fishing activity patterns for Portuguese seiners based on VMS data analysis

A. Campos & P. Fonseca
IPMA, Portuguese Institute for the Ocean and the Atmosphere, Lisbon, Portugal
CCMAR, Universidade do Algarve, Faro, Portugal

P. Lopes & J. Parente
IPMA, Portuguese Institute for the Ocean and the Atmosphere, Lisbon, Portugal

N. Antunes & P. Lousã
XSealence, Sea Technologies SA, TagusPark, Edifício Ciência II, Lisbon, Portugal

ABSTRACT: VMS (Vessel Monitoring System) data are increasingly being viewed as essential in commercial stock research and management. However, the usefulness of these data for the estimation of the fishing effort is highly dependent on the correct identification of the VMS registers associated to fishing operations. In this study, a VMS sample of unprocessed data with an average resolution of 10 minutes, belonging to four vessels licensed for purse seining, was used to analyze the fishing operations, starting with the identification of fishing trips and then the identification of hauls within trips. The methodological approach was based on the in-depth knowledge of the typical fishing operations carried out by the purse-seine fleet, thus allowing for the correct interpretation of the information on position, course and speed contained in the VMS data. An algorithm was developed incorporating that knowledge, resulting in a first approach to the automatic delimitation of the different phases of a single trip, basically by differentiating between fishing and non-fishing VMS points. The data time resolution was not enough to enable the identification of the gear deployment phase, which is an extremely fast operation. Haul back, carried out at very low speeds and extending over time, was found to be the best candidate to be a marker of the fishing activity.

1 INTRODUCTION

In 2007, the Portuguese purse seine coastal fleet operating from the mainland comprised 151 active vessels with mean age over 20 years. The dimensions of a typical vessel are 20 m overall length, 5 m beam, gross tonnage of 50 GRT and engine power of 300 HP (Hough et al., 2009). This fleet targets mainly sardine (*Sardina pilchardus*) and operates nets whose size can reach 1000 m in length by 120 m in height (Parente, 2001). Fishing trips have maximum duration of one day (Parente, 2003), starting at a port and ending at the same or another port. In each trip the vessel may carry out one or more hauls, each haul comprising several main operational phases including net deployment (setting and encircling); net closing, by hauling up the purse line; drying-up the net, gradually reducing the volume of fish encircled; and hauling the net up. It is preceded by a period of fish detection using electronic fish finding equipment. In some trips, there may be an extra phase (Feijó, 2013) in which the vessels are floating or steaming at very low speeds, waiting for the best moment to launch the seine gear.

The characterization of the different operational phases during a fishing trip was carried out in a previous study (Parente, 2001). Steaming from port to fishing areas is performed at maximum engine power, corresponding to the maximum vessel speed in navigation; during the detection phase the vessel generally operates at speeds slightly lower; setting is an extremely fast operation carried out near the maximum speed threshold, and finally, hauling up the net and the catch on board may take up to two hours at speeds close to zero. Floating, when it occurs, may depend on information on the activity of other vessels, the opening time for the auction (fish market), and the possible need to fish transhipment to another vessel.

Vessel Monitoring System (VMS) is mandatory for most vessels in this fleet, reporting vessel position, speed and heading over time. This information, used by fisheries control bodies in Europe for monitoring and control of fisheries, has a great potential for research, allowing the mapping and quantification of fishing activities. It has been used in this field since the last decade, in the analysis of fleet dynamics, as well as in the estimation of the fishing effort.

Underlying this analysis is the identification of the most important phases of fishing operations. The use of VMS points for this purpose entails a detailed knowledge of the operations on board the different types of vessels.

However, high-quality information regarding the frequency of records relating to the position of vessels is crucial. Faced with a high frequency of records, an experienced observer will be able to recognize spatial "signatures" corresponding to the different phases of the fishing operations.

At present, VMS data are received in the control centre every two hours, the minimum rate set by EU regulations. VMS registers obtained with this frequency do not match the spatiotemporal scales of the different phases of the fishing operations (Katara and Silva, 2017), thus compromising the identification of hauls. However, the blue box registers occur every 10 minutes and so this possibility was explored throughout this work in support to a first approach to the automatic detection of fishing operations and the identification of VMS fishing and non-fishing registers.

2 DATA ANALYSIS

The data analyzed is unprocessed VMS data with an average resolution of 10 minutes, for a total of four vessels licensed for seining and belonging to the period from 2009 to 2016. The data were obtained at the scope of the project SeaITAll – "Sistema para Gestão Integrada de Pescas".

The area of activity is represented in Figure 1, where all available VMS points are mapped for each vessel during the period under analysis. Three out of the four vessels operated off the south (Algarve) coast, and one at the southwest coast between Lisbon and Sines.

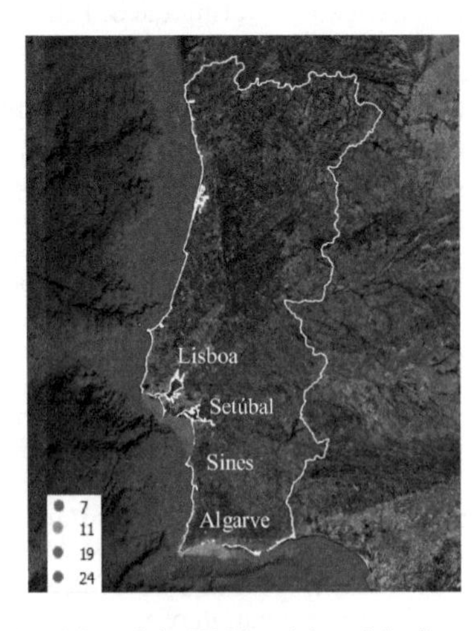

Figure 1. Map of the VMS activity of the four seiners.

2.1 *Identification of fishing trips*

The limits (beginning and end) of the fishing trips were defined by using an automatic procedure, considering the entry in port as the end of a trip and the departure, immediately afterwards, as the beginning of the next trip. Port entering and exiting is interpreted as the entry and exit of an area of influence calculated from a circumference radius centred at the port.

The data were processed for the identification of the fishing trips. For each trip, a trajectory was produced, starting at a port and ending at the same or another port. These trajectories, consisting of a succession of points 10-minutes apart, define patterns that were analysed and interpreted to identify and characterize the main operational phases of the trip.

2.2 *Identification of fishing hauls within trips*

To identify hauls within the previously defined trips there is a need to differentiate between fishing and non-fishing points. To this end, the defined procedure combines knowledge on fishing operations carried out by purse seine vessels with VMS information on position, course and speed. An algorithm was developed allowing to take the first steps in automatic fishing identification.

The different operational phases in a fishing trip have been typified in terms of vessel speed and heading, over time, at which they are carried out. The periods with higher speeds correspond to steaming and detection and to net deployment. The latter is an extremely fast operation, lasting 2 to 5 minutes, and consequently is not detected even in the presence of high frequency registers such as those analyzed here. Consequently, fishing hauls could only be identified by analyzing those periods where the speed remains approximately constant and close to zero, corresponding to the hauling operations. These periods include the haul-up of the purse line, net hauling and catch loading on board.

Based on the existing knowledge on these operations, none of these periods, individually considered, is sufficiently distinct to be reliably identified with a frequency of records every 10 minutes. However, in a "standard" haul, when pooled together the total duration of the haul-up operations allows for its detection.

In this analysis, haul identification was based on the detection of the hauling periods within each trip, using an automatic procedure based on the differences in speed between steaming, detection, deployment and the hauling operations. The minimum time interval at reduced speeds for an operation to be identified as a hauling operation is between 1 and 2 hours. This information is used in the algorithm for the identification of hauls, along with the speed pattern identified during the hauling operations:

speed approximately constant and close to zero during this phase and marked variations of speed at the beginning and at the end. The automatic process consists in the implementation of the algorithm that analyses each fishing trip individually aiming at the determination of the number of hauls.

3 RESULTS

A total of 652 fishing trips corresponding to 597 identified fishing operations could be identified for the 4 vessels in study (Table 1), 518 of which belonging to a single vessel.

Trip duration was generally between 3 and 14 hours, seldom surpassing 14 hours (Figure 2a).

The number of hauls per trip varied from 0 up to 2 for most trips (Figure 3). In a reduced number of trips the vessel carried out 3 and 4 hauls. It was not possible to identify hauls on a large number of trips. Trips without hauls had very short duration, mostly between 1 and 4 hours (Figure 2b).

Frequencies for the trip and haul-up starting hour are shown in Figure 4. In most trips, vessels leave the port early in the evening or in the early hours of the day (Figure 4a), with most hauls starting a couple of hours following departure (Figure 4b). This pattern was observed in all vessels except one, which fished mainly during the last hours of the day.

Finally, the speed profile corresponding to VMS registers was analysed for all the identified trips. Two modes are evidenced (Figure 5), the first around zero knots, corresponding predominantly to the hauling period and possibly to floating, and another around 9 knots, corresponding to steaming, detection and net deployment.

Table 1. Number of fishing trips identified for each vessel.

Vessel	N° trips	N° trips < 1 day	N° trips with identified hauls
11	518	518	336
7	41	41	30
19	64	64	57
24	54	29	24

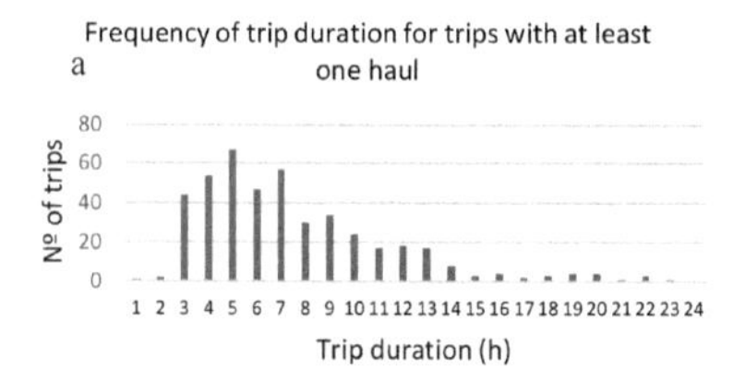

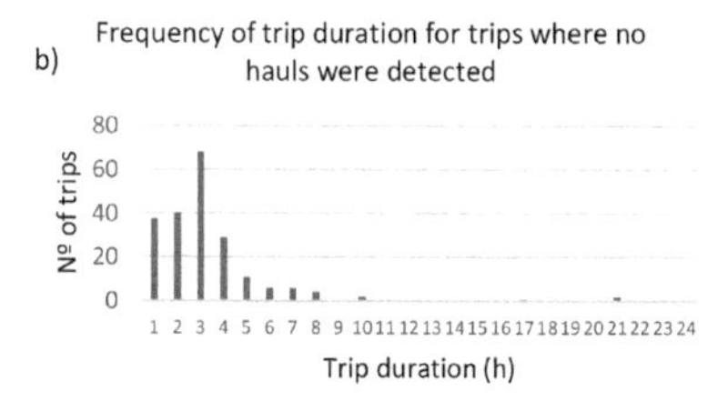

Figure 2. Frequency of fishing trip (FT) durations. a) FT where it was possible to identify at least 1 haul; b) FT in which no hauls were identified.

3.1 *Analysis of trips with no identified hauls*

In an attempt to explain the reason for the existence of trips without identified hauls (31% of the total

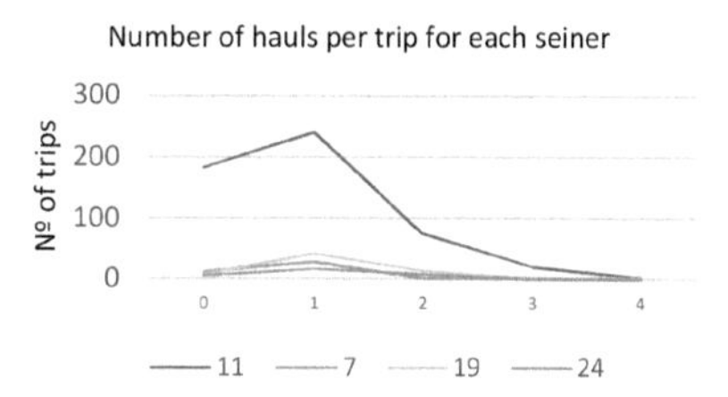

Figure 3. Frequency of the number of hauls per trip for the vessels under analysis.

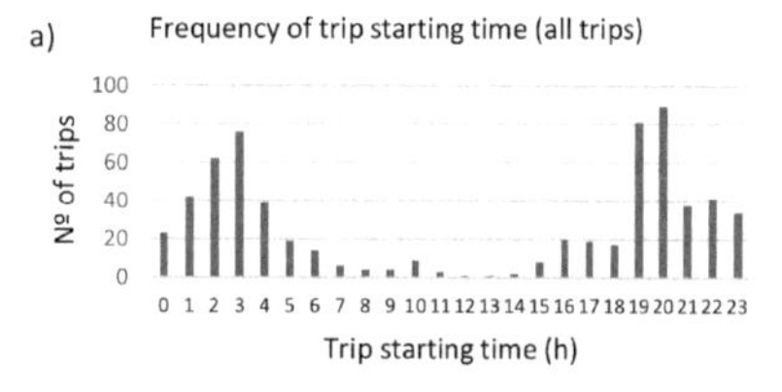

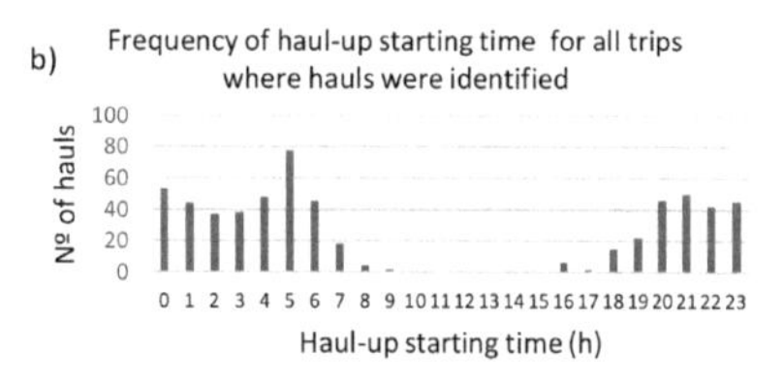

Figure 4. a) Frequency of the trip starting hour for all vessels under analysis; b) frequency of haul-up starting hour for trips where hauls were identified.

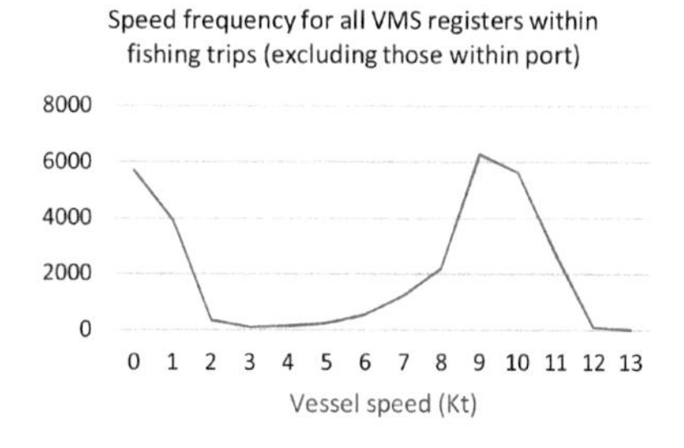

Figure 5. Vessel speed frequency analysed during fishing trips.

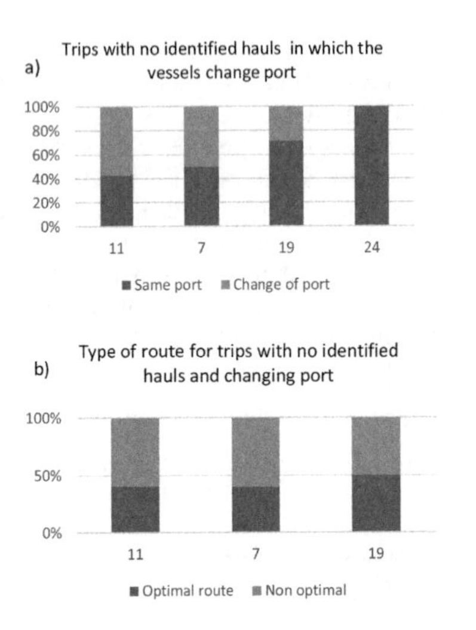

Figure 6. a: Percentage of trips with non-identified hauls in which vessels change port; b: Type of route for trips with non-identified hauls and changing port.

number of trips), the trajectories for these trips were analysed, allowing to conclude that in many of these trips the port of arrival is not the same as the port of departure (Figure 6a). For trips without hauls taking place between different ports, the route between ports is expected to be an approximately rectilinear path (optimal route). However, for most trips where vessels change port this optimal route has not been observed (Figure 6b).

In a significant fraction of these trips the trajectories suggest that other activities took place besides steaming. Fish detection is one of these activities, and the failure to detect hauls may indicate that fish detection was not successful, i.e., that did not end up in a haul. Additionally, hauls might have occurred during a time interval too narrow to be classified as such.

This last situation may correspond to aborted hauls or hauls where only part of the net was deployed, implying a very significant reduction in the total time for fishing operations.

4 DISCUSSION

In this study, high-frequency VMS data (10-minute interval between registers) obtained through the MONICAP (Continuous Monitoring System for Fishing Activity) were analysed for four Portuguese seiners. In a first phase, this information was used in the recognition of the detailed routes associated to fishing trips and in the segmentation of the fishing trip in its various components, characterizing the duration, speed and course in each operational phase. The objective was to identify the fishing behaviour patterns allowing the automatic detection of fishing operations.

A trip begins with the departure of the vessel from a port and ends with its return to the same or a different port, usually to sell the fish caught within that trip. As such, collating information on landings with available geo-referenced information can help to identifying fishing trips with effective fishing activity, also contributing to clarifying the reason for the existence of trips where no hauls were identified.

This analysis provided baseline information for identifying and characterizing the different phases of fishing operations. Furthermore, it contributed for the development of fishing rules for mapping fishing effort for seiners using high discrimination georeferenced data.

The current results must be considered preliminary, given the very small sample, with only four vessels involved in the analysis and 79% of the number of trips belonging to a single vessel. In future studies, the entire fleet should be considered in the analysis.

ACKNOWLEDGEMENTS

This study was carried out at the scope of the project SeaITAll "Sistema para Gestão Integrada de Pescas", project SI COP N° 017693 PT2020, and received national funds from FCT—Foundatin for Science and Technology through ject UID/Multi/04326/2013.

REFERENCES

Feijó, D, (2013), "Caracterização da pesca do Cerco na Costa Portuguesa". Dissertação de Mestrado apresentada na Faculdade de Ciências da Universidade do Porto, Instituto Português do Mar e da Atmosfera. Mestrado em Recursos de Biológicos Aquáticos. Departamento de Biologia.

Hough, A., Nichols, J., Scott, I., Vingada, J., (2009), "Portuguese Sardine Purse Seine Fishery". Public Comment Draft Report. Moody Marine Ltd. Derby, UK.

Katara, I. & Silva, A., (2017), "Mismatch between VMS data temporal resolution and fishing activity time scales", *Fisheries Research* 188:1–5. doi.org/10.1016/j.fishres.2016.11.023.

Parente, J., (2001), Frota de cerco costeira. Tipologia das embarcações e das redes de cerco. Relat. Cient. Téc. Inst. Invest. Pescas Mar nº 74, pp. 50.

Parente, J., (2003), "Caracterização da frota de cerco costeira e perspectivas de modernização". Dissertação original apresentada para Provas de acesso à categoria de Investigador Auxiliar, no Instituto de Investigação das Pescas e do Mar, pp. 216.

Progress in Maritime Technology and Engineering – Guedes Soares & Santos (Eds)

Characterizing container ship traffic along the Portuguese coast using big data

R.C. Botter
Universidade de São Paulo, São Paulo, Brasil

T.A. Santos & C. Guedes Soares
Centre for Marine Technology and Ocean Engineering (CENTEC), Instituto Superior Técnico, Universidade de Lisboa, Lisbon, Portugal

ABSTRACT: Container ship traffic has increased consistently during the last decades, but detailed information about it is not always available. An overall analysis of the main routes is annually published by the United Nations Conference on Trade and Development (UNCTAD), containing data lagged typically 1–2 years in relation to the year of its publication. This paper presents research carried out using a database containing information from September 2016 to September 2017 on container ship traffic throughout the world. This database contains a very significant amount of data, requiring big data analysis techniques to be used in the extraction of information. Such information enables the detailed characterization of deep sea world routes, their particularities such as frequency in the ports, capacity of ships used, among other information. The analysis presented in this paper examines in detail all the routes that pass along the Portuguese coast or call some national port, with the purpose of evaluating the number of services, the total transportation capacity made available and characterize the size of the ships deployed in each route. Conclusions regarding the container vessel traffic in this geographical area are drawn.

1 INTRODUCTION

In order to characterize the traffic of container ships on the Portuguese coast, specific data is required on the routes and ships passing along this coast, whether calling in Portuguese ports or not. One of the possible sources of information on this topic was the Review of Maritime Transport (2016), which presents data in summary form relating to the year of 2014. In that yearbook it is possible to identify the major routes (most notably the East-West trades), but without the number of ships, their main characteristics and which ports were called. Another possible source of information is the National Center for Ecological Analysis and Synthesis, which published a global map of human impacts to marine ecosystems containing information on the traffic density in world routes, however without differentiating between type of cargo. Still another possibility is to use AIS data, which has been used for various purposes (eg. Silveira et al. 2013) and incorporates enough information to identify containerships (Santos & Guedes Soares 2018).

In order to cover this gap in information, a more recent data source, provided by LINESCAPE (2017), with information from August 1, 2016 to August 15, 2017, was obtained, with information on the global traffic of container ships. This database contains 414,587 rows and 12 columns. The objective of this research is to analyze this updated database and characterize the routes, ports and container ships that pass through the Portuguese coast and also that use the ports of the country, characterizing the capacity of transport and the area of influence of traffic.

Based on the literature review, Big Data, Data Mining and Business Intelligence analysis techniques were applied. The statistics were then obtained by route and by vessels, in order to assess the capacity of the system.

The paper is structured as follows. Section 2 provides a general overview of the methodology followed for analyzing the database. Section 3 presents some of the information contained in the database. Section 4 analyzes the information and extracts results for the Portuguese coast. Section 5 presents some conclusions.

2 METHODOLOGY FOR DATA ANALYSIS

Data analysis has been based on the concepts presented by Norman (2001):

Dematerialisation: highlights the ability to separate the informational aspect of an asset/resource and its use in context from the physical world.

Liquification: highlights the point that, once dematerialised, information can be easily manipulated and moved around (given a suitable infrastructure), allowing resources and activity sets that were closely linked physically to be unbundled and 'rebundled' – in ways that may have traditionally been difficult, overly time-consuming or expensive.

Density: is the best (re)combination of resources, mobilised for a particular context, at a given time and place—it is the outcome of the value creation process.

In order to extract information from the database, several algorithms were used for the data analysis. These algorithms have been applied following the work phases indicated in Slavin (2011), namely:

- the cleaning of data,
- the analysis and collection of statistics
- calculation of transport capacity and analysis of the results obtained.

3 DATABASE DESCRIPTION

3.1 *General*

The contents of the columns contained in the database is shown in Table 1.

Table 1. Database description and preliminary statistics.

Column	Description	Statistics listed in the data base
Carrier_code Carrier_name	Name and code of Carrier name. The top 20 based on the number of ports attended are shown in the Table 2	58 carriers
Service_code Service_name	Name and code of the service. The top 20 based on the number of ports attended are shown in the Table 3	1050 Services
Vessel_imo Vessel_name Vessel_voyage	Vessel Name, Imo code and number of voyage	2140 ships
Port_code Port_name Port_eta Port_etd	Port Code, Name, For each vessel the eta and etd date. The 20 portes more called are shown in the Table 4	467 ports
Date off	Date of last voyage announcement	From August 1st, 2016 to September, 7th, 2017

Data cleaning is inevitable when manipulating a database of that size. Repeating lines is the most frequent error. In the case of this database, the last column of Table 1 indicates the correct route followed by the vessel, since it indicates the last publication of the dates that the ship will arrive in the ports. Thus, it was necessary to analyze together the ETA port, the ETD port and Date Off for the perfect characterization of the route. After this step the database was reduced to 319,000 lines and statistics are shown in Tables 2 to 5.

3.2 *Top carriers, services and vessels*

It has been noted that the Carrier Name is not associated with the ship owner. Many Carriers may appear on an MSC ship, for example. The Ship MSC Amsterdam carries cargo from Alianca, Hamburg Sud, Hyundai, Maersk, MSC and Safmarine, indentified by different service names.

This shows that companies operate on a Code Share.

Table 2 shows a top 20 Carrier based on counting of services in all ports.

Table 3 shows the top 20 containers services. The NEMO service performed by the ship AEGIALI for a year, call the following ports on the route:

Tilbury, Hamburg, Rotterdam, Le Havre, Fossur-Mer, Genova, Dumyat, As Suways, Port de Pointe des Galets, Fremantle, Melbourne, Sydney, Adelaide, Singapore, Port Klang, Chennai, Colombo, Cochin, As Suways, Dumyat, Malta Freeport Distripark, Salerno, London Gateway Port.

Table 2. Top 20 carriers.

Carrier name	Number of services in all ports
CMA-CGM	50954
UASC	25207
Maersk	22927
MSC	20856
Evergreen	18215
Hamburg SUD	17493
Alianca	16923
Yang Ming	12831
NYK Line	10594
OOCL	9634
US Lines	9257
ANL	8830
Hyundai	8570
Safmarine	7979
Zim	6635
UFS	6163
Arkas Line	5499
OPDR	5216
K-Line	4970
SM Line	4939

Table 3. Top 20 containers services.

Top 20 services	Number of port calls
NEMO	3290
SOUTH AMERICA EAST COAST – EUROPE	3043
EUROPE – SOUTH AMERICA EAST COAST	2982
Mediterranean Club Express	2855
ASIA – NORTH EUROPE SERVICE	2409
FAL1	1919
Bosphorus Express	1839
EPIC1	1706
MEX2	1683
New North Europe Med Oceania	1681
ASIA – NORTH EUROPE	1462
MEDFDR	1416
NEW SERVICE CCWMS	1402
MINA	1392
NEW SERVICE MESAN	1375
South Mediterranean – North Europe	1280
MED AMERICAS	1266
BLX	1264
FRENCH ASIA LINE 8	1261
ASIA SUEZ EXPRESS	1257

Table 4. Count of container services by port.

Top 20 ports	Count of container services in each port
Rotterdam	12316
As Suways	11324
Antwerpen	9590
Shanghai	8273
Le Havre	8063
Hamburg	7437
Ningbo	7389
Singapore	7200
Valencia	6729
Algeciras	6433
Yantian	6263
Genova	6094
Piraeus	5610
Port Said	4759
Barcelona	4699
Jeddah	4647
Felixstowe	4351
Port Klang	4221
Bremerhaven	3830
Busan	3780

The AEGIALI is a container ship built in 2002 with container capacity at 14t (TEU): 4170. Its gross tonnage is 66573 tonnes, L = 280 m and B = 40 m.

However, the sequence of ports can be changed from a complete voyage to the next. For example, the port of Tilbury is called in a voyage and the port London Gateway is called in the next one.

It is observed in the database that the service name is directly connected to the carrier name (that is, the same service has different names for different carriers). The carriers that offer the most services, as reported in this database, are CMA-CGM (199), UASC (81), Evergreen (71), Maersk (71) and CSCL (69).

For each port in database the number of services were counted and are shown in Table 4. It should be mentioned that American ports do not appear in Table 4 because they are generally attended by direct scales and there are fewer movements in American ports compared to the European Union.

3.3 *Characterization of routes*

For the analysis of routes that call Portuguese ports or navigate across the Portuguese coast it was necessary to identify the routes of interest. A total of 14 routes were selected and are shown in Table 5, which also shows the main ports in each route.

Based on these routes, new algorithms were developed to separate each route, indicating each ship that used it and determining the sequence of ports visited, thus determining the travel time, the number of trips per year and the carrying capacity. Table 6 shows for the route with the highest ship traffic, that is the North Europe-East Asia route, the main individual services which exist in the route and the carrier name. A total of 143 services run along this route and 470 ships were used. Table 6 also shows that among the top 20 services, 4 belong to Maersk, 3 to CMA-CGM, 3 to MSC and 3 to Evergreen. This could be expected as the first three are the largest carriers in the world.

To find the ship sequence in the route, it was first analysed how much time each ship operates in the route and results are shown in Figure 1. It may be seen that 179 ship spent the year in the route. However, surprisingly, many other ships (approximately 300) stay in the route less than one year.

Ships that sailed on the route for up to 180 days completed between 1 and 2 trips between some ports of the route. These voyages appear to be spot ones.

A sample of the sequence of ports for a few examples of such ships is shown in Table 7. Note that for the last vessel, between September and December the vessel did not participate in this route and also after 27 January 2017, having been moved to another route.

In the analysis carried out in this work, we considered only the ships that remained on the route for a further 300 days and an example of the travel sequence is shown in Appendix 1.

Table 5. Routes selected that passing through the Portuguese coast.

Origin	Barcelona	Valencia	Fos-sur-Mer	Genova
Destination	Guayaquil	Callao	Valparaiso	Balboa
Route Name	**West Med – WCSA (west coast south America)**			
Origin	Barcelona	Valencia	Fos-sur-Mer	Genova
Destination	Salvador	Rio de Janeiro	Santos	Paranagua
Route Name	**West Med – ECSA**			
Origin	Barcelona	Valencia	Fos-sur-Mer	Genova
Destination	New York	Newark	Montreal	Savannah
Route Name	**West Med – USEC (US east coast)**			
Origin	Barcelona	Valencia	Fos-sur-Mer	Genova
Destination	Los Angeles	Long Beach	Oakland	Seattle
Route Name	**West Med – USWC (US west coast)**			
Origin	Barcelona	Valencia	Fos-sur-Mer	Genova
Destination	New Orleans	Houston	Veracruz	Altamira
Route Name	**West Med – US Gulf Coast**			
Origin	Barcelona	Valencia	Fos-sur-Mer	Genova
Destination	Salvador	Rio de Janeiro	Santos	Paranagua
Route Name	**North Europe – ECSA (east coast south America)**			
Origin	Shanghai	Yantian	Ningbo	Hong Kong
Destination	Rotterdam	Antwerpen	Hamburg	Bremerhaven
Route Name	**North Europe – East Asia**			
Origin	Rotterdam	Antwerpen	Hamburg	Bremerhaven
Destination	Jebel Ali	Khor al Fakkan	Muhammad Bin Qasim	Mundra
Route Name	**North Europe – South Asia**			
Origin	Felixstowe	Rotterdam	Antwerpen	Bremerhaven
Destination	Haifa	Limassol	Alexandria	Mersin
Route Name	**North Europe – East Med – Variant 1**			
Origin	Felixstowe	Rotterdam	Antwerpen	Bremerhaven
Destination	Gebze	Kumport Terminal	Istanbul	Aliaga
Route Name	**North Europe – East Med – Variant 2**			
Origin	London	Felixstowe	Hamburg	Antwerpen
Destination	Dakar	Abidjan	Takoradi	Tema
Route Name	**North Europe – West Africa**			
Origin	London	Felixstowe	Hamburg	Antwerpen
Destination	Cape Town	Coega	Durban	Port Elizabeth
Route Name	**North Europe – South Africa**			
Origin	Shanghai	Yantian	Ningbo	Hong Kong
Destination	New York	Newark	Montreal	Savannah
Route Name	**East Asia – USEC**			
Origin	Jebel Ali	Khor al Fakkan	Karachi	Muhammad Bin Qasim
Destination	New York	Newark	Montreal	Savannah
Route Name	**South Asia – USEC**			

4 FLEET CAPACITY ESTIMATION FOR SHIPS CROSSING PORTUGUESE COAST

4.1 *Ship characteristics*

Based on Figure 1, a set of 236 ships was analyzed as shown in Table 8 that corresponds to ships with more than 300 days in the route North Europe-East Asia.

Figure 2 shows a histogram of the ships nominal TEU capacity. From Table 8 and Figure 2 it can be concluded that small vessels are still used but that last generation containerships are being introduced in the routes, as is evident from the peaks for 13000–14000 TEU ships and 18000–19000 TEU ships.

Figure 3 shows the histogram of age of ships. The world fleet of container ships has an average age of approximately 11.0 years in late 2016.

Table 6. Services deployed in the North Europe – East Asia route.

North Europe – East Asia	
# Services = 143	
# Ships = 470	
Top 20 Services	
ASIA – NORTH EUROPE	APL
ASIA – NORTH EUROPE SERVICE	OOCL
NORTH EUROPE – ASIA	APL
NEMO	ANL
New North Europe Med Oceania	CMA-CGM
Europe to Asia (AE6)	MAERSK
Asia to Europe (AE6)	MAERSK
Europe to Asia (AE1)	MAERSK
LION WESTBOUND	MSC
LION EASTBOUND	MSC
Asia to Europe (AE2)	MAERSK
FRENCH ASIA LINE 7 (FAL 7)	CMA-CGM
SHOGUN EASTBOUND	MSC
FRENCH ASIA LINE 1 (FAL 1)	EVERGREEN
LOOP 5	NYK Line
Evergreen Europe – French Asia Line (FAL3)	EVERGREEN
VESPUCCI SERVICE	CMA-CGM
ASIA – EUROPE CONTAINER SERVICE 5 EAST	UASC
ASIA – EUROPE CONTAINER SERVICE 5 WEST	UASC
EVERGREEN ASIA/NORTH EUROPE (NE7)	EVERGREEN

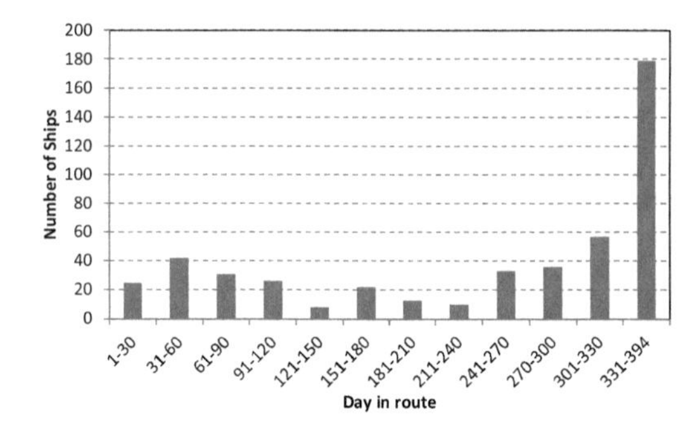

Figure 1. Days in the route.

Table 7. Port sequence for vessel under 180 days in the route.

Vessel 1 – 10 days in the route	
Initial Date	Port Call
3-jun-17	Rotterdam
5-jun-17	Hamburg
9-jun-17	Felixstowe
11-jun-17	Antwerpen
12-jun-17	Shanghai
12-jun-17	Ningbo
13-jun-17	Shanghai
13-jun-17	Ningbo
13-jun-17	Le Havre
Vessel 2 – 30 days in the route	
Initial Date	Port Call
10-fev-17	Rotterdam
11-fev-17	Hamburg
13-fev-17	Bremerhaven
15-fev-17	Wilhelmshaven
16-fev-17	Felixstowe
18-fev-17	Antwerpen
20-fev-17	Le Havre
25-fev-17	Tanger Med
8-mar-17	Salalah
12-mar-17	Jebel Ali
Vessel 3 – 60 days in the route	
Initial Date	Port Call
26-abr-17	London Gateway Port
28-abr-17	Hamburg
29-abr-17	Rotterdam
1-mai-17	Le Havre
7-mai-17	Fos-sur-Mer
8-mai-17	Genova
13-mai-17	Dumyat
15-mai-17	As Suways
24-mai-17	Port de Pointe des Galets
3-jun-17	Fremantle
8-jun-17	Melbourne
11-jun-17	Sydney
15-jun-17	Adelaide
25-jun-17	Singapore
Vessel 4 – 180 days in the route	
Initial Date	Port Call
3-ago-16	Ningbo
4-ago-16	Shanghai
7-ago-16	Chiwan
9-ago-16	Yantian
13-ago-16	Tanjung Pelepas
29-ago-16	Sines
2-set-16	Antwerpen
4-set-16	Le Havre
6-set-16	Southampton
8-set-16	Felixstowe
17-dez-16	Yokohama
22-dez-16	Ningbo
23-dez-16	Shanghai
26-dez-16	Chiwan
28-dez-16	Yantian
1-jan-17	Tanjung Pelepas
17-jan-17	Sines
21-jan-17	Antwerpen
23-jan-17	Le Havre
25-jan-17	Southampton
27-jan-17	Felixstowe

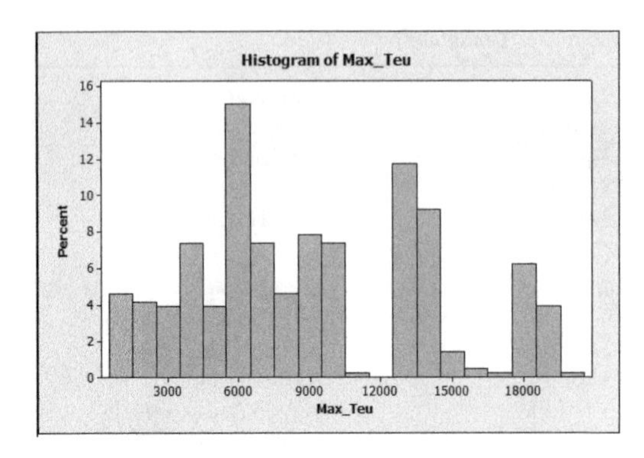

Figure 2. Nominal TEU capacity for sample of 236 ships.

Table 8. Descriptive statistics: DWT, Nominal TEU and Ship Age.

Variable	Mean	StDev	Minimum	Maximum
DWT	101998	50280	8622	202347
Nominal TEU	9039	5063	700	19870
Age	8.8	5.4	1.0	31.0

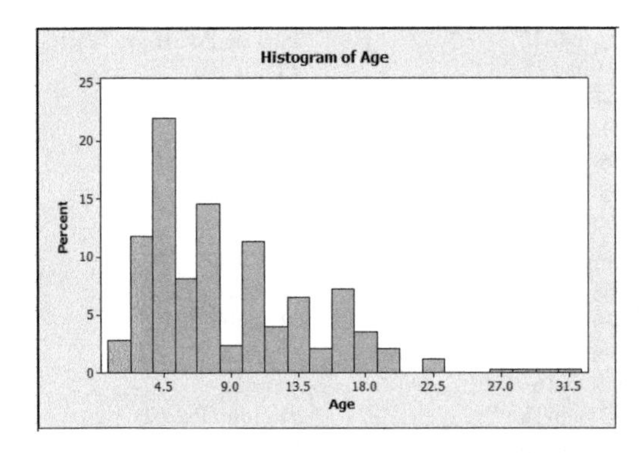

Figure 3. Age for sample of 236 ships.

The average of ships age in the route is therefore lower than the average for the world fleet. Ships such as BOUNDARY, CHARLOTTE BORCHARD, RUTH BORCHARD, RACHEL BORCHARD, GRANDE ATLANTICO, GRANDE MEDITERRANEO, GRANDE ARGENTINA and GRANDE AFRICA represent the older ships in the route.

4.2 *Fleet capacity*

For all ships in Figure 1 an algorithm counted the number of complete voyages that appears in data base. After that, for each ship was considered that the average number of TEU was 75% of nominal TEU capacity. The results are shown in Table 9. The value found for the capacity of the fleet that sailed during the period from August 2016 to August 2017 is compatible with the volume of containers published by UNCTAD for the year 2015.

Table 9. Fleet capacity.

Number of days in the route	Number of ships	Number of voyages	TEU capacity
0–30	25	1	206 250
31–60	42	1	346 500
61–90	31	1.5	383 625
91–120	26	1.8	386 100
121–150	8	2	132 000
151–180	22	3.5	635 250
181–210	13	4	429 000
211–240	10	5	412 500
241–270	33	5.8	1 579 050
270–300	36	6	1 782 000
301–330	57	6.7	3 150 675
331–394	179	7	10 337 250
	Fleet Capacity (TEU's)		19 780 200

5 CONCLUSIONS

The database used in this research is complex and its analysis used techniques for cleaning and classification of information to extract the routes and the ships that travel along the Portuguese coast. The selection of the routes of interest for this research using the database made possible a more detailed description of the containerized maritime transport system.

The major results found are listed below:

- the verification that ship routes over time may change;
- that the sequence of ports served within a route by a ship may also change from one voyage to the next;
- that ships can be moved from one route to another;
- that the age of the fleet in the routes that pass by the Portuguese coast is low, evidencing the use of ships increasingly larger;
- the shipping companies operate in code share, where several services of different carriers are transported by the ships, trying to maximize the occupation of the ships;

It is recommended that the analysis of this database be improved and extended to all major routes.

It would also be interesting to study in detail each route executed by each ship. Through this study, it could be analyzed whether new transshipment arrangements and new port sequences could minimize the cost of the global logistics chain. It could also be studied how to avoid that exports and imports from countries like Portugal use large road transport distances to access the main ports of Northern Europe.

ACKNOWLEDGEMENTS

This work was performed within the scope of the Strategic Research Plan of the Centre for Marine Technology and Ocean Engineering (CENTEC), which is financed by the Portuguese Foundation for Science and Technology (Fundação para a Ciência e Tecnologia—FCT). The first author would like to acknowledge the financial support of the Brazilian Navy. The authors would like to acknowledge the LINESCAPE for permission to use this database.

REFERENCES

Halpern, B.S. et al (2008), A Global Map of Human Impact on Marine Ecosystems, Science 319, 948 (2008), DOI: 10.1126/science.1149345.

Linescape (2017), [WWW document] https://www.linescape.com/.

National Center for Ecological Analysis and Synthesis (2011). The Global Map of Human Impacts to Marine Ecosystems.

Normann R (2001), Reframing Business: When the Map Changes the Landscape. John Wiley & Sons, Chichester, Sussex.

Review of Maritime Transport (2016), UNCTAD/RMT/2016, UNITED NATIONS PUBLICATION Sales no. E.16.II.D.7, ISBN 978-92-1-112904-,5 eISBN: 978-92-1-058462-3, ISSN 0566-7682.

Santos, T.A. & Guedes Soares, C. 2018; Methodology for estimating technical characteristics of container ships from AIS data, Santos, T.A. & Guedes Soares, C. (Eds) Progress iin Maritime Technology and Engineering, UK: London, Taylor & Francis.

Silveira, P.A.M.; Teixeira, A.P., & Guedes Soares, C. 2013; Use of AIS Data to Characterise Marine Traffic Patterns and Ship Collision Risk off the Coast of Portugal. Journal of Navigation. 66(6):879–898.

Slavin K. (2011), How algorithms shape our world. [WWW document] http://www.ted.com/talks/kevin_slavin_how_algorithms_shape_our_world.html (accessed 11 April 2013).

APPENDIX

Appendix 1. Sequence of ports for vessel with more than 365 days in the route.

379 days in the route	
Initial Date	Port Call
04/ago/16	Le Havre
07/ago/16	Rotterdam
08/ago/16	Hamburg
10/ago/16	Antwerpen
13/ago/16	Hamburg
13/ago/16	Felixstowe
16/ago/16	Rotterdam
18/ago/16	Le Havre
17/set/16	Port Klang
22/set/16	Chiwan
25/set/16	Hong Kong
27/set/16	Shanghai
29/set/16	Ningbo
02/out/16	Vostochnyy Port
03/out/16	Yantian
05/out/16	Vung Tau
08/out/16	Port Klang
19/out/16	As Suways
24/out/16	Magadansky, Port
27/out/16	Le Havre
30/out/16	Rotterdam
31/out/16	Hamburg
03/nov/16	Antwerpen
05/nov/16	Felixstowe
07/nov/16	Rotterdam
08/nov/16	Le Havre
20/nov/16	As Suways
23/nov/16	Jeddah
10/dez/16	Port Klang
15/dez/16	Chiwan
17/dez/16	Hong Kong
20/dez/16	Shanghai
23/dez/16	Ningbo
26/dez/16	Yantian
28/dez/16	Vung Tau
29/dez/16	Cai Mep International Terminal
30/dez/16	Vung Tau
31/dez/16	Port Klang
11/jan/17	As Suways
14/jan/17	Petropavlovsk-Kamchatskiy
19/jan/17	Magadansky, Port
20/jan/17	Le Havre
22/jan/17	Rotterdam
25/jan/17	Hamburg
27/jan/17	Antwerpen
28/jan/17	Felixstowe
29/jan/17	Antwerpen
30/jan/17	Rotterdam
31/jan/17	Felixstowe
02/fev/17	Rotterdam
04/fev/17	Le Havre
12/fev/17	As Suways
19/fev/17	Jeddah
04/mar/17	Port Klang
10/mar/17	Hong Kong
13/mar/17	Chiwan
14/mar/17	Shanghai
14/mar/17	Hong Kong
15/mar/17	Shekou
16/mar/17	Chiwan
17/mar/17	Hong Kong
20/mar/17	Yantian
15/mar/17	Shekou
16/mar/17	Chiwan
17/mar/17	Hong Kong
20/mar/17	Yantian
21/mar/17	Shanghai
22/mar/17	Vung Tau
23/mar/17	Shanghai
24/mar/17	Ningbo
27/mar/17	Yantian
29/mar/17	Cai Mep International Terminal
30/mar/17	Vung Tau
01/abr/17	Port Klang
05/abr/17	As Suways
13/abr/17	Le Havre
14/abr/17	Rotterdam
15/abr/17	As Suways
16/abr/17	Antwerpen
19/abr/17	Hamburg
20/abr/17	Le Havre
21/abr/17	Rotterdam
22/abr/17	Le Havre
24/abr/17	Rotterdam
26/abr/17	Hamburg
28/abr/17	Antwerpen
01/mai/17	Rotterdam
02/mai/17	Felixstowe
04/mai/17	Le Havre
07/mai/17	As Suways
10/mai/17	Jeddah
27/mai/17	Port Klang
01/jun/17	Chiwan
08/jun/17	Ningbo
09/jun/17	Shanghai
10/jun/17	Shekou
11/jun/17	Ningbo
11/jun/17	Shanghai
12/jun/17	Ningbo
13/jun/17	Shanghai
15/jun/17	Hong Kong
16/jun/17	Ningbo
17/jun/17	Shekou
19/jun/17	Yantian
21/jun/17	Cai Mep International Terminal
22/jun/17	Tanjung Pelepas
23/jun/17	Port Klang
24/jun/17	Tanjung Pelepas
25/jun/17	Port Klang
04/jul/17	As Suways
07/jul/17	Piraeus
13/jul/17	Le Havre
14/jul/17	Rotterdam
15/jul/17	Antwerpen
18/jul/17	Hamburg
19/jul/17	Antwerpen
20/jul/17	Hamburg
21/jul/17	Rotterdam
22/jul/17	Felixstowe
23/jul/17	Rotterdam
24/jul/17	Felixstowe
26/jul/17	Southampton
18/ago/17	Singapore

Progress in Maritime Technology and Engineering – Guedes Soares & Santos (Eds)
 ISBN 978-1-138-58539-3

Methodology for estimating technical characteristics of container ships from AIS data

T.A. Santos & C. Guedes Soares
Centre for Marine Technology and Ocean Engineering (CENTEC), Instituto Superior Técnico, Universidade de Lisboa, Lisbon, Portugal

ABSTRACT: A methodology to estimate technical characteristics of container ships using the limited information provided by the Automatic Identification System (AIS) is presented. This methodology uses the information on length and breadth of the ship to estimate its nominal TEU capacity and then combines such information with a comprehensive database of container ships to estimate other ship characteristics such as draught, depth, gross tonnage, TEU capacity (14t), propulsion power, electrical installation power, service speed, newbuilding price. These technical characteristics may be used to support economic studies of new liner routes and studies of container ship emissions in coastal and port waters. This approach is tested for the fleet of container ships calling in key Portuguese ports, aiming at evaluating the accuracy of the produced estimates, allowing conclusions to be drawn on the feasibility of its application in the mentioned studies.

1 INTRODUCTION

Globalization has implied in the shipping industry an increased importance for the container ship segment of the world fleet. The number and average size of these ships has grown continuously, particularly since 1990, as a result of the growth in maritime flows of containerized cargo. In parallel with these industry developments, research has focused on developing methods and tools for planning and optimizing container ship liner services, both deep-sea routes and short sea ones. Furthermore, more complicated voyage patterns have evolved, namely using the hub and spoke paradigm, allowing substantial economies of scale, which have been the subject of many research works.

In economic studies underlying problems such as ship sizing, fleet sizing, fleet mix, ship routing, it is often necessary to determine the technical parameters of ships found to be suitable for the intended operations. The information which generally arises from optimization studies is the TEU capacity of the ships necessary for carrying the cargo, as anticipated from demand forecasting studies. It is then necessary to be able to estimate the technical parameters of the ship based on this cargo required capacity. It is necessary to take into account that in container ships the nominal capacity is a theoretical geometrical capacity, but that ships are not able to carry the same number of fully loaded containers. It is then important to be able to estimate the number of TEUs at 14 tons which the ship is able to carry. Furthermore, the demand for transportation of refrigerated cargo will drive the number of reefer slots (and plugs) needed in the ship.

Once the nominal and real (at 14 tons) TEU capacities of the ship have been estimated, other technical parameters, including the length over all, breadth, maximum draught and air draught, that is the main dimensions of the container ship may be estimated. These dimensions need to be taken into account, as restrictions, in such problems as ship deployment, optimization of ship routes and in fleet sizing. The characteristics of ports and canals which are included in the envisaged voyage are then to be compared with ship's characteristics in these studies.

Subsequently, in studying the economics of the services arising within optimization studies, as explained by Santos & Guedes Soares (2017), it is also important to be able to estimate other technical parameters of importance for calculating ship's capital costs, operating costs and voyage costs. These parameters include the main dimensions, but also the newbuilding price, gross tonnage, service speed, propulsion power (MCR), auxiliary power, crew number. Cargo handling equipment is also of importance as it will influence costs in port and productivity rates. This information will allow the calculation of all components of ship costs and the economic evaluation of the envisaged route.

The ability to estimate container ship's characteristics from a small number of technical parameters such as the main dimensions (length over all and breadth), is also of interest in other studies. These two main dimensions are generally widely available in automatic identification systems (AIS) messages. Therefore, it is important to be able to estimate ship's technical characteristics from such simple and widely available information within the scope of, for example, studies on ship emissions in ports and coastal waters.

Another possible application of these estimation methods is in the characterization of liner services passing along certain coastlines, aiming at assessing the size and capacity of ships deployed in such lines and its fuel and ballast capacities. These capacities may also be of interest in case of accident involving oil/fuel spill, providing readily available estimates of probable fuel content in ship's tanks.

This paper proposes a methodology for estimating technical parameters of container ships using AIS data or TEU nominal capacity. The scarce information provided by this system is first used to determine the nominal and real capacity of the ship and subsequently used to derive many other technical parameters. These estimations are mainly carried out using a database of container ships. This database is combined with some other information for estimating certain economic parameters (newbuilding price). The identification of ship type in AIS is not very detailed, namely containerships are taken within a broad type of cargo ships. However, it is today possible to promptly obtain from the MMSI number (also provided by AIS) the specific type of the ship, thus allowing the identification of container ships as such.

This paper consists of a section 2 devoted to the presentation of the container ship database, followed by a section 3 detailing the methodology used for estimating the technical characteristics of ships. Section 4 presents a numerical example of application of the methodology, consisting in estimating the parameters of sets of container ships calling in key Portuguese ports. Three sets of ships have been considered: container ships calling in the ports of Leixões, Lisbon and Sines in the month of October 2017. The estimated characteristics are then compared with the real characteristics of the ships in order to evaluate the precision of the methodology. Conclusions regarding the feasibility of this approach are indicated in section 5.

2 CONTAINERSHIP DATABASE

As mentioned in the preceding section, it is useful to have a container ship database in order to support different studies. A brief survey in literature and websites has shown that such databases are not freely available today. Consultancies such as Lloyd's List Maritime Intelligence Unit, HIS Fairplay or Alphaliner provide a substantial number of payed services including tracking ships, ship technical details and capacity analysis. These services are generally individualized per market segment (container ships, tankers, bulk carriers, gas carriers).

The academic literature on this topic shows a few examples of databases of container ships, mainly used for two purposes: supporting basic ship design; providing information for determining the size of design ship to be used in port design. Kristensen (2013) provides a statistical analysis of container ship main dimensions based on a database from IHS Fairplay. In addition, this study contains information on the block coefficient and lightweight of ships. However, no information is reported on other ship technical characteristics. This IHS database has also been used to support the studies on ship emissions by IMO (2014).

Takahashi *et al.* (2006) show the results of a statistical study, based on the Lloyd's List Maritime Intelligence Unit database, of the main dimensions of different ship types, including an analysis of container ships discriminated in different classes. The results are used to obtain tables relating deadweight to ship dimensions. However, again, characteristics such as propulsion power, auxiliary power, speed, and others of importance for economic analysis are not considered in this study.

Another interesting source of information is the report by MAN B&W (2015) which includes a significant body of data on typical container ship main dimensions but also on propulsion power and speed for different classes. This data is presented under the form of graphs with regression lines, with no further details (points) on the underlying database.

In any of these cases, the databases used for these studies are not freely available to the public. Additionally, the databases do not cover the whole range of technical characteristics required for the studies indicated in the introduction to this paper. This has led to the development of a tailor-made database of container ships using technical magazines (RINA (2000–2013)) and information extracted from various websites/databases, including Equasis, Scheepvaartwest and Classification Societies. Technical information for container ships of all dimensions (300 TEU to 21.000 TEU) was collected and cross checked using the different sources.

This verification has shown that some technical characteristics are more prone to incongruencies between different sources than others. One example is ship speed, which rarely states whether it is maximum speed or service speed. Draught is also quite frequently unclear, as in most cases it is not stated whether it is summer draught, design draught or scantling draught. Another frequent issue is the fact that very frequently a significant part of ship's data is not available, even upon cross-checking of different sources. Typically, number and power of generators and capacity of boilers is difficult to obtain. Displacement and lightweight are also seldom indicated. The number of reefer containers which may be taken on board is also generally unclear as to whether these are TEUs or FEUs. It is also often unclear if indicated numbers for crew refer to the maximum capacity of the ship (cabins) or the actual normal crew of the ship.

Taking in consideration these issues, ships have been selected for inclusion in the database based on the amount of information available and the quality of this information. As a result, a total of 265 ships are currently inserted in the database and the fleet composition has been tuned to the actual composition of the world fleet, as shown in Figure 1. The size categories of ships adopted in this study have been taken from Alphaliner (2017).

The sample of 265 ships represents approximately 5.2% of the world containership fleet on 31 December 2016, which consisted of a total of 5112 ships. The average age of the ships in the database is 11 years. In 2015 the average age of the world containership fleet was 11.5 years, according with Sea Europe (2017). This implies that both the fleet composition (by size categories) and the average age of ships included in the database are in line with similar parameters for the world fleet. It is assumed, therefore, that the sample is representative of the world fleet.

The database includes a large number of characteristics of the ships, as shown in Table 1. These include some commercial characteristics of qualitative nature but also a significant number of technical characteristics related with propulsion, auxiliary power generation, main dimensions, tank capacities and cargo capacities.

Figures 2, 3 and 4 show the length over all, the breadth of ships and depth of ships contained in the database as a function of the TEU nominal capacity of the ships. It may be seen that the coefficient of determination (R^2) values are generally very good, with that of the draught being slightly

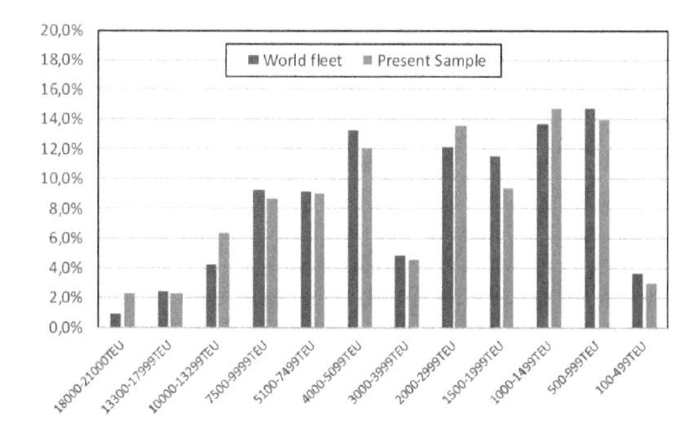

Figure 1. World fleet and database compositions.

Table 1. Ship characteristics included in database.

Characteristics	
IMO Number	Service Speed
MMSI number	Length over all
Call Sign	Length between perpendiculars
Ship Name	Breadth
Flag	T max
Owner	T scantling
Operator	Depth
Year of building	Freeboard
Shipyard	Displacement
Yard number	Lightweight
Engine maker	Fuel capacity
Engine model	Ballast capacity
Number of propulsion engines	Nominal TEU capacity
Propulsion engines power	Nominal TEU capacity (holds)
Length of engine room	Nominal TEU capacity (deck)
Number of propellers	TEU 14t capacity
Diameter of propellers	Reefer TEU capacity
Type of propellers	Deadweight
Number shaft generators	Cargo deadweight
Shaft generator power	Gross Tonnage
Total number of generators	Net Tonnage
Total generators power	Number of cranes
Emergency generator power	Crew number
Bow thrusters power	Newbuilding price
Stern thrusters power	

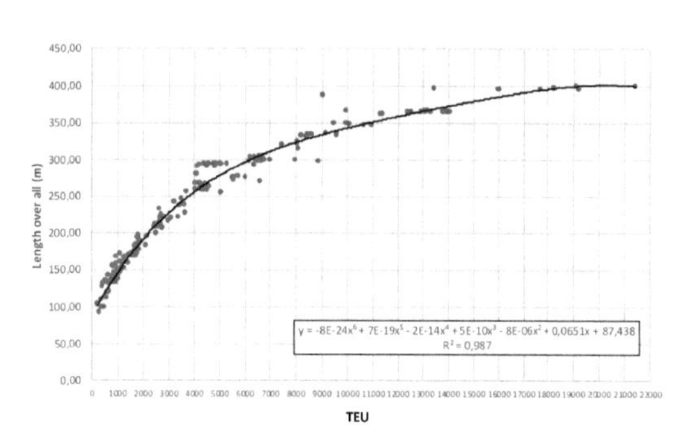

Figure 2. Length over all—TEU nominal capacity.

lower. Regarding the breadth of the ships, it is clear the grouping of ships in steps (approximately 2.6 m spaced) as the number of rows of containers increases with the size of the ship. This is particularly evident for the old panamax ships, all with breadths around 32.2 m. The same is visible for the depth of ships in Figure 5, which also present the typical grouping in steps of approximately 2.6 m.

Apart from the main dimensions of the ships, other technical characteristics of interest are included in the database. Figures 5 and 6 show some of these characteristics, namely the deadweight

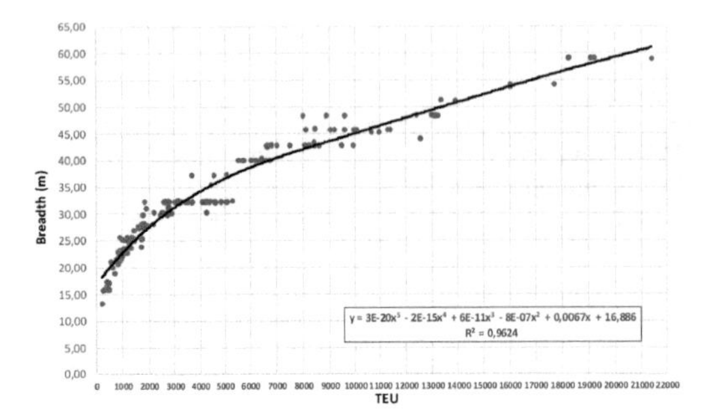

Figure 3. Breadth—TEU nominal capacity.

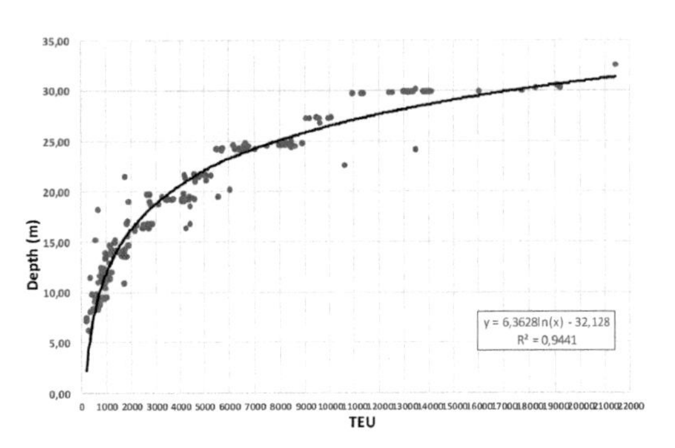

Figure 4. Depth—TEU nominal capacity.

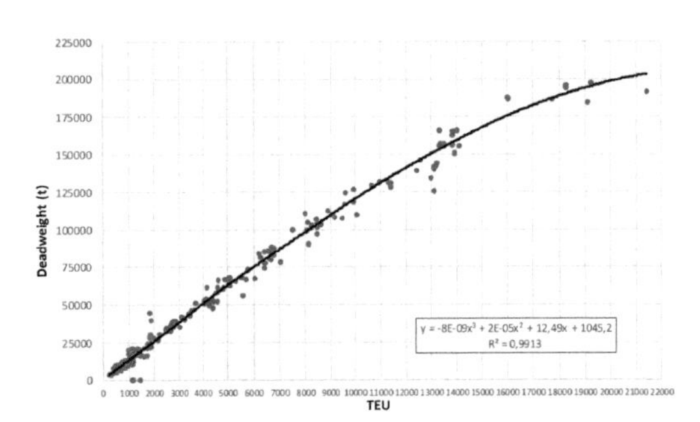

Figure 5. Deadweight—TEU nominal capacity.

and gross tonnage. As can be seen, in general, the coefficient of determination R^2 of regressions of these variables with TEU nominal capacity are very good. The graphs exhibit an almost linear trend up to 10.000 TEU capacity ships. Upwards, deadweight and gross tonnage appear to grow less with increasing TEU capacity. However, this conclusion must be taken with precaution as there is a limited number of ships available in the database and dispersion appears to be larger.

Figures 7 and 8 show the installed propulsion power and speed as a function of TEU nominal capacity of the ships. As may be seen, the dispersion of the values is now much wider and simple regression formulae with good R^2 coefficients of

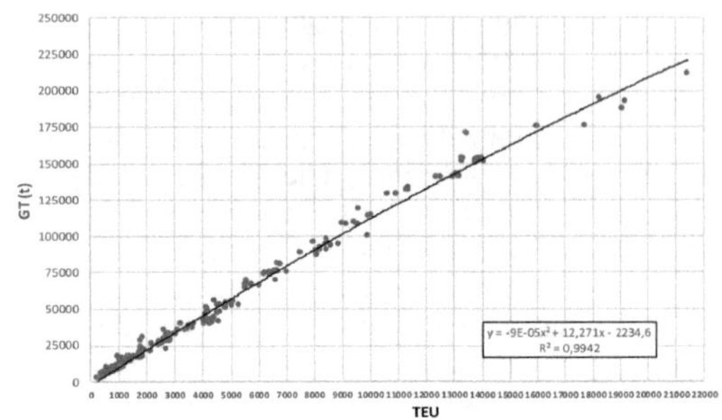

Figure 6. Gross tonnage—TEU nominal capacity.

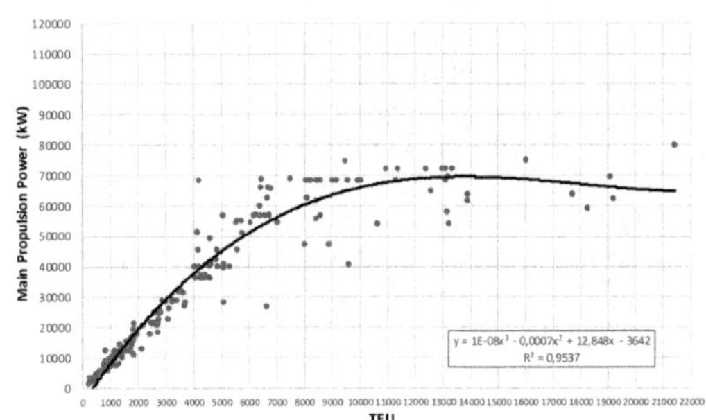

Figure 7. Main propulsion power—TEU nominal capacity.

determination are much more difficult to obtain. This may be explained by the fact that main propulsion power also depends on the design speed of the ship which is largely a decision of the shipowner and is known to have decreased in recent years due to high fuel prices. Main propulsion power appears to be relatively well defined when plotted against nominal TEU up to ships of 4000 TEU capacity, but upwards a significant dispersion exists. Most notably, a ship of 4.200 TEU with almost 70.000 kW has been found to correspond with a specific Maersk fast containership (service speed 29 knots).

Ship speeds are shown in Figure 8 and also show some dispersion. For ships below 6.000 TEU the speed tends to increase with TEU nominal capacity, but upwards the trend is for speed to remain constant. Very large ships (above 12.000 TEU) seam to have slightly lower speeds, perhaps because these are the newest vessels, designed and built in an age of high fuel prices. The main propulsion power, shown in Figure 7, exhibits a similar trend for larger ships.

Auxiliary power, shown in Figure 9, shows an even wider dispersion and consequently the coefficient of determination is smaller. Here, other variables which might explain this dispersion would be the number of reefer containers that the ships can take. For example, the ship of 9.500 TEU with an auxiliary power of 25.000 kW corresponds to a Hapag-Lloyd ship built for the Europe-South

America trade lane, which is characterized by high volumes of reefer cargo.

In addition to the information presented and discussed in the previous figures, the database also contains information on newbuilding prices. Figure 10 shows the newbuilding prices for various classes of container ships between 1999 and 2017.

Data for this Figure has been gathered from Clarkson Research, Danish Ship Finance and reports of different shipping companies. Therefore, this data represents generic values for certain sizes of ships rather than individual prices for specific ships, which are generally not disclosed by shipping companies. The variety of sources of information explains the presence of superimposed series. The most recent numbers allow estimates of container ship newbuilding prices, which are necessary for economic studies.

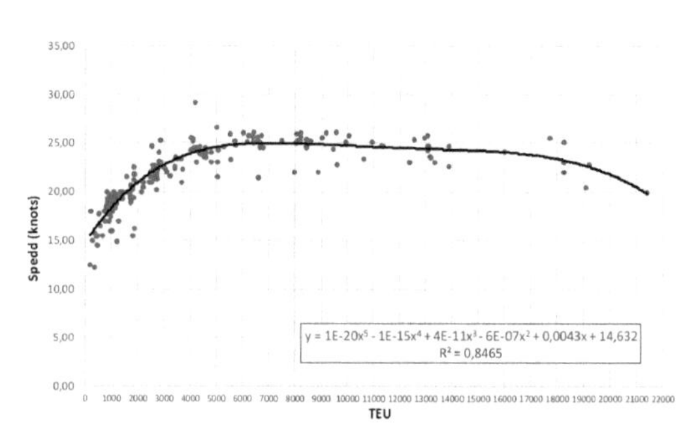

Figure 8. Speed—TEU nominal capacity.

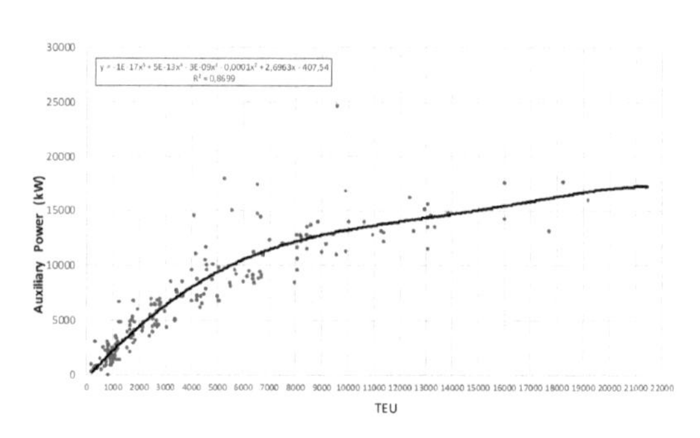

Figure 9. Electrical power—TEU nominal capacity.

3 METHODOLOGY

The database described above may be used for deriving containership technical characteristics from a limited number of parameters. One application case is the one involving economic studies of liner services, which produce the required TEU capacity of the ships, but lack other details of the ships. The database above may be used to estimate other technical parameters required for economic studies of the feasibility of the liner service. A second case application is the one relating to the estimation of containership parameters from the length and breadth of the ship, taken from AIS information or from port's I&T systems. This might have application in estimating ship emissions and characterizing the maritime traffic along a coastline or in a port area.

Figure 11 shows the flowchart of the estimation process. The process may start from the nominal TEU capacity previously calculated from a transportation problem. A second option is to start from knowledge of the ship's length over all and breadth (from AIS). These two options correspond to the two application cases mentioned above. In any case, following the determination of nominal TEU capacity, it is possible to obtain by interpolation the ship's depth, summer draught, scantling draught, gross tonnage and net tonnage. Apart from these geometrical parameters, it is possible to obtain the newbuilding price through the use of statistical data from shipbrokers and shipping consultancies, information which has been collected

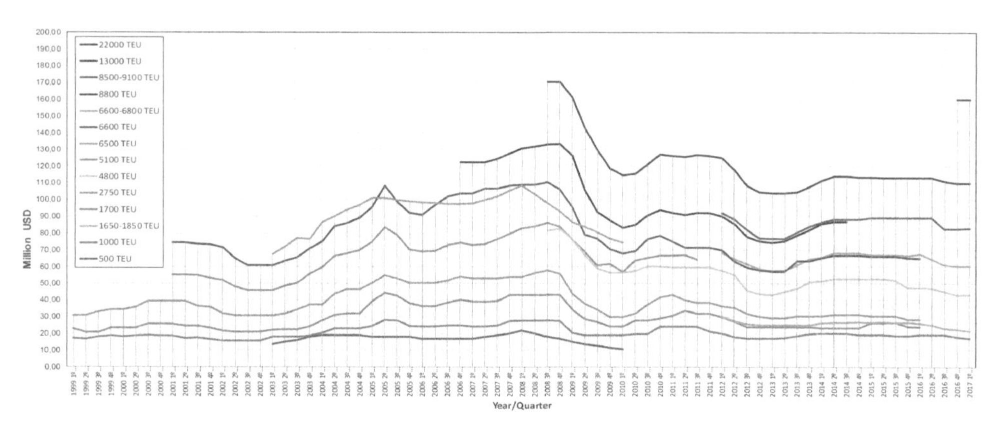

Figure 10. Newbuilding prices.

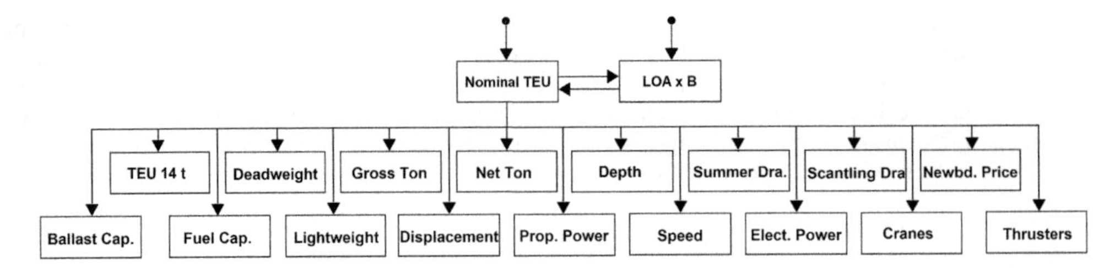

Figure 11. Methodology for estimation of containership technical characteristics.

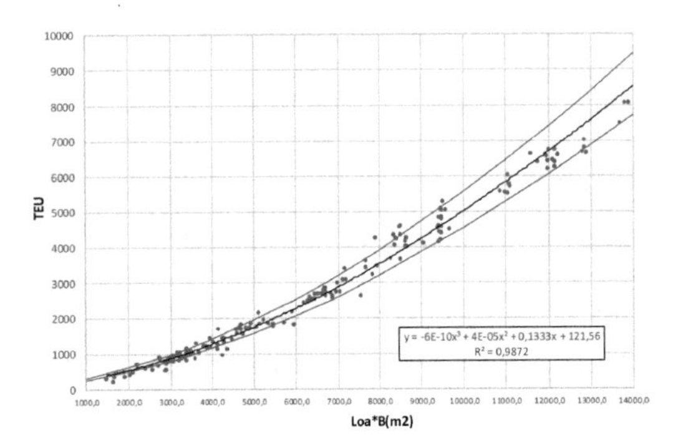

Figure 12. TEU nominal capacity plotted against LOA × B.

in separate and added to the fleet database. Other parameters related with the ship's machinery may also be estimated from the database such as: maximum continuous rating (propulsion power), auxiliary power, speed, power of bow and stern thrusters, number and capacity of cranes. Finally, displacement, deadweight, fuel capacity, ballast capacity and lightweight may also be estimated.

The initial parameters which are assumed to be known are the length over all (m) and the breadth (m). These allow the calculation of LOA × B and this allows the estimate of the nominal TEU capacity of the ship, using the regression formula in Figure 12.

Having obtained the nominal TEU capacity of the ship, it is then possible to use the regression formulae in the figures above to estimate the different technical characteristics shown in Figure 11.

4 APPLICATION TO CONTAINER SHIPS CALLING IN PORTUGUESE PORTS

4.1 *Database of containerships calling in Portuguese ports*

Following the same lines of the containership database described above, three databases were developed containing ships which called in three major Portuguese ports: Leixões, Lisbon and Sines. The ships listed in these three databases are not included in the previous database. The database of ships calling in ports relates to the month of October 2017 and was built using information provided in port authorities websites. It was noted that some ships were calling in the same port several times throughout this month and were also calling in one or the two other ports considered in this study. The information provided in these websites contains a few technical characteristics of ships, including IMO number, name and main dimensions, apart from estimated and actual times of arrival and departure from the relevant port terminal.

Table 2 shows the number of ships, average age, average TEU nominal capacity and total nominal capacity of ships calling in each of the three Portuguese ports. Ships calling in Leixões and Lisbon are slightly older, on average, than those coming to Sines and in the main database (and in world fleet). Ships calling in Leixões are smaller than those in Lisbon and even more than those calling Sines.

Figure 13 shows the distribution of ships per size category (in percentage of total number of containerships) for the three ports and for the main database, whose distribution closely matches that of the world fleet. It may be seen that containership fleets calling in Leixões and Lisbon comprise ships of much lower capacity than the fleet calling in Sines. Even in Sines, in the month of October 2017, only one ship with capacity equal or above 13.300 TEU called in this port. Leixões and Lisbon receive mainly ships with capacity below 3.000 TEU. Lisbon receives a few ships with capacities up to 5.100 TEU. Both ports receive predominantly ships in the 500 TEU to 1.000 TEU range. Ships which are smaller than 4.000 TEU are severily underrepresented in Sines (with an exception for feeder ships between 500 TEU and 1.000 TEU), while all size ranges from 4.000 TEU to 13.300 TEU are overrepresented in this port. These conclusions confirm the status of Sines as a transshipment hub for containerized cargo.

4.2 *Estimates of ship technical parameters*

In order to test the methodology for the estimation of ship technical parameters, regression formulae such as those presented in Figures 2 to 9 have been applied to the three samples of ships calling in Portuguese ports. However, since ships calling in

Table 2. Characteristics of container ship fleets calling in Portuguese ports during October 2017.

Parameter	Leixões	Lisbon	Sines	Main database
Number of ships	48	50	66	265
Average age	13,7	12,3	11,3	11,0
Average TEU	1149	1777	6336	–
Total capacity (TEU)	83.063	113.357	459.747	–

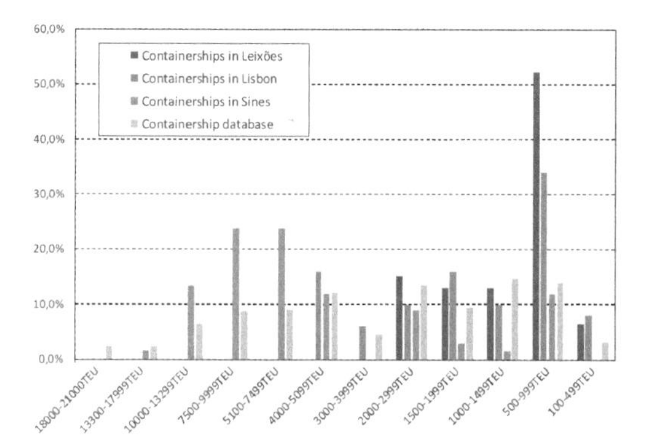

Figure 13. Distribution of ships per size category in the Portuguese ports (October 2017) and in the main database.

Leixões and Lisbon are generally below 5.000 TEU in capacity, regressions have been carried out on the main database restricted to ships of that maximum capacity. For Sines, where ships of all sizes call, the regressions are based on the full database.

The first step has been to use Figure 12 to obtain the nominal TEU capacity, since the length over all and breadth of the ships calling each port are known. With the TEU capacity of each ship, estimates of draught, depth, gross tonnage, speed, deadweight, ballast capacity and fuel capacity have been obtained using regression formulae. Experience has shown that regression formulae for propulsion power and auxiliary power yield very poor results, as might be expected considering data dispersion. Therefore, a multiple regression analysis has been carried out resulting in the following formula:

$$P_B = 74.V_s + 7,9.TEU \quad (1)$$

where P_B is the propulsion power (MCR), V_s is the ship speed and TEU is the nominal capacity.

The same approach for the auxiliary power resulted in the following formula:

$$P_E = 0,55.TEU + 0,1.P_B \quad (2)$$

where P_E is the auxiliary power.

Table 3. Average errors (%) in estimates of various ship technical characteristics for the three fleets.

Parameter	Leixões	Lisbon	Sines
TEU	6.1	12.9	7.4
Draught	4.9	6.0	8.6
Depth	8.1	8.8	7.3
Gross Tonnage	7.8	8.2	8.3
Speed	4.4	4.3	10.7
Propulsion power	9.2	8.9	17.7
Auxiliary power	35.9	39.0	25.6
Deadweight	9.8	9.6	10.2
Ballast capacity	16.5	16.9	20.3
Fuel capacity	15.9	20.9	17.6

With the estimates of different technical characteristics and the actual ship characteristics it is possible to calculate the errors associated with the estimates. Table 3 shows the average errors associated with these estimates. Separate values are shown for the three ports. Estimates are produced from length over all and breadth only. Unfortunately, like in the main database, also in the fleets calling in the different ports, there are characteristics of the ships which could not been found using the same data sources as for the main database. The errors reported in Table 3 refer only to ships for which the characteristic under consideration is known.

The overall conclusion is that most errors (17 in 30) are smaller than 10%, while in 10 cases out of 30 the errors are between 10% and 20%. However, in general, auxiliary power is very poorly estimated (errors between 25% and 40%). Also, ballast and fuel capacity are poorly estimated (errors between 16% and 21%). Results for the port of Sines are in general worse than for the other two ports. This could probably be overcome in Sines by estimating separately for ships under 5.000 TEU and then for ships above this capacity, instead of using the regressions for the full database.

5 CONCLUSIONS

This paper has presented a database of container ships containing a significant number of technical characteristics of ships in the world fleet. The database contains 265 ships, which represent 5.2% of the world fleet (by number of ships). The database has been found to be representative of the world fleet as regards average age and number of ships per dimension class. This database has been developed to allow further studies on the economics of new liner routes and studies of container ship emissions in coastal and port waters.

A methodology for estimating ship technical characteristics from AIS parameters, which typically include only length over all and breadth of the ship, has been presented. Its application in a simple numerical example to estimate the technical characteristics of container ships indicates that errors of 10% may be expected in many cases. The most difficult parameter to estimate has been found to be the auxiliary power, followed by tank capacities (ballast and fuel). In general, for preliminary economic studies of new liner services and estimates of ship emissions, these errors may be acceptable.

Future research will be devoted to ensuring the completeness of the database. It also appears feasible to apply more sophisticated analysis techniques to obtain improved estimates of the ship's characteristics.

ACKNOWLEDGEMENTS

This work was performed within the scope of the Strategic Research Plan of the Centre for Marine Technology and Ocean Engineering (CENTEC), which is financed by the Portuguese Foundation for Science and Technology (Fundação para a Ciência e Tecnologia—FCT). The authors would like to acknowledge, in the development of this database, the valuable contribution of Pedro Martins and the information kindly made available by Manuel Ventura.

REFERENCES

Alphaliner 2017. Monthly monitor.

IMO 2014. Reduction of GHG emissions from ships—Third IMO GHG Study 2014, Report MEPC 67/INF.3.

Kristensen, H.O. 2013. Statistical Analysis and Determination of Regression Formulas for Main Dimensions of Container Ships based on IHS Fairplay Data. Project no. 2010–56, Emissionsbeslutningsstøttesystem, Technical University of Denmark.

MAN B&W 2015. Propulsion Trends in Container Vessels, Copenhagen, Denmark.

RINA 2000–2013. Significant Ships of the Year, London, United-Kingdom.

Santos, T.A., Guedes Soares, C. 2017. Ship and fleet sizing in short sea shipping, Maritime Policy and Management, Vol. 47, Issue 7, pp. 859–881.

Sea Europe 2017. Market forecast report.

Takahashi, H., Goto, A., Abe, M. 2006. Study on Standards for Main Dimensions of the Design Ship, Technical Note N° 309, National Institute for Land and Infrastructure Management, Japan.

Intelligent ship navigation

Progress in Maritime Technology and Engineering – Guedes Soares & Santos (Eds)
 ISBN 978-1-138-58539-3

Challenges and developments of water transport safety under intelligent environment

H.B. Tian, B. Wu & X.P. Yan
Intelligent Transportation Systems Center (ITS Center), Wuhan University of Technology, Wuhan, China
National Engineering Research Center for Water Transportation Safety (WTS Center), Wuhan University of Technology, Wuhan, China

ABSTRACT: As intelligence technology has developed very quickly, which makes the autonomous ships also develop fast, the challenges and developments of water transportation safety is analyzed in this paper. First, the four components of future water transportation system, which are shore-based power supply system, marine electromechanical system, navigation brain system, shipping network information system are discussed. Second, the five safety challenges, including reliability of the shaft-less rim-driven thruster system, reliability of marine electromechanical system, security of the intelligent navigation system, information security of shipping network system, emergency rescue in the future shipping system are analyzed in detail and some cases are used to demonstrate the potential safety challenges. Third, some suggestions are given to enhance the water transportation safety under intelligent environment. It should be admitted that the sculpture of the future water transportation should be improved and more attention should be drawn to such research area in the future.

1 INTRODUCTION

Intelligent vehicle becomes a hot topic in both research and industry area recently. For example, Google driverless car uses cameras, radar sensors to navigate and has safely driven 480.000 kilometers (Luo et al. 2017). The same trend appears in maritime transportation. In recent years, with the development of marine radar, external sensor, control algorithms and so on (Ma et al. 2015, 2016), intelligent ships are expected to be available to use in the near future, and several projects has been conducted to develop such autonomous merchant ships.

The Maritime Unmanned Navigation through Intelligence in Networks (MUNIN) project, which is co-funded by the European Commissions, manages to identify and verify the concept of autonomous ships (Wahlström et al. 2015). In this project, the future autonomous ships is described as a vessel primarily guided by automated on-board decision systems but controlled by a remote operator in a shore side control station (Burmeister et al. 2014). Moreover, another project, Advanced Autonomous Waterborne Applications Initiative (AAWA), is conducted to produce the specification and preliminary designs for the next-generation advanced ship solutions by exploring economic, social, legal, regulatory and technological factors.

The researches of intelligent ships stimulated the development of the ship propulsion technology, and these technologies will be applied to the future autonomous ships. Specifically, the shore-based energy propulsion technology, which supplies the power from shore side, can largely reduce the CO_2 emission (Yan & Wan 2015). Moreover, the shaft-less rim-driven thruster (RDT), which can promote the energy efficiency, is also developed fast (Yan et al. 2017).

As both the navigation and propulsion systems will be changed a lot in the future, several new challenges of such ships may exist and many attention has been paid to this field recently. In the MUNIN project, the different risk levels especially the unacceptable of autonomous ships is identified by using the formal safety assessment method (Rødseth & Burmeister 2015), and the associated risk control options are also proposed in this project. Based on this research work, the safety of autonomous ships in the AAWA project is further analyzed. From the perspective of human reliability by using human factors classification and analysis system method, the likelihood of navigation accidents are discovered to be reduced while the consequences of non-navigation accidents can be reduced (Wróbel et al. 2016, 2017).

To analyze the future challenges maritime safety under intelligent environment, the remainder of thepaper is organized as follows. Section 2 presents the sculpture of the future intelligent navigation system. Section 3 proposes the six challenges of water transportation safety. Five suggestions are given in Section 4, and conclusions are drawn in Section 5.

2 DEVELOPMENT OF FUTURE WATER TRANSPORTATION SYSTEM

As intelligent and green ship technology develops fast, four associated technologies will be used in the future water transportation system, which are shore-based power supply system, marine electromechanical system, navigation brain system and shipping network information system. The sculpture of the future water transportation system is shown in Figure 1.

A shore-based power supply system, which uses electricity as the main power, will replace the diesel engine in the ship. Moreover, the future ship will use shaft-less rim-driven thruster.

A marine electromechanical system will be used to discover the fault of electromechanical system.

A navigation brain system, which give instructions to make the ship follow a predefined voyage plan within certain degrees of freedom to adjust route, can achieve perception of the navigational environment, cognition of the navigation situation and decision and control of the ship.

A shipping network information system, which connects all the entities of the water transportation system, can communicate and share information with the involved entities.

2.1 *Shore-based power supply system*

In order to reduce the environmental pollution, the shore-based power supply system will provide the electricity as power of the ship rather than diesel engine using shore-based energy. The system is consisted by three components, which are shore-based power supply system, ship power supply system and ship propulsion system. The shore-based power supply system includes power plant, power battery charging station, high voltage transmission lines and electricity substation. The ship power system is combined of shore equipment, ship power station and ship battery supply systems. Moreover, the ship propulsion system is combined by ship control system and electric propulsion system. The shore-based power supply system is shown in Figure 2. Compared with traditional power system, there are three advantages of using such system.

1. Diesel engine room is not needed so that the cabin capacity can be saved.
2. As the ship is powered by electricity, the energy can be saved by using power management system.
3. The pollution of air water and noise can be reduced by using shaft-less rim-driven thruster.

From the previous works, three types of the shore–based power supply systems can be discovered. The first type is the contact type, the principle of this type is that the ship power station contacts with electricity catenary to charge the powered the ship. The second type is power battery type, which uses battery as power and it needs to be replaced when the batteries is run out of energy. The third type is wireless power supply types, which transmits the electricity through wireless energy launcher and receiving device (Yan & Wan 2015).

It should be mentioned that in this shore-based power supply system, the propulsion system uses shaft-less rim-driven thruster. The merit of this system is that it can eliminate ship shafting and simplify the structure; as a result, the system can reduce the cost of ship design and manufacture. The shaft-less rim-driven thruster (RDT) is composed of rotor bearing, fixed bearing, multipole stator, shell (Tan et al. 2015). Compared with traditional diesel propulsion system, there are many advantages such as in ship design, propulsion performance, vibration, reduction, manufacturing and maintenance (Yan & Liang 2017).

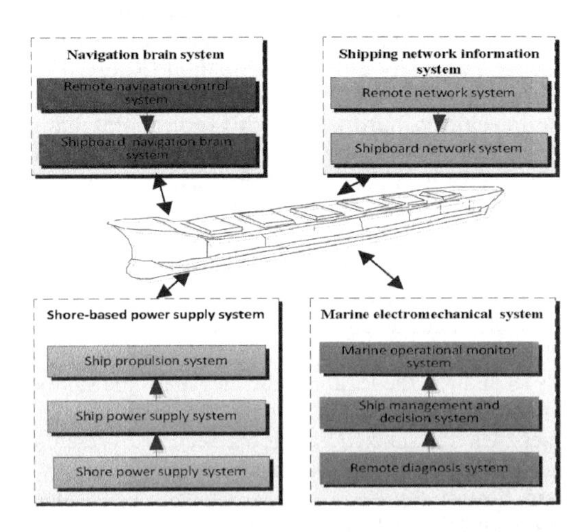

Figure 1. Sculpture of future water transportation system.

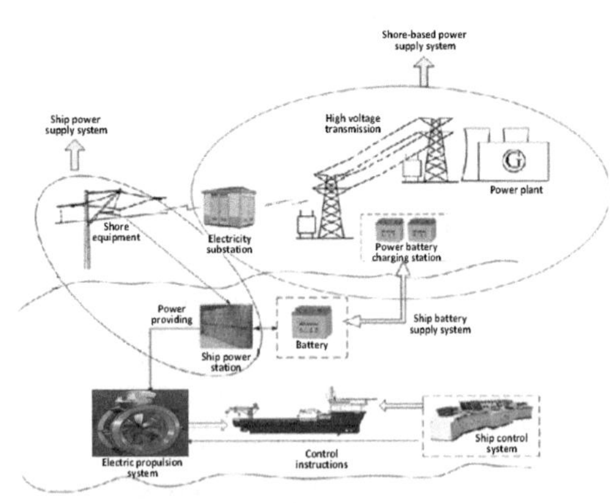

Figure 2. Shore-based power supply system.

2.2 *Marine electromechanical system*

Marine electromechanical system is used for diagnosis of fault of the operational systems. The system includes three components, which are operational monitor system, ship management and maintenance decision system.

The principle of the marine electromechanical system is as follows. Operational monitor system and ship management and maintenance decision system pass short messages through GSM and transmit data and instructions regularly through GPRS. Ship management and maintenance decision system and marine power remote diagnosis system transmits message through the internet. The system model is presented in Figure 3.

However, the system also includes an intelligent ship network system, which can realize ship to shore information interaction and ship to ship information interaction. The intelligent network system can help to accomplish some intelligent activities, such as navigation, fire alarm, loading and unloading. Furthermore, remote operation and intelligent maintenance service could realize.

2.3 *Navigation brain system*

Navigation brain system includes three components, which are perception systems, cognition systems, and decision and control systems. First, perception system receives the cargo information, power system condition navigation environment information. Then, cognition system uses information fusion technology to analyze navigation situation. Last, decision and control systems conduct navigation control and give the right order to the bridge room to control the ships automatically. The navigation brain system is presented in the Figure 4.

Specifically, perceptual space is equipped with millimeter radar, laser radar, cameras, AIS (automatic information system), forward-looking sonar and other sensors to derive navigational environment information. Moreover, this space is equipped with GPS (global positioning system), gyrocompass

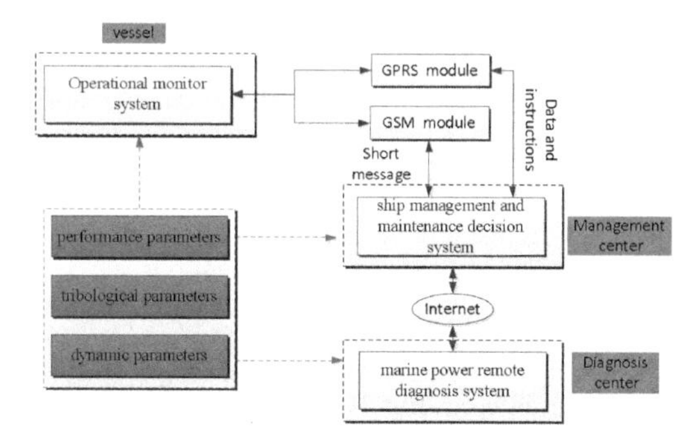

Figure 3. Marine electromechanical system.

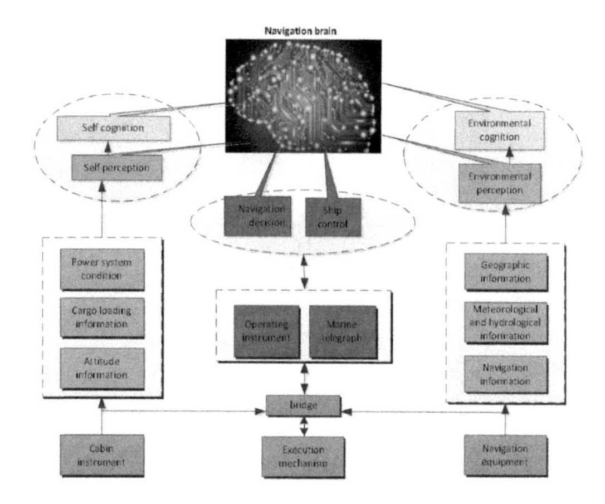

Figure 4. Framework of navigation brain system.

and other sensors to obtain the information of the ship's own motion. It is also equipped with oil consumption sensor and shaft power sensor and speed sensor to obtain the monitoring condition information of ship engine room. The cognitive space is equipped with computing equipment. The navigation situation analysis algorithm and the learning algorithm of ship behavior are completed on the computing equipment. Decision making space is equipped with computing equipment, controllers and so on. The navigation decision algorithm and the ship navigation control algorithm should also be developed. Perception space is the basic of cognitive space. The feedback from cognitive space will affect the preprocessing process of perception space. Decision making space uses the feedback information of perceptual space to influence the situation cognition of cognitive space.

2.4 *Shipping network information system*

Shipping network information system includes onboard ship network system and remote network system, the inland river shipping network information system in Figure 5 (Yan 2010, Sun 2010). Recently, the security of shipping network becomes a concern especially when all the shipping entities are connected by using internet. When the hacker attacks the shipping network, these hackers may control the ships and robbery and hijacking incidents may be occurred.

Five basic functions are required for shipping network information system, which are identification, protection, detection, responding and recovering. The identification function means the system can identify hazards to system, data, and assets. The protection function means the system has the ability to limit or contain the impact of a potential cyber incident. Detection function means to identify the occurrence of a network incident. The detection function enables timely discovery of

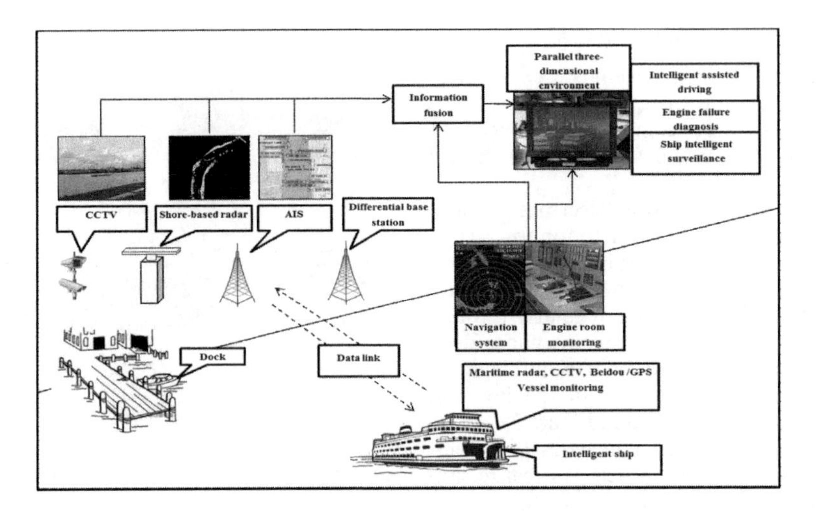

Figure 5. Inland river network information system.

a network incident. The respond function owns the ability to contain the impact of a potential cyber safety and security incident. Recovering function means to maintain resilience and to restore any capabilities or services that were impaired due to a network incident.

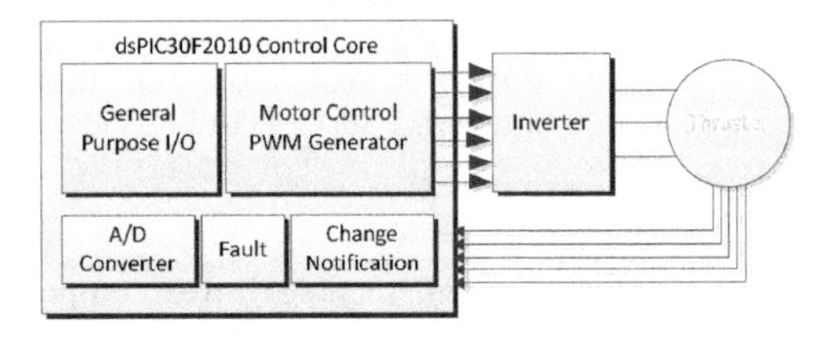

Figure 6. RDT control system.

3 CHALLENGES OF WATER TRANSPORTATION SAFETY

3.1 *Reliability of the shaft-less rim-driven thruster system*

The RDT control system driver architecture is presented in Figure 6. With the increase of the power of the motor, the size of the non-axial propulsion machinery also increases, the basic structure is more complicated, and the integration design is more difficult. Non-axial propulsion motor has high reliability and high efficiency. In spite of the rapid development of the speed control technology of the no-position sensing system, the problem of instability in the start-up and low-speed operation is very important. The research results show that it is difficult to control the permanent magnet motor with the existing position sensing technology. With the continuous development and improvement of this technology, RDTs can be reasonably expected to play a critical role in ship propulsion systems in the future, and may have a profound influence on the green shipping industry as well.

3.2 *Reliability of marine electromechanical system*

Accidents caused by the failure of marine power equipment occur frequently, and this type of accidents has caused large economic losses and fatalities. Ship reliability analysis indicated that 60%~80% of the whole ship all fault occurred in power plant, therefore, researching fault identification, prediction and mitigation of marine electromechanical system can help us eliminate equipment problems in a timely manner, as a result, it can guarantee the safe operation of the machinery equipment.

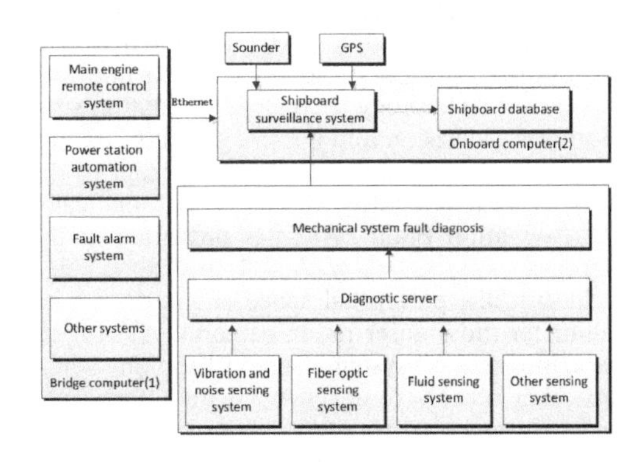

Figure 7. Monitoring and diagnosis system for ship engine room.

Ship engine room monitoring and diagnosis system is presented in Figure 7 (Yan & Li 2013). The system integrates multiple sources information and can make timely judgment on the potential failure of the main engine and other mechanical equipment, which is beneficial for avoiding the occurrence of major accidents. Bridge computer includes main engine remote control system, power station automation system, fault alarm system. Through diagnostic server, four types of mechanical system fault diagnosis can be achieved, which are vibration and noise sensing system, fiber optic sensing system, fluid sensing system and other sensing system. Onboard computer includes shipboard surveillance system and shipboard database. Shipboard surveillance system obtains data through sounder, GPS (Hsieh et al. 2007).

However, existing monitor and diagnosis system can help to ship engine room. It is necessary to develop a new generation of marine electromechanical system to improve the safety.

1. Construction three stage management system of ship, diagnosis center and management center
2. Researching intelligent operation and maintenance system software for remote diagnosis and maintenance decision.

3.3 *Security of the intelligent navigation system*

Different from the smart cars, which is a two dimensional motion system, intelligent ship is three dimensional, including yaw, roll and pitch. The ships features, cargo fuel, as well as the navigation environment and climate conditions, speed, and many factors forming dynamics model (Hinnenthal & Clauss 2010). And artificial intelligence in the navigation of ships use is facing the shortage of samples. Other simple movement model based on the parameters including depth study is difficult to directly apply to the process of driving behavior and realize intelligent driving.

For realizing the cargo ship unmanned and intelligence, a set of "the brain" function of "artificial intelligence" system is being developed (Ma et al. 2015). The system is called brain, because the system could complete the voyage by him with the functions of perception, cognition and decision-making. It is able to complete all kinds of ships sailing with the unmanned driving system and unmanned sailing brain work on ships and autonomous navigation realization. The function of navigation brain is to perceive space, cognitive space and decision-making space. In the process of these three parts, the most middle one is to strengthen perception, learn the captain's driving experience, and finally realize the autonomous decision-making and control of the ship. We also made a description, began to realize dynamic system and external environment perception, by this time most is to rely on people sail, finally steered the ship one hundred percent of the realization of artificial intelligence. Second, the situation cognition including historical data, experience to work on the deep processing, the description to the navigation environment, the situation to make the judgment of safety or not. The third is decision making, including how to avoid collision, especially in narrow waters, such as how to dock and offshore when entering or leaving Hong Kong. At present, I have such a three-body ship in the east lake as an experiment, which can perceive the external information. After the perception and decision, through learning with driving experience, the construction of operation model is realized (Ma et al. 2016).

3.4 *Information security of shipping network system*

With the development of technology, the ships information technology (IT) and operation technology (OT) are more and more connected by the Internet. This increases the risk of unauthorized access or malicious attacks on ship systems and networks. Risk can also occur on entering the system; through the mobile media malicious software can invasion the shipping network. The security, environmental, and commercial consequences of not preparing for a network event can be significant. Here we list some typical cases of network security of shipping enterprises in Table 1.

Table 1. Shipping information network accidents.

Time	Country	Event	Consequence
2017	British	Clark shipping company was hacked by hackers	Company data leakage
2017	United States	Danish Petya virus attacks Maersk group IT system	System crippled, losing $300 million
2017	China/ Russia/ Britain	WannaCry virus transmission	300,000 users were affected, losing $8 billion
2013	Belgium	Drug smuggling case at Antwerp	Dealer infiltrate the harbor system and hide the drug in the cargo
2012	China/ Japan/ Korea	Cefog cyber attacks	Military ship building and maritime sectors were attacked, 4,000 IP cases infected.
2011	Iran	National shipping company was being attacked	cargo details were damaged

Many international maritime related organizations have taken measures for the safety of ship network information. In 2016, IMO incorporated the network security of ships into the agenda and formulated the voluntary guidelines for cyber security cooperation. The Baltic and international maritime commission issued guidelines on ship network safety. In 2017, China classification society issued the guidance on the requirements and safety assessment of ship network system. The international association of classification societies established a joint working group to assess the development of standardized industry norms that could be used to combat the threat of maritime network threats. The international maritime insurance alliance issued a statement to strengthen cooperation with international telecommunication union, IMO and other organizations.

In general, there are two types of shipping network attacks that affect companies and ships. One is untargeted attacks, systems and data from companies or ships are one of many potential targets. Another is target attack, company or ship system and data is the intended target. Untargeted attacks may use tools and technologies available on the Internet that can be used to locate companies and ship known vulnerabilities. Examples of tools and techniques that might be used in these situations include: targeted attacks can be more complex and use tools and technologies specifically targeted at specific companies or ships. Examples of tools and techniques that can be used in these situations include: the above example is not exhaustive. The number and complexity of the tools and techniques used in network attacks continues to grow, and only those organizations and individuals are limited in their intelligence. Crucially, It system users on board are aware of potential network security risks and are trained to identify and mitigate such risks.

3.5 *Emergency rescue in the future shipping system*

From analysis of Eastern star accident, the existing Yangtze River emergency rescue forces cannot meet the requirement of the Yangtze River waterway construction. The emergency rescue ability cannot meet the need of emergency access to the main channel of the Yangtze River. New rescue and salvage technologies should be proposed (Wu et al. 2015). For example, distress searching technology and information transmission technology should be developed to locate the target accurately in the bad weather conditions. Moreover, the marine emergency rescue and decision-making support system should also be developed.

There are three aspects to focus on salvage technologies, which are large tonnage wreck salvage technology, deep diving technique, underwater operation technique at large velocity. Through the research of large tonnage wreck salvage technology, strengthen the reserve of advanced and prospective salvage technology, and improve the comprehensive technical ability of wreck salvage. It mainly includes large tonnage underwater lifting technology, underwater plugging technology, fishing auxiliary decision support system, underwater search and salvage information transmission technology. Through the research on the complete set of saturation diving and the research and development of diving equipment, we can improve the salvage ability of deep underwater rescue. It mainly consists of 300 meters deep saturation diving complete technology; large depth salvage robot (ROV), submarine can operate manned submersible application technology. Through the research and development of diving equipment and speed reduction equipment, the ability of underwater diving under great velocity is improved. It mainly includes the energy dissipation breakwater and the large velocity diving operation training pool.

In terms of ecological environment protection, China's ministry of transport has formulated a special plan of action for the prevention of pollution by the Yangtze River economic belt. To strength the management of pollution sources, we will strengthen on-site supervision over the prevention and control of pollution. The establishment of a Marine emission control zone prohibits the development of the Yangtze River basin.

4 SUGGESTIONS FOR ENHANCEMENT OF MARITIME SAFETY

4.1 *Establish ships safety management framework*

Together with the present situation of ship safety management system and the weak link, ships safety management system including three levels, which are the safety management system, security technology, ship safety laws and regulations (Wu et al. 2015). Among them, ship safety protection technology includes ship design, ship construction, ship operation, and ship accident rescue and ship personnel. Ship design has stability, fire, life, strength, turbine, electric and other aspects. The construction of ships consists of craft, materials, outfitting, hull and inspection. Ship operation is divided into passenger ship navigation, support guarantee (navigation facilities, navigation environment, port anchorage, and communication facilities), cargo management, passenger management, maintenance and so on. The ship accident rescue depends on contingency plan, system, mechanism, and legal system – "one case three system". The crew includes crew, passengers and tour guides. The ship safety management system involves the use of departments and agencies, such as shipping companies, trade associations, local governments, maritime agencies, travel agencies. Ship safety laws and regulations including all aspects of the legal norms and regulations mentioned above. The ship safety system is presented at Figure 8.

4.2 *Safety administration of unmanned ship in the future*

The existing maritime administration methods and technology may not be useful for the future safety administration, and it is necessary to take measures to study new supervision methods under intelligent navigation environment. Here are suggestions on safety management of unmanned ships.

1. Identification of the differences between traditional ships supervision and unmanned ships supervision is necessary, which is also the initial step for administration.
2. Further research on the key technologies for unmanned ship administration, including promote the function of VTS and AIS equipment.
3. New rules or laws between unmanned ships and conventional ships should be set for water traffic safety management.
4. Improvement of the crew ability of ship maneuvering should be carried out under intelligent environment.

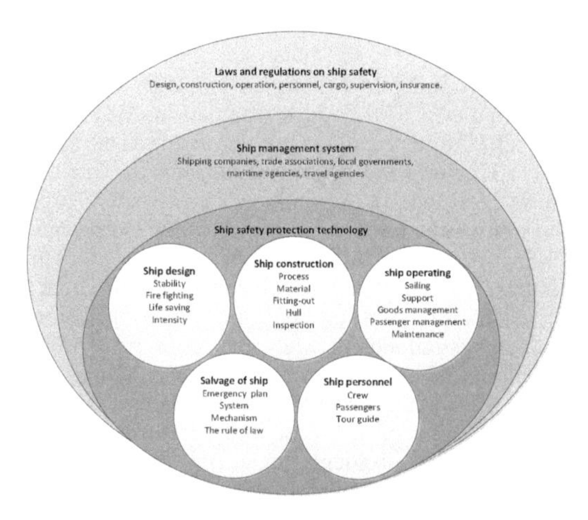

Figure 8. Ship safety management framework.

4.3 *Use of big data for safety management*

Big data is a novel technology for maritime safety management as there are different types of data in maritime transportation, such as accident data, AIS data. Moreover, some works have been done.

The big data framework for safety management is showed in Figure 9. Three types of safety management can be carried out. First, the ship motion trajectory data of the ship can be obtained through the VTS AIS LRIT device. Then, the multi-source location data is integrated to achieve the recognition, track the target, and realize maritime traffic supervision. Through the establishment of historical track database and combining with navigation infrastructure data, data mining and knowledge discovery are carried out to realize water traffic situation assessment. Identification of ship activity can help to research early risk warning of navigation and identify ship emissions and pollutants diffusion; as a result, the monitoring level of ship pollution can be improved. According to the regional emission inventory and the spatial and temporal distribution characteristics of pollutant

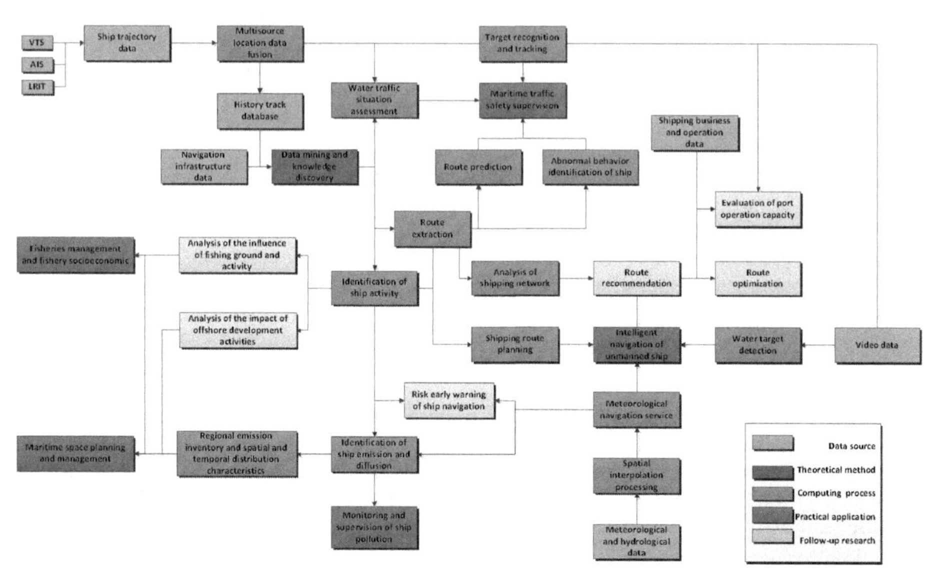

Figure 9. Big data framework for safety management.

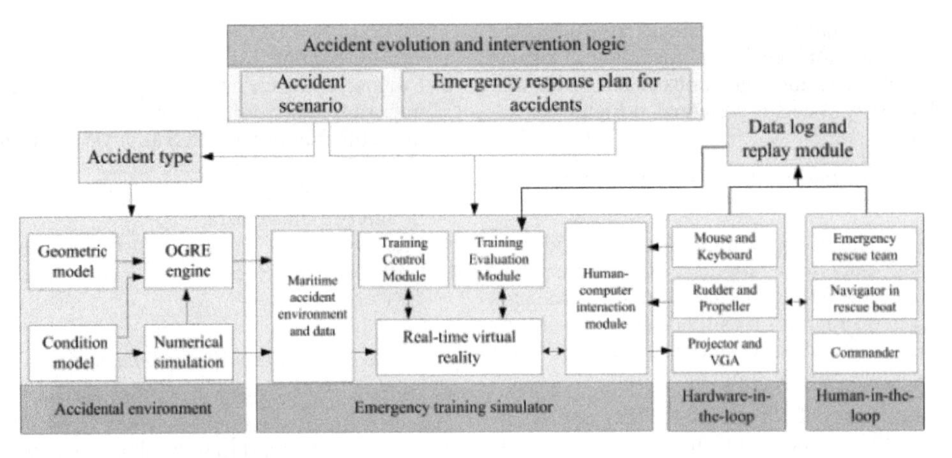

Figure 10. System architecture of MESS.

emission, assistance for maritime space planning and management can be provided. Identification of ship activity could have an impact on the analysis for the study of offshore development activities and the analysis of fisheries and fishing activities. As a result, it can provide guidance for fisheries management and fishing social economy. In addition, data mining and knowledge discovery can be used to calculate and extract shipping information. Route extraction can help to predict the route of the ship, identify the abnormal navigation behavior of the ship, and conduct maritime traffic safety supervision. The route information extracted can be used in the analysis of shipping network, and then the safe route of ship is recommended. Finally, route optimization configuration is carried out. Shipping business and operation data, ship trajectory data and video data provide support for the research of port operation capability evaluation. Shipping route is planned by the identification of the ship's activities with the extracted information. The spatial interpolation processing of meteorological moisture data could provide the meteorological navigation service, which could guarantee for the ship emission and navigation risk warning. Recommended routes planned shipping routes, weather navigation services, water target detection data through video data could provide support and services for ship intelligent navigation (Wu et al. 2016).

4.4 *Develop maritime emergency simulation system*

As virtual simulation system owns the characteristics of interactivity, immersion, low cost and low risk, it has been proved to be a practical tool for improving the skill of participants. This advanced technology has been applied in many fields. For example, the simulation system (MSS) is used to train crew and pilots (Feng et al. 2012), and auto pilot (Yu et al. 2013), train simulation (Watanabe et al. 2011) and also in the military field (Rizzo et al. 2011). However, few systems focuses on training for crews in Rescue Coordination Centre (RCC), so the skillsof the rescuers need a maritime emergency simulation system. MESS is different from MSS in several ways. First, the participants were different, the former focuses on multi-person cooperation and process modeling while the latter paid more attention to the accuracy of ship motion. Moreover, this system could provide different scenarios with the advantage of low-cost and random-scenario analysis. The accident logic and intervention method is based on the evolution and development of ship accidents, thus the emergency process for participants would effective and useful. Furthermore, the traffic organization and rescue of man overboard are also in line with reality, so this could be applied to the actual SAR in future. This system could also be used for accident investigation, adaptive decision-making and human reliability. For further research, the accident such as stranding should be carried out (Wu et al. 2014). The system architecture for MESS is shown in Figure 10.

5 CONCLUSION

This paper proposed the development and challenges of the water transportation system in the future. The developments are shore-based energy, shaft-less rim-driven thruster system, fault diagnosis of ship electromechanical system, ship navigation with navigation brain system and network information system. Afterwards, in view of the future developments direction of water transport system, this paper proposed challenges to the future water transportation system, including reliability of the shaft-less rim-driven thruster, Reliability of marine electromechanical system, security of navigation brain system, information security of shipping network system, emergency rescue. Furthermore, some suggestions are given to enhance the maritime safety, including ship safety management, unmanned ships management, shipping large data application, emergency rescue research.

However the future description of waterway transportation system should be improved and more researches should be focused to explore new way to solve problems owing to intelligent technology. In this context, more work should be done to improve the safety and efficient of water transportation system. The opportunities and challenges should be further analyzed with the development of shipping industry.

ACKNOWLEDGEMENTS

The research presented in this paper was sponsored by a grant from National Science Foundation of China (Grant No. 51609194), a grant from the Hubei Natural Science Foundation (Grant No. 2017CFB202) and Marie Curie RISE RESET (Grant No. 730888).

REFERENCES

Burmeister H C, Bruhn W, Rødseth Ø.J., et al. Autonomous Unmanned Merchant Vessel and its Contribution towards the e-Navigation Implementation: The MUNIN Perspective 1. *Int. J. e-Navigation & Maritime Economy*, 2014, 1(2): 1–13.

Feng M, Li Y. Ship intelligent collision avoidance based on maritime police warships simulation system. Electrical & Electronics Engineering (EEESYM), 2012 *Symposium on IEEE*, 2012: 293–296.

Hinnenthal J, Clauss G. Robust Pareto-optimum routing of ships utilising deterministic and ensemble weather forecasts. *Ships and Offshore Structures*, 2010, 5(2): 105–114.

Hsieh M F, Chen J H, Yeh Y H, et al. Integrated design and realization of a hubless rim-driven thruster. Industrial Electronics Society (IECON), 2007 *33rd Annual Conference on IEEE*, 2007: 3033–3038.

Luo J, Yan B, Wood K. InnoGPS for Data-Driven Exploration of Design Opportunities and Directions: The Case of Google Driverless Car Project. *Journal of Mechanical Design*, 2017, 139(11): 111–416.

Ma F, Chen Y W, Huang Z C, et al. A novel approach of collision assessment for coastal radar surveillance. *Reliability Engineering & System Safety*, 2016, 155: 179–195.

Ma F, Chen Y, Yan X, et al. A novel marine radar targets extraction approach based on sequential images and Bayesian Network. *Ocean Engineering*, 2016, 120: 64–77.

Ma F, Wu Q, Yan X, et al. Classification of Automatic Radar Plotting Aid targets based on improved Fuzzy C-Means. *Transportation Research Part C Emerging Technologies*, 2015, 51: 180–195.

Rizzo A, Parsons T D, Lange B, et al. Virtual reality goes to war: a brief review of the future of military behavioral healthcare. *Journal of Clinical Psychology in Medical Settings*, 2011, 18(2): 176–187.

Rødseth Ø J, Burmeister H C. Risk assessment for an unmanned merchant ship. *TransNav: Int. J. Marine Navigation & Safety of Sea Transportation*, 2015, 9(3): 357–364.

Sun X, Wu Y, Chu, X, Intelligent Yangtze River Shipping and Its Prospects, *Journal Transport Information and Safety*, 2010, 28 (6): 48–52. (in Chinese).

Tan W, Yan X, Liu Z, et al. Technology Development and Prospect of Shaftless Rim-driven Propulsion System. *Journal of Wuhan University of Technology*, 2015, 39 (3), 601–605. (in Chinese).

Wahlström M, Hakulinen J, Karvonen H, et al. Human factors challenges in unmanned ship operations–insights from other domains. *Procedia Manufacturing*, 2015, 3: 1038–1045.

Watanabe N, Shimomura T, Sasaki K, et al. Hardware-in-the-loop simulation system for duplication of actual running conditions of a multiple-car train consist. Quarterly Report of RTRI, 2011, 52(1): 1–6.

Wrobel K, Krata P, Montewka J, et al. Towards the Development of a Risk Model for Unmanned Vessels Design and Operations. *Transnav Int. Journal on Marine Navigation & Safety of Sea Transportation*, 2016, 10(2): 267–274.

Wrobel K, Montewka J, Kujala P. Towards the assessment of potential impact of unmanned vessels on maritime transportation safety. *Reliability Engineering & System Safety*, 2017, 165: 155–169.

Wu B, Wang Y, Zhang J, et al. Effectiveness of maritime safety control in different navigation zones using a spatial sequential DEA model: Yangtze River case. *Accident analysis & prevention*, 2015, 81: 232–242.

Wu B, Yan X, Wang Y, et al. Maritime emergency simulation system (MESS)–a virtual decision support platform for emergency response of maritime accidents. Simulation and Modeling Methodologies, *Technologies and Applications* (SIMULTECH), 2014: 155–162.

Wu B, Yan X, Wang Y, et al. Selection of maritime safety control options for NUC ships using a hybrid group decision-making approach. *Safety Science*, 2016, 88: 108–122.

Yan X, Li Z, Zhang Y, et al. Study on key techniques of wear monitoring and fault diagnosis for marine diesel engines: a review. *China Mechanical Engineering*, 2013, 24(10): 1413–1419. (in Chinese).

Yan X, Liang X, Ouyang W, et al. A review of progress and applications of ship shaft-less rim-driven thrusters. *Ocean Engineering*, 2017, 144: 142–156.

Yan X, Wan J, System Construction of Canal Ship Propulsion Technology Based on Shore Power. *Ship & Ocean Engineering*, 2015, 44(3): 159–168. (in Chinese).

Yan Z, Yan X, Ma F, et al. Green Yangtze River Intelligent Shipping Information. *Journal Transport Information and Safety*, 2010, 6(29): 76–81. (in Chinese).

Yu Y, El Kamel A, Gong G. Modeling intelligent vehicle agent in virtual reality traffic simulation system. *Systems and Computer Science* (ICSCS), 2013 2nd International Conference on IEEE, 2013: 274–279.

Progress in Maritime Technology and Engineering – Guedes Soares & Santos (Eds)
© 2018 Taylor & Francis Group, London, ISBN 978-1-138-58539-3

Collision avoidance, guidance and control system for autonomous surface vehicles in complex navigation conditions

M.A. Hinostroza & C. Guedes Soares
Centre of Marine Technology and Ocean Engineering (CENTEC), Instituto Superior Técnico, Universidade de Lisboa, Lisbon, Portugal

ABSTRACT: This paper presents a collision avoidance, guidance and control system for operation of autonomous surface vehicles in critical navigation conditions. The collision avoidance unit is based on fuzzy logic intelligent decision-making algorithm, the guidance unit uses the line-of-sight algorithm and, the control unit, is composed by a PID heading controller and a speed controller. A set of numerical simulations were carried out for different collision scenarios to validate the effectiveness and feasibility of the system. Also a case study involving ship traffic in the estuary of Tagus river, Lisbon was performed and a good performance of the system was found.

1 INTRODUCTION

Human error contributes to more than 80% of ship collision accidents, which pose a rather severe threat at sea. Research on autonomous collision avoidance at open sea is important for mitigating the overall risk of collisions in intelligent navigation, (Perera et al. 2011). These systems are designed to support decision-making for navigators and isolate human errors during collision avoidance. However, in near future, autonomous surface vessels will enter in maritime transportation, and it will needs to incorporate an automatic collision avoidance system to ensure safe navigation.

With the increasing number of successful projects in autonomous surface vehicles (ASVs), there is an increased interest in its concept application in the maritime industry. This paper refers to the application of the concept of unmanned navigation, involving the reduction and minimization of crew on-board vessels, with the ultimate goal in mind of a ship carrying out its functions without man on-board overtime. In this scenario, Rolls-Royce (2014) has announced that remotely operated local vessels are expected by 2020 and ocean-going ships for maritime transport by 2030. These unmanned ships aim to increase safety of operations at the sea, reduce fuel consumption, and transform the work roles in the maritime domain. The Defense Advanced Research Project Agency (DARPA) a developed Sea Hunter, built as part of its Anti-Submarine Warfare Continuous Trail Unmanned Vessel (ACTUV) program, began sea trials in April of 2015. Sea Hunter is a 132 feet long trimaran (a central hull with two outriggers), manufactured deep in the Silicon Forest of Portland, Oregon, (Njus, 2016). It is expected to undergo two years of testing before being in service with the U.S. Navy. If tests are successful, future such craft may be armed and used for anti-submarine and counter-mine duties, operating at a small fraction of the cost of operating a destroyer.

Recently, the World's first autonomous cargo vessel, was announced, YARA Birkeland is planned to sail in the Norwegian fjords, and is scheduled for fully autonomous operation in 2020 (KongsbergMaritime, 2017). This will be a giant step for maritime traffic, and sets the start for unmanned marine vehicles. Nevertheless, there are still challenges related to these operations.

Other examples is small scale the family of ASVs which includes scaled fishing trawler type vessel, ARTEMIS, the catamaran models, ACES and AutoCat, and the SCOUT vessels. Of all these prototypes, the kayak type, SCOUT vessels have successfully implemented COLREGs at a basic level for head-on situations whilst maintaining wireless communication (Benjamin & Curcio, 2004).

Moreira et al. (2008) performed successfully model tests using a scaled model of the tanker "Esso Osaka", which was instrumented for autonomous operation and different guidance and control approaches. Perera et al. (2015), Ferrari et al. (2015), Hinostroza et al. (2017) and Xu et al. (2018) have conducted model tests with an autonomous surface vehicle of 2.5 m for intelligent navigation and collision avoidance, manoeuvring tests and marine exploration and system identification, respectively.

Many techniques have been proposed for avoidance of collision situations, Sato & Ishii (1998); Statheros et al. (2008), but in general those techniques ignore the law of the sea as formulated by the International Maritime Organization (IMO) in 1972. These rules and regulations are expressed in the Convention on the International Regulations for Preventing Collisions at Sea (COLREGs). In recent literature new formulations including COLREGs regulation were addressed, Zhang et al. (2015) has presented a distributed anti-collision decision support formulation in multi-ship encounter situations under COLREGs. He et al. (2017) has performed a quantitative analysis of COLREGs rules and seamanship for autonomous collision avoidance at open sea. The present convention was designed to update and replace the Collision Regulations of 1960, which were adopted at the same time as the International Convention for Safety of Life at Sea (SOLAS) Convention.

Fuzzy-logic based systems, which are formulated for human type thinking, facilitate a human friendly environment during the decision making process. Hence, several decision making systems in research and commercial applications have been presented before. Automatic collision avoidance systems for ship systems using fuzzy logic based control systems have been proposed by Hasegawa (1987). The conjunction of human behavior and the decision making process has been formulated by various fuzzy functions in several works. Fuzzy logic based decision making system for collision avoidance of ocean navigation under critical collision conditions was presented in Perera et al. (2011). However, the simulation results are limited to the two-vessel collision avoidance situations.

This paper focuses on a development of a novel complete navigation system for ASVs, including a collision avoidance unit, guidance and control system, Figure 1. This system combines a practical collision avoidance unit based on fuzzy logic intelligent decisions and a guidance and control unit based in Line-of-sight algorithm and heading and speed controllers. It is a work specifically solving the collision avoidance problem navigating in complex environments, including dynamic obstacles. The algorithm designed in this paper is able to extract information from a real navigation map to construct a synthetic grid map, where both static and dynamic obstacles are well represented. By using such a map, a collision free path may be generated which can be directly used as a guidance trajectory for practical navigation.

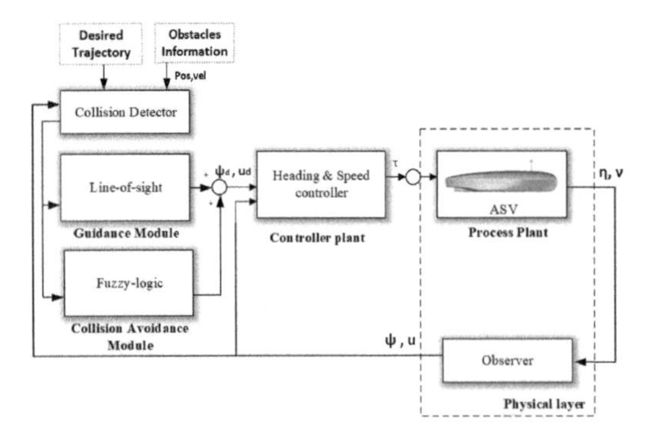

Figure 1. Block diagram of collision avoidance, guidance and control system.

2 MATHEMATICAL FORMULATION

In this section, the mathematical formulation of collision avoidance and guidance system for ASVs, showed in Figure 1, is presented. The collision detection block uses the close distance of approach between vessels to determine the collision risk from COLREGs rules. The collision avoidance unit uses the fuzzy logic algorithm to compute the desired heading and speed in order to avoid the obstacles. The Guidance and Control block is where the trajectory tracking is performed based in a desired heading and ASV speed. Notice that the collision avoidance unit is based in the work proposed by Perera et al. (2011). However, this paper includes a collision detection module, which is based in the distance to closest point of approach and the time to the closest point of approach detects potential conflicts; an iterative fuzzy membership function in order to reduce the number of fuzzy rules, and the ship motions are simulated using a realistic 3DOF nonlinear mathematical model. The present system also incorporates a path-planning system and controller plant.

2.1 *COLREGs rules and regulations*

The COLREGs, IMO (1972), include 38 rules that have been divided into Part A (General), Part B (Steering and Sailing), Part C (Lights and Shapes), Part D (Sound and Light signals), and Part E (Exemptions). In this study the COLREGs Part B, concerning Steering and Sailing rules are considered.

It is a fact that the COLREGs rules and regulations regarding collision situations in ocean navigation have been ignored in most of the recent literature. The negligence of the IMO rules may lead to conflicts during ocean navigation. As for the reported data of maritime accidents, 56% of major maritime collisions include violations of the COLREGs rules and regulations. Therefore, the methods proposed by the literature ignoring the COLREGs rules and regulations should not be implemented in ocean navigation. On the other hand, there are some practical issues regarding implementation of the COLREGs rules and regulations during ocean navigation. The Own vessel Head-on and Overtake situations and crossing are presented in Figure 2. In Crossing situations where the Own vessel is in "Give way" situations or, "Stand on" situations where there are velocity constrains in implementing COLREGs rules and regulations of the "Give way" and "Stand on" vessel collision situations when the Target vessel has very low or very high speed compared to the Own vessel.

2.2 *Collision detection module*

The collision detection module is responsible to determine if there is a potential conflict in route as shown in Figure 2, where two vessels have the possibility of collision at the trajectory crossing point. To effectively eliminate the conflict, it is necessary to compute the distance to the closest point of approach (DCPA) and the Time to the closest point of approach (TCPA). As shown in Fig. 3, vessel 1 is travelling with the velocity V1 and vessel 2 has the velocity of V2. V12 is the relative velocity of the vessel1 with the respect to vessel 2, D is the distance between two vessels and γ is the angle between the relative motion line and the bearing angle of vessel 1.

The collision detection module uses this DCPA and the TCPA to determine the collision risk (CR),

$$CR = \begin{cases} DCPA \leq d_{threshold} \\ TCPA \leq t_{threshold} \end{cases} \quad (1)$$

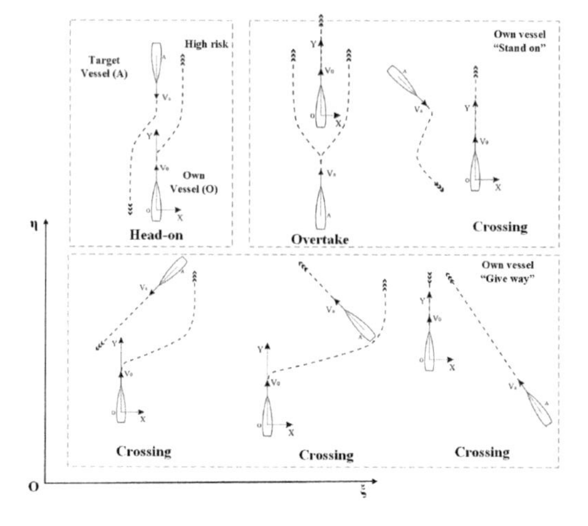

Figure 2. COLREGs rules and regulations, Perera et al. (2011).

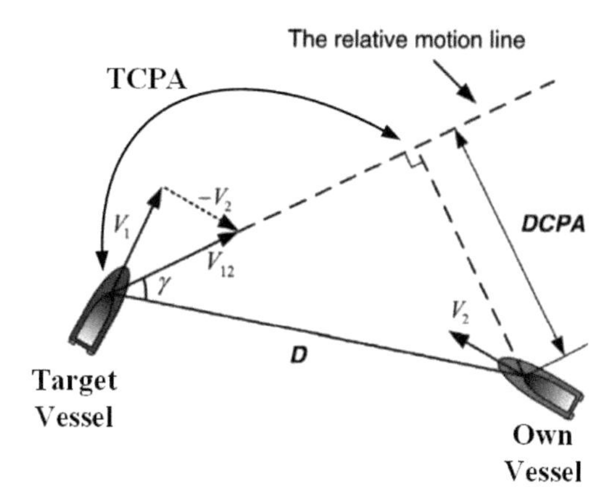

Figure 3. Calculation of DCPA and TCPA.

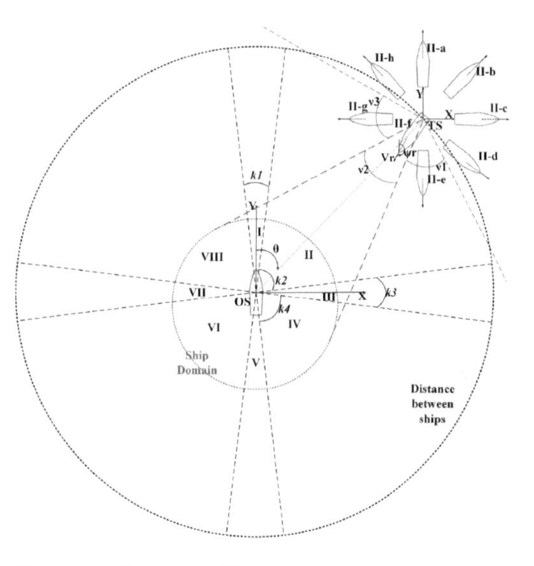

Figure 4. Relative collision situation for two vessels, Perera et al. (2011).

where $d_{threshold}$ and $t_{threshold}$ are defined by COLREGs rules and regulations, i.e. $d_{threshold}$ = 6 miles and $t_{threshold}$ = 20 min (He et al., 2017).

2.3 *Fuzzy-logic collision avoidance unit*

The fuzzy logic collision avoidance unit is responsible to compute the desired heading and velocity of own vessel in order to avoid potential conflicts in a desired trajectory. This study uses the Mandami formulation, the relative collision situation for two vessels are shown in Figure 4.

Table 1. Rules for fuzzy logic algorithm.

Region	Div.	ψ rel.	DCPA	TCPA	Decision	Decision
II	f	v_2	medium	Moderate	$\delta\psi > 0$	$\delta V_0 < 0$
II	f	v_1	medium	Moderate	$\delta\psi > 0$	N.A
II	f	v_3	medium	Moderate	$\delta\psi < 0$	N.A
I	e	v_1	medium	Moderate	$\delta\psi > 0$	N.A
I	e	v_2	medium	Moderate	$\delta\psi > 0$	$\delta V_0 < 0$
I	e	v_3	medium	Moderate	$\delta\psi < 0$	N.A
VIII	d	v_1	medium	Moderate	$\delta\psi > 0$	N.A
VIII	d	v_2	medium	Moderate	$\delta\psi < 0$	$\delta V_0 < 0$
VIII	d	v_3	medium	Moderate	$\delta\psi < 0$	N.A
II	f	v_2	Small	Short	$\delta\psi > 0$	$\delta V_0 < 0$
II	f	v_1	Small	Short	$\delta\psi > 0$	$\delta V_0 < 0$
II	f	v_3	Small	Short	$\delta\psi < 0$	$\delta V_0 < 0$
I	e	v_1	Small	Short	$\delta\psi > 0$	$\delta V_0 < 0$
I	e	v_2	Small	Short	$\delta\psi > 0$	$\delta V_0 < 0$
I	e	v_3	Small	Short	$\delta\psi < 0$	$\delta V_0 < 0$
IV	h	v_1	medium	Moderate	$\delta\psi > 0$	N.A
IV	h	v_2	medium	Moderate	$\delta\psi > 0$	$\delta V_0 < 0$
IV	h	v_3	medium	Moderate	$\delta\psi < 0$	N.A

*N.A, no action.

2.3.1 *Collision conditions*

Figure 4 presents a relative collision situation in ocean navigation that is similar to a Radar plot. The Own vessel ocean domain is divided into two circular sections, the ship domain and the actual approximate distance to the target vessel.

The Own vessel Collision Regions are divided into eight regions from I to VIII. These regions are separated by dotted lines that are coincident with the Collision Regions as formulated in the Fuzzy membership function (FMF), presented in subsection 2.3.2. It is assumed that the Target vessel will be located within one of these eight regions and the collision avoidance decisions are formulated in accordance to each region.

Target vessel positions have been divided into eight divisions of vessel orientations regarding the relative course (II-a, II-b, II-c, II-d, II-e, II-f, II-g and II-h). These divisions are separated by dotted lines that are coincident with the Relative Collision Angle FMF.

Table 1 present the summarized collision risk assessments and decisions of the two-vessel collision situation in Figure 3. The first column represents the Collision Regions (Reg.) with respect to the Own vessel, and the second column represents the Divisions (Div.) of the Target vessel orientations. The third column ψ rel represents relative collision angle, fourth column is the DCPA and fifth is TCPA, using these inputs is possible to assessment the Collision Risk (Risk) with respect to each of the Collision Regions. From this collision risk the actions to avoid collision are in column sixth and seventh, according COLREGs rules and regulations.

2.3.2 *Fuzzy membership function (FMF)*

FMF describes fuzzy sets that map from one given universe of discourse to a unit interval. This is conceptually and formally different from the fundamental concept of probability, (Pedrycz & Gomide, 2007). The core of the fuzzy set A is defined as the set of all elements of the universe typical to A that are associated with the membership value of 1.

The support of the fuzzy set is defined as the set of all elements of X that have nonzero membership degree in A. The FMF for inputs, collision distance (R), DCPA, TCPA and relative collision angle (v), are presented in Figs 5,6, and 7 respectively. Figure 8 are formulated for the output FMFs of speed (dψ) and course (dV) change of the Own vessel. The core and support variables are listed on the respective figures of inputs and outputs FMFs.

In this work the FFM relate to the relative regions v1, v2, v3 are dependent on the relative region value, θ, Figure 7 present the FMF for $\theta = 0$;

2.4 *Guidance and control module*

2.4.1 *3DOF Mathematical model*

In order to simulate the 3DOF planar ship motions. The Abkowitz (1980) model is modified in order to make the modelling more flexible and realistic physically. In this study, the current effect is considered as the main external excitation, because the ship model has a small above water structure. Figure 9 presents the coordinate frames for 3DOF ship motions.

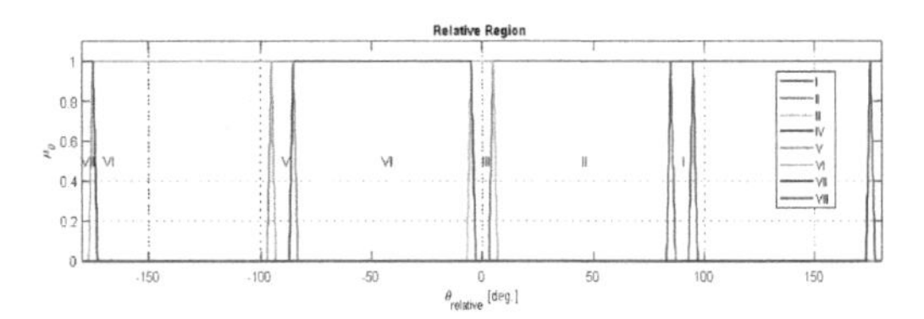

Figure 5. Relative region input fuzzy function.

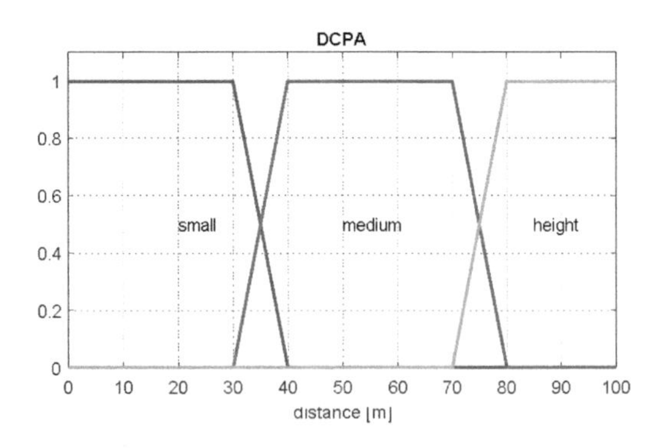

Figure 6. Distance to closest point of approach input FMF.

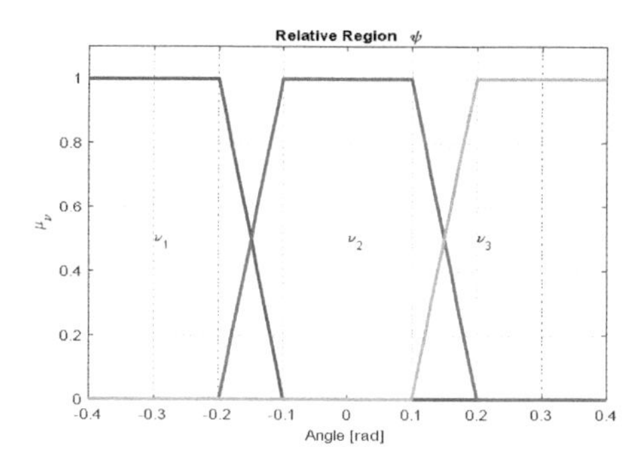

Figure 7. Variable fuzzy membership function.

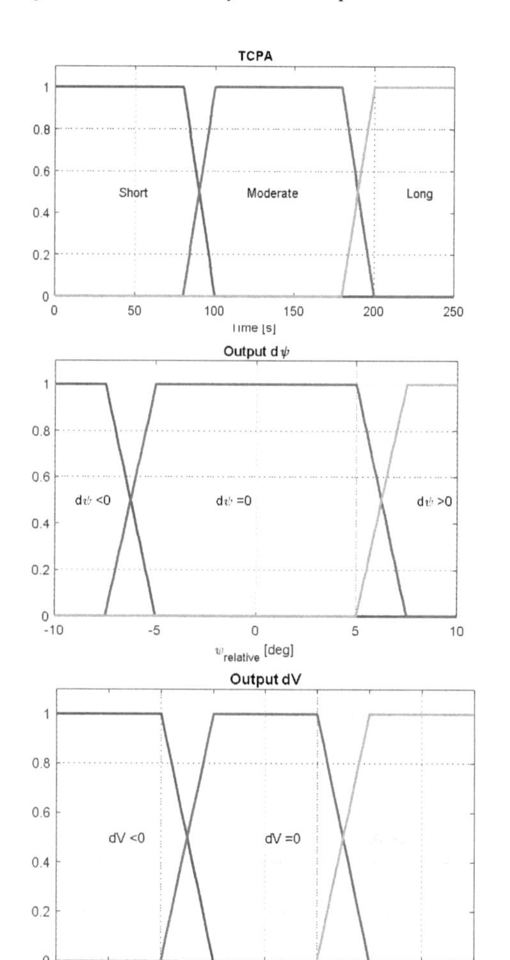

Figure 8. Fuzzy membership function.

As presented in Figure 9, u_c is the current's magnitude, α is the current's direction, ψ is the ship's heading angle, u is the forward component of velocity over ground, and v is the transverse

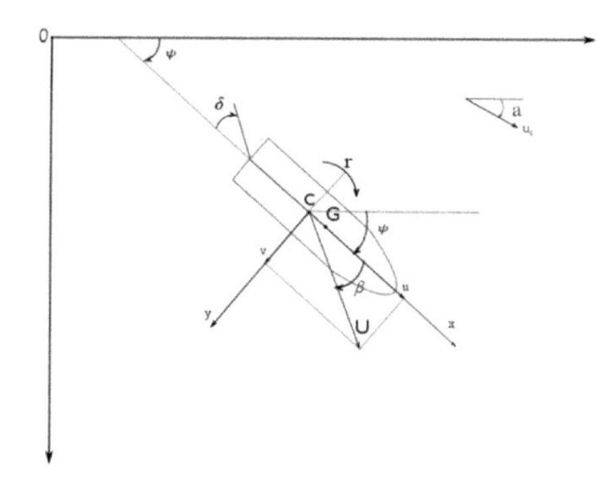

Figure 9. Coordinate frames for 3DOF marine surface vehicle.

component of velocity, the relative forward velocity and transverse velocity are given by

$$\begin{aligned} u_r &= u - u_c \cos(\psi - \alpha) \\ v_r &= v + u_c \sin(\psi - \alpha) \end{aligned} \tag{2}$$

The time derivatives of u and v are given:

$$\begin{aligned} \dot{u} &= \dot{u}_r - u_c r \sin(\psi - \alpha) \\ \dot{v} &= \dot{v}_r - u_c r \cos(\psi - \alpha) \end{aligned} \tag{3}$$

where the accelerations of the motion in 3 degree of freedom (surge, sway and yaw) are given by

$$\begin{aligned} &(m - X_{\dot{u}_r})\dot{u}_r - m v_r r - m x_G r^2 = f_1 \\ &(m - Y_{\dot{v}_r})\dot{v}_r + (m x_G - Y_{\dot{r}})\dot{r} + m u_r r = f_2 \\ &(m x_G - N_{\dot{v}_r})\dot{v} + (I_z - N_{\dot{r}})\dot{r} + m x_G u_r r = f_3 \end{aligned} \tag{4}$$

where m the mass of the ship is, $-X_{\dot{u}_r} - Y_{\dot{v}_r} - Y_{\dot{r}} - N_{\dot{v}_r} - N_{\dot{r}}$ are the added mass and moment, respectively. The dimensionless forces are defined as multi-variety third-order regression polynomials depending on the non-dimensional velocities.

$$\begin{aligned} f_1' = {} & \kappa_{\eta_1} \eta_1' u_r'^{\,2} + \kappa_{\eta_2} \eta_2' n' u_r' + \kappa_{\eta_3} \eta_3' n'^{\,2} \\ & - \kappa_{C_R} C_R' + \kappa_{X_{v_r v_r}} + X'_{v_r^2} v_r'^{\,2} \\ & + \kappa_{X_{ee}} X'_{e^2} e^2 + \kappa_{X_{rr}} X'_{r^2} r'^{\,2} \\ & + \kappa_{X_{v_r r}} X'_{v_r r} v_r' r' + \kappa_{v_r^2 r^2} X'_{v_r^2 r^2} v_r'^{\,2} r'^{\,2} \end{aligned} \tag{5}$$

$$\begin{aligned} f_2' = {} & \left\{ \kappa_{v_r} Y'_{v_r} v_r' + \kappa_{Y_\delta} Y_\delta' (c - c_0) v_r' \right\} \\ & + \left\{ \kappa_{Y_r} Y_r' r' - \kappa_{Y_\delta} \frac{Y_\delta'}{2} (c - c_0) r' \right\} \\ & + \kappa_{Y_\delta} Y_\delta' \delta + \kappa_{Y_{rrv_r}} Y'_{r^2 v_r} r'^{\,2} v_r' \\ & + \kappa_{Y_{eee}} Y'_{e^3} e^3 + \kappa_{Y_0} Y_0' \end{aligned} \tag{6}$$

$$\begin{aligned} f_3' = {} & \kappa_{N_0} N_0' + \left\{ \kappa_{N_{v_r}} N'_{v_r} v_r' - \kappa_{N_\delta} N_\delta' (c - c_0) v_r' \right\} \\ & + \kappa_{N_\delta} N_\delta' \delta + \kappa_{N_{rrv_r}} N'_{r^2 v_r} r'^{\,2} v_r' + \\ & + \kappa_{N_{eee}} N'_{e^3} e^3 + \left\{ \kappa_{N_r} N_r' r' + \frac{1}{2} \kappa_{N_\delta} N_\delta' (c - c_0) r' \right\} \end{aligned} \tag{7}$$

Detailed information about symbols used in equations (4), (5), (6) and (7) can be found in Moreira et al. (2007).

2.4.2 *Line-of-sight algorithm*

According to the approach presented in Caccia (2006), the basic automatic guidance capabilities, i.e. PID auto heading and line-of-sight (LOS) guidance, is proved to be sufficient for an unmanned surface vehicle to satisfactorily accomplish its operational goal.

In the LOS algorithm the desired geometric path is composed by a collection of way-points p_k in a way-point table. The LOS position is located somewhere along the straight-line segment connecting the previous p_{k-1} and current p_k way-points as shown in Moreira et al. (2007). Thus, let the ship's current horizontal position $p = [x,y]$ be the centre of a circle with radius of n ship lengths ($nLpp$). This circle will intersect the current straight-line segment at two points where $p_{LOS} = [x_{LOS}, y_{LOS}]$ is selected as the point closest to the next way-point. To calculate p_{LOS}, two equations with two unknowns must be solved online. These are:

$$(y_{LOS} - y)^2 + (x_{LOS} - x)^2 = (nL_{pp})^2 \tag{8}$$

$$\frac{y_{los} - y_{k-1}}{x_{los} - x_{k-1}} = \frac{y_k - y_{k-1}}{x_k - x_{k-1}} = \tan(\alpha_{k-1}) \tag{9}$$

The Eq. (8) is recognized as the theorem of Pythagoras, while the second equation states that the slope of the path between the previous and current way-point is constant.

Selecting way-points in the way-point table relies on a switching algorithm. A criterion for selecting the next way-point, located at $p_{k+1} = [x_{k+1}, y_{k+1}]$, is for the ship to be within a circle of acceptance of the current way-point pk. Hence, if at some instant of time t the ship position p(t) satisfies,

$$(x_k - x(t))^2 + (y_k - y(t))^2 \leq R_k^2 \tag{10}$$

the next way-point is selected from the way-point table. R_k denotes the radius of the circle of acceptance for the current way-point. It is imperative that the circle enclosing the ship has a sufficient radius such that the solutions of Eq. (10) exist. Therefore, $nL_{pp} \geq R_k$, for all k is a necessary bound.

Two independent PID controllers with the gains obtained with the pole placement method were used for controlling the speed and the heading, as was presented in Moreira et al. (2007).

3 NUMERICAL SIMULATIONS

This section presents results of numerical simulations for the collision avoidance system including the guidance and control for the own and target vessels. The own and target ships are simulated considering the same vessel, based in particularities of the "Esso-Osaka" tanker ship.

3.1 *"Esso Osaka" ship model*

This sub-section describes the model of the "Esso Osaka" ship. For the verification of the guidance and control designs, a good mathematical model of the ship is required to generate typical input/output data. The dynamics of the "Esso Osaka" tanker ship is described based on the horizontal motion with the variables of surge, sway and yaw. The model was scaled 1:100 from the real "Esso Osaka" ship. The vehicle main characteristics are listed in Table 2. The non-dimensional hydrodynamics coefficients are presented in Moreira et al. (2007).

3.2 *Numerical simulations*

In this, sub-section numerical simulations of the collision avoidance system for two different collision scenarios. The grid map is a rectangular map of 500 × 700 [m], where 1 pixel is equal to 1 m, It is assumed that the position and speed of the target vessel is always available.

Table 3 shows the model configuration for the collision scenarios. The own vessel starts at the origin with a constant heading pointed the North. The targets vessel starts at an arbitrary position in order to simulate collision, with a constant speed and heading.

Fig. 10 presents the trajectories of the own vessel and target vessel during the execution of the collision avoidance scenario in case A. In this figure, the own vessel trajectory is on blue line and the target vessel in orange. The desired trajectories are plotted in red dotted lines, for each vessel.

Table 2. "Esso Osaka" model particulars.

"Esso Osaka" Model	
Overall Length (mm)	3430
Length between perp.	3250
Breadth (mm)	530
Draught (estimated at the tests) (mm)	217
Displacement (estimated at trials) (kg)	319.4
Rudder area (m²)	0.0120
Propeller area (m²)	0.0065
Scaling coefficient	100

Table 3. Model configurations for collision avoidance tests.

Case		Vessel 1	Vessel 2
A	Start point (m)	(0,0)	(250,550)
	Initial heading (°)	0	225
	Des. target (m)	(0,500)	(–100,100)
	Speed (m/s)	0.5	0.6
B	Start point (m)	(0,0)	(–350,350)
	Initial heading (°)	0	90
	Des. target (m)	(0,500)	(250,350)
	Speed (m/s)	0.5	0.6

From the plot is possible to see the modification of the ship heading of vessel 1 when a collision situation is detected. According to COLREGs rules and regulation, the own vessel pass through the portside.

Figure 11a presents the time series of distance between own and target vessel during the execution of the collision avoidance task in case A. From this plot is possible to see a constant decrease of distance between ships at begin of simulation, it is because the trajectories are in an imminent collision situation, However, after the collision situation is detected, the ship changes in order to avoid collision and the distance increases. Figure 11b, presents speeds of each vessel, the speed of the own vessel varies in order to avoid a potential collision, according to the COLREGs rules and regulations. The target vessel has a constant speed during whole simulations.

Fig. 12 presents the trajectories of two vessel during the execution of the collision avoidance scenario in case B. In this figure, the own vessel trajectory is on blue line and the target vessel in orange. The desired trajectories are plotted in red dotted lines, for each vessel. From the plot is possible to see the modification of the ship heading of vessel 1 when a collision situation is detected. According to COLREGs rules and regulation, the own vessel pass through the portside.

Figure 13a presents the time series of distance between own and target vessel during the execution of the collision avoidance task in case B. From this plot is possible to see a constant decrease of distance between ships at begin of simulation, it is because the trajectories are in an imminent collision situation,

However, after the collision situation is detected, the ship changes in order to avoid collision and the distance increases. Figure 13b, presents speeds of each vessel, the speed of the own vessel varies in order to avoid a potential collision, according to the COLREGs rules and regulations. The target vessel has a constant speed during whole simulations.

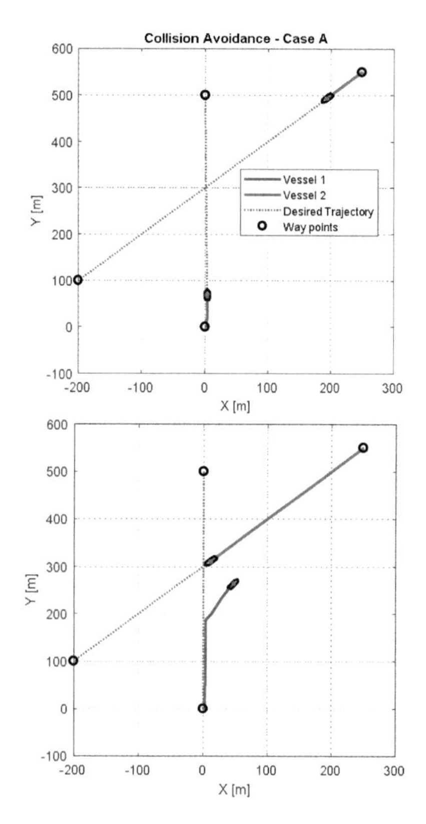

Figure 10. Simulation of collision situation A.

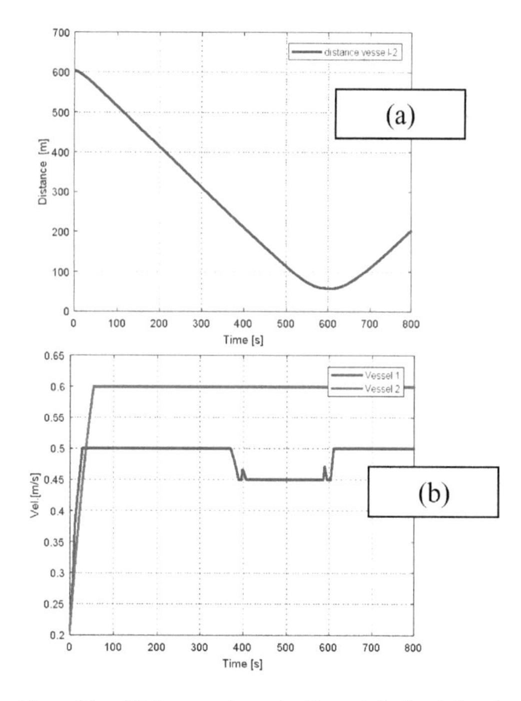

Figure 11. Distance and speeds of vessels in simulation A.

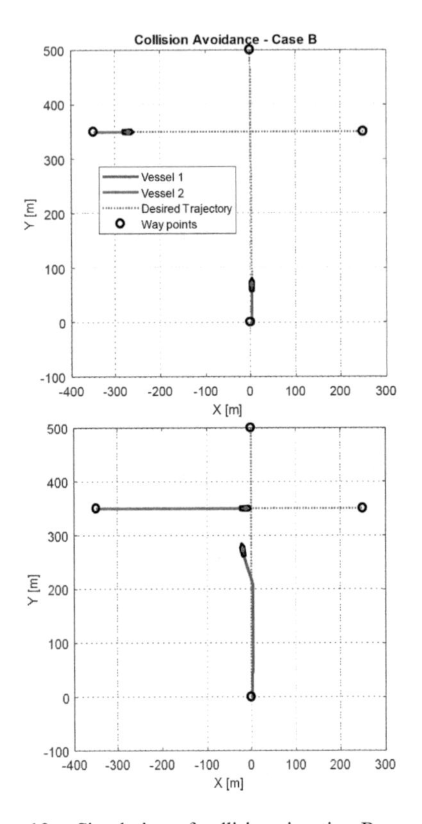

Figure 12. Simulation of collision situation B.

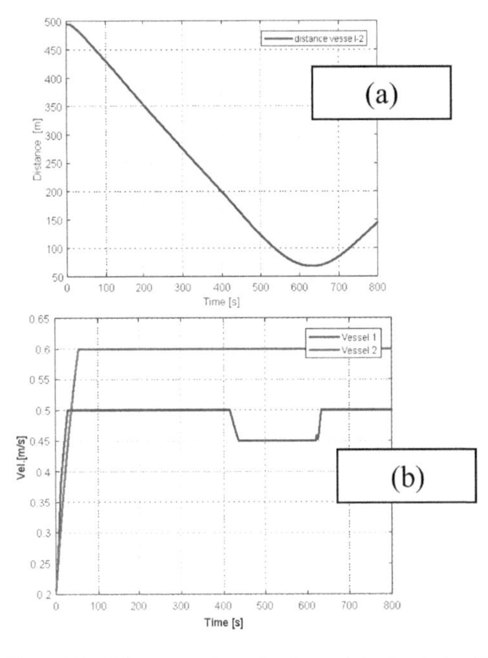

Figure 13. Distance and speeds of vessels in simulation B.

4 CASE STUDY

In this case study, the marine traffic in estuary of Tagus river in Lisbon is chose as place for simulations. The objective is test the capabilities of the collision avoidance and guidance system in a grid map with real dimensions, considering a real marine environment, For this purpose the marine routes of two "Transtejo" vessels are simulated.

4.1 *Place of tests*

The place chosen for the numerical simulation is the estuary of the Tagus river in Lisbon. The area chosen for the simulation is a square area of 4 Km side length, Figure 14a,b. This area includes the "Cais do Sodré" dock in Lisbon, Portugal and two routes of river transport, i.e. "Cacilhas-Cais do Sodré" and "Seixal-Cais do Sodré".

Figure 15a,b shows the "Esso-Osaka" model, used in for numerical simulations and the "Transtejo" vessel, this vessel is used for river transport.

Figure 14 (a,b) Place for simulations.

Figure 15. (a) Esso-osaka model (b) "Transtejo" vessel.

In this study the "Transtejo" vessels are modelled as "Esso Osaka" scaled model

Notice that in this case study the influence of wind speed, river waves and others environmental disturbances are neglected.

4.2 *Results*

In order to study the performance of collision avoidance, guidance and control system in a real collision scenario, an aerial picture of the estuary of Tagus river is pre-processing to create the grid map. Figure16 presents a bit map of the grid map in Rhinoceros software. Figure 16 also shows the scaled map for the simulations. In the grid map 1 [pixel] is equal to 10 [m].

Table 4 presents the model configuration of the numerical simulation. In this configuration, the starting point of the "Transtejo" vessel is located in "Cais do Sodré" dock, Lisbon and the target point are a located Dock in "Cacilhas" and "Seixal" docks.

The starting point of the own vessel is an arbitrary point in the middle of "Tagus" river and the target point is in a Navy dock in Lisbon.

Figure 17 presents the sequence motions of collision avoidance manoeuvres of an ASV in a complex marine environment including static and dynamic obstacles. In this plots the trajectory of ASV is plot in red, the "Transtejo" 1 ship in brown and "Transtejo" 2 ship in orange. The desired trajectories are plotted in red dotted lines, the desired waypoints are represented by a black circle, and the static environment is plotted in blue asterisks. From this sequence it is clear to see the heading angle changes in order to avoid potential collision

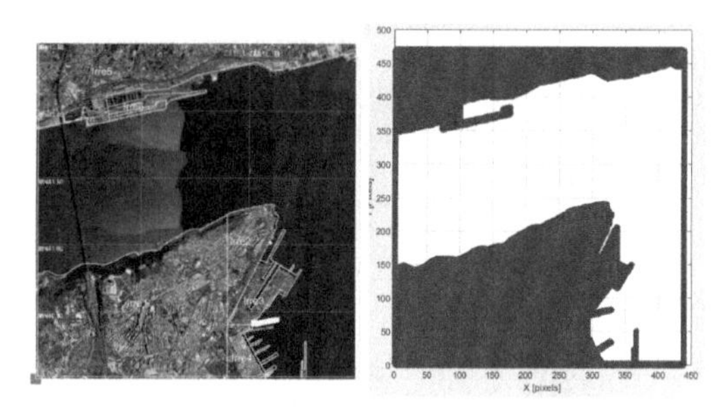

Figure 16. Generation of the grid map from aerial picture.

Table 4. Model configurations for case study simulation.

	ASV	"Transtejo" 1	"Transtejo" 2
Start point (pixel)	(25,250)	(235,418)	(436,44)
Initial heading (°)	60	100	340
Des. target (pixel)	(425,436)	(180,190)	(235,418)
Des speed (pix./s)	0.5	0.6	0.6

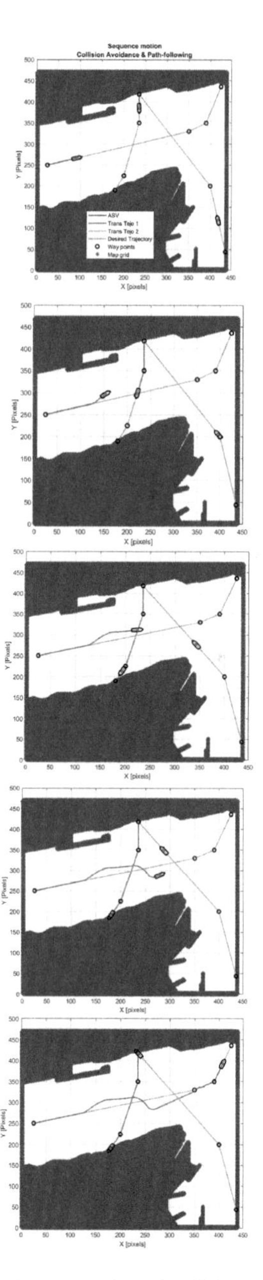

Figure 17. Sequence motion of collision avoidance and path-following algorithm.

situation. At the begin of simulation the ASV is tracking the desired path, however, after some minutes the first "Transtejo" 1 ship is detected and a potential collision is identified, Thus, according to the COLREGs rules and regulations, the ASVs starts the collision avoidance manoeuvre. Later, a second collision situation is detected, with the "Transtejo" 2 ship and the ASV, once more; modify his heading angle in order to avoid the imminent collision. Finally, when the collision risk disappears, the ASVs continues to tracking the desired path.

Figure 18a presents the time series of distance between ASV and "Transtejo" 1, in blue, ship and ASV and "Transtejo" 2 ship, in green, during the execution of the collision avoidance task. From this plot is possible to see a constant decrease of distance between ships at begin of simulation, it is because the trajectories are in an imminent collision situation, However, after the collision situation is detected, the ship changes in order to avoid collision and the distance increases. Figure 18b, presents speeds of each vessel, the speed of the own vessel varies in order to avoid a potential collision, according to the COLREGs rules and regulations. The target vessel has a constant speed during whole simulations.

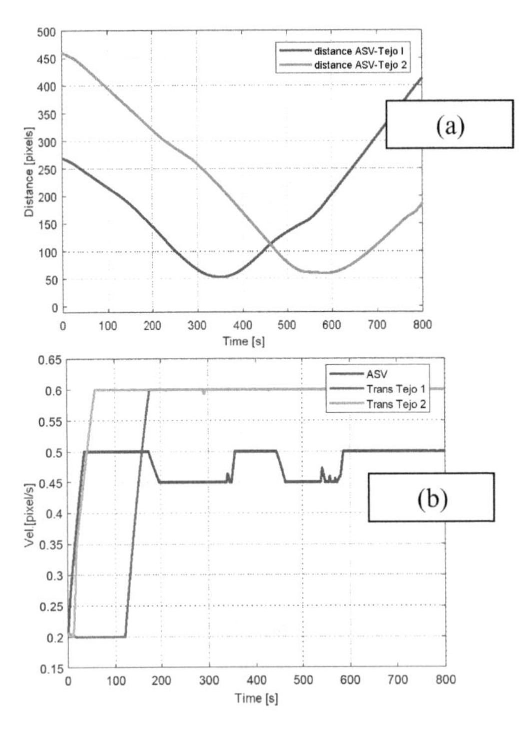

Figure 18. Distances between AVS and "Trantejo" vessels and speeds.

5 CONCLUSIONS

A system for collision avoidance, guidance and control for operation of ASVs in a complex marine environment was presented. The fuzzy logic based intelligent decision-making algorithm was used for collision avoidance. The guidance unit employs the LOS algorithm and, the control unit is composed by a PID heading controller plus a speed controller.

Two different collision scenarios were simulated in order to study the collision avoidance system performance and a good results were found.

From the case-study simulation, a good performance of the system was found in real marine environment.

Future improvements are to increase the number of collision scenarios and more vessels. Expand the collision avoidance, guidance and control unit for a fleet of ships. Introduce disturbances i.e. wind and waves.

ACKNOWLEDGEMENTS

The first author has been funded by a PhD scholarship from the University of Lisbon and from Centre for Marine Technology and Ocean Engineering (CENTEC).

This work was performed within the Strategic Research Plan of the Centre for Marine Technology and Ocean Engineering (CENTEC), which is financed by Portuguese Foundation for Science and Technology (Fundação para a Ciência e Tecnologia—FCT). The authors are grateful to H. Xu for his help and advises in guidance of vessels.

REFERENCES

Abkowitz, M.A. (1980). Measurement of hydrodynamic characteristics from ship maneuvring trials by system identification. Sname Transactions, *88*, 283–318.

Benjamin, M.R., & Curcio, J.A. (2004). COLREGs-based navigation of autonomous marine vehicles. Autonomous Underwater Vehicles, *2004 IEEE/OES* (pp. 32–39). IEEE.

Caccia, M., 2006, "Autonomous surface crafts: prototypes and basic research issues", in Proc. 14th Mediterranean Conference on Control and Automation, Ancona, Italy, pp. 1–6.

Ferrari, V., Perera, L.P., Santos, F.P., Hinostroza, M.A., Sutulo, S., & Guedes Soares, C. (2015). Initial experimental tests of a research-oriented self-running ship model. Guedes Soares, C. & Santos T.A. (Eds.) Maritime Technology and Engineering, Taylor & Francis Group, London, UK, 913–918.

Fossen, T.I. (2011). Handbook of Marine Craft Hydrodynamics and Motion Control. John Wiley & Sons Ltd.

Hasegawa K (1987) Automatic collision avoidance system for ship using fuzzy control. In: Proceedings of 8th ship control system symposium, pp 234–258.
He Y., Jin Y., Huang L., Xiong Y., Chen P., Mou J., (2017), Quantitative analysis of COLREG rules and seamanship for autonomous collision avoidance at open sea, Ocean Engineering, V. 140, pp. 281–291.
Hinostroza M.A., Xu, H. & Guedes Soares C. (2017), Path-planning and path-following control system for autonomous surface vessel, Maritime Transportation and Harvesting of Sea Resources, Taylor & Francis Group, London, pp. 991–998.
IMO (1972) Convention on the international regulations for preventing collisions at sea (COLREGs). http://www.imo.org/conventions/.
KongsbergMaritime, 2017. Autonomous ship project, key facts about YARA Birkeland. URL, https://www.km.kongs berg.com/ks/web/nokbg0240.nsf/AllWeb/4B8113B707A50A4FC125811D00407045?OpenDocument.
Moreira, L., Fossen, T.I., and Guedes Soares, C. 2007, "Path Following Control System for a Tanker Ship Model". Ocean Engineering. 34:2074–2085.
Moreira, L., Santos, F.J., Mocanu, A., Liberato, M., Pascoal, R., and Guedes Soares, C., 2008, "Instrumentation used in guidance, control and navigation of a ship model", 8th Portuguese Conference on Automatic Control, Vila Real, Portugal, pp. 530–535.
Njus E. (2016), The military's Oregon-built drone ship is headed to California, Oregon Business News, 7 April 2016.
Pedrycz W, Gomide E (2007) Fuzzy systems engineering toward human centric computing. Wiley, Hoboken.
Perera L.P., Carvalho J. and Guedes Soares C. (2011). Fuzzy-logic based decision making system for collision avoidance of ocean navigation under critical collision conditions. Journal of Marine Science and Technology, 16(1), 84–99.
Perera, L.P., Ferrari, V., Santos, F.P., Hinostroza, M.A., and Guedes Soares, C. (2015). Experimental Evaluations on Ship Autonomous Navigation and Collision Avoidance by Intelligent Guidance. IEEE Journal of Oceanic Engineering, *40*(2), 374–387.
Rolls-Royce, "Autonomous ships: The next step," *White Paper*, Available: http://www.rolls-royce.com/~/media/Files/R/Rolls-Royce/documents/customers/marine/ship-intel/rr-ship-intel-aawa-8pg.pdf.
Sato Y., Ishii H., (1998), Study of a collision-avoidance system for ships, Control Engineering Practice, Vol. 6, pag. 1141–1149.
Statheros, T., Howells, G., and McDonald-Maier K., (2008), "Autonomous ship collision avoidance navigation concepts, technologies and techniques," The Journal of Navigation, Vol. 61, pp. 129–142.
Xu, H., Hinostroza M.A, & Guedes Soares, C. (2018). Identification of hydrodynamic coefficients of ship nonlinear manoeuvring mathematical model with free running model tests., International Journal of Maritime Engineering, Accepted for publication.
Zhang J., Zhang D., Yan X., Haugen S. and Guedes Soares C., (2015) A distributed anti-collision decision support formulation in multi-ship encounter situations under COLREGs, Ocean Engineering, Vol. 105, Pag. 336–348.

Progress in Maritime Technology and Engineering – Guedes Soares & Santos (Eds)
© 2018 Taylor & Francis Group, London, ISBN 978-1-138-58539-3

A framework of network marine meteorological information processing and visualization for ship navigation

Xin Peng, Yuanqiao Wen & Chunhui Zhou
School of Navigation, Wuhan University of Technology, Wuhan, Hubei, China

Liang Huang
National Engineering Research Center for Water Transport Safety and School of Navigation, Wuhan University of Technology, Wuhan, Hubei, China

ABSTRACT: With the rapid development of information technology in shipping industry, massive network meteorological datasets come available and valuable. Intuitive visualization of surrounding marine meteorological data is helpful to improve the safety and reliability of ship navigation. In this article, build a system framework of network marine meteorological information processing and visualization that can automatically obtain, store and process these network meteorological information and provide well-designed and content-rich meteorological thematic map for ship navigation. Tested, the system can effectively provide the sharing service of marine meteorological information.

1 INTRODUCTION

The hydrological and meteorological environment of ship navigation is complex and changeable, so master meteorological information is one of most important guarantee to ensure the safety and reduce the fuel consumption of ships. The traditional marine meteorological service is difficult to meet user's requirements for two reasons. (1) Most systems are performance of a single type of meteorological information (Liu, 2014; Holmukhe, 2010). And it is mostly manual input information. (2) The meteorological elements differences are difficult to reflect in service areas, low degree of Information visualization, and user experience is poor (Tang, 2011; Khotimah, 2011). Obviously, the traditional marine meteorological information system has been difficult to meet the development needs of the modern shipping business.

With the rapid development of information technology in shipping industry, massive network meteorological datasets come available and valuable. The contents and formats of these datasets vary with types and sources of weather data and are difficult for operators to recognize and understand. Therefore, it is essential and useful to build a framework that can automatically obtain, store and process these network meteorological information and provide well-designed and content-rich meteorological thematic map for ship navigation. The system is B/S (Browser/Server) mode, and supports users to instantly get all kinds of meteorological information through a web browser. as simple as possible. Avoid excessive notes and designations.

2 SYSTEM DESIGN

2.1 *System requirements analysis*

In this article, a framework of network marine meteorological information processing and visualization has been developed to integrate multisource network meteorological datasets and offer global marine weather information perception and visualization, including wave height, sea temperature, swell, ocean current, air pressure, high altitude 500 Mb isobar, visibility, and ocean wind. The system contributes to grasp the meteorological condition for ship and its surrounding area during the voyage, in order to prevent danger and play a supporting decision-making of the voyage effectively. Based on investigation and analysis for ships navigation and the system have the following requirements.

1. The system should support acquisition and updating the original marine meteorological data periodically and automatically, and storage of data.
2. The system should have the function of classify and process the meteorological data automatically, according to the types and characteristics of all kinds of meteorological data.
3. The system shall visualize the eight categories of meteorological information, including creating iso-line and thematic map remarks. The air pressure map should show cold front or warm front, the ocean pressure map includes wind speed and wind direction, the visibility map should highlighting the areas with low visibility, the wave map includes wave height and wave

direction, the ocean current map includes the flow of ocean currents and judgment the type of warm and cold currents.
4. The system should support the publication of maps, and user get services through a browser.
5. To achieve function of map zoom and roaming, to meet the needs of different users.

2.2 *Design of system function module*

The function of network marine meteorological information processing and visualization system showing marine meteorological data about the ships and specific areas in the course of ships navigation accurately and directly. Establish an efficient, open and interactive visualization service system of integrates multi-source global ocean meteorological information. The system realized the network data source acquisition, analysis and processing of various types of meteorological information periodically and automatically, and provide multi-scale ocean meteorological thematic map visualization services for different types of locations such as ship locations. Analysis of the requirements of the system, the framework consists of four layers that are data collection layer, data parsing layer, data processing layer and data sharing layer, as shown Figure 1.

2.2.1 *Data acquisition module*

The reliability and accuracy of original marine meteorological data plays a key role in system. The sub-module is the basis on realizing the automation service of system. This sub-module can automatically classify the raw meteorological data and write to database without human intervention, and this module is the basis of realizing intelligent management and network sharing of marine meteorological information.

This module is to periodically collect weather datasets twice a day (0:00/12:00) from APPLIED WEATHER TECHNOLOGY (AWT) and

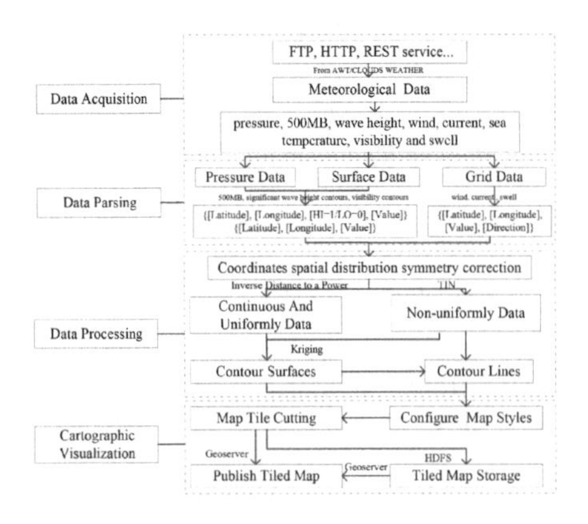

Figure 1. The framework of network meteorological information processing and visualization.

CLOUDS WEATHER. When the system is started and a timer is triggered, this system based on FTP standard network protocol to automatically download eight types of obtained meteorological data are pressure, 500 MB, wave height, wind, current, sea temperature, visibility and swell. For each data class, an individual work thread is designed to automatically retrieve datasets and multiple access ways are supported to compatible with download requirements of different data resources, such as FTP, HTTP, REST service, etc. If we use AWT to obtain meteorological data, first need to apply for an FTP account. We can download eight kinds of Weather data, which are ZIP files of TXT format. Still another technique, we use HTTP or REST request server to download meteorological data by CLOUDS WEATHER.

2.2.2 *Data parsing module*

This module defines several parsers to extract detailed content from various network data files, including pressure data parser, grid data parser and surface data parser according to the characteristics of different data.

1. Pressure data parser: The pressure data parser is mainly responsible for reading pressure contours and similar high altitude 500 MB, significant wave height contours, visibility contours, because these three types of weather elements have same data format. The format of such data consists of two parts that are triples data {[Latitude], [Longitude], [Value]} denoting observed pressure values and quadruples data {[Latitude], [Longitude], [HI = 1/LO = 0], [Value]} indicating pressure centers, as shown Figure 2 and Figure 3.
2. Grid data parser: The grid data parser is used to extract grid point sampling data in the form of quadruples {[Latitude], [Longitude], [Value], [Direction]}. The parser applies to wind, current

```
[Latitude],[Longitude],[Value]
-75.50,-55.12,952
-75.20,-54.00,952
-74.97,-53.00,952
-74.69,-51.50,952
-74.50,-50.29,952
-74.32,-49.00,952
-74.14,-47.50,952
-74.00,-46.17,952
-73.89,-45.00,952
-73.80,-43.50,952
-73.74,-42.00,952
-73.74,-40.50,952
-73.77,-39.00,952
-73.85,-37.50,952
-73.96,-36.00,952
-74.14,-34.50,952
-74.42,-33.00,952
-74.77,-32.00,952
-75.50,-31.15,952
```

Figure 2. Example triples Pressure data.

```
LF
[Latitude],[Longitude],[HI=1/LO=0],[Value]LF
-56.56,109.57,0,951LF
-58.20,39.21,0,951LF
-68.70,-167.05,0,959LF
-62.55,-83.57,0,975LF
-70.03,-75.11,0,975LF
-70.49,-41.51,0,979LF
47.78,123.77,0,983LF
73.06,17.16,0,991LF
58.14,-34.60,0,995LF
31.74,72.27,0,995LF
51.53,-66.63,0,999LF
42.67,97.67,0,999LF
39.83,82.95,0,999LF
32.44,86.97,0,999LF
66.33,-116.80,0,1003LF
39.72,96.39,1,1005LF
32.54,-80.24,0,1003LF
```

Figure 3. Example quadruples pressure data.

```
[Latitude],[Longitude],[Value],[Direction]LF
75.00,-179.00,0.327,69LF
75.00,-178.00,0.272,84LF
75.00,-177.00,0.270,79LF
75.00,-176.00,0.316,71LF
75.00,-175.00,0.238,67LF
75.00,-174.00,0.080,58LF
75.00,-173.00,0.065,37LF
75.00,-172.00,0.053,332LF
75.00,-171.00,0.071,331LF
75.00,-170.00,0.196,344LF
75.00,-169.00,0.145,332LF
75.00,-168.00,0.214,306LF
75.00,-167.00,0.325,305LF
75.00,-166.00,0.200,251LF
75.00,-165.00,0.220,268LF
75.00,-164.00,0.215,221LF
75.00,-163.00,0.298,229LF
75.00,-162.00,0.268,231LF
```

Figure 4. Example grid data.

and swells data files. Example quadruples pressure data as shown in Figure 4.

3. The surface data parser is specially designed to analyze surface front data files. The data structure of each front observation is composed of a group of coordinates {[Latitude], [Longitude]}, an end-of-segment indicator and a type-of-front indicator.

2.2.3 *Data processing module*

The function of sub-module is to display original marine weather data graphically on the base map, and display meteorological elements, including revises meteorology elements by spatial distribution symmetry correction, generate pressure iso-surface, the pressure iso-line tracing (take the air pressure data for example).

1. Coordinate spatial distribution symmetry correction

Firstly, The function StreamReader sr = new StreamReader (pcfile) can read the sampling weather data in turn, and to blank lines as the same iso-line of demarcation line in the process of reading. Three types of data would be recorded, including the value of air pressure iso-line, the coordinate latitude and longitude, the data of boundary points of the iso-line.

Secondly, Create a string object (LineString ll = new LineString (cc);) according to the coordinate point data, and determine the coordinate is located in the eastern hemisphere or the western hemisphere based on longitude value. We determine string object start in western hemisphere when the longitude in data less than –180°, and based on this function (cc2[kk] = new Coordinate(cc[kk].X + 360, cc[kk].Y)) to calculate coordinate of eastern hemisphere located in the same circle of longitude. Otherwise, determine string object start in Eastern hemisphere when the longitude in data more than +180°, and based on this function (cc2[kk] = new Coordinate(cc[kk].X –360, cc[kk].Y)) to calculate coordinate of Western hemisphere located in the same circle of longitude. This process is not necessary when a contour located in the same hemisphere.

Finally, Using the ship format of the world map boundary, the data of the first two processing steps have been finished to cut off data points on land.

2. Generate pressure iso-surface

Based on the difference of continuity and homogeneity in distribution of various original meteorology data, this paper disposes the original data points by interpolation, using anti-distance weighted average algorithm which is a kind of grid interpolation. First gains boundary of data, divides the Earth into m × n grids (In this paper, the size of the grid is 0.01° × 0.01°), and then disposes data by interpolation. To heterogeneous point data, using irregular triangle mesh interpolation method. Using two interpolation methods process data, and form a continuous data set. Finally, generate continuous iso-surface by Kriging Spatial Interpolation technology.

3. Pressure iso-line tracing

This sub-module calls pressure contours dynamic link library in C#, which encapsulates the completion of the contour generated by a variety of structures and functions, including the equivalent point search, iso-line search and others. The pressure contours dynamic link library entrance is (ReadContourFromDB(_RecordsetPtr Record, layerName);), and return a type structure (ContourFeatureGraphicsLayer). This structure is used to store the name of the contour layer (string layerName), data point attribute information, and labeling information: (labelLayer).

2.2.4 *Data cartographic visualization module*

The module is divided into two sub-modules: Thematic map visualization, Meteorological data sharing.

1. Thematic map visualization
Take into account the visual content of the various thematic maps, the richness of the map elements, and the different display configurations. Configuring map style for different types and level of thematic maps based on general marine meteorological visual configuration requirements. And all kinds of thematic maps are cut into six levels in total from level 0 to level 5.
2. Meteorological data sharing
Based on HDFS, the module realizes the storage of vast tile data and coding the tile maps, in order to save and load the tile maps, which is beneficial for spatial server Geoserver to issue the tile maps.

2.3 *The overall system architecture design*

We analyze the function requirements of system, designed the overall architecture of the system. The system is consisting of Web client tier, cloud server tier and cloud database tier. As shown in Figure 5.

1. Client tier
Client is a presentation tier in the form of system services. The client of this system uses standard Web browser, any Web browser is allowed to use this system, and increases the degree of convenience for users.

2. Cloud server tier
This tier is a bridge between the users and the application server. The user sends a request to the server through Web browser, and server returns the processing result to the users, to complete the response for the user access request.

3. Cloud database tier
The application database layer is the storage of the system database, which stores the original data of the AWT marine meteorological data, the various types of meteorological data and the thematic map of the tiles generated through the system processing.

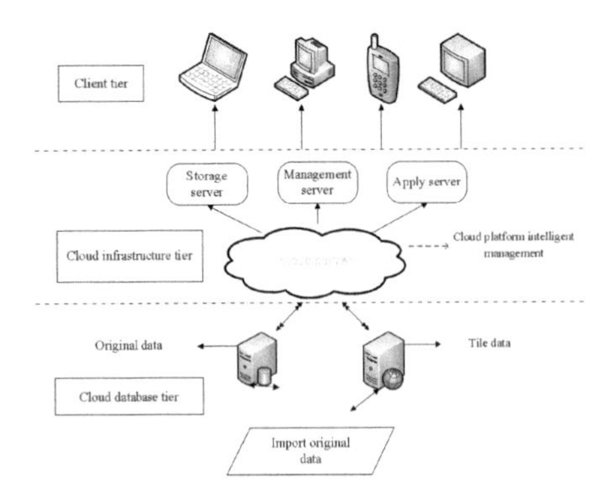

Figure 5. The system overall structure diagram.

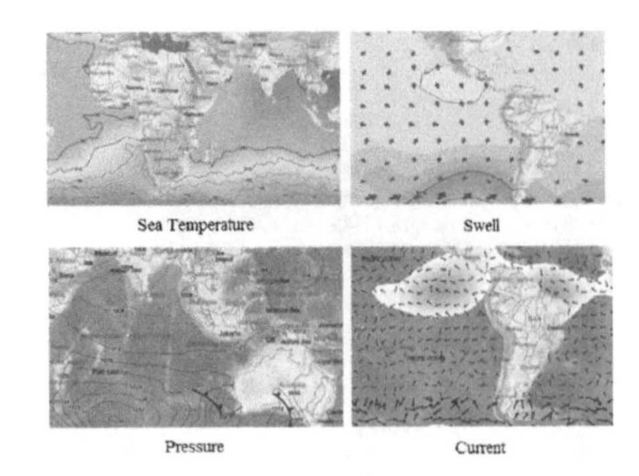

Figure 6. Integrated visualization of ECDIS and various meteorological information.

3 SYSTEM IMPLEMENTATION EFFECT

The system is a B/S (Browser/server) architecture whose front stage is written by HTML5, JQuery and JSP and the back stage is established by Java, with the application released via Tomcat, and the database is Oracle 11 g. The system fully utilizes the convenience brought by the cloud computing and the hardware flexible extension of the cloud platform allows the system applications to cope with the calculation no matter how large it is, which significantly lowers the probability of malfunction and increases the interaction and immediacy of the system.

After the experiment, Users are allowed to obtain eight kinds of global meteorological information. The visualization thematic maps as shown in Figure 6.

REFERENCES

Holmukhe R M, Chaudhari P S, Kulkarni P P, et al. (2010), "Measurement of Weather Parameters via Transmission Line Monitoring System for Load Forecasting". *International Conference on Emerging Trends in Engineering and Technology*. IEEE Computer Society: 298–303.

Khotimah P H, Krisnandi D, Sugiarto B. (2011), "Design and implementation of Remote Terminal Unit on Mini Monitoring Weather Station Based on Microcontroller". *International Conference on Telecommunication Systems, Services, and Applications*. IEEE: 186–190.

Liu X X, Sun Y Q, Xiang L, et al. 2014. Design and implemen-tation of meteorological service information system based on WebGIS. *Computer Engineering & Design*. 35(01): 322–326.

Tang Z Y, Hui L S, Xiao X L, et al. (2011), "The system of automatic observing present weather based on kinds of technology". *International Conference on Electronics*, Communications and Control. IEEE: 2971–2974.

Progress in Maritime Technology and Engineering – Guedes Soares & Santos (Eds)
ISBN 978-1-138-58539-3

Role assignment and conflict identification for the encounter of ships under COLREGs

Y. Zeng & J.F. Zhang
Intelligent Transportation Systems Center (ITS Center), Wuhan University of Technology, Wuhan, China
National Engineering Research Center for Water Transportation Safety (WTS Center), Wuhan University of Technology, Wuhan, China

A.P. Teixeira & C. Guedes Soares
Centre for Marine Technology and Ocean Engineering (CENTEC), Instituto Superior Técnico, Universidade de Lisboa, Lisbon, Portugal

ABSTRACT: Collision avoidance is one of the most important operations in ship navigation. A fast and effective assessment of the navigating situation and decision on adequate actions are essential for the safety of the ship itself and all the surrounding ships. The regulations on the role assignment introduced in the International Regulations for Preventing Collisions at Sea (COLREGs) are studied. The paper addresses the possible conflicts between Rules 12 and 15 when information about the ship types is not complete or is assessed differently by the two ships involved in the encounter situation. With respect to the Rule 12, algorithms of determining the wind direction on ships and the ship to windward are introduced, so that the role assignments using Rules 12 and 15 can be stablished automatically. The results derived from the two rules are compared and the possible conflicts are demonstrated. Possible ways of dealing with such conflicts are also discussed.

1 INTRODUCTION

With the increase of the ships in ports and waterways, the water areas are becoming increasingly crowded. The ship traffic is also becoming much more complex, resulting in more difficult collision avoidance operations. The risks of collision, grounding and other navigational accidents need to be dealt with properly (Zhang et al. 2016; Liu et al. 2016).

Collision between ships is one of the most common types of accident in many water areas. Once a collision accident occurs, it may cause great economic loss, environmental damages, and casualties. Consequently, the reduction of the occurrence of collision accident has become one of the greatest concerns.

Collision avoidance is one of the most important operations for navigating ships. A fast and effective detection, assessment, decision making and operation are essential for the safety of the ship itself and the surrounding ships, especially in complex encounter situations. For any sailing ships, when making the collision avoidance decisions, they have to comply with the International Regulations for Preventing Collisions at Sea (COLREGs), which was formulated by the International Maritime Organization (IMO, 1972).

In COLREGs, the geographical relationship between two ships is classified into three types, which are head-on, crossing or overtaking. The ships need to assess their roles (either give way to another ships or keep course and speed) before making collision avoidance operations.

Many approaches have been introduced to make collision avoidance decisions for ships, like Ant Colony Algorithm (Tsou et al. 2010, Liu et al. 2007), Evolution Algorithm, Genetic Algorithm (GA) (Smierzchalski, et al. 2000; Tsou et al. 2010), Fuzzy Logic approach (Tam et al. 2013; Perera et al. 2011; Hwang et al. 2010) and some other deterministic algorithms. The main idea of the suggested approaches is to find the shortest path that can pass through both moving and static objects safely.

Smierzchalski et al. (2000) proposed an evolution algorithm under multiple moving and static obstacles. The environment is modelled as a set of polygons considering their shapes. The trajectory is expressed by a set of maneuverers (e.g. course alteration, speed change). The algorithm can obtain the result with a computation time between 15 s and 4 min. Ant Colony Algorithm that was investigated by Tsou et al. (2010) is another similar path planning algorithm, in which a set of trajectories from origin to destination are first constructed randomly. The limitations on safety distance between ships, the degree of course and speed alterations are used to evaluate the effectiveness of each trajectory. The algorithm would gradually converge to the optimum solution after enough iterations. The simulation results indicate that the approach converge a little bit faster than GA algorithm. The study did not consider static obstacles.

Genetic Algorithm (GA) is essentially a random algorithm. Therefore, there will be violation of the COLREGs when making collision avoidance decisions. Ant Colony Algorithm is mostly applied to collision avoidance decision between two ships, which will not work well for multi-ship encounter situation.

Unlike the above heuristic algorithms, some deterministic algorithms have also been proposed, in which the optimum path is generated using analytical solutions. Tam et al. (2013) proposed a cooperative path planning by considering all the surrounding ships. The turning angle is set to be 30° and return to initial course after some distance (D). The value of D would increase gradually until the path is safe enough. The heading angle is set to be constant and the procedure was simplified into determining the distance that it should navigate on the new course to keep clear of the ships with higher priority by linear extension.

Benjamin et al. (2006a,b) proposed a protocol-based programme for unmanned marine vehicles. Anti-collision was transformed into an interval programming model for multi-objective decision making. The model was based on 'IF-THEN' rules derived from the Coast Guard Collision Regulations of the USA. Similar research work has been carried out by Perera et al. (2012). The decisions were made in two steps. In the first step, fuzzy 'IF-THEN' rules were constructed for some typical encounter situations based on COLREGs and expert navigation knowledge. In the second step, a Bayesian network was used to decide the actions in a quantitative way.

One problem for the above model is that the decision may fail to avoid collision due to the counteraction of the rules when making defuzzification. In order to overcome the problem, some further improvements have been proposed by Perera et al. (2014), including adding fuzzy smooth bearing regions for contradictory decision boundaries, multilevel decision action formulations.

Zhang et al. (2012) proposed an anti-collision decision approach on the basis of COLREGs. The minimum distances required to make collision avoidance under both normal and critical situations are calculated for typical encounter situation considering ship manoeuvrability. Deterministic algorithms have advantage in computation efficiency. However, their effectiveness in dealing with complex situations still need to be further validated.

Path planning for ships that have collision risk with other ships have been widely investigated by several researchers. However, there are only a few research works focusing on the requirements of COLREGs, especially the formalizations of some specific requirements on the ships' roles and on the actions they should take under complex encounter situations. Kreutzmann et al. (2013) presented such research by formalizing the navigational rules. The compliance of COLREGs is considered from the perspective of software verification. The focus of the research work was on the effect of wind on the ships. With the proposed formalized rules is very easy to determine which ship should give way and which should keep its course and speed. However, it should be noted that the effect of wind is considered only when the encounter ships are sailing vessels. If not, it should be determined according to their relative position and bearings, which was not taken into fully consideration. Similar research work has been performed by Banas & Breitsprecher (2011), in which the good seamanship from seafarers' experience was integrated in the decision support system derived from COLREGs under collision situations.

When the ships make collision avoidance decisions, the first step is to determine their roles according to the dynamic information of the ships and the meteorological conditions, on the basis of COLREGs requirements. Then they can make specific decisions accordingly. However, conflicts between different rules may occur when the information on the ships or on the environment is not available or is assessed differently by the two ships involved in the scenario. As a result, the formalizations of the rules and the possible conflicts from Rules 12 and 15 are investigated in this paper. The possible measurements that can deal with such conflict is also discussed.

The remainder of the paper is organized as follows: The main procedure of collision avoidance decision making and some specific requirements in COLREGs are introduced in section 2. Section 3 mainly demonstrates how the results from Rule 12 and 15 may conflict under some specific encounter situations. The algorithm on how to express the relationship between ship' heading and wind direction is also introduced. Discussions on how to deal with such conflict situations are given in section 4. Finally, conclusions are drawn in section 5.

2 COLLISION AVOIDANCE

2.1 *Collision avoidance procedure*

When a ship is navigating in any water area, it assesses the risk of collision for every encounter situation and makes anti-collision decisions and

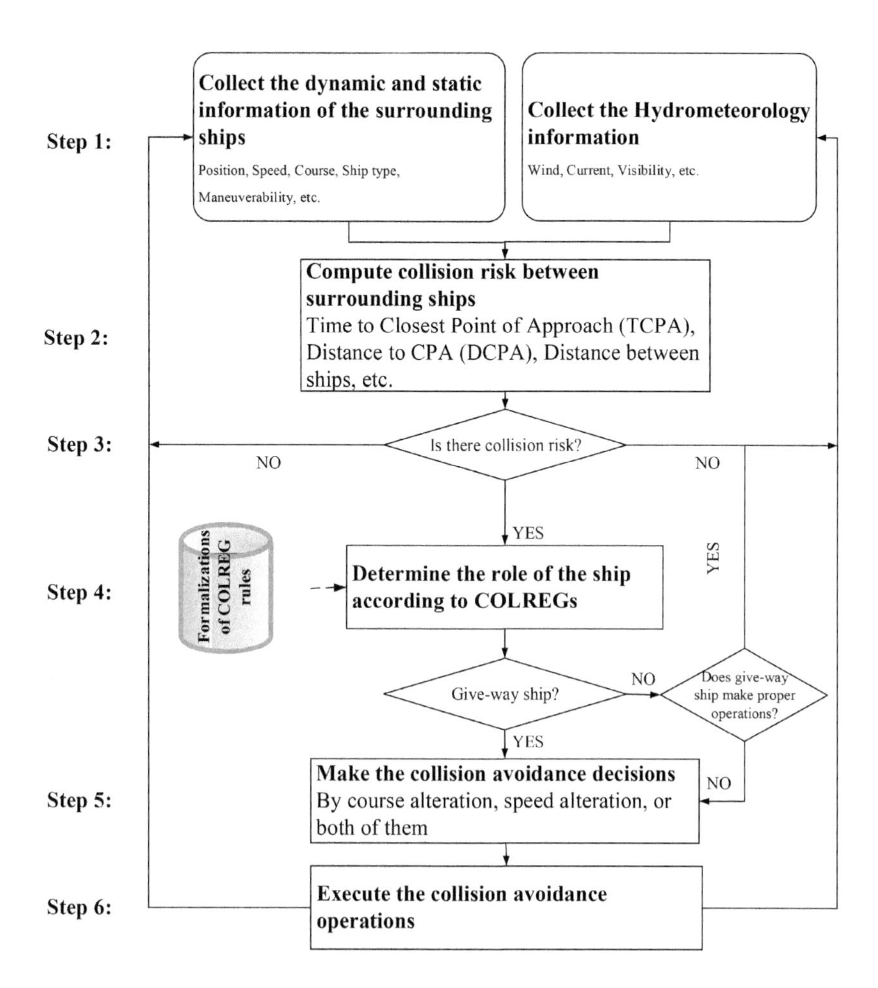

Figure 1. Anti-collision decision-making procedure.

actions when necessary. Fig. 1 presents the procedure of collision avoidance for ships, in which the decisions are made according to the following steps:

STEP 1: Data collection. There are mainly two types of information. One is the information on the surrounding ships, which includes the positions, courses, velocities of the ships. Moreover, the manoeuvrability of the target ships is also necessary for some special ships. The other type is the hydrometeorology information, including wind and current speed and directions. Visibility is also useful to measure the degree of collision risk.

STEP 2: Collision risk assessment. The possibilities of collision with all the surrounding ships are assessed by synthesizing the available data. The most important indicators are the Time to Closest Point of Approach (TCPA), Distance to CPA (DCPA), and the distance between ships. The seafarers make the judgment on whether collision avoidance manoeuvre is necessary according to these indicators.

STEP 3: Role assignments. The ships need to determine their roles under specific encounter situations according to COLREGs. The key issue in this step is that the seafarers in different ships have to make unanimous judgements in order to avoid conflicts. It should be noted that if the ship is assigned to be stand-on, it is still possible to take collision avoidance actions if the give-way ship fails to take actions. This step is the focus of this paper.

STEP 4: Collision avoidance decision making. The give-way ship should take decisions to avoid collision with other ships by course/speed alterations, or by both of them. It is important to compare the effectiveness of avoiding collision by steering and by speed changing. This issue is discussed in detail by Zhang et al. (2015).

STEP 5: Collision avoidance execution. The ships need to navigate following the planned safe path in the previous step. During the execution, the seafarers are required to maintain continuous watch keeping on the other ships and make adjustments if new collision risk is arising.

2.2 *Role assignment based on COLREGs*

In COLREGs, many of the requirements are expressed in qualitative ways, such as the definition of good seamanship, safe speed, action to avoid collision should be early and substantial, etc. Besides these, there are also some specific requirements that can be analysed quantitatively. One of them is about how to assign the roles of two encountered ships, which are mainly introduced in Rule 15 and 12.

Rule 15 of COLREGs states that "When two power-driven vessels are crossing so as to involve risk of collision, the vessel which has the other on her own starboard side shall keep out of the way and shall, if the circumstances of the case admit, avoid crossing ahead of the other vessel".

According to this rule, the roles of ships can be determined based on the relative bearings of the ships. In general, the encounter between two ships can be categorized into three types, which are crossing, overtaking and head-on. If the own ship is placed at the center of a coordination system with its course heading to north, the role of own ship can be determined using the graph presented in Fig. 2.

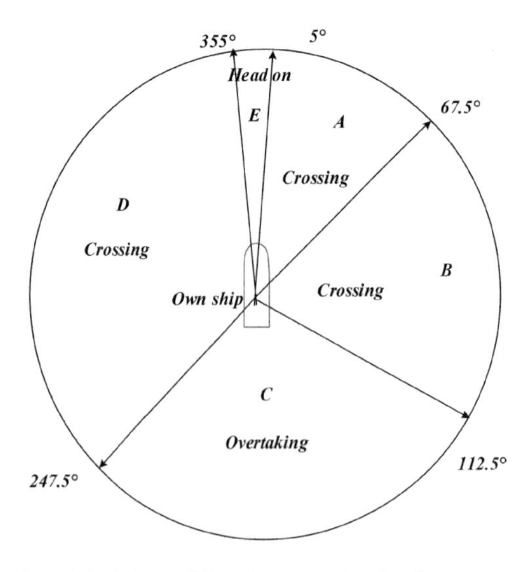

Figure 2. Types of the ship encounter situations.

In many real applications, the own ship is not always located in the origin. As a result, the algorithms of coordination translations and rotations are necessary. The details of them are introduced by Zhang et al. (2012).

Another rule for determining the roles of two encountered ships is Rule 12, which states that "When two sailing vessels are approaching one another, so as to involve risk of collision, one of them shall keep out of the way of the other as follows: (*i*) when each has the wind on a different side, the vessel which has the wind on the port side shall keep out of the way of the other; (*ii*) when both have the wind on the same side, the vessel which is to windward shall keep out of the way of the vessel which is to leeward; (*iii*) if a vessel with the wind on the port side sees a vessel to windward and cannot determine with certainty whether the other vessel has the wind on the port or on the starboard side, she shall keep out of the way of the other". The effect of wind needs to be considered under such circumstance, rather than the relative bearing between the two ships.

It should be noted that the two rules cannot be used simultaneously. Rule 12 takes effect only when they are sailing vessels. Moreover, according to Rule 11, Rules 12 and 15 apply to vessels in sight of one another, and therefore if all the relevant information, especially the ship types and the wind direction, is available to the seafarers, there would be no conflict between these rules. However, such information is sometimes not available due to some practical or technical reasons. For example, the static data in Automatic Identification System (AIS) may not be available or is wrongly transmitted due to technical or communication failures. Also, the environmental conditions may affect differently the identification of the ship type by the two ships, which poses difficulties in the role assignment according to COLREG and, therefore, increases the collision risk.

Regarding Rule 12, there are two algorithms that need to be developed before determining the roles of ships automatically. The first is how to calculate the wind speed and direction relative to the vessel according to the ship velocity and wind direction. As shown in Fig. 3, $V_1 = (V_x, V_y)$ is the vector of ship's velocity and $V_W = (V_{wx}, V_{wy})$ is the vector of wind direction. If V_1 is rotated clockwise for 90° to transform into $V_3 = (V_y, -V_x)$. As can be seen in Fig. 3, if the intersection angle between V_3 and $V_W\,(\alpha)$ is less than 90° the wind is blowing on the port side of the ship, that is:

$$\mathbf{V_3} \cdot \mathbf{V_W} = V_y V_{wx} + (-V_x V_{wy}) > 0 \tag{1}$$

Otherwise, the wind is blowing on the starboard of the ship.

If both ships have wind on the same side, further judgement on which ship is to windward is necessary. This can be calculated by comparing the relationship between the relative position between two ships and the wind direction. As can be seen in Fig. 4, the vector representing the relative bearing of target ship can be expressed as $P = (x_1 - x_2, y_1 - y_2)$. If the intersection angle between P and $V_W = (V_{wx}, V_{wy})$ (θ) is less than 90°, that is,

$$P \cdot V_w = (x_1 - x_2)V_{wx} + (y_1 - y_2)V_{wy} > 0 \tag{2}$$

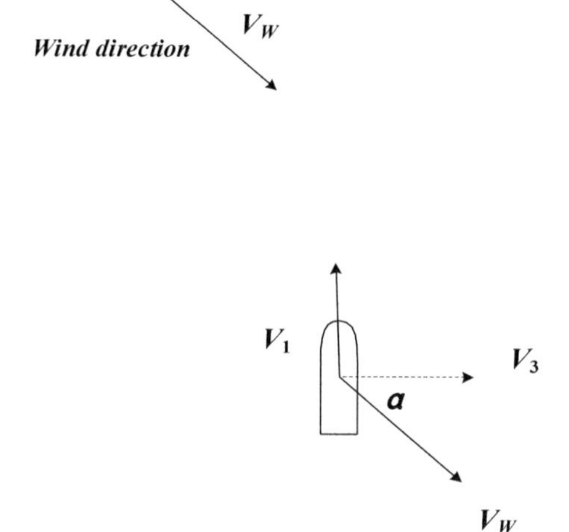

Figure 3. Wind speed and direction relative to the vessel.

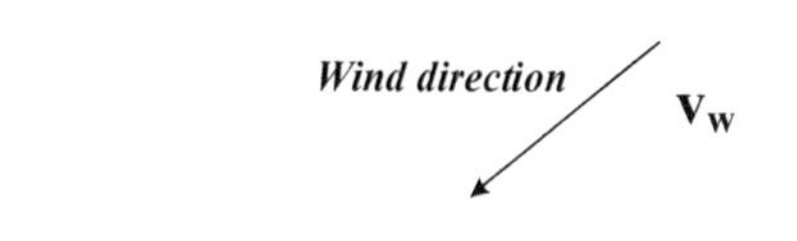

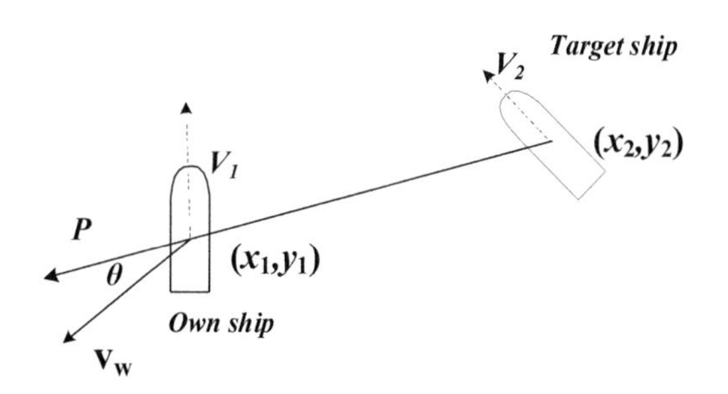

Figure 4. Ship to windward assessment.

the target ship is to the windward direction, which means that it should give way to the own ship. Otherwise, the own ship should take collision avoidance operations.

3 CASE STUDY

In this section, some typical encounter situations between two ships are evaluated using the two rules introduced in the previous section. Noting that the encounter situations of head-on and overtaking are simple and easy for seafarers to make decisions, only crossing situations are considered. With respect to this, two types of crossings with large and small crossing angle are analysed.

In the first case, the own ship is set to be heading to the north and the target ship is located on the starboard of the own ship, which is presented in Fig. 5. According to Rule 15 and Fig. 2, the target ship is within the region A (Fig. 2). Therefore, the own ship should be the give-way ship, which is shown in red in Fig. 5(a). Moreover, considering the effective of collision avoidance operations, steering is usually much more effective than changing speed in a large crossing angle situation. This issue is analysed in detail in Zhang et al. (2015). Noting that the own ship is supposed to cross from behind of the target, the give way ship should turn to starboard to avoid collision.

When considering such situation with Rule 12, it can be inferred from Fig. 5(b) that the wind is blowing to the port of own ship and to the starboard of the target ship. That is to say, the own ship is also the give-way ship, which is the same with the result from Rule 15 and there is no conflict between them. On the contrary, if the wind direction is reversed, the target ship becomes the give-way ship and the conflict arises.

Another case is an encounter situation with small crossing angle. As can be seen in Fig. 6 (a), the target ship is within region B, indicating that the own ship is the give way ship using Rule 15. Under such situation, the own ship usually reduces its velocity to cross from behind the target ship. When considering Rule 12, the wind is blowing from the starboard of both ships according to Eq. (1). The ship to the windward needs to be further identified using Eq. (2). It is obvious from Fig. 6 (b) that the target ship is to the windward, thus it is the give way ship. Such conflict between different rules would make the collision avoidance decision making much more difficult. The coordination and communication between the ships becomes vital to deal with the collision risk in a proper and effective way. Theoretically, there should be no conflict between Rule 15 and 12 if all the relevant navigational information is available. The reason is that

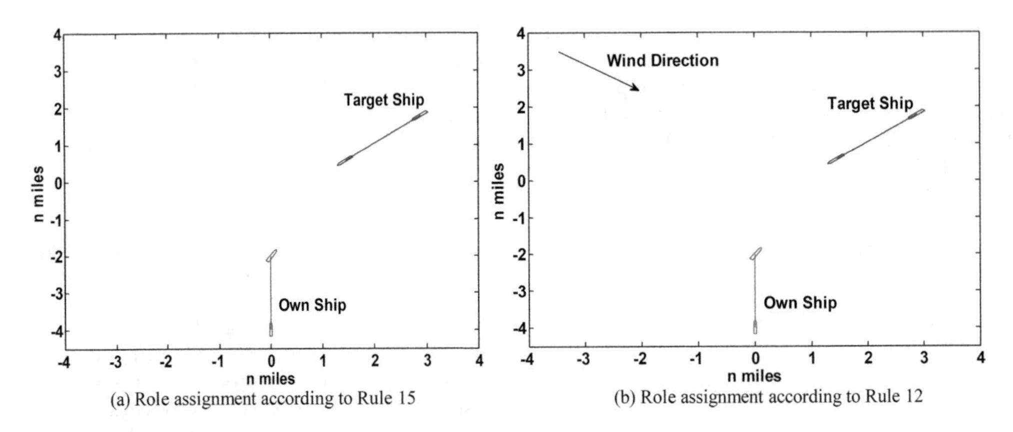

(a) Role assignment according to Rule 15

(b) Role assignment according to Rule 12

Figure 5. Encounter situation with large crossing angle.

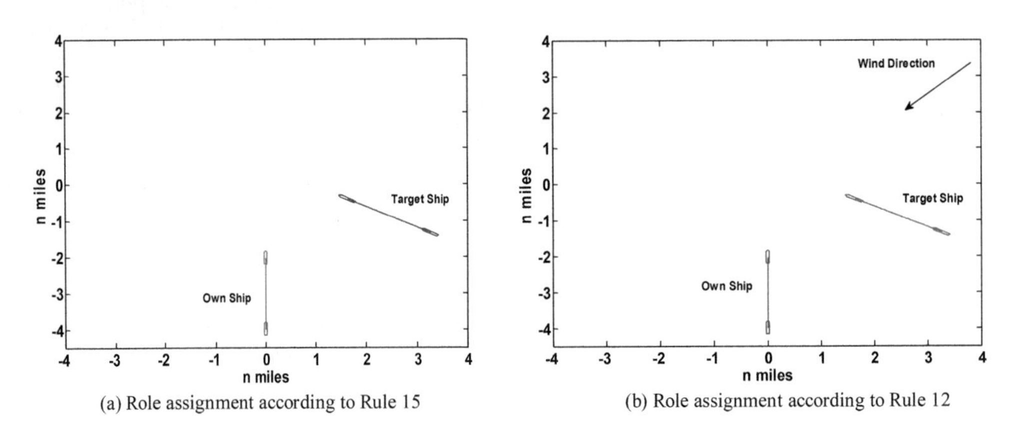

(a) Role assignment according to Rule 15

(b) Role assignment according to Rule 12

Figure 6. Encounter situation with small crossing angle.

Rule 12 is only valid when any of the encounter ships are sailing vessels. However, such conflict demonstrated above often occurs in real situations due to not complete and asymmetric information among ships. The seafarers may make decisions based on different rules.

Compliance with COLREGs is one of the most important requirements for safety navigation. However, the communication between the ships is also important, especially under uncertain situations. Collision would not be avoided if both of them believe that they are head-on ship. According to the statement of good seamanship in COLREGs, the ships need to make early and substantial operations, and the head-on ship should make continuous lookout and take actions if the give-way ship fails to reduce the collision risk. Another possibility is that both ships believe they are the give-way ship. According to COLREGs, the give-way ship should cross behind of the stand-on ship. The effectiveness of the actions taken by the two ships may be offset with each other and the collision risk may not reduce with time. If it increases until a near miss situation, it may not be possible for them to make quick and proper adjustments on their decisions in such a short time.

Besides the discussed conflict between Rule 12 and 15, there are several other ambiguities between different rules when navigational information is not complete. For example, the definitions on some parameters, such as safe distance between ships, safe speed, visibility, vary a lot in different types of sea areas. Seafarers' subjective opinions on them are not identical because of different experience and knowledge. This would influence their subjective assessment of the collision risk, and therefore have impact on the collision avoidance decision making process. Therefore, the Human Reliability Analysis (HRA) during the anti-collision process needs to be investigated in more detail by focusing on how they deal with such conflicts between the COLREGs requirements.

4 CONCLUSIONS

In this paper, the issue of collision avoidance decision making is studied based on the requirements of COLREGs. The general process of anti-collision decision and operation is first introduced. The focus of this paper is on the role assignment of the encountered ships and the possible conflicts between Rules 12 and 15 are discussed in detail. The results indicate that the conflicts may occur in real applications due to lack or incomplete information about the ships or the situation is assessed by the two ships in different ways in an encounter situation. Moreover, the possibility of conflict between other rules, such as rule 15 and 18 (or 19) should also be investigated in more detail in future work. The priorities of different ship types in collision avoidance operation should be involved. Therefore, a more formalized and clear expression on COLREGs is useful for constructing a real-time decision support system for collision avoidance, especially under critical situations. Moreover, with respect to the situations where there may be conflicting judgements, the communications and coordination between the ships play a vital role in avoiding collision. A further investigation on the human errors in dealing with such complex situations using bridge simulators would be helpful for identifying the key factors in ship collisions.

ACKNOWLEDGEMENTS

The research was supported by the National Science Foundation of China (NSFC) under grant No. 51609194, Double First-rate Project of WUT, REliability and the National Key Technologies Research & Development Program (2017YFC 080490, 2017YFC 0804904), as well as the EU project "Safety Engineering and Technology for large maritime engineering systems", GA nr. 730888 – RESET—H2020-MSCA-RISE–2016.

REFERENCES

Banas, P. & Breitsprecher, M. (2011). Knowledge base in the interpretation process of the collision regulations at sea. *TransNav-International Journal on Marine Navigation and Safety of Sea Transportation*, 5(3): 359–364, Gdynia: Poland.

Benjamin, M.R., Curcio, J.A., Leonard, J.J. & Newman, P.M. (2006). Navigation of unmanned marine vehicles in accordance with the rules of the road. *IEEE International Conference on Robotics and Automation.* vol. 2006, pp. 3581–3587, IEEE.

Benjamin, M.R., Leonard, J.J., Curcio, J.A. & Newman, P.M. (2010). A method for protocol-based collision avoidance between autonomous marine surface craft. *Journal of Field Robotics*, 23(5), 333–346.

Cheng, M. & Kuang, T.C. (2010). The study of ship collision avoidance route planning by ant colony algorithm. *Journal of Marine Science and Technology*, 18(5), 746–756.

Collisions at Sea. Journal of Engineering for Maritime Environment 226 (3), 250–259.

Conventions on the International Regulations for Preventing Collision at Sea (COLREGs). 1972. The International Maritime Organization (IMO).

Hwang, C.N., Yang, J.M. & Chiang, C.Y. (2001). The design of fuzzy collision-avoidance expert system implemented by h∞-autopilot. *Journal of Marine Science & Technology*, 9(1), 25–37.

Kreutzmann, A., Wolter, D., Dylla, F., et al. Towards Safe Navigation by Formalizing Navigation Rules. *International Journal on Marine Navigation and Safety of Sea Transportation* (TransNav), 2013, 7(2): 161–168.

Liu, L.Q., Dai, Y.T., Wang, L.H. & Gan, X.L. (2007). Res earch on global path planning of underwater vehicle based on ant colony algorithm. *Journal of System Simulation*, 19(18), 4174–4177.

Perera, L.P., Carvalho, J.P., Guedes Soares, C. (2012). Intelligent ocean navigation and fuzzy-Bayesian decision/action formulation. *IEEE J. Ocean. Eng.*, 37 (2), 204–219.

Perera, L.P., Carvalho, J.P., Guedes Soares, C. (2014). Solutions to the failure and limitations of mamdani fuzzy inference in ship navigation. *IEEE Trans. Veh. Technol.*, 63 (4), 1539–1554.

Smierzchalski, R. & Michalewicz, Z. (2000). Modeling of ship trajectory in collision situations by an evolutionary algorithm. *IEEE Transactions on Evolutionary Computation*, 4(3), 227–241.

Tam, C., Bucknall, R. Cooperative path planning algorithm for marine surface vessels. *Ocean Engineering*, 2013, 57: 25–33.

Tsou, M., Hsueh, C. The study of ship collision avoidance route planning by ant colony algorithm. *Journal of Marine Science and Technology*, 2010, 18(5): 746–756.

Zhang, J.F., Teixeira, A.P. & Guedes Soares, C., Yan, X. & Liu, K. Maritime Transportation Risk Assessment of Tianjin Port with Bayesian Belief Networks. *Risk Analysis*, 2016, 36(6):1171.

Zhang, J.F, Yan, X.P. A distributed anti-collision decision support formulation in multi-ship encounter situations under COLREGs. *Ocean Engineering*, 105(2015):336–348.

Zhang, J.F., Xin, L. & Peng, J. (2012). A novel approach for assistance with anti-collision decision making based on the international regulations for preventing collisions at sea. *Proceedings of the Institution of Mechanical Engineers Part M, Journal of Engineering for the Maritime Environment*, 226(3), 250–259.

Ship performance

Progress in Maritime Technology and Engineering – Guedes Soares & Santos (Eds)
© 2018 Taylor & Francis Group, London, ISBN 978-1-138-58539-3

Design related speed loss and fuel consumption of ships in seaways

M. Riesner, O. el Moctar & T.E. Schellin
Institute of Ship Technology, Ocean Engineering and Transport Systems (ISMT),
University of Duisburg-Essen, Duisburg, Germany

ABSTRACT: Fuel consumption is directly related to ship resistance. The design of ships is usually performed on the basis of powering requirements in calm water without considering actual operating conditions. The effect of the seaway is included using an experience-based allowance on the required power, the so-called sea margin. This practice can lead to either unnecessary excessive power reserves or to underpowered ships that cannot sail against wind and waves in heavy weather. Furthermore, a ship may rarely ever experience calm water conditions it was optimized for. To fulfill requirements of limited carbon emissions and to optimize fuel consumption, ships should be optimized for sea condition in which most of the fuel is consumed. To reliably predict a ship's fuel consumption in seaways, a velocity prediction program accounting for calm water resistance, wave resistance, wind drag, and propulsion characteristics, we calculated ship speed and engine output in seaways. Using a scatter table based on encountered sea states during worldwide service, we calculated fuel consumption for a post-Panamax containership, a VLCC tanker, and a cruise ship.

1 INTRODUCTION

The increasing competitiveness imposed by the shipping market and new international regulations on environmental impact caused fuel consumption to become an important issue. In general, ship owner and shipyard define several contractually listed requirements for newly built ships, some of which have to be fulfilled during sea trials. One important part of the contract is the definition of minimum speed to be reached under defined environmental conditions during sea trials. The ship's design as well as its optimization of hull lines and propulsion components is based on these conditions, which are usually adapted for calm water conditions. The disadvantage of this process is that these mild weather conditions rarely occur in the course of a ship's operating time. Consequently, the ship may not be optimized for realistic operational conditions, resulting in higher fuel consumption, environmental pollution, and operating costs. Recently, fuel consumption has become an important issue because of increasing competitiveness imposed by the shipping market and new international regulations on environmental impact.

To ensure that the ship can retrieve the additional power required due to waves and wind, an experience based allowance, the so-called sea margin, is added to the calm water power requirements. In general, the calm water power is increased by 15%, independently of ship type. Here we showed that the sea margin should be chosen ship dependently, e.g., the wave added resistance of a tanker is relatively large compared to that of a slender containership. Consequently, the added power in waves needs to be greater. The added power of a post-Panamax containership and a cruise ship was investigated numerically using the RANS solver of el Moctar et al. (2017). RANS simulations allowed precise calculations of wave added resistance.

We introduced an approach to efficiently calculate a ship's speed loss in waves, based on assessing the installed power as proposed by the Marine Environment Protection Committee of the International Maritime Organization (IMO, 2013). Our approach solved the equilibrium of longitudinal and transverse forces and yaw moment and considered force components in calm water (including drift), in waves, and in wind while accounting for the effects of the rudder and the propulsion system. Shigunov (2017) used a similar approach to investigate the added power of another post-Panamax containership in different sea states.

We computed speed loss, engine output, and fuel consumption of a post-Panamax container ship, a VLCC tanker, and a cruise ship in sea states these ships will statistically be exposed to during worldwide service. Based on fuel consumption in each sea state and the sea state's probability of occurrence (scatter table), we obtained a distribution of fuel consumption associated with all sea states. The resulting speed losses and fuel consumptions represented relevant factors for the optimization of ship lines and the design of propulsion systems, here summarized as effective weather routing in so-called design related sea states.

2 WORLDWIDE SERVICE ROUTES

The encountered sea states depended on the route the ships sailed, the season, and the wind direction. To rely on seasonally independent wave statistics, we assumed a worldwide service route based on long-terms statistics. This route comprised the shortest sea connections between the four selected ports of Rotterdam, Singapore, Tokyo, and Vancouver. The chronological connection of these ports crossed the following sea areas: English Channel, Bay of Biscay, Strait of Gibraltar, Mediterranean Sea, Suez Canal, Red Sea, northern Indian Ocean, Malacca Strait, Singapore Strait, South China Sea, Philippine Sea, North Pacific Ocean, Caribbean Sea, and North Atlantic Ocean.

Generally, seaway statistics divide the world's oceans into a number of sea areas and include data for all different sea states in each sea area. Normally, these statistics comprise joint occurrence probabilities of significant wave height, H_S, and zero up-crossing wave period, T_Z. Different statistics exist; e.g., Söding (2001), Hogben et al. (1986), and BMT Agross BV (2013). The main differences between these statistics are the methodic procedure of collecting data and the sea areas covered. Söding's data are based on comparing and correlating wave measurements obtained from the Geosat satellite of the US Navy and from seaway calculations based on wind field data. The global wave statistics of Hogben et al. originated from visual assessments of significant wave heights and zero up-crossing wave periods. They divided the world's oceans and sea areas into 104 sub areas, and their data recordings began in 1967. The data from BMT ARGOSS BV comprise digital wave statistics based on current satellite measurements and from a mathematical model. We relied on the well known global wave statistics of Hogben et al. for our worldwide service route. Table 1 lists the percentage distribution of seaways. Columns represent different significant wave heights, $H_S[m]$; rows, zero up-crossing wave periods, $T_z[s]$.

3 VELOCITY PREDICTIONS

Maximum ship speed, engine output, and fuel consumption in waves were calculated using a velocity prediction program to assess the installed power of a ship's propulsion provided by the Marine Environment Protection Committee of the International Maritime Organization (IMO, 2013). Its original aim was to assess the maneuverability of ships in adverse sea conditions and to identify the minimum necessary engine power for safe manoeuvring.

To compute speed loss and engine output, we solved for the force equilibrium in surge and sway and the moment equilibrium in yaw:

$$\vec{X}_G + \vec{X}_S + \vec{X}_W + \vec{X}_R + \vec{X}_T = \vec{0} \quad (1)$$

where index G designates calm water forces, index S wave forces, W wind forces, R ruder forces, and T propeller trust. All force components are velocity dependant and were obtained from an external data base. As the ship was free to rotate about its vertical axis, it was able to execute stationary drift motion. Therefore, the calm water forces comprised two components, namely, a calm water resistance due to the ship's forward motion and a change of calm water resistance due to the ship's drift motion. This resulted in a longitudinal force, a transverse force, and a yaw moment.

Table 1. Distribution of seaways for the selected worldwide service route.

H_S/T_Z	4	4–5	5–6	6–7	7–8	8–9	9–10	10–11	11–12	12–13	13–14	Total
0–1	2.03	5.79	6.23	3.60	1.33	0.33	0.06	–	–	–	–	19.37
1–2	0.54	4.09	9.03	9.77	6.54	2.91	0.91	0.24	0.04	–	–	34.09
2–3	0.10	1.17	3.86	6.07	6.02	3.96	1.75	0.61	0.15	0.04	–	23.73
3–4	0.03	0.29	1.21	2.41	3.14	2.72	1.58	0.68	0.22	0.05	–	12.33
4–5	–	0.08	0.38	0.86	1.31	1.41	1.02	0.53	0.18	0.05	0.01	5.83
5–6	–	0.02	0.10	0.31	0.54	0.66	0.54	0.33	0.13	0.05	0.01	2.69
6–7	–	0.01	0.04	0.13	0.21	0.30	0.28	0.18	0.08	0.03	–	1.27
7–8	–	–	0.02	0.05	0.08	0.13	0.13	0.10	0.06	0.03	–	0.60
8–9	–	–	–	0.02	0.04	0.06	0.07	0.05	0.04	–	–	0.28
9–10	–	–	–	–	0.01	0.03	0.04	0.03	0.03	–	–	0.14
10–11	–	–	–	–	–	0.01	0.02	0.01	–	–	–	0.04
11–12	–	–	–	–	–	–	–	–	–	–	–	0.00
Total	2.70	11.45	20.87	23.87	19.24	12.53	6.41	2.77	0.93	0.25	0.01	100

In general, wave forces oscillate in time. However, our approach yielded a stationary solution. Therefore, $\vec{X}_S$ included only the mean second order components of wave induced forces. Seaways were modeled by superimposing single regular waves of different heights, phases, and periods. Consequently, drift forces in a seaway were calculated by superimposing contributions from regular waves:

$$\vec{X}_S = 2\int_0^{\infty}\int_0^{2\pi} \frac{\vec{R}_{AW}}{\zeta^2} S(\omega) D(\mu) d\omega d\mu \tag{2}$$

where $\vec{R}_{AW}/\zeta^2$ is the quadratic transfer function of the time average drift forces and moments, ζ is the wave amplitude, $S(\omega)$ is the seaway's energy spectrum, ω is wave frequency, and $D(\mu)$ is a spreading function which ensures that the encounter angle of single waves are distributed about the wave encounter angle μ. The available wave statistics did not differentiate between swells and wind seas. Therefore, we used the modified Pierson-Moskowitz wave spectral density function (Lloyd, 1998) for all seaways:

$$S(\omega) = 123\, H_S^2 T_Z^{-4} \omega^{-5} e^{\left(-495 T_Z^{-4} \omega^{-4}\right)} \tag{3}$$

$\vec{X}_W$ is the sum of the relative wind forces due to the forward velocity of the ship and wind. The resulting apparent wind speed and apparent wind angle considered the geometric relation between the true wind angle, true wind speed, and ship speed. Here, wind direction was always parallel to wave propagation, thus the true wind angle and the wave encounter angle were equal. The true wind speed (TWS) was related to the significant wave height, HS, using the well known formula of Bretschneider (Michel, 1999):

$$H_S = 0.0248\, TWS^2 \tag{4}$$

Rudder forces were calculated using a simplified ruder model. The model considered a lift force, L_R, perpendicular to the ruder inflow and a drag force, D_R, parallel to the ruder inflow:

$$L_R = c_L(\alpha)\frac{1}{2}\rho A_R v_R^2 \tag{5}$$

$$D_R = c_D(\alpha)\frac{1}{2}\rho A_R v_R^2 \tag{6}$$

where c_L is the lift coefficient, c_D is the drag coefficient, α is rudder angle of attack, ρ is density of water, A_R is ruder area, and v_R is mean flow velocity at the ruder. Here v_R depended on the geometry at the stern and the effects of the propeller stream. The flow velocity was higher in the region of the propeller stream and lower above and below the propeller stream. The mean flow velocity $\bar{v}_R$ was calculated according to Brix (1993), using propeller stream theory:

$$\bar{v}_R = \sqrt{\left(1-\frac{A_{R,p}}{A_R}\right)v_a^2 + \left(\frac{A_{R,p}}{A_R}\right)v^2(x)} \tag{7}$$

where $A_{R,p}/A_R$ is the ratio of ruder area, A_R, and the area inside the propeller stream, $A_{R,p}$, and $v_a = v_S(1-w)$ is the inflow velocity at the propeller plane. It was calculated from ship speed v_s and the wake number w, where v is flow velocity of the rudder inside the propeller stream.To satisfy the equilibrium condition in the longitudinal direction, we determined the propeller trust $\vec{X}_T$ required to maintain a specific ship speed. We used the free running propeller curve and relied on a simplified approach for the dynamics of diesel engines proposed by IMO (2013). If the propulsion system was not able to deliver the required propeller trust, the ship speed was iteratively corrected to achieve equilibrium.

4 FUEL CONSUMPTION

Generally, fuel consumption depends on the fuel's calorific value, the powering system's effectiveness, and the main engine's required power output. We assumed that all ships used the same fuel and that their specific fuel consumption was a constant 180 grams per kWh. Although system effectiveness normally changes with power output of the main engine, we ignored this and kept system effectiveness constant. We calculated fuel consumption, B, as follows:

$$B = b_e P_B \tag{8}$$

where b_e is the constant specific fuel consumption, and P_B is the engine brake power. Hence, engine break power, P_B, and power delivered by the propeller, P_D, were related by shaft efficiency, η_S:

$$P_B = P_D \eta_S \tag{9}$$

Normally, shaft efficiency is high. Here we assumed that η_S is a constant 0.99. Thus, P_D was calculated using the open water characteristics of the propeller multiplied by its relative rotative efficiency, η_R, which considered that the propeller, situated behind the ship, does not operating in open water conditions. However, in waves rotative efficiency and open water characteristics of the propeller may differ from those in calm water conditions. El Moctar et al. (2017) investigated experimentally and numerically the propulsion characteristics of the subject containership and the subject cruise ship in head waves. They found that, compared to calm water conditions, rotative efficiency, η_R, and hull efficiency, η_H, change only minimally in waves (less than 1%), i.e., they were unable to determine the effects of waves on rotative efficiency. Moor and Murdey (1970) and Nakamura and Naito (1977) reached similar conclusions.

Hull efficiency of the containership seems to be more effected by waves. However, regarding the containership, el Moctar et al. (2017) investigated only two different wave lengths. Thus, they were unable to determine effects of hull efficiency in different wave lengths.

Wu et al. (2014) and Taskar and Steen (2015), for example, studied propulsion characteristics of the VLCC tanker in waves. Among others, they showed that rotative efficiency and open water propeller characteristics are only slightly affected by waves. However, it was difficult to establish a clear tendency even in head waves rather than in oblique waves. Usually, in calm water η_R for a conventional ship with two propellers is approximately 0.98, and for a conventional ship with one propeller it is slightly higher. We assumed η_R to be 1.0 and constant for the subject ships and sea states. We assumed also that open water propeller characteristics remain the same in waves. Therefore, P_D was directly calculated as follows:

$$P_D = 2\pi\rho n^3 D_P^5 K_Q(J) \tag{10}$$

Here ρ is water density, n is propelle revolution, D_P is propeller diameter, and K_Q is the non-dimensional torque coefficient that depends on propeller advance number, J. Open water characteristics were obtained from available model test measurements documented by el Moctar et al. (2017) for the containership and for the cruise ship and by Taskar and Steen (2015) for the tanker.

5 RESULTS

The sea states a subject ship encountered during its voyage were directly related to fuel consumption; specifically, to added resistance in waves and wind in severe seaways encountered while underway. In which sea state a ship consumed most fuel depended not only on its operational time in a particular sea state, but also on its characteristics.

Therefore, we investigated three different kinds of ships, namely, a 355 m long generic post-Panamax containership, known as the Duisburg Test Case (DTC) (el Moctar et al., 2012), a 220 m long cruise ship, and a 320 m long tanker, known as KVLCC2 (Larsson et al., 2010). Table 2 lists their principal particulars. Here Lpp denotes length between perpendiculars; B, molded breadth; D, scantling draft; GM, transverse metacentric height; Δ, displaced volume; CB, block coefficient; SW, hull wetted surface, VS, service speed; t, thrust deduction coefficient; w, wake fraction; and Pmax, maximum engine output. We considered these subject ships traveling the same worldwide service route.

Table 2. Principal particulars of subject ships.

	DTC	KVLCC2	Cruise ship
Lpp [m]	355.0	320.0	220.2
B [m]	51.0	58.0	32.2
D [m]	14.5	20.8	7.2
GM	1.50	5.71	2.754
Δ [m³]	173470	312622	33229
C_B	0.661	0.810	0.651
SW [m²]	22032	27194	7823
V_S [knots]	25.0	15.5	21.0
t [–]	0.09	0.11	0.18
w [–]	0.27	0.12	0.31
P_{max} [kW]	67000	29500	17000

Solving the stationary equilibrium equation (1) yielded the attainable ship speed in waves. Especially calm water forces $(\vec{X}_G)$, drift forces in waves $(\vec{X}_S)$, and wind forces $(\vec{X}_W)$ were difficult to compute and required extensive model tests or numerical calculations for validation. To shorten computation time, our method extracted the required force coefficient from previously calculated data bases. Data bases for $\vec{X}_G, \vec{X}_S$, and $\vec{X}_W$ were obtained numerically, whereby $\vec{X}_S$ depended on ship speed and wave encounter angle. To account for varying ship speeds and wave encounter angles accurately, wave drift forces were calculated for ten different velocities and 14 different wave encounter angles, ranging from head waves to following waves. Wave drift forces were computed using the frequency domain boundary element method of Söding (2014), a code relyings on Rankine sources as basic flow potentials and accounting for the fully nonlinear steady flow caused by the ship's forward speed in calm water. This method yielded results that compared favorably to model test measurements and CFD simulations; see, e.g., Lyu et al. (2017) and Ley et al. (2014).

Figure 1 plots exemplarily the non-dimensional quadratic transfer function of wave added resistance of Lyu et al. (2017), here expressed as the added resistance coefficient Cxadd, for the cruise

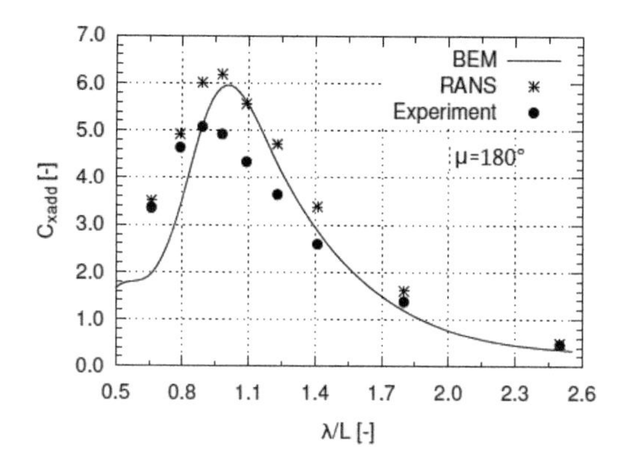

Figure 1. Wave added resistance coefficients of the cruise ship advancing at 15kn in regular head waves (Lyu et al., 2017).

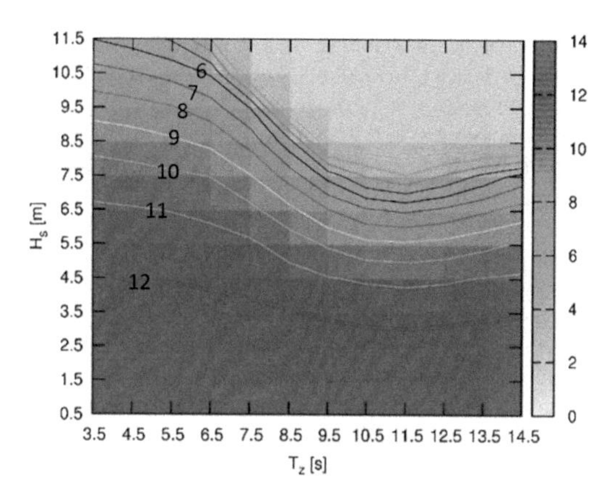

Figure 2. Attainable speed of the containership for a wave encounter angle of 150 deg.

ship advancing at 15 kn forward speed in regular head waves. Symbol BEM denotes results obtained from the frequency domain boundary element method; Symbol RANS, results from a field method solving the Reynolds-averaged Navier-Stokes equations; and Experiment, results from model test measurements.

The data base of calm water drift and wind forces was obtained from CFD simulations that solved the Reynolds-averaged Navier-Stokes equations (el Moctar et al., 2009). Drift angle and wind encounter angle were varied systematically, resulting in non-dimensional coefficients of longitudinal and transverse forces and yaw moments. Alternatively, available empirical data could be used to estimate aerodynamic forces on ships; see, e.g., Blendermann (1993, 2014).

Solving equations (5) and (10) yielded ships speed in waves and the respective engine output. For a wave heading angle of 150 deg, Figures 2, 3, and 4 plot maximum attainable ship speeds of the containership, the VLCC tanker, and the cruise ship, respectively, attained in sea states encountered during the worldwide route. The ordinate specifies significant wave height; the abscissa, zero up-crossing wave period. The continuous lines designate associated isotachs. The coloured scale denotes the attainable ship speed. Red areas designate sea states in which the ship is able to advance at relatively higher (service) speeds; yellow areas, sea states in which the ship is nearly standing still.

Figure 2 shows that speed loss of the containership is greatest in sea states with zero up-crossing wave periods of around $T_z \approx 11.5s$. We would have expected maximum wave added resistance to occur in sea states with wave length nearly equal to ship length, i.e., at a wave length to ship length ratio of 1.0. The container ship has a length of 355 m, and

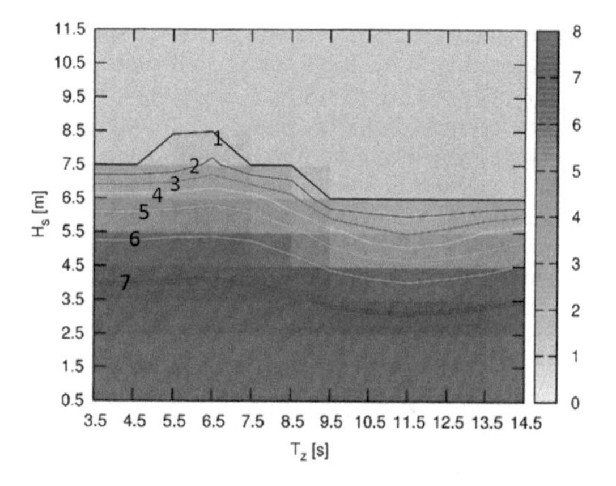

Figure 3. Attainable speed of the tanker for a wave encounter angle of 150 deg.

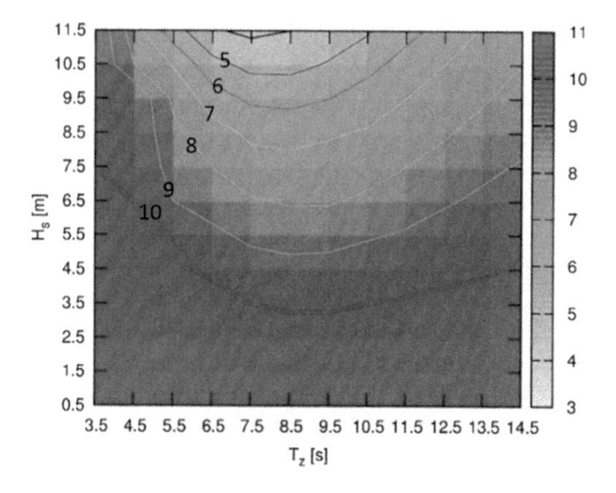

Figure 4. Attainable speed of the cruise ship at wave encounter angle of 150 deg.

a wave of similar length has a period of 15.1 s. Therefore, wave added resistance increased in sea states with longer periods and, consequently, the maximum ship speed was more affected. The highest isotach designates a ship speed of 12.0 m/s, and the attainable ship speed drops below this value in sea states with $H_s \approx 4.5s$ and $T_Z \leq 5.5s$ as well as in sea states with $H_s \approx 3.0s$ and $T_Z = 11.5s$. With increasing significant wave height, the isotachs are closer together, which is reasonable because wave added resistance as well as wind force were proportional to wave height squared (see equations (6) and (4)). Consequently, the ship speed drops below 2.0 m/s in sea states with a significant wave height of H_S ≈ 7.8 m and a zero up-crossing period of T_Z ≈ 11.5 s.

The plot of attainable ship speed of the tanker (Figure 3) differs from results for the containership. The design speed of the tanker was only 15.5 kn compared to 25.0 kn for the containership. As a result, the installed engine power was considerable smaller (approximately 2.3 times smaller). However, due to the large block coefficient of the tanker, its wave added resistance was significantly higher compared to the slender container ship. Consequently, the speed loss in waves was clearly more pronounced. The ship speed dropped below 7.0 m/s for $H_S \geq 3m$, and the tanker was not able to move forward when the significant wave height was above 6.5 m and for $9.5 \leq T_Z \leq 14.5s$.

Furthermore, the velocity drop between small and large zero up-crossing periods is less pronounced than for the containership. This was due to the tanker's relatively greater wave added resistance also in short waves, caused by high wave diffraction at its bow. Furthermore, in short (head) waves the frequency domain panel method slightly underestimated wave added resistance of the containership. However, this behavior was less pronounced for the tanker. Consequently, wave added resistance of the tanker was substantial also in short waves.

Results for the cruise ship (Figure 4) again differ from previous plots. In contrast to the containership and the tanker, which are ships equipped with internal combustion engines, the cruise ship is powered by two electrically driven steerable rudder propellers. Electrical engines react less critically to sudden power reductions at reduced engine speeds. We assumed that the electrical engines were able to deliver maximum output also at reduced engine revolutions. Consequently, the speed loss in waves of increasing height was less compared to that of the containership and the tanker.

Cruise ships tend to reduce ship speed to avoid slamming loads and to reduce accelerations. Furthermore, our worldwide service route is not typical for a cruise ship. Cruise ships try to avoid critical sea states more than cargo ships. However, our aim was to investigate fuel consumption of different ships for the same route. Therefore, we did not consider a weather-induced speed reduction.

An additional characteristic shown in Figure 4 is that the maximum speed loss occurs in sea states with a zero up-crossing wave period of 9.0 s instead of 11.5 s as was the case for the tanker and the containership. This was attributed to the smaller ship length of the cruise ship. A wave with a length equal to the ship length (wave added resistance was assumed to be maximum in that case) has a period of 11.8 s.

Compared to the period of a wave with a length equal to the ship length of the containership (this period would be 15.2 s), it seemed reasonable that the maximum speed loss occurred in sea states with lower zero up-crossing periods.

Due to different ship characteristics (e.g., wave added resistance or engine dynamics) the speed loss in waves seemed to be different. This speed loss was less pronounced in sea states with lower significant wave height, but it increased in sea states with increasing wave height.

In a sea state with significant wave height of 2.5 m, the containership was able to attain 96.1%, the cruise ship 95.4%, and the tanker 93.7% of design speed. This indicates that the container ship's engine was equipped with a relatively larger amount of reserve power. In sea states with higher significant wave heights, the influence of engine dynamics was greater regarding the deliverable engine output and, consequently, the attainable ship speed was more effected. For a significant wave height of 6.5 m, the container ship was able to attain 53.3% of its design speed. However, the attainable speed of the tanker dropped to below 10% of its design speed. In contrast, the cruise ship was able to reach 73% of its design speed.

Based on engine output and fuel consumption for every single sea state and the distribution of the sea state occurrence (Table 1), we calculated long-term distributions of fuel consumption. We assumed that wave encounter angles were evenly distributed. Figure 5 plots the distribution of fuel consumption for the container ship. The red colored areas denote sea states where the ship consumed a high percentage of total fuel while underway; the green and yellow colered areas, sea states where the ship consumed less fuel. The plot includes three isolines marked 80%, 50%, and 17.5%, denoting the percentage fuel consumption in sea states surrounded by these isolines. Consequently, most of the fuel was consumed in moderate sea states, with the largest contribution in sea states with H_S ≈ 1.5 m and T_Z ≈ 6.5 s.

Comparative distributions of fuel consumption for the containership, plotted in Figure 5, indicate that fuel consumption was dominated by the probability of sea state occurrences and not by the specified fuel consumption of the ship while sailing in each sea state, as this fuel consumption depended on wind, wave added resistance, attainable ship speed, engine output, etc.

This finding was reinforced in that fuel consumption distributions for the tanker, plotted in Figure 6, and (although not plotted here) for the cruise ship, were almost identical to the fuel consumption distribution for the containership. This was so although installed engine power, wave added resistance, and attainable ship speed differed.

To visualize the correlation between wave added resistance and percentage contribution of fuel consumption of each sea state, we calculated the associated Pearson correlation coefficients. This coefficient, r_{xy}, relates the dependency of two pairs of assigned values, x and y, as follows:

$$r_{xy} = \frac{\sum_{i=1}^{n} x_i y_i - n\bar{x}\bar{y}}{\sqrt{\sum_{i=1}^{n} x_i^2 - n\bar{x}^2}\sqrt{\sum_{i=1}^{n} y_i^2 - n\bar{y}^2}}. \quad (11)$$

where the bar indicates average values. This coefficient has a value between +1 and −1, where 1 is total positive linear correlation, 0 is no linear correlation, and −1 is total negative linear correlation.

Figure 7 visualizes exemplarily the linear correlation between wave added resistance (abscissa) of the container ship for every sea state (represented by each dot) and the percentage contribution to fuel consumption (ordinate). The straight line representing the linear regression function indicates that the correlation between the wave added resistance and fuel consumption contribution was low, as confirmed by the associated Pearson correlation coefficient of $r_{R,C} = -0.436$. Exemplary, sea states causing a high wave added resistance are plotted at the left side of Figure 7 and it can be seen that these sea states did almost not contribute to the total fuel consumption. Contrary sea states of small and medium wave added resistance had the main contribution.

Figure 8 visualizes the correlation between sea state occurrence based on the scatter table

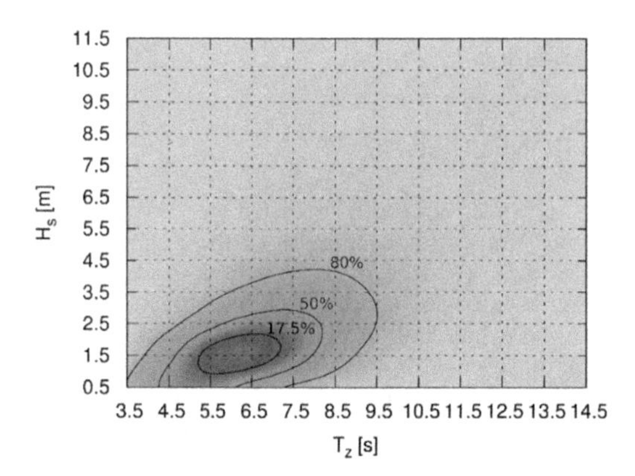

Figure 5. Distribution of fuel consumption for the containership.

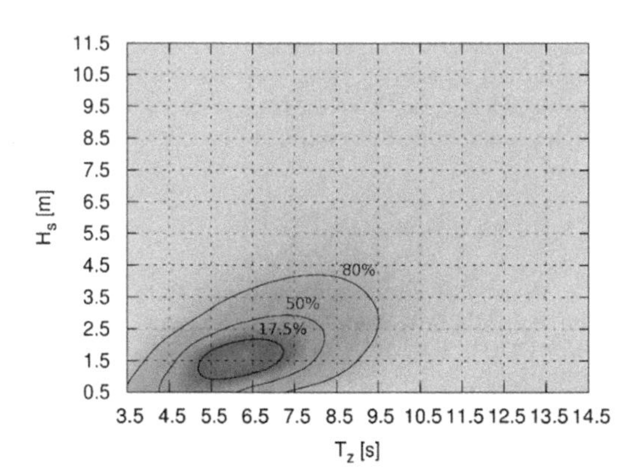

Figure 6. Distribution of fuel consumption for the tanker.

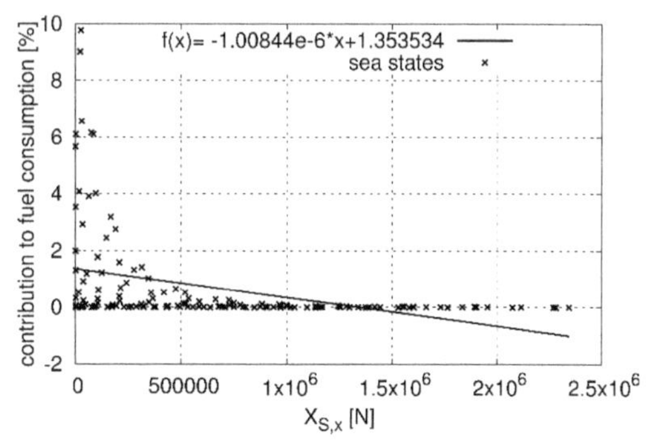

Figure 7. Correlation of mean added wave resistance and its contribution to total fuel consumption of each sea state.

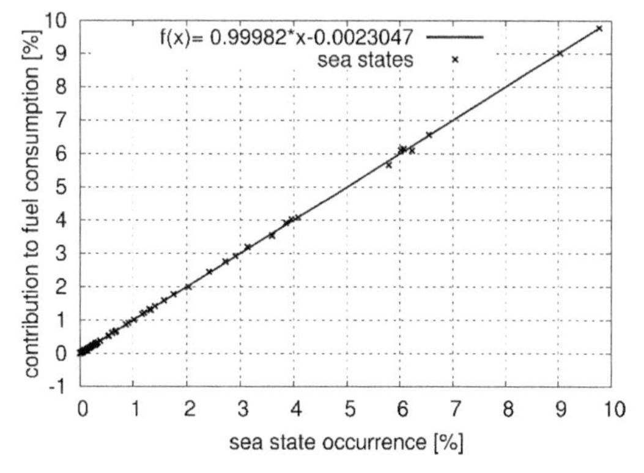

Figure 8. Correlation of frequency of occurrence and percentage contribution to total fuel consumption of each sea state.

(abscissa) and the percentage contribution of total fuel consumption (ordinate). Both values correlate favorably, which was confirmed by the associated correlation coefficient of $r_{O,C} = 0.9991$. That means that the distribution of the fuel consumption did not depend on the ship characteristic (calm water, wind and wave forces or propulsion system). The fuel consumption distribution did almost depend on the probability of occurrence of each sea state.

We obtained similar results for the tanker and the cruise ship. For the three subject ships, Table 3 lists the linear correlation coefficients between wave added resistance and fuel consumption distribution, $r_{R,C}$, and between sea state occurrence and fuel consumption distribution, $r_{O,C}$. For the containership and the tanker, these coefficients turned out to be almost identical. For the cruise ship, correlation coefficient $r_{O,C}$ was almost identical to that of the other ships; however, correlation coefficient $r_{R,C}$ is slightly smaller.

Our computed distributions of fuel consumptions for all three ships depended almost completely on the occurrence of each sea state. Consequently, design related sea states could have been determined during the early design process, based on the operational sea areas and the expected sea states. However, our primarily concern was the high correlation of fuel consumption and sea state occurrence.

Our computed distribution of fuel consumption was based on the required engine output for each sea state and the frequency occurrence of that sea state. For the three subject ships, this frequency of occurrence varied from 0.010 to 9.770%, with a mean value of 1.153%. For the containership, the required engine output varied from 51 to 100%, with a mean value of 81.8%. This suggests that the variance of engine output was much lower than the variance of seaway occurrence probability. For a correct comparison of these variances, mean values should have been identical. Therefore, we multiplied the required engine output by an occurrence factor, and these mean variances were determined as follows:

$$Var(X) = \frac{1}{n}\sum_{i=1}^{n}\left(x_i - \bar{x}\right)^2 \quad (12)$$

Table 3. Linear correlation coefficients $r_{R,C}$ and $r_{O,C}$ for subject ships.

	$r_{R,C}$	$r_{O,C}$
Containership	–0.436	0.999912
Tanker	–0.435	0.999905
Cruise ship	–0.393	0.999880

where n is the number of values in the sample, x_i is a single sample value, and $\bar{x}$ is the mean value of all sample values. The variance was 4.25 for the distribution of sea state occurrence. In contrast, the variances of main engine power output were 0.01701, 0.01128, and 0.001589 for the containership, the tanker, and the cruise ship, respectively. Thus, the variance of the sea state occurrence distribution was at least 250 times higher than the variance of the engine output distribution. Therefore, the distribution of fuel consumption depended almost entirely on the distribution of sea state occurrence.

The variance of engine power output distribution for the cruise ship was about ten times smaller than for the containership. This difference was related to the associated engine dynamics. In contrast to the electrical engines that power the cruise ship, the diesel engine installed on the containership and the tanker was not able to deliver full power at lower engine revolutions. Consequently, the delivered engine power dropped, depending on the speed loss of the ship in higher waves. In turn, this increased the variance of the engine output distribution.

3 CONCLUSIONS

Our approach efficiently predicted speed loss, engine output, and fuel consumption of three ships in sea states and we developed a method to determine design related sea states. First, we obtained speed loss and engine output in sea states typically occurring during worldwide service. We demonstrated that speed loss in sea states with high waves was considerable and that engine dynamics played an important role for the attainable ship speed. Furthermore, speed loss of the tanker was higher compared to speed loss of the containership and the cruise ship.

The distribution of the fuel consumption was almost independent of ship properties. Pragmatically, frequently occurring milder sea states contributed most to overall fuel consumption in contrast to the rarely occurring severe sea states, which caused added resistance to be high.

ACKNOWLEDGEMENTS

This work was funded by the German Federal Ministry for Economic Affairs and Energy under Grant FK03SX339B.

REFERENCES

BMT Agross BV (2013). www.waveclimate.com, Marknesse, The Netherlands.

el Moctar, O., Müller, S.-B., Neugebauer, J. (2009). Schleppkräfte, Internal Report, University of Duisburg-Essen, Institute of Ship Technology, Ocean Engineering and Transport Systems.
el Moctar, O., Müller, S.-B., Neugebauer, J. (2009). Schleppkräfte, Internal Report, University of Duisburg-Essen, Institute of Ship Technology, Ocean Engineering and Transport Systems.
el Moctar, O., Sigmund, S., Ley, J., and Schellin, T.E. (2017). Numerical and Experimental Analysis of Added Resistance in Waves. J Offshore Mechanics and Arctic Engg. Vol. 139, DOI: 10.1115/1.4034205.

Hogben, N., Da Cunha, N.M., and Oliver, G.F. (1986). Global Wave Statistics. British Maritime Technology Ltd., Feltham, Middlesex, UK.

Larsson, L., Stern, F., and Visonneau, M. (ed.) (2010). Proc. Gothenburg Workshop on Numerical Ship Hydrodynamics, Vol II, Gothenburg.

Ley, J., Sigmund, S., and el Moctar, O. (2014). Numerical Prediction of the Added Resistance of Ships in Waves. Proc. 33rd Int. Conf. on Ocean, Offshore, and Arctic Engg., San Francisco, OMAE2014-24216.

Michel, WH. (1999) Sea spectra revisited. Marine Technol. 36(4): 211–227.

Moor, DI. and Murdey, DC. (1970). Motions and propulsion of single screw models in head seas, Part II, The Royal Institution of Naval Architects, Transactions Vol. 112(2).

Nakamura, S. and Naito, S. (1977). Propulsive performance of a container ship in waves. J Soc Naval Architects Japan. Vol. 15, pp. 24–48.

Shigunov, V. (2017). Added power in seaway. Ship Technology Research, Vol. 64, pp. 65–75, DOI: 10.1080/09377255.2017.1331953.

Söding, H. (2001). Global Seaway Statistics. Schriftenreihe Schiffbau, Report No. 610, Technical University Hamburg-Harburg, Hamburg.

Söding, H., Shigunov, V., Schellin, T.E., and el Moctar, O. (2014). A Rankine Panel Method for Added Resistance of Ships in Waves. J. Offshore Mechanics and Arctic Engineering, Vol. 136, 031601, DOI: 10.1115/1.4026847.

Taskar, B., Steen, S. (2015), Analysis of Propulsion Performance of KVLCC2 in Waves, Fourth International Symposium on Marine Propulsors, Austin, Texas, USA.

Valanto, P, Hong, Y. (2017), Wave added resistance and pro-pulsive performance of a cruise ship in waves, Proceedings of the 27th International Offshore andPolar Engineering (ISOPE) Conference, San Francisco, USA.

Wu, P-C., Okawa, H., Kim, H., Akamatsu, K., Sadat-Hosseini, H., Stern, F., Toda, Y. (2014), Added Resistance and Nominal Wake in Waves of KVLCC2 Model Ship in Ballast Condition, 30th Symposium on Naval Hydrodynamics Hobart, Tasmania, Australia.

Progress in Maritime Technology and Engineering – Guedes Soares & Santos (Eds)
© 2018 Taylor & Francis Group, London, ISBN 978-1-138-58539-3

Influence of main engine control strategies on fuel consumption and emissions

R. Vettor, M. Tadros, M. Ventura & C. Guedes Soares
Centre for Marine Technology and Ocean Engineering (CENTEC), Instituto Superior Técnico, Universidade de Lisboa, Lisbon, Portugal

ABSTRACT: Whether the main engine is controlled by keeping constant the engine speed, the torque, the brake power or the ship speed, variations on ship resistance, typically due to evolving environmental conditions, result in a different working points of the engine and, consequently, affect in different ways the fuel consumptions, emissions and duration of the navigation in a seaway. In this work, a qualitative analysis is conducted to evaluate the effects of the different control strategies and some examples are performed providing some quantitative assessment by considering two different ships and properly modelling their behaviour in a seaway and their main engine performance.

1 INTRODUCTION

In navigation, it is common practice to control the main engine operation by defining some variables to be maintained constant up to the next waypoint in order to match the propulsion output with the propeller requirements (Stapersma and Woud, 2005). The engine speed (RPM) is a value that is easy to set and control, thus it has been traditionally chosen as a practical control parameter, allowing the torque and the ship speed to adapt accordingly in case the ship resistance should vary, typically due to evolving environmental conditions. The dynamics of the propulsion system in changing marine environment is a complex process, in which the control variables are in transient conditions rather than in the actual set stage (Yum et al., 2017), these transients may significantly influence the overall fuel consumption and emissions, especially when larger variations are required to maintain the configuration.

Nowadays electronically controlled marine diesel engines combined with ship propulsion models (Mizuno, 2009; Papalambrou and Kyrtatos, 2006; Zhao et al., 2015) make it easier to automatically define any required configuration, thus one of the following three strategies can be adopted: the traditional fixed RPM, fixed engine brake power (BP), fixed torque and fixed ship speed.

In this paper, the effects of the four strategies in ship operations and engine performance will be discussed. A propulsion system model has been developed (Tadros et al., 2015) in order to relate any weather condition in which the ship is required to operate to a specific engine configuration, accordingly with the defined control parameter. In the following the mean value of the added resistance is considered, thus transients are neglected, nevertheless useful general considerations can be done on the differences among the different strategies.

Finally, two simulations in a real weather scenario will be performed by considering the navigation of a containership in the open ocean between Azores Islands and continental Portugal and the transfer of a small fishing vessel in the partially sheltered North Sea. Besides added resistance and propeller settings, also the influence of involuntary speed reduction is assessed by imposing safety constraints to a number of commonly used seakeeping responses. Different strategies may be suitable to different ship type and mission requirements, thus analysing their effects in the realistic voyages will offer a practical meter of comparison.

2 PROPULSION SYSTEM MODEL

A consistent model for the whole propulsion system chain from the total resistance to the prime mover performance has been used. The method allows to relate the environmental conditions in the location where the ship is operating, with the characteristics of the main engine, identifying the exact working point, accordingly with the specific load diagram. Such an accurate model is fundamental for predicting attainable ship speed, fuel consumption and emissions, and to allow the system to identify the most efficient and less pollutant routes.

2.1 *Total hull resistance*

The total hull resistance (R_{tot}), is calculated as the sum of hull resistance in still water (R_{still}), and the added resistance due to the effect of environmental factors (R_{env}), according to the equation:

$$R_{tot} = R_{still} + R_{env} \tag{1}$$

The still water resistance is calculated with the Holtrop and Mannen's method (Holtrop, 1984). Developed at the Netherlands Model Basin, the model is based on regression analysis of several model experiments and full-scale data and is widely considered an approximate procedure for a preliminary assessment of ship resistance in calm water for regular hull forms.Relatively to the environmental loads, only the effect of waves has been considered here. A procedure developed by Faltinsen et al. (1980), based on the direct pressure integration, is used for the calculation of added resistance in waves. The method provides information for added resistance, transverse drift force and mean yaw moment on a ship in regular waves of any direction. The velocity potential is calculated by the strip theory proposed by Salvesen et al. (1970). When the ratio between wavelength and ship length is lower than 0.5, the results are corrected according to the asymptotic expressions presented by Faltinsen et al. (1980).

2.2 *Propeller operation*

Once the total resistance has been calculated, the propeller settings required to maintain a given speed (V) can be identified by knowing the corresponding open water K_T–K_Q curves. The possible effects of cavitation or ventilation, which generally require empirical approaches or tests, are not taken into account.

First of all, the effective thrust (T_E) and advanced speed (V_A), must be calculated taking into account the effects of the presence of hull on the propeller operation, namely the wake fraction (w) and the thrust deduction (t), as:

$$T_E = R_T/(1-t) \tag{2}$$

$$V_A = V/(1-w) \tag{3}$$

Then, from the definition of the thrust coefficient (K_T) and advance coefficient (J), it is possible to identify the operation point of the propeller as the one that equals the coefficient K_1 defined as following:

$$K_1 = \frac{K_T}{J^2} = \frac{T_E}{\rho \cdot V_A^2 \cdot D^2} \tag{4}$$

where D is the propeller diameter and ρ is the water density. Once the operation point is identified, the required torque and RPM can be readily computed.

2.3 *Full engine model*

The two engines considered in this paper are simulated using 1D engine simulation software (Ricardo Wave Software, 2016) according to the methods presented by Watson and Janota (1982) taking into account the different parts of the engine as presented in Figure 1.

This 1D model is an extension of the 0D model presented in (Benvenuto et al., 2016; Morsy El Gohary and Abdou, 2011; Tadros et al., 2015), where the fluid behaviour inside the engine and its effect on the combustion process and exhaust emissions is taking into consideration.

This model helps to minimize the brake Specific Fuel Oil Consumption (SFOC), thus the carbon dioxide (CO_2) emissions and to achieve the limitation of the nitrogen oxides (NO_X) emissions according to the IMO restrictions (http://www.imo.org).

The atmospheric initial conditions, type of fuel, turbocharger, intercooler, intake and exhaust systems and injectors are the main input data required for the numerical model. The different processes inside the cylinders are calculated according to the first law of thermodynamics (Heywood, 1988), taking into account the heat transfer suggested by Woschni (1967). The heat release rate of the combustion process is modelled using the Wiebe function (Watson et al., 1980). Furthermore, while the CO_2 emissions depend on the amount of carbon for each type of fuel used, they are calculated as functions of SFOC using the emission factor presented by Kristensen (2012) and the NO_X emissions are calculated using the extended Zeldovich mechanism (Heywood, 1988).

2.4 *Surrogate engine model*

A surrogate model is also made available to facilitate the implementation of the engine model in the simulations (Tadros et al., 2018). This is based

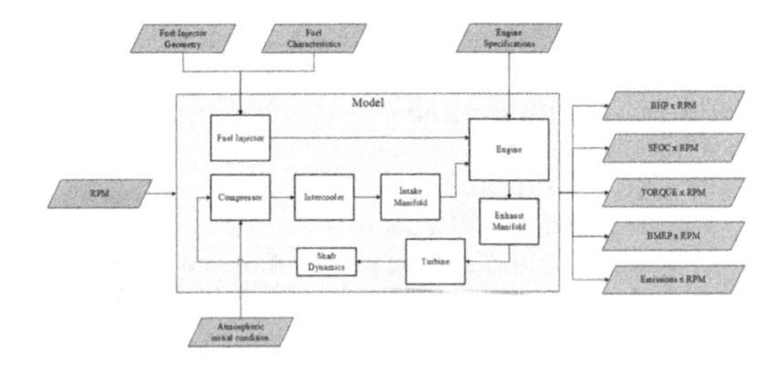

Figure 1. Schematic diagram of diesel engine (Tadros et al., 2016).

Table 1. Seakeeping responses and corresponding limits.

Seakeeping responses	Limit
Slamming probability	3%
Green water probability	5%
RMS of vertical acceleration at the bridge	0.2 g

on the statistical analysis of the results provided by the full model and provides a surface response function whose coefficients have to be calculated for the specific engine installed on-board.

By means of the surrogate model, one can readily estimate fuel consumption and exhaust gases emissions, by knowing the required brake power and engine speed (RPM).

2.5 *Voluntary speed reduction*

Other than for limitations in the operability of the main engine, ship speed in rough weather conditions may be influenced by the choice of the shipmaster to slow down aiming at reducing ship motions and thus their effect on safety of the ship, crew, cargo and passenger on-board. This results in a lower attainable ship speed, especially for severe sea-states, characterized by a significant wave height (H_S) higher than 3 m, as shown, for instance, in Prpić-Oršić et al. (2016).

The actual occurrence of voluntary speed reduction is difficult to estimate, indeed it depends on factors not easy to be controlled or known a-priori, as the attitude of the Shipmaster to risk. Nevertheless, several studies can be found in the literature where the ship responses are analysed and appropriate limits are proposed (e.g. Dubrovskiy, 2000; NORDFORSK, 1987) and can be used as a reference when no other information is available.

For the scope of this work, the ship responses taken into account are slamming, green water and vertical acceleration at the bridge. The corresponding chosen limits are listed in Table 1.

3 ENGINE MANAGEMENT STRATEGIES

In navigation, the engine operation mostly depends on the required speed, the dynamic environmental loads, and the strategy chosen for its control. In this work engine strategies differentiate depending on the parameter that is chosen as control variable, as for instance ship speed, RPM, torque or brake power. Indeed, due to the continuously variating environmental loads, the actual working point of the engine cannot be constant along the route,

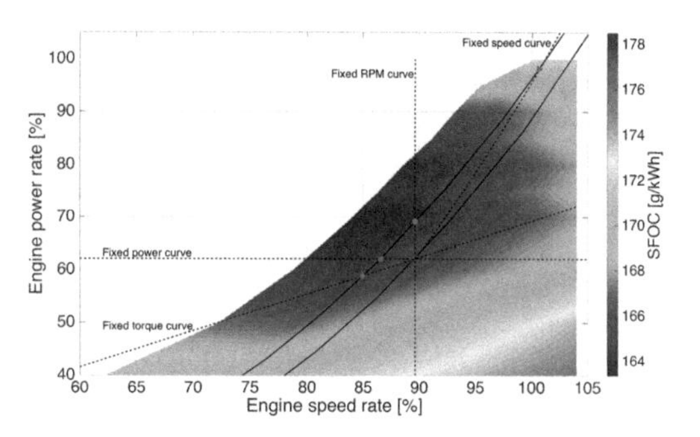

Figure 2. Example of the effect of different strategies in the working point of the main engine.

but has to contrast the effect of this variation by changing its configuration.

While sailing between two consecutive waypoints, the Shipmaster typically impose one of the previously mentioned variables to be maintained constant, in terms of mean value, until next order. Historically, the common practice was to contrast an increased resistance by operating on the engine control maintaining a constant value of RPM, hence varying the ship speed and the brake power accordingly. Nowadays, automatic controls of the main engine allow easier regulations, and the choice of a particular strategy can have a different impact on engine performance and ship operations.

Figure 2 shows the load diagram of a marine diesel engine, with the corresponding fuel consumptions in the background. The considered ship for this example is the test containership S175. The lower solid line represents the propeller curve in calm water, and the red star indicates the corresponding working point for a ship speed of 18.5kn. The upper solid line represents, instead, the propeller curve when the ship is operating in head sea, with H_s equal to 4.0 m and peak period (T_P) of 10s. Facing such a sea-state, may be a hazard for some ships, however, in the following example, only the engine performance is considered, thus no restrictions are made regarding safety and seakeeping responses, which will be included in the simulations presented in section 4.

The four dashed lines, refer, as indicated in Figure 2, to the variation in the engine configuration when the added resistance increases:

- if the ship speed is required to be constant, the new working point will be far from the original one (red triangle in Figure 2), in particular with a significant higher power required (about 36% more in this case) as well as a higher RPM (about 10%), due to the cubic behaviour of the propeller curve. Moreover, although the specific fuel

consumptions and CO_2 emissions, in terms of mass per unit power-time, increase just slightly, the consumption per nautical mile is about 50% higher. Even worse the case of the NOX emission;
- if the fixed RPM strategy is imposed, the additional torque required to oppose the incremented resistance, is provided by increasing the brake power only the amount needed to cross the new propeller curve (red dot in Figure 2). Nevertheless, this causes a significant reduction in the ship speed. Even in this case, the consumption per nautical mile increases, due to the combined effect of slightly higher brake power and longer time required to sail the track;
- when an additional power is not allowed, the extra torque causes the propeller, thus the engine, to lower its RPM (red square in Figure 2). Although both specific fuel consumption and brake power are the same as in calm water, the lower attainable ship speed causes the ship to take a longer time to sail the same track (for instance in this case about 45 minutes more, every 100 nautical miles), thus causing an increase in the overall consumption and emissions.
- being the torque directly proportional to the power and inversely proportional to the RPM, both have to decrease when a constant torque strategy is chosen and the total resistance increases (red diamond in Figure 2). This is the case resulting in the lowest attainable speed, but at the same time, the only one in which power decreases, so the opposing effects on consumption and emission almost nullify reciprocally.

In Table 2, the data corresponding to the above discussed configurations are listed.

Although the previous considerations are specific for the given example and engine, it is clear that the variation in the consumption and emissions per distance sailed are more strongly governed by the dual relation between attainable ship speed and required brake power, than by the variations in the specific consumption and emissions. For this reason, the results of the previous analysis can be considered qualitatively valid, even for any regular hull ship and diesel engine.

It is also worth to consider that constant ship speed control imposes more significant variations in the configuration of the main engine, thus resulting in increased loads due to the transition phases and may affect the long-term life-time of the engine.

Table 2. Route followed from Azores Islands to Lisbon.

	Ship Speed	RPM Rate	BP Rate	SFOC	
	[kn]	[%]	[%]	g/kWh	kg/nmi
Calm Water	18.5	89.6	62.1	165	160
Fixed Speed	–	99.3	98.2	169	245
Fixed RPM	16.8	–	69.1	164	192
Fixed Power	16.3	86.9	–	165	181
Fixed Torque	16.1	85.6	58.8	165	176

4 SIMULATIONS IN REAL WEATHER SCENARIO

In order to assess the effect of the different strategies in realistic operations, two simulations have been carried out. The first reproduces the transit of a containership between Lisbon in continental Portugal and Ponta Delgada in Azores Islands, while the second refers to a small fishing vessel crossing the North Sea from Aberdeen in UK to Bergen in Norway.

In both cases, real weather conditions are considered as deriving from the ERA-interim reanalysis database.

4.1 *Containership in a west North Atlantic route*

The containership MV Adee, with length 117.60 m and breadth 20.20 m, operating the shipping between continental Portugal and Azores Islands as shown in Figure 3, has been considered for evaluating the effects of applying the different strategy in an open ocean navigation.

The ship is equipped with a four-strokes 18-cylinders medium speed marine diesel engine with a maximum power of 9,180 kW. The baseline engine configuration corresponds to a ship speed of 15 knots in calm water, which is pointed by the green dot in the engine load diagram in Figure 5.

The navigation starts from São Miguel Island on January 27th at midnight in a typical North Atlantic winter sea-state characterised by long

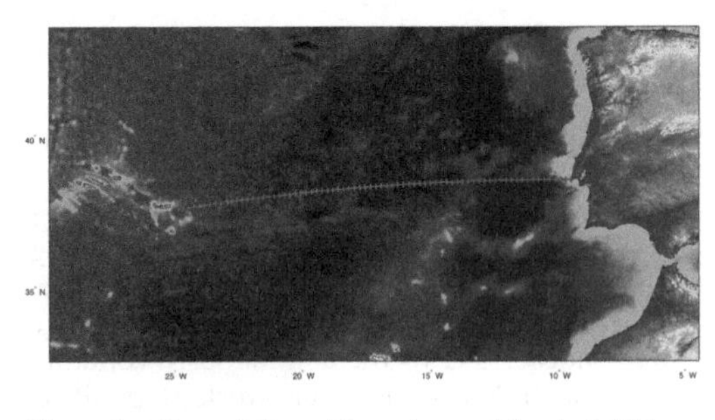

Figure 3. Route followed from Azores Islands to Lisbon.

waves with a significant wave height of 3.5 m. Progressing eastwards, the weather conditions become more severe due to the development of an extra-tropical storm that reaches the western Iberian coast on January 28th.

Figure 4 shows the operational profiles accordingly to the four different strategies, as well as the one corresponding with cam water condition. In particular: the encountered significant wave height, the ship speed, the engine speed, torque and power, and the fuel consumption per nautical mile are reported.

When an engine related parameter (namely RPM, torque or brake power) is maintained constant, the engine configurations are constrained to a small range of operational conditions for the entire voyage, as also shown in Figure 5.

On the other hand, when ship speed is required to be constant, the engine has to significantly variate its operating point, especially with regards to the brake power that increases up to about 20%, reflecting in higher fuel consumptions.

Nevertheless, the effects on ship speed reduction are considerable, with a maximum of almost 10% in case of fixed torque strategy.

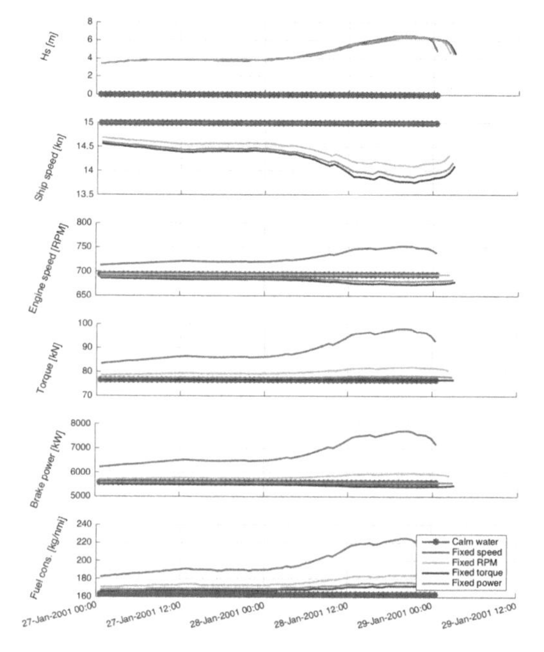

Figure 4. Operational profiles of the MV Adee containership depending on the required strategy.

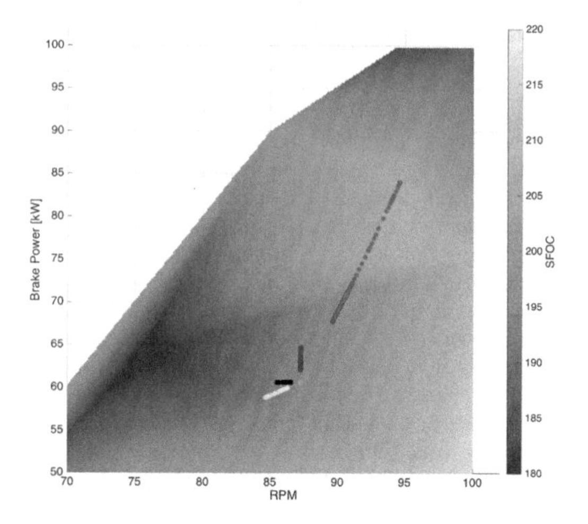

Figure 5. Load diagram of the engine and working points during the journey for (red) calm water, (green) constant speed, (blue) constant RPM, (black) constant power and (yellow) constant torque.

4.2 *Fishing vessel in a North Sea route*

A 29 m trawler vessel, whose body plan and main dimensions are shown in Figure 6, operating in the North Sea has been selected for the second simulation, in order to also analyse the case of small engines and a wave climate typical of sheltered seas.

Departing from the port of Aberdeen in UK on January 10th 2001 at 15:00, the vessel heads towards the port of Bergen in the Norwegian west coast, at a distance of about 272 nmi, following the route shown in Figure 7.

The vessel is imposed to navigate in the engine settings corresponding, in calm water, to ship speed of 11 knots, thus to complete the journey about one day will be required. When the vessel leaves the port, weather conditions are characterised by head sea with 1.3 m of H_s and T_P of about 9.7 s. Along the route the weather intensifies, waves become shorter with a T_P of 8.1 s and a H_s that almost reaches 3 m when approaching the Norwegian coast.

The operational profiles in navigation are plotted in Figure 8. In this case, the variations are less pronounced than in the previous simulation, mainly due to the less severe weather conditions,

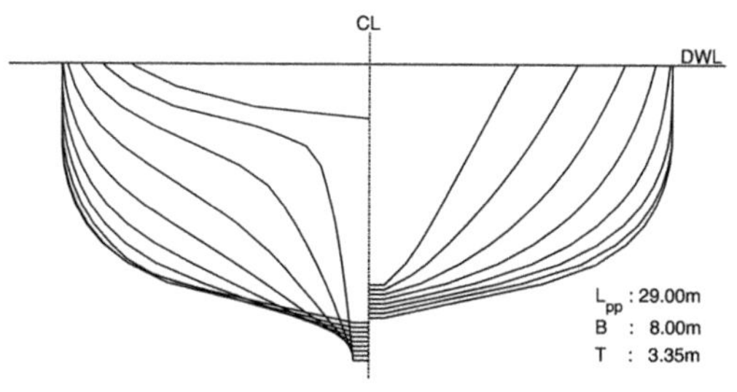

Figure 6. Body plan of the fishing vessel.

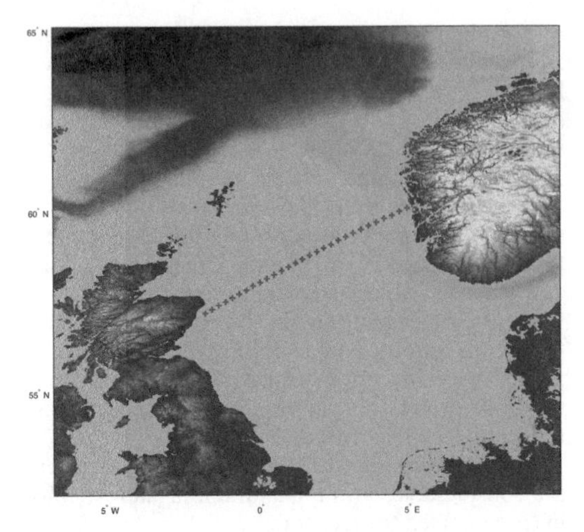

Figure 7. Route followed between Aberdeen and Bergen.

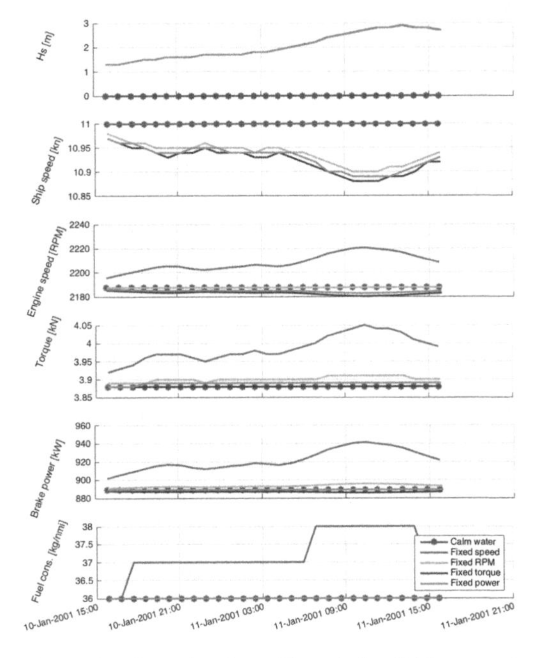

Figure 8. Operational profiles of the fishing vessel depending on the required strategy.

however a marked difference can be notice when adopting a constant ship speed on respect to the other strategies.

Figure 9 shows the working point experienced by the main engine accordingly to the control strategies adopted. In this case, the variation in the brake power is limited to about 5%, while in the engine speed to less than 3%.

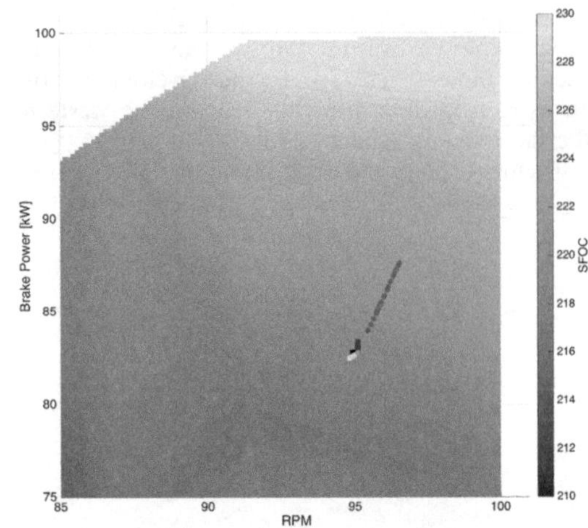

Figure 9. Load diagram of the engine and working points during the journey for (red) calm water, (green) constant speed, (blue) constant RPM, (black) constant power and (yellow) constant torque.

5 CONCLUSIONS

In the presented work, it is analysed the effect of different engine control strategies in the ship operations and performance of the main mover. Although controlling the operation by imposing an engine related parameter, as traditionally done for the RPM, ensures smoother variations in the engine configuration, it may cause a significant drop in the attainable ship speed, thus resulting inappropriate if a rigorous schedule must be followed. On the other hand, requiring the ship to navigate at a constant speed may result in a considerable increase of the fuel consumption, affecting the operational costs of the voyage.

It must be considered that noticing that, due to the wider variation of the engine configuration, when a constant ship speed strategy is adopted, the required power is more prone to exceed the engine operability, thus requiring intervention for re-defining the settings. Differently, with the other strategies, alterations in the defined operations can be expected only if navigating close to the maximum speed or for safety reason due to excessive ship motions. This may also have an effect when the strategies are implemented in navigation software, such as weather routing, influencing the computational time.

Finally, it is worth noticing that, although a reliable modelling of engine performance is fundamental, the overall quantities of fuel and exhaust gases are strongly governed by the variations in the required power, rather than by the differences in the specific consumptions and emissions (g/kWh) between the engine operating points. As an example, by re-computing route of the MV Adee at fixed speed with a constant value of the specific fuel consumption corresponding to the calm water condition, the variation on the overall consumption resulted in about 2.5%.

ACKNOWLEDGEMENTS

This work was performed within the Strategic Research Plan of the Centre for Marine Technology and Ocean Engineering (CENTEC), which is financed by Portuguese Foundation for Science and Technology (Fundação para a Ciência e Tecnologia—FCT).

REFERENCES

Benvenuto, G., Lavitola, M., Zaccone, R., Campora, U., 2016. Comparison of a natural gas engine with a diesel engine for marine propulsion, in: Guedes Soares, Santos (Eds.), Maritime Technology and Engineering 3. Taylor & Francis Group, London, pp. 725–734.

Dubrovskiy, V.A., 2000. Complex Comparison of Seakeeping: Method and Example. Mar. Technol. 37, 223–229.

Faltinsen, O.M., Minsaas, K.J., Liapis, N., Skjordal, S.O., 1980. Prediction of Resistance and Propulsion of a Ship in a Seaway, in: Proceedings of the 13th Symposium on Naval Hydrodynamics. Tokyo, Japan, pp. 505–529.

Holtrop, J., 1984. A statistical re-analysis of resistance and propulsion data. Int. Shipbuild. Prog.

Mizuno, N., 2009. Marine Main Engine Control with Adaptive Extremum Control Scheme, IFAC Proceedings Volumes. IFAC.

Morsy El Gohary, M., Abdou, K.M., 2011. Computer based selection and performance analysis of marine diesel engine. Alexandria Eng. J. 50, 1–11.

NORDFORSK, 1987. Assessment of Ship Performance in a Seaway: The Nordic Co-operative Project: "Seakeeping Performance of Ships."

Papalambrou, G., Kyrtatos, N.P., 2006. Robust Control of Marine Diesel Engine Equipped with Power-take-in System, IFAC Proceedings Volumes. IFAC.

Prpić-Oršić, J., Vettor, R., Faltinsen, O.M., Guedes Soares, C., 2016. The influence of route choice and operating conditions on fuel consumption and CO_2 emission of ships. J. Mar. Sci. Technol. 1–24.

Salvesen, N., Tuck, E., Faltinsen, O., 1970. Ship motions and sea loads. Trans. SNAME.

Stapersma, D., Woud, H., 2005. Matching propulsion engine with propulsor. J. Mar. Eng. Technol. 25–32.

Tadros, M., Ventura, M., Guedes Soares, C., 2018. Surrogate models of the performance and exhaust emissions of marine diesel engines for ship conceptual design, in: Guedes Soares, Teixeira (Eds.), Maritime Transportation and Harvesting of Sea Resources. Taylor & Francis Group, London, pp. 105–112.

Tadros, M., Ventura, M., Soares, C.G., 2016. Assessment of the performance and the exhaust emissions of a marine diesel engine for different start angles of combustion, in: Guedes Soares, Santos (Eds.), Maritime Technology and Engineering 3. Taylor & Francis Group, London, pp. 769–775.

Tadros, M., Ventura, M., Soares, C.G., 2015. Numerical simulation of a two- stroke marine diesel engine, in: Guedes Soares, C., Dejhalla, R., Pavletic, D. (Eds.), Towards Green Marine Technology and Transport. Taylor & Francis Group, London, pp. 609–617.

Yum, K.K., Taskar, B., Pedersen, E., Steen, S., 2017. Simulation of a two-stroke diesel engine for propulsion in waves. Int. J. Nav. Archit. Ocean Eng. 9, 351–372.

Zhao, F., Yang, W., Wan, W., Kiang, S., 2015. An Overall Ship Propulsion Model for Fuel Efficiency Study. Energy Procedia 75, 813–818.

Progress in Maritime Technology and Engineering – Guedes Soares & Santos (Eds)
 ISBN 978-1-138-58539-3

Analysis of multipurpose ship performance accounting for SME shipyard building limitations

Y. Denev & P. Georgiev
Technical University of Varna, Varna, Bulgaria

Y. Garbatov
Centre for Marine Technology and Ocean Engineering (CENTEC), Instituto Superior Técnico, Universidade de Lisboa, Lisbon, Portugal

ABSTRACT: This study analyses the performance of new design multipurpose ships, accounting for the constraints of a small and medium sized shipyard during the building process of new ships. The conceptual design of ships with deadweight in the range of 5,000 to 8,000 tons with and without shipyard building constraints is performed. The ship hulls, for all studied ships, are developed based on the combined linear scale and Lackenby transformation approach using 3D models of existing already built ships. The evaluation of the performance includes the total resistance of ship at three different speeds, intact stability, cargo volume and number of containers on the deck. The impact of the shipyard building constraints is evaluated by comparing the performance index of studied ships with the same deadweight with and without shipbuilding restrictions. Based on the present analysis, several important conclusions are derived.

1 INTRODUCTION

Nowadays, the European maritime industry relies on small and medium-sized enterprises, SME to restore traditional European shipbuilding industry while ensuring the youth employment. According to the European Association of Craft, Small And Medium-sized Enterprises, (UEAPME, 2014), 99.8% of the more than 20 million enterprises in the EU are SMEs. The average European enterprise provides a job for six employees, including the owner-manager, and SMEs count for 2/3 of the private employment and produces about 60% of the added value in the European economy. In the last decade, SMEs created 80% of the new jobs.

According to the EU Craft and SME Barometer (http://www.ueapme.com), for the second half of 2017, the SME Climate Index reached 80.2 ppts and it is the highest score ever achieved since the outbreak of the global financial crisis in 2007–08. The SME Climate Index is calculated as the average of the companies that have reported positive or stable business growth and expect a positive or stable development for the next period. Therefore, the index can vary from 100 (all positive or neutral) to 0 (all negative).

In this respect, the EU funded Project Shiplys main objectives are to respond to the needs of the SME shipyard designers, shipbuilders and ship-owners (Bharadwaj et al., 2017) in the development of a ship risk-based design framework a framework tool will be developed to support the competitiveness of SMEs in design, shipbuilding and retrofitting.

The shipbuilding is strongly related to the transportation of cargoes. For the general/dry cargo shipping in the Black Sea and Mediterranean regions, the Istanbul Freight Index, ISTFIX (http://en.istfix.com) provides information that can be used for any type of shipping analysis. ISTFIX is an internet reference website that contains statistical information derived from various sources.

The analysed routes, taken by the ship's operation are:

- Route 1: Black Sea – Marmara;
- Route 2: Black Sea – East Mediterranean;
- Route 3: Black Sea – Central Mediterranean;
- Route 4: Black Sea – West Mediterranean;
- Route 5: Black Sea – Continent.

Four groups of ships are analysed:

- Group 1: 2,000–4,000 DWT;
- Group 2: 4,000–6,000 DWT;
- Group 3: 6,000–8,000 DWT;
- Group 4: 8,000–12,000 DWT.

The ISTFIX provides a unique freight index in the short sea coaster shipping, starting from

January 1st, 2008. A study shows that the market risk in the ISTFIX shipping area is much lower than in the international Baltic handy size index, BHSI (Ünal & Derindere, 2014).

The shipping in Route 1 to 5 are a part of the Short Sea Shipping that operates in the EU coastline of about 70,000 km.

According to a recent study, the current short sea transportation capacity is inefficient, especially in the dry bulk and general cargo segments (Gustafsson et al., 2016).

This study highlighted five directions for improvement of the competitiveness:

- increase the freight market efficiency through transparency,
- dynamic and integrated production and logistics planning,
- efficient cargo handling,
- performance-driven shipbuilding and operation, and
- sustainable investment and governance models for the system-wide transition.

The goal of the present study is to analyse the importance of the SME shipyard building limitations on the performance of new design ships built in the conditions of SME shipyard. The study considers the shipbuilding capacity of an SME shipyard and the demand of the ship-owners for an efficient ship, operating in the Black sea – Mediterranean region. The building limitations considered in the present study are as follows (Atanasova et al., 2018):

- maximum docking capacity – 1,800 t;
- maximum dock dimensions that allow a ship to be built with a length not greater than 135.8 meters and a breadth not greater than 16 meters;
- the depth of the fairway determines the draft of the ship to be not more than 8 meters.

The software tool "Expert" has been used in defining design solutions for multipurpose ships subjected to shipbuilding, operational and functional constraints (Damyanliev et al., 2017, Garbatov et al., 2017a). This software is structured as an open system allowing the search design solution by Sequential Unconstrained Minimization Technique (SUMT) for different types of ships for which a suitable mathematical model can be generated. Different mathematical models can be employed in identifying the main dimensions of ship, ship hull form, mass and volume distributions, general arrangement, ship hull structures and equipment; propulsion complex; freeboard requirements; stability; sea-keeping; manoeuvrability etc. The conceptual framework is capable of accounting for series of constraints.

Table 1. Main ship dimensions of non-restricted ships.

DWT	5,000	6,000	7,000	8,000
Ship	S1	S2	S3	S4
L, m	86.79	93.21	96.71	99.98
B, m	16.17	16.87	18.60	19.23
d, m	7.18	7.33	7.98	8.53
D, m	8.90	9.32	10.06	10.81
L/B	5.37	5.52	5.20	5.20
B/d	2.25	2.30	2.33	2.26
L/D	9.76	10.00	9.61	9.25
Cb	0.685	0.76	0,67	0.669
Δ, t	7,068	7,981	9,853	11,241

Table 2. Main ship dimensions of restricted breadth B = 16 m.

DWT	5,000	6,000	7,000	8,000
Ship	$S1_R$	$S2_R$	$S3_R$	$S4_R$
L, m	88.63	106.60	120.62	135.06
B, m	16.00	16.00	16.00	16.00
d, m	7.08	6.88	6.67	6.57
D, m	8.81	8.93	9.03	9.17
L/B	5.54	6.66	7.54	8.44
B/d	2.26	2.33	2.40	2.44
L/D	10.06	11.93	13.36	14.73
Cb	0.69	0.721	0.772	0.812
Δ, t	7,091	8,665	10,181	11,809

Analysing the SME ship repair yard capacity in building new ships (Atanasova, et al., 2018) the authors considered two groups of ships in the range of 5,000–8,000 DWT developed by the software tool "Expert". The same ships are used in this study too. The main dimensions of the ships are obtained using CAPEX as an optimization criterion. The first group consists of ships without shipbuilding restrictions and the second one is for ships with a 16 m restricted breadth that is related to a specific SME shipyard (Atanasova, et al., 2018).

The length of the second group of ships increases to compensate the restricted breadth and at the same time the draught decreases, which can be explained by the condition of reaching the minimum required ship stability criterion.

2 SHIP HULL FORM

2.1 *Hull form transformation*

The evaluation of the ship performance is based on the hull form obtained by a transformation of an existing parent ship hull, PSH.

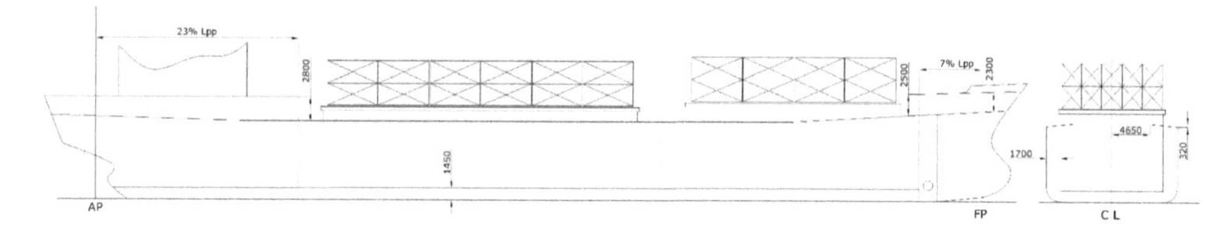

Figure 1. Multipurpose ship typical profile and cross section.

Figure 2. 5,000 DWT ship hull with restricted breadth, PSH 1.

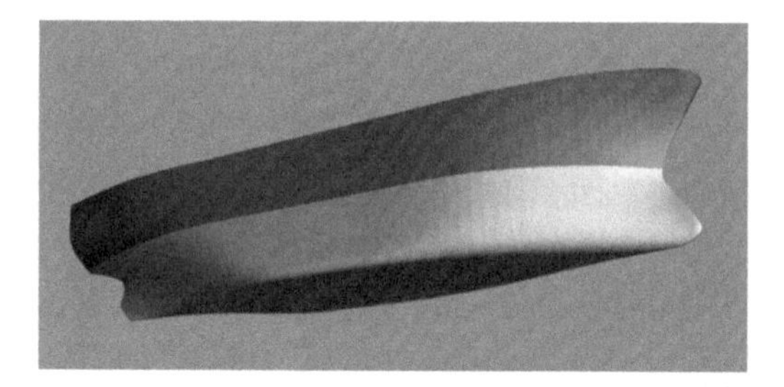

Figure 3. 8,000 DWT ship hull without restrictions, PSH 2.

For the hull transformation a combined linear scaling and Lackenby transformation approach in two steps is used. The first step consists of a scale transformation of the longitudinal, transverse and vertical coordinates multiplying them by a factor. The linear scaling keeps hull coefficients unchanged.

The second step implements the Lackenby (1950) method to estimate the desired displacement (block coefficient, C_B). The limits of the changes in parameters, which still lead to decent hull forms, depend on the hull form. Due to the wide range of C_B for all eight ships, from 0.67 to 0.812, two different parent hulls are used taken from the database of Freeship Plus software (https://freeship-plus.en.softo nic.com). The first one is for the ships of 5,000 and 6,000 DWT and the second one for the rest of the ships. Figure 2 and Figure 3 present the PSH1 and PSH2.

2.2 *General arrangement of ships*

The analysed multipurpose ships are with a forecastle and poop and the superstructure is in the aft. The length of the forecastle and the poop is 7% and 23% of the length between the perpendiculars of the ship respectively.

There are several constant dimensions like, the height of the superstructures and double bottom, breadth of the double side, chamber, etc. The side view and cross section of the ship are shown in Figure 1.

The ship is intended only to carry containers on the deck, but the breadth of the cargo holds is consistent with 5 rows of containers.

3 SHIP HULL PERFORMANCE

The overall performance of both groups of ships is evaluated by the total resistance of the ship for 3 different speeds, intact stability, according to the IS Code, and the cargo capacity measured by the total volume of the cargo holds, excluding hatches and the number of containers on the deck.

3.1 *Ship resistance*

The total resistance, R_T, kN is calculated by the Holtrop & Mennen (1982) method, where only, three speeds are considered, i.e. 10.5 kn; 12.5 kn and 15.0 kn. The relative total resistance is defined as:

$$R_{RT} = R_T/(\gamma\nabla) \tag{1}$$

where R_{RT} is the relative total resistance, γ is the specific weight of the sea water, 10.055 kN/m^3, and ∇ is the immersed volume, m^3. The total resistance is shown in Table 3.

3.2 *Intact stability*

The intact stability is evaluated by the maximum permissible vertical centre of gravity, KG_{max}, related to the ship depth, calculated according to the Part A, Chapter 2.2.2 of the Intact Stability Code (MSC.267(85), 2008) for several draughts in the range of 0.6 to 1.05 of the summer draught.

The relative intact stability, R_{IS} is defined as:

$$R_{IS} = KG_{max}/D \tag{2}$$

Table 3. Total resistance of ships, kN.

Ship	DWT	Speed, kn		
		10.5	12.5	15.0
$S1_R$	5,000	92.1	142.4	251.0
$S2_R$	6,000	105.3	157.5	261.3
$S3_R$	7,000	114.6	176.4	304.2
$S4_R$	8,000	129.1	196.0	328.0
S1	5,000	91.3	142.0	253.0
S2	6,000	95.6	145.8	252.9
S3	7,000	105.4	164.4	303.4
S4	8,000	113.3	175.0	320.3

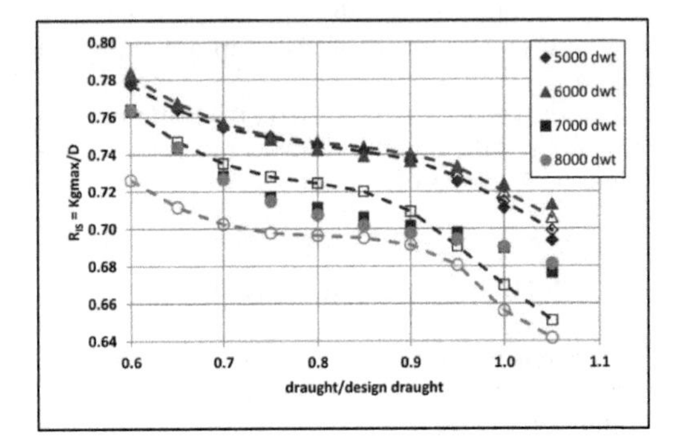

Figure 4. Relative intact stability vs. relative draught.

Figure 4 presents R_{IS} for all ships, Si_R and Si. The non-restricted ships, Si are presented by a dot curve.

3.3 *Cargo volume and storage*

The cargo volume is calculated based on several assumptions:

- the breadth of the double side of the midship section is constant for all ship equals to 1.70 m;
- the cargo space is located from the ER bulkhead to the most forward bulkhead that coincides with the forecastle wall;
- the total length of the cargo holds is 70% of the length between the perpendiculars of the ship;
- there is not double side at the forward and aft ends of the cargo space. The shape in this area needs especial design;
- the volume of hatches is not considered, which will compensate the fact that the double side at the ends is not accounted for.

The relative value of the cargo volume, RCV is estimated as:

$$R_{CV} = CV/LBD \quad (3)$$

where CV is the cargo volume, m^3.

It is envisaged to transport containers only on the hatch covers. The longest ships may carry containers on the poop deck too.

The relative container capacity, R_{CC}, $-/m^2$ is defined as:

$$R_{CC} = Nc/LB, -/m^2 \quad (4)$$

where Nc is the number of containers, TEU.

The estimation of the number of containers considers the shape of the deck and hatch cover at the forward end and the reduced number of tiers in the first or second bays. The cargo volume and container capacity are presented in Table 5.

4 PERFORMANCE ANALYSIS

The performance evaluation for all analysed ships is based on the results presented in Table 4.

The comparison of the analysed ships is made, for the same deadweight using the following criterion:

$$PI_{ij} = (V_R/V_{NR})_j \text{ if the performance } i \rightarrow \max \quad (5)$$

$$PI_{ij} = (V_{NR}/V_R)_j \text{ if the performance } i \rightarrow \min \quad (6)$$

where PI is the performance index, $i \in [1, 4]$ (resistance, intact stability, cargo volume, container capacity), $j \in [1, 4]$ (5,000, 6,000, 7,000 and 8,000 DWT), V_R is the relative performance of the ship with a restricted breadth, V_{NR} is the relative performance of the ship with a non-restricted breadth.

The maximum is looked for the intact stability, cargo volume and container capacity and the minimum is looked for the relative total resistance.

According to Eqn (5) and (6), the performance index, PI will be greater than 1.0 if the performance of the ship with a restricted breadth is superior to the non-restricted ship.

The relation of the relative total resistance R_{RT} of the non-restricted to the restricted ships, $(R_{RT})_{NR}/(R_{RT})_R$, for the analysed range of speeds is presented in Figure 5.

The non-parallel grow-up of the relative total resistance, as has been shown in Figure 5, can be explained by the existence of different block coefficients and slenderness of the ships. For the ships of 5,000 and 6,000 DWT, the C_B is relatively low and comparable with the ones of the non-restricted ships and are combined with the higher values of the slenderness as can be seen in Table 4.

The high slenderness leads to a reduction of the intensity of the generated by the ship waves, and the wave resistance. That is why the restricted ship poses a higher PI in the total resistance for a speed greater than 12 kn. In the contrary, the higher block coefficient, which is about 20% for

Table 4. Ships' performance.

	Restricted breadth				Non-restricted breadth			
Ship	$S1_R$	$S2_R$	$S3_R$	$S4_R$	S1	S2	S3	S4
DWT	5,000	6,000	7,000	8,000	5,000	6,000	7,000	8,000
L, m	88.63	106.60	120.62	135.06	86.79	93.21	96.71	99.98
B, m	16.00	16.00	16.00	16.00	16.17	16.87	18.60	19.23
d, m	7.08	6.88	6.67	6.57	7.18	7.33	7.98	8.53
D, m	8.81	8.93	9.03	9.17	8.90	9.32	10.06	10.81
Cb	0.690	0.721	0.772	0.812	0.685	0.676	0.670	0.669
Δ, t	7,090	8,665	10,181	11,809	7,068	7,981	9,853	11,241
L/B	5.539	6.663	7.539	8.441	5.367	5.525	5.199	5.199
B/d	2.260	2.326	2.399	2.435	2.252	2.302	2.331	2.254
L/D	10.060	11.937	13.358	14.728	9.752	10.001	9.613	9.249
Cm	0.974	0.974	0.994	0.994	0.974	0.974	0.994	0.994
Cw	0.844	0.867	0.861	0.894	0.841	0.831	0.778	0.778
Cp	0.708	0.740	0.776	0.816	0.703	0.694	0.674	0.672
LCB, m	43.832	52.741	64.313	72.016	42.933	46.112	51.554	53.295
LCB/Lpp	0.4946	0.4948	0.5332	0.5332	0.4947	0.4947	0.5331	0.5331
Slenderness, –	4.6515	5.2330	5.6114	5.9799	4.5597	4.7028	4.5484	4.5001
Resistance								
R_T(Vs = 10.5), kN	92.1	105.3	114.6	129.1	91.3	95.6	105.4	113.3
R_T(Vs = 12.5),kN	142.4	157.5	176.4	196.0	142.0	145.8	164.4	175.0
R_T(Vs = 15), kN	251.0	261.3	304.2	328.0	253.0	252.9	303.4	320.3
Δ, kN	69,558	85,001	99,872	115,847	69,341	78,290	96,656	110,270
$R_{RT}*10^3$, (10.5 kn)	1.3241	1.2388	1.1475	1.1144	1.3167	1.2211	1.0905	1.0275
$R_{RT}*10^3$, (12.5 kn)	2.0472	1.8529	1.7663	1.6919	2.0479	1.8623	1.7009	1.5870
$R_{RT}*10^3$, (15.0 kn)	3.6085	3.0741	3.0459	2.8313	3.6486	3.2303	3.1390	2.9047
Stability								
R_{IS}	0.7109	0.7145	0.7239	0.7201	0.6895	0.6698	0.6899	0.6561
Cargo capacity								
CV, m^3	5,828	6,996	8,235	9,440	5,616	6,668	8,568	9,966
R_{CV}	0.4665	0.4594	0.4726	0.4764	0.4496	0.4550	0.4735	0.4795
Nc	82	92	116	126	82	88	108	108
$R_{CC}*100$, $1/m^2$	5.7825	5.3940	6.0106	5.8307	5.8430	5.5964	6.0040	5.6173

Table 5. Cargo volume and container capacity.

Ship	CV, m^3	$R_{CV,-}$	Nc	$R_{CC}*100$, $-/m^2$
$S1_R$	5,829	0.4665	82	5.7825
$S2_R$	6,996	0.4594	92	5.3940
$S3_R$	8,235	0.4726	116	6.0106
$S4_R$	9,440	0.4764	126	5.8307
S1	5,616	0.4496	82	5.8430
S2	6,668	0.4550	88	5.5964
S3	8,568	0.4735	108	6.0040
S4	9,966	0.4795	108	5.6173

the 8,000 DWT ships, leads to a higher total resistance in the case of the 7,000 and 8,000 DWT ships, almost in the whole speed range.

The performance index for the relative intact stability, R_{IS} versus draughts in the range of 0.6 to 1.05 of the summer draught is shown in Figure 6.

For the draughts close to the design one, the both groups of ships with deadweight of 5,000 and 6,000 tons the PI is about 1.0. For ships of 7,000 and 8,000 DWT, the PI with respect to the intact stability is 3–5% better for the restricted ships, which can be explained with the higher values of B/d and Cw (see Table 4).

The performance indices for the four values of the deadweight are presented in Table 6 and Figure 7. Bold type of numbers (green colour) indicates the indices where the ship with a restricted breadth has a better performance index with respect to the ships without a restriction with respect to the breadth.

The last two rows of Table 6 include the performance index for the cargo capacity. The higher relative cargo volume of 5,000 and 6,000 DWT ships with a restricted breadth is due to the greater ship length. For 7,000 and 8,000 DWT ships the

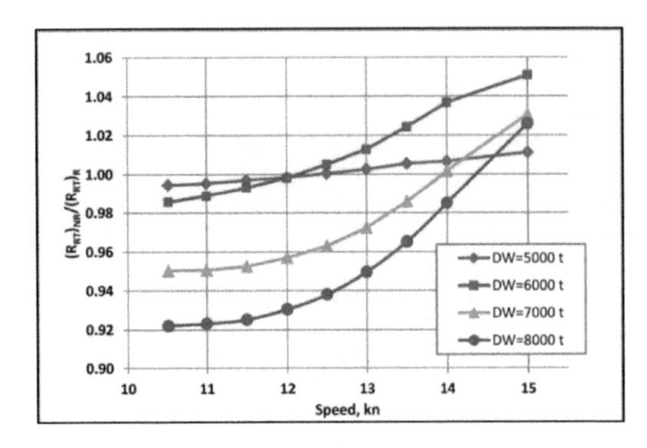

Figure 5. Relative total resistance.

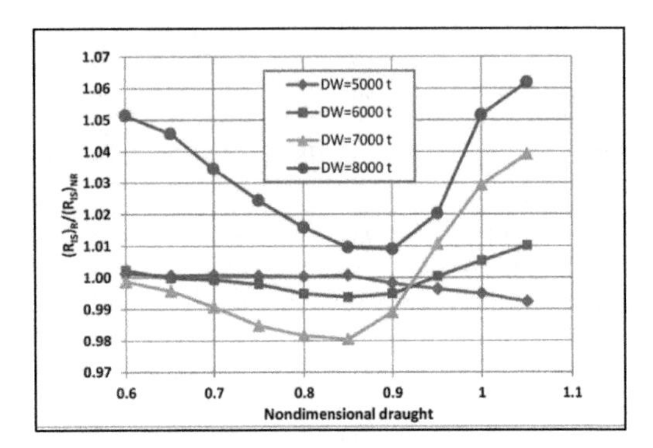

Figure 6. Relative intact stability R_{IS}.

Table 6. Performance index, PI.

	Deadweight, t			
Performance index	5,000	6,000	7,000	8,000
PI_{RRT} Vs = 10.5 kn	0.9944	0.9857	0.9503	0.9220
PI_{RRT} Vs = 12.5 kn	**1.0003**	**1.0051**	0.9630	0.9380
PI_{RRT} Vs = 15.0 kn	**1.0111**	**1.0508**	**1.0306**	**1.0259**
PI_{RIS}	0.9950	**1.0053**	**1.0293**	**1.0515**
PI_{RCV}	**1.0376**	**1.0096**	0.9980	0.9935
PI_{RCC}	0.9896	0.9638	**1.0011**	**1.0380**

breadth increases considerably and this leads to a higher cargo volume for the non-restricted ships.

The better performance index of the container capacity of the restricted ships with a 7,000 and 8,000 DWT is explained by the ship length. A bigger poop length permits stowage of one or two bays of containers on the poop deck.

One can see from Table 6 that the performance of the ships with a restricted breadth for some of the operational characteristics is better than the corresponding ship with the same deadweight, but without a breadth restriction.

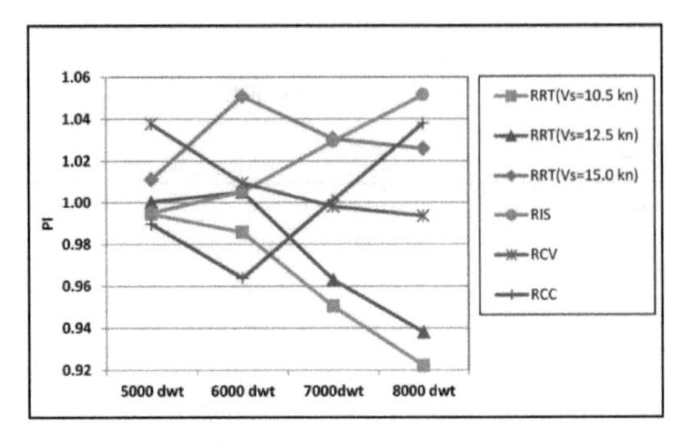

Figure 7. Performance index.

For all deadweights studied, the restricted ships have up to 5% lower relative total resistance at a speed of 15 kn. The all restricted ships, except the one of 5,000 DWT possess a belter intact stability. The cargo volume is relatively higher for smaller restricted ships of 5,000 and 6,000 DWT while the container capacity is better for the 7,000 and 8,000 DWT restricted ships.

5 CONCLUSIONS

The presented study analysed the impact of the specific SME shipyard limitations on the ship performance in building new ships with respect to the breadth of the ship.

The evaluation of the total ship resistance, intact stability and cargo capacity is estimated based on the transformation of two parent hulls. The comparison between the ships with the same deadweight with or without restrictions is based on the estimated performance index for the acceptable ship operational characteristics.

It was concluded that the designed ships with a constrained breadth, due to shipyard building limitations, doesn't lead to a considerable reduction of the ship performance, and in some cases the ship performance may be even better than the one of the ships with unrestricted breadth.

The analyses are based on 3D ship hull models of existing, already built ships. The conclusions may be influenced by other local parameters of the chosen parent hull like the U-V shape of the frames, location of LCB, LCF etc.

The obtained results may be confirmed when the conceptual design of ships is performed by minimizing the required freight rate, including the fast hull geometry prototyping as it was stipulated in the Shiplys Scenario 2 (Garbatov et al., 2017b).

ACKNOWLEDGEMENTS

This paper reports a work developed in the project "Ship Lifecycle Software Solutions", (SHIPLYS), which was partially financed by the European Union through the Contract No 690770 – SHIPLYS – H2020-MG-2014-2015.

REFERENCES

Atanasova, I., Damyanliev, T.P., Georgiev, P. & Garbatov, Y. 2018. Analysis of SME ship repair yard capacity in building new ships. *Proceedings of MARTECH*. London: Taylor & Francis Group.

Bharadwaj, U.R., Koch, T., Milat, A., Herrera, L., Randall, G., Volbeda, C., Garbatov, Y., Hirdaris, S., Tsouvalis, N., Carneros, A., Zhou, P. & Atanasova, I. 2017. Ship Lifecycle Software Solutions (SHIPLYS) — an overview of the project, its first phase of development and challenges. *Maritime Transportation and Harvesting of Sea Resources*. 889–897.

Damyanliev, T.P., Georgiev, P. & Garbatov, Y. 2017. Conceptual ship design framework for designing new commercial ships. *In:* Guedes Soares, C. & Garbatov, Y. (eds.) *Progress in the Analysis and Design of Marine Structures*. London: Taylor & Francis Group, 183–191.

Garbatov, Y., Ventura, M., Georgiev, P., Damyanliev, T.P. & Atanasova, I. 2017a. Investment cost estimate accounting for shipbuilding constraints. *In:* Guedes Soares, C. & Teixeira, A. (eds.) *Maritime Transportation and Harvesting of Sea Resources*. London: Taylor & Francis Group, 913–921.

Garbatov, Y., Ventura, M., Guedes Soares, C., Georgiev, P., Koch, T. & Atanasova, I. 2017b. Framework for conceptual ship design accounting for risk-based life cycle assessment. *In:* Guedes Soares, C. & Teixeira, A. (eds.) *Maritime Transportation and Harvesting of Sea Resources*. London: Taylor & Francis Group, 921–931.

Gustafsson, M., Nokelainen, T., Tsvetkova, A. & Wikstrom, K. 2016. Revolutionizing short sea shipping. *Positioning Report*. Åbo Akademi University.

Holtrop, J. & Mennen, G.G.J. 1982. An approximate power prediction. *International Shipbuilding Progress*, 29, 166–170.

Lackenby, H. 1950. On the Systematic Geometrical Variation of Ship Forms. *Transactions of INA*, 92, 289–315.

MSC.267(85) 2008. Adoption of the International Code on Intact Stability. London, UK: IMO.

UEAPME 2014. SMEs mean jobs and growth—Crafts and SMEs 2020. Brussels.

Ünal, G. & Derindere, S. 2014. Revealing the freight market risk in Istfix shipping area. *International Journal of Shipping and Transport Logistics*, 6, 593–610.

Computational fluid dynamics

Progress in Maritime Technology and Engineering – Guedes Soares & Santos (Eds)
 ISBN 978-1-138-58539-3

Wake of a catamaran navigating in restricted waters

G.T.P. McSullea, J.M. Rodrigues & C. Guedes Soares
Centre for Marine Technology and Ocean Engineering (CENTEC), Instituto Superior Técnico, Universidade de Lisboa, Lisbon, Portugal

ABSTRACT: High speed catamaran ferries are used throughout the world in confined waters, such as rivers and ports. Wake from these and other small vessels has a variety of effects on the waterways, including deterioration of banks, damage to structures and other vessels, and danger to people close to the shore. A CFD study of the effects of hull characteristics on the wave height and wave energy of the wake wash was undertaken. The hulls were varied using the design of experiments methodology so that the main effects of each hull factor could be analysed with a reduced number of tests. A secondary study was also undertaken focussing on the effect of vessel speed and water depth on wave height and wave energy. Design of experiments analysis highlighted that of the four factors (beam, demi hull beam, bow keel rake and bow entry angle) demi hull beam had the largest effect on the wave heights and wave energy. It was also shown that increasing the vessel speed resulted in increased wave heights and wavelengths. Furthermore, the reduction in water depth resulted in an increase in wave heights.

1 INTRODUCTION

Maritime mass transportation has shown to be a feasible commuting solution for cities having access to interior waterways, with catamaran designs providing a good speed-comfort-economy relation. However, there can be negative side-effects to this solution due to the effect of wake wash of these vessels on the, typically unprotected, shores. Issues related to the safety of people and property in coastal environments and the effect of wake wash on the coastal ecosystems need to be addressed. Predicting accurately the wake wash allows for designing better vessels and establishing operational limits aimed at reducing those negative side-effects.

The use of CFD codes to study wake wash has grown in the last decades. In many studies the aim is to predict the wave making resistance of the ship, which has led to studies on wake wash. Tarafder & Suzuki (2007) used a potential based panel method to study the effects of wave interference and hull separation on the wave making resistance of a Wigley hull catamaran in deep water. From this study, they noted that the magnitude of the wave profile on the inner side of the catamaran is much larger due to wave interference, and that at higher speeds and hull separations in excess of 40% of the hulls length interference effects were negligible. Echoing the results found by Tarafder & Suzuki, (2007) Nizam et al. (2013) found, in their paper on a numerical study of wave making resistance of a pentamaran in unbound water using a surface panel method, that at higher speeds ($Fr > 0.8$) interference between hulls was greatly reduced. He et al. (2015) also considered the wave interference effects on the far field wake characteristics of both monohull and catamarans, noting that the hull shape has a larger impact on the wave heights, while the kinematics of the ship has more influence on the shape of the wake field.

The main objectives of this study are to numerically predict the wakes generated by a set of catamarans as well as analysing the main and interacting effects of the *Beam* (B, the overall width of the ship), *Demi-Beam* (DB, the width of each hull), *Bow Keel Rake* (BKR, the angle the bow stem makes with the vertical axis) and *½ Entry Angle* (½ EA, the angle the bow makes with the longitudinal axis) on the wave height and wave energy of the wake wash. These are identified by completing a Design of Experiments analysis in conjunction with Computational Fluid Dynamics (CFD) studies.

The study is structured into four phases:

Phase 1: CFD model validation
CFD studies of Incat Crowther's latest 29.6 m hull in a deep-water channel as a catamaran are carried out. This phase is used to validate the CFD model, by comparing it with experimental results (Macfarlane, et al., 2012) (Macfarlane, 2012) of similar hulls.
Phase 2: Design main and interaction effects
Eight CFD simulations of catamarans in a deep-water channel are carried out. The hull

characteristics are varied according to a half factorial study of four factors mentioned above, each with two levels. The results of the CFD study are analysed using the DOE methodology and the main and interacting effects are determined. The hull models were provided by Incat Crowther (IncatCrowther, 2018).

Phase 3: Effect of advance speed

This phase considers the effect of varying hull speed on the wake wash of a specific hull. Vessel speed is a key factor, as it greatly affects the characteristics of wake wash of catamarans, such as propagation angle, wave height and wavelength. For this study a lower speed of 8 m/s and a higher speed of 15 m/s are considered.

Phase 4: Effect of water depth

The water depth is reduced to make the $Fr_h > 1$ resulting in a super-critical regime. Water depth affects both the form and characteristics of the wake such as wave height and wavelength. Simulating a hull moving at super critical regime speed allows comparison between the effects of water depth, vessel speed and hull characteristics.

2 BACKGROUND THEORY

2.1 *Wake wash*

Today, wake wash is well understood, and the form can be predicted with accuracy. In general, the wave pattern generated is more affected by the vessel's speed/depth ratio than the vessel's form. The *Length Froude Number* and *Depth Froude Number* are both useful dimensionless relations which can be used to predict the wave wash form.

Length Froude Number,

$$Fr_L = \frac{u}{\sqrt{gL}} \tag{1}$$

Depth Froude Number,

$$Fr_h = \frac{u}{\sqrt{gh}} \tag{2}$$

When the vessel speed corresponds to a $Fr_h < 0.75$, the speed is said to be sub-critical. As the Fr_h approaches one ($0.75 \leq Fr_h < 1.0$) the speed is said to be trans-critical and when $Fr_h = 1$it is called critical. A speed corresponding to $Fr_h > 1$ is termed super-critical.

For vessel speeds within the sub-critical region, all vessels produce the wake pattern called a Kelvin Wave Pattern. It consists of two wave systems: transverse and divergent. Transverse waves propagate parallel to the sailing line behind the ship; the height of the waves is heavily dependent on the vessel's length-displacement ratio. Divergent waves propagate at around 35 degrees from the sailing line, from the bow and stern of the ship. The divergent wave trains form a Kelvin Wedge with an angle of 19 degrees at each side of the sailing line. This angle, termed the wave angle, and the propagation angle (19 and 35 degrees, respectively) are dependent on Fr_h. Havelock (1908) showed that the wave angle is around 19 degrees until the Fr_h approaches 0.75, where it rises sharply, peaking at 90 degrees for $Fr_h = 1$. At this point the transverse waves merge with the divergent waves. Once the Fr_h becomes super critical the transverse waves disappear and the divergent waves take on a more acute angle (dependent on the ship's velocity).

Wave energy is a key factor that can be used to compare the wake wash of vessels, particularly with respect to it's potential to be destructive to coastal environments.

The wave energy per wavelength per unit width of wave crest (E_w) in deep water is given by USACERC (1977),

$$E_w = \frac{\rho g^2 H^2 T^2}{16\pi} \tag{3}$$

2.2 *Design of experiments methodology*

Successfully implemented Design of Experiments (DOE) methodology can yield large gains in terms of time and effort saved by reducing the number of experiments required to obtain results for multiple factors. Factorial studies relate to experiments where more than one factor may be important and Factorial Experimental Design (FED) is applied when each combination of factor levels must be tested (Montgomery & Runger, 2003). When there are multiple factors to be considered, engineers will usually use a fractional factorial study. A fractional factorial study is one where only some combinations of the factorial study are tested. The combinations are chosen specifically to allow the main and interacting effects to be separated and studied. This means that the engineer only needs half or even a quarter of the tests usually required for the full factorial study, while still obtaining an accurate perspective of which factors have the largest effects.

It is evident that this can be applied to the area of naval architecture where there are always multiple factors affecting any situation.

3 CFD MODEL VALIDATION

This preliminary study includes CFD simulations of Incat Crowther's latest 29.6 m catamaran. This

Table 1. Incat Crowther 29.6 m catamaran characteristics.

Characteristic	Value (s)
Length [m]	29.6
Beam [m]	8.8
Draft [m]	1.3
Demi Beam [m]	1.2
Vessel Speed [kt]	19

study is used for validation, benchmark and convergence of the numerical model. The hull has the characteristics shown in Table 1.

3.1 *CFD model setup*

The CFD simulations were made using the Reynolds-averaged Navier-Stokes equations (RANSE) solver STAR CCM+. The model includes the boundaries: inlet, outlet, side walls, top, bottom and hull surfaces. The side walls, top and bottom boundaries use the wall type boundary condition with the slip Shear Stress Specification. The hull also utilises the wall type condition but uses the non-slip Shear Stress Specification and a smooth wall surface specification. The inlet uses a velocity inlet type condition with a current of 10 m/s for both air and water in the phase 1study and 12.86 m/s in the phase 2 study. The outlet uses a pressure outlet type condition, utilising the hydrostatic pressure from the Volume of Fluid model. The Boundary conditions are summarised in.

The simulation uses the Volume of Fluid model with the FlatVOFWave option selected, creating a flat initial water surface with inlet and outlet boundary conditions as mentioned above. The simulations are solved using the Implicit Unsteady Solver, due to the Segregated Flow model also being used. Turbulence is accounted for using the Realizable K epsilon model, which is the improved version of the standard K epsilon model and includes a new transport equation for the turbulent dissipation rate epsilon (CD Adapco, 2006). The hull is fixed in all axis of rotation and translation, this means that trim and sinkage effects are not included in the simulation. Convergence is measured by plots of the x and y forces and the z moments over time. Measurements are taken in the form of surface elevations and longitudinal wave cuts. The domain size is chosen based on multiples of the ships length. The domain extends from four ship lengths aft of the ship, to 1.5 ship lengths forward of the bow. The domain width extends 1.5 ship lengths to each side of the sailing line, the water depth is 2 ship lengths below the water surface and the top of the domain is 0.5 ship lengths above.

The mesh used was a uniform prism with custom surface mesh sizes at the boundaries and two volumetric controls at the water surface. The hulls used prism layers and near wall modelling method, Two Layer all y+ Wall Treatment. The grid size at the domain boundary surfaces is a 5 m by 5 m square, while the smallest grid size on the surface of the hull was 0.05 m. The mesh growth rate was set to slow, which yielded a growth pattern shown in Figure 1 below. The near wall prism layer thickness was set to 0.01 m and the prism layer thickness was set to 0.04 m, where 3 prism layers are applied.

A convergence test of the mesh was carried out in this phase. Three simulations were run with different numbers of cells: 10 million, 4 million cells and 2.8 million cells. Analysis of these simulations revealed that a 4 million cells mesh provided results very close to those of 10 million cells, suggesting that 4 million cell simulations is sufficiently accurate to use.

3.2 *CFD validation*

The verification and validation of results are important parts of any study based on numerical methods. In this study, experimental testing was outside the scope of work due to the extent of time and resources it would require. Therefore, experimental results were sourced from other projects to provide a validation reference. The validation of the CFD results was completed using experimental data for

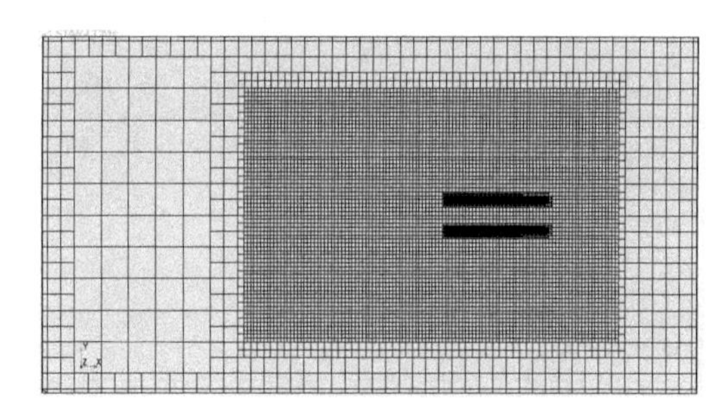

Figure 1. CFD grid growth pattern.

Table 2. CFD boundary condition.

Boundary	Type	Specifications
Side Walls, Top & Bottom	Wall	Slip
Hulls	Wall	Non-Slip
Inlet	Velocity Inlet	10 m/s – Phase 1 12.86 m/s – Phase 2
Outlet	Pressure Outlet	Hydrostatic Pressure of VOF

shallow water gathered from Macfarlane (2012) and data extrapolated from experimental results for deep water from Macfarlane et al., (2014). These experimental results were gathered at the test basins in the Australian Maritime College (AMC) in Tasmania, Australia. The difference between the two data sets is that the experimental results from Macfarlane (2012) are for a 24 m catamaran in shallow water (h/L = 0.55), where h is the water depth and L is the ship length, while the results from Macfarlane et al., were extrapolated from experimental data including the same 24 m catamaran and a 30 m catamaran in deep water, using the Wave Predictor tool developed by Macfarlane et al. (2014).

It was impossible to replicate the experiments exactly in CFD, because the time required to model the new hulls was not available. However, since the hull shapes tested by Macfarlane (2012) and Macfarlane et al. (2014) are of a similar shape and have similar characteristics, as shown in Table 3, it was decided that their results are appropriate for a high-level validation of the CFD model. Table 3 compares the main characteristics of the hulls used for CFD validation, this includes water depth (h), ship length (L), vessel speed (V), ship displacement (Δ), depth/speed ration (Fr_h), length/speed ratio (Fr_L) and depth/length ratio (h/L). Comparison of the three sets of results, shown in Table 4, reveals

Table 3. CFD validation hull characteristics.

HULLS	h (m)	L (m)	V (kt)	Δ (t)	Fr_h	Fr_L	h/L
Deep water data	60	29.6	19	31	0.4	0.57	2.03
Hull used in CFD simulation	60	29.6	19	28	0.4	0.57	2.03
Shallow water data	12	24	19	55	0.92	0.65	0.55

Table 4. CFD validation results.

Results	H_A [mm]	T_A [s]	H_B [mm]	T_B [s]
Wave Predictor 1	171	5	289	4.5
Hull used in CFD Simulation	175	4.5	275	3.9
Exp. Results of 24 m cat at 10 m/s	250–450*	5.75–6.5*	550–850*	3.5–4.75*

*Range of experimental results with error bars included.

that the CFD results are within a similar range with only a small margin of error.

To compare the wave heights and wave periods of the experimental results and the CFD results, wave-cuts were made using STAR CCM+ at the lateral distance from the sailing line of 41.4 m (corresponding to the y/L ratio of 1.38, where y is the lateral distance, used in the experimental results of Macfarlane (2012) and Macfarlane et al. (2014)). These wave heights could then be compared with the wave heights and periods (Macfarlane, 2012) and (Macfarlane, et al., 2014) for the corresponding Froude length number of 0.75.

The leading waves of a wake are described by Macfarlane (2012) and Robbins (2013) as two separate but equally important waves. The first wave (termed wave A) is characterized by a large period and medium to high wave height. Adding these two characteristics together means that in many instances this first wave can have the largest total wave energy. The second wave, named wave B, is characterized by the highest wave height of the wave group following wave A and usually has a larger wave height but a lower period than wave A. In many situations this is the wave with the highest wave energy. Therefore, these waves are important to analyze, as they stand out because of their potential for damage in coastal areas due to their high wave energies.

The results of the CFD validation are shown in Table 4, where H_A, T_A, H_B, T_B are the wave heights and periods of waves A and B[1], respectively. These values are calculated from wave-cuts taken 45 m laterally from the sailing line. These wave-cuts show wave heights and wavelengths, from the wavelength the wave period can be calculated by dividing it by the wave speed. Wave A corresponds to the first wave in the wake wash and wave B corresponds to the highest wave in the following wave group after wave A, shown in Figure 2.

It can be seen in the surface plot, Figure 3, that the wake pattern shares the general form of the standard Kelvin wake pattern for sub critical flow, as expected.

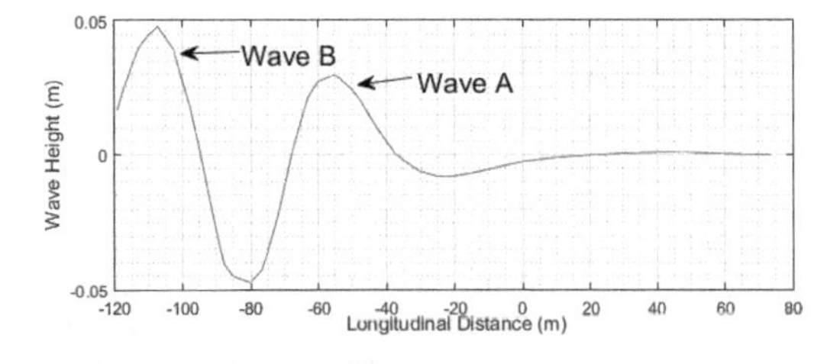

Figure 2. Definition of Wave A & B.

[1]Wave A and B are defined in detail in the following section.

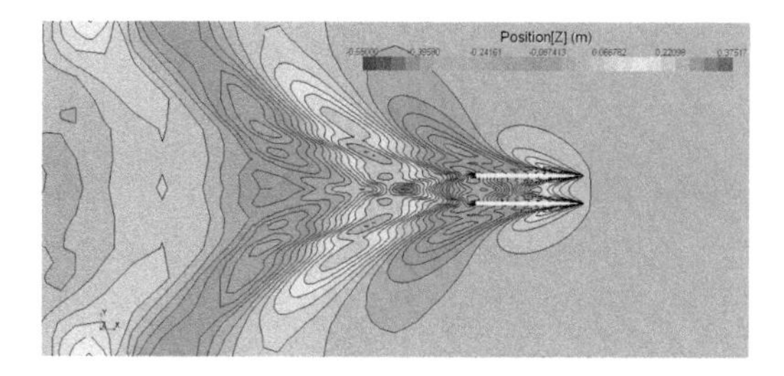

Figure 3. Preliminary Study CFD Simulation Results.

Furthermore Figure 3 shows the wake has multiple waves following in quick succession, this is because each hull has two main wave trains: one that propagates from the bow and one that propagates from the stern.

4 DESIGN MAIN AND INTERACTION EFFECTS

The main study focuses on the 8 combinations of hull characteristics created using the DOE methodology. Eight different combinations of hull characteristics (B, DB, BKR & ½ EA) of the "Incat Crowther 29.6 m" reference design were simulated to assess their main and interacting effects.

4.1 *Simulation setup*

The simulations for the main study use a similar range of hull characteristics to the preliminary study so it was considered acceptable to use the same model, since the model was shown to be accurate and validated. The range of values used in the main study for the CFD simulations are summarized in Table 5. Measurements are taken at the lateral line 45 m from the sailing line, inside the Near Field Region, and at the bow. The Near Field Region is defined as the space within one ship length each side of the sailing line and from the stern to two ship lengths aft.

4.2 *Surface plots*

The surface elevation plots highlight how the variation in hull characteristics affect the wake wash. Although the kelvin wave pattern is similar for each pattern due to the sub critical depth Froude regime, the wave heights and wave periods vary. This is due to the varying hull characteristics, shown in Table 6.

4.2.1 *Hull 1*

This catamaran shares the smallest B and DB with Hull 4, which is combined with a smaller ½ EA, making each hull extremely thin. The surface plot shown in Figure 4 displays a well-defined wake wash of the sub-critical Kelvin form, including interaction between the bow and stern waves from each hull. In contrast to the preliminary hull, the largest wave heights are located at the bow. The largest troughs are located just aft of the ship's stern where the inner stern waves interact.

Table 5. Main study hull characteristics.

Characteristic	Value (s)
Length [m]	30
B [m]	8 & 10
Draft [m]	1.3
DB [m]	1.5 & 2.5
½ EA [deg]	14 & 24
BKR [deg]	0 & 50
Vessel Speed [kt]	25

Table 6. Main study hull characteristics.

Hull	B [m]	DB [m]	1/2 E A [°]	BKR [°]
1	8	1.5	14	0
2	10	1.5	24	0
3	10	1.5	14	50
4	8	1.5	24	50
5	10	2.5	14	0
6	8	2.5	24	0
7	8	2.5	14	50
8	10	2.5	24	50

4.2.2 *Hull 2*

The hull has a small DB but a larger B, which means that the separation between the hulls is larger. Comparing the surface plot of Hull 1 with that of Hull 2 the more complex and turbulent wake of Hull is emphasized. In Figure 4, showing the surface plot of Hull 1, three defined wave peaks can be identified in the wake, whilst in Figure 5, for Hull 2, there are only two wave peaks. Hull 2 also shows milder wave heights and a cleaner and calmer wake wash. Again, the maximum surface elevation is at the bow and the minimum elevation is located at an interaction point between the inner bow/stern waves.

4.2.3 *Hull 3*

Hull 3 has a wide beam and thin hulls, lower entry angle and demi beam. The wake has many similarities to the wake of Hull 2. Effectively the only difference between the two catamarans is in the bow: the bow entry angle of Hull 2 is larger than that of Hull 3 and the bow keel rake is vertical for Hull 2 and angled at 50 degrees for Hull 3. The difference in the maximum surface elevation shows the effect

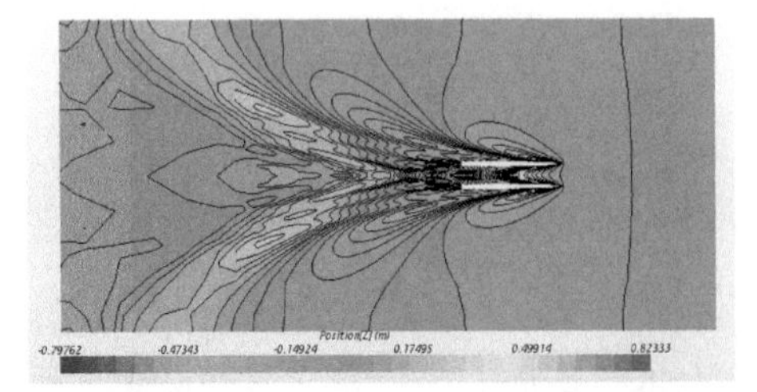

Figure 4. Hull 1 V = 12.86 m/s.

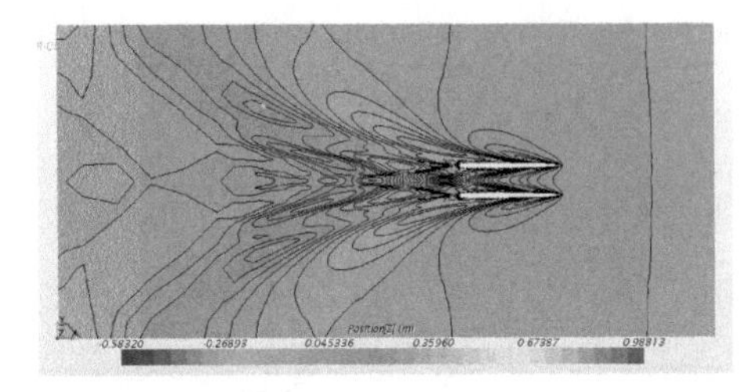

Figure 5. Hull 2 V = 12.86 m/s.

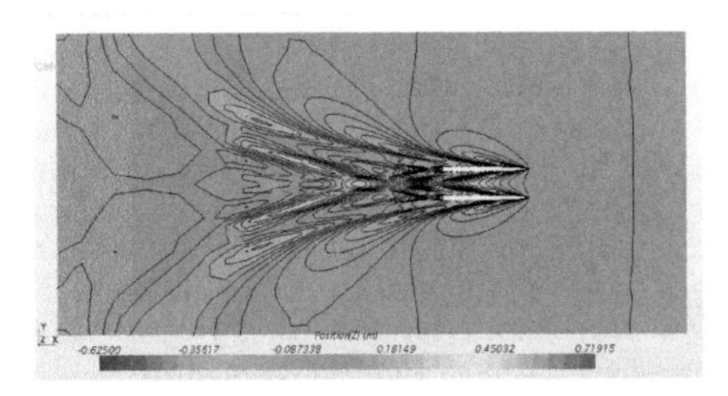

Figure 6. Hull 3 V = 12.86 m/s.

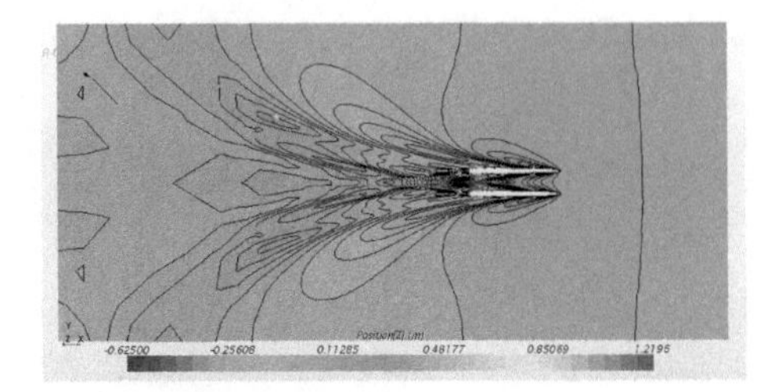

Figure 7. Hull 4 V =12.86 m/s.

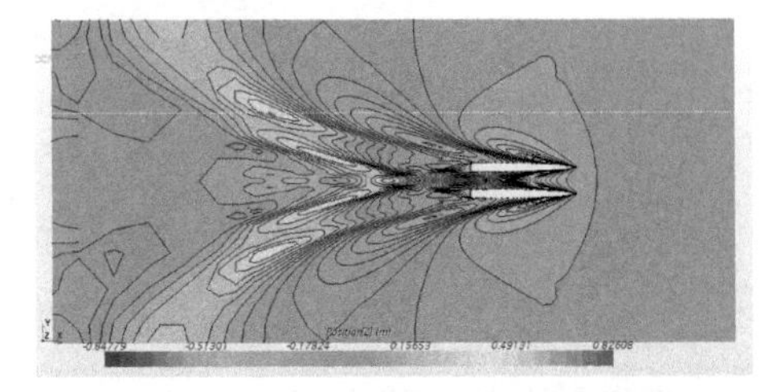

Figure 8. Hull 5 V = 12.86 m/s.

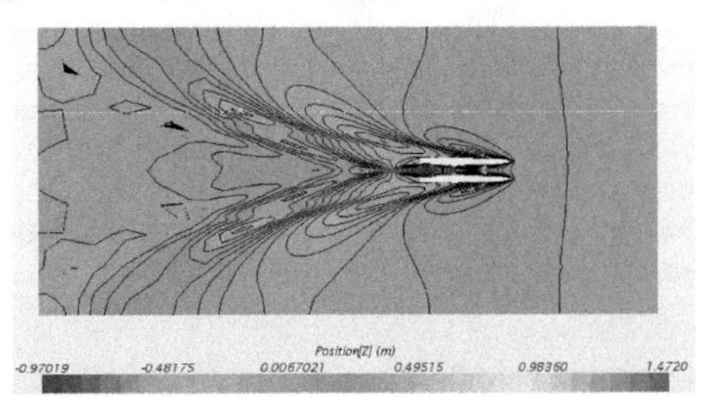

Figure 9. Hull 6 V = 12.86 m/s.

of these minor differences in the bows. It can also be seen that the wave heights for Hull 2 are higher than Hull 3 and so the effect of the bow is carried through the wave trains.

4.2.4 *Hull 4*

Hull 4 shares similar characteristics to Hull 1, equal beam and demi beam and different bow characteristics. Comparing the two surface plots shown in Figure 4 and 7 highlights the effect of the combination of a wider entry angle and a larger bow rake verses a thinner and vertical bow. In the surface plot of Hull 4, the maximum surface elevation is over 1.2 m, 50% larger than that of Hull 1, while the wave heights in the wake are also slightly higher. In contrast, the maximum trough depression is lower for Hull 4 than for Hull 1, suggesting the bow characteristics effect the interactions in some way.

4.2.5 *Hull 5*

Hull 5 is the first of the hulls tested with a wider demi beam. It is evident that this characteristic is important because it has a large effect on the wave heights of the wake wash. The wake wash pattern shows wave peaks in the wake between 1–2 boat lengths aft of the ship. Also, the trough caused by the interaction of the inner bow and stern wakes is larger than any of the catamarans with a 1.5 m Demi Beam.

4.2.6 *Hull 6*

Hull 6 shares the smallest separation distance with Hull 7, due to the smaller Beam and larger Demi Beam values. This is coupled with a larger entry angle, which causes very large bow waves and large interactions at the stern. The wake pattern is also unique as the two waves following in close succession in the 1–2 ship lengths aft region have almost joined, forming very long crested waves. This is a dangerous possibility to be avoided, as the merging of these waves would increase the wave height substantially.

4.2.7 *Hull 7*

Hull 7 shows a very interesting wake pattern, characterized by very steep waves in the near field,

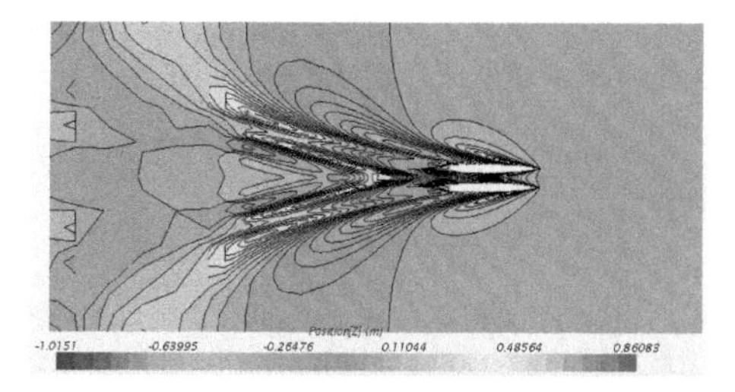

Figure 10. Hull 7 V = 12.86 m/s.

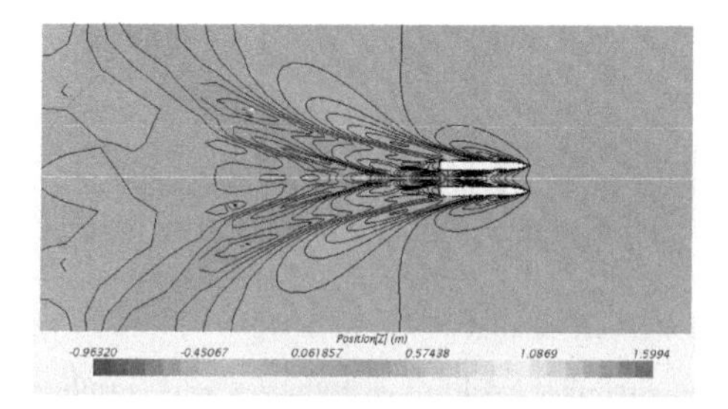

Figure 11. Hull 8 V = 12.86 m/s.

which then show the effects of attenuation in the mid-field region closer to the domain boundaries. The steepness of the waves and the narrow initial propagation angle could be a symptom of the bow entry angle changes. It is most likely more connected with this factor than with the bow keel rake. The surface plot also shows large interactions at the aft of the ship causing a deep trough at the stern. This could also be linked to the bow characteristics which may be affecting how and where the waves interact. Lastly, it is clear that the bow characteristics again have affected the maximum surface elevation at the bow, this time a thinner raked bow has created lower elevations than Hull 6.

4.2.8 *Hull 8*

The surface plot for Hull 8 shown in Figure 11 is again very interesting. The waves in the wake wash are less steep, but also extremely high. This combination of factors also produces the largest surface elevations at the bow. The stern trough caused by the interactions of the inner bow and stern waves is larger.

4.3 *Analysis of main and interacting effects*

The results taken from these measurements are shown in Table 7. The wave heights are measured in meters and are measured from the height of the wave peak to the bottom of the next trough, except for at the bow where the height is just the surface elevation from the undisturbed free surface. The different combinations of hull characteristics have

Table 7. Main study CFD results.

	Measurements taken at 45 m			Near field Region	At Bow
Hull	Max wave height [m]	Period [s]	Energy [kJ/m/s]	Max wave height [m]	Max wave height [m]
1	0.350	5.8	0.981	0.781	0.823
2	0.248	6.6	0.515	0.469	0.988
3	0.207	5.8	0.358	0.469	0.719
4	0.219	6.2	0.402	0.547	1.220
5	0.396	5.8	1.287	0.781	0.826
6	0.409	6	1.373	0.938	1.472
7	0.350	5.6	1.030	1.094	0.861
8	0.374	5.8	1.173	0.938	1.599
AVG	0.319	6.0	0.890	0.752	1.064

a significant impact on the wave heights in the near-field region and at the bow, however these effects become less apparent at the 45 m lateral line. This is due to wave height attenuation and at a distance far from the ship (in deep water) it would not matter which hull was chosen, as the waves heights would all be extremely similar. However, in congested waterways or where the bathymetry is varied, this type of data is very important for understanding what effects each characteristic has on the wave heights and wave energy.

For simplicity, each factor is given a letter and the high and low levels are represented by a + and– sign, respectively. The factorial has a specific order such that every combination of factors A, B and C are tested. The value D is calculated by taking the product of the high/low values of A, B and C. To calculate the effect (E) of a factor, the product of the corresponding high/low value $((+/-)_i)$ and result (H_i) for each hull is summed and divided by the number of Factors (4), as shown equation (4).

$$E = \frac{1}{4}\sum_{i=1}^{8}\left(\frac{+}{-}\right)_i H_i \quad (4)$$

The results of this analysis are shown in Table 8 and Table 9. To make the table easier to understand, the absolute values (ABS) were normalized by the average value (AVG) for each output shown in the last row of Table 7, the normalized values are indicated by a %. In Table 8 and Table 9, the larger the value in absolute terms, the larger the effect that factor had on the output. A positive value means that the trend is increasing, that is, an increase in the value of the hull characteristic will increase the output. Inversely, a negative value suggests that the opposite is true. The results highlight that of the 4 factors, *Demi Hull Beam* is the key

Table 8. Main study main effects.

		½ EA [°]	BKR [°]	DB [m]	B [m]
DOE analysis		A	B	C	D
Hw 45 m	ABS	–0.014	–0.063	0.126	–0.026
Hw NFR	ABS	–0.059	0.020	0.371	–0.176
Hw Bow	ABS	0.513	0.072	0.252	–0.061
Ew 45 m	ABS	–0.010	–0.242	0.429	–0.080
Hw 45 m	%	–4%	–20%	40%	–8%
Hw NFR	%	–8%	3%	49%	–23%
Hw Bow	%	48%	7%	24%	–6%
Ew 45 m	%	–2%	–39%	69%	–13%

Table 9. Main study interacting effects.

		2nd Order Interactions		
DOE analysis		AB=CD	AC=BD	AD=BC
Hw 45 m	ABS	0.031	0.032	0.023
Hw NFR	ABS	0.020	0.059	0.137
Hw Bow	ABS	0.107	0.180	0.009
Ew 45 m	ABS	0.094	0.116	0.038
Hw 45 m	%	10%	10%	7%
Hw NFR	%	3%	8%	18%
Hw Bow	%	10%	17%	1%
Ew 45 m	%	15%	19%	6%

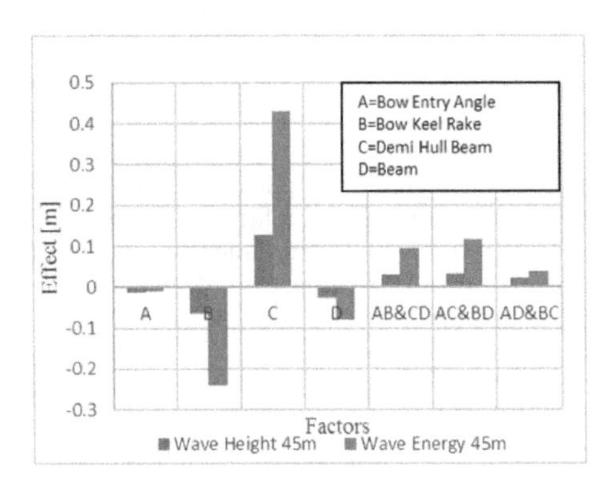

Figure 12. DOE analysis of results.

factor in terms of reducing the wave heights that propagate from the catamaran. This is most likely because this factor strongly affects the displacement of each hull, which according to Macfarlane (2012), is one of the crucial factors which affects wake wash. For the waves propagating into the far field, the *Bow Keel Rake* is also important. This is most likely because in this study this factor affects the waterline length, which has been linked to wake wash wave heights by Fox et al. (1993). In the Near Field Region, the Beam also has a significant effect and Bow Wave is most affected by the *Bow Entry Angle* and *Demi Hull Beam*.

5 EFFECT OF ADVANCE SPEED

This section discusses the effects of vessel speed on the wake wash of a catamaran hull. The study was carried out using Hull 1, which has all factors at their lower value. Using this hull allows a comparison of three different vessel speeds. This study includes a low level of 8 m/s, a medium level of 12.86 m/s and a high level of 15 m/s. Vessel speed affects the wake wash characteristics in several ways. The propagation angles change, as does the wave heights and wavelengths. The CFD model used for the main study was maintained for these tests.

5.1 *Study results*

The surface plot for the highest speed simulation is shown in the Figure 13. The wake wash retains the same shape and similar wave heights to the 12.86 m/s simulation, however the wavelength increased. Since wavelength, in deep water, is proportional to vessel speed this was expected. Due to the increase in wavelength, the wave energy will also increase (since the wave heights stayed practically the same as those of the mid speed simulation).

The surface plot for the low speed simulation is shown in Figure 14. Although the wake wash pattern

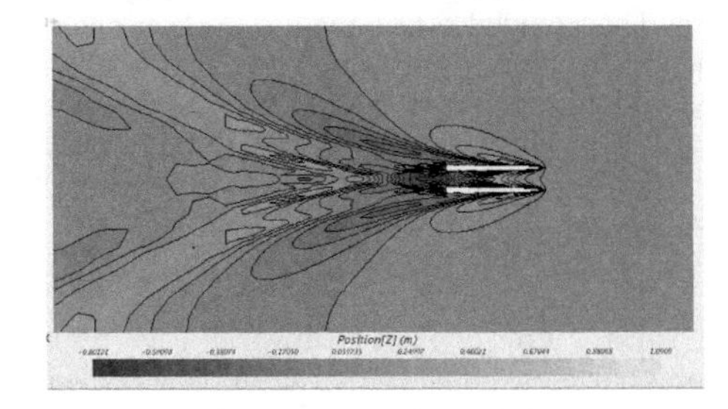

Figure 13. Surface elevation Hull 1, V = 15 m/s.

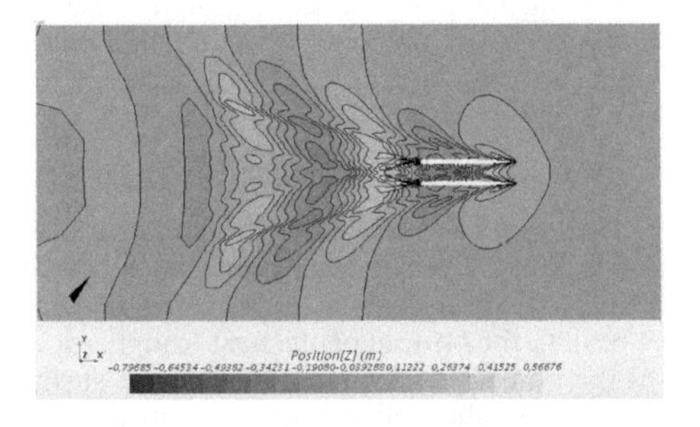

Figure 14. Surface elevation Hull 1, V = 8 m/s.

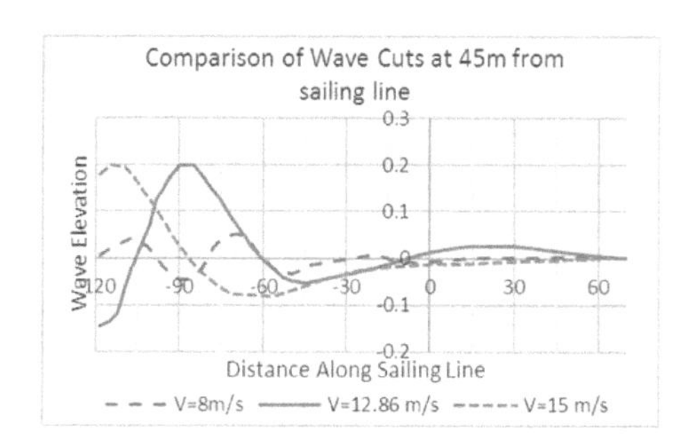

Figure 15. Comparison of wave heights for various vessel speeds.

is maintained, the wave heights and wavelengths are reduced. This highlights the effect of vessel speed on the wake wash of the catamaran hull. It also shows that in situations where wave height and wave energy is a critical factor, reducing the vessel speed may be the most effective response.

A comparison of the wave heights along a wave cut at 45 m from the sailing line are shown in Figure 15. It is highlighted in this graph how the wave heights and wave lengths are affected by vessel speed. The wave heights and wave lengths at a lower speed are greatly reduced and, as the vessel speed increases, both wave height and wave length increase. At the highest vessel speed, the wave height remains the same however the wave length increases again, which leads to a higher wave energy.

Although the wave characteristics will undoubtedly change as the waves propagate into shallow water, the higher wave energy will cause increased risk of damage to the coastal environment. Furthermore, in congested waterways the increase in wave height can cause more dangerous situations for other vessels and can put increased loads on moorings and jetties, so reducing these two factors is of the upmost importance.

6 EFFECT OF WATER DEPTH

This section discusses the effect of water depth on the wake of a catamaran travelling in a channel. This section uses the simulation for Hull 1 made in Phase 2 of the study and compares it to a new simulation made with a water depth of 10 m. This water depth corresponds to a $Fr_h \cong 1.3$ (super critical regime), as mentioned previously, the relationship between vessel speed and water depth has a large impact on the form and characteristics of the wake. Using this study, it is possible to compare the effects of water depth, vessel speed and hull characteristics on the wake of a catamaran.

The simulation is made using the model of Hull 1. The domain size is reduced so that the water depth is only 10 m. Reducing the water depth, will increase the turbulence of the water flowing around the hull. To accommodate this, the number of cells is increased to approximately 9 million. This change included an increase (of one) in the number of prism layers around the hull as well as reduction in the mesh size on the bottom exterior boundary. Due to time constraints a new mesh convergence study was not possible.

6.1 *Study results*

The surface plot for Hull 1 with h = 10 m, shown in Figure 16, highlights the effect of shallow water on the wake wash form and characteristics. Although the wake form is similar to the wake of the same hull in deep water (Hull 1 from Phase 2 study), there are some differences. As expected, there is an absence of transverse waves, this is highlighted by the variations in surface elevation laterally within the wake. The wave heights are also much larger than those from the phase 2 study.

Comparing wave cuts made 45 m laterally from the sailing line of Hull 1 in shallow and deep water, shown in Figure 17, this change in wave heights is evident. Although the wave lengths are similar, the phase of the waves is different, this is most likely due to a change in wave angle. For deep water (Fr_h < 0.75) the wave angle is approximately 19 degrees,

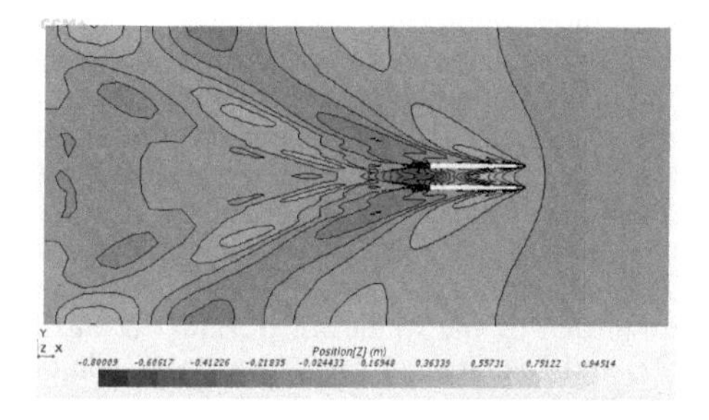

Figure 16. Surface elevations for Hull 1, V = 12.86 m/s, h = 10 m.

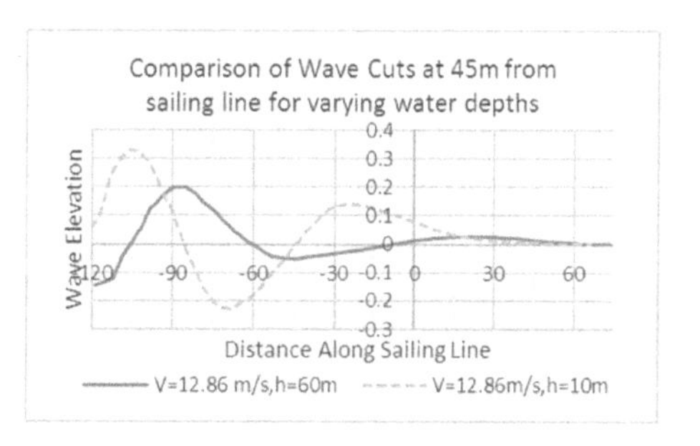

Figure 17. Comparison of wave cuts for Hull 1 in shallow and deep water.

while for shallow water ($Fr_h > 1$) the angle is dependent on the vessel speed (Havelock, 1908).

7 DISCUSSIONS AND CONCLUSIONS

This paper utilized a commercial CFD software to study the effect of hull characteristics and vessel speed on the wake wash of catamarans. These variables greatly affect the wake wash of ships and from this study it was shown that a combination of factors can be used to greatly reduce the wake wash effects. By improving the hull design to reduce the negative wake wash characteristics, catamarans can travel at greater speeds without creating as large or dangerous wake wash.

This research studied the effect of four different hull characteristics on the wake wash of a catamaran. These four characteristics were: *Beam*, *Demi Beam*, *Bow Entry Angle* and *Bow Keel Rake*. The results showed that, of these four characteristics, Demi Beam was the most influential factor in affecting maximum wave height and wave energy. The results also showed that the Bow Keel Rake was influential. These results agree well with conclusions by Macfarlane et al. (2014), who suggest that the *Demi Beam* and the *Waterline Length* greatly affect the maximum wave height and wave energy of a ship's wake wash. Analysis of the interacting effects did not show that any particular interaction was of more significant importance. This could be due to the small number factor levels studied or the range. Although it makes sense that certain combinations of factors will inevitably create large constructive or deconstructive wave interactions, it is also most likely that the main effects of the hull characteristics will have the largest effect on wave height and wave energy overall. This theory is supported by the studies of Tarafder & Suzuki (2007) and Nizam et al. (2013) who conclude that for higher Froude numbers and hull spacings the interference effects are less distinguishable.

For assessing the effect of vessel speed, two extra simulations for a higher and lower vessel speed were carried out. These speeds were 8 m/s and 15 m/s. The results showed that increasing the vessel speed (from 12.86 m/s to 15 m/s) resulted in an increase in wave length and a reduction in the vessel speed resulted in a reduced wave height and wave length. These conclusions are important when considering congested waterways or sensitive ecosystems where lower wave heights and wave energy are required.

To assess the effects of water depth on the wake wash a simulation was made with a reduced water depth of 10 m. This resulted in a $Fr_h \cong 1.3$, making the vessel speed regime super critical. This resulted in an increase in wave heights and a disappearance of transverse waves. This highlights the importance of water depth when addressing the safety concerns around wake wash.

ACKNOWLEDGEMENTS

The authors would like to thank Incat Crowther for their technical support in the form of models of catamarans they provided, to be use in the CFD simulations.

REFERENCES

CD Adapco, 2006. *Star CCM+ User Guide.* [Online] Available at: http://www.cd-adapco.com/products/star-ccm/documentation. [Accessed 1 November 2017].

Fox, K., Gornstein, R. & Stumbo, S., 1993. Wake Wash Issues and Answers, Pacific Northwest Section: *Society of Naval Architects and Marine Engineers.*

Havelock, T., 1908. The Propagation of Groups of Waves in Dispersive Media, with application to Waves produced by a Travelling Disturbance. London, *Proceedings of the Royal Society*, pp. 398–430.

He, J. et al., 2015. Interference effects on the Kelvin wake of a catamaran represented via a hull-surface distribution of sources. *European Journal of Mechanics B/Fluids,* 56(1), pp. 1–12.

IncatCrowther, 2018. *IncatCrowther.* [Online] Available at: http://www.incatcrowther.com. [Accessed 19 Febuary 2018].

Macfarlane, G., 2012. *Marine Vessel Wave Wake: Focus on Vessel Operations within Sheltered Waterways (Doctors dissertation).* Hobart, Tasmania: Australian Maritime College Press.

Macfarlane, G., Bose, N. & Duffy, J., 2012. *Wave Wake: Focus on vessel operations within sheltered waterways,* s.l.: SNAME.

Macfarlane, G., Bose, N. & Duffy, J., 2014. Wave Wake: Focus on vessel operations within sheltered waterways. *Journal of Ship Production and Design,* Volume 30, pp. 109–125.

Montgomery, D. & Runger, G., 2003. *Applied Statistics and Probability for Engineers.* 3rd ed. Arizona: John Wiley and Sons.

Nizam, M., Ali, M. & Tarafder, M., 2013. Numerical prediction of wave-making resistance of pentamaran in unbounded water using a surface panel method. *Procedia Engineering,* Volume 56, pp. 287–296.

Robbins, A., 2013. *Shallow Water Catamaran Wash - Simple Characteristics for a Complex Phenomenon (Doctors dissertation),* Hobart, Tasmania: Australian Maritime College Press.

Tarafder, M. & Suzuki, K., 2007. Computation of wave-making resistance of a catamaran in deep water using a potential-based panel method. *Ocean Engineering,* Volume 34, pp. 1892–1900.

USACERC, 1977. *Shore Protection Manual.* Fort Belvoir, Virginia: U.S. Army Coastal Engineering Research Center.

Progress in Maritime Technology and Engineering – Guedes Soares & Santos (Eds)
 ISBN 978-1-138-58539-3

A CFD study of a ship moving with constant drift angle in calm water and waves

H. Islam & C. Guedes Soares
Centre for Marine Technology and Ocean Engineering (CENTEC), Instituto Superior Técnico, Universidade de Lisboa, Lisbon, Portugal

ABSTRACT: Prediction of forces and moments on hull form is essential to determine ship's stability and maneuvering capabilities during voyage. The presented study provides a comparison among the non-dimensional lateral forces and yaw moments encountered by a ship in calm water and in waves. A very large crude carrier model, KVLCC2, is simulated with static drift motion using OpenFOAM, both in calm water and in waves. The calm water simulation results are compared with experimental data and linear hydrodynamic derivatives are derived from the results and are compared. Next, static drift simulations were performed in waves, with three different wave lengths, and the lateral force and yaw moment results were compared with calm water simulation results.

1 INTRODUCTION

The common practice for determining ship's stability and maneuverability characteristics is to determine its hydrodynamic derivatives using forced motion experiments or simulations. Maneuverability studies are mostly performed in calm water with PMM or CMT tests. However, such tests are quite expensive, since they require good test facilities and proper ship model. Furthermore, ships in their regular voyage mostly maneuver in water with waves. Thus, the hydrodynamic derivatives derived from calm water studies might not properly represent the ship's maneuvering behaviors at open sea.

Although experimental studies are most popular for determining ship's maneuvering capabilities, with recent developments in computing resources and CFD (Computational Fluid Dynamics) tools, such studies through CFD have become very efficient and economic. Thus, shipbuilding industry and researchers are slowly moving to CFD tools for ship maneuvering tests. However, the test practices remain the same as established in 1960s, captive maneuvering test in calm water. Whereas, in practice, ships mostly maneuver at seas with waves and currents, not in calm water. Thus, present practices might be questionable.

Studies related to application of CFD in ship maneuverability prediction has been relatively recent. Most of the early works related to maneuvering were focused on planar motion mechanism (PMM) simulations. PMM simulations were first widely discussed in SIMMAN 2008 workshop (2008), where different research groups presented static drift, pure sway and pure yaw simulation results. Broglia et al. (2008) showed pure sway and pure yaw motion results for KVLCC1 and 2 models with propeller and rudder, simulated using a solver developed by INSEAN. Hochbaum et al. (2008) simulated static drift, pure sway and pure yaw case for the two tanker models with propeller and rudder, using a self-developed code. Ghullmineau et al. (2008) provided PMM results for US navy frigate using ISIS-CFD solver. Miller (2008) provided PMM calculation for DTMB 5415 using CFDShip-Iowa. Wang et al. (2011) simulated oblique motion for KVLCC2 in deep and shallow water using commercial code FLUENT. Simonsen et al. (2012) presented zig-zag, turning circle and PMM results for an appended KCS model using STAR-CCM+ and compared with experimental data. Lee & Kim (2015) performed PMM simulation for a wind turbine installation vessel using OpenFOAM, ignoring free surface calculation. Later, Kim et. al. (2015) presented PMM simulation results for KCS model using in-house code SHIP_Motion and predicted hydrodynamic derivatives from simulation results. Hajivand & Mousavizadegan (2015, 2015a) also performed PMM simulation using OpenFOAM and STAR-CCM+ for DTMB 5512 model and predicted hydrodynamic derivatives from the simulation results.

Recently, a lot of the work related to maneuvering in waves was performed under the SHOPERA (Papanikolaou et al., 2015) project. Uharek & Hochbaum (2015) developed a mathematical model for approximating mean forces and moments due to waves based on a double parametric approach (encounter angle and wave length) using hydrodynamic coefficients obtained from RANS computations. Sprenger et al. (2016) presented experimental data from four European leading research institutes to establish a benchmark and validation database that addresses seakeeping and maneuvering in waves in different environmental conditions and water depths. Moctar et al. (2016) investigated second order wave-induced forces and moments, along with maneuvering motions in calm water and in waves for a container and a tanker ship. Fournaraskis et al. (2017) explored different techniques for estimating the drift forces, yaw moment and added resistance of ships in waves.

This paper presents results for static drift simulations for a KVLCC2 model, both in calm water and in wave. Initially, calm water drift results are validated with experimental data. Linear hydrodynamic derivatives are predicted from the calm water simulated results and compared with experimental data. Next, static drift simulations are performed at three different wave lengths Finally, the drift forces and moments for both calm water and waves were compared. The paper highlights the effect of waves on the drift forces and moments through static drift simulations.

2 METHOD

2.1 *Simulation solver*

OpenFOAM (Open Field Operation and Manipulation) is an open source library, written in C++ language following object-oriented paradigm. It can be used to numerically solve a wide range of problems in fluid dynamics, from laminar to turbulent flows, with single and multi-phases. It has several packages to perform multiphase turbulent flow simulation for floating objects. OpenFOAM also allows relatively easy customization and modification of solvers, because of its modular design. The solver has been elaborately described by Jasak (1996, 2009).

The OpenFOAM solver used to perform ship hydrodynamic simulations for this paper simulates incompressible, two-phase flow. The governing equations for the solver are the Navier-Stokes equation (1) and continuity equation (2) for an incompressible laminar flow of a Newtonian fluid. The Navier-Stokes and Continuity equation are given by:

$$\rho(\partial v/\partial t + v\cdot\nabla v) = -\nabla p + \mu\nabla\hat{}2v + \rho g \tag{1}$$

$$\nabla v = 0. \tag{2}$$

where v the velocity, p is the pressure, μ is the dynamic viscosity, g is acceleration due to gravity, and ∇^2 is the Laplace operator. Further, the continuity equation is of the form

The Volume of Fluid (VOF) method is used to model fluid as one continuum of mixed properties. This VOF method determines the fraction of each fluid that exists in each cell, thus tracks the free surface elevation. The equation for the volume fraction is obtained as

$$\frac{\partial \alpha}{\partial t} + \nabla(\alpha U) = 0, \tag{3}$$

where U is the velocity field, α is the volume fraction of water in the cell and varies from 0 to 1, full of air to full of water, respectively.

The unstructured collocated Finite Volume Method (FVM) using Gauss theorem together with user-defined and implemented solution algorithm and time-integration schemes (Drikakis, et al., 2007) is used to discretize the governing equations. Time integration is performed by a semi-implicit second-order, two-point, backward-differencing scheme. Pressure-velocity coupling is obtained through PIMPLE algorithm (Ferziger & Peric, 2008), a combination of SIMPLE and PISO. OpenFOAM follows a Cartesian coordinate system, if not specified otherwise. All systems are based on an origin point and coordinate rotation. The solver has a local and a global coordinate system. The solver incorporates three different turbulence models, k–ε, k–ω and SST k–ω. Turbulence is discretized using a 2nd order upwind difference. Turbulence for the presented simulations was modeled with the Reynolds-averaged stress (RAS) SST k–ω two-equation model. The parameters were calculated using the guidelines that resulted from a recent study (Labanti et al., 2016).

2.2 *Ship model*

The model ship used in this research for simulations, is the KRISO Very Large Crude Carrier 2 (KVLCC2). Table 1 provides the specifications of the KVLCC2 model.

All simulations were performed with the 7 m model ship. The model is widely popular among researchers and has been elaborately discussed at Gothenburg, SIMMAN and Tokyo workshops.

2.3 *Simulation mesh*

The domain size (blockMesh) for simulations was set following general ITTC (2011) guidelines; the inlet was placed one and a half ship length windward the bow, the outlet three and a half ship length downstream the stern, each lateral boundary was one and a half ship lengths away from the ship's symmetry plane, the depth or bottom of

Table 1. Specifications of the oil tanker ship model KVLCC2.

Specification		KVLCC2 (full scale)	KVLCC2 (model scale)
Length between perpendiculars	Lpp (m)	320.0	7.0
Breadth	B (m)	58.0	1.2688
Depth	D (m)	30.0	0.6563
Draft	T (m)	20.8	0.4550
Wetted surface area	S (m^2)	27194.0	13.0129
Displacement volume	V (m^3)	312622	3.2724
LCB from mid-ship	LCB (m)	11.136	0.2436
Kyy	Kyy (m)	0.25 Lpp	0.25 Lpp

domain was set at one ship length and the atmosphere was at half ship length from free surface.

The hull form was integrated to the blockMesh by using snappyHexMesh utility, which created a "body fitted" hexahedral mesh around the hull surface from the specified STL file. The drift angles for simulations were applied to STL files and then snappyHexMesh was executed. Two separate mesh resolution was used for the calm water simulations and for simulations in waves. In order to properly capture the pressure on the hull form and also the free surface elevation, three successive refinements were performed near the hull surface. The average mesh resolution used for the calm water simulations with varying drift angles was 4.1 million. The number of cells varied slightly depending on the drift angle, as the refinement area was adjusted depending on the drift angle.

In case of static drift simulations in waves, similar mesh topology was used, with a reversed domain. This is done to avoid some settings issues in OpenFOAM. However, all images for wave cases were flipped to avoid confusion. Furthermore, for ensuring proper and even propagation of incoming waves, the refinements were performed covering a larger domain area, extending to the wave inlet. Mesh resolution in z-axis was also increased to properly capture the incoming wave steepness. The average mesh resolution used for the static drift cases in waves was 8.1 million. General mesh assembly for both the simulations are shown in Figure 1.

2.4 *Computational resource*

The simulations were performed in an Intel(R) Core i7 CPU with 8 cores, clock speed 3.60 GHz and 16 GB of physical memory. The average time step used was 0.02 second and for simulating each case with stable output, the required physical time was about 20 hours per calm water case, and around 48 hours per wave simulation case. All the simulations were run up to 50 seconds (simulation time) for attaining stable results.

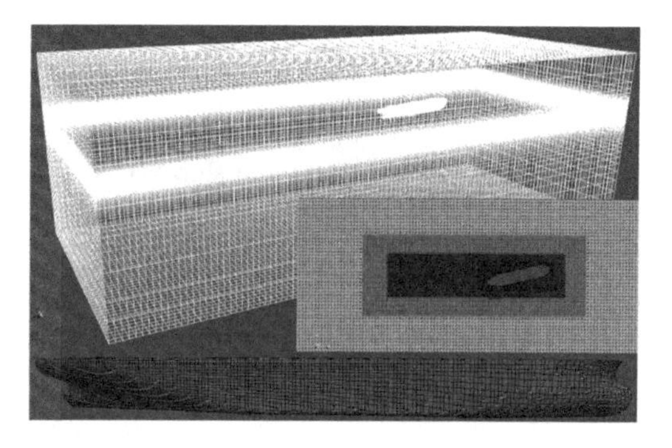

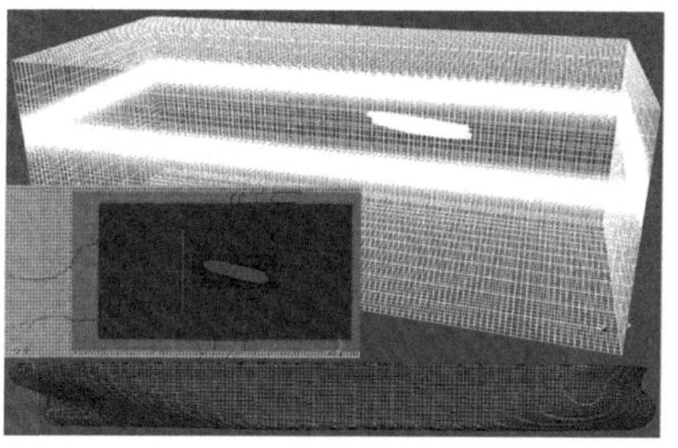

Figure 1. Simulation mesh; overall simulation domain, mesh distribution on the free surface (drifted hull) and mesh distribution on hull form, for calm water on top and for waves at bottom.

3 RESULTS

For static drift simulation, a setup similar to experimental facility was created through CFD medium. Static drift motion is asymmetric, thus full hull was simulated. All simulations were performed at the design speed of the ship ($Fr = 0.142$, $Rn = 4.6 \times 10^6$), and all ship motions were restricted (0 DoF). Total six static drift cases were simulated in calm water. As for wave cases, simulations were performed for five drifting angles at three different wave lengths.

3.1 *Static drift simulation in calm water*

Static drift test is the towing of ship in a tank in oblique condition, as shown in Figure 2, where β represents drift angle. For the simulations, the same condition was reproduced in CFD environment using OpenFOAM. The interFoam solver was used to simulate the static drift cases with a static mesh. The total drag resistance encountered by the ship during forward motion with zero drift angle was 3.81×10^{-3}, whereas, for experimental data, the value is 4.056×10^{-3} (Larsson, 2011). The deviation observed here is mostly because, the ship in experiment was free to heave and pitch. Whereas in simulation cases, the ship was static. A grid convergence study was not reported here since direct validation with experimental data was performed. In order to show simulation convergence, time history for forces and moments are shown in Figure 3, for drift angles 0, 6 and 12 degrees. The figure shows that the moment is the last to reach stability and that happens after roughly 70 seconds of simulation.

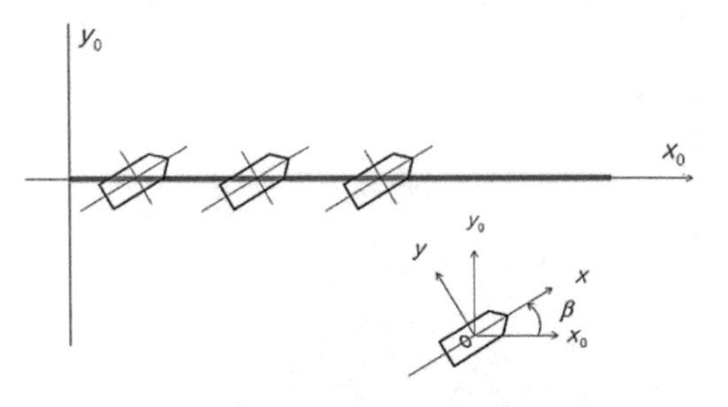

Figure 2. Schematic diagram representing static drift motion.

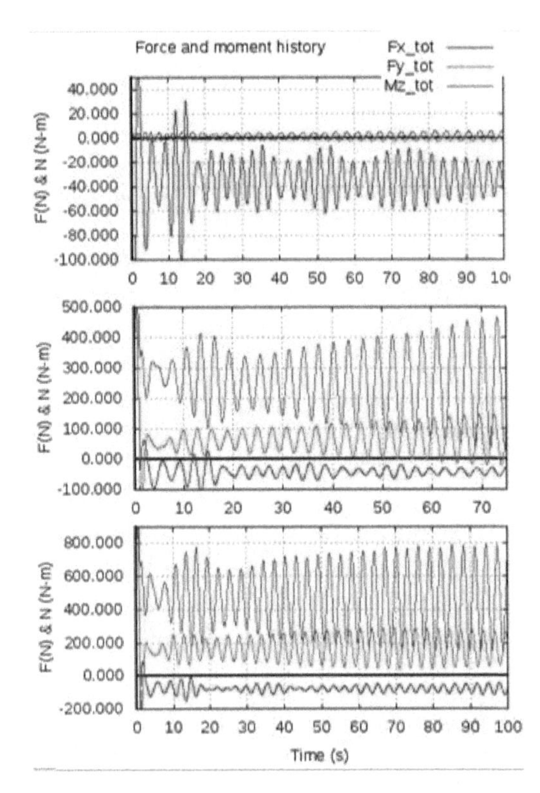

Figure 3. Time history for forces and moment in calm water static drift simulation, for drift angle 0°, 6° and 12°, from top to bottom.

For static drift test, simulations were performed for seven drift angles, −2°, 0°, 2°, 6°, 8°, 10° and 12°. The simulation results for drag and lateral force coefficients, and yaw moment coefficients are shown in Figures 4, 5 and 6. The comparison data for lateral force and yaw moment was taken from Wang et al. (2011). Experimental data for the drag force is not available. The figures also show Experimental Fluid Dynamics (EFD) data produced by National Maritime Research Institute (NMRI) and Shanghai Jiao Tong University (SJTU), and simulation data produced using commercial code FLUENT. The results are shown in non-dimensional scale. For conversion, following equations (4, 5 and 6) were used. In the equations, ρ is the fluid density, v is ship speed, L_{PP} is the ship length between perpendiculars and T is the ship draft.

$$F_X' = \frac{\bar{F}_X(N)}{0.5 \times \rho \times v^2 \times A_{ws}} \tag{4}$$

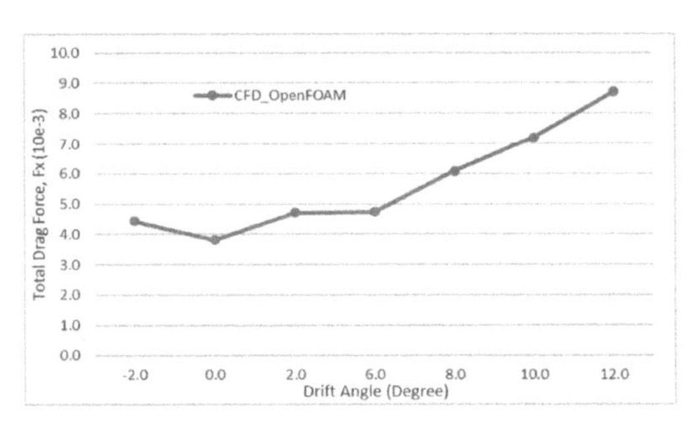

Figure 4. Total drag force coefficients experienced by KVLCC2 model at different drift angles in calm water.

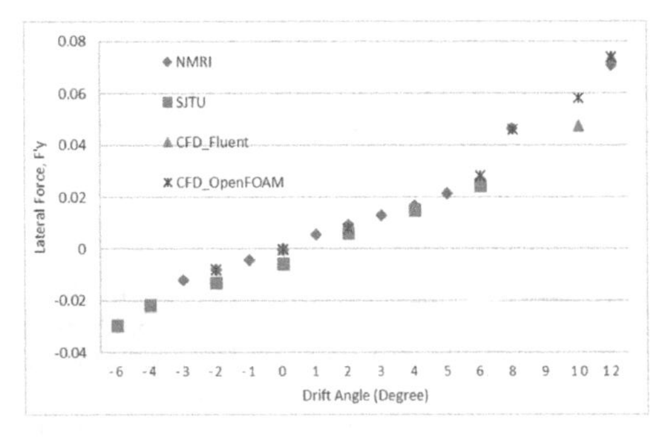

Figure 5. Lateral forces experienced by KVLCC2 model at different drift angles in calm water simulation.

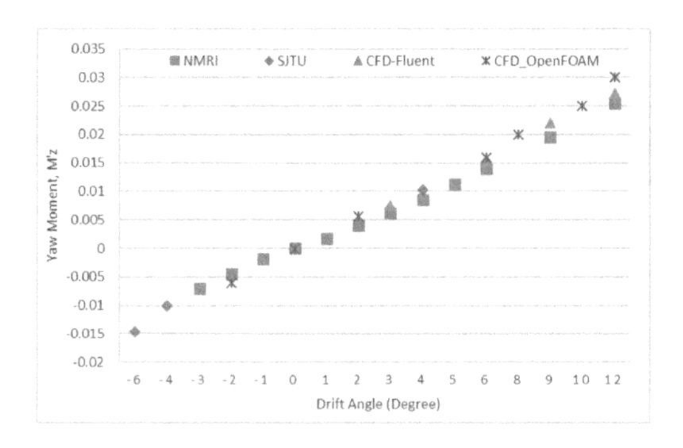

Figure 6. Yaw moments experienced by KVLCC2 model at different drift angles in calm water simulation.

$$F_Y' = \frac{\bar{F}_y(N)}{0.5 \times \rho \times v^2 \times L_{PP} \times T} \tag{5}$$

$$M_Z' = \frac{\bar{M}_z(N-m)}{0.5 \times \rho \times v^2 \times L_{pp}^2 \times T} \tag{6}$$

As can be seen from the results, the simulation results agree very well with the experimental data and properly captures the trend of lateral force and yaw moment. In case of static drift, as ship is moved forward with a drift angle, the drifted side of the bow encounters higher resistance than the other. Overall, with the increase in drift angle, the encountered drag resistance for the ship also increases. The free surface elevation and pressure distribution at hull surface for a drift angle of 12° are shown in Figure 7.

3.2 *Hydrodynamic derivatives prediction*

Linear hydrodynamic derivatives are predicted from lateral forces and yaw moments encountered by the ship at drifting conditions. By using curve fitting to the data for forces and moments as a function of drift angle, the hydrodynamic derivatives or coefficients, X_{vv}, Y_v, Y_{vvv}, N_v and N_{vvv} are predicted. Among the derivatives, Y_v and N_v are linear, and rest are non-linear. Although non-linear derivatives can be predicted from static drift results, their reliability can be limited. Thus, only linear derivatives were calculated. Y_v is predicted from the slope of non-dimensional lateral force F'y and N_v is predicted from the slope of yaw moment M'z, against changing drift angle. For calculating the slopes from the predicted results, linear trend lines were generated on the data and slopes were calculated.

Initially, the linear hydrodynamic derivatives were calculated for calm water static drift conditions and results were compared with available

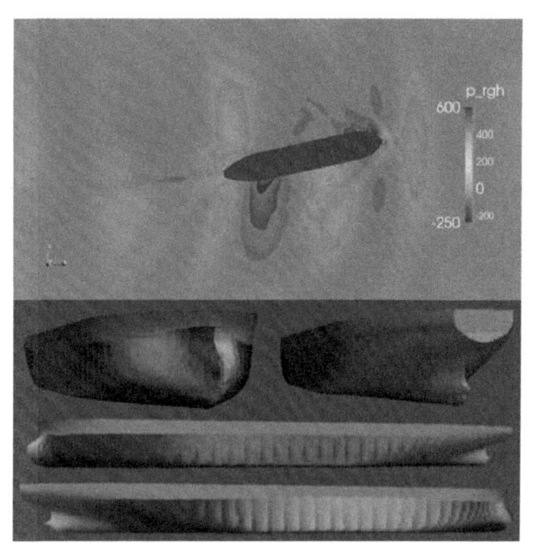

Figure 7. Hydrodynamic pressure distribution at zero water level during drift motion of KVLCC2 at 12 degree; and pressure distribution at bow, stern and mid-plane during drifted motion.

experimental data. The trend lines are shown in Figure 8 and 9, and the predicted derivatives are shown in Table 2.

The relative comparison among derivatives show reasonable agreement among each other. Although the deviation in percentage among CFD and EFD data is large, the difference in physical value is minor. Furthermore, the measured values (force and moment), both for EFD and CFD cases are so minor that, it is impossible to say that the results are devoid of uncertainties. Overall, it can be concluded that the CFD solver was well capable in predicting the linear hydrodynamic derivatives of the ship model.

3.3 *Static drift simulation in waves*

The case setup for static drift simulation in waves was similar to the setup for simulation in calm water, except that there was wave propagation through the field. For simulating wave cases, waveDyMFoam solver was used. The solver is coupling between waves2Foam and interDyMFoam. For simulating static drift cases in waves, three wave lengths were chosen, $\lambda/L = 0.3$, 0.5 and 0.7; where λ is the wave length and L is the ship length between perpendiculars. The waves were generated using Stokes second order wave equation and the wave height was set at 0.08 meter for $\lambda/L = 0.3$, and 0.1 meter for rest of the cases. In each wavelength five drifted conditions were simulated, −2°, 0°, 2°, 6° and 12°. The wave lengths were limited to relatively

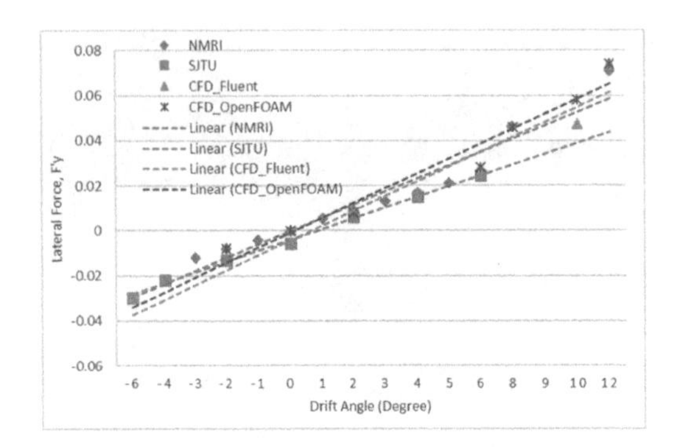

Figure 8. Results for lateral force prediction for KVLCC2 in calm water with linear trend lines.

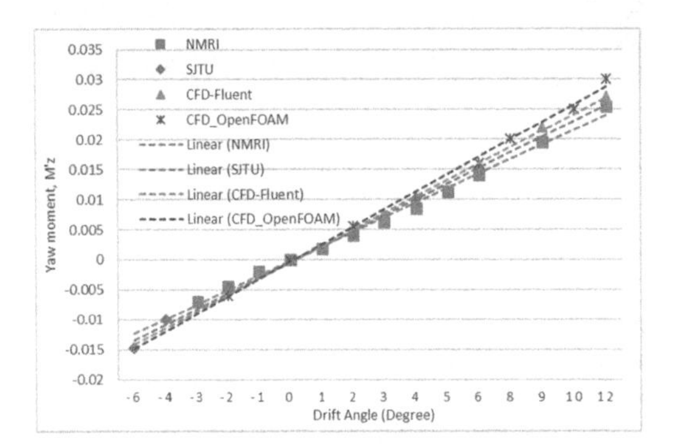

Figure 9. Results for yaw moment prediction for KVLCC2 in calm water with linear trend lines.

Table 2. Prediction of linear hydrodynamic derivatives from calm water simulation results.

Derivatives	CFD OpenFOAM	EFD NMRI	EFD SJTU	CFD Fluent
Y_v	0.0066	0.0059	0.0048	0.0066
N_v	0.0029	0.0024	0.0026	0.0027

low wavelength cases since large ships like KVLCC2 are rarely exposed to sea environment with λ/L greater than 0.8. The drift angles were limited to five cases to save computational time. For establishing reliability of the head wave simulation cases, initially added resistance values for the three wave lengths at zero drift angle were computed and compared with experimental data (Kim et al., 2013). The results are shown in Table 3.

For added resistance calculation, equation 7 was used. Although the results show deviation with experimental data, this is mostly because of the restricted motion of the ship during simulation. As for the case at λ/L = 0.3, perhaps second order Stokes wave theory wasn't sufficient for this case and higher order wave theory should have been used. A separate grid convergence study was not performed for the wave cases, since applying a different mesh resolution comparing the calm water cases leaves the results incomparable. Furthermore, mesh dependency in lateral force and yaw moment prediction is far less comparing to added resistance cases. A time history for the forces and moment prediction for the wave lengths at three drift angles are shown in Appendix.

For conversion of the encountered resistance and moment to non-dimensional form, equations 8 and 9 were used. In the equations, g is the gravitational constant and ζ_H is the wave amplitude.

$$C_{AW} = \frac{\bar{F}_X(N) - R_T(N)}{\rho g \zeta_H^2 \frac{B^2}{L_{PP}}} \tag{7}$$

$$F'_Y = \frac{\bar{F}_y(N)}{\rho \times g \times \zeta_H^2 \times L_{PP}} \tag{8}$$

$$M'_Z = \frac{\bar{M}_z(N-m)}{\rho \times g \times \zeta_H^2 \times L_{PP}^2} \tag{9}$$

The simulation results for lateral force coefficient and yaw moment coefficient are shown in Figures 10 and 11. The figures include results for

Table 3. Added resistance results for KVLCC2 in head waves, at zero drift angle.

λ/L	Caw (EFD_MOERI)	Caw (CFD_OpenFOAM)
0.3	3.5	6.90
0.5	3.6	3.26
0.7	3.2	3.55

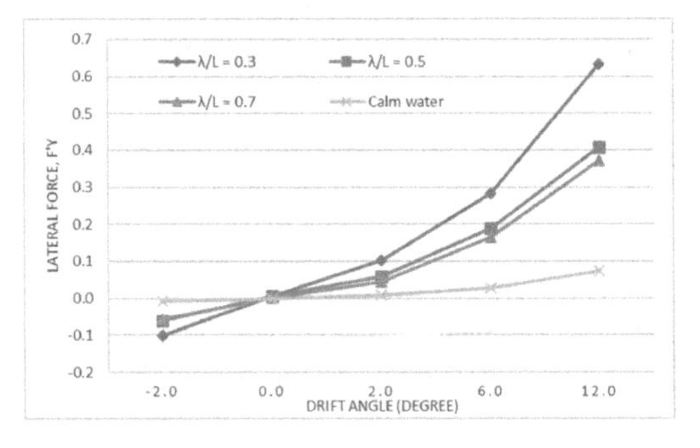

Figure 10. Lateral forces experienced by kvlcc2 model at different drift angles, for different incoming wave lengths.

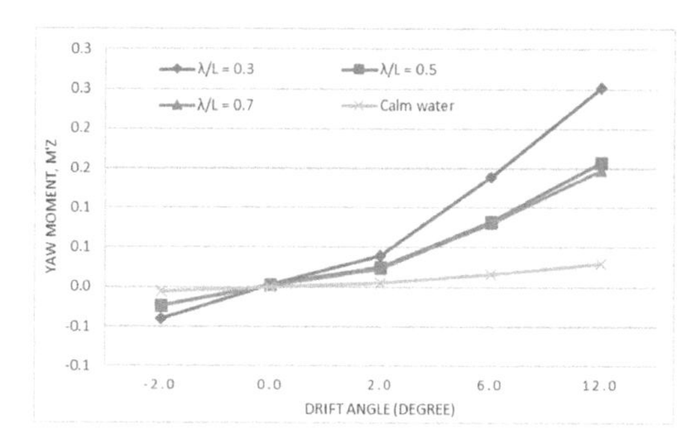

Figure 11. Yaw moment experienced by KVLCC2 model at different drift angles, for different incoming wave lengths.

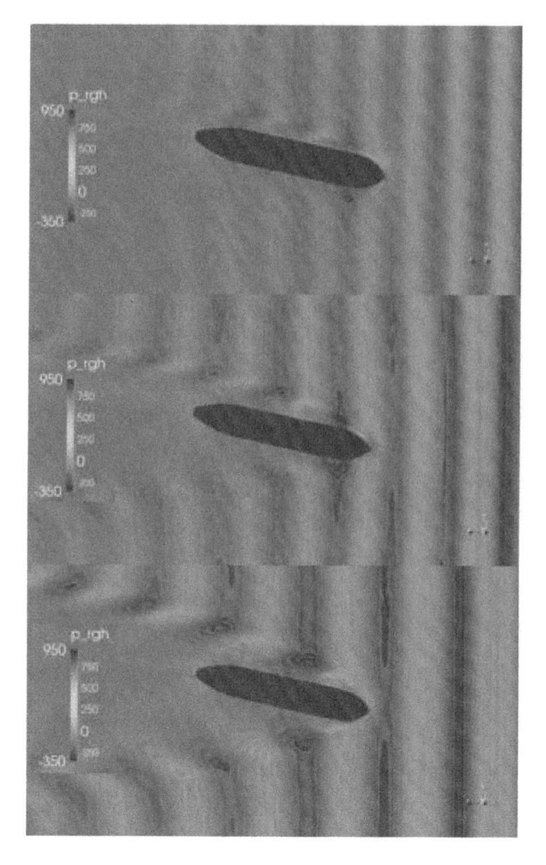

Figure 12. Pressure distribution at zero water level for drift angle 12° at λ/L = 0.3, 0.5 and 0.7, from top to bottom.

the three incoming wave lengths, and also for calm water for easier comparison.

As can be seen from the results, there is a significant difference among the force and moment values with and without waves. Not just the values, even the trend is significantly different. This is natural since the drifting hull would encounter higher resistance in waves, comparing to calm water. Furthermore, the non-dimensionalization process for calm water simulation and for waves is different. In wave simulation, the ship also encounters green water, which also influence the results. Nevertheless, the results are a good indicator to realize that the maneuvering characteristics predicted in calm water may not be applicable in sea environment.

The pressure distribution at zero water level for a drift angle of 12° at wave length/ship length = 0.3, 0.5 and 0.7 are shown in Figure 12.

4 CONCLUSIONS

In the paper, static drift simulations were performed for a KVLCC2 model using OpenFOAM, and the results were compared with experimental data. After validation, static drift simulations were performed for the same ship model in waves.

The lateral forces and yaw moments predicted form the static drift simulations in calm water showed good agreement with experimental data. The hydrodynamic derivatives calculated from the results also agreed very well with experimental results. The static drift simulations performed in waves showed significant difference in force and moment prediction, comparing the calm water simulation cases, signifying that ship has comparatively lower stability while encountering waves, comparing to calm water cases.

Further investigation is needed to better understand ship's stability in waves, as ships do not voyage through the sea with zero degrees of freedom. Hence, the cases should be repeated with at least heave and pitch free conditions to better represent the actual voyage conditions. A more elaborate study with more wavelengths and all PMM motions is also necessary to better understand how ship's stability is affected in short and long wave lengths.

ACKNOWLEDGMENTS

The work was a part of SHOPERA (Energy Efficient Safe SHip OPERAtion) Collaborative Project which was co-funded by the Research DG of the European Commission within the RTD activities of the FP7 Thematic Priority Transport/FP7-SST-2013-RTD-1/Activity 7.2.4 Improving Safety and Security/SST.2013.4-1: Ships in operation. http://shopera.org/

REFERENCES

Broglia R., Muscari R. and Mascio A.D., 2008. Numerical simulations of the pure sway and pure yaw motion of the KVLCC1 and 2 tankers. *SIMMAN 2008 Proceedings*.

Drikakis, D., Fureby, C., Grinstein, F. and Liefendahl, M., 2007. ILES with limiting algorithms. *In: In Implicit Large Eddy Simulation: Computing Turbulent Fluid Dynamics*. s.l.:Cambridge University Press, pp. 94–129.

Ferziger J.H. & Milovan P., 2008. Numerische Stromungsmechanik. *SpringerVerlag*, Heidelberg, Berlin, second edition.

Fournarakis N, Papanikolaou A, Liu S, 2017. Estimation of the drift forces and added resistance in waves of the KVLCC2 tanker. *J. Ocean Eng. Mar. Energy*, DOI 10.1007/s40722-017-0077-7.

Gullmineau E., Queutey P., Visonneau M., Leroyer A. and Deng G., 2008, RANS simumotion of a US NAVY frigate with PMM motions. *SIMMAN 2008 Proceedings*.

Hajivand A. & Mousavizadegan S.H., 2015. Virtual simulation of maneuvering captive tests for a surface vessel. *International Journal of Naval Architecture and Ocean Engineering*, vol. 7, pp. 848–872.

Hajivand A. and Mousavizadegan S.H., 2015a. Virtual maneuvering test in CFD media in presence of free surface. *International Journal of Naval Architecture and Ocean Engineering*, vol. 7, pp. 540–558.

Hochbaum A.C., Vogt M. and Gatchell S., 2008. Maneoeuvering prediction for two tankers based on RaNS simulations. *SIMMAN 2008 Proceedings*.

Jasak, H., 1996. "Error Analysis and Estimation for the Finite Volume Method with Applications to Fluid Flows", *Ph.D. thesis, Imperial College of Science, Technology & Medicine*, London, UK.

Jasak, H., 2009. OpenFOAM: Open Source CFD in research and industry. *International Journal of Naval Architecture and Ocean Engineering,* Volume 1(2), pp. 89–94.

Kim H., Akimoto H. and Islam H., 2015. Estimation of the hydrodynamic derivatives by RaNS simulation of planar motion mechanism test. *Ocean Engineering*, vol. 108, pp. 129–139.

Kim, J., Park, I.-R., Kim, K.-S., Kim, Y.-C., Sik Kim, Y. and Van, S.-H., 2013. Numerical Towing Tank Application to the Prediction of Added Resistance Performance of KVLCC2 in Regular Waves. *Proceedings of the Twentythird (2013) International Offshore and Polar Engineering (ISOPE) Anchorage*, Alaska, USA, June 30-July 5, 2013.

Labanti, J., Islam, H. and Guedes Soares, C., 2016. CFD assessment of Ropax hull resistance with various initial drafts and trim angles. Guedes Soares, C. and Santos T.A., (Eds.), *Maritime Technology and Engineering 3,* London, UK: Taylor & Francis Group; pp. 325–332.

Larsson L., Stern F. & Visonneau M., 2011. CFD in Ship Hydrodynamics- Results of the Gothenburg 2010 Workshop. *MARINE 2011, IV International Conference on Computational Methods in Marine Engineering, Computational Methods in Applied Sciences.*

Lee S. & Kim B., 2015. A numerical study on manoeuvrability of wind turbine installation vessel using OpenFOAM. *International Journal of Naval Architecture and Ocean Engineering*, vol. 7, pp. 466–477.

Lewis, E., 1988. *Principles of Naval Architecture.* Jersey City, NJ: The Society of Naval Architects and Marine Engineers.

Miller R.W., 2008. PMM calculation for the bare and appended DTMB 5415 using the RaNS solver CFDSHIP-IOWA. *SIMMAN 2008 Proceedings*.

Papanikolaou, A.; Zaraphonitis, G.; Bitner-Gregersen, E.; Shigunov, V.; El Moctar, O.; Guedes Soares, C.; Reddy, D.N., and Sprenger, F., 2016, Energy efficient safe ship operation (SHOPERA)., *Transportation Research Procedia* 14:820–829.

Simonsen C.D., Otzen J.F., Klimt C., Larsen N.L. and Stern F., 2012. Maneuvering predictions in the early design phase using CFD generated PMM data. *29th Symposium on Naval Hydrodynamics*, Gothenburg.

SIMMAN 2008. [Online] Available at: http://www.simman2008.dk/ [Accessed 2016].

Sprenger, F; Maron, A; Delefortrie, G; Cura-Hochbaum, A; Lengwinat, A; Papanikolaou, A; 2016. *Experimental Studies on Seakeeping and Manoeuvrability in Adverse Weather Conditions*. SHOPERA Technical report.

Uharek S. and Hochbaum C.A., 2015. Modelling mean forces and moments due to waves based on RANS simulations. *Proceedings of the Twenty-fifth International Ocean and Polar Engineering Conference Kona*, Big Island, Hawaii, USA, June 21–26, 2015.

Wang H M, Tian X, Zou Z., and Wu B, 2008. Experimental and numerical researches on the viscosity hydrodynamics of hydrodynamic forces acting on a KVLCC2 model in oblique motion. *SIMMAN 2008 Proeedings.*

APPENDIX

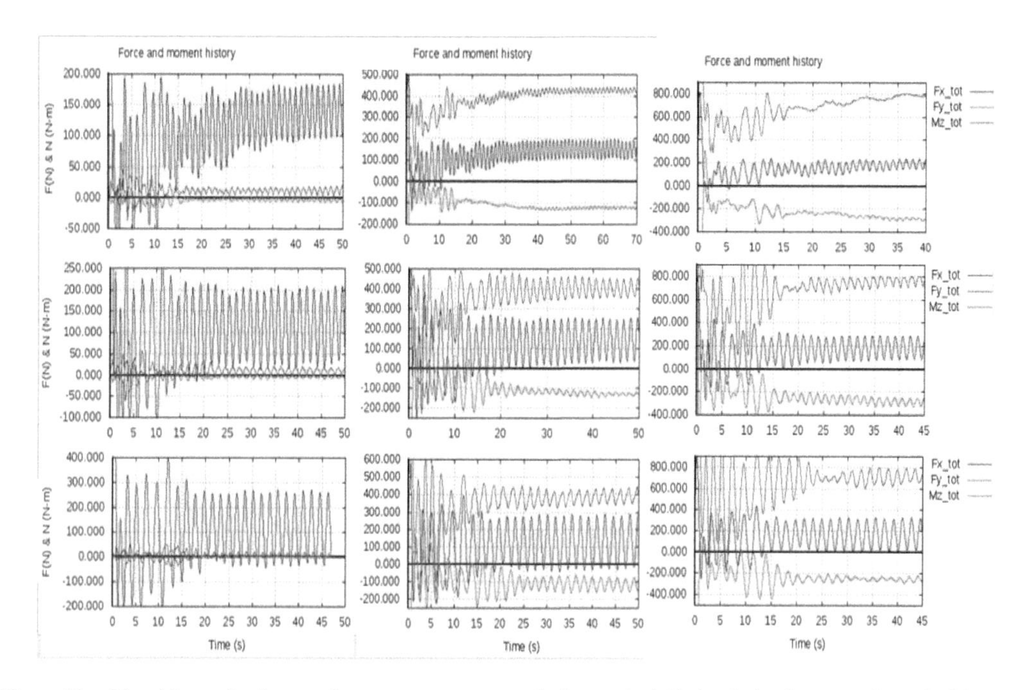

Figure 13. Time history for force and moment convergence during static drift simulation in waves. Cases for $\lambda/L = 0.3$ at left, $\lambda/L = 0.5$ at middle and $\lambda/L = 0.7$ at right. From top to bottom, drift angle 0°, 6° and 12°.

Progress in Maritime Technology and Engineering – Guedes Soares & Santos (Eds)
© 2018 Taylor & Francis Group, London, ISBN 978-1-138-58539-3

Ship self-propulsion performance prediction by using OpenFOAM and different simplified propeller models

S. Gaggero, T. Gaggero, G. Tani, G. Vernengo, M. Viviani & D. Villa
Department of Electrical, Electronic, Telecommunications Engineering and Naval Architecture, University of Genoa, Genoa, Italy

ABSTRACT: Classic hydrodynamics-related ship design problems can nowadays be approached by CFD viscous solvers. Ship self-propulsion performance prediction represents one of the most interesting problems in this framework. The capabilities of CFD codes to resolve accurately the separate problems (open water propeller performance and hull resistance) have been demonstrated over the last decades. The complexity of the combined problem (and, in turn, the required computational time) has restricted its solution to research applications still far from everyday industrial practice. Some approaches have been developed to reduce the computational burden, based e.g. on simple actuator-disk theory or, recently, on BEM/RANS coupled solvers. In this respect, different approaches exploiting the open-source solver OpenFOAM are presented, focusing on the main self-propulsion parameters. In addition, a new numerical strategy able to provide more information compared to classical simplified approaches, is herein presented and validated against experimental measurements on the well-known Kriso Container Ship (KCS) test case.

1 INTRODUCTION

Over the last decades, Computation Fluid Dynamics (CFD) methods became a mature research tool to solve ship hydrodynamics problem and they also started to be considered for industrial applications at different stages of the design. This was possible thanks to the easier access to computational resources and to the theoretic improvements to the numerical solution of always more complex physical problems such as multi-phase and free surface flows. Considering these progress in hardware solutions and CFD techniques, some standard hydrodynamic problems in naval architecture can now be quite easily resolved in the framework of the so-called Virtual Towing Tank. This is the case for instance of performance prediction of the hull resistance advancing in calm water or of the propeller efficiency.

Both commercial and open-source software reach similar confidence levels when compared to experimental measurements (as shown in Larsson et al. 2015). All the applied solvers demonstrate to be able to predict the total hull drag in calm water within an error of 1–2% maximum. This is an encouraging result considering that the same accuracy has been shown in the experimental procedures reported by the International Toking Tank Conference (ITTC 2017), with different facilities performing the same test and cross-comparing each other measurements.

Recently, both modern commercial and open-source CFD software have been demonstrated to accurately predict e.g. both the planning hulls performance (Ferrando et al. 2015) as well as that of displacing ships (Gaggero et al. 2015). Moreover, it was possible to achieve these results at a relatively low computational cost. This is particularly important in order to allow to consider these type of simulations in early ship design stages and as viable alternative and complement (e.g. for typical medium size shipyards) to towing tank measurements. These modeling techniques, indeed, can also be used to support experimental campaigns, improving the knowledge of the flow field around the ship hull or for predicting the full-scale hull wake to the propeller to be further used as input for cavitation tunnel experiments (Tani et al. 2017) or final propeller design.

Accordingly, also propeller performance prediction by CFD methods has significantly grown (see, among the other, Gaggero et al. 2017a, Gaggero et al. 2010b, Gaggero et al. 2014b Gaggero et al. 2017d and Gaggero and Villa 2017b). The PPTC 2015 test case has been widely solved by using the open-source OpenFOAM libraries showing high accuracy both in cavitating and sub-cavitating regimes. Propeller design by optimization has been extensively and successfully completed by using both viscous unsteady Reynolds Averaged Navier Stokes (URANSE) solvers and ad-hoc developed Boundary Element Methods (BEM) (Gaggero et al. 2017c).

Despite the aforementioned solutions are nowadays considered as everyday practice at least by large and medium-size companies, there are still several problems that are the object of research.

This is mainly due to the relatively high computational effort required for those solutions, as the case of seakeeping (Grasso et al. 2010, Guo et al. 2012 and Deng et al. 2010, Bonfiglio et al. 2016) or ship manoeuvring in in calm water and waves (Sung and Park 2015, Ferrant et al. 2008).

Another complex problem is represented by self-propulsion performance prediction, i.e. the evaluation of the propeller revolution rate when operating in the hull wake considering all the mutual interactions between the ship and the propeller (and possibly other appendages and/or the rudder). In addition, the equilibrium between the propeller thrust and the hull resistance at the design speed need to be reached. In fact, one of the main issues related to this condition is the increase in computational time. This is due to the very different time scales between the free surface flow around the hull and the flow closed to the propeller. Usually, the solver is constrained to impose a time-step defined by the faster flow dynamics, usually related to the propeller, while the total computational time is correlated to the hull flow, i.e. the one with the slower dynamics. The overall computational time can be much higher than those required to solve both problems separately. Simplified numerical strategies designed to reduce the overall computational effort, possibly limiting the loss of accuracy, have been recently proposed, ranging from simple actuator disk models (Carrica et al. 2013 and Fu et al. 2015) up to BEM/RANS coupled approaches (Villa et al. 2011, Gaggero et al. 2017c, Kim et al. 2006 and Alexander et al. 2009). Despite the propeller effects are included in the hull performance evaluation, the simplest methods based on actuator disks are not able to provide the complete set of data usually derived from self-propulsion tests (i.e. the propeller rate of revolution at equilibrium).

The proposed research focuses on the systematic comparison of different strategies, already successfully applied to investigate the rudder/propeller system (Bruzzone et al. 2014, Villa et al. 2017), to tackle the self-propulsion problem. In addition, a new procedure able to provide a wider set of useful information also from simplified actuator disk based calculations is presented and validated by comparison against available experimental measurements and higher fidelity (BEM/RANS coupling and full RANS analyses) calculations.

2 TEST CASE

The well-known KRISO Container Ship (KCS) is used as a numerical benchmark in the present study. Table 1 reports the main model data and Figure 1 shows a profile view of the underwater hull shape (Larsson et al. 2015). A scale ratio of

Table 1. KCS main data in model scale.

Ship data in model scale	Symbol	Value
Length between perp.	L_{PP} [m]	7.2786
Length of waterline	L_{WL} [m]	7.3577
Max. beam at waterline	B_{WL} [m]	1.0190
Draft	T [m]	0.3418
Displacement volume	∇ [m^3]	1.6490
Wetted surf. area w/o rudder	S_W [m^2]	9.4379
Propeller center, (from FP)	x/L_{PP}	0.9825
Propeller center, (below WL)	z/L_{PP}	–0.02913
Prop. rotation dir.		clockwise

Figure 1. Sketch of the KCS test case hull shape.

1/31.6 has been used. All the simulations have been performed at the ship design speed equal to 24 knots, corresponding to a Froude number of 0.26. The self-propulsion prediction has been carried out considering the hull with its propeller but without the rudder to reproduce the same configuration adopted during the experimental measurements (as reported in Larsson et al. 2015). As requested by the ITTC'78 procedure, a Skin Friction Correction (SFC) equal to 30.3 N is included. The towing tank test has been performed with the ship free to sink and trim. The dynamic attitude computed in towing conditions has been kept fixed during the self-propulsion numerical simulations.

3 NUMERICAL MODEL: APPROXIMATION OF THE PROPELLER EFFECT

The open source OpenFOAM libraries have been used to investigate the KCS self-propulsion condition. This suite provides a wide type of solvers and pre/post processors designed to solve most of the flow problem in the engineering field. One of the goals of this work is, in fact, to test a full license-free approach on a particular and complex naval hydrodynamic problem. In this light, both the built-in Cartesian hex-dominant mesher *snappyHexMesh* and the *interDyMFoam* solver have been applied. The necessary tuning of this meshing tool often represents the bottleneck of such kind of applications but its rich features, which can be freely customized and parameterized, ensure the generation of high-quality meshes, mandatory for the so-called "Virtual toking tank" computations.

Figure 2 shows an overview and a close-up of the generated mesh. The figure highlights the free-surface vertical refinement, essential to properly capture the hull generated wave pattern, and the

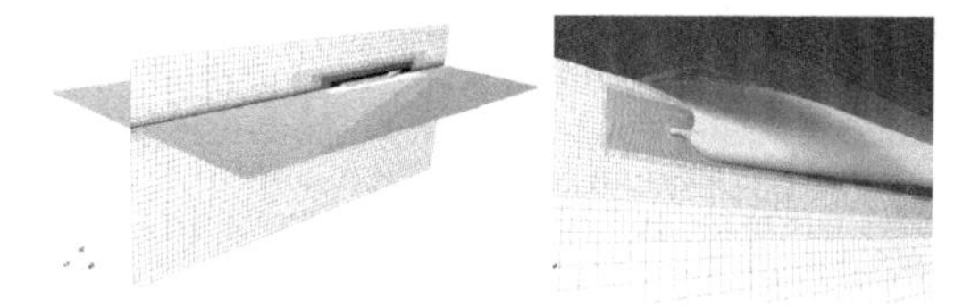

Figure 2. KCS mesh layout with hull wake detail (right).

horizontal refinements inside the Kelvin region (which extends for about ±20° from the bow of the hull). In addition, due to the need of resolving the hull-propeller interaction, a finer mesh has been clustered on this region. A final mesh of about 3 million cells is used for the whole computational domain.

Regarding the solver set-up, two approaches have been used to address time discretization: the Local Time Step (LTS) approach for steady solutions and an implicit time marching scheme for transient solutions. The two approaches need different computational resources: the first is faster but can be only used if a steady free surface solution is expected while the latter, that is the slower, requires higher computational resources because the solution time-step is constrained by the maximum admissible Courant number. Turbulence has been modeled using the *SST k-ω* formulation. A more detailed description of the mesh, domain and of solvers characteristics can be found in Gaggero et al. (2015) and Gaggero et al. (2017a).

As mentioned, the exploration of the cost/benefit balance resulting from different approaches when including the effects of the propeller working behind the hull is one of the main goals of this research. Several approaches have been considered with different levels of complexity:

- The simpler constant translational actuator disk which includes only the thrust force (CD_T), uniformly distributed over the entire propeller disk (Figure 3a),
- The rotor-translational disk, which includes also the torque moment (CD_{TQ}), uniformly distributed over the entire disk (Figure 3b),
- The radially varying actuator disk, which considers the radial distribution of propeller force and moment (RD_{TQ}), as per Figure 3c,
- The radial-tangential varying actuator disk which considers also the circumferential variation of forces generated by propeller in a non-uniform wake (RTD_{TQ}), as per Figure 3d,
- The BEM/RANS coupled approach (Figure 3e), originally developed in Villa et al 2011.
- The full-RANS approach (Figure 3f) where the propeller is fully resolved in the RANS solver. This is used as the reference.

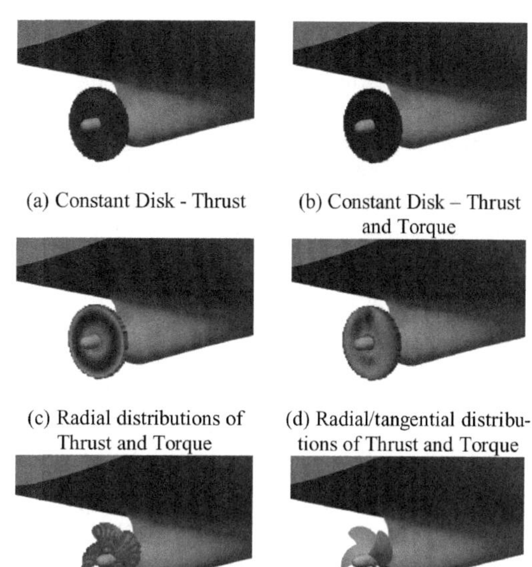

(a) Constant Disk - Thrust
(b) Constant Disk – Thrust and Torque
(c) Radial distributions of Thrust and Torque
(d) Radial/tangential distributions of Thrust and Torque
(e) Coupled BEM/RANS
(f) full RANS

Figure 3. Distribution of forces imposed in the simplified models and propeller geometry.

The comparison of the costs/benefits of different propeller approximations has been previously analyzed in Bruzzone et al. (2014). Those methods have been used to resolve the propeller/rudder interaction, comparing the generated flow fields and focusing on the accelerated flow field and its interaction with the rudder. In the present paper, due to the absence of the rudder, the focus will be on the model interaction with a body (the hull) located in front of the propeller region.

A dedicated implementation of these actuator disks in the OpenFOAM framework has been developed adding a new *fvOption* class method. Each of the proposed models requires a different level of knowledge of the propeller geometry and functioning. This makes them suitable at different ship design stages. For example, the BEM/RANS coupled approach requires the detailed propeller geometry knowledge and the availability of an unsteady BEM solver, restricting its application only at an advanced design stage. On the contrary, the CD_T model requires only the propeller diameter, commonly known since the preliminary design stages.

Figure 3 displays some perspective views of the described models with their own forces distributions.

The two simpler models, CD_T and CD_{TQ}, distribute uniformly the thrust force on the whole propeller disk region; the latter adds also a homogeneous tangential force proportional to the imposed torque. These models require only the knowledge of the propeller diameter and, for the second one, a preliminary estimation of the propeller efficiency, being the thrust always equal to the ship drag. These data commonly are known or can be estimated, since from the preliminary ship design stages. As demonstrated in Bruzzone et al. (2014), these simple and efficient models are not able to generate a reliable propeller wake, which can be obtained instead by the radial varying actuator disk model. The RD_{TQ} approach distributes both axial and tangential forces following expected blade force distributions along the radius (those computed from a lifting line design approach, for instance), therefore increasing the axial force at r/R about equal to 0.7. This model, which has been previously demonstrated to be able to reasonably predict the propeller wake field, is not able to take into account the blade force variation during a revolution due to the presence of the non-uniform hull wake. The influence of this unsteadiness is, instead, partially included with the RTD_{TQ} model, which shows, as in Figure 3, the typically increased distribution of forces on the upper-left part of the disk where a slower wake interacts with the propeller blade, increasing its load. Both RD_{TQ} and RTD_{TQ} models, compared to the constant disk ones, require more detailed propeller information (propeller force radial and/or tangential distributions and unsteady functioning), which can be provided, for example, by a simplified potential method as lifting line/lifting surface, BEMT or BEM codes. In the present work, these data have been computed by a dedicated BEM code (Gaggero 2010a), which has been also used to compute the body force distributions necessary for the BEM/RANS coupling. Behind the BEM/RANS coupling, indeed, there is the prediction of unsteady blade forces with the Panel Method, which turn into the unsteady body forces to be included into the RANS domain, and the iterative prediction of the effective wake by subtracting the propeller self-induced velocities (BEM) from the total velocity field (RANS) in front of the propeller.

3.1 *Self-propulsion procedure*

One of the most challenging aspects related to the self-propulsion simulation is the evaluation of the self-propulsion parameters (thrust deduction factor and wake factor) especially when simplified models, like actuator disks, are used.

When a full-RANS approach is used, the same procedure adopted in the experimental activities can be implemented. Setting the desired ship velocity, an initial propeller revolution rate is imposed (commonly defined by means of the nominal wake and the open water propeller curves), then, computing both the hull actual drag and the provided propeller thrust, the revolution rate is varied until an equilibrium condition is obtained (considering also the SFC factor). As a result, the equilibrium thrust (T), the resulting thrust deduction factor (t), the torque (Q) and propeller RPS are found and, by means of the propeller open water curves, as prescribed by ITTC'78 procedures, the effective wake fraction (w) can be computed.

When any actuator disk model is used, differently, only the equilibrium thrust and the consequently thrust deduction factor can be directly determined by the simulation. The effective wake fraction, on the contrary, needs a dedicated analysis. Considering that the propeller open water curves cannot be used because the propeller rate of rotation is unknown, the wake fraction can be directly evaluated by the integral of the velocity field in the propeller plane, as preliminarily shown in Villa et al. (2011). Unlikely, this data, due to the presence of the body force region, is affected also by the self-induced velocities generated by the momentum source represented by the actuator disk. The proposed procedure intends to evaluate this latest contribution by means of a dedicated open water simulation where the same actuator disk used for self-propulsion operates in a homogeneous mean nominal wake. In such way, an approximated self-induced velocity field in the propeller plane (inside the body force disturbed region) can be directly evaluated and by subtracting it (Equation 1) from the total velocity field, the effective wake can be determined:

$$V_{effective} = V_{total} - \left(V_{actuator\ disk\ in\ nominal\ wake} - V_{ship} \cdot (1-w)_{nominal}\right) \quad (1)$$

where V_{total} is the "total" velocity (Carlton, 2007) field computed by RANS on the propeller plane with the influence of the actuator disk and $V_{actuator\ disk\ in\ nominal\ wake}$ is the "total" velocity field on the propeller plane (including the influence of the actuator disk) of a simplified analysis where the actuator disk, operating in open water condition, is subjected to an uniform inflow equivalent to the nominal wake of the ship in towing condition $(V_{ship} \cdot (1 - W)_{nominal})$.

Once the effective wake (by Equation 1) and the thrust deduction factor are known, also the propeller revolution rate can be estimated by means of the numerical (or experimental if available) open water propeller curves.

This procedure for the calculation of the effective wake, at the cost of a simple and trivial additional calculation (actuator disk in uniform inflow equivalent to nominal wake), has been adopted for all the actuator disk models. Differently, in the BEM/RANS coupled approach, the effective wake has been estimated by directly computing the unsteady self-induced velocities in front of the propeller using the Panel Method and iteratively exchanging body forces at a given revolution rate (from BEM to RANS) and total velocity fields (from RANS to BEM) as described in Gaggero et al. (2017c). The computed effective wake in one or more plane in front of the propeller are then extrapolated to the propeller plane, to have a more reliable estimation of the effective wake. The functioning conditions (thrust and the relative propeller rate of revolution) are consequently directly provided by the unsteady BEM calculations.

An example of the application of Equation 1 is presented in Figure 4, where the mean wake fractions on a region equivalent to the propeller disk are computed at different longitudinal positions, being x/D = 0 the propeller plane (negative forward). In particular, the comparison is among the nominal wake (ship in towing condition), the "total" wake in self-propulsion condition (RANS with the CD_T actuator disk model) and the "total" wake of the same actuator disk model subjected to a uniform open water inflow equivalent to the nominal wake of towing condition ($V_{ship} \cdot (1 - W)_{nominal}$). The resulting effective wake does not show anymore the "jump" due to the momentum sources representing the action of the propeller across x/D = 0 and, as expected, it is possible to appreciate that it is "faster" than the nominal wake while maintaining a very similar behavior.

This analysis, which has been applied to the wake fraction seen as the average of the velocities on the propeller plane, can be identically applied point by point on the propeller plane, opening (as shown in the next section) the way to the adoption of this procedure also for non-equilibrium conditions, as maneuvering ones.

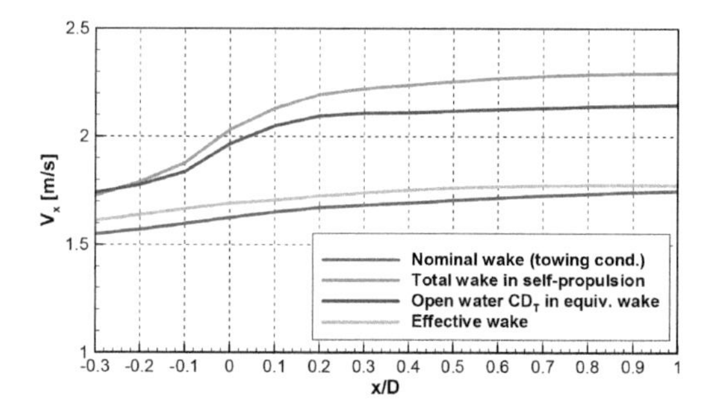

Figure 4. Example (model scale with CD_T) of the wake fraction variation along the longitudinal direction.

4 RESULTS

Preliminarily to self-propulsion analyses, towing tank tests have been carried out and compared with literature data. Results are summarized in Table. A discrepancy lower than 1% has been found for the total drag coefficient and even lower discrepancies are present for the ship attitude. Results of present analyses, which are within the numerical accuracy reported in the Tokyo 2015 Workshop where several solvers and users experience have been compared, can be considered acceptable. To have a complete insight into the reliability of the simplified procedure based on actuator disk calculations proposed to estimate the self-propulsion propeller/hull interactions, it is important to verify also the accuracy in predicting the nominal wake, both in terms of average and local values. Results of this analysis are summarized in Table 2 and in Figure 5.

Differences in terms of average values are less than 3%. The computed nominal wake is slightly faster than the measured wake and this is mainly due to the underestimation of the velocity reduction in correspondence of the hull stern while maintaining an overall good similarity in terms of local distribution of velocity.

Self-propulsion simulations have been carried out using all the propeller approaches outlined in Section 3. When simplified propeller methods (CD_T to BEM/RANS coupling) have been adopted, calculations have been carried out exactly with the same mesh of the towing test analyses. On the contrary, for full RANS analyses, sliding meshes

Table 2. CFD and EFD comparison of the KCS towing tank test with model free to sink and trim.

	Sinkage	Trim			
KCS ship	[cm]	[deg]	$C_T \cdot 10^3$	$C_F \cdot 10^3$	1–w
CFD	1.391	–0.170	3.674	2.897	0.739
EFD	1.395	–0.169	3.711	2.883	0.719

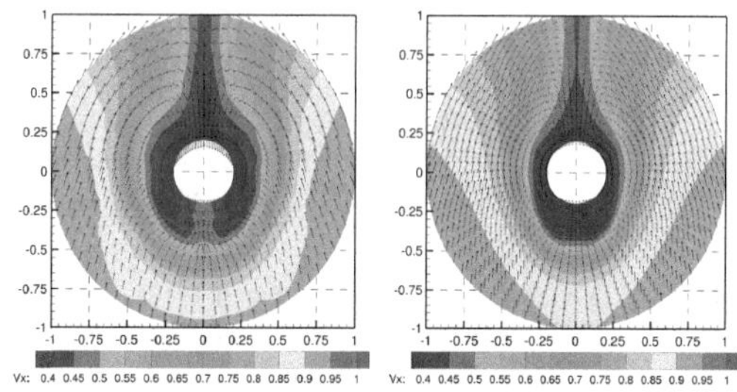

Figure 5. Comparison between EFD (left) and CFD (right) nominal wakes for the KCS in model scale at 0.26 Froude number.

have been employed by using a dedicated inner, rotating, region where the propeller has been modelled. The exact representation of the propeller required 1 Million of additional cells, which significantly contributes to the total cells count.

Furthermore, when the BEM/RANS coupling or the full-RANS approach is used, an unsteady solver is necessary. This aspect further contributes to the computational weighting of the calculations with respect to the actuator disk models that, by representing the propeller as its average action on the flow, operate in steady mode. The increase of the computational time is particularly significant for the full-RANS analyses, being the simulation time step (about 0.5° of propeller rotation) strictly constrained by sufficiently low Courant numbers. Table 3 reports a summary of the computed results for each propeller model compared to the experimental measures. The first four models (CD_T to RTD_{TQ}) use Equation 1 for the estimation of the effective wake and, coherently with their application (radial and tangential distribution of forces are obtained from initial estimation of propeller performances using BEM), make use of the open water propeller curves from BEM calculations to estimate the propeller rate of revolution. For the BEM/RANS coupling, the propeller rate of revolution is a direct result of the analysis and the wake fraction is the spatial average of the distribution of velocities on the propeller disk once propeller self-induction is subtracted, at convergence, from total velocities. For full RANS, instead, the wake fraction is obtained, in analogy to ITTC'78 experimental procedures, from the open water propeller curves estimated with the RANS itself.

According to the reported results, it is evident that the interaction between the hull and the simplified propeller models is only slightly influenced by the adopted approach. The thrust at equilibrium (and, in turn, the thrust deduction factor) varies less than 4%, with a discrepancy in terms of predicted

Table 3. Self-propulsion parameters comparison varying the actuator disk with the experimental data as reference.

Model	Prop. Rev. [RPS]	Thrust [N]	1–w	1–t
CD_T	9.533	61.50	0.7695	0.8871
CD_{TQ}	9.581	62.50	0.7713	0.8710
RD_{TQ}	9.581	62.66	0.7704	0.8701
RTD_{TQ}	9.637	62.89	0.7778	0.8667
BEM/RANS	9.570	64.20	0.7816	0.8494
Full-RANS	9.656	62.77	0.7618	0.8688
Exp.	9.500	65.15	0.7860	0.8460

propeller rate of revolution lower than 1%. In addition, a certain convergence trend can be appreciated by looking at the results provided by the actuator disks models. As the model increases its complexity (from constant to radial and to radial/tangential distribution of forces), the predicted rate of revolution tends to the most complete, full RANS, computation (even if discrepancies between open water propeller curves by BEM and RANS should be taken into account). BEM/RANS coupling, instead, shows slightly different results which, in the light of the analyses carried in Gaggero et al. (2014a), can be ascribed to the differences between equivalent open water calculations and unsteady analyses with BEM when significantly radially varying wakes are considered. Overall, the agreement with experiments is good, also in the light of the numerical results submitted to the Tokyo 2015 Workshop (Larsson et al. 2015). In this Workshop, the average predicted propeller rate of revolution was equal to 9.576 with a quite large dispersion of the data (±0.2 RPS): current results, also those obtained with the simplified models, are exactly within the confidence reported in the Workshop.

Figure 6 reports the velocity flow field computed in the propeller plane, then subjected to the acceleration provided by the actuator disk models. Several aspects can be highlighted.

The CD_{TQ} model, compared to the CD_T one, shows the rotational component of the velocity field related to the inclusion of the tangential momentum sources proportional to the torque.

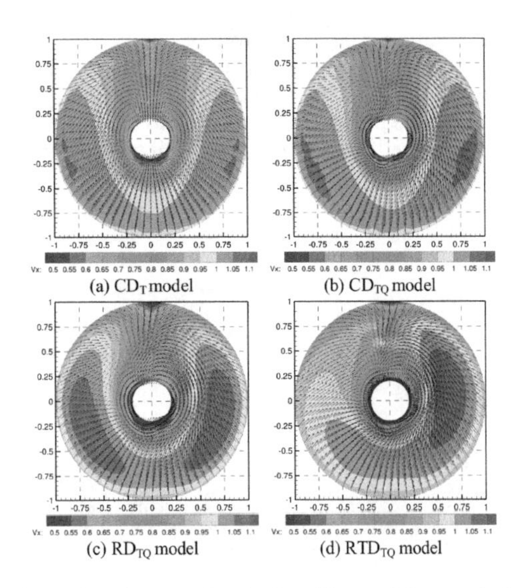

Figure 6. Computed wake in self-propulsion mode in the propeller plane (seen from aft). (a) CD_T, (b) CD_{TQ}, (c) RD_{TQ} and (d) RTD_{TQ}.

The distribution of axial velocity is not significantly affected by the inclusion of tangential forces and the negligible differences at outer radii can be ascribed mainly to the slightly different equilibrium point (i.e. higher momentum sources) predicted by the actuator disk accounting also for torque.

Differently, the RD_{TQ} model shows, as expected, an increased axial velocity in correspondence of r/R = 0.7, where the propeller load, and in turn, the momentum sources, computed by the BEM, is higher. All these models (CD_T, CD_{TQ}, and RD_{TQ}), as previously mentioned, are axially symmetrical, therefore do not include the different load experienced by the blades during a single revolution. The RTD_{TQ} model, on the contrary, adds this effect by including the actual distribution of momentum sources at any angular position based on unsteady calculations by the Panel Method with the preliminary nominal wake as the spatial non-homogeneous inflow. This peculiarity is well observable in Figure 6d where the higher axial velocities on the starboard side are representative of the higher blade load when it operates into the tangential flow of Figure 5. These differences (and peculiarities) among the various actuator disk models are furthermore evidenced in Figure 7, where the propeller induced velocities are extracted from the equivalent open water tests needed by Equation 1.

Almost uniform distribution of axial self-induced velocities when constant actuator disks (CD_T and CD_{TQ}) are used, axisymmetric radial and tangential components when torque is added to the models (CD_{TQ}), higher self-induced velocities in correspondence of higher local loads (RD_{TQ} and RTD_{TQ} respectively for radial and radial/circumferential distributions of body forces) can be highlighted as well.

Figure 8, finally, shows the comparison of the effective wakes obtained by applying Equation 1 with the four proposed actuator disk models and with the extrapolated effective wake on the propeller plane from the BEM/RANS coupling.

Regardless the actuator disk model, predictions are very similar, both in terms of wake fractions (Table 3) and local distribution of velocities, to the effective wake by the BEM/RANS coupling which, among the approaches for the prediction of the effective wake as the difference between the total and the self-induced velocity fields (Carlton 2017), can be considered the most accurate implementation. This confirms the consistency and the reliability of the simplified approach, which results consequently useful also for an initial estimation of the flow field for wake adapted propeller design.

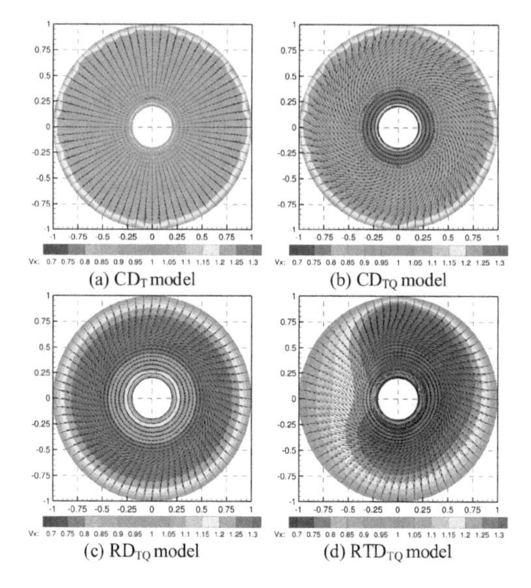

Figure 7. Induced velocity fraction during open water test in the propeller plane. (a) CD_T, (b) CD_{TQ}, (c) RD_{TQ} and (d) RTD_{TQ}.

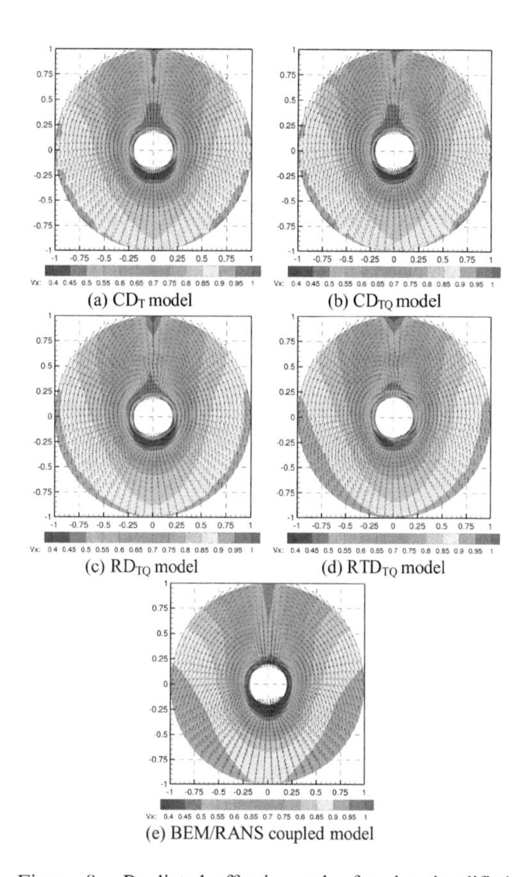

Figure 8. Predicted effective wake for the simplified propeller models. (a) CD_T, (b) CD_{TQ}, (c) RD_{TQ}, (d) RTD_{TQ} and (e) BEM/RANS coupled model.

5 CONCLUSIONS

A systematic analysis of different methods to include the propeller effect in a viscous RANS solver to simulate a ship in self-propulsion condition has been carried out. Four simplified actuator disks models have been explored, comparing the results, both in terms of global self-propulsion parameters and local flow field variables, with the BEM/RANS coupling approach, with full RANS calculations, and with the available experimental measurements.

Results from any of the proposed methods are good, especially in the light of the computational efficiency provided by the simplified approaches, which, consequently, could be applied during the preliminary design phases with an already satisfactory level of confidence. The propeller rate of revolution is overestimated by all the methods but the highest difference (with the full RANS approach) is, however, lower than 1.5%. In this particular case, a certain underprediction (also in open water condition) of the propeller forces at moderate/high advance coefficients by RANS, widely evidenced in literature, could explain the highest predicted propeller rate of revolution.

For the thrust deduction factor and the wake fraction, the differences between measurements and predictions with the simplified actuator disk models or with the BEM/RANS coupling are almost negligible. The highest differences (when the simplest model, CD_T, is applied), for both the quantities, are respectively of 5% and 2%. As the complexity of the model increases, discrepancies are between 1 and 2.5%, which can be considered more than satisfactory in the light of simplification introduced in the model. As in the case of the prediction of the propeller rate of revolution, full RANS calculations, especially for the prediction of the wake fraction, are significantly influenced by the inaccurate prediction of the open water propeller curves.

Also locally, the simplified approaches predict, almost regardless the considered model, very similar distributions of the effective velocity field. Together with the computational efficiency of the models (even higher than that of the BEM/RANS coupling which is usually considered a standard for an estimation of the effective wake field) this is a key point of the proposed analysis, opening the opportunity to have, in a very preliminary design phase, a reliable estimation of all the quantities necessary for a wake adapted propeller design.

The required steps for an even more robust approach consist of an extensive validation of these simplified models, including in the analyses configurations with shaft brackets, rudders and severe hull sterns generating strongly disturbed wakes.

REFERENCES

Alexander, B., Phillips, S., Turnock, R. and Furlong, M., 2009: Evaluation of manoeuvring coefficients of a self-propelled ship using a blade element momentum propeller model coupled to a Reynolds averaged Navier Stokes flow solver, *Ocean Engineering*, Volume 36, Issues 15–16, 2009, pp. 1217–1225.

Bonfiglio, L., Vernengo, G., Brizzolara, S. and Bruzzone, D. 2016: A hybrid RANSE – strip theory method for prediction of ship motions. *Proceedings of the 3rd International Conference on Maritime Technology and Engineering*, MARTECH 2016, in: Marine Technology and Engineering, 3, CRC Press, Editors: Guedes Soares & Santos, Volume 1, pp. 241–250.

Bruzzone, D., Gaggero, S., Podenzana Bonvino, C., Villa, D. and Viviani, M., 2014: Rudder-Propeller Interaction: analysis of different approximation techniques, *Proceedings of the 11th International Conference on Hydrodynamics* ICHD 2014, Singapore, October 19–24 2014, pp. 230–239.

Carlton, J.S., (2007) Marine Propellers and Propulsion, 2nd edition, Butterworth-Heinemann, 2007.

Carrica, P.M., Ismail, F., Hyman, M., Bhushan, S. and Stern F. 2013: Turn and zigzag maneuvers of a surface combatant using a URANS approach with dynamic overset grids. *J Mar Sci Technol* 18:166.

Deng, G.B., Queutey, P. and Visonneau, M., 2010: RANS prediction of the KVLCC2 tanker in head waves, *Journal of Hydrodynamics*, Ser. B, Volume 22, Issue 5, Supplement 1, 2010, pp. 476–481.

Ferrando, M., Gaggero, S. and Villa, D. 2015: Open Source Computational of Planing Hull Resistance, *Transactions of the Royal Institution of Naval Architects Part B: International Journal of Small Craft Technology*, Vol 157, Issue Jul-Dec 2015, pp. 83.98,, Royal Institution of Naval Architects.

Ferrant, P., Gentaz, L., Monroy, C., Luquet, R., Ducrozet, G., Alessandrini, B., Jacquin, E. and Drouet, A., 2008: Recent advances towards the viscous flow simulation of ships manoeuvring in waves, *Proceedingd of 23rd International Workshop on Water Waves and Floating Bodies*, Jeju, Korea.

Fu, H., Michael, T.J. and Carrica, P.M., 2015: A method to perform self-propulsion computations with a simplified body-force propeller model, *The Twenty-fifth International Ocean and Polar Engineering Conference*, International Society of Offshore and Polar Engineers.

Gaggero S., 2010a: Development of a potential panel method for the analysis of propellers performances in cavitating and supercavitating conditions, *Ph.D. Thesis*, University of Genoa, Italy, 2010 (in Italian).

Gaggero, S., Villa, D. and Brizzolara, S., 2010b: RANS and PANEL method for unsteady flow propeller analysis, *Journal of Hydrodynamics*, 22 (5 SUPPL. 1), pp. 547–552.

Gaggero, S., Villa, D. and Viviani, M., 2014a: An investigation on the discrepancies between RANS and BEM approaches for the prediction of marine propeller unsteady performances in strongly non-homogeneous wakes, *Proceedings of the 33rd International Conference on Ocean, Offshore and Artic Engineering*, OMAE 2014, San Francisco, USA, June 201, pp 1–13.

Gaggero, S., Villa, D., Viviani, M. and Rizzuto, E., 2014b: Ship wake scaling and effect on propeller performances. *Proceedings of IMAM 2013, 15th International Congress of the International Maritime Association of the Mediterranean,* in: Developments in Maritime Transportation and Exploitation of Sea Resources, CRC Press, Editors: Guedes Soares & Lopez-Pena, Volume 1, pp. 13–21.

Gaggero, S., Villa, D. and Viviani, M., 2015: The Kriso Container Ship (KCS) test case: an open source overview, *Proceedings of VI International Conference on Computational Methods in Marine Engineering*, MARINE 2015, June 15–17, Rome, Italy pp. 735–749.

Gaggero, S., Tani, G., Villa, D., Viviani, M., Ausonio, P., Travi, P., Bizzarri, G. and Serra, F., 2017a: Efficient and multi-objective cavitating propeller optimization: An application to a high-speed craft, *Applied Ocean Research*, Volume 64, pp. 31–57.

Gaggero, S. and Villa, D., 2017b: Steady cavitating propeller performance by using OpenFOAM, StarCCM+ and a boundary element method, *Proceedings of the Institution of Mechanical Engineers Part M: Journal of Engineering for the Maritime Environment*, 231 (2), pp. 411–440.

Gaggero, S., Villa, D. and Viviani, M., 2017c: An extensive analysis of numerical ship self-propulsion prediction via a coupled BEM/RANS approach, *Applied Ocean Research*, Volume 66, pp. 55–78.

Gaggero, S. and Villa, D., 2017d: Cavitating Propeller Performance in Inclined-Shaft Conditions with OpenFOAM: the PPTC 2015 test case, Accepted for publication on *Journal of Marine Science and Applications* (JMSA-2016-09-0083).

Grasso, A., Villa, D., Brizzolara, S. and Bruzzone, D., 2010: Nonlinear motions in head waves with a RANS and a potential code, *Journal of Hydrodynamics*, Volume 22, Issue 5, Supplement 1, pp. 10–15.

Guo, B.J., Steen, S. and Deng, G.B., 2012: Seakeeping prediction of KVLCC2 in head waves with RANS, *Applied Ocean Research*, Volume 35, 2012, Pages 56–67.

Kim, J., Kim, K.S., Kim, G.D., Park, I.R. and Van S.H., 2006: Hybrid RANS and potential based numerical simulation for self-propulsion performances of the practical container ship *J Ship Ocean Technol*, 10 (4), pp. 1–11.

Larsson, L., Stern, F., Visonneau, M., Hirata, N., Hino, T., and Kim, J., 2015: Tokyo 2015: A workshop on CFD in ship hydrodynamics. In Proceedings (Vol. 2).

Resistance Committee, 2017: Final Report and Recommendations to the 28th ITTC, *Proceedings of 28th International Towing Tank Conference*, Vol 1, Wuxi, China.

Sung, Y.J., and Park, S.H., 2015: Prediction of ship manoeuvring performance based on virtual captive model tests, *Journal of the Society of Naval Architects of Korea*, 52(5), 407–417.

Tani, G., Viviani, M., Villa, D. and Ferrando, M., 2017: A study on the influence of hull wake on model scale cavitation and noise tests for a fast twin-screw vessel with inclined shaft, *Proceedings of the Institution of Mechanical Engineers Part M: Journal of Engineering for the Maritime Environment*, pp. 1–24.

Villa, D., Gaggero S. and Brizzolara, S., 2011: Simulation of ship in self propulsion with different CFD methods: from actuator disk to potential flow/RANS coupled solvers, *Proceedings of International Conference-Developments in Marine CFD* RINACFD2011, London, England, 22–23 March 2011.

Villa, D., Viviani, M., Tani, G., Gaggero, S., Bruzzone, D. and Podenzana Bonvino, C., 2017: Numerical Evaluation of Rudder Performance Behind a Propeller in Bollard Pull Condition, *Accepted for publication on Journal of Marine Science and Applications.*

Resistance and propulsion

Progress in Maritime Technology and Engineering – Guedes Soares & Santos (Eds)
 ISBN 978-1-138-58539-3

Experimental study of frictional drag reduction on a hull model by air-bubbling

E. Ravina & S. Guidomei
DITEN, Polytechnic School, Genoa, Italy

ABSTRACT: The paper refers on a research activity developed at DREAMS Lab of the University of Genoa (Italy), focused on experimental application of air-bubbling technique on a hull model. In this study the reduction in the frictional resistance by air bubbling is tested with the injection of compressed air on the bottom of the model tested in the towing tank. The injection of compressed air is managed by a customized pneumatic circuit, designed in according to the model shape and size. The pneumatic unit is designed to allow a flexible distribution of air in different areas of the hull and to organize flexible tests usually not implemented in standard test facilities.

1 INTRODUCTION

As well known, one the most strategical problems is the air pollution and a lot of researches are oriented to propose solutions to reduce it. About 30% of the total use of energy, mainly given by oil, is spent in the transports areas and 80% of the global trade is represented by the maritime one. Consequently there is a wise interest to improve ship energy efficiencies to control the emissions.

Ships efficiency can be improved reducing the loss of energy or using new kind of alternative energies. One of the most important causes of energy lost is the hull friction drag, representing 60÷70% of the total drag. Different approaches and techniques are proposed and applied to reduce this kind of resistance, in particular modifying the hydrodynamic conditions of the hull-water interface.

One of most promising techniques modifying the boundary layer structure is the air-bubbling, subject of this experimental study, implemented on a hull model. The main goal is the measurement of the changes in the local frictional drag at different levels of flow rate and pressure of injected air. The experiments in the towing investigate on the differences in terms of drag reduction between hull without and with holes on the bottom, modifying the characteristics of speed, pressure, flow rate and areas interested to the air injection.

In the paper are collected the results corresponding to different working conditions, showing the advantages related to air bubbling.

The hull equipped with pneumatic circuit is still used to arrange new systematic unconventional experiments oriented to optimize the gain related to air bubbling technology.

2 STATE OF THE ART

In the last few years a lot of studies and the experimentations have been done about air lubrication to have more information about its application and benefits.

From the experimental point of view Mitsubishi has developed this technique on real ship, with the installation of an air-lubrication system achieving an energy saving of 12% (Mizokami et al. 2010).

Istanbul Technical University proposes experimental tests on vertically and horizontally flat plates with holes powered by compressed air, showing gains about 5% (Gökçay 2012).

Samsung Ship Model Basin has tested in 2014 a ship model with 6 injection units which were independent one to each others and characterized by a matrix with a lot of small holes. Resistance gains of 8÷10% and power gains of 10÷12% are obtained (Jang et al. 2014).

From the numerical point of view recent progresses in CFD for naval architecture and ocean engineering are synthesized by specific analyses (Stern et al. 2015). In particular, CFD air lubrication methods and effects of air lubrication on resistance of chemical tankers are numerically investigated (Dogrul et al. 2010).

Many others studies and detailed deepening are available in literature, showing a diffuse interest on this topic.

3 EXPERIMENTS ON AN HULL MODEL

The study is implemented on a model with a large flat bottom having main dimensions L = 1.8 m, B = 0.3 m and D = 0.13 m (Fig. 1). The sinkage is

105 mm and the trim is equal to zero, reached adding balance weights inside the model.

3.1 *Air distribution circuit*

The pneumatic unit is designed to allow a flexible distribution of air in different areas of the hull: an original pneumatic distribution circuit is designed and realized for this model, customized for the hull shape. The presence of two collectors that feed the holes at the stern and two collectors for the holes at the bow gives the possibility to select the zone for the injection of the air (Fig. 2).

Because of the connection at non-return valves, these 4 collectors are independent one from each other: this solution allows to make different tests with the injection of air in different zones of the hull and each collector is equipped with a flow regulator. From each collector 10 pneumatic hoses allow the distribution of the compressed air in each one of the 40 holes made in the bottom of the hull (Fig. 3).

The unit is arranged in advance to install flow rate proportional valves: this configuration, at the moment under development, will allow a flexible modulation of the air injection during the navigation.

Figure 1. Model of hull.

Figure 2. Detail of air collector.

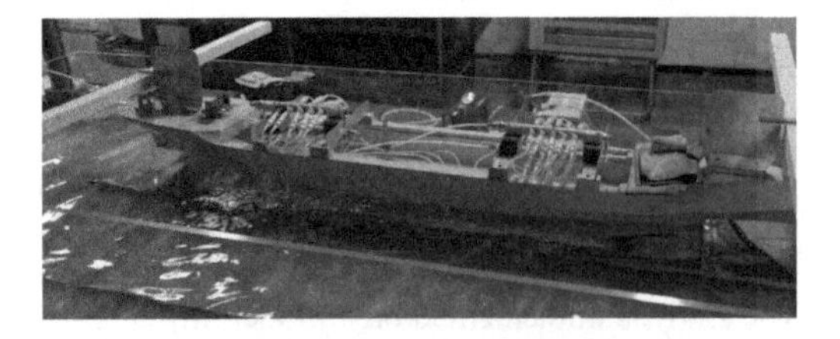

Figure 3. Hull equipped with the pneumatic circuit.

3.2 *Test at the towing tank*

Seven different operating conditions at three different levels of speed were measured at the towing tank with the goal to estimate the changes in local frictional drag at different levels of flow rate and pressure of injected air in different areas of the bottom (Fig. 4)

In particular a pressure regulator allows feed the circuit from the compressor at 0.7 bar, 0.5 bar and 1.5 bar: for these three levels of pressure the test was made feeding all the holes after the test made with the original hull and with the hull equipped but without the injection of the air.

All the tests are made respectively at 0.66 m/s, 0.83 m/s and 1.01 m/s.

After preliminary tests the air pressure is selected to 0.7 bar, and in this condition tests with the injection of air only in the holes of the bow and only in the holes of the stern are implemented. For each operative condition the results were analysed and the average value of the resistance at different speeds was calculated to estimate the drag's law.

Significant differences on the draft and on the wetted surface comparing the original hull and the hull with air injection are not detected. All resistance measurements concern the total resistance of the hull under different working conditions.

3.2.1 *First condition: Original hull*

The results of the first condition representing the original hull without holes are reported in the Table 1.

Figure 4. Hull assembled in the towing tank.

Table 1. Results first condition.

Speed	Resistance
m/s	g
0.66	63.26
0.83	109.46
1.01	170.14

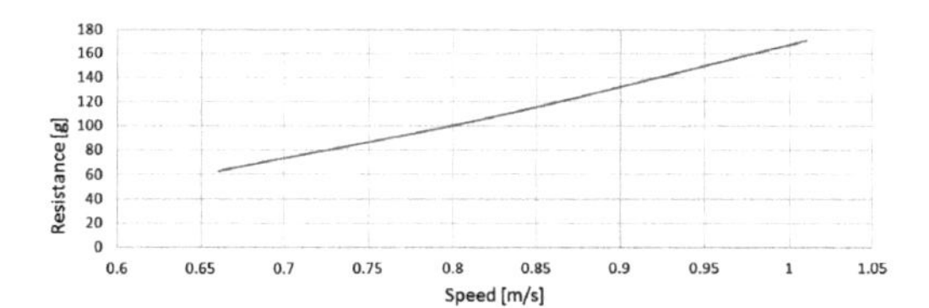

Figure 5. Resistance law: first condition.

Table 2. Results second condition.

Speed	Resistance
m/s	g
0.66	59.43
0.83	106.45
1.01	170.93

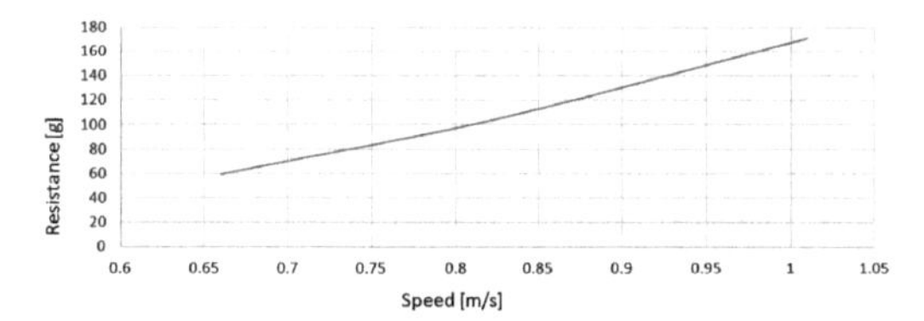

Figure 6. Resistance law: second condition.

The corresponding resistance law is shown in Fig. 5. The resistance values are expressed in grams, in according to the calibration of the load cell installed on the towing tank.

3.2.2 *Second condition: Hull modified without air injection*

The results of the second condition representing the hull modified with holes on the bottom but without the air injection are reported in Table 2 with the corresponding resistance law in Fig. 6.

The presence of the matrix of holes generates a gain of resistance that decreases increasing the speed (from 6% at 0.66 m/s to –0.5% at 1.01 m/s). Anyway these differences are comparable with the measure uncertainty of the load cell (sensitivity tolerance ± 0.1%, nominal load 10 kg).

3.2.3 *Third condition: Hull modified with air injection at 0.7, 1.5 and 0.5 bar in all the holes*

The results reported in the Table 3 correspond to the third condition representing the hull modified with air injection in all 40 holes at three different values of pressure: 0.7, 1.5 and 0.5 bar.

The corresponding resistance laws are collected in Fig. 7.

Table 3. Results third condition.

Speed [m/s]	Resistance [g]		
Pressure [bar]	0.7	1.5	0.5
0.66	53.47	53.76	56.29
0.83	96.25	96.50	97.25
1.01	161.45	149.78	167.58

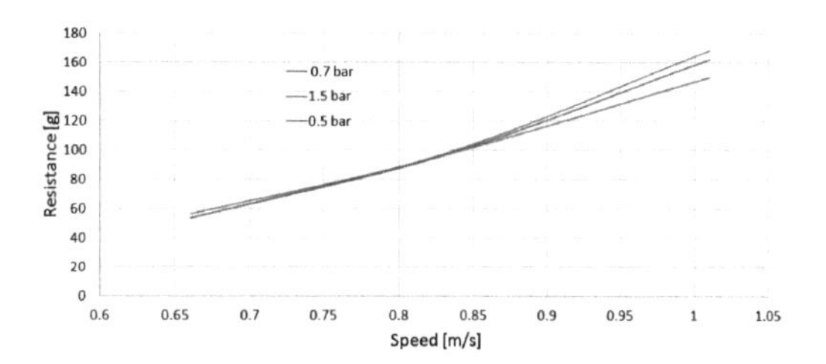

Figure 7. Resistance law: third condition.

Table 4. Results fourth condition.

Speed	Resistance
m/s	g
0.66	55.77
0.83	98.11
1.01	164.33

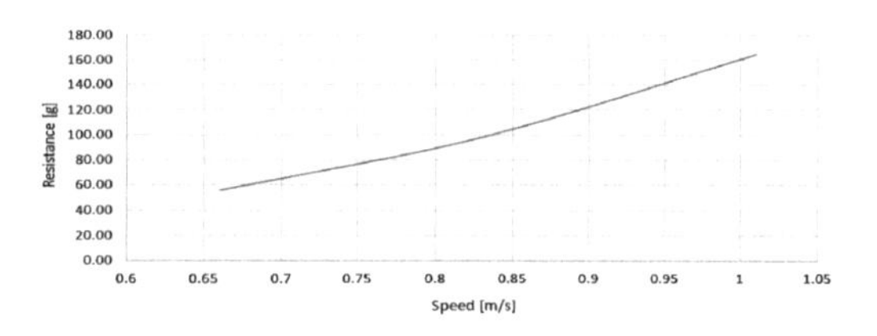

Figure 8. Resistance law: fourth condition.

The advantage of the air bubbling is shown at higher speeds, increasing the air pressure.

On the contrary, with reference to the hull under test high pressure levels coupled to high speed generate significant turbulence and the hull tends to swamp water.

For this reason the further experiments are implemented to 0.7 bar of pressure. In particular, hereafter injections only to bow and to stern are discussed.

3.2.4 *Fourth condition: Hull modified with air injection at 0.7 bar in holes at the bow*

The results of the fourth condition representing the hull modified with air injection in only holes at the bow at 0.7 bar are reported in Table 4.

The corresponding resistance law is shown in Fig. 8.

In comparison with the previous condition, at the same pressure, the levels of resistance are a little in In comparison with the previous condition, at the same pressure, the levels of resistance are little higher, showing the positive contribution of the other areas of bubbling.

Then the opposite bubbling configuration is tested, as described hereafter.

3.2.5 *Fifth condition: Hull modified with air injection at 0.7 bar in holes at the stern*

The results of the condition representing the hull modified with air injection in only holes at the stern at 0.7 bar are reported in the Table 5.

The corresponding resistance law is shown in Fig. 9.

These tests show the relevant effects of the air injection at the stern of the hull on the resistance values, with enhancements respect to the bubbling at the bow.

Table 5. Results fifth condition.

Speed	Resistance
m/s	g
0.66	53.02
0.83	91.70
1.01	159.13

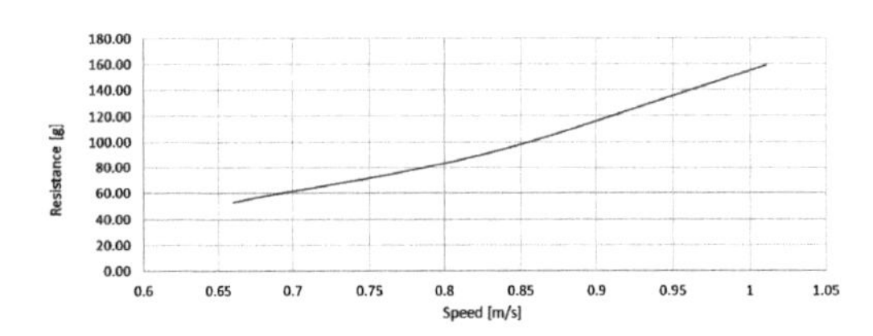

Figure 9. Resistance law: fifth condition.

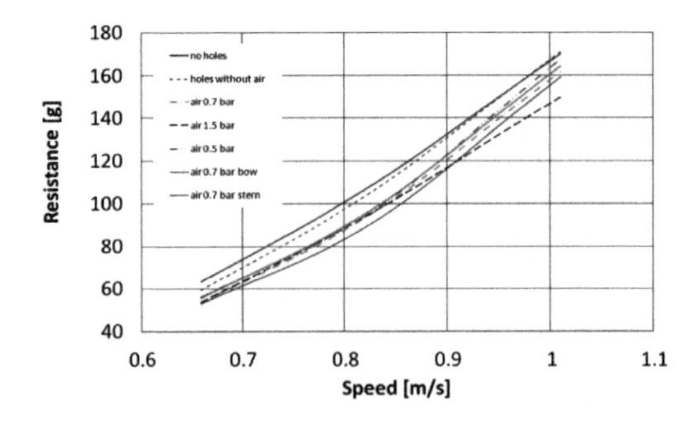

Figure 10. Resistance law: comparison.

3.3 *Final results*

All experiments are compared to understand which operative condition gives more benefit. (Fig. 10).

The corresponding gains are evaluated under the following working conditions:

1. hull modified with air injection at 0.7 bar in all the holes
2. hull modified with air injection at 1.5 bar in all the holes
3. hull modified with air injection at 0.5 bar in all the holes
4. hull modified with air injection at 0.7 bar in holes at the bow
5. hull modified with air injection at 0.7 bar in holes at the stern

and are compared to the condition of the hull modified with the matrix of holes but without air injection.

The corresponding results are collected in Table 6.

In general the injection of compressed air shows a positive influence in terms of gain of friction resistance; in particular the best condition is given by a medium level of pressure for a medium level of speed with the injection of the air in only zone of the hull's stern with a gain that is more than the 13% (value comparable with the results available in literature).

Finally the percentage of the resistance gain is evaluated (Table 7).

Table 6. Resistance gains.

Speed	Resistance gain				
m/s	%				
	1	2	3	4	5
0.66	9.57	9.54	5.28	6.16	10.79
0.83	9.58	9.35	8.64	7.83	13.85
1.01	5.38	12.37	1.96	3.86	6.90

Table 7. Resistance gain with air injection in stern holes at 0.7 bar in comparison with hull equipped but without air injection.

Speed	Resistance gain
m/s	%
0.66	10.79
0.83	13.85
1.01	6.90

For the higher speed of 1.01 m/s the more relevant differences from the operative conditions are detected and so also the benefits given by air injection are more evident.

4 TRANSPOSITION AT FULL SCALE

A possible transposition of the results obtained to value the effective efficiency of this solution in terms of gain in fuel consumption e gas emissions is estimated.

Unfortunately, this transposition is influenced by the towing tank dimensions ($60 \times 2.5 \times 3$ m) and by the maximum speed reachable (max. 2 m/s). That limits the scale of the models under test. Consequently is very difficult to discuss observed trends when considering the air-bubbling application at full scale, in terms of resistance.

In particular, the resistance gain achieved with the injection of air at 0.7 bar in all the holes in comparison with the condition of hull modified but without are injection is analyzed. Supposing the hydrostatic pressure and estimating the pressure drops inside the pneumatic circuit at full scale the air injection pressure is estimated around 4 bar.

In this case the resistance gain is 9.58%: the same gain is assumed for the real ship corresponding at the model used.

The resistance of the ship in the two operative conditions is evaluated and consequently the power of the engines with the corresponding fuel consumption that allows to estimate the gain in terms of fuel cost.

Then the cost of the compressed air is estimated and compared with the previous one gain.

Finally a net profit is estimated in around 17 Euros per hour, corresponding to 6.5% of the costs. It is aligned with the economical results available in technical literature and could be considered satisfactory in long terms of time during a travel of the ship. It is important to note that these estimations are deduced by experiments on model in calm water. The towing tank used for experiments is not equipped with wave making facility: consequently the profit estimation cited above is certainly different in presence of waves. The effect generated by waves should be negative on the benefits related to air-bubbling but in order to confirm this opinion additional tests into towing tanks equipped with wave generators could be necessary.

5 CONCLUSIONS

The experimental activity on plates and on model of hull show very interesting results related to the application of the air bubbling technique.

The advantage of air bubbling is more evident to high speed, with different gain related to the pressure level.

With reference to tests on the plates grater resistance gain occur in presence of matrix of a lower number of greater holes with respect to a greater number of smaller holes.

With reference to the test on the model interesting different performances are detected varying the zone of the air injection, in particular, from stern to bow.

The pneumatic unit is designed to allow a flexible distribution of air in different areas of the hull and to organize flexible tests usually not implemented in standard test facilities.

In general the injection of compressed air shows a positive influence in terms of gain of friction resistance; in particular the best condition is given by a medium level of pressure for a medium level of speed with the injection of the air in only zone of the hull's stern with a gain that is more than the 13% (value comparable with the results available in literature). Actually the realized test benches can be considered as pilot test facilities able to arrange new systematic unconventional experiments oriented to optimize the gain related to air bubbling technology.

ACKNOWLEDGEMENTS

The authors thank Professor Carlo Podenzana to make available the model of hull and Dr. Alberto Ferrari and Mr. Sergio Talocchi for the support given during the tests at towing tank.

REFERENCES

Blaine W. Andersen (1976), "The Analysis and Design of Pneumatic Systems", Robert E. Krieger Publishing Company, Malabar, Florida, USA.

Dogrul A, Alikan Y. and Celik F. (2010), "A numerical investigation of air lubrication effect on ship resistance", *Intl. Conf. on Ship Drag Reduction (SMOOTH-Ships)*, 20–21 May, Istanbul, Turkey.

Gokcay S. (2012), "Ship drag reduction through air injection to boundary layer", May, Istanbul Technical University.

Jang J., Ho Choi S., Ahn S., Kim B., Seo J, (2014), "Experimental investigation of frictional resistance reduction with air layer on the hull bottom of a ship", Marine Research Institute, Samsung Heavy Industries, Korea.

Kawabuchi M., Kawakita C., Mizokami S., Higasa S., Kodan Y., (2011), Takano S., *CFD* "Prediction of Bubbly Flow around an Energy-saving Ship with Mitsubishi Air Lubrication System", *Mitsubishi Heavy Industries Technical Review* Vol. 48 No. 1.

Kumagai I., Nakamura N., Murai Y., Tasaka Y., Takeda Y., Takahashi Y., (2010), "A New Power-saving Device for Air Bubble Generation: Hydrofoil Air Pump for Ship drag Reduction", *International Conference on Ship Drag Reduction*, May, Istanbul, Turkey.

Lyu X., Tang H., Sun J., Wu X., Chen X., (2014), "Simulation of microbubble resistance reduction on a suboff model", *Brodogradnja/Shipbuilding*, Volume 65, Number 2.

Mizokami S., Kawakita C., Kodan Y., Takano S., Higasa S., Shigenaga R., September (2010), "Experimental study of air lubrication method and verification of effects on actual Hull by means of sea trials", *Mitsubishi Heavy Industries Technical Review*, Vol. 47 n°3.

Stern F. et al., (2015), "Recent progress in CFD for naval architecture and ocean engineering", *Journal of Hydrodynamics*, Ser. B. Vol. 17, Issue 1, pp. 1–23.

Progress in Maritime Technology and Engineering – Guedes Soares & Santos (Eds)
 ISBN 978-1-138-58539-3

Procedure for production of scaled ship models for towing tank testing

K.D. Giannisi, D.E. Liarokapis, J.P. Trachanas, G.P. Milonas & G.D. Tzabiras
National Technical University of Athens, Athens, Greece

ABSTRACT: Ship owners traditionally use the results of towing tank tests to evaluate ship performance. Scaled ship models must be created for this purpose. According to ITTC instructions the models dimensions must not exceed 1mm tolerance in all directions. The accuracy of the design and manufacture process is crucial to obtain the desired result. In other ways, the reliability of the towing tank tests may be questionable. In this respect, the Laboratory for Ship and Marine Hydrodynamics (LSMH) of NTUA propose a design method to ensure the reliability of the procedure. The Laboratory has recently revised a two and half axis CNC machine for producing scaled ship models, based on a traditional twin spindle 2½ axis hand driven ship model milling machine. This machine consists of four servo motors in conjunction with Mach3 controller that cooperates with CAD/CAM software. In our case, the commercial 3D software Rhinoceros 5 was used. The process used to build the model from the ship's construction plans will be presented in detail. Areas were special attention is needed will be remarked.

1 INTRODUCTION

1.1 *Ship models*

Towing tank testing is often used to evaluate the performance of ships. In that order, scaled ship models are created for testing. The process of the construction strongly affects the total accuracy of the measured results. For that reason, ITTC (ITTC, 2002) has published limits that the constructor must take into account, to obtain the required accuracy. These limits are very hard to achieve, even with modern tools. Especially in the stage of line fairing it is very hard to produce a smooth surface without deforming the initial lines.

The more problematic areas are those in the regions of stern and stem. These areas have either intense curvatures or sharp edges. In order to avoid unwanted fluctuations, the user must fair the lines to such extend that in some cases the deviation from the preliminary design exceeds the recommended limits.

Another restrain concerns the initial lines that are usually taken from the construction plans. In many cases, these plans differ from the design plans. Moreover, in some cases the designs are scanned and are not in digital form, which introduces additional inaccuracies.

Concerning all the above, without the use of modern CNC machines, ITTC guidelines are hard to achieve.

1.2 *Traditional model building methods*

Until recent times, model boat building for testing purposes was created by hand (Davis, 1989). Expert wood working personnel needed a great amount of effort and time to produce an accurate scaled model of the real ship. Moreover, the latest designs for the stern and the stem increased the demands for producing the models. Especially the geometry of the bulb needs special care to create and was usually produced as a separate part.

There are two main techniques of ship model producing, similar to traditional wooden boat building (Chapelle, 1994). The first one is plank on frame models. The frames of the hull are accurate assembled together and then small stripes of wood are fastened from edge to edge (carvel planking), gaining support from the frame and forming a smooth surface (Fig. 1). This procedure has the advantage of low weight model production, which is required for 'power boats' testing. In the second technique, which is traditionally called "bread and butter" (Kempf & Remmers, 1954), the user must initially build a hollow block, made of lumber wood. Then, by using CAD software, the model is divided by height in several horizontal planes (Fig. 2). Consequently, by using a hand-driven milling machine, waterlines are formed on the wooden block. The inner edges between the planes define the outside surface of the hull. Based on these traces, the user

Figure 1 Carvel method in boat building.

Figure 2. "Bread and butter" method in boat building.

has to remove the excess wood in order to create a smooth surface.

The height of the planes strongly affects the accuracy of the method. Usually, for optimum result and minimum working time, the user chooses to use a sufficient number of waterlines.

Both techniques require highly experienced working crew and the model production takes at least a month, including design and construction. However, the most detrimental about these techniques is the dependency of the final results on the user. Moreover, in order to ensure that the model dimensions are within the ITTC recommendations, the tested model must be accurately measured. Concerning the above, it is obvious that a new, automated procedure is needed to reduce the working time and inaccuracies.

In this manner, the current work is based on proposing the steps needed to design and create a ship model, suitable for towing tank testing. These include the drawing of the ship lines from the construction plans, the creation of the hull surface, as well as the hardware equipment needed to transform the Rhino commands into machine code. Finally, we will analyze how to retain the outcome under the acceptable limits.

1.3 *Modern model building methods*

Nowadays, most of the towing tanks use highly sophisticated CAD-CAM software in conjunction with CNC milling machines. These consist of up to five axis motion, robotic arms, rotating support tables etc. Correspondingly, the Laboratory for Ship and Marine Hydrodynamics, recently updated an old hand-driven milling machine to a 2½ axis, symmetrical CNC machine. The machine consists of four servo motors matched with position encoders, Mach3 controller and the input surface is produced by Rhino5 CAD-CAM software.

2 DESIGN OF MODEL'S HULL

2.1 *Lines insertion*

As it was mentioned above, the line plans are usually obtained from the construction plans given from the ship company. In this way, the experimentalist can produce a model from the actual ship. These plans are usually found in printed versions and cannot be directly imported to the design software. There are three methodologies to overcome this problem (Kostas, 2014). The first one is to transfer the offsets from the plans and import the data to the software manually. The second one is to scan the plans and import them as images to CAD. But the most efficient method is to use digitizing software to scan the plans. This procedure, although is time efficient, the quality of the scanning affects the final result. In all cases, the user must correct the inaccuracies, resulting from the data transfer. In order to create a faired surface of the ship's hull, the obtained lines must be faired. When the information from the construction plans is inadequate, further ship sections are needed to describe the hull geometry. In cases where there are intense alternations in the geometry, a denser grid of curves is required.

2.2 *Surface creation*

In our case, the commercial software Rhinoceros 5 was used. Before inserting the line plans, we started by creating a frame spaced baseline in the workspace, based on the ships dimensions. This defines our basis and corrects small misalignments during the lines' insertion (Fig. 3). A duplicate of the inserted curves provides the prospect to check the deviation from the faired lines.

The next step is the fairing of the line plans, starting from the profile curve and proceeding to the sections. At this point, we perform an initial check of the faired lines, so as not to exceed the outer dimensions. That can be performed by applying the List command, which provides us the coordinates of the curve points. The process is terminated when the curvature analysis gives a fairly smooth result (Robert Mc Neel, 2015).

Flat of Side, Flat of Bottom and Deck curves are then created, based on the faired profile lines and sections. Following, we repeat the previous procedure of fairing the new lines and checking. To reinsure the validity of our progress an additional measure is performed on the construction plans.

The next step is the creation of several waterlines, based on the information derived from previous analysis. The step between the waterlines depends on the judgment of the designer. Areas of intense alternations in the geometry usually requires denser grid of waterlines. Again the lines need to be faired. Important issue in this stage is not to lose the conjunctions between the lines. In other case the surfaces cannot be created. The designer must rebuild and re-fair the sections from the dense grid of the waterlines. This is an iterative process which is terminated when all the curves are properly jointed, beneath the given deviation. The aim of this process is to capture the real geometry of the hull.

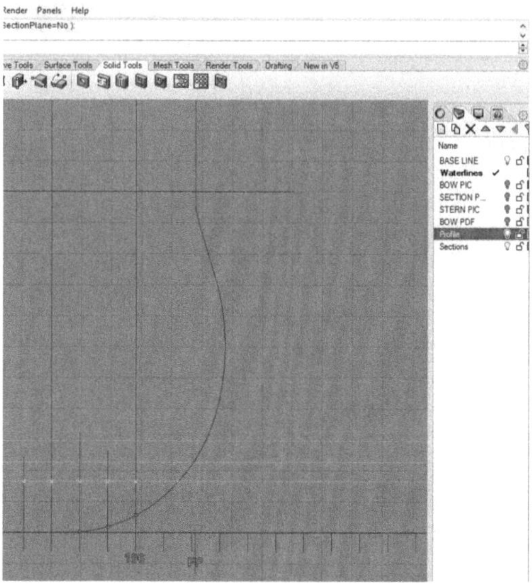

Figure 3. Alignment of bow profile with baseline.

At Figure 4, we notice the insertion of the surfaces. In some cases, likewise the bulb and the stern region, the designer must use a different approach. More precisely, new sections must be added where given information is inadequate. This addition will not alternate the hulls geometry, as they derive from the profile, FOS and the waterlines. It must be mentioned that in some cases is unavoidable to create the surface partially. To ensure the integrity of the final surface, we join the smaller surfaces with G1 continuity (Kostas, 2014). After visual inspection, we proceed with surface's curvature analysis. The result we obtained can be observed at the following Figure (Fig. 5).

At this point, the design process is finished. The maximum deviation from the initial lines was estimated at 101 mm at the area of the deck, close to the bow. This refers to full scale dimensions. The entire wetted surface's deviation was under 35 mm. The model's length is about 4.5 m, which corresponds to a scale ratio of 1:40. Thus, the ITTC guidelines were accomplished.

A schematic illustration of the hull's frame and surface is presented in the following figure (Fig. 6).

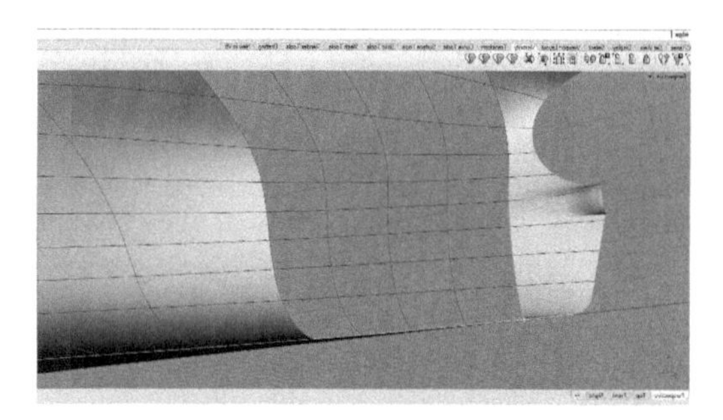

Figure 4. Surface creation.

Figure 5. Curvature Analysis.

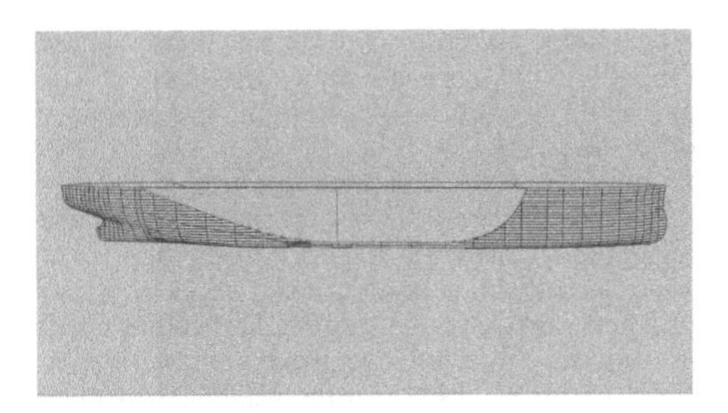

Figure 6. Final design.

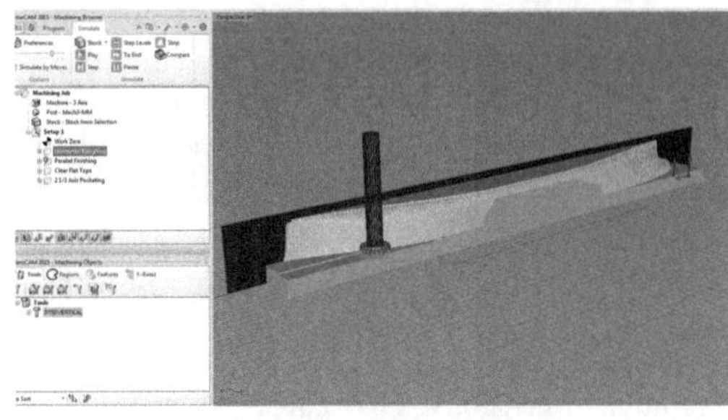

Figure 7. Simulation of roughing process.

3 CUTTING PROCESS

3.1 *Computer aid design to manufacturing*

As it was mentioned before, the milling machine utilizes two spindles that allow the symmetrical cutting, which greatly reduces the cutting time. The vertical axis utilizes two moving parts, which move identically. Typical cutting time is estimated to about twenty hours for a model of 3.5 meters. Thus, only the starboard side of the model is used. Again, the commercial software Rhino 5 CAM was used.

Usually, the geometry is imported in the CAM workspace with the deck facing down. At that point, the orientation of the model must be specified. The next step is to create the stock (bounding box), which defines the dimensions of the block to be cut (Robert Mc Neel, 2015). The stock must encloses the model and its dimensions must not exceed the workspace of the milling machine. Usually, in order to avoid winnowing any wastage of material, the stock's dimensions are slightly bigger from the model's. Furthermore, the stock material must be selected. The user chooses the cutting tool and sets the vertical step. Depending on the type of the material, the cutting tool characteristics (diameter and number of blades) and the spindle speed, the software adjusts the cutting speed. The cutting path is automatically produced from the Rhino CAM, based on the design created in CAD. Afterwards, the work zero of the machine must be determined and defined.

For optimum results, the cutting process can be divided in two or more processes. Usually, the first one consists of the roughing of the bounding box (Fig. 7). At that stage, it is proposed to set up an offset of few millimeters, to increase the surface's smoothness. In our case, the offset was chosen to 2mm. The second process (that is referred as finishing), combines a concurrent movement of all axis (Fig. 8). For an even better result, the user can divide these processes in more stages. However, this is not suggested for wooden constructions, as it is time-consuming.

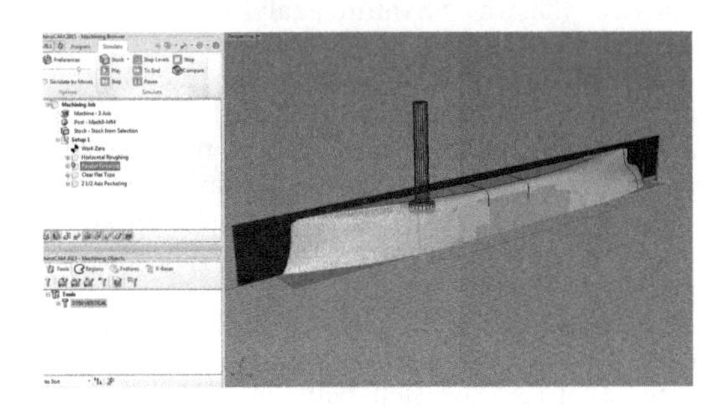

Figure 8. Simulation of finishing process.

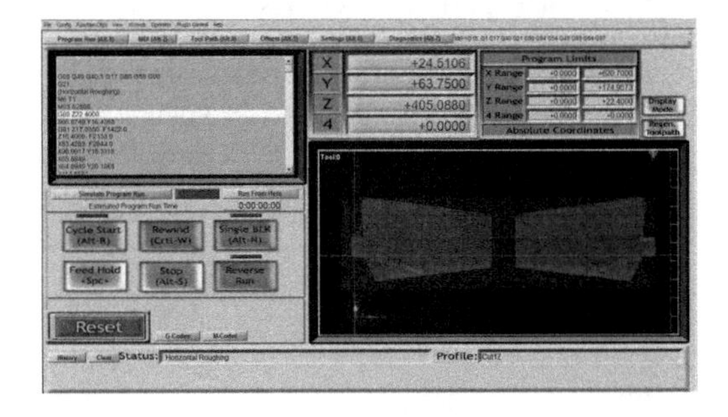

Figure 9. G-code in Mach 3 Workspace.

Before continuing to the cutting procedure, it is highly recommended to run a process simulation. Any conflicts will be noticed in this stage. Finally, by using the post analysis, the software generates the G-code that is transferred to the Mach 3 controller (Fig. 9).

3.2 *Model making*

The cutting procedure begins with the alignment of the bounding box (stock) with the center of the cutting table. The use of a base is suggested, to avoid any accidental contact between the cutting tools and the working table. The stock must be firmly fixed up to the table, to prevent any unwanted vibrations during cutting. Then, the cutting tools are

Figure 10. Finishing process.

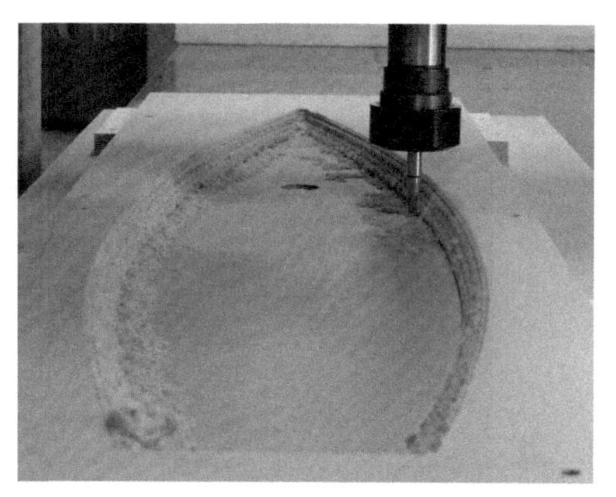

Figure 11. Internal roughing of horizontal plane.

Figure 12. Assembly of horizontal planes.

set to zero point position (ArtSoft Software Inc., 2005), matching to the work zero defined in CAM. It is suggested to define the reference point in every process, to reduce machine's digitizing error.

Depending on the processes defined in CAM, the user must begin implementing the G-codes, from roughing to finishing (Fig. 10). To improve the quality of cutting, the cutting speed can be adjusted. In any case, any intervention in the position of the tool is feasible, either to introduce new coordinates or to interrupt the procedure. Concerning the deviation, after calibrating the CNC machine, we measured a maximum tolerance of 0.2 mm per m.

3.3 *Alternative solutions for light weight and accurate weight distributed models (MDF and composites)*

In some cases, model weight or strength is crucial in the production process. Typical techniques are not sufficient for such creations.

Different types of material or even different approach must be used. In our case, the aim was to keep the overall weight as low as possible. Thus, composite material was used (in this example, MDF was used), combined with the forming of the internal hull. Each horizontal plane was designed and cut separately (Fig. 11). The deviation is considered the same as previously.

Then follows the assembly of the formed planes (Fig. 12). After creating the hollow block, the procedure referred at model making is repeated. With this procedure, the hull width can be altered, depending on the needs of experiment. The hull's external surface is designed and by setting an offset, the hull's width is produced. At Figure 13, the finishing process of the model is presented.

3.4 *Rudder*

A part that is commonly used in self-propulsion and resistance tests is the ship's rudder. The rudder is manufactured in two separate parts that were afterwards joined together. Guidelines are used to ensure that the two parts will precisely be assembled. To provide a light weight and rigid

Figure 13. External finishing.

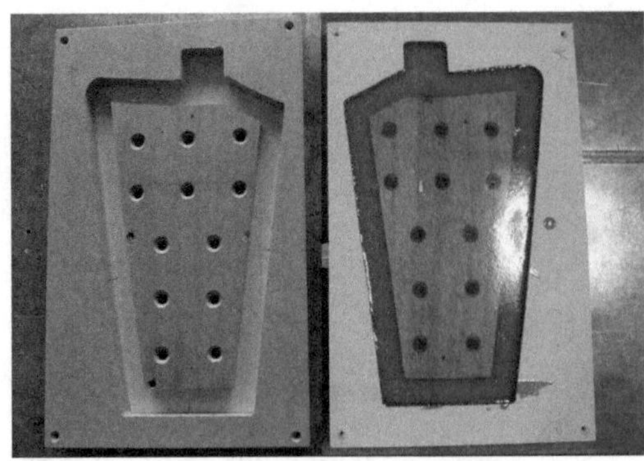

Figure 15. Epoxy rudder after first process.

Figure 14. Stock creation for rudder production.

Figure 16. Rudder ready to use.

construction, composite materials were used. After the forming of the rudder, an epoxy coat with fiberglass covered the surface. To avoid exceeding the recommended limits, the rudder's external surface was designed 1 mm smaller than the preliminary design. Each layer of the fiberglass is estimated about 0.3 mm in width. By using three layers of fiberglass, we assume that the total added width is about 1 mm.

An alternative approach is to create the stock with composite materials. The CNC machine is used to create a hollow container, which gives shape to the epoxy resin when it cools and hardens. Below, an illustration of the process is presented (Fig. 14).

After the creation of the stock, the procedure described in model making is repeated (Fig. 15).

The painted rudder with its axis fitted is displayed below, at Figure 16.

3.5 *Intake mould creation*

Except from the model creation, the CNC machine can be used for producing several parts needed for the experiments. Below, the mould creation of an intake manifold is presented. In this example, the stock was created with green MDF, which is easy to machine and is completely moisture resistant.

Once again, the mould was created in two parts that were later assembled. At Figures 17–18, we can see the shaping of the mould and the final result.

Figure 17. Shaping of the intake's mould.

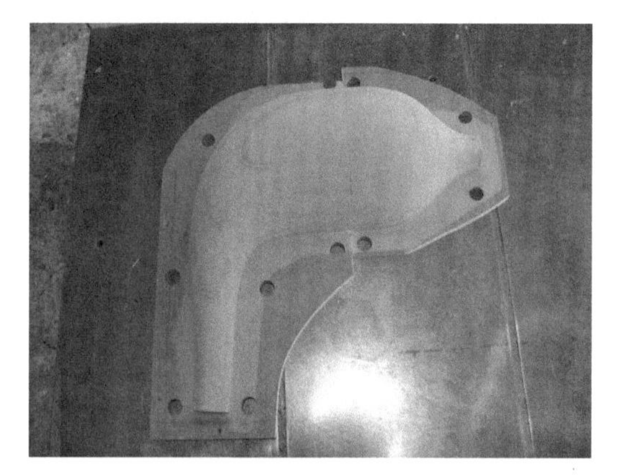

Figure 18 Intake's mould.

4 FUTURE WORK

The LSMH is currently working on the implementation of measuring device to the CNC machine. More specifically, the Mach 3 controller has an option to connect a measuring devise so as to perform an automated scanning procedure. Both contact probes and laser measuring devises can be used for this purpose.

5 CONCLUSIONS

The LSMH has recently updated a traditional milling machine to a modern CNC for model making purposes. The procedure followed to produce a model was described in detail. Alternative techniques to match the needs of experiments were presented.

As it is obvious from the manuscript, trying to minimize the deviation from the original dimensions is a difficult task to achieve. That applies even from the design stage. Considering that the final surface needs further treatment (primer, paint), the model must be accurately measured in order to ensure that is beneath the requested deviation.

Finally, we suggest that the guidelines for the recommended tolerances must be revised. The deviation limit should be a function of the scale ratio. Higher scale ratios should refer to stricter tolerances.

ACKNOWLEGMENTS

Acknowledgments must be given to laboratory faculty, who did their best to get this work completed.

REFERENCES

ArtSoft software incorporated, 2005. *Using Mach3 Mill A user's guide to installation, configuration and operation*. ArtSoft software incorporated.

Chapelle, I.H. 1994. *Boatbuilding: A complete handbook of wooden boat construction*. New York: W.W. Norton & Company.

Davis, G.C. 1989. *The built-up ship model*. New York: Dover Editions. International Towing Tank Conference, (01). 2002. *ITTC – Recommended Procedures and Guidelines: Model Manufacture Ship Models*.

Kempf & Remmers, 1954. *Making wooden ship models*. Hamburg: Kempf & Remmers.

Kostas, K.B. 2014. *3D Design and calculations on Rhino 3D*. Athens: Da Vinci.

Robert McNeel & Associates, 2015. *Rhino for Windows: User Guide*. Barcelona: Robert McNeel & Associates.

Progress in Maritime Technology and Engineering – Guedes Soares & Santos (Eds)
 ISBN 978-1-138-58539-3

A benchmark test of ship resistance in extremely shallow water

Q. Zeng, C. Thill & R. Hekkenberg
Delft University of Technology, Delft, The Netherlands

ABSTRACT: CFD (Computational Fluid Dynamics) calculations have been making dramatic contributions to ship resistance prediction. Since the virtual models used are always simplifications of the real ships and fluid, results need to be validated by experiments. However, although a certain number of model tests can be found, the publicly available resistance tests in shallow to extremely shallow water (water depth-to-draft ratio $h/T \leq 2.0$) are rare. Therefore, to provide data for validation, a test of an inland ship model in shallow water is performed in a towing tank. Four shallow water depths are applied and one deep water case is added for comparison. The uncertainties in this test are analyzed for the measuring instruments as well as the resistance, trim and sinkage. This test is motivated to provide the path for the ongoing research by the authors on the improved prediction of ship resistance in shallow water and enables the benchmark for other researchers, who investigate ship resistance in extremely shallow water, with experimental data to validate CFD calculations.

1 INTRODUCTION

Inland shipping is—in terms of CO_2 efficiency—regard as a "green" alternative for inland transportation due to its high energy efficiency. It is also capable of transporting much larger volume and weights of cargos at relatively lower costs compared to rail and road transport (van der Laan et al., 2010). Therefore, inland shipping plays an irreplaceable role in intracontinental transportation. Therefore, understanding ships' behavior in inland (usually shallow and restricted) waterways is important as is the corresponding hull optimization.

Predicting ship resistance accurately in shallow and/or confined water is crucial to design a resistance-optimized hull form and estimate the minimum required power. Investigations have been done and quantitatively indicated the speed reduction due to shallow water effects, e.g. Schlichting (1934) (translated by Roemer (1940)). With the development of computer hardware and software, more hydrodynamic details of ship resistance in shallow water can be understood through Computational Fluid Dynamics (CFD) calculations (Raven, 2012, Terziev et al., 2018). As CFD methods are more affordable than model tests, they have become popular tools and have been making dramatic contributions to ship resistance prediction.

However, the virtual models used in CFD computations are approximations of real ships. Simplifications are frequently used to grab the most important parts concern, e.g., using a half model in a symmetrical condition. In addition, the precision of CFD method depends on the mathematical way of solving the Navier-Stokes equations, the method of space discretization, the turbulence models, etc. For instance, as shown by Eça and Hoekstra (2008), results of the friction on a flat plate varied significantly when different turbulence models applied. As a consequence, results from CFD computations should be validated by experiments.

For sea-going ships in the deep water, a plenty of experimental data can be found, especially for those "standard" ship types like the Kriso Container Ship (KCS) (Moctar et al., 2012). However, publicly available experiments for ships sailing in shallow water are insufficient. Schlichting (1934) performed a number of ship resistance tests in shallow water, but the ships he used are cruisers, which is less applicable for fuller inland ships. Jiang (2001) did some tests with a typical inland ship and his method is only applicable for moderately shallow water (water depth-to-draft ratio $h/T \geq 1.5$). A recent test with the h/T up to 1.2 was performed by Mucha et al. (2017), but a limited number of results can be found for the resistance in different water depths.

In this study, tests of an inland ship model (bare hull) in shallow and calm water are performed in a towing tank. Four different shallow water cases are implemented and one deepwater case is added for comparison. The depth Froude number (Fn_h) varies from 0.1 to 0.75 (subcritical speed range), which is common for inland vessels. Additionally, the uncertainties of the measuring instruments, as well as the results of resistance, trim and sinkage, are evaluated according to the guidelines of ITTC (2014b).

The benchmark tests in this study can extend the experimental database of ship resistance, trim, and sinkage in shallow water and will provide researchers who investigate ship resistance in extremely shallow water with experimental data to validate CFD calculations.

2 TEST MODEL AND TEST SCHEME

2.1 *Test model*

A bare-hull wood-made ship model (Figure 1) is applied. This is a 1/30 scaled model of an inland ship with the main parameters shown in Table 1.

Figure 1. The ship model in towing tank.

Table 1. Main parameters of both the model and full-scale ship.

	symbol	unit	model	full-scale
Scale	λ	–	1/30	1
Length	L	m	2.867	86.0
Beam	B	m	0.380	11.4
Draft	T	m	0.117	3.5
Mass	M	kg	109.19	2948118
Wetted surface	S	m^2	1.575	1417.8
Block coefficient	C_B	–	0.864	0.864

Sand strips are used to stimulate a turbulent flow at three positions: x = 2.787 m (close to the end of the bow), x = 2.487 m (transition point to the parallel body), and x = 2.187 (an appropriate place to re-stimulate the flow). The strips are 40 mm wide and have an experimentally validated resistance coefficient of 0.01.

The ship model is mounted in the small towing tank of TU Delft, which is 85 m long, 2.75 m wide and with the maximum carriage speed of 3 m/s. The water depth of the tank can be adjusted within the range from 0 m to 1.25 m.

2.2 *Test scheme*

In this test, a waterway for which the depth-to-draft ratio (h/T) is less than 2.0 is assumed to be extremely shallow. Four shallow water cases with an h/T of 2.0, 1.8, 1.5 and 1.2 are applied. One case of deep water (h/T = 10.71, the deepest condition for this towing tank) is applied for comparison.

In general, the design speed of an inland ship is mostly in the range of 10 ~ 15 km/h. In this study, this range is expanded to 6 ~ 22 km/h not only to include more navigation conditions but also to cover a large enough part of the subcritical speed range. The values of depth Froude number ($Fn_h = V/\sqrt{gh}$, where V is ship's speed) for each case are shown in Table 2.

Table 2. The depth Froude number for each case.

V (m/s)	V (km/h)	Depth Froude number (Fn_h)				
1/30 model	full-scale	h/T = 10.71	h/T = 2.01	h/T = 1.80	h/T = 1.50	h/T = 1.20
0.3	5.92	0.086	0.198	0.209	0.229	0.256
0.4	7.89	0.114	0.263	0.279	0.305	0.341
0.5	9.86	0.143	0.329	0.348	0.382	0.427
0.6	11.83	0.171	0.395	0.418	0.458	0.512
0.7	13.80	0.200	0.461	0.488	0.534	0.597
0.8	15.77	0.228	0.527	0.557	0.611	–*
0.9	17.75	0.257	0.593	0.627	0.687	–
1.0	19.72	0.286	0.659	0.697	0.763	–
1.1	21.69	0.314	0.724	0.766	–	–

*To avoid grounding, cases marked "–" were not performed.

3 SOURCES AND PROPAGATION OF UNCERTAINTIES

For ship resistance tests, uncertainties may be generated throughout the process and will propagate into the final results of resistance, trim and sinkage. Based on the description of ITTC (2014b), the sources of uncertainty can be grouped into five types: geometry, installation, calibration, repeat measurement and data reduction. Most items in each group are listed in Figure 2.

3.1 *Hull geometry*

The manufacture of the ship model is based on the provided lines plan. Uncertainty might be generated if deviations happen between the physical model and the lines plan. Those deviations may result in a different displacement and/or different wetted surface, which will propagate into the final results (e.g. the resistance). In this study, the model used was well constructed by trained engineering staff for the parallel midship segments and NC milled at the bow and the stern. Therefore, the deviation of the shape is believed being quite small and the corresponding uncertainty can be ignored.

Additionally, deformations due to a different temperature condition, material property, etc. are minor and also hard to measure. Consequently, the corresponding uncertainty can also be ignored.

3.2 *Test installation*

This group of uncertainty source relates the drift angle, the alignment of the direction of the dynamometer with the ship model's centerline, and other aspects of the test mounting.

During ship resistance tests, the model should be ballasted to its designed waterline. A deviation can cause an uncertainty into the displacement ($u(\Delta)$). Consequently, an uncertainty is generated into the wetted surface and then propagates into the value of resistance. The wetted surface is assumed to be a two-thirds power of the volume ($S \propto \Delta^{2/3}$). Therefore, the relative uncertainty of the wetted surface ($u'(S)$) can be represented as

$$u'(S) = \frac{u(S)}{S} = \left(\frac{u(\Delta)}{\Delta} + 1\right)^{2/3} - 1. \quad (1)$$

As the wetted surface area has linear effects on ship resistance, the influence of the deviation of the displacement on the resistance can be easily derived.

For the tests in shallow water, if the model is not ballasted to the designed waterline, the water depth will change accordingly. The resistance, trim, and sinkage in shallow water, which depend on the water depth, will be affected by the mismatched draught. In this test, the ship model and the measuring instruments are carefully mounted and the quality of the installation is guaranteed (e.g. the drift angle is within 0.1 degrees, the initial trim is less than 0.1 degrees). Therefore, the uncertainty caused by the test installation into the final results is deemed negligible.

3.3 *Instrument calibration*

In this section, the calibrations of the instruments for measuring forces, trim, sinkage and temperature are illustrated sequentially.

3.3.1 *Dynamometer for resistance*

Two dynamometers were chosen. One was for the towing point in the front of the model, which undertook the majority of force; the other was at the back and connected with a slider (free to move back and forth). Meanwhile, the predicted maximum drag of the ship in shallow water is 30 N. Considering a three to four times larger starting force is needed, two dynamometers with a range of 100 N and 50 N were used for front and the back, respectively.

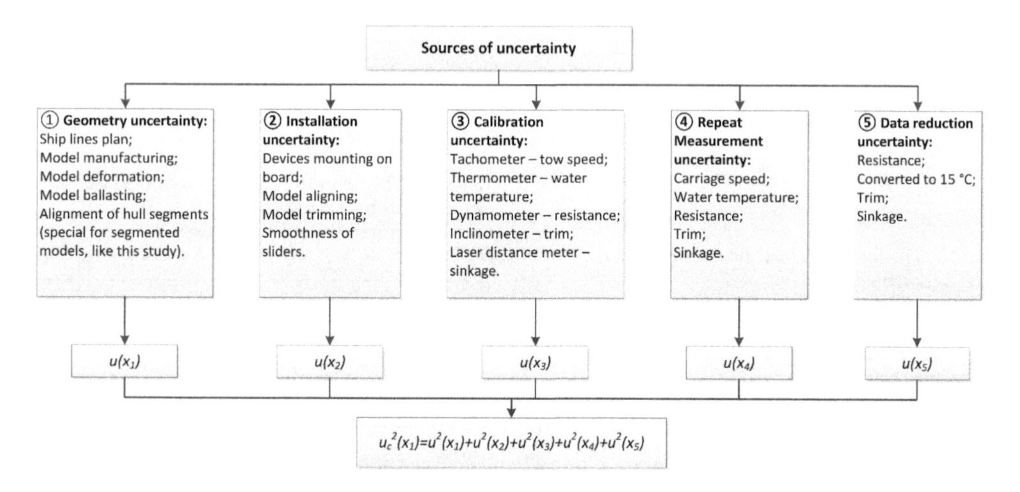

Figure 2. Sources of uncertainty in ship resistance test.

Based on the ITTC (2014d), end-to-end calibrations were performed for both two dynamometers, in both the positive and the negative direction. By regularly adding the masses to the maximum and then reducing them to zero, changes of forces against the voltage are shown in Figure 3.

The calibration factor k_R for resistance is defined as

$$Force(\mathrm{N}) = Voltage(\mathrm{V}) \times k_R. \tag{2}$$

The value of k_R and its standard uncertainty ($u(k_R)$) and relative uncertainty ($u'(k_R)$) are shown in Table 3.

The uncertainties of the dynamometers will directly propagate into the measurements of the resistance.

3.3.2 *Inclinometer for trim*

An inclinometer was applied to measure the trim angle which was mounted at the center of gravity. Errors and uncertainty caused by the deviation from the center of gravity during the test are ignored.

The calibration factor k_{Tr} is defined based on the average angle (θ_A, in degree) and the measured angle (θ_M, in degree):

$$\theta_A = \theta_M \times k_{Tr}. \tag{3}$$

The k_{Tr} and its standard uncertainty ($u(k_{Tr})$) and relative uncertainty ($u'(k_{Tr})$) are shown in Table 4.

The uncertainty of the inclinometer will directly propagate into the measurements of the trim angle.

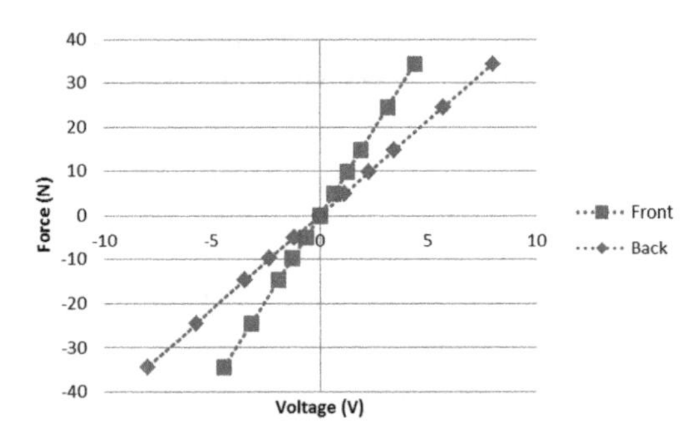

Figure 3. The forces against the voltages for the front and the back dynamometers.

Table 3. The calibration factors k_R of the dynamometers and their uncertainties.

Dynamometer position	k_R	$u(k_R)$	$u'(k_R)$
Front	7.787	0.013	0.17%
Back	4.273	0.007	0.16%

3.3.3 *Laser distance meter for sinkage*

The sinkage is measured by a laser distance meter, which is fixed at the same longitudinal position as the center of gravity of the model. The resolution of this device is sufficiently high (0.1 mm).

The calibration factor k_{Si} for the sinkage measurements is defined as follows:

$$Distance(mm) = Voltage(\mathrm{V}) \times k_{Si}. \tag{4}$$

The k_{Si} and its standard uncertainty ($u(k_{Si})$) and relative uncertainty ($u'(k_{Si})$) are shown in Table 5.

The uncertainty of the laser distance meter will directly propagate into the measurements of sinkage.

3.3.4 *Thermometer for water temperature*

The temperature in the water was regularly recorded with a thermometer, which has a resolution of 0.1°C. Five positions chosen for recording distributed evenly along the towing tank. As the test was performed in a short time (around one week), the effects of temperature among different days on the results of the test were assumed to be negligible.

The calibration factor k_{Te} for temperature measuring is defined based on the average temperature (T_A) and the measured temperature (T_M, in °C):

$$T_A = T_M \times k_{Te}. \tag{5}$$

The k_{Te} and its standard uncertainty ($u(k_{Te})$) and relative uncertainty ($u'(k_{Te})$) are shown in Table 6.

The uncertainty of the thermometer will directly propagate into the measurements of temperature,

Table 4. The calibration factor k_{Tr} of the inclinometer and its uncertainties.

Device	k_{Tr}	$u(k_{Tr})$	$u'(k_{Tr})$
Inclinometer	1.00	0.043	4.26%

Table 5. The calibration factors k_S of the laser distance meter and its uncertainties.

Device	k_{Si}	$u(k_{Si})$	$u'(k_{Si})$
Laser distance meter	6.601	0.056	0.85%

Table 6. The calibration factor k_{Te} of the thermometer and its uncertainties.

Device	k_{Te}	$u(k_{Te})$	$u'(k_{Te})$
Thermometer	1.00	0.0042	0.42%

which affects the density and viscosity of the water. As discussed in ITTC (2011), when the temperature is changed by 0.1 °C (at 20°C), the change of density and viscosity can be estimated as ± 0.0021% and ± 1%, respectively. Therefore, the relative uncertainties of density ($u'(\rho)$) and viscosity ($u'(\mu)$) in this test, considering the uncertainty of thermometer, can be estimated as (T_A is 19.5°C in this test)

$$u'(\rho)=\frac{u(\rho)}{\rho}=\pm\frac{19.5\times 0.42\%}{0.1}\times 0.0021\%=0.0017\%$$
$$u'(\mu)=\frac{u(\mu)}{\mu}=\pm\frac{19.5\times 0.42\%}{0.1}\times 1\%=0.819\%. \quad (6)$$

3.3.5 *Tachometer for carriage velocity*

A tachometer is installed on the carriage indicating the velocity. The accuracy of this tachometer is sufficiently high.

The calibration factor k_V for the tachometer is defined based on the average velocity (V_A, in m/s) and the measured velocity (V_M, in m/s):

$$V_A=V_M\times k_V. \quad (7)$$

The k_V and its standard uncertainty ($u(k_V)$) and relative uncertainty ($u'(k_V)$) are shown in Table 7.

The uncertainty of the tachometer will propagate into the calculation of the total resistance coefficient.

3.4 *Repeating tests*

If an average of the results is used for the final measurand, the standard uncertainty and the relative standard uncertainty can be represented as follows.

- For single measurement

1000 samples were recorded within a unit time (1 second). A moving average $(\hat{x}_i)$ with the interval of 1000 was taken for the results of resistance, trim and sinkage. The number of time units during the effective time range (the recording value is stable) was expressed with the symbol n. The measured value (x_i) and its uncertainty can be derived as

$$x_i=\frac{1}{n}\sum_{i=1}^{n}\hat{x}_i,\ u(x_i)=\sqrt{\frac{1}{n}\sum_{i=1}^{n}(\hat{x}_i-x_i)},$$
$$u'(x_i)=\frac{u(x_i)}{x_i}. \quad (8)$$

Table 7. The calibration factor k_V of the tachometer and its uncertainties.

Device	k_V	$u(k_V)$	$u'(k_V)$
Tachometer	1.00	0.0004	0.04%

- For repeating tests (repeat N times)

$$\overline{x}_i=\frac{1}{N}\sum_{i=1}^{N}x_i,\ u(\overline{x}_i)=\sqrt{\frac{1}{N}\sum_{i=1}^{N}(x_i-\overline{x}_i)},$$
$$u'(\overline{x}_i)=\frac{u(\overline{x}_i)}{\overline{x}_i}. \quad (9)$$

3.5 *Data reduction*

Normally, the results of the total resistance need to be converted into the condition that the temperature is 15°C. The following relation is usually applied (ITTC, 2014a):

$$\frac{R_t(15°C)}{R_t(t_i)}=\frac{\rho(15°C)}{\rho(t_i)}\left[1+\frac{C_f(15°C)-C_f(t_i)}{C_t(t_i)}\right]. \quad (10)$$

where the t_i is the temperature during the test, the C_f can be estimated by ITTC correlation line for deep water. However, the ITTC 57 line is not applicable in shallow water (Zeng et al., 2017), and the exact expression for friction prediction in shallow water is temporarily not available. Therefore, all the results of the total resistance in this test will be expressed under the exact temperature.

In this study, the data reduction for the coefficient of ship resistance can be represented as

$$C_t=\frac{R_t}{0.5\cdot\rho V^2 S}. \quad (11)$$

The measurements of resistance (R_t), water density (ρ), ship velocity (V) and wetted surface (S) can be seen as independent. Therefore, the uncertainty of C_t can be written as follows (ITTC, 2014c):

$$u^2(C_t)=\left(\frac{\partial C_t}{\partial R_t}\right)^2u^2(R_t)+\left(\frac{\partial C_t}{\partial\rho}\right)^2$$
$$u^2(\rho)+\left(\frac{\partial C_t}{\partial V}\right)^2u^2(V)+\left(\frac{\partial C_t}{\partial S}\right)^2u^2(S). \quad (12)$$

Based on the analyses in the above section, the standard uncertainty of C_t can be written as

$$u(C_t)=$$
$$\sqrt{\left(\frac{2}{\rho V^2S}\right)^2\cdot u(k_R)+\left(\frac{2R_t}{\rho^2V^2S}\right)^2\cdot u(\rho)+\left(\frac{4R_t}{\rho V^3S}\right)^2\cdot u(k_V)+\left(\frac{2R_t}{\rho V^2S^2}\right)^2\cdot u(S)}. \quad (13)$$

As discussed in Section 3.2, $u(S)\approx 0$. Therefore, the $u(C_t)$ can be rewritten as

$$u(C_t)=\sqrt{\left(\frac{2}{\rho V^2 S}\right)^2\cdot u(k_R)+\left(\frac{2R_t}{\rho^2 V^2 S}\right)^2\cdot u(\rho)+\left(\frac{4R_t}{\rho V^3 S}\right)^2\cdot u(k_V).} \tag{14}$$

The above equation is the uncertainty of a single test on the measuring instrument, and the uncertainty of repeating tests can follow the method introduced in section 3.4.

For trim and sinkage, they are measured directly. Therefore, the uncertainties of trim ($u(trim)$) and sinkage ($u(sinkage)$) can be derived from the calibration of the inclinometer and laser distance meter:

$$u(trim)=u(k_{Tr});\quad u(sinkage)=u(k_S). \tag{15}$$

4 RESULTS AND UNCERTAINTY EVALUATIONS

During the test (18–28 July 2017), the average temperature in the water in the towing tank is 19.5°C. Therefore, the corresponding density is 998.3091 kg/m^3 and the dynamic viscosity is 0.001014 Pa·s.

The signs of the resistance, trim, and sinkage and are defined as follows:

- resistance is positive pointing to the stern;
- trim is positive with bow up;
- sinkage is positive downwards.

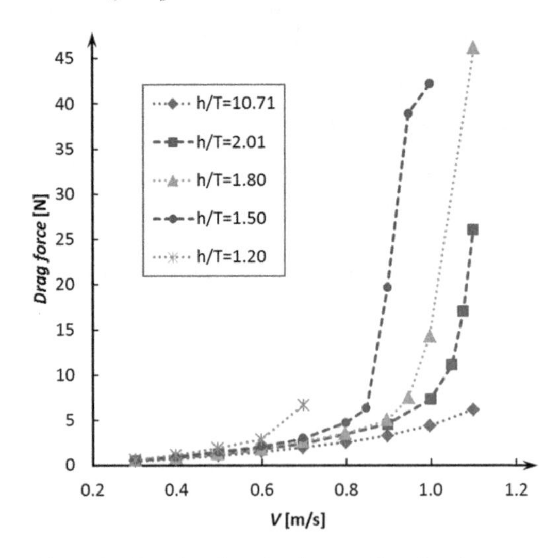

Figure 4. Results of total resistance (drag force) against the velocities in the model tests.

The results of the total resistance and its coefficient are shown in Figure 4 and Figure 5, respectively. The relative uncertainty (in percentage) of the total resistance coefficient is shown in Table 8. Some extra towing speeds were added where necessary (e.g. the range where the trim changes reversely).

Compared with that in deep water, the drag force on ship hull increases in shallow water. A faster "back flow" might be a reason for this. When the depth Froude number (Fr_h) is around 0.7, significant increases of the drag force are observed. Furthermore, a smaller h/T will lead to an earlier change of the drag.

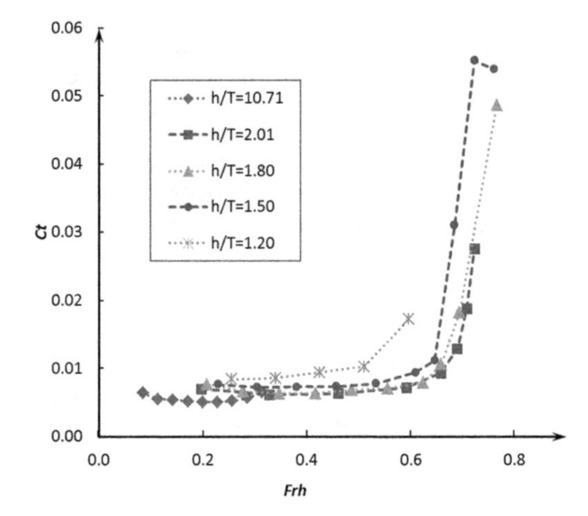

Figure 5. Results of total resistance coefficient against the depth Froude number (Fr_h) in the model tests.

Table 8. The uncertainty of the total resistance coefficient (C_t) (the sign is ±).

	h/T				
V(m/s)	10.71	2.01	1.80	1.50	1.20
0.30	2.316%*	14.223%	14.875%	13.840%	25.578%
0.40	1.717%*	–	10.353%	17.086%	7.347%
0.50	3.139%*	7.477%	5.721%	6.975%	5.311%
0.60	6.590%	–	6.339%	3.673%	4.721%
0.70	2.723%**	5.089%	3.574%	3.807%	7.672%
0.80	4.418%	–	3.509%	3.898%	–
0.85	–	–	–	2.393%	–
0.90	4.400%*	3.683%	2.428%	0.708%	–
0.95	–	–	3.167%	0.604%	–
1.00	5.575%*	1.838%	2.103%	0.618%	–
1.05	–	1.949%	–	–	–
1.075	–	1.432%	–	–	–
1.10	2.746%*	1.470%	0.658%	–	–

*Repeated twice.
**Repeated thrice.

It should be pointed out that a "turning point" ($V = 1.0$ m/s at $h/T = 1.80$) is observed in Figure 5, i.e. when the speed goes up to a certain value, a revered trend happens suddenly. It would be interesting to add more towing speeds around this point to clearly illustrate the physics in this velocity range.

Those cases that were tested multiple time are marked in Table 8. Details of the results are shown in the appendix.

Results of the trim and sinkage are shown from Figure 6 to Figure 9. The relative uncertainty (in percentage) for trim and sinkage are shown in Table 9 and Table 10, respectively.

For the same depth Froude number (Fr_h), trim and sinkage in shallow water are smaller than those in a deep one. This is probably due to a higher

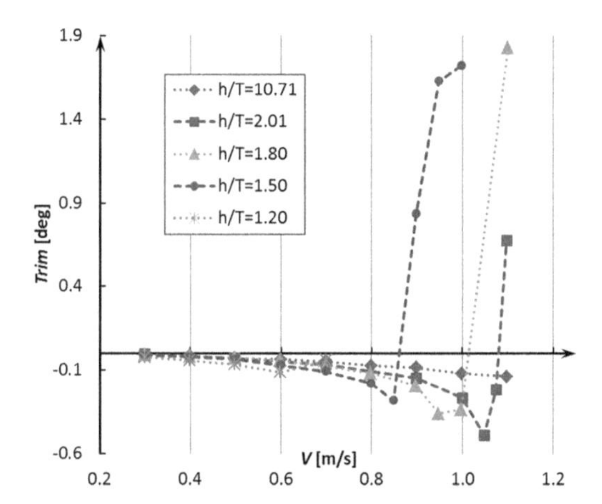

Figure 6. Results of trim against the velocities in the model tests.

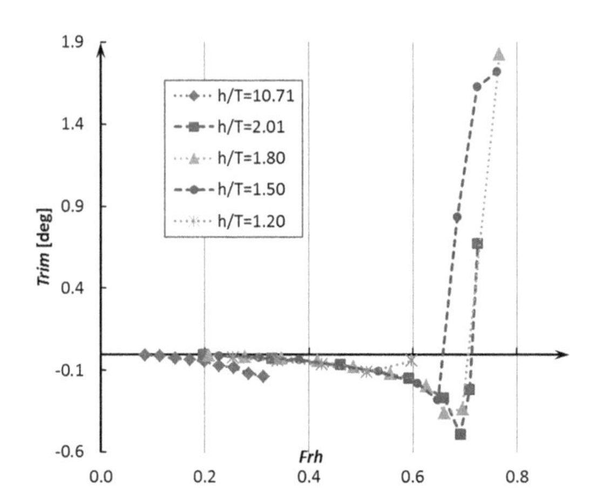

Figure 7. Results of trim against the depth Froude number (Fr_h) in the model tests.

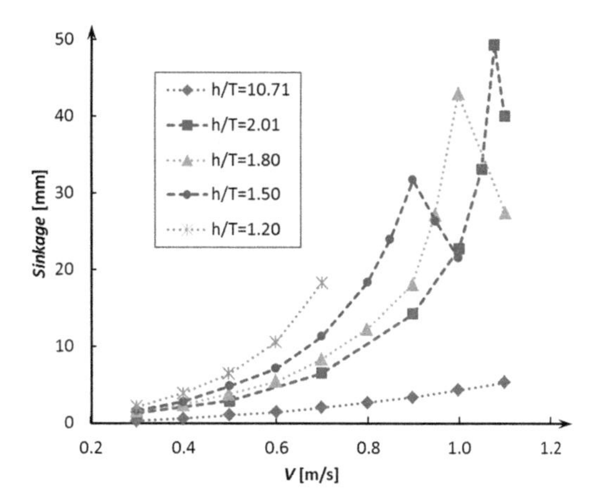

Figure 8. Results of sinkage against the velocities in the model tests.

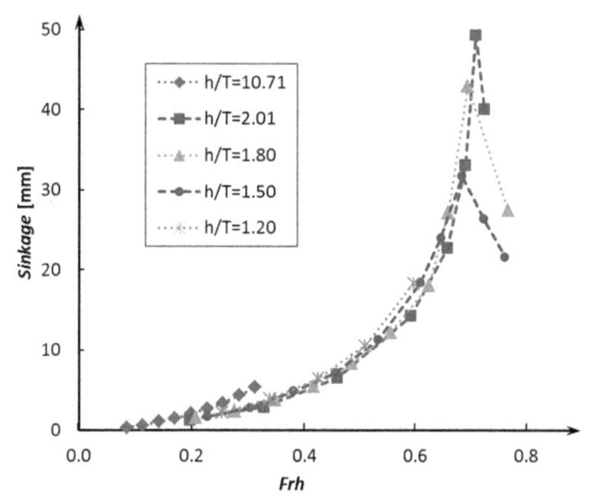

Figure 9. Results of sinkage against the depth Froude number (Fr_h) in the model tests.

Table 9. The uncertainty of trim in the test (the sign is ±).

	h/T				
V(m/s)	10.71	2.01	1.80	1.50	1.20
0.30	42.02%	104.64%	83.35%	49.69%	43.89%
0.40	29.09%	–	45.83%	32.89%	17.89%
0.50	19.16%	26.53%	27.97%	26.00%	13.62%
0.60	20.40%	–	20.83%	13.43%	16.27%
0.70	5.44%	15.85%	16.35%	10.70%	42.13%
0.80	14.94%	–	13.63%	14.57%	–
0.85	–	–	–	10.58%	–
0.90	11.90%	11.54%	9.26%	5.86%	–
0.95	–	–	7.83%	4.86%	–
1.00	9.13%	7.36%	11.12%	4.82%	–
1.05	–	9.36%	–	–	–
1.075	–	12.16%	–	–	–
1.10	6.97%	6.31%	5.20%	–	–

Table 10. The uncertainty of sinkage in the test (the sign is ±).

	h/T				
V(m/s)	10.71	2.01	1.80	1.50	1.20
0.30	13.53%	2.09%	7.83%	10.62%	3.70%
0.40	4.03%	–	3.27%	2.81%	4.87%
0.50	2.62%	5.28%	2.17%	1.82%	2.19%
0.60	9.26%	–	2.60%	1.67%	2.50%
0.70	12.22%	2.06%	2.57%	2.04%	1.46%
0.80	7.82%	–	1.48%	2.55%	–
0.85	–	–	–	1.61%	–
0.90	3.76%	3.40%	1.57%	1.38%	–
0.95	–	–	1.44%	1.39%	–
1.00	2.82%	1.83%	2.80%	2.43%	–
1.05	–	1.21%	–	–	–
1.075	–	0.90%	–	–	–
1.10	3.40%	1.53%	2.24%	–	–

dynamic pressure on the ship hull. When $Fr_h \leq 0.6$, the trim and sinkage seem to be independent on h/T; however, when $Fr_h > 0.6$, a smaller h/T can trigger an earlier inverse of the trend of the trim and sinkage. After this inverse, the dynamic pressure begins to play an important role in ship's behavior in shallow water.

5 REMARKS

In this study, four shallow water cases ($h/T = 1.2$, 1.5, 1.8 and 2.0) were used to do benchmark tests of an inland ship model. One deep-water case ($h/T = 10.71$) was added for comparison. The depth Froude number (Fn_h) varies from 0.1 to 0.75, which covers most of the subcritical speed range. It was confirmed that ship's resistance, trim, and sinkage deviate significantly from those in deep water.

The uncertainties of the measuring instruments and the resistance, trim and sinkage were evaluated. Results of the resistance, trim, and sinkage in different shallow water conditions were expressed with graphs, which build a database for this benchmark test. This test showed the path for the ongoing research by the authors on improving the prediction method of ship resistance in the shallow water. It also enables the benchmark for other researchers, who investigate ship resistance in shallow water, with experimental data to validate CFD calculations.

REFERENCES

Eça, L. & Hoekstra, M. 2008. The Numerical Friction Line. *Journal Of Marine Science And Technology,* 13, 328–345.

Ittc 2011. Recommended Procedures-Fresh Water And Seawater Properties. *Proceedings Of The 26th International Towing Tank Conference*, 7.5–02–01–03.

Ittc 2014a. Recommended Procedures-Example For Uncertainty Analysis Of Resistance Tests In Towing Tank. *Proceedings Of The 27th International Towing Tank Conference*, 7.5–02–02–02.1.

Ittc 2014b. Recommended Procedures-General Guideline For Uncertainty Analysis In Resistance Tests. *Proceedings Of The 27th International Towing Tank Conference*, 7.5–02–02–02.

Ittc 2014c. Recommended Procedures-Guide To The Expression Of Uncertainty In Experimental Hydrodynamics. *Proceedings Of The 27th International Towing Tank Conference*, 7.5–02–01–01.

Ittc 2014d. Recommended Procedures-Uncertainty Analysis Instrument Calibration. *Proceedings Of The 27th International Towing Tank Conference*, 7.5–01–03–01.

Jiang, T. A New Method For Resistance And Propulsion Prediction Of Ship Performance In Shallow Water. Proceedings Of The 8th International Symposium On Practical Design Of Ships And Other Floating Structures, 2001 Shanghai, China.

Moctar, O.E., Shigunov, V. & Zorn, T. 2012. Duisburg Test Case: Post-Panamax Container Ship For Benchmarking. *Ship Technology Research,* 59, 50–64.

Mucha, P., El Moctar, O., Dettmann, T. & Tenzer, M. 2017. Inland Waterway Ship Test Case For Resistance And Propulsion Prediction In Shallow Water. *Ship Technology Research,* 64, 106–113.

Raven, H. A Computational Study Of Shallow-Water Effects On Ship Viscous Resistance. 29th Symposium On Naval Hydrodynamics, Gothenburg, 2012.

Roemer, M.C. 1940. Translation: Ship Resistance In Water Of Limited Depth-Resistance Of Sea-Going Vessels In Shallow Water.

Schlichting, O. 1934. Schiffswiderstand Auf Beschränkter Wassertiefe: Widerstand Von Seeschiffen Auf Flachem Wasser. Jahrbuch Der Stg.

Terziev, M., Tezdogan, T., Oguz, E., Gourlay, T., Demirel, Y.K. & Incecik, A. 2018. Numerical Investigation Of The Behaviour And Performance Of Ships Advancing Through Restricted Shallow Waters. *Journal Of Fluids And Structures,* 76, 185–215.

Van Der Laan, A., Bloemhof, J. & Beijer, C. 2010. Sustainable Inland Transportation.

Zeng, Q., Hekkenberg, R., Thill, C. & Rotteveel, E. 2017. Numerical And Experimental Study Of Resistance, Trim And Sinkage Of An Inland Ship Model In Extremely Shallow Water. *International Conference On Computer Applications In Shipbuilding (Iccas2017).* Singapore: Rina.

APPENDIX

Table A. The values of resistance (N) in different cases.

V(m/s)	h/T 10.71	2.01	1.80	1.50	1.20
0.30	0.449	0.487	0.540	0.538	0.586
0.40	0.685	–	0.820	0.904	1.068
0.50	1.048	1.193	1.228	1.421	1.836
0.60	1.450	–	1.774	2.069	2.869
0.70	1.933	2.407	2.596	2.971	6.660
0.80	2.521	–	3.534	4.697	–
0.85	–	–	–	6.266	–
0.90	3.288	4.544	5.004	19.629	–
0.95	–	–	7.460	39.195	–
1.00	4.385	7.287	14.260	42.165	–
1.05	–	11.093	–	–	–
1.075	–	17.041	–	–	–
1.10	6.136	26.063	46.200	–	–

Table B. The values of trim (°) in different cases.

V(m/s)	h/T 10.71	2.01	1.80	1.50	1.20
0.30	–0.007	–0.005	–0.008	–0.012	–0.022
0.40	–0.015	–	–0.017	–0.022	–0.043
0.50	–0.028	–0.029	–0.026	–0.036	–0.066
0.60	–0.034	–	–0.043	–0.069	–0.111
0.70	–0.048	–0.065	–0.078	–0.107	–0.041
0.80	–0.073	–	–0.119	–0.180	–
0.85	–	–	–	–0.284	–
0.90	–0.083	–0.148	–0.194	0.833	–
0.95	–	–	–0.360	1.627	–
1.00	–0.119	–0.267	–0.336	1.722	–
1.05	–	–0.489	–	–	–
1.075	–	–0.218	–	–	–
1.10	–0.138	0.673	1.828	–	–

Table C. The values of sinkage (mm) in different cases.

V(m/s)	h/T 10.71	2.01	1.80	1.50	1.20
0.30	0.293	1.289	1.540	1.656	2.138
0.40	0.612	–	2.393	2.775	3.858
0.50	1.025	2.880	3.768	4.832	6.407
0.60	1.445	–	5.440	7.120	10.549
0.70	2.018	6.530	8.331	11.273	18.329
0.80	2.671	–	12.256	18.360	–
0.85	–	–	–	23.984	–
0.90	3.368	14.296	18.103	31.714	–
0.95	–	–	27.177	26.393	–
1.00	4.323	22.762	42.936	21.587	–
1.05	–	33.092	–	–	–
1.075	–	49.334	–	–	–
1.10	5.332	40.096	27.456	–	–

Ship propulsion

Progress in Maritime Technology and Engineering – Guedes Soares & Santos (Eds)
 ISBN 978-1-138-58539-3

Optimization scheme for the selection of the propeller in ship concept design

M. Tadros, M. Ventura & C. Guedes Soares
Centre for Marine Technology and Ocean Engineering (CENTEC), Instituto Superior Técnico, Universidade de Lisboa, Lisbon, Portugal

ABSTRACT: In this study, an optimization scheme is presented coupling an open-source parametric design and analysis tool for propellers, Open Prop, and a nonlinear optimizer. This developed model, implemented in Matlab, is used to optimize different parameters of the geometry of the propeller to maximize the propeller efficiency and to verify the limits of cavitation. Propeller diameter, rotation speed, number of blades, expanded area ratio, skew and the hub diameter are the parameters optimized. The optimization procedure was tested as follows: first, a container ship was selected and the resistance of the ship was computed, then the propeller was optimized for the specified ship design speed, taking into account the power losses from the engine to the propeller, wake fraction and thrust deduction fraction. The optimization model is constructed, including the objective and the boundary conditions, while the fitness function is proposed employing penalty functions to express constraints. The model developed can be further used to optimize the parameters of any fixed pitch propeller for a given vessel.

1 INTRODUCTION

A ship is a very complex system, which requires different optimization methods during the design stage to control the huge amount of variables involved. These methods require a large number of numerical iterations to identify the optimal solution which become more easily with the aid of advanced computers to achieve (Tadros et al., 2017). Ship design deals with different fields and each one has its methods of calculation (Watson, 1998). These methods are based on real data collected from existing ships.

Marine propulsion systems are one of these complex systems that are responsible to propel the ship at the desired speed and to provide sufficient thrust to support maneuvering, stopping and backing. It is composed of four primary components; the prime mover, the transmission shaft, the propulsor and a control system. The prime mover must supply efficient power to rotate the propeller to achieve the required speed under any load. The power is transmitted to the propeller from the prime mover passes through the transmission shaft, then the thrust is transferred from the propeller to the main thrust bearing of the ship. A control system is used to convey the operator's desired actions to the main propulsion plant.

1.1 *Propeller selection*

Prior of propeller selection, the hull of the ship must be generated to fit the owner requirements, then the resistance in calm water must be computed for different speeds. Watson (1998) presented different methods to generate hull surface using know design parameters or geometric modifications of a basis ship or a particularly suitable methodical series data. Each method has its own calculations to estimate the resistance and propulsion power.

After the establishment of the ship hull, the type of propeller is chosen taking into account the propeller geometry and the blade section. Also, the interaction between the propeller and the ship must be considered presented in the propulsive coefficient as the wake fraction, the thrust deduction factor and relative rotative efficiency. The propeller must be efficiently designed to improve the energy efficiency design index (EEDI) to maximize the deadweight of the ship and to minimize the fuel consumption and the CO_2 emissions. These are the main guidelines to choose a propeller from the point of view of ship design.

1.2 *Literature review*

Rankine (1865) calculated the propeller performance using the momentum theory. This theory was very simple and did not considered the geometry of the blade (Carlton, 2012). Then, Froude (1878) extended the work done by Rankine and developed the blade element theory. This method took into account the geometry of the propeller blade without considering inflowing water from its far upstream value relative to the propeller disc.

Recently, the lifting line method have been considered during the design process for propellers as preliminary design tools (Epps & Kimball, 2013b). It can further applied in lifting surface blade design (Lee et al., 2014), panel method analysis (Peterson, 2008) and computational fluid dynamics (CFD) analysis (Stanway & Stefanov-Wagner, 2006).

In addition to the theories of propeller, different optimization methods are used to optimize the identified parameters in order to select the suitable propeller for a given ship. These optimization methods show a good ability to deal with complex and nonlinear system in different applications (Hillier and Lieberman, 1980).

Radojčić (1985) used sequential unconstrained minimization technique (SUMT) to optimize propeller design at the preliminary design stage. The genetic algorithms (GA) technique is also used to maximize efficiency of B-series propeller with nonlinear constraints (Suen & Kouh, 1999). Karim & Ikehata (2000) used some additional parameters than Suen & Kouh including Reynolds number correction of design parameters and material strength. Benini (2003) used multi-objective optimization methods to maximize the efficiency of B-screw propeller and thrust coefficient taking into account the cavitation constraints. Chen & Shih (2007) used genetic algorithm to obtain an optimal set of parameters leading to efficient propeller performance for both hydrodynamic efficiency and vibration consideration, while, Gaafary et al. (2011) took into account cavitation, material strength and required propeller thrust.

In this study, a model based optimization scheme is developed coupling OpenProp (Epps & Kimball, 2013a) a software tool for designing and analyzing the marine propeller and a nonlinear optimization method. Due to the large number of parameters that affect the propeller performance, the model, implemented in Matlab, shows a good flexibility to select and design a propeller for a given vessel by optimizing the parameters of propeller according to the lifting line method. This model helps the designer to maximize the propeller efficiency and to verify the limits of cavitation as suggested by Burrill & Emerson (1978). Propeller diameter, rotation speed, number of blades, expanded area ratio, skew and the hub diameter are the parameters optimized at the design speed of a container ship. The losses along the gearbox and transmission shaft are considered, also, the wake fraction and thrust deduction fraction are calculated. The optimization model is constructed, including the objective of the study, the boundary conditions and the penalty functions to express constraints.

This model can be used to select and design a fixed pitch propeller during the preliminary stage of ship design. Also, it can be coupled with an engine model as presented in (Maftei et al., 2009, Morsy El Gohary & Abdou, 2011, Welaya et al., 2013, Tadros et al., 2015, Theotokatos et al., 2016, Martelli & Figari, 2017) to control the behaviour of the propeller with the operating range of the engine and to be connected to a ship routing code to determine the behavior of the ship and to analyse the amount of fuel consumption and exhaust emissions during the voyage (Vettor & Guedes Soares, 2016, Prpić-Oršić et al., 2016, Vettor et al., 2016).

The rest of the paper is organized as follows: section 2 presents a general presentation of ship data and engine characteristics; section 3 gives an overview of the methodology and software used to optimize the propeller taking into account the cavitation problem; the simulation and results are presented in section 4, while conclusion and are discussed in section 5.

2 GENERAL PRESENTATION OF THE SHIP AND THE ENGINE

2.1 *Ship data*

A container ship is selected to apply the optimization model to design and select the suitable propeller. Table 1 shows the main characteristics of the ship.

The total ship resistance, RT, is calculated using numerical equations suggested by Holtrop & Mennen (1982) and Holtrop (1984) for different ship speeds as pre-sented in Figure 1.

The resistance is divided into two main components, the viscous resistance, R_V, and the wave making resistance, R_W, as presented in following equation.

$$R_T = R_V + R_W \tag{1}$$

Table 1. Ship data.

Parameter	Value
Length water line (m)	111.187
Breadth (m)	19.5
Draft (m)	7.24
Displacement (m³)	10,894
Block Coefficient	0.694
Design speed (knots)	18
Number of propellers	1

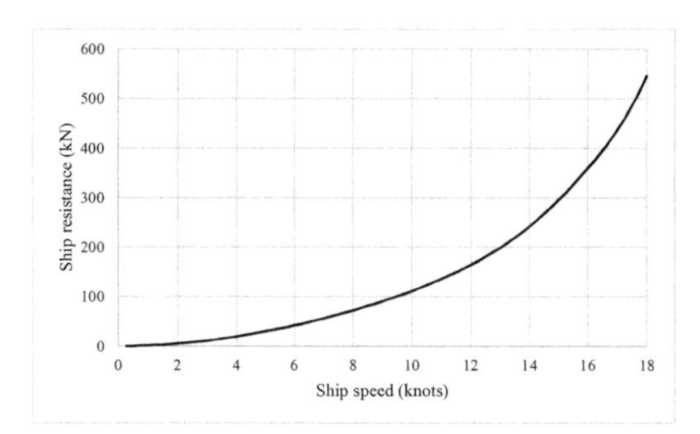

Figure 1. Total resistance of container ship.

Table 2. Technical data of MAN 18V32–44CR.

Parameter	Value
Bore (mm)	320
Stroke (mm)	440
No. of cylinders	18
Displacement (liter)	640
BMEP (bar)	23.06
Piston speed (m/s)	11
Rated speed (RPM)	750
Maximum torque (Nm)	116,886
Engine speed range (RPM)	450–750
Specific fuel consumption (g/kW.h)	190
Power-to-weight ratio (kW/kg)	0.1
Exhaust gas status	IMO Tier 2

The advance speed, VA, is calculated as function of ship speed, VS, taking into account the wake fraction, w, as in the following expression.

$$V_A = V_S(1 - w) \tag{2}$$

While the thrust deduction fraction, t, is used to estimate the thrust force, T, from the total resistance.

$$T = \frac{R_T}{1 - t} \tag{3}$$

2.2 *Engine characteristics*

The engine selected is the MAN 18V32/44CR. It is a large marine four-stroke turbocharged diesel engine.

It is applied for mechanical or diesel-electric propul-sion drives in container vessel.

The maximum power of this engine is 9,180 at 750 RPM and Table 2 shows the main characteristics of the engine (MAN Diesel & Turbo, 2014).

3 METHODOLOGY

3.1 *Propeller model*

The performance of the propeller is calculated using OpenProp as a suite of open source code implement-ed in Matlab used to design both, propeller and tur-bine developed by students and professors from MIT (Epps et al., 2009) which shows good results in comparison with experimental data (Yum et al., 2017).

This model is based on the lifting line propeller design theory using the propeller vortex lattice (PVL) where the propeller is represented by a set of lifting lines partitioned into panels (Kobayakawa & Onuma, 1985). It has the ability to consider the cavi-tation and the blade stress analysis.

The input parameters required for simulation of the propeller model as shown in Figure 2 are divided into four groups:

1. General parameters such as number of blades (Z), propeller speed (N), propeller diameter (D), required thrust (T), ship speed (V_S), hub diameter (D_{hub}) and water density (ρ).
2. Airfoil type defined by mean line type (i.e. NACA = 0.8) and thickness form (i.e. NACA 65 A010)
3. Blade design values such as r/R, c/D, tmax/D, VA/Vs, skew and rake, where r is the blade element position, R is the propeller radius, c is the chord length and tmax is maximum thickness at each blade element.
4. The characteristics of the duct.

3.2 *Cavitation method*

The cavitation is calculated using Burrill's method (Carlton, 2012). This method is suitable for the fixed pitch and conventional propellers. It has been used with considerable success by propeller

Figure 2. Graphical User Interface (GUI) for OpenProp v3.3.4.

designers to estimate the basic blade area ratio associated with a propeller design.

The mean cavitation number ($\sigma_{0.7R}$) is calculated based on the static head relative to the center-line of the shaft, while, the dynamic head is referred to the 0.7R blade section. The propeller cavitation limit is defined by following expression:

$$\frac{T / EAR}{0.5 \rho V_{0.7R}^2} \leq \left[0.2533 + 0.03892 \ln(\sigma_{0.7R}) \right] \tag{4}$$

where,

$$Burrel\ Cavitation_{limit} = \left[0.2533 + 0.03892 \ln(\sigma_{0.7R}) \right]^2 \tag{5}$$

$$Burrel\ Cavitation_{actual} = \frac{T / EAR}{0.5 \rho V_{0.7R}^2} \tag{6}$$

$$\sigma_{0.7R} = \frac{P_{atm} + \gamma(h - 0.7R) - P_v}{0.5 \rho V_{0.7R}^2} \tag{7}$$

where, P_{atm} is the atmospheric pressure, P_v is the vapor pressure, γ is the specific weight, h is the propeller center-line immersion and $V_{0.7R}$ is the flow velocity at $0.7R$ blade section.

3.3 *Nonlinear programming method*

In a Matlab environment, an optimization scheme is used to couple the propeller model with a constrained nonlinear optimization model as shown in Figure 3. The optimization model is the in Matlab function fmincon. This function uses gradient-based optimization algorithms to find the minimum or maximum value of the objective function using interior-point algorithm. It shows a good computational timing in comparison with other optimization methods (MathWorks, 2017).

In order to handle constraints, most of nonlinear programming problems have used the penalty function approach which was proposed by Homaifar et al. (1994) in an application of genetic algorithms (GAs). Deb (2000) suggested to set right the penalty function to control the constraints and to obtain feasible solutions. Michalewicz (1995) analysed the performance and the difficulties of different methods used describing the penalty functions on a number of test problems. However, the static penalty function method shows a good ability to deal better with all complicated systems than the sophisticated methods (Michalewicz & Schoenauer, 1996).

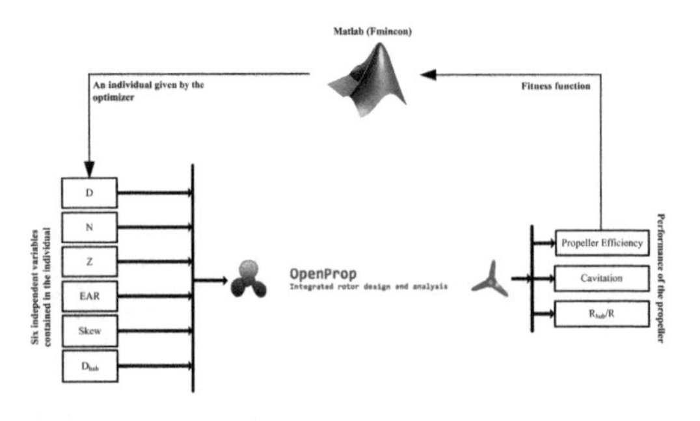

Figure 3. Matlab/OpenProp coupling scheme.

So, the static penalty function in form of violation degree is used in this study in the created nonlinear model to find the optimal solution of the different input parameters of the propeller geometry to maximize the propeller efficiency while keeping the cavitation limit in the tolerated scope. The input vector that must be optimized is defined by:

$$x = (D, N, Z, EAR, Skew, D_{hub}) \tag{8}$$

and the boundary limits of each input are expressed as:

$$3 \leq D \leq 5 \tag{9}$$

$$150 \leq N \leq 300 \tag{10}$$

$$3 \leq Z \leq 5 \tag{11}$$

$$0.3 \leq EAR \leq 1.05 \tag{12}$$

$$0 \leq Skew \leq 108 \tag{13}$$

$$0.15 \times min(D) \leq D_{hub} \leq 0.3 \times max(D) \tag{14}$$

where N is in RPM, D and D_{hub} in meters and *Skew* in degrees. The range of values of propeller diameter is suggested according to the ship draft. The boundary limits of D_{hub} is a function of the propeller diameter and the values of the limits are taken from different propellers data collected from (Tachmindji & Milam, 1957, Cummings, 1973). The range of skew angles are also collected from the data presented in Boswell (1971). The range of values of the expanded area ratio (EAR) and the values of the blade design are taking according to Wageningen B-screw series (van Lammeren et al., 1969).

Then, two restriction conditions, equality and inequality, are described and the static penalty function presented in the form of violation degree. Equality restriction condition is normalized as in the following equation:

$$h(x) = \frac{P_D}{P_{D-obj}} - 1 \tag{15}$$

where P_D is the delivered power and must subjected to:

$$Delivered\ power(x) = Delivered\ power_{obj} \tag{16}$$

While the inequality restriction conditions equations (17)–(19) are normalized as:

$$g_1(x) = \frac{Burrel\ Cavitation_{actual}}{Burrel\ Cavitation_{limit}} - 1 < 0 \tag{17}$$

$$g_2(x) = -\frac{R_{hub}/R}{0.15} + 1 < 0 \tag{18}$$

$$g_3(x) = \frac{R_{hub}/R}{0.3} - 1 < 0 \tag{19}$$

where, R_{hub} is the radius of the hub and is equal to the half of D_{hub}.

Therefore, the fitness function for the whole nonlinear model may be written as:

$$Fitness\ function = -Prop_{Eff} + R\left[\sum_{i=1}^{3} max(g_i(x),0) + |h(x)|\right] \tag{20}$$

where, the value of the penalty parameter R is very important and equal to 1,000 as suggested by Zhao and Xu (2013).

4 SIMULATION AND RESULTS

Based on the developed optimization scheme, the parameters of the geometry of the propeller are optimized for a given container ship to maximize the propeller efficiency and verify the limits of cavitation according to Burrill's method.

The propeller is designed for 18 knots ship speed and at 9,180 kW maximum engine power. The effi-ciency of the gear box ($\eta GB = 99\%$) and the efficiency of the transmission shaft ($\eta T = 99\%$) are considered as losses between the brake power, PB, and the deliv-ered power, PD, which is equal to 8,997 kW as pre-sented in equation (21). The wake fraction and thrust deduction fraction are calculated based on the em-pirical formulas suggested by Holtrop (1984). Figure 4 shows a general overview of the power train of the propulsion system between the marine diesel en-gine and the propeller.

$$P_D = P_B \times \eta_{GB} \times \eta_T \tag{21}$$

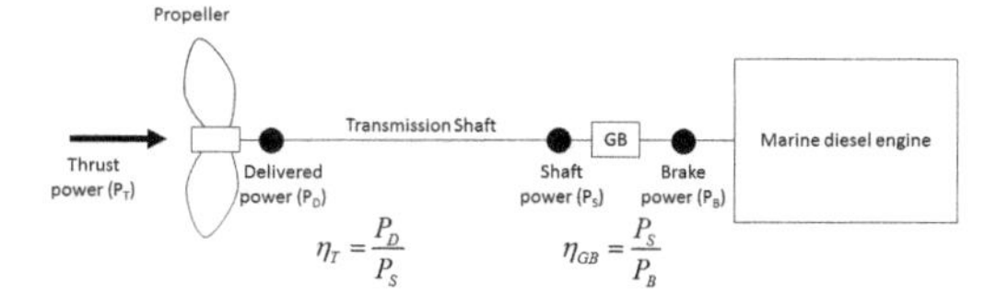

Figure 4. Schematic diagram of the power train.

The optimized parameters are presented in Table 3 and the cavitation does not exceed the limit re-quired, while the variation of c/D, t_{max}/D and pitch diameter ratio (P/D) along the propeller radius are presented in Table 4. While the rake is assumed to be zero. Also, the 3D geometry of the propeller is shown sin Figure 5 for a better over-view of the propeller shape. The reduction ratio of the gearbox can be 3.13:1 when comparing the maximum rotational speed of the engine with the rotation speed of the propeller.

According to the non-dimensional parameters ex-pressed in equations (22) – (25), the performance curve of the propeller as shown in Figure 6 is com-puted for different advance coefficients.

$$K_T = \frac{T}{\rho N^2 D^4} \tag{22}$$

$$K_Q = \frac{Q}{\rho N^2 D^5} \tag{23}$$

Table 3. Calculated results of the optimized input parameters.

Optimized parameter	Value
Z	4
N (rpm)	253.7
D (m)	3.59
EAR	0.6512
D_{hub} (m)	1.06
Skew (degree)	1
Cavitation limit	0.0294
Cavitation actual	1.82e-05

Table 4. Calculated results of the blade design parameters.

r/R	c/D	t_{max}/c	P/D
0.296	0.356	0.091	0.702
0.351	0.375	0.081	0.718
0.406	0.391	0.071	0.735
0.513	0.411	0.057	0.750
0.565	0.415	0.051	0.753
0.664	0.413	0.041	0.759
0.710	0.405	0.037	0.761
0.794	0.377	0.031	0.764
0.831	0.358	0.028	0.765
0.896	0.305	0.024	0.766
0.923	0.268	0.022	0.767
0.966	0.182	0.016	0.768
0.981	0.136	0.012	0.769
0.991	0.090	0.008	0.769
0.998	0.044	0.004	0.769
1.000	0.000	0.004	0.769

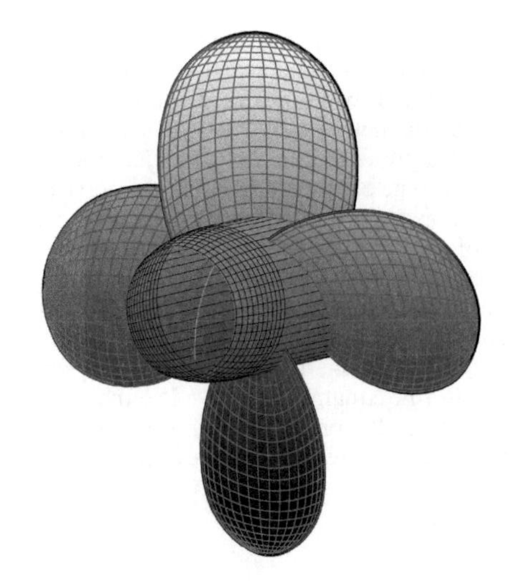

Figure 5. 3D image of optimized propeller.

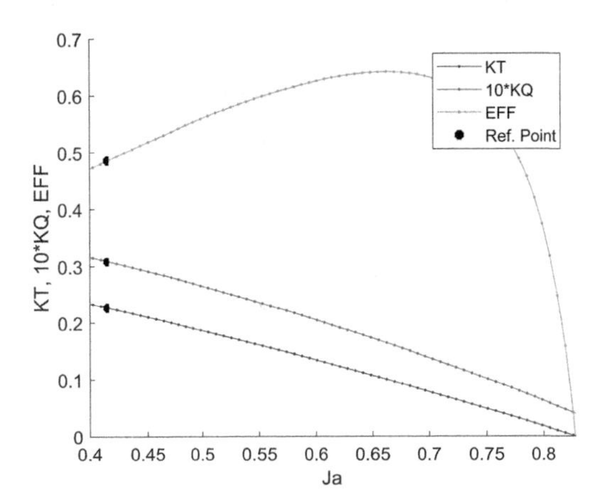

Figure 6. Propeller performance curve.

$$\eta_o = \frac{J_A}{2\pi} \frac{K_T}{K_Q} \tag{24}$$

$$J_A = \frac{V_A}{ND} \tag{25}$$

where, K_T is the thrust coefficient, K_Q is the torque coefficient, η_o is the open water efficiency and J_A is the advance coefficient.

5 CONCLUSIONS

A model based optimization scheme was developed to optimize the selection of a fixed pitch propeller for a given ship. This code is implemented in Matlab coupling OpenProp, a design and analysis software of marine propellers and a nonlinear optimizer. The optimization model is constructed, including the objective, the boundary conditions and the penalty functions to express constraints. Different propeller geometries are optimized to maximize the propeller efficiency and to verify the limits of cavitation.

The losses along the gearbox and the transmission shaft are considered. Also, the wave fraction and the thrust deduction fraction are calculated.

This code can be improved by adding different mean line type, thickness type and blade design values for a better choice from a wide range of propellers. It can be further used during the preliminary stage of ship design in order to select an optimum and suitable propeller. Also, it can be connected to ship weather routing to optimize the voyage of the ship where both the still water resistance and the wave effects are taking into consideration.

ACKNOWLEDGEMENTS

This work was performed within the scope of the Strategic Research Plan of the Centre for Marine Technology and Ocean Engineering (CENTEC), which is financed by the Portuguese Foundation for Science and Technology (Fundação para a Ciência e Tecnologia—FCT).

REFERENCES

Benini, E. 2003. Multiobjective design optimization of B-screw series propellers using evolutionary algorithms. *Mar Technol,* 40, 229–238.

Boswell, R.J. 1971. Design, Cavitation Performance, and Open Water Performance of a Series of Research Skewed Propellers.

Burrill, L.C. & Emerson, A. 1978. Propeller cavitation: further tests on 16 in. propeller models in the King's College Cavitation Tunnel. *Trans. NECIES,* 195.

Carlton, J. 2012. *Marine Propellers and Propulsion*, Butterworth-Heinemann.

Chen, J.-H. & Shih, Y.-S. 2007. Basic design of a series propeller with vibration consideration by genetic algorithm. *Journal of Marine Science and Technology,* 12, 119–129.

Cummings, D.E. 1973. Numerical Prediction of Propeller Characteristics. *Journal of Ship Research,* 17, 12–18.

Deb, K. 2000. An efficient constraint handling method for genetic algorithms. *Computer Methods in Applied Mechanics and Engineering,* 186, 311–338.

Epps, B.P. & Kimball, R.W. 2013a. *OpenProp v3: Open-source software for the design and analysis of marine propellers and horizontal-axis turbines.* [Online]. Available: http://engineering.dartmouth.edu/epps/openprop [Accessed 01.10.2016].

Epps, B.P. & Kimball, R.W. 2013b. Unified Rotor Lifting Line Theory. *Journal of Ship Research,* 57.

Epps, B.P., Stanway, M.J. & Kimball, R.W. 2009. OpenProp: An Open-source Design Tool for Propellers and Turbines. *SNAME Propellers and Shafting*.
Froude, W. 1878. On the elementary relation between pitch, slip and propulsive efficiency. *Trans. RINA*, 19.
Gaafary, M.M., El-Kilani, H.S. & Moustafa, M.M. 2011. Optimum design of B-series marine propellers. *Alexandria Engineering Journal*, 50, 13–18.
Hillier, F.S. & Lieberman, G.J. 1980. *Introduction to operations research*, Holden-Day.
Holtrop, J. & Mennen, G.G.J. 1982. An approximate power prediction method. *International Shipbuilding Progress*, 29, 166–170.
Holtrop, J. 1984. Statistical re-analysis of resistance and propulsion data. *International Shipbuilding Progress*, 31, 272–276.
Homaifar, A., Qi, C.X. & Lai, S.H. 1994. Constrained Optimization Via Genetic Algorithms. *SIMULATION*, 62, 242–253.
Karim, M. & Ikehata, M. A genetic algorithm (GA)-based optimization technique for the design of marine propellers. Proceedings of the propeller/shafting 2000symposium, 2000 Virginia Beach, USA.
Kobayakawa, M. & Onuma, H. 1985. Propeller aerodynamic performance by vortex-lattice method. *Journal of Aircraft*, 22, 649–654.
Lee, K.-J., Hoshino, T. & Lee, J.-H. 2014. A lifting surface optimization method for the design of marine propeller blades. *Ocean Engineering*, 88, 472–479.
Maftei, C., Moreira, L. & Guedes Soares, C. 2009. Simulation of the dynamics of a marine diesel engine. *Journal of Marine Engineering & Technology*, 8, 29–43.
MAN Diesel & Turbo. 2014. *32/44CR Project Guide—Marine* [Online]. Available: https://www.engines.man.eu [Accessed 31.08.2017].
Martelli, M. & Figari, M. 2017. Numerical and experimental investigation for the performance assessment of full electric marine propulsion plant. *In:* Guedes Soares & Teixeira (eds.) *Maritime Transportation and Harvesting of Sea Resources*. Taylor & Francis Group, London, pp. 87–93.
MathWorks. 2017. *Comparison of Five Solvers* [Online]. Available: https://www.mathworks.com/help/gads/example-comparing-several-solvers.html [Accessed 02.06.2017].
Michalewicz, Z. & Schoenauer, M. 1996. Evolutionary algorithms for constrained parameter optimization problems. *Evol. Comput.*, 4, 1–32.
Michalewicz, Z. Genetic algorithms, numerical optimization, and constraints. *In:* Eshelman, L., ed. Proceedings of the Sixth International Conference on Genetic Algorithms, 1995 Morgan Kauffman, San Mateo. 151–158.
Morsy El Gohary, M. & Abdou, K.M. 2011. Computer based selection and performance analysis of marine diesel engine. *Alexandria Engineering Journal*, 50, 1–11.
Peterson, C. 2008. Minimum Pressure Envelope Cavitation Analysis Using Two-Dimensional Panel Method. Msc MS thesis, Massachusetts Institute of Technology.
Prpić-Oršić, J., Vettor, R., Faltinsen, O.M. & Guedes Soares, C. 2016. The influence of route choice and operating conditions on fuel consumption and CO2 emission of ships. *J Mar Sci Technol* 21, 434–457.
Radojčić, D. 1985. Optimal preliminary propeller design using nonlinear constrained mathematical programming technique. *Ship Science Reports*. Southampton: UK University of Southampton
Rankine, W.J.M. 1865. On the Mechanical Principles of the Action of Propellers. *Trans. RINA*, 6.
Stanway, M.J. & Stefanov-Wagner, T. Small-diameter ducted contrarotating propulsors for marine robots. OCEANS 2006, 18–21 Sept. 2006 2006. 1–6.
Suen, J.-b. & Kouh, J.-s. Genetic algorithms for optimal series propeller design. Proceedings of the third international conference on marine technology, 1999 Szczecin, Poland.
Tachmindji, A.I. & Milam, A.B. 1957. The Calculation of the Circulation for Propellers with Finite Hub Having 3, 4, 5 and 6 Blades.
Tadros, M., Ventura, M. & Guedes Soares, C. 2015. Numerical simulation of a two-stroke marine diesel engine. *In:* Guedes Soares, Dejhalla & Pavleti (eds.) *Towards Green Marine Technology and Transport*. Taylor & Francis Group, London, pp. 609–617.
Tadros, M., Ventura, M. & Guedes Soares, C. 2017. Surrogate models of the performance and exhaust emissions of marine diesel engines for ship conceptual design. *In:* Guedes Soares & Teixeira (eds.) *Maritime Transportation and Harvesting of Sea Resources*. Taylor & Francis Group, London, pp. 105–112.
Theotokatos, G., Stoumpos, S., Lazakis, I. & Livanos, G. 2016. Numerical study of a marine dual-fuel four-stroke engine. *In:* Guedes Soares & Santos (eds.) *Maritime Technology and Engineering 3*. Taylor & Francis Group, London, pp. 777–783.
van Lammeren, W.P.A., van Manen, J.D. & Oosterveld, M.W.C. 1969. The Wageningen B-screw series. *Trans. SNAME*, 77.
Vettor, R. & Guedes Soares, C. 2016. Development of a ship weather routing system. *Ocean Engineering*, 123, 1–14.
Vettor, R., Tadros, M., Ventura, M. & Guedes Soares, C. 2016. Route planning of a fishing vessel in coastal waters with fuel consumption restraint. *In:* Guedes Soares & Santos (eds.) *Maritime Technology and Engineering 3*. Taylor & Francis Group, London, pp. 167–173.
Watson, D.G.M. 1998. *Practical Ship Design*, Elsevier.
Welaya, Y.M.A., Mosleh, M. & Ammar, N.R. 2013. Thermodynamic analysis of a combined gas turbine power plant with a solid oxide fuel cell for marine applications. *International Journal of Naval Architecture and Ocean Engineering*, 5, 529–545.
Yum, K.K., Taskar, B., Pedersen, E. & Steen, S. 2017. Simulation of a two-stroke diesel engine for propulsion in waves. *International Journal of Naval Architecture and Ocean Engineering*, 9, 351–372.
Zhao, J. & Xu, M. 2013. Fuel economy optimization of an Atkinson cycle engine using genetic algorithm. *Applied Energy*, 105, 335–348.

Progress in Maritime Technology and Engineering – Guedes Soares & Santos (Eds)
 ISBN 978-1-138-58539-3

Marine propulsion shafting: A study of whirling vibrations

S. Busquier & S. Martínez
Polytechnic University of Cartagena, Cartagena, Spain

M.J. Legaz
University of Cádiz, Cádiz, Spain

ABSTRACT: Whirling vibration is an important part of the calculations of design of a marine shaft. In fact, all Classification Societies require a propulsion shafting whirling vibration calculation in the scope of the critical speeds i.e. free whirling vibration calculation. However, whirling vibration is a source of fatigue failure of the bracket and aft stern tube bearings, destruction of high-speed shafts with universal joints, noise and hull vibrations. There are a lot of uncertainties in the calculation of whirling vibration, namely in shafting system modeling, in determination of the excitation and damping forces. Moreover, whirling vibration calculation mathematics is much more complex than torsional or axial. The marine propulsion shaft can be studied as a self-sustained vibration system which can be modeling with de Van der pol equation. In this document, a fashion way to solve the Van der pol equation is presented in a theoretical way.

1 INTRODUCTION

The denomination 'whirling' was introduced into mechanical engineering by W.J.M. Rankine in the year of 1869. The main goal of whirling vibration calculations of rotating machinery was in this year and is now to determine the critical speeds of the shaft. The whirling vibration resonance cause by residual unbalance of the fast rotating turbine rotor might result in a catastrophic failure.

The theory of whirling vibration was applied to the marine propulsion shafting. Nowadays, all Classification Societies require a propulsion shafting whirling vibration calculation named in some Class Rules as bending or lateral vibration calculation.

The classification societies make reference to whirling vibration calculation, requiring the calculus of critical speeds. In the reference to forced whirling vibration, the classification societies only say that this calculation could be request.

The classification societies have clear requirements for shaft modelling and acceptance criteria in the case of torsional vibration calculation. However, in the case of whirling vibrations the criteria are not so specific as the torsional vibration calculation.

It can be found various causes due to which the calculus of whirling amplitudes is still not well developed. Firstly, it is referring to a fatal damage, whirling vibration hardly ever provoke one compared with torsional vibration. Secondly, regarding to mathematics, whirling vibration calculation is much more complex. Thirdly, in the whirling calculation there are a lot of uncertainties particularly in determination of excitation and damping forces in the shafting system modelling.

Nonetheless, whirling vibration is a source of fatigue failures in aft stern tube bearings as it is shown in the Figure 1. They also cause destructions of high-speed shafts with universal joints, it is shown in the Figure 2. They can provoke excessive hull and superstructure vibrations. All this can lead to failure of seals which can cause the leakage of lubrication oil into the sea. All these failures show the importance of whirlin vibration calculations.

Figure 1. Fatigue failures in universal joints.

Figure 2. Fatigue failures in aft stern tube bearings.

With the see in the development and improvement of marine propulsion systems and prevention of sea pollution, a complete analysis has to be done. In a complete analysis of marine shaft system, the whirling vibrations play an important role. The shaft designers must have taking into account all the aspect of the vibrations and the new tools of calculation such as softwares and numerical methods of resolution.

In spite of the fact that marine propulsion shafting whirling vibration studies started at the middle of the last century very few publications concerning of whirling vibration calculation can be found, e.g. (Jasper, 1954), (Murawski, 2003, 2005), (Šverko and Šestan, 2010), (Zhou et al., 2014).

2 THE THEORICAL BACKGROUND

The definition of whirling motion can be done as any motion of a rotating shaft in which its center line points move in a transverse direction along some trajectories called whirling orbits.

There are many factors, which can lead to a rotary machine to continuous whirling (Gunter, 1972). For a propulsion shafting they may be listed as the followings.

1. Residual unbalance.
2. Alternating propeller hydrodynamic loads
3. Alternating motor electromagnetic loads.
4. Internal-combustion engine radial forces.
5. Gear imperfections and hammering.
6. Gyroscopic moments.
7. Asymmetric shafting properties.
8. Hydrodynamic forces in lightly loaded bearings.
9. Hysteretic internal friction damping (shaft internal damping, friction in couplings and shrink fits, etc.).
10. Friction in journal bearings.

Propulsion shafting whirling vibration may be excited by one factor or by a combination of the factors listed above.

Propulsion shafting is a complex continuous body with variable mass and elastic properties, characterized by an infinite number of degrees of freedom.

Taking into account only one free degree, we can modelling the shaft as a supported-supported beam. The supported are the bearing and the load is the weigh of the shaft and the couplings. We can take the load as uniformly distributed load between two supports.

The vibration equation used in calculation is as follows:

$$Mx''(t) + Cx'(t) + Kx(t) = F(t) \tag{1}$$

where: $X(t)$ – solution of the equation (deflection and slopes at the system nodes);

M – mass matrix;
C – damping matrix;
K – stiffness matrix;
F – excitation forces vector.

In whirling vibration every point of a flexible rotating shaft moves with angular speed (ω) around the tangent to deflection curve of the shaft.

At the same time there are no obstacles for rotating shaft to vibrate as above-mentioned beam in two orthogonally related planes if correspondent alternating forces are applied. As a result, the shaft rotation axis will move in the space along the beam vibration orbit.

Change in the position of the rotation axis of a rotating body in classical mechanics is known as a ***precession*** motion. Conventional name for the precession motion of the rotating shafts is '***whirling***'.

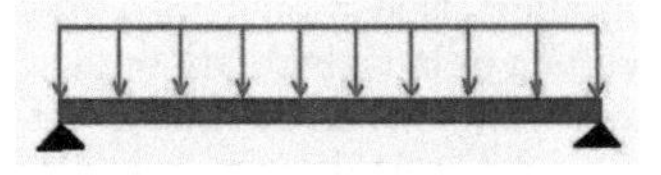

Figure 3. Supported-supported beam.

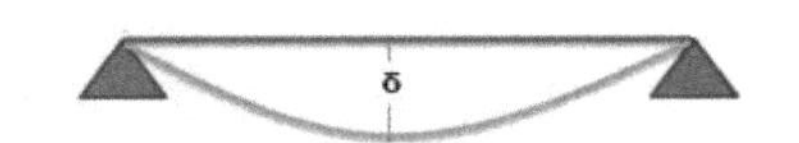

Figure 4. Deformation of the beam.

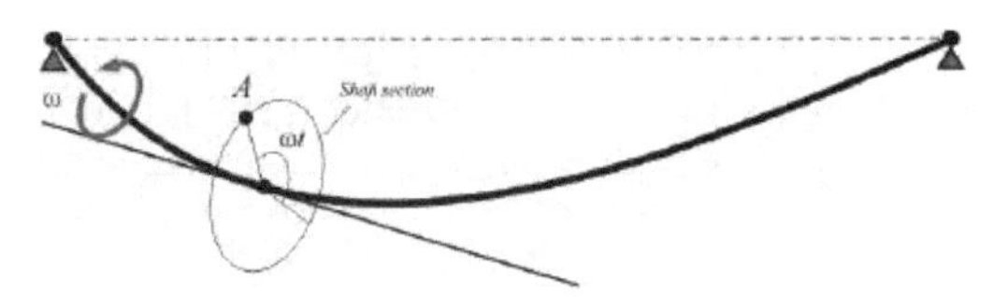

Figure 5. Rotation of a shaft.

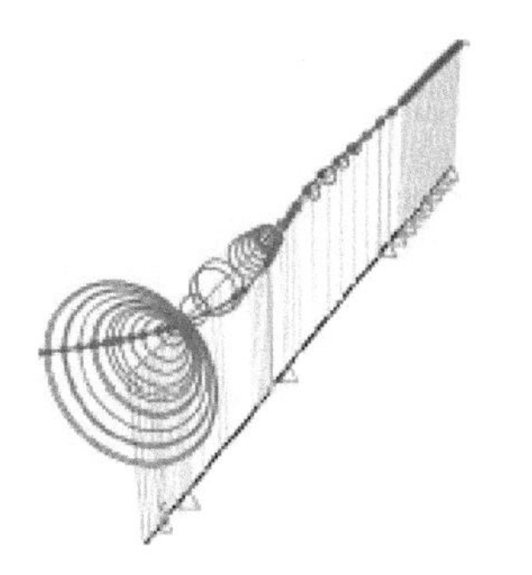

Figure 6. Deflection shapes diagrams of the first mode.

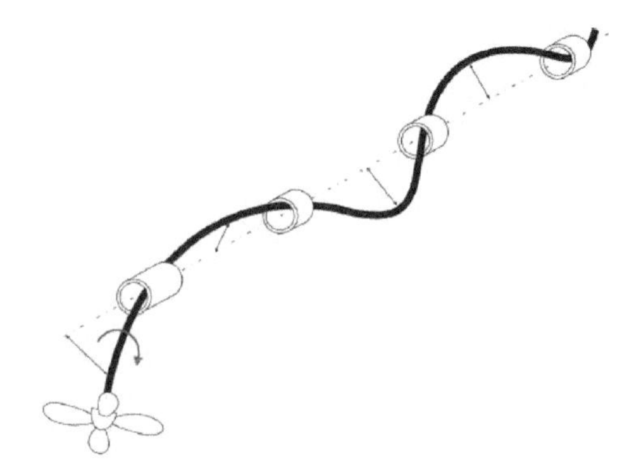

Figure 7. Shaft line whirling vibration motion.

When precession orbit shape is circular or elliptical we have the case of a circular or elliptical whirling. Very often term 'whirling' associate with a circular or elliptical orbit shape exclusively, but it is not quite correct from general point of view.

In the next figure, it is shown the deflection shape diagrams of the first mode for a shafting.

The whirling vibration can cause deformation in several planes and movements within the bearing clearance as it is shown in the Figure 7.

3 SHAFT VERSUS ROTORS

It could be interesting to compare the marine shafting with rotors with the objective of studying if some results of rotors can be use in the analysis of whirling vibration in propulsion shafting. Whirling vibration theory in rotor is well developed nowadays. There has been a huge amount of publications along of his history.

The first and most important difference is in spin speed: hundreds rpm for propulsion shafting versus thousands or even tens of thousands for rotors. Quantitative difference results in qualitative differences in phenomena to be accounted for in the analyses. For fast rotating rotors the residual unbalance is a main source of whirling vibration this is not the case for propulsion shafting. There are no huge centrifugal forces in propulsion shafting. The instability issues that are widely discussed in rotor domain publications are not inherent to marine shafts.

The second are structure and operating condition differences: propulsion shafting are multi-bearings, multi-shafts and include various joints and interfaces such as shrink-fitted connections, flange couplings, gearing, flexible couplings. Propulsion shafting having a massive propeller at the cantilevered end are affected by fluid environment.

The rotor whirling principal concern of single shaft rotor systems (Jeffcott rotors) and hardly ever it is found multi-bearing rotors. On the other hand, propulsion shafting are statically indeterminate multi-bearing systems. In the propulsion shafting the dynamic motions in whirling vibration depend on bearings' spatial positions.

Regarding to place, high speed rotors are usually place on massive concrete or stiff steel supports while propulsion shafting bearings are located on flexible hull structures.

4 SELF-EXCITED VIBRATION

Self-excited systems begin to vibrate of their own accord spontaneously, the amplitude increasing until some nonlinear effect limits any further increase. The energy supplying these vibrations is obtained from a uniform source of power associated with the system which, due to some mechanism inherent in the system, gives rise to oscillating forces. The nature of self-excited vibration compared to forced vibration is:

In self-excited vibration the alternating force that sustains the motion is created or controlled by the motion itself; when the motion stops, the alternating force disappears.

In a forced vibration the sustaining alternating force exists independent of the motion and persists when the vibratory motion is stopped.

The occurrence of self-excited vibration in a physical system is intimately associated with the stability of equilibrium positions of the system. If the system is disturbed from a position of equilibrium, forces generally appear which cause the system to move either toward the equilibrium position or away from it.

In the latter case the equilibrium position is said to be unstable; then the system may either oscillate with increasing amplitude or monotonically recede from the equilibrium position until nonlinear or limiting restraints appear. The equilibrium position is said to be stable if the disturbed system approaches the equilibrium position either in a damped oscillatory fashion or asymptotically.

The forces which appear as the system is displaced from its equilibrium position may depend on the displacement or the velocity, or both. If displacement-dependent forces appear and cause the system to move away from the equilibrium position, the system is said to be statically unstable. For example, an inverted pendulum is statically unstable. Velocity-dependent forces which cause the system to recede from a statically stable equilibrium position lead to dynamic instability.

Self-excited vibrations are characterized by the presence of a mechanism whereby a system will vibrate at its own natural or critical frequency, essentially *independent* of the *frequency* of any external stimulus.

Self-excited vibrations pervade all areas of design and operations of physical systems where motion or time-variant parameters are involved—aeromechanical systems (flutter, aircraft flight dynamics), aerodynamics (separation, stall, musical wind instruments, diffuser and inlet chugging), aerothermodynamics (flame instability, combustor screech), mechanical systems (machine-tool chatter), and feedback networks (pneumatic, hydraulic, and electromechanical servomechanisms), marine (shaft line).

The mechanisms of self-excitation which have been identified can be categorized as follows:

Whirling or Whipping
Hysteretic whirl
Fluid trapped in the rotor
Dry friction whip
Fluid bearing whip
Seal and blade-tip-clearance effect in turbomachinery
Propeller and turbomachinery whirl
Parametric Instability
Asymmetric shafting
Pulsating torque
Pulsating longitudinal loading
Stick-Slip Rubs and Chatter
Instabilities in Forced Vibrations
Bistable vibration
Unstable imbalance

5 WHIRLING. ANALITIC MODELING

As source of whirling vibration two classes forces are found. The two classes of forces that excite whirling vibration in propulsion system are internal and external forces. In the case that whirling vibration in propulsion system is excited by internal forces there is no direct dependence between vibration intensity and transmitted power.

The alternating forces of both classes excite two kinds of whirling vibrations: forced vibration and self-excited vibration.

Focus on self-excited vibration, due to whirling phenomena. The whirling or whipping phenomena is the most important subcategory of instabilities origin self-excited.

The way in what whirling produces self-excited phenomena is the following. A generation of a tangential force is provoked, normal to an arbitrary radial deflection of a rotating shaft, whose magnitude is proportional to (or varies monotonically with) that deflection.

At some "onset" rotational speed, such a force system will overcome the stabilizing external damping forces which are generally present and induce a whirling motion of ever-increasing amplitude, limited only by nonlinearities which ultimately limit deflections.

The vibration itself takes place by excitation coming from within the system itself. Due to some transient disturbance, small amplitudes of vibration that may initiate grow until the restoring forces in the system are sufficient to limit the resulting vibratory motion.

Therefore, these vibrations are called self-excited vibrations. A close mathematical analogy to this class of phenomena is the concept of "negative damping" A convenient way of modelling such negative damping is by the Van der pol equation (Rao and Gupta, 1999). Van de pol used a damping model in which the viscous damping coefficient itself is a function of square amplitude, i.e.

$$F_d = \mu(1-x^2)x' \tag{2}$$

A simple mathematical representation of a self-excited vibration may be found in the concept of negative damping. Consider the differential equation for a damped, free vibration:

$$M\frac{dx^2}{dt} + C\frac{dx}{dt} + Kx = 0 \tag{3}$$

In the case of a whirling or whipping shaft, the equations of motion (for an idealized shaft with a single lumped mass m) and considering only viscous damping.

For self-excited vibration this equation can be written:

$$M\frac{dx^2}{dt} + \mu(1-x^2)\frac{dx}{dt} + Kx = 0 \tag{4}$$

6 VAN DER POL EQUATION

The Van der Pol oscillator is a dynamical system which includes positive feedback and a nonlinear damping term. In its original application, at the turn of the 20th century, it was used in the electrical field as an electrical oscillator with a nonlinear element was the forerunner of the early commercial radios. A circuit of this type helps small oscillations and causes the large ones to damp.

The analysis of behaviour of the Van der Pol oscillator with a damping coefficient ε is:

When ε = 0 the system becomes a linear oscillator without damping.

As ε increases, the system reaches a nonlinear state.

The system only has one equilibrium point, which is P(0,0) and is always unstable. However, all trajectories in phase space tend to a stable limit cycle.

The equation seen in the previous section is the equation of Van der pol oscillator.

$$M\frac{dx^2}{dt}+C\frac{dx}{dt}+Kx=0 \tag{5}$$

This equation can become to have a chaotic behaviour as the parameter increases. If the parameter is very large, the system becomes stiff.

The stiff problems are a particular type of problems in which some classical method fail in approximate their solution. Especially for the approximation of stiff system of ordinary differential equations, explicit Runge-Kutta methods are not suitable (Hairer and Wanner, 1991).

The stiffness property of a system of ordinary differential equations cannot be defined in precise mathematical terms. Following Lambert (Lamber, 1991) stiffness occurs when stability requirements, rather than those of accuracy, constrain the steplength, or when some components of the solution decay much more rapidly than others.

For these and others related statements, a classical recommendation is to use, in general, implicit schemes when we are interested in stiff problems.

In this document we present a fashion method to solve the Van de pol equation which it is able to found the solution whatever the value of nu parameter.

7 FASHION WAY TO SOLVE VAN DER POL EQUATION

The Van der Pol equation can be written as:

$$\frac{dx}{dt}=x' \tag{6}$$

$$\frac{dx'}{dt}=\mu(1-x^2)\frac{dx}{dt}+x \tag{7}$$

Basically, this equation can be represented as a system of non-linear equations with damping.

In order to solve this system of equations a theoretical fashion approach is used.

For simplicity, our attention is focused on a system of non-linear equations of the form

$$Mx'(t)=f(x(t)) \quad in\,(0,T),\ x(0)=x_0 \tag{8}$$

where M is a given, eventually singular, matrix depending on t.

The original system of non-linear equations is linearized obtaining an iterative scheme. Our ideas are based on the analysis of a certain error functional of the form

$$E(x)=\frac{1}{2}\int_0^T\left|Mx'(t)-f(x(t))\right|^2 dt \tag{9}$$

to be minimized among the absolutely continuous paths $x:(0;T)\rightarrow R^N$ with $x(0)=x_0$.

In (Amat et al., 2012) it is demonstrated that the existence of solutions for the system of non- linear equations is equivalent to the existence of minimizers for $E(x)$ with vanishing minimum value. The typical minimization schemes like (steepest) descent methods will work fine as they can never get stuck in local minima, and converge steadily to the solution of the problem, no matter what the initialization is.

Since our optimization approach is really constructive, iterative numerical procedures are easily implementable.

1. Start with an initial approximation $x^{(0)}(t)$ compatible with the initial conditions (ex. $x^{(0)}(t)=x_0+tf(x_0)$).
2. Assume the approximation $x^{(j)}(t)$ in [0,T], it is known.
3. Compute its derivative $M(x^{(j)})(t)$ in [0, T].
4. Compute the auxiliary function $y^{(j)}(t)$ as the numerical solution of the problem

$$My'(t)-\nabla f\left(x^{(j)}(t)\right)y(t) = f\left(x^{(j)}(t)\right)-M\left(x^{(j)}\right)'(t) \ \ in\,(0,T), \quad y(0)=0 \tag{10}$$

by making use of a numerical scheme for algebraic equations with dense output (like collocation methods).

5. Change $x^{(j)}$ to $x^{(j+1)}$ by using the update formula

$$x^{(j+1)}(t)=x^{(j)}(t)+y^{(j)}(t) \tag{11}$$

6. Iterate (3), (4) and (5), until numerical convergence. In practice, the stopping criterium used is

$$max\left\{\left\|y^{(j)}\right\|_\infty,\sqrt{2E\left(x^{(j)}\right)}\right\}\leq TOL \tag{12}$$

In particular, this numerical procedure can be implemented, in a very easy way, using a problem-solving environment like MATLAB (MathWorks, 2017).

Some other references about studies in marine shafting systems can be found in Gurrr and Rulfs, 2008. And in Carlton, 2007.

8 CONCLUSIONS

The importance of analysis of whirling vibration in marine shafting has been highlighted.

A revision of mathematical model of vibrating shaft have been made. The focus has been put in the self-excited vibration.

The Van der Pol equation has been introduced as a way to model the behaviour of a self-excited shaft with negative damping.

A fashion way to solve the Van der Pol equation is presented. This method is able to find the solution of equation for high values of parameter.

Moreover, this method never gets stuck in a local minimum and always converges to the solution sought regardless of the initialization.

This is an important improvement over the methods that need a close initialization to the solution sought.

ACKNOWLEDGEMENTS

Research supported in part by Programa de Apoyo a la investigacion de la fundacion Seneca-Agencia de Ciencia y Tecnología de la Region de Murcia 19374/PI/14 and MTM2015-64382-P (MINECO/FEDER).

REFERENCES

Amat S., Legaz M.J. and Pedregal P., 2012. On a Newton-type method for differential algebraic equations. Journal of applied mathematics 2012: 1–15.

Carlton, J. S., 2007. Marine Propellers and Propulsion, Oxford. Butterworth–Heinemanns.

Gunter, E. J., 1966. Dynamic Stability of Rotor-Bearing Systems. NASA SP-113.

Gurr, C. and Rulfs H., 2008. Influence of transient operating conditions on propeller shaft bearings. Journal of Marine Engineering and Technology.

Hairer E. and Wanner G., 1991. Solving Ordinary Differential Equations II: Stiff and Differential Algebraic Problems, Berlin, Germany. Springer-Verlag.

Jasper, N., 1954. A Design Approach to the Problem of Critical Whirling Speeds of Shaft-Disk Systems, Navy Department the David Taylor Model Basin, December, Report 890.

Lambert J.D., 1991. Numerical Methods for Ordinary Differential Systems: The initial value problem. John Wiley and Sons.

Murawski, L., 2003. Static and Dynamic Analyses of Marine Propulsion Systems, Warszawa. Oficyna Wydawnicza Politechniki Warszawskiej.

Murawski, L., 2005. Shaft Line Whirling Vibrations: Effects of Numerical Assumptions on Analysis Results. Marine Technology and SNAME News. Vol. 42, no. 2.

Rao, J. and Gupta, K., 1999. Introductory course on theory and practice of mechanical vibrations, India. New age international.

Šverko, D. and Šestan, A., 2010. Experimental Determination of Stern Tube Journal Bearing Behaviour. Brodogradnja, 61, 2.

Zhou, Xincong., and Qin, Li., Chen, Kai., Niu, Wanying., Jiang, Xinchen., 2014. Vibrational Characteristics of a Marine Shaft Coupling System Excited by Propeller Force, The 21st International Congress on Sound and Vibration. Beijing.

Dynamics and control

Progress in Maritime Technology and Engineering – Guedes Soares & Santos (Eds)
© 2018 Taylor & Francis Group, London, ISBN 978-1-138-58539-3

Assessment of the electric propulsion motor controller for the Colombian offshore patrol vessel

C. Morales, E. Insignares, B. Verma, D. Fuentes & M. Ruiz
COTECMAR, Cartagena, Colombia

ABSTRACT: Recent initiative in the preliminary design of the OPV93 led to the assessment of various propulsive arrangements, including the CODELOD propulsion system. As the electric propulsion set of the CODELOD arrangement could be fitted with either a soft starter or variable frequency drive, the two options are required to be evaluated so as to define the final arrangement for CODELOD as it will be used for life cycle costs calculations. For the evaluation, the propeller-gearbox-motor integration was performed, discussing the performance of each alternative considering the propeller curve and ship speed. Additionally, analysis of harmonic distortion, efficiency and power factor correction was performed along with the comparison of the main characteristics for both options. The soft starter-motor set come up as the appropriate solution to the electric propulsion train of the CODELOD arrangement, however, additional analysis on life cycle costs are recommended to define the impact of power savings of the VFD-motor set.

1 INTRODUCTION

Recently, COTECMAR has been working in the design of a new generation of OPVs—Offshore Patrol Vessel-. The project, named OPV93, is 100% Colombian design and is currently in the basic design stage with its construction expected to begin at 2019.

As part of the design spiral, several propulsion systems and manufacturer combinations are under evaluation to define the ship propulsive arrangement. The alternative to be selected is the one that meets the Colombian Navy requirements with the lowest life cycle costs.

Among the options, the CODELOD—COmbined Diesel ELectric or Diesel—propulsion system is included. As this hybrid propulsion system is fitted with electric motors to thrust the ship within a speed range, the life cycle costs depends whether the electric propulsion motors will be speed controlled. The above emerges as result of two points to consider: 1. the ship is to be fitted with CPP—Controllable Pitch Propeller—propellers, meaning that ship speed can be varied without motor speed control only requiring inrush current reduction by soft starter and 2. In maritime propulsion literature as well manufacturer recommendations, electric propulsion systems are often associated with speed control of electric propulsion motors (Woud & Stapersma 2003, Patel 2012).

In this context, the aim of the present study is to evaluate the use of variable frequency drives—VFD—or soft starters for propulsion motor control within the CODELOD arrangement for the OPV93 to use as input for the life cycle cost estimation of the propulsive arrangement.

2 THE OPV93

The new generation of Colombian OPVs were conceived to meet the Colombian Navy operational requirements for patrol and security within the Colombian Exclusive Economic Zone, exceeding the capabilities of current OPVs. The Table 1 displays the main characteristics of the ship, while the Figure 1 shows the operational profile for 2500 hours of annual operation.

Since during 90% of the operational time the ship is expected to travel at speeds up to 12 knots, but full power will be required during remaining operational time, a hybrid propulsion system may be an appropriate solution (Woud & Stapersma

Table 1. OPV93 Main characteristics.

Characteristic	Value
Length overall	93 m
Beam, molded	14.0 m
Depth	7 m
Design draft	4.1 m
Displacement	2250 Ton
Maximum speed	20 knots

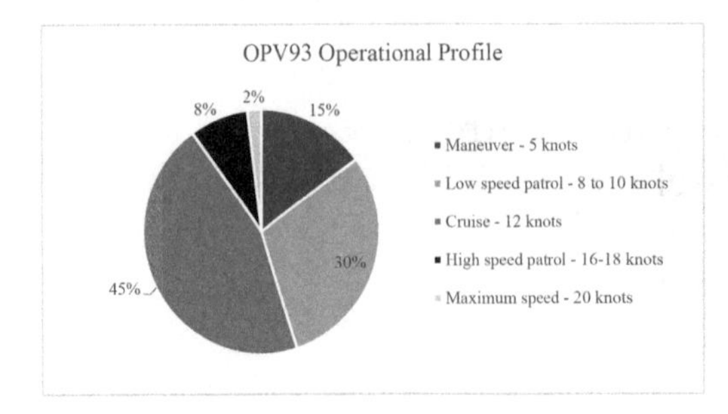

Figure 1. OPV93 design operational profile.

2003 pp. 120, Watson 2002 pp. 275, Lamb 2003 pp. 55–25). In this context, the CODELOD propulsion system emerges as an interesting option, running in electric propulsion mode at speeds up to 12 knots while in Diesel mode for greater speeds.

2.1 *The CODELOD propulsion system*

A typical electromechanical arrangement for CODELOD propulsion system in naval ships is illustrated in Figure. As shown in the Figure, the Diesel generator sets and the main Diesel engines are the sources of power, the former delivering electrical power to ship consumers and the latter delivering power to the propeller.

In electric propulsion mode, the service generators produce electric power to either, hotel and electric propulsion to thrust the ship, while in Diesel propulsion mode, the generators deliver power only to hotel loads. The distribution of electrical power for hotel and propulsion loads is known as Integrated Power System – IPS- (Doerry 2015). Depending on the total generating power level and consumers, the distribution voltage could be low or medium voltage – up to 13.8 kV- in alternating current or direct current high voltage – exceeding thousands of kV- (Doerry 2015, Woud & Stapersma 2003 pp. 120).

The propeller receives power from the main Diesel engine or electric propulsion motor through a gearbox, with the electric motor controlled normally by power electronics. The type of motor control significantly influences the life cycle costs of the CODELOD arrangement.

For the OPV93, the design constraints set the propeller type as CPP. Two 2.6 m diameter and 5 blades propellers were selected for maximum efficiency at 20 knots. Consequently, two 600 kW induction motors and two 4800 kW four stroke Diesel engines were selected as propulsion engines; the main generators are 4 units with 740 kW electric rated power according to the n-1 philosophy. The main characteristics of the selected motors are detailed in Table 2.

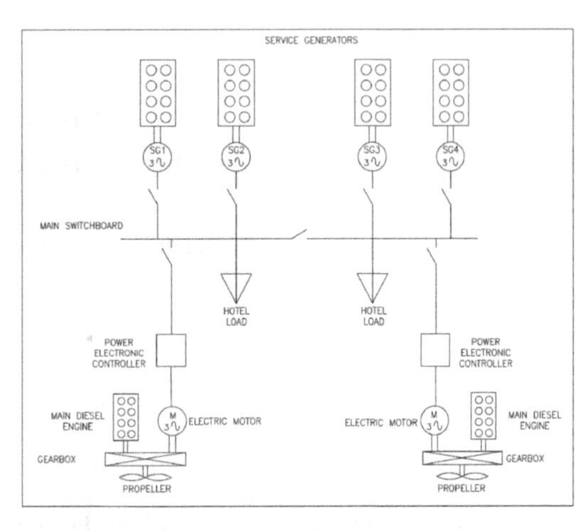

Figure 2. Main electromechanical arrangement for CODELOD propulsion system.

Table 2. Main characteristics of the OPV93 electric propulsion motors.

Characteristic	Value	Characteristic	Value
Rated output	600 kW	Rated voltage	440 V
Frequency	60 Hz	Speed	1190 rpm
Starting current	530%	Rated torque	4815 Nm
Power factor	100% load 0.82	75% load 0.79	50% load 0.7
Efficiency	100% load 96.2%	75% load 96.5%	50% load 96.2%

2.2 *Fixed or variable speed motor control*

In literature as well as ship references, the electric propulsion motors within hybrid or all-electric propulsion systems are typically associated with variable frequency control and fixed pitch propellers—FPP (Adnanes 2003, Woud & Stapersma 2003 pp. 118–120, Lamb 2003 pp. 55–25, Watson 2002 pp. 275). The above to change ship speed by motor rotational speed variation according to the FPP behavior.

Unlike FPP, electric motor operation along with CPP is not extensively discussed in literature; as the CPP allows ship speed variation only by changing the pitch-to-diameter ratio, electric propulsion motor speed control is not required (Adnanes, 2003). Accordingly, electric propulsion motors along with CPP can operate over the entire range of the electric propulsion mode at fixed rotational speed. In this case a control device is needed to reduce the impact of the motor inrush current since motor power is large compared with generator power (Woud & Stapersma 2003, pp. 357).

Nevertheless, Radan (2004) addresses that electric propulsion motors above the high-power range—assumed as 900 kW for electric propulsion motors—are recommended to operate with VFD even along with CPP. The above mainly due to the power factor correction by the VFD, which compensates the initial investment as well as the additional weight and volume of the unit.

As the OPV93 is fitted with CPP and the electric propulsion motor is below the high power range, VFD control will be discarded. Instead soft starter seems as the most appropriate option for starting current reduction requirement due to the rated power of the electric propulsion motors, which exceeds 20% of the generator rated power as recommended by Woud & Stapersma (2003 pp. 357).

In spite of that, some proposals received for the OPV93 CODELOD propulsion system disapproved the operation of the electric propulsion motors at fixed speed, pointing mainly at the ship electrical power system instability and the reduction in the quality of power supply. Furthermore, Woud & Stapersma (2003 pp. 358) and Patel (2012) states the drawbacks of soft starters, emphasizing in the power factor distortion and lack of power factor correction. Their solution lead to use of VFD to control the electric propulsion motors.

The aforementioned discussion demands the execution of a study to define whether soft starters or variable frequency drives should be used for motor control in the CODELOD arrangement, to be used as an input for the life cycle study of the propulsion arrangement.

3 ASSESSMENT OF SOFT STARTER AND VFD FOR ELECTRIC PROPULSION MOTOR

3.1 *Electric motor operation*

The current and torque behavior of the induction motor for the CODELOD arrangement, is depicted in Figure 3. As is widely explained in Chapman (2012), during starting the current has its maximum value and the torque is high, depending of motor construction, the starting current exceeds 5 to 7 times the rated current and torque can exceed 200% the rated torque. As the motor accelerates, the current decreases; the torque drops shortly and later grows as speed increases. Near synchronous speed, the motor reaches its rated operational zone with torque behavior becoming linear above 95% of synchronous speed, tending to zero at synchronous speed; simultaneously the current decreases, attaining rated values.

3.2 *Propeller-motor integration*

The preliminary design of the ship propeller sets the pitch-to-diameter (P/D) ratio as 1.25 (P/Dc) for maximum efficiency at 20 knots. Besides, the propeller blade angle of attack range was set from P/Dc+3° to P/Dc–15° to fit the fixed speed operation of the electric propulsion motors, as will be explained below. The Figure 4 shows the CPP power-speed curves for each blade angle variation, with isolines indicating ship speeds from 8 to 13 knots. Calculated torque and power data for propeller includes 15% sea margin and 5% additional operation margin for the electric motors.

To thrust the ship at 12 knots with the lowest power demand (correspondent to P/Dc + 3°), each propeller requires almost 478 kW of input power. Accordingly, to avoid cavitation effects for high rotational speeds, the maximum rotational speed of the propulsion motor at the input of the gearbox was calculated as 1200 rpm.

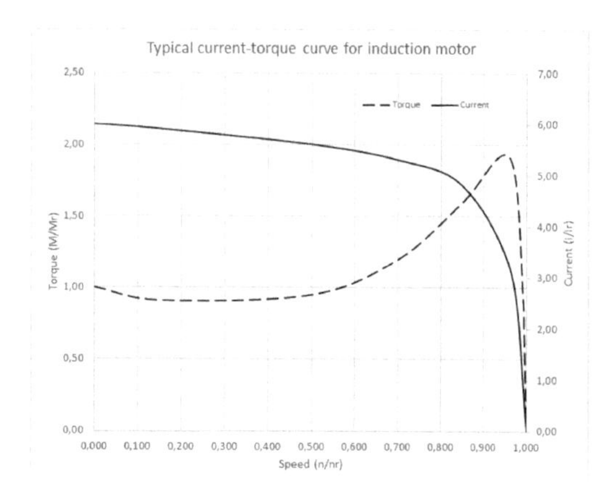

Figure 3. Current-torque curve for the OPV93 electric propulsion motor.

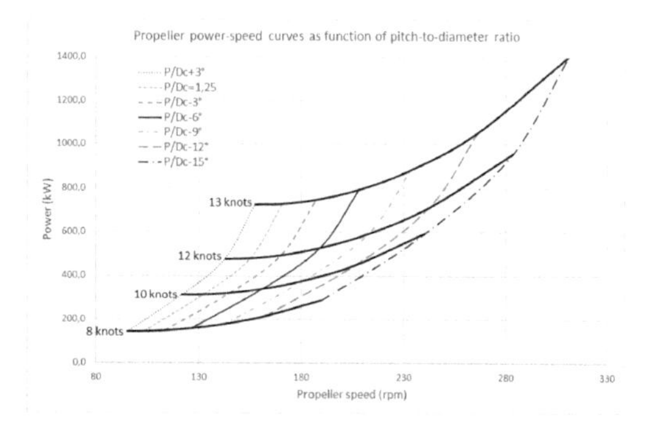

Figure 4. CPP power-speed curves as function of P/Dc.

3.2.1 *Soft starter control*

Considering the power range of the selected motors and the current behavior in Figure 3, direct on line starting will produce significant voltage drop in the ship power system, leading to black-out. The soft starters are devices which reduce the motor starting current to less than 200% (Rockwell Automation 2014), by voltage reduction with power electronics. The voltage reduction is maintained about 20 seconds after motor starting, until the motor attains its rated speed; later, the semi-conductor devices are bypassed and the motor is direct on-line connected.

At steady state, the propulsion motor with soft starter will operate at almost fixed speed following its rated operational curve. Hence, in the CODELOD arrangement the propeller-motor integration occurs when the propeller operational point (torque/power at certain rotational speed) matches the motor rated operational curve (torque/power-speed characteristic) in the same rotational speed.

The first step to the integration is to set the gearbox reduction ratio, to reduce the motor shaft speed to the propeller speed range. As the propeller demands 478 kW at 12 knots with blade angle curve P/Dc+3°, the correspondent propeller speed is 142.7 rpm. Considering a linear trend in the propulsion motor power-speed characteristic as informed by the manufacturer, the motor will provide the required power at 1192 rpm. Hence, the speed reduction ratio should be approximately 8.35.

With the constant reduction ratio defined, a decrease in the propeller blade angle should cause the ship speed to drop with the consequent change in the power and speed delivered by the motor. This scenario is illustrated in Figure 5, as the blade angle decreases to P/Dc-3° the speed is reduced to 10 knots with the propeller demanding about

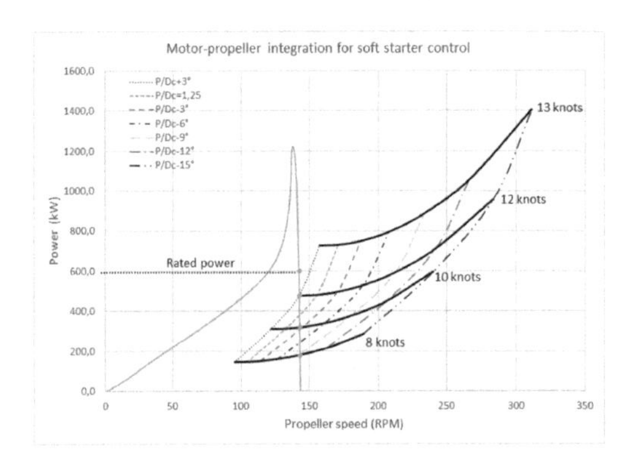

Figure 5. Motor-propeller integration for soft starter control.

320 kW at 143.0 rpm. If blade angle continues decreasing to P/Dc-9°, will cause speed change to 8 knots, demanding 177 kW at 143.3 rpm. Note that propeller rotational speed varies less than one rpm through ship speed change.

Therefore, soft starter control of the OPV93 propulsion motors allows to direct on-line connection of motors reducing its starting current; however, since motor induced torque is proportional to square of the supply voltage, it is recommended to start the motor without shaft load.

3.2.2 *VFD control*

The VFD is an equipment fitted with controlled semiconductor devices, receiving at the input electrical power with fixed voltage and frequency and delivering it with variable voltage and frequency at the output. This allows torque and rotational speed control of induction motors, since motor synchronous speed depends on motor pole number and supply frequency, as defined in Equation 1:

$$\eta_s = \left(-\frac{120 \times f_e}{p} \right) \tag{1}$$

where η_S is the synchronous speed, f_e is the electrical frequency and p is the pole number.

With VFDs, motor speed can be reduced below or increased above the rated speed.

In electric propulsion systems, VFDs are normally used along with fixed pitch propellers (FPP) to control ship speed (Adnanes, 2003). In addition, they are used normally with CPP to control motors in the high-power range (Radan, 2004) owing to the power factor compensation, which permits the reduction of cross section of motor cable and avoids the transport of reactive power in the power system. As well as the soft starters, VFDs can reduce the motor starting current also by voltage control.

The propulsion motor and propeller integration is made starting from the same conditions as before: 8.35 gearbox reduction ratio and motor delivering power at rated conditions to thrust the ship at 12 knots, matching with propeller curve correspondent to blade angle P/Dc+3°. Since variation of the pitch-to-diameter ratio is not required to change ship speed, the motor-propeller integration will follow the power curve correspondent to blade angle P/Dc+3°.

At rated voltage and frequency, the propulsion motor power-speed characteristic is the same of Figure 5. If the frequency supplied to the motor is decreased, then, the synchronous speed is also reduced, displacing the power-speed characteristic along the propeller speed axis; besides, as the voltage should be also decreased in the same proportion, torque also drops. The above is depicted

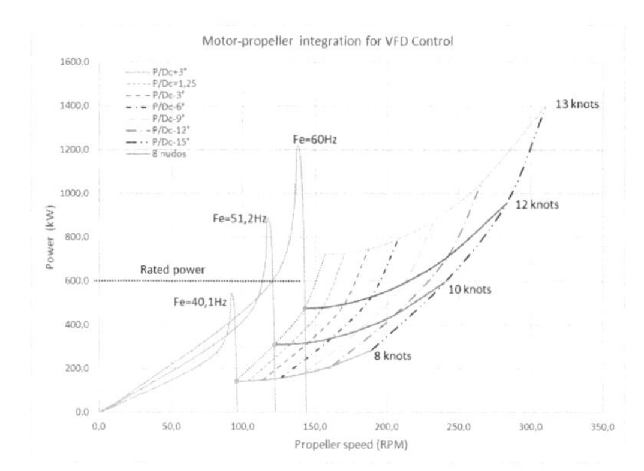

Figure 6. Motor-propeller integration for variable speed control.

in Figure 6, to thrust the ship at 10 knots without change in the propeller blade angle of attack the electrical frequency supplied to the motor is reduced from 60 Hz to 51.2 Hz; the new motor power-speed characteristic matches the P/Dc+3° propeller curve at 10 knots. If the electrical frequency is set to 40.1 Hz, the new motor-speed characteristic matches the propeller P/Dc+3° curve at 8 knots and the ship can travel at this speed.

Compared with the soft starter control, it can be noted that for 10 and 8 knots, the propeller required power is lower. The difference is 2.6% for 10 knots and 19.2% for 8 knots; according to the trend, for 5 knots a greater difference is expected.

Although the speed variation of the OPV93 with propulsion motor speed control does not require pitch-to-diameter change, it is also possible to vary it while the electrical frequency of propulsion motors is kept constant. In this case the system will have the same behavior the soft starter control.

3.3 *Harmonic distortion*

The harmonics are frequency components in the voltage or current signal that are integer multiples of the frequency of the main signal. Harmonics cause distortion of current and voltages, producing undesirable effects in ship power systems.

The Total Harmonic Distortion (THD), which is a measure of a signal distortion, is set by the IEC (1995) as 5% maximum for ship power systems. Soft starters and VFD are sources of harmonic components, producing harmonic distortion in the electric system.

3.3.1 *Soft starter*

When a motor is controlled by soft starter, its supply voltage is highly distorted by the firing control performed by the thyristors. At the instant which the motor is started, the thyristors fire at large angles, decreasing to zero as the motor speed grows. The variable firing angle produces partial voltage amplitudes leading to harmonic distortion as high as 10% (Rockell Automation 2014).

However, as the soft starter operates only during motor starting (<20 s) harmonic distortion is not relevant for the ship power system, because in steady state the propulsion motors are directly connected to the network and no harmonic components are produced.

3.3.2 *VFD*

In the VFDs the AC input power is converted to DC power and filtered, later the signal is inverted back to AC with voltage and frequency control. Most VFD are fitted with diode bridge or SCR rectifier at the AC-DC converter, while the inverter section is normally equipped with IGBT.

As the rectifier transforms the AC sinusoidal wave in a DC signal by uncontrolled (diode) or controlled (SCR) firing, the input current signal is distorted to a partial sinusoidal signal caused by the switching of the semiconductor devices. To size the harmonic contribution of this type of VFD in the ship power system, the harmonic spectrum ship service generators were calculated for the two 600 kW induction motors supplied by a 6-pulse, 600 kW VFD with the results shown in Figure 7.

As seen in the figure, the amplitude of the harmonic components in the current and voltage signals of the generator sets is high, especially for the current. In both cases, the THD exceeds the maximum values allowed by the IEC (1995).

To mitigate the harmonic component contribution on ship electric power system, some measures must be applied, as the use of 18-pulse or 24-pulse AC-DC converters within the VFDs along with phase shift transformers, as well as harmonic filters at the input. Another option is the use of Active Front-End (AFE) converters, fitted with IGBTs

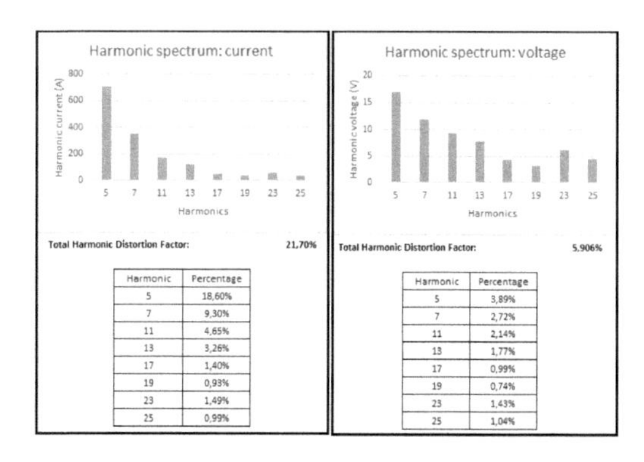

Harmonic	Percentage
5	18,60%
7	9,30%
11	4,65%
13	3,26%
17	1,40%
19	0,93%
23	1,49%
25	0,99%

Harmonic	Percentage
5	3,89%
7	2,72%
11	2,14%
13	1,77%
17	0,99%
19	0,74%
23	1,43%
25	1,04%

Figure 7. Calculated harmonic spectrum and THD for voltage and current signals of a 600 kW, 6-pulse VFD.

as the inverter section of the VFDs; this type of converter rectifies the AC power to DC producing low harmonic components thanks to the switching control of IGBTs (Rockwell automation 2014, Woud & Stapersma 2003, Patel 2012).

For the CODELOD arrangement with speed control, a 750 kW AFE drive was proposed to perform speed control of propulsion motors with low harmonic distortion of voltage and current, fulfilling with the IEC limit.

3.4 *Efficiency*

3.4.1 *Soft starter*

In the soft starters, as the thyristors performs the voltage control at the start, less than 1V drops across each device. For instance, while in operation, a soft starter can attain an efficiency near to 99.5% (Rockell Automation, 2014). However, after the motor starting process is complete and the soft starter is bypassed, the efficiency is almost 100% because the motor is directly connected to the switchboard.

3.4.2 *VFD*

The variable frequency drives fitted with IGBT devices for rectifier and inverter have typical efficiencies between 95–98% owing to the IGBTs voltage drop. As the motor is running, switching devices are on, with the consequent voltage drop across them. The preliminary efficiency of the 750 kW AFE drive proposed for the OPV93, as reported by the manufacturer is 96.1%.

3.5 *Power factor and apparent power demand*

Induction motors, as the main propulsion motors for the CODELOD arrangement, typically consumes reactive power because of the lagging power factor. For the propulsion motors mentioned before, the power factor varies between 0.7 and 0.82, lagging, as shown in Table 2.

Additionally, the distortion power factor produced by non-linear loads is another issue to consider. The distortion power factor is the decrease of the average power transferred to the load caused by harmonic components in the current signal.

3.5.1 *Soft starter*

During motor starting, while the soft starter thyristors are on, harmonic distortion is produced in the current signal, causing distortion power factor which becomes 0% as the thyristors are bypassed and the motor is directly connected to the network. Consequently, in steady state, the OPV93 service generators must supply the propulsion motors reactive demand.

In line with the abovementioned idea, at 12 knots each propulsion motor must deliver 478 kW to the propeller. Since the efficiency and power factor of the motor at this loading are approximately 96.5% and 0.79, respectively, each soft starter-propulsion motor set will require about 630kVA from the ship service generators. The reactive power demand can be supplied with power factor correction capacitors to reduce the magnitude of the current delivered by the generators to the propulsion motors.

3.5.2 *VFD*

Unlike the soft starter, the AFE driver proposed for the CODELOD propulsion motors performs power factor correction thanks to the switching control of the rectifier IGBTs and the harmonic filters at the input. As consequence, by using the AFE driver, the power factor is corrected to 1 as reported by the manufacturer. According to this, at 12 knots the VFD-propulsion motor set will demand 515kVA, including both efficiencies, VFD and motor.

The difference between apparent power consumption between the VFD-motor set and soft starter-motor set exceeds 22% for VFD-motor set.

3.6 *Comparison of soft starter-motor set and VFD-motor set*

The Table 3 summarizes the main characteristics of soft starter and VFD along with propulsion motors as discussed before. Additionally, the table includes data as dimensions and weight of the proposed equipment for the OPV93 within the pre-

Table 3. Comparison of soft starter-motor and VFD-motor sets.

Characteristic	Soft starter	AFE VFD
Dimensions (L × W × H)	1200 mm × 2000 mm × 800 mm	2400 mm × 1955 mm × 600 mm
Weight	500 kg	1450 kg
Harmonic distortion	0% (steady state)	<5% THD
P.F. correction	No, equals motor	Yes, ≈1
Electrical Efficiency	100%	96.1%
Motor starting	Without shaft load	Low shaft load
Ship speed variation	Changing the pitch-to-diameter ratio	Motor frequency control
Power delivered to CPP	Up to 1.19	1
Normalized acquisition cost	1	>2.5

liminary stage. The cost is also addressed assuming the comparison made by Rockwell automation (2014) and confirmed with the received proposals.

Comparing the main features of each set, the soft starter-motor set initially appears to be the most appropriate for the CODELOD arrangement, particularly due its dimensions, weight, harmonic distortion and cost. Moreover, the electrical efficiency of the soft starter is almost 100% in steady state operation, while the maximum efficiency expected for the VFD is 96.1%.

However, compared to the VFD-motor set, there is an increase in power consumption owing to the lack of power factor compensation and the power demanded by the CPP at low speeds. The former can be compensated by power factor capacitors at the load, while the latter occurs during 45% of the total operational time, which is the same time that the ship is expected to be in cruise mode. Accordingly, this can lead to additional fuel consumption of the service generators.

According to the stated by Radan (2004) it is not expected that the benefits of using VFDs with the electric propulsion motors for the OPV93 exceed the investment. Still, further analysis to determine the effects of the additional power demand of the soft starter-motor set over the life cycle costs of the ship compared with the AFE VFD-motor set is recommended.

4 CONCLUSIONS

The use of soft starter or variable frequency drive along with electric propulsion motors was assessed to define the most suitable option for OPV93 CODELOD arrangement. The ship is expected to operate 90% of its operational time in electric propulsion mode and will be fitted with CPP propeller.

The evaluation was performed by motor-propeller integration and comparison of the electric power system performance for each method of propulsion motor control.

For the soft starter-motor set, the motor-propeller integration demonstrated that the ship can change its travel speed by variation of propeller blade angle of attack while the propulsion motors runs at fixed speed. Besides, harmonics and efficiency analysis shown that the soft starter-motor set does not contribute to harmonic distortion in the electric power system and has zero power loss during normal steady state operation of propulsion motors. However, the propulsion motor power factor is not corrected, requiring measures to reduce the reactive power demanded from generator sets.

The analysis of the AFE VFD-motor and CPP integration exposed that vessel speed change could be performed by either blade angle and motor electrical frequency variation. As the VFD controls the motor supply frequency, the CPP blade angle is not required to change, leading to a reduction in propeller power demand at low speeds. But, the VFD produces harmonic distortion even with the AFE rectifier and has power losses owing to converter efficiency. Finally, the acquisition cost is significantly high.

Consequently, the soft starter-motor set stands as the better alternative for the CODELOD propulsion system. Further analysis and simulation regarding to life cycle costs is required to define its benefits over VFD control.

REFERENCES

Adnanes, A.K. 2003. Maritime Electrical Installations and Diesel Electric Propulsion.

Chapman, S. 2012. *Electric Machinery Fundamentals 5th Edition*. New York: McGraw-Hill.

Doerry, N. 2015. Naval Power Systems. *IEEE Electrification, Vol. 3, No. 2. June 2015: pp. 12–21.*

IEC 1995. *IEC 60092-101:1994 + AMD1: 1995 Electrical installations in ships – Part 101: Definitions and general requirements*. Geneva: The International Electrotechnical Commission.

Lamb, T. 2003. *Ship Design and Construction*. Jersey City: Society of Naval Architects and Marine Engineers.

Patel, M. 2012. *Shipboard Propulsion, Power Electronics and Ocean Energy*. Boca Raton: CRC Press.

Radan, D. 2004. *Power Electronic Converters for Ship Propulsion Electric Motors* Trondheim, Norway: NTNU.

Rockwell Automation 2014. When to use a soft starter or an AC variable frequency drive. *Rockwell Automation Publication 150-WP007A-EN-P.*

Watson, D. 2002. *Practical Ship Design.* Oxford: Elsevier.

Woud, H. & Stapersma D. 2003. *Design of Propulsion and Electric Power Generation Systems*. London: IMarEST.

Progress in Maritime Technology and Engineering – Guedes Soares & Santos (Eds)
© 2018 Taylor & Francis Group, London, ISBN 978-1-138-58539-3

Simulation of a marine dynamic positioning system equipped with cycloidal propellers

M. Altosole, S. Donnarumma, V. Spagnolo & S. Vignolo
Polytechnic School of Genoa University, Genoa, Italy

ABSTRACT: Marine cycloidal propellers represent a good alternative to traditional propellers especially in Dynamic Positioning (DP) applications, since they can generate almost the same thrust in all directions. The present study aims to combine a simulation platform, previously developed by some of the authors for DP systems, with the performance modelling of an epicycloidal propeller. The latter bases the propeller thrust and torque evaluation on the kinematics of the blades, taking into account suitable correction factors in order to consider the interference phenomena among blades. As a case study, the control and allocation logics of a DP system are analyzed for a surface vessel, equipped with a single bow thruster and two epicycloidal propellers at stern. The examined ship is the same for which a DP system, characterized by a conventional twin-screw propulsion, was already studied and installed on board. A performance comparison between the two distinct propulsion configurations is then carried out by dynamic simulation.

1 INTRODUCTION

In order to automatically maintain position and heading of a vessel, Dynamic Positioning (DP) systems employ in general waterjets or azimuth thrusters; in case, rudders and propellers (together with bow and stern thrusters) can be used too (Alessandri et al. 2014). Marine cycloidal propellers (Jürgens et al. 2002, Taniguchi 1962, Esmailian 2014), usually driven by diesel-electric propulsion to better handle the large changes in power demand (typical during DP operations), can represent a good alternative to traditional propellers since they can generate almost the same thrust in all directions. They are classified into true cycloidal, epicycloidal (e.g. Voith Schneider Propeller) and trochoidal propellers on the basis of their eccentricity value e, namely the ratio between the distance of the steering center from the propeller axis and the radius of the circular orbit described by the blade axes (the rotor radius): a true cycloidal propeller is characterized by $e = 1$, while the conditions $e < 1$ and $e > 1$ distinguish epicycloidal and throcoidal propellers, respectively (Bose 2008). In the present study, the performance of an epicycloidal propeller is modelled within a DP propulsion simulator, already developed by some of the authors for a surface vessel equipped with two conventional twin-screw propellers and a bow thruster. This kind of configuration is not very suitable for station-keeping and DP applications (Sørensen 1996, Sørensen 2011, Fossen 1996, Fossen 2002), nevertheless, a conventional propulsion configuration could be requested for specific operations characterized by limited DP capabilities. For instance, the mentioned simulator was developed for a patrol vessel designed with a twin propeller-rudder configuration and a single bow-thruster, which were requested to provide a certain dynamic positioning performance at zero-speed with moderate weather conditions. The main purpose of the DP simulation model was to validate the Force and Thrust Allocation Logic (FAL, TAL, Johansen 2013), specifically designed for such propulsion configuration (Donnarumma et al. 2015).

In this new work, the same vessel, but supposed equipped with a single bow thruster and two epicycloidal propellers at stern, is simulated in order to analyze the main differences during DP operations, in terms of general performance and control system behavior. This kind of simulation involves a reliable representation of the epicycloidal propellers, whose manufacturers unfortunately do not publicly share their performance maps for confidential reasons. Therefore, simplified simulation approaches, as possible for traditional propellers (Altosole et al. 2012, Martelli 2015) or waterjets (Altosole et al. 2005), are quite difficult to be developed. The present numerical modelling is based on a mixture of theoretical and empirical considerations: in particular, the propeller thrust and torque evaluation is based on the kinematics of the blades, taking into account suitable correction factors in order to consider the interference phenomena among blades. The result is a simulation approach able to predict the performance of an epicycloidal propeller, avoiding demanding computations (e.g. CFD methods) that would not allow an effective simulation of the whole DP system.

2 DP SIMULATION MODEL

Figure 1 provides a sketch of the devised DP-logic circuit, specific for the target ship. The system block simulates, through the equations of motion, the presence of a Positioning Reference System (PRS), composed by a DGPS and a Fiber Optic Gyro (FOG), which picks up the instantaneous position and velocity of the vessel. Such measurements are compared with the corresponding desired quantities in order to compute the position and velocity errors.

The circuit extrapolates the low-frequency (LF) components of the errors and send them to the regulator, where the required forces and moments are firstly evaluated and subsequently allocated to the actuators. Environment action and delivered forces are then used within the equations of motion to obtain the new position and velocity.

The mathematical models adopted for ship motions and for wind, wave, and current forces, as well as for the controller, are briefly illustrated below.

2.1 *Ship motions*

Setting $\eta := [x, y, \psi]^T \in \mathbb{R}^3$, the array of the position (longitudinal and lateral position and orientation) of the vessel w.r.t. the Earth-fixed frame, and $\nu := [u, v, r]^T \in \mathbb{R}^3$, the array of the components of velocity (linear and angular) expressed in the body-fixed basis, the ship kinematics is described by the relations:

$$\dot{\eta} = R(\psi)\,\nu, \quad R(\psi) = \begin{pmatrix} \cos\psi & -\sin\psi & 0 \\ \sin\psi & \cos\psi & 0 \\ 0 & 0 & 1 \end{pmatrix} \tag{1}$$

The ship motion equations are given by:

$$M\dot{\nu} + C(\nu)\,\nu + D_0\nu + D(\nu)\,\nu = \tau_D + \tau_E \tag{2}$$

where M, C and D are mass-inertia and added mass, Coriolis and damping matrices respectively, the array $\tau := [X, Y, N]^T \in \mathbb{R}^3$ represents the components of the resultant force and moment (τ_D for delivered and τ_E for environmental forces and moments), expressed in the body-fixed basis.

2.2 *Environmental forces and moments*

Environmental disturbances are evaluated as the sum of forces and moments due to wind, current and wave respectively. Forces and moments are expressed making use of the well-known resistance form, depending on non-dimensional coefficients C_X, C_Y, and C_N, related respectively to the longitudinal force, the lateral force and the moment. In order to consider the occurring worst condition, all environmental disturbances are supposed to be aligned in the same incoming direction. The current and the wind speeds are assumed constant and wave drift forces are modelled as proportional to the square of the significant height H_s. Collecting all the (body-fixed basis) components of the force and moment in a unique 3-dimensional array τ, we have:

$$\tau_E = \tau_{\text{current}} + \tau_{\text{waves}} + \tau_{\text{wind}} \tag{3}$$

2.3 *Controller*

The controller consists of a PD (proportional and derivative controller), a wind forces reconstruction, a sea force estimation and a block for allocation logic (Figure 2).

The controller law is given by:

$$\tau_R = K_P\tilde{\eta} + K_D\dot{\tilde{\eta}} + \bar{\tau}_{PD} - \tau_W \tag{4}$$

where the output τ_R represents the required force and moment; in Eq. (4), K_P and K_D are constant matrices, $\bar{\tau}_{PD}$ and τ_W are contributions which compensate the environmental disturbances (Donnarumma et al. in press) and the quantities $\tilde{\eta} := \eta - \eta_d$ and $\dot{\tilde{\eta}} := \dot{\eta} - \dot{\eta}_d$ are controller input errors, η_d and $\dot{\eta}_d$ denoting the desired position and velocity respectively.

2.4 *Allocation*

The adopted thrust allocation logic (TAL) is based on a constrained minimum problem. The idea is to minimize a cost function of the seven variables

$$\underline{x} = [T_{pt}, T_{sb}, T_{bow}, X_{pt}, Y_{pt}, X_{sb}, Y_{sb}] \in \mathbb{R}^7 \tag{5}$$

subjected to some suitable constraints. In particular, denoting by T_{pt} and T_{sb} the portside and starboard thrusts respectively, T_{bow} the thrust of the bow thruster, (X_{pt}, Y_{pt}) and (X_{sb}, Y_{sb}) the components of the portside and starboard thrust forces in the body-fixed basis, the constrained minimum problem is formulated as

$$\min_{\underline{x}} f(\underline{x}) \text{ with } h_i(\underline{x}) = 0;\ g_j(\underline{x}) > 0 \tag{6}$$

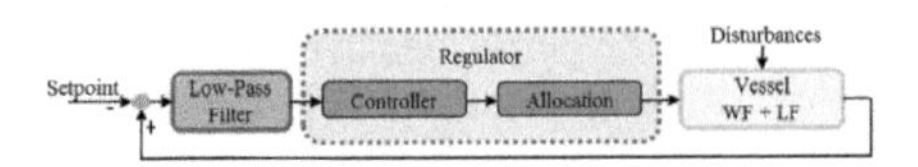

Figure 1. DP simulation model.

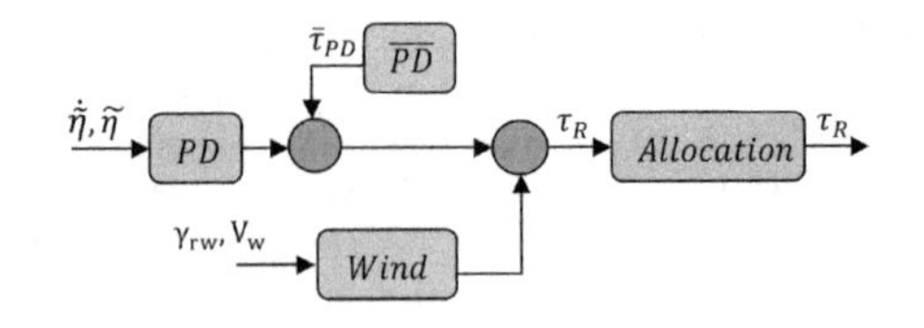

Figure 2. Controller layout.

where

$$f(\underline{x}) = \left(\frac{T_{\mathrm{pt}}}{T^{\mathrm{tot}}_{\mathrm{max}}}\right)^2 + \left(\frac{T_{\mathrm{sb}}}{T^{\mathrm{tot}}_{\mathrm{max}}}\right)^2 + \left(\frac{T_{\mathrm{bow}}}{T^{\mathrm{tot}}_{\mathrm{max}}}\right)^2 \tag{7}$$

is the cost function to be minimized, and

$$\begin{cases} h_1(\underline{x}) = X_{env} - X_{pt} - X_{sb} = 0 \\ h_2(\underline{x}) = Y_{env} - Y_{pt} - Y_{sb} - T_{bow} = 0 \\ h_3(\underline{x}) = N_{env} - x_{bow}\, T_{bow} - x_{pt}\, Y_{pt} + y_{pt}\, X_{pt} + - x_{sb} \\ \quad Y_{sb} + y_{sb}\, X_{sb} = 0 \\ h_4(\underline{x}) = T^2_{pt} - X^2_{pt} - Y^2_{pt} = 0 \\ h_5(\underline{x}) = T^2_{sb} - X^2_{sb} - Y^2_{sb} = 0 \end{cases} \tag{8}$$

are the constraints to be satisfied; in Eq. (8), $\{X_{env}, Y_{env}, N_{env}\}$ are the components of the force and the moment due to environmental disturbances, (x_{pt}, y_{pt}) and (x_{sb}, y_{sb}) are the coordinates of the propellers thrust centres and x_{bow} is the longitudinal coordinate of the bow thruster. Moreover, we have

$$g_1(\underline{x}) = T_{pt} > 0, \; g_2(\underline{x}) = T_{sb} > 0 \tag{9}$$

Eq. (6) and (7) require that the sum of the squared desired thrusts is minimum. Eq. (8) details the constraints: the first three represent the equilibrium between the environmental disturbances and the delivered force and moment; the last two correlate the modulus of the portside and starboard thrust forces with their longitudinal and lateral components. Finally, Eq. (9) ensures that the modulus of the two aft thrusts is positive.

3 EPICYCLOIDAL PROPELLER MODEL

3.1 *Kinematics*

In this subsection, we sketch the kinematical model adopted to describe the motion of the blades of a given epicycloidal propeller. Such 2-dimensional plane model makes use of two distinguished reference frames (see Figure 3): the first one $(O, \underline{b}_1, \underline{b}_2, \underline{b}_3)$ is fixed to the hull and it has its origin O at the center of the rotor, the unit vector $\underline{b}_1$ points towards the bow, the unit vector $\underline{b}_2$ points towards starboard and the unit vector $\underline{b}_3 = \underline{b}_1 \wedge \underline{b}_2$ points downwards; the second one $(O, \underline{e}_1, \underline{e}_2, \underline{e}_3)$ rotates clockwise about the vertical axis passing through O and parallel to $\underline{b}_3 = \underline{e}_3$, by an angle $\beta \epsilon [0, 2\pi]$ which determines (the perpendicular of) the steering force direction. The angle β is related to the rudder pitch of the epicycloidal propeller. The steering centre C lies on the straight line passing through O and parallel to $\underline{e}_2$. During the revolution motion, the projection P of the blade shaft on the plane $\langle O, \underline{b}_1, \underline{b}_2 \rangle$ describes a circumference having centre O and radius R, coinciding with the rotor radius. We parameterize such a circumference by an angle θ (function of time), in such a way that the unit tangent vector $\underline{t}$ is expressed as $\underline{t}(\theta) = -\sin\theta \underline{b}_1 + \cos\theta \underline{b}_2$. Introducing the vector $(C - O) = s\underline{e}_2 = -s\sin\beta\, \underline{b}_1 + s\cos\beta \underline{b}_2$ (with $s \in [0, 0.8R]$), the vector joining the steering centre C with the point P can be expressed as $(P - C) = (R\cos\theta + s\sin\beta)\, \underline{b}_1 + (R\sin\theta - s\cos\beta)\, \underline{b}_2$. The variable s is usually called driving pitch and it controls the magnitude of the thrust. The unit vector $\frac{(P-C)^{\perp}}{|(P-C)^{\perp}|}$, orthogonal to $(P - C)$ and belonging to the plane $\langle O, \underline{b}_1, \underline{b}_2 \rangle$, identifies with the unit vector of the blade chord. Therefore, the pivoting motion of the blade around its shaft can be described by the angle α (function of time) between the unit vectors $\underline{t}$ and $\frac{(P-C)^{\perp}}{|(P-C)^{\perp}|}$. Choosing anticlockwise the positive direction of rotation around the blade shaft, the pivoting angle α can be defined as

$$\alpha = \begin{cases} \cos^{-1}\left(\dfrac{(P-C)^{\perp}}{|(P-C)^{\perp}|} \cdot \underline{t}\right) & \text{if } \cos(\theta - \beta) \geq 0 \\ -\cos^{-1}\left(\dfrac{(P-C)^{\perp}}{|(P-C)^{\perp}|} \cdot \underline{t}\right) & \text{otherwise} \end{cases} \tag{10}$$

where

$$\frac{(P-C)^{\perp}}{|(P-C)^{\perp}|} \cdot \underline{t} = \frac{R + s\sin(\beta - \theta)}{\sqrt{(-R\sin\theta + s\cos\beta)^2 + (R\cos\theta + s\sin\beta)^2}} = \cos\alpha \tag{11}$$

Figure 3 summarizes the above outlined kinematical scheme.

Supposing now that the vessel is moving, let $\underline{v}_O = \hat{u}\underline{b}_1 + \hat{v}\underline{b}_2$ be the velocity of O (w.r.t. the Earth-fixed frame) expressed in the hull-fixed basis. Denoting by $\underline{v}'_P = -R\dot{\theta}\sin\theta \underline{b}_1 + R\dot{\theta}\cos\theta \underline{b}_2$ the velocity of the point P w.r.t. the body-fixed frame, the velocity of P w.r.t. the Earth-fixed frame is given by

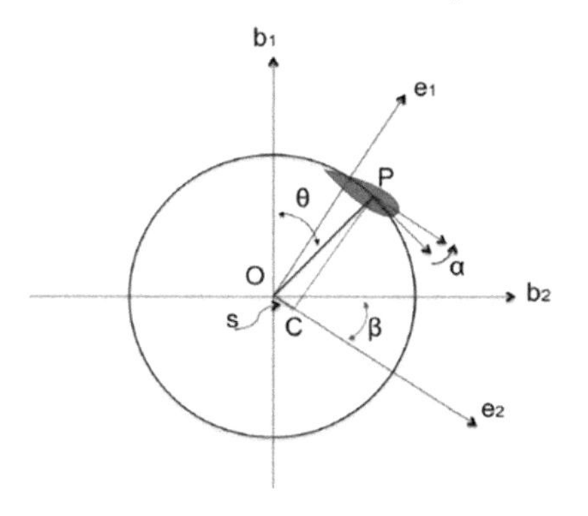

Figure 3. Kinematics of the blade.

$$\begin{aligned}\underline{v}_P &= \underline{v}_P' + \underline{v}_o + \underline{\omega} \wedge (P-O) \\ &= [\hat{u} - R(\dot{\theta}+r)\sin\theta]\,\underline{b}_1 + [\hat{v} + R(\dot{\theta}+r)\cos\theta]\,\underline{b}_2\end{aligned} \quad (12)$$

where $\underline{\omega} = r\underline{b}_3$ is the angular velocity of the vessel. The velocity of the incoming flow experienced at P by a blade-fixed observer is then $-\underline{v}_P$; its unit vector $\hat{\underline{t}}$ is expressed as

$$\begin{aligned}\hat{\underline{t}} &= -\frac{\underline{v}_P}{|\underline{v}_P|} = \\ &-\frac{[\hat{u} - R(\dot{\theta}+r)\sin\theta]\,\underline{b}_1 + [\hat{v} + R(\dot{\theta}+r)\cos\theta]\,\underline{b}_2}{\sqrt{[\hat{u} - R(\dot{\theta}+r)\sin\theta]^2 + [\hat{v} + R(\dot{\theta}+r)\cos\theta]^2}}\end{aligned} \quad (13)$$

Making use of the unit vector $\hat{\underline{t}}$, it is possible to characterize the attack angle of the incident flow as

$$\hat{\alpha} = \pi - \cos^{-1}\left[\frac{(P-C)^{\perp}}{|(P-C)^{\perp}|} \cdot \hat{\underline{t}}\right] \quad (14)$$

according to Figure 4.

3.2 *Hydrodynamic forces*

In this subsection, making use of some simplifying assumptions, we present a simple model for evaluating the hydrodynamic forces generated by each blade. It is supposed that the velocity of the incident flow be the same over the entire surface of the blade and coincide with $-\underline{v}_P$. Under such a condition, the lift and drag produced by each blade can be expressed as

$$\underline{L} = c_L \frac{1}{2} \rho_w A |\underline{v}_P|^2 \,\hat{\underline{n}} \quad (15)$$

$$\underline{D} = c_D \frac{1}{2} \rho_w A |\underline{v}_P|^2 \,\hat{\underline{t}} \quad (16)$$

where c_L is the lift coefficient, c_D is the drag coefficient, ρ_w is sea water density, A is the blade lateral area, $|\underline{v}_P|$ is the incoming flow speed, $\hat{\underline{t}}$ is the unit vector of the lift force (unit vector of the incoming flow at P), and $\hat{\underline{n}}$ is the unit vector of the drag force (perpendicular to $\hat{\underline{t}}$).The unit vector $\hat{\underline{n}}$ can be determined by the following procedure, in which two main scenarios are distinguished:

- the attack angle $\hat{\alpha}$ belongs to the interval $]0, \frac{\pi}{2}[$, namely the incoming flow hits the blade from the front. In such a circumstance, the unit vector $\hat{\underline{n}}$ is determined according to the requirements:

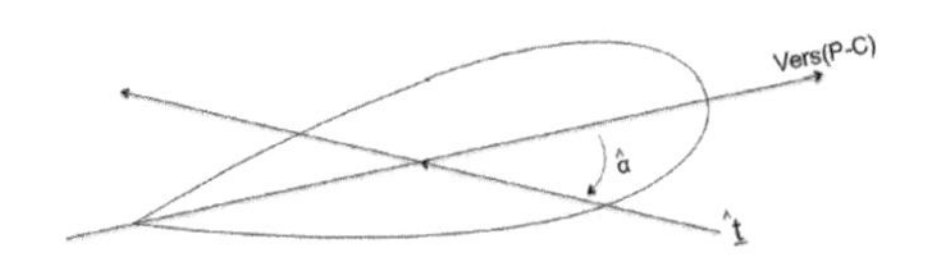

Figure 4. Angle of attack.

$$\hat{\underline{n}} = \begin{cases} \underline{b}_3 \wedge \hat{\underline{t}} & \text{when } \hat{\underline{t}} \wedge \frac{(P-C)^{\perp}}{|(P-C)^{\perp}|} \cdot \underline{b}_3 > 0 \\ -\underline{b}_3 \wedge \hat{\underline{t}} & \text{when } \hat{\underline{t}} \wedge \frac{(P-C)^{\perp}}{|(P-C)^{\perp}|} \cdot \underline{b}_3 < 0 \end{cases} \quad (17)$$

- $\hat{\alpha} \in]\frac{\pi}{2}, \pi[$, the incoming flow hits the blade from the back. In this case, $\hat{\underline{n}}$ is singled out by the requests:

$$\hat{\underline{n}} = \begin{cases} -\underline{b}_3 \wedge \hat{\underline{t}} & \text{when } \hat{\underline{t}} \wedge \frac{(P-C)^{\perp}}{|(P-C)^{\perp}|} \cdot \underline{b}_3 > 0 \\ \underline{b}_3 \wedge \hat{\underline{t}} & \text{when } \hat{\underline{t}} \wedge \frac{(P-C)^{\perp}}{|(P-C)^{\perp}|} \cdot \underline{b}_3 < 0 \end{cases} \quad (18)$$

As remaining particular cases, if $\hat{\alpha} = 0$ or $\hat{\alpha} = \pi$ there is no lift while if $\hat{\alpha} = \frac{\pi}{2}$ then $\hat{\underline{n}} = \hat{\underline{t}}$. The above described procedure allows to determine the lift and drag provided by each single blade. The resultant hydrodynamic force generated by the epicycloidal propeller can be computed as the sum of all contributions given by each blade.

3.3 *Torque acting on the rotor*

In order to calculate the torque acting on the rotor, the Newton-Euler moment equations for each single blade and for the rotor are considered separately. Developed in the hull-fixed reference frame and with respect to the point O (center of the rotor), the Newton-Euler moments equation for each blade can be expressed as

$$\begin{aligned}\underline{M}_O^H + \underline{M}_O^G + \underline{M}_O^R + \underline{M}_O^I &= I_G(\dot{\underline{\omega}}) + \underline{\omega} \wedge I_G(\underline{\omega}) + \\ &+ m(G-O) \wedge \underline{a}_G\end{aligned} \quad (19)$$

where, $\underline{M}_O^G, \underline{M}_O^H, \underline{M}_O^R$ and $\underline{M}_O^I$ are respectively the hydrodynamic, weight force, reactive force, and inertial force torques w.r.t. O acting on the blade; I_G is the inertia tensor w.r.t. the center gravity G of the blade; $\underline{\omega} = (\dot{\theta} - \dot{\alpha})\underline{b}_3$ is the blade angular velocity w.r.t. the hull-fixed frame; $\underline{a}_G$ is the acceleration of G w.r.t. the hull-fixed frame; and m is the blade mass.

Knowing the revolution velocity of the rotor and the position of the steering center as well as the velocity of the incoming flow, the consequent motion of the blade is known from kinematics; at the same time, the knowledge of the hydrodynamic forces allows the evaluation of their moment. Evaluating the reactive torques from Eq. (19) and inserting them in the moment equation for the rotor, the (scalar value of) engine torque can be calculated as

$$M_O^E = \sum_{i=1}^{n} \left(\underline{M}_O^R \right)_i \cdot \underline{b}_3 - \underline{M}_O^I \cdot \underline{b}_3 + I_O \left(\dot{\underline{\omega}}_r \right) \cdot \underline{b}_3 \qquad (20)$$

where $\underline{M}_O^I$ is the inertial forces torque acting on the rotor, I_O is the inertia tensor of the rotor and $\underline{\omega}_r = \dot{\theta}\underline{b}_3$ is the angular velocity of the rotor, n is the number of blades.

3.4 *Validation and thrust generation*

The main features and the validation of the simulator based on the mathematical model illustrated above have been presented in Altosole et al. (2017). For sake of shortness, we only recall that the interference among the blades is taken into account by means of three correction factors validated with the open water diagram of an existing propeller: shielding correction, referring to the shielding of the blades that are in the half circumference not directly exposed to the incoming flow (in the model, the correction factor, depending on driving pitch values, reduces the right velocity of the incoming water flow); interference correction, modeled by reducing the attack angle of the incoming flow with respect to the chord of the blade section (the correction depends on the advance coefficient and pitch values); reverse thrust correction, representing the reduction of the reverse thrust (comparing butterfly diagrams found in open source with those obtained by simulation, we found out that for advance coefficient more than 0.4 there was a reduction of the thrust when the steering pitch β was between $\pi/2$ and $3\pi/2$: we introduced this coefficient to take into account this further phenomenon).

As we have illustrated in the description of the epicycloidal propeller model, there are two different pitches (s and β) that control the magnitude and the direction of the thrust, together with the choice of a suitable rpm. In this case study, simulations have been made by keeping rpm constant, so modifying the thrusts only by means of the two geometric pitches. When the simulator runs, the required thrusts are translated in terms of corresponding control pitches and thus the delivered thrusts are generated: the steering pitch β is strictly linked to direction of the required thrust (aligned along the unit vector $\underline{e}_1$, see Figure 3), while the other pitch s is uniquely determined by a matching algorithm that combines a given required thrust (at fixed rpm) to a predetermined geometric pitch.

4 SIMULATION RESULTS

In this section, some simulation results concerning the vessel equipped with epicycloidal propellers are presented and compared with those obtained by previous simulations of the same vessel, equipped with a conventional twin screw propulsion system.

The environmental disturbances have been modeled as detailed in Alessandri et al. (2014) and briefly recalled in Subsection 2.2. In order to consider the worst environmental condition, the disturbances (sea, wind and current) are considered aligned and coming from the same direction. For this work we maintained Mediterranean SS 4 (significant wave height of 1.8 m, wave period of 8.8s) and a constant current speed of 1 kn as for the DP capability plots presented in (Donnarumma et al., in press), where a static analysis of the ship performance has been presented.

We show the simulation results in the presence of environmental disturbances coming from an angle of 30° with respect to the desired heading, in two distinct cases: 10 kn and 30 kn wind speed. Consistently with what proposed in Donnarumma et al. (in press), the evaluation of the environmental disturbances mean components $\overline{\tau}_{PD}$ and τ_W requires some minutes of transient that are not relevant for the station keeping performances evaluation. For such a reason, first few minutes of simulation have been neglected. In Figures 5, 6 and 7 the variations of the ship position and heading are shown in the two different environmental conditions. As we can see, for a wind speed of 10 kn both the propulsion configurations are able to keep the desired position and heading; however, different amplitudes of the oscillations around the desired set-point and then different performances of the two propulsion systems are evident. For a wind speed of 30 kn instead, the conventional propulsion cannot perform the desired DP maneuver (see Figure 6). The same conclusions are reflected in the Figures from 8 to 13,

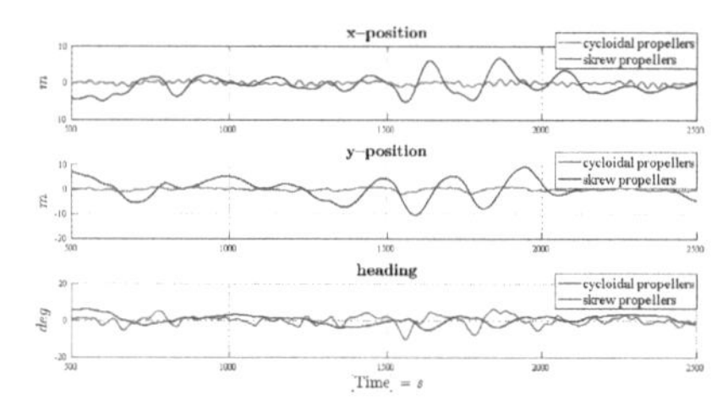

Figure 5. Motions time history for wind speed of 10 kn.

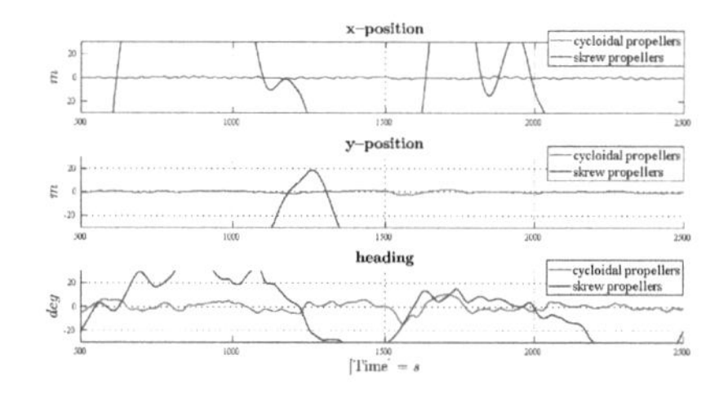

Figure 6. Motions time history for wind speed of 30 kn.

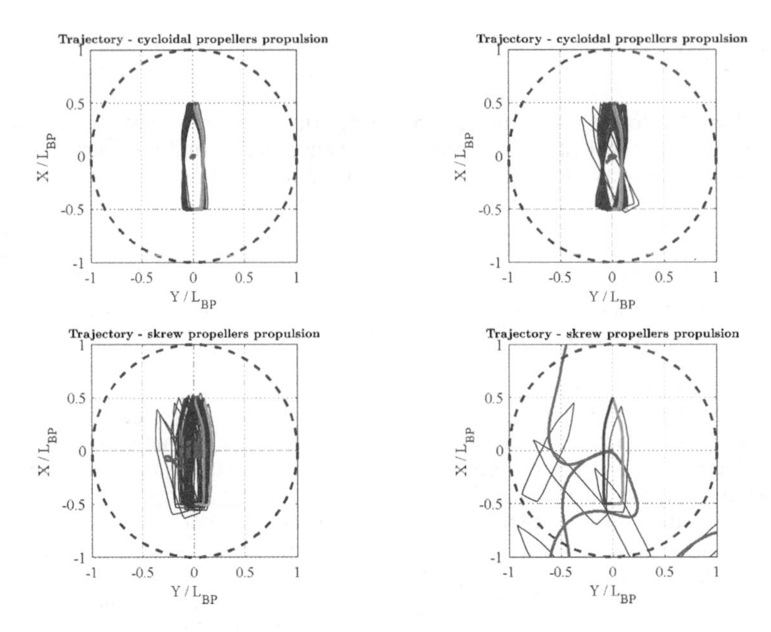

Figure 7. Ship position (trajectory of the origin of the body-fixed frame) and orientation variations for wind 10 kn (on the left) and 30 kn (on the right).

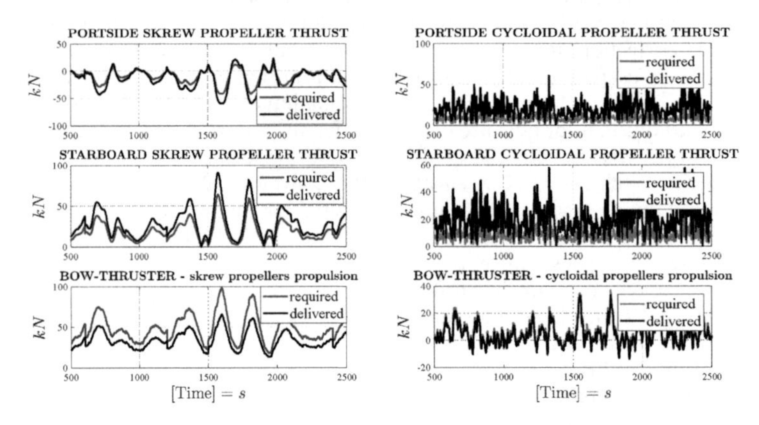

Figure 8. Time history of required and delivered thrust for wind speed of 10 kn.

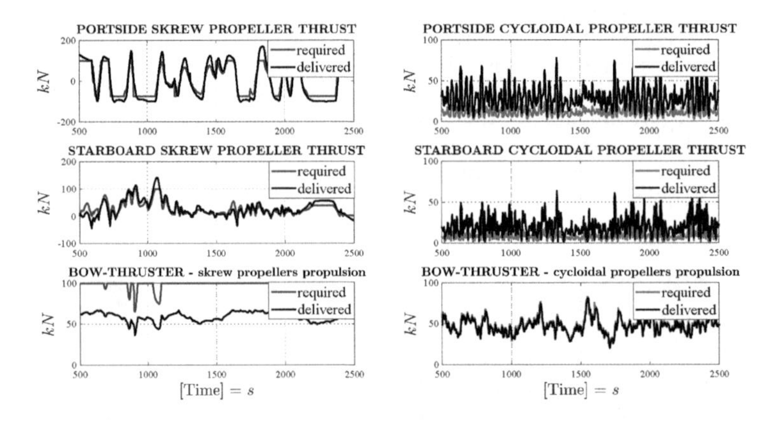

Figure 9. Time history of required and delivered thrust for wind speed of 30 kn.

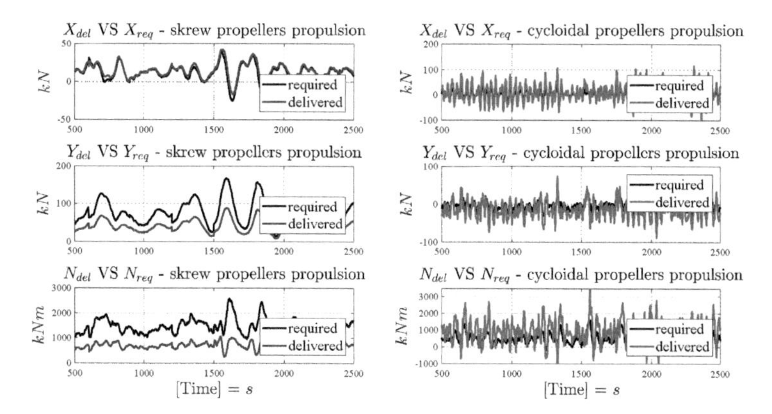

Figure 10. Time history of required and delivered force and moment for wind speed of 10 kn.

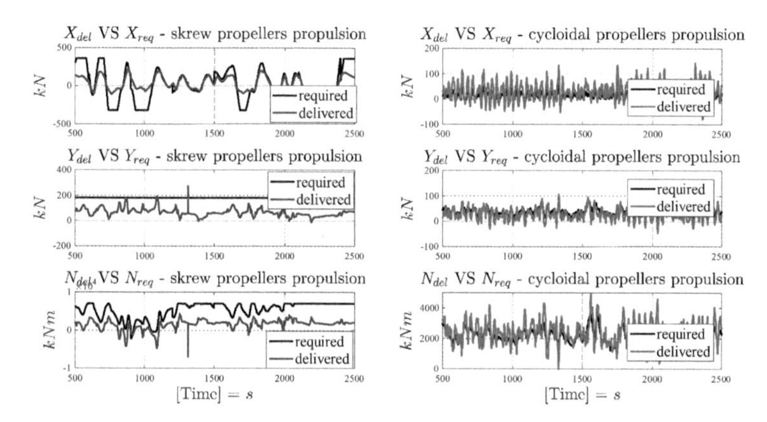

Figure 11. Time history of required and delivered force and moment for wind speed of 30 kn.

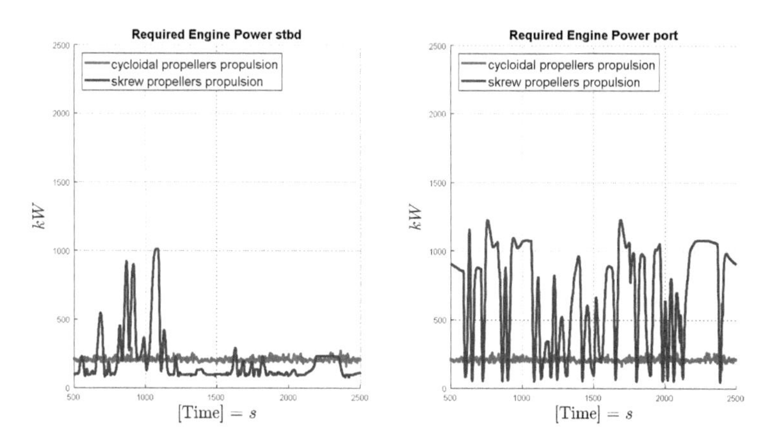

Figure 12. Time history of required engine power for wind speed of 10 kn.

where differences between the two propulsion plants are underlined. The thrusts required to the actuators and thus delivered by the two propulsion plants are very far from each other. Also the required and delivered force and moment are deeply unequal, since the deviation from the desired set-point is very different. These results are reflected in Figures 12 and 13, where the power required to the engines for the two propulsion configurations is shown.

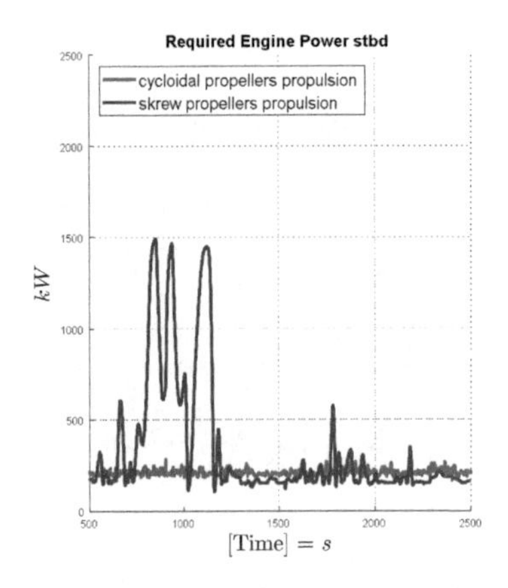

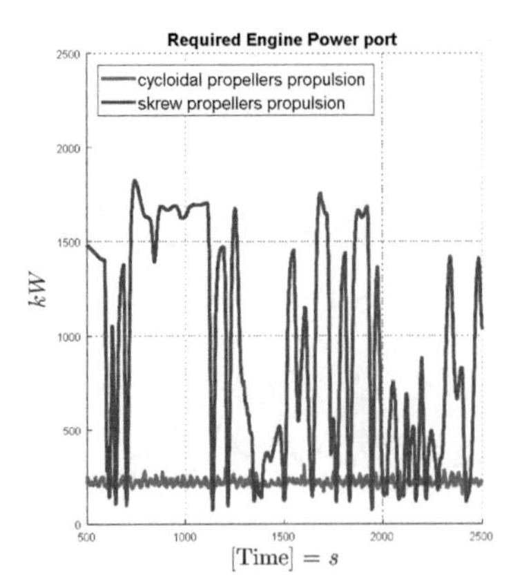

Figure 13. Time history of required engine power for wind speed of 30 kn.

5 CONCLUSIONS

A simulation model for dynamic positioning of a vessel equipped with cycloidal propellers has been presented. Dynamic simulations have been carried out and compared with those regarding the same ship equipped with conventional propellers. The obtained results confirm the conclusions of a previous work (Altosole et al. 2017), where simulations in static conditions were performed. As it was expected, it is shown that cycloidal propellers can be a valid alternative to traditional propellers in case of strong DP requirements. Simulation results have been provided also to illustrate the effectiveness of the proposed propulsion plant and the corresponding thrust allocation, as well as the reliability of the mathematical and numerical model implemented for cycloidal propellers. Future works will concern dynamic simulations of maneuvering at design speed.

REFERENCES

Alessandri, A., Chiti, R., Donnarumma, S., Luria, G., Martelli, M., Sebastiani, L., Vignolo S. 2014. Dynamic Positioning system of a vessel with conventional propulsion configuration: Modeling and Simulation. *Proceedings of MARTECH 2014, 2th International Conference on Maritime Technology and Engineering*, Lisbon, Portugal, 15–17 October 2014, 725–733.

Altosole, M., Benvenuto, G., Figari, M. 2005. Performance prediction of a planing craft by dynamic numerical simulation. *Proceedings of HSMV 2005, 7th Symposium on High Speed Marine Vehicles*, Naples, Italy, 21–23 September 2005, pp. 105–111.

Altosole, M., Donnarumma, S., Spagnolo, V., Vignolo, S. 2017. Marine cycloidal propulsion modelling for DP applications. *Proceedings of Marine 2017*, 7th *International Conference on Computational Methods in Marine Engineering*, Nantes, France, 15–17 May 2017, 206–219.

Altosole, M., Figari, M, Martelli, M. 2012. Time-domain simulation for marine propulsion applications. *Proceedings of SCSC 2012, 2012 Summer Computer Simulation Conference*, Genoa, Italy, 8–11 July 2012, 36–43.

Bose, N. 2008. Marine Powering Prediction and Propulsors. *The Society of Naval Architects and Marine Engineers*.

Donnarumma, S., Figari, M., Martelli, M., Vignolo, S., Viviani, M. Design and Validation of Dynamic Positioning for Marine Systems: A Case Study. *IEEE Journal of Oceanic Engineering*, DOI: 10.1109/JOE.2017.2732298, in press.

Donnarumma, S., Martelli M. and Vignolo S. 2015. Numerical models for ship dynamic positioning. *Proceedings of Marine 2015, 6th International Conference on Computational Methods in Marine Engineering*, Rome, Italy, 15–17 June 2015, 1078–1088.

Esmailian, E., Ghassemi, H., Heidari, S.A., 2014. Numerical investigation of the performance of voith schneider propulsion, *American Journal of Marine Science*, Vol. 2, No. 3, 58–62.

Fossen, T.I., 2002. Marine Control System. Marine Cybernetics. Norway: Trondheim.

Fossen, T.I., Sagatun, S.I., Sørensen, A.J. 1996. Identification of dynamically positioned ships, *Control Engineering Practice*, Vol. 4, Issue 3, 369–376.

Johansen, T.A. & Fossen, T.I., 2013. Control Allocation-A Survey. Automatica Vol. 49, 1087–1103.

Jürgens, D. & Moltrecht, T. 2002. Enhanced cycloidal propulsion. *The International Workboat Show*, New Orleans.

Martelli, M. 2015. *Marine propulsion simulation.* Berlin: Walter de Gruyter GmbH.

Sørensen, A.J. 2011. A survey of dynamic positioning control systems, *Annual Reviews in Control*, Vol. 35, Issue 1, 123–136.

Sørensen, A.J., Sagatun, S.I., Fossen, T.I. 1996. Design of a dynamic positioning system using model-based control, *Control Engineering Practice*, Vol. 4, Issue 3, 359–368.

Taniguchi, K. 1962. Sea Analysis of the vertical axis propeller. *Proc. of the 4th symposium on Naval Hydrodynamics*, Office of Naval Research, Washington.

Progress in Maritime Technology and Engineering – Guedes Soares & Santos (Eds)

ISBN 978-1-138-58539-3

Reliability analysis of dynamic positioning systems

M.V. Clavijo & M.R. Martins
Analysis, Evaluation and Risk Management Laboratory—LabRisco, Naval Architecture and Ocean Engineering Department, University of Sao Paulo, Sao Paulo, Brazil

A.M. Schleder
Department of Production Engineering, Sao Paulo State University—UNESP, Sao Paulo, Brazil

ABSTRACT: This paper presents the quantitative reliability assessment for two typical configurations (a DP Class 2 semi-submersible platform and a DP Class 3 drillship); based on the present technological maturity and the offshore industry-reported component failure rates. In the analyzed DP2 and DP3 configurations, the control and thruster subsystems contribute to 1%, and 7% of the total DP system failures, while the power subsystem contribute to 10% and 6% of the total DP2 and DP3 system failure. Regarding the MTTF analysis, the drillship presents a Mean Time to Failure of 4.65 years, a 1.41 fold increase over semisubmersible platform. The results presented could be used for reliability-centered design and maintenance planning of DP systems.

1 INTRODUCTION

Nowadays, most of the operations related to offshore oil exploration are performed by Dynamically Positioned (DP) units; in Brazil, the use of those units is usual to perform drilling operations. The DP ships or DP semisubmersible platforms must keep their position by computer-assisted thrusters that are constantly opposed to the environmental forces, (IMO, 1994). The advent of DP systems, which allows the vessel to maintain a position without anchor lines, has brought great flexibility to oil exploration field; however, all systems are subject to failure including the DP system.

Usually, in the drilling areas, there are numerous simultaneous complex operations being performed by several operators subject to the most diverse climatic conditions; a failure in the DP system may cause vessel drift, which may occasionally result in a collision with other field equipment, causing material, personnel and environment damage that can be catastrophic, (Patino, 2012).

Currently, there are only few studies that actually present a quantitative evaluation of DP systems failures, although, the information about the DP system is essential to guarantee operations at tolerable risk levels; (Hauff, 2014), (Ebrahimi, 2010) and (Ferreira, 2016).

Some authors, as Vedachalam & Ramadass (2017), argue that reliability is the key requirement for DP systems to ensure safe and successful completion of critical offshore operations. According to International Maritime Contractors Association (IMCA) database with 71 DP-related incidents reported from 54 vessels during 2014, 54% of the failures were due to technical failures in the DP system and the rest were due to human error, procedural and environment related, (IMCA, 2016). Taking into consideration the increasing number of DP vessels and the demand on risk reduction with increased HSE regulations; IMO insist fault-tolerant DP systems using redundant architectures, Det Norske Veritas—DNV (2015) & American Bureau of Shipping—ABS (2013). Hence, studies on reliability centered DP system design based on the recently reported offshore component failure rates are inevitable to determine the trade off in redundancy, cost, weight and footprint.

In this context, this paper presents a reliability analysis of a DP system class 2 and a DP system class 3 (International Maritime Organization Classification according to redundancy levels presented at the system) used in the drilling units; it will be described typical system configurations, presented a database-driven survey, and it will be discussed differences between reliability results of each DP class.

The authors use Fault Tree diagrams (FT) as technique for reliability and safety analysis in order to calculate the probability of failure of each class of DP system (DP2 and DP3).

The reliability analysis will provide crucial information to future risk analysis involving DP drilling units.

2 GENERAL ASPECTS OF DP SYSTEMS

Dynamic positioning (DP) systems are used for the station keeping of marine vessels. In this section is presented the general structure of DP system.

According to IMO (1994), ABS (2013) and DNV (2013), the DP systems comprise three subsystems: power, thruster and DP control.

Power subsystem means all components and systems necessary to supply the DP system with power. The power subsystem includes: mover-generators with auxiliary subsystems, switchboards, electrical distribution system including cabling and cable routing and power management system—PMS (if applicable).

Thruster subsystem means all components and systems necessary to supply the DP system with thrust force and direction. The thruster subsystem includes: thruster with drive units and auxiliary subsystems, main propellers and rudders (if these are under the control of the DP system), thruster control electronics, manual thruster controls and associated cabling and cable routing.

DP control subsystem means all control components and systems, hardware and software necessary to dynamically position the vessel. The DP control subsystem consists of the following: computer system, Position Reference Systems—PRS, Uninterruptible Power Supply Systems—UPS, environment sensors, Independent Joystick System—IJS, associated cabling and cable routing.

Additionally, there is a set of auxiliary subsystems whose functions have a direct effect on the thrusters and mover-generators; these are ventilation, lubrication oil, cooling water, compressed air and fuel oil subsystem.

Figure 1 show the general configuration of the DP system considering the three main subsystems described above (power (red), thruster (yellow) and control DP (green)). Notice that the auxiliary subsystems does not appear in this figure, since its operation is the support of the main components (mover-generators and thrusters), and here it is only seeing a general configuration.

As can be seen in Figure 1, the DP system power is produced in mover-generators and then delivered to the switchboard, where it is redirected to the control and the thruster subsystems.

To supply power to the DP control subsystem equipment, the switchboard filters power through UPSs to absorb current spikes (Rappini et al., 2003). If the mover-generators fail, there is an emergency power source that will provide power to the UPSs in order to maintain control of vital components. This measure excludes the thrusters which will lose power because the emergency mover-generator cannot supply the amount of power that the thrusters need.

Regarding the flow of information, the control computers receive information from two sources: the environment sensors (Gyro, Wind and Vertical Reference Sensor—VRS) and PRS. The computer processes all the information and sends signals to the electronic controls of the thruster subsystem.

The DP vessels need to engage the alternative manual control of the thrusters, to ensure this, there is the IJS which is directly connected to the electronic controls of the thruster subsystem, also powered by the UPS, and receives information from some of the environment sensors and PRS.

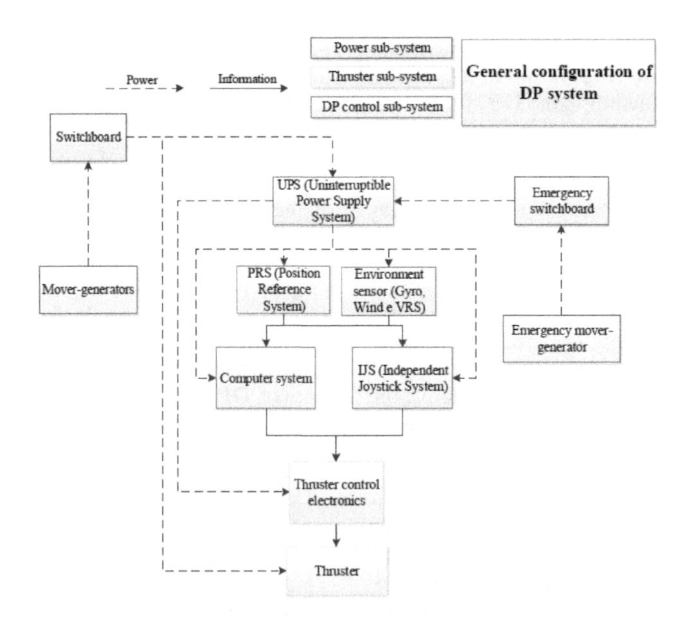

Figure 1. General configuration of DP system.

Finally, once the information and power are transmitted to the electronic controls of the thruster subsystem, these equipment sends the information to the thrusters to reach the directions and commands previously established, and to obtain control of the position of vessel.

3 TYPICAL SYSTEMS CONFIGURATION

To evaluate the reliability of the DP system, two typical configurations are selected: a DP Class 2 semi-submersible platform and a DP Class 3 drillship. It is worth noting that the general safety requirement for classes 2 and 3 is that any single fault should not propagate into a loss of position; and in class 3 the single fault including complete loss of a compartment due to fire or flood (IMO, 1994).

3.1 *Configuration 1: Semisubmersible platform*

The first case study is a semisubmersible platform with eight thrusters arranged in pairs at the ends of the port and starboard pontoons. The vessel has a Diesel electric propulsion system with six mover-generators providing power to variable speed, fixed pitch thrusters. Figure 2 shows the configuration of the DP system (class 2).

How it can spot in Figure 2, the power is generated by six mover-generators and passes through circuit breakers before powering the main switchboard (two high voltage); such the platform has two 11kV switchboard, the DP system electrical circuit is divided in two (side A and side B).

Usually the platform operates with closed bus, in other words, the two switchboard sections of high voltage are connected (even in low-voltage switchboards); that is why, the case study was developed assuming this operating condition.

The two High Voltage switchboards are able to provide power to the thruster subsystem and the DP control subsystem. The power supplied to the thruster subsystem feeds the Siplinks (functions as a DC network coupling, interconnecting the electrically isolated parts of the on-board medium voltage systems) of each pair of thrusters, after that the transformers decrease the voltage; and the power supplied to the DP control subsystem throughout three phase transformers (two) and the Low Voltage switchboards distribute power to the UPSs.

Finally, all environment sensors and PRSs send information to the three control computers and some duplicate this information to the IJS (enabling the system to operate as a simple DP system). The three control computers operate in parallel, each receiving the same input from the sensors and PRSs; all controllers independently

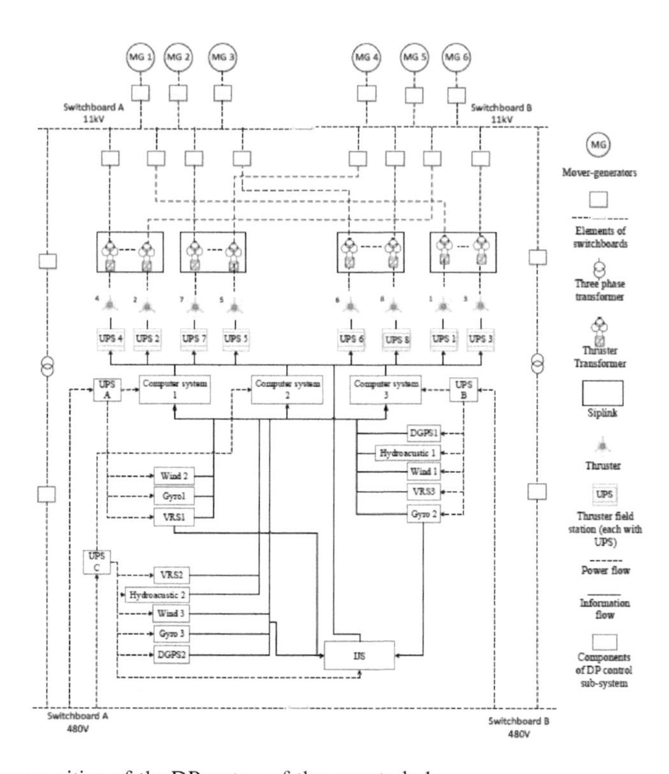

Figure 2. General composition of the DP system of the case study 1.

compute required information to maintain station and pass this information to the field station of the thrusters.

The field stations use the median value as the commands signal and feed this torque and azimth commands to the Aquamaster units (interface between the DP control subsystem and the thruster); while the IJS in case of being enabled as DP system, sends isolated analog signals to the field stations of the thrusters in order not to share connection with the control computers.

In addition to what is shown in Figure 2, there is the emergency switchboard and its mover-generator whose their function is to ensure the power supply (480V) for all UPSs of the DP system (UPSs of the control subsystem and the UPSs of the thruster subsystem), in case of Low Voltage switchboards failure.

In addition to the three main subsystems (power, thruster and DP control) there are five other subsystems that support the operation of the DPS-2 (ventilation, compressed air, cooling water, fuel oil and lubrication). According to failure modes and effects analysis—FMEA of the configuration 1, failure modes of the ventilation and compressed air subsystem will not cause any transient effect on DP positioning. However, the failure of the cooling water, fuel oil and lubrication subsystem can lead to loss of mover-generators or thrusters, and consequently leading to loss of controlled propulsion, (TRANSOCEAN, 2009).

3.2 *Configuration 2: Drillship*

The second case study is a drillship with six thrusters arranged in pairs. The vessel has a Diesel electric propulsion system with six mover-generators. Figure 3 shows the configuration of the DP system class 3.

As can be seen in Figure 3, the power is generated by six mover-generators and passes through circuit breakers before powering the main switchboard (three high voltage); such the platform has three 11kV switchboard, the DP system electrical circuit is divided in three (port, cent and starboard).

The three High Voltage switchboards provide power to the thruster subsystem while only the starboard and port switchboard provide power to the DP control subsystem.

The power supplied to the thruster subsystem is filtered by the thruster transformers while the power supplied to the DP control subsystem is filtered by two three-phase transformers that convert the high voltage (11kV) into a low voltage (440V). The Low Voltage switchboards are responsible power to all DP system UPSs.

Finally, all environment sensors and PRSs send information to the three control computers and

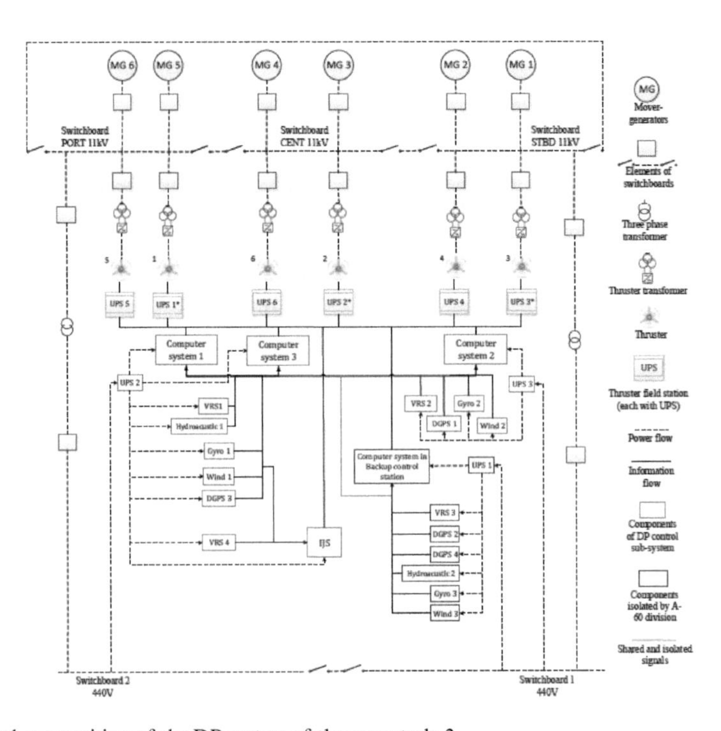

Figure 3. General composition of the DP system of the case study 2.

some duplicate this information to the IJS (enabling the system to operate as a simple DP system).

Notice that the environment sensors (Wind 3, Gyro 3 and VRS 3), and the PRSs (DGPS 2 and 4, Hydroacustic 2) of the Backup control station, send signals to the control computers through an isolating box.

As in the previous case, in addition to what is shown in Figure 3, there is an emergency switchboard and its mover-generator whose their function is to ensure the power supply (440V) for all UPSs of the DP system (UPSs of the control subsystem and the UPSs of the thruster subsystem) in case of Low Voltage switchboards failure.

Also as in the previous case, the failure of the cooling water, fuel oil and lubrication subsystem can lead to loss of mover-generators or thrusters, overcoming the worst-case projected failure and consequently leading to loss of controlled propulsion, (Samsung Heavy Industries, 2012).

4 RELIABILITY DETERMINATION

Reliability is the capability of the system to perform its intended function during the defined period under pre-established conditions (Smith & Simpson, 2004). The methodology adopted for reliability determination involving multiple systems is described in Figure 4. The failure rates for the components and subsystems are obtained from the relevant failure data base and the FT are developed based on the bottom-top approach and system functional architecture (Vedachalam & Ramadass, 2017).

Component failure rates published by DNVs Handbook for Offshore Reliability Data (OREDA), IEEE 493-IEEE Recommended practice for the design of reliable industrial and commercial power systems, US Naval Surface Warfare Center (NSWC) handbook of reliability prediction procedures for mechanical equipment, report of the 20th International Conference: Computer Safety, Reliability and Security—SAFECOMP, and India's National Institute of Ocean Technology—

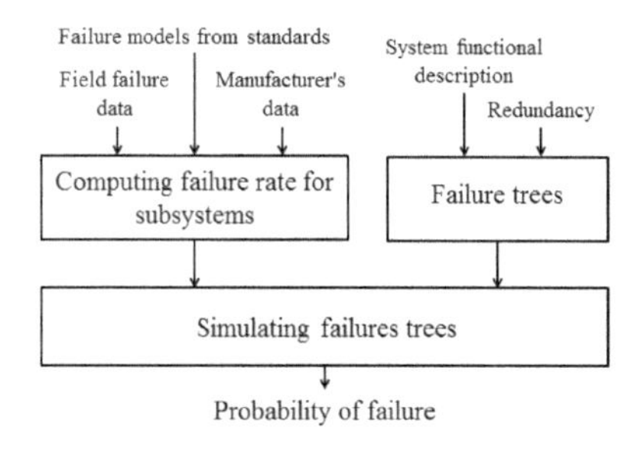

Figure 4. Reliability determination methodology; Vedachalam & Ramadas (2017).

Failure Reporting, Analysis and Corrective Actions System (NIOT—FRACAS) database are used for the reliability analysis. Table 1 details the failure rate data of the components/subsystems used in the analysis for configuration 2.

Failure rate of a component/subsystem is defined in Failure-In-Time (FIT), and represented as, expressed in failures per hour, OREDA (2015). Given the number of failures and the total cumulative hours of operation, λ is calculated as represented in Equation 1.

$$\lambda = \left(\frac{Number\ failures}{Total\ operating\ time\ in\ hours} \right) \quad (1)$$

Table 1. Failure rate data used for the analysis.

Component	Failure rate [failures/h]	Source of data
Engine	5,49E-05	OREDA (2015)
Generator	3,00E-05	OREDA (2015)
Circuit Breaker	4,79E-07	IEEE (2007)
PMS	1,74E-05	Vedachalam et al. (2013)
Switchboard [11kV]	4,57E-06	EPRI (2001)
Bus-Bar	1,08E-06	IEEE (2007)
High Voltage Cable	1,61E-07	IEEE (2007)
Transformer	6,74E-07	IEEE (2007)
Lubrication pump	3,57E-05	OREDA (2015)
Gravity and inspection tank	3,01E-05	OREDA (2015)
Rexpeller Control Unit	1,14E-05	SAFECOMP (2001)
Drive Control Unit	1,14E-05	SAFECOMP (2001)
Variable Frequency Drive	8,80E-06	Vedachalam et al. (2013)
Field Station	2,20E-06	Blischke & Murthy (2003)
Hydraulic motor	2,50E-05	OREDA (2015)
Hydraulic pump	3,57E-05	OREDA (2015)
Gear box	1,61E-06	NSWC (2011)
Motor	2,50E-05	OREDA (2015)
VRS	1,13E-05	NIOT (2015)
Wind	5,15E-05	NIOT (2015)
Gyro	3,33E-05	NIOT (2015)
UPS	8,95E-06	OREDA (2015)
DGPS	1,02E-05	Novatel (2017)
Hydro acoustic	2,85E-05	NIOT (2015)
Control computers	1,74E-05	OREDA (2015)
IJS	1,74E-05	OREDA (2015)
FW Cooler	2,54E-05	OREDA (2015)
SW pumps	6,15E-06	OREDA (2015)
FW pumps	6,15E-06	OREDA (2015)
Transfer pumps (fuel oil)	1,18E-04	OREDA (2015)
Transfer pumps (lubrication)	3,57E-05	OREDA (2015)

From λ, the Mean Time to Fail (MTTF) is computed as,

$$MTTF = \left(\frac{1}{\lambda} \right) \quad (2)$$

For a component/system with a failure rate of λ, the Probability of Failure (PoF) in a continuously operating period t, Q(t) (which is between 0 and 1), is computed based on Equation 3,

$$Q(t) = 1 - e^{-\lambda t} \quad (3)$$

Assuming the failures rate of component/subsystem as a constant failure rate (λ) and using the Fault Tree (FT) module of Shapire8 software were build the FTs for each configuration (DP2 and DP3), in order to know the probability of DP system failure (SHAPIRE, 1987).

The following subsections present the fault trees for the power subsystem in each of the configuration; to show with an example how the failure trees of the subsystems of each configuration were analyzed.

4.1 *Fault tree of power subsystem (configuration 1)*

Figure shows the failure tree of power subsystem, in which it is identified that the subsystem presents a total failure if any of the following failures occur:

- Failure to generate power: occurs when all the mover-generators fail on the sides of electric circuit (side A and B). This failure results in the loss of position of the vessel once the thrusters will lose because the UPSs cannot supply the amount of power that the thrusters need, although it is possible to operate some equipment of the DP control subsystem and electronic components of the thrusters with the UPSs.
- PMS failure: even though in the case PMS failure the switchboards can be active manually, in the fault tree in question it was considered PMS failure that generate false demands on both sides of the electrical circuit.
- Failure of both the high voltage switchboards (11kV): the power subsystem fails because the propulsion is loss due to lack of power supply to the thrusters and the main power source of the control devices is lost.
- Bus-bar failure; i.e. failures like short circuit, earth fault, etc.

4.2 *Fault tree of power subsystem (configuration 2)*

The drillship's power subsystem fails completely when at least one of the following faults occurs:

- Failure of all mover-generators, resulting in total loss of power supply.
- Three high-voltage switchboards fail (11kV); several electrical faults such as earth fault, over current, etc. could cause the failure of HV switchboard. Since each HV switchboard is independently operated with open bus tie breaker the failure of HV switchboard will affect dedicated generators and consumers, not other HV switchboards. For this reason, a DP3 failure occurs when the three HV switchboards fail.
- PMS failure sending wrong signals to all three sides of circuit electric of the vessel (starboard, center and port).

Figure 6 shows the failure tree according to the previous descriptions of failures.

4.3 *Differences between fault trees*

To obtain the failure of each subsystem in both fault trees, it is necessary to model the loss of all

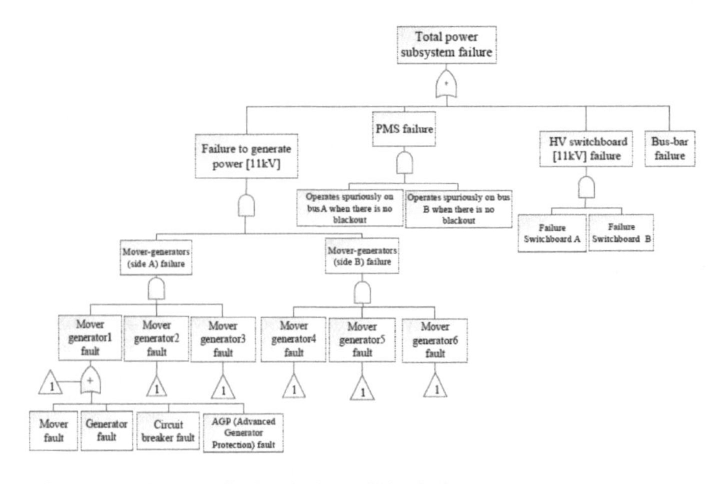

Figure 5. Fault tree for power subsystem for Semisubmersible platform.

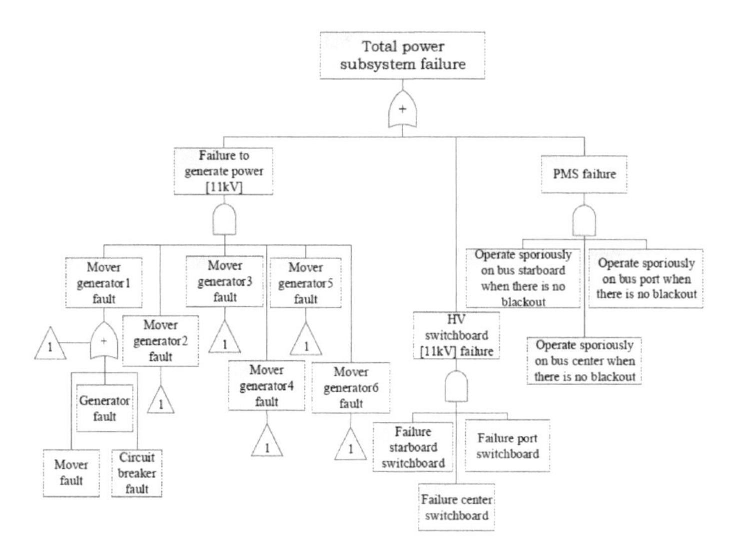

Figure 6. Fault tree for power subsystem for Drillship.

Table 2. PoF of subsystems for the two configurations.

	Probability of Failure (PoF)	
Subsystem/Classification	DP2	DP3
Power subsystem	0,0961	0,0565
Thruster subsystem	0,0137	0,0134
Control subsystem	0,0697	0,0665
Critical auxiliary subsystems	0,1094	0,0717
PoF of total DP system	0,2614	0,1934
MTTF of the DP system	3.30 years	4.65 years

the components and their redundancies. But there is a significant difference between the two trees.

As shown in Figure 2 the platform has two switchboard of high voltage (11kV) that operate with closed bus, that is, the two switchboard sections of high voltage are connected together; while the drillship has three switchboards of high voltage (11kV) that operate with open bus, i.e. each side of circuit electric operate independently (Figure 3). This is modeled on the fault trees as follows:

- The fault tree of configuration 1 considers the failure of two switchboards, while the fault tree of configuration 2 considers threes switchboards; and the same goes for the PMS because the PMS is divided for each circuit electric side of vessel, (MTS, 2007).
- While the DP2 power subsystem fault tree considers the failure of bus-bar that can generate a total loss of energy, the DP3 power subsystem fault tree does not consider the failure of this component because the drillship operates with open bus, so failures like short circuit, earth fault, etc.; in bus-bar could not spread between switchboards.

4.4 *DP2 and DP3 systems*

Based on the similar methodology, reliability computations are carried out for the DP2 and DP3 configurations and the identified results are shown in Table 2; the Probability of Failure was calculated for one year of operation.

5 SUMMARY AND CONCLUSION

The paper presents the methodology for reliability assessment of two typical configurations of DP systems.

The power subsystem and critical auxiliary subsystems reflect differences between the evaluated configurations, while the thruster and control subsystem show similar probability of failure between the semisubmersible platform and drillship.

This analysis allows concluding that incorporating additional redundancies control computers, sensor control power supplies and thruster control electronics in DP3 could result only in marginal increase in the system probability of failure.

While some authors assert that with the advancing technologies in sensors, power electronics, computer processors, condition monitoring systems, best application and maintenance practices; it could be possible to attain a higher reliability/ MTTF in future (Vedachalam & Ramadass, 2017); it is worth noting that this study was developed considering a semisubmersible platform with Siplink, an electronic component that increases the reliability of the DP system because even when one 11-kV switchboard is faulty, the Siplink is able to provide the required power of thrusters from other 11kV switchboard (Siemens, 2010); and because of this it is possible that the probability of failure of thruster subsystem is similar in both configurations.

Regarding the MTTF analysis, the drillship presents a Mean Time to Failure of 4.65 years, a 1.41 fold increase over semisubmersible platform.

The data and the results presented could be used as a guideline by DP system engineers to realize reliability-centered designs and for maintenance planning of higher capacity DP systems.

REFERENCES

ABS. 2013. *Guide for dynamic positioning system.* Houston: American Bureau of Shipping—ABS.

Blischke, R. & Murthy, D. 2003. *Case studies in reliability and maintenance.* Stockholm.

DNV. 2013. *Dynamic Positioning Systems Rules for classification of Ships, Part 6 chapiter 7.* (pp. 29–42). Noruega: DET NORSKE VERITAS AS—DNV.

DNV. 2015. *Dynamic Positioning Systems—operation. Guidance, Recommended practice DNVGL-RP-E307.* Edition July 2015.

Ebrahimi, A. 2010. *Effect analysis of Reliability, Availability, Maintainability and Safety (RAMS) Parameters in design and operation of Dynamic Positioning (DP) systems in floating offshore structures.* Stockholm.

EPRI. 2001. *A review of the Reliability of Electric Distribution System Components.* California: Electric Power Research Institute.

Ferreira, D. 2016. *Confiabilidade de sistemas de geração de energia em sondas de posicionamento dinâmico por simulação Monte Carlo.* Rio de Janeiro: UFRJ.

Hauff, K. 2014. *Analysis of Loss of Position Incidents for Dynamically Operated Vessels.* Marine Engineer, Marine Technology, Norwegian University Of Science and Technology, Trondheim.

IEEE. 2007. *Design of Reliable Industrial and Commercial Power Systems.* New York: Institute of Electrical and Electronics Engineers.

IMCA. *DP station keeping incident reports,* http://www.imca-int.com/marine-division/dynamic-positioning.aspx, browsed on 01.09.2016.

IMO. 1994. *Guidelines for vessels with dynamic positioning systems.* London: International Maritime Organization—IMO.

MTS. 2007. *Power management control of electrical propulsion systems.* Houston: Marine Technology Society—MTS.

NIOT. 2015. *Assessment of the reliability of the Indian tsunami buoy system,* Underwater Technol. 32 (2015) 255–270. India: National Institute of Ocean Technology.

NOVATEL. 2017. MTBF specifications of GPS receivers, http://www.novatel.com/support/known-solutions/mtbf-specifications-for-gps-700-series/.v.

NSWC. 2011. Handbook for Reliability Prediction of Mechanical Equipment, Caderock Division. Naval Surface Warfare Center.

OREDA. Offshore reliability dat handbook by DNV, SINTEF and group of oil companies 2015.

Patino, C. 2012. Avaliação probabilística do risco. *Análise de risco em operações de "Offloading" – Um modelo de avaliação propbabilística dinâmica para a tomada de decisão* (pp. 8–36). São Paulo: Universidade de São Paulo.

Rappini, S. Pallaoro, A. & Heringer, M. 2003. Fundamentos de Posicionamento Dinâmico. Sao Paulo: PETROBRAS.

SAFECOMP. 2001. *Computer Safety, Reliability and Security: 20th International Conference,* Budapest, Hungary, September 26–28, 2001 Proceedings.

SAPHIRE (Systems Analysis Programs for Hands-on Integrated Reliability Evaluations). Version 8.x. "Software" (1987): U.S. Nuclear Regulatory Commission—NRC.

SAMSUNG. 2012. *Preliminary FMEA report of the DP system.* Panama: Samsung Heavy Industries Co., Ltd.

SIEMENS. 2010. *Siplink—reliable and economical,* https://w3.siemens.com/powerdistribution/global/SiteCollectionDocuments/en/mv/power-supply-solutions/onshore-power-supply/brochure-SIPLINK_Drilling-Ships_en.pdf, browsed on 08.10.2017.

Smith, J. & Simpson, K. 2004. *Functional Safety—A Straight Forward Guide to Applying IEC 61508 and Related Standards,* Elsevier, Burlington MA, Butterworth, Heinemann, 280.

TRANSOCEAN. 2009. *Sedco 706 Failure modes, effects and criticality analysis of the DP system.* Brasilia: TRANSOCEAN.

Vedachalam, N.; Ramesh, R.; Muthukumaran, D. & Subramanian, A. 2013. Reliability centered development of deep water ROV ROSUB 6000. In *Marine Technology Society Journal,* 47(3): 55–71.

Vedachalam, N., & Ramadass, G. 2017. *Reliability assessment of multi-megawatt capacity offshore dynamic positioning systems.* Chennai: National Institute of Ocean Technology—NIOT.

Marine pollution and sustainability

Progress in Maritime Technology and Engineering – Guedes Soares & Santos (Eds)
 ISBN 978-1-138-58539-3

Sustainability in fishing vessel design process 1988–2018

F.A. Veenstra
TU Delft, Delft, The Netherlands
VFC, Haarlem, The Netherlands

J.A.A.M. Stoop
TU Delft, Delft, The Netherlands
Kindunos, Dordrecht, The Netherlands

J.J. Hopman
3mE, TU Delft, Delft, The Netherlands

ABSTRACT: Since 1988, the Dutch flatfish fisheries are dealing with sustainability aspects in the vessel design process. In the first place at a time when social and political trends call for more attention to personal safety and working conditions and later extended with emphasis on a greener image and the triple P aspects (People, Planet, Profit). Nowadays due to the Paris climate agreement, every country has set their goals for radical emission reductions, even zero-emissions in 2050. This means that the CO2 targets are also becoming a (scientific) challenge developing new fishing vessel design methodologies including eco-technical solutions (ß-design), personal safety and working conditions(α-design) and socio-technical strategies (α-design). An effective combined ß-α design methodology does not exist yet, let alone also integrating the new circular economy (ʏ-) requirements. For such a new fishing vessel design process the evolving circular economy, principles (CE) are challenging with the ultimate sustainability goals: zero-emissions, zero-waste, zero-accidents on-board. The design experiences of the safety-integrated Beamer 2000 redesign (1990) and the sustainability-integrated MDV-1 new design (2015) are stimulating starting points.

1 INTRODUCTION

1.1 *Dutch beamtrawl bottom fisheries in sustainable transition*

The North Sea flatfish fisheries can be characterized by small, family owned enterprises (SME), where the fishermen have a decisive voice in (re) designing and in anticipating on the nowadays sustainable requirements and corporate social responsibilities (CSR). Their approach is always practically and cost-efficiently. So-far the SME's applied technical innovations through a derivative approach: step-by-step, ad-hoc energy saving; safety-integrated Beamer 2000 re-design (1990).

But in 2015 2 SME's chose for the first time for a disruptive approach: out-of-the-box, prospective energy-efficient; sustainability-integrated MDV-1 new design.

For fishery SME's the ultimate chosen innovations are based on a realistic return of investments by substantially decreasing the operational costs (investment, fuel) and strongly increasing the earnings (landed fresh fish). However after the economic crisis (2008), aging fishing fleet and more stringent sustainability requirements, the sector is urgently looking for innovative new-buildings with year-round positive business models. Besides the SME's must increasingly deal with the new political issues and marine strategies, e.g. Brexit, Natura 2000, landing undersized species, work-rest cycle, pulse dossier (transition from beamtrawl to electric pulse fishing to make that the bottom flatfish selectively leap into the fishnet). Thereby skipper-owners are becoming more eager to keep good fishermen onboard, e.g. when reducing the fishermen workload by further automation of fresh flatfish processing and more autonomous fish hold storage.

Figure 1. North sea traditional beam trawler 1988 present.

Figure 2. North Sea MDV-1 twinrigger 2015-present.

1.2 *Sustainable fishing vessel design drivers*

Through the past decennia the sustainable fishery design drivers have been evolved from improving the safety- and working conditions (beamer 2000 redesign) towards energy-efficient new designs (MDV-1 twinrigger), while the politics/NGO's are now requiring more energy-neutral designs and new business models. For this the triple P and the circular economy principles (CE) are nowadays becoming the new design drivers for ultra-sustainable fishing vessels (Table 1). For the beamtrawl fishing vessels, the so-called beamers the overall safety regulations are well institutionalized and are applied on the existing fishing vessels and new builds since the 90's. However when it comes to personal safety, in the sense of working conditions and well-being, the improvements are often a matter of personal experiences and preferences. It depends on the skill of the skipper/crew but also on the age of the vessel and good maintenance. Besides, because of more eco-system rules, high seas fishery inspections and decreasing North Sea fishing grounds the fishermen workload (and stress) is increasing considerably. Such personal safety aspects (People) were already a political issue in the 88's and for this the Beamer 2000 projects were conducted by RIVO fishery researchers (Naval architecture/engineering; a.o. Frans Veenstra,ref.) and the TU Delft (Safety Sciences; a.o. John Stoop, ref.) always in close cooperation with the fishing sector. Firstly the design drivers were based on the safety and occupational accidents (analysis), thereafter extended with the quality aspects in the 2000's. This research resulted in a practical safety-integrated redesign of existing beamers with proven solutions as well as the development of a scientific design methodology dealing with personal safety in the redesign process of fishing vessels (Kindunos: safety integrated redesign process reducing occupational accidents and improving onboard personal safety/working conditions/well-being without introducing new hazards or safety side effects). Because of the economic crisis (2008; high fuel prices, low fish prices) and aging beamer fleet, the sector was urgently looking for innovative new-buildings, while the SME's couldn't afford it. For this reason the Masterplan Sustainable Fisheries (MDV in Dutch) started in 2010, firstly creating sector-and public support where after a feasibility study was conducted to design and build an innovative-sustainable pilot vessel. Where in the Beamer 2000 design the technical solutions were leading, in the MDV-1 new design the main design driver was a year-round positive business model through very low operational costs and higher landed fresh fish prices. The applied innovations were green shipping proven and for personal safety reasons the 30 m pilot vessel also must be an excellent North Sea working platform. From a vessel designers point of view the applied methodology was disruptive and for sector support more than 1 SME should participate. To keep the financial risks for the SME's acceptable, the EU and Dutch ministry granted some innovation-subsidies. The MDV-1 is already 2 years fishing and the technical design targets have been fully realized: 80% FO $\downarrow$, multi-sustainable innovations, seaworthy platform. The design requirements for the MDV follow-up fishing vessels have already been drafted

Table 1. Fishing vessel design drivers 1988–2018.

Pre-designs	Beamer 2000 1990s	MDV-1 twinrigger 2015	CE–vessel design 2018
Focus	Safety and working conditions (People)	Business model fish2dish (Profit, Planet)	Value chain (Planet Profit, People)
Design	Product	Product & fishprocess	Product & fishprocess & value chain
Goals	Improved safety and working conditions	Proven sustainable/ future-proof business model/happy crew	Circular economy principles
Classi-fication	Safety Design Index	Eco-Economy Design Index	CE Design Index
Tools	Solution matrix	Innovation pillars	Re-use
Methodo-logy	Safety-integrated redesign	Sustainability integrated new design	Circular fishing vessel (pre) design

in discussion with the MDV-1 crew. In the Netherlands a number of new fishing vessels are being ordered now with many MDV innovations followed suit. Because the new Dutch cabinet (2017) has set new climate targets the fishery researchers are working on additional MDV requirements. In which also the evolving circular economy principles (CE) will be taken in consideration as a new set of design drivers. Ultimately resulting in an intrinsic sustainable CE fishing vessel design, what is good for the fishermen (People), the (shipbuilding)sector (Profit) and the ecosystems (Planet).

In the 1988–2014 period second to none multicriteria design methodologies were available. For the first time the Kindunos-method applied a dual-criteria approach (ß-α aspects) in redesigning fishing vessels. Regarding the evolving sustainable design drivers, the TNO-organization has developed a Life Cycle Analysis based design model (LCA) to look after the eco-impact of the used construction materials, equipment and installations in the vessel design process and lifetime (LCA: life cycle analysis, a step by step method to map the complete environmental impact of products and services). For the first time such a LCA approach has been fully applied in the designing and building of the Rainbow Warrior II (2009). Based on these experiences the MDV team designed and launched the MDV-1 pilot vessel (2015), the first sustainability-integrated fishing vessel. Depending on the realistic return of costs and the required short building time, the ß-α design aspects as well as the new ɣ-design aspects were incorporated as much as possible.

2 DESIGN METHODOLOGIES

2.1 *Beamer 2000 redesign methodology in practice*

In the 1990s a number of Beamer 2000 design studies were conducted by RIVO, TU Delft and Shipbuilding College Haarlem, whereby Frans Veenstra (fishery designs, ref.) and John Stoop (safety sciences, ref.) were the leading researchers. The aim was to reconsider the existing beamer designs from an integral safety point of view and good working conditions. The researchers have always had a close cooperation with the Dutch fishing sector, fishermen/SME's and Dutch shipyards and suppliers. During the project phases many publications and discussions were held as well as contributions given on (safety) symposiums. Finally the studies resulted in practical solutions of which many have already been implemented since. From a scientific point of view there were stimulating exchanges of the (intermediate) results, many master-students involved and regularly scientific cooperation with colleague researchers abroad, in particular with Norway (Marintek), France (Ifremer/CEASM) and UK (Seafish Authority). On the basis of accident and labor analysis a set of explicit safety- and working requirements were formulated. Where-after the fishery designers came up with many technical solutions avoiding new hazards or side effects in the (re)design process. Through this designer & scientific safety methodology the Beamer 2000 approach could be expanded in the Beamer 2000 projects, a set of sub-projects reducing the shortcomings of the North Sea beam trawling in the fields of safety and working conditions (People) as well as the fish quality (Profit) and the eco-system aspects (Planet). This resulted in the sustainable design requirements and some conceptual redesigns, the sustainable Beamer2000 and Trawler 2000 (ref.). With these designs and knowledge transfer via discussions, journals, papers and fishery exhibitions the fishery safety, quality and eco—awareness of het Dutch fishing sector has been increased considerably since. And in the safety science disciplines the Kindunos methodology is a successful example of a design approach whereby the technical aspects (beta) as well as the 'vague' aspects as health and well-being (alpha) were integrated. However nowadays in safety science circles a more sociological approach like Resilience is predominantly (social-technical system capacity to adjust/adapt to safety-wellbeing innovations and/or new organization, ref.). From the Beamer 2000 projects many lessons have been learned, such as that for an thoroughly analysis of the occupational accidents and working conditions the data from different sources must be looked at. One should not only analyze data regarding the occupational accidents but also on absenteeism and working conditions as registered by the Ministries, Red Cross, Shipping Council, Coastguard and Fishermen Organizations. For the analysis the most important indicators are related to the fishing process (capsizing, collisions, heavy occupational accidents), the working process (light accidents, rejected employees, sickness leave) as well as the fishing vessel working platform (sea performance, noise, vibration, climate). Especially in a predominantly SME sector it is important that the beamer 2000 researchers went onboard testing their findings and possible solutions (technical, operational, educational). Based on the functional redesigns of the different working places, 5 safety-integrated solution matrices were developed with a realistic return of costs (overall design (Table 2), fish processing deck, wheelhouse, engine-room, fishhold).

At the beamer 2000 time with many new fishing vessels and good earnings the fishing sector and stakeholders were not interested in the overall

Table 2. Beamer safety-integration matrix overall design.

Safety-integration matrix	Envisaged effects	Envisaged effects	Envisaged effects
Overall Design	Accidents	Workload	Cross linkages
Implementation of sound-proofing packages	Improved commu-nication	Less stress and risk of noise induced hearing loss	Less fatigue, more attention to the quality of fish and environ-ment
Winch (amidship) to change places with fish-processing (under forecastle)	No warps/lines across deck, less movement on deck	Minimum accel-eration levels; less strain on the back and joints	Less fatigue, more attention to the quality of fish and environ-ment
Raising wheelhouse	Improved vision lines to working deck and horizon	Less strain on watchman	Less risk of collisions and environ mental pollution

Figure 3. Beamer 2000 fish processing amid ships.

values and new business-models. While with fluctuating fuel prices every SME came up with own ad-hoc energy-saving solutions. Since the 2008 crisis and aging fishing fleet the most important focus is nowadays the earning capacity of the vessels and also on the operational costs, in particular less energy but also costs regarding selectivity, by-catch and maintenance. Besides additional financial aspects are also becoming stricter design drivers, such as how to finance the vessel, the end-of-life value and the fish-chain dependency (fish2dish). The Beamer 2000 lessons were a good start-up for the MDV team to design an innovative and safe fishing vessel with a year round positive business model and proven sustainability.

2.2 *MDV-1 twinrigger design methodology in practice*

After the economic crisis 2008 the sector urgently needed an innovative pilot fishing vessel with substantial low operational costs (investment ↓, fuel ↓) and a greener image (↑). To realize the design and building of such an innovative vessel a foundation MDV was established with a multi-disciplinary team and a good network. Based on his Beamer 2000 experiences Frans Veenstra (ref.) was asked to become the MDV innovation manager: “to stimulate and guard the MDV goals and observe the innovations in practice”. The main MDV objective was to develop a positive business model for sustainable North sea flatfishing by designing and launching of an appealing pilot fishing vessel; not only from a technical point of view (ß-aspects) but also from a personal safety point of view (α-aspects) and anticipating the socio-technical business approaches (ϒ-aspects). To decrease the investments one could decide for a smaller fishing vessel, however from the overall safety point of view (Beamer 2000) the pilot vessel must be an excellent North sea working platform. For this and to substantially reduce the fuel costs (80% ↓) a revolutionary 30 m hullform was designed with input from MARIN (Wageningen) and D3 (Spain). The innovative hull is a combination of a low resistance yachting hullform and stable working vessel with an optimal length/displacement ratio. From the sustainability point of view many multi-green shipping techniques and hybrid constructions have been applied, more or less according the LCA approach for ship lifecycle perspective, however excluding the fishing gears. For recycling reasons and also to decrease the shipweight, hybrid construction materials have been considered. Even a biobased composite wheelhouse was looked into, but because of still lacking of Dutch fishing vessel building rules (Dutch Shipping Inspectorate (IL&T)) in this regard, it could not applied on the MDV-1 or the strict building period of 12 months would extend too much. To realize the MDV required business model the maximum vessel investment was 4 million euro, excluding the fishing gears. To become proven sustainable 5 MDV innovation pillars were addressed and applied (Table 3).

In the MDV design approach 3 decisive phases were crucial, firstly to enhance governmental and sector support, secondly to clarify the design- and financial feasibility and at last not at least the draft of 3 innovative pre-designs. After these phases one of the predesigns was chosen and 2 SME’s joined the MDV team to further design, build and launch the pilot vessel and operate the MDV-1 in practice (2015-present). Already during the first trials the

Table 3. MDV-1 sustainable innovation pillars.

MDV-1 five Innovation Pillars (IP)	Triple P
I.P.1 hullform, hull weight	People (working conditions) Profit (operational costs)
I.P.2 energy-efficent engineroom	Profit (fuel costs), Planet (CO2 reductions)
I.P.3 hybrid construction materials	Profit (operational costs), Planet (recycling)
I.P.4 fishhandling, fishprocessing	People (automation), Profit (quality, fish2dish)
I.P.5 eco-friendly fishing gears	People (save hganbdling0, Planet (selectivity), Profit (quality)
MDV-1 IMMANUEL (2015-present, Fig. 2)	Length 30.15 m Beam 8.60 m Depth to maindeck 5.87 m Depth to shelterdeck 8.54 m E-motor 400 kW Propellor 3 m Generator 1 × 500 ekW Harborgenerator 1 × 117ekW Fishing boxes 800–900 Fishprocessing deck 100 m^2 Twinrigging astern flatfish

shipyards, SME's and MDV learned that the shipbuilding objectives were fully realized: a seaworthy platform, 80% energy saving, low noise levels, good vision lines. However the operational targets were not reached yet, esp. the innovative fishing gears (twinrigpulse) and 2 prototypes of fishprocessing equipment (flatfish gutting, -sorting). These are being further developed in the MDV-1 phase 4 research period, (2016–2018). From the Beamer 2000 point of view the MDV-1 is the ultimate Beamer as Frans Veenstra and John Stoop were at the time aiming at. With this layout the main safety shortcomings were remedied, in particular the choice for fish gear handling over the stern with no fishing lines at deck-level and excellent vision lines from the wheelhouse to the working deck-areas. For MDV-1 the operational excellence was leading (positive business model) with off-course optimal working- and safety conditions. The safety/well-being aspects and sustainability were all incorporated as much as possible through application of a new layout and with multi green shipping innovations. The MDV multidisciplinary cooperation (SME, government, research) was the important success factor and also commitment of at least 2 family enterprises, who are also delivering the MDV-1 skipper(s) and crew for the MDV phase 4 research period. The cooperation of 2 Dutch shipyards (Hoekman, Padmos), with an excellent track-record in the fisheries and constructive interaction with vessel supplying companies lead to the revolutionary new fishing pilot vessel MDV-1, the MDV starting point for eco-friendly transition of the Dutch beamer fleet towards sustainable flatfish fisheries. The fact that MDV-1 became the KNVTS "Innovative Ship of the Year 2016" and the MDV foundation became the most innovative SME on Urk (2016), have proven that the MDV design process was very successfully. Also the multi-green shipping approach was absolutely decisive and instrumental. In the Maritme Holland the MDV-1 has been extensively described under the title of "A sneak peek at the future of fishing vessels" (ref.). After 2 years fishing the MDV-1 and innovative equipment got a lot of attention and many multi-innovations are already going to be applied in the upcoming newbuilding plans.

2.3 *Circular fishing vessel design methodology in theory*

For ultra-sustainable fishing vessel designs the derivative Beamer 2000 safety-redesign- and the disruptive MDV-1 sustainable new design experiences are a challenging start-up. For this the α-, ß and ʏ design aspects must be fully incorporated as well as a customization to the evolving circular economy requirements. The concept of a 'Circular Economy (CE)' is currently high on the political agenda. The necessity of a circular economy has been highlighted by many, but perhaps most convincingly by the Ellen McArthur foundation. The foundation describes the circular economy as a system that is restorative or regenerative by intention and the outcome must be achieved by eliminating waste through the superior design of materials, products and systems. Remanufacturing is a key strategy and an important methodology for substantially reducing the CO2 emissions. Besides new business- and value chain models must be introduced. With ultra-sustainable fishing vessels and new fishery value chains the SME's will actually become future proof (30–40 yrs.). As learned by the MDV approach with the CE-design one must firstly define the ultimate horizon 2050 goals (Table 4). Originating from these goals one can draft a new pre-design, which are feasible for the SME's, firstly financially with realistic return of investments but secondly fully triple-P sustainable. Also with this CE design one can cost-efficiently anticipate on future regulations. Then the materials, products and value chains will last longer than today. If making use of bio-based materials in a never-ending cycle, the SME's get the most out of their assets, the ultra-sustainable fishing vessel.

The actors for future-proof eco-designs are mainly related to the transition from fossil fuels to

Table 4. Circular economy principles, horizon 2050 goals.

Circular Economy principles 2018	Horizon 2050 ultimate Triple P (triple Zero)
Planet (CO2-emissions)	0%
Profit (Waste)	0%
People (Accidents onboard)	0%
Modular & flexible design (re-use)	100%
Recycling	100%
End-of-life value	50%
Longevity	50 yrs.

alternatives (LNG, CNG) and re-use of raw materials and equipment anticipating world-wide shortages of materials. This means that energy-efficient and climate neutral drivers must become the CE-design approach. Instead of 'made-to-be' vessels it must become 'made-again' through value creation (extended life-cycle and good end-of-life value). For such a circular fishing vessel design a SMART, modular flexible approach is needed, whereby (raw) materials and equipment must be re-used as much as feasible from a business point of view. For the SME is a higher end-of-life value a stimulating target. The current end-of-life value of the aging beamers is next to nothing. In the first place because of the tailor-made North sea fishing method and the strict Dutch North Sea classification rules. For this reasons the aging fleet isn't direct usable for other fishing methods abroad or a costly refit is necessary. Regarding new fishery business models a start-up discussion is beginning for further flatfish chain integration, in which fishermen & the processing industry are working closely together regarding good quality fresh fish (from fish2dish). With such an approach the sector also can anticipate on new EU eco-system regulations and EU political changes (Brexit, Marine Strategy; Landing obligation undersized species). From the SME perspective all the α-, ß and ʏ design aspects can be visualized in the so-called fishing vessel design spiral with the realistic sets of design objectives (Table 5).

In the CE-fishing vessel design process a helpful and transparent design tool can become a Circular Economy Design Index (CEDI). This index must be developed yet, so that already in the pre-designing process the degree of sustainability will be classified for ultra-sustainable fishing vessels complying with the future proof fisheries environment. Since the 90's there is an Energy Efficient Design Index (EEDI) in the merchant marine industry, but the focus here is only on energy-saving and energy-efficient operation, while the CEDI design index must also deal with the fishing method and value chains.

Table 5. Fishing vessel design spiral and goals/objectives.

CE Design Aspects	CE goals in 5–10 yrs.	CE objectives
CE fishing vessel design spiral (Fig. 4)	Incorporating technical (ß), operational (α) and societal (ʏ) aspects	Remanufacturing and reconditioning
ßèta	Energy-neutral fishing vessels	Green shipping/ fishing techniques; proven sustainable, modular and energy-efficient; innovative materials and equipment to reduce their environmental footprint; increased flexibility in design reducing weight, fuel demand and maintenance costs
αlpha	SME-perspective and fishermen wellbeing	Socio-technical system-capacity to adapt/support onboard safety, wellbeing and organizational innovations; participative and prospective designing
ʏ gamma	New fishery business models	Smart chain management anticipating (fuel/ material) shortages and securing value chains; restorative and regenerative by design

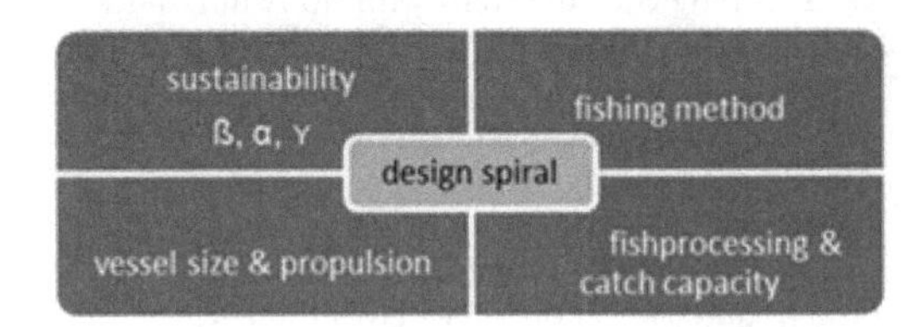

Figure 4. Design spiral including sustainability aspects.

3 CONCLUSIONS AND RECOMMENDATIONS

Since 1988, the Dutch fishery sector and fishery research are familiar with sustainability in the design process. There has always been a good cooperation between the SME and researchers. The design drivers evolved from the personal and vessel safety aspects into the triple P aspects. There is an open mind for the next sustainability step, even incorporating the CE-principles and CE business models. With the design and launching of MDV-1 the first disruptive design approach has successfully been applied. The pilot vessel is already fishing for 2 ½ yrs. and the MDV success factors are already well-known and are incorporated in various new design plans. After the proven derivative Beamer 2000 redesign approach (optimal fishery management, adaptive design, integration (partial)solution aspects) and the disruptive MDV-1 new design approach (multi-innovative, proven sustainable, revised fishery value chain) the Kindunos/MDV-1 approach is an excellent starting point for the next step. Not only for the academic world with the development of a circular economy design index (CEDI), but also to focus on the additional MDV CE-goals, finally resulting in ultra-sustainable North Sea and Wadden Sea fishing vessels and new flatfish business models. During three decades the Dutch fishing vessels had to adapt to three major socio-political changes: from traditional to a Safe & Health Environment (SHE), from SHE to Sustainable and from Sustainable to Circular Economy (re)designs. These adaptations grew in magnitude from a derivative (ad-hoc) to a disruptive (prospective) to an innovative (foreseeable) level. These adaptations came with very high transition costs, especially under conditions of the 2008 economic crisis. This severely hampered the ability of individual SME's to invest in midlife upgrades, let alone in new builds. In order to remain flexible and adaptive to external political and socio-economical demands (ß-α-γ-aspects), the fishing vessel design process has to become more transparent. Such a transparency should already start at the conceptual level with identification of design values and business models (CEDI), rather than a further restrictive detailing and quantification of performance indicators at a detailing level. Such a starting point facilitates foresight on future use and exploitation of fishing vessels in a life cycle approach, independent of political, business or socio-economic changes on the short and long term. With the CE-transition the Dutch fisheries/SME's do take their corporate social responsibilities, substantially increasing their already green image and at the same time keeping a 'happy crew' onboard. Actually giving fully substance to the ultimate-CE sustainability requirements: zero-emissions, zero-waste, zero-accidents and the here intended new fishing vessel CE-design process.

REFERENCES

Bates, Quentin 2015. Innovative IMMANUEL joins Dutch fleet. *Fishing News*, July 30; pg. 10–14.

Brouckaert, Bruno 2016. MDV-1IMMANUEL, a sneak peek at the future of fishing vessels. *Maritime Holland*. vol. 1, 2016, pg 46–54.

De Vos-Effting 2008. *A LCA based eco-design considerations for the Rainbow Warrior III*. International Hiswa symposium Yacht design and- construction. 17–18 nov. 2008. Amsterdam

Hale, A, Heijer, T. 2006. Defining resilience. Resilience engineering: concept and precepts. *Ashgate publishing*; pg 115–137.

Kimura, N, van Drieën, J, Veenstra, F.A. 1996. Simulation study on effect on human response to the motion of Dutch beamers *RIVO-DLO report 96.003*. IJmuiden.

Morel, Gael, Chauvin, Christine 2006. Towards a new state of resilience for the socio-technical system of the sea fishing industry. *Safety Science*. 44; pg 599–619.

Stoop J.A. 1993. *Human factors in bridge operations: decision-support at future bridges*. Proceedings Tenth Ship Control System Symposium, 25–29 october, 1993, National Defence Headquarters DGMEM/DMEE, Ottawa.

Stoop J.A, Veenstra F.A. 1992. *Information technology in the fishing industry*. Science Publishers, Computer applications in ergonomics, occupational safety and health; pp 219–226. Elsevier. Amsterdam.

Stoop J.A., 1997. *Design for safety, a new trend?* Proceedings of the International Conference on Engineering Design ICED 1997 Tampere, Finland, august 19–21, 1997.

Stoop J.A. 1993, *Towards a safety integrated design method*. Proceedings 9th International Conference on Engineering Design ICED 93, August 17–19, 1993, the Hague.

Stoop, J.A. (1990). Safety and the Design Process. *Doctoral Thesis TU Delft. ISBN 90-9003301-*. Delft

Stoop, J.A, Veenstra, F.A. 1990. Safety integrated redesign of Dutch beamtrawlers *Schip en werf*, 57e jaargang, no.5, mei 1990.

Stoop, J.A. Veenstra, F.A. 1993. *Towards a safety integrated design method.* Proceedings 9th International Conference on Engineering Design ICED 93, August 17–19, The Hague.

Veenstra, F.A, Hopman, J.J, Stoop, J.A.A.M. (2017). Multicriteria Fishing Vessel Design Methodology. *USA Journal of Fisheries and Aquaculture;* Volume 2017, Issue Oct. 06.

Veenstra, F.A. 2002. Dutch newbuildings after the 2000 re-design requirements. *Fishing News International, Good Gear Guide 2002-new vessels*. ISBN 0-9518579-9-1; pages 25–31.

Veenstra, F.A., Mul, N. (1991). *IQAS flatfish processing on board beamer 2000*. ICES meeting, 22–24 april. Ancona

Veenstra, F.A. 1989. Noise levels and noise control onboard beamtrawlers. *Schip en werf*, nr. 7, pg 237–241. Rotterdam.

Veenstra, Frans 1992. *Integrated quality to improve the onboard safety, fresh fish handling and marine environment*.: Safety and working conditions onboard fishing vessels. 15–17 sept. 92. Villagarcia de Arosa.

Veenstra, Frans 1999. *Accident prevention onboard Dutch Fishing vessels.* Dutch-France workshop.CEASM,23–25 April. Lorient.

Veenstra, Frans. 2017. *MDV-1,from innovative IDEA (2006) to ShipoftheYear(2017)(inDutch).* https://masterplanduurzamevisserij.nl/nl/kennisbank/onderzoek/publicaties.Urk.

Veenstra, Frans. Brinkman, Rick. 1995. Stern Trawler 2000, a new approach designing on points of sustainable aspects: working conditions, HACCP quality control, environment. *HSB International;* VOL 44-No1; pages: 51–53.

Veenstra, Frans. Stoop, John. 1990. *Beamer 2000*; Safety integrated (re)designing, the Kindunos method. CIP-DATA Royal Library, ISBN 90-74549-02-0. the Hague.

Progress in Maritime Technology and Engineering – Guedes Soares & Santos (Eds)
© 2018 Taylor & Francis Group, London, ISBN 978-1-138-58539-3

Ballast water management: And now, what to do?

L. Guerrero, J. Pancorbo & J.A. Arias
Bureau Veritas Marine and Offshore Division Iberia

ABSTRACT: After the Ballast Water and Sediments Convention was ratified in IMO on 8th September, 2016, it came into force one year later. This entry into force creates a decision making process for all the shipowners of new and existing ships, deciding what system to apply and when. It is not an easy task, and even more taking into account the two different sets of rules that coexis in the marine world: IMO and USCG standards. This paper discusses the different existing systems, their main principles and advantages. It also discusses the practical steps in a retrofitting process.

1 INTRODUCTION

Ballast water can contain aquatic and marine microbes, animals and plants, which will be carried to the place of deballasting, thus creating a potential risk of the local ecosystem. The International Convention for the control and Management of Ship's Ballast Water and Sediments was adopted in 2004, to create safety barriers to this problem of the invasive species, and after a long period of ratifications and recounting of percentages, was finally ratified on September 8th 2016 with the signature of Finland, achieving the needed conditions for ratification (35% of the world fleet by gross tonnage and 30 states).

After the Convention has entered into force, every ship over 400 GT will be required to undergo an initial survey and be issued with an International Ballast Water Management Certificate, which will be valid for five years subject to annual/intermediate surveys and subsequent 5 yearly renewal surveys. Initial survey will confirm that:

- An approved Ballast Water Management (BWM) Plan and ballast water record book are on board and the arrangements for ballast water management are as noted within the plan.
- Verify that the ballast water treatment system is approved, fitted according to class and statutory requirements and operational (unless vessel is exempted according to regulation A-4 of the convention).

Although the provisions of the Convention apply to all ships operating in the aquatic environment designed to carry ballast water, the survey and certification requirements apply only to ships of 400 GT and above.

The application of this Convention will be done in different phases:

1. Phase 1. This phase corresponds to D-1standard, as an "interim" standard until phase 2 comes into force.
2. Phase 2. This phase corresponds to D-2 standard (performance standard).

The standard to be achieved, when the Convention was created, was D-2, but, at that time, no treatment system plant was existing, so a period of time that could allow other method (D-1) was agreed. This D-1 method includes the 3 methods we are using at present in most of the ships, i.e. flow-through, sequential and dilution.

Until the final application of the Convention to a certain ship, she will have to use the system D-1, until her renewal of IOPP, when system D-2 is to be working, thus BWTS (Ballast Water Treatment System) will have to be installed on board.

After a long period of time, lasting more than 10 years, the Ballast Water Convention has been ratified in IMO. This means the starting shot for the regulation to be complied with and a headache for many shipowners that will have to update their vessels to these regulation requirements, not only in terms of costs but also in terms of selection, loss of profit, uncertainty in the disposal of both equipment fitted for the ship, trade concerns and the availability of a shipyard which can do the necessary works.

The set of regulations concerning ballast water exchange that affect the world fleet is the IMO standard. In addition, USCG/EPA (Environmental Protection Agency) standard ("final rule") applies to the fleet in the US waters and is applicable today (not as IMO standard, which will be applicable 8th September 2017). In this paper, we will introduce briefly the approval procedure (for USCG and IMO), as well as the drivers in retrofitting an existing vessel, and the news about this issue.

2 TYPE APPROVAL SYSTEMS

The regulations (USCG and IMO) defer basically in their testing standards as well as the chronogram of application dates. These standards for testing are defined in G8/G9 for IMO and in ETV protocol. Bear in mind that the testing standard in IMO has been modified in the last MEPC.70 (October 2016), which adopted a new set of regulations for that purpose, which superseded the current program for TA (Type Approval).

This means that, under IMO standard, the 69 already approved systems would lose their Certification, and new one should be applied for. But these systems can still be used, in accordance to the phase-out calendar agreed in the same meeting, as follows:

1. It was agreed that the guidelines should be used from adoption and that **all new Type Approvals from 28 October 2018 should meet the new requirements**.
2. I was also agreed that Ballast Water Systems **installed** on ships after 28 October 2020should be approved under the revised G8 guidelines.

This means an existing period for all the already approved systems, but also means that they will have to update to the new testing standard.

Regarding US Type Approval systems, USCG accepts systems which are Type Approved by other Administrations under G8/G9 procedures, as Alternative Management Systems—AMS. On December 2nd 2016, the first TA was granted by USCG to Optimarin Ballast System, which is a combined filtration/ultraviolet ballast water management system ranging capacities from 167 m^3/h to 3000 m^3/h. Does this mean that USCG will not accept IMO approved systems anymore? The answer is clearly "NO" since IMO systems will still be valid under Alternate Management System. In any case, there are already 6 systems approved by USCG and 2 more under approval.

As an example, see in Figure 2 a typical sequence for a system of UV+filtration.

USCG provides 4 ways for complying with the BW Standard in addition to the use of a USCG TA system:

1. Alternate Management Systems (AMS). This allows that an IMO type approved system can be accepted for the purpose of compliance of USCG regulation, up to a period of 5 years.
2. Use of ballast from US public water system.
3. Discharge of water to a reception facility.
4. No discharge of unmanaged ballast water inside 12 nm (in USA waters).

In addition, an extension to a vessel's compliance date can be granted when all the efforts have

Figure 1. Final phase of ballasting.

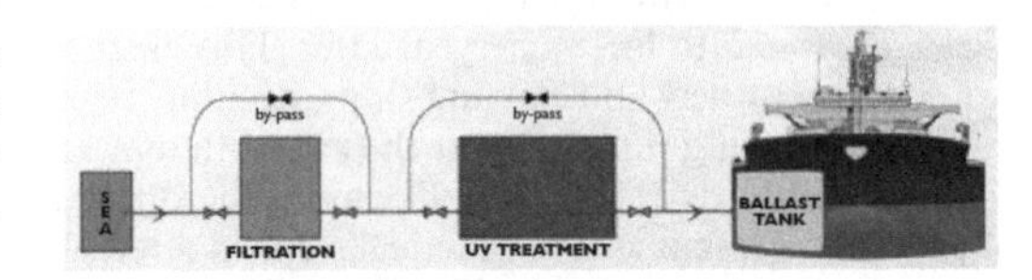

Figure 2. Ballasting: Filtration + UV disinfection.

been made to comply with the standard, and it is demonstrated that has not been possible. In such a case, USCG can grant this extension. This request is to be done, since now a (Ballast Water Management System) BWMS is already available, through an explicit statement supported by documentary evidences.

The main additional requirements that USCG would place compared to IMO regulations are:

- Clean ballast tanks regularly to remove sediments.
- Rinse anchors and chains when the anchor is retrieved.
- Remove fouling from the hull, piping and tanks on a regular basis.
- Maintain a BWM Plan that includes the above in addition to ballast water management (no requirement that the BWM Plan must be approved).
- Maintain records of ballast and fouling management.
- Submit a report form 24 hours before calling at a US port.

In addition, some US states, such as California and New-York, have different regulations. Bear in mind that California is the "entry point" for 79% of Invasive Species found on the North American west coast and 89% (257 out of 290) of known west coast Invasive Species are established in California coastal waters.

At this moment, 6 systems have already been granted with the USA TA in accordance with their specific procedure. The type approval programs are really not the subject of this paper, thus we will leave it for the reader to deepen their knowledge on this subject through the reading of G8/G9 IMO standards as well as USCG/EPA test protocols.

3 BALLAST WATER MAIN SYSTEMS

The system more commonly used at this moment is the UV type. Anyway, this system is not approved by USCG unless in conjuntion with other system.

The systems can be divided roughly in 3 technologies and their sub-types:

1. Mechanical
   a. Cyclonic separation (hydroclone)
   b. Filtration
2. Chemical treatment and biocides
   a. Clorination
   b. Chlorine dioxide
   c. Advanced oxidation
   d. Residual control (sulphur/bisulphate)
   e. Peraclean Ocean
3. Physical desinfection
   a. Coagulation/Floculation
   b. Ultrasound (US)
   c. Ultraviolet (UV)
   d. Heat
   e. Cavitation
   f. Deoxygenation
   g. Electro-chlorination/electrlosysis
   h. Electro-catalysis
   i. Ozonation

Between these types, the distribution per technology would be as shown in Figure 4. The figure shows BV data from 2016 based on 64 approved systems and 26 under process).

We can easily see from the above that the most common technologies are based on UV/US and electrochemical systems and chemical injection (being the two last ones using active substances). We will focus in this paper on the UV systems systems only.

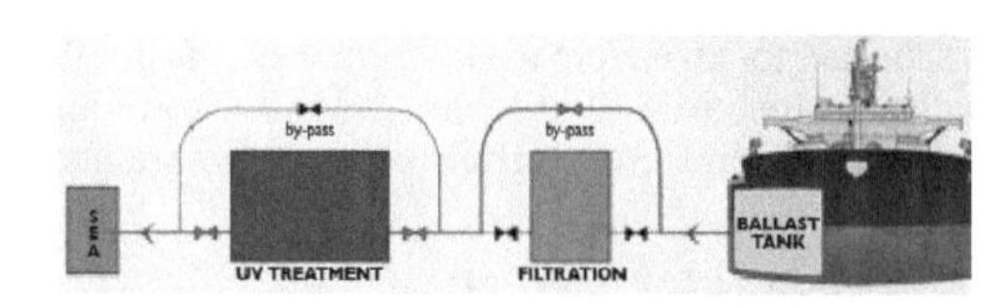

Figure 3. Deballasting: Filtration + UV bypassed.

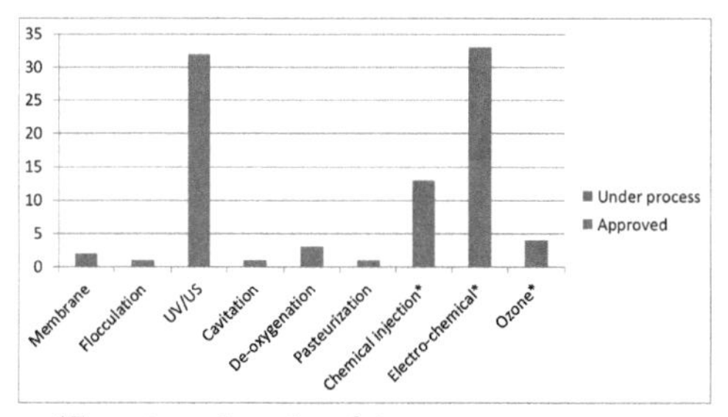

Figure 4. Distribution by type.

3.1 *UV treatment systems*

In these systems, UV radiation is used to attack and break down the cell membrane killing the organism outright or destroying its ability to reproduce. The effectiveness highly depends on the turbidity of the ballast water as this could limit the transmission of the UV radiation. We have to take into account that UV lights are required to be maintained and power consumption needs to be considered. An example of a UV system would be as shown in Figure 5.

The typical functioning system (in this case in conjuntion with a filtration system) would be as shown in Figure 6.

Normally, in many systems, the treatment is done when ballasting, and in many cases not necessary while unballasting (or partially).

3.2 *Deoxygenation*

This method reduces pressure of oxygen in space above the water with inert gas injection or by means

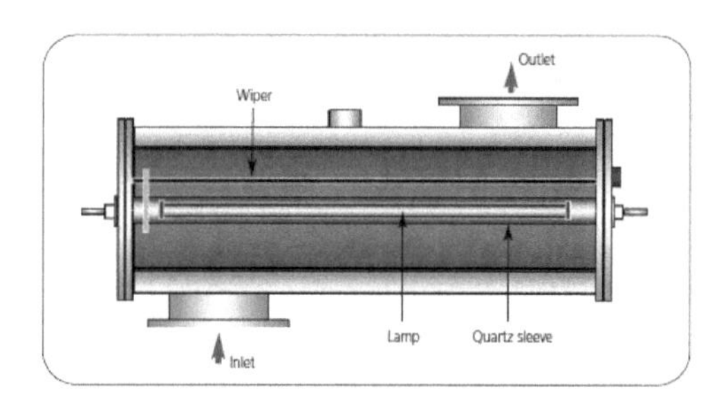

Figure 5. UV system.

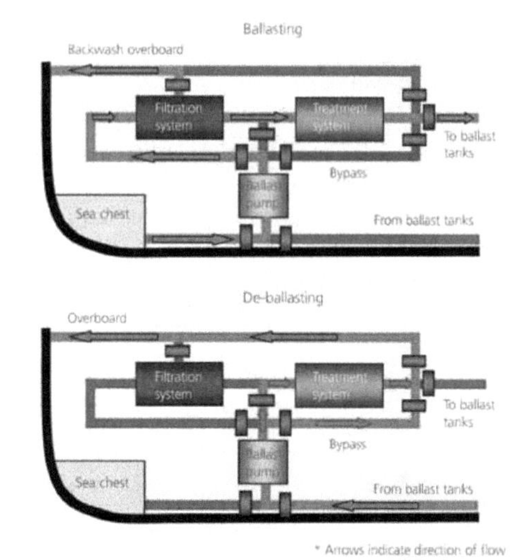

Figure 6. Functioning principles of treatment system.

of a vacuum to asphyxiate the micro-organisms. Typically, the time required for organisms to be asphyxiated is between 1 and 4 days. Removal of oxygen may result in a decrease in corrosion propensity. If an inert gas generator is already installed on the ship then deoxygenation plant would take up little additional space. This process has been developed specifically for ballast water treatment whereby the deaerated water is stored in sealed ballast tanks.

3.3 *Cavitation*

This is induced by ultrasonic energy or gas injection and disrupts the cell wall of organisms. Such high pressure ballast water cavitation techniques are generally used in combination with other technologies. Ballast water treatment systems may combine cavitation with filtration, UV, ozone or deoxygenation.

3.4 *Chemical disinfectation*

The chemical inactivation of the micro-organisms is produced through either:

- Oxidizing biocides – general disinfectants which act by destroying organic structures, such as cell membranes, or nucleic acids; or
- non-oxidizing biocides – these interfere with reproductive, neural, or metabolic functions of the organisms.

Efficiency of these processes can vary according to conditions of the water such as PH, temperature and type of organism. Systems in which chemicals are added normally need to be neutralised prior to discharge to avoid environmental damage in the ballast water area of discharge. Chemical products are injected into the system in order to "eliminate" the invasive species. Figure 7 shows an example of this arrrangement.

This chemical disinfectation type may be divide in several types. Between those, we will focus on the most common types:

a. Chlorination/Chorine dioxide/Electrochlorination

In this case the water chlorination is produced through chemical process (Cl_2, ClO_2) or electrolysis.

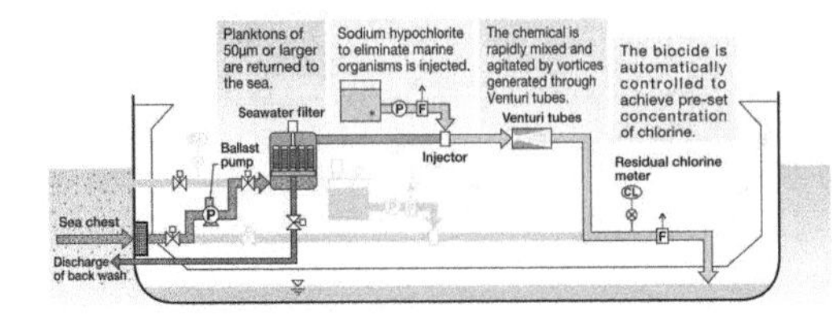

Figure 7. Chemical desinfectation.

This system creates a oxidizing solution that destroys the cell walls of micro-organisms. This system is well established and used in municipal and industrial water disinfection applications.

Most ozone and chlorine systems are neutralized but some are not. Chlorine dioxide has a half life in the region of 6–12 hours, according to suppliers, but at the concentrations at which it is typically employed it can be safely discharged after a maximum of 24 hours. Reagents used can be chemically hazardous.

When this system is selected, special attention is to be paid to the fact that it may lead to by-products (e.g. chlorinated hydrocarbons/trihalomethanes) or gas release (H_2).

b. Peracetic acid and hydrogen peroxyde

Those products are classed as an oxidizing biocide that, when diluted in water, destroys cell walls of micro-organisms. It is infinitely soluble in water. Produces few harmful by-products and is relatively stable. It is important to bear in mind that reagent is typically dosed at high levels, requires suitable storage facilities and can be relatively expensive.

3.5 *Ozonation*

In this system, ozone gas (1–2 mg/l) is bubbled into the water which decomposes and reacts with other chemicals to kill micro-organisms. It is specially effective at killing micro-organisms, but it is not as effective at killing larger organisms. It produces bromate as a by-product. One of the things to be accounted for is the fact that ozone generators are required in order to treat large volumes of ballast water. These may be expensive and require sufficient installation space. Systems in which chemicals are added normally need to be neutralised prior to discharge to avoid environmental damage in the ballast water area of discharge. Most ozone and chlorine systems are neutralised but some are not.

4 IMPLEMENTATION OF BALLAST WATER CONVENTION: APPLICATION DATES

There has been no modification in the applicability date of ballast water systems required by IMO, therefore A.1088(28) (see Table 2) is still the calendar to be used for IMO (as already said, USCG is requesting TA or acceptable protocol). This means that all ships would have to comply with D-2 standard in the date of the renewal of the IOPP Certificate.

This calendar would make (and already made) some shipowners decide whether to fit a BWMS in the IOPP renewal date or pay in advance the visit for the IOPP renewal, before the entry into force

Table 1. Implementation schedule according with USCG.

Category	Ballast Water Capacity	Compliance Date
New Vessels (Constructed on or after December 1, 2013)	All	On delivery
Existing Vessels (Constructed before December 1, 2013)	1,500–5,000 m³	First scheduled drydocking after January 1, 2014
	Less than 1,500 m³ orgreater than 5,000 m³	First scheduled drydocking after January 1, 2016

Table 2. Adjusted implementation according with IMO Resolution A.1088(28).

Resolution A.1088(28) · Adjusted implementation of Regulation D – 2			
BW capacity (m³)	Ship constructed < 2009	Ship constructed ≥ 2009 < 2012	Ship constructed ≥ 2012 <EIF
< 1500	Compliance by first IOPP renewal survey following anniversary date of delivery in 2016 If EIF > 2016 : Compliance by first IOPP renewal survey following EIF	Compliance by first IOPP renewal survey following EIF	
1500 – 5000	Compliance by first IOPP renewal survey following anniversary date of delivery in 2014 If EIF > 2014 : Compliance by first IOPP renewal survey following EIF	Compliance by first IOPP renewal survey following EIF	
> 5000	Compliance by first IOPP renewal survey following anniversary date of delivery in 2016 If EIF > 2016 : Compliance by first IOPP renewal survey following EIF		Compliance by first IOPP renewal survey following EIF

date (i.e. 8th September 2017), so they can take advantage of the 5 years period, which is the IOPP renewal date period. This would mean breaking the harmonization between the IOPP (statutory certificate) and the class certificate. This is allowed, but not recommended by Administrations.

In the last MEPC.70, IMO received a request of modification of the dates for the applicability of the requirement, which consists essentially, on the possibility that the rule will not require ships whose IOPP certificate falls due prior to 8th September 2019 (2 years after Entry into Force of the Convention), to fit BWMS until the second IOPP Renewal whereas ships with an IOPP renewal date after 8th September 2019 will be required to maintain the first renewal date for fitting of BWMS.

The proposal was led by Liberia, India and World Shipping Council (WSC) and is supported by Intercargo. This proposal was further discussed during the MEPC.71, which was scheduled mid-2017. This means that, if this proposal was not approved, shipowners that did not advance their IOPP renewal would have to update their ships to D-2 standard almost immediately. If this proposal went on, those owners who advanced their visit for IOPP renewal before 8th September 2017, would lose a maximum of 2 years plus that period between the renewal date of IOPP and 8th September 2017. So the decision was not simple.

Table 3. New scheme to revoke A.1088(28).

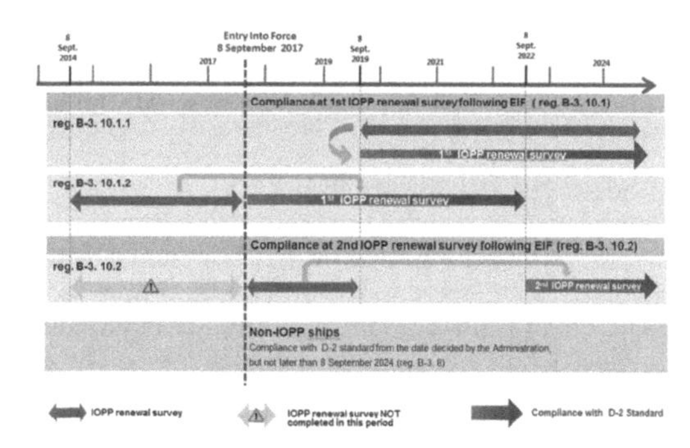

In MEPC.71, the Committee approved a resolution stating this new scheme, which will have to be ratified by the member states in MEPC.72.

This IMO scheme has been circulated to member states for foreseen adoption during MEPC 72 scheduled to be held from April 9–13, 2018.

Under the proposal, ships constructed on or after September 8, 2017 are to comply with the D-2 standard on or after that date. Vessels built before September 8, 2017, are to comply with the D-2 standard at the first MARPOL IOPP renewal survey completed on or after:

- September 8th, 2019 (Reg B-3/10.1.1); or
- September 8th, 2017, only if a MARPOL IOPP renewal survey is completed on or after September 8th, 2014, but prior to September 8th, 2017 (Reg B-3/10.1.2).

If the survey as per Reg B-3/10.1.2 is not completed, then compliance with the D-2 standard is required at the second MARPOL IOPP renewal survey after 8th September, 2017, only if the first MARPOL IOPP renewal survey after 8th September 2017 is completed prior to 8th September 2019 and a MARPOL IOPP renewal survey was not completed on or after 8th September 2014, but prior to 8th September 2017 (Reg B-3/10.2). For ships constructed before 8th September 2017 and which are not subject to the MARPOL IOPP renewal survey, compliance with the D-2 standard is required not later than 8th September 2024 (Reg B-3/5).

This scheme can be graphically summarized as shown below in Table 3.

5 RETROFITTING TO EXISTING SHIPS

Once owners have decided to install a BWMS in their vessel, they have to study several issues with the manufacturer and yard. The model to be followed to choose and install the system is the following:

a. Selection/Design feasibility.
b. Detailed design.
c. Installation on board.

A more detailed analysis of these would be as follows:

a. Selection/Design feasibility
Both phases are carried out almost simultaneously in order to combine the different systems with the pattern of operation, ship type and arrangement, etc of our ship. Many questions will have to be placed to the equipment manufacturers. These questions will reduce the options:

- Is there enough capacity in the BWTS? Some vessel types will need as much as 6.000 m^3/h or more, which not many BWTS can provide.
- Is there enough power? Any BWTS will increase the power demand, which will have to be satisfied in a new balance.
- What about Safety? The introduction of BWTS can introduce new Gas Dangerous Zones that would affect the electrical equipment around.
- Do we need hydrogen removal? Some BWTS produce hydrogen, and it has to be removed.
- Training. A new BWTS needs additional training to deal with it, and also to know when, where and how it has to be used.
- Risk assessment/HAZID study has to be done.
- Holding time, dosage, discharge limits. In some systems we need to inject products within the piping. This injection will have its dosage, holding time, that will have to be known in advance in order to create the requested tanks of this products.

Some other considerations that we will have to take into account for the selection would be:

- We will need to find a suitable location (to minimize pipe and electrical runs). Some equipment will not be fitted for the location we have chosen.
- We need to integrate electrically/hydraulically the BWTS into vessel system.
- Special requirements exist for the installation in Oil & Chemical Tankers.
- Availability of equipment in the shipyard. We will have limited options due to this fact, and depending on the time window we are working with, we will have or not a certain system available for our use.
- Treatment process is limited by water temperature and salinity to achieve optimum performance. A heat exchanger might be used to raise the water temperature.

For the selection of the BWTS, we should also take into account commercial and temporary aspects:

- The whole process can take as little as 9 but up to 18 months. Therefore, and to reduce off-hire times, a logistics plan is to be drawn and followed.
- Availability of shipyards and workforce with relevant skills. It is foreseen that around 40 retrofits per day are to be made in till 2021. This means a overloading of yards and equipment manufacturers which can provoke undesired delays in our retrofit.
- Fleet operators will be able to negotiate deals for multi-ship installations but maybe a single manufacturer does not have suitable systems for all vessels in a fleet, in terms of capacity, etc.

b. Detailed design
Once the system has been selected and a feasibility study has been carried out, the detailed design should take place, in order to "accommodate" or chosen BWTS into our vessel. When we are planning a retrofit in our vessel, we should be able to prepare new isometric drawings of piping. When the ship is an old existing ship, drawings regarding the structure of the engine room are usually not available. In this case, a 3-D scanning of the engine room would be an asset and help us prepare the piping to and from the BWMS, within the existing limitation of the actual space. These systems can create a cloud of points from the actual scanning of the engine room and create a whole image from them. The amount of points per second can be around 1 million per second. There are quite a lot of these systems, one example of which is shown in Figure 8, working in different countries.

c. Installation on board
Once the detailed design is finished (with the uncertainties that a vessel in service can bring), there are some installation pointers that should be considered:

- Installation surveys to ensure that the BWTS is installed in accordance with the approved drawings and class requirements.
- Makers must provide information on whether system can generate hydrogen or other hazardous gases affecting ballast water tank coatings.

Figure 8. Example of 3D-scanner.

- Plastic pipes are to meet the fire endurance requirements of Bureau Veritas (2018) and IMO (1993) regulations.
- Transfer of ballast water from non-hazardous to hazardous area may be accepted but not vice versa (very important in chemical carriers and oil tankers).
- Penetrations of engine/pump room bulkheads are in general not permitted.
- Spool pieces are removed after ballast transfer and the open pipe flanges are to be covered by blanking plates adequate for the service pressure.

6 CONCLUSIONS

After a long period of ratification process, the BW Convention has entered into force on the 8th September 2017. This guarantees a common standard for all the fleet throughout the world, and another proof of the commitment that the marine community has with the environment by promoting eco-friendliness and sustainability. This willingness comprises not only BWTS, but also the marine compromise with the reduction of polluting emissions of CO_2, NO_x and SO_x, by different strategies that include creation of Energy Efficiency Indexes, creation of Emission Control Areas (SECA's for SO_x and NECA's for NO_x) and the use of alternative fuels (like LNG/CNG).

One of the main problems that industry will have to face will be the availability of shipyards and manufacturers to install in such a short period of time (at present 5 years starting in the entry date), such a large number of systems. It is estimated that around 70.000 ships will have to update their ballast water standard from D-1 to D-2 in that period, so 40 retrofits per day are foreseen.

This is a major step forward for our industry, but also a big challenge, that would need a major effort of logistics to cope with this new situation. If we also take into account the USCG foreseeable new certifications of BWTS on the market, more variables would be in the equation. Therefore, a good schedule for complying with D-2 standard is a must, combining yard, manufacturer and charterer needs together with the Owners fleet in order to avoid unnecessary off-hire times and thus reducing the economic impact for the company in terms of their loss of revenue. A correct selection of the BWTS that is best fitted for each of the ships of their fleet will reduce also the CAPEX/OPEX and make operation still viable and more sustainable.

REFERENCES

Bureau Veritas 2018. NR 467 Rules for the Classification of Steel Ships. Paris, France.

IMO 1993. Assembly Resolution A.753 (18) - Guidelines for the application of plastic pipes on ships, London, United Kingdom.

IMO 2004. International Convention for the control and Management of Ship's Ballast Water and Sediments. London, United Kingdom.

IMO 2013. Resolution A.1088(28), Application of the International Convention for the control and Management of Ship's Ballast Water and Sediments, 2004. London, United Kingdom.

IMO 2016. Resolution MEPC.279(70) - 2016 Guidelines for approval of Ballast Water Management Sstems (G8). adopted on 28 October 2016, London, United Kingdom.

US 2003. Code of Federal Regulations 33 CFR 151.1510 – Ballast water management requirements.

Progress in Maritime Technology and Engineering – Guedes Soares & Santos (Eds)
© 2018 Taylor & Francis Group, London, ISBN 978-1-138-58539-3

Persistent organic pollutants in Baltic herring in the Gulf of Riga and Gulf of Finland (north-eastern Baltic Sea)

L. Järv
Estonian Maritime Academy of Tallinn University of Technology, Tallinn, Estonia

T. Raid & M. Simm
Estonian Marine Institute, University of Tartu, Tallinn, Estonia

M. Radin
Ministry of Rural Affairs, Republic of Estonia, Tallinn, Estonia

H. Kiviranta & P. Ruokojärvi
National Institute for Health and Welfare, Kuopio, Finland

ABSTRACT: Baltic herring (*Clupea harengus membras*) is one of the most abundant and economically valuable fish species in Estonian waters dominating both in offshore and coastal sea fishery. Herring has been historically one of the most important species both on the domestic market and for export. The aim of the present study is to get an overview of the occurrence of some organic pollutants in the Baltic herring caught in Estonian Exclusive Economic Zone (EEZ). The samples of Baltic herring were collected from the Gulf of Riga and from the Gulf of Finland in 2013–4 with the aim to assess its suitability for human consumption standards in the Estonian EEZ. The observed content of dioxins (PCDD, PCDF, dioxins), polychlorinated biphenyls (PCB), polybrominated diphenyl ethers (PBDE), organic tin (OT), perfluorocompounds (PFAS) are examined and discussed in the study. The results revealed that higher content of all studied organic pollutants were observed in older Baltic herring and in fish from semi-enclosed gulfs and these contents may cause concern with respect to human health.

1 INTRODUCTION

The Baltic Sea environment of the is susceptible to pollution by hazardous substances due to its natural features such as long water residence period, shallowness and the large catchment area. There are about 85 million people living in the Baltic Sea catchment area, where also various types of industrial activities, busy sea traffic and intensive agriculture are taking place. Hazardous substances, emitted by the human activities are transported to the sea *via* water and air and therefore their sources can be found even far away from the Baltic Sea area. As a result, the Baltic Sea has been quoted as one of the most polluted seas in the world (Schultz-Bull et al. 1995). The HELCOM Baltic Sea Action Plan (BSAP) Reaching Good Environmental Status for the Baltic Sea: hazardous substances (Anon 2007) and EU Water Framework Directive (EC 2010) set strategic goals related to hazardous substances: (1) Baltic Sea with life undisturbed by hazardous substances; (2) identification of ecological objectives corresponding to the "good environmental" status to minimize the concentration of hazardous substances close to zero or to the natural background levels, (3) all fish in the Baltic Sea should be safe for human consumption (the content of organochlorines retained in fish should not exceed the safety limits established by the EU).

However, regardless to the actuality of the issue, monitoring of hazardous substances, with an ambitious zero-emission target for all man-made hazardous substances in the whole Baltic Sea catchment area by 2021 (HELCOM 2010a) as one of the four elements of the ecosystem health, targeted by the HELCOM BSAP (eutrophication, biodiversity, hazardous substances and maritime activities (Anon 2007), has gained a widespread attention.

The aim of this study is to provide an overview of the occurrence of some organic pollutants in Baltic herring (*Clupea harengus membras*) caught in Estonian Exclusive Economic Zone of the Gulf of Riga and of the Gulf of Finland in 2013–4 with the aim to assess its suitability for human consumption standards.

2 MATERIAL AND METHODS

2.1 *Study area*

The Gulf of Riga and the Gulf of Finland are located in the north-eastern part of the Baltic Sea. Both gulfs are under extensive anthropogenic impact (Lilja et al. 2009).

The Gulf of Riga (GoR) is a semi-enclosed sub-basin connected to the Baltic Proper *via* Irbe Strait (Fig. 1). The shallowness of the gulf (the average depth is only 23 m by Omstedt et al. 2009), results in complete vertical water mixing during winter. The salinity varies in wide range from 0.5 psu in coastal surface layers to 7 psu close to the Irbe Stait (Berzinsh 1995). Due to limited water exchange, the GoR is more euthrophicated than the Baltic Proper. The outflow of nutrients is higher than inflow through the straits (Jurkovskis, P. & Poikane 2008). The biggest bay of the area is semi-enclosed Pärnu Bay which is regarded by HELCOM (2002) as one of the most euthrophicated part of the Baltic Sea.

The Gulf of Finland (GoF) is located in the north-eastern part of the Baltic Sea and is the easternmost arm of the Baltic Sea (Fig. 1). The gulf is a continuation of the Baltic Proper without having any separating sill and can be characterized as a transition zone between brackish coastal sea and open sea conditions. The hydrological conditions are variable depending on the distance from the Baltic Proper. The small bays of the southern coast have good connection to the open GoF and are strongly influenced by the main Baltic Sea current, upwelling and river inflow (Soomere et al. 2009). Due to the large influx of fresh water from rivers, the gulf water has very low salinity—between 0.2 and 5.8 PSU at the surface and 0.3–8.5 PSU near the bottom (ESI 2016).

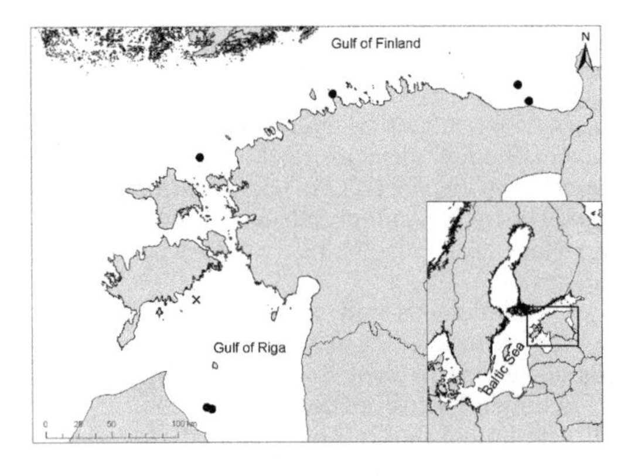

Figure 1. Study area and sampling sites: • spring spawning Baltic herring and **x** autumn spawning Baltic herring sampling sites.

2.2 *Fish sampling*

Sampling process, analyses of samples and compiling of biological material for chemical analyses followed requirements of Manual KJ I/16 (Biological tissue sampling from fish and molluscs for chemical analysis). The method has been accredited according to the ISO 17025 standard by Estonian Accreditation Centre (testing laboratory L179).

The fish samples were collected from four sampling sites in the GoF and from three sampling sites in the GoR in 2013–14 (Fig. 1) from Estonian commercial pelagic trawl fishery in western and eastern parts of the GoF in October and in the GoR in April. In 2014, one additional sample from traditional gillnet-fishery of the autumn spawning of Baltic herring was taken from the northern part of the GoR. Since the fish catches sold on Estonian market for human consumption are not sorted, the biological material, used as a basis of present study, was collected randomly in order to characterise the composition of consumable fish.

The total length and weight of sampled fish was measured. Additionally, age, sex and maturity stage using six-grade scale (ICES 2010) was assessed. The Fulton's body condition index (F) was calculated as $F = TW \times 100/SL^3$, were TW is total body weight (g) and SL is standard length of fish (mm). Counting of the growth zones on otoliths was used to age Baltic herring. The average biological parameters (±SE) of studied fish are presented in Table 1.

To find the correlation (R^2) between age of fish and dioxins content data, altogether of 112 samples, of Baltic herring had been analysed.

2.3 *Chemical analyses*

The chemical analyses were carried out at the Environmental Health Unit of the National Institute for Health and Welfare (THL), which has been accredited according to the ISO 17025 standard by FINAS (testing laboratory T077). In addition to the internal quality control, the laboratory participates in several annual inter-laboratory comparisons e.g. for food, feed, and environmental samples organized by Folkhelseinstitutet in Norway (PCDDs, PCDFs, PCBs and PBDEs), EU-Central Reference Laboratory in Germany (PCDDs, PCDFs, PCBs and PBDEs), Centre de Toxicologie in Canada (PFAAs), and Quasimeme intercalibration trial (OTC).

2.3.1 *Dioxins (PCDD), furans (PCDF), dioxin-like PCB, non-dioxin-like PCBs (ndl-PCBs) and polybrominated diphenyl ethers (PBDEs)*

Fish samples were spiked with corresponding ^{13}C-labelled standards. After pooling and

Table 1. Biological parameters (avg ± SE), the content of dry matter and lipids in Baltic herring of the Gulf of Riga (GoR) and of the Gulf of Finland (GoF) in 2013–14.

	n	TL*, cm	Tw**, g	F***	Age	Dry matter, %	Lipids, %
GoR							
Spring spawners	6	13.9 ± 0.2	15.8 ± 0.12	0.6 ± 0.01	2.7 ± 0.01	23.7	6.1
Autumn spawners	5	19.5 ± 2.6	53.4 ± 2.7	0.62 ± 0.02	6.8 ± 0.3	22.8	7.3
GoF							
Eastern part	6	16.6 ± 0.1	32.0 ± 0.6	0.70 ± 0.01	4.0 ± 0.1	23.5	5.1
Western part	6	13.4 ± 0.1	14.4 ± 0.01	0.64 ± 0.01	2.5 ± 0.1	22.9	4.6

*Total length; **Total weight; ***Fulton's body condition index.

homogenization the samples were freeze-dried, and fat was extracted with ethanol-toluene (15%/85% v/v) in Accelerated Solvent Extractor (Dionex ASE 300). After that solvent was exchanged to hexane and the fat content was determined gravimetrically. Further the samples were defatted on an acidic multi-layer silica column and purified and fractionated on alumina and carbon columns. PCDDs/PCDFs, PCBs and PBDEs were analyzed with gas chromatography/high-resolution mass spectrometry—GC/HRMS (Autospec Ultima) using selected ion monitoring mode with a 10,000 resolution. PCDDs/PCDFs, PCBs and PBDEs were separated on a DB5 MS column (60 m × 0.25 mm × 0.25 μm; for BDE209 the column length was 5 m). For quantification of these compounds ^{13}C-labelled PCDDs/PCDFs, PCBs and PBDEs were used. The recoveries of the individual internal standards were determined by adding the recovery standards just before mass spectral analysis, and were 50–120%. The limits of quantitation (LOQ) for individual congeners corresponded to a signal-to-noise ratio of 3:1 (based on EU Commission Regulation No 252/2012) or concentration of the analyte in blank sample. Two blank samples (a reagent blank and a methodological blank sample handled as the other samples) and one laboratory control sample of Baltic herring was run with every analysis batch with an allowable variation of ±20%.

2.3.2 *Organotin compounds (OT)*

The OT compounds analysed were mono-, di-, and tributyltin (MBT, DBT, TBT) and mono, di- and triphenyltin (MPhT, DPhT, TPhT) and dioctyltin (DOT). The analysis of freeze -dried fish samples (0.25 g) was performed by solvent extraction, ethylation with sodium tetraethylborate and high resolution, GC/HRMS analysis according to the tissue method developed by Ikonomou et al. (2002) with slight modifications. Details of the method used have been described previously (Rantakokko et al. 2008, Rantakokko et al. 2010). Summary concentration of measured OTs as cation/g wet weight (ng/g w.w.) is denoted as ΣOTs. In the calculations of ΣOTs, results less than the limit of quantification (<LOQ) were treated as zero. Two laboratory reagent blank samples were treated and analysed as the actual samples in each series of samples. Certified mussel tissue CRM 477 was analysed in every series of samples. It has certified concentrations for MBT, DBT and TBT and indicative concentrations for MPhT, DPhT, and TPhT, respectively (Pellegrino et al. 2000).

2.3.3 *Perfluorinated Alkylated Substances (PFAS)*

The content of 13 analogues of PFAS—PFHpA (perfluoroheptanoic acid), PFOA (perfluorooctanoic acid), PFNA (perfluorononanoic acid), PFDA (perfluorodecanoic acid), PFUnA (perfluoroundecanoic acid), PFDoA (perfluorododecanoic acid), PFTrA (perfluorotridecanoic acid), PFTeA (perfluorotetradecanoic acid), PFHxS (perfluorohexane sulfonate), PFHpS (perfluoroheptane sulfonate), PFOS (perfluorooctane sulfonate), and PFDS (perfluorodecane sulfonate), PFHxA (perfluorohexanoic acid), was examined. For quantitation prior to an extraction procedure mass labelled internal standards were added into freeze-dried fish samples. The samples were extracted with ammonium acetate in methanol. and centrifuged. The supernatants were collected, extracts were evaporated to dryness and filtered The PFAS were analysed using liquid chromatography negative ion electrospray tandem mass spectrometry (LC–ESI–MS/MS). Details of the LC–ESI–MS/MS parameters and quantitation have been presented earlier (Koponen et al. 2013). Measurement uncertainty of PFAS was 30%.

3 RESULTS AND DISCUSSION

Altogether 23 samples of Baltic herring were collected and analysed. The mean biological parameters of Baltic herring samples on chemical substances are presented in Table 1. The average

Table 2. The average content (avg ± SE) of persistent organic pollutants the Baltic herring (w.w.) of the Gulf of Riga (GoR) and of the Gulf of Finland (GoF) in 2013–14.

	dioxins*, pg/g	indPCB, ng/g	PBDE, ng/g	OT, ng/g	PFAS, ng/g
GoR					
Spring spawners	5.65 ± 0.65	19.2 ± 3.0	0.64 ± 0.007	4.76 ± 0.25	6.43 ± 0.55
Autumn spawners	9.61 ± 0.79	32.6.0 ± 2.3			
GoF					
Eastern part	5.72 ± 0.41	17.8.3 ± 1.2	0.63 ± 0.12	10.9 ± 0.83	5.01 ± 0.28
Western part	2.39 ± 0.08	9.1 ± 0.1	0.37 ± 0.02	7.85 ± 0.84	5.33 ± 0.49

* Dioxins = PCDDs+PCDFs+dl-PCBs.

content of persistent organic pollutants in the Baltic herring by sampling sites in 2013–14 are presented in Table 2.

3.1 *Dioxins (PCDD, PCDF, dioxin-likePCB)*

The concentrations of the dioxins: polychlorinated dibenzo-*p*-dioxins (PCDDs), polychlorinated dibenzofurans (PCDFs) and dioxin-like polychlorinated biphenyls (dl-PCBs) were determined in 23 pooled samples of Baltic herring in 2013–14 (Table 1). The average dioxins content in the Baltic herring was 6.38 pg WHO_{2005}-TEQ/g in w.w. Three samples of 2–4 years old spring spawning herring and all five samples of 5+ years old autumn spawning herring exceeded allowable EU concentration threshold of dioxins 6.5 pg WHO_{2005}-TEQ/g in w.w. (Table 2). The following congeners of PCDD/F: 2,3,4,7,8-PeCDF (44%) and 2,3,7,8-TCDF (29%) and of dl-PCB: 118 (60%) and 105 (20%) dominated the TEQ in Baltic herring. The dl-PCBs prevailed over the PCDD/F in all samples in 2013–14.

However, in all cases the content of dioxins was higher in the fish from the semi-enclosed GoR (Fig. 2) and in older fish and in fish with higher fat content (Table 1). The higher fat content was caused by the good feeding condition (Table 1: F) of Baltic herring in autumn and from another hand by the food composition of herring consisting mostly of zooplankton with high content of dioxins (Hallikainen et al. 2011). In the Baltic herring collected in spring from the western part of the GoF the dioxin content was low and did not exceed the allowable EU concentration threshold (Fig. 2).

The higher levels of dioxins are legitimate in older fish, but the relatively high dioxins values in younger herring is alarming. Since the content of dioxins exceeded allowable EU concentration threshold not only in older (7+ years), but also in younger (2–4 years) fish in the GoR, the dioxin monitoring should be continued.

Altogether in 112 samples of Baltic herring had been analysed on the content of dioxins during

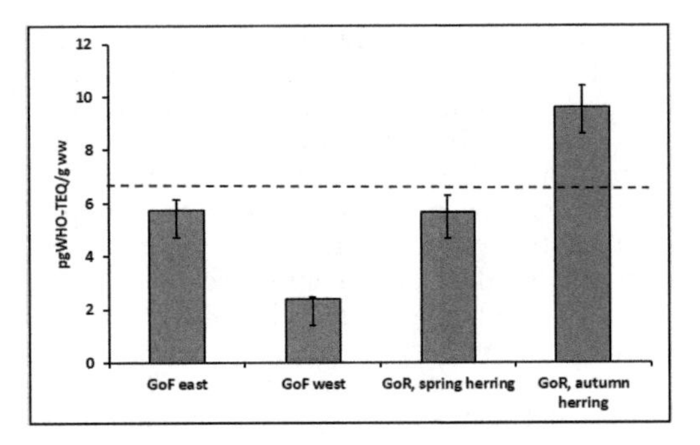

Figure 2. The PCDD/Fs and dl-PCBs content in the Baltic herring in 2013–14: GoF – the Gulf of Finland, GoR – the Gulf of Riga. ---- maximally allowable EU concentration values of PCDD/F + dioxin-like PCB.

the last decades in Estonia (Pandelova et al. 2008, Roots et al. 2010, Järv et al. 2014). The results indicate that the dioxin concentration grows linear with age of the Baltic herring. The positive correlation ($R^2 = 0.566$) between the dioxins concentration and age was found to be statistically reliable ($p < 0.05$). According on our data the concentration of dioxins exceeds the maximum allowable EU concentration threshold in age of 8 years and of summarised PCDD/F and dl-PCB in age of 7.5 years old B. herring (Fig. 3).

The studies made recently in Finland support our results and demonstrate that the content of dioxins increase linear until 5 years old herring and after that accelerates in semi-enclosed sea areas (Bothnia Bay) and exceed the EU concentration threshold already in age of 5+ years old herring (Hallikainen et al. 2011). It can be partly explained with the food composition of herring consisting mostly of zooplankton and nectobenthos which content of dioxins is high (Hallikainen et al. 2011).

In Estonian commercial catches the herring at age 7 years and older constituted about 4% in the eastern part of the GoF and of the GoR, and up to 10% in the western part of the GoF in 2004–2013.

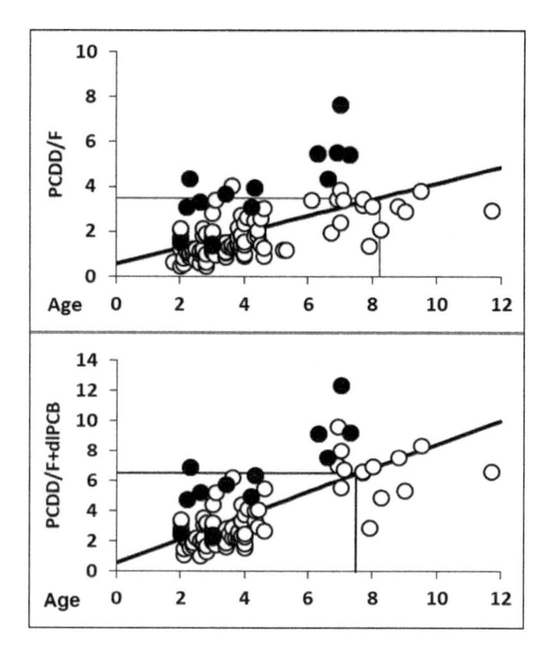

Figure 3. The summarised dioxins content ($pgWHO_{2005}TEQ/g$ w.w.) dependency on the age of Baltic herring.

That means, that some of the Baltic herring caught from these areas have dioxin concentrations above maximum allowable concentrations set by EU and therefore ineligible to be placed in the market. However, it is very important to link the dioxins content to lipid content of fish and follow their seasonal variation in different age classes.

Due to high concentration of dioxins the dietary recommendations were developed by National Institute for Health Development in Estonia (Pitsi et al. 2017). The annual consumption of fish is 11.6 kg per person in Estonia. Dietary recommendations were developed originally for coastal fishermen to reduce the health risks associated with the high dioxins concentrations in fish. According on this recommendation there is no dioxin risk if people consume up to 12 servings (900 g) per week or up to 130 g per day of fish with low (perch, pike-perch) or medium fat content (B. herring, flounder). At the same time, it was recommended not to eat Baltic herring with total length ≥ 22 cm (Roots 2011).

3.2 *Non dioxin-like polychlorinated biphenyls (PCB)*

Altogether six congeners (indicator PCBs) of ndl-PCB: CB 28, 52, 101, 138, 153 and 180, constitute proximately 50% of PCBs in fish studied. The maximum allowable EU concentration threshold

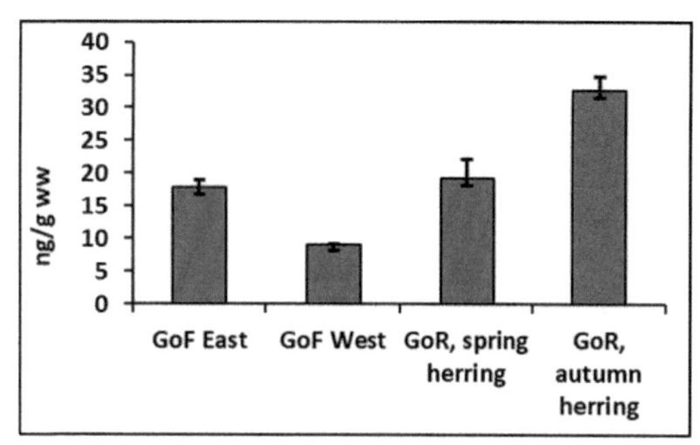

Figure 4. The ind-PCBs content in the Baltic herring in 2013–14: GoF – the Gulf of Finland, GoR – the Gulf of Riga.

of ind-PCBs in fish is 75 ng/g w.w. (Commission regulation EU No 1881/2006).

However, the higher concentrations of ind-PCBs were observed in the GoR and in the eastern part of the GoF (Fig. 4). Since PCBs are lipid-soluble compounds the highest concentration of ind-PCB (*ca* 32.6 ng/g w.w.) were found in the fatter and older (7+ years) autumn spawning herring (Table 1) of the GoR (Table 2). Despite it the mean content of PCB was well below the allowed EU limits.

3.3 *Organotin compounds (OT)*

To characterise the content of organotin compounds (OT) the total content of four compounds TBT, DBT, TPhT and DOT was calculated. The maximum allowable concentration value of organotins 0.25 μg/kg body weight/day for TBT, DBT, TPT and DOT compounds was established by EFSA (2004).

The average DBT content in Baltic herring was 0.61 ± 0.06 ng/g (w.w.) and it was a bit higher in the Baltic herring from the eastern part of the GoF. The lowest content of DBT was observed in the GoR where it was? below the quantification range. The average content of OT analogues: TBT, DPhT and TPhT, in B. herring were 4.92 ± 0.61, 0.17 ± 0.01 and 1.20 ± 0.11 ng/g (w.w.), respectively.

In all cases the content of congeners was lower in the fish from the GoR and higher in the fish from the eastern part of the GoF (Fig. 5). The OT content in Estonia was also investigated within the HELCOM Project "Screening of hazardous substances in the eastern Baltic Sea". According to the results of that project the content of TBT in the Baltic herring varied from 3 to 8 ng/g (w.w.) except one sample from the impact area of the Port of Sillamäe (34 ng/g w.w.) in 2008. The high content of TBT observed in some samples can probably be explained with the location of sampling sites in the

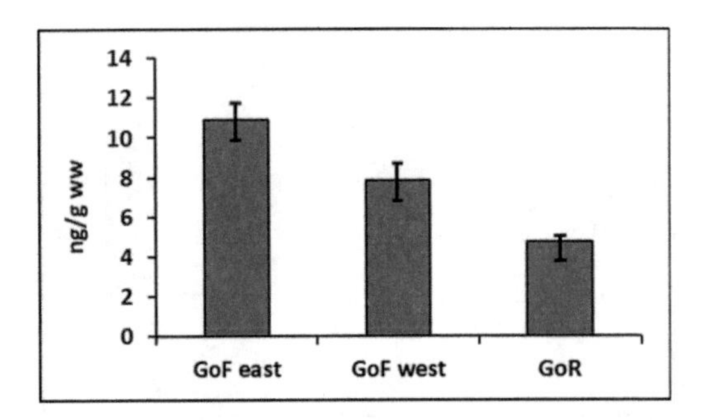

Figure 5. The content of four OT analogues in the Baltic herring in 2013–14: GoF – the Gulf of Finland, GoR – the Gulf of Riga.

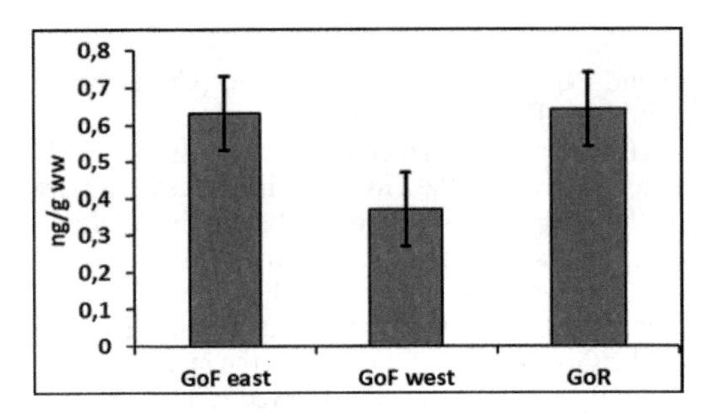

Figure 6. The total content of eight of PBDE analogues in the Baltic herring in 2013–14: GoF – the Gulf of Finland, GoR – the Gulf of Riga.

impact areas of ports characterised by the elevated concentration of OT compounds in sediments and fishes including Baltic herring (Lilja *et al.* 2009).

3.4 *Polybrominated diphenyl ethers (PBDE)*

According to the European Food Safety Authority (ESFA) regulation (2011) eight analogues of PBDE (BDE-28, –47, –99, –100, –153, –154, –183, –209) should be monitored in the marine environment. Since the food quality limit is not set for PBDE, we based our assessment on the EQS—European Quality Standards given for living resources (EC 2008/105).

Our results revealed that the content of three analogues: BDE-85, –138 ja –183, was below the quantification range and the analogue BDE-47 clearly dominated (53% of summarised 15 studied congeners) in the Baltic herring. Since PBDEs are lipid-soluble compounds, the total content was always considerably higher in fatter and older autumn spawning Baltic herring in the GoR (Table 1, Fig. 6). The mean total of eight analogues of PBDEs ranged from 0.37 ± 0.02 to 0.63 ± 0.12 to 0.67 ± 0.04 ng/g w.w. in Western and Eastern GoF and in the GoR, respectively. Such a wide variation: 0.46–0.71 ng/g (w.w.) was observed also in 2006–2011 (Roots *et al.* 2010).

By EQS Directive (EC 2008/105) the summary content of six congeners of PBDE cannot exceed 0.0085 µg/kg (w.w.) in fish. By our results, the concentration of one congener BDE-47 exceeded this value in all samples. According to the HELCOM (2017) the status of the ∑BDE congeners 28, 47, 99, 100, 153, and 154 in fish during the period up to 2015 shows that the threshold is exceeded at every monitoring site in the Baltic Sea. Therefore, the all monitoring sites have been classified as a "areas of not good state".

3.5 *Perfluorinated alkylated substances (PFAS)*

The content of 13 different PFAS compounds: PFHpA, PFOA, PFNA, PFDA, PFUnA, PFDoA, PFTrA, PFTeA, PFHxS, PFHpS, PFOS and PFDS, were studied in the Baltic herring of the GoF and of the GoR in 2013–2014.

In all cases the concentrations of PFHxA, PFHpA, PFTeA and PFHpS remained below the LOQ (the limit of quantitation). Two congeners PFOS and PFDS dominated in the Baltic herring in both studied sea areas. The total content of 13 PFAS compounds in Baltic herring studied was somewhat higher in the western part of the GoF compared to the eastern part of the GoF (Fig. 7). The congener PFDA exceeded the LOQ in one sample from the GoR.

The content of PFOA remained under the LOQ in all samples collected during present study in 2013–2014. In this respect our results were similar to those of reported for the Finnish coastal waters in 2009–2010. The content of PFOS in Baltic herring generally remained below 1–5 ng/g w.w. in the both studies (Hallikainen et al. 2011). However, the content of PFOS found in more polluted areas in the vicinity of the ports of Helsinki were substantially

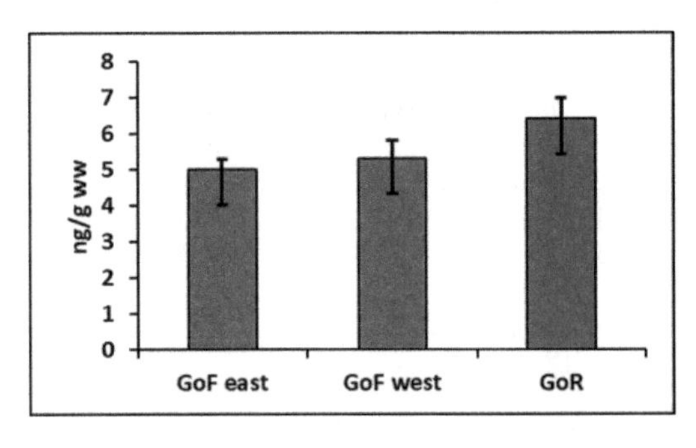

Figure 7. The total content of thirteen PFAS analogs in the Baltic herring in 2013–14: GoF – the Gulf of Finland, GoR – the Gulf of Riga.

higher, reaching 16–40 ng/g w.w. (Hallikainen et al. 2011). The overall average content of PFOS in Baltic Sea fish muscle tissues remains below the limit of quantification (HELCOM 2010b). The maximum allowable concentration limits of PFAS for fish are not established.

4 CONCLUSIONS

1. The content of dioxins, ind-PCBs and PFAS was always higher in semi-enclosed Gulf of Riga and in the eastern part of the Gulf of Finland.
2. The content of ind-PCB is generally low in Baltic herring and the fish is suitable for human consumption.
3. The concentration of one congener—BDE-47 of PBDE exceeded the value of EQS in all samples and therefore the further study of PBDE in fish is necessary.
4. While the content of dioxins is increasing not only in older (7+ years), but also in younger (2–4 years) Baltic herring it should be continuously monitored in the future and it is important to link the dioxins content to lipid content of fish and follow their seasonal variation in different age classes.
5. Maybe a common estimate with which analyte groups the exposure via consumption of Baltic herring are the most relevant when respective ADI/TDI is estimated.

ACKNOWLEDGEMENTS

The work presented in this paper was funded by the Estonian Ministry of Rural Affairs, research project 3.4–29/334. We thank the laboratory technicians in the Chemicals and Health Unit at the National Institute for Health and Welfare, Finland for an excellent analytical work performed.

REFERENCES

Airaksinen, R., Hallikainen, A., Rantakokko, P., Ruokojärvi, P., Vuorinen, P.J., Parmanne, R., Verta, M., Mannio, J. & Kiviranta, H. 2014. Time trends and congener profiles of PCDD/Fs, PCBs, and PBDEs in Baltic herring off the coast of Finland during 1978–2009. Chemosphere, 42: 165–171.

Airaksinen, R., Hallikainen, A., Rantakokko, R., Ruokojaärvi, P., Vuorinen, P.J., Mannio, J., Kiviranta, H. 2015. Levels and Congener Profiles of PBDEs in Edible Baltic, Freshwater and Farmed Fish in Finland. Environ. Sci. Technol., 49: 3851–3859.

Anon2007. http://www.helcom.fi/BSAP/ActionPlan/en_GB/ ActionPlan/ 12.12.2017.

Berzinsh. V. 1995. Dynamics of hydrological parameters of the Gulf of Riga. In Ecosystem of the Gulf of Riga between 1920 and 1990 (Ojaveer. E. ed.), Estonian Academy Publishers, Tallinn: 8–31.

Commission Regulation (EC) No 1881/2006 of 19 December 2006 setting maximum levels for certain contaminants in foodstuffs (Text with EEA relevance) OJ L 364, 20.12.2006: 5–24.

EC 2008/105. Environmental Quality Standards Directive: http://ec.europa.eu/environment/water/water-dangersub/pri_substances.htm.

EC 2010. Common Implementation Strategy for the Water Framework Directive (2000/60/EC): Guidance Document No. 25 On chemical monitoring of sediment and biota under the Water Framework Directive Luxembourg: Office for Official Publications of the European Communities, 1–74.

EFSA 2004. Opinion of the Scientific Panel on Contaminants in the Food Chain on a request from the Commission to assess the health risks to consumers associated with exposure to organotins in foodstuffs. EFSA Journal, 102: 1–119.

EFSA 2011. Scientific Opinion on Polybrominated Diphenyl Ethers (PBDEs) in Food. EFSA Journal, 9: 1–274. ESI 2016. http://www.europeanstraits.eu/Partners/Gulf-of-Finland 15.12.17.

Hallikainen, A., Airaksinen, P., Rantakokko, J., Koponen, J., Mannio, P.J., Vuorinen, T., Jääskeläinen, H. and Kiviranta, H. 2011 Environmental pollutants in Baltic sea fish and other domestic fish: PCDD/F, PCB, PBDE, PFC and OT compounds, Evira Research reports, 2/2011: 1–101.

HELCOM 2002. Environment of the Baltic Sea Area 1994–1998. Balt. Sea Environ. Proc. 82B: 3–216.

HELCOM 2010a. Ecosystem Health of the Baltic Sea 2003–2007: HELCOM Initial Holistic Assessment, Balt. Sea Environ. Proc. No. 122.

HELCOM 2010b. Hazardous substances in the Baltic Sea. An integrated thematic assessment of hazardous substances in the Baltic Sea. Balt. Sea Environ. Proc. No. 120 A.

HELCOM 2017. Polybrominated diphenyl ethers (PBDE). HELCOM core indicator report, www.helcom.fi/baltic-sea-trends/indicators/polybrominated-diphenyl-ethers-(pbde).

ICES 2010. Report of the Advisory Committee. Book 8. The Baltic Sea. ICES Adv. 2010, 8, 118 p.

Ikonomou, M.G., Fernandez, M.P., He, T. and Cullon, D. 2002. Gas chromatography-high-resolution mass spectrometry based method for the simultaneous determination of nine organotin compounds in water, sediment and tissue. Journal of Chromatography, 975: 319–333.

Jurkovskis, A. & Poikane 2008. Biogeochemical, physical and anthropogenic transformations in the Daugava River estuary and plume, and the open Gulf of Riga (Baltic Sea) indicated by major and trace elements. J. of Mar. Syst., 8: 32–48.

Järv, L., Simm, M., Raid, T., Järvik, A. 2014. Environmental status of the North-eastern Baltic Sea: the results of long-term monitoring of organochlorine compounds. 15th International Congress of the International Maritime Association of the Mediterranean IMAM 2013, Developments in Maritime Transportation and Exploitation of Sea Reasources, A Coruna, Spain, 14.–17. October 2013, Vol. 2: 819–824.

Koponen, J., Rantakokko, P., Airaksinen, R. and Kiviranta, H. 2013. Determination of selected perfluorinated alkyl acids and persistent organic pollutants from a small volume human serum sample relevant for epidemiological studies. J. Chromatogr. A, 1309, 48–55.

Lilja, K., Norström, K., Remberger, M., Kaj, L., Egelrud, L., Junedahl, E., Brorström-Lunden, E., Ghebremeskel, M. & Schlabach, M. 2009. The Screening of Selected Hazardous Substances in the Eastern Baltic Marine Environment. Swedish Environmental Research Institute Ltd. Report B1874. 1–59.

Omstedt, A., Gustafsson, E. & Wesslander, K. 2009. Modelling the uptake and release of carbon dioxide in the Baltic Sea surface water. Cont. Shelf Res., 29: 870–885.

Pandelova, M., Henkelmann, B., Roots, O., Simm, M., Jäev, L., Benfenati, E., Schramm, K.-W. 2008. Levels of PCDD/F and dioxin-like PCB in Baltic fish of different age and gender. Chemosphere, 71(2), 369–378.

Pellegrino, C., Massanisso, P. and Morabito, R. 2000. Comparison of twelve selected extraction methods for the determination of butyl- and phenyltin compounds in mussel samples. TrAC Trends in Analytical Chemistry, 19: 97–106.

Peltonen, H., Roukojarvi, P., Korhonen, H., Kiviranta, H., Flinkman, J., Verta, M. 2014. PCDD/Fs, PCB and PBDEs in zooplankton in the Baltic Sea – spatial and temporal shifts in the congener specific concentrations. Chemospere, 114: 172–180.

Pitsi, T., Zilmer, M., Vaask, S., Ehala-Aleksejev, K. 2017. Eesti toitumis- ja liikumissoovitused 2015. Tervise Arengu Instituut. Tallinn: 284. (https://intra.tai.ee/images/prints/documents/149019033869_eesti%20toitumis-%20ja%20liikumissoovitused.pdf 14.12.17.)

Rantakokko, P., Hallikainen, A., Airaksinen, R., Vuorinen, P.J., Lappalainen, A., Mannio, J. and Vartiainen, T., 2010. Concentrations of organotin compounds in various fish species in the Finnish lake waters and Finnish coast of the Baltic Sea. Science of the Total Environment, 408: 2474–2481.

Rantakokko, P., Turunen, A., Verkasalo, P.K., Kiviranta, H., Männistö, S. and Vartiainen, T. 2008. Blood levels of organotin compounds and their relation to fish consumption in Finland. Science of the Total Environment, 399: 90–95.

Roots, O. & Nõmmsalu, H. 2011. Report on hazardous substances screening in the aquatic environment in Estonia. In: Viisimaa, M., (Ed.), Baltic Environmental Forum, Tallinn, 1–97.

Roots, O., Zitko, V., Kiviranta, H., Rantakokko, P. and Ruokojärvi, P. 2010. Polybrominated diphenyl ethers in Baltic herring from Estonian waters, 2006–2008. Ecological Chemistry, 19 (1): 14–23.

Schultz-Bull, D.E., Petrick, G., Kannan, N. & Duinker, J.C. 1995. Distribution of individual chlorobiphenyls in solution and suspension in the Baltic Sea. Mar Chem 48: 245–270.

Soomere, T., Leppäranta, M. & Myrberg, K. 2009. Highlights of physical oceanography of the Gulf of Finland reflecting potential climate changes. Boreal. Env. Res., 14(1): 152–165.

Vuorinen, P.J., Myllylä, T., Keinänen, M., Pönni, J., Kiviranta, H., Peltonen, H., Verta, M., Koistinen, J., Karjalainen, J., Kiljunen, M. 2012. Biomagnification of organohalogens in Atlantic salmon (Salmo salar) from its main prey species in three areas of the Baltic Sea. Science of the Total Environment, 421–422, 129–143.

Ship design

Progress in Maritime Technology and Engineering – Guedes Soares & Santos (Eds)
 ISBN 978-1-138-58539-3

Critical wind velocity for harbor container stability

A. Balbi, M.P. Repetto, G. Solari, A. Freda & G. Riotto
Department of Civil, Chemical and Environmental Engineering (DICCA), University of Genova, Genova, Italy

ABSTRACT: The seaport areas are very exposed to wind action, which affects all the terminal activities. Piled containers can fall and get damaged under extreme wind actions, causing serious safety problems for port operators and economic damage. For this reason, the management of the companies usually adopts safety systems furnishing alert and regulating the terminal operability on the basis of the real time wind velocity monitored in the terminal areas. However, the study of wind stability of the piled containers has not reported in the literature yet, and the critical wind velocity values at the basis of the safety systems are still affected by many uncertainties. This is a preliminary paper concerning in situ and laboratory measures whose results are preparatory for the formulation of the critical wind velocity for the containers stability. The friction coefficient between the containers surface and the piled containers aerodynamic coefficients are the input parameters of critical speeds causing container's sliding, tilting and lifting. The friction coefficient has been analyzed by means of in situ tests on the piled containers. The aerodynamic coefficients have been studied by means of wind tunnel tests.

1 INTRODUCTION

The existence of seaport infrastructure is essential in the marine transportation system. The movement of goods is crucial to the development of economic and social communities. Due to their natural location, ports are continuously threatened by natural hazards. This in turn necessitates continuous monitoring their operation, in case of extreme weather events. The piled containers can fall under strong winds, causing serious safety problems for port operators and economic losses. For this reason, the management of the companies usually adopts safety systems that furnish alert and regulate the terminal operability on the basis of the real time wind velocity monitored in the terminal areas. University of Genova (Italy) developed a wide research activity in this field, with reference to the European Projects "Wind and Ports" (Solari et al. 2012) and "Wind, Ports and Sea" (Repetto et al, 2017), handling the problem of the wind safe management and risk assessment of the North Tyrrhenian seaports. The projects developed a web-based GIS platform, integrated into the operational management of the involved ports, making directly available in real time in situ monitoring wind velocities, wind and wave medium- and short-term forecasting, statistical analyses and wind climatological mapping (Repetto et al., 2018). However, the study of wind stability of the piled containers has not reported in the literature yet (Zhen et al. 2011), and the critical wind velocity values at the basis of the safety systems are still based on experience and affected by many uncertainties.

This paper presents a further step towards a complete formulation of the safety management procedure against extreme wind in containers terminals. It reports a preliminary work on in situ and laboratory measures, whose results characterize statistically two fundamental parameters for the formulation of the critical wind velocity for the containers stability. The first input parameter is the friction coefficient that has been analyzed by means of in situ tests on the piled containers. On-site sliding tests were carried out by stacking several containers and applying an horizontal force to one of them, measuring for which force, conditions and configurations the sliding occurs. The second input parameter is the aerodynamic coefficient that has been studied by means of wind tunnel test on scaled models. Several set-up conditions have been reproduced, considering seven different container storage configurations, two different incoming wind profiles and two different container's tier, obtaining as many aerodynamic coefficient sets of values. The conclusions summarize the obtained results and describe the perspectives of the analysis.

2 METHODOLOGY

Considering the container as a rigid body, the critical conditions of sliding, tilting and lifting can be defined by imposing the wind-induced forces and

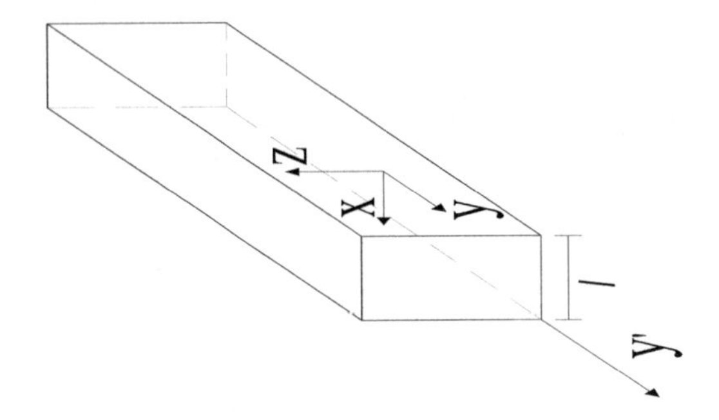

Figure 1. Cartesian axes system.

moment to the body and applying the equilibrium equations. In particular, we define a local element reference system xyz, with the x axis orthogonal to the container, the y axis aligned with the container and the z axis vertical (Figure 1); the origin of the reference system is placed in the center of the area of the container basis. F_{xy}, F_z and M'_y are the wind-induced horizontal, vertical force and moment on the container, where M_y is calculated with respect to the y'–y'axis, placed at the basis longitudinal corner of the container (Figure 1). The sliding, tilting and lifting equilibrium conditions can be expressed as

$$F_{xy} = (P - F_z)C_a \quad (1)$$

$$F_z = P \quad (2)$$

$$M_y = P\frac{l}{2} \quad (3)$$

where P is container's weight force, C_a is the friction coefficient between the containers' surface and l is the transverse width of the container

The forces and moments that appear in Eqs. (1)–(3) are originated from the peak wind velocity, which takes into account the average wind speed over 10 minutes and the gust effect caused by atmospheric turbulence. The resulting wind loads acting on the container's surface can be expressed by the relationship (Simiu & Scanlan 1986):

$$F_i = \frac{l}{2}\rho\hat{U}^2\bar{C}_{Fi}A \quad (4)$$

where $i = x, y, z$, $\hat{U}$ is the peak wind velocity at the site, $\bar{C}_{Fi}$ is the aerodynamic force coefficients, A is the area of the container. The peak wind velocity at the site can be defined according to the relation as:

$$\hat{U} = \bar{U}(1 + g_U I_U) \quad (5)$$

where I_U is the site turbulence intensity, g_U is the gust coefficient. Expressing the global wind loads on the basis of Eq. (4) and substituting them into Eq. (1)–(3), the critical velocity can be obtained in explicit form. As it can be shown, it depends on the wind direction, turbulence intensity, aerodynamic force coefficients and friction coefficients of the piled containers. These two last families of parameters are strongly uncertain, as they are not furnished in literature for this kind of structure. Therefore, a wide experimental campaign has been developed in order to characterize statistically these parameters.

2.1 *Static friction coefficient*

The static friction coefficient between two stacked containers is a key parameter for assessing the stability of the container; however, the data provided in the literature present a considerable dispersion, inducing an important uncertainty in the analyses.

Therefore, an on-site experimental campaign has been carried out to experimentally measure the friction coefficient between stacked containers. Figure 2 represents a scheme of the set-up test.

It consists in – six full containers placed at the base, a full container with the role of matching part for the force actuator, an empty container, object of the test, connected to the force actuators by means of ropes. Two ropes are anchored to the lower end of the corner blocks (Figure 3), by two symmetrical metal restraints, and connected to a centered galvanized metallic shackle with a diameter of 32 mm, from which a ratchet strap provides a preload to the rope system. The force transmission system is then connected, by means of a second belt, to the hydraulic jack, placed behind the full container, by means of a system designed and built by the DICCA laboratory. The hydraulic jack, with a maximum capacity of 20 kN, is connected to load cells with a maximum load capacity of 25 kN.

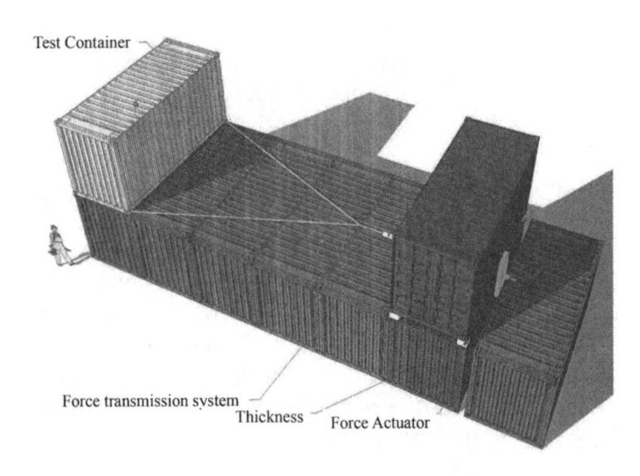

Figure 2. Scheme of experimental test for the measurement of the friction coefficient.

Figure 3. Photographs of the experimental test for the measurement of the friction coefficient: force transmission system.

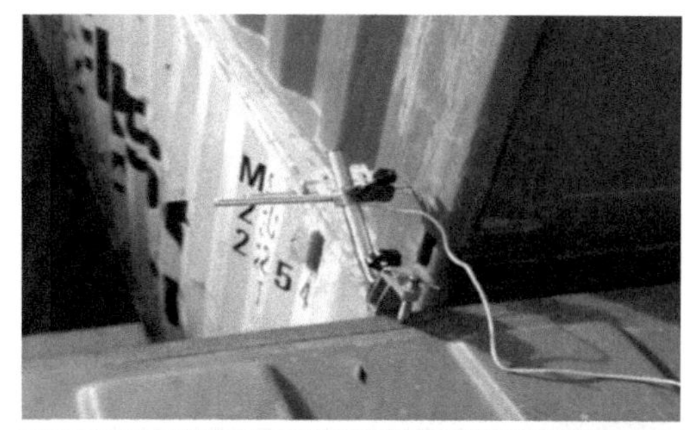

Figure 4. Photographs of the experimental test for: a) measure of displacement through LVDT b) hydraulic jack, load cell and acquisition system.

Figure 5. Test operations in order to obtain wet surface conditions on the test container.

On the lateral faces of the test container, two LVDT displacement sensors were positioned at the lower corner blocks (Figure 4a). The load signals, transmitted by the load cell, and displacement signals, transmitted by the LVDTs, were acquired simultaneously by means of a PC acquisition system placed of the recording station (Figure 4b). Tests were carried out on 5 different containers, characterized by different wear and surface conditions.

Tests were repeated on each container, varying the position of the corner blocks (precision and staggered support) and the surface conditions: wet or dry (Figure 5). For each test, the displacement at the two ends of the container and the force impressed were measured simultaneously with the variation of time. During the first test, a sampling frequency of 20 Hz was used, while for the subsequent a sampling frequency of 1000 Hz was used. In post-processing, the signals were filtered by means of a moving average operation in order to purify the signal from environmental fluctuations.

Figure 6 shows, as example, the results of the measurements carried out on the first test container, in conditions of wet surfaces. The diagrams show moving average signals. Figure 6a shows the time-history of the force measured by the load cell. The diagram shows a first part with irregular growth, which corresponds to the manual application of the pre-tension to the rope system; a second section with regular growth, corresponding to the implementation of the hydraulic jack pull; a third part with a sudden decrease of the force, corresponding to the overcoming of the static friction and the sliding of the container. Figure 6b shows the time-histories of the displacement measurements carried out at both ends of the container. In this case, the two diagrams are almost perfectly coincident, indicating a pure rigid translation of the container. A first portion of zero displacement is detected, followed by a sudden increase in the displacement due to the sliding of the container

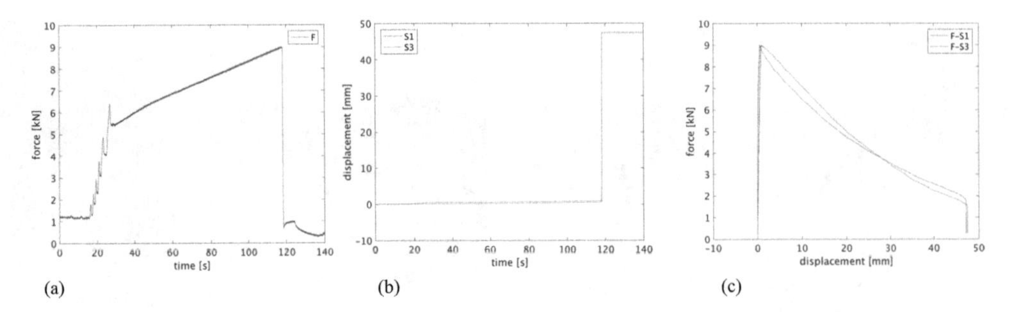

Figure 6. Diagrams of the recordings made on the container 1, wet surface condition: a) recording of force as time changes; b) recording of movements as time changes; c) force-displacement diagram.

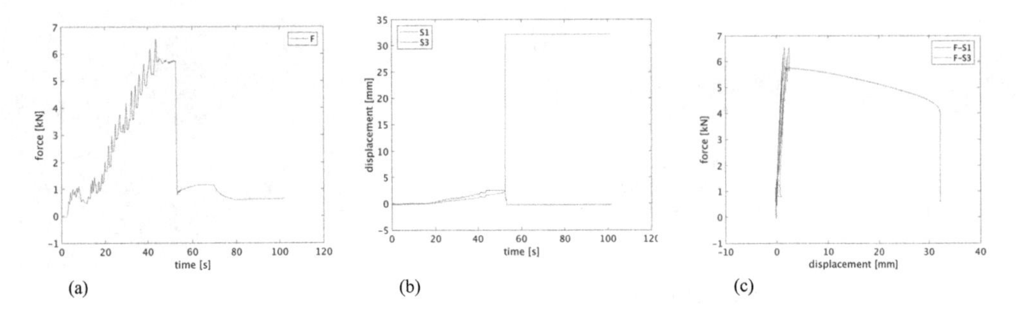

Figure 7. Diagrams of the discarded recordings made on the container 1 roto-translation case with wet surface condition: a) recording of force as time changes; b) recording of movements as time changes; c) force-displacement diagram.

to overcome static friction. The starting moment of the translation coincides with the one of a fall in force, shown in Figure 5a. Figure 5c shows the force-displacement diagram obtained from the simultaneous recordings. We note a first stretch of force displacement with zero displacement, a second section of decrease of force and increase of displacement. Not all the tests gave a clear result as in the case shown in Figure 6.

In some cases, roto-translations of the container have been highlighted, due to positions not perfectly centered by the container itself with respect to the force actuator. These conditions are revealed by a different trend ofthe displacement measures at the two ends of the container (Figure 7b). The diagrams of the records were carefully analyzed and, in the case of excessive rotations, the test was discarded

From these data it is possible to identify the force corresponding to the overcoming of the static friction and derive the value of the static friction coefficient by means of the relation:

$$C_a = \frac{F_{\max}}{P} \tag{6}$$

where F_{max} is the force corresponding to the sliding of the test container, P is the weight of the test container. In the overall, 40 tests were considered valid; Figure 8 shows the histogram of the friction coefficient values evaluated by the tests.

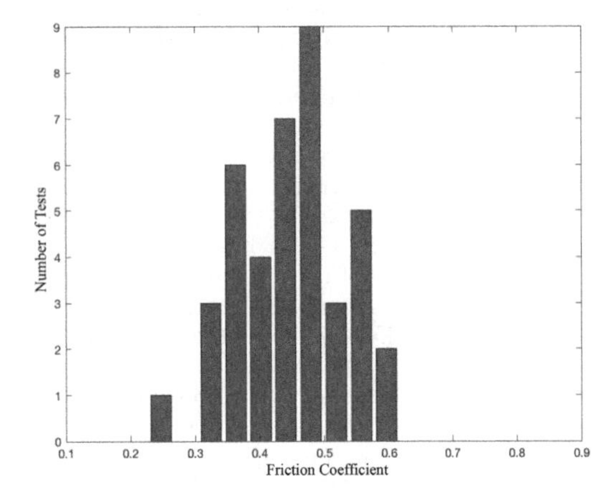

Figure 8. Histogram of the values of the friction coefficient.

2.2 *Aerodynamic coefficient*

The value of the container aerodynamic coefficients are difficult to assess using current literature and standards, given the peculiarities of the shape and the surface of the container, and the variability of the configurations that the stacked containers can take. The aerodynamic coefficients have been studied by means of wind tunnel test on 1:50 scaled models. The Aerodynamic tests, aimed at evaluating the pressure coefficients on the external surface of the scaled container model, have been performed in the wind tunnel (WT) of the Department of Civil, Chemical and Environmental Engineering (DICCA).

The wind tunnel has two different test sections, with cross section of 1.7×1.35 m, (width $\times$ height). The wind tunnel test concerned the 40' high cube container, the characteristics are shown in Table 1. The model is equipped with 225 external pressure taps (Figure 9b): 15 are placed on each of the lateral smaller faces B and D, 65 are placed on each of lateral faces R and L, 65 are placed on the top face T (Figure 10).

Seven set-up conditions have been reproduced, considering seven different container storage configurations (Figure 11) which correspond to different stacking methods on other containers of the same properties. The isolated pile of containers is the standard reference case, which allows a qualitative comparison. For each configuration the instrumented container was placed in 4th and 5th tier, always considering it as the top container of the pile and removing the lower row of the configuration.

Two different wind profiles were considered for the wind tunnel test, corresponding to a less intense but more turbulent (*Urban wind*) and a more intense but less turbulent (*Sea wind*). The wind characteristics were reproduced in the wind tunnel through the insertion of passive devices at the entrance and on the floor of the test chamber (Figure 9a). The container symmetry allows to repeat the pressure measurements just for 9 wind directions, starting from 0° up to 180° with steps of 22.5° (0°, 22.5°, 45°, 67.5°, 90°, 112.5°, 135°,

Table 1. Characteristics of the examined containers.

40' high cube box dimensions	
Width b (m)	12,192
Height h (m)	2.896
Depth d (m)	2.438

	L (left)	
D (door)	T (top)	B (back)
	R (right)	

Figure 10. Scheme and denomination of the sides of the container.

(a)

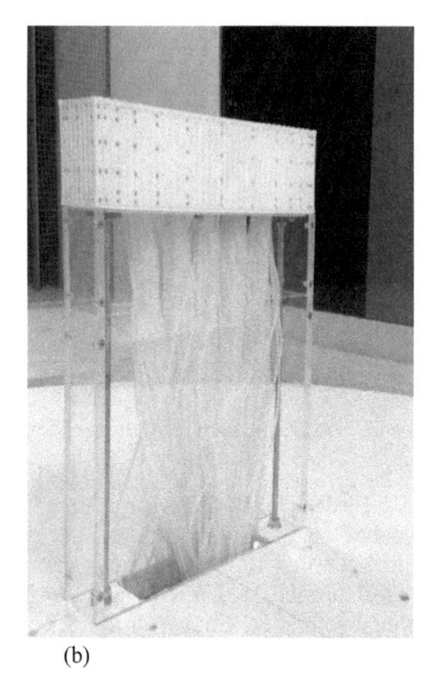

(b)

Figure 9. a) Test chamber set-up; b) Instrumented container model.

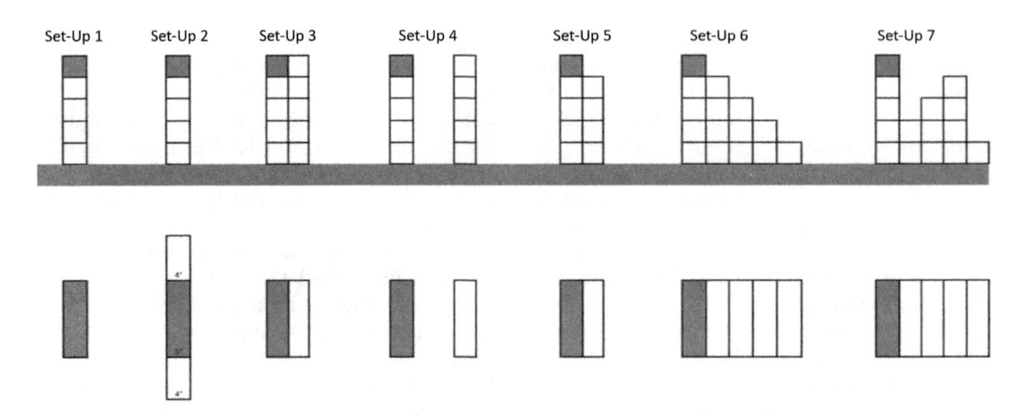

Figure 11. Test set-up in 5th tier, the instrumented container is shown in blue.

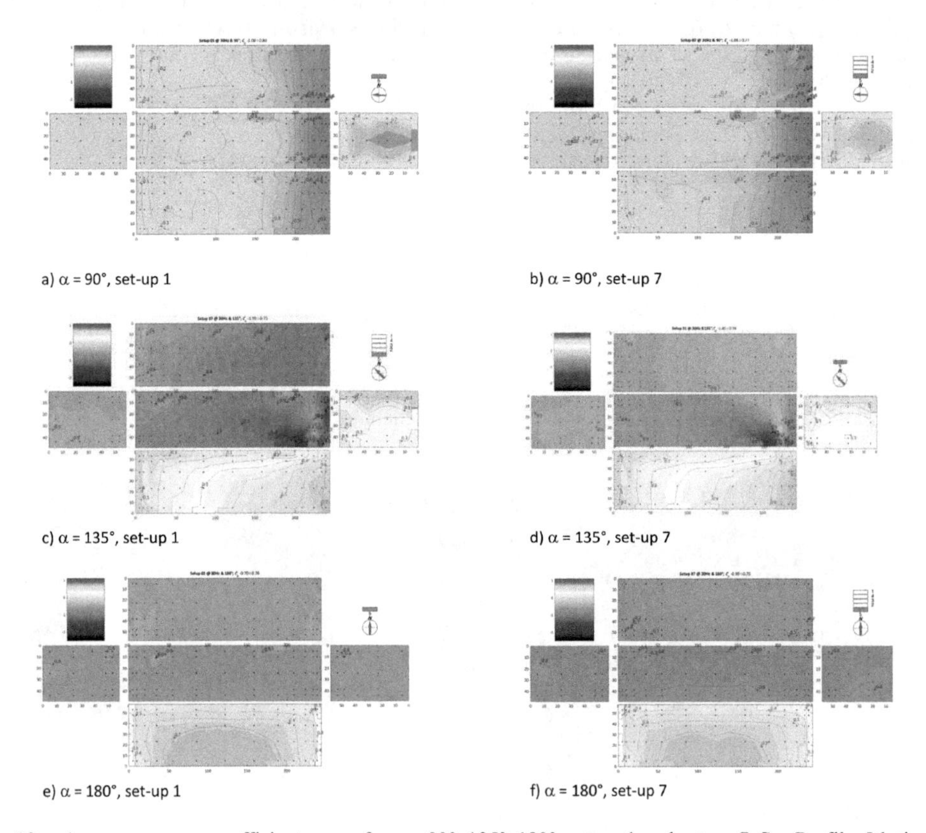

Figure 12. Average pressure coefficients map, for $\alpha = 90°$, 135°, 180°, set-up 1 and set-up 7, Sea Profile, 5th tier.

157.5°, 180°). The 0° direction represents the incident wind on the longer side of the container. It should be noted that the directional reference in the wind tunnel does not coincide with the geographical reference. Each test condition was analyzed in steady state, for a measurement period of 120 s (wind tunnel scale). The tests have been carried out for all the set-ups and for all the angles with the *Sea profile* and containers in the 5th shot. After a post-processing phase, identified the most penalizing cases, these were repeated with containers in the 4th tier both with *Sea profile and Urban profile*. The pressure coefficient c_p (Figure 12), at the point x of the test body surface at time t, is defined (Holmes 2001, Simiu & Scanlan 1986) as the dimensionless value:

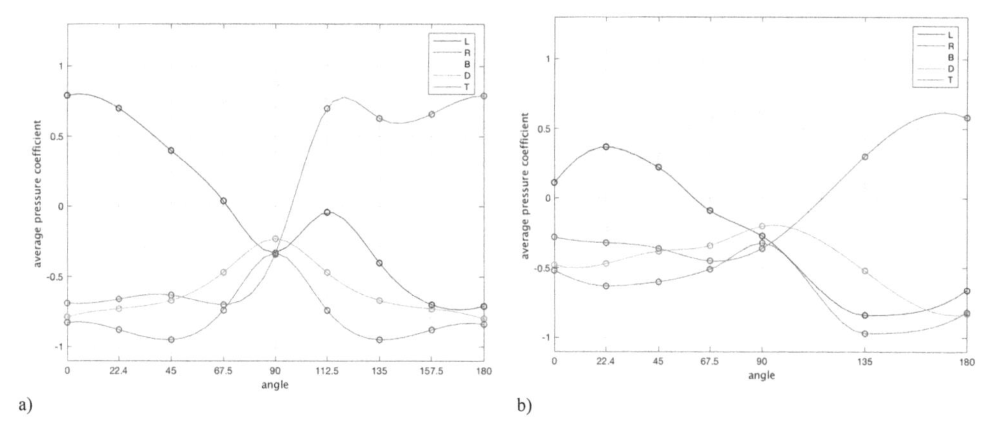

Figure 13. Average pressure coefficient on the face δ = L, R, B, D, T, Sea profile, 5th tier, set-up 1 (a) and set-up 7 (b).

$$c_p(x,t) = \frac{p(x,t) - p_0}{\frac{1}{2}\rho U^2} \tag{7}$$

where $p(x, t)$ is the pressure measured at point x at time t, p_0 is the static pressure in the test chamber, ρ the air density based on local test chamber conditions and U the undisturbed average velocity at the reference height found from the undisturbed artificial boundary layer used for the test. By averaging over time the pressure coefficient the average pressure map is obtained for each test configuration (Figure 13). By averaging these values over pressure taps mesh, the average pressure coefficient on the container faces are obtained.

3 CONCLUSIONS

This paper presents the results of two experimental campaigns aimed at quantifying the friction coefficient and the aerodynamic coefficients of piled containers, respectively. The friction coefficient has been characterized statistically on the basis of 40 measured values. On-site sliding tests were carried out on 5 different containers, characterized by different wear and surface conditions, precision and staggered support, in order to catch the variability of operational conditions. The aerodynamic coefficients has been studied by means of wind tunnel test on scaled models, reproducing seven different container storage configurations, two different incoming wind profiles and two different containers tier, obtaining as many aerodynamic coefficient sets of values. The coefficients presented in this work are general and valid for any terminal location. They represents key parameters to obtain containers critical wind velocity values. Future work will develop the analytical formulation of the critical wind velocity, and will derive its probability distribution, considering the variability of the input parameters and taking into account local turbulence conditions and the actual orientation of the containers at the site.

REFERENCES

Repetto, M.P., Burlando, M., Solari, G., De Gaetano, P., Pizzo, M. (2017). "Integrated tools for improving the resilience of seaports under extreme wind events", Sust. Cities Soc., 32, pp. 277–294.

Repetto, M.P., Burlando, M., Solari, G., De Gaetano, P., Pizzo, M., Tizzi, M. (2018)," A web-based GIS platform for the safe management and risk assessment of complex structural and infrastructural systems exposed to wind.", Adv. Eng. Soft., 117, pp. 29–45.

Simiu, E., & Scanlan, R. H. (1986). "Wind effects on structures", an introduction to wind engineering. John Wiley.

Solari, G., Repetto, M.P., Burlando, M., De Gaetano, P., Pizzo, M., Tizzi, M., Parodi, M. (2012). The wind forecast for safety management of port areas. J. Wind Eng. Ind. Aerodyn. 104–106, pp. 266–277.

Zhen, W., Xingqian, P., & Chunhui, Z. (2011). "Overturning analysis of Harbor Containers based on wind tunnel test of rigid models", In *Electric Technology and Civil Engineering (ICETCE), 2011 International Conference on* (pp. 544–548). IEEE.

Progress in Maritime Technology and Engineering – Guedes Soares & Santos (Eds)
© 2018 Taylor & Francis Group, London, ISBN 978-1-138-58539-3

Tool for initial hull structure dimensioning at ship concept design

F. Sisci & M. Ventura
Centre for Marine Technology and Ocean Engineering (CENTEC), Instituto Superior Técnico, Universidade de Lisboa, Lisbon, Portugal

ABSTRACT During the ship concept design, the hull weight estimates are generally based on top-down methods such as empirical formulas resulting from statistical analysis of data from existing ships and using as variables the main dimensions and some hull form coefficients. These formulas are generally outdated, and do not reflect the actual configuration of the hull structure. In this work, a tool for the initial dimensioning of the hull structure by class rules was developed as an additional module of a ship numeric model for the ship concept design. A non-linear constrained optimization procedure is used to obtain the scantlings of plates and stiffeners that minimize a measure of the structural weight per unit length. The methodology adopted and the data models are presented as well as the results of a case study application.

1 INTRODUCTION

During the concept stage of ship design, due to the reduced amount of information available, a number of tasks are carried out in a simplified way and using mostly data estimated from empirical methods. This is the case of the hull structure weight that is generally estimated by some empirical expression as a function of the ship's length, breadth, depth and block coefficient (Schneekluth and Bertram, 1998). In merchant ships, the hull weight accounts for about 70% of the lightship weight, and in ships with the engine room located aft, the weight of the cargo area can sum up to 70% of the total hull weight.

Taking this into consideration, is developed a concept that allows to improve the estimation of the weight of the cargo area by producing a midship section configuration (considering plates and stiffeners) and estimating the total weight as function of it. In order to achieve this, many considerations and limitations have to be taken into account, as the produced structural configuration at midship shall be realistic (e.g. few standard elements), feasible (considered economically and for production) and compliant with the rule sets defined by the classification societies.

In this work is developed a tool for the initial dimensioning of the scantlings in the midship section of a steel hull, which, through an optimization procedure, allows to comply with specified requirements. Although in the tool this optimization procedure is applied exclusively on longitudinal elements, in a second stage also the transversal elements are taken into consideration, as their specifications depend on the main dimensions and their effect over the total structural weight can't be neglected to achieve an accurate estimation.

Objective of the presented tool is to define a feasible structural configuration of the considered section based on few initial data (mostly main dimensions and ship design parameters), which in a quick but still accurate way, estimates the section modulus and the weight per unit length of the cargo area with more precision as using only empirical methods. Thus, the challenge is to provide a procedure that allows to define the structural configuration of a cross-section in the early stage of a conceptual design when main dimensions are known, and the structural arrangement has to be estimated. Having as input the main dimensions, the task is to create a cross-section typical for the specified ship type, defining a feasible plates and stiffeners arrangement that fulfills the requirements imposed by the classification societies over the local and global minimum section modulus and on the minimum thickness of the structural elements.

A number of parameters allows the user to define the profile type of the stiffeners, the maximum width of plates, height of double bottom and width of the side tanks. Moreover, the algorithm used considers the midship section formed by stiffened panels, structural elements that define a number of plates and stiffeners with constant dimensions, material, stiffener's profile and spacing. This concept allows to choose in a flexible way the most convenient arrangement of structural components, but still being realistic regarding the variety of the elements.

The concept consents to dimension the structural configuration based on the typical arrangement for the considered ship type with symmetric midship section. In the presented tool the case study is a multi-purpose/container vessel of small dimensions (feeder ship).

2 STATE OF THE ART

Over the years the structure of a vessel has been analyzed with the objective to get more precise estimations and eventually to propose improvements to the ship design process by optimizing the total weight or costs of the vessel.

Rigo (2001a, 2001b, 2003) presented an optimization procedure for the ship hull to be used in the preliminary design, in which the scantlings are computed based on first principles and the minimizations of the weight and of the normalized building cost were used as objectives, either individually or combined. The design variables are the thicknesses of the plates. For the longitudinal stiffeners and the transverse frames the variables are the spacing, the height and thickness of the web and the width of the flange.

Andrews (2006) applied the computer-aided environment on the preliminary ship design stage on single structural modules, in order to analyze and compare outputs deriving from different solutions, being in this way possible to evaluate different solutions and produce a more detailed preliminary design.

Ehlers et al. (2010) presented an optimization procedure for the concept design of the structure of a chemical product tanker. Only the longitudinal structure is considered for the ship weight, hull production cost evaluation and its fatigue life. The analyzed tanker structure is composed by one corrugated bulkhead and 22 stiffened panels. The bulkhead is defined by the corrugation geometry and the plate thickness. Each of the stiffened panels is defined by three design variables: plate thickness, type and number of stiffeners.

Sun and Wang (2012) applied on the midship section of a VLCC a modelling procedure combined with a GA-based (Genetic Algorithm) optimization procedure, which has a set of 35 design variables relative to the plates thickness and beams dimensions. As the design variables are related exclusively to longitudinal elements, the chosen objective function is the minimization of the mass of all longitudinal structures in the central body of the VLCC.

Ma et al. (2014) also aimed the optimization of the scantlings in the cross section by applying the finite element method on every plate of the structure, analyzed individually. It was adopted a GA-based multi-objective optimization to obtain the minimum weight and costs, and the maximum value for the safety measure.

Andric et al. (2017) proposed a multi-objective scantling optimization applied on the structure of a cruise vessel with the aim to minimize the structural weight and the vertical center of gravity, having as design variable the scantling of the plates. The set of optimal solutions has been further analyzed with a Pareto frontier.

Gaspar et al. (2012) decompose the structure in subcomponents with defined characteristics, to reduce the complexity of the problem and at the same time producing information about the structural subcategory, which is more detailed and accurate. Similarly, the structural module taken into consideration for this tool is the midship section, that is the most relevant section in order to estimate weights, structural strength and costs of the total vessel.

The previous work done shows how through different methods is aimed the optimization of weights and costs through the modification of selected design variables. While Rigo (2001a) optimizes weights and costs adopting as design variables the plate thickness, number of stiffeners, stiffeners' profile and spacing, Ehlers et al. (2010) define a model where the spacing is not taken into consideration, but the variables are associated to one stiffened panel in order to get minimum structure weight, cost and fatigue life. Ehlers (2010) applies the same procedure over the side structure of the hull in order to define the most convenient structural configuration in function of the crashworthiness, having as design variables the plate thickness, stiffener types and spacing.

Being the objective of the presented approach to define initial hull structure scantlings through an optimization process, rather than the optimization of such scantlings, it is aimed just the minimization of the weights of the structure (that is important to increase service speed, cargo carried or to minimize the building costs), having as design variables the thickness of the plates in the deck, from which depend respective stiffeners dimensions and spacing of longitudinal and transversal elements.

In recent years the interest over a software that could support the engineer in the design process has grown, and some of the classification societies have developed software tools able to perform different levels of strength assessment based on a variable number of inputs. For example, Lloyd's Register and Bureau Veritas have developed respectively the software "RulesCalc" and "MARS2000", which are able to perform an analysis over a structure given by the user and check the rules compliance. In addition to rules compliance checking, the "Leonardo" system by RINA implements a FE analysis of the given structure. DNV-GL developed the "Poseidon" system, which also allows to perform an automatic initial scantling determination.

3 MIDSHIP SECTION LONGITUDINAL STRUCTURE

In this work the hull structure is represented by 2D representations of cross sections in the cargo area. The procedure developed is a compromise between accuracy and the level of detail. A number of simplifications were assumed to reduce the computational time so that it can be included in the ship optimization process. The geometry of the hull molded lines is described by a number of cross sections represented by polylines. The scantlings of plates and stiffeners are determined based on the requirements of a classification society (DNV GL, 2016; DNV GL, 2016a).

Main objective of the work is to define the structural configuration and the initial dimensioning of the scantlings in the selected section and to perform over these a strength, economic and reliability analysis. In the algorithm presented in Figure 1 are used as inputs the vessel's main dimensions, ship type and, if specified by the user, hull compartment parameters and specific shipyard's constraints.

3.1 *Methodology*

Firstly, the geometry of the section and the structural configuration are defined in accordance with the global parameters associated with the specific ship type Figure 1. The dimensions that are used as input in the tool are ship's main dimensions (L_{pp}, B, D, T, C_b and C_m), structural specifications (maximum plate width and minimum structural thickness) and the hull compartment parameters (double bottom height and side tanks width), which can be defined by the user out of a defined range of dimensions depending on the ship type.

The continuous longitudinal material is then decomposed in stiffened panels (*Bottom, Bilge, Side, Deck, DoubleBottom, BottomGirders, LongitudinalBulkhead, SideGirders* and *Hatch*) with defined plates and stiffeners.

For this purpose, for each panel is specified a *PlateSet* and a *StiffenersSet*. A *PlateSet* is an array of groups of identical plates, specifying for each of them quantity, width and thickness. A *StiffenerSet* is an array of identical stiffeners belonging to the same plate, specifying for each a quantity, the profile section type, the orientation, the dimensions and the spacing.

In order to define the scantling for each of the plate sets, rules of the DNV-GL classification society are taken into consideration, which constraint the minimum thickness based on two methods. The chosen thickness is the highest of the values resulting from the two methods. Other rules of the same classification society are also applied to calculate the local minimum section modulus respective to

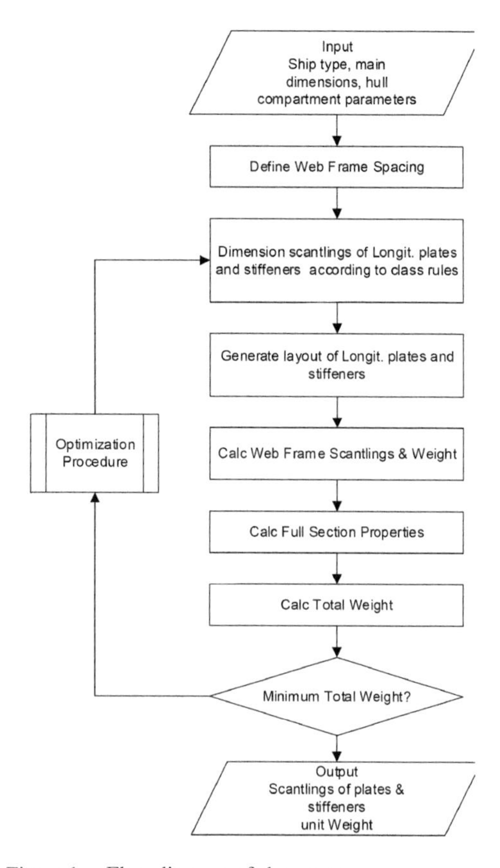

Figure 1. Flow diagram of the process.

a specific stiffener and its minimum web thickness (that takes also into account the ratio between web of stiffeners and thickness of the plate at which the stiffener is attached to). In this way, having defined as default the profile section type of the stiffeners for each of the stiffened panels, it is possible to consult the embedded catalog of standardized stiffener profiles and chose the one with the lowest sectional area that complies with minimum local section modulus and minimum web thickness.

Rules of the classification society are also applied on the whole structure, to define the associated wave coefficient (C_w) and the minimum section modulus relative to deck and to bottom, which will be used in the optimization tool as constraint.

As the midship section structure needs to be computed geometrically in order to be evaluated, polylines describing the shape are assigned to the stiffened panels. Both the elements stored in the stiffened panel (plates and stiffeners) are in this way associated to a position. With this process it is possible to obtain the complete structure layout. The number of stiffeners associated to each panel is obtained by dividing the available space by the stiffeners spacing, similar approach adopted to get the number of vertical girders in the double bottom and of horizontal girders in the side tanks, where the available space is divided by the girder spacing, in this case chosen to be three times the stiffener spacing.

Stiffeners and girders are initially created independently from each other, so that these elements can be equally distributed along the structure. In a second stage the position of the stiffeners is checked again, to erase the stiffeners that are too close to a girder or to another stiffener, as it would be impossible to weld these (chosen tolerance for this purpose is 0.1 m). Similarly, also plates dimensions need to be constrained, by avoiding designing plates with width smaller than 1 m (with exception of the sheer strake) and bigger than the maximum width imposed at the beginning of the tool (2.5 m, with exception of the bilge plate), with a thickness not smaller than 7 mm, imposed as the minimum feasible scantling for strength requirements.

In the horizontal stiffened panels *(Bottom, DoubleBottom, Deck)* the stiffeners are located starting from the longitudinal center plane to the sides in order to maintain the vertical alignment even when the hull shape changes.

Being at this point the sectional structure fully described, it is possible to compute geometrical properties like the sectional area, moments of inertia, neutral axis and resulting section modulus. Having sectional area and material properties, the longitudinal contribution to the weight per unit length of the parallel body of the vessel can be calculated.

The produced midship section is limited to symmetric shapes, as just half of the section is designed and mirrored to create the complete section. This tool is designed with the objective to analyze multipurpose/container vessels with typical dimensions of a feeder ship.

3.2 *Data model*

The methodology described above was implemented in an object-oriented code whose main classes are presented in Figure 2.

The central class *CrossSection*, the basis of the data structure, contains the description of a single selected section of the vessel, which is associated to ship characteristics through the class *Ship* (main dimensions, type, etc.) and for which rules of the classification society are computed in *ClassRules*, in order to establish constraints regarding the design of the structure.

The class of *StiffenedPanel* consists in the description of panels that form the whole cross-section of the vessel. A *StiffenedPanel* is composed by a set of plates welded side by side and reinforced by stiffeners. It is assumed that within a panel the thickness of the plates and the profile of the stiffeners are constant, reason for which the classes *PlateSet* and *StiffenerSet* are adopted, in order to store the qualitative description of the element type relative to a panel. The *PlateSet* stores information about the quantity, width, thickness and material of a group of plates, the *StiffenerSet* class has information about the quantity, section profile type and dimensions, spacing, web and flange direction and material of stiffeners associated to a panel.

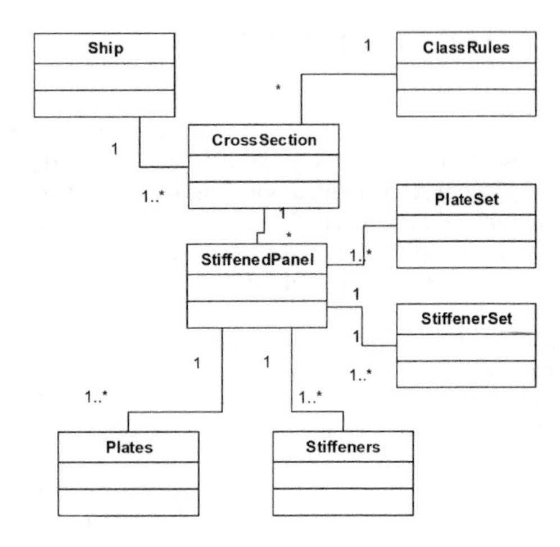

Figure 2. Main data classes developed for the procedure.

While the classes *PlateSet* and *StiffenerSet* are associated to a group of elements, the classes *Plates* and *Stiffeners* are the actual description of these elements, being associated individually to each of these.

After that the required information is stored in the class CrossSection, it is possible to calculate the weight per unit length based on the analyzed section, being given as objective function of the tool.

4 TRANSVERSE STRUCTURE

In this work the transverse structure is not optimized, being dimensioned in accordance to the classification society rules, in order to complement the longitudinal weight and obtain in this way the total unit weight estimate.

A single typical web frame in the midship section area is dimensioned both for the double-bottom and the side-shell regions. The weight of the complete frame is then computed and divided by the web frame spacing as a contribution to the hull weight per unit length.

The dimensioning of the typical transverse bulkhead is not yet considered in this work. The contribution of transversal material to the total weight is obtained by multiplying the total number of bulkheads in the cargo area by the individual weight which is computed by multiplying the cross-sectional area by an empirical index in ton/m^2.

5 OPTIMIZATION PROCEDURE

In this tool the optimization process is adopted with the aim to find a feasible solution for the initial dimensioning of the stiffened panels in the midship cross section. This solution is not to be intended as the best structural configuration for the given section as the optimization algorithm is applied exclusively to guarantee the compliance with selected requirements, as described in section 5.2.

The optimization procedure adopted is based on the genetic algorithm, but the convergence criteria has not been applied as, as said, objective of the tool is to produce feasible outputs.

5.1 *Design variables*

The design variables for the model of the multi-purpose/container vessel are the thicknesses of the plates in the panels in the upper part of the section, as being more distant from the neutral axis have a more critical and sensitive role in the compliance to the overall minimum section modulus imposed by the classification societies. Indeed, the upper part of the structure in such type of vessels presents a section modulus that is lower than the one in the lower part, as the material in there is less and thus the neutral axis is lower. The analyzed thicknesses are identified with the upper longitudinal bulkhead panel and the deck panel. The selection of the design variables is limited to the upper panels as these have the biggest influence over the section modulus. Moreover, the number of these is chosen to be low as, having more design variables, would lead to higher computational time without improving the output quality.

The design variables are assumed to have an initial value equal to the minimum local scantling defined by the classification societies, also considered to be the lower bound of the variable during the optimization process.

5.2 *Constraints*

The bounds applied over the design variables used for the single-objective genetic algorithm, as described in the previous chapter, are the minimum thickness imposed by the classification society as lower bound and the double of such value as upper bound:

$$t_{CSR} \leq t_i \leq 2 * t_{CSR}$$

where t_{CSR} is the minimum thickness for a specified plate and t_i is the actual plate thickness.

As the algorithm adopted for the optimization process does not deal with discrete values and thus it is not possible to create a catalog of available plate thicknesses, the obtained feasible distribution of the scantlings has been rounded up to the next integer number. This process shall be improved in a second moment by adopting a different algorithm, which enables to select the plate scantling from a given catalog specific to the shipyard.

Constraint applied over the optimization process is the section modulus at deck and bottom that must be bigger than the minimum section modulus defined by the classification societies:

$$W_{bottom} \geq W_{min,CSR}$$
$$W_{deck} \geq W_{min,CSR}$$

where W_{bottom} and W_{deck} are the calculated section moduli relative to the bottom and deck, and $W_{min,CSR}$ is the minimum section modulus imposed by the classification societies.

Other types of constraints applied in the tool are indirect, as they are not part of the optimization process, but the algorithm of the tool constraints the possible range of application. This is the case of stiffener selection, which is done by computing the required local section modulus as given by the classification societies and choosing out of a catalog the stiffener with the lowest sectional area that complies with such requirement. Another constraint considered for the selection of stiffeners is the minimum web thickness, which must be higher than 90% of the scantling of the plate to which the stiffener is attached. Output from the stiffener selection is the stiffener type and its web and, eventually, flange dimensions.

5.3 *Objective function*

Although Winkle & Baird (1986) investigated the fabrication cost of stiffened ship structures and came to the conclusion that relative cost optima rarely bear any relation to traditional minimum weight criteria, being the aim of the tool to find a quick and feasible solution, it is enough to minimize the weight per unit length. Moreover, the objective of the presented algorithm is to estimate initial dimensions for the structural configuration through an iteration procedure defined by the optimization process, rather than to find the minimum for weights or costs. The total unit weight is calculated as follows:

$$W_{tot} = A_{section}\, \gamma_{steel} + (V_{web}\, \gamma_{steel}) / S_{web}$$

where W_{tot} is the total unit weight in [t/m], $A_{section}$ is the area of the longitudinal structural components, γ_{steel} is the specific weight of steel assumed to be 7.85 [t/m³], V_{web} is the volume of the transverse structural elements in one frame and S_{web} is the web frame spacing.

In order to keep the initial dimensioning process fast, the total weight is split into longitudinal and transversal components, where the first depends on the optimization process, and the second on the stiffeners spacing, which allows the program to be fast, at the expense of the level of detail of the transverse components.

5.4 *Optimization algorithm*

A non-linear constrained single-objective algorithm is adopted. The algorithm aims the minimization of the objective function, the weight per unit length, by increasing or decreasing the variables associated to the scantling of the plates in the upper part of the external side shell, in the upper part of the longitudinal bulkhead, used as double side shell, and in the deck.

6 RESULTS AND VALIDATION

The design of a small multi-purpose ship is adopted as a case study for the developed procedure.

Main dimensions of the considered vessel are shown in Table 1.

In addition to these values, the double bottom height is assumed to be 1.2 m, the side tank width to be 1.6 m, the maximum plate width to be 2.5 m, to comply with requirements of the shipyard, and the overall minimum plate thickness is assumed to be 7 mm.

For this case are chosen the flat bar stiffeners in the bottom, the double bottom and girders' panels, while in the side, longitudinal bulkhead and deck panels are adopted stiffeners with the bulb profile.

The produced midship section is shown in Figure 3 and consists of a set of panels with associated plates and stiffeners. The section before the optimization presents values and characteristics as shown in Table.

As it can be seen in the previous table, the values produced from the given initial dimensions deliver a section modulus at the bottom that does not comply with the requirements imposed on the minimum section modulus by the classification society.

The tool, in order to verify the imposed constraints on the section modulus, modifies the scantling of the plates in the mentioned panels until all the requirements are fulfilled. After the optimization process is concluded, the constraints are verified and a local minimum for the weight is found, the tool presents as output the values listed in Table 2.

The final arrangement of the midship section consists of a total of 22 plates (with scantling range between 7 mm and 18 mm) and 45 longitudinal stiffeners, subdivided in 24 with a flat bar profile and 21 with bulb profile.

Table 1. Main dimensions of the considered example.

	Value	Unit
L_{pp}	115.0	m
B	20.00	m
D	10.00	m
T	8.30	m
C_b	0.72	
C_m	0.99	

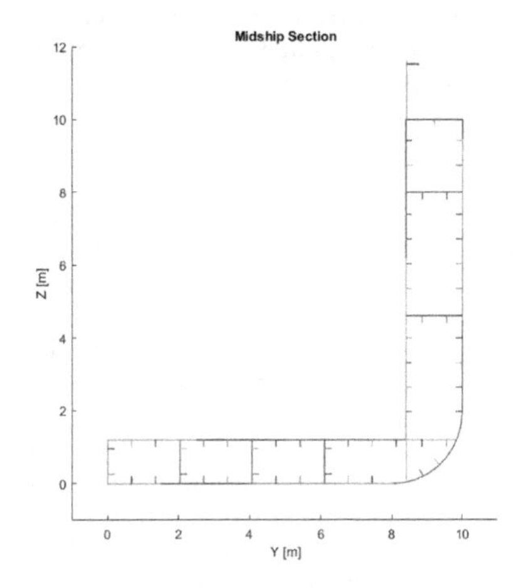

Figure 3. Produced midship section.

Table 2. Output values before optimization.

Definition	Value	Unit
Total sect. area	1.1	m²
Neutral axis	3.16	m
Unit weight	9.0	t/m
Sect. modulus deck	4.534	m³
Sect. modulus bottom	2.095	m³
Min. sect. modulus	3.09	m³

Table 3. Output value after optimization.

_Definition	Value	Unit
Total sect. area	1.3	m²
Neutral axis	3.83	m
Unit weight	10.3	t/m
Sect. modulus deck	4.973	m³
Sect. modulus bottom	3.09	m³
Min. sect. modulus	3.09	m³

7 CONCLUSIONS

The concepts and the methodology developed for the task of the initial structural dimensioning of the hull were presented. This methodology was implemented as a computational tool to be used as a new module of a system for the determination of the ship main characteristics at the early concept design stage.

This concept has been driven by the need of a practical and fast algorithm to provide an initial structural arrangement of the cross-section, to analyze the influence of different design factors on ship design, to allow the user to check what is the effect of different structural configurations on the unit weight and to check its feasibility. Different structural configurations can be obtained by changing stiffeners spacing and profile, double bottom height, width of side tanks, number of bottom and side girders. Main challenge of this procedure is to define an algorithm which estimates the initial structural arrangement of the typical cross-section and is adaptable to different ship types and ship dimensions.

The work presents a method to design a feasible structure configuration, where the scantlings in the defined section are obtained through an optimization procedure. The single-objective optimization procedure is carried minimizing the weight, having as design variables the scantlings of the panels in the upper part of the section (deck and upper plates of side and bulkhead), which allows the program to be fast and to present a feasible arrangement that complies with classification societies rules. The algorithm is able to select the stiffener dimensions out of an imbedded catalog with standard elements, to ensure that variety of components is limited. Same procedure shall be implemented in the future over the plate thickness, with an imbedded catalog of available standard thicknesses, which, up to this point, has been approximated by rounding up the value obtained from the initial dimensioning.

Profile type of the stiffeners, maximum width of plates, height of double bottom and width of the side tanks can be chosen by the user, in order to create a case-specific model that takes into account the shipyard specifications and shipowner preferences, besides the structural configuration specific to the ship type. Moreover, the section is formed by panels, which are structural elements grouping plates and stiffeners with constant dimensions, material, stiffener's type and spacing. This tool structure allows a flexible selection of the most convenient arrangement of structural components, but still being realistic regarding the variety of the elements, as the properties are associated to the stiffened panels.

Although this process has general applicability independent from ship type or dimensions, in this work is shown the case applied on a feeder multipurpose/container vessel.

ACKNOWLEDGEMENTS

This paper reports a work developed in the project "Ship Lifecycle Software Solutions", (SHIPLYS), which was partially financed by the European Union through the Contract No 690770 – SHIPLYS – H2020-MG-2014–2015.

REFERENCES

Andrews, D.J. (2006). Simulation and the design building block approach in the design of ships and other complex systems. Proceedings of the Royal Society A. Vol. 462, No. 2075.

Andric, J.; Prebeg, P. and Stipic, T. (2017). Multi-objective scantling optimization of a passenger ship structure.

DNV-GL (2016). Rules for Classification of Ships, Part 3 Hull, Chapter 5 Hull girder strength.

DNV-GL (2016a). Rules for Classification of Ships, Part 3 Hull, Chapter 6 Hull local scantling.

Ehlers, S.; Remes, H.; Klanac, A. and Naar, H. (2010). A Multi-Objective Optimisation-Based Structural Design Procedure for the Concept Stage – A Chemical Product Tanker Case Study. Ship Technology Research, Vol. 57, No. 3, pp. 182–196.

Ehlers, S. (2010). A procedure to optimize ship side structures for crashworthiness. Proceedings of the Institution of Mechanical Engineers. Journal of Engineering for the Maritime Environment, Vol. 224, Part M.

Gaspar, H.M.; Ross, A.M.; Rhodes, D.H. and Erikstad, S.O. (2012). Handling Complexity Aspects in Conceptual Ship Design. International Marine Design Conference (IMDC), Glasgow, UK, 11–14 June 2012.

Ma, M.; Freimuth, J.; Hays, B. and Danese, N. (2014). Hull Girder Cross Section Structural Design using Ultimate Limit States (ULS) Based Multi-Objective Optimization. COMPIT 2014, Redworth, UK, 12–14 May 2014, pp. 511–520.

Rigo, P. (2001a). A module-oriented tool for optimum design of stiffened structures. Marine Structures, Vol. 14, No. 6, pp. 611–629.

Rigo, P. (2001b). Least-Cost Optimization Oriented Preliminary Design. Journal of Ship Production, Vol. 17, No. 4, pp. 202–215.

Rigo, P. (2003). An Integrated Software for Scantling Optimization and Least Production Cost. Ship Technology Research, 50(4):126–141.

Schneekluth, H.; Bertram, V. (1998). Ship design for efficiency and economy. Butterworth-Heinemann, Oxford.

Sun, L. and Wang, D. (2012). Optimal Structural Design of the Midship of a VLCC Based on the Strategy Integrating SVM and GA. J. Marine Sci. Appl., No. 11, pp. 59–67.

Winkle, I. E. and Baird, D. (1986). Towards more effective structural design through synthesis and optimisation of relative fabrication costs. Transactions of RINA, Vol. 128, pp. 313–336.

Progress in Maritime Technology and Engineering – Guedes Soares & Santos (Eds)
© 2018 Taylor & Francis Group, London, ISBN 978-1-138-58539-3

Conceptual design of multipurpose ship and fleet accounting for SME shipyard building limitations

T. Damyanliev & P. Georgiev
Technical University of Varna, Varna, Bulgaria

I. Atanasova
Varna Maritime Ltd., Varna, Bulgaria

Y. Garbatov
Centre for Marine Technology and Ocean Engineering (CENTEC), Instituto Superior Técnico, Universidade de Lisboa, Lisbon, Portugal

ABSTRACT: This work deals with the design and optimization of a new multipurpose ship and fleet that will be built in a small and medium sized shipyard accounting for the constraints related to the shipbuilding limitations. The problem is solved in two stages, where in the first one, based on the cargo flows, the number of ships to be built and speed is defined. At the second stage, the completing of the fleet leads to defining the technical specification for the individual ship as a part of the fleet. A constraint associated to the maximum breadth of the ship is used as an additional limitation in defining the basic characteristics of the ship. The design solutions comply with the shipyard construction and navigation restrictions and the CAPEX and OPEX costs are used as a part of the optimization function employing the specialized software "Expert". Based on the identified design solution several conclusions are derived.

1 INTRODUCTION

The technological development of the small and median enterprises, SMEs in Europe with respect to the economic growth and employment is of a great importance nowadays. According to LeaderSHIP strategy (ec.europe.eu/DocsRoom/documents/10504/) the maritime technology industry is of strategic importance for the EU and in the RDI area one of the objectives is "encouraging open innovation in clusters to enhance participation of maritime technology SMEs in RDI projects and access to RDI results".

In this respect, one of the main goals of the EU Project Shiplys is to respond to the needs of the SME shipyard designers, shipbuilders and shipowners (Bharadwaj et al., 2017) in the development of a ship risk-based design framework accounting for the life cycle cost assessments. The framework deals with the conceptual ship design, risk-based target structural reliability assessment, risk-based maintenance, and fast hull geometry prototyping and shipbuilding management (Garbatov et al., 2017b). This framework will enable the SMEs shipyards to make more reliable estimates, given the client requirements, in the early stages of the inquiry and the existing shipyard's shipbuilding capacity.

The shipbuilding capacity of one specific SME ship repair yard with respect to building new ships was analysed by Atanasova et al. (2018). The main conclusion from the analysis is that it is possible to build new ships of different types with deadweight up to 7,000 tons with a limitation of the ship breadth due to the existing dock capacity. Considering the existing shipyard facilities and implemented shipbuilding technology and equipment, it was concluded that it is possible to build a new multipurpose ship in a shipbuilding period of eight months.

During the shipbuilding process, a limiting factor for the main dimensions of the ship is the width of the dock of 16 m and the docking weight capacity permits building of vessels with restricted breadth up to 6,800 tons.

The initial ship design is normally split into two stages – "Fleet Composition" and "Conceptual Design", "external" and "internal" design tasks. The expedience of jointly tackling the two tasks was originally formulated by Gallin (1973) and Pashin (1983).

The main characteristics of the subsystem "fleet", are the parameters that ensure the economic efficiency of a group of ships operating according to a predefined transportation scenario, where the most important output parameter the ship speed, load capacity (deadweight, number of containers, cargo volume etc.) and the number of ships. The main task of the subsystem "ship" is defining the main dimensions of the ship, hull form coefficients, etc. that would provide the best economic performance during the ship operation. The design solution considers all conditions formulated by the "fleet" subsystem, which are part of the design specification.

In this respect, Wagner et al. (2014) presented a scenario-based optimization procedure of the KRISO container ship, using a statistically developed operational profile generated from an existing container vessel. The main conclusion was that the usage of scenarios within the optimization process has a strong impact on the hull form.

Ventura and Guedes Soares (2015) integrated a voyage model in to a ship design optimization procedure, where the voyage scenario allowed an estimation of the sailing and port times and operational costs. Two objective functions were established in minimizing of the required freight rate and attained EEDI.

The present work deals with the design and optimization of a new multipurpose ship and fleet that will be built in a small and medium sized shipyard accounting for the constraints related to the constructional limitations. The problem is solved in two stages, where in the first one, based on the cargo flows, the number of ships to be built and ship speed are defined. At the second stage, the completing of the fleet leads to defining the technical specification for the individual ship as a part of the fleet.

The design solution of the two tasks is performed by employing the software Expert (Damyanliev & Nikolov, 2002), which was recently used by Damyanliev et al. (2017) for designing new commercial ships. The fleet composition task considers a given distance between ports, total amount of transported cargo and Panama Canal restrictions.

Two transportation scenarios with a different total amount of cargo and distances between the ports are analysed. The second scenario is most suitable for a ship with deadweight up to 6,000 tons.

2 DESIGN DEFINITION

The "Fleet composition" and "Conceptual design" tasks are defined for a specific transportation conditions of a cargo flow, where the optimal design solution estimates the number of ships, speed and deadweight of required ships (external task) and the main dimensions and ship hull form coefficients (internal task).

The optimization of the object function, F(**X, Q**) is formulated as (Damyanliev et al., 2017):

$$F(\mathbf{X}^*) = minF(\mathbf{X},\mathbf{Q}), \quad \mathbf{X} \in \mathbf{E}^n \tag{1}$$

which is subjected to design constraints:

$$\mathbf{H}\{h_i(\mathbf{X},\mathbf{Q})\} > 0, \quad i = 1,2,\ldots,m \tag{2}$$

where **X** is the vector of design variables x_1, x_2, ..., x_n, $\mathbf{X}^*(x_1^*, x_2^*, \ldots, x_n^*)$ is the vector optimum design solution, $\mathbf{h}_i(\mathbf{X}, \mathbf{Q})$ are the inequality constraints as a function of design variables **X** and uncontrollable parameters **Q**.

The components of the vectors of the design variables **X**, constraints, $\mathbf{h}_i$, and uncontrollable parameters, **Q** are part of the external and internal tasks.

The vector of design variables, **X** includes:

- number and speed of the ships, $\mathbf{X}_E$ (external task);
- main dimensions and ship hull form coefficients, $\mathbf{X}_I$.(internal task);

Uncontrollable parameters in most cases are input variables in the mathematical model and are defined as:

- descriptors of the transportation scenario and cargo flow (characteristics of the cargo, voyage distance, port performance, crew number, etc.);
- descriptors of the ship (coefficient of structures etc.);
- descriptors of the economic performance (normative and statistical coefficients etc.).

Similarly, the vector of constraints includes:

- constraints related to the external task, $\mathbf{H}_E$;
- constraints related to the internal task, $\mathbf{H}_I$.

The optimal solution is obtained by employing the Sequential Unconstrained Minimization Technique, SUMT as defined by Fiacco and McCormick (1968), Himmelblau (1972).

This algorithm is formulated in using nonlinear programming (1) and (2) without constraints by introducing a penalty parameter. The solution is based on a sequential unconstrained minimization of the transformed objective functions $\mathbf{P}(\mathbf{X}, \mathbf{Q}, r_k)$ in the following form:

$$\mathbf{P}(\mathbf{X},\mathbf{Q},r_k) = \mathbf{F}(\mathbf{X},\mathbf{Q}) + 1/r_k \sum\{min[0;\mathbf{H}(\mathbf{X},\mathbf{Q})]\}^2 \tag{3}$$

$$\mathbf{F}(\mathbf{X}^*) = \lim\{min\mathbf{P}(\mathbf{X},\mathbf{Q},r_k\}, \quad r_k \to 0 \tag{4}$$

where r_k is the penalty parameter, $r_k > 0$.

This algorithm allows eliminating the intermediate checks for the compatibility of the design solution with the constraints. Employing this algorithm, a universal ship conceptual design framework was developed in (Damyanliev et al., 2017). The developed framework can solve the external and internal ship design tasks, subjected to different initial conditions and constraints and will be used in the present study.

3 BASE CASE STUDY

A case study in in defining a design solution of the "Fleet composition" and "Conceptual design" tasks is presented here.

3.1 *Transportation scenario*

The transportation scenario involves a transportation of cargo, mainly containers, from the terminal, T to Port 1, P1 and Port 2, P2 and return as can be seen in Figure 1.

The amount of transported cargoes is as follows:

- Total amount of cargo from Terminal to Port 1 and Port 2 and vice versa per year is Q_{sum} = 1,000,000 tons;
- Cargo from Terminal to Port 1 and vice versa is $Q_{t1} = Q_{1t} = k_{t1}.Q_{sum}$;
- Cargo from Terminal to Port 2 and vice versa is $Q_{t2} = Q_{2t} = k_{t2}.\ Q_{sum}$;
- Cargo from Port 1 to Port 2 and vice versa is $Q_{12} = Q_{21} = k_{12}.Q_{sum}$.

It is assumed that the cargo consists of 16-ton TEU. It is assumed 10% void space in the transported containers resulting in the average weight of one container of 14.65 tons.

The distances between the ports and terminal are:

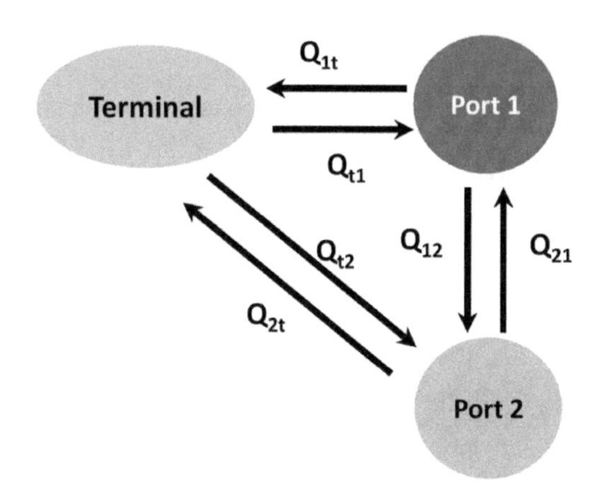

Figure 1. Transportation scenario.

- Terminal – Port 1 = 1161 nm;
- Port 1– Port 2 = 339 nm.

The cargo handling time is:

- Terminal 630 TEU/day;
- Port 1 570 TEU/day;
- Port 2 520 TEU/day;

The freight rate per ton of cargo is:

- Terminal – Port 1 = 30 USD/ton;
- Terminal – Port 2 = 40 USD/ton;
- Port 1 – Port 2 = 10 USD/ton;

3.2 *Ship definition*

The type of ships is multi-purpose, intended for transport of bulk and other dry cargoes. The ships are equipped with cranes for loading and loading of containers.

The ships are single-decked, with an engine room located aft, single propeller with a slow-speed diesel engine, and a superstructure located extremely aft. There is a bulb bow and transom stern.

3.3 *Design parameters*

The design parameters are defined as:

- Number of ships Ns;
- Speed, kn Vs;
- Length between perpendiculars, m L_{pp};
- Breadth, m B;
- Draught, m d;
- Depth, m D;
- Block coefficient $C_{B.}$

There are no formal constraints to the design variables. The design solution of the transportation of cargo is controlled by an indicator, P_{Qsum}, which is defined as:

$$P_{Qsum} = TC_{sum}/Q_{sum} \quad (5)$$

where:

$$TC_{sum} = Ns\ Nv\ TCsv \quad (6)$$

where Nv is the number of voyages per year and TCsv is the transported cargo per ship per voyage.

The condition when $P_{Qsum} = 1$ indicates that the total amount of cargo is transported during the year.

The required deadweight of the ships is provided by the condition when $P_{Dw} = 1$ defined as:

$$PD_{w} = DW/\ DWr \quad (7)$$

where DW is the estimated deadweight and DWr is the required one.

In the cases where the deadweight is a resultant value, the buoyancy index, P_{FL} is defined as:

$$P_{FL} = \Delta/(LW+DW) \quad (8)$$

where Δ is the weight displacement, tons, LW is the lightweight, tons and DW is the deadweight, tons.

The condition when $P_{Qsum} = 1$ represent the case where the buoyancy equilibrium is satisfied.

Additionally, some functional constraints are also satisfied including:

- Summer free board, P_{FB};
- Minimum stability with containers, P_{GMc};
- Sufficient cargo volume, P_v.

The objective function may use one of the following economic indicators:

- Required Freight Rate, RFR;
- Profit, Pr;
- Profitability, Re.

The required freight rate is defined as:

$$RFR = (OPEX + CFR.CAPEX)/Q, \text{ USD/ton} \quad (9)$$

where OPEX is the operational cost per year, USD, CFR is the capital recovery factor, CAPEX is the capital expenditure, USD and Q is the transported cargo per year, tons. A recent analysis about a CAPEX estimation in the condition of a SME shipyard was presented in (Garbatov et al., 2017a)

The profit is defined as:

$$Pr = (Rev - OPEX)/Q, \text{ USD/ton} \quad (10)$$

where Rev = Q.FR is the revenue per year, USD, FR is the market freight rate, USD/ton and Q is the amount of transported cargo, tons.

The profitability is defined by:

$$Re = (Rev - OPEX)/CAPEX, \% \quad (11)$$

The above economic indicators are of a universal nature and are often used in assessing the economic efficiency of complex technical systems.

The required freight rate assesses the rate of return of the initial investments; the profit includes only the revenues from the shipping activity.

Through the profitability, the effectiveness of the investments, accounting for the operating costs and revenues from the shipping may be controlled.

3.4 *Design solution*

The defined design tasks were solved by using the software Expert, considering the three economic indicators RFR, Pr and Re.

The design solution of the optimized design variables is presented in Table.

Two of the economic indicators involved in the optimisation procedure, defining the design solution, RFR and Pr, lead to similar optimal ships with similar main dimensions and deadweight.

According to the profitability criterion, Re, the ship has a larger deadweight. For the three indicators, the L_{pp}/B ratio, which is associated with the ship propulsion and seakeeping performance, is close to the lover limit of 5.2. The ratio B/d is higher, which can be explained by the P_{GMc} limitation, which determines the minimum stability in the load cargo condition with containers.

For the assumed transportation scenario, the number of ships needed to transport the cargo in one year is tree units.

A more detailed analysis is needed to explain the relatively low optimum speed of the ship, which are close to the minimum one of 10 kn as a limit.

Figure 2 shows that the design speed for ships of deadweight between 8,000 and 12,000 tons is in the range of 15–17 kn for the analysed 32 multipurpose vessels. The reason for the lower speed can be related to the assumed economic conditions and transportation scenario.

In fact, it is a current practice to reduce the speed for relatively short voyages using so-called "economical speed". The reduction in the design speed results in a lowering in fuel and oil consumption, which may reduce the OPEX up to 30%.

Table 1. Design parameters.

	Indicators	RFR (min)	Pr (max)	Re (max)
	Design variables			
1	Ns	3.078	3.072	2.561
2	Vs, kn	10.411	11.549	10.592
3	L_{pp}, m	123.734	126.436	129.811
4	B, m	23.796	23.711	24.961
5	d, m	7.156	7.108	7.322
6	D, m	9.639	10.181	10.181
7	C_B	0.728	0.700	0.813
	"Active" constraints			
1	P_{Qsum}	1.00	1.00	1.00
2	P_{Fl}	1.00	1.00	1.00
3	P_{FB}	1.00	1.00	1.02
4	P_v	1.09	1.03	1.09
5	P_{GMc}	1.00	1.02	1.04
6	L_{pp}/B	5.20	5.33	5.20
	Output			
	DW, tons	11050	10500	14300
	L_{pp}/B	5.20	5.33	5.20
	B/d	3.33	3.34	3.41
	L_{pp}/D	12.84	12.42	12.75

The optimum speed is influenced by the relation between the travel time and time for cargo handling. The change in the voyage descriptors: the voyage duration, Ts, cargo handling time, Th, the total time for one voyage operation, Tv and the number of voyages per year, Nv, for the assumed transportation scenario as a function of the deadweight and speed is presented in Figure 3.

As the speed of the ship increases, the voyage time decreases. For the ship with grater deadweight the time for handling the cargo also increases, which leads to an increase the total voyage time. In the case of a relatively short operational distance between the ports, the cargo handling time may be synchronised with the voyage time by reducing the higher ship speed.

In practice, the ship can operate in different operational conditions and to be effective the speed may need to be reduced. In this respect, a power margin that is related to the need to provide a higher speed to deliver the cargo on time and the use of controllable pitch propeller, CPP that may allow effective load of the main engine at speed different of the design one is analysed.

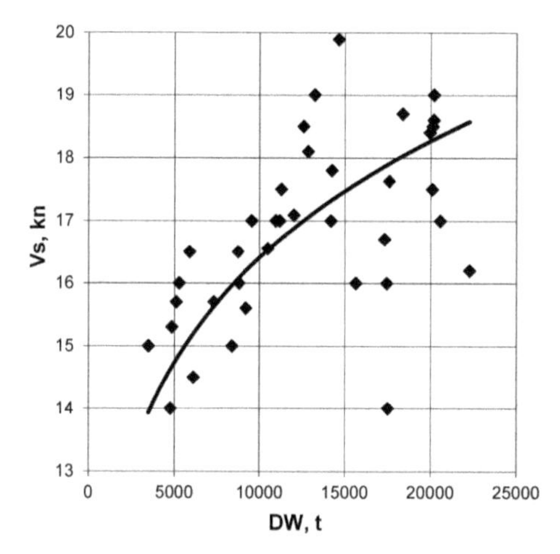

Figure 2. Speed as a function of DW.

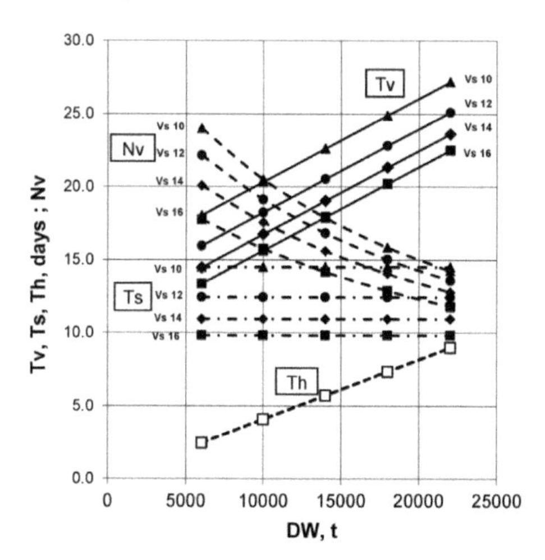

Figure 3. Voyage descriptors as a function of DW.

One can see from Table 1 that the design solution depends on the chosen criterion as an objective function. For the deadweight range from 6,000 to 22,000 tons with a fixed speed of 15 kn the normalized economic indicators RRFR, RPR and RRE are presented in Figure 4.

It is commonly accepted that with increasing of the deadweight, the economic efficiency of the ship improves—initially sharply, and then smoothly to reach asymptotic (constant values).

In the case of RFR and Pr, the optimum ship deadweight is between 10,000 and 12,000 tons, and after that one can see a slight decrease in the efficiency. Profitability increases rapidly, reaching a clearly defined optimum of DW between 14,000 and 16,000 tons, followed by a decrease in the efficiency.

Figure 5 presents the required number of ships for transportation of total amount of cargo $Q_{sum} = 1,000,000$ tons per year. For the deadweight in the range of 10,000–14,000 tons and speed Vs = 15 kn, the number of ships is 2.5–3.

The optimal length between the perpendiculars does not differ significantly for the presented economic indicators as can be seen in Figure 6.

In the case of the profit indicator, for deadweight bigger than 15,000 tons, there is a significant increase in the optimal length between the perpendiculars.

The reason for this is that the profit indicator does not consider the increasing of CAPEX due to increasing of the ship length.

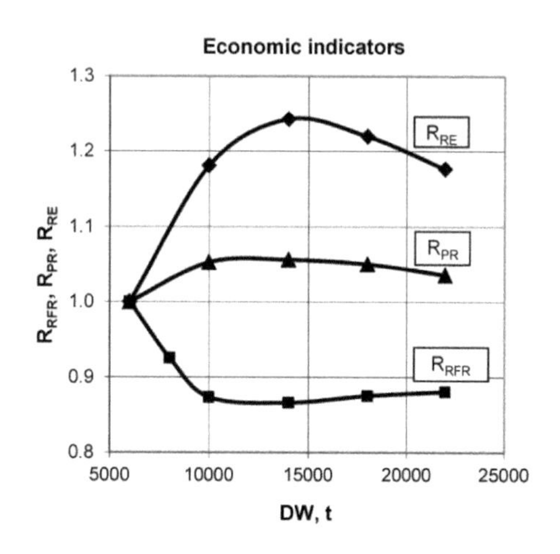

Figure 4. Relative economic indicators as a function of deadweight.

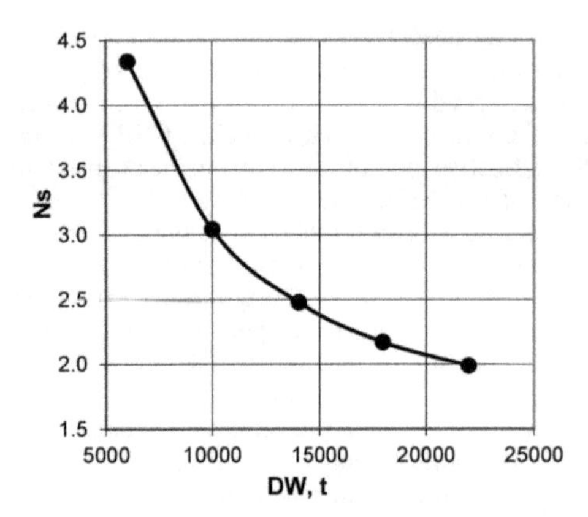

Figure 5. Number of ships, Ns as a function of DW.

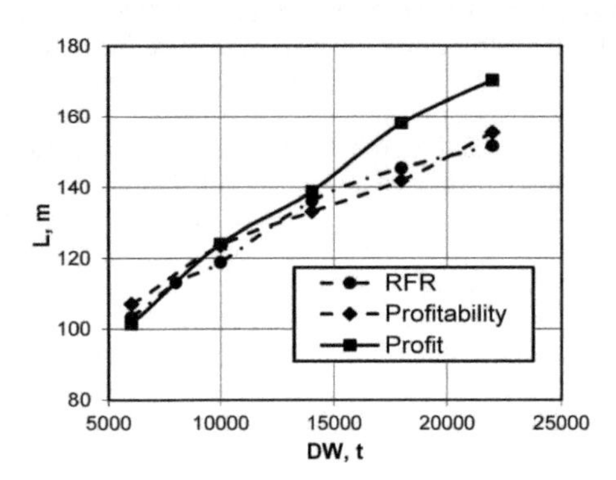

Figure 6. Ship length as a function of economic indicators.

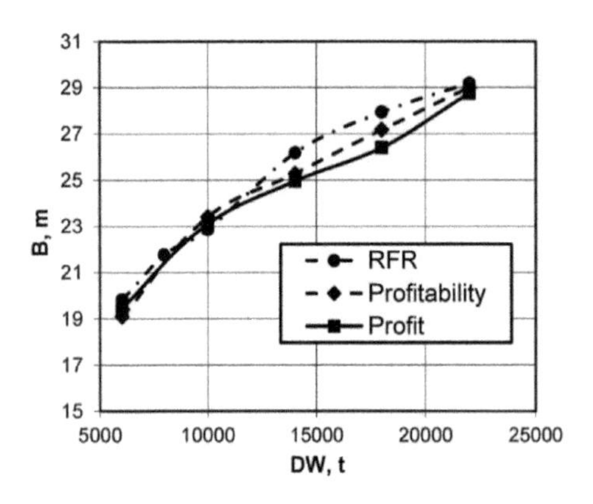

Figure 7. Ship breadth as a function of economic indicators.

The breadth of the vessel varies in narrow ranges for the three indicators as can be seen in Figure 7.

Table 2 presents some results of the analysed ship the main dimension ratios in the deadweight range of 14,000 to 16,000 tons.

Table 2. Dimension ratios and economic indicators.

Indicator	L_{pp}/B	B/d	L_{pp}/D
Statistical data			
min	5.10	2.42	11.25
max	6.30	3.12	15.32
Constraints			
min	5.20	2.00	8.00
max	12.00	4.00	18.00

The L_{pp}/B ratio, which is commonly referred to as an indicator of the ship propulsion and seakeeping is at or close to the minimum values, typical for wider ships. The B/d ratio, which influences the stability, is close to its upper limit. The L_{pp}/D ratio as an indirect indicator of the stiffness of the ship structure, takes values close to the average one.

4 SHIPS DESIGN OF DW UP TO 6000 TONS

To investigate the economic efficiency of cargo transportation with a ship built under the constraints of SME shipyard, ships with a DW range from 4,000 to 5,500 tons are analysed. Two case studies will be analysed accounting of the SME constraint. Case Study 1, CS1, the transportation scenario is the same as in the previous section and Case study 2, CS2, where the transportation scenario is defined as a cargo volume of Q_{sum} = 500,000 tons.

Distance b/w ports:

- Terminal – Port 1: 340 nm
- Port 1 – Port 2: 420 nm

Freight rate:

- Terminal – Port 1 10 USD/ton
- Terminal – Port 2 10 USD/ton
- Port 1 – Port 2 12 USD/ton

For both case studies, the constraints are related to the ship hull constructional capacity of the facilities of a SME shipyard (Garbatov et al., 2017a, Atanasova et al., 2018), where the breadth of the ships cannot be bigger than 16 m

The profitability, Re is considered as an objective function and a speed of 14 kn is adopted.

4.1 *Case study 1*

Table 3 and Table 4 present the output design parameters in the case of restriction and without restriction with respect to the breadth of the ships.

The constraints that set up the optimum solution are related to the requirements of transportation of the cargo volume, minimum intact stability

Table 3. Output design parameters, Case study 1, without restriction.

DW, tons	4,000	4,500	5,000	5,500
Relative values of Re (RRe)				
RRe	1.000	1.058	1.107	1.168
Design variables				
Ns	5.841	5.295	4.869	4.520
L_{pp}, m	93.576	103.342	106.187	114.988
B, m	17.73	17.385	17.837	18.002
d, m	5.567	5.818	6.069	6.184
D, m	6.979	7.418	7.786	8.057
C_B	0.650	0.650	0.656	0.656
Main dimensions ratio				
L_{pp}/B	5.278	5.944	5.953	6.388
B/d	3.185	2.988	2.939	2.911
L_{pp}/D	13.408	13.931	13.638	14.272

Table 4. Output design parameters, Case study 1, with restriction.

DW, tons	4,000	4,500	5,000	5,500
Relative values of Re (RRe)				
RRe	0.993	1.052	1.096	1.123
Design variables				
Ns	5.882	5.297	4.874	4.517
L_{pp}. m	96.49	105.923	113.199	119.477
B. m	16.001	16.005	16.001	16.001
d. m	5.662	5.716	5.553	5.406
D, m	7.171	7.392	7.423	7.564
C_B	0.678	0.695	0.743	0.792
Main dimensions ratio				
L_{pp}/B	6.030	6.618	7.074	7.467
B/d	2.826	2.800	2.882	2.960
L_{pp}/D	13.456	14.329	15.250	15.795

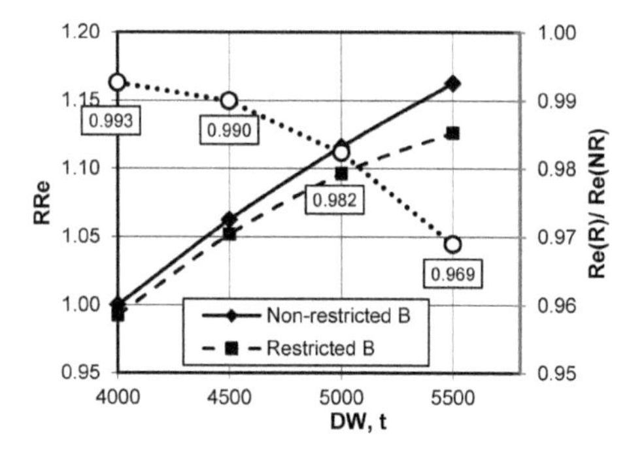

Figure 8. Relative profitability as a function of DW. CS1.

and summer free board waterline. The imposed constraint in the breadth of the ship is active in the investigated range of the deadweight and leads to an increase of the length and block coefficient of the ship as can be seen in Table 4.

The relationship between the profitability, in the case of non-restricted design, and the deadweight is presented in Figure 8. The effectiveness of the ship with a restricted breadth decreases with increasing the length of the ship.

The relation between the relative profitability for restricted Re(R) and non-restricted Re(NR) ships is presented in Figure 8 as a dotted line. The decreasing of Re due to the constraint related to the breadth varies from 0.7–3.1%.

The profitability of ships with a deadweight in the range from 4,500 to 5,500 tons without restriction in the breadth is about 4 times lower than for the ships with deadweight around 14,000 tons. With a restriction of the breadth, the profitability additionally drops down by about 4%.

4.2 *Case study 2*

The output design parameters for the Case study 2 are presented in Table 5 and Table 6.

The impact of the restricted breadth leads to a relative lengthening of the ship and increasing the block coefficient, which may explain the reduction of the efficiency (see Figure 9).

The relatively short voyages and associated lower freight rate, in a comparison to Case study 1, which reduces the profitability about two to three times.

However, in the case of a ship with a design constraint due to the SME construction limitation and without shipbuilding restriction in the cargo transportation condition of Case study 2, the effectiveness of the two design ships is not very different, which is in the range of 2% (see Figure 9).

Table 5. Output design parameters, Case study 2, without restriction.

DW, tons	4,000	4,500	5,000	5,500
Relative values of Re (RRe)				
RRe	1.000	1.112	1.219	1.307
Design variables				
Ns	2.186	1.997	1.844	1.720
L_{pp}. m	98.809	101.332	104.215	109.985
B. m	16.145	16.878	17.229	18.075
d. m	5.74	5.859	5.861	5.858
D. m	7.258	7.461	7.551	7.633
C_B	0.65	0.662	0.693	0.695
Main dimensions ratio				
L_{pp}/B	6.120	6.004	6.049	6.085
B/d	2.813	2.881	2.940	3.086
L_{pp}/D	13.614	13.582	13.801	14.409

Table 6. Output design parameters, Case study 2, with restriction.

DW. tons	4,000	4,500	5,000	5,500
Relative values of Re (RRe)				
RRe	0.983	1.094	1.188	1.220
Design variables				
Ns	2.193	2.002	1.854	1.737
L_{pp}. m	102.379	108.310	115.114	121.491
B. m	16.001	16.000	16.001	16.001
d. m	5.484	5.594	5.471	5.294
D, m	7.010	7.265	7.194	7.535
C_B	0.669	0.693	0.738	0.794
Main dimensions ratio				
L_{pp}/B	6.398	6.769	7.194	7.593
B/d	2.918	2.860	2.925	3.022
L_{pp}/D	14.605	14.908	16.001	16.124

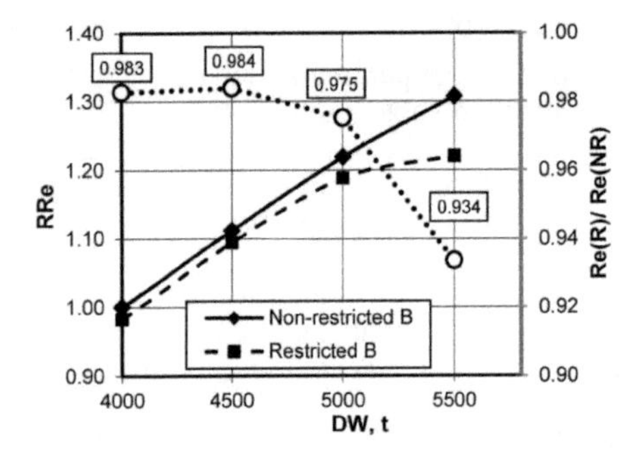

Figure 9. Relative profitability as a function of DW for restricted and non-restricted breadth, CS2.

5 CONCLUSIONS

This work performed a concept design and optimization of a new multipurpose ship and fleet that can be built in the condition of a small and medium sized shipyard accounting for the existing building limitations.

The analyses demonstrated that ships with a deadweight range of 4,000 to 5,500 tons can be built in the condition of SME and efficiently used for transportation of cargo with varying voyages specifications, especially in small consignments.

Even in a relatively short voyage, with an expected lower freight rate, the designed small ship with a maximum breadth of 16 m has a positive economic performance in transporting a cargo in both directions in absence of a ballast passage.

ACKNOWLEDGEMENTS

This paper reports a work developed in the project "Ship Lifecycle Software Solutions", (SHIPLYS), which was partially financed by the European Union through the Contract No 690770 – SHIPLYS-H2020-MG-2014-2015.

REFERENCES

Atanasova, I., Damyanliev, T.P., Georgiev, P. & Garbatov, Y., (2018), "Analysis of SME ship repair yard capacity in building new ships", *Progress in Maritime Transportation and Engineering.* C. Guedes Soares & T.A. Santos (Eds.), Taylor & Francis Group, London, UK.

Bharadwaj, U.R., Koch, T., Milat, A., Herrera, L., Randall, G., Volbeda, C., Garbatov, Y., Hirdaris, S., Tsouvalis, N., Carneros, A., Zhou, P. & Atanasova, I., (2017), "Ship Lifecycle Software Solutions (SHIPLYS) – an overview of the project, its first phase of development and challenges", *Maritime Transportation and Harvesting of Sea Resources.*, C. Guedes Soares & A.P. Teixeira (Eds.), Taylor & Francis Group, London, UK, pp. 889–897.

Damyanliev, T. P., Georgiev, P. & Garbatov, Y. (2017). "Conceptual ship design framework for designing new commercial ships", *In:* Guedes Soares, C. & Garbatov, Y. (eds.) *Progress in the Analysis and Design of Marine Structures.* London: Taylor & Francis Group, 183–191.

Damyanliev, T.P. & Nikolov, N.A., (2002), "Computer system "Expert_ SRS" – valuations and analyses in the ship repair", Proceedings of the 6th International Conference on Marine Science and Technology (Black Sea, 2002), Varna, Bulgaria., Union of Scientists of Varna, 14–18.

Fiacco, A.V. & McCormick, G.P., (1968). *Nonlinear Programming: Sequential Unconstrained Minimization Techniques*, Chichester, England, Wiley.

Gallin, C., (1973), Which way computer aided preliminary ship design and optimization, Proceeding of ICCAS, Tokyo.

Garbatov, Y., Ventura, M., Georgiev, P., Damyanliev, T.P. & Atanasova, I. (2017a). "Investment cost estimate accounting for shipbuilding constraints", *In:* Guedes Soares, C. & Teixeira, A. (eds.) *Maritime Transportation and Harvesting of Sea Resources.* London: Taylor & Francis Group, 913–921.

Garbatov, Y., Ventura, M., Guedes Soares, C., Georgiev, P., Koch, T. & Atanasova, I., (2017b). "Framework for conceptual ship design accounting for risk-based life cycle assessment". *In:* Guedes Soares, C. & Teixeira, A. (eds.) *Maritime Transportation and Harvesting of Sea Resources.* London: Taylor & Francis Group, 921–931.

Himmelblau, D. 1972. *Applied Nonlinear Programing,* New York, McGraw – Hill.

Pashin, V., M. 1983. *Ship optimization system approach and mathematical models*, St Petersburg, Sudostroenie.

Ventura, M. & Guedes Soares, C., (2015). "Integration of a voyage model concept into a ship design optimization procedure". *In:* Guedes Soares, D. P. (ed.) *Towards Green Maritime Technology and Transport.* London: Taylor & Francis, 539–548.

Wagner, J., Binkowski, E. & Bronsart, R., (2014), "Scenario based optimization of a container vessel with respect to its projected operating conditions". *Int. J. Nav. Archit. Ocean Eng.*, 496–506.

Ship structures I

Progress in Maritime Technology and Engineering – Guedes Soares & Santos (Eds)
© 2018 Taylor & Francis Group, London, ISBN 978-1-138-58539-3

Analysis of the ultimate strength of corroded ships involved in collision accidents and subjected to biaxial bending

J.W. Ringsberg, Z. Li & A. Kuznecovs
Department of Mechanics and Maritime Sciences, Chalmers University of Technology, Gothenburg, Sweden

E. Johnson
Department of Safety—Mechanics Research, RISE Research Institutes of Sweden, Borås, Sweden

ABSTRACT: This study presents an analysis of the effects of sudden damage, and progressive deterioration due to corrosion, on the ultimate strength of a ship which has been collided by another vessel. Finite Element Analyses (FEA) of collision scenarios are presented where factors are varied e.g. the vessels involved in the collision, and consideration of corroded ship structure elements and their material characteristics in the model. The striking ship is a coastal tanker, the struck ship is either a RoPax ship, or, a coastal oil tanker vessel. The ultimate strength analysis of the struck vessel accounts for the shape and size of the damage opening from the FEA. The Smith method is used to calculate the ultimate strength of intact and damaged ship structures during biaxial bending. The study shows how corroded, collision-damaged ship structures suffer from a reduction in crashworthiness and ultimate strength, how this should be considered and modelled in FEA.

1 INTRODUCTION

With an ever-increasing worldwide ship traffic, larger ship sizes, and increased number and different types of marine structures offshore, there is an enhanced risk for collision accidents with, for example, wind/wave/tidal energy farms, structures for oil and gas extraction or other ships. Many of these marine structures, especially the ships, have been in operation for many years. Corrosion, permanent deformations and accumulated fatigue damage are some factors that influence their residual strength and structural integrity. In this study, the structural integrity analysis that refers to the accidental and ultimate limit states of corroded ship structures that are damaged by another vessel from a collision event is investigated by nonlinear explicit finite element analyses (FEA) in a parametric study.

In conventional ship design practice, it is required to carry out an assessment of a ship's structural and ultimate strength under intact conditions (i.e. no damage opening in the side-shell). In case of a collision accident, where a damage opening occurs in the struck vessel's side-shell structure, the safety margin against the ultimate limit strength is greatly reduced due to the damaged condition. An ultimate strength analysis of a damaged ship from a collision accident needs a description of the collision location, shape and size of the ship structure's damages in order to make a reliable estimation of the ship's reserve strength.

Considerable research efforts have been spent on the structural response during collision as well as on the residual ultimate strength of struck ships. Examples of such investigations with the representation of the material and its characteristics can be found in AbuBakar & Dow (2016), Ehlers (2010), Ehlers & Østby (2012), Hogström & Ringsberg (2012), Hogström et al. (2009), Marinatos & Samuelides (2015), Samuelides (2015), Storheim et al. (2015a,b), Yamada (2014), and Zhang & Pedersen (2016). Faisal et al. (2016) studied the hull collapse strength of double hull oil tankers after collisions using a statistical approach. Several parameters were considered such as impact location, extent of damage (represented by penetration depth in the struck ship), and the collision scenario. Simplified shapes of the structural damages were used in the study.

The ship's physical condition due to e.g. corrosion is also important to consider in this regard. Campanile et al. (2015) present a study on the same topic and type of damage but for bulk carriers including the effect from corrosion using a corrosion model proposed by Paik et al. (2003). The results show how the influence from corrosion of the material leads to a significant decrease in the residual ultimate strength index (RSI). There are several investigations on the buckling ultimate strength that support this finding for intact structures that suffer from either minor or major corrosion wastage; see e.g. Paik et al. (2009) and Saad-Eldeen et al. (2011).

There are few studies in the literature which systematically present the consequences of corrosion on the collision resistance and the ultimate strength together. The influence from corrosion is typically simplified by removing the extra corrosion margin in the damage assessment, see Ringsberg et al. (2017a) and Parunov et al. (2017). The objective of the current investigation is to study the effects of sudden structural damage, and progressive deterioration due to corrosion, on the ultimate strength of a ship which has been collided by another vessel. The influence from corrosion is considered here by three factors:

- Reduction of the thickness of the ship structure's elements due to corrosion.
- Increased friction coefficient on corroded surfaces.
- Change of the material characteristics of a corroded material as compared with a non-corroded material using the approach presented in Garbatov et al. (2014, 2016).

The study is a continuation of previous work reported in Ringsberg et al. (2017b). The methodology has been improved and extended to enable biaxial loading conditions for both intact and damaged ship structures using the Smith method. Section 2 presents the methodology adopted in the study to calculate the ultimate strength capacity of intact, damaged, non-corroded and corroded ship structures. Section 3 presents two case study vessels, their FE models, the stress-strain curves (constitutive models) of the non-corroded and corroded materials, and other model details. In the following Section 4, the results from analyses in a parametric study are presented with emphasis on the influence from corrosion on a damage opening's shape and size, and the struck vessel's residual strength. Results from analyses of biaxial loading conditions are highlighted using the coastal oil tanker as the case study vessel. The conclusions of the study are presented in Section 5.

2 METHODOLOGY

A methodology has been developed for the assessment of the structural integrity of non-corroded/corroded ships that have been damaged during a ship-ship collision accident, see Figure 1. The collision scenario must first be identified, together with all data needed to carry out a nonlinear finite element analysis (FEA) with the vessels involved. The FEA will with this set-up and for the collision location on the hull give the shape and size of the struck ship's structure damages. The information is used to make an estimation of the ship's reserve strength in a subsequent ultimate strength analysis. An analysis can be carried out with either of the following two options, which both include biaxial loading conditions (horizontal and vertical bending moment): an in-house MATLAB code of the Smith method, or, continued FEA based on the damaged geometry from the former FEA.

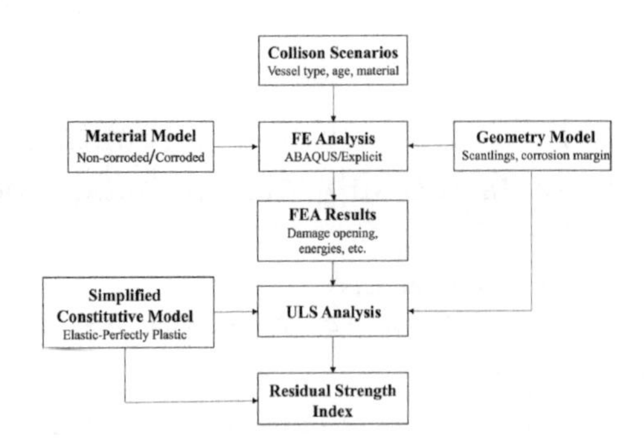

Figure 1. Flowchart of the analysis procedure of the current study.

This study focuses on the presentation of results using the Smith method in the ultimate strength analyses. It is an established method, it is fast to use and is well-suited for parametric studies which are presented in this investigation. Ultimate strength analyses using FEA are more cumbersome and require detailed modelling of structures, materials etc., and they are not presented in this study.

The methodology is used in a parametric study of collision scenarios and ship conditions for two case study vessels (see Section 3). Geometrically, the two ship types have different requirements for the allowed loss of material due to corrosion in the cross-section related to the age of the ship. Corrosion status modelling is considered in the analyses by adjusting the remaining corrosion margin and the material's stress-strain curve as proposed by Garbatov et al. (2014, 2016) for non-corroded and corroded materials; both effects are accounted for in the nonlinear FEA of the ship-ship collision scenarios. In addition, the ultimate strength and reserve strength calculations of the collision-damaged ship cases are compared with results from calculations for intact ship structures; these are based on the same simulation models except for that the collision event is not simulated.

3 SIMULATION MODELS AND ANALYSES

Explicit FEA are presented where the ship type, damaged/intact vessel, the corrosion margin thickness and material characteristics are varied systematically in a parametric study. The striking ship is represented by a coastal tanker while the struck ship is either a RoPax ship, or, a coastal oil tanker vessel.

A detailed description of the FE models and analyses have been presented in Ringsberg et al. (2017a, 2017b), hence, only a brief summary is presented in this section with an emphasis on corroded ship structures and modelling details related to that.

3.1 *Case study vessels and finite element models*

The collision scenario in the study was a collision between two similar-sized vessels. The striking ship was a coastal product/chemical tanker with a total displacement of 10,800 metric tons, in the following referred to as the striking tanker. Two different struck ships were studied for comparison: one RoPax ship and one coastal oil tanker. Figure 2 presents the mid-ship section scantlings of the struck vessels which are collided amidships by the striking tanker. The design draft for the RoPax ship is 4 m and 7.4 m for the coastal oil tanker.

The RoPax ship is a coastal RoRo cargo vessel with a typical side-shell structure with small distance between the inner and the outer side-shells, which makes the ship sensitive to collision damage (Karlsson 2009). It has three RoRo decks, one on the tank top, one on the main deck and one outdoors on the upper deck. The ship is longitudinally stiffened above the main deck and in the double bottom, and it has a transversely stiffened double side-shell.

The coastal oil tanker has longitudinally stiffened double bottom and weather deck, while the double side-shell structure is transversely stiffened. It has a corrugated longitudinal bulkhead in the centre plane of the cross-section. The corrugation is vertical; hence, the bulkhead may be omitted in the ULS calculations since it does not contribute effectively to the hull girder longitudinal strength (IACS 2017).

The FE analyses were carried out using the software Abaqus/Explicit ver. 6.13–3 (Abaqus 2016). A thorough presentation of the design of

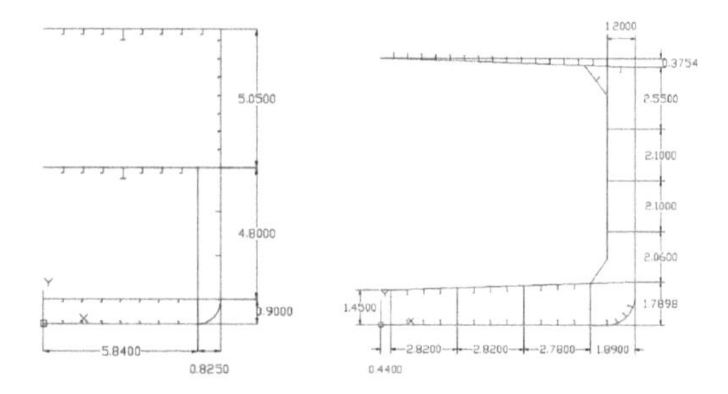

Figure 2. The mid-ship section of the struck vessels (unit: meters): (left) the RoPax ship, and (right) the coastal oil tanker.

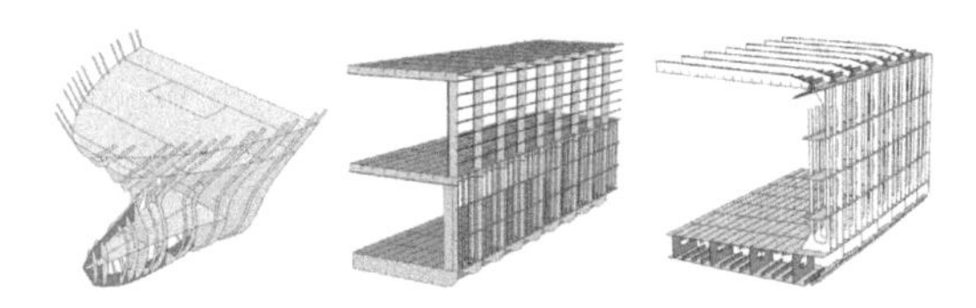

Figure 3. The geometry of the inner structure of: (left) the bulbous bow section [i.e. the striking tanker ship], the outer shell of the side-shell structure of (middle) the RoPax ship, and (right) the coastal oil tanker.

the FE models for collision simulations, description of material characteristics and damage modelling can be found in previous work by the authors; see Hogström & Ringsberg (2012, 2013) and Ringsberg et al. (2017a) for detailed descriptions. Figure 3 presents the geometry of the inner structure of the bulbous bow section of the striking tanker, and the geometries of the struck RoPax and coastal oil tanker ships.

The FE models of the struck ships were made sufficiently large to avoid influence from the boundary conditions. The collision impact was amidships and between bulkheads and web frames. The bow section of the striking tanker was modelled as deformable and restricted to only move in a prescribed and right-angle collision direction. (Other collision angles have been investigated previously by Ringsberg et al. (2017a)). The striking ship was given an initial forward velocity of seven knots, while the side-shell structure of the struck ships was held fixed along its circumference. The velocity of the striking bow gradually decreased to zero knots during the collision event as energy was dissipated through deformations and fracture in the structures.

The FE meshes were made of four-node shell elements with reduced integration (S4R and some three-node S3R in Abaqus/Explicit) and five section points through the thickness. A mesh convergence analysis resulted in an element size of 60 mm. The element length/thickness ratio was 5 in the part of the model with the largest sheet thickness. Explicit time stepping was used with an automatic choice of time step. The general contact condition criterion in Abaqus/Explicit was used in conjunction with a friction coefficient of 0.3 or 0.5 (non-corroded or corroded surface) to model the contacts between surfaces that occur in the collision.

3.2 *Corroded ship structure: Corrosion margin and material modelling*

The two most important factors that have a big impact on strength reduction in addition to the net section area loss are, according to Garbatov et al. (2016), the change in material parameters caused by corrosion and stress concentration due to local corrosion pits. Several investigations in the literature have studied how the strength of corroded metal structures depends on factors such as the degree of degradation, geometric modelling of pit density and initial imperfections in simulation models used for nonlinear FE analysis; see e.g. Paik et al. (2008) and Paik & Melchers (2008).

In the current study, another approach presented by Garbatov (2014, 2016) is used. The degree of corrosion is taken into account by simply varying the parameters in the stress-strain law, the structural thickness and the coefficient of friction and can therefore easily be implemented in nonlinear FE analyses of the marine structures. Three different representations of an NVA shipbuilding mild steel were used depending on the grade of corrosion (ship's age and severity of corrosion): NVA virgin (non-corroded), NVA minorly corroded and NVA severely corroded. A detailed description of the material models used in this study are presented in Ringsberg et al. (2017b) and only a brief summary is given below. The material parameters for the three materials are summarized in Table 1 and illustrated in Figure 4.

The virgin NVA material was represented by a nonlinear elastic-plastic power law constitutive material model with isotropic hardening. The influence from strain rate effects was considered using the Cowper-Symonds (CS) relationship, including two constants C and P. Degradation leading to failure was modelled with one model for onset of failure based on the shear criterion in Abaqus/Explicit (damage initiation, DI) and one model for damage degradation (damage evolution, DE). In the DE-model the length dependence between element size and fracture strain was accounted for through Barba's law; see Hogström et al. (2009) for details.

The constitutive material parameters for the minorly and severely corroded NVA steels were obtained from Garbatov et al. (2014). Both materials were represented by a bilinear elastic-plastic constitutive material model, with linear isotropic hardening between the yield and ultimate tensile stresses. As found in Garbatov et al. (2014), a corroded material is less ductile compared to a non-corroded material, and the necking point is not easily observed during tensile tests. Hence, the damage model of the corroded materials was represented solely by the shear failure DI criterion without any DE law.

Nevertheless, due to lack of material data for the corroded materials, a sensitivity study was carried out. The influence from taking into consideration a Cowper-Symonds model and a simplified DE model (activated after the fracture strain is reached) were investigated. The results show that both issues had only minor influence on the size of the damage opening in corroded ship structure as compared to analyses when no strain rate dependence or DE model were included. Hence, the latter were used i.e. no Cowper-Symonds or DE models were included in the FEA with corroded material properties.

During the lifetime of a ship, the hull is subjected to corrosion which results in reduced corrosion margin of its structural members. The corrosion margins of the struck coastal oil tanker, estimated from the Common Structural Rules for

Table 1. Material parameters used in the constitutive material and damage models.

Parameter	NVA virgin (non-corroded)	NVA minorly corroded	NVA severely corroded
Young's modulus, E (GPa)	210	179	158
Poisson's ratio, υ (–)	0.3	0.3	0.3
(Static) Yield stress, $\sigma_{y,s}$ (MPa)	310	310	291
Ultimate tensile strength (MPa)	579	518	440
Hardening coefficient, K (MPa)	616	845	752
Hardening exponent, n	0.23	1.00	1.00
Necking strain, ε_n (%)	23.0	–	–
Fracture strain, ε_f (%)	35.1	24.8	20.0
Cowper-Symonds constant, C (–)	40.4	–	–
Cowper-Symonds constant, P (–)	5	–	–
DE parameters, bilinear model; see Abaqus (2016) for details	(0, 0), (0.02, 0.00458), (1, 0.01832)		

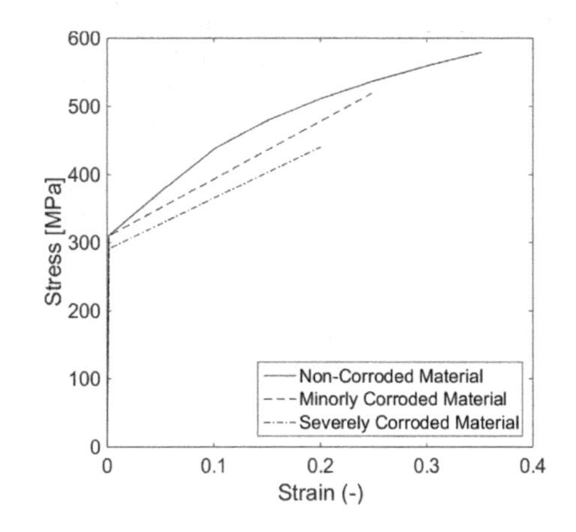

Figure 4. True stress-strain curves for the non-corroded, the minorly, and the severely corroded materials.

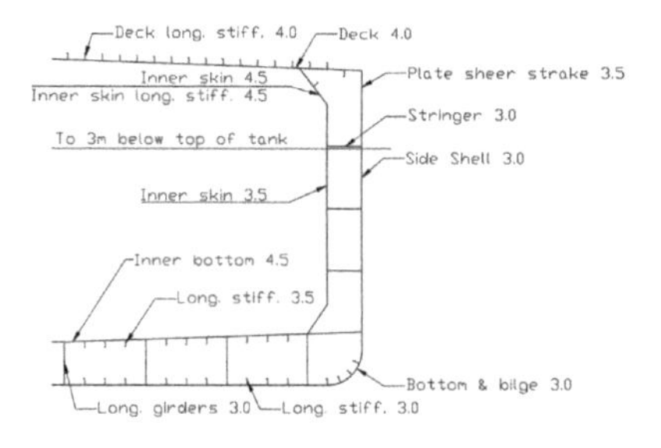

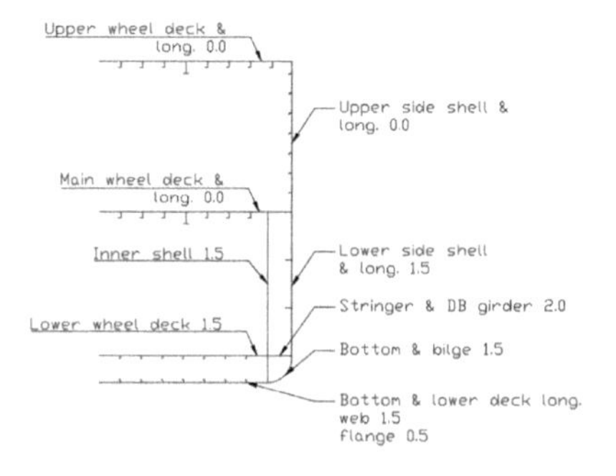

Figure 5. Corrosion margins (unit: mm) for (upper) the coastal oil tanker, and (lower) the RoPax ship.

Bulk Carriers and Oil Tankers (CSR-BC&OT) (IACS 2017), are shown in Figure 5. For RoPax vessels, there are no clear guidelines for corrosion margins, and a corrosion margin of 20% has been added on all parts in the double bottom and side-shell structures in accordance with Ringsberg et al. (2017a). Full reduction of a corrosion margin could represent 25 years of operation for a vessel and a 50% reduction of the corrosion margin around 16 years of operation.

The friction coefficient of corroded surfaces in contact was altered, from 0.3 to 0.5. Table 2 presents the relationship between the remaining corrosion margin in intervals and how they correlate with the material model used in the FE analyses. The intervals are relevant for the coastal oil tanker where different parts of the structure have various percentages of loss of material due to corrosion, whereas for the RoPax vessel, with uniform corrosion margin, the minorly corroded material had a reduction of 50% and the severely corroded material 100% of this margin. Figure 6 presents where in the ships' cross-section different material properties where used in the simulation models; see Kuznecovs & Shafieisabet (2017) for details.

3.3 *Parametric study*

The parametric study was designed to enable systematic analysis of several factors and their influence on the shape and size of the damage opening of the struck vessels, and how the ultimate strength was

Table 2. Corroded material model assignment based on different percentage of corrosion.

Remaining corrosion margin	Material model
100%	Virgin (non-corroded) material
50% – <100%	Minorly corroded
0% – <50%	Severely corroded

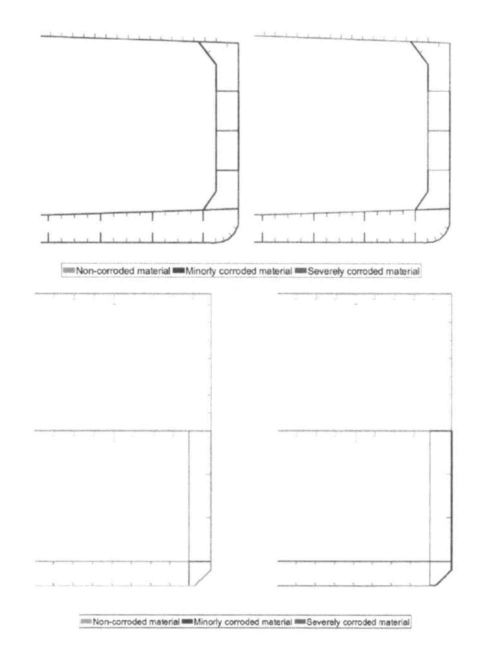

Figure 6. Illustrations of where in the ships' cross-sections different material models were used for (left) "minorly" and (right) "severely" corroded ship cases, for the (upper) coastal oil tanker ship and (lower) the RoPax ship; note: the RoPax did not reach a state with severly corroded material.

affected by these factors; see Table 3 for the analysis matrix. Status "intact" refers to ultimate strength analyses of cases where the RoPax and coastal oil tanker were assessed without any damage opening, while status "damaged" refers to ultimate strength analyses of cases where the vessels have damage openings as calculated by the FE analyses.

Thickness reductions due to corrosion of the ship structures were considered in the three corrosion margin cases shown in Table 2 and the material behaviour was varied between the three material models in Table 1. A "reference" value of the friction coefficient was set to 0.3 for all non-corroded surfaces and a value of 0.5 was used to represent a corroded surface (see Ringsberg et al. (2017a), where the influence from friction alone as a parameter was studied in detail). Moreover, the striking tanker ship speed was 7 knots, and it was always assumed to

Table 3. Analysis matrix in the parametric study.

Case	Status (I/D)	Ship[1] (T/R)	Material model	Remaining corrosion margin (%)	Friction coefficient (–)
T1-I	Intact	Tanker	Virgin	100	–
T2-I	Intact	Tanker	Virgin	50	–
T3-I	Intact	Tanker	Virgin	0	–
T4-I	Intact	Tanker	Minor	50	–
T5-I	Intact	Tanker	Severe, Minor	0, 50	–
T1-D	Damaged	Tanker	Virgin	100	0.3
T2-D	Damaged	Tanker	Virgin	50	0.5
T3-D	Damaged	Tanker	Virgin	0	0.5
T4-D	Damaged	Tanker	Minor	50	0.5
T5-D	Damaged	Tanker	Severe, Minor	0, 50	0.5
R1-I	Intact	RoPax	Virgin	100	–
R2-I	Intact	RoPax	Virgin	50	–
R3-I	Intact	RoPax	Virgin	0	–
R4-I	Intact	RoPax	Minor	50	–
R5-I	Intact	RoPax	Minor	0	–
R1-D	Damaged	RoPax	Virgin	100	0.3
R2-D	Damaged	RoPax	Virgin	50	0.5
R3-I	Damaged	RoPax	Virgin	0	0.5
R4-I	Damaged	RoPax	Minor	50	0.5
R5-D	Damaged	RoPax	Minor	0	0.5

[1]Tanker refers to the struck coastal oil tanker.

have full corrosion margin and material characteristics according to the NVA virgin material.

4 RESULTS

4.1 *Shape and size of damage, structural deformation*

Figure 7 presents the projected shape and size of damage openings of the RoPax ship and the coastal oil tanker from the FE analyses in Table 3. Figure 8 presents the deformed and damaged cross-sections where the damage openings are the largest.

For the coastal oil tanker, the damage shapes on the inner side-shell differ compared to the outer side-shell because the bilge hopper and the inner bottom are damaged in addition to the inner side-shell. The sizes of the damage openings for the coastal oil tanker cases are, for all of the cases except for T5-D, smaller compared to the corresponding RoPax ship cases. It was expected because the ships were impacted by the same striking vessel and the coastal oil tanker has more structural elements with larger dimensions compared with the RoPax ship. Hence, the kinetic energy is dissipated more efficiently in this ship before fracture occurs. For the T5-D case, the influence from corrosion degraded the structure even more giving a significantly lower resistance to the impact load. Further, except for these observations, the same trend was found as for the RoPax ship: a reduction of the corrosion margin

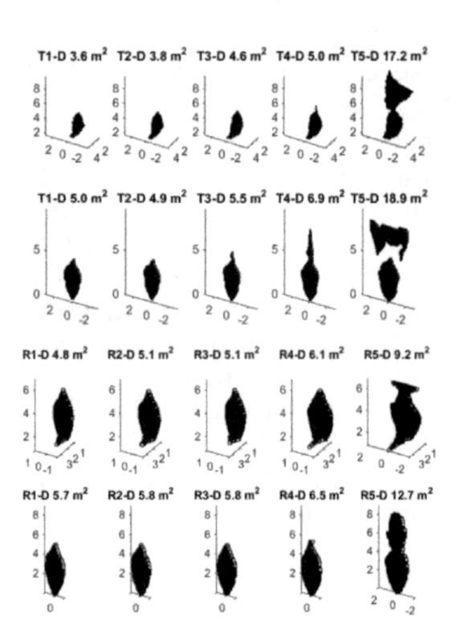

Figure 7. Shape and size of the damage openings for the coastal oil tanker (T) and the RoPax (R) cases: (upper) inner side-shell and (lower) outer side-shell.

results in a larger size of the damage opening, cf. T1-D and T3-D. The material models for minorly and severely corroded materials (T5-D) resulted in larger damage opening compared with the virgin material model (T3-D). The results show that the increase in damage opening size in the inner side-shell is 378% between a new-built ship (T1-D) and a minorly/severely corroded ship (T5-D).

For the RoPax ship, the sizes of the damage openings in the inner side-shell are smaller com-

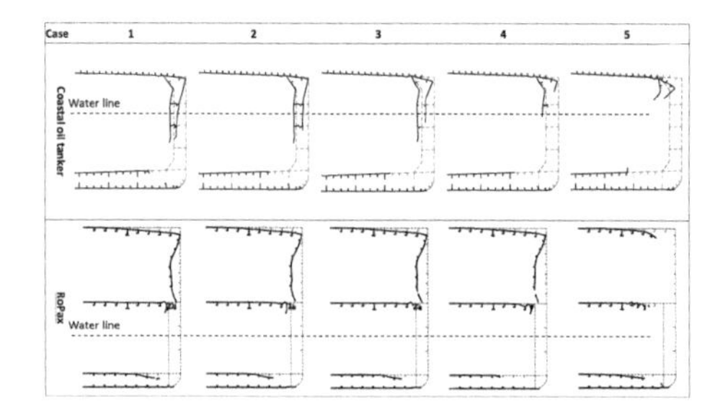

Figure 8. Deformed and damaged cross-sections for the RoPax and costal oil tanker cases.

pared to the outer side-shell. It is also found that there is a small increase in size of the damage opening when the corrosion margin is reduced, cf. R1-D and R3-D. There is an influence from the material model which shows that the damage opening is larger when the minorly corroded material model (case R5-D) is used compared with the virgin material model (R3-D). The results show that the increase in damage opening size in the inner side-shell is 92% between a new-built ship (R1-D) and a minorly corroded ship (R5-D).

4.2 *Ultimate and residual strength analyses*

The ultimate strength analyses were carried out using the Smith method in Fujikubo et al. (2012) with biaxial loading conditions, i.e., combined vertical and horizontal bending moment. This Smith method was implemented in an in-house MATLAB code. It was developed for the analysis of both undamaged and damaged ship hulls, where the latter required that the code must take into account for the rotation and the displacement of the neutral axis (NA) during the analysis. The code was verified and the results were compared with good agreement for undamaged ship hull cross-sections against a large number of cases taken from the literature, such as from Nishihara (1984) who carried out experiments, and Downes et al. (2016) who carried out numerical simulations using the FE method; see details for vertical bending conditions in Kuznecovs & Shafieisabet (2017). All verification cases were also modelled and analysed using the commercial software MARS2000 (Bureau Veritas 2015). Results from the ultimate strength analyses of undamaged ships (i.e. intact conditions in Table 5) are presented in Section 4.2.1.

Yamada (2014) presented a residual strength index, RSI2 = M_U/M_P, to show the dependence of ultimate strength due to age for intact corroded vessels without collision damage; M_U is the ultimate vertical bending moment, M_P is the ultimate fully plastic bending moment. In the current study, the calculation of the RSI2 index for different material conditions (virgin and corroded) and corrosion margin reduction models were used to show how the RSI2 decreases as the corrosion margin decreases and the material's properties degrade. This means that the loss in structural integrity was assumed to be proportional to the age of the ship structure. The results for vertical bending moement only are presented in Section 4.2.2.

4.2.1 *Vertical bending moment: Intact and damaged conditions*

The ultimate strength for intact conditions of the coastal oil tanker and the RoPax ship were calculated for all "intact" cases presented in Table 3. The results are presented in Figure 9 for only vertical bending as the ultimate vertical bending moment, M_U, and the vertical position of the NA, versus the curvature, χ. Detailed analyses of these results have been presented in Ringsberg et al. (2017b).

The M_U for the coastal oil tanker in hogging and sagging was calculated to 1.42 GNm and −1.02 GNm for T1-I, respectively. The higher ultimate vertical bending moment in hogging condition was mainly due to the stiffness contribution from the double bottom structure. The reduction in ultimate load capacity during sagging condition at the curvature $\chi = -0.20 \times 10^{-3}$ m^{-1}, was caused by buckling of the strength deck; this resulted in a rapid shift of the neutral axis towards the baseline. In hogging condition at the curvature $\chi = 0.26 \times 10^{-3}$ m^{-1}, the reduction in ultimate strength occurred due to buckling of the double bottom stiffener elements.

For the RoPax ship, the ultimate vertical bending moment for R1-I was calculated to 615 MNm and −563 MNm in hogging and sagging, respectively. The ultimate strength in sagging condition was governed by buckling collapse of the upper and main decks, while in the hogging condition it was determined by buckling of the double bottom structure. The other curves in the figure show how the M_U in hogging and sagging are reduced due to corrosion.

Figure 10 presents the $M_U - \chi$ results for the coastal oil tanker and the RoPax ship for the "damaged" cases in Table 3 under vertical bending conditions: see Ringsberg et al. (2017b) for more details. For the RoPax ship, the M_U reduction in hogging and sagging, between R1-D and R5-D, is 21% and 15% respectively. For the coastal oil tanker, the same analysis between T1-D and T5-D shows a reduction by 41% and 55% in hogging and sagging, respectively. A larger reduction in hogging was expected since the damage opening is on the compression side. Figure 11 shows examples of stress distributions in the coastal oil tanker and Ropax ship cross-sections for the cases T1-D and R1-D at the ultimate stress for these cases.

The residual ultimate strength in damaged condition was calculated in the cross-section that had the largest damage expansion (see Figure 7 and

8) in the FE analysis. The residual strength index RSI2 according to Yamada (2014) was used in the assessment which compared all the damaged and the intact cases in Table 3. Figure 12 presents the results in hogging and sagging conditions. For the coastal oil tanker, the RSI2 value was always lower in sagging compared to hogging, independent of if the ship was intact or damaged. For the RoPax ship, the index was almost the same in hogging and sagging for the intact cases, but for all damaged cases a larger value was found in sagging compared to hogging. Thus, this ship type suffers more from the collision damage and corrosion with respect to the relationship between hogging and sagging RSI2 values, even if it is relatively constant between the cases R1-D to R5-D. For the coastal oil tanker, there is a significant reduction in the residual ultimate strength index when comparing the T3-D and T5-D cases with most of the other coastal oil tanker cases. It is also notable that the coastal oil tanker has significantly larger reduction in RSI2 values between its cases compared to the RoPax ship's cases.

The results in Section 4.1 show that the damage opening of the coastal oil tanker was smaller for all cases but one (T5-D) compared to the RoPax ship. The larger percentage in reduced ultimate strength between the two vessels is because of two interacting reasons. Firstly, the tanker's ultimate strength is affected more by corrosion compared to the RoPax ship, cf. Figure 9 for intact conditions. Secondly, the Smith method in Fujikubo et al. (2012) shall be applied in the cross-section where most of the structure's elements are removed because of the collision damage. This means that the shape of the damage opening and its location are important. A comparison of the damage shapes in the outer and inner side-shells in Figure 8 shows that the expansion or "lengths" of the damages, i.e. along the side-shell, bilge corner and double bottom, are quite similar. However, the damage openings in the coastal oil tanker are in locations where the structural elements contribute significantly to the moment of inertia and structural strength; the RoPax ship is obviously less sensitive to this in its structural design.

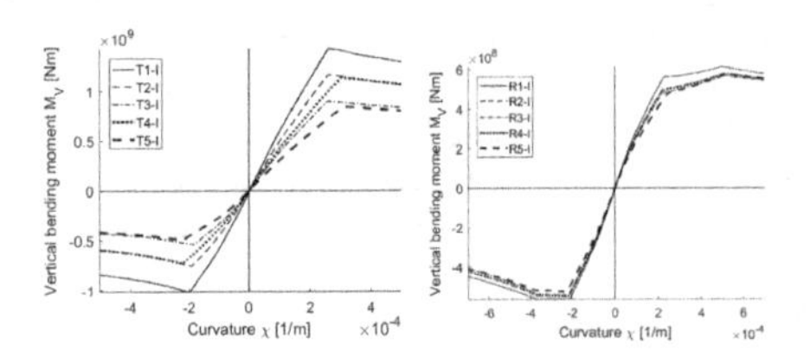

Figure 9. Ultimate strength analysis of the undamaged (intact) coastal oil tanker (left) and the RoPax ship (right).

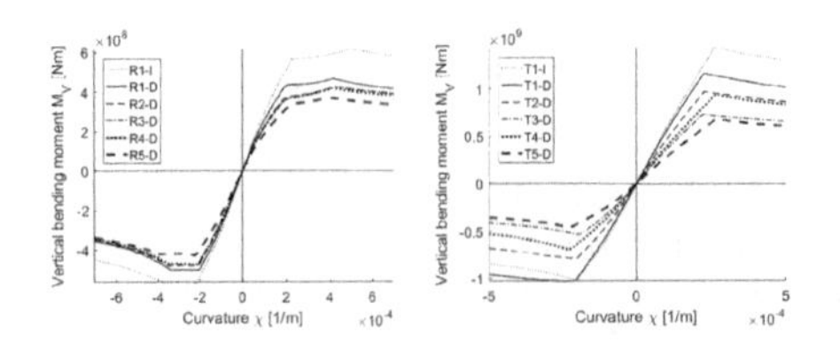

Figure 10. Ultimate strength analysis of damaged vessels presented as vertical bending moment versus curvature: (left) the RoPax ship, and (right) the coastal oil tanker.

4.2.2 *Biaxial bending moments: Intact and damaged conditions*

Section 4.2.1 presented the results from ultimate strength analyses when the ships where subjected only to vertical bending conditions. The Smith method in the MATLAB code was further developed to enable ultimate strength analyses during biaxial bending loading conditions. The results for the coastal oil tanker are presented with the vertical and horizontal bending moments at the ultimate stress plotted against each other in Figure 13.

The results for the intact cases show that the reduction in ultimate strength between pure vertical and horizontal bending moments is about the same, and that the biaxial loading plot is symmetric around the *y*-axis. It is very clear that corrosion has a negative effect in the ultimate bending strength, and also that the degree of corrosion and how it is represented in the model has great influence, cf. T3-I and T5-I.

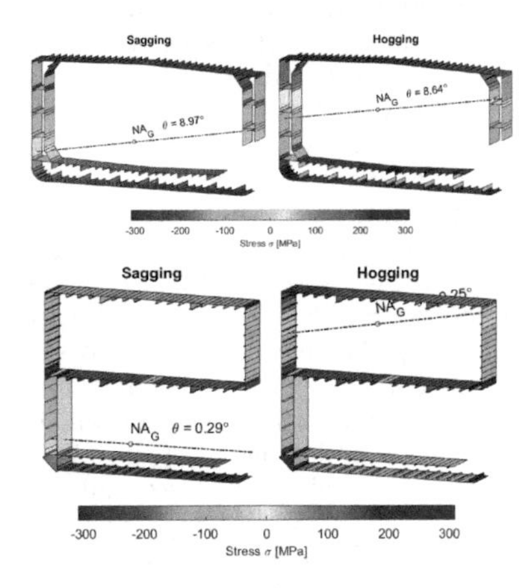

Figure 11. Axial average-stress distribution of T1-D and R1-D at ultimate strength point.

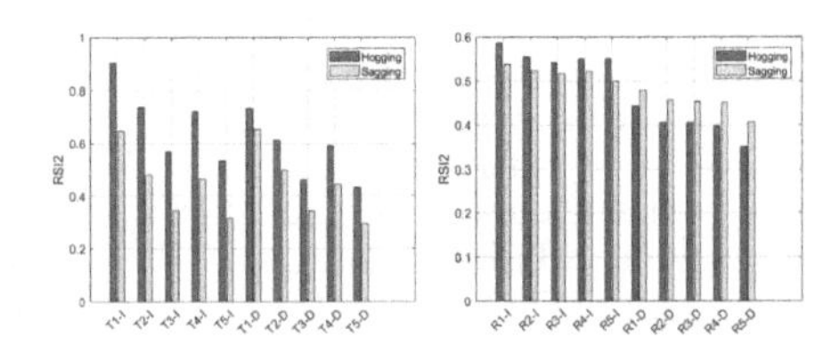

Figure 12. RSI2 for the (left) coastal oil tanker, and (right) the RoPax ship.

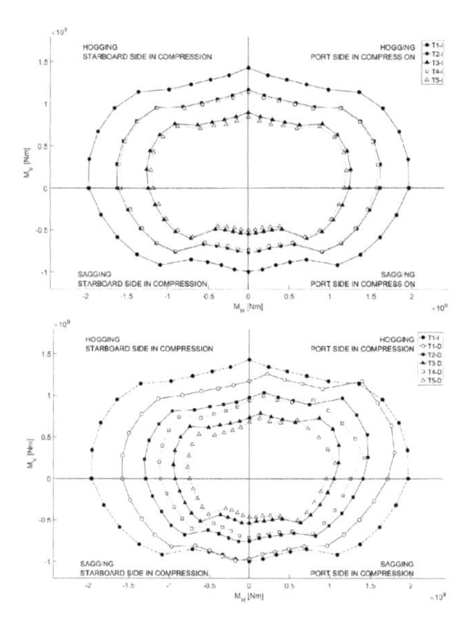

Figure 13. Ultimate bending moment diagrams in the vertical and horizontal directions at the ultimate stress for the coastal oil tanker. The upper figure shows results from the intact cases and the lower figure shows the results for the damaged cases.

For the damaged cases, the same trend regarding the influence from corrosion and how it is modelled can be seen. In addition to that, the biaxial loading plot is no longer symmetric around the *y*-axis because of the damage opening on the port side of the ship structure. The results show that the coastal oil tanker has lower ultimate strength capacity when the starboard side is in compression, i.e. the damage opening is subjected to tension loads. Overall, the biaxial ultimate strength bending moment results show that the damaged ship structure has lower ultimate strength capacity compared to the intact ship structure.

5 CONCLUSIONS

This study presented a methodology which was used in numerical analyses of the ultimate strength of intact and ship-collision-damaged ships, with different degrees and status of metal corrosion. Two case study vessels, a RoPax ship and a coastal oil tanker, were investigated when they were struck by a coastal tanker. Explicit FE analyses of collision events were presented where the consideration of corroded ship structure elements, their material characteristics in the model and vessel type were varied systematically in a parametric study. The ultimate strength of the struck vessel, for each collision event, was calculated using a verified in-house MATLAB code of the Smith method which used the shape and size of the damage openings from the ship-ship collision FE analyses.

The crashworthiness of the side-shell structures was reduced when the corrosion margin was reduced. The size of the damage opening increased even more when corroded material properties were considered in the analysis. Hence, an assessment of corroded ships involved in collision accidents should be carried out considering a reduction of the corrosion margin and modelling of corroded material properties, otherwise the damage opening may be greatly underestimated.

For a collision-damaged vessel, a prediction of the reduction in ultimate strength capacity must be carried out taking into account the status of its corrosion margin and the corroded material properties of the structure's material. It was found in the ultimate strength analyses that not only the size of the damage opening is important, but also its location and shape (here, largest expansion in a cross-section) are very important. The results limited to vertical bending moment conditions show that this is also dependent on the ship type: the coastal oil tanker suffered from a larger reduction in ultimate strength compared to the RoPax ship due to its structural design. Thus, in ultimate strength analyses of collision-damaged and corroded ships, it is recommended to carry out explicit nonlinear FE analyses of the ship collision scenarios in order to as realistically as possible estimate the damage openings' characteristics which are needed in e.g. the Smith method.

Ultimate strength analyses using the Smith method for biaxial bending moment conditions were carried out on the coastal oil tanker vessel. The results follow the same trends as for the vertical bending moment analyses regarding influence from corrosion and how it is accounted for in the numerical models. Howver, the collision-damaged cases show larger reduction in ultimate strength during biaxial loading conditions as compared with the intact cases, and also in comparison with the results for vertical bending moment only. Another observation was in the asymmetry in results around the vertical bending moment axis due to the asymmetry in the cross-section caused by the damage opening. The results show that when the damaged side of the ship is subjected to tensile loading conditions, it has a lower ultimate strength capacity compared to when it is subjected to compression loads.

ACKNOWLEDGEMENTS

The FE analyses carried out in this study were partly performed on resources at Chalmers Centre for Computational Science and Engineering (C3SE www.c3 se.chalmers.se) provided by the Swedish National Infrastructure for Computing (SNIC).

REFERENCES

Abaqus. 2016. Dassault Systemes Simulia, Abaqus version 6.13–3. [Available online: http://www.3ds.com/products-services/simulia/products/abaqus/; December 2017].

AbuBakar A, Dow RS. 2016. The impact analysis characteristics of a ship's bow during collisions. In: Proceedings of the Seventh International Conference on Collision and Grounding of Ships and Offshore Structures (ICCGS 2016); Ulsan, Korea, 15–18 June 2016.

Bureau Veritas. 2015. MARS2000. [Available online: http://www.veristar.com/; December 2017].

Campanile A, Piscopo V, Scamardella A. 2015. Statistical properties of bulk carrier residual strength. Ocean Engineering. 106(1): 47–67.

Downes J, Tayyar GT, Kvan I, Choung J. 2016. A new procedure for load-shortening and - elongation data for progressive collapse method. International Journal of Naval Architecture and Ocean Engineering. 9(6): 705–719.

Ehlers S. 2010. The influence of the material relation on the accuracy of collision simulations. Marine Structures. 23(4): 462–474.

Ehlers S, Østby E. 2012. Increased crashworthiness due to arctic conditions—The influence of sub-zero temperature. Marine Structures. 28(1): 86–100.

Faisal M, Noh SH, Kawsar MRU, Youssef SAM, Seo JK, Ha YC, Paik JK. 2016. Rapid hull collapse strength calculations of double hull oil tankers after collisions. Ships and Offshore Structures. 12(5): 624–639.

Fujikubo M, Zubair MA, Takemura K, Iijima K, Oka S. 2012. Residual hull girder strength of asymmetrically damaged ships. Journal of the Japan Society of Naval Architects and Ocean Engineers. 16: 131–140.

Garbatov Y, Guedes Soares C, Parunov J, Kodvanj J. 2014. Tensile strength assessment of corroded small scale specimens. Corrosion Science. 85(1): 296–303.

Garbatov Y, Parunov J, Kodvanj J, Saad-Eldeen S, Guedes Soares C. 2016. Experimental assessment of tensile strength of corroded steel specimen subjected to sandblast and sandpaper cleaning. Marine Structures. 49(1): 18–30.

Hogström P, Ringsberg JW, Johnson E. 2009. An experimental and numerical study of the effects of length scale and strain state on the necking and fracture behaviours in sheet metals. International Journal of Impact Engineering. 36(10–11): 1194–1203.

Hogström P, Ringsberg JW. 2012. An extensive study of a ship's survivability after collision—A parameter study of material characteristics, non-linear FEA and damage stability analyses. Marine Structures. 27(1): 1–28.

Hogström P, Ringsberg JW. 2013. Assessment of the crashworthiness of a selection of innovative ship structures. Ocean Engineering. 59(1): 58–72.

IACS. 2017. Common structural rules for bulk carriers and oil tankers. Version: 1 January, 2017. International Association of Classification Societies (IACS), London, U.K.

Karlsson U. 2009. Improved collision safety of ships by an intrusion-tolerant inner side-shell. Marine Technology. 46(3): 165–173.

Kuznecovs A, Shafieisabet R. 2017. Analysis of the ultimate limit state of corroded ships after collision. MSc thesis Report X-17/376, Department of Mechanics and Maritime Sciences, Chalmers University of Technology, Gothenburg, Sweden.

Marinatos JN, Samuelides MS. 2015. Towards a unified methodology for the simulation of rupture in collision and grounding of ships. Marine Structures. 42(1): 1–32.

Nishihara S. 1984. Ultimate longitudinal strength of mid-ship cross section. Naval Architecture and Ocean Engineering (SNAJ Publication). 22: 200–214.

Paik JK, Lee JM, Park YI, Hwang JS, Kim CW. 2003. Time-variant ultimate longitudinal strength of corroded bulk carriers. Marine Structures. 16(8): 567–600.

Paik JK, Melchers RE (2008). Condition assessment of aged structures. Cambridge (England): Woodhead Publishing Limited and CRC Press LLC.

Paik JK, Kim BJ, Seo JK. 2008. Methods for ultimate limit state assessment of ships and ship-shaped offshore structures: Part II stiffened panels. Ocean Engineering. 35(2): 271–280.

Paik JK, Kim DK, Kim M-S. 2009. Ultimate strength performance of Suezmax tanker structures: Pre-CSR versus CSR designs. The International Journal of Maritime Engineering. 151(A2): 39–58.

Parunov J, Rudan S, Gledić I, Bužančić Primorac B. 2017. Finite element study of residual ultimate strength of a double hull oil tanker damaged in collision and subjected to bi-axial bending. In: Proceedings of the International Conference on Ships and Offshore Structures (ICSOS 2017); Shenzhen, China, 11–13 September 2017.

Ringsberg JW, Li Z, Johnson E. 2017a. Performance assessment of crashworthiness of corroded ship hulls. In: Proceedings of the Sixth International Conference on Marine Structures; Lisbon, Portugal, 8–10 May 2017.

Ringsberg JW, Li Z, Johnson E, Kuznecovs, A, Shafieisabet, R. 2017b. Reduction in ultimate strength capacity of corroded ships involved in collision accidents. In: Proceedings of the International Conference on Ships and Offshore Structures (ICSOS 2017); Shenzhen, China, 11–13 September 2017.

Saad-Eldeen S, Garbatov Y, Guedes Soares C. 2011. Experimental assessment of the ultimate strength of a box girder subjected to severe corrosion. Marine Structures. 24(4): 338–357.

Samuelides MS. 2015. Recent advances and future trends in structural crashworthiness of ship structures subjected to impact loads. Ships and Offshore Structures. 10(5): 488–497.

Storheim M, Amdahl J, Martens I. 2015a. On the accuracy of fracture estimation in collision analysis of ship and offshore structures. Marine Structures. 44(1): 254–287.

Storheim M, Alsos HS, Hopperstad OS, Amdahl J. 2015b. A damage-based failure model for coarsely meshed shell structures. International Journal of Impact Engineering. 83(1): 59–75.

Yamada Y. 2014. Numerical study on the residual ultimate strength of hull girder of a bulk carrier after ship-ship collision. In: Proceedings of the ASME Thirty-Third International Conference on Ocean, Offshore and Arctic Engineering (OMAE2014); San Francisco, California, USA, 8–13 June 2014.

Zhang S, Pedersen PT. 2016. A method for ship collision damage and energy absorption analysis and its validation. In: Proceedings of the First International Conference on Ships and Offshore Structures (ICSOS 2016); Hamburg, Germany, 31 August–2 September 2016.

Progress in Maritime Technology and Engineering – Guedes Soares & Santos (Eds)
 ISBN 978-1-138-58539-3

Residual strength assessment of a grounded container ship subjected to asymmetrical bending loads

M. Tekgoz, Y. Garbatov & C. Guedes Soares
Centre for Marine Technology and Ocean Engineering (CENTEC), Instituto Superior Técnico, Universidade de Lisboa, Lisbon, Portugal

ABSTRACT: The objective of this work is to analyze the effect of grounding related damage and associated neutral axis rotation on the residual load carrying capacity of a container ship hull subjected to asymmetrical bending. The assessment is performed by the finite element method and a formulation based on the Common Structural Rules (CSR). An update to the progressive collapse approach stipulated by CSR is proposed and compared with the finite element solution. Finally, several conclusions are presented.

1 INTRODUCTION

Structural damage is a very important phenomenon to be investigated since it may lead to a major reduction in the structural loading capacity and poses a danger in seafarer lives, marine environment and the cargo being carried on board. Therefore, damaged ship structural strength assessment is of utmost importance from a safety point of view. The longitudinal strength, which is related to ship hull girder strength, is the most important strength to ensure the safety of ship structures.

Guedes Soares et al. (2008) evaluated the ability of using simplified methods based on the progressive collapse to predict the ultimate strength of a damaged ship. The results of the approximate methods agreed well with each other for both conditions, damaged and intact. The simplified methods were more conservative than the finite element analysis in hogging while it seems to give a good approximation to the result of sagging with some overestimation.

Yoshikawa et al. (2008) studied the damage strength for bulk carriers after the ship grounding using a different approach in that the damage parts were not removed but considered in the load-displacement relation of the damaged panel, independently of the fact that the ultimate strength of the damaged panel is insignificant when the damage occurs between two neighbouring transverse frames.

Tekgoz et al. (2015a) studied the compressive strength of a single hull damaged tanker ship subjected to asymmetrical bending loading. It was concluded that the structural damage leads to the maximum bending capacity shift and reduction as a function of the heeling angle degree.

When the damage occurs, the ship strength is reduced, which leads to changes in the ship cross-section structural descriptors. Therefore, as a result of the created asymmetrical bending loadings, the neutral axis translates and rotates.

Fujikubo et al. (2012) discussed the influence of the rotation of the neutral axis on the residual strength of bulk carriers and double-hull hankers having collision damages at the side structures. The progressive collapse analysis was applied, employing the Smith's approach for a biaxial bending problem. The reduction of the ultimate strength was investigated for different damage locations and extends.

Choung et al. (2012) studied the residual strength of an asymmetrically damaged tanker considering rotational and translational shifts of the neutral axis plane. A new criterion defined as a force vector equilibrium condition was proposed for defining the translational and rotational location of the neutral axis plane.

Tekgoz et al. (2018) assessed the ultimate strength capacity of an intact and damage container ship under asymmetrical bending loadings taken into the account of the influencing factors, that may have significant impact on the residual and intact ultimate loading capacity, such as the neutral axis plane mobility, the external boundary conditions and the structural definition of a model to be analyzed in the finite element and closed-form solutions.

The objective of this work is to analyze the effect of structural damage and the neutral axis rotation and translation and the resulting asymmetrical bending load on the ultimate strength assessment of a grounded container ship. The numerical evaluation of the ultimate strength of a container ship hull girder subjected to asymmetrical bending is performed based on the common structural rules, CSR, IACS (2012) and advanced nonlinear finite element method, ANSYS (2009).

2 THE DESCRIPTION OF THE DAMAGE CASES AND FINITE ELEMENT MODELLING

2.1 *Structural description*

The containership is presented in Figure 1 and its stiffener descriptors are given in Table 1.

2.2 *Finite element modelling*

The ultimate strength is analyzed by a finite element analysis (FEA), using the commercial software ANSYS (2009), which enables modelling of the elastic plastic material properties and large deflections. Four-node quadrilateral shell element, Shell 181 has been used to model the plates and stiffeners. The kinematic assumption of the finite element analysis is a large displacement and rotation. The material is assumed to be bilinear elastic-perfectly-plastic without hardening.

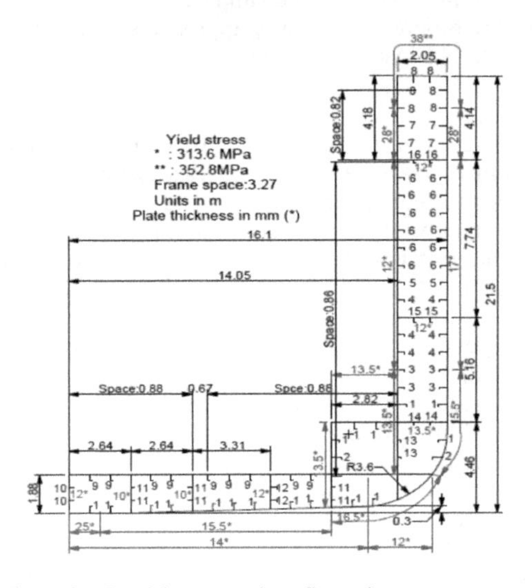

Figure 1. Model structural configuration.

Table 1. Container ship stiffener specifications.

Stiff. No	Dimensions (mm)	Type
1	350 × 100 × 12/17	Angle-bar
2	400 × 100 × 11.5/16	Angle-bar
3	300 × 90 × 13/17	Angle-bar
4	300 × 90 × 11/16	Angle-bar
5	250 × 90 × 12/16	Angle-bar
6	250 × 90 × 10/15	Angle-bar
7	300 × 28	Flat-Bar
8	300 × 38	Flat-Bar
9	300 × 90 × 13/17	Angle-bar
10	230 × 10	Flat-Bar
11	150 × 10	Flat-Bar
12	300 × 90 × 11/16	Angle-bar
13	150 × 90 × 9/9	Angle-bar
14	150 × 12	Flat-Bar
15	250 × 90 × 12/15	Angle-bar
16	150 × 90 × 12/12	Angle-bar

The initial geometry imperfection of plates and stiffeners of the container ship is generated by a pre-deformed surface (see Figure 2).

The initial geometry surface imperfection is modelled as proposed by Paik et al. (2012):

$$w_L(x,y) = w_{L\max}\sin(\pi\frac{mx}{l})\sin(\pi\frac{ny}{b_l}) \quad (1)$$

$$w_G(x,y) = w_{G\max}\sin(\pi\frac{mx}{l})\sin(\pi\frac{ny}{b_G}) \quad (2)$$

$$w_s(x,z) = C_o\frac{z}{h_w}\sin\frac{\pi x}{l} \quad (3)$$

where h_w is the stiffener web height, l is the length of the panel and b_l and b_G are the breath between stiffeners and the girder respectively, x and y are the Cartesian coordinates of any location on the plate and m and n are the number of half waves assumed in the x and y directions. In all studied cases, n has been set to be 1, $w_{G\max}$ has been taken identical to C_o and the imperfection magnitude between longitudinals has been defined as:

$$w_{L\max} = b/200 \quad (4)$$

$$C_o = 0.0015l \quad (5)$$

For the finite element model studied herein, four finite elements for the webs, two elements for the

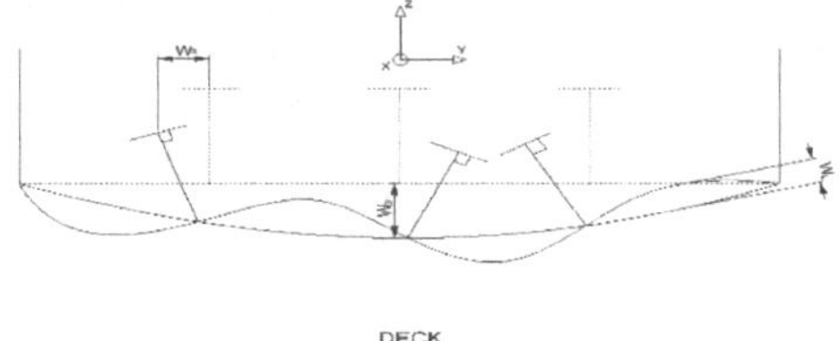

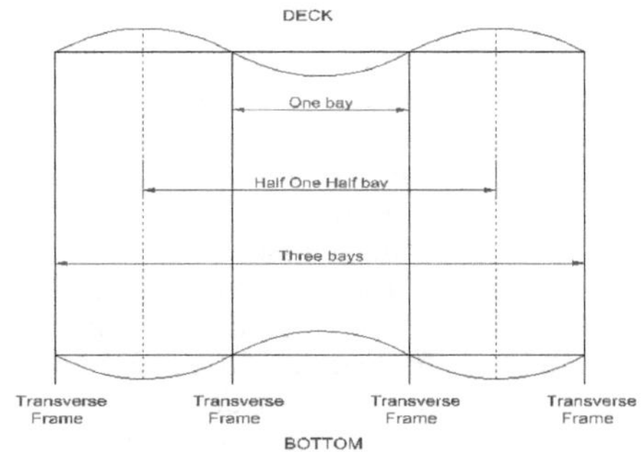

Figure 2. Initial imperfection, cross sectional, up and global longitudinal, down view.

flanges and ten elements for the plates between longitudinal stiffeners have been generated. It is a common practice, but also more importantly, due to the initial imperfection imposed on the stiffener, the stiffener flanges are modelled by shell elements.

When the structure is modelled with big elements, it decreases the structural strain energy potential and leads to a more optimistic strength capacity estimate. In the case of a model with a smaller element size, the modelled structure presents a more strain energy potential and in turn more degrees of freedom, which subsequently lead to a less capacity. This is also true for the element with higher order shape functions. In order to establish the optimum finite element mesh size, an analysis is performed to identify the minimum resistance to the applied external load.

To keep the end cross-section plane, stiff beam elements have been implemented, generating umbrella boundary conditions. The boundary conditions are defined using master nodes located at the end of the crossing point between the central line and the initial elastic neutral axis in the fore and aft net sections of the studied segment. The master nodes are connected by stiff beams to the element nodes of the two extreme edges of the segment, preventing any local corrugation and keeping the two net-section planes during the step loading (see Figure 3).

To calculate the curvature at the collapsing bay, the so-called coupling equations in the finite element solution is implemented, which allows the nodes in the collapsing bay transverse frames to act as one unit. For this, additional master nodes are created at the transverse frames of the middle bay, which is in line with the initial neutral axis position. The nodes at the collapse bay transverse frames are linked to their corresponding master nodes, created through the coupling equations, which allows to estimate the curvature.

This method has also been implemented in order to estimate the progressive neutral axis rotations in the case of the finite element solution (Tekgoz et al., 2012, 2015b, a, Tekgoz et al., 2018).

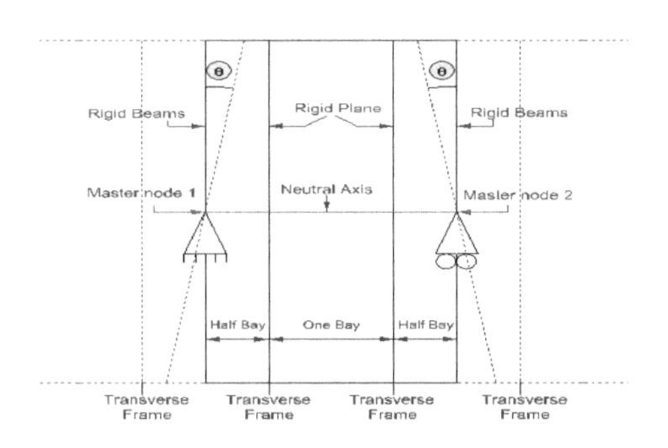

Figure 3. Boundary conditions and Finite element model extent.

2.3 *Description of damage cases*

Three damage scenarios, based on the harmonized common structural rules, IACS (2014) have been studied here (see Figure 4 and Table 2).

Where b is the damage breath, h is the damage height and B is the ship breath. The total area reductions are kept close as much as possible. The damage is shifted transversely in order to find the most unfavorable location in terms of the ultimate residual loading capacity.

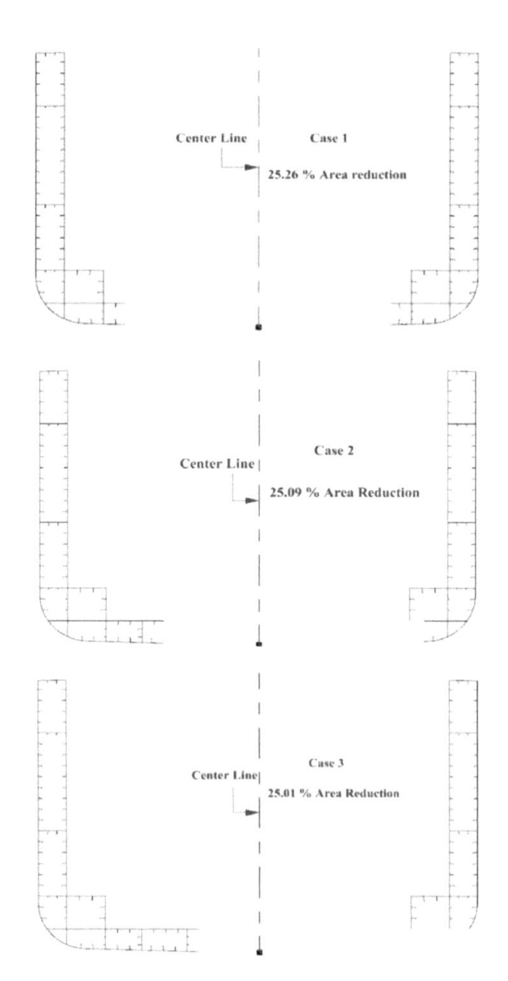

Figure 4. Damage cases.

Table 2. Damage area for bulk carrier, H-CSR.

Damage penetration, in m	
Height, h	Min (B/20, 2)
Breadth, b	0.6B

3 ULTIMATE STRENGTH ESTIMATION

When the structure is progressively and asymmetrically loaded to determine the ultimate limit (collapse), some of its structural components will progressively lose their strength, introducing additional asymmetry in the global structural stiffness. As a result of that at the ultimate limit state of the loading, as long as the external load does not act on one of the principle planes, the neutral axis will not only translate but also rotate with respect to its initial position as can be seen in Figure 5.

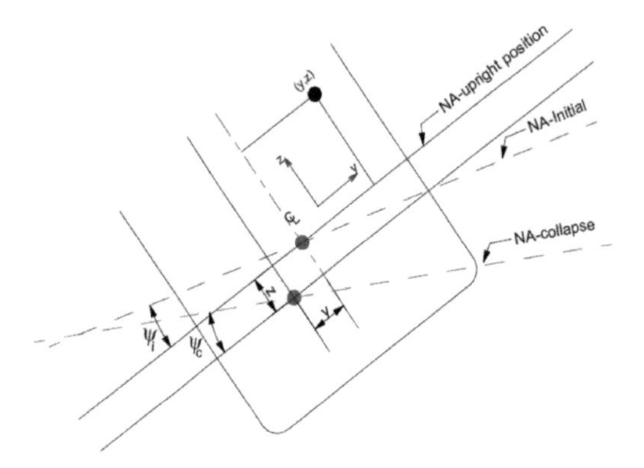

Figure 5. Neutral axis transition and rotation.

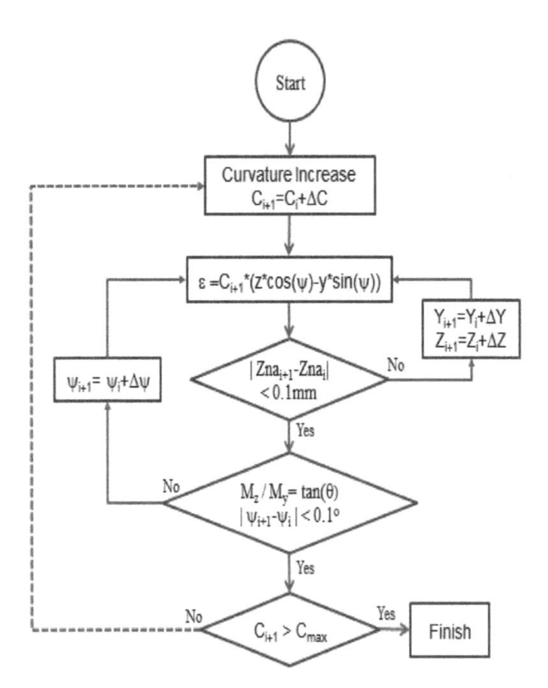

Figure 6. Ultimate strength assessment iteration procedure with the newly proposed convergence method.

The ultimate strength estimations are based on the solution of the common structural rules, CSR and the finite element solution, FEM. The model, CSR is the solution of the Common Structural Rules accounting for only the neutral axis translations. The model, CSR-U is the updated solution of the common structural solution with the neutral axis translations and plus rotations. The model, FEM is the solution of the finite element analysis considering both the neutral axis translations and rotations.

4 DISCUSSION

Three damage scenarios have been studied here, based on the Common Structural Rules solution under the asymmetrical bending loadings in the sagging bending conditions and are compared with the solution of the finite element method at 0 degree of heeling.

Figure 7 shows the interaction bending moment curve under the asymmetrical bending load and it has been broken down into four regions. Regions A and D have only been studied. The results are shown in Table 3 and Table 4.

The model CSR-U, which accounts for the neutral axis translations and rotations, is the update of the Common Structural Rules solution, CSR that accounts for only the neutral axis translations under the asymmetrical bending load. The strength reductions have been calculated between CSR-U intact and CSR-U damage and between CSR intact and CSR damage. It turns out that the neutral axis rotation effect on the residual ultimate strength capacity reduction becomes more significant in 0 degrees of heeling for Case 3 in this particular damage and ship.

Figure 8 shows the ultimate strength reductions under the asymmetrical bending loading in Region

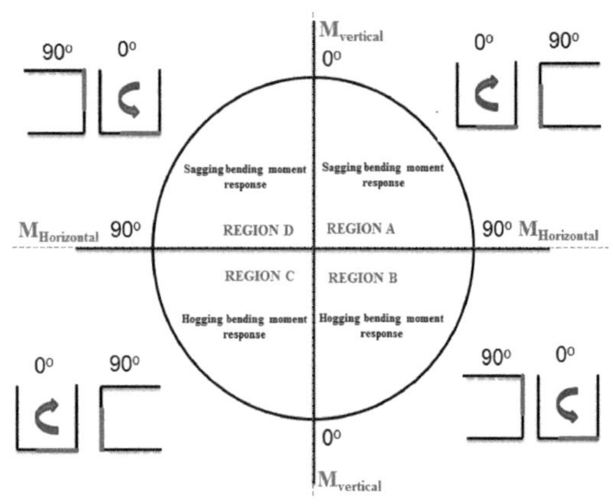

Figure 7. Interaction bending moment curve.

Table 3. Ultimate strength, Region A.

Model	Case	Ship Heeling-Region A, Ultimate strength, GNm				
		0°	30°	45°	60°	75°
CSR-U	Intact	6.37	7.15	7.62	8.21	9.31
	Case 1	5.56	5.73	6.05	6.72	8.09
	Case 2	5.46	5.46	5.69	6.29	7.52
	Case 3	5.23	5.16	5.34	5.87	6.97
CSR	Intact	6.37	7.01	7.31	8.03	9.86
	Case 1	5.56	5.59	5.72	6.51	8.77
	Case 2	5.57	5.43	5.43	6.07	8.19
	Case 3	5.53	5.29	5.16	5.68	7.54
CSR-U, Strength reduction, %	Case 1	12.77	19.86	20.57	18.10	13.08
	Case 2	14.34	23.66	25.29	23.38	19.21
	Case 3	17.92	27.82	29.89	28.41	25.14
CSR, Strength reduction, %	Case 1	12.77	20.17	21.82	18.84	11.03
	Case 2	12.48	22.50	25.71	24.33	16.93
	Case 3	13.12	24.46	29.44	29.25	23.55

Table 4. Ultimate strength, Region D.

Model	Case	Ship Heeling-Region D, Ultimate strength, GNm				
		0°	30°	45°	60°	75°
CSR-U	Intact	6.37	7.15	7.62	8.21	9.31
	Case 1	5.56	5.73	6.05	6.72	8.09
	Case 2	5.46	5.99	6.43	7.18	8.66
	Case 3	5.23	6.18	6.74	7.58	8.62
CSR	Intact	6.37	7.01	7.31	8.03	9.86
	Case 1	5.56	5.59	5.72	6.51	8.77
	Case 2	5.57	5.79	6.05	6.98	8.99
	Case 3	5.53	5.97	6.36	7.35	8.62
CSR-U, Strength reduction, %	Case 1	12.77	19.86	20.57	18.10	13.08
	Case 2	14.34	16.30	15.68	12.44	6.95
	Case 3	17.92	13.67	11.53	7.62	7.37
CSR, Strength reduction, %	Case 1	12.77	20.17	21.82	18.84	11.03
	Case 2	12.48	17.36	17.32	13.04	8.78
	Case 3	13.12	14.75	12.97	8.42	12.58

A. The ultimate strength calculation is based on the CSR-U that is the update of the Common Structural Rules solution accounting for the neutral axis translations and rotations and based on the CSR, which only accounts for the neutral axis translations. It shows that, Case 3 is the most significant damage case for this particular container ship and its impact is most pronounced in 45 degree of ship heeling. In fact, in this case if the progressive neutral axis rotation is not considered, the residual ultimate strength prediction may lead to an optimistic one at 0 degree of heeling.

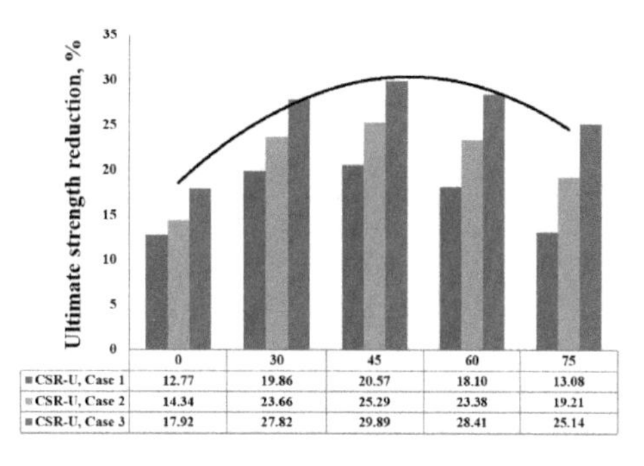

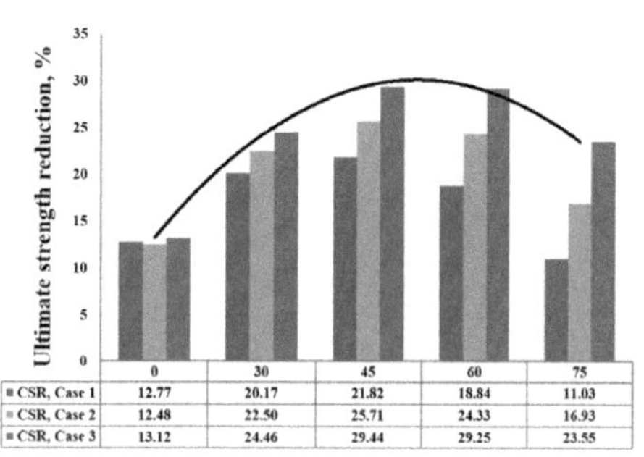

Figure 8. Region A, Ultimate strength reductions under asymmetrical bending loading, CSR-U (up), CSR (down).

Figure 9 shows the ultimate strength reductions under the asymmetrical bending loading in Region D. The ultimate strength calculation is based on the CSR-U that is the update of the Common Structural Rules solution accounting for the neutral axis translations and rotations and based on the CSR, which only accounts for the neutral axis translations. As it turns out that, Case 1 is the most significant damage case for this particular container ship and its impact is most pronounced in 45 degree of ship heeling as it is also the Case 3 in Region A. In addition to that, in the Case 3 and partially in the Case 2, the damage impact decreases as the ship heeling angle increases.

When the load or geometrical asymmetry is present, this leads to an asymmetrical bending. In this case, the neutral axis does not only translate, but also rotate due to the fact that the external bending moment is being subjected to both principal axes whose curvature difference leads to a neutral axis rotation.

To find the effect of the neutral axis rotation on the residual ultimate strength capacity in-between the CSR-U that accounts for the neutral axis translations and rotations and the model of CSR, which only accounts for the neutral axis translations, Region A and D have been compared. The results have been shown in Table 5 and Table 6.

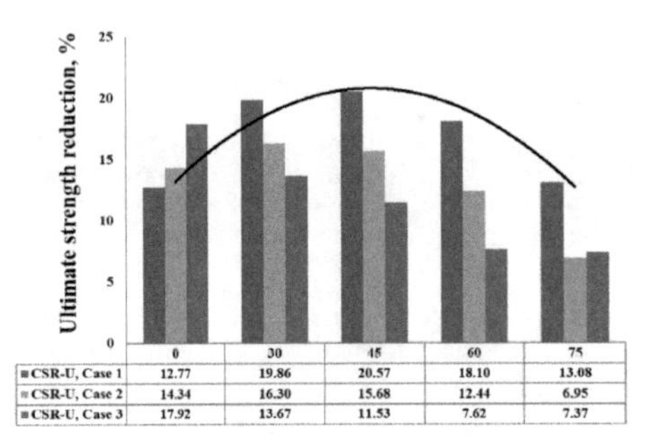

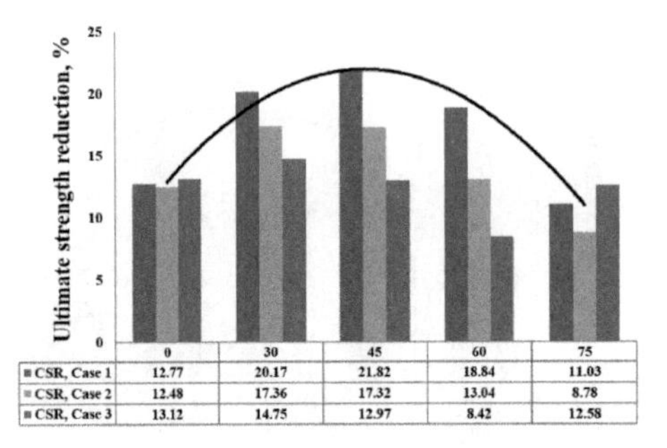

Figure 9. Region D, Ultimate strength reductions under asymmetrical bending loading, CSR-U (up), CSR (down).

Table 5. Ultimate strength difference, Region A.

	Ship Heeling-Region A, Ultimate strength, GNm					
Model	Case	0°	30°	45°	60°	75°
CSR-U	Case 1	5.56	5.73	6.05	6.72	8.09
	Case 2	5.46	5.46	5.69	6.29	7.52
	Case 3	5.23	5.16	5.34	5.87	6.97
CSR	Case 1	5.56	5.59	5.72	6.51	8.77
	Case 2	5.57	5.43	5.43	6.07	8.19
	Case 3	5.53	5.29	5.16	5.68	7.54
Strength difference, %	Case 1	0.0	−2.5	−5.9	−3.2	7.8
	Case 2	2.1	−0.6	−4.8	−3.5	8.2
	Case 3	5.5	2.4	−3.5	−3.5	7.5

ultimate strength prediction, which may arrive as far as 8% in the case of 0 and 75 degrees of ship heeling or some optimistic estimations that may arrive as far as 6% as in the case of 30, 45 and 60 degrees of heeling.

Figure 11 shows the ultimate residual strength differences using both the models of CSR-U and CSR. The results are quite similar to the ones derived from Region D. As it suggests that unlike the neutral axis translations, the neutral axis rotation has a different effect on the ultimate strength capacity since in certain degrees of heeling, when

Figure 10 shows the ultimate residual strength differences using both the models of CSR-U and CSR. As it suggests that unlike the neutral axis translations, the neutral axis rotation has a different effect on the ultimate strength capacity since in certain degrees of heeling, when the neutral axis rotation is considered it may lead to a pessimistic

Table 6. Ultimate strength difference, Region D.

	Ship Heeling-Region D, Ultimate strength, GNm					
Model	Case	0°	30°	45°	60°	75°
CSR-U	Case 1	5.56	5.73	6.05	6.72	8.09
	Case 2	5.46	5.99	6.43	7.18	8.66
	Case 3	5.23	6.18	6.74	7.58	8.62
CSR	Case 1	5.56	5.59	5.72	6.51	8.77
	Case 2	5.57	5.79	6.05	6.98	8.99
	Case 3	5.53	5.97	6.36	7.35	8.62
Strength difference, %	Case 1	0.0	−2.5	−5.9	−3.2	7.8
	Case 2	2.1	−3.4	−6.3	−2.9	3.7
	Case 3	5.5	−3.4	−5.9	−3.1	0.0

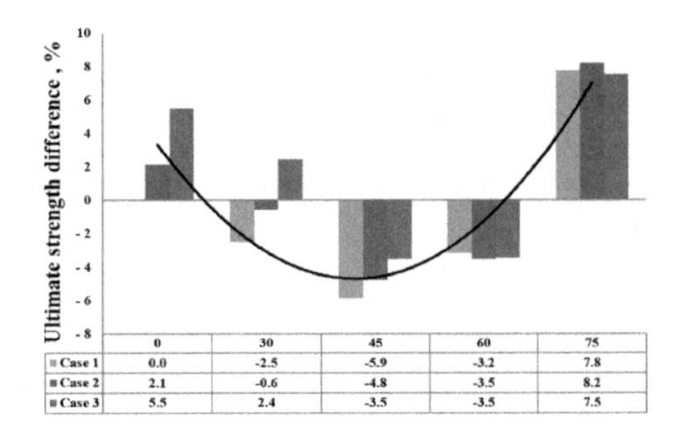

Figure 10. Region A, Ultimate strength differences, CSR-U and CSR.

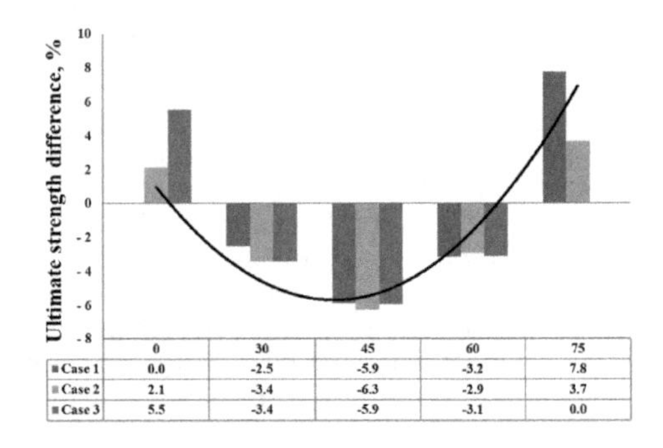

Figure 11. Region D, The ultimate strength prediction differences, CSR-U and CSR.

the neutral axis rotation is considered it may lead to a pessimistic ultimate strength prediction which may arrive as far as 8% in the case of 0 and 75 degrees of ship heeling or some optimistic predictions that may arrive as far as 6% as in the case of 30, 45 and 60 degrees of heeling.

Figure 12 shows the bending moment—curvature relationship under 0 degree of heeling for intact and damage cases studied here. The damage leads to the structural bending rigidity and ultimate load capacity reductions.

Figure 13 and Figure 14 show the bending moment—curvature relationships of Case 2 and Case 3 under 0 degree of heeling along with the progressive neutral axis rotations respectively. The model, CSR-U accounts for the neutral axis translations and rotations, the model CSR accounts for only the neutral axis translations.

Similar results have been presented in Figure 15 to Figure 17 where the bending moment—curvature relationships have been shown along with the progressive neutral axis rotations under 45 degrees of heeling for Case 1, Case 2 and Case 3.

Lastly, the CSR, CSR-U and FEM solutions have been compared in terms of the bending moment—curvature relationship along with the neutral axis

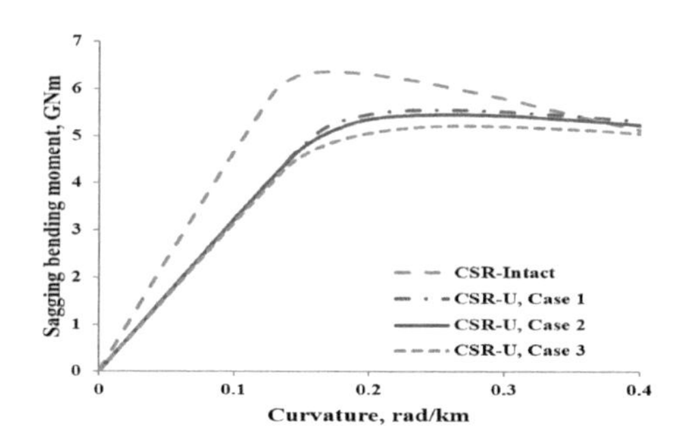

Figure 12. Residual sagging bending moment—curvature relationship, 0° of heeling.

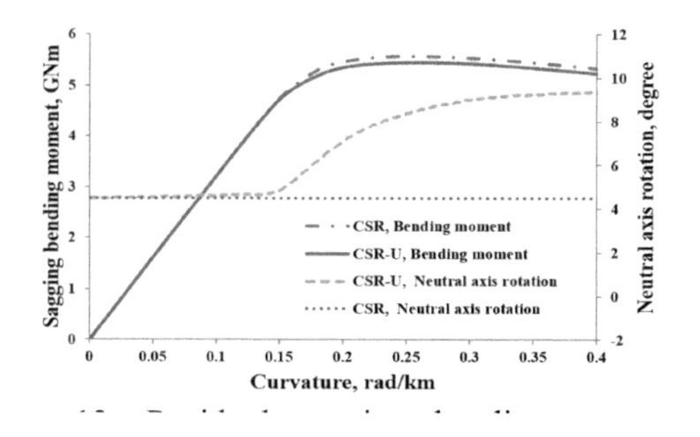

Figure 13. Residual sagging bending moment—curvature relationship under 0° of heeling, Case 2.

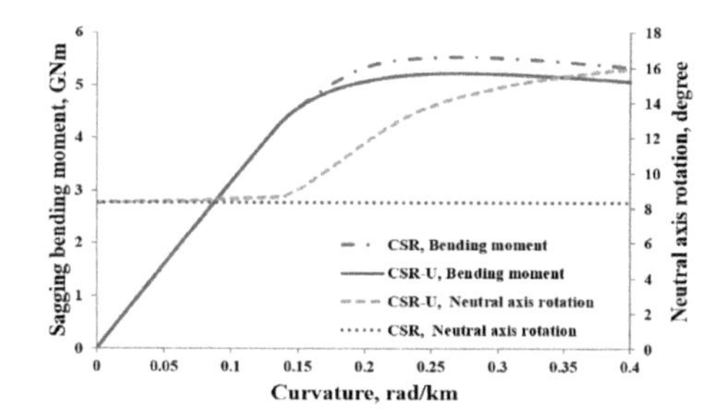

Figure 14. Residual sagging bending moment—curvature relationship under 0° of heeling, Case 3.

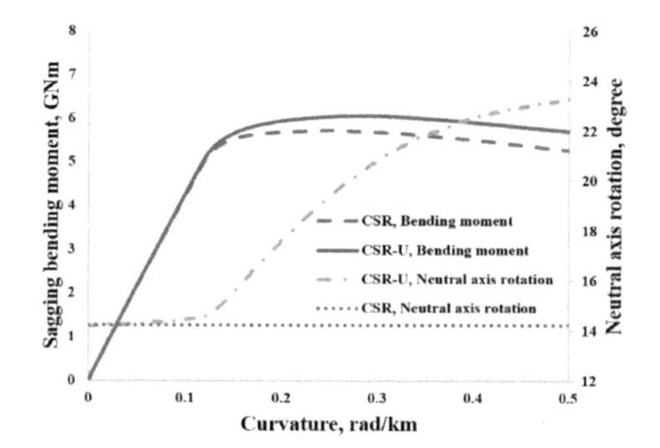

Figure 15. Residual sagging bending moment—curvature relationship under 45° of heeling, Region A, Case 1.

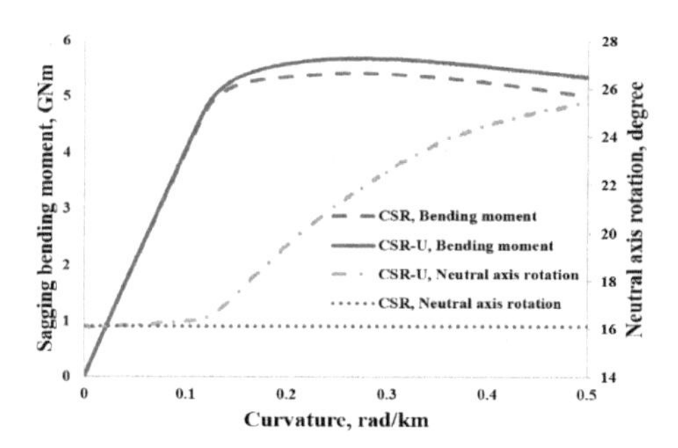

Figure 16. Residual sagging bending moment—curvature relationship under 45° of heeling, Region A, Case 2.

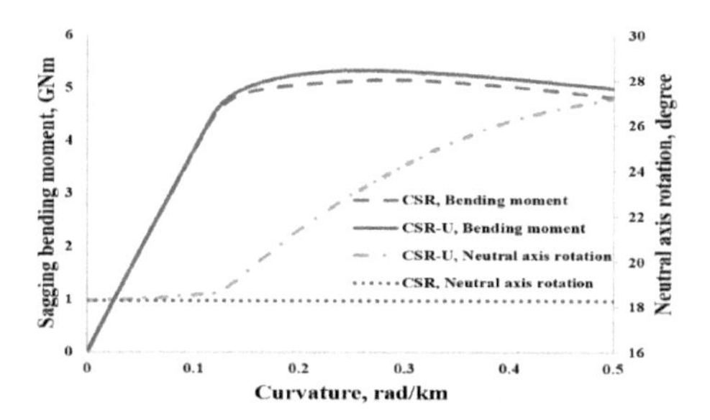

Figure 17. Residual sagging bending moment—curvature relationship under 45° of heeling, Region A, Case 3.

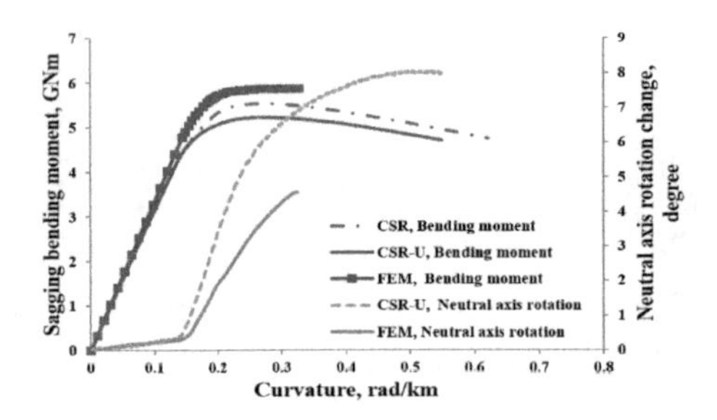

Figure 18. Residual sagging bending moment—curvature relationship under 0° of heeling, Case 3, CSR-U, CSR and FEM.

rotation variations as shown in Figure 18. As can be seen the neutral axis rotation variations are quite similar up to the non-linear range and the bending moment curvature relationship results of CSR are closer to the one of FEM solution.

5 CONCLUSIONS

Grounding related three damage cases have been studied in terms of the residual ultimate strength for a container ship under the asymmetrical bending load based on the Common Structural Rules and the finite element method accounting for the neutral axis translations and rotations.

The residual strength reduction is more pronounced at 45 degree of ship heeling for each region studied here that may result in 30% strength reduction in the case of 25.01% area reduction for this particular grounding related damage. In addition, depending on the ship heeling region, the damage case significance may differ. It is important to point out that the damage shape and accompanying residual stresses were not taken into account, which should decrease the strength reduction.

As the grounding related damage shifts transversely from the center line, its impact becomes more significant on the strength reduction.

The neutral axis rotation consideration on the residual strength estimate may lead to a lower strength is the case of 0 and 75 degrees of heeling or upper residual strength in the cases of 30, 45 and 60 degrees of ship heeling.

ACKNOWLEDGEMENTS

This work was performed within the Strategic Research Plan of the Centre for Marine Technology and Ocean Engineering (CENTEC), which is financed by Portuguese Foundation for Science and Technology (Fundação para a Ciência e Tecnologia-FCT).

REFERENCES

ANSYS 2009. Online Manuals, Release 12.

Choung, J., Nama, J.-M. & Ha, T.-B. 2012. Assessment of residual ultimate strength of an asymmetrically damaged tanker considering rotational and translational shifts of neutral axis plane. *Marine Structures,* 25, 71–84.

Fujikubo, M., Takemura, K., Oka, S., Alie, A.Z.M. & Ijima, K. 2012. Residual Hull Girder Strength of Asymmetrically Damaged Ships -Influence of Rotation of Neutral Axis due to Damages. *Journal of Japan Society of Naval Architects and Ocean Engineers,* 16, 131–140.

Guedes Soares, C., Luís, R.M., Nikolov, P., Dowes, J., Taczala, M., Modiga, M., Quesnel, T., Toderan, C. & Samuelides, M. 2008. Benchmark study on the use of simplified structural codes to predict the ultimate strength of a damaged ship hull. *International Shipbuilding Progress,* 55, 87–107.

IACS 2012. Common Structure Rules for Double Hull Oil Tankers, Consolidated version, July 2012.

IACS 2014. Common Structural Rules for Bulk Carriers and Oil Tankers.

Paik, J.K., Amlashi, H., Boon, B., Branner, K., Caridis, P., Das, P., Fujikubo, M., Huang, C.H., Josefson, L., Kaeding, P., Kim, C.W., Parmentier, G., Pasqualino, I.P., Rizzo, C.M., Vhanmane, S., Wang, X. & Yang, P. 2012. ISSC Commitie III.1 Ultimate Strength. *In:* Fricke, W. & Bronsart, R. (eds.) *18th International ship and offshore structures congress.* Hamburg: Schiffbautechnische Gesellschaft, 285–364.

Tekgoz, M., Garbatov, Y. & Guedes Soares, C. 2012. Ultimate strength assessment accounting for the effect of finite element modelling. *In:* Guedes Soares, C., Garbatov, Y., Sutulo, S. & Santos, T. (eds.) *Maritime Engineering and Technology.* London, UK: Taylor & Francis Group, 353–362.

Tekgoz, M., Garbatov, Y. & Guedes Soares, C. 2015a. Strength assessment of a single hull damaged tanker ship subjected to asymmetrical bending loading. *In:* Guedes Soares, C., Dejhalla, R. & Pavletic, D. (eds.) *Towards Green Marine Technology and Transport.* London: Taylor & Francis Group, 327–334.

Tekgoz, M., Garbatov, Y. & Guedes Soares, C. 2015b. Ultimate strength assessment of a container ship accounting for the effect of neutral axis movement. *In:* Guedes Soares, C. & Santos, T.A. (eds.) *Marine Technology and Engineering.* London, UK: Taylor & Francis Group, 417–425.

Tekgoz, M., Garbatov, Y. & Guedes Soares, C. 2018. Strength assessment of an intact and damaged container ship subjected to asymmetrical bending loadings *Marine Structures,* 58, 172–198.

Yoshikawa, T., Maeda, M. & Inoue, A. 2008. A study on the residual strength of bulk carriers after impact loading. *Proceedings of the Japan Society of Naval Architects and Ocean Engineers,* 7W.

Progress in Maritime Technology and Engineering – Guedes Soares & Santos (Eds)
© 2018 Taylor & Francis Group, London, ISBN 978-1-138-58539-3

Strength assessment of an aged single hull tanker grounded in mud and used as port oil storage

N. Vladimir, I. Senjanović & N. Alujević
University of Zagreb, Zagreb, Croatia

S. Tomašević
Adriatic Tank Terminals d.o.o., Ploče, Croatia

D.S. Cho
Pusan National University, Busan, Republic of Korea

ABSTRACT: This paper deals with a practical problem of structural integrity evaluation of a single hull tanker which, after its regular service, was grounded in mud and used as port oil storage. The fore and the aft part of the ship were removed, whereas its middle module was kept to be used as a liquid cargo storage and a permanent ballast that ensures the ship continuous contact with the sea bottom. Therefore, the ship can be considered as supported in part by the mud and in the other part by the surrounding water. Although the ship is classed as a stationary object, the reliable strength assessment is still necessary to check its structural integrity in order to avoid potential environmental issues. The analysis was done in two steps. At first, the mathematical model of ship static equilibrium has been formulated, consistently taking into account the supporting mud reaction and buoyancy. Secondly, the developed physically consistent mathematical model is used as a basis for a detailed 3D FEM analysis, where the 3D FEM model is generated based on the available technical documentation and thickness measurements of the whole ship structure, which was performed in order to directly account for the effect of ageing (corrosion). Representative stresses in all structural elements are determined and compared with the permissible ones. Conclusions on ship suitability to serve as port oil storage are drawn.

1 INTRODUCTION

Ship hull inspection is a very important task to ensure its structural integrity during the whole lifetime. It is well-known that corrosion affects both local and global ship strength so that a number of models have been developed in order to improve structural analysis, inspection and maintenance procedures as for instance (Wirsching et al., 1997; Saad-Eldeen et al., 2014; Ventikos et al., 2017). Moreover, corrosion is one of the time-dependent detrimental phenomena which can lead to catastrophic failures (Rahbar-Ranji, 2012). On the other hand, grounding of ships is a special problem in maritime sector that has also been extensively considered in the relevant literature (Prestileo et al., 2013; Heinvee and Tabri, 2015).

This paper is devoted to the strength assessment of a grounded aged single-hull oil tanker,

Figure 1, which is being used as port oil storage, (Vladimir and Senjanović, 2017a,b). After its regular service, the ship was converted in such a way that its cargo area module can be used as a

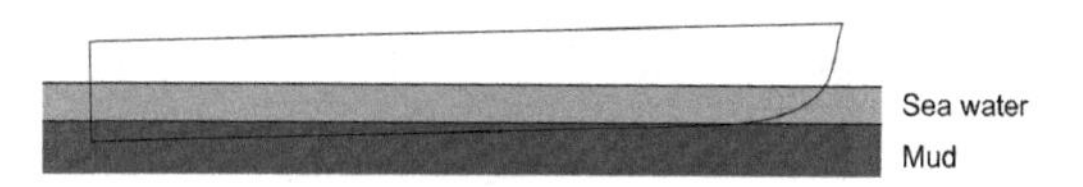

Figure 1. Schematic presentation of ship grounded in mud.

stationary object with the aim of increasing the port storage capacities. (aft structure from A.P. to FR60 was removed). However, bearing in mind the age of the ship, an extensive structural integrity evaluation was necessary in order to ensure that the risk of the oil loss and the consequent pollution is minimised.

For this purpose, a fully consistent mathematical model is derived, taking into account the fact that the ship is supported by both the mud and the sea water. The ship immersion in the mud is determined by the in-situ water depth readings and the ship dimensions. The developed theoretical model is based on static equilibrium and geometric relations, where the mud mass density is adjusted in order to properly take into account the sea bottom stiffness, i.e. the bottom reaction force. This is because it was confirmed by observations that the exact ship position is independent on the loading condition variation. Therefore it must be that the mud reaction continuously changes depending on the cargo filling levels.

A detailed strength assessment is performed by 3D Finite Element Method (FEM), (Senjanović, 2002). As a basis for the calculation, the above mentioned consistent global ship static equilibrium model is used. Accordingly, besides imposing the ship and cargo weight and surrounding water pressures on the generated 3D FEM structural model, additional bottom pressures are calculated and applied to simulate the effect of the supporting mud. The analysis was performed by means of general finite element software NASTRAN (MSC, 2005). Two loading conditions, covering the most critical cases are identified and considered in detail. The ageing (corrosion) effect is directly included in the calculation by taking into account the exact plating and stiffener thicknesses. These were obtained through a thickness measurement campaign of the complete ship structure (Vladimir and Senjanović, 2017a,b).

2 MATHEMATICAL MODEL

2.1 *Ship static equilibrium*

The weight of the grounded ship is equilibrated partly by the water buoyancy and partly by the sea bottom reaction. Hence, the total reaction $R = Q\text{-}B$ is known, but its distribution along the ship, which is important for the longitudinal strength analysis, is unknown. To the best of the authors' knowledge, a directly applicable procedure for determining the distribution of the reaction force has not been previously developed, neither by calculation nor empirically. In the procedure proposed in this paper, it is assumed that the ship hull freely floats in both water and mud. Since the ship does not change her draught at different loading conditions, the sea bottom reaction is simulated by buoyancy of a particularly high-density mud. The ship immersed into the water and mud is shown in Figure 2.

Ship weight, Q, has to be equilibrated by the water and mud buoyancies, respectively:

$$Q = \rho_w g \int_0^L A_w(x)\,\mathrm{d}x + \rho_m g \int_0^L A_m(x)\,\mathrm{d}x, \tag{1}$$

that gives:

$$Q = \rho_w g V_w + \rho_m g V_m, \tag{2}$$

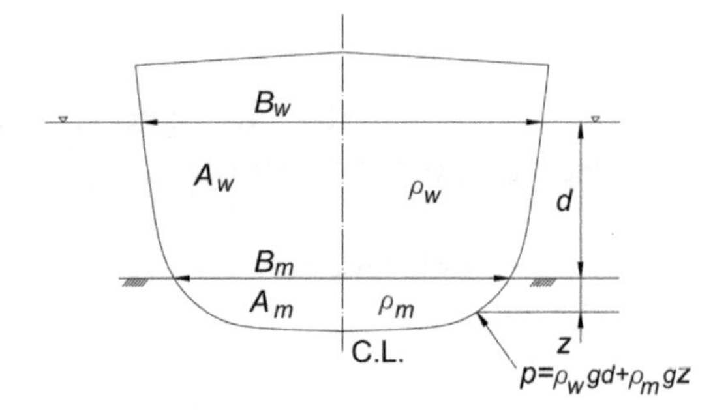

Figure 2. Schematic presentation of ship cross-section immersed in water and mud.

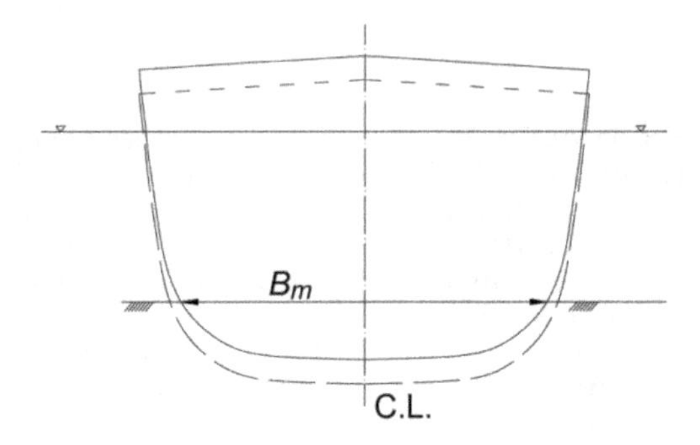

Figure 3. Unit immersion of ship in water and mud.

where:

ρ_w – water density,
ρ_m – mud density,
V_w – volume of ship immersed in water,
V_m – volume of ship immersed in mud,
A_w – cross-section area immersed in water,
A_m – cross-section area immersed in mud, and
g – gravity constant.

From Eq. (2) one can calculate the mud density necessary to maintain the static equilibrium:

$$\rho_m = \frac{Q - \rho_w g V_w}{g V_m}. \tag{3}$$

Therefore the mud density is determined for each loading condition separately, by assuming that the ship immersion in mud is constant (as confirmed by the in-situ observations).

In case of a unit immersion, Figure 3, additional buoyancy represents the sea bottom (mud) stiffness, and yields:

$$k = \rho_m g B_m. \tag{4}$$

where B_m represents the cross-section breadth at the borderline between the mud and water, Figure 2.

2.2 *Determination of the mud pressure*

As mentioned above, the ship is in a static equilibrium, where the hull and cargo weights are partially equilibrated by the supporting mud pressure. Also, the hull sides are exposed to the hydrostatic pressure of the surrounding water. For the purpose of the analysis, the 3D FEM model weight should be adapted to real ship weight by the material density adjustments. In order to eliminate rigid body motions in the FE model, the ship is supported at three selected nodes for easier modelling. However, it is controlled that reaction forces at these nodes are zero. The boundary conditions are shown in Figure 4.

The positions of the supports are shown in Figure 5, where it is indicated that the aft supports are at the cross-sections of the transverse bulkhead at FR91 and longitudinal bulkheads, whereas the fore support is on transverse bulkhead at FR187 in the centreline. The supports are located below very stiff structural elements in order to minimize stress concentrations due to a possibly slightly unbalanced FE model.

The distribution of the mud reaction along the ship hull is shown in Figure 6. It can be calculated by knowing the total ship and cargo weight, F, and the longitudinal position of the centre of gravity (CG), x_{CG}, which are here taken from the FEM software output file (MSC, 2005). However, based on the static equilibrium of forces shown in Figure 5, the expression for determination of the CG position can be derived:

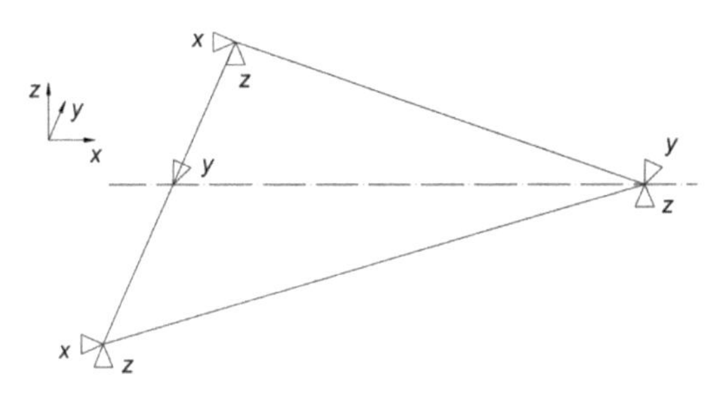

Figure 4. Boundary conditions of FE model in the supported nodes.

$$Fx_{CG}^* = F_2 a, \tag{5}$$

where

$$F = 2F_1 + F_2. \tag{6}$$

The distance of CG from aft supports, x_{CG}^*, reads:

$$x_{CG}^* = \frac{F_2}{F} a. \tag{7}$$

Before calculating mud pressures on the hull, the mud continuous reaction defined by q_1 and q_2, Figure 6, is expressed via the ship total weight F and the longitudinal position of CG, x_{CG}. According to Figure 6, the force and moment equilibrium can be written in the following form:

$$\sum F_i = 0: \quad \frac{1}{2}(q_1 + q_2) l_1 + \frac{1}{2} q_2 l_2 = F, \tag{8}$$

which gives:

$$q_1 + q_2 \left(1 + \frac{l_2}{l_1}\right) = \frac{2F}{l_1}, \tag{9}$$

$$\sum M_i = 0: \quad \frac{1}{2}(q_1 + q_2) l_1 \frac{1}{2} l_1 + \frac{1}{2} q_2 l_2 \left(l_1 + \frac{1}{3} l_2 \right) = F x_{CG}, \tag{10}$$

$$q_1 + q_2 + 2q_2 \frac{l_2}{l_1^2}\left(l_1 + \frac{1}{3} l_2 \right) = \frac{4}{l_1^2} F x_{CG}, \tag{11}$$

and finally leading to:

$$q_1 + q_2 \left[1 + 2\frac{l_2}{l_1}\left(1 + \frac{1}{3}\frac{l_2}{l_1} \right) \right] = \frac{4}{l_1^2} F x_{CG}. \tag{12}$$

After substituting (9) into (12), one can write:

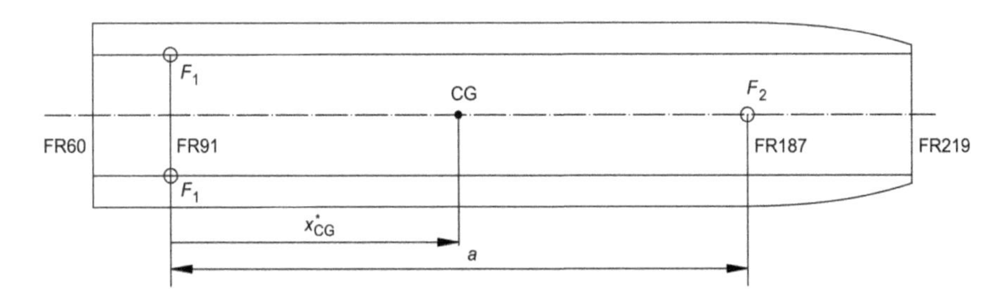

Figure 5. Positions of supports, i.e. reaction forces along the 3D FE model.

$$q_2 \frac{l_2}{l_1}\left(1+\frac{2}{3}\frac{l_2}{l_1}\right)=\frac{2F}{l_1}\left(2\frac{x_{CG}}{l_1}-1\right). \quad (13)$$

Based on the above derivation, q_2 and q_1 can be expressed as:

$$q_2 = \frac{\frac{2F}{l_1}\left(2\frac{x_{CG}}{l_1}-1\right)}{\frac{l_2}{l_1}\left(1+\frac{2}{3}\frac{l_2}{l_1}\right)}, \quad (14)$$

$$q_1 = \frac{2F}{l_1} - q_2\left(1+\frac{l_2}{l_1}\right). \quad (15)$$

For the pressure calculation, the width of the mud waterline, $B_{WLMUD,}$ is relevant, Figure 6. According to Figure 7 one can write:

$$p_1 = \frac{q_1}{B_{WLMUD}}, \quad p_2 = \frac{q_2}{B_{WLMUD}}, \quad p_p = \frac{q_p}{B_p}. \quad (16)$$

Along a single cargo hold, the mud pressure is assumed to be constant, i.e. a step-like distribution is used for simplicity.

3 SHIP DATA, DESCRIPTION OF FEM MODEL AND CALCULATION SETUP

3.1 *Ship data*

The considered single hull oil tanker was built in 1956, with the following main particulars:

Length between perpendiculars: L_{PP} = 149.07 m
Breadth: B = 25.72 m
Depth to main deck: H = 14.1 m

The tank layout is shown in Figure 8. The central cargo tanks 4C-9C are intended to be used as oil storages. The side tanks 1 L, 1D, 4 L and 4D are aimed to be empty, whereas tanks 2 L, 2D, 3 L, 3D, 5 L, 5D, 6 L and 6D contain permanent ballast (sea water, 95% of tank capacity), that together with the ballast in the fore peak (20% of FP capacity) and the lightship mass, ensure a permanent contact between the ship structure and the bottom. This tank filling strategy is summarized in Table 1 together with the tank capacities.

An extensive thickness measurement campaign has been performed in order to accurately take into account the corrosion effect, i.e. to reliably simulate a realistic ship condition. It should be mentioned that condition based maintenance is regularly carried out in this case, meaning that if excessive

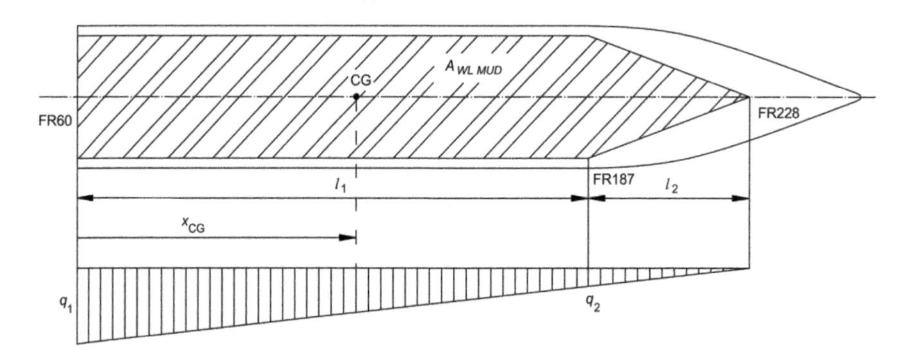

Figure 6. Longitudinal distribution of mud reaction.

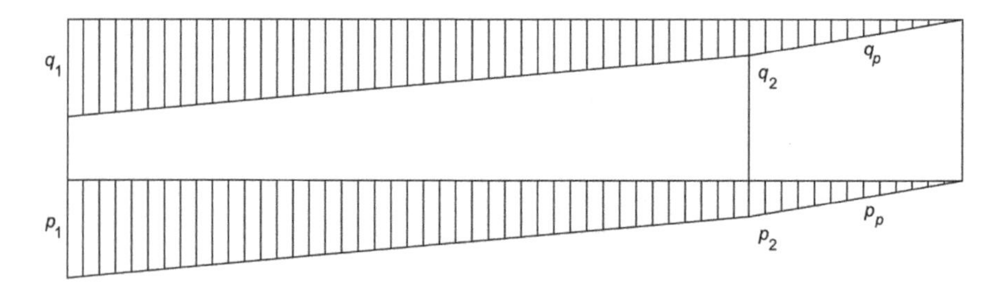

Figure 7. Longitudinal mud pressure distribution.

corrosion or pitting is noticed, Figure 9, the structure is locally remedied, Figure 10.

As claimed by the ship-owner, ship was made of steel, and for the needs of the calculation the values of Young's modulus, material density and Poisson's ratio are set at 2.1×10^{11} N/m^2, 7850 kg/m^3 and 0.3, respectively.

3.2 *FEM model*

A fine mesh 3D FEM model of the tanker is generated in order to calculate local stresses. The

Figure 8. Ship general arrangement with tank disposition.

Table 1. Containment of ship tanks and tank capacities.

Tank designation	Filling	Capacity, m^3
1C	None	N/A
2C	None	N/A
3C	Pump room	N/A
4C	Oil	3000
5C	Oil	3000
6C	Oil	3000
7C	Oil	3000
8C	Oil	3000
9C	Oil	3000
10C	None	N/A
1 L/D	None	N/A
2 L/D	Ballast (95%)	1500
3 L/D	Ballast (95%)	750
4 L/D	None	N/A
5 L/D	Ballast (95%)	1500
6 L/D	Ballast (95%)	703

Figure 9. Corroded tank bottom plating.

Figure 10. Remedied tank bottom plating.

axonometric view on the FEM model (stretching from FR60 (in this case AP, x = 0 m) to FR219 (x = 124.179 m)) is shown in Figure 11. In Figure 12 the model is shown without deck plating and beam elements in order to ensure better visibility of the internal structure.

Before going into details on the calculation setup, a more detailed overview of the FEM model is provided. In Figure 13 a typical cargo tank structure together with the side (ballast) tank structure is presented in an axonometric view.

Besides ordinary frames, the analysed ship structure is characterized by both semi-web frames, Figure 14, and web frames, Figure 15.

In total, the FE model used for the analysis has 30762 nodes and consists of 64733 elements, whereas 34434 are the plate elements and 30299 are the beam elements. The mesh density in the longitudinal direction equals to the ordinary frame spacing (781 mm). The only exception is the fore part of tanks 1 L/1D where the mesh density is two times coarser. The mesh density in the transverse direction is equal to the spacing of the longitudinal stiffeners.

In Figure 16 a view of a side tank without the side shell is given, so that the internal tank framing can be seen. Also, one can see the vertical stiffeners of the transverse bulkhead.

The complete FEM model was updated with thickness measurement data, as for instance shown in case of a typical transverse bulkhead,

Figure 17, and a typical cargo and side tank structure, Figure 18.

3.3 *Calculation setup*

Two representative load cases are defined next, i.e. LC1 and LC2. Load case LC1 is characterized by full cargo tanks 4C, 6C and 8C, Figure 19, whereas in load case LC2 cargo tanks 5C and 8C are empty, and the other cargo tanks are full, Figure 20.

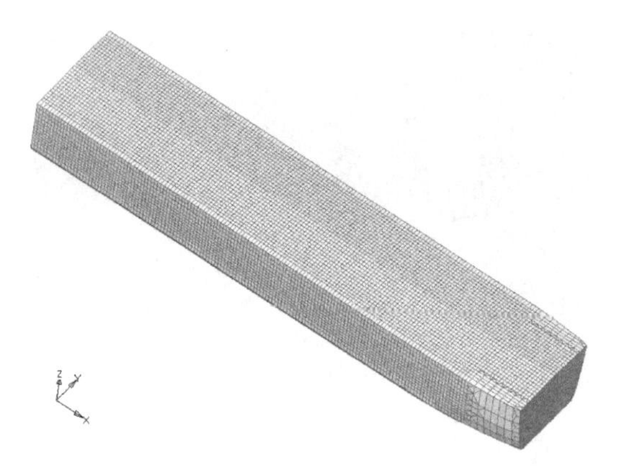

Figure 11. 3D FEM model used in the structural analysis.

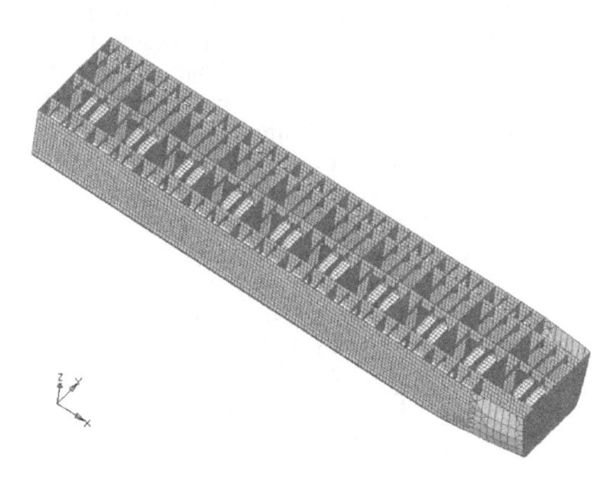

Figure 12. 3D FEM model used in structural analysis without deck plating and without beam elements.

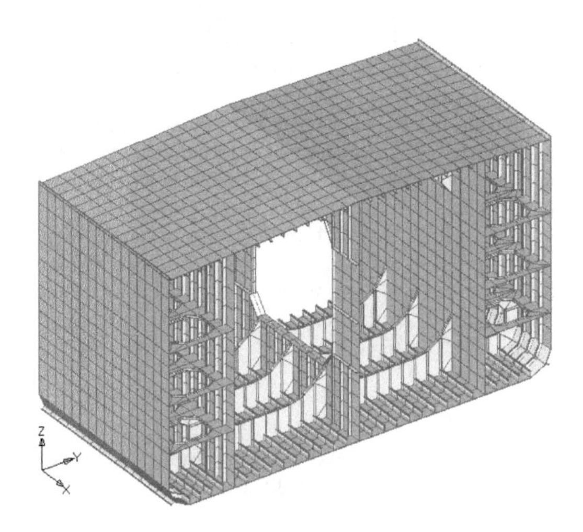

Figure 13. Typical cargo and side (ballast) tank structure.

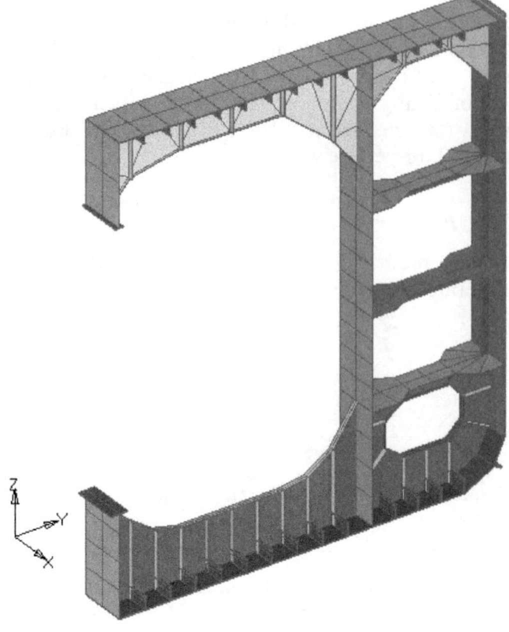

Figure 14. Typical view on semi-web frame.

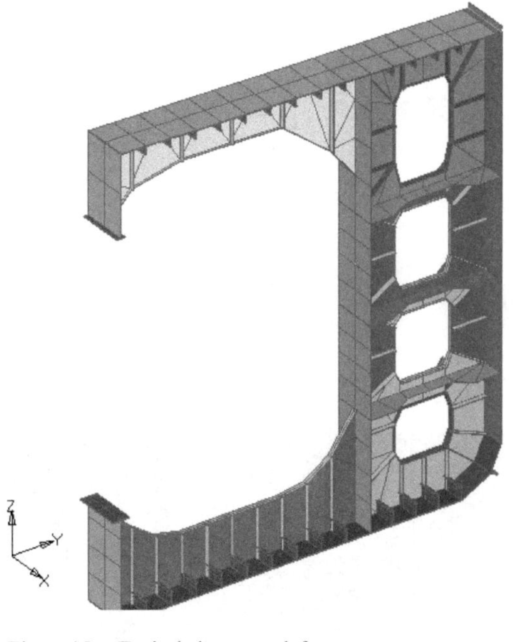

Figure 15. Typical view on web frame.

For both of the above defined load cases, the total weight and the centre of gravity are determined, and the mud pressure for each tank is calculated.

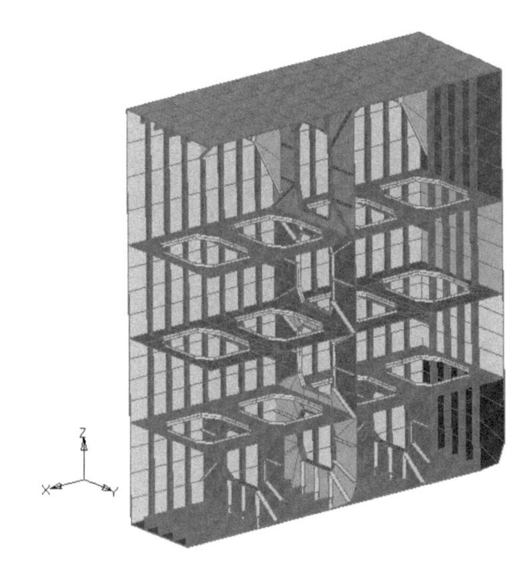

Figure 16. Side tank internal structure and stiffeners of longitudinal and transverse bulkhead.

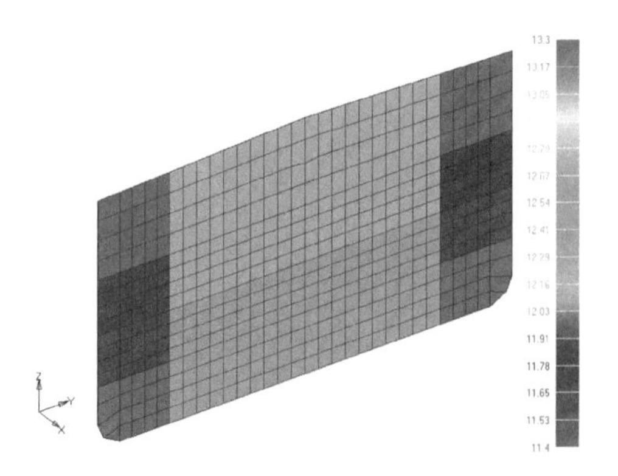

Figure 17. Plate thicknesses of transverse bulkhead at FR139 [mm].

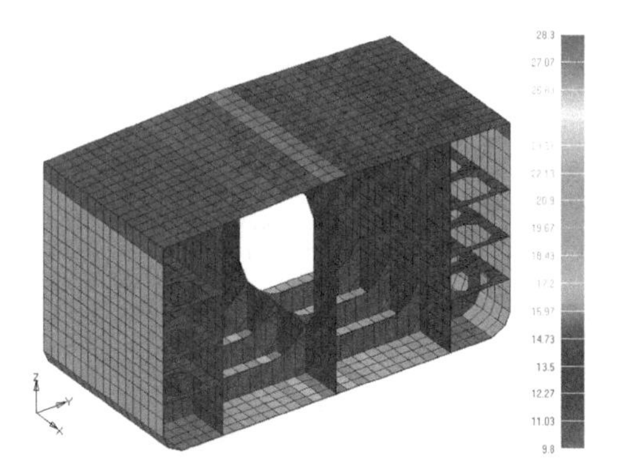

Figure 18. Plate thicknesses in Tank 5C and corresponding side tanks [mm].

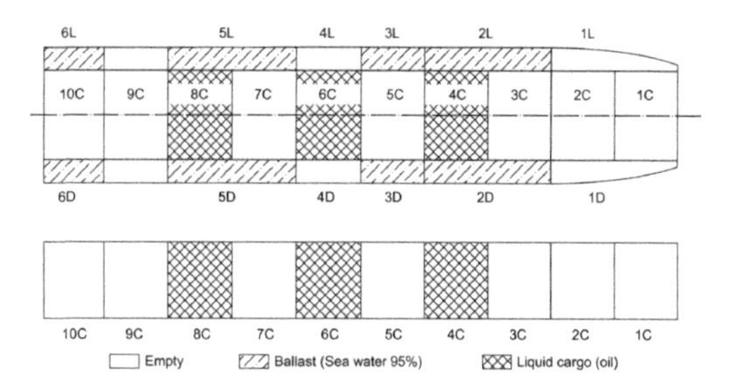

Figure 19. Schematic representation of LC1.

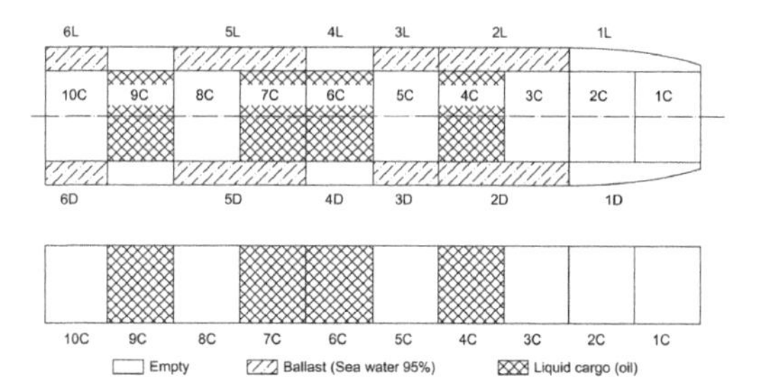

Figure 20. Schematic representation of LC2.

It has been found that the variation of sea level due to tide and ebb has no significant influence the results. Therefore the average value of sea level (water depth) is assumed. Also, some local deteriorations (like pitting) are not considered in this calculation. After imposing the mud pressures on the bottom, the above-mentioned reactions in the three supporting nodes are checked. Each load case is characterised by the same filling of ballast tanks, i.e. only the filling of the cargo tanks is variable.

4 RESULTS

The static analysis is performed by NASTRAN software (MSC, 2005) and von Mises stresses are obtained for LC1 and LC2 in all elements of the structural model. They are compared with the maximum allowable stress which reads 230 N/mm^2 (CRS, 2016). It is found that the calculated stresses are below the allowable stresses for both load cases within the complete FE model. Next, an overview of the results is given where stress distributions are shown for characteristic structural members within the cargo area. The maximum calculated stresses are tabulated in Tables 2 and 3.

Von Mises stress levels in the complete ship structure for LC1 and LC2 are illustrated in Figures 21 and 22, respectively. Stress units in all figures below are N/m^2.

Table 2. Von Mises stresses in representative structural elements for LC1.

Name	Tank	Position along the ship	Maximum von Mises stress (N/mm²)
Transverse bulkhead	10C	FR60	65.2
Transverse bulkhead	9C, 10C	FR75	35.6
Transverse bulkhead	8C, 9C	FR91	190.7
Transverse bulkhead	7C, 8C	FR107	196.7
Transverse bulkhead	6C, 7C	FR123	189.9
Transverse bulkhead	5C, 6C	FR139	208.3
Transverse bulkhead	4C, 5C	FR155	200.1
Transverse bulkhead	3C, 4C	FR171	206.3
Bottom plating	10C	FR60-FR75	123.9
Bottom plating	9C	FR75-FR91	41.1
Bottom plating	8C	FR91-FR107	36.2
Bottom plating	7C	FR107-FR123	52.9
Bottom plating	6C	FR123-FR139	41.0
Bottom plating	5C	FR139-FR155	64.1
Bottom plating	4C	FR155-FR171	45.5
Bottom plating	3C	FR171-FR187	53.8
Tank internal structure (central bottom girder, deck girder, bulkhead central web, web frames, floor plates)	10C	FR60-FR75	97.5 (floor plate at FR67)
Tank internal structure (central bottom girder, deck girder, bulkhead central web, web frames, floor plates)	9C	FR75-FR91	158.2 (connection of bulkhead web and bulkhead at FR91)
Tank internal structure (central bottom girder, deck girder, bulkhead central web, web frames, floor plates)	8C	FR91-FR107	124.7 (connection of bulkhead web and bulkhead at FR107)
Tank internal structure (central bottom girder, deck girder, bulkhead central web, web frames, floor plates)	7C	FR107-FR123	139.3 (connection of bulkhead web and bulkhead at FR123)
Tank internal structure (central bottom girder, deck girder, bulkhead central web, web frames, floor plates)	6C	FR123-FR139	132.6 (connection of bulkhead web and bulkhead at FR139)
Tank internal structure (central bottom girder, deck girder, bulkhead central web, web frames, floor plates)	5C	FR139-FR155	141.0 (connection of bulkhead web and bulkhead at FR155)
Tank internal structure (central bottom girder, deck girder, bulkhead central web, web frames, floor plates)	4C	FR155-FR171	155.4 (connection of bulkhead web and bulkhead at FR155)
Tank internal structure (central bottom girder, deck girder, bulkhead central web, web frames, floor plates)	3C	FR171-FR187	152.4 (connection of bulkhead web and central bottom girder at FR174)
Longitudinal bulkhead, port side	10C	FR60-FR75	42.0
Longitudinal bulkhead, port side	9C	FR75-FR91	82.1
Longitudinal bulkhead, port side	8C	FR91-FR107	61.2
Longitudinal bulkhead, port side	7C	FR107-FR123	65.4
Longitudinal bulkhead, port side	6C	FR123-FR139	53.7
Longitudinal bulkhead, port side	5C	FR139-FR155	66.2
Longitudinal bulkhead, port side	4C	FR155-FR171	52.6
Longitudinal bulkhead, port side	3C	FR171-FR187	77.0

Table 3. Von Mises stresses in representative structural elements for LC2.

Name	Tank	Position along the ship	Maximum von Mises stress (N/mm²)
Transverse bulkhead	10C	FR60	82.4
Transverse bulkhead	9C, 10C	FR75	189.0
Transverse bulkhead	8C, 9C	FR91	208.9
Transverse bulkhead	7C, 8C	FR107	174.9
Transverse bulkhead	6C, 7C	FR123	49.0
Transverse bulkhead	5C, 6C	FR139	208.0
Transverse bulkhead	4C, 5C	FR155	199.9
Transverse bulkhead	3C, 4C	FR171	202.6
Bottom plating	10C	FR60-FR75	184.8
Bottom plating	9C	FR75-FR91	38.0
Bottom plating	8C	FR91-FR107	57.4
Bottom plating	7C	FR107-FR123	50.1
Bottom plating	6C	FR123-FR139	59.3
Bottom plating	5C	FR139-FR155	76.3
Bottom plating	4C	FR155-FR171	57.9
Bottom plating	3C	FR171-FR187	62.0
Tank internal structure (central bottom girder, deck girder, bulkhead central web, web frames, floor plates)	10C	FR60-FR75	138.4 (bulkhead web at FR72)
Tank internal structure (central bottom girder, deck girder, bulkhead central web, web frames, floor plates)	9C	FR75-FR91	121.2 (connection of bulkhead web and bulkhead at FR91)
Tank internal structure (central bottom girder, deck girder, bulkhead central web, web frames, floor plates)	8C	FR91-FR107	169.3 (connection of bulkhead web and bulkhead at FR107)
Tank internal structure (central bottom girder, deck girder, bulkhead central web, web frames, floor plates)	7C	FR107-FR123	133.6 (connection of bulkhead web and bulkhead at FR107)
Tank internal structure (central bottom girder, deck girder, bulkhead central web, web frames, floor plates)	6C	FR123-FR139	142.4 (connection of bulkhead web and bulkhead at FR139)
Tank internal structure (central bottom girder, deck girder, bulkhead central web, web frames, floor plates)	5C	FR139-FR155	134.9 (connection of bulkhead web and bulkhead at FR155)
Tank internal structure (central bottom girder, deck girder, bulkhead central web, web frames, floor plates)	4C	FR155-FR171	163.1 (connection of bulkhead web and bulkhead at FR155)
Tank internal structure (central bottom girder, deck girder, bulkhead central web, web frames, floor plates)	3C	FR171-FR187	171.7 (connection of bulkhead web and central bottom girder at FR174)
Longitudinal bulkhead, port side	10C	FR60-FR75	64.4
Longitudinal bulkhead, port side	9C	FR75-FR91	59.4
Longitudinal bulkhead, port side	8C	FR91-FR107	82.2
Longitudinal bulkhead, port side	7C	FR107-FR123	67.8
Longitudinal bulkhead, port side	6C	FR123-FR139	64.8
Longitudinal bulkhead, port side	5C	FR139-FR155	78.7
Longitudinal bulkhead, port side	4C	FR155-FR171	59.8
Longitudinal bulkhead, port side	3C	FR171-FR187	83.6

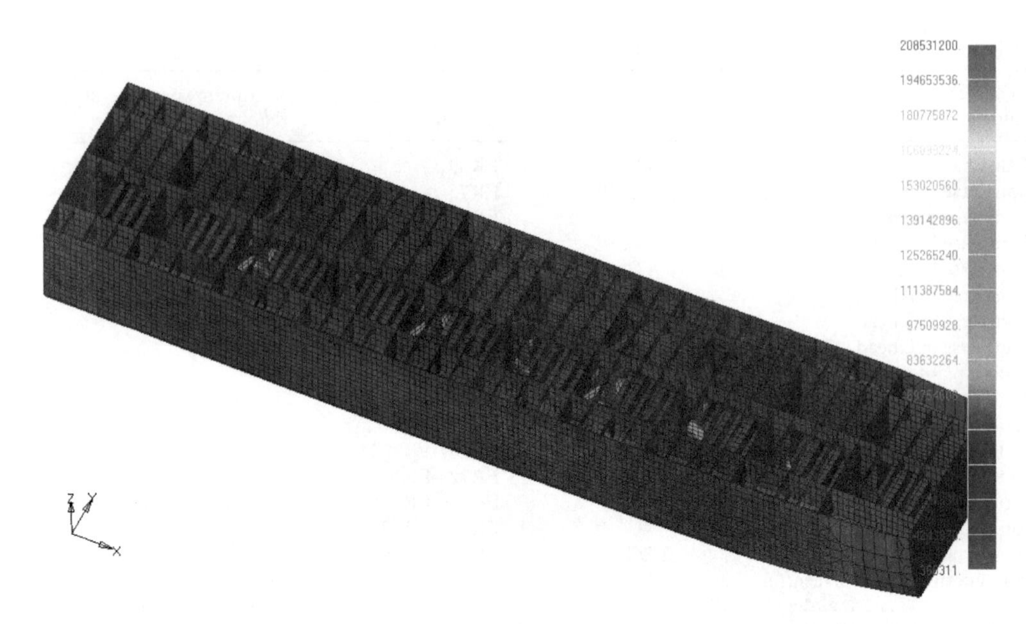

Figure 21. Von Mises stresses in plate elements of ship hull, LC1.

Figure 22. Von Mises stresses in plate elements of ship hull, LC2.

Besides the von Mises stresses in the transverse bulkheads, Figure 23, shear stresses are also checked, Figure 24.

Typical von Mises stress distribution in a cargo tank area is shown in Figure 25.

In Figure 26 the stress distribution in the typical ballast tank structure is presented, while Figure 27 shows characteristic structural elements with the highest stress levels for LC2 (transverse bulkhead at FR91).

The results in Tables 2 and 3 clearly indicate that the maximum stresses are obtained in the transverse bulkheads of central tanks, and their maximum levels are obtained if two consecutive

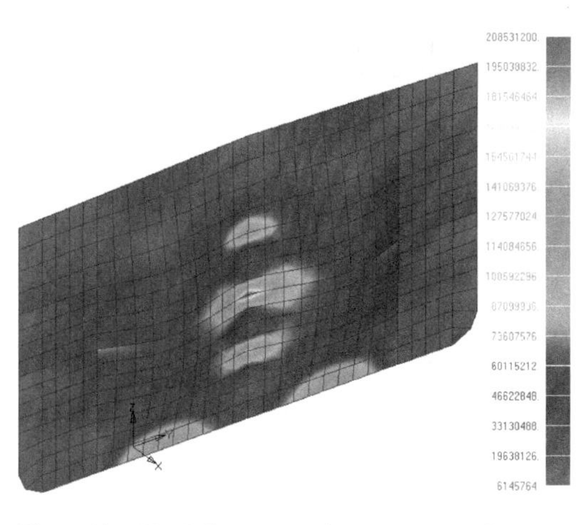

Figure 23. Von Mises stresses in transverse bulkhead at FR139, LC1.

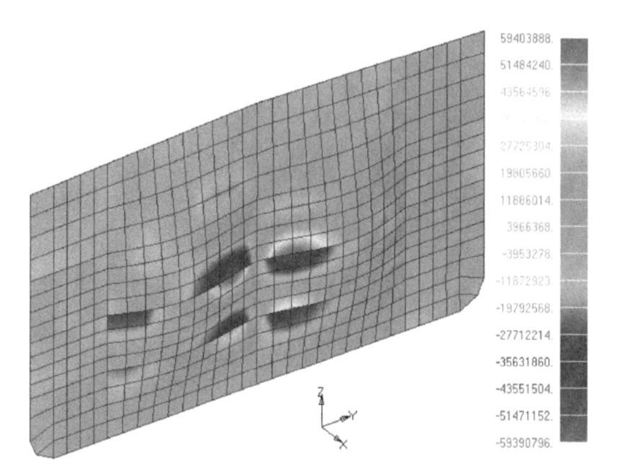

Figure 24. Shear stresses in transverse bulkhead at FR139, LC1.

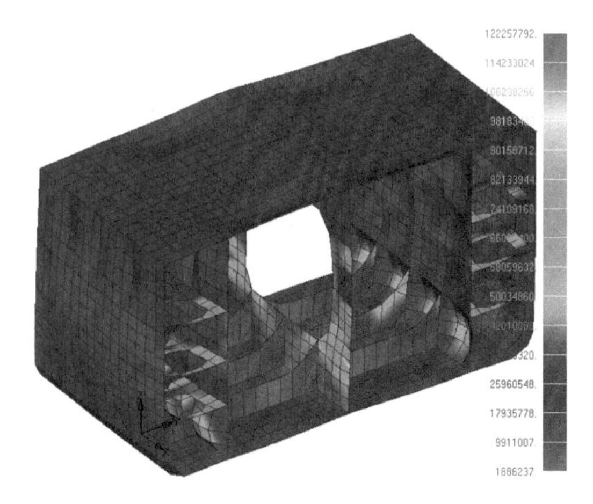

Figure 25. Von Mises stresses in tank 6C, LC1.

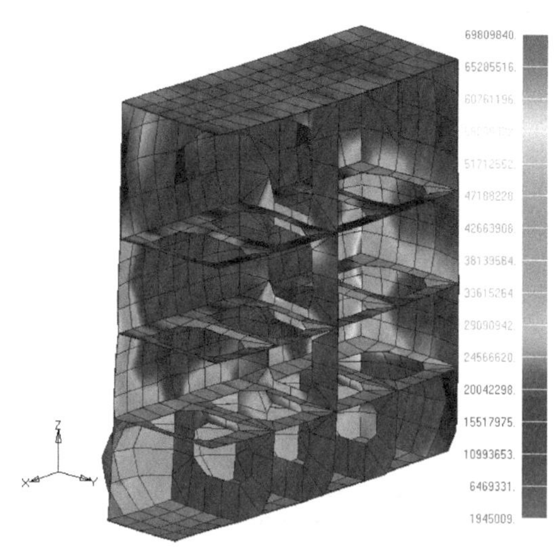

Figure 26. Von Mises stresses in stringers and internal walls of tank 3 L, LC1.

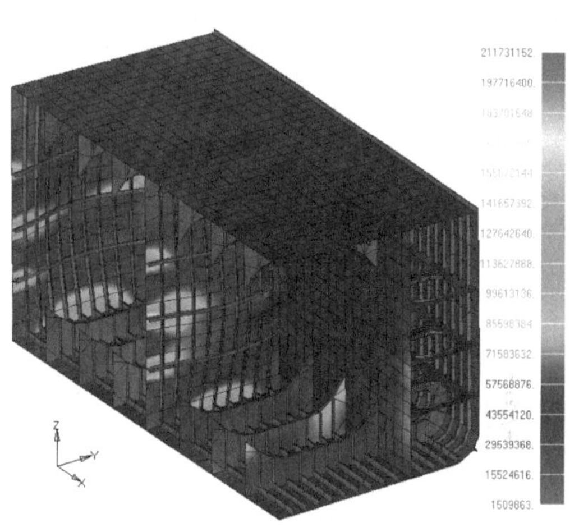

Figure 27. Tanks 8C and 9C with transverse bulkheads at FR75 and FR91, one-half of the model, LC2.

tanks have significantly different filling ratios (i.e. full and empty tanks are altering). Therefore, the stresses can be minimized if similar filling level in adjacent tanks is kept.

5 CONCLUDING REMARKS

The strength assessment of a single hull tanker, converted to port oil storage and grounded in mud, has been carried out by 3D FEM. A 3D finite element model of the ship structure is generated. A special mathematical model for loading of the ship grounded in the mud has been developed in order to take into account the external loading by both the mud and the surrounding water. In such a model the total loading of the FEM ship structure consists of 1) ship weight, 2) liquid cargo pressure, 3) surrounding water pressure on ship sides and 4) mud buoyancy. The mud buoyancy is imposed as a hydrostatic pressure on the ship bottom. Von Mises stresses and shear stresses are taken as the representative ones for the evaluation of structural integrity.

Locations with the highest stress levels are identified for selected load cases. Despite the selection of the most severe loading conditions, represented by different combinations of successive empty and full cargo tanks, the calculated stresses are below permissible values for all structural members. Maximum stress levels are regularly obtained in transverse bulkheads of central tanks. Therefore, if the cargo amount in the adjacent tanks is similar, stresses in bulkheads between them will be minimized.

Moreover, the obtained results show that the ship in its current static equilibrium position can be exploited with arbitrary combinations of cargo tank fillings. However, due to safety reasons and bearing in mind the age of the ship, the structural elements should be continuously surveyed.

ACKNOWLEDGEMENTS

The investigation presented in this paper is funded by the Adriatic Tank Terminals d.o.o. and the European Union's Horizon 2020 research and innovation programme under the Marie Sklodowska-Curie grant agreement no. 657539 STARMAS. Publication of this paper is also supported by the National Research Foundation of Korea (NRF) grant funded by the Korea government (MSIT) through GCRC-SOP (No. 2011-0030013), and within the project Green Modular Passenger Vessel for Mediterranean (GRiMM), funded by the Croatian Science Foundation (Project No. 2017-05-UIP-1253).

REFERENCES

CRS. 2016. *Rules for Classification of Ships, Part 2 – Hull*, Croatian Register of Shipping, Split, Croatia.

Heinvee, M., Tabri, K. 2015. A simplified method to predict grounding damage of double bottom tankers. *Marine Structures*, 43:22–43.

MSC. 2005. *MSC.NASTRAN2005: Installation and Operations Guide*, MSC Software.

Prestileo, A., Rizzuto, E., Teixeira, A.P., Guedes Soares, C. 2013. Bottom damage scenarios for the hull girder structural assessment. *Marine Structures*, 33:33–55.

Rahbar-Ranji, A. 2012. Ultimate strength of corroded steel plates with irregular surfaces under in-plane compression. *Ocean Engineering*, 54:261–269.

Saad-Eldeen, S., Garbatov, Y., Guedes Soares, C. 2014. Strength assessment of a severely corroded box girder subjected to bending moment. *Journal of Constructional Steel Research*, 92:90–102.

Senjanović, I. 2002. *Finite element method in analysis of ship structures*, University of Zagreb, Zagreb. (textbook, in Croatian).

Ventikos, N., Sotiralis, P., Drakakis, M. 2018. A dynamic model for the hull inspection of ships: The analysis and results. *Ocean Engineering*, Volume 151, Pages 355–365.

Vladimir, N., Senjanović, I. 2017a. Evaluation of structural integrity of a ship hull used as port oil storage, Part I: 1D FEM analysis, University of Zagreb, Faculty of Mechanical Engineering and Naval Architecture, (internal report).

Vladimir, N., Senjanović, I. 2017b. Evaluation of structural integrity of a ship hull used as port oil storage, Part II: 3D FEM analysis, University of Zagreb, Faculty of Mechanical Engineering and Naval Architecture, (internal report).

Wirsching, P.H., Ferensic, J., Thayamballi, A. 1997. Reliability with respect to ultimate strength of a corroding ship hull. *Marine Structures*, 10(7):501–518.

Ship structures II

Progress in Maritime Technology and Engineering – Guedes Soares & Santos (Eds)
© 2018 Taylor & Francis Group, London, ISBN 978-1-138-58539-3

Failure assessment of transition piece of jacket offshore wind turbine

B. Yeter, Y. Garbatov & C. Guedes Soares
Centre for Marine Technology and Ocean Engineering (CENTEC), Instituto Superior Técnico, Universidade de Lisboa, Lisbon, Portugal

ABSTRACT: The present study performs a failure assessment of the transition piece of a jacket offshore wind turbine based on a nonlinear finite element analysis. The extreme and fatigue loads acting on the offshore wind turbines increases since the size of the wind turbine increases. Therefore, it is aimed here to develop a multi-dimensional failure criterion to be used in the early stages of the design. Once the structure is subjected above its design loads, it is expected to fail either by tension failure (rupture), buckling, or both. High amplitude plastic strains occurred at the welded-tubular connections, because of the extreme loads and high stress concentration, are dealt with the low-cycle fatigue damage approaches. Using the Neuber's rule, the global stress range is transformed into a true level of stress-strain, where there is a high-stress concentration located. Moreover, the nonlinear structural behaviour under compression is assessed by performing a FE analysis accounting for the nonlinearities related to geometry and material properties.

1 INTRODUCTION

The continuous technological development in wind turbines in terms of a capacity has motivated the offshore wind industry sector to move towards deeper waters to extract more power from a less turbulent wind. Consequently, the offshore wind turbines are subjected to a significantly higher overturning moment due to the perpetually increasing wind-induced loads and the height of the support structure and tower. It is therefore of interest to study the structural response of the support structure and the possible failure mechanisms.

For the structures such as an offshore wind turbine support structure, fatigue damage has always been a vital issue to deal with. The low-cycle fatigue load regime occurs because of a technical problem associated with the operation of a wind turbine if not due to environmental conditions. Due to the high local stress concentration, the structure may experience significant localised plastic deformation even after few cycles of a low-cycle fatigue load regime.

The localised plastic strain, accumulated over the course of the service life of an offshore wind turbine, may give rise to a crack initiation, which affects the structural integrity directly. Moreover, the accumulated plastic strain may also cause a localised geometrical imperfection on the structure, which affects the ultimate load-carrying capacity of the structure.

Both low-cycle fatigue and the ultimate strength are essential failure mechanism for offshore wind turbine support structures, and therefore have to be studied in the design stage. To this end, the present study assesses the design of a transition piece of a jacket support structure, which is an essential part of the support structure, based on both failure mechanisms.

The low-cycle fatigue has been investigated in the search of an explanation for the rapid crack initiation at the marine structures after few years spent in service such as in (Kim et al., 2005, Wang et al., 2006, Garbatov et al., 2011). A generic approach adopted by the standards to deal with this phenomenon at the design stage, the studies conducted the LCF assessment based upon the pseudo hotspot stress range that is the linearly equivalent of the actual strain.

In addition, NORSOK N-006 (2009) also presented the S-N curves developed mainly to address the low cycle fatigue regime for tubular joints in air and seawater with a cathodic protection to check the welded tubular joints that are subjected to cyclic loads such as abnormal waves due to the danger of a crack growing up to a certain size that leads to sudden failure.

The hotspot stress range derived from the linear elastic FE analysis may indicate a local yielding. Therefore, one must consider using the cyclic stress-strain relation combined with the Neuber (1961)'s rule for a derivation of the actual strain. Subsequently, the accumulated fatigue damage is calculated in the same fashion done for the high-cycle fatigue using the linear Palmgren—Miner (1945)'s rule.

As far as design requirements for offshore structures are concerned, the ultimate limit state analysis is performed in such way that the structural response remains within the elastic limit for the specific storm conditions, which is leading the structure to be designed with very thick structural components. However, this is merely a consequence of the safety concerns for the offshore structure in the Oil & Gas industry.

Available documented publications on the ultimate strength of offshore structures are relatively limited compared to the considerable number of publication on ship structures. The study performed by Kallaby and Millman (1975) was one of the initial works to assess the ultimate strength of the offshore structure, including beyond the elastic limit. The study aimed to investigate the energy absorption capacity of a jacket offshore platform under seismic loading. Following this, Marshall (1975) showed that offshore structures might have the ultimate strength twice as the design capacity.

As far as the structural behaviour of a circular cylindrical shell is concerned, Brazier (1927) concluded that the ovalisation of the circular cross-section has a direct effect on the ultimate strength of the structure subjected to bending. Later on, Sherman (1976) conducted a series of test for a broad range of slenderness ratio D/t, from 18 to 102. The study concluded that the tubular structures with slenderness ratio greater than 50 do not have sufficient plastic rotation capacity to develop the classical ultimate strength.

Fabian (1977) studied the long cylindrical elastic shells subjected to axial, bending and internal pressure and found out that the circumferential flattening causes an ultimate load and compression wrinkles generate the bifurcation buckling axially.

The ultimate strength assessment of the cylindrical shells accounting for the initial imperfection and different slenderness ratios are also studied by Gellin (1980), Kyriakides and Ju (1992) and Ju and Kyriakides (1992), Dimopoulos and Gantes (2012), Guo et al. (2013).

In the scope of the present study, the transition piece is subjected to a bending moment transferred from the wind turbine, the low cycle fatigue assessment is performed based on the strain-based approach accounting for the localised plastic deformation, and the ultimate strength analysis is carried out by a nonlinear finite element analysis accounting for the material and geometrical nonlinearity.

2 STRUCTURAL DESCRIPTION

The jacket support structures are found to be suitable for the intermediate water depth, and the support structure can be adaptable to greater water depths (Yeter et al., 2017). Like the other OWT support structures, the wind-induced loads are transferred to the jacket support structure through a transition piece. The jacket offshore wind turbine and the transition piece are illustrated in Figure 1.

The transition piece is an integral part of the whole offshore wind turbine, and it needs a robust design, which stands against the cyclic bending moment caused by the wind-induced loads. The strength of the transition piece has to be analysed in terms of ultimate strength and fatigue capacities.

The transition piece consists of a column, four brace elements, and four pile elements. The column element is directly connected to the tower and transmits the load coming from the tower to the braces, whereas the brace elements support the column and connect the column to the pile elements. The piles are connected to the jacket support structure.

Moreover, the design of transition piece is constrained by the dimensions of the tower and jacket support structure, especially in terms of the diam-

Figure 1. A typical jacket offshore wind turbine.

eter. However, by varying the slenderness ratio of the tubular elements, in turn, their thicknesses, a robust design can be achieved. In addition, the position of the braces can also be another design variable. The geometrical specification of the transition piece is given in Figure 2, and the values are presented in Table 1.

The present study analyses a wide range of different designs of the transition piece. The designs of the transitional pieces are varied based on the slenderness ratio of the column γ_c, and the ratio of the slenderness of the column to the slenderness of the brace/pile, μ. As the diameters of the tubular components are fixed, the associated thickness varied. The transition pieces designed to be analysed here are presented in Table 2.

The slenderness ratio of the column is denoted as $\gamma_c = D/(2T)$, and it ranges from 10 to 160. The parameter associated with the ratio of the slenderness of the components μ ranges from 0.25 to 1. By employing this parameter, more degree of freedom is given to create more models in the feasible design space.

It is worth mentioning that the feasible design space is modelled based on manufacturing allowance rather than the typical range used in the offshore industry. By doing so, it is aimed here to investigate prevailing failure mechanism for different regions of the design space.

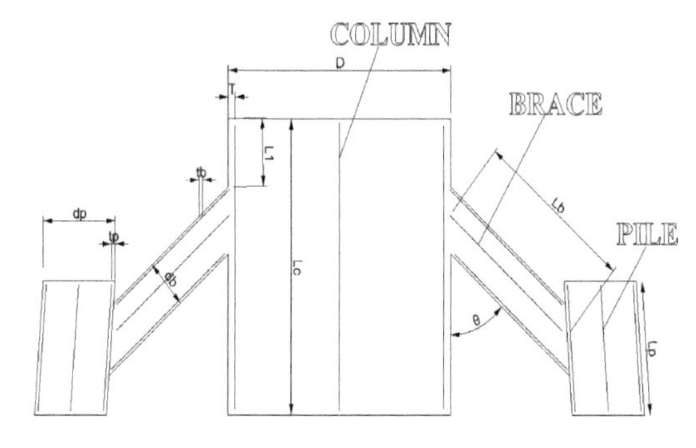

Figure 2. Configuration of the transition piece.

Table 1. Geometrical properties.

Description, parameter	Value
Diameter of column, D	4.000 m
Diameter of brace, db	0.800 m
Diameter of pile, dp	1.200 m
Thickness of column, T (M7)	40 mm
Thickness of brace, tb (M7)	24 mm
Thickness of pile, tp (M7)	32 mm
Length of column, Lc	4.500 m
Length of brace, Lb	5.260 m
Length of pile, Lp	2.000 m
Gap between brace and tower, L1	0.672 m
Angle between column and brace, θ	30°

Table 2. Designs of transition pieces.

			Column		Brace		Pile	
Model	μ	T	T (mm)	γ_c	tb (mm)	γ_b	tp (mm)	γ_p
M1	0.50	0.40	12.5	160	5.0	80.0	7.5	80.0
M2	0.50	0.40	14.3	140	5.7	70.0	8.6	70.0
M3	0.50	0.40	16.7	120	6.7	60.0	10.0	60.0
M4	0.50	0.40	20.0	100	8.0	50.0	12.0	50.0
M5	0.50	0.40	25.0	80	10.0	40.0	15.0	40.0
M6	0.50	0.40	33.3	60	13.3	30.0	20.0	30.0
M7	0.50	0.40	40.0	50	16.0	25.0	24.0	25.0
M8	0.50	0.40	50.0	40	20.0	20.0	30.0	20.0
M9	0.50	0.40	66.7	30	26.7	15.0	40.0	15.0
M10	0.50	0.40	80.0	25	32.0	12.5	48.0	12.5
M11	0.50	0.40	100.0	20	40.0	10.0	60.0	10.0
M12	0.50	0.40	133.3	15	53.3	7.5	80.0	7.5
M13	0.50	0.40	166.7	12	66.7	6.0	100.0	6.0
M14	0.50	0.40	200.0	10	80.0	5.0	120.0	5.0
M15	0.50	0.80	133.3	15	106.7	3.8	160.0	3.8
M16	0.50	0.80	166.7	12	133.3	3.0	200.0	3.0
M17	0.50	0.80	200.0	10	160.0	2.5	240.0	2.5
M18	0.50	0.63	133.3	15	83.3	4.8	125.0	4.8
M19	0.50	0.63	166.7	12	104.2	3.8	156.3	3.8
M20	0.50	0.63	200.0	10	125.0	3.2	187.5	3.2
M21	0.50	0.50	133.3	15	66.7	6.0	100.0	6.0
M22	0.50	0.50	66.7	12	83.3	4.8	125.0	4.8
M23	0.50	0.50	200.0	10	100.0	4.0	150.0	4.0
M24	0.75	0.27	20.0	100	5.3	75.0	8.0	75.0
M25	0.75	0.27	25.0	80	6.7	60.0	10.0	60.0
M26	0.75	0.27	33.3	60	8.9	45.0	13.3	45.0
M27	0.75	0.27	40.0	50	10.7	37.5	16.0	37.5
M28	0.75	0.27	50.0	40	13.3	30.0	20.0	30.0
M29	0.75	0.27	66.7	30	17.8	22.5	26.7	22.5
M30	0.75	0.27	80.0	25	21.3	18.8	32.0	18.8
M31	0.75	0.27	100.0	20	26.7	15.0	40.0	15.0
M32	1.00	0.20	20.0	100	4.0	100.0	6.0	100.0
M33	1.00	0.20	25.0	80	5.0	80.0	7.5	80.0
M34	1.00	0.20	33.3	60	6.7	60.0	10.0	60.0
M35	1.00	0.20	40.0	50	8.0	50.0	12.0	50.0
M36	1.00	0.20	50.0	40	10.0	40.0	15.0	40.0
M37	1.00	0.20	66.7	30	13.3	30.0	20.0	30.0
M38	1.00	0.20	80.0	25	16.0	25.0	24.0	25.0
M39	1.00	0.20	100.0	20	20.0	20.0	30.0	20.0

3 FINITE ELEMENT ANALYSIS

The nonlinear finite element analyses are carried out using the commercial software ANSYS to analyse the ultimate strength of the transitional piece. The nonlinearities that are associated with the material and structural geometry are accounted for, and the large deformation option is activated. Moreover, the arch-length method is adopted to solve the convergence problem by introducing an arc instead of line to converge the Newton-Raphson equilibrium, and following the post-collapse behaviour is obtained.

For the low-cycle fatigue damage assessment, the pseudo-hot spot stress is calculated by performing a linear FEM analysis. Here the objective is to obtain the stress occurred because of the maximum load that the structure can carry by using the linear stress-strain relationship. To this end, the maximum principal stress is calculated and combined with the appropriate stress concentration factor to estimate the pseudo-hot spot stress, which is used to be translated to the true (actual) strain by employing strain-based approach and the Nueber's rule in tandem.

3.1 *FEM modelling*

The finite element type SHELL181 is used to model the tubular monopile structure. The element has four nodes with six degrees of freedom at each node: translations in the X, Y, and Z-axes, and rotations about the X, Y, and Z-axes. The element type includes the stress stiffness terms by default and supports the nonlinear material models, which makes the element type well-suited for a linear, large rotation, and large strain nonlinear applications. Furthermore, the material model used in the nonlinear FE analysis is a bilinear elastic-perfectly plastic stress-strain relationship where the yield stress is 355 MPa.

The finite element model is using a coarse mesh density, which may overestimate the ultimate strength due to the over-stiffening. On the other hand, selecting a refined mesh density not only increases the computational time immensely but also cause a convergence problem. Thus, a mesh sensitivity analysis is carried out based on the relation between the gradient of the ultimate bending moment and the finite element size to identify the optimal finite element size. The result of the sensitivity analysis and the appropriate finite element size is found to be 0.1 m.

The present study introduces the geometrical imperfection into the FEM model of the transition piece to model the imperfection that may be occurred before and during the manufacturing process. The imperfections are generated by modifying the vertical and horizontal position of the nodes, and they are applied based on a function generating superimposed periodic buckling wave shapes onto a 2-D surface throughout the structure. The amplitude of a half wave is defined as a function of the length and thickness of the column component, and the number of the half wave is defined as 8.

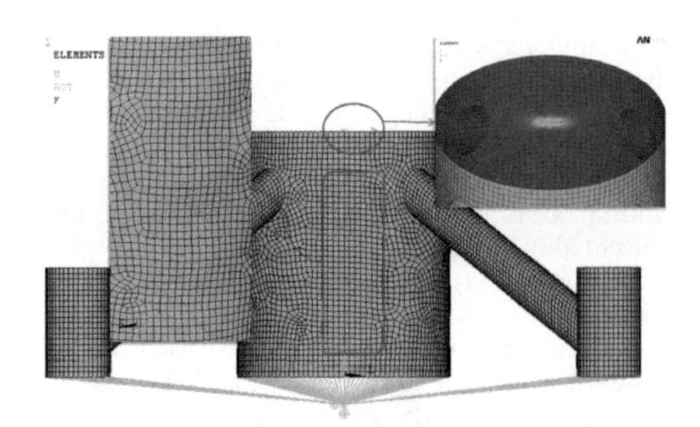

Figure 3. Finite element modelling of the transition piece.

3.2 *Loading and boundary conditions*

For the application regarding the load and boundary conditions, two master nodes are defined at both ends of the FEM model of the tubular monopile structure. These master nodes are connected to the nodes at the bottom and top of the structure through rigid beams, which allow the structure to follow the behaviour of the master nodes. Figure 3 demonstrates the load and boundary conditions applied in the nonlinear FE analysis.

A concentrated load is applied to the master node on the top, while the master node at the bottom is connected to three nodes at X, Y and Z-axes by springs.

4 ULTIMATE STRENGTH ASSESSMENT

The transition piece is a redundant frame system, and when a local failure occurs because of an overloading, the load can be carried by an alternative load path until the global collapse of the support structure occurs.

In the present study, the structural response of the transition piece is discussed at four points. The first point is associated with the proportional limit where the structure follows a linear force-displacement relationship. The second point is where the deformation becomes obvious. The point 3 is associated with the ultimate strength of the transition piece, and the last point accounts for the failure. These points are illustrated in Figure 4 on a force-displacement relationship of a transition piece design, M6.

At the point of the proportional limit, the deformation occurred at the structure is elastic except for the column-brace connection where the structure experiences a slight plastic deformation. Figure 5 shows that the von Mises elastic and plastic strain distribution of the studied structure. As mentioned, the elastic strain is distributed throughout the column structure; the concentration can be observed between the column and braces, as well as the circumferences of the braces. Also, the bottom of the column has shown high elastic strain due the subjected bending moment. The proportional limit is estimated to be approximately 75 percent of the ultimate strength of the transition piece.

After the proportional limit, the structure commences having a visible plastic strain, in turn, permanent deformation. The plastic strain emerges between the saddle and the crown toe of the tubular joint, and gradually grows on to the column. Figure 6 shows the von Mises plastic strain distribution, which is found to be concentrated between the column and braces. In addition to this, the structure has started to experience plastic strain at a location close to the bottom of the column side under compression. The point 2 occurs approximately 94 percent of the ultimate strength.

The ultimate strength is reached at the point 3 from this point on the structure do not have the capacity to carry the load anymore. Figure 7 shows

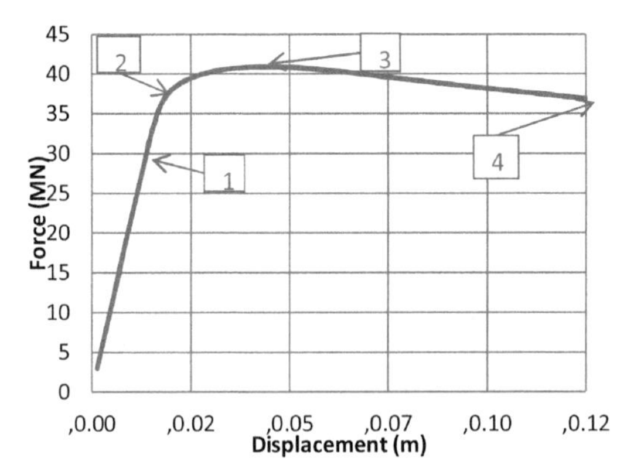

Figure 4. Force-displacement diagram of M6.

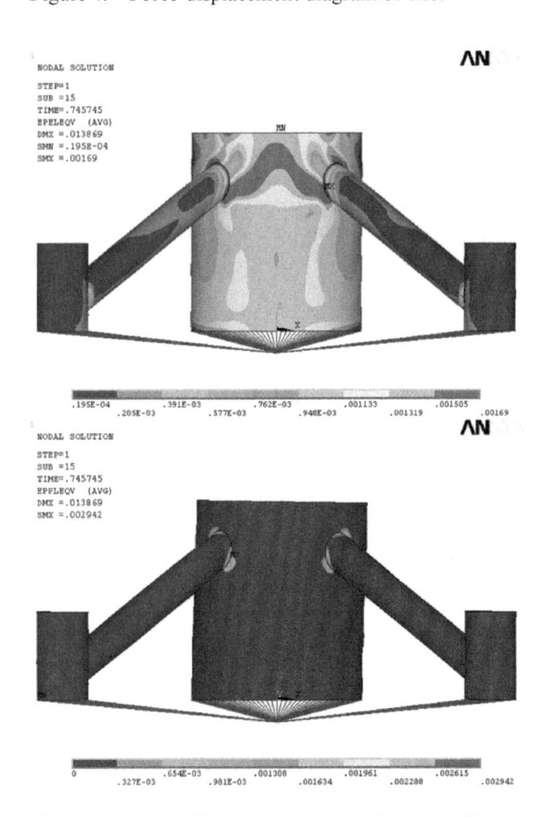

Figure 5. Von Mises elastic (up) and plastic (down) strain at the proportional limit.

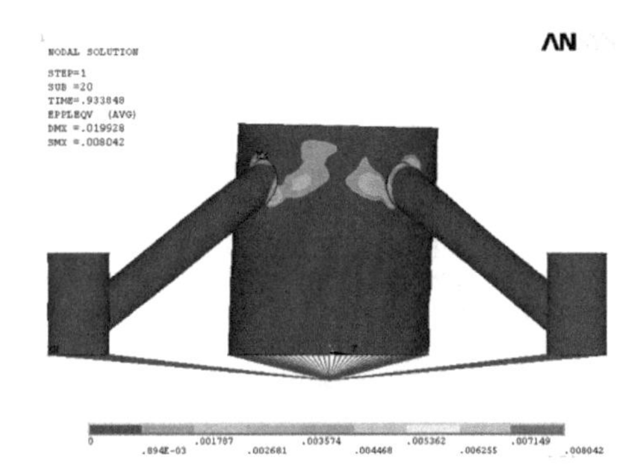

Figure 6. Von Mises plastic strain at point 2.

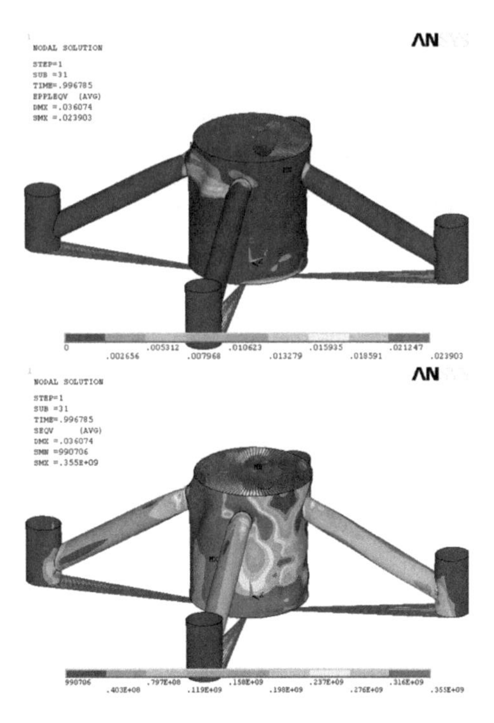

Figure 7. Von Mises plastic strain (up) and total stress at the ultimate strength.

the von Mises plastic strain as well as the von Mises stress distribution of the structure.

The plastic strain distribution points out that the deformation is spread between the braces at side-to-side with reference to the applied load in the direction of the X-axis. On the sides, deformation takes a somehow a wave shape, which seems to be the prevailing indicator of the buckled column, consequently the capacity reduction. The deformation that has started to be visible at the point 2 becomes obvious at the ultimate strength point.

Moreover, a significant part of the column structure goes up to the yield stress. Apart from the tubular joint, along the braces at both the tensile and compressive sides it seems nominal stress remain well below the yield stress.

Figure 8 shows the deformation shape at the failure, which is represented as point 4. The aim of studying the point 4 is to detect the critical location of the transition piece as the deformation shape becomes quite apparent. There seem to be three locations, where the significant deformation happens, namely "a", "b" and "c". The most visible deformation is "a", and it is associated with a shape similar to a wave with a high crest, whereas "b" is an ovalised-type deformation shape happening around the lower portion of the column under the compressive side. Furthermore, "c" accounts for the ovalisation at one of the braces at the under compressive side of the transition piece.

It is worth mentioning that the structural responses illustrated here represent the structural responses of the transition piece design with a relatively lower thickness. It can be expected that the structural response of the thicker structure has a higher capacity, and the stresses are even more distributed. For the structure with thicker plates, it is plausible to observe not only the almost whole column structure but also the brace components reach the yield stress. The given assertion can also be drawn from the force-displacement (see Figure 9).

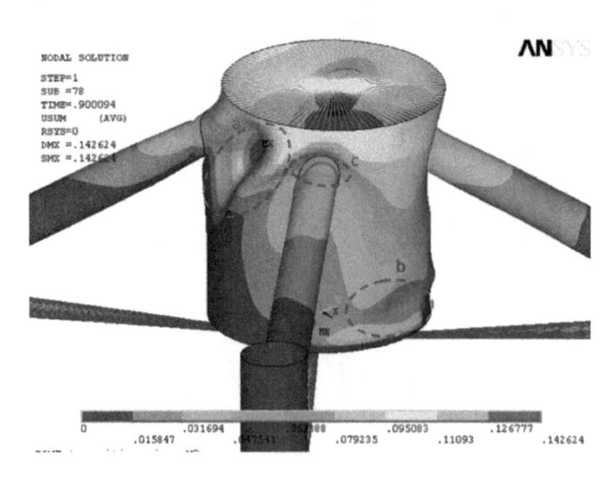

Figure 8. Deformation shape at failure.

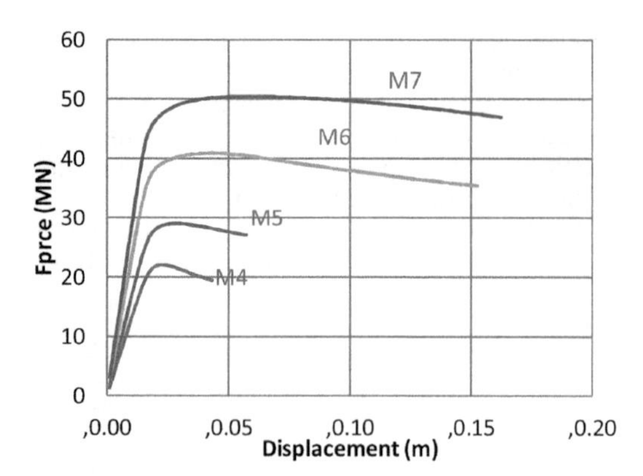

Figure 9. Force-displacement diagram of different designs.

The force-displacement diagram demonstrates that when the structure gets thicker, it goes through relatively higher plastic deformation before reaching the ultimate strength.

5 LOW-CYCLE FATIGUE

When the structure is subjected to the cyclic loads beyond the elastic region, the strain-based approach, which accounts for the plastic deformation is more suitable to be employed for the fatigue damage assessment. The true (actual) strain accounts for the elastic-plastic mechanism that covers the interaction between the notch plastic zone and the surrounding elastic material.

$$\Delta\varepsilon_t = \Delta\varepsilon_{el} + \Delta\varepsilon_{pl} \tag{1}$$

where $\Delta\varepsilon_{el}$ and $\Delta\varepsilon_{pl}$ are the elastic and plastic strain range respectively. The total local strain range can be described as a function of the local stress range and the material properties (Yeter et al., 2016).

The Ramberg and Osgood (1943) model express the cyclic elastic-plastic stress-strain behaviour of the material as given below:

$$\Delta\varepsilon_{loc,t} = \frac{\Delta\sigma_{loc}}{E} + 2\left(\frac{\Delta\sigma_{loc}}{2\mathrm{K}'}\right)^{\frac{1}{n'}} \tag{2}$$

where K′ and n' are the material parameters and can be described in terms of the fatigue strength and ductility coefficients and exponents. The stress concentration factor (SCF) calculated by the non-dimension parametric equations for the offshore tubular joints is used to estimate the global hotspot stress. Using the Neuber's rule, the global stress range is transformed into an actual level of strain.

The Ramberg–Osgood equation for the stress-strain relation to the Neuber's rule leads to:

$$\frac{\Delta\sigma_{nom}{}^{2}K_{t}{}^{2}}{E}=\frac{\Delta\sigma_{loc}{}^{2}}{E}+\left(2\Delta\sigma_{loc}\right)\left(\frac{\Delta\sigma_{loc}}{2K'}\right)^{\frac{1}{n'}} \tag{3}$$

The given relation can be solved through an iteration approach for the local stress range.

6 RESULTS AND DISCUSSIONS

The nonlinear finite element analyses are carried out for all the designs of the transition piece as given in Table 2. Thus, the ultimate load-carrying capacity of the designs is calculated. Afterwards, using the maximum load that the structure may carry, linear FE analyses are carried out to calculate the (pseudo) hot spot stress, which is translated to the true strain by using the local-strain approach. By doing so, the transition pieces are evaluated from both ultimate strength and low-cycle fatigue standpoint.

Figure 10 shows the relation between the thickness of the column and the ultimate strength. A strong correlation is observed here. Although the braces are integral parts of the transition piece, the component that carries the load most is the column itself. This observation is valid for different ratios of the thickness studied here; however, it must be mentioned that for the structures with thicker plates there seems to be a little contribution from the braces to the maximum load carrying capacity. Especially at 100 mm of a column thickness, the differences of the ultimate strength for different brace thicknesses can be noticed, whereas the difference below 50 mm can hardly be found.

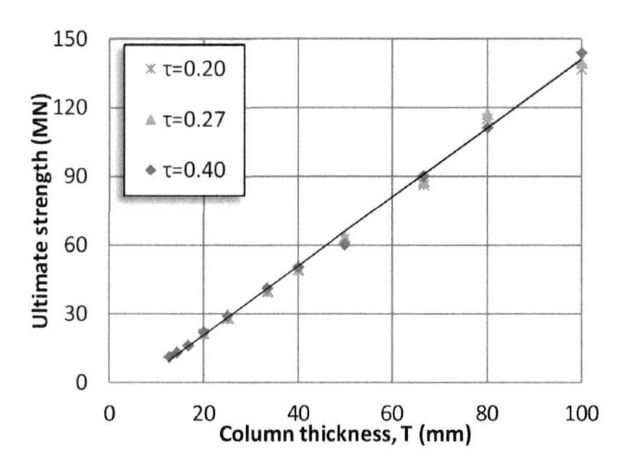

Figure 10. Ultimate strength and column thickness relation.

The ultimate strength assessment can be evaluated from the column slenderness perspective as well since it is an essential parameter for the design purposes to define design constraints. The results shown in Figure 11 indicate that there is a negative correlation between the ultimate strength and the column slenderness ratio, which can be explained with a power function. The ultimate strength increases dramatically once going towards to lower slenderness ratios.

The slenderness ratios lower than 20 are also studied. The dramatic increase in the ultimate strength still continues; however, as mentioned before, as the column gets thicker the contribution of the brace components to the ultimate strength is increasing as seen in Figure 12.

As far as the low-cycle fatigue damage is concerned, differentiation of the nominal stress and localised stress is quite essential. Thus, the nominal and localised stress ranges with respect to the brace thickness are given in Figure 13.

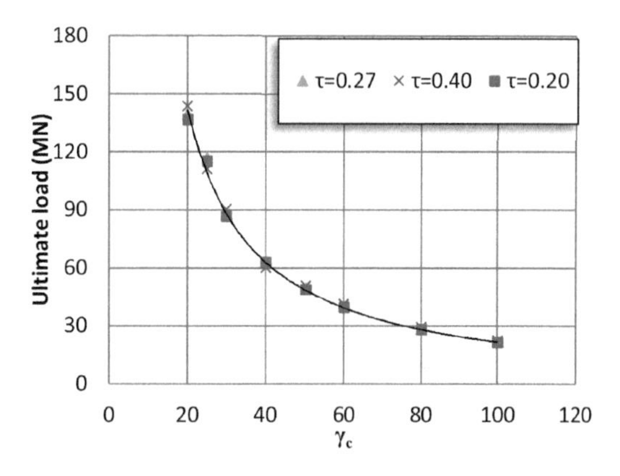

Figure 11. Ultimate strength and slenderness ratio relation.

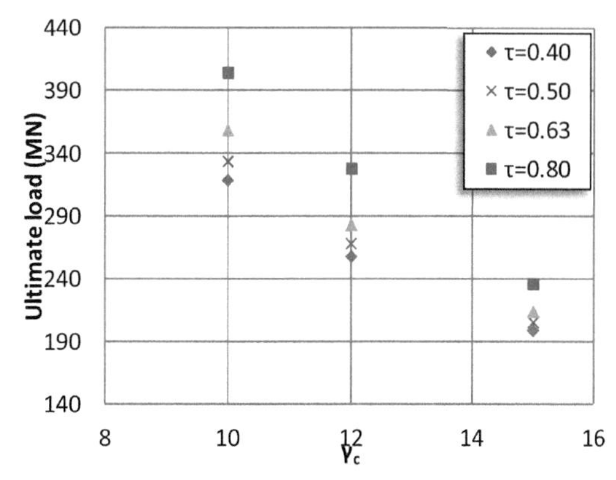

Figure 12. Ultimate strength at lower slenderness ratio.

It is because the structure is thick enough to carry a higher load by using all part of its components, the stresses occurred at the failure found to be higher and more distributed. Because of this, the nominal stresses at the brace component seem to increase as the thickness of both column and brace components increase (see Figure 13).

The increase in localised stress is also observed, however, on a smaller scale, which is a result of the pseudo hot spot stress translated into the localised true stress based on the Neuber's rule.

In addition, the stress concentration factor is reducing as the structure is designed to be thicker, as seen in Figure 14. Therefore, even though the nominal stress is higher for the thick structures because of SCF, the localised stress does not increase in the same proportion.

Moreover, the thickness ratio between the brace and column has a significant influence on the stress concentration factor; therefore, it is expected to have a lower pseudo hot spot stress, in turn, localised stress and strain. For the transition

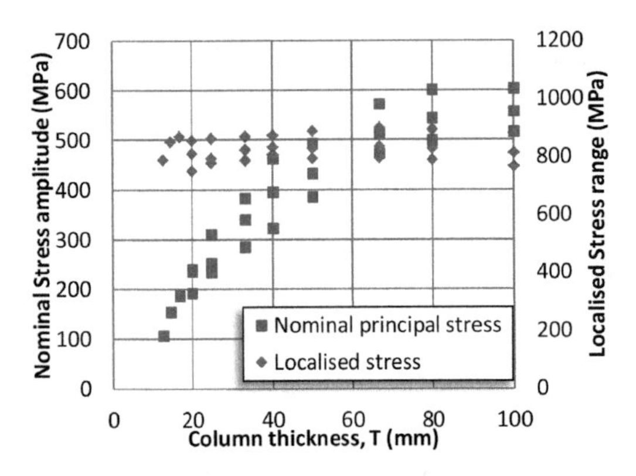

Figure 13. Nominal and localised stresses.

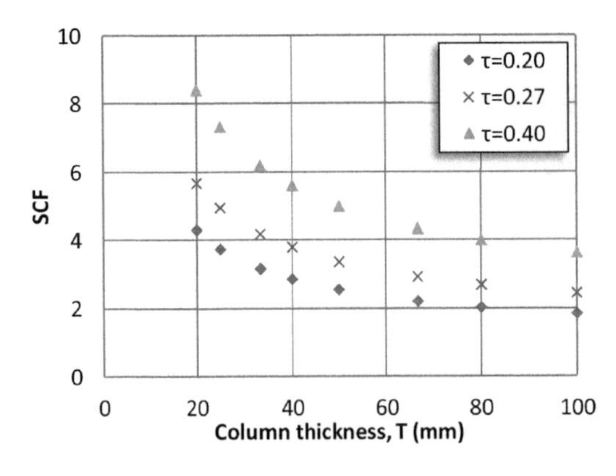

Figure 14. Stress concentration factor for different thickness.

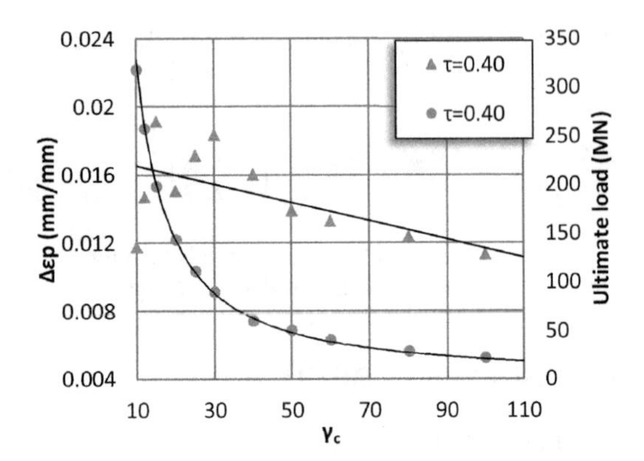

Figure 15. Ultimate strength and low-cycle damage together.

models with less than $\tau = 0.40$, low-cycle fatigue is expected to lose the significance as a failure mode.

As the primary objective of the present study, the structural capacity of different transition pieces from the ultimate strength and the low-cycle fatigue are analysed together. It is desired to have a lower low-cycle fatigue damage on one hand; on the other hand, it is desirable to achieve a higher ultimate strength level. Figure 15 shows the coupled assessment of designs of the transition piece.

It is seen that as the slenderness of the column increases the ultimate strength is reduced. However, the fatigue damage is also reduced. Therefore,a column slenderness ratio below 40 is found to be recommendable since the increase of the ultimate strength more dramatic than the increase of the low-cycle fatigue damage.

7 CONCLUSIONS

In the present study, the failure assessment of the transition piece of a jacket offshore wind turbine to extreme loadings was performed. The structural response is analysed by performing nonlinear finite element analysis, and the local-strain approach is adopted to estimate the true strain as a response to a low-cycle fatigue load.

It is found that the applied load is mostly carried by the column component. The brace components had a slight effect on the ultimate strength of the transition piece only in the models with very thick plates. Moreover, a negative correlation between the ultimate strength and the column slenderness ratio was defined with a power function.

The nominal principal stresses at the brace components were increased as the structures were designed with thicker plates. However, because of the stress concentration factor, the localised stress did not increase in the same proportion.

Finally, the column slenderness ratio below 40 was found recommendable after taking into consideration the ultimate strength and the low-cycle fatigue together.

ACKNOWLEDGEMENTS

This work was performed within the Strategic Research Plan of the Centre for Marine Technology and Ocean Engineering (CENTEC), which is financed by Portuguese Foundation for Science and Technology (Fundação para a Ciência e Tecnologia-FCT).

REFERENCES

Brazier, L.G. 1927. On the flexure of thin cylindrical shells and other "thin" sections. *Proceedings of the Royal Society of London. Series A, containing papers of a mathematical and physical character*, 116, 104–114.

Dimopoulos, C.A. & Gantes, C.J. 2012. Experimental investigation of buckling of wind turbine tower cylindrical shells with opening and stiffening under bending. *Thin-Walled Structures,* 54, 140–155.

Fabian, O. 1977. Collapse of cylindrical, elastic tubes under combined bending, pressure and axial loads. *International Journal of Solids and Structures,* 13, 1257–1270.

Garbatov, Y., Rudan, S., Gaspar, B. & Guedes Soares, C. 2011. Fatigue assessment of marine structures. *In:* Guedes Soares Et Al (ed.) *Marine Technology and Engineering.* London: Taylor & Francis Group, 876–881.

Gellin, S. 1980. The plastic buckling of long cylindrical shells under pure bending. *International Journal of Solids and Structures,* 16, 397–407.

Guo, L., Yang, S. & Jiao, H. 2013. Behavior of thin-walled circular hollow section tubes subjected to bending. *Thin-Walled Structures,* 73, 281–289.

Ju, G.T. & Kyriakides, S. 1992. Bifurcation and localization instabilities in cylindrical shells under bending—II. Predictions. *International journal of solids and structures,* 29, 1143–1171.

Kallaby, J. & Millman, D.N., 1975, Inelastic analysis of fixed offshore platforms for earthquake loading, Offshore Technology Conference (OTC), Houston.

Kim, K.S., Kim, B.O., Kim, Y.K., Lee, C.H. & Lee, S.W., 2005, A study on the low cycle fatigue behavior of the steel for shipbuilding industry, Key Engineering Materials, Trans Tech Publ, 10–15.

Kyriakides, S. & Ju, G.T. 1992. Bifurcation and localization instabilities in cylindrical shells under bending—I. Experiments. *International Journal of Solids and Structures,* 29, 1117–1142.

Marshall, P.W. 1975. Failure modes for offshore platforms. *Methods of Structural Analysis.* ASCE, 309–323.

Miner, M.A. 1945. Cumulative Damage in Fatigue. *Journal of Applied Mechanics -Transactions of the ASME,* 12, 159–164.

Neuber, H. 1961. Theory of stress concentration for shear-strained prismatical bodies with arbitrary nonlinear stress-strain law. *Journal of Applied Mechanics,* 28, 544–550.

NORSOK N-006 2009. Assessment of structural integrity for existing offshore load-bearing structures. Norway.

Ramberg, W. & Osgood, W.R. 1943. Description of stress-strain curves by three parameters. *Technical Note No. 902.,* Washington DC: National Advisory Committee For Aeronautics.

Sherman, D.R. 1976. Test of circular steel tubes in bending. *Journal of the structural division,* 102, 2181–2195.

Wang, X., Kang, J.-K., Kim, Y. & Wirsching, P.H., 2006, Low cycle fatigue analysis of marine structures, 25th International Conference on Offshore Mechanics and Arctic Engineering, American Society of Mechanical Engineers, 523–527.

Yeter, B., Garbatov, Y. & Guedes Soares, C., June 2016, Reliability of Offshore Wind Turbine Support Structures Subjected to Extreme Wave-Induced Loads and Defects, Proceedings of The 35th International Conference on Ocean, Offshore and Arctic Engineering, OMAE16, Busan, South Korea, American Society of Mechanical Engineers.

Yeter, B., Garbatov, Y. & Guedes Soares, C. 2017 System reliability of a jacket offshore wind turbine subjected to fatigue. *In:* C., G.S. & Garbatov, Y. (eds.) *Progress in the Analysis and Design of Marine Structures.* London, UK: Taylor & Francis Group, 939–950.

Progress in Maritime Technology and Engineering – Guedes Soares & Santos (Eds)
© 2018 Taylor & Francis Group, London, ISBN 978-1-138-58539-3

Low-cycle fatigue of damaged stiffened panel in ship structures

I. Gledić & J. Parunov
Faculty of Mechanical Engineering and Naval Architecture, University of Zagreb, Zagreb, Croatia

ABSTRACT: This work aims to extend the conventional post-accidental structural safety assessment by investigating possibility of damage propagation in stiffened panels during salvage period as low-cycle high stress fatigue process caused by fluctuating wave loads. Low-cycle loading can cause high-amplitude stresses at the edge of damage opening, which consequently lead to high stress concentrations. In the present study, intended for conceptual considerations, damage of stiffened panel is idealized as rectangular, diamond and circular shape. A finite element model is created for intact stiffened panel and for each idealized damage shape for which the stress concentration factor is calculated. The number of constant-amplitude wave load cycles to initiate a fatigue crack is calculated using the strain-life method defined according to Det Norske Veritas classification notes. Accumulated low-cycle fatigue damage is then estimated by Monte Carlo simulation, where individual stress amplitudes are drawn as random numbers according to Rayleigh distribution. The analysis is performed for three assumed types of short-term random sea state: calm, moderate and rough seas Conclusions of the study are related to the conditions that could cause low-cycle fatigue and to the importance of assumed idealized damage shapes.

1 INTRODUCTION

The increase in the number of ships in the global fleet necessarily implies the increase of collision and grounding risk. During safety assessment of a damaged ship hull it is generally considered that the damage is time invariant (e.g. Prestileo et al. 2013, Youssef et al. 2016) However, the initial damage caused by collision or grounding can further propagate during the ship salvage operations, in a similar way as large fatigue cracks.

The propagation of damage during towing period is being discussed in some studies (Kwon et al. 2010, Sasa & Incecik 2009), attempting to define in a conceptualized manner a methodology adequate for calculation of the residual ultimate strength of a damaged ship by taking into account increase of the damage during towing period. The physical background of the damage propagation problem is somewhat similar to the issue of large fatigue crack propagation through the ship structure (Ayala-Uraga & Moan 2007).

The study by Bardetsky (2013) has pointed out the importance of the assessment of the crack propagation as well as the importance of the residual strength reduction of a damaged ship. Study showed the significance of the development of the computational model for rapid crack growth assessment in a damaged ship structure, while some modelling options for such assessment were proposed.

A simplified method for studying propagation of the damage during towing period after collision or grounding is proposed by Bužančić Primorac & Parunov (2013). A preliminary conclusion was that a damaged ship would be able to withstand a towing period of 7 days before crack propagates to the critical size.

Gledić & Parunov (2015) examined whether the weight function method could be used for rapid stress intensity factor calculation in case when a ship is damaged in collision or grounding. It was shown by the weight function method that a stiffener causes only small reduction of the stress intensity factor and that the weight function method may be considered as rapid stress intensity factor assessment method for damaged ship, but it would be necessary to perform further verification using finite element method (FEM) and fracture mechanics (FM).

Most of the research so far regarding the fatigue life of an intact or damaged ship is done assuming that a ship is experiencing only high-cycle fatigue (HCF) caused by a high number of stress cycles with small amplitudes. In some cases, however, a marine structure is subjected to stress cycles of such a large magnitude that small but significant parts of the structure in question experience cyclic plasticity. Fatigue caused by a low number of cycles with high amplitude is known as the low-cycle fatigue (LCF). In a scenario where a damaged ship e.g. encounters a severe storm during the salvage period, the damage accumulation could lead to the LCF. Importance of the LCF is such that it should be considered as principal failure mode, associated with ultimate limit state (ULS) or accidental limit state (ALS) (DNVGL-RP-C208 2016). NORSOK (2009) prescribes a cyclic check for marine structures during storm actions in ULS and ALS, and provides the acceptance criteria for LCF.

The aim of this study is to determine the LCF damage accumulation in a damaged stiffened panel. The effect of the damage shape is examined through three types of idealized damage shapes: diamond, circular and rectangular. Fluctuating wave–induced stresses in short-term sea state are assumed to follow Rayleigh distribution. LCF damage is determined by strain-based approach for calculation of the fatigue damage accumulation. At the end of this study, uncertainties regarding parameter definition and material properties are discussed and conclusions based on results are provided.

2 LOW-CYCLE FATIGUE

During service life, a ship is subjected to fluctuating wave-induced stresses which can be divided into two main groups: high-cycle low-amplitude stresses, causing HCF and low-cycle high-amplitude stresses leading to LCF. HCF is caused by high-cycle wave-induced stresses, considered to be in the range between 10^4–10^8 numbers of cycles. Low-cycle stresses are the consequence of e.g. loading and unloading operations, temperature fluctuation or extreme waves developed in the storm. Low-cycle stresses leading to LCF are considered to be in the range between 10 and 10^4 numbers of cycles. The classification rules traditionally have been prescribing simplified fatigue analysis of structural details exposed to high-cycle stresses (DNVGL-CG-0129 2015). Recently, however, more attention is given to low-cycle stresses, mostly resulting from loading and unloading operations of oil tankers or FPSO vessels. The classification rules developed a procedure to assess LCF of ship structures under frequent loading and unloading cycles (DNVGL-CG-0129 2015). The procedure defines the minimum requirement to LCF strength. It also provides expressions for calculation of combined fatigue damage due to HCF and LCF.

Currently, LCF calculation procedures may fit into one of the two groups: methods using range of local deformations and methods of pseudo-elastic stresses. The method of local deformations aims to define the deformation and stresses at highly stressed micro-locations as a function of global deformation and stresses of structural details (Yeter et al. 2015). The latter method is compatible with hot-spot strain range approach where total strain is converted to pseudo-elastic stress range by using a plasticity correction factor (DNVGL-CG-0129 2015, Urm et al. 2004).

Wang et al. (2006) proposed the extended S-N curve in order to define the fatigue life of ship structural details in the low-cycle regime. They also developed a method for fatigue damage calculation based on a hot-spot stress approach.

Urm et al. (2004) developed a simple procedure to assess LCF of ship structures by using pseudo elastic stress range and plasticity correction factor. For calculation of LCF, static loads are presumed to be due to loading and unloading of cargo. The procedure for calculation of combined damage due to LCF and HCF is also developed.

In severe sea conditions, the fluctuating stresses due to high short-term loading like waves with high amplitudes may exceed the yield stress of material and lead to the structural failure. In this case, in order to predict the time of a new crack initiation, LCF analysis and the short-term statistics should be used. Opposed to the high-cycle regime, the fatigue life in the low-cycle regime is normally expressed in terms of the total strain range rather than the stress range (Heo et al. 2004).

The strain-life approach, used for calculation of LCF, requires the knowledge of stress-strain behavior of material, which is presented by cyclic stress-strain curve. The use of monotonic stress-strain curve must be avoided since it may provide non-conservative fatigue life estimates, which is particularly important in case of high strength steels (DNVGL-CG-0129 2015). In comparison with the monotonic stress-strain material curve, the cyclic stress-strain material curve captures a hardening effect of a notch root area. The elasto-plastic stress state at notch root is expressed by Ramberg-Osgood function:

$$\varepsilon = \frac{\sigma}{E} + \left(\frac{\sigma}{K'}\right)^{\frac{1}{n'}} \tag{1}$$

where: σ and ε are local stress and strain amplitudes respectively; E is the Young's modulus of elasticity, while K' and n' are coefficients of the material.

In order to calculate the local stress and strain it is necessary to apply analytical model such as, e.g., Neuber's rule to relate nominal stresses and strains to local stresses and strains. Neuber's rule, Equation 2, is the most widely used notch stress-strain analytical model.

$$\sigma\varepsilon = \frac{(K_t \sigma_n)^2}{E} \tag{2}$$

Here, σ and ε are local stress and strain amplitudes respectively while σ_n is nominal stress and K_t is stress concentration factor obtained from the linear finite element analysis.

Heo et al. (2004) performed fatigue testing of base metal and welded joints in low-cycle high stress regime and compared it to a design guidance for LCF in ship structure proposed by classification rules. They concluded that Neuber's rule gives conservative results in estimation of elasto-plastic notch stress-strain state.

Chen (2016) proposed the procedure for calculation of appropriate size of a stop-hole for cracked marine structures. The proposed method incorporates both HCF and LCF analysis into remain service life prediction, taking into account the long-term and short-term wave induced loading.

Yeter et al. (2015) studied the influence of abnormal waves on wind turbine supporting structures. It was found that abnormal waves can lead to LCF.

Brief literature overview shows the relevance of this topic for different types of marine structures.

3 COMPUTATIONAL METHODOLOGY FOR LCF

In this section a procedure of low-cycle fatigue calculation is explained in detail.

3.1 *Wave-induced loading*

The wave-induced stress amplitudes are obtained by using the Monte Carlo (MC) simulation (Chen 2016). The random amplitude of wave induced stresses for a short-term duration is assumed to follow Rayleigh distribution and may be simulated using the following expression:

$$y = \sigma\sqrt{-2lnc} \tag{3}$$

where: σ is scale parameter of the Rayleigh distribution while c is the random number with uniform distribution [0,1].

3.2 *Neuber's Rule and Ramberg-Osgood relationship*

The fluctuating wave loading generates the fictitious, perfectly elastic stress at the tip of the damage that could exceed the actual local yield stress. In the same time, the actual plastic strain could be much higher than the fictitious elastic strain. The calculation of local stress and strain is described in Figure 1. Three curves are presented: the straight line represents the perfectly elastic stress-strain relationship, the dashed curve is the Neuber's Rule and the curved line represents the Ramberg-Osgood cyclic stress-strain. If elastic stresses are lower than yield stress, there is no difference between local and elastic stress and strain values. Once the elastic stress exceeds the value of yield strength of the material the relationship between the local stress and strain becomes non-linear and there is a difference between actual stress and strain and those predicted by the linear analysis. At this point, with previously calculated nominal stress and stress concentration factor (K_t), it is necessary to apply Neuber's Rule to calculate the associated local non-linear stress and strain. Combining the Ramberg-Osgood equation (Equation 1) and Neuber's rule (Equation 2) leads to:

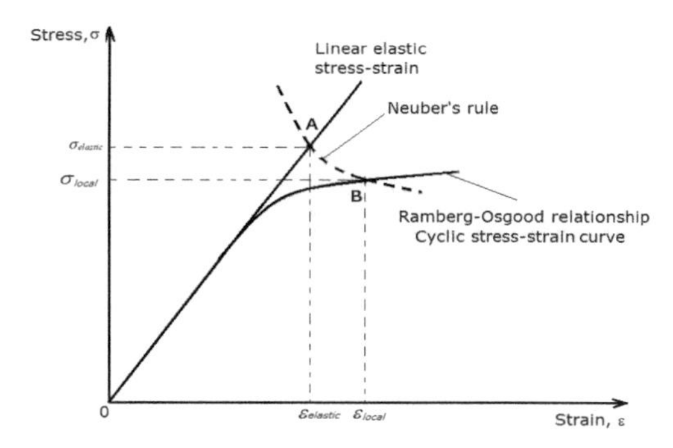

Figure 1. Neuber-Ramberg-Osgood (Heo et al. 2004).

$$\frac{(\sigma \cdot K_t)^2}{E} = \frac{\sigma}{E} + \sigma\left(\frac{\sigma}{K'}\right)^{\frac{1}{n'}} \tag{4}$$

Solution of this $\sigma \cdot E$ loop expression is found by iterative process using the Newton-Raphson method.

In graphic terms, the Neuber's curve intersects both linear elastic stress-strain curve and cyclic stress-strain curve. The elastic stress and strain are solution of the left part of the Equation 4 (point A) which is also the first intersection of the Neuber's curve and linear elastic stress-strain curve. For the particular elastic stress, the local stress and strain are found at the intersection of the same Neuber's curve and Ramberg-Osgood cyclic stress-strain curve, i.e. at point B. The local stress value is lower than the stress predicted by the linear analysis, but the local strain is larger than the strain predicted by the linear analysis.

3.3 *Strain-life method*

Once the value of the local stress and strain is determined, the number of cycles to failure is calculated by the strain-life method. There are different versions of the strain-life method. Some of them take into the account the mean stress effect such as: "Morrow-elastic", "Morrow elastic-plastic"

Table 1. Cyclic stress-strain characteristics for base metal material (DNV-RP-C208 2016.) (Air conditions).

Parameter, Symbol	Value, Units
Cyclic strength coefficient, K'	410 MPa
Fatigue strength coefficient, σ_f'	175 MPa
Fatigue ductility coefficient, ε_f'	0.091
Fatigue strength exponent, b	–0.1
Strain hardening exponent, n'	0.1
Fatigue ductility exponent, c	–0.43

and "Smith-Watson-Topper" strain-life relationship (Stephens et al. 2001). In this article the mean stress effect is not considered and therefore "Manson-Coffin" relationship is used for calculation of strain-life, according to the guidance of DNVGL-RP-C208 (2016):

$$\frac{\Delta\varepsilon_l}{2} = \frac{\sigma_f'}{E}\left(2N_f\right)^b + \varepsilon_f'\left(2N_f\right)^c \tag{5}$$

where: $\Delta\varepsilon_l$ is local stress range, N_f number of cycles to failure; σ_f', ε_f', b and c are cyclic stress-strain characteristics for base metal material in air conditions, provided in Table 1.

Number of cycles for given local stress is calculated by the iterative process using the Newton-Raphson method.

3.4 *Low-cycle fatigue damage*

The low-cycle accumulated fatigue damage induced within a short-term sea condition is given by adopting the Palmgren-Miner rule to LCF:

$$D_l = \sum_{j=1}^{n_s} \frac{1}{N_j} \tag{6}$$

where: n_s is the total number of wave-induced stress cycles in the short-term condition; N_j is the low-cycle fatigue life predicted by the "Manson-Coffin" strain life method.

Herein, it is assumed short-term sea state duration of three hours. The number of wave-induced stress cycles, n_s, within this short–term sea state is assumed to be 1500 cycles, meaning that one cycle corresponds to about 7 seconds. For each stress cycle, partial damage is calculated, and then summation across all cycles is performed using Equation 6.

4 LCF OF DAMAGED STIFFENED PANEL

Damage shape is usually very complex with many irregular notches and the idealization is necessary. For conceptual considerations, three idealized damage models of a damaged stiffened panel are presented in this work: diamond, circular and rectangular shape (Underwood et al. 2012).

The finite element model of the stiffened panel was created using FEMAP software (Fig. 2). The unit panel is consisting of a portion of the plate of width b with a stiffener centered on the plate strip. The stiffener 5030 6 is chosen from ISA (Unequal Indian Standard Angles), according to Underwood et al. (2012) and Suneel Kumar et al. (2009). It denotes an unequal Indian standard angle of flange width 30 mm, with overall web depth of 50 mm and uniform thickness of section 6 mm. Length of the panel and the thickness of the plate is taken as 1500 mm and 6 mm, respectively. In the present study, the width of the plate is equal to the stiffener spacing, i.e. 510 mm.

The damage opening is assumed to be in the center of the panel. As it is symmetrical with respect to the unit panel, only a half portion of the damage opening is considered in the analysis. The width of the opening equals half the size of plate width, i.e. 255 mm. In the region of the stress concentration, a more refined mesh, txt mm, was used in all three models (DNVGL-CG-0129 2015). Outside of the region of stress concentration was used coarser mesh, 25 × 25 mm, according to Suneel et al. (2009).

Firstly, the intact model was used to calculate the distribution of uniform nominal stress across the entire panel. The intact model is subjected to forced unit displacement in positive direction of axis x. The application of the forced displacement was necessary in order to have "pure" axial force and to avoid an undesired bending of the loaded edges of the plate-stiffener combination. The element has three translational (U_X, U_Y, U_Z) and three rotational (R_X, R_Y, R_Z) degrees of freedom at each node. Along the unloaded edges i.e. between the adjacent stiffened plates, translation along axis y is prevented as well as the rotation around axis x and axis z. At the ends of stiffened plate all degrees of freedom are constrained. The material properties

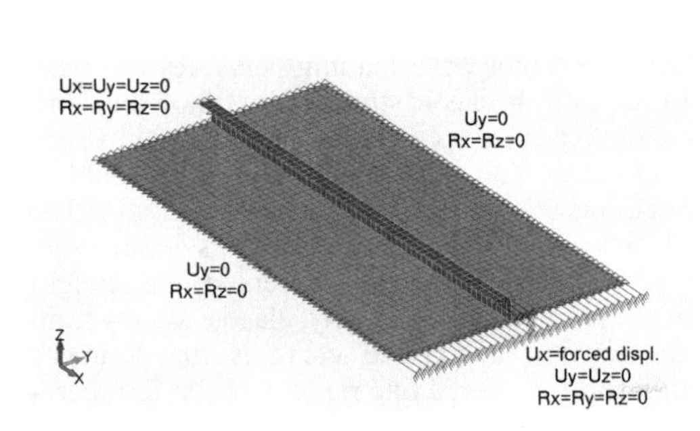

Figure 2. Intact finite element model with constrains.

for the finite element model are provided in the Table 2.

Secondly, the same type of constraints and the same loading is applied on finite element models with diamond, circular and rectangular damage shape (Fig. 3). There are no boundary conditions applied in the region of the damage. Maximum hot-spot stresses are calculated by the finite element analysis using FEMAP.

Table 2. Material properties of damaged stiffened panel.

Parameter, Symbol	Value, Units
Yield stress, σ_y	235 MPa
Young Modulus, E	206 000 MPa
Poisson's ratio, ν	0.3

a)

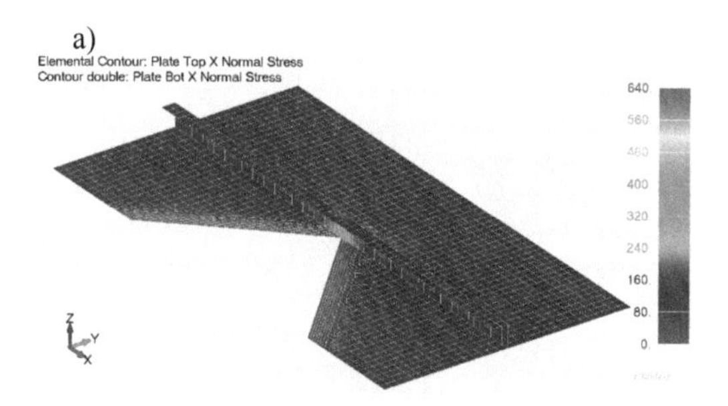

b)

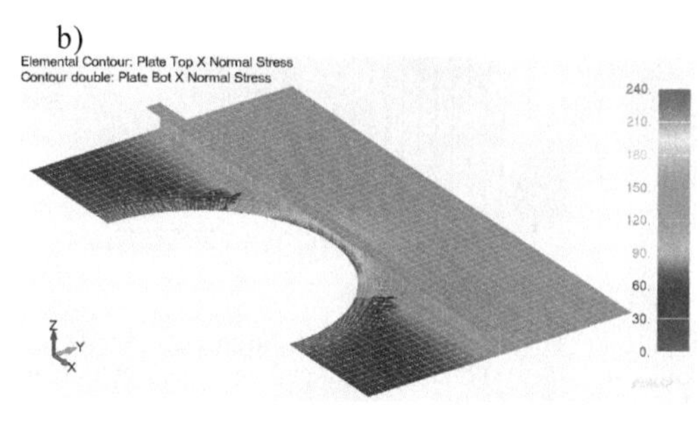

c)

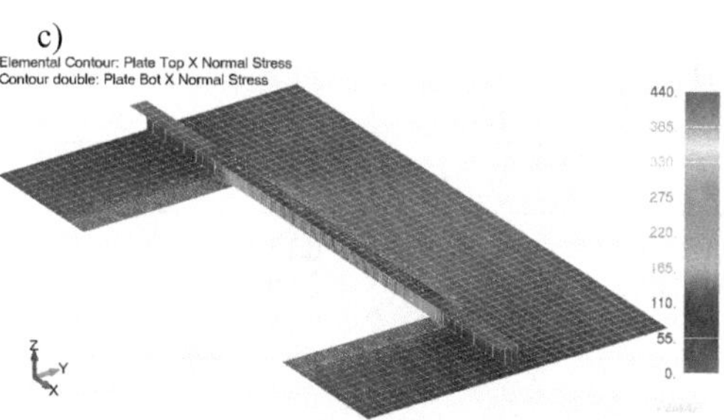

Figure 3. Damaged stiffened panel: a) Diamond shaped damage; b) Circular shaped damage; c) Rectangular shaped damage.

Once the nominal stress and the maximum hot-spot stress are known, K_t may be determined as:

$$K_t = \frac{\sigma_{\max}}{\sigma_n} \tag{7}$$

where: σ_{max} is the maximum hot-spot stress at the damage tip, while σ_n is the nominal stress, calculated for the intact panel.

As expected, the maximum obtained value of K_t is in the case of diamond damage shape (Table 3). It should be pointed out that the stress components considered in the Equation 7 are the major principal stresses parallel to the applied load.

After the calculation of K_t, the final step is to calculate the LCF damage accumulation. Three sea states are considered: calm, moderate and rough.

According to Hoffman & Karst (1975), the statistical most probable value of the amplitude of the highest wave, $\sigma_e{}^*$, given by the maximum value (or the mode) of the extreme probability distribution is given as follows:

$$\sigma_e{}^* = \sqrt{2m_0\, ln(n)} \tag{8}$$

where n is the number of cycles within the short-term sea state; m_0 is the variance of stress amplitude. Other statistical parameters are calculated as:

$$STDEV = \sqrt{m_0} \tag{9}$$

$$\sigma_{1/3} = 2STDEV \tag{10}$$

Main nominal stress parameters, provided in Table 4, which are assumed as representative for each sea state, are calculated by using the Equations (8), (9) and (10). Values from the Table 4. are rounded to the nearest value. It should be pointed out that sea states used in the present study are hypothetical sea states, assuming that they result in wave-induced stresses specified in Table 4.

The wave-induced nominal stress amplitudes are herein generated by MC simulation, as described in Section 3.1, with parameters specified in Table 4. The ideal elastic stresses at hot-spot locations are obtained by multiplying nominal stress amplitudes by stress concentration factor K_t.

Table 3. K_t for three models of damaged stiffened panel.

Damage shape	K_t
Diamond	3.47
Circular	1.61
Rectangular	1.79

Table 4. Sea state.

Sea state	σ_e*	VAR	STDEV	$\sigma_{1/3}$
Calm	50	171	13	26
Moderate	100	685	26	52
Rough	150	1538	39	78

*σ_e is the most probable wave-induced stress amplitude for hypothetical sea-state; **VAR is the variance of stress amplitude; ***STDEV is the standard deviation of stress amplitude; ****$\sigma_{1/3}$ is the significant wave-induced stress amplitude (mean of the highest one third of the stresses).

The actual local stress and strain are calculated from ideal elastic stresses by the strain-life approach and iterative process using the Newton-Raphson method, as described previously in Section 3.2.

Once the local stress and strain are determined, it is possible to calculate the number of cycles by using the "Manson-Coffin" relationship (Eq. 5).

In order to calculate histogram of accumulated damage, D_l (Eq. 6) a 5000 three-hour MC simulations are performed (Chen 2016). In each MC simulation, different random numbers for calculating the Rayleigh-distributed nominal stresses are drawn. Such repeated MC simulation is necessary as the procedure for calculating LCF is non-linear and each short-term simulation results in different D_l. This procedure is done for each sea state and for each type of damage.

As a result, histograms of LCF damage D_l are generated for each damage type. It was found that LCF damage accumulation for calm and moderate sea states is negligible because accumulated damage D_l in all cases is well below 0.01 while the obtained histograms have rather narrow shapes. The histograms for rough sea state and three cases of damage shape are presented in Figure 4.

In case of the circular and rectangular damage shape, the accumulated LCF is very low, even for rough seas. The damage accumulation is in the range between 0.04 to 0.06 for the rectangular shape and 0.03 to 0.04 for the circular shape. The results are much more important for the case of the diamond shape damage, where the accumulated LCF reads between 0.5 and 0.8.

The arithmetic mean, standard deviation and D_l corresponding to 1% probability of exceedance for diamond shape damage are presented in Table 5.

According to DNVGL-CG-0129 (2015), if the contribution of the LCF damage accumulation is equal or above 0.25, then LCF is considered as an important contributor to the total fatigue damage and it is recommended to calculate the combined fatigue damage accumulation due to HCF and LCF. In this particular example, the diamond shape damage in rough seas is worth of further investigation, as D_l is rather large.

a)

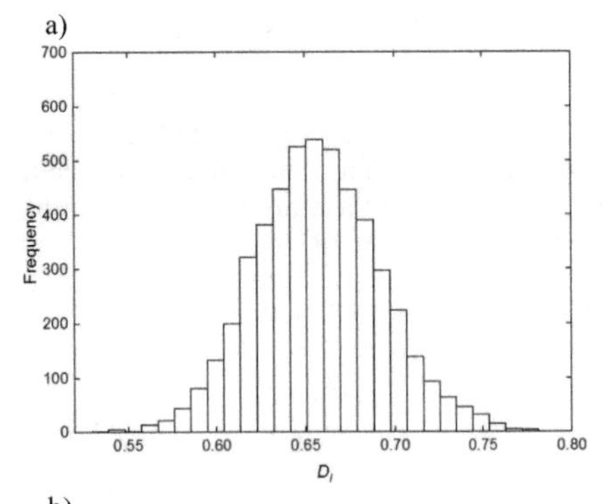

b)

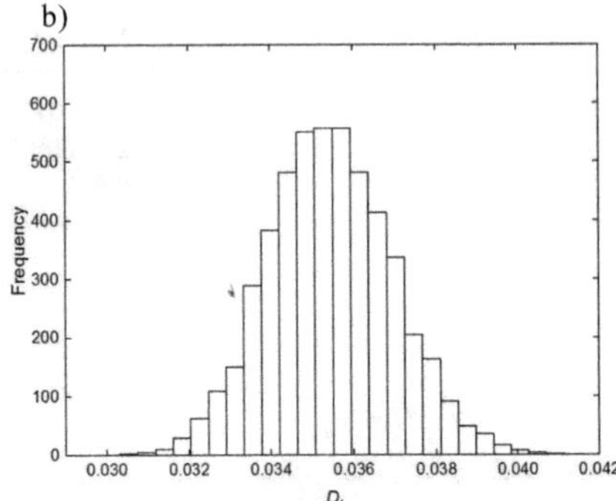

c)

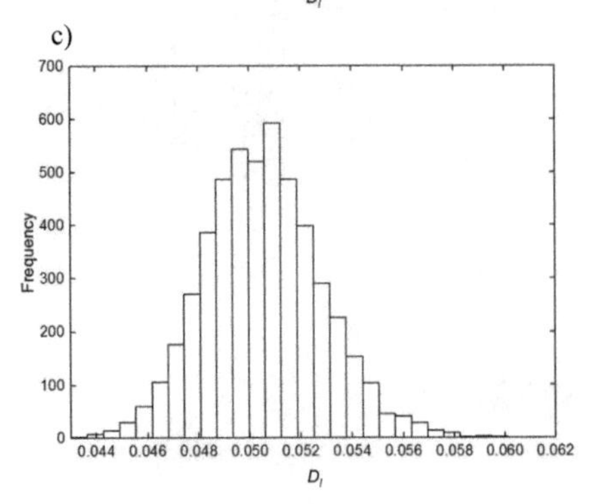

Figure 4. Histogram of low-cycle fatigue damage D_l for rough sea state: a) Diamond shaped damage; b) Circular shaped damage; c) Rectangular shaped damage.

Table 5. Mean value and standard deviation for diamond shape damage.

	Rough	*Moderate*	*Calm*
MEAN	0.66	0.13	0.01
STDEV	0.035	0.007	0.0004
$D_{1\%}$	0.75	0.15	0.01

5 DISCUSSION

The cyclic material properties and parameters in the "Coffin-Manson" equation vary in the literature. Lassen & Recho (2006) proposed calculating these parameters by assuming the dependency of various criteria on the Brinell hardness as the key parameter. However, they found that the results can lead to an overestimation of the damage initiation time. Wang et al. (2006) proposed the same empirical relationships as Lassen & Recho (2006) for calculating σ_f', ε_f', b and c, but for the calculation of K' and n' they suggested certain parameters based on the DSME testing of four types of ship steel. DNVGL-RP-C208 (2016) prescribed the values of the σ_f', ε_f', b and c parameters for air conditions and seawater with cathodic protection. Value of K' and n' are different for base material and for welded joint. Further research needs to be done in order to examine the effect of the parameters variation on the LCF.

The strain-life method chosen for this study assumes that there is no mean stress effect. This assumption is in the agreement with Stephens et al. (2001). It is found that in the LCF region the mean stress effect is smaller due to more stress relaxation at high strain amplitudes. The mean stress effect should be examined more as a part of the further research using some of the proposed models with mean stress effect included.

The assessment of the effect of the damage shape on the damage accumulation of a damaged stiffened panel has shown that, in some cases, the damage shape has a significant effect but further work is needed in order to analyze realistic damage shape on a damaged ship structure. The intention of the authors is to make research efforts in that direction. Representation of service loads in realistic manner is also essential for successful fatigue analysis or design.

According to Stephens et al. (2001), load sequence effect is important and it does effect fatigue. They also concluded that this effect depends on the exact details of the load history and not on the number of cycles. So far, there are no quantitative rules that tell when sequence effects must be considered in predicting fatigue life. Based on the experience it is stated that if the sequence of service loads is completely unknown, one must decide whether to assume significant sequence effects or not.

Zakaria et al. (2016) did a review of the previous literature on the effects of loading sequence on fatigue life behavior. They concluded that through load interaction, load sequences can significantly affect fatigue life. Although there are many fatigue damage accumulation rules used throughout the literature, because of its simplicity, the most commonly used rule is the linear damage accumulation, also known as the Palmgren-Miner's rule. The major drawback of this rule is the lack of consideration of the load sequence effect, that is, the interaction between higher and lower stress loads. For this reason, the Palmgren-Miner's rule may under—or over—estimate fatigue damage.

Unlike the literature regarding load sequence effect on HCF, the literature regarding load sequence effect on LCF is rather scarce and therefore it cannot be precisely concluded if the load sequence effect has the same effect on LCF as it does on HCF. It can only be assumed that load sequence has effect on LCF, but further research is needed to investigate to what extent.

6 CONCLUSIONS

The aim of this work was to investigate the possibility of damage propagation in damaged stiffened panels as a low-cycle high stress fatigue process caused by fluctuating wave loads.

For conceptual considerations, the damage shape of the damaged stiffened panel was idealized as: diamond, circular and rectangular (Figure 3). According to these idealizations, a finite element study of linear elastic stress concentration is performed (Table 3). The assumption was that the wave-induced stresses within short-term sea state follow the Rayleigh distribution. Individual wave-induced stresses are then generated by MC simulation. Using the Ramberg-Osgood relationship and the Neuber's rule, the local stress and strain were calculated. The strain-life method is used to calculate the accumulated damage induced by each stress cycle.

For the calculation of LCF damage, calm, moderate and rough short-term sea states are considered. The LCF accumulated damage histograms are calculated by performing 5000 three-hour MC simulations for each sea state and type of damage (Figure 4).

It was found that both the damage shape and severity of sea state have effect on LCF. It was found that LCF damage accumulation for calm and moderate sea states is negligible because accumulated damage in all cases is well below 0.01. In case of the rough sea and the rectangular shape the damage accumulation is in the range between 0.04 to 0.06 and 0.03 to 0.04 for the circular shape. For the case of the diamond shape damage, the results are much more significant and the accumulated LCF reads between 0.5 and 0.8. This shows that in some unfavorable circumstance such as rough sea and diamond shaped damage (Table 5), LCF could occur at the damage tip and the damage could start propagating. In such case, the crack propagation across stiffened panels should be further analyzed.

ACKNOWLEDGEMENTS

This work has been fully supported by Croatian Science Foundation under the project 8658. The first author has received Mobility Grant for PhD students of University of Zagreb.

REFERENCES

Ayala-Uraga, E. & Moan, T. (2007). "Time-variant reliability assessment of FPSO hull girder with long cracks". *Journal of Offshore Mechanics and Arctic Engineering*. 129: 81–89.

Bardetsky, A. (2013). "Fracture mechanics approach to assess the progressive structural failure of a damaged ship". *Collision and Grounding of Ships and Offshore Structures*. (ed.): Amdhal, J., Ehlers, S., J. Leira, B. Taylor & Francis Group. London. 77–84.

Bužančić Primorac, B. & Parunov, J., (2013). "Reduction of the ultimate strength due to crack propagation in damaged ship structure", *Developments in Maritime Transportation and Exploitation of Sea Resources*, Guedes Soares, C. & López Peña, F. (Eds.), Taylor & Francis Group. London. 365–371.

Chen N. Z, (2016). "A stop-hole method for marine and offshore structures". *International Journal of Fatigue*. 88: 49–57.

DNVGL-CG-0129. (2015). Fatigue assessment of ship structures.

DNVGL-RP-C208. (2016). Determination of structural capacity by non-linear finite element analysis methods.

Gledić, I. & Parunov, J., (2015). "Application of weight function method in the assessment of crack propagation through stiffened panel", *Towards Green Marine Technology and Transport*, Guedes Soares, C., Dejhalla, R. & Pavletic, D. (Eds.), Taylor & Francis Group, London, pp. 247–252.

Heo, J.S., Kang, J.K., Kim, Y., Yoo, I.S., Kim, K.S., Urm, H.S., (2004). "A study on the design guidance for low cycle fatigue in ship structures". *9th Symposium on Practical Design of Ships and Other Floating Structures (PRADS),* Luebeck-Travemuende, 12–17 September 2004. Germany. 782–789.

Hoffman, D. & J. Karst, O. (1975). "The theory of Rayleigh distribution and some of its Applications". *Journal of Ship research. 19*(3):172–191.

Kwon, S., Vassalos, D., Mermiris, G., (2010) "Adopting a risk-based design methodology for flooding survivability and structural integrity in collision/grounding accidents". *Proceedings of the 11th International Ship Stability Workshop*, Wageningen, The Netherlands.

Lassen T. & Recho N., (2006). "Fatigue life analyses of welded structures (FLAWS)". ISTE Ltd. London, United Kingdom.

Norsok N-006., (2009). Assessment of structural integrity for existing offshore load-bearing structures. Norway.

Prestileo, A., Rizzuto, E., Teixeira, A.P., Guedes Soares, C., (2013). "Bottom damage scenarios or the hull girder structural assessment". *Marine Structures*. 33: 33–35.

Sasa, K. & Incecik, A. (2009). "New Evaluation on ship strength from the view point of stranded casualties in coastal areas under rough water". *Proceedings of the ASME 2009 28th International Conference on Ocean, Offshore and Arctic Engineering, OMAE 2009*. Honolulu, Hawaii. 1–8.

Stephens, R.I., Fatemi, A., Stephens R.R., Fuchs, H.O. (2001), Metal fatigue in engineering. Second Edition. *John Wiley & Sons*, Inc. Hoboken, New Jersey, United States of America.

Suneel Kumar, M., Alagusundaramoorthy, P., Sundaravadivelu, R. (2009). "Interaction curves for stiffened panel with circular opening under axial and lateral loads". *Ships and Offshore Structures*. 4 (2): 133–143.

Underwood, J.M., Sobey, A.J., Blake, I.R.J. & Ajit Shenoi, R.,(2012). "Ultimate collapse strength assessment of damaged steel-plated structures". *Engineering Structures*. 38: 1–10.

Urm, H.S., Yoo, I.S., Heo, J.H., Kim, S.C., Lotsberg, I. (2004). "Low cycle fatigue assessment for ship structures." *9th Symposium on Practical Design of Ships and Other Floating Structures (PRADS),* Luebeck-Travemuende, 12–17 September 2004. Germany. 774–781.

Wang, X., Kang, J.K., Kim, Y., Wirsching, P.H., (2006). "Low cycle fatigue analysis of marine structures." *25th International Conference on Offshore Mechanics and Artic Engineering, OMAE 2006*, Hamburg, Germany. 523–527.

Yeter, B., Garbatov, Y., Guedes Soares, C. (2015). "Low cycle fatigue assessment of aging offshore wind turbine supporting structure subjected to abnormal wave loads". *Towards Green Marine Technology and Transport.*, Guedes Soares, C., Dejhalla, R. & Pavletic, D. (Eds.), Taylor & Francis Group, London, pp. 287–294.

Youssef, S.A.M., Faisal, M., Seo, J.K., Kim, B.J., Ha, Y.C., Kim, D.K., Paik, J.K., Cheng, F., Kim, M.S. (2016). Assessing the risk of ship hull collapse due to collision. *Ship and Offshore Structures*. 11(4): 335–350.

Zakaria, K.A., Abdullah, S., Ghazali, M.J. (2016). A Review of the Loading Sequence Effects on the Fatigue Life Behaviour of Metallic Materials. *Journal of Engineering Science and Technology Review.* 9(5): 189–200.

Progress in Maritime Technology and Engineering – Guedes Soares & Santos (Eds)
© 2018 Taylor & Francis Group, London, ISBN 978-1-138-58539-3

Failure assessment of wash plates with different degree of openings

S. Saad-Eldeen
Centre for Marine Technology and Engineering (CENTEC), Instituto Superior Técnico, Universidade de Lisboa, Lisbon, Portugal
(On leave from the Naval Architecture and Marine Engineering Department, Faculty of Engineering, Port Said University, Port Fouad, Egypt)

Y. Garbatov & C. Guedes Soares
Centre for Marine Technology and Engineering (CENTEC), Instituto Superior Técnico, Universidade de Lisboa, Lisbon, Portugal

ABSTRACT: Wash plates are used to minimize the impact load resulting from the sloshing. The design of wash plates, to withstand both lateral and axial loads, is an important issue to be satisfied and the presence of multiple openings affect directly the strength of the wash plates. Therefore, the aim of the present study is to investigate the effect of the number of opening sets on the strength capacity of wash steel plates and to find out the appropriate number of opening sets to optimize the structural capacity. A series of finite element analyses are performed, considering wash plates with a different degree of openings represented by a different number of opening sets. The number of openings sets, as well as the plate thickness, varies. Structural assessment of wash plates with a different degree of openings is carried out, aiming to identify a failure criterion.

1 INTRODUCTION

Large experience has been accumulated during the last decades about the behaviour of marine structural components such as plates. The plates are used in several manners either intact or with a single opening, even with multiple openings, to satisfy the design condition. For plate with a single opening, several research analyses have been performed as the work done by Moen and Schafer (2009) who studied the influence of one opening on the critical elastic buckling stress, considering compressive and bending loadings. The ultimate strength of perforated steel plates under edge shear is investigated by Paik (2008).

Experimentally, Kim et al. (2009) carried out buckling and ultimate strength tests of plates and stiffened panels with one opening, subjected to an axial compressive load, varying several parameters as the aspect ratio, plate slenderness, opening size, shape and location.

Saad-Eldeen et al. (2016a, c) carried out a series of experimental compressive tests on steel plates with one central elongated circular opening of different opening sizes. Considering the initiation of cracks around the with the opening due to complex loading, the ultimate strength is investigated experimentally with and without the locked crack of different crack lengths. The effect of initial imperfections, corrosion degradation and cracks on the strength capacity of steel plate with one circular opening have been studied experimentally by Saad-Eldeen et al. (2016b).

DNV (2008) gives some notes regarding the design of wash plates, which are fitted in a ship's ballast tanks to prevent sloshing due to the free surface of the water when the ship is rolling or pitching. The minimum plate thickness of wash plates is a function of the reaction forces imposed on the neighbour structures and the fitting of a wash plate is mainly based on the tank configurations, i.e. if the breadth of the peak tank is bigger than 2/3 of the moulded breadth of the ship, a centre wash bulkhead shall be fitted.

Regarding the formation of openings, IACS (2010) stated that the total area of the openings shall be greater than 10% of the total area of the original plate. For wash plates, which are normally used in closed tanks to minimize the rising dynamic stresses from large liquid movement. SSC (1990) reported twelve case studies of sloshing damage from several types of ships. Coefficients for sloshing forces, pressures were determined using model tests.

Saad-Eldeen et al. (2015) analysed the effect of combined loading, with full and partial lateral loads on strength of wash plates with different degree of openings.

From design point of view, there are no specific rules, which defines the number of opening in a wash plate, this shortage of information regarding the design of wash plates motivates the authors to carry out a series of nonlinear finite element analysis for different plate thicknesses, to identify the optimum number of openings for different degree of openings that show higher strength.

2 FINITE ELEMENT MODELLING OF WASH PLATES

Generally, the plate slenderness, β is the governing parameter of the plate buckling strength, which is a function of the plate geometry configurations; plate breadth and plate thickness as well as material properties; yield stress, σ_y and modulus of elasticity, E. The thickness of the analysed plate varies from 4 mm to 10 mm with an interval of 2 mm, with a plate length, a and breath, b of 800 and 400 mm, resulting in a plate slenderness range of 1.35 to 3.38.

To introduce a set of openings with a specific degree of openings, DOO, the generalized expression of DOO is defined as:

$$DOO = \frac{1}{ab}\sum_{i=1}^{N} A_O 100\% \tag{1}$$

where N is the number of openings, A_0 is the surface area of the ith opening. In the present analysis, three-degree of openings are considered as 16.80%, 28.30% and 35.5%, which are represented by three sets of opening numbers 4, 8 and 24, respectively, as shown in Figure 1.

The current analyses of intact and wash steel unstiffened plates are carried out using the nonlinear finite element commercial code ANSYS. The finite element analysis uses the full Newton–Raphson equilibrium iteration scheme. Having activated the large strain option, it encompasses the shape change of the elements (i.e., strains are finite), rigid-body effects (e.g., large deformation), stress stiffening and regular linear analysis to solve the geometric and material nonlinearities and to pass through the extreme points. The automatic time stepping features are also employed, allowing to determine the appropriate load steps.

Shell element 181 is used to generate the entire finite element model, defined by four nodes, with six degrees of freedom at each node. Shell181 is well-suited for linear, large rotation, and/or large strain nonlinear applications and suitable for analysing thin to moderately-thick shell structures.

The wash plate is made of low carbon steel with a yield stress of 235 MPa and the modulus of elasticity of 206 GPa. The stress-strain curve is assumed to be elastic-perfectly plastic. The wash plates are subjected to a combined load; uniaxial compressive and lateral pressure, with boundary conditions at the four edges as shown in Figure 1.

The initial imperfection is modelled as one half downward wave, based on the Fourier series and the amplitude of the initial imperfection for each plate is different, which is calculated according to the equation developed by Faulkner (1975) and Smith et al. (1988), considering α of 0.15:

$$\omega_0 = \alpha \beta^2 t \tag{2}$$

The intact model used in the analysis is a master model used by the authors in several analyses by Saad-Eldeen et al. (2014), in which was proved that the best element size is 5 mm.

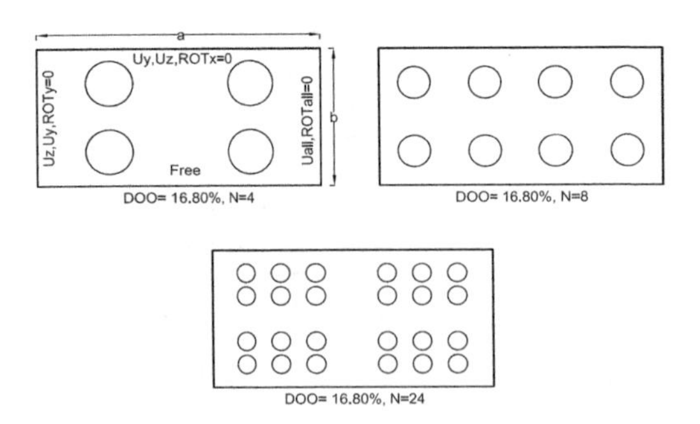

Figure 1. Wash plate boundary conditions and degree of opening.

3 ANALYSIS RESULTS

For plates without openings, intact plates, the relationship between the normalized strength and normalized strain is presented in Figure 2. The normalized strain is calculated as a ratio between the ultimate strength and the yield stress of the material, where the normalized strain is the ratio between the strain corresponding to ultimate strength and the yield strain of the material.

As may be seen from Figure 2, with increasing the plate thickness, the normalized strength is increasing and the normalized strain at the ultimate loading is decreasing. The relationship between the normalized ultimate strength and the plate slenderness is shown in Figure 3 and given in Table 1, in which with increasing the plate slenderness, the normalized ultimate strength decreases nonlinearly.

In order to assess the strength of wash plates, three degrees of the opening are considered as 16.80%, 28.30% and 35.5%, using the recommendation given in IACS (2010), where the minimum

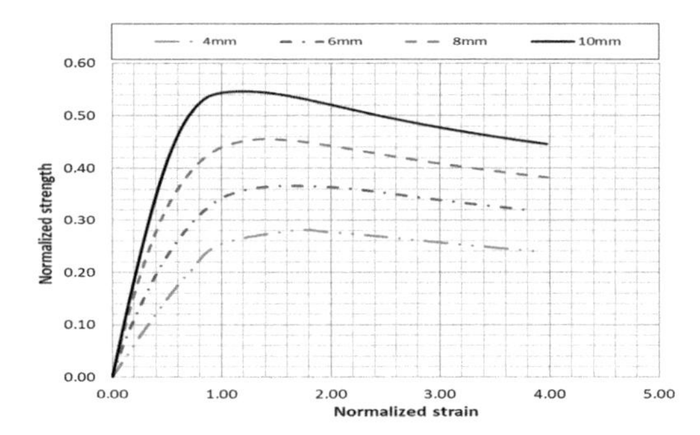

Figure 2. Normalized strength vs. normalized strain for different plate thicknesses.

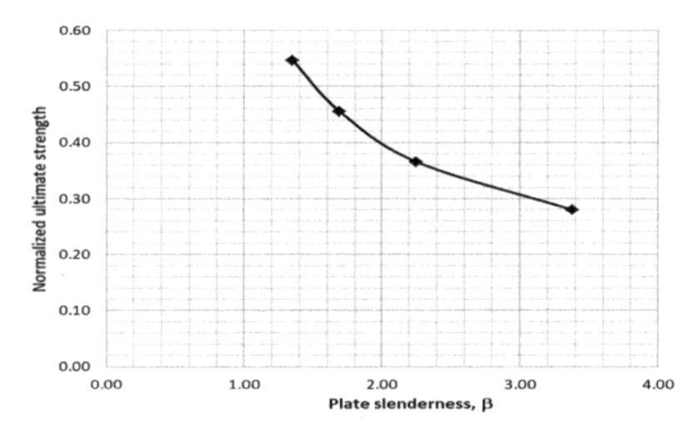

Figure 3. Normalized ultimate strength as a function of plate slenderness.

Table 1. Normalized ultimate strength for intact and wash plates, DOO = 16.80%.

	Intact	DOO 16.80%		
t, mm	N = 0	N = 4	N = 8	N = 24
4	0.28	0.16	0.19	0.18
6	0.37	0.22	0.25	0.23
8	0.46	0.29	0.31	0.29
10	0.55	0.34	0.38	0.29

opening ratio with respect to the original area of the plate does not to be less than 10%. For each DOO, three opening numbers are used as 4, 8 and 24.

For a plate thickness of 4 mm, the normalized strength versus normalized strain for an intact and with DOO of 16.80% plate is presented in Figure 4. As may be seen, the slope of the relationship is different between the intact one and the ones with DOO of 16.80%, which shows the effect of the residual breadth due to the openings on the developed strain at the same load, where the developed strain for wash plates with DOO 16.80% is bigger than the one of the intact plates within the linear

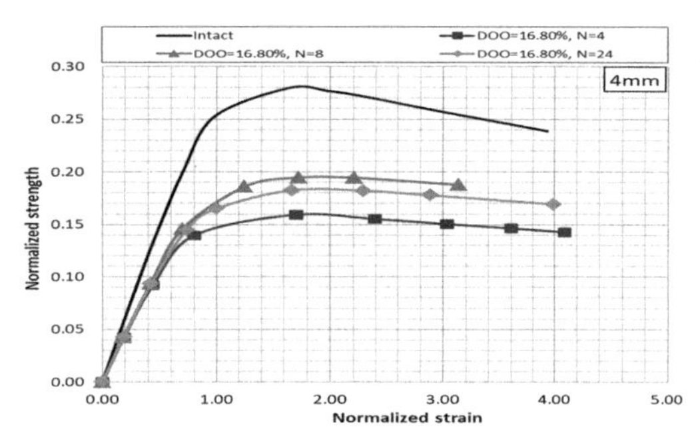

Figure 4. Normalized strength vs. normalized strain, DOO = 16.80%, t = 4 mm.

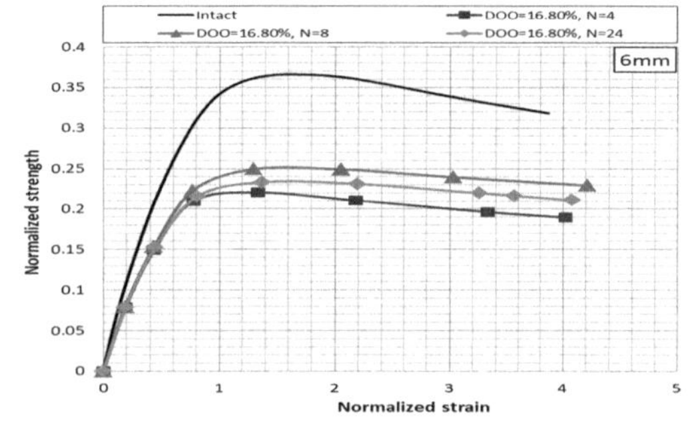

Figure 5. Normalized strength vs. normalized strain, DOO = 16.80%, t = 6 mm.

region. From a strength point of view, it is obvious that the lowest strength is developed by a number of opening numbers 4 and the highest strength is for the set number of 8, where a number of openings are 24 in between, as shown in Figure 4 and tabulated in Table 1. Form Table 1, the developed strength of N = 24 is bigger than the one developed by N = 4 with 12.74%. This indicates that for a plate thickness of 4 mm, the relationship between the number of openings and the normalized ultimate strength is not systematic.

For a plate thickness of 6 mm, the relationship between the normalized strength and strain is shown in Figure 5, in which the same observation of the deviation in the slope exists as for a thickness of 4 mm. Also, it may be noticed that the higher strength is for N = 8 and the lowest is for N = 4, but the strength of N = 24 is bigger than the one of N = 4 with 5.45%, which is less than the one registered for a thickness of 4 mm, see Table 1.

With increasing the plate thickness from 6 mm to 8 mm, the relationship of the normalized strength and normalized strain is quite different, where the strength developed by N = 4 and N = 24 is the same, keeping the higher strength for a number of openings of N = 8. This means that with increasing

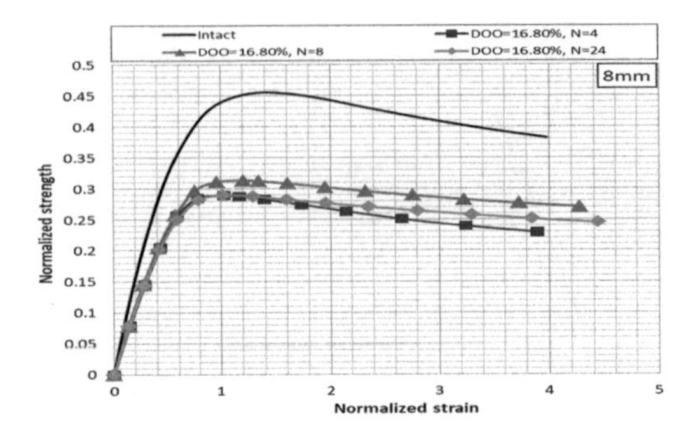

Figure 6. Normalized strength vs. normalized strain DOO = 16.80%, t = 8 mm.

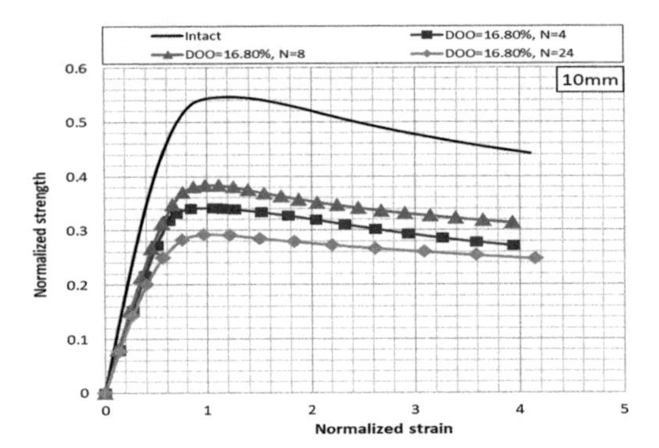

Figure 7. Normalized strength vs. normalized strain, DOO = 16.80%, t = 10 mm.

the plate thickness the benefit of using N = 24 is decreasing, see Figure 6 and Table 1.

For a thicker plate of 10 mm, the trend changes, where the number of openings N = 4 shows higher strength than the one of N = 24, which is on the contrary to the plate thickness of 4 mm, 6 mm and 8 mm, respectively.

This indicates that for DOO = 16.80%, there is an inflection point at which the trend of the opening sets changes and this point is between plate thickness 6 mm and 8 mm.

A plot of the plate thickness and the difference between the normalized ultimate strength of the opening numbers 24 and 4 is shown in Figure 8. With increasing the plate thickness, the sign of the difference changes showing an inflection of the behaviour around 8 mm, after which the opening sets 4 shows higher strength than N = 24, represented by a negative difference between N = 24 and N = 4.

The shear stress distributions for N = 4 and N = 24 are presented in Figure 10, as may be seen, the locations with high stresses for N = 4 are concentrated around the lower opening, but for N = 24 the stresses are concentrated between the first and

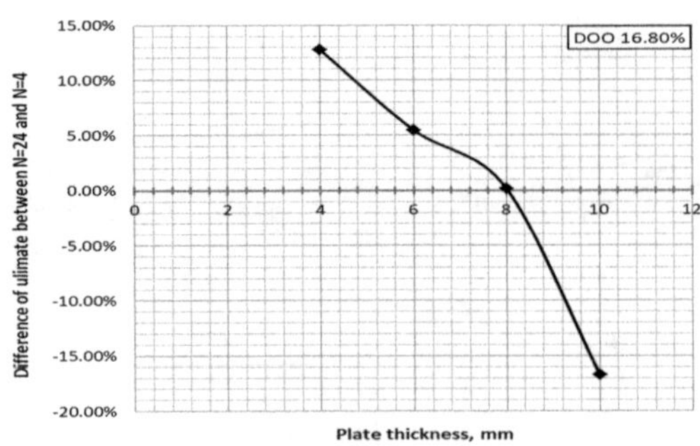

Figure 8. Normalized ultimate strength between N = 24 and N = 4 as a function of plate thickness, DOO = 16.80%.

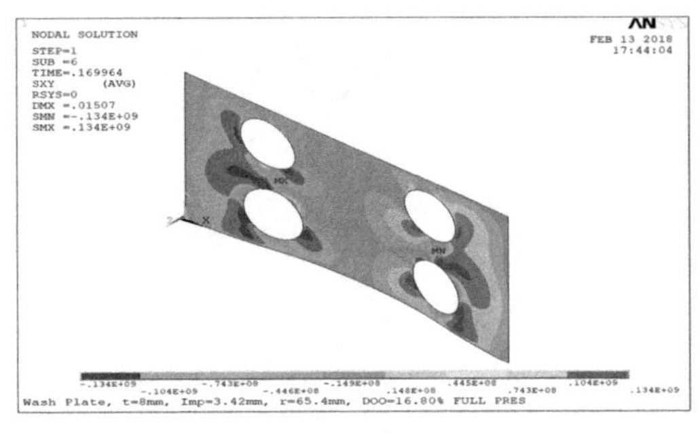

Figure 9. Shear stress distribution of N = 4 (up) for t = 8 mm and DOO 16.80%.

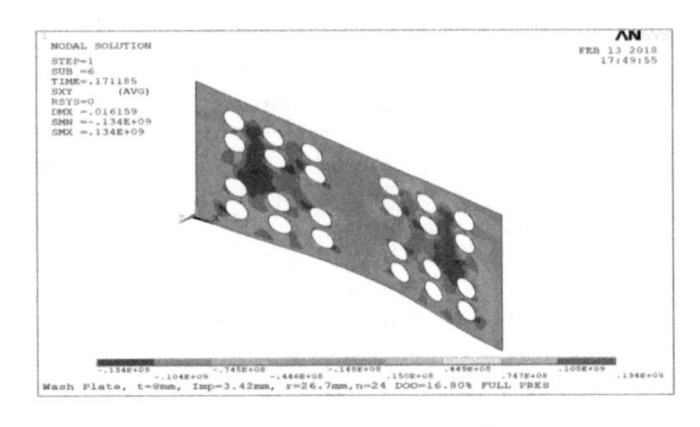

Figure 10. Shear stress distribution of N = 24 for t = 8 mm and DOO 16.80%.

the second set of openings and the higher stress are with 134 MPa for both cases.

In the case of DOO = 28.30%, four plate thicknesses are used in the analysis as the one of DOO = 16.30%. For the first plate thickness of 4 mm and with a number of openings 4, the ultimate normalized strength decreases to 0.08 with a reduction of 71.34% with respect to the intact one. With increasing the opening numbers to 8, the plate shows higher strength than the one of 4 mm, as may be seen from Figure 11 and Table 2.

For the biggest number of openings, N = 24, the ultimate strength decreases again, conceding with the observation occurs for the plate thickness of 4 mm and DOO = 16.80%. Therefore, the higher strength is related to N = 8 and the lowest one is for N = 4 and the difference between the ultimate normalized strength of N = 24 and N = 4 is 36.44%. This difference is higher than the one of DOO = 16.80%.

By increasing the plate thickness to 6 mm, the relationship between the normalized strength and strain is shown in Figure 12, in which the number of openings 8 keeps showing the higher strength and N = 4 shows the lower strength. But the difference between N = 24 and N = 4 is 32.52% which is lower than the one for the plate thickness of 4 mm. This observation coincides also with the one for

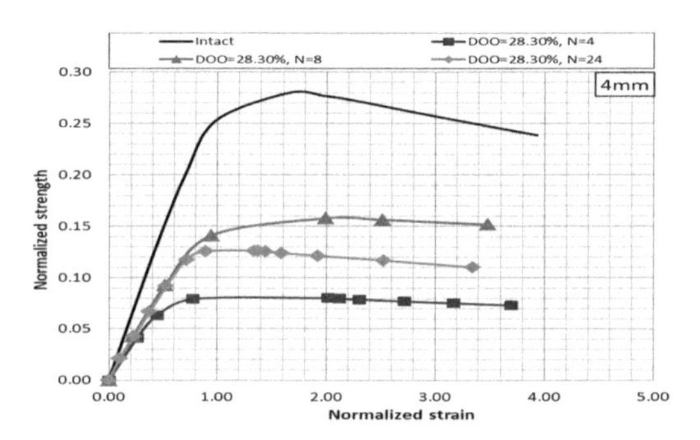

Figure 11. Normalized strength vs. normalized strain, DOO = 28.30%, t = 4 mm.

Table 2. Normalized ultimate strength for intact and wash plates of DOO 28.30%.

	Intact	DOO 28.30%		
t, mm	N = 0	N = 4	N = 8	N = 24
4	0.28	0.08	0.16	0.13
6	0.37	0.12	0.20	0.18
8	0.46	0.15	0.25	0.22
10	0.55	0.16	0.31	0.22

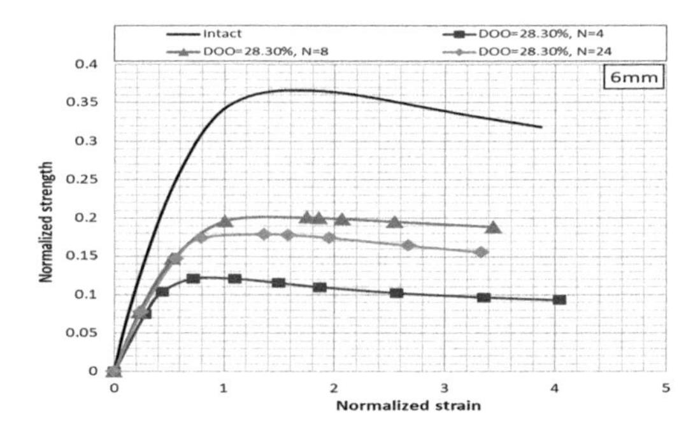

Figure 12. Normalized strength vs. normalized strain, DOO = 28.30%, t = 6 mm.

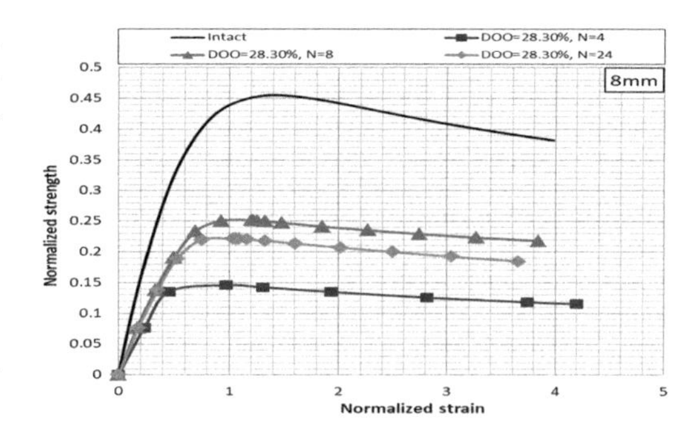

Figure 13. Normalized strength vs. normalized strain, DOO = 28.30%, t = 8 mm.

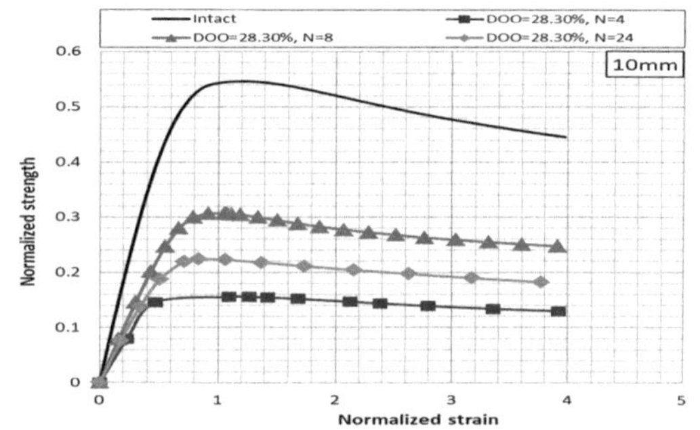

Figure 14. Normalized strength vs. normalized strain, DOO = 28.30%, t = 10 mm.

DOO = 16.80%. For a plate thickness of 8 mm, the difference between the ultimate normalized strength of n = 24 and N = 4 increases to 34.19%, which is on the contrary to the one that occurs at the same thickness but with DOO = 16.80%.

For a thicker plate of 10 mm, the relationship of the normalized strength and strain is presented in Figure 14 and the difference of the normalized ultimate strength between the two openings 24 and 4 decreases to 30.50%. A relationship between the plate thickness and the difference of the normalized ultimate strength between the opening numbers 24 and 4 for DOO = 28.30% is shown in Figure 15, in which there is a disturbance in the trend at the plate thickness of 8 mm, which is on the contrary to the one with DOO = 16.80% at which the difference is almost zero, see Figure 8.

The third DOO used in the present analysis is 35.5%, and it is represented by three numbers of openings. The resultant normalized strength and strain for a plate thickness of 4 mm are presented in Figure 16 and the normalized ultimate strength is given in Table 3 for different opening numbers. It is clear that for the number of openings 8 it shows the higher strength and the lowest strength is represented by the number 4. The difference of the ultimate strength between N = 24 and N = 4 is

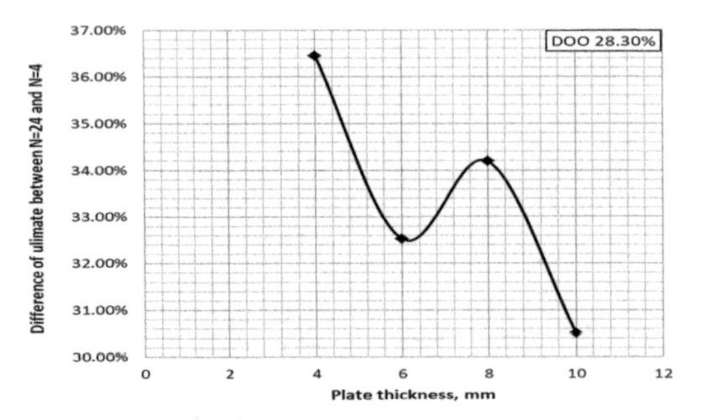

Figure 15. Difference of normalized ultimate strength between N = 24 and N = 4, DOO = 28.30%.

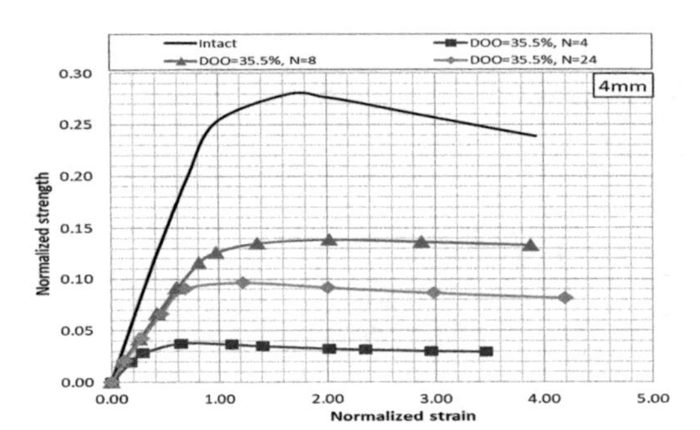

Figure 16. Normalized strength vs. normalized strain, DOO = 35.5%, t = 4 mm.

Table 3. Normalized ultimate strength for intact and wash plates of DOO = 35.5%.

	Intact	DOO 35.5%		
t, mm	N = 0	N = 4	N = 8	N = 24
4	0.28	0.04	0.14	0.10
6	0.37	0.05	0.18	0.14
8	0.46	0.05	0.22	0.16
10	0.55	0.05	0.26	0.20

61.09% which is higher than of both differences for DOO 16.80 and % and 28.30%. This indicates the increase of the difference of the normalized ultimate strength between the two opening numbers as the degree of openings increases.

For a plate thickness of 6 mm, the relationship between the normalized strength and strain is presented in Figure 17, where it may be noticed that the lowest strength response is for N = 4 and the difference between N = 24 and N = 4 increases to 64.44%, which is on the contrary to DOO = 16.80% and 28.30%.

By increasing the plate thickness to 8 mm, the obtained normalized ultimate strength for N = 4 is the same as the one of the plate thickness of 6 mm. On the contrary, the obtained ultimate strength for

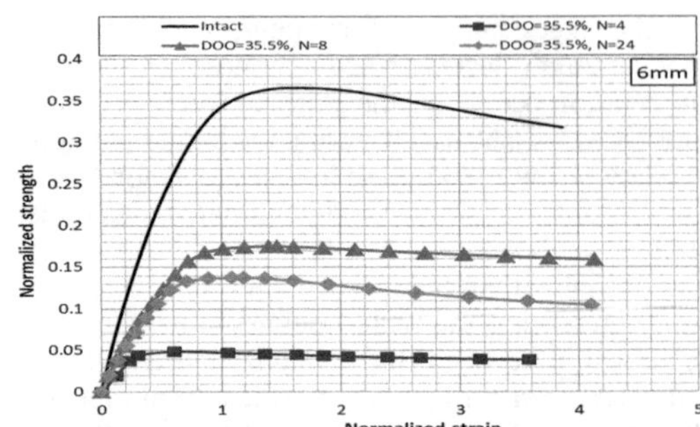

Figure 17. Normalized strength vs. normalized strain, DOO = 35.5%, t = 6 mm.

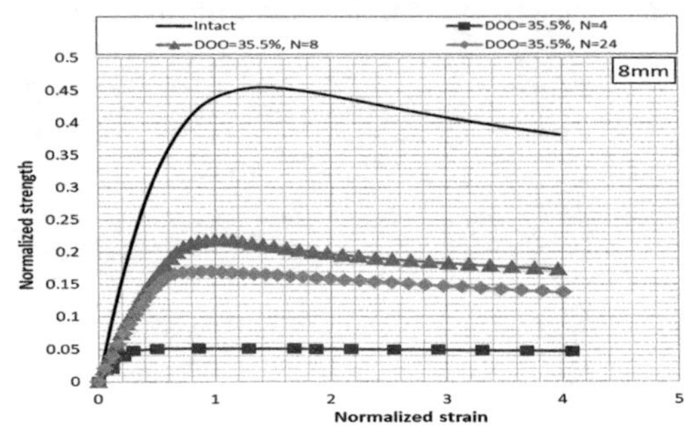

Figure 18. Normalized strength vs. normalized strain, DOO = 35.5%, t = 8 mm.

N = 24 is higher than the one of a thickness of 6 mm, which arises as an increase of the difference between N = 24 and N = 4, as may be noticed from Figure 18 and Table 3.

For the thicker plate of 10 mm, see Figure 19, it is obvious that the normalized ultimate strength developed by N = 4 is the same as for the thickness of 6 mm and 8 mm, see Table 3, but the normalized ultimate strength for N = 24 increases registering a difference of 73.04% with respect to N = 4. This indicates the difference of the plate response for a higher degree of openings.

The relationship between the plate thickness and the difference of the normalized ultimate strength between N = 24 and 8, for DOO = 35.5% is shown in Figure 20. As the plate thickness increases, the difference of the normalized ultimate strength between the two opening numbers increases, which are on the contrary to DOO = 16.80% and 28.30%. Therefore, it may be concluded that for a higher degree of opening, the number of the opening may be ordered according to the ultimate strength from the best to the worst as 8, 24 and 4.

The shear stress distributions for a plate thickness of t = 6 and 8 mm with N = 4 are shown in Figure 21 and Figure 22. At the ultimate load step for both cases, a tripping of the residual strip near the upper edges and between the two openings

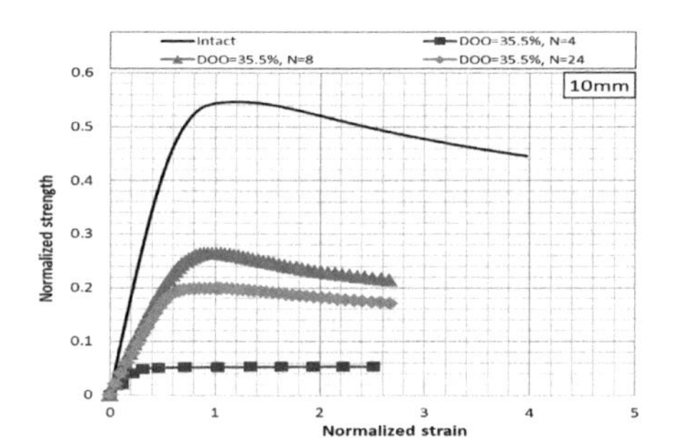

Figure 19. Normalized strength vs. normalized strain, DOO = 35.5%, t = 10 mm.

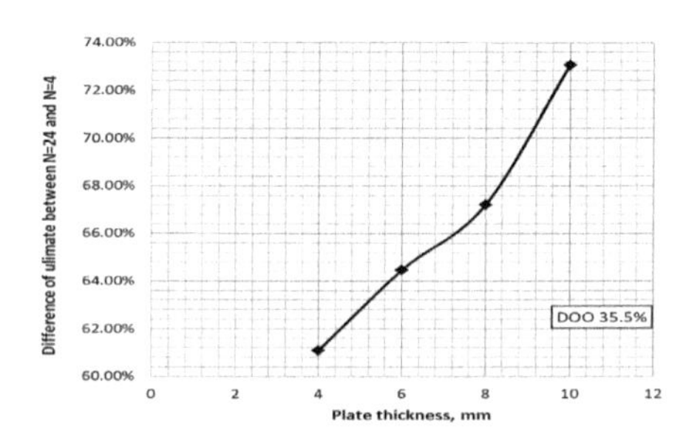

Figure 20. Difference of normalized ultimate strength between N = 24 and N = 4, DOO = 35.5%.

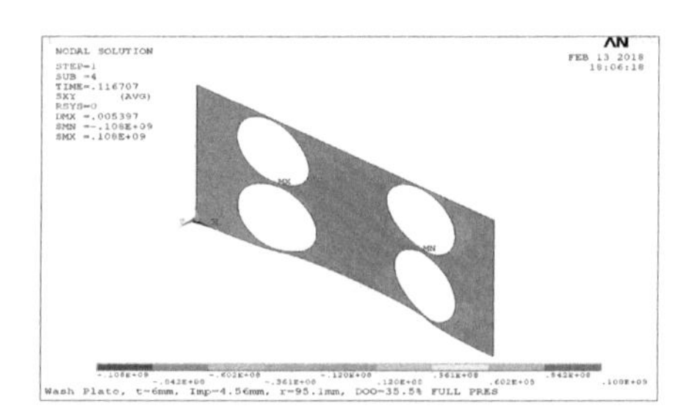

Figure 21. Shear stress distribution of N = 4 for t = 6 mm for DOO 35.5%.

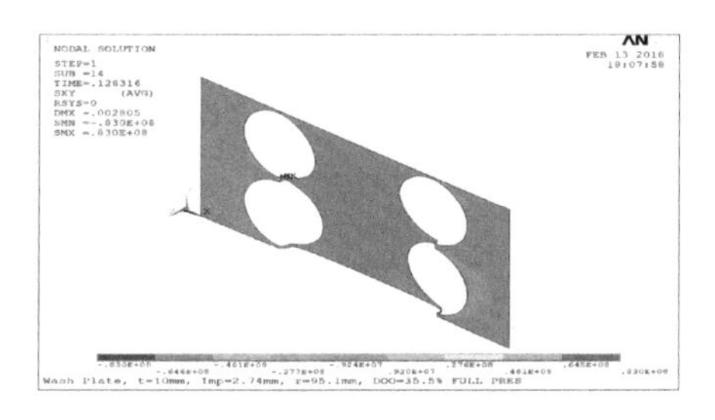

Figure 22. Shear stress distribution of N = 4 for t = 10 mm (down) for DOO 35.5%.

occurs with higher deformation for t = 8 mm rather than t = 6 mm, which facilitate the occurrence of a failure.

4 FAILURE ASSESSMENT

To identify the failure of wash plates, a criterion should be defined. A relationship between plate slenderness and normalized ultimate strength for a different degree of openings are presented in Figure 23, to Figure 25.

As may be seen in Figure 23, the trends for the three opening numbers for DOO = 16.80% are not systematic where this is an overlap between N = 24 and N = 4 at a plate slenderness of 2, which is represented by the zero difference of the normalized ultimate strength between N = 24 and N = 4, see Figure 8. For DOO = 28.30%, see Figure 24, the trends of N = 24 and N = 8 are almost the same with an increase in the difference between the normalized ultimate strength. As the degree of openings increases to 35.5%, see Figure 25 the difference between N = 24 and N = 4 increases,

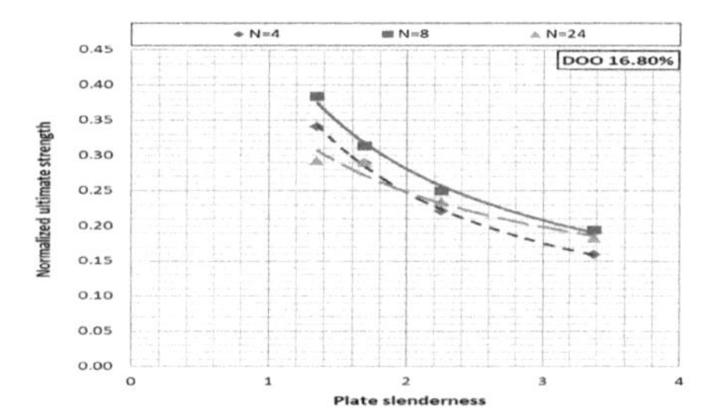

Figure 23. Normalized ultimate strength as a function of plate slenderness, DOO = 16.80%.

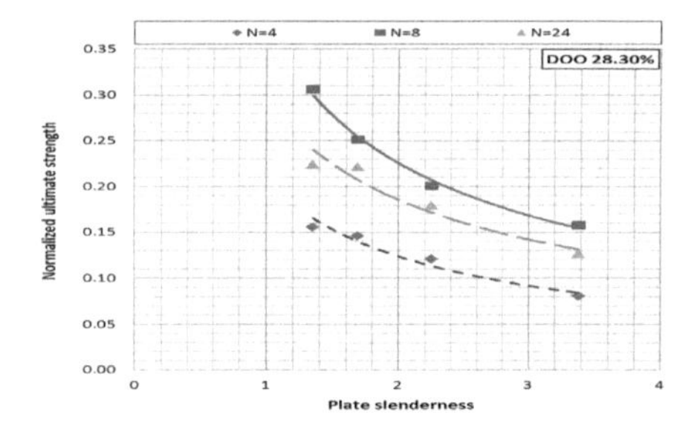

Figure 24. Normalized ultimate strength, DOO = 28.30%.

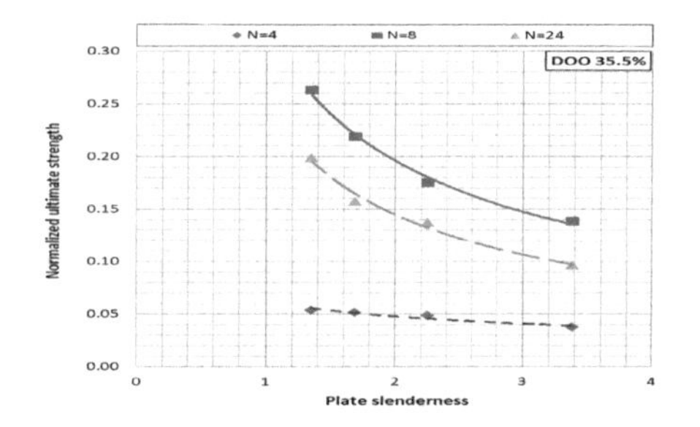

Figure 25. Normalized ultimate strength, DOO = 35.5%.

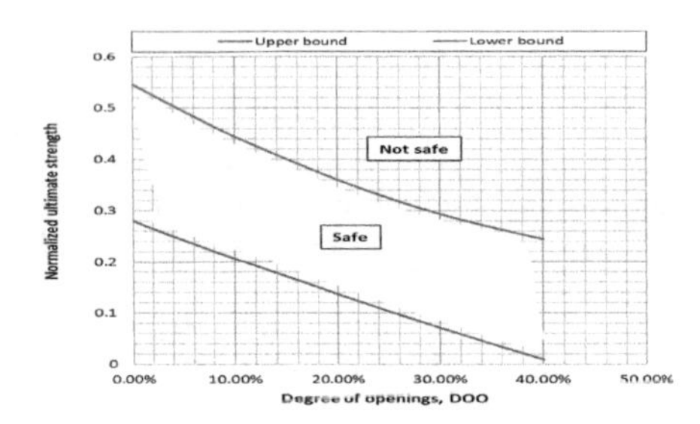

Figure 26. Failure assessment diagram.

with a different trend of N = 4, compared to N = 8 and N = 24, which is due to the almost constant normalized ultimate strength for $\beta \leq 2.25$. This indicates that for a higher degree of openings, DOO = 35.5% and N = 4, the benefit of having better ultimate strength due to increasing the plate thickness is vanishing at $\beta \leq 2.25$.

Based on the results of the current analysis for wash plates under combined loading; uniaxial compressive and lateral load, a failure strength assessment diagram is developed as presented in Figure 26, which is a relationship between the normalized ultimate strength and the degree of openings. The upper bound represents the obtained normalized ultimate strength for a number of opening N = 8 and the lower bound is the lower strength for a number of openings N = 4. Based on that, the wash plate will be safe if the normalized ultimate strength at any degree of openings is within the borders of the safe zone, which is represented by the upper and lower bounds, as shown in Figure 26.

5 CONCLUSIONS

A series of static nonlinear finite element analyses have been performed for wash plates with three degrees of openings and three numbers of openings subjected to a combined load. Based on the current analysis, it was concluded that as the degree of openings increases, the difference of the normalized ultimate strength between the two opening numbers 24 and 4 increases.

From the ultimate strength point of view, the number of openings, which represents a specific degree of opening may be ordered from the best to the worst as 8, 24 and 4.

It was observed that for DOO = 35.5% and $\beta \leq 2.25$ the normalized ultimate strength is almost constant, which indicates the diminishing of the benefit of having better ultimate strength due to the increase of the plate thickness at $\beta \leq 2.25$.

A failure strength assessment diagram is developed with upper bound which represents the obtained normalized ultimate strength for a number of opening N = 8 and the lower bound is the lower strength for a number of openings N = 4.

ACKNOWLEDGMENTS

The first author has been funded by the Portuguese Foundation for Science and Technology (Fundação para a Ciência e Tecnologia—FCT) under contract SFRH/BPD/84823/2012.

REFERENCES

DNV 2008. Hull Structural Design, Ships with Length 100 metres and above. *Rules for Classification of Ships, Det Norske Veritas.*

Faulkner, D. 1975. A review of Effective Plating for Use in the Analysis of Stiffened Plating in Bending and Compression. *Journal of Ship Research,* 19, 1–17.

IACS 2010. Common structural rules for double hull oil tankers. *IACS Common Structural Rules.* Det Norske Veritas.

Kim, U.N., Choe, I.H. & Paik, J.K. 2009. Buckling and Ultimate Strength of Perforated Plate Panels subject to Axial Compression: Experimental and Numerical Investigations with Design Formulations. *Ships and Offshore Structures,* 4, 337–361.

Moen, C.D. & Schafer, B.W. 2009. Elastic Buckling of Thin Plates with Holes in Compression or Bending. *Thin-Walled Structures,* 47, 1597–1607.

Paik, J.K. 2008. Ultimate strength of perforated steel plates under combined biaxial compression and edge shear loads. *Thin-Walled Structures,* 46, 207–213.

Saad-Eldeen, S., Garbatov, Y. & Guedes Soares, C. 2014. Ultimate strength assessment of steel plates with a large opening. *In:* Guedes Soares, C. & Peña, L. (eds.) *Developments in Maritime Transportation and Exploitation of Sea Resources.* Taylor & Francis Group, London, UK.

Saad-Eldeen, S., Garbatov, Y. & Guedes Soares, C. 2015. Strength assessment of wash plates subjected to combined lateral and axial loading. *In:* Guedes Soares, C. & Santos, T. a. R. (eds.) *Maritime Technology and Engineering.* Taylor & Francis Group, London, UK.

Saad-Eldeen, S., Garbatov, Y. & Guedes Soares, C. 2016a. Experimental investigation on the residual strength of thin steel plates with a central elliptic opening and locked cracks. *Ocean Engineering,* 115, 19–29.

Saad-Eldeen, S., Garbatov, Y. & Guedes Soares, C. 2016b. Experimental strength analysis of steel plates with a large circular opening accounting for corrosion degradation and cracks subjected to compressive load along the short edges. *Marine Structures,* 48, 52–67.

Saad-Eldeen, S., Garbatov, Y. & Guedes Soares, C. 2016c. Experimental strength assessment of thin steel plates with a central elongated circular opening. *Journal of Constructional Steel Research,* 118, 135–144.

Smith, C.S., Davidson, P.C., Chapman, J.C. & Dowling, J.P. 1988. Strength and Stiffness of Ships' Plating under In-plane Compression and Tension. *Transactions RINA,* 130, 277–296.

SSC 1990. Ship Structure Committee: Liquid sloshing in cargo tanks. *SSC-336.*

Structures in composite materials

Progress in Maritime Technology and Engineering – Guedes Soares & Santos (Eds)
© 2018 Taylor & Francis Group, London, ISBN 978-1-138-58539-3

Experimental and numerical structural analysis of a windsurf fin

F. Nascimento, L.S. Sutherland & Y. Garbatov
Centre for Marine Technology and Engineering (CENTEC), Instituto Superior Técnico, Universidade de Lisboa, Lisbon, Portugal

ABSTRACT: Structural analyses, both numerical and experimental, have been made of a composite windsurf fin. Structural analyses, both numerical and experimental, have been made of a composite windsurf fin. An FE model was developed employing commercial software ANSYS using both linear and nonlinear analyses of the structural response due to external loads (selected to give an approximation of the in-service longitudinal deformation) giving the resulting stresses and deflections. To calibrate the numerical model, experimental tests using a servo-hydraulic mechanical test machine was performed. To try to estimate the failure load of the fin Tsai-Wu, Maximum Stress and Maximum Strain failure criteria were applied.

1 INTRODUCTION

Fibre reinforced plastic (FRP) composite materials are now ubiquitous in the marine industry, offering many advantages over the use of steel, aluminium or wood, such as resistance to corrosion and rot, ease of forming complex seamless shapes, and high specific material properties due to such advantages (Sutherland, 2018). For cases with a highly loaded structure that also requires thinner sections due to hydrodynamic considerations such as windsurfer fins (or hydrofoils) composites are often the only feasible solution. Since these fins must be both thin for efficient hydrodynamic performance and strong enough not to break under the often very demanding loadings asked of them it is important to be able to predict their structural response.

There is little previous work concerning windsurfer fins, although there were a few scientific studies on windsurfer fin performance, mostly as the sport increased rapidly in popularity in the 1990's (Broers et al., 1992, Sutherland, 1993, Sutherland & Wilson, 1994, Kunoth et al., 2007, Gourlay & Martellotta, 2011) and there are various articles of varying technical depth in the windsurfing press and online, e.g. (Fagg, 1997, Drake, 2005).

The complex nature of both composite materials and the internal structural arrangement of a fin mean that a correspondingly complex numerical analysis such as FEM is required to analyse structural responses. Also, very small changes in deflections can result in important consequences in terms of hydrodynamic efficiency and control, especially in a competitive environment. The fins studied here are fabricated by F-Hot Fins and are used by top international sailors and even world champions. Hence the scope for obtaining a structural model that will be developed to allow faster and easier evaluation of future improvements to the foil sections, planforms and lay-ups of these fins is of great interest.

Hence, the aim here is to develop an initial FE model of an actual windsurfer fin and to both calibrate and validate the model in terms of stiffness and deflection responses with full-scale mechanical tests on the fabricated fin itself. Finally, Letters f, m and c are used to indicate fibre, matrix and composite respectively. Volume and weight fractions are denoted by V and W respectively estimates of the ultimate strength of the fin will be made using various failure criteria. The FE model obtained will be the first stage in an overall work incorporating parallel computational fluid dynamics (CFD) and fluid structure interaction (FSI) studies to obtain a design tool for the improvement of windsurfer fin design, or in fact for other underwater foils.

2 MATERIAL PROPERTIES

The material property inputs to an FE model are essential in ensuring that correct results are obtained, but this is often the most difficult part of this type of numerical modelling due to the large number of properties required for these anisotropic materials. Further, this is usually even more challenging since the composites used in the marine industry are hand produced, as they are in this case, leading not only to inherent variability, but also to very non-standardised materials (Sutherland, 2018).

One of the most basic properties which will both depend on the exact production method and conditions (and even laminator) considered and affect many of the strength and especially stiffness material properties is the relative proportions of fibre and matrix, expressed as either weight or volume fractions. For theoretical analysis, volume fractions are more helpful, but weight fractions are easier to obtain experimentally, but they are easily converted knowing the raw fibre and matrix densities (Shenoi & Wellicome, 1993).

Fibre weight fraction (FWF) is easily measured from mass measurements taken during fabrication:

$$FWF = \frac{m_f}{m_f + m_m} \tag{1}$$

where the subscripts f and m refer to fibre and matrix, respectively.

And this is then converted to fibre volume fraction (FVF) via

$$FVF = \frac{1}{\left[1 + \frac{\rho_f}{\rho_m}\left(\frac{1}{FWF} - 1\right)\right]} \tag{2}$$

where ρ is density.

The thickness of each ply is also of great importance in developing the geometrical model, and this is given by Shenoi and Wellicome (1993):

$$t = \frac{W_F}{\rho_f \,.\, FVF \,.\, 1000}\,[mm] \tag{3}$$

where WF is the 'fibre areal weight' in g/m^2.

The exact amounts of materials used in the present study were measured by the producer of the wind-surf fin and Eqn (1) presented an average FWF of about 0.41. The densities of the epoxy resin, and the E-glass and carbon reinforcements were 1.2, 2.6 and 1.5 g/cm^3 respectively (Gurit-Holding, 2000, Exel, 2016), applying Eqn (2) the FVF values are estimated approximately as 0.3 and 0.4 for E-glass and carbon plies, respectively.

Knowing the fibre and matrix materials, the approximate FVF values and the production process used, initial estimates of mechanical properties from the test data reported in (Miller, 1991) and the ANSYS materials property library were defined as can be seen in Table 1.

Knowing the fibre and matrix materials, the approximate FVF values and the production process used, initial estimates of mechanical properties from the test data reported in (Miller, 1991)

Table 1. Composite material descriptors.

Description	Symbol	Value	Unit
Matrix Density	ρ_m	1.2	g/cm^3
E-Glass Density	$\rho_{E\text{-}Glass}$	2.580	g/cm^3
Density Carbon UD	$\rho_{Carbon\text{-}UD}$	1.5003	g/cm^3
Density Carbon Woven	$\rho_{Carbon\text{-}Woven}$	1.470	g/cm^3
Mass of Fiber	m_f	97	g
Mass of Matrix	m_m	140	g
Total Fiber Weight Fraction	FWF	40.92	%
Fiber Weight Fraction Carbon UD	$FVF_{carbon\ UD}$	39.56	%
Fiber Weight Fraction Carbon Woven	$FVF_{carbon\ woven}$	40.04	%
Fiber Weight Fraction E-Glass	$FVF_{e\text{-}glass}$	27.57	%
Thickness of Carbon-Woven (100 g/m^2)	$t_{carbon\text{-}woven\ (100g/m^2)}$	0.170	mm
Thickness of Carbon-Woven (200 g/m^2)	$t_{carbon\text{-}woven\ (200g/m^2)}$	0.340	mm
Thickness of Carbon-UD	$t_{carbon\text{-}UD}$	0.169	mm
Thickness of E-Glass	$t_{E\text{-}Glass}$	0.281	mm

and the ANSYS materials property library were defined as can be seen in Table 1.

3 FINITE ELEMENT MODEL

A finite element model is created using the commercial software ANSYS (2009). The model consists of 2597 nodes and 2610 elements. Most of the generated finite elements are quadrilateral, of a size from 2 to 10 mm. Shell element, type SHELL181, is used. This element is a four-node finite element with six degrees of freedom at each node, including translations and rotations in the x, y, and z axes. SHELL181 element is well-suited for linear and nonlinear large deflection finite element analyses (see Figure 1).

Figure 1 shows a thickness distribution over the fin. The root of the fin, located in the fixed support, is thicker and hence has more plies and it is the most highly stressed area. The lay-up includes woven and unidirectional carbon and E-glass reinforcements, as can be seen in Figure 2. The fin is symmetrical about the centre line and Figure 2 only shows the half of the fin and the ply identification numbers used. The two identical halves of the fin were identified as 'TOP' and 'DOWN' referring to their positions in the mechanical tests.

The boundary conditions of the finite element model are shown in Figure 3, fully representing the set-up of the experimental test. In this case, the

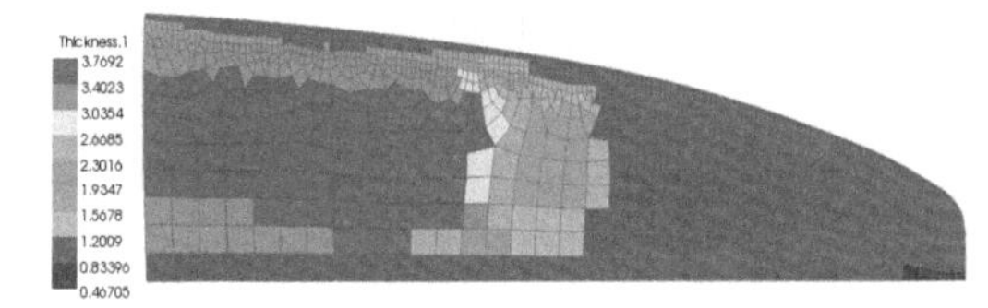

Figure 1. Finite element thickness.

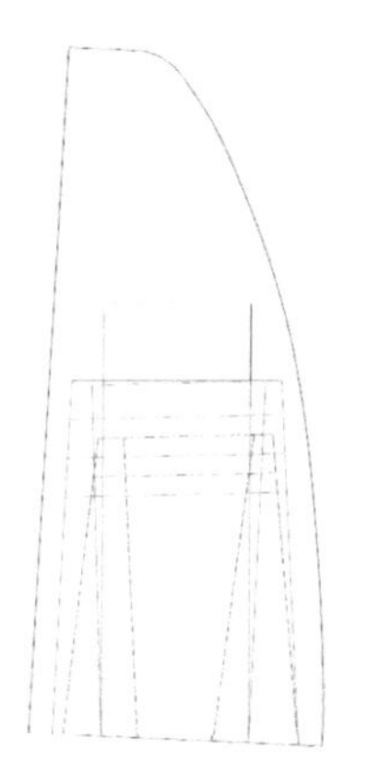

Figure 2. Lay-up plan.

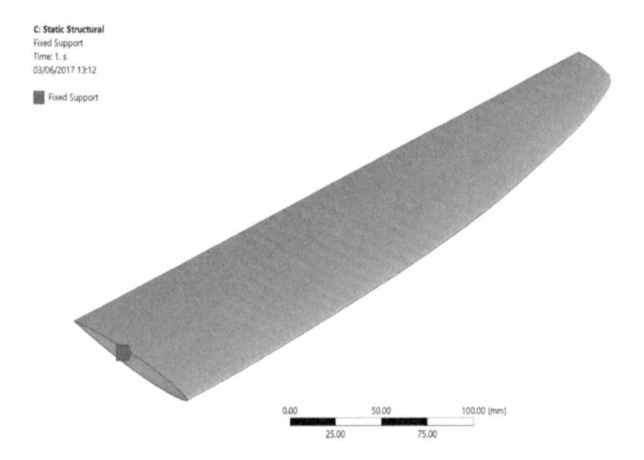

Figure 3. Boundary conditions.

boundary conditions are set to zero displacement and rotation at the support, where y = 0.

4 MECHANICAL TESTS

A calibrated computer-controlled servo-hydraulic machine was used to perform the mechanical tests, recording force and displacement with time. In terms of supports, it was a simple matter to clamp the fin base to avoid any rotation or translations (see Figure 4) as would be the case in-service. However, it was not feasible to replicate the varying distributed hydrodynamic loading that the fin would be subjected to in the water and hence a simple point load, applied via a 10 mm diameter hemispherical steel load applicator was used (see Figure 4).

The application point of the point load was, however, selected to give the best approximation of the in-service loading. The load was applied at the quarter chord distance from the leading edge, corresponding to the effective position of the hydrodynamic loading. Also, the centre of pressure of a semi-elliptic loading is at approximately 40% span, and so the fin was loaded at this point (see Figure 5).

Further, since the in-service loading is distributed, applying the load at twice this distance gives a reasonable approximation to the working lengthwise-deformed shape, and hence the fin was also tested with the load at 80% span (see Figure 6).

Loading was applied under displacement control at a speed of 0.1 mm/s up to approximately 40% of the estimated failure load to ensure that the fin was not damaged as it would be needed for

Figure 4. Experimental set-up.

Figure 5. 40% span test.

Figure 6. 80% span test.

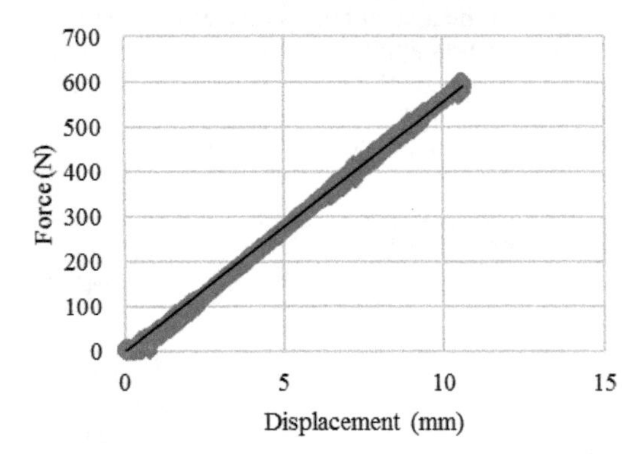

Figure 7. Force-Displacement, 40% span.

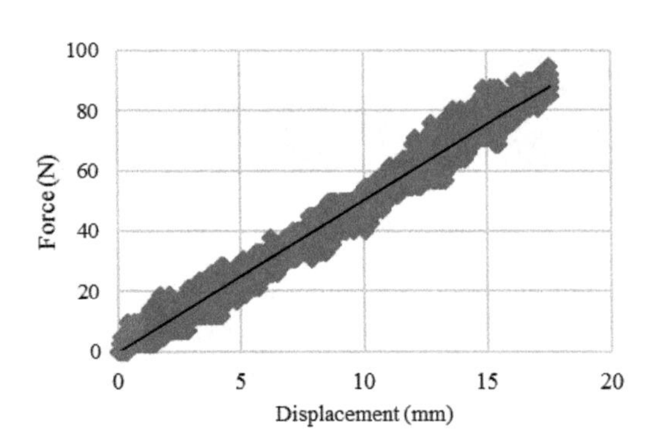

Figure 8. Force-Displacement, 80% span.

later mechanical and hydrodynamic testing; the main objectives of these tests was to calibrate the material property inputs to the FE model.

The resultant force-deflection results from both tests are shown in Figure 7 and Figure 8. The 'Unfiltered data' is the raw data from the sensors, but the very high R2 values of the fitted lines show that once electrical and other noise has been removed, the mechanical behaviour is very linear. This indicates that, up to the loadings used in these tests, no significant damage had occurred.

5 FEA MODEL CALIBRATION

To calibrate the developed finite element model, an idealized scenario is assumed, where, the report of Miller (1991), related to the mechanical properties of tested fin, is used to perform an initial estimation of the force-displacement relationship and the final calibration of the mechanical properties is performed using the result of the experimental tests.

The force-displacement relationship of the initial numerical estimate and experimental ones are shown in Figure 9. The first estimation of the force-displacement relationship, in the case of 40% of span, shows that the first assumption about the mechanical properties leads to an overestimation of the strength.

To adjust the numerical model to the experimental results the Young moduli were calibrated, and the final values are presented in Table 2 and the resulting force-displacement relationship in Figure 10.

The moment-curvature, M = (k) is presented in Figure 11, demonstrating a strong linear relationship between the moment and curvature.

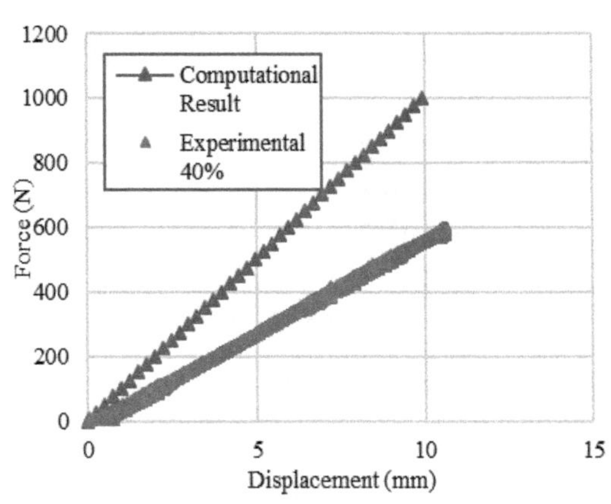

Figure 9. Force-displacement relationship, 40% span.

Table 2. Young moduli of fibres.

	Units	Epoxy Carbon UD	Epoxy Carbon Woven	Epoxy E-Glass UD
E_x	MPa	58606	30670	12000
E_y	MPa	2413	30670	2668
E_z	MPa	2413	3450	2668

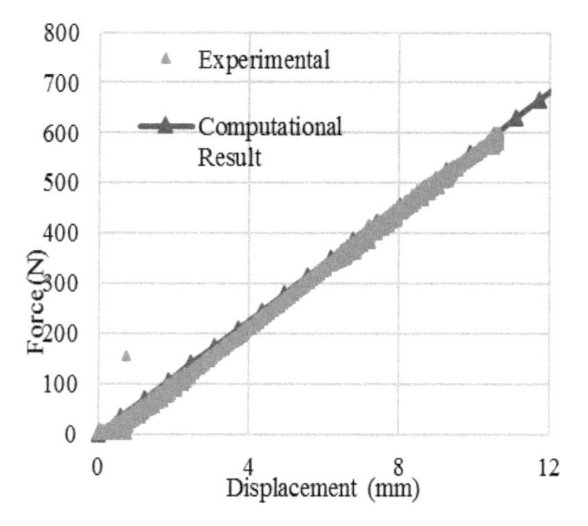

Figure 10. Force-displacement, 40% span.

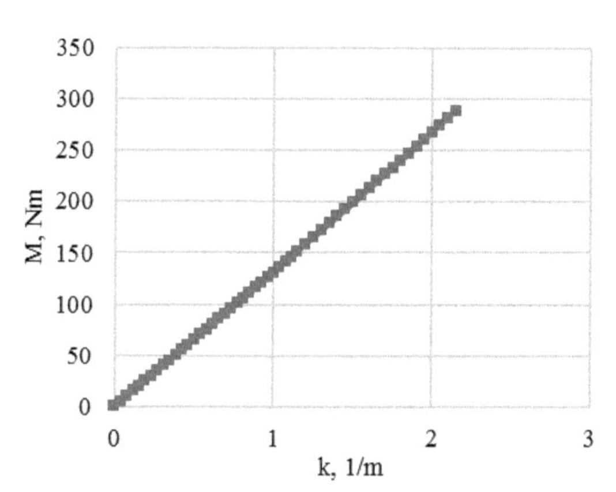

Figure 11. Moment-curvature relationship.

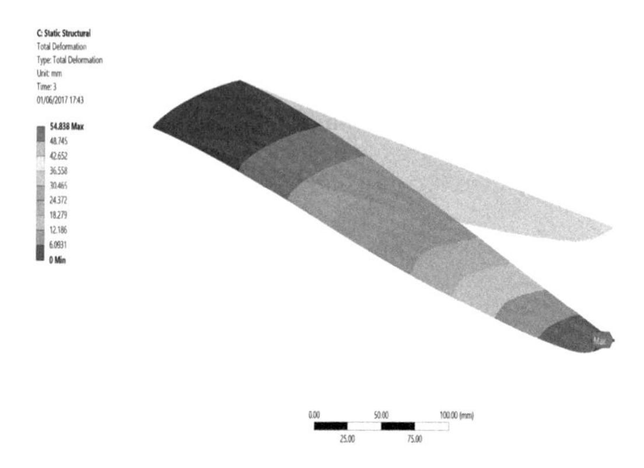

Figure 12. Displacement (z), 1,000 N, maximum displacement of 54.838 mm.

In terms of displacement, when the fin is subjected to a concentrated force of 1,000 N, the fin behaviour is shown in Figure 12, where the

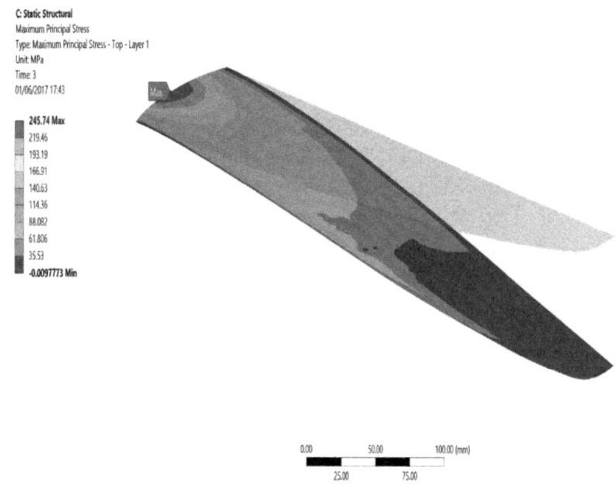

Figure 13. First principal stresses, 1,000 N, maximum stress of 245.74 MPa.

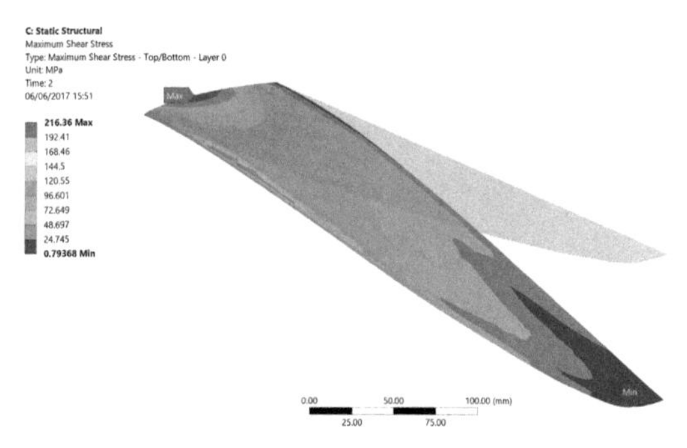

Figure 14. Shear stresses, 1,000 N, maximum shear stress of 216.36 MPa.

deformed and un-deformed shapes of the fin can be seen.

The distribution of the first principal stresses as a result of the loading force of 1,000 N is shown in Figure 13, where can be seen that the maximum stresses are located closely to the clamped support area.

The distribution of shear stresses, when applying a concentrated force of 1,000 N is presented in Figure 14.

6 FAILURE ANALYSIS

The critical failure load is estimated by using the Tsai and Wu (1971) failure criterion. Unlike the conventional isotropic materials, where one critical stress constant will suffice for the failure assessment, the composite materials require more elaborate methods to establish a failure criterion. The strength of the composite can be defined based on the strength of the individual plies. In addition, the failure of the plies occurs consequently as the applied load increases. It means that there may be a sequence of the first ply failure followed by other ply failures until the last one fails, denoting the ultimate failure. The progressive failure is therefore a quite complex in composite structures.

The Tsai-Wu failure criterion is used here to identify the failure based on the stresses and material properties. A failure index of 1 denotes the onset of failure, and a value less than 1 denotes no failure. The Tsai-Wu failure criterion is commonly used for orthotropic materials with unequal tensile and compressive strengths. The failure index is defined as:

$$\text{F.I.} = F_1\sigma_1 + F_2\sigma_2 + F_{11}\sigma_1^2 + F_{22}\sigma_2^2 + F_{33}\tau_{12}^2 + 2F_{12}\sigma_1\sigma_2 \tag{4}$$

where

$$F_1 = \frac{1}{X_t} - \frac{1}{X_c};\quad F_2 = \frac{1}{Y_t} - \frac{1}{Y_c};\quad F_{11} = \frac{1}{X_t X_c}; \tag{5}$$

$$F_{22} = \frac{1}{Y_t Y_c}; F_{33} = \frac{1}{\tau_{12^2}}; F_{12} = \frac{-1}{2\sqrt{Y_t Y_c X_t X_c}} \tag{6}$$

where the coefficient F12 represents the interaction between σ1 and σ2, and Xc, Yc are the compressive strength and Xt, Yt are the tensile strength of the material in the longitudinal (X) and transversal direction (Y). The parameter τ12 is the in-plane shear strength of the material.

A sample of the resulting Tsai-Wu failure surface is shown in Figure 15:

Applying the Tsai-Wu criterion, the critical load that produces the first ply failure is identified as 910 N in the case when the 1st ply is under compression as can be seen in Figure 16.

Higher stress values are concentrated near the clamped support (the left hand side of the fin in Figure 16). It is also observed some higher stresses in a small region near the leading edge around the half of the span, which may be developed because of the denser finite element mesh.

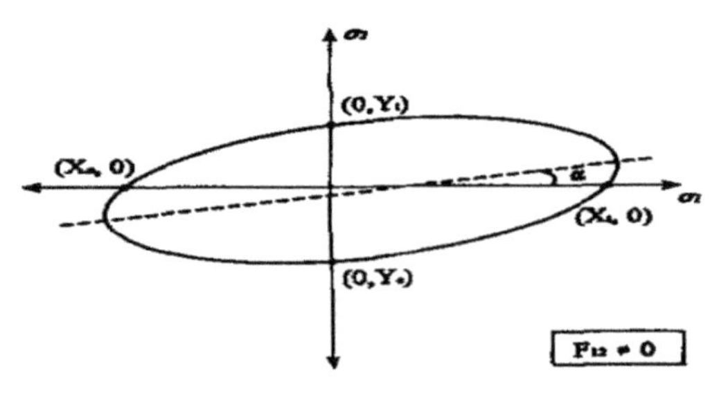

Figure 15. Tsai-Wu failure surface.

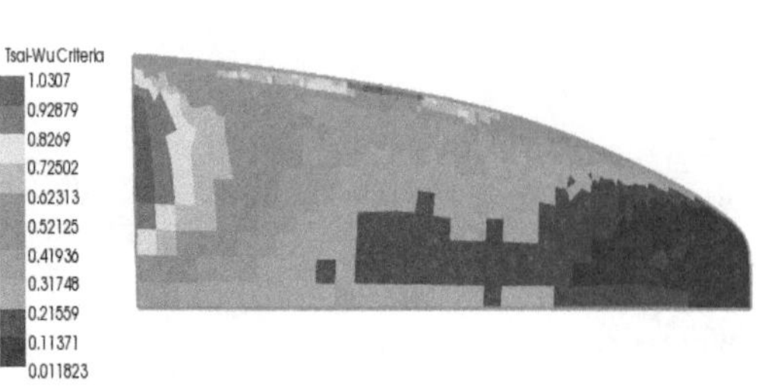

Figure 16. Tsai-Wu failure.

Figure 17. Fibre failure-maximum stress.

Although, the Tsai-Wu criterion identifies the presence of failure, it does not specify the type of failure. Two additional criteria for verifying the stress-strain behaviour of the fin with respect to the maximum stress and strain are applied here, considering three different failure conditions (Camanho et al., 2015):

$$Fibre:\quad \sigma_1 \geq \sigma_{1T}^u \quad or \quad |\sigma_1| \geq \sigma_{1C}^u \tag{7}$$

$$Matrix:\quad \sigma_2 \geq \sigma_{2T}^u \quad or \quad |\sigma_2| \geq \sigma_{2C}^u \tag{8}$$

$$Shear:\quad |\sigma_{12}| \geq \sigma_{12}^u \tag{9}$$

Verifying the fibre failure, the required force to achieve this failure mode has to be above 2,050 N, and the area near the clamped support firstly demonstrate this failure mode, which is associated with the compressed part of the fin related to 4.DOWN ply.

Three different failure conditions are considered in respect to the maximum strain of fibre direction, matrix or transversal direction and for shear strains:

$$Fibre:\quad \varepsilon_1 \geq \varepsilon_{1T}^u \quad or \quad |\varepsilon_1| \geq \varepsilon_{1C}^u \tag{10}$$

$$Matrix:\quad \varepsilon_2 \geq \varepsilon_{2T}^u \quad or \quad |\varepsilon_2| \geq \varepsilon_{2C}^u \tag{11}$$

$$Shear:\quad |\sigma_{12}| \geq \sigma_{12}^u \tag{12}$$

Figure 18 shows the failure with respect to the maximum strain that occurs in the compressive side, ply 4. DOWN, when the loading force reaches

Table 3. Failure criteria.

Failure criterion	Force (N)	Ply failure	Deflection (mm)	Stresses (MPa)
Tsai-Wu	910	1.DOWN	49.902	223.62
Max Stress	2050	4.DOWN	109.68	393.18
Max Strain	825	4.DOWN	40.031	201.51

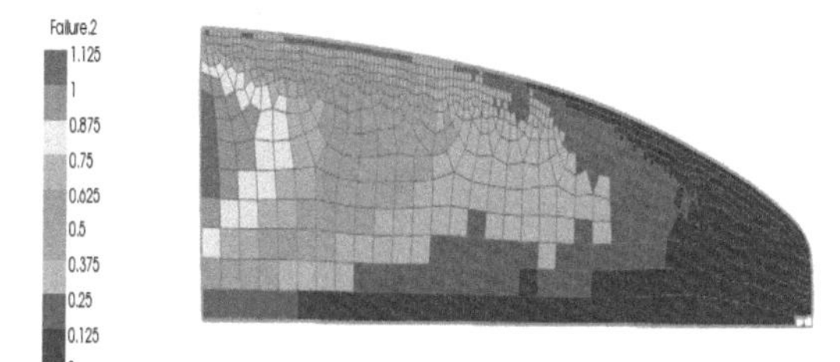

Figure 18. Fibre failure-maximum strain.

a value of 825 N. The failure verifications are shown in Table 3.

Table 3 also shows that the first failure criterion is in a very close agreement with the one of the maximum strain, indicating that the fibres stretch more. The Tsai-Wu failure criterion indicates a relatively low force of failure, which is associated with the matrix. All criteria show failure in plies that are subjected to compressive loading.

7 CONCLUSIONS

An initial FE model of a real built windsurfer fin has been developed and calibrated experimentally in terms of stiffness and deflection responses with full-scale mechanical tests.

The tracking down of relevant material property input data was a problem, but suitable initial values were found and then calibration of these values with the force-deflection data from the mechanical tests enabled a good fit of the predicted numerical results with the actual experimental behaviour.

The FE model gave a highly linear behaviour, with only a very short 'pseudo-plastic' due to failure mechanisms just before final failure. Initial failure load estimations have been made using the Tsai-Wu, Maximum Stress and Maximum Strain criteria, but further work should include experimental testing of the fin to failure to verify these values.

The current FE model will now be developed to allow exploration and optimisation of the laminate lay-up schedule and will be combined with a parallel computational fluid dynamics (CFD) work to finally give a working fluid structure interaction (FSI) model.

REFERENCES

ANSYS 2009. Online Manuals, Release 12.

Broers, A.M., Chiu, T.W., Pourzanjani, M.M.A., Buckingham, B.J. & van den Bersselaar, T. 1992. The Effects of Tip Flexibility on the Performance of a Blade-Type Windsurfer Fin. *Manouevring and Control of Marine Craft.* Comput. Mech. Publ, 261–273.

Camanho, P.P., Arteiro, A., Melro, A.R., Catalanotti, G. & Vogler, M. 2015. Three-dimensional invariant-based failure criteria for fibre-reinforced composites. *Int. J. Solids Struct.*, 55, 92–107.

Drake, J. 2005. *An Introduction to the Physics of Windsurfing.*

Exel. 2016. *Exel-Raw Materials-Reinforcements* [Online]. http://www.exelcomposites.com/en-us/english/composites/rawmaterials/reinforcements.aspx.

Fagg, S. 1997. *The development of a reversible and finitely variable camber windsurf fin.*

Gourlay, T. & Martellotta, J. 2011. *Aero-Hydrodynamics of an RS:X Olympic Racing Sailboard.*

Gurit-Holding, A. 2000. Guide to Composites.

Kunoth, A., Schlichtenmayer, M. & Schneider, C. 2007. Speed Windsurfing: Modelling and Numeric. *Int. J. Numer. Anal. Model,* 4, 548–558.

Miller, P. 1991. NNS composite materials properties database, unpublished composite test program report.

Shenoi, R.A. & Wellicome, J.F. 1993. *Composite Materials in Maritime Structures,* Southampton.

Sutherland, L.S. 1993. *Windsurfer Fin Hydrodynamics*, University of Southampton.

Sutherland, L.S. 2018. A review of impact testing on marine composite materials: Part I – Marine impacts on marine composites. *Compos. Struct,* (accepted for publication).

Sutherland, L.S. & Wilson, P.A. 1994. *Fin Hydrodynamics of a Windsurfer,* Southampton.

Tsai, S.W. & Wu, E.M. 1971. A General Theory of Strength for Anisotropic Materials. *J. Compos. Mater.*, 5, 58–80.

Progress in Maritime Technology and Engineering – Guedes Soares & Santos (Eds)
 ISBN 978-1-138-58539-3

Uncertainty propagation and sensitivity analysis of a laminated composite beam

M. Calvário, A.P. Teixeira & C. Guedes Soares
Centre for Marine Technology and Ocean Engineering (CENTEC), Instituto Superior Técnico, Universidade de Lisboa, Lisbon, Portugal

ABSTRACT: This paper aims at performing an uncertainty propagation and sensitivity analysis of the Tsai-Will and Tsai-Wu failure criteria for simply supported laminated beam subjected to uniform load. Four elastic moduli and five ultimate stresses are modeled as random variables. The correlation between the mechanical properties is accounted and assessed by Monte Carlo Simulation based on a micromechanics model. Monte Carlo Simulation methods are used for uncertainty propagation analysis using the Nataf transformation to simulate the non-normal correlated input random variables. Sensitivity analyses of the composite failure criteria are conducted based on the standardized regression coefficients of the input parameters. It is found that the transverse tensile strength for a unidirectional lamina is the most important parameter.

1 INTRODUCTION

Composite materials have been used in different industrial areas, resulting in a significant weight reduction and higher resistance to corrosion relatively to the traditional steel construction. A composite laminate is an assembly of fiber-reinforced plies (made off fiber and matrix) with uncertain properties due to the uncertainty on the layup assembly, manufacturing process, etc. (Lin 2000).

A research effort has been dedicated to the reliability of composite structures, including probabilistic approaches concerning the strength of composite materials (Guedes Soares 1997, Sutherland & Guedes Soares 1997) and their assessment in structures (e.g. plates). Lin (2000) has assessed the reliability of laminated plates under transverse loads with different methods (Monte Carlo simulation and first order second moment method), while Chen & Guedes Soares (2007) have proposed a method for reliability assessment of the post-buckling compressive strength of laminated composite plates and stiffened panels under axial compression. Also, Zhao et al. (2016) have studied the reliability of stiffened panels with compressive loads.

An important issue is the effect of statistical correlation between material properties and their effect on the reliability of composite plates, highlighted by Shaw et al. (2010), Zhang et al. (2015, 2016). Also, sensitivity analyses in terms of parameters role on the reliability of composite structures were carried out in different studies (Thomas & Wetherhold 1991, Zhou et al 2016 and Wang et al. (2017).

The objective of this paper is to study the uncertainty and sensitivity of Tsai-Will and Tsai-Wu failure criteria on a simply supported laminated beam. The Monte Carlo simulation method is adopted to evaluate the correlation between the mechanical properties, as well as for uncertainty propagation analysis, using the Nataf transformation to simulate the non-normal correlated input random variables. Standardized regression coefficients (SRC) are used as sensitivity measures to assess the most important mechanical properties on the failure criteria.

The paper is organized in five sections. In Section 2, the main equations related with the micro/macro mechanics of a lamina, as well the bending of a laminated composite beam, are presented. The Nataf transformation model is highlighted in Section 3 and in Section 4 the case study is presented. Results are presented and discussed in Section 5.

2 COMPOSITE MECHANICS

In this section, an overview of composite materials mechanics is presented. In section 2.1, the main constitutive laws based on the macro-mechanical of an orthotropic composite lamina are described. Sections 2.2 and 2.3 present the micro-mechanical analysis regarding the estimation of mechanical properties and the bending of a composite laminated beam, respectively. The contents of this chapter are based on model and assumptions of Kaw (2006).

2.1 *Macro-mechanical analysis*

For a unidirectional lamina it is assumed that the layer of the laminate is orthotropic, axis 1 is the axis along the fiber direction, axis 2 is transverse to axis 1 and 3 is perpendicular to lamina plane. Also, it is considered that the lamina is thin, which allows the plane stress assumption. The stresses in referential (local) are given by (Kaw 2006):

$$\begin{bmatrix} \sigma_1 \\ \sigma_2 \\ \tau_{12} \end{bmatrix} = \begin{bmatrix} Q_{11} & Q_{12} & 0 \\ & Q_{22} & 0 \\ sim. & & Q_{66} \end{bmatrix} \begin{bmatrix} \varepsilon_1 \\ \varepsilon_2 \\ \varepsilon_{12} \end{bmatrix} \tag{1}$$

where: σ_1 = Normal stress along the material direction 1; σ_2 = Normal stress along the material direction 2; τ_{12} = Shear stress in the pane 12; Q_{ij} = reduced stiffness coefficients; ε_1 = strain along the material direction 1; ε_2 = strain along the material direction 2; and ε_{12} = Shear strain in the 1–2 plane.

The coefficients Q_{ij} are obtained by:

$$\begin{aligned} Q_{11} &= \frac{E_1}{1-\nu_{12}^2 E_2/E_1} \\ Q_{12} &= \frac{\nu_{12}E_2}{1-\nu_{12}^2 E_2/E_1} \\ Q_{22} &= \frac{E_2}{1-\nu_{12}^2 E_2/E_1} \\ Q_{66} &= G_{12} \end{aligned} \tag{2}$$

where: E_1 = Young's modulus in direction 1; E_2 = Young's modulus in direction 2; ν_{12} = major Poisson's ratio and G_{12} = in-plane shear modulus.

The local stresses in an angle lamina can be obtained from the global stress:

$$\begin{bmatrix} \sigma_x \\ \sigma_y \\ \tau_{xy} \end{bmatrix} = [T]^{-1} \begin{bmatrix} \sigma_1 \\ \sigma_2 \\ \tau_{12} \end{bmatrix} \tag{3}$$

where: $[T]$ = transformation matrix, obtained by:

$$[T] = \begin{bmatrix} c^2 & s^2 & 2cs \\ s^2 & c^2 & -2cs \\ -cs & cs & c^2-s^2 \end{bmatrix} \tag{4}$$

where: $c = \cos(\theta)$; $s = \sin(\theta)$ and (θ) = angle lamina.

Also, the local and global strains are related with each other (Kaw 2006):

$$\begin{bmatrix} \varepsilon_1 \\ \varepsilon_2 \\ \gamma_{12} \end{bmatrix} = [R][T]\,[R]^{-1} \begin{bmatrix} \varepsilon_x \\ \varepsilon_y \\ \gamma_{xy} \end{bmatrix} \tag{5}$$

where: R = Reuter matrix; ε_x = normal strain in x direction; ε_y = normal strain in y direction; and γ_{xy} = shear strain in the xy plane.

The matrix R is:

$$[R] = \begin{bmatrix} 1 & 0 & 0 \\ 0 & 1 & 0 \\ 0 & 0 & 2 \end{bmatrix} \tag{6}$$

The global stresses are as:

$$\begin{bmatrix} \sigma_x \\ \sigma_y \\ \tau_{xy} \end{bmatrix} = [T]^{-1}[Q]\,[R][T][R]^{-1} \begin{bmatrix} \varepsilon_x \\ \varepsilon_y \\ \gamma_{xy} \end{bmatrix} \tag{7}$$

Equation (7) can be written as:

$$\begin{Bmatrix} \sigma_x \\ \sigma_y \\ \tau_{xy} \end{Bmatrix} = \begin{bmatrix} \overline{Q}_{11} & \overline{Q}_{12} & \overline{Q}_{16} \\ & \overline{Q}_{22} & \overline{Q}_{26} \\ sim. & & \overline{Q}_{66} \end{bmatrix} \begin{Bmatrix} \varepsilon_x \\ \varepsilon_y \\ \varepsilon_{xy} \end{Bmatrix} \tag{8}$$

where: $\overline{Q}_{ij}$ = Elements of the transformed reduces stiffness matrix.

The coefficients $\overline{Q}_{ij}$ are obtained by:

$$\begin{aligned} \overline{Q}_{11} &= Q_{11}c^4 + Q_{22}s^4 + 2(Q_{12}+2Q_{66})s^2c^2 \\ \overline{Q}_{12} &= (Q_{11}+Q_{22}-4Q_{66})s^2c^2 + Q_{12}(c^4+s^2) \\ \overline{Q}_{22} &= Q_{11}s^4 + Q_{22}c^4 + 2(Q_{12}+2Q_{66})s^2c^2 \\ \overline{Q}_{16} &= (Q_{11}-Q_{12}-2Q_{66})c^3s - (Q_{22}-Q_{12}-2Q_{66})s^3c \\ \overline{Q}_{26} &= (Q_{11}-Q_{12}-2Q_{66})cs^3 - (Q_{22}-Q_{12}-2Q_{66})c^3s \\ \overline{Q}_{66} &= (Q_{11}+Q_{12}-2Q_{12}-2Q_{66})s^2c^2 + Q_{66}(s^4+c^4) \end{aligned} \tag{9}$$

In case of strength failure of an angle lamina, two criteria are studied: Tsai-Hill and Tsai-Wu given respectively by (Kaw 2006):

$$\frac{\sigma_1^2}{\sigma_{1\,ult}^{T\,2}} - \frac{\sigma_1\sigma_2}{\sigma_{1\,ult}^{T\,2}} + \frac{\sigma_2^2}{\sigma_{2\,ult}^{T\,2}} + \frac{\tau_{12}^2}{\tau_{12ult}^2} < 1 \tag{10}$$

$$H_1\sigma_1 + H_2\sigma_2 + H_6\tau_{12} + H_{11}\sigma_1^2 + H_{22}\sigma_2^2 + H_{66}\tau_{12}^2 + 2H_{12}\sigma_1\sigma_2 < 1 \quad (11)$$

where: σ_{1ult}^T = Longitudinal tensile strength; σ_{2ult}^T = transverse tensile strength; τ_{12ult} = in-plane shear strength; and H_1, H_2, H_{11}, H_{22}, H_6, H_{66}, H_{12} = Tsai-Wu criteria coefficients.

The coefficients H_1, H_2, H_{11}, H_{22}, H_6, H_{66} and H_{12} are given by:

$$H_1 = \frac{1}{\sigma_{1ult}^T} - \frac{1}{\sigma_{1ult}^C} \quad (12)$$

$$H_{11} = \frac{1}{\sigma_{1ult}^T \sigma_{1ult}^C} \quad (13)$$

$$H_2 = \frac{1}{\sigma_{2ult}^T} - \frac{1}{\sigma_{2ult}^C} \quad (14)$$

$$H_{22} = \frac{1}{\sigma_{2ult}^T \sigma_{2ult}^C} \quad (15)$$

$$H_6 = 0 \quad (16)$$

$$H_{66} = \frac{1}{\tau_{12ult}^2} \quad (17)$$

$$H_{12} = -\frac{1}{2}\sqrt{\frac{1}{\sigma_{1ult}^T \sigma_{1ult}^C \sigma_{2ult}^T \sigma_{2ult}^C}} \quad (18)$$

where: σ_{1ult}^C = Longitudinal compressive strength; and σ_{2ult}^C = transverse compressive strength.

2.2 *Micro-mechanical analysis*

In the present study, the material properties of the composite structure consist of four elastic moduli and the five strength parameters. The constants E_{11}, ν_{12} and G_{12} are obtained by (Huang, 2001):

$$E_{11} = V_f E_{f11} + V_m E_m \quad (19)$$

$$\nu_{12} = V_f \nu_{f12} + V_m \nu_m \quad (20)$$

$$G_{12} = G^m \frac{(G_{f12}+G_m)+V_f(G_{f12}-G_m)}{(G_{f12}+G_m)-V_f(G_{f12}-G_m)} \quad (21)$$

where: V_f = fiber volume ratio; E_{f11} = fiber modulus in longitudinal direction; V_m = matrix volume ratio; E_m = matrix elastic modulus; ν_{f12} = in-plane fiber Poisson ratio; ν_m = matrix Poisson ratio; G_{f12} = in-plane fiber shear modulus; and G_m = matrix shear modulus.

The E_{22} is determined by (Morais, 2000):

$$E_{22} = \frac{\sqrt{V_f}}{\frac{\sqrt{V_f}}{E_{22f}} + \left(1-\sqrt{V_f}\right)\frac{1-\nu_m^2}{E_m}} + \left(1-\sqrt{V_f}\right)\frac{E_m}{1-\nu_m^2} \quad (22)$$

where: E_{22f} = fiber modulus in transverse direction.

The ultimate stresses are obtained with (Kaw, 2006):

$$\sigma_{1ult}^T = V_f \sigma_{fult} + \frac{\sigma_{fult}}{E_{f11}}\left(1-V_f\right)E_m \quad (23)$$

$$\sigma_{2ult}^T = E_2 \varepsilon_{mult}^T \left[\frac{d}{s}\frac{E_m}{E_f} + \left(1-\frac{d}{s}\right)\right] \quad (24)$$

$$\tau_{12ult} = G_{12}\gamma_{12mult}\left[\frac{d}{s}\left(\frac{E_m}{E_f}-1\right)+1\right] \quad (25)$$

$$\sigma_{1ult}^C = \frac{E_1}{\nu_{12}}\varepsilon_{mult}^T\left[\frac{d}{s}\left(\frac{E_m}{E_f}-1\right)+1\right] \quad (26)$$

$$\sigma_{2ult}^C = E_2 \varepsilon_{mult}^C \left[\frac{d}{s}\frac{E_m}{E_f} + \left(1-\frac{d}{s}\right)\right] \quad (27)$$

where: σ_{fult} = fiber tensile strength; σ_{mult} = matrix tensile strength; ε_{mult}^T = ultimate tensile strain of the matrix; d = diameter of the fibers; s = center-to-center spacing between the fibers; ε_{mult}^C = ultimate compressive failure strain of matrix; γ_{12mult} = ultimate shearing strain of the matrix.

The values of $\varepsilon_{mult}^T, \varepsilon_{mult}^C$ *and* γ_{12mult} are obtained with (Kaw, 2006):

$$\varepsilon_{mult}^T = \frac{\sigma_{mult}^T}{E_m} \quad (28)$$

$$\varepsilon_{mult}^C = \frac{\sigma_{mult}^C}{E_m} \quad (29)$$

$$\gamma_{12mult} = \frac{\tau_{12mult}}{G_{12}} \quad (30)$$

2.3 *Bending of a laminated beam*

For the present study, a simply supported laminated symmetric rectangular beam with uniform load is considered (Fig. 1). The determination of

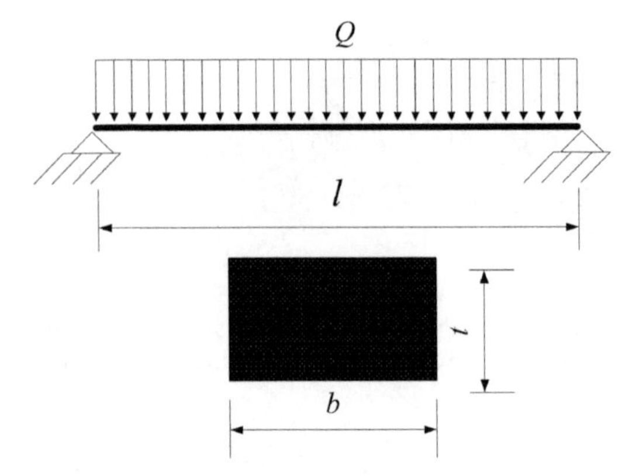

Figure 1. Laminated beam. Adapted from Kaw (2006).

the stresses (local referential) at the top the plies is based on Equations (1) to (9). For this structure and with the assumptions based on Kaw (2006), the global strains are obtained by:

$$\varepsilon_x = z\kappa_x \tag{31}$$

$$\varepsilon_y = z\kappa_y \tag{32}$$

$$\varepsilon_{xy} = z\kappa_{xy} \tag{33}$$

where κ_x, κ_y, κ_{xy} = midplane curvatures and z = coordinate.

The midplane curvatures are obtained by:

$$\kappa_x = D^{-1}_{11} M_x \tag{34}$$

$$\kappa_y = D_{12} M_x \tag{35}$$

$$\kappa_{xy} = D^{-1}_{16} M_x \tag{36}$$

where $[D_{ij}]$ = Bending stiffness matrix; and M_x = moment.

The values of $[D_{ij}]$ and M_x are obtained by:

$$D_{ij} = \frac{1}{3}\sum_{k=1}^{n}\left[\bar{Q}_{ij}\right]_k \left(h^3_k - h^3_{k-1}\right),\ i = 1,2,6; j = 1,2,6 \tag{37}$$

$$M_x = \frac{M}{w} \tag{38}$$

where: h = coordinate location of plies; M = bending moment at the center of the beam; w = section with.

The value of M_x is obtained by:

$$M = \frac{Ql^2}{8} \tag{39}$$

where Q = uniform load; l = beam length.

3 NATAF TRANSFORMATION

In engineering problems with uncertainty, it is common to model the input variables as a random vector and is preferable to transform this vector into an independent standard normal space, being the Nataf transformation commonly used (e.g. Liu & Kiureghian 1986, Xiao 2014). The Nataf transformation allows the simulation of a correlated random vector (X) with a specified correlation matrix from a vector of independent standard normal variables (Xiao, 2014). The method is based on the transformation from standard normal deviates to random variables x, in which the cumulative distribution function $F(x)$ is determined by (e.g. Liu & Kiureghian 1986, Xiao 2014):

$$\mathrm{F}(x) = \Phi(z) \tag{40}$$

$$x = F^{-1}\left[\Phi(z)\right] \tag{41}$$

$$X = \left(x_1,\ldots,x_i,\ldots,x_m\right)^T \tag{42}$$

where $\Phi(\cdot)$ = cumulative distribution function of the standard normal variables z; $F^{-1}(\cdot)$ = inverse of the cumulative distribution function of x; and X = random vector.

The correlation matrix of X and the correlation matrix of a standard normal vector $Z = (z_1, \ldots z_i, \ldots z_m)^T$ are given with:

$$R_X = \begin{pmatrix} \rho^x_{1,1} & \cdots & \rho^x_{1,j} & \cdots & \rho^x_{1,m} \\ \vdots & \vdots & \vdots & \vdots & \vdots \\ \rho^x_{i,1} & \cdots & \rho^x_{i,j} & \cdots & \rho^x_{i,m} \\ \vdots & \vdots & \vdots & \vdots & \vdots \\ \rho^x_{m,1} & \cdots & \rho^x_{m,j} & \cdots & \rho^x_{m,m} \end{pmatrix} \tag{43}$$

$$R_Z = \begin{pmatrix} \rho^z_{1,1} & \cdots & \rho^z_{1,j} & \cdots & \rho^z_{1,m} \\ \vdots & \vdots & \vdots & \vdots & \vdots \\ \rho^z_{i,1} & \cdots & \rho^z_{i,j} & \cdots & \rho^z_{i,m} \\ \vdots & \vdots & \vdots & \vdots & \vdots \\ \rho^z_{m,1} & \cdots & \rho^z_{m,j} & \cdots & \rho^z_{m,m} \end{pmatrix} \tag{44}$$

where: ρ_x = correlation coefficient of two random variables; ρ_z = equivalent correlation in the standard normal space, R_X = correlation matrix; R_Z = correlation matrix in the normal space.

The matrix R_z can be generated with a linear transformation, which is given by:

$$U = L^{-1}Z \leftrightarrow Z = LU \quad (45)$$

$$R_Z = LL^T \quad (46)$$

where U = $(u_1,\ldots u_i,\ldots u_m)^T$ independent standard normal vector; and L = lower triangular matrix from the Cholesky decomposition.

The relationship of $\rho_{xi,xj}$ and $\rho_{zi,zj}$ is expressed by (Liu & Kiureghian, 1986):

$$\rho_{x_i,x_j} = \int_{-\infty}^{+\infty}\int_{-\infty}^{+\infty}\left(\frac{x_i-\mu_i}{\sigma_i}\right)\left(\frac{x_j-\mu_j}{\sigma_j}\right) \varphi_2\left(z_i,z_j,\rho_{z_i,z_j}\right)dz_i dz_j \quad (47)$$

where: μ_i, μ_j = mean of X_i and X_j respectively; σ_i, σ_j = standard deviation of X_i and X_j respectively; and φ_2 = two dimensional normal probability density function of zero means, unit standard deviations and correlation matrix R_{zz}.

Empirical formulae expressed by the function F relating ρ_{xx} and ρ_{zz} have been proposed by Liu & Kiureghian (1986):

$$\rho_{x_i,x_j} = F\rho_{z_i,z_j} \quad (48)$$

4 CASE STUDY

Based on assumptions and methods presented in the previous sections, uncertainty propagation and sensitivity analyses are conducted in order to evaluate the variables importance on the Tsai-Hill and Tsai-Wu failure criteria. The measure of sensitivity adopted is the standard regression coefficient, given by (Gan, 2014):

$$SRC = \frac{b_i \hat{s}_i}{\hat{s}} \quad (49)$$

where: b_i = ith regression coefficient of the random variable X_i, $\hat{s}$ = standard deviation of y (model ouput); $\hat{s}_i$ = standard deviation of X_i.

The standard deviations are obtained with:

$$\hat{s} = \sqrt{\frac{1}{N-1}\sum_{k=1}^{N}\left(y_k - \bar{y}\right)^2} \quad (50)$$

$$\hat{s}_i = \sqrt{\frac{1}{N-1}\sum_{k=1}^{N}\left(X_i^k - \bar{X}_i\right)^2} \quad (51)$$

A measure of the quality of the regression model relatively to observed data is given by the coefficient of determination, as (Helton et al. 2006):

$$R^2 = \frac{\sum_{i=1}^{N}\left(\hat{y}_i - \bar{y}\right)^2}{\sum_{i=1}^{N}\left(y_i - \bar{y}\right)^2} \quad (52)$$

The composite structure is a simply supported laminated beam made of graphite fiber and epoxy matrix (Shiao & Chamis 1999) with an orientation $[0/90/45/0/\text{-}45]_s$ (Chen & Guedes Soares 2007). The dimensions and load values are presented in Table 1, while the material properties related with fiber and matrix are presented in Table 2.

Based on the information of Table 2 and Equations (19) to (30), 10^5 independent Monte Carlo simulations were conducted in order to determine the correlation between the 9 system random variables (presented in Tables 3 and 4) following the approaches of Shaw et al (2010), Zhang et al (2015, 2016). Normal, Lognormal and Weibull distributions were fitted to the parameters and then assessed in terms of Chi-square goodness-of-fit test. The results of probability characteristics of the random variables are presented in Table 3 while the linear correlation coefficients are presented in Table 4.

Table 1. Load and dimensions values on a simply supported beam.

Q [N/m]	l [m]	b [mm]	d [mm]
1000	0.5	30	6

Table 2. Material properties for fiber and matrix components. Adapted from Shiao & Chamis (1999).

Variable	Unit	Mean	Cov	Distribution
E_{f11}	[GPa]	213.74	0.05	Normal
E_{f22}	[GPa]	13.79	0.05	Normal
ν_{f12}		0.2	0.05	Normal
G_{f12}	[GPa]	13.79	0.05	Normal
E_m	[GPa]	3.45	0.05	Normal
ν_m		0.35	0.05	Normal
V_f		0.66	0.05	Normal
σ_{fult}	[MPa]	2757.90	0.05	Weibull
σ_{mult}	[MPa]	103.42	0.05	Weibull
σ^C_{mult}	[MPa]	241.32	0.05	Weibull
τ_{12mult}	[MPa]	89.63	0.05	Weibull
G_m	[GPa]	1.28	0.05	Normal

Table 3. Probabilistic characteristics of the input random variables.

Variable	Unit	Mean	Cov	Distribution
E_{11}	[GPa]	142.267	0.07	Normal
E_{22}	[GPa]	8.361	0.058	Lognormal
ν_{12}		0.251	0.041	Normal
G_{12}	[GPa]	4.395	0.089	Lognormal
σ^T_{1ult}	[GPa]	1.835	0.07	Normal
σ^T_{2ult}	[MPa]	24.667	0.197	Normal
τ_{12ult}	[MPa]	51.421	0.07	Weibull
σ^C_{1ult}	[MPa]	1.668	0.187	Normal
σ^C_{2ult}	[MPa]	56.954	0.197	Normal

Table 4. Linear correlation coefficients of the input random variables.

	E_{11}	E_{22}	ν_{12}	G_{12}	$\sigma_{1ult}{}^T$	$\sigma_{2ult}{}^T$	τ_{12ult}	$\sigma_{1ult}{}^C$	$\sigma_{2ult}{}^C$
E_{11}	1								
E_{22}	0.55	1							
ν_{12}	−0.34	−0.30	1						
G_{12}	0.63	0.70	−0.43	1					
σ^T_{1ult}	0.49	0.55	−0.34	0.63	1				
σ^T_{2ult}	−0.7	−0.72	0.48	−0.86	−0.67	1			
τ_{12ult}	−0.46	−0.51	0.32	−0.56	−0.46	0.63	1		
σ^C_{1ult}	−0.46	−0.80	0.26	−0.8	−0.62	0.92	0.59	1	
σ^C_{2ult}	−0.69	−0.72	0.48	−0.86	−0.67	0.93	0.63	0.85	1

The probabilistic models of the random variables and their correlation matrix (Tables 3 and 4) are then used for uncertainty propagation of the failure criteria by Monte Carlo simulation (10^4 samples). The Nataf transformation, Equations (40) to (48), is used to simulate the non-normal correlated random variables. The values of F (Equation 48) for marginal distributions: Normal-Normal, Normal-Lognormal, Normal-Weibull, and Lognormal-Weibull are given by Liu & Kiureghian (1986).

5 RESULTS

In this section, the sensitivity measures of the parameters of the Tsai-Hill and Tsai-Wu failure criteria are shown. In Tables 5 and 6 the values of SRC are presented for the Tsai-Hill and Tsai-Wu, respectively, and the corresponding coefficients of determination are presented in Table 7 and Table 8.

Analysing Table 5 for ply number 1, 2, 3 and 4, it can be seen that σ2Tult is the parameter with more influence on the Tsai-Hill criterion (with negative influence) while for the ply 5 is E11 the most influential parameter (also with negative influence). The parameter E11 show high values of sensitivity in all plies, while σ1ultT has influence in plies number 1 and 4 and in ply 5 G12 and τ12ult are relevant (high SRC values).

Table 5. SRC of random parameters affecting uncertainty of Tsai-Hill criteria.

Ply	0 (1)	90 (2)	45 (3)	0 (4)	−45 (5)
SRC	[°]	[°]	[°]	[°]	[°]
E_{11}	−0.52	−0.79	−0.77	−0.52	-1.39
E_{22}	0.23	0.49	0.46	0.22	0.16
ν_{12}	0.29	0	0.02	0.29	0.05
G_{12}	0.10	0	0.02	0.10	0.71
σ^T_{1ult}	−0.58	0	0	−0.58	−0.02
σ^T_{2ult}	−0.77	−0.95	−0.95	−0.77	−0.31
τ_{12ult}	−0.14	0	−0.04	−0.14	−0.47

Table 6. SRC of random parameters affecting uncertainty of Tsai-Wu criteria.

ply	0 (1)	90 (2)	45 (3)	0 (4)	−45 (5)
SRC	[°]	[°]	[°]	[°]	[°]
E_{11}	0.37	0.09	0.19	0.37	0.15
E_{22}	−0.27	−0.07	−0.20	−0.25	−0.34
ν_{12}	−0.26	0	0	−0.25	−0.06
G_{12}	0.04	0	0.07	−0.02	0.24
σ^T_{1ult}	0.54	0	0.04	0.57	−0.18
σ^T_{2ult}	1.12	1.06	1.10	1.09	1.15
τ_{12ult}	−0.05	−0.01	−0.04	−0.02	−0.21
σ^C_{1ult}	−0.23	0	0	−0.20	−0.06
σ^C_{2ult}	−0.21	−0.42	−0.31	−0.21	−0.23

Table 7 shows a coefficient of determination R value close to 1 in all plies for the Tsai-Hill criterion, indicating that the linear regression model accounts for most of the uncertainty.

Concerning the sensitivity on Tsai-Wu failure criterion, $\sigma_{2ult}{}^T$ is also the most important parameter in all plies (positive influence, Table 6). The parameter $\sigma_1{}^T$ is relevant in plies 1 and 4 (positive influence), $\sigma_{2ult}{}^C$ in plies 2 and 3 (negative influence) and in ply 5 E_{22} an important parameter. Also, τ_{12ult} has shown to be less important in this criterion with exception of ply 5. The coefficient of determination R (Table 8) is also close to 1 in all plies, showing that the linear regression model provides a satisfactory approximation.

It should be noted that the formulation to model the ultimate strength of unidirectional lamina is based on a simplified formulation and, therefore, different sensitivity results would be obtained for other micro-mechanics models. Also, beam thickness, width, load and the fiber angle alignment were not modelled as random variables, which may also have impact on sensitivity results. Finally, strength parameters of a unidirectional lamina are particular sensitive to material and geometric non homogeneity, fiber-matrix interface, manufacturing process, etc. and so more difficult to predict their quantities than the stiffness parameters (Kaw 2006). So,

Table 7. Coefficient of determination considering the Tsai-Hill linear regression model.

Ply	0 [°]	90 [°]	45 [°]	0 [°]	–45 [°]
R	0.97	0.95	0.95	0.97	0.98

Table 8. Coefficient of determination considering the Tsai-Wu linear regression model.

Ply	0 [°]	90 [°]	45 [°]	0 [°]	–45 [°]
R	0.98	0.98	0.98	0.98	0.97

experimental techniques play an important role to validate the micro-mechanics model and would contribute to more accurate sensitivity measures.

6 CONCLUSIONS

Uncertainty propagation and sensitive analyses on the Tsai-Hill and Tsai-Wu failure criteria were conducted for a laminated simply supported beam with random mechanical properties. Four elastic moduli and five strength parameters were considered as random variables and for this case study σ_{2ult}^{T} was the most important parameter. The sensitivity analysis provided information about the importance of a certain parameter, being this useful to characterize composite materials, allowing for customization of each ply.

It should be noted that the strength parameters of unidirectional are very sensitive to several issues such as the quality of the manufacturing process, fiber-matrix interface and so different results of sensitivity may be obtained considering other micromechanics models or data from experimental tests.

ACKNOWLEDGEMENTS

This work was performed within the Strategic Research Plan of the Centre for Marine Technology and Ocean Engineering, which is financed by Portuguese Foundation for Science and Technology (Fundação para a Ciência e Tecnologia—FCT).

REFERENCES

Chen, N.Z. & Guedes Soares, C., (2007), "Reliability Assessment of Post-Buckling Compressive Strength of Laminated Composite Plates and Stiffened Panels under Axial Compression", *Int. J. Solids Struct.* 44:(22–23), 7167–7182.

Gan, Y., Duan, Q., Gong, W., Tong, C., Sun, Y., Chu, W., Ye, A., Miao, C., Di, Z. (2014). "A comprehensive evaluation of various sensitivity analysis methods: A case study with a hydrological model". *Environmental Modelling & Software* (51): 269–285.

Guedes Soares, C. (1997). "Reliability of components in composite materials". *Reliability Engineering and System Safety* (55) 171–177.

Helton, J., Johnson, J., Sallaberry, C., Storlie, C. (2006). "Survey of sampling-based methods for uncertainty and sensitivity analysis". *Reliability Engineering and System Safety* (91): 1175–1209.

Huang, Z. (2001). "Micromechanical prediction of ultimate strength of transversely isotropic fibrous composites" *International Journal of Solids and Structures* (38): 4147–4172.

Kaw, A. (2006). *Mechanics of Composite Materials*. Boca Raton: CRC Pres Taylor & Francis Group.

Lin, S. (2000). "Reliability predictions of laminated composite plates with random system parameters". *Probabilistic Engineering Mechanics* (15):327–338.

Liu, P. & Kiureghian, A. (1986). "Multivariate distribution models with prescribed marginal and covariances". *Probabilistic Engineering Mechanics* (1): 105–112.

Morais, A. (2000). "Transverse moduli of continuous-fibre-reinforced polymers". *Composites Science and Technology* (60): 997–1002.

Shaw, A., Sriramula, S., Gosling, P., Chryssanthopoulos, K. (2010). "A critical reliability evaluation of fibre reinforced composite materials based on probabilistic micro and macro—mechanical analysis". *Composites: Part B* (41) 446–453.

Shiao, M. & Chamis, C. (1999). "Probabilistic evaluation of the fuselage-type composite structures". Probabilistic *Engineering Mechanics* (14): 179–187.

Sutherland, L. & Guedes Soares, C. (1997). "Review of probabilistic models of the strength of composite materials". *Reliability Engineering and Systems Safety* (56) 183–196.

Thomas, D. & Wetherhold, R. (1991). "Reliability analysis of continuous fiber composite laminates". *Composite structures* (17): 277–293.

Wang, X., Ma, Y., Wang, L., Geng, X., Wu, Di. (2017). "Composite laminate orientated reliability analysis for fatigue life under non-probabilistic time-dependent method". *Comput. Methods Appl. Mech. Engrg.* (326): 1–19.

Xiao, Q. (2014). "Evaluating correlation coefficient for Nataf transformation". *Probabilistic Engineering Mechanics* (37) 1–6.

Zhang, S., Zhang, C., Chen, X. (2015). "Effect of statistical correlation between ply mechanical properties on reliability of fibre reinforced plastic composite structures". *Journal of Composite Materials* (49): 2935–2945.

Zhang, S., Zhang, L., Wang, Y., Tao, J., Chen, X. (2016). "Effect of ply level thickness uncertainty on reliability of laminated composite panels". *Journal of Composite Materials* (35): 1387–1400.

Zhao,W., Liu, W., Yang, Q. (2016). "Reliability analysis of ultimate compressive strength for stiffened composite panels". *Journal of Reinforced Plastics & Composites* (35): 902–914.

Zhou, X., Gosling, P., Ullah, Z., Kaczmarczyk, Ł., Pearce, C. (2016). "Exploiting the benefits of multiscale analysis in reliability analysis for composite structures". *Composite Structures* (155): 197–212.

Progress in Maritime Technology and Engineering – Guedes Soares & Santos (Eds)
© 2018 Taylor & Francis Group, London, ISBN 978-1-138-58539-3

Experimental study of the residual strength of damaged hybrid steel-FRP balcony overhangs of ships

N. Kharghani & C. Guedes Soares
Centre for Marine Technology and Ocean Engineering (CENTEC), Instituto Superior Técnico, Universidade de Lisboa, Lisbon, Portugal

ABSTRACT: A composite-to-steel hybrid balcony overhang is tested in order to find strength and stiffness under imposed bending load at the second load cycle after initial failure. The specimens are loaded under shear at the first cycle until the failure occurs. The configuration tested is representative of a solution being considered for the balconies of cruise ships, where the substitution of steel by composites aims at weight saving. The steel component is a channel type of structure made of two plates that serve as external supports of the sandwich plate. The purpose of the current study is evaluating the reduction of ultimate strength and stiffness of the damaged structure and identifying the critical locations and failure progress in bending condition.

1 INTRODUCTION

The increasing demand for stronger, lighter and more energy efficient materials has accelerated in recent years the transfer of composite technology into commercial and consumer sectors. It has been driven by the need for increased performance properties in high volume such as the structures in marine industry (Kharghani & Guedes Soares, 2016). In ship structures, there are many applications for composites, including the current and potential use in hulls and superstructures, decks, bulkheads, advanced mast systems, propellers and other equipment (Mouritz et al. 2001). One of the parts which is applied as a superstructure is the balcony. The current study concentrates on a damaged hybrid ship balcony, consisting of a steel support, to which a composite material plate (a sandwich panel) is attached. This part was tested primarily in shear and bending conditions as the first cycle of loading (Kharghani & Guedes Soares, 2018). Although the core shear took place earlier than steel yielding and debonding initiated at very low values of load, the structure showed a significant post-collapse strength. In the current paper, the failed specimen is imposed under bending as the second load cycle.

There are remarkable published studies, dealing with steel-to-composite joints: Chen et al. (2014) prepared four identical specimens of FRP/steel composite plates to conduct their test. Two of the specimens were under uniaxial loading, and the other two were under cyclic uniaxial loading. During the test, they monitored the specimens by Acoustic Emission (AE). Clifford et al. (2002) examined the mechanical response of a prototype joint between a glass-fibre reinforced polymer superstructure and a steel hull formed via a resin infusion process and subsequently modified to improve performance through a combined program of modelling and mechanical testing. Cao et al. (2003) replicated the joint design type A from Clifford et al. (2002), with different glass fibre, vinylester and core material.

Kharghani et al. (2015) performed a numerical analysis of a composite to steel joint in order to determine the stress distributions on the joint as a function of the imposed bending and torsional loads. The steel component was a channel type of structure made of two plates that serve as external supports of the sandwich plate. The attachment of the steel joint to the structure was supported by a bracket. Finite element analysis was conducted with 3D models in order to determine the most highly stressed zones where failure can be expected. The above-mentioned study addressed the extent of the steel composite overlap and bracket angle as the main parameters. It was concluded that the most appropriate geometries are 45 or 63 degrees for the bracket angle, 6 mm thickness for the steel (at channel and bracket parts) and 238 mm steel-composite overlap. It was observed that the initiation of debonding depends on the length of the overlap between steel and composite. For the minimum overlap of 38 mm debonding initiated at lower values of loads than for overlaps of 138 mm and 238 mm. But for the last two overlaps the difference was negligible. These results served as the basis to design test specimens and to plan the current experimental investigation.

Jiang et al. (2014) studied the adhesively-bonded joint under shear loading experimentally and numerically. Also, Jiang et al. (2015) focused on mechanical behaviours of adhesively—bonded joints between FRP sandwich decks and steel girders. A specific tensile–shear loading device was designed with the capacity to provide the combination of tensile and shear loads in six different ratios. Ultimate failure loads, load-deformation behaviours and failure modes of adhesively-bonded joints were investigated experimentally and compared regarding to different loading angles. Li et al. (2015) examined the mechanical response of the joint between a Glass-fibre Reinforced Polymer (GRP) superstructure and a steel hull formed and subsequently modified to improve performance through a combined program of modelling and testing.

Kotsidis et al. (2015) studied an adhesively bonded butt-joint, comprised of a double lap steel-GFRP joint and a GFRP sandwich composite part. In order to simulate the mechanical behaviour of the joint subjected to tensile and bending loading, a two-dimensional finite element model was developed. Various design parameters were examined in order to evaluate their effects on the joint load bearing capacity and stiffness. Hentinen et al. (1997) presented several joint elements developed for joining large FRP sandwich panels to ships. Metal profiles were adhesively bonded to the panels in the prefabrication phase of the sandwich. This made it possible to weld the FRP part directly to the metal structure in the shipyard. Cao et al. (2004) concentrated on hybrid ships, consisting of an advanced double hull stainless steel center section, to which a composite material bow and/or stern was attached. Two concepts of joints, a bonded-bolted joint and a co-infused perforated joint, were evaluated.

According to the above-mentioned studies which mostly investigate different types of hybrid joints up to initial failure, it is of vital importance to improve the researches on the damage propagation of a channel type of joints including a conservative overlap experimentally and in a more realistic condition. In the current paper some large-scale experiments are carried out on a composite-to-steel hybrid balcony to analyse the behaviour of the structure in bending condition after failure due to shear and bending.

2 EXPERIMENTS

2.1 *Specimens*

Three specimens were tested with the given unique identities: Sp1, Sp2 and Sp3. The steel component of the specimens is a channel type of structure made of two plates that serve as external supports of the sandwich plate. The attachment of the steel joint to the structure is supported by two brackets (see Figure 1). The technical drawing of the joint is presented in Figures 2 and 3 and it is the result of a previous initial numerical 3D parametric study (Kharghani et al. 2015). The thicknesses of the core, steel and skins are 30, 6 and 2.5 mm, respectively. The thickness of the steel base-plate is 20 mm (see Figure 3), whereas the angles of the steel brackets

Figure 1. 3D view of the specimen.

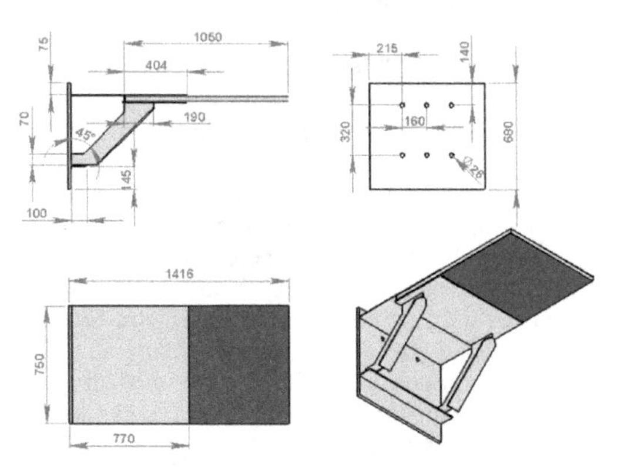

Figure 2. Technical drawing of the specimen.

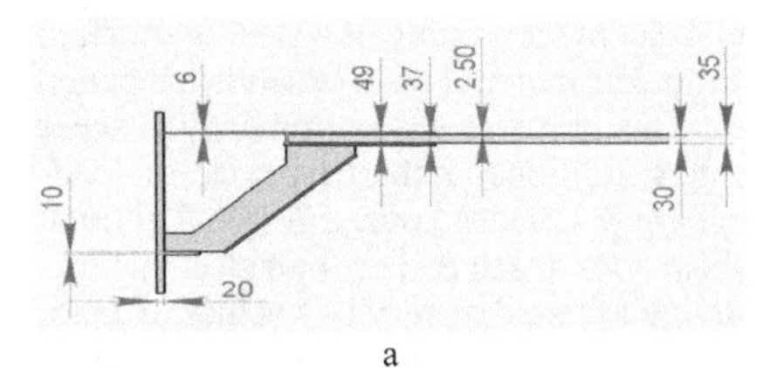

a

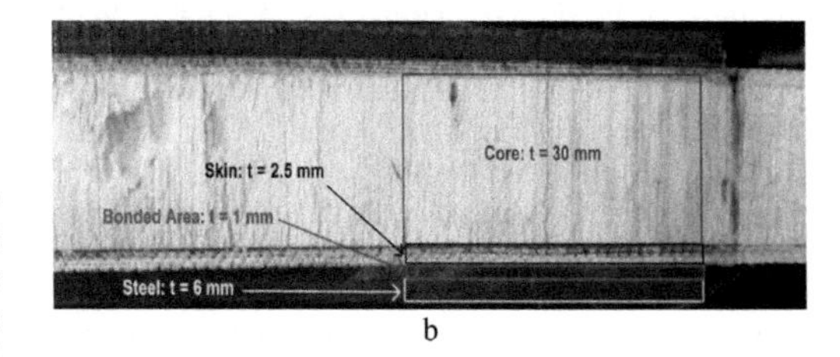

b

Figure 3. Thicknesses in detail: a) front view, b) lay-up.

are 45° and the width of all specimens is equal to 750 mm. Finally, the overlap length between the steel parts and the sandwich panels are 404 mm for all of the specimens (see Figure 2).

The composite system of the sandwich skins consisted of four layers of 813 gr/m^2 biaxial stitched E-glass fabric from Metyx Composites together with a Scott Bader Crystic VE679PA vinylester resin fabric was a bi-axially stitched E-glass and a Vinylester resin system, Crystic VE 679 PA, was used to impregnate the laminated fabrics. The ProBalsa Standard wood 155 kg/m^3 from DIAB Balsa wood has been used as the core of the sandwich plate, whereas the steel used was of AH36 grade with density of 7860 kg/m^3 with isotropic, elastic-plastic characteristics.

Adhesion between all parts of the joint (steel/composite, steel/core and composite/core) was achieved only with the aid of the vinylester resin during manufacturing, i.e. there was no use of any structural adhesive for bonding pre-fabricated parts.

The weight of each specimen was measured using an overhead crane and was observed 150 kg (10% of the total weight is related to the sandwich panel). It would be about 365 kg, in case of full-steel specimens. Thus, the hybrid ones are about 60% lighter than the conventional full-steel specimens.

2.2 *Loading conditions*

A calibrated servo-hydraulic mechanical test rig with a 5-tone load cell was used (see Figure 4). The load roller was made of steel with the length and the diameter of 800 and 60 mm respectively. Two types of loading condition (LC) were considered in this experiment: Shear (LC1) and bending (LC2). Figures 5 and 6 demonstrate the details of loading conditions and the exact position of the load roller. Although both loading conditions include shear and bending simultaneously, but LC1 indicates that can impose the maximum shear load to the steel-composite overlap whereas LC2 provides the maximum possible bending load. Specimens 1 and 2 were tested under shear load and specimen 3 was tested under bending for the first cycle of loading. In reality both loading conditions were combined shear and bending although one case was dominated by shear and the other by bending. After failure of the specimens in the first cycle, all of them were tested under bending (LC2- see Figure 6) for the second cycle (see Table 1) loading was applied along the line indicated in Figures 5 and 6. As loading is linearly increasing the vertical displacement with a rate of 2 mm/min, i.e. this procedure was very slow. Also, the period of data capturing was 0.02 second.

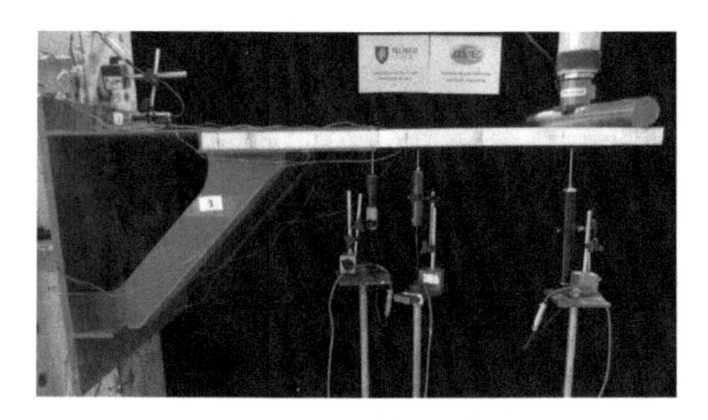
Figure 4. Specimen in loading condition.

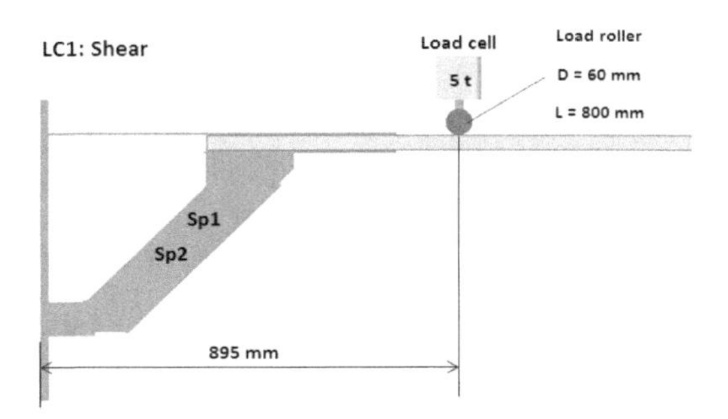

Figure 5. Details of the shear loading condition (LC1) (Sp1 and Sp2).

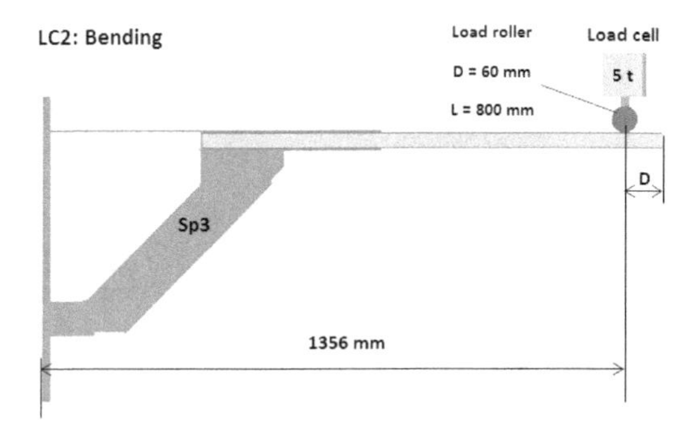

Figure 6. Details of the bending loading condition (LC2) (Sp3).

The specimens were bolted to one of the vertical columns of the main frame. Six bolts with 26 mm diameter were used (see Figure 7). The quality of clamping was checked before and during the test. Before the test, the slopes of the top surfaces were measured. The maximum observed value was about 0.3°. Also, a LVDT was used horizontally to measure the displacement of the base plate in the longitudinal direction during the tests (see Figure 8a). The maximum recorded displacement was 0.45 mm.

The deflection of the specimens in three points A, B and C (see Figures 8b and 9) was measured

Table 1. Load condition.

Specimen	1st load cycle	2nd load cycle
Sp1	Shear (LC1)	Bending (LC2)
Sp2	Shear (LC1)	Bending (LC2)
Sp3	Bending (LC2)	–

Figure 7. Connection method of the specimen to the main frame.

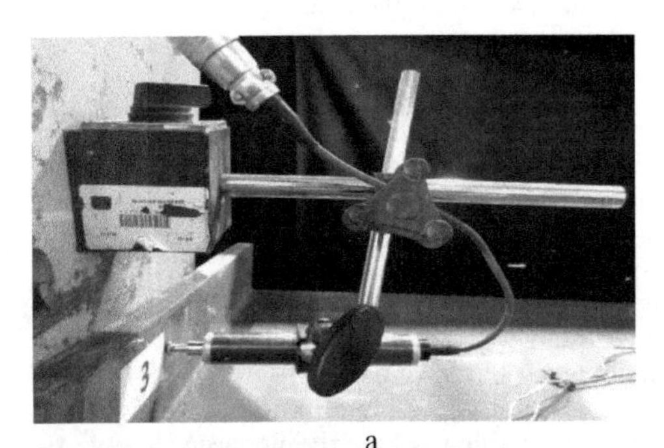

a

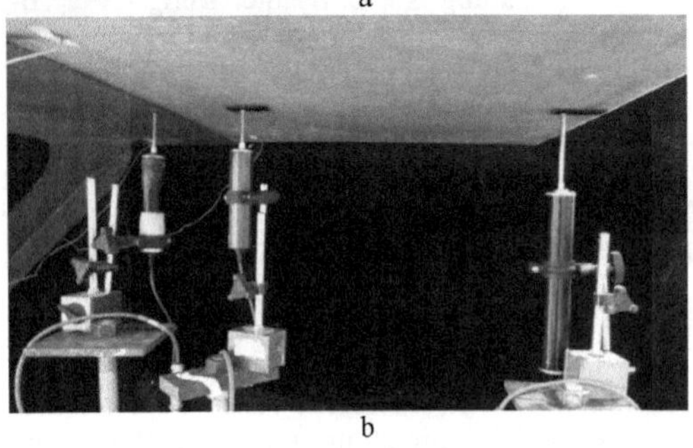

b

Figure 8. LVDTs to measure: a) Horizontal displacement of the specimen in the connection point to the main frame (Point H), b) Vertical displacements of the steel and sandwich panel (Points A, B and C).

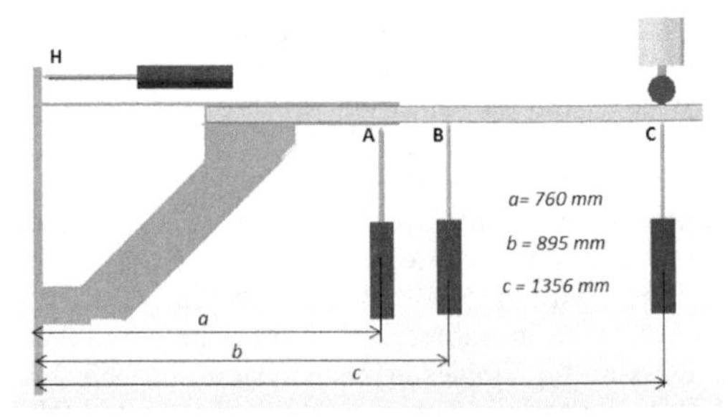

Figure 9. LVDTs positions.

using LVDTs. All of these three points are located along the symmetry line of the specimen. Moreover, smooth thin plates were connected to the composite surfaces to prevent any instability in the results (see Figure 8b).

3 RESULTS AND DISCUSSION

This section concentrates on the analysis of the experimental results at the second load cycle after initial failure and compared them with each other. Furthermore, the locations of the failure regions are shown and the behaviour of the structure is described in detail for both cycles. According to Table 1, for the first cycle two specimens are loaded under shear and the other one under bending. Then the specimens that were tested under shear in the first cycle are loaded under bending at the second one (see Figures 5 and 6).

Figure 10a shows the failed specimen Sp1 (12.8 mm–26.0 kN) under shear load (LC1). Clearly, debonding in the upper skin-steel interface can be observed. At the outer surface of the core, inside the steel channels, there are several obvious cracks. The longest crack has a length of 30 mm along the block bond of the core and penetrate the whole thickness completely. Finally, the specimen fails as demonstrated in Figure 10b. For Sp2, it can be seen that the crack at the core occurs outside the steel channel (see Figure 11).

There are some local failures of the specimen Sp3 during the bending test (LC2). The test was stopped at around 170 mm deflection and 18.0 kN load at point C, when a sudden core failure occurred. The force keeps raising until the bonding of steel-upper skin fails at the backside. Through a combined action the debonding of steel-lower skin at the front side was in progress. It is followed by a progressive steel-upper skin debonding along the width of the specimen (right and top side). A local core shear failure occurs, and it consists in a core crack, aligned with the core discontinuities, followed by upper skin-core debonding at the front side.

Once the steel and sandwich panel are no longer fully-jointed, their effectiveness in carrying lateral loads drops. It should be mentioned that Balsa core panels are manufactured with small rectangular pieces of balsa, glued together. When testing a panel, there will inevitably exists some discontinuities in the core, along its length. These discontinuities are weak points, and will induce shear failure (Castilho et al. 2015). The location of the core shear at the end of the test is around the corner of the steel channel (see Figure 12).

Sp1 and Sp2, which had been tested and failed under shear loading (see Figure 5) were tested under bending load in another cycle (see Figure 6). Their stiffness is approximately identical but the ultimate loads are different. The maximum force for both was about 27 kN at the first cycle (shear loading) (see Table 2). But the ultimate value for Sp2 is about twice Sp1 at the second cycle (see Figures 13 and 14).

However, the failure is progressive in Sp2 and it deflects more than Sp1. It seems that the location of the core shear in Sp2 which is out of the steel-composite overlap region causes in increasing the ultimate strength for Sp2 (see Figures. 15 and 16).

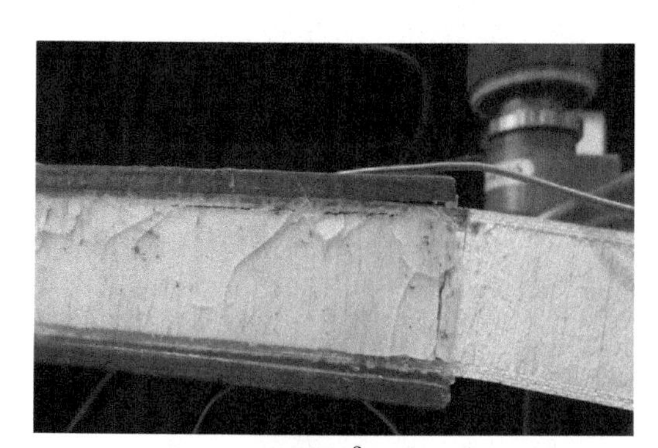

a

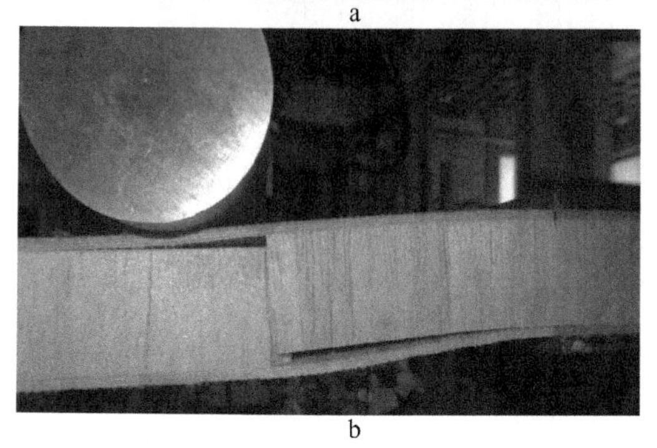

b

Figure 10. Sp1under shear load in the 1st cycle: a) core shear and debonding (front view), b) core shear (back view).

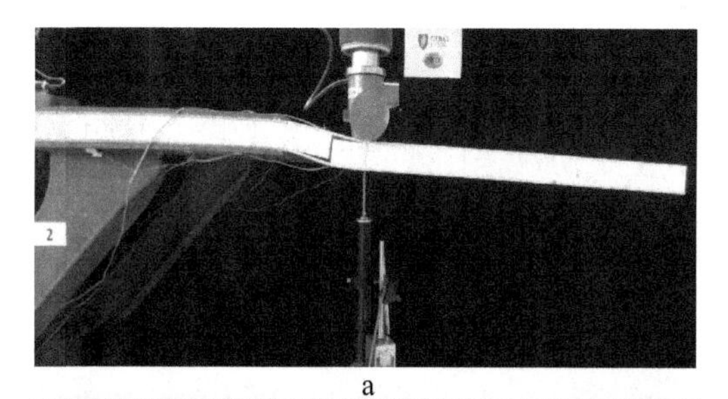

a

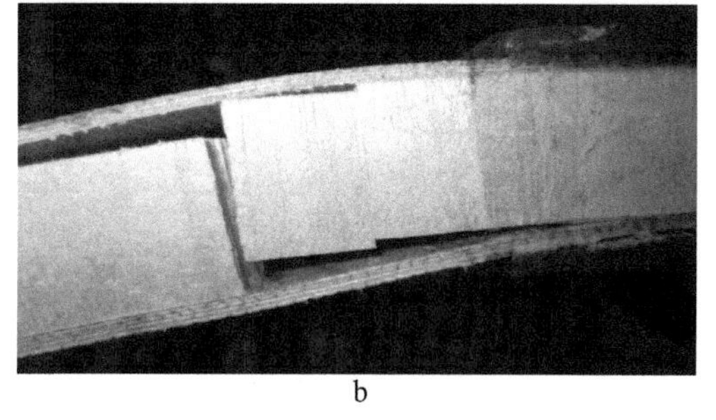

b

Figure 11. Sp2 under shear load in the 1st cycle: a) core shear and debonding (front view), b) core shear (back view).

a

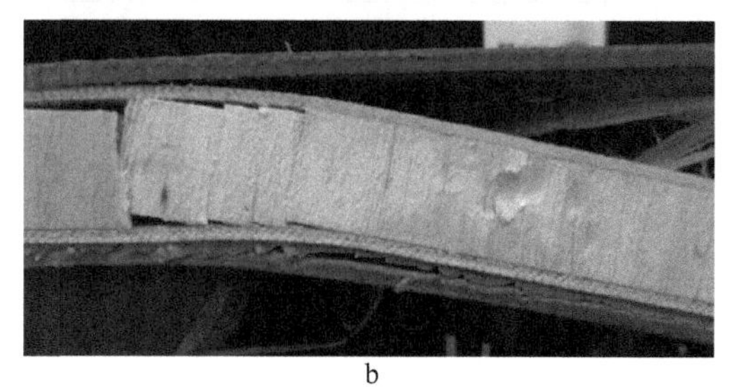

b

Figure 12. Sp3 under the ultimate bending load: a) core shear (front view), b) core shear and debonding (front view).

Figure 16 demonstrates the progressive failure in Sp2 at the second load cycle. Core shear results in skin-core delamination. This delamination develops until the whole of the skin separates from the core (see Figure 17). Figure 18 shows the path of the shear progress along the width of the core. As it can be seen this path is not regular and not aligned with the core block bond.

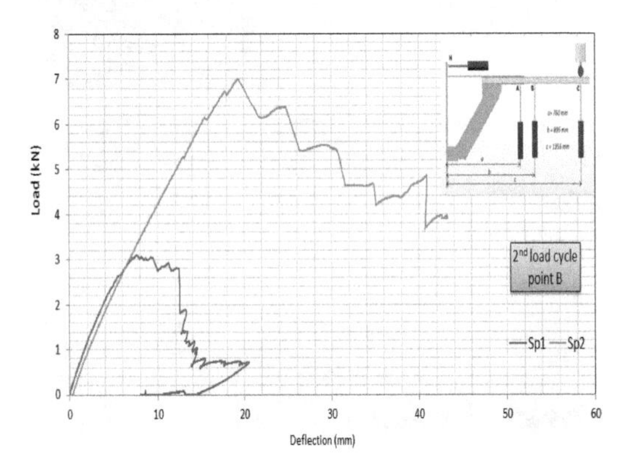

Figure 13. Load-deflection graph at the second load cycle for all of the specimens (point B).

Table 2. Comparison of the stiffness, maximum force and the maximum deflection of all of the specimens at the 1st load cycle.

	Stiffness [kN/mm]	Max. load [kN]	Max. deflection [mm]
Position	B		B
Sp1	2.3	26	13
Sp2	3.1	27	25
Sp3	–	18	55

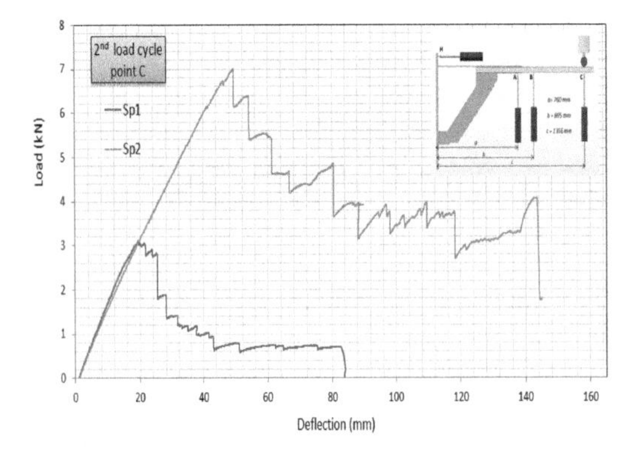

Figure 14. Load-deflection graph at the second load cycle for all of the specimens (point C).

a

b

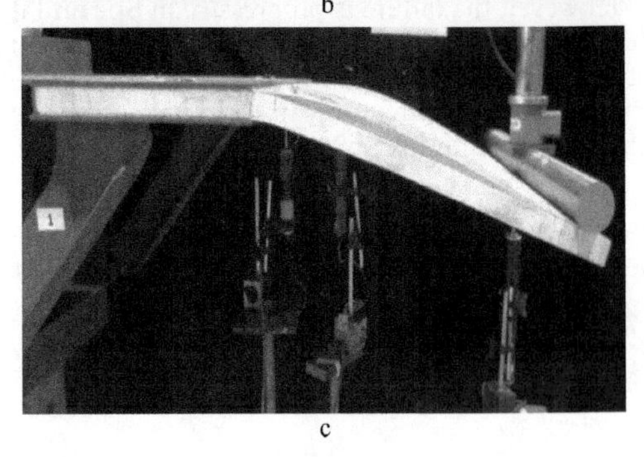

c

Figure 15. Sp1under bending load at the 2nd cycle: a) core shear and debonding (front view), b) debonding (back view), c) fully debonded under maximum deflection (front view).

The ratio of the ultimate load in shear condition to the bending one at the first load cycle is 1.5. Also, the ultimate strength in bending condition at the second cycle decreases about 70 percent for the failed specimen (see Table 3). Whereas the stiffness reduces only 16 percent. The maximum deflection value is 166 mm in bending around the width edge of the balcony (point C). This value for the steel support is 31 mm and it should be mentioned that the plastic deformation of the steel support in point "A" is 0.18 mm for Sp1 and Sp2 after the second load cycle. The specimens were failed under shear at the first load cycle have the same stiffness about 0.17 kN/mm at the second one (see Table 3).

Figure 16. Progressive core shear and debonding in front view for Sp2 under bending load at the 2nd cycle.

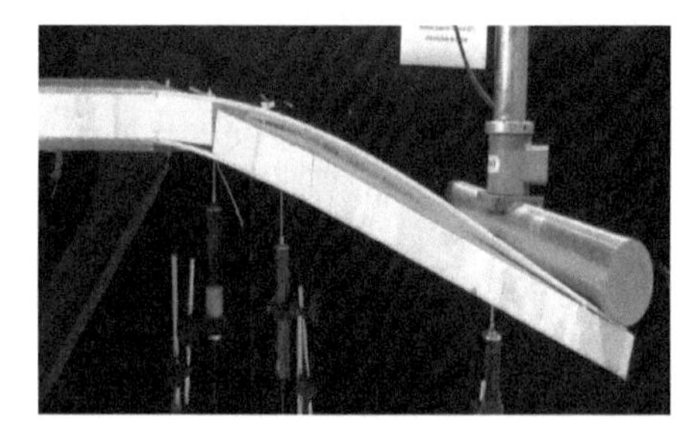

Figure 17. Core shear and debonding for Sp2 under bending load at the 2nd cycle (front view).

Table 3. Comparison of the stiffness, maximum force and maximum deflection in bending condition of Sp1 and Sp2 at the 2nd load cycle and Sp3 at the 1st load cycle.

	Stiffness [kN/mm]	Max. load [kN]	Max. deflection [mm]		
Position	C	C	A	B	C
Sp1	0.18	3	1.6	7	20
Sp2	0.16	7	8	19	49
Sp3	0.23	18	31	55	166

Figure 18. Core shear progress and debonding for Sp2 under bending load in the 2nd cycle (back view).

4 CONCLUSIONS

A damaged hybrid steel-FRP balcony overhang of ship was investigated experimentally to determine the post collapse residual strength. It was demonstrated that at the second cycle, the balcony has only one-third of its bending load capacity of the first cycle due to core shear failure at the end of the first one. Although the stiffness remains unchanged, 70 percent reduction of ultimate strength can be observed. Also, the location of the core shear failure at the first load cycle plays a significant role in determination of the residual ultimate strength of the balcony after initial failure. If the crack is located outside the steel channel the failure can be progressive due to skin-core failure. It should be mentioned that the crack of core shear does not necessarily occur along the block bond of Balsa. Finally, the stiffness of the structure is identical at both cycles (shear and bending) because it essentially depends on the steel channel elasticity and its brackets which have a negligible maximum deflection due to plasticity at the end of the second cycle.

ACKNOWLEDGEMENTS

This work has been done within the project 'Materials Onboard: Steel Advancement and Integrated Composites—MOSAIC', (www.mosaicships.com), which was partially funded by the European Community's Seventh Framework Program under grant No. 314037.

REFERENCES

Cao, J. & Grenestedt, J.L. 2003. Test of a redesigned glass-fiber reinforced vinyl ester to steel joint for use between a naval GRP superstructure and a steel hull. *Composite Structures* 60: 439–445.

Cao, J. & Grenestedt, J.L. 2004. Design and testing of joints for composite sandwich/steel hybrid ship hulls. *Composites: Part A* 35: 1091–1105.
Castilho, T., Sutherland, L.S. & Guedes Soares, C. 2015. Impact resistance of marine sandwich composites. *Maritime Technology and Engineering 3*, Guedes Soares, C. & Santos, T.A. (Eds) Taylor & Francis Group, London, 607–617.
Chen, Z., Li, D., Li, Y. & Feng, Q. 2014. Damage analysis of FRP/steel composite plates using acoustic emission. *Pacific Science Review* 16: 193–200.
Clifford, S.M., Manger, C.I.C & Clyne, T.W. 2002. Characterisation of a glass-fibre reinforced vinylester to steel joint for use between a naval GRP superstructure and a steel hull. *Composite Structures* 57: 59–66.
Hentinen, Markku, Hildebrand, Martin, Visuri & Maunu. 1997. Adhesively bonded joints between FRP sandwich and metal. Different concepts and their strength behaviour. Espoo Technical Research Centre of Finland, VTT Tiedotteita—Meddelanden—Research Notes 1862: 44.
Jiang, X., Kolstein, M.H. & Bijlaard, F.S.K. 2014. Experimental and numerical study on mechanical behavior of an adhesively-bonded joint of FRP–steel composite bridge under shear loading. *Composite Structures* 108: 387–399.
Jiang, X., Kolstein, M.H., Bijlaard, F.S.K. & Qiang, X. 2015. Experimental investigation on mechanical behavior of FRP-to-steel adhesively-bonded joint under combined loading-part 1: Before hygrothermal aging. *Composite Structures* 125: 672–686.
Kharghani, N., Guedes Soares, C. & Milat, A. 2015. Analysis of the stress distribution in a composite to steel joint. *Maritime Technology and Engineering 3*, Guedes Soares, C. & Santos, T.A. (Eds) Taylor & Francis Group, London, 619–626.
Kharghani, N. & Guedes Soares, C. 2016. Effect of uncertainty in the geometry and material properties on the post-buckling behaviour of a composite laminate. *Maritime Technology and Engineering 3*, Guedes Soares, C. & Santos, T.A. (Eds) Taylor & Francis Group, London, 497–503.
Kharghani, N. & Guedes Soares, C. 2018. Experimental and numerical study of hybrid steel-FRP balcony overhang of ships under shear and bending. Submitted to *Marine Structures*.
Kotsidis, E.A., Kouloukouras, I.G. & Tsouvalis, N.G. 2015. Finite element parametric study of a composite-to-steel-joint. *Maritime Technology and Engineering*, Guedes Soares, C. & Santos, T.A. (Eds), Taylor & Francis Group, London, 627–635.
Li, X., Li, P., Lin, Z. & Yang, D. 2015. Mechanical Behavior of a Glass-fiber Reinforced Composite to Steel Joint for Ships. *J. Marine Sci*. Appl. 14: 39–45.
Mouritz, A.P., Gellert, E., Burchill, P. & Challis K. 2001. Review of advanced composite structures for naval ships and submarines. *Composite Structure* 53:21–41.

Shipyard technology

Progress in Maritime Technology and Engineering – Guedes Soares & Santos (Eds)
© 2018 Taylor & Francis Group, London, ISBN 978-1-138-58539-3

Model to forecast times and costs of cutting, assembling and welding stages of construction of ship blocks

A. Oliveira & J.M. Gordo
Centre for Marine Technology and Ocean Engineering (CENTEC), Instituto Superior Técnico, Universidade de Lisboa, Lisbon, Portugal

ABSTRACT: The international competition in shipbuilding is a subject that weighs heavily in new ships orders and leads to the study about the feasibility of adopting new technologies and processes in the production flow, in order to respond to the current challenges. Thus, a program was developed to forecast times and costs in the construction processes stages of ship blocks in a shipbuilding yard, allowing the simulation of implementation of alternative cutting and welding technologies. The main goal of this study is to understand the relation between operational and labor costs in various types of cutting and welding technologies, and the potential earnings related to cost savings in downstream stages of the production flow due to the application of higher quality technologies in upstream processes. The times and cost values computed by the developed algorithm grant a deeper understanding of the consequences of the adoption of alternative shipbuilding technologies in the productive process.

1 INTRODUCTION

The construction by blocks is the most reliable production scheme in a construction shipyard in order to achieve a more cost-effective production, with simultaneous increase on the quality of the processes, and is vastly accepted that is undoubtedly the today's mainstream scheme of ship construction (Storch, et al., 2007).

The block construction is today a well-defined sequence of stages, according with the type and characteristics of the block, as illustrated in Figure 1.

It is important to stress that in the present study the pre-outfitting activities of the block were not considered, although they are an important strategy to contribute to a more cost-effective production process.

For each stage of the sequence shown in Figure 1 there is a different set of available technologies and techniques. One can exemplify with the current available technologies for the cutting process. Either for the steel plate or for the frames cutting stage, different possibilities are currently available: Oxy-fuel cutting, plasma cutting, laser cutting and abrasive water-jet cutting (Oliveira & Gordo, 2018). The same principles apply for example on the several structural levels, from small complex pieces to final block construction stage, where many different welding techniques can be used, from electrodes to the newest welding technologies, such as laser beam welding or plasma welding (Gordo, et al., 2006).

In way to understand the implications of the implementation of different options of cutting and welding technologies in the block construction flow process, several studies were conducted

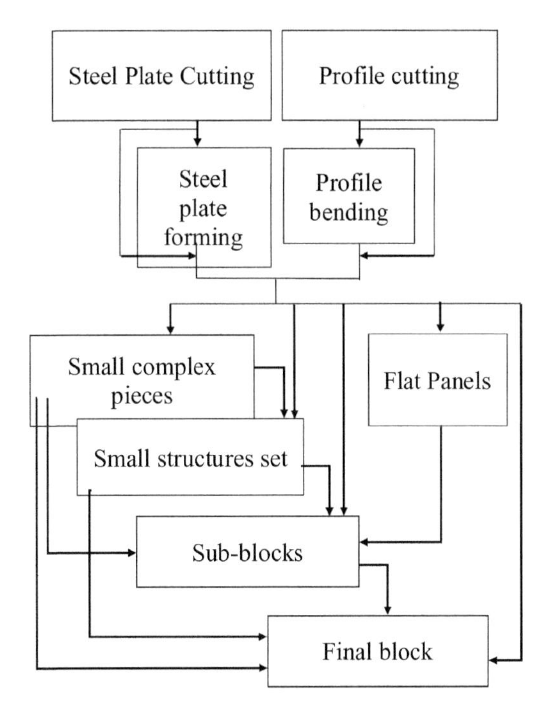

Figure 1. Block production sequence main stages scheme.

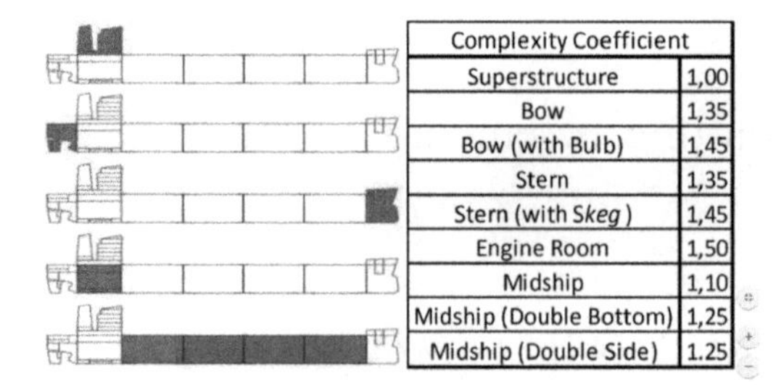

Complexity Coefficient	
Superstructure	1,00
Bow	1,35
Bow (with Bulb)	1,45
Stern	1,35
Stern (with *Skeg*)	1,45
Engine Room	1,50
Midship	1,10
Midship (Double Bottom)	1,25
Midship (Double Side)	1.25

Figure 2. Block complexity coefficient (Leal & Gordo, 2017).

to analyze the consequences of the variation of the production's time and cost parameters (Leal & Gordo, 2017). Also, by implementing and developing simulation tools, other set of studies were conducted to obtain a better understanding of the production flow when faced with different production options (Ljubenkov, et al., 2008) (Oliveira & Gordo, 2018). In this way, several studies have proved gains in the construction process, for example in double bottom blocks (Ozkok & Helvacioglu, 2013). Also, by applying lean tools, several studies had proven positive results on the manufacturing processes (Kolich, et al., 2016), hence stressing the importance for a careful manufacturing planning.

In way to obtain a reliable set of results on the construction process, it is important to specify as well as possible, not only the production process, like cutting or welding processes, but also the block that is being analyzed. For the understanding of one of the present study's main goals, it is key to realize that in many past parametric studies on the block production process, the block is characterized only by a small set of values, according with the block type and dimensions, as shown in the Figure 2:

The present study aims to avoid the use of the above values shown in Figure 2, by developing a model where the user set the characteristics of the production process and a deeper characterization of the block which production cost and times one wants to analyze.

2 PRODUCTION ANALYSIS MODEL

2.1 *Data flow*

The developed model aims to conduct a more reliable analysis of the production process of the block in the shipbuilding yard, and for that one of the key features is the consideration of all the steel pieces which form the block.

Through a set of graphic interfaces, the user should be able to define the block's pieces, as also the shipyard processes specifications. Hence, as shown in Figure 3, the model's input arguments are the block's pieces characteristics and the production process specification. Through a series of computations, the model creates automatically a set of PDF files with the times and costs of the main construction stages, as well as a Microsoft Project file with the flow production.

The characterization stage of the block's pieces is realized through the Rhinoceros CAD program. According with a standardized way of definition of the characteristics of each block piece, the user defines those characteristics in the Rhinoceros program. The characteristics defined by the user are divided in several values, which comprehend values that deal with:

- Piece type;
- Piece dimensions;
- Lengths of cutting and welding;
- Level of possible bending;
- Stage of block construction to which belongs.

The main menu of the developed model presents an option that allows to update the Rhinoceros file with a different block or with an actualized block pieces' characteristic.

The shipyard block construction processes specifications are defined by the user through a set of graphic interfaces of the developed software.

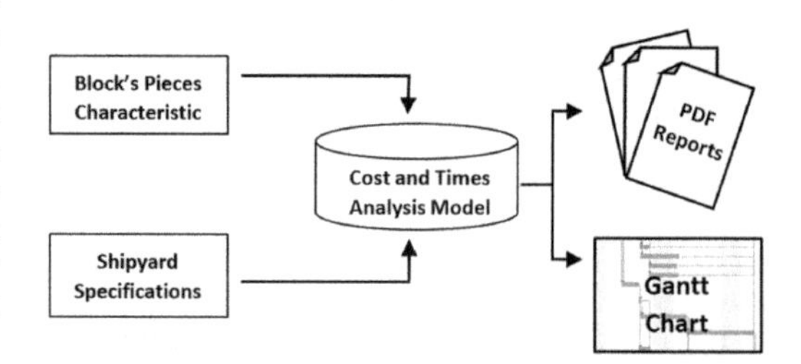

Figure 3. Model data flow.

Figure 4. Main menu model's graphic interface.

The production specifications are divided in the following sets of data input, each one with their own graphic interfaces:

Cutting Processes { Profiles Cutting; Steel Plates Cutting }

Forming Processes { Profiles Bending; Steel Plates Bending; Steel Plates Forming }

Assembly and Welding Processes { Small Complex Pieces; Small Structures; Flat Panels; Sub-blocks; Final Block }

The data defined in each of the production stages presented above is related with the process type, but can be summarized in the following set of values:

- Process speeds and times;
- Technologies used;
- Number of active workstations;
- Number of workers needed.

In the shipyard specification phase of the values input activities is also included the specification of the costs of the several processes, according with the type of technology and equipment used, as shown in the Figure 5.

The costs definition also depends strongly on the type of technology and equipment one is defining, but can also be summarized in the following set of values:

- Consumables flow rate;
- Electricity consumption;
- Equipment Depreciation;
- Wage of the technician worker.

For a more reliable construction process analysis, the user should also define the production flow of the several structures levels that made up the block, as shown in Figure 6.

The flow production sequence definition is defined through a graphic interface, as shown in Figure 7.

In the flow sequence definition, as shown above, the program presents, in the two columns of the left, a certain structure A, and, in the two columns of the right, the destination structure to which the structure A will be joined.

After defining the block and shipyard characteristics, as displayed in the present paper chapter, the needed data is completed and the computations of the cost and time analysis can be initiated.

2.2 *Model algorithm computations*

Considering the specifications set by the user presented in the previous chapter, the developed model computes the cost and times analysis of each one of the main stages of the steel block construction. The sequence in which the algorithm run can be illustrated in the flow shown in Figure 8:

Although it is not feasible to present in this paper all the formulas used for the computations of the several block construction stages, attending that each main construction stage got his own sequence of activities, treated each one individually, is useful for a better comprehension to exemplify with one of the several activities of one given stage, for example, the automatic cutting activity of the profile cutting stage.

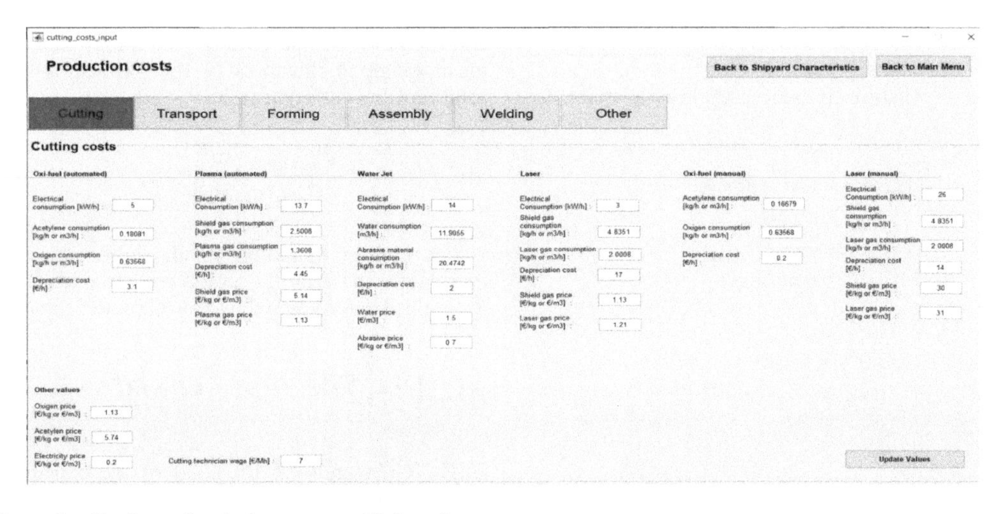

Figure 5. Cutting technologies costs graphic interface.

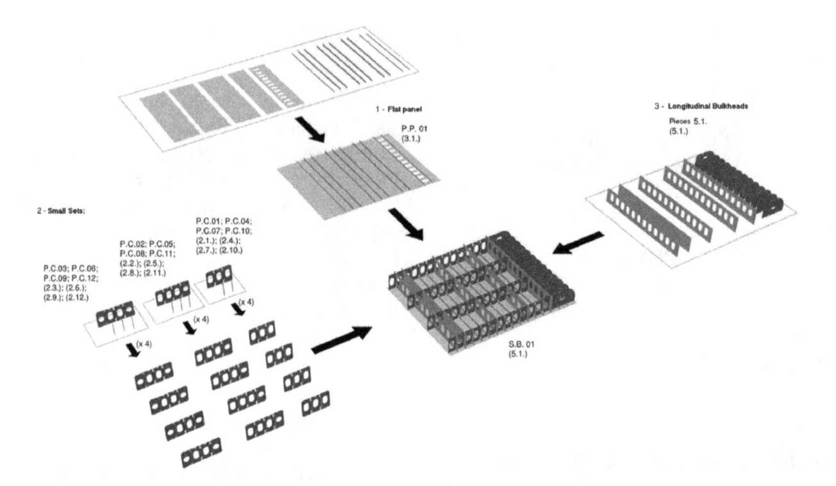

Figure 6. Block construction sequence.

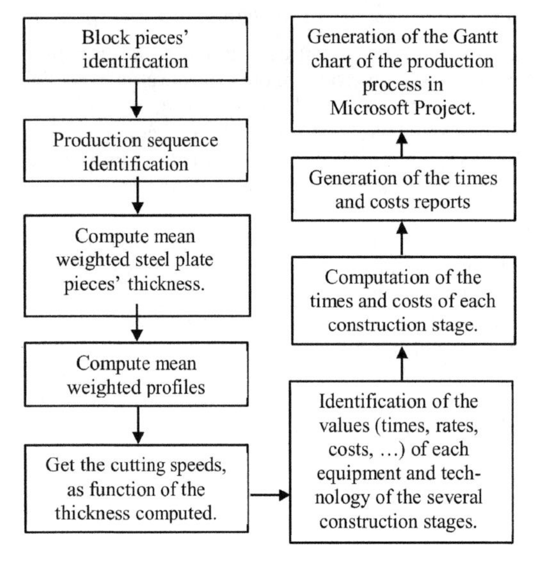

Figure 7. Construction sequence graphic interface.

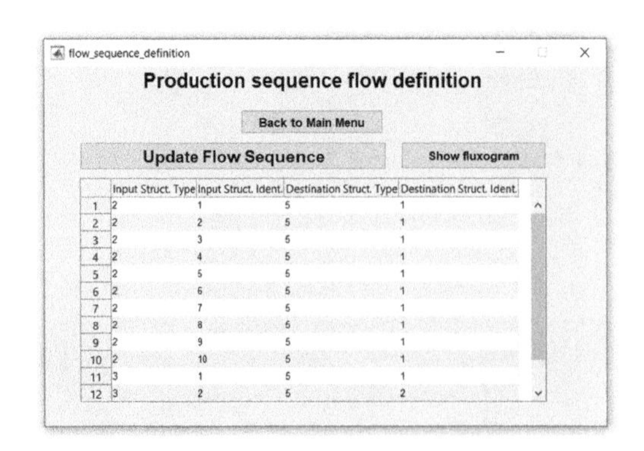

Figure 8. Algorithm flow chart.

After the work distribution of the profiles by each active profile cutting equipment is computed so that the stage is performed in the minimum possible time, the following set of activities are considered and analyzed:

- Location of the profile to cut;
- Transport and positioning of the profile before cut;
- Cut preparation;
- Automatic marking;
- Automatic cutting;
- Manual Cutting
- Dimensional control;
- Manual marking;
- Transport of the profile after cut.

The present example deals with the automatic cutting stage. The time need for this stage is computed simply applying the cutting speed, AC_{speed} [m/min], which was previous calculated through consideration of the cutting technology and the mean weighted thickness of the profiles, on the total profile's cutting length, $P_{cutting\ length}$[m]:

$$AC_{time}[\text{min}] = \frac{P_{cutting\ length}[\text{m}]}{AC_{speed}[\text{m/min}]} \tag{1}$$

The cost computation of the automatic cutting activity in the profiles cutting construction stage if obtained considering the following sum of items:

$$AC_{cost}[€] = \left(\sum_{1}^{n} CC_n\right) + DC + LC + EC \tag{2}$$

The CC stands for the costs of the consumables, according with the type of cutting technology defined, and is computed in the following way:

$$CC_n[€] = CR_n[\text{un}/\text{h}] \times \frac{AC_{time}[\text{min}]}{60} \times CR_n[€/\text{un}] \quad (3)$$

where:
CR_n – Consumable n flow rate [un/h];
CR_n – Consumable n specific price [€/un].

The unit [un] can stand usually for [m^3] or [l], according with the type of consumable.

The DC value of formula (2) stands for the depreciation cost of the automatic cutting equipment and is easily computed through:

$$CD[€] = ACE_{dep}[€/\text{h}] \times \frac{AC_{time}[\text{min}]}{60} \quad (4)$$

where
ACE_{dep} – Automatic cutting equipment depreciation rate cost [€/h].

The LC value of the formula () stands for the labor costs and is obtained by applying:

$$LC[€] = NW \times WW[€/\text{h}] \times \frac{AC_{time}[\text{min}]}{60} \quad (5)$$

where
NW – Number of workers needed in this cutting activity phase;
WW – Worker wage [€/h].

The last item of the formula (2) concerns the electricity cost and is computed through the following formula:

$$EC[€] = PC[\text{W}] \times \frac{AC_{time}[\text{min}]}{60} \times PP[€/\text{W.h}] \quad (6)$$

where
PC – Power of the equipment [W], which can be specified also with the aim of a power efficiency value, P_{eff}[%];
PP – Power specific price $[€/\text{W.h}]$.

Is important to stress that all the above values are defined by the user either in the block pieces' definitions stage or in the shipyard characteristics set stage.

2.3 *Model validation*

The validation process of the developed model was conducted in two distinct fronts: The first concerning the cost values computed and the second concerning the time values.

For both validation processes were used two blocks, as shown in the Figures 9 and 10:

The Block A, shown in Figure 4, belongs to a pontoon with a length of 20.78 m, breadth of 9.86 m and depth of 1.42 m, with 41 ton. The Block B, illustrated in Figure 5, is a mid-ship double bottom block of a chemical tanker, with a length of 10 m, breadth of 13 m and depth of 1.5 m, with 47 ton.

The construction sequence of Block A was performed through the assembling of 8 small structures and 2 flat panels. The combination of one sub-block and the second flat panel finish the block construction sequence. The Block A is made up of 277 steel plate pieces and 192 steel profiles. There is a total of 959 meters of steel cutting work.

Figure 9. Block A – pontoon block.

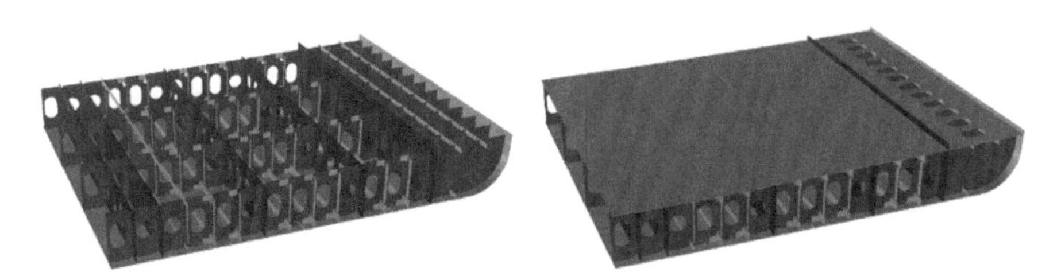

Figure 10. Block B – double bottom block.

The steel plate pieces present a range of thickness 8 mm to 6 mm.

The construction sequence of Block B was performing through the assembling of 12 small structures and 2 flat panels. In a similar way of the Block A, the combination of one sub-block and the second one flat panel finish the block construction sequence. The Block B is made up of 138 steel plate pieces and 19 steel profiles. There is a total of 1132 meters of steel cutting work. The steel plate pieces present a range of thickness 10 mm to 15 mm.

2.3.1 *Cost analysis validation*

The cost analysis computed values were validated through the application and comparison with the model developed by Gordo & Leal (2018).

Both blocks A and B were run in the Leal & Gordo model, as well as a production characteristics set as similar as possible to the one characterized in the model developed in the present study when running the production process also for both blocks. The values obtained in both models are presented in Table 1.

Although a different way of study approach, where Leal & Gordo's model specify the block through a set of values, like the ones presented in Figure 2 and the developed model of the present paper requires a deeper specification of the block, it is acceptable to compare the results obtained.

Although some disparity on the values of the Block B, which can be explained by the lower structural complexity it would be expected for a double bottom block, the values presented in the Table 1 allow to validate the cost analysis values of the developed model.

2.3.2 *Time analysis validation*

To conduct the validation process of the time values computed by the developed model, was used real case data given by a construction shipyard in Portugal, WestSea S.A., hereby called WS, of four blocks (i, j, l, k) of a ship construction project.

Table 1. Cost analysis validation Values. values in [€].

		Leal and Gordo's model	Present model
Block A	Cutting	2387	2820
	Assembly and welding	10776	9949
Block B	Cutting	5066	11549
	Forming	570	1214
	Assembly and welding	1265	3388

The four blocks manufactured in the WestSea Shipyard (Block i, j, k and l) are blocks of military ships, hence it was not feasible for the Shipyard to give more detailed information rather than the total weight of the block's steel plate pieces and profiles, and the main block construction stages times.

Several block construction stages were studied separately, and the ratios [ton/day] were computed and compared with the obtained applying Blocks A and B in the developed model, as shown in Tables 2 to 6.

The validation process through the comparing method of the ratio values is done only with situations where there is a similar block characteristic, hence not all the blocks appear in all tables. Is also important to stress that the shipyard characteristics defined in the developed program were set to be as similar as possible to the ones of the WS shipyard during the construction process of the four blocks. For example, regarding the welding technologies, the butt welds performed in the plate blanket construction stage of the panel line are performed through one side automated submerged

Table 2. Profiles cutting time validation values.

Block	$\frac{\textit{Profiles weigth}\,[ton]}{\textit{Cutting time}\,[days]}$
Block *l* (WS)	0.62
Block *k* (WS)	0.64
Block A (model)	0.58
Block B (model)	3.89

Table 3. Steel plate cutting time validation values.

Block	$\frac{\textit{Steel plate pieces weigth}\,[ton]}{\textit{Cutting time}\,[days]}$
Block *i* (WS)	8.46
Block *j* (WS)	2.38
Block *k* (WS)	3.05
Block *l* (WS)	6.67
Block A (model)	4.31
Block B (model)	8.30

Table 4. Plate forming time validation values.

Block	$\frac{\textit{Block weigth}\,[ton]}{\textit{Plate forming time}\,[days]}$
Block *i* (WS)	6.4
Block *j* (WS)	6.4
Block B (model)	11.9

Table 5. Flat panels construction time validation values.

Block	$\frac{Block\ weigth\ [ton]}{Flat\ panels\ time\ [days]}$
Block *i* (WS)	1.75
Block *j* (WS)	2.35
Block *k* (WS)	2.97
Block A (model)	2.27
Block B (model)	7.97

Table 6. Global block construction time validation values.

Block	$\frac{Block\ weigth\ [ton]}{Block\ construction\ time\ [days]}$
Block *i* (WS)	0.28
Block *j* (WS)	0.44
Block *k* (WS)	0.61
Block *l* (WS)	0.16
Block A (model)	0.41
Block B (model)	0.67

arc welding, and all the other welding works are performed through manual flux cored arc welding.

The time ratios obtained allow to validate the algorithms of the time analysis conducted by the developed model. Hence, validated the time and cost analysis, the model is validated.

2.4 *Alternative cutting technologies implementation study*

Validated the developed program is possible to undergo on a simple simulation of alternative production processes implementations. The chosen process to study on the simulation hereby presented relates to the study of implementation of alternative automatic cutting technologies. The four situations studied were:

- Situation 1 – All the cutting processes are performed through oxy-fuel cutting;
- Situation 2 – The panel line cutting stage and profiles cutting are execute with oxy-fuel, and the steel plates cutting process, to generate pieces, is performed by plasma cutting. This is actually the most similar situation when compared to the actual WesSea Shipyards S.A. production process.
- Situation 3 – All the steel cutting processes are performed through laser technology;
- Situation 4 – All the steel cutting processes are performed through abrasive water jet technology.

The time and cost values obtained by running the developed program are presented in Table 7.

It is important to state that the calculations were conducted in a way that the differences present in the time and cost values of the various assembly and welding phases, are only justified due to the decrease of the gridding stage, resulting from the increase of cutting quality that each cutting technology allows. The analysis of the results allows to comprehend that although some decrease in the assembly and welding cost values is obtained, mainly due to cost savings by the reduction of the man-hours needed in the gridding process, the global saving is not so large as one would expect, reaching, at most 5%. This cost saving due to the reduction of work in the gridding phase do not justify per himself the increase of the cost of the cutting technologies with better cutting quality. However, is important to stress that a better cutting quality also allows important improvements in the dimensional control, decreasing possible reworks or corrections in the assembly and welding stages, although those savings are hard to estimate and, by that reason, were not consider in the developed software tool.

If one considers only the analysis on the cutting costs, the values obtained are in line with the actual shipbuilding industry, where the plasma cutting is the most attractive technology. The cutting speed of the plasma saves precious man-hours, hence balancing its higher operational costs when compared to the oxy-fuel technology and even obtaining cost savings.

As expected, the high operational costs of the laser and the low cutting speed of the waterjet cutting do not yet allow to implement economically that technologies on the ship production process at large-scale.

Although assembling and welding values are here shown together, it is possible and interesting to exemplify its time ratio, i.e., the time ratio of the assembling related works *vs* the time ratio

Table 7. Alternative cutting technologies scenarios.

		Cutting processes		Assembly and welding processes	
Block	Situation	Time [days]	Cost [€]	Time [days]	Cost [€]
A	1	24.5	3017	109	9310
	2	21.0	2656	108	9079
	3	19.9	4469	107	8964
	4	29.0	8371	107	8906
B	1	11.4	1836	81	5414
	2	7.3	1428	81	5268
	3	7.1	1965	81	5181
	4	16.9	6917	78	5142

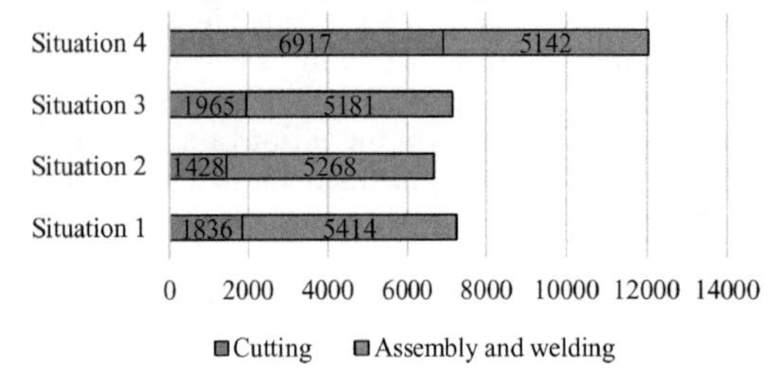

Figure 11. Cutting costs, in [€], of the implementation of different cutting technologies in block B.

Table 8. Assembling and welding related works ratios, of Block A, for oxy-fuel cutting and water Jet cutting situations.

	Oxy-fuel cutting		Water jet cutting	
	Assembling related works	Welding related works	Assembling related works	Welding related works
[min]	21202	18539	19346	18539
[%]	53.4	46.6	51.0	49.0

of the welding related works. Illustrated those ratios to the Block A by applying the developed model, with similar shipyard characteristics to the WestSea Shipyard's, one has the results presented on Table 8.

As expected, the ratio gap between the assembling related works and the welding related works increases with a higher cutting quality.

3 CONCLUSIONS

The main goal of the developed model was to prove the reliability of the implementation of an alternative approach in the block construction process study. Through a validation process was considered that such approach, implemented in the developed program, is reliable. Further studies are needed to prove higher quality level of the values computed considering the approach here chose when compared to the more classical and less detailed approach, like the ones illustrated in Figure 2.

To achieve more reliable results, further work is needed to be conducted in the developed software, mainly in the assembling and welding activities. The consideration of the weld dimensions according with the base metal thickness, as well as a deeper characterization of the consumable rates of the welding processes are some examples of future works to accomplish and implement in a better way in the developed model.

In addition to the goal of proving the reliability of this type of block production analysis, the model aims also to serve as a tool to understand the consequences of the implementation of alternative and more recent process technologies. Although the cutting technologies, even the newer, are quite well recorded in existing studies, the more recent welding technologies have not yet well defined published parameters of consumes rates. Such data is fundamental to perform a reliable analysis, and that was the reason why the authors only simulated alternative scenarios with different cutting technologies.

ACKNOWLEDGEMENTS

This paper reports a work partially developed in the project "Ship Lifecycle Software Solutions", (SHIPLYS), which was partially financed by the European Union through the Contract No 690770 - SHIPLYS—H2020-MG-2014–2015.

REFERENCES

Gordo, J. & Leal, M., 2018. A tool for analysis of costs on the manufacturing of the hull. In: G. Soares & A. Teixeira, eds. *Maritime Transportation and Harvesting of Sea Resources.* Lisbon: Taylor & Francis Group, pp. 743–748.

Gordo, J.M., Carvalho, I. & Guedes Soares, C., 2006. Potencialidades de processos tecnológicos avançados de corte e união de aço em reparação naval. In: C. Guedes Soares & V. Gonçalves Brito, eds. *Inovação e Desenvolvimento nas Actividades Marítimas.* Lisboa: s.n., pp. 877–890.

Kolich, D., Storch, R.L. & Fafandjel, N., 2016. *Optimizing Shipyard Interim Product Assembly Using a Value Stream Mapping Methodology.* Rhode Island, USA, World Maritime Technology Conference.

Leal, M. & Gordo, J.M., 2017. Hull's manufacturing cost structure. *Shipbuilding,* 68(3), pp. 1–24.

Ljubenkov, B., Dukié, G. & Kuzmanié, M., 2008. Simulation Methods in Shipbuilding Process Design. *Journal of Mechanical Engineering,* Volume 54, pp. 131–139.

Oliveira, A. & Gordo, J., 2018. Cutting processes in shipbuilding – a case study. In: G. Soares & A. Teixeira, eds. *Maritime Transportation and Harvesting of Sea Resources.* Lisbon: Taylor & Francis Group, pp. 757–762.

Oliveira, A. & Gordo, J., 2018. Implementation of new production processes in panel's line. In: G. Soares & A. Teixeira, eds. *Maritime Transportation and Harvesting of Sea Resources.* Lisbon: Taylor & Francis Group, pp. 763–773.

Ozkok, M. & Helvacioglu, I.H., 2013. A Continuous Process Improvement Application in Shipbuilding. *Brodogradnja,* 64(1), pp. 31–39.

Storch, R.L., Hammon, C.P., Bunch, H.M. & Moore, R.C., 2007. *Ship Production.* 2nd ed. s.l.:SNAME.

Progress in Maritime Technology and Engineering – Guedes Soares & Santos (Eds)
 ISBN 978-1-138-58539-3

Causal analysis of accidents at work in a shipyard complemented with Bayesian nets modelling

B. Costa & C. Jacinto
UNIDEMI, Research Unit of Mechanical and Industrial Engineering, Universidade Nova de Lisboa, Lisbon, Portugal

A.P. Teixeira & C. Guedes Soares
Centre for Marine Technology and Ocean Engineering (CENTEC), Instituto Superior Técnico, Universidade de Lisboa, Lisbon, Portugal

ABSTRACT: This paper analyses the causality of occupational accidents occurred in a Portuguese shipyard during two years, 2014 and 2015. The objective is to identify not only the immediate causes of the accidents, but also their underlying causal factors at the organizational level, aiming to identify opportunities for improvement and to reduce accident rates in the shipyard. A second objective of the paper is to explore a way of modelling, quantitatively, the system studied from the causation point of view (cause-to-effect relationships). The method for Recording, Investigation and Analysis of Accidents at Work (RIAAT) is adopted to achieve the first goal. A Bayesian Network model of the causal relationships and consequences of the occupational accidents is then developed and used to simulate and analyse quantitatively different scenarios. A tangible plan of action is proposed, in which three categories of preventive measures are suggested: Engineering, Training & Awareness and Management & Control measures.

1 INTRODUCTION

This work was developed in a traditional Portuguese shipyard environment. It belongs to the general manufacturing industry and its principal activities are the construction, repair and maintenance of ships. The Economic Activity of this shipyard is classified in Section C, Division 33.15 according to code NACE Rev.2 (Eurostat, 2008).

The activities carried out in any shipbuilding industry require a great control of accident risks as they involve several high-risk tasks and hazardous maintenance operations. Such hazardous operations have been reported, and their risks evaluated by means of different techniques (Jacinto & Silva, 2010; Fragiadakis et al., 2014; Fernandes & Crispim, 2016; Tsoukalas & Fragiadakis, 2016).

In developed countries, and within the European Union (EU), every employer has a legal duty to assess workplace risks, aiming at prevention planning and risk reduction. In addition, employers also have the obligation to record (and analyse) accidents that caused absence from work, as well as incidents (or dangerous occurrences) that could have been particularly serious from a safety perspective (e.g.: former EU-Directive 89/391/EEC, amended by EU-Directive 2007/30/CE and subsequent). Company records are particularly useful since they provide the primary information for analysing, studying and understanding accident main causes and causal relationships at different levels. This build-up of knowledge from accident information has been progressively recognised to have a fundamental impact on designing safety measures and stimulating organisational learning (Hovden et al., 2011; Lukic et al., 2012; Drupsteen & Guldenmund, 2014; Silva et al., 2017).

Accidents can be analysed individually, usually through an in-depth case-to-case approach, or by statistical analysis of aggregate results. In the first case, Reason's (1997) theory of accident causation is still quite popular. According to Reason (1997, p.1), "there are two types of accidents: those that happen to individuals and those that happen to organisations". The frequency of occurrence of individual accidents is much higher because there are many hazards in the workplace and also due to specific factors of the humans that promote human error, and hence promote the occurrence of active failures. One should bear in mind, however, that human factors are not only characteristic of the individual occurrence of accidents at work, since they are also very important in the complex organizational accidents.

The fallibility (potential to fail) is a human "unavoidable" feature but it is, nowadays, recognised that anyone at work can make mistakes, or violate procedures, for reasons that often go beyond the simple elementary logic (Reason 1997, p.10). In his accident causation model, Reason makes a clear distinction between *active failures* (which are directly associated with the accident) and *latent conditions* (or dormant failures), which are more difficult to identify for being hidden in the system. Reason's causation theory proposed two decades ago is presently still very popular with a widespread use.

Another way of studying accident causation is through the use of large accident datasets. In this case "accident rates" are frequently used to assess the damaging effects of "non-safety" and to search for patterns and trends. Such indicators can be interpreted as the numerical result of the consequences inherent to the risk of workplace activities. In a shipyard, these measures, e.g., "frequency", "severity" and "incidence" rates, tend to be high (Jacinto & Silva, 2010; Fragiadakis et al., 2014; Tsoukalas & Fragiadakis, 2016). The mentioned indicators are often used to characterise the situation at a given moment and allow managers to monitor the evolution of accident rates over time.

More recently, there have been a few attempts to use Bayesian Networks (BN) for modelling workplace accidents. A Bayesian analysis of accidents involving falls from height (stairs and scaffolds) has been proposed by Martín et al. (2009), who related fall accidents with factors such as, experience, task duration, training, knowledge of rules, or hazard perception. Likewise a Bayesian network analysis has been applied for modelling occupational risk in harsh offshore environments (Song et al., 2016). Other examples come, for instance, from Rivas et al. (2011) who used BN to explain and predict accidental events in mining environments, or from Abdat et al. (2014) who used Bayesian Networks in the metalwork industry for mapping accident scenarios.

Authors using Bayesian Networks frequently highlight the ability of such technique to show causal links between multiple variables (as opposed to most common traditional techniques), thus allowing a more realistic modelling. The information obtained gives a valuable input for risk management and accident prevention.

The available data in the shipyard under observation shows high accident rates, as it happens with many other shipyards. This led the authors to formulate the *key question* in this study: what are the particular causal factors (active failures and latent conditions) that most contribute to the accidents? This study aims at providing an answer to this question, focusing on causal analysis and accident modelling. It involved a detailed analysis of accidents occurring in a period of two years (2014 and 2015).

The objective is to identify ways of improving safety and, in particular, to reduce accident rates in the three most critical services (departments) of the company.

2 METHODOLOGY

As mentioned, the study covered a time span of two consecutive years (2014–2015). The methodology adopted was designed into three stages, depending on the specific purpose, as follows:

1. *Preliminary Diagnosis.* It involved colleting general information (number of accidents, hours worked, number of employees, total days lost per accident). This was used to calculate accident rates and to make a general diagnosis of the organisation as a whole. It also allowed pinpointing the most three critical Services (departments);
2. *Detailed causal analysis.* It involved gathering additional in-depth information on each accident individually (description of the accident, type of injury, workplace factors, etc.). This was carried out by application of RIAAT (Registration, Investigation and Analysis of Accidents) (Jacinto et al., 2011). This stage allowed to make an in-depth observational study following a case study approach (case-by-case analysis). Cause-effect relationships were established at this point between two variables: Deviation (immediate cause) versus Contact (type of accident, the effect).
3. *Modelling multivariate causal relationships.* This was achieved by application of Bayesian Networks to the previous sets of accident data from stage 2.

Stage 1 (Diagnosis) started by considering all accidents occurring in the company. Accident rates for frequency (I_f), gravity/severity (I_g) and incidence (I_i) were calculated according to the ILO (16th resolution of 1996), as follows:

$$I_f = \frac{\text{Num. of accid. involving 1 or more days lost}}{\text{Man-hours actually worked}} \times 10^6 \quad (1)$$

$$I_g = \frac{\text{Num. of lost days (every calendar day)}}{\text{Man-hours actually worked}} \times 10^6 \quad (2)$$

$$I_i = \frac{\text{Num. of accid. involving 1 or more days lost}}{\text{Nimber of workers}} \times 10^3 \quad (3)$$

Stage 2 (Detailed causal analysis) focused only on three critical Services identified in stage 1. This second analysis was further subdivided into two sets: the first part of RIAAT (elementary characterization of accident) was applied to all accidents occurred in the three Services selected (N_1 = 113 cases), while the remaining steps of RIAAT were applied only to a restricted sub-set (individual case studies) ($N_2 = 23$). This selection was based on the high frequency and(or) the severity of each accident, i.e., those which represented more lost days for the company.

Finally, in Stage 3 (Modelling), a specialised software (GeNie®) was used for modelling accidents through Bayesian Networks. This modelling tool may be seen as a practical way to assist managers to monitor accidents' causality in the company.

3 SHIPYARD AND GENERAL DIAGNOSIS

3.1 *The shipyard*

The company studied is a large shipyard in Portugal. Its main activity deals with repair and maintenance of vessels, i.e., large, medium and small ships (activity code NACE 33.15).

The company employed around 500 people in 2015 and a little more in 2014. The working force is composed mostly by male workers (94.4%) and the average age is 47.5 years. The age classes that include more workers are 40–44 years (21.8% of workforce) and 55–59 years (23.6% of workforce). Most workers have qualifications equal to or higher than the third cycle of basic education (≥ 9 years of school).

The organizational structure of the company is divided into four main Departments (D): Commercial (DC), Production (DP), Resources (DR) and Technical & Customer (DT). They report directly to the management board, the Council of Directors.

The Services (term used internally to designate each of the technical/operational business units) under DP (18 in total) are the basic units of the structure where most of the productive activities take place. Each Service is internally identified by a 4-letter code, as follows:

SVMT – Manoeuvring and Transportation (land);
SVCN – Metalwork & ship Construction;
SVSC – Civil Locksmiths;
SVCA – Carpentry;
SVSO – Welding;
SVME – Mechanics;
SVCT – Piping Workshop;
SVEA – Electronic and Automation;
SVMF – Machinery Tools;
SVTS – Surface Treatment;
SVRM – Maintenance;
SVTM – Torpedos, Missiles & Mines;
SVCO – Telecommunications, Giroelectrics, Radar;
SVEL – Electrical Repairs and Constructions;
SELQ – Quality control laboratories;
SVSA – Stocks & Warehouses.

3.2 *Accidents and rates*

The number of accidents, hours actually worked, number of workers and days lost in 2014 and 2015, in each Department, are shown in Table 1. There were no fatal accidents in the last 15 years.

Table 1 shows that the Production Department (DP) holds the highest values in terms of both frequency of accidents and number of days lost in the period. Within DP, there were 175 accidents leading to 751 lost days in 2014, and 114 accidents originating 956 days lost in 2015. This is explained by the nature of the activities carried out in DP that cover all activities of manufacturing and productive nature and, naturally, by the higher number of workers exposed to risk. The frequency and severity rates calculated for DP Services in 2014, are shown in Figure 1.

Figure 1 indicates that the services that present simultaneously the higher accident rates, I_f and I_g, are the SVMT, SVCA, SVCN, SVSC and SVSO. From a prevention strategic point of view, these five Services were apparently the best "candidates" for applying the RIAAT process, in order to better characterise the various causal factors of the accidents.

Table 1. Hours actually worked; number of workers; number of accidents; working days lost—per Department in 2014/15.

Year	Department	Hours (actually worked)	Number of workers	Number of accidents	Working days lost
2014	DC	1774	29	1	0
	DP	693499	400	175	751
	DR	85259	154	2	85
	DT	70950	59	0	0
2015	DC	5254	3	0	0
	DP	679554	397	114	956
	DR	102839	63	1	71
	DT	69835	41	2	11

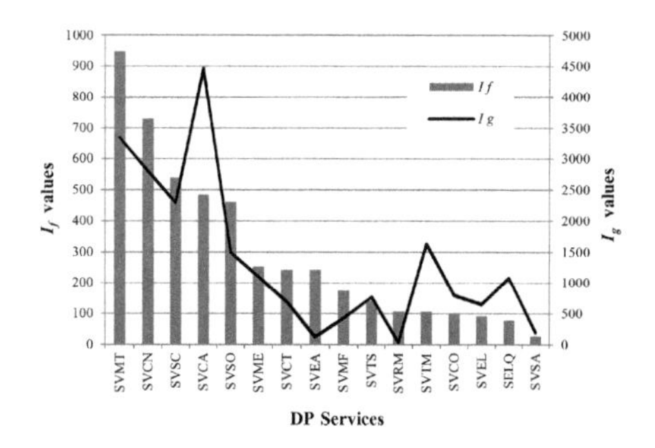

Figure 1. Frequency (If) and Severity rates (Ig) by DP Services in 2014.

However, the subsequent calculation of the Incidence rate (I_i), demonstrated that the SVCA (Carpentry Service) loses importance in terms of incidence. By contrast, SVSO (Welding Service) gained importance by this indicator, but the low severity rate reduces its priority. The 2015 results were fairly similar.

Considering all three indicators simultaneously, for both years, it was concluded that the most critical (Criticality criteria considered the most prominent "peaks" of each indicator ($I_f \geq 300$ accidents per million hours effectively worked, $I_g \geq 1000$ days lost per million hours actually worked and $I_i \geq 300$ accidents per thousand workers) production units to study were the three Services: SVMT, SVCN and SVSC. Therefore, these were the three Services selected for the causality analysis.

4 CAUSALITY ANALYSIS

4.1 *Coding accidents according to ESAW*

The first part of RIAAT process, for Recording accidents (Jacinto et al. 2011), uses the ESAW (European Statistics of Accidents at Work) coding system (Eurostat, 2013), which specifies the different explanatory variables of the accident.

As such, the first step was to code accordingly the relevant variables for all accidents occurred in 2014–2015, in the three targeted Services ($N_1 = 113$).

Of the 113 occurrences, SVCN recorded 39 accidents, SVMT 31 accidents and SVSC 43 accidents.

Figure 2 shows the relative distribution for the variable Contact (main classes) occurred in 2014 and 2015. This variable identifies the modality of the injury, i.e., it represents the type of accident. Contact is subdivided into 10 classes (main categories), defined in ESAW (Eurostat, 2013).

Particularly, it was found that the most frequent classes, or types of accident, were:

- C30 (~21%) – vertical movements, crushes on, or against something, i.e., result of a fall;
- C40 (~25%) – struck by object in motion, projected;
- C70 (~20%) – physical constraint of the musculoskeletal system.

Logically, the type of Contact is a direct result of the "deviation" occurred (last immediate cause leading to the contact itself), rendering the variable Deviation equally important. Its relative distribution is shown in Figure 3.

The relevant classes of Deviations (causes) were:

- D20 (~19.5%) – overflow, overturn, (…), projection, emission of material agent;
- D40 (~18%) – loss of control of something, machine or object;
- D60 (~26%) – body movement (bad movement, uncoordinated) without physical stress (generally leading to external injury);
- D70 (~22%) – body movement with physical stress; overexertion of force (generally leading to musculoskeletal injuries).

Noteworthy saying that categories D40, D60 and D70 are all deviations of human nature. Together with D50 (slipping, stumbling and falling), they sum up ~70%, showing the very high contribution of human failures in the genesis of these accidents.

These two distributions (Contact & Deviation) from the shipyard under study are corroborated by the national statistics for the entire Portuguese activity sector "repair and maintenance of ships" (code NACE 33.15). Evidence is given in Figures 4–5, showing very similar patterns for both data sets: the shipyard under study versus the National data provided directly by GEP (Office for Strategy and Planning, Portuguese Ministry of Labour). The period is different, only because 2012

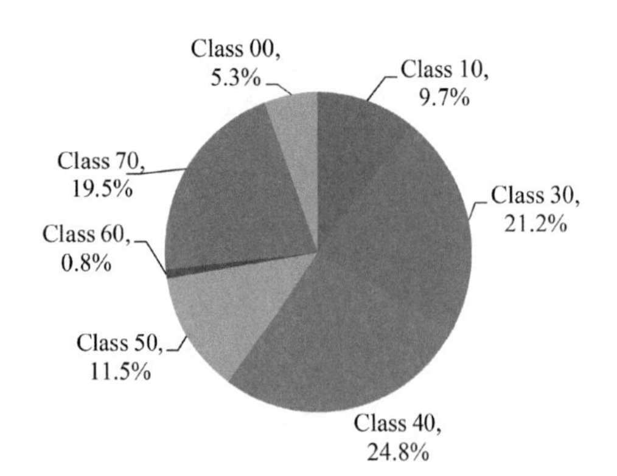

Figure 2. Relative distribution of accidents, by Contact, in 2014–2015 (SVMT, SVCN and SVSC) ($N_1 = 113$).

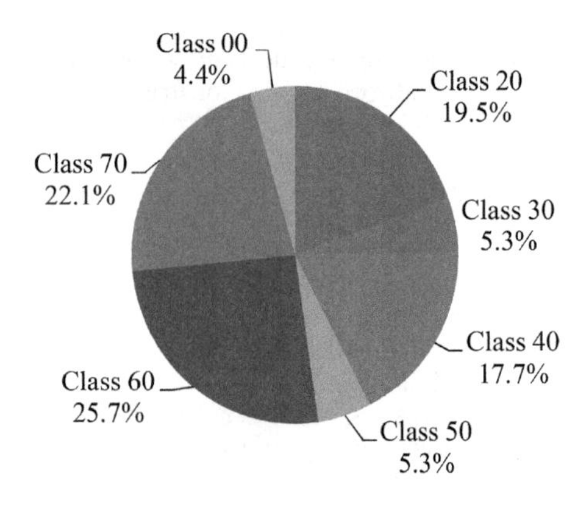

Figure 3. Relative distribution of accidents, by Deviation in 2014–2015 (SVMT, SVCN and SVSC) ($N_1 = 113$).

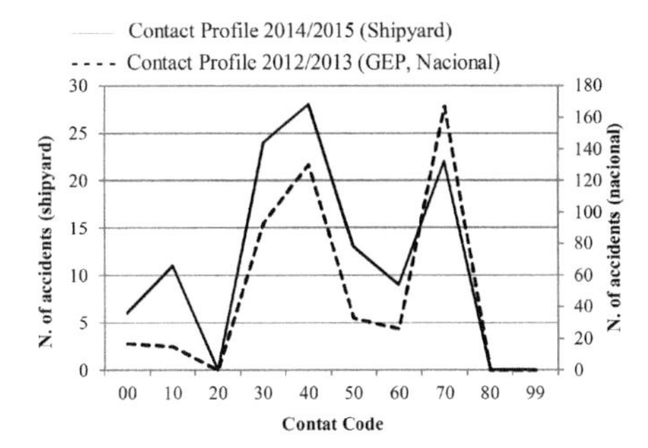

Figure 4. Statistical comparison (Shipyard vs National): Variable Contact.

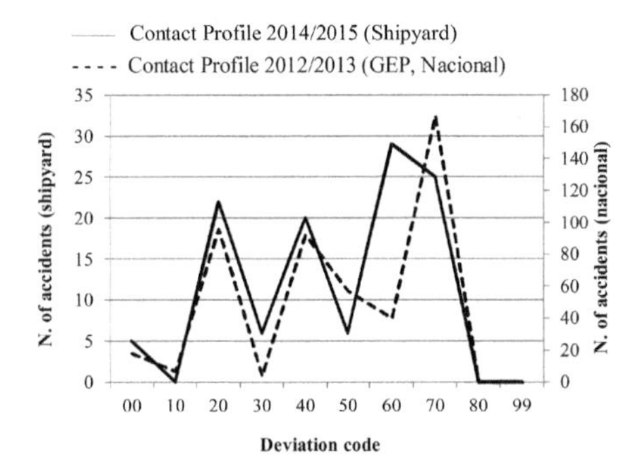

Figure 5. Statistical comparison (Shipyard vs National): Variable Deviation.

and 2013 were the last available national data at the time of the study.

The similarity of the patterns indicates that the shipyard analysed here is representative of its specific industrial setting, at national level.

Several other ESAW harmonised variables were included in this analysis, namely: Age of Worker, Type of Injury and Days Lost. The univariate analysis (relative distribution) on this set of accidents ($N_1 = 113$) revealed that the prevailing categories for these variables were, respectively: Age of Worker (victim) in the range 45–54 years old (~38%), Type of Injury was "superficial injuries" (cod.010, ~32%), and Days Lost in the range 4–30 days (36%).

4.2 *Cause-effect relationships*

Once the dominant modalities of Deviation (*cause*) and Contact (*effect*) had been found, a statistical test was conducted to establish whether such modalities of the two variables were statistically associated with each other. The method is already thoroughly explained by others authors (Chauvin & Le Bouar, 2007; Jacinto & Guedes Soares, 2008). The test is an adaptation of the independence test of Chi-square (χ^2), having the advantage of dealing with dependency relationships between modalities of variables (X_i, Y_i) rather than just between their main variables (X, Y). It allows the calculation of an "R" ratio for grading the importance of the relationship; for instance, R > 1.5 denotes strong positive relationship between the two modes, and R > 2 indicates very strong positive relationship. The resulting R ratios for all pairs of Deviation × Contact (D × C) are shown in Table 2.

"Typical accidents" of this activity were then defined as those complying with two requirements: high frequency and strong or very strong "*cause-effect*" relationship. The results in Table 2 show what the "typical accidents" in this shipyard are:

1. Pair D20 × C40; R = 2.94 (very strong positive relationship). Accident 1: *victim was struck by the projection or emission of something*. D30 was also very strongly associated with struck/collision, but the frequency was low (Fig. 3, ~5%).
2. Pair D60 × C30; R = 1.85 (strong positive relationship). Accident 2: *victim falls after a bad movement*. D50 was also very strongly associated with falls, but the frequency was low (Fig. 3, ~5%).
3. Pair D70 × C70; R = 3.90 (very strong positive relationship). Accidentv3: victim suffers a musculoskeletal injury, caused by overexertion of force or bad movement with physical stresss.

The above relationships disclose the relevant causal links between the most frequent types of Deviations (direct causes) and each type of Contact (accident). However, the limitation of this approach is that it only allows bivariate analysis (pairwise correlations), leaving aside the potential relationships with other variables.

The findings at this stage, combined with traditional accident pyramids build for each relevant Contact C30, C40 and C70 (not included in this article), allowed to identify the most critical individual accidents, in terms of both frequency and consequence (days lost). From this, a restricted number of 22 individual accidents (out of 113) were

Table 2. R ratios between Contacts and Deviations.

Deviation (*cause*)	Contact (*type of accident; the effect of cause*)									
	C10	C20	C30	C40	C50	C60	C70	C80	C99	C00
D10	–	–	–	–	–	–	–	–	–	–
D20	2.33	–	–	**2.94**	–	–	0.23	–	–	–
D30	–	–	–	4.04	–	–	–	–	–	–
D40	–	–	0.90	0.58	2.07	4.78	0.24	–	–	–
D50	–	–	4.71	–	–	–	–	–	–	–
D60	1.83	–	**1.85**	0.29	2.17	0.45	0.18	–	–	0.67
D70	0.41	–	0.57	0.16	0.35	–	**3.90**	–	–	–
D80	–	–	–	–	–	–	–	–	–	–
D99	–	–	–	–	–	–	–	–	–	–
D00	–	–	–	–	–	–	–	–	–	–

selected for the subsequent in-depth study, aiming at analysing the latent causal factors as well.

5 IN-DEPTH ACCIDENT INVESTIGATION

An in-depth analysis of the 22 most critical individual accidents was carried out through interviews, aiming to identify latent conditions that somehow facilitated the occurrence of the active failures. This corresponds to the second phase of the RIAAT process.

5.1 *Active failures*

The first step was to identify other active failures in addition to the Deviation itself, i.e., the last event before the accident (also an active failure). As proposed by the WAIT method (Jacinto & Aspinwall, 2003), active failures can be classified into 5 groups. In the 22 accidents studied in depth, the analysis revealed the results summarised in Table 3.

The distribution found, in particular the large proportion of human failures, is similar to that observed in other studies carried out in several sectors of activity, namely in the extractive sector (Jacinto & Guedes Soares, 2008), the fishing sector (Antão et al., 2008) and the food industry (Jacinto et al., 2009).

5.2 *Latent conditions*

According to the Reason's model embedded in the RIAAT process (Jacinto et al., 2011), latent conditions (indirect causes) can be structured in three main categories, namely: (1) Individual Contributing Factors (ICF), which are personal conditions that may trigger or influence human errors and behaviour; 2) Workplace Factors (WPF) that directly or indirectly contributed negatively to the event under analysis, and could offer future opportunities for prevention; and finally 3) Organizational and Management Factors (OMF) that represent organizational and management weaknesses that may have facilitated previous events and unsatisfactory working conditions.

In the 22 accidents selected for the in-depth investigation, 100 latent conditions (contributing factors) were identified during the interviews and field visits. The results are summarised in Table 4.

Table 3. Classification of Active Failures (N_2 = 22 accidents).

Code	Meaning / Nature of failure	Number (%)
HUM	Human failures	20 (69%)
E&B	Equipment & Buildings	4 (14%)
HAZ	Hazards (chemicals, debris, etc.)	3 (10%)
LOR	Living Organisms (fungi, animals,)	0 —
NAT	Natural phenomena (ice, rain, etc.)	2 (7%)

During the interviews it was possible to learn that several Work Place and a few Management factors, had an influence in the occurrence of safety violations (human failures). The problems identified in the analysis were subjected to internal discussion for creating a hierarchized and detailed plan of action.

5.3 *Action plan / recommendations*

The plan was firstly developed with specific actions to prevent or control the risks and failures identified in the 22 most critical accidents analysed. At a second stage, the action plan was extended by including suggestions of improvement proposed by the workers themselves during the interviews. To a certain extent, costs were also considered when establishing the plan of action, in an attempt to conform to the ALARP principle - "As Low as Reasonably Practicable", i.e., "as far as possible".

The corrective and improvement measures were structured into three main categories: (1) engineering, (2) training and awareness, and (3) management and control measures.

In this particular case, it was considered that the priority measures, likely to have the greatest short-term impact, would be **training and awareness** (2), as listed below. The purpose of these is to discourage workers from engaging in risky behaviour that may jeopardize their integrity or that of those around them.

Table 4. Distribution of Latent Conditions (N2 = 22 acci.).

Code	Meaning / Nature of failure (relevant examples found)	N° (%)
ICF	Individual (*most frequent were disattention, fatigue and natural human variability*)	15 (15%)
WPF	Work Place (*unpleasant environment by fumes/dust, inadequate equipment or tools and task related problems, namely handling of heavy/large objects and lack of training and competence*)	37 (37%)
OMF	Management (*poor communication, poor workers involvement and safety related factors, namely insufficient risk assessment and lack of safety representatives*)	48 (48%)
	Total	100 items

- Visual risk management ("advertisement") related to material handling, equipment and machinery (welding machines, grinders, drills, hammers, etc.);
- Tools for preventing accidents: APPT (Pre-Task Hazard Analysis). It consists basically of a "quick standard form" that needs to be filled in, by workers, on their workplace, at the beginning of shifts. The objective is to increase safety awareness by "pushing" workers to assess the environment around them;
- New rules for the systemic storage of materials, and for the correct use of PPE (protective equipment);
- Training on ergonomics and handling of equipment and machinery.

With respect to **management and control** (3) the proposed measures were as follows:

- Hiring more safety technicians;
- Increasing supervision in the workplaces;
- Carrying out more frequent risk assessments;
- Creating a Safety Commission;
- Stimulating Organizational Learning, namely by discussing results of investigations;
- Improving maintenance plans for machinery and equipment.

Finally, the **engineering measures** (1) were designed to impact directly on the sources of risk, either by elimination or reduction; they were as follows:

- Modifications to equipment: addition of steps to mobile cranes, equip rectifiers with protections against projections, use springs or elastics to tie-up the bottom of the pants when riding bicycles;
- Modifications to workplaces: increase lighting and ventilation inside docked ships (especially in confined spaces), apply anti-slippery coatings on floors;
- Area delimitation and risk signs: paint the first steps of scaffolding, paint salient structures in workshops and ships, use lightning cords (Led).

6 BAYESIAN NETWORK MODELLING

Bayesian Networks (BN) are an interpretive and analytical approach to probabilistic reasoning and offer a decision-support modelling method. A BN consists of a directed acyclic graph that includes nodes and relations of probabilistic nature that show their reciprocal influence (Heckerman 1996, p.11).

The BN modelling was applied in the third and last stage of this case-study. The aim was to demonstrate the applicability of BNs in this specific context, and the method's ability to deliver a quantified multivariate analysis.

To this purpose, a dedicated software, GeNIe® (Druzdzel, 1999) was used. This tool offers a graphical user interface (GUI) that allows interactive modelling of the domain and also algorithms for parameter learning, i.e. for stabilising the model correlation structure from data (Bayesian learning). In particular, GeNIe® uses the EM (Expectation Maximization) algorithm that is adopted in various computational applications for data modelling involving probabilistic models such as Markov chains and Bayesian networks. The EM method belongs to a class of statistical techniques for the estimation of parameters in statistical models when latent (or hidden) variables exist. The parameters found are maximum likelihood estimates (MLE).

It is known that a certain type of accident may cause a certain type of injury to a worker, but nothing can be taken for granted and probabilities must be used in a way to minimize the uncertainty associated with this type of problem. Thus, GeNIe® is useful to model uncertainty, i.e., Bayesian networks are especially useful for risk analysis and to model relations between causal factors and consequences.

Using BN models it possible to perform a multivariate probabilistic study that interrelates all the explanatory variables of the accidents considered in the ESAW methodology, or in the RIAAT method and others variables inherent to the company itself.

Figure 6 shows the developed BN model that includes the main variables describing the cause-effect relationships of the accidents. The independent variables (i.e., without parent nodes) in this model, are Services (the 3 operational areas analysed before) and Age of Worker (victim).

Figure 7 shows the probability distribution of each model variable obtained by parameter learning from data. For this purpose the real 113 accidents recorded and analysed before, were "expanded" into a larger database, rendering more than 200 possible accident scenarios, or possible

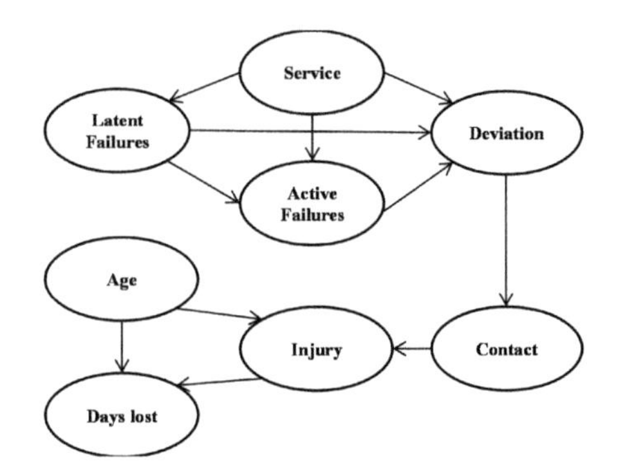

Figure 6. BN model of accidents.

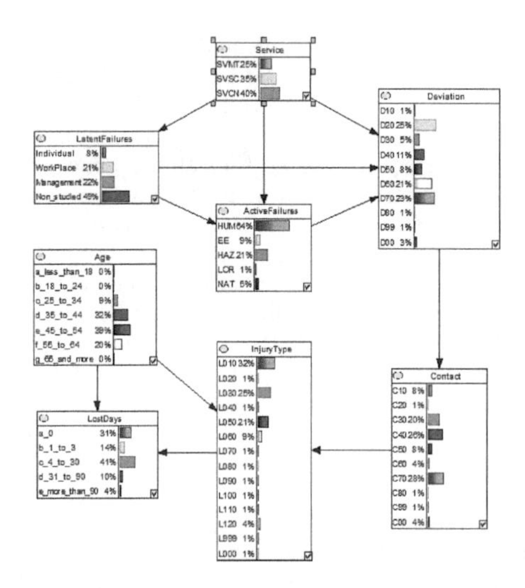

Figure 7. BN model of accidents in the shipyard (probability distribution of the BN model variables).

combinations. This expansion of the original database happens because *one single real accident*, characterised by the basic ESAW variables may be, or not, associated with *other causation variables* that were identified in the 22 cases analysed in-depth, such as, one or more human error(s), one or more local workplace factor(s) and one or more management factor(s). As a result, each event, corresponding to a single line in the original database, needs to be replicated in more lines, each representing a "possible" combination of latent causes. This expanded dataset was then used by the software for parameter learning, i.e., to derive the probability distributions of all marginal (independent) random variables and the correlation structure among all dependent variables of the BN model.

It can be observed from Figure 7 that the most frequent modalities of Deviation and Contact are D20, D60, D70, and C30, C40, C70, respectively, which is in line with the results of the univariate and bivariate analysis. Human Failures (HUM ~64%) also shows good adherence to the reality found in the shipyard. Moreover, the category "Non_Studied" (49%) within Latent Failures shows that many of the 113 accidents included in the dataset were not subjected to in-depth investigation.

Therefore, a limitation of this study is that there are "missing variables", since not all accidents (113 – 22) were analysed for latent failures. This explains the different frequencies (%) in the model, as compared with those in the previous sections of this paper. Despite this limitation, the BN model shows how particular scenarios can be analysed by propagating the effect of "evidences" on specific variables.

In BN analysis, an "evidence" is defined as a known value of a particular random variable of the Bayesian network model. It is related to the concept of "inference" and it can be defined as a "belief updating". This is used to analyse scenarios by updating probabilities (i.e. by assessing posterior probabilities) across the whole Bayesian network given a set of selected "evidences" on particular variables.

For instance, if the SVMT Service (manoeuvring & transportation) is defined as statistical "evidence" (SVMT appears underlined in box), i.e., if it is hypothesized that 100% of cases occur in this Service, the BN model returns, by probabilistic inference, the posterior probability distribution of each variable (Fig. 8).

Another example is given in Figure 9, in which Contact C30 (crash on/over; falls) is defined as "evidence" (C30 is underlined), i.e., considering that this happens in 100% of cases, it is observed that it is caused mostly by D50 (slips and trips, 38%), D60 (bad body movements, 37%) and D40 (loss of control, 13%). The previous bivariate analysis had already revealed the statistically significant cause-effect relationship between D60 × C30 (R = 1.85)

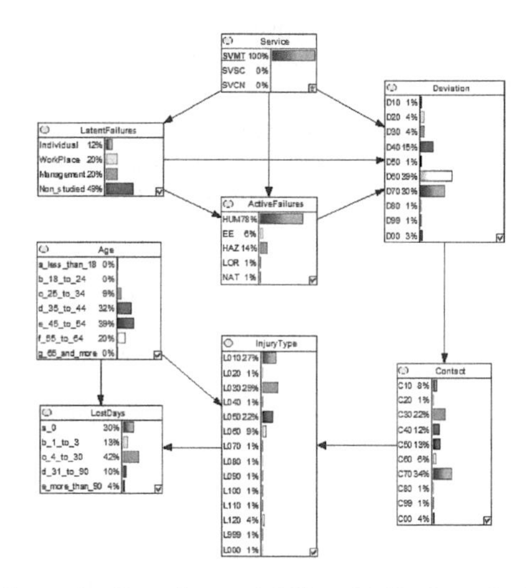

Figure 8. Posterior probability distribution of the model variables given evidence: Service = SVMT.

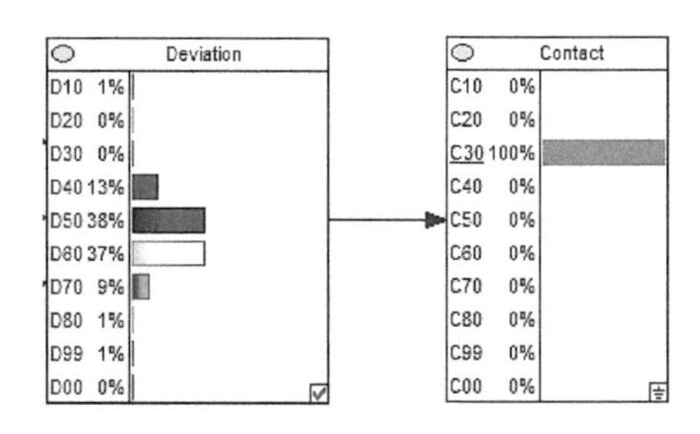

Figure 9. Posterior distribution of Deviation given evidence: Contact = C30 (crash on/over; fall).

and also D50 × C30 (R = 4.71) (*c.f.* Table 2). The first was highlighted as one of "typical accidents" due to its real high frequency.

In another scenario shown in Fig. 10, Contact C40 (being struck by something) was defined as "evidence" (C40 is underlined), i.e., considering that this happens in 100% of cases, the BN model shows that its most probable causes are D20 (emission, projection of something, 70%) and D30 (breakage, bursting, fall, collapse of something, 18%). Once again this corresponds to one of the typical accidents already identified by the notation D20 × C40 (R = 2.94).

Finally, Figure 11 depicts the result of a scenario in which Contact C70 (physical stress on the musculoskeletal system) is defined as "evidence" (C70 is underlined). In this case, the most probable cause is D70 (body movement under or with physical stress / overexertion of force, 71%). This evidence is in agreement with the previous findings pinpointing accident D70 × C70 (R = 3.90) as one of the "typical accidents" as well, since it is very frequent and holds very strong positive cause-effect correlation.

All cases analysed have shown that the results of the BN model are representative of the accident scenarios studied. Moreover, it was found that the use of this tool constitutes a good complement to RIAAT as a practical analysis tool. Summing up, the Bayesian Networks have proved to be a versatile tool for modelling occupational accidents,

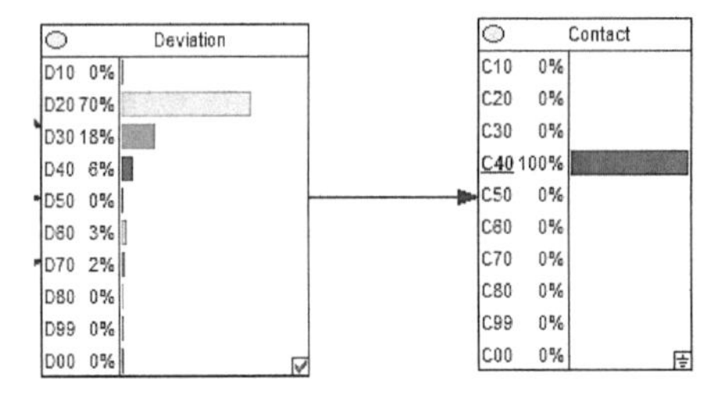

Figure 10. Posterior distribution of Deviation given evidence: Contact = C40 (struck by, collision).

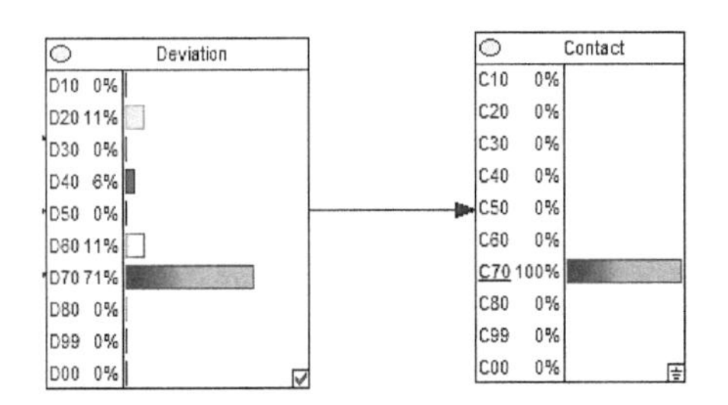

Figure 11. Posterior distribution of Deviation given evidence: Contact = C70 (physical stress on the musculoskeletal system).

capable of analysing various relevant scenarios using data collect by an organisation.

7 CONCLUSIONS

The paper presented a detailed analysis of accidents at work occurred in a large shipyard (~500 employees), in which *ship repair and maintenance* are the main activities.

The analysis was carried out in three phases that has included: 1) a general characterization of accident indicators, 2) a thorough causal analysis of the most serious and frequent accidents occurred in the three critical Services (areas of work) in the period 2014–2015, using the RIAAT method, and finally, 3) a causality modelling and analysis using Bayesian Networks.

The first part of the study has identified the Services (areas of work) with the highest accident rates, which were the focus of the subsequent analysis, namely:

1. SVMT (Manoeuvring and Transportation (land)), with incidence rates of 1615 and 769 accidents per 1000 workers in 2014 and 2015 respectively;
2. SVSC (Civil Locksmiths), with incidence rates of 961 and 692 accidents per 1000 workers in 2014 and 2015 respectively;
3. SVCN (Metalwork and Ship Construction), with incidence rates of 1300 and 650 accidents per 1000 workers in 2014 and 2015 respectively.

These three Services were further investigated and their "typical accidents" were analysed in more detail. The study covered both active and latent causes, which allowed establishing the most pertinent control measures.

Three typical "accidents" (the most frequent ones) were characterised as being:

1. "Crushes in vertical or horizontal movements against immobile objects; fall of victim" (cod. C30, ~21%), caused by "bad movements of the body not subjected to physical constraints" (deviation cod. D60)
2. "Strikes / collisions by moving objects" (cod. C40, ~25%), caused by "emission/projection of something, namely particles and dust" (deviation cod. D20);
3. "Physical constraints of the body leading to musculoskeletal injuries"(cod. C70, ~20%), caused by "movements of the body subject to physical constraints; overexertion of force" (deviation cod. D70).

The study revealed that active failures were mostly human erroneous actions (69%), or related to equipment and buildings (14%), diverse hazards (10%) and natural causes (7%).

On the other hand, the latent failures at the workplace level calling for more attention were related to "unpleasant and unhealthy physical environments", "task related" and "inadequate equipment or tools".

Finally, at the management level, the three main weaknesses identified were related to "specific safety factors", "top management communication" and "procedures and rules". Based on this analysis, 18 tangible risk control and improvement measures were defined within three main categories: training and awareness-raising measures, management and control measures and engineering measures.

Finally, the accident dataset was used to develop a Bayesian Network model that allowed the probabilistic analysis of several scenarios by means of evidence propagation. Broadly speaking, the results of BN analysis of particular "typical accidents" confirmed the conclusions of the in-depth analysis, showing that Bayesian Networks provide a very realistic representation of accident causes and of their relationships. The BNs have the advantage of modelling the multivariate causal relations among all variables and they provide a good complement to the in-depth investigation of accidents.

ACKNOWLEDGEMENTS

The authors are grateful to CENTEC for funding the study (Research Grant Ref. BL 342/2016), to GEP (the Cabinet of Strategy and Planning, of the Portuguese Ministry of Labour) for supplying national accident data, and to Paulo Martins and Vera Silva (Production and H&S Managers, respectively) for their collaboration and helpful support in the hosting institution.

REFERENCES

Abdat, F., Leclercq, S., Cuny, X., & Tissot, C. (2014). Extracting recurrent scenarios from narrative texts using a Bayesian network: Application to serious occupational accidents with movement disturbance. Accident Analysis and Prevention, 70, 155–166.

Antão, P., Almeida, T., Jacinto, C., Guedes Soares, C. 2008. Causes of occupational accidents in the fishing sector in Portugal. Safety Science, 46(6), 885–899.

Drupsteen, L., Guldenmund, F.W. 2014. What is learning? A review of the safety literature to define learning from incidents, accidents and disasters. Journal of Contingencies Crisis Management, 22(2), 81–96.

Druzdzel, M. 1999. SMILE: Structural Modelling, Inference, and Learning Engine and GeNIe: a development environment for graphical decision-theoretic models. In: Proceedings of the Sixteenth National Conference on Artificial Intelligence (AAAI-99), pp 342–343, Orlando, Florida.

Eurostat. 2008. NACE Rev.2 – Statistical classification of economic activities in the European Community. Eurostat, European Commission, Luxembourg.

Eurostat. 2013. European Statistics on Accidents at Work (ESAW) – Summary methodology. Edition 2013. Eurostat, European Commission, Luxembourg.

Fernandes, J.D. & Crispim, J. 2016. The Construction Process of the Synthetic Risk Model for Military Shipbuilding Projects in Brazil. Procedia Computer Science, 100, 796–803.

Fragiadakis, N.G., Tsoukalas, V.D., Papazoglou, V.J. 2014. An adaptive neuro-fuzzy inference system (anfis) model for assessing occupational risk in the shipbuilding industry. Safety Science, 63, 226–235.

Heckerman, D. 1996. A Tutorial on Learning with Bayesian Networks. Innovations in Bayesian Networks, 1995 (November), 33–82.

Hovden, J., Størseth, F., Tinmannsvik, R.K. 2011. Multilevel learning from accidents—case studies in transport. Safety Science, 49(1), 98–105.

Jacinto, C. & Aspinwall, E. 2003. Work Accidents Investigation Technique (WAIT) – Part I. Safety Science Monitor, Vol. 7 (1), Article IV-2, 17p.

Jacinto, C. & Guedes Soares, C. (2008). The added value of the new ESAW/Eurostat variables in accident analysis in the mining and quarrying industry. Journal of Safety Research, 39(6), pp. 631–644.

Jacinto, C. & Silva, C. 2010. A semi-quantitative assessment of occupational risks using bow-tie representation, Safety Science, 48(8), 973–979.

Jacinto, C., Canoa, M., Guedes Soares, C. (2009). Workplace and organisational factors in accident analysis within the food industry. Safety Science, 47(5), pp. 626–635.

Jacinto, C., Guedes Soares, C., Fialho, T., Silva, A.S. 2011. The Recording, Investigation and Analysis of Accidents at Work (RIAAT) process. IOSH Publications, UK (ISSN: 1477-3996), Policy and Practice in Health and Safety, 9(1), pp. 57–77 (user's manual of RIAAT (2010) available at http://www.mar.ist.utl.pt/captar/en/riaat.aspx).

Lukic, D., Littlejohn, A., Margaryan, A. 2012. A framework for learning from incidents in the workplace. Safety Science 50(4), 950–957.

Martin, J.E., Rivas, T., Matias, J.M., Taboada, J., Arguelles, A. 2009. A Bayesian network analysis of workplace accidents caused by falls from height. Safety Science, 47, 206–214.

Reason, J. 1997. Managing the risks of organizational accidents. Ashgate Publishing Ltd, Aldershot Hants.

Rivas, T., Paz, M., Martín, J.E., Matías, J.M., Garcia, J.F., Taboada, J. (2011). Explaining and predicting workplace accidents using data-mining techniques. Reliability Engineering and System Safety, 96(7), 739–747.

Silva, A.S., Carvalho, H., Oliveira, M.J., Fialho, T., Guedes Soares, C., Jacinto, C. (2017). Organisational practices for learning from work accidents throughout their information cycle. Safety Science, 99, 102–114.

Song, G., Khan, F., Wang, H., Leighton, S., Yuan, Z. 2016. Dynamic occupational risk model for offshore operations in harsh environments. Reliability Engineering & System Safety, 150, 58–64.

Tsoukalas, V.D. & Fragiadakis, N.G., 2016. Prediction of occupational risk in the shipbuilding industry using multivariable linear regression and genetic algorithm 42/42 analysis. Safety Science, 83, 12–22.

Progress in Maritime Technology and Engineering – Guedes Soares & Santos (Eds)
 ISBN 978-1-138-58539-3

Analysis of SME ship repair yard capacity in building new ships

I. Atanasova
Varna Maritime Ltd., Varna, Bulgaria

T. Damyanliev & P. Georgiev
Technical University of Varna, Varna, Bulgaria

Y. Garbatov
Centre for Marine Technology and Ocean Engineering (CENTEC), Instituto Superior Técnico, Universidade de Lisboa, Lisbon, Portugal

ABSTRACT: The objective of this work is to analyse the existing capacities of a small sized ship repair yard in building new ships, accounting for the existing constraints raised from the implemented ship repair technology, facilities, equipment and human resources. The output of the study is to identify if the already implemented technology and infrastructure in one specific SME ship repair yard can build new ships and what the limitations are. Two optional solutions for building new ships are proposed, including some measures in the enhancement of the equipment and shipyard technological processes. The main conclusion is that the analysed SME shipyard has sufficient workshop space and floating dock capacity that may be employed in building new ships of different type with a dead weight up to 7,000 tons.

1 INTRODUCTION

The small and medium sized enterprises (SME) are the backbone of the industry and the European governments are taking care of their development. The European shipbuilding and ship repair industry is made up of around 300 yards and more than 80% are considered to be 'small to medium' enterprises, SME (LeaderSHIP2020, 2013). On the other side, the European marine equipment manufacturing and industry (propulsion, cargo handling, communication, automation, integrated systems, etc.) is made up of around 7,500 companies, where the clear majority is also considered SMEs.

According to Unit_E.4. (2009), 99% of the European SMEs generate about 58% of the EU's turnover, are responsible for two thirds of the total private employment and in the last 5 years, 80 percent of the new jobs were created by them.

However, Parc and Normand (2016) claimed that more effective policies for enhancing the competitiveness of European shipbuilders are necessary by taking a holistic approach to solve core problems. The reason for that is the simultaneous implementation of two competitive strategies i.e. 1) cost leadership and 2) differentiation. With respect to the first strategy, the companies try to ensure competitive advantage over costs and according to the second one it is seeking to be unique in its industry. Considering the advantages of the European shipbuilding industry, attention should be paid to constructing expensive and complex vessels that yield a high added-value.

Lee (2013) analysed the efficiency, productivity, growth, and stability of Korean SMS shipyards and suggested some directions for improvements. One alternative for SMS shipyards is to switch to the marine equipment industry or the maintenance and repair of ships. Third alternative is the ship breaking, although as a labour-intensive industry it is developing in China, India, Bangladesh, and Pakistan. At the end of 2016 the European Commission adopted the first version of the European List of ship recycling facilities. The first 18 shipyards included in the List are all located in the EU.

Fourth direction suggested is establishing the subcontract structure as a value chain among large shipyards and SMEs. Finally, the possible solution is building clusters with governmental support in many ways.

The Maritime Administration, Department of Transportation (USA) has provided the Small Shipyard Grant Program, with $9,800,000 available for grants for capital and related improvements to qualified shipyard facilities that will be effective in fostering efficiency, competitive operations, and ship construction, repair, and reconfiguration. (https: //www.marad.dot.gov).

There is an increasing number of scientific publications devoted to the problems of a small shipyard. Song et al. (2009) presented a simulation-based support system for a ship production management that can be applied in SMEs for different processes. The simulation includes layout optimization, load balancing, work stage planning, block logistics and material management.

Typically, the basic planning and initial design in SMEs are done by design agents, outside the shipyards, and outfitting and detailed designs for construction are done in the shipyards. Shin et al. (2012) proposed a prototype of ship basic planning system for SME based on the internet technology and concurrent engineering concept. The used internet environment enables remote design and information exchange between shipyards and design agents.

The use of modern CAD tools is of particular importance for the competitiveness of SMEs. Paine et al. (2013) shared the experience gained from implementing many commercially available design software packages with vessel models for automatic generation of bill of materials (BOM) combined with an in house developed material requirements planning (MRP) system. The basis of the concept is to develop a low budget integrated design and production system suitable for a small shipyard.

In the last decade, various "Design for X" methods have been developed. The goal of "design for production" is reducing production cost without sacrificing the design performance or product quality (Misra, 2016). Some of the considered aspects include:

- Simplicity in design: minimum number of parts; reduction in part variability; reduction in welding joint length; standardization of parts; integration of structure and outfit etc.;
- Design based on shipyard facilities: limits on ship dimensions; maximum weight and size of blocks; maximum size of panels—panel line turning and rotating capabilities; maximum berth loading; launching limitations;
- Other production considerations: simplified hull forms; avoidance of double curvature and large single curvature; developable surfaces; constant hold or tank length, constant hatch size, etc.

The goal of the presented study is to analyse the existing capacities of a small sized ship repair yard to build new ships, accounting for the existing constraints. This analysis is a part of the Shiplys project (Bharadwaj et al., 2017) Scenario 2 related to the development of a software tool for a conceptual ship design accounting for the risk-based life cycle assessment (Garbatov et al., 2017b).

The output of the study is to identify if the already implemented technology and infrastructure in one specific SME ship repair can build new ships and what the limitations are.

2 CURRENT SHIPYARD BUILDING CAPACITY

The existing production areas and facilities, implemented technology and human resources of the ship repair yard, allow partial and class repairs of small and middle-tonnage ships of different types. Shipyard's building berth area of analysed SME ship repair yard consists of four ship building berths equipped with one floating dock, No 2, with a floating system for vessel lifting and shifting. The floating dock can accommodate vessels with length over-all/breadth of 136/16 meters with a maximum launching weight of 1,800 tons. These restrictions limit the build or repair of the vessel deadweight of about 6,000 to 7,000 tons.

The existing larger floating dock, No 1 (not presented in the shipyard plan), can accommodate vessels with length overall/breadth of 155/22.8 meters and with a maximum launching weight of 7,500 tons, which corresponds to ships with a maximum deadweight of about 20,000 tons.

The existing production hall was built around the launching complex (Figure 1, position 1). The hall is with the following dimensions—a total length of 192 meters (8 sections, each of 24 meters), a width of 96 meters and a length of 10.8 meters. The hall has not been equipped yet. Two of the sections have specialized areas for the processing of constructions, intended to be used for ship hull repairs.

Several conclusions may be derived with respect to the fact that currently, the shipyard does not have the capacity to build new ships due to the lack of ad-equate facilities and equipment, and to correct this deficiency in the case of building new vessels the following needs to be fulfilled:

- to complete the production hall (Figure 1, position 1) and to be equipped with the appropriate machines, crane and other equipment, required for the processing of plates, profiles and production of welded joints and sections;
- a pipe preparation shop has to be set up in the production hall;
- in the eastern part of the production hall a site for the pre-slipway assembly must be established; it may be equipped with appropriate production systems, including gantry cranes with a lifting capacity of 30 to 40 tons;
- the new ship building may also be performed employing the existing building berth in the eastern part of the launching complex. This area to be equipped with cranes with a lifting capacity

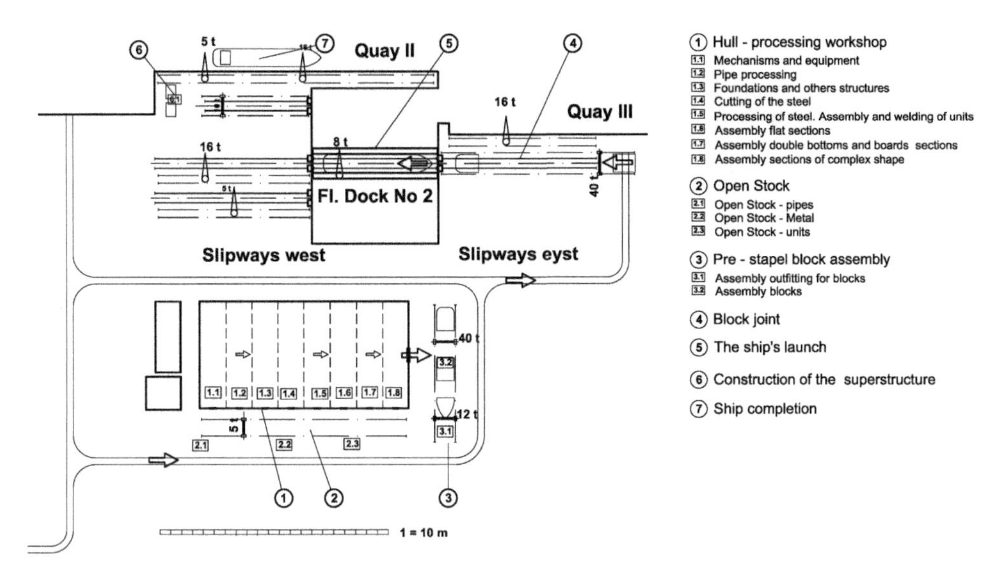

Figure 1. Shipyard building berths and facilities.

not less than those on the site for a pre-slipway fitting;
- the ship superstructure assembly site has to be established in the vicinity of the Berth Wall II.

3 SHIPBUILDING TECHNOLOGY

Assessing the currently existing capacity and if the above described shipyard retrofitting is implemented, the yard will beze to perform new shipbuilding along with its ship repair program. To construct new ships, two options are possible, which can be defined as a function of the location of the building berth:

- Option 1 – the hull assembly may be performed on the building berth of the Floating dock No 1;
- Option 2 – the hull assembly may be performed on the building berth, which is a part of the launching complex.

In the case of Option 1, the Floating dock No 1 has to be equipped with an appropriate crane with a lifting capacity not less than 50 tons, otherwise the existing crane of 5 tons will limit the assembly capacity of the building berth.

Option 2 considers the maximum load carrying capacity of the Floating dock No2, which is about 1,800 tons. In the case of relatively large ships, the building berth may assemble the ship hull, ship systems and equipment, main engine and propulsion system and if it is possible may include some large-sized machinery and equipment in the engine room. The ship construction afloat includes the assembly of the superstructure and all remaining completion works.

In the following tables, the technological process for a new ship building in the shipyard's building berth complex is described. The general layout of the production process and associated production areas of a hull construction is shown in Figure 1. The remaining technological shipyard equipment will ensure the implementation of the mechanical, metal, electrical and other works. The mounting and testing of the equipment and gears is carried out by the manufacturers.

Currently, the shipbuilding process may be performed following these stages:

- the zero processing of the base material: plates and rolled profiles may be performed in the existing facilities or may be subcontracted to the existing neighbor's shipyards (Table 1 position 1).
- cleaning and cutting of plates and rolled profiles, including bending of rolled profiles may be performed in the hull workshop as can be seen in Figure 1, position 1, where a specialized area will be established. Table 1 position 2 shows the equipment needed to perform cleaning, cutting and bending of plates and profiles.
- the process of fabrication of sections—plane and curved will be performed in the Hull workshop as can be seen in Figure 1, position 1, where a specialized area in the section 1.5 will be established. Table 1 position 3 shows the required equipment.

3.1 *Assembly and welding of sections*

This stage of the production covers the following processes:

- the assembly and welding of plane sections—forming plates and fitting the frames and reinforcements in the case of flat sections. This process will take place in the Hull workshop; see position 1 in Figure 1, where a specialized area in the Section 1.6 will be established. Table 2 position 1 shows the required equipment.
- the assembly and welding of the double bottom, side shells and curved sections will be performed in the hull workshop, see the position 1 in Figure 1, where a specialized area in the section 1.7 will be established. Table 2 position 2 shows the required equipment.
- the assembly and welding of the 3-D sections will be performed in the hull workshop as can be seen from the position 1 in Figure 1, where a specialized area in the section 1.8 will be established. Table 2 position 3 shows the required equipment.

3.2 *Pre-slipway assembly*

In the eastern part of the production hall a specific site for a pre-slipway assembly will be established, as can be seen in the position 3 in Figure 1. The required equipment is shown in Table 3.

At this production stage, the following processes will be carried out:

- outfitting of the constructions;
- consolidation of volumetric sections—the constructed sections are consolidated in blocks and (or) conglomerations in the pre-slipway assembly area.

3.3 *Assembly on building berth*

The 3-D assembled structures, mechanisms and equipment will be mounted using the building berth, which has a capacity of the admissible launching weight of 1,800 tons.

The ship assembly will be performed on the eastern building berth. The required equipment is shown in Table 4. At this production stage, the following processes should be carried out:

- placing/positioning of the building berth support–slipway cars;
- positioning and aligning of blocks—ship hull blocks are assembled consecutively;
- hull blocks are welded;
- mounting of the main engine;
- fitting of gears and installations;
- hull testing before launching;
- painting;
- positioning on a dock and launching afloat.

Table 1. Stages of the technological process.

Processing	Equipment	Characteristics
1 Zero processing of plates and profiles	Subcontracted service	
2 Cleaning, cutting and bending of plates and profiles	ZINSER Gas Cutting Machine Plasma cutting machine Vertical hydraulic press Profile bending hydraulic press Stacking crane	4 × 12 m 2,2 × 12 m 250 t; 800 mm 8 t; jib with electrical magnets 5 t;
3 Fabrication of welded joints, assemblies and sections	Assembly pallets Three rolled bending machines Hydraulic profile bending machine Mechanical cutting machine Stand for T-shaped profile Stacking crane	S = 20 mm; L = 800 mm S = 16 mm; L = 2,5 m 8 t; 5 t;

Table 2. Assembly and welding of sections.

Processing	Equipment	Characteristics
1 Assembly and welding of flat sections—Section 1.6	Welding portal for sheets Hydraulic press for Direction I frames Stacking crane	12.5 t; 12.5 t;
2 Assembly and welding of double bottom, boards. Outfitting—Section 1.7	Tooling for French curve templates Stacking crane	12.5 t; 12.5 t;
3 Assembly and welding of curved and volumetric sections. Outfitting—Section 1.8	Tooling for French curve templates Telescopic stand for French curve template Stacking crane	12.5 t; 12,5 t;

Table 3. Pre-slipway assembly.

Process	Equipment	Characteristics
1 Outfitting	Gantry crane	12.5 t;
2 Consolidation in blocks	Gantry crane	12.5 t;
and conglomerations	Gantry crane	40.0 t;
3 Consolidation in volumetric sections	Gantry crane	12.5 t;
	Gantry crane	40.0 t;

Table 4. Assembly on a building berth.

Processes	Equipment	Characteristics
1 Positioning of the building berth support	Slipway cars	Slipway system
2 Positioning and aligning of blocks	Gantry crane, Bridge crane	40.0 t; 16.0 t;
3 Hull welding		
4 Main engine mounting	Floating (mobile) crane	100.0 t;
5 Gears and installations fitting	Bridge crane	16.0 t;
6 Hull testing before launch		
7 Painting		
8 Positioning on a dock and launching afloat	Floating dock No2	Slipway system

3.4 *Completion afloat, testing and trials*

The ship's completion is carried out on the completion berth as can be seen from the position 7 in Figure 1. Table 5 lists the required equipment.

At this production stage, the following processes will be carried out:

- fitting of machinery and systems.
- mounting of the superstructure—consolidation of superstructure constructions is carried out on a specialized site, which is located at the position 6 in Figure 1.
- mounting of deck constructions, machinery, gears, systems, electrical, navigational and other equipment;
- superstructure outfitting;
- completion works;
- tests on a mooring station and sea trials.

3.5 *Transportation*

The transportation of the sections made in the construction workshop to the assembly site and from there to the building berth will be done by a transport platform with a carrying capacity of over 50 tons. In-house transport between the workshops consists of non-self-propelled rail trolleys.

Table 5. Completion afloat, testing and trials.

Processes	Equipment	Characteristics
1 Fitting of machinery and systems in the main engine room	Bridge crane	16 t.
2 Superstructure mounting	Floating (mobile) crane	100 t.
3 Mounting of deck constructions, machinery, gears	Bridge crane	16 t.
4 Superstructure outfitting	Bridge crane	16 t.
5 Completion works.	Bridge crane	16 t.
6 Tests on a mooring post		
7 Sea trials		
8 Delivery		

3.6 *Delivery of resources and equipment, warehousing and depot*

The plates and profiles will be transported after the "zero" treatment of the material by vehicles or by sea barges. An open warehouse for base materials and intermediate assembly of units is to be set up near the hull workshop (Figure 1, position 2).

4 SHIP DESIGN FOR BUILDING IN SME SHIP REPAIR SHIPYARD

As the multi-purpose vessel's capacity increases, the efficiency improves, and the objective here is to assess the impact of the restrictive production conditions of an SME ship repair shipyard.

In this case, the restrictions are determined by the capacity of the launching complex of the yard, which has the following limitations:

- maximum docking capacity – 1,800 t. This condition defines the maximum weight of the ship upon the launching—the lightweight (LW_1), which includes ship hull and all units that need to be mounted before the ship launching in the water;
- maximum dock dimensions that allow a ship to be built with a length not greater than 135.8 meters and a breadth not greater than 16 meters;
- the depth of the fairway determines the draft of the ship to be no more than 8 meters.

Based on the numerical analysis, employing concept design software tool "Expert" (Damyanliev & Nikolov, 2002), the investment costs, CAPEX is adopted as a criterion for evaluation of the design solutions. The software tool "Expert" has been also recently employed in defining design solutions subjected to different constraints. Details for the structure of the software framework and used algorithms can be found in Damyanliev et al. (2017), Garbatov et al. (2017a).

Table 6 shows the optimum main dimensions for ships with deadweight from 5,000 to 10,000 t obtained by "Expert" system. Here the main dimensions do not have any restrictions.

Table 7 shows the optimum main dimensions for ships with deadweight from 5,000 to 8,000 t, accounting for a constraint arising from the vessel's breadth of 16 m.

All vessel designs shown in Table 6 have a breadth of more than 16 meters. To fulfil the sets of deadweight accounting for the constraint of the breadth, the length is increased. Because of that, the draft is reduced, which can be explained by the condition of reaching the required minimum of the stability.

Based on the chosen optimization criterion, CAPEX, the designed ships without breadth limitations are relatively short with low ratios of L/B and L/D.These results are shown in Figure 2, Figure 3 and Figure 4.

Table 6. Ship design solutions for different DW.

DW, t	5,000	6,000	7,000	8,000	9,000	10,000
L, m	86.79	93.21	96.71	99.98	109.81	115.21
B, m	16.17	16.7	18.60	19.23	19.77	19.75
d, m	7.18	7.33	7.98	8.53	7.99	8.34
D, m	8.90	9.32	10.06	10.81	10.44	11.03
L/B	5.37	5.2	5.20	5.20	5.56	5.83
B/d	2.25	2.30	2.33	2.26	2.47	2.37
L/D	9.6	10.00	9.61	9.25	10.52	10.45
Cb	0.685	0.676	0.67	0.669	0.714	0.721

Table 7. Ship design solutions for different DW, B = 16 m.

DW, t	5,000	6,000	7,000	8,000
L, m	88,63	106.60	120.62	135.06
B, m	16.00	16.00	16.00	16.00
d, m	7.08	6.88	6.67	6.57
D, m	8.81	8.93	9.03	9.17
L/B	5.54	6.66	7.54	8.44
B/d	2.26	2.33	2.40	2.44
L/D	10.06	11.93	13.36	14.73
Cb	0.69	0.721	0.772	0.812

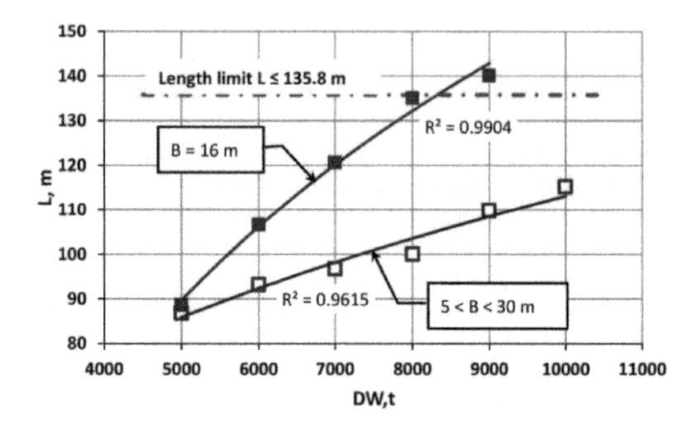

Figure 2. Length of the ship as a function of DW (L < 135.8 m).

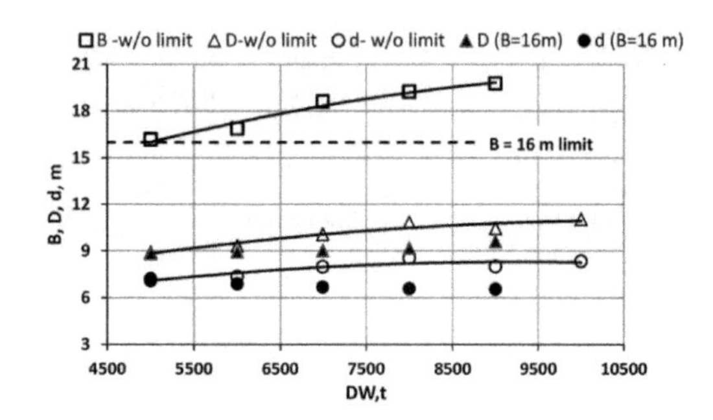

Figure 3. Main dimensions B, D, d as a function, of DW.

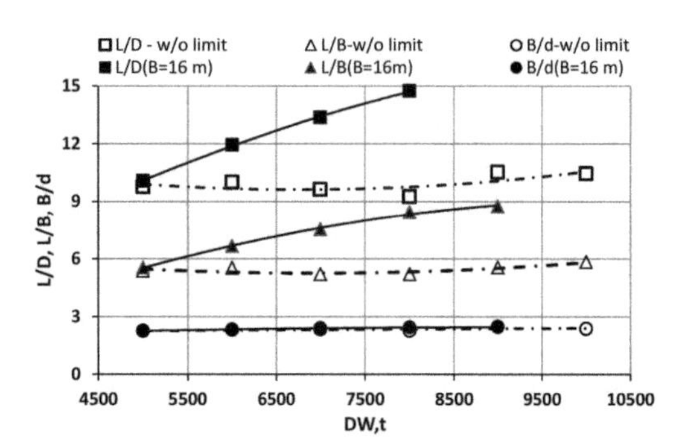

Figure 4. Main dimensions ratios as a function of DW.

The lightweight, LW_1, (as opposed to LW) includes the weight of the main hull, without 20% of the deck plating, part of the gears, including the steering gear, main engine and propulsion system.

The weight characteristic of the vessels, LW and LW_1 are presented as a function of the deadweight, DW in Table 8 and Table 9.

The LW1 includes the systems without which the ship cannot be launched into the water. In terms of the maximum docking capacity, the maximum deadweight of the ship will be around 6,700 t with

Table 8. Light weight as a function of DW, B = 16 m.

DW, t	5,000	6,000	7,000	8,000
LW_1, m	1,153	1,566	1,948	2,429
LW, m	2,104	2,678	3,550	3,831

Table 9. Light weight as a function of DW, without restriction.

DW, t	5,000	6,000	7,000	8,000	9,000	10,000
LW_1, t	1,127	1,438	1,745	1,835	2,030	2,331
LW, t	2,079	2,542	2,997	3,243	3,400	3,445

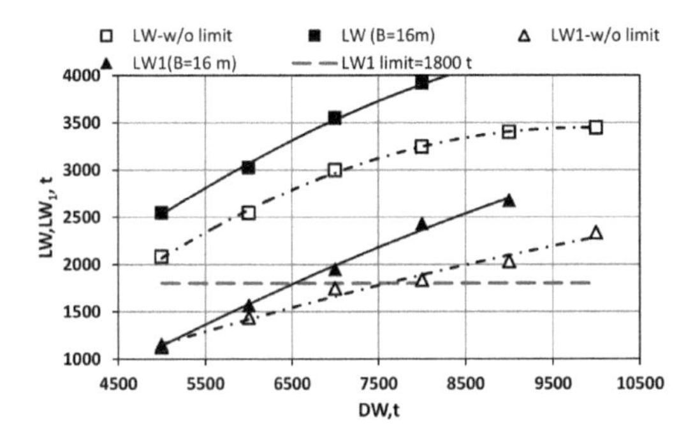

Figure 5. Light weight LW and LW_1 as a function of DW.

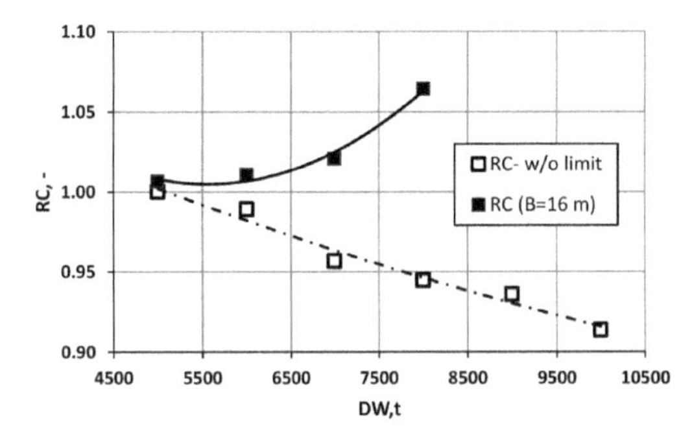

Figure 6. Relative CAPEX (RC) as a function of DW.

vessels of a limited breadth (Table 8) and about 7,600 t for those without a restriction (Table 9).

Figure 5 shows how LW1 changes as a function of B and DW and Figure 6 presents CAPEX, in a relative form (RC) related to the value at unrestricted 5,000 DWt ship. The imposed limitation on the breadth of the ship leads to a significant increase in the relative capital with a deadweight of the designed ships over 6,000 t. The optimal vessel with a limited breadth up to 16 meters is with deadweight of 5,500 t, where after this deadweight CAPEX increases essentially.

5 CONCLUSIONS

The objective of the present study was to identify if the already implemented technology and infrastructure in one specific SME ship repair yard can build new ships and what the limitations are. The conclusion derived is that the new building of ships of different types with deadweight up to 7,000 t is possible to be carried out alongside the existing repair capacity in the analysed SME ship repair yard. Considering the shipyard's capacity and its technological equipment, it is possible to build new multipurpose ships up to 7,000 t in a building cycle of eight months.

From the results shown, the following conclusions may be derived:

- a determining factor in defining the main characteristics of the ship is the width of the dock;
- the dock length affects only vessels with a restricted breadth and with deadweight of above 8,000 t;
- the docking capacity is a limitation for vessels with an unrestricted and limited breadth, with a deadweight of 7,600 t and 6,700 t, respectively.

ACKNOWLEDGEMENTS

This paper reports a work developed in the project "Ship Lifecycle Software Solutions", (SHIPLYS), which was partially financed by the European Union through the Contract No 690770 – SHIPLYS – H2020-MG-2014–2015.

REFERENCES

Bharadwaj, U.R., Koch, T., Frank, D., Herrera, L., Randall, G., Volbeda, C., Garbatov, Y., Hirdaris, S., Tsouvalis, N., Carneros, A., Zhou, P. & Atanasova, I. 2017. Ship Lifecycle Software Solutions (SHIPLYS) – an overview of the project, its first phase of development and challenges. *In:* Guedes Soares, C. & Teixeira, A. (eds.) *Maritime Transportation and Harvesting of Sea Resources.* London: Taylor & Francis Group.

Damyanliev, T., Georgiev, P. & Garbatov, Y. 2017. Conceptual ship design framework for designing new commercial ships. *In:* Guedes Soares, C. & Garbatov, Y. (eds.) *Progress in the Analysis and Design of Marine Structures.* London: Taylor & Francis Group, 183–191.

Damyanliev, T.P. & Nikolov, N.A., 2002, Computer system "Expert_ SRS" – valuations and analyses in the ship repair, Proceedings of the 6th International Conference on Marine Science and Technology (Black Sea 2002), Varna, Bulgaria., Union of Scientists of Varna, 14–18.

Garbatov, Y., Ventura, M., Georgiev, P., Damyanliev, T.P. & Atanasova, I. 2017a. Investment cost estimate accounting for shipbuilding constraints. *In:* Guedes Soares, C. & Teixeira, A. (eds.) *Maritime Transportation and Harvesting of Sea Resources.* London: Taylor & Francis Group, 913–921.

Garbatov, Y., Ventura, M., Guedes Soares, C., Georgiev, P., Koch, T. & Atanasova, I. 2017b. Framework for conceptual ship design accounting for risk-based life cycle assessment. *In:* Guedes Soares, C. & Teixeira, A. (eds.) *Maritime Transportation and Harvesting of Sea Resources.* London: Taylor & Francis Group, 921–931.

LeaderSHIP2020 2013. The Sea, New Opportunities for the Future. Brussels.

Lee, J. 2013. Directions for the Sustainable Development of Korean Small and Medium Sized Shipyards. *The Asian Journal of Shipping and Logistics,* 29, 335–360.

Misra, S.C. 2016. Design Principles of Ships and Marine Structures, Taylor & Francis group.

Paine, A.N.O., Ransing, R.S., Gethin, D.T., Sims, G.J., Richley, J.M., Boissevain, L.J. & Lewis, M.W. 2013. Challenges faced by a small shipyard in integrating computer aided design and production processes - A real life case study. *Proceedings of ICCAS.*

Parc, J. & Normand, M. 2016. Enhancing the Competitiveness of the European Shipbuilding Industry: A Critical Review of its Industrial Policies. *Asia-Pacific Journal of EU Studies,* 14, 73–92.

Shin, S., Lee, S., Kang, D. & K., L. 2012. The development of internet based ship design support system for small and medium sized shipyards. *International Journal of Naval Architecture and Ocean Engineering,* 4, 33–43.

Song, Y.J., Woo, J.H. & Shin, J.G. 2009. Research on a simulation-based production support system for middle-sized shipbuilding companies. *International Journal of Naval Architecture and Ocean Engineering,* 1, 70–77.

Unit_E.4. 2009. Think Small First—Considering SME interests in policy-making including the application of an 'SME Test'. European Commission, Enterprise and Industry Directorate General.

Progress in Maritime Technology and Engineering – Guedes Soares & Santos (Eds)
© 2018 Taylor & Francis Group, London, ISBN 978-1-138-58539-3

Shipyards of the 21st century: Industrial internet of things on site

Vicente Díaz-Casas, Alicia Munin Doce, Pedro Trueba Martinez & Sara Ferreño Gonzalez
UDC—Navantia Research Unit, Universidade da Coruña, Ferrol, Spain

M. Vilar
UDC—Navantia Research Unit, Navantia, Ferrol, Spain

ABSTRACT: The range of possibilities offered by the IoT in sectors such as shipbuilding, still unexplored, is very interesting. In a work environment where the flow of information is complicated and information is disaggregated at different points, the aggregation of information and connectivity between systems, jobs, suppliers, etc., will allow improving the efficiency inside a shipyard. In this paper, the challenges of the IIOT in the shipyard that have been addressed in the work carried out by the Research Unit of the University of A Coruña—Navantia are presented. Thus, a process to study the applicability of IoT platforms to ship manufacturing has been started: to test different IoT platforms and find the one that best suits a particular use case. The case chosen is the Preoufitting Workshop.

1 INTRODUCTION

The technology known as "Internet of Things" (IoT) is a new paradigm that encompasses, in turn, different technologies of ubiquitous computing, sensors wired or not (Wireless), RFID cards, networks, embedded systems and mobile devices. So that physical world and available resources have their replica in the digital world.

Thus, the IoT sector is in the middle of an important boom and in a high demanding process. Due to its transversal and integrating behaviour, it has led to the appearance of hundreds of IoT platforms and it is expected that this number will continue to grow. But many of these solutions are specific for each company. According to the IoT Forrester report of the last quarter of 2016 the sector's leading platforms are:

- PTC Thingworx
- IBM Watson
- SAP Leonardo
- GE Predix
- Microsoft Azure

Recently, the Industrial IoT (IIoT) has emerged as a sub-paradigm of the previous one, meaning the connection of machines in a factory. In this type of communication, the versatile software layer that characterizes the IoT is of great importance, and which makes it applicable to all types of sectors. Therefore, an IIoT environment consists of several layers of hardware and software, which leads to the installation of sensors in the factories, which combined with a large computational capacity, leads to lay the foundations for a new generation of intelligent factories (Ahmed, 2016).

In the maritime sector there are not many examples of application of this type of technology. Hyundai Heavy Industries started in 2015 (Hyun, 2015) a project to incorporate an IoT platform in its newly built vessels. It is about getting "smart and connected" ships that allow shipowners to better manage their fleets and improve their efficiency. The project consists in installing a sensor network in the vessels that capture information during the ship's navigation. The type of information is location, marine currents, weather, as well as, data of the equipment installed on board. By making use of the real-time analysis capabilities of the IoT platforms, the shipping companies can monitor the status of their ships in real time and make decisions for a more efficient operation.

Another example is the IoT application carried out in the Møre cluster (Norway). This cluster includes 200 companies dedicated to providing support services to offshore platforms. Its business ranges from the construction of vessels to the management of the service of these vessels (Wang et al., 2016). In this case, they have opted for a combination of Big Data and IoT. Because the analysis of large volumes of data requires a large computational capacity, the OSV ship would install a hybrid CPU platform with the capacity to perform a descriptive and real-time predictive analysis of the ship's critical equipment. The data for this analysis would come from the IoT part, whose components would include RFID readers, sensors, actuators, cameras and GPS. The analysis of the data with the Big Data methodology allows an ad hoc maintenance of the vessels and it obtains information to improve future designs.

2 WHAT IS THE IIOT IN THE SHIPYARD?

There is a gap in the application of IoT technologies in the shipbuilding sector, that is, the application of these new tools to the shipyard itself and its production methods.

There are fundamentally three perspectives (Atzori et al., 2010) in the adoption of these technologies in the industry:

1. The object-oriented, which focuses on the visibility of the machines
2. The internet-oriented, which seeks to improve the protocols of the network
3. The communication-oriented semantics, which focuses on topics of how to represent, store, interconnect, search and organize all the information generated by the "intelligent objects", increasingly numerous

In our approach to the introduction of the IIoT in a shipyard we have opted for this last perspective. In a first analysis, the diversity of sources of information existing in the shipyard was detected. In a second analysis we focused on a workshop verifying the complexity to access usefull information, that could facilitate the daily work and planning of the workshop.

The application of the IIoT in the shipyard supposes a change of paradigm regarding the management of the information in the shipyard. This change implies the expansion of what is traditionally understood as a shipyard, since it involves the idea of shipbuilding process from a holistic point of view. This encompasses not only the processes that are carried out in the shipyard itself, but each and every one of the actions/processes that leads to the achievement of the final product, the ship.

So the goal of this technology in the shipyard is to achieve a connected manufacturing, sharing the information generated throughout the process to create a collaborative environment, so that different processes can be synchronized for a global optimization of the ship construction.

This change of vision implies a conceptual change in the management of all the information inherent to the construction process. All the information is interconnected and is available in real time for each position. Thus, both, external personnel of suppliers and subcontractors and internal, from the operator to the director of the shipyard have access to the information of the system.

Figure 1 shows how the IIoT layer is located within the information system in the plant, directly

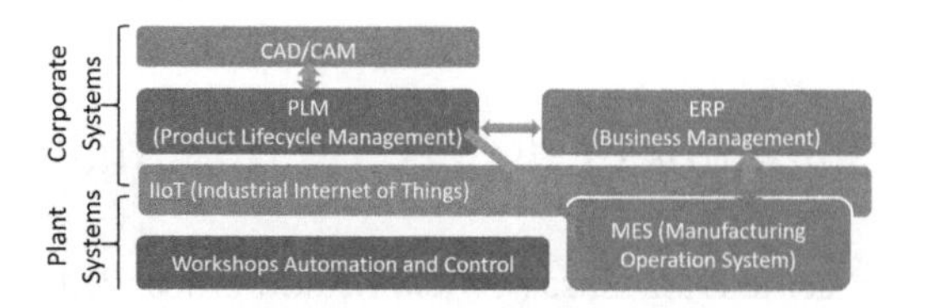

Figure 1. Information structure inside the shipyard.

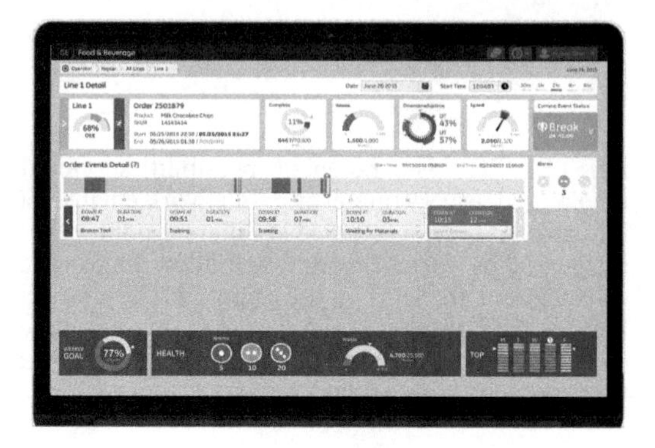

Figure 2. Visualization of an IIoT app in a mobile device.

linked to the production process and the corporate systems of design, management and planning. Positioning itself as the connecting link that enables the interconnection of both fields.

The implementation of this technology implies the use of mobile tools. In such a way that they have to adapt to each task. In the case of shipbuilding, the operator does not have a fixed position, but rather works inside a block or ship. So the system must accompany him in a transparent way for the user.

3 INTEGRATION OF THE IIOT IN SHIPBUILDING

Therefore, the initial objective is to find a platform that allows designing a viable application in a shipyard, wich responds to the problems of the different workshops, offering the services and functionalities that each one requires. This platform must allow connectivity with the rest of the information systems already implemented in the shipyard (ERP, PLM, MONTH, etc.). Access to information and its treatment in a user-oriented manner (operator, workshop manager, director) results in time savings, a reduction in errors and a faster identification of possible problems (bottlenecks, mismatches, etc.). In addition, an IoT platform applied to industry (IIoT) allows managing a "sensorized" installation that provides information on consumption, product movements, etc.

In addition, all this data flow can provide very valuable conclusions for the management of any industry or workshops.

The IoT platform must necessarily connect with the ERP, since that is where all the information related to the products to be manufactured is stored. Due to the access and the dump of the information in a ERP can be different according to what one wants to consult, it is laborious and costly in time to obtain the data sought. That is why the first functionality is defined as the query of information about MBOM in the ERP and the availability of materials in the warehouse.

The query of the Manufacturing Orders (MO) of the workshop itself would allow obtaining information about the operations to be carried out, the status of the order, assigned work position, associated materials, start and end date of work, etc.

The access to the MO's of the workshop suppliers will allow to know the status of the pending material. That is, the degree of progress, notifications made, the expected dates of end of work, etc.

The same happens with the access to the transactions of the warehouse since it would allow knowing what materials are discharged, pending quality approval or ready to be delivered.

These last two actions would involve starting the traceability operation of the intermediate products, equipment or other elements that the workshop will need in the near future.

The combination of queries is necessary for the control of the works in the workshop, and the crossing of data that can be done with them, fundamental to organize the programming of the workshop not only day to day, but practically in real time.

Another of the basic functionalities of the application is the notification of finished works. This can mean ending the work of equipment assembling (or installation of an element) and, therefore, give a certain degree of progress within the Manufacturing Order, or close the entire Manufacturing Order.

These notifications would be made by the operator himself when finishing the work from a mobile device. Doing so speeds up the updating of information, without detriment to a manager verifying that the work has been completed correctly and rectifying, if necessary, the degree of progress of the MO or its closure.

This updated information on the progress of the work can be incorporated into the PLM with which the work is carried out in the shipyard. That is, in 3D CAD, for example stored in the PLM, could be visualized what is installed or manufactured and what is not. The differentiation would be done through colors. Therefore, the application of IoT would allow real-time update of what is already mounted / installed in the 3D visualization of a block, where it would be appreciated graphically what is the situation of this block in terms of pending work, possible interferences, etc.

4 THE CHALLENGES OF THE IIOT FOR ITS REAL IMPLEMENTATION

In this section, the challenges that have been addressed in the work carried out by the UDC-Navantia Joint Research Unit (JRU) are presented. Thus, a process to study the applicability of IoT platforms to ship manufacturing has been started: to test various IoT platforms and find the one that best suits a particular use case. The case chosen is the Preoutifitting Workshop.

This workshop presents a bunch of particularities that differentiate it from the rest of the workshops. It is not a production workshop proper, but assembly. It does not work with raw material (steel plates) but with finished products that must be incorporated in that step of construction. These products can be intermediate products made within the shipyard (pipes, foundations, HVAC ducts, etc.) or products from external suppliers (electrical and mechanical equipment, connection elements, etc.).

These circumstances make this workshop the most dependent on external agents for the smooth running of the work (compliance with planning, management of waiting times due to lack of material, errors in the delivery of products, etc.). Therefore, the definition and development of an application based on an IoT platform can integrate all the information stored in the information systems, but not intelligently connected to this workshop, solving some of the problems posed.

On the other hand, in a workshop such as Preoutfitting, entirely oriented to the assembly of different elements (steel parts, electrical elements, pipes, compartment elements, etc.), the progress of the work is totally dependent on the availability of those elements. Therefore, its management is vital for the coordination in the workshop.

In this sense, the functionalities sought are:

1. Show the list of materials necessary to perform the work indicated in the MO and the associated information (drawings, specifications, etc.) through the selection of a specific material/equipment.
2. Perform the traceability of the components, materials or equipment to be installed/assembled. This traceability would begin in the warehouse, as already mentioned. Therefore, the IoT application must access information on whether the material is registered in the system. Subsequently, it will be followed up whether it is in the warehouse or an external provider.

Knowing if transport's material has been requested, the expected date of delivery by the supplier (warehouse or external supplier) and the margin of uncertainty with which you work for the delivery of the material would allow the application to estimate when the materials will arrive and if it will be on the scheduled dates. The difference of days between the estimate and the planned dates would be able to assess if it is necessary or not to modify the workshop schedule. In addition, in the application it could be posible to Access to the material movements (origin, in transit, etc.), the type of stock (no restrictions, pending quality control, blocked, etc.), estimated time for the reception process or others.

3. Perform the traceability of the pipes. These pipes represent around 60% of the total components to be installed in the Preoutfitting workshop. Inside the JRU a Research Line is underway focused precisely on the traceability of the pipes through RFID tags, conecting with this work the IoT application can make use of RFID readers to record the entrance of the pipe pallets in the workshop of Preoufitting. This allows to continue with the traceability started in the Pipes workshop and to automatically performe the notification of material delivery. Each pipe has a tag that contains the basic information that identifies it, so the application would inform the person in charge which layer, area and block corresponds to each pallet, and to which cell it should be moved from the reception area.
4. Show the materials to be placed on the "edge of the line". This concept gives an overview of whether the next scheduled MO can be started, indicating whether the components are available in the workshop or not. In the case of having been received by the workshop, they will be assigned a stowage place associated with the functionality of "material traceability". Thanks to this their location will be automatic.
5. Address the necessary information to perform the work from the materials located in the "edge of line". This applies to components that are located at one end of the cell. The incorporation of a barcode reader or QR, in a mobile device, would allow to identify the material, the MO to which it belongs and the information associated with it immediately.
6. Confirm the consumption of materials in the ERP. In the same way that the application will allow to notify the progress of the works corresponding to an MO, the same will be done with the installed/mounted components.

Another of the functionalities is the activation of a series of alarms in different degrees, identified by colors (for example, yellow, orange, red), for the identification of abnormal situations in the Workshop. These alarms can arise as anomaly detection in the data taken by the sensors installed in the workshop. Intermediate and high limit values will be established which, once overcome, activate visual alarms on the general screen of the workshop and send warning messages to workshop managers to their mobile devices.

It will also be necessary to manage notifications that arise due to incidents that are recorded, or notices associated with any other relevant information.

As a result of the above, and the use of mobile devices, four types of users are defined: operator, supervisor, workshop manager and director. Each of them will access a different level of aggregated information. For this purpose, it is necessary to design different "control panels", that is, screens that offer the necessary data for the development of each one's work, in a way geared to their needs, contributing at each level to making the most appropriate decisions.

Other relevant information for works control in the workshop are the management indicators (KPI's). This concept references to a set of parameters, expressed graphically, that give a general view of the state of the Workshop and the progress of the work. These indicators are prepared based on the performance statistics of the workshop, and this information is accessed by the IoT platform due to the previously defined functionalities.

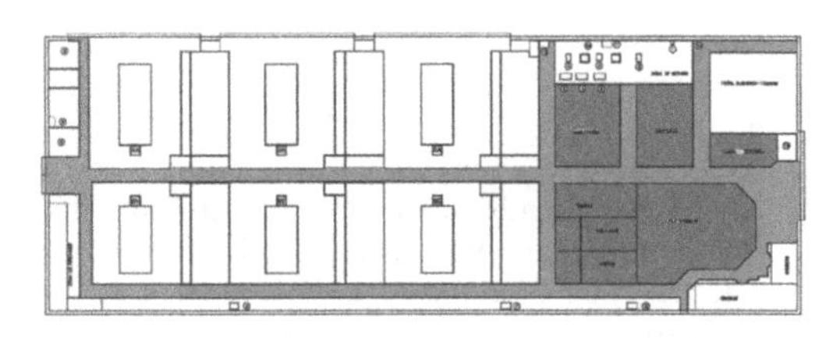

Figure 3. Preoutfitting plant.

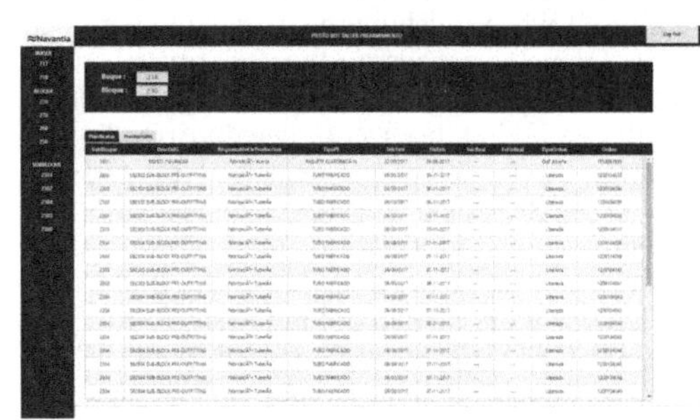

Figure 4. List of manufacturing orders sort by ship and block.

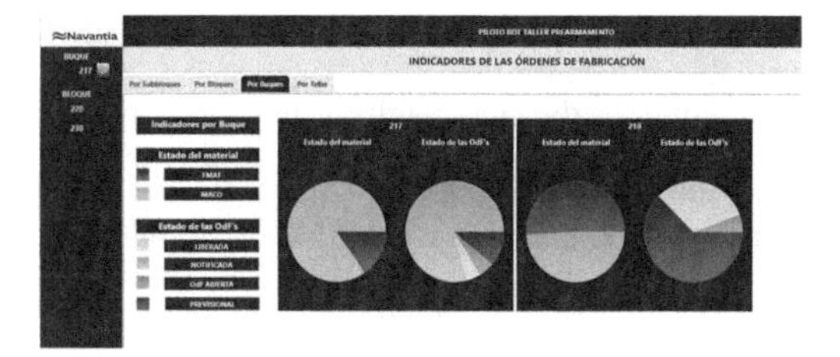

Figure 5. Indicators of workshop performance by ship.

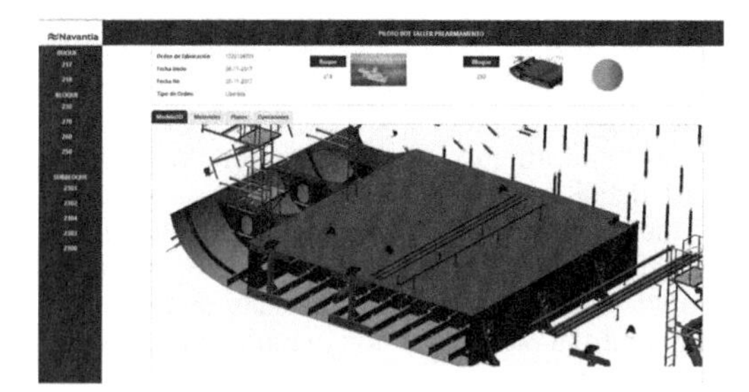

Figure 6. 3D Model and detail information for a specific manufacturing order.

On one hand, there are the indicators of Earned Value Management (EVM) such as the SPI and the CPI. These indicators give a macro view of the situation of the work carried out. These are shown in the Control Panel of type of users such as workshop management or director.

And, on the other hand, a group of specific indicators of the Preoutfitting Workshop are developed, which would contribute to the management people assessment about for the situation of the Workshop every day.

These indicators can be displayed graphically (pie charts, bar diagrams, etc.) and their updating is immediatly, as relevant notifications are registered in the systems. This type of indicators would be shown in the Control Panel of the supervisors and the person in charge of the Workshop.

As a support to the work developed by the workshop personnel, the possibility of viewing CAD files with 3D and 2D models is fundamental, according to their needs. In the case of the 3D model there must be the possibility of moving this model to obtain different views or to zoom in certain areas. This forces the platform to have the capacity to manage files that may be "heavy", resulting in slowed motion, or the collapse of the device.

It is also important the visualization of drawings, procedures, specifications, etc., as a complement to the information of the manufacturing orders.

All these files are usually posted in a PLM application, so the IoT application has to Access this system too for a query.

Figure 7. Status of the manufacturing orders and work progress by ship.

Regarding the programming of the workshop, a compact view of the same is looked for where the orders that can be started in the next days can be seen, as well as notifications of possible overloads of the workshop according to the degree of progress of the manufacturing orders opened. The application has the capacity to perform a reprogramming of works according to the casuistry of the workshop at that moment (availability of materials, production orders released, etc.)

For this, it is essential the visualization (with a color code, statistics or similar) of the workshop load according to the completed MO, the degree of progress of the open MO and the delays recorded in the workshop schedule. This would allow decisions to be made on the balancing of the workload through the reprogramming of tasks.

As for the hardware part of the project, it is necessary to sensorise certain elements in the workshop (electrical consumption, movement of cranes, etc.), which will provide complementary information to that obtained from the management systems.

The crossing of both sources of data will allow to draw useful conclusions that contribute to the optimal perfoming of the workshop. Therefore, the consultation of the information obtained from the smart sensors installed in the workshop must be online, in order to have an immediate view of the workshop's status. But it is also necessary ti view the historical data that have been recorded, expressed through graphs or statistics.

Finally, one of the potential of the IoT platforms is the integration of the entire logistics chain. The flow of information among the members of that chain results in mutual benefit, since it is possible to optimize processes time that are dependent on each other.

The Preoutfitting workshop is the clearest case of dependence on external suppliers and their deadlines for delivery of materials. Therefore, access to external suppliers is considered for actions such as forward notification of the order and expected delivery date. This information is useful both to

Figure 8. Arduino board to be used in the workshop test.

the Preoutfitting workshop and to the Warehouse, which is the one that manages the material order.

5 CONCLUSIONS

The possibility of incorporating new technologies into productive processes approaches to the present something that seemed a challenge for the future. This supposes a new industrial revolution (the fourth one) since it implies an important qualitative change in the way of realizing, controlling and obtaining points of improvement in the productive chain. Access and management of relevant information is simplified and oriented to obtain a more effective decision support tool.

The range of possibilities offered by the IoT in sectors such as shipbuilding, still unexplored, is very interesting. In a work environment where the flow of information is complicated and information is disaggregated at different points, the aggregation of information and connectivity between systems, tasks, suppliers, etc., will allow improving the efficiency of the work of manufacture inside a shipyard.

One of the problems of the industry that is not 4.0, is that there is a lot of information about machines and processes. But this information is difficult to access, stored in different isolated points and no data performance is obtained. What is offered by an IoT platform is that each type of user will be able to quickly and intuitively see the information that corresponds to the performance of their work in a web interface accessible from the browser of any device, including personal computers, tablets or mobile phones. The platform adds different sources of information (ERP, PLM, machines) in a single point to make decisions and obtain performance from the data. In this way, the user does not have to spend time looking for information in any type of application, but has it available at a glance.

To analyze the possibilities offered by an IoT platform, a proof of concept has been proposed in a specific workshop, the Preoutfitting workshop, which is the point at which workflows converge from the other workshops and because, for its day-to-day work, It depends on more information than the rest of the workshops. The selection of this workshop was based on the particularities of the work system, that made it more suitable to test the capabilities (connectivity, information management, user differentiation, uses of web frames, etc.) that this type of platform offers. Within this workshop, the platform is accessible by different user roles with their corresponding information. For an operator, the platform must detail the work to be done, showing details of the manufacturing orders. On the other hand, the information presented to a workshop manager should offer an overview of the state of the workshop and the progress of the manufacturing orders with respect to the planning.

In short, the IoT platform offers the opportunity to convert the data into useful knowledge for the operation of the workshop, so that decisions can be made to avoid problems, delays, and improve the production process.

ACKNOWLEDGEMENTS

The results presented in the paper include the results of the work carried out within the IIoT research line of the UDC—Navantia Joint Research Unit. This project is carried out thanks to the support of the Galician Innovation Agency (GAIN) of the Xunta de Galicia and the Ministry of Economy, Industry and Competitiveness (Spain Government).

REFERENCES

Ahmed, Ejaz, et al. Internet-of-things-based smart environments: state of the art, taxonomy, and open research challenges. *IEEE Wireless Communications*, 2016, vol. 23, no. 5, p. 10–16.

Hyun, Lee. Strategies for improving the competitiveness of the Korean shipbuilding industry: Case study of Hyundai Heavy Industries. 2015.

Wang, Shiyong, et al. Implementing smart factory of industrie 4.0: an outlook. *International Journal of Distributed Sensor Networks*, 2016, vol. 12, no. 1, p. 3159805.

Coating and corrosion

Progress in Maritime Technology and Engineering – Guedes Soares & Santos (Eds)
© 2018 Taylor & Francis Group, London, ISBN 978-1-138-58539-3

Internal corrosion simulation of long distance sandwich pipe

Cheng Hong, Yuxi Wang, Jiankun Yang, Segen F. Estefen & Marcelo Igor Lourenço
Subsea Technology Laboratory, COPPE, Federal University of Rio de Janeiro, Rio de Janeiro, Brazil

ABSTRACT: In many deep water oil field developments, oil and gas products are transported from the wellhead to the topside in multiphase flow without separation. As a result, the presence of carbon dioxide (CO_2), hydrogen sulfide (H_2S), and free water of the production fluid can cause internal corrosion in the subsea pipeline. In this paper, the internal corrosion process is numerically simulated for a long distance sandwich pipe which transports oil with high CO_2 content. Parametric study is conducted to find out the effect of different factors on the corrosion rate, including fluid pressure, temperature, flow rate, water cut and flowline profile. Based on the results, qualitative recommendations are proposed for reducing the corrosion damage.

1 INTRODUCTION

Deepwater activities in offshore Brazil mainly centered on Santos, Campos and Espírito Santo Basins. The first two basins occupy 90% of the deep water discoveries in Brazil. Exploration activities in Santos Basin began in 2002, encompassing an area of approximately 11600 km^2, which is known as the Pre-Salt Cluster. Water depths in this area vary from 2000 to 2500 m (Costa Fraga et al. 2014, Neto et al. 2009).

The oil produced from Santos Basin is with hight content of CO_2 (e.g. 8%–12% in Lula oil field) (Beltrao et al. 2009, de Abreu Campos et al. 2017, Neto et al. 2009). The occurrence of CO_2, in the presence of water, produces carbonic acid (H_2CO_3) which reduces the pH of the environment and causes uniform and localized corrosion in carbon steel (Beltrao et al. 2009). Therefore, the subsea flowlines in Santos Basin area are facing the risk of internal CO_2 corrosion damage and needs the corresponding monitoring and control measures (Beltrao et al. 2009, Filho et al. 2009).

Besides, considering that the ultra-deep water spreads over very large area, high pressure and low temperature also challenges the oil field development. New concepts for subsea pipelines and risers have been proposed to achieve flow assurance and structural reliability in the deep water environment. Sandwich pipe (SP) has been considered as one of the alternatives for the single wall pipe (SW). Numerical simulations and experimental collapse tests indicate high structural capacity for SP compared to SW (Estefen et al. 2005, Kyriakides and Netto 2000). Flow assurance issues were numerically simulated, indicating a good insulation performance of SP compared with SW (Estefen et al. 2016). However, the internal CO_2 corrosion for SP has not been discussed in detail yet, which is an important issue for scenarios such as the Santos Basin area.

Therefore, this paper aims at analyzing the internal CO_2 corrosion for SP through numerical simulation. The effect of several production aspects is discussed, including flow pressure, fluid temperature, water cut, flow rate and flow line profile. Based on the simulation results, suggestions of measures to reduce internal CO_2 corrosion are proposed.

2 BASIC INFORMATION

2.1 *Configuration of sandwich pipe*

The concept of sandwich pipe (SP) is relatively recent, and several studies are under development to support its implementation. Typically, the SP contains three layers, including two internal and external thin layers and a thick central core as the annular, as shown in Figure 1. The central layer characteristics satisfy simultaneously mechanical and thermal requirements. Therefore, structural strength combined with adequate flow assurance can be obtained.

In this paper, a SP with 203.2 mm (8 in) internal diameter is assumed. Internal and external layers

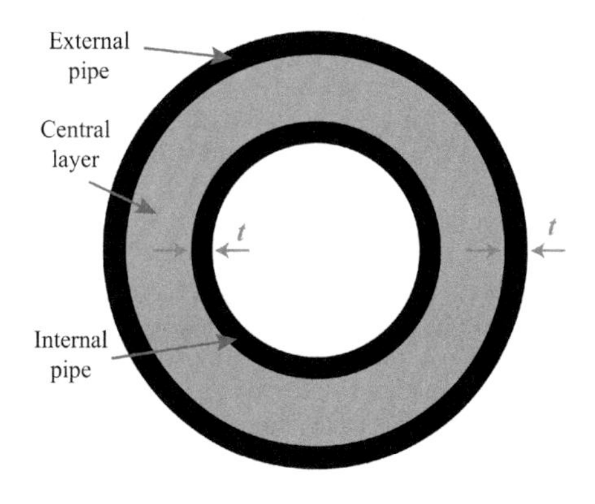

Figure 1. Typical section of a sandwich pipe (SP).

Table 1. Sandwich pipe configuration.

Layer	Thickness (mm)	Material	Density (kg/m^3)	Heat Capacity ($J/kg \cdot K^{-1}$)	Thermal Conductivity ($W/kg \cdot K^{-1}$)
Internal	3	Carbon steel	7850	500	45
Central	31	SHCC	1473	880	0.28
External	3	Carbon steel	7850	500	45

are 3 mm thick carbon steel, and the central core is a strain hardening cementitious composite (SHCC), which have both suitable strength and insulation performance (An et al. 2012, Estefen et al. 2016, Su et al. 2005). The thickness of SHCC is 31 mm to meet the strength requirement. Table 1 presents the material and related properties for each layer.

2.2 *Oil properties*

The oil employed in this paper is API grade 18, and viscoisty 15cp at standard condition. Related oil properties are listed in Table 2.

2.3 *Internal CO_2 corrosion simulation models*

Many different mathematical models for CO_2 corrosion prediction have been developed based on the theory and extensive corrosion data obtained from experiments conducted in laboratories and field monitoring.These models can be divided into threetypes: empirical, semi-empirical and mechanism models (Kahyarian et al. 2016, Nešić 2007, Peng and Zeng 2015).

The corrosion simulation is conducted by the software OLGA. Currently, OLGA has three corrosion models implemented, the NORSOK M-506 model (Norsork 2005), the de Waard 95 model (De Waard et al. 1995) and the IFE top-of-line corrosion model (Nyborg and Dugstad 2007). Each model haas its feasible scope of application and limitations. Detailed mathematical models are not presented here.

The top-of-line model focuses on the corrosion at the top of the pipe line and is only fit for the situation of stratified flow. Therefore, the application of this model is limited. The NORSOK and de Waard 95 models are both regarded as conservative models as they include only limited effects of protective corrosion films OLGA. Besides, many other factors are considered, such as CO_2 partial pressure (which is the product of system pressure and CO_2 mole fraction), pH value, andso on. The NORSOK model is more accurate when the fluid temperature is 100–150°C compared with other temperature condition, while the de Waard 95 model can provide relatively reliable results when the temperature is lower than 80°C (Norsork 2005, De Waard, Lotz, & Dugstad 1995). Besides, in software OLGA, the largest CO_2 mole fraction that are acceptable is 10%.

Table 2. Oil properties.

API grade	18
GOR, Sm^3/Sm^3	150
Viscosity at standard condition, ($mPa \cdot s$)	15
CO_2 mole fraction	0.1
Gas specific gravity	0.64

In practical situation, the fluid temperature inside the subsea pipe is usually lower than 100°C, therefore, the de Waard 95 model is more applicable for our work. Besides, according to the fluid properties provided in Table 2, the CO_2 mole fraction is 10%, which does not exceed the upper limit set by OLGA. As a results, de Waard 95 model is selected to study the sandwich pipe internal corrosion simulation.

3 CORROSION PREDICTION COMPARISON BETWEEN SANDWICH PIPE AND SINGLE WALL PIPE

The CO_2 internal corrosion rate of sandwich pipe and single wall pipe are compared. The single wall pipe is made of carbon steel, which is the same material as the internal layer of the sandwich pipe. The two types of pipe have the same profile and internal diameter. The flowline length is 20.07 km, and the profile is presented in Figure 2. The internal diameter is 203.2 mm, as already mentioned in section 2.

The seawater temperature is 2.2°C. Polypropylene foam is used as external insulation layer for single wall pipe, with thickness of 20 mm, to obtain the same U-value as the sandwich pipe, causing a similar temperature drop along the flowline. The U-value is assumed as 10.5 $W/m^2 \cdot K^{-1}$. The liquid flow rate is 9000 sm^3/d and the inlet fluid temperature 70°C. The simulated corrosion rates of the two types of pipe are shown in Figure 3.

Figure 3 indicates that under the given pressure and temperature condition, the corrosion rates of the two types of pipe are similar. The average corrosion rate for sandwich pipe is 1.25 mm/year, and for single wall pipe, the value is 1.21 mm/year. However, it should be note that to keep the fluid temperature, the single wall pipe needs extra external insulation layer, which increases the pipe cost. As a result, for similar corrosion rates, sandwich pipe is more economical. Besides, as shown in Figure 4,

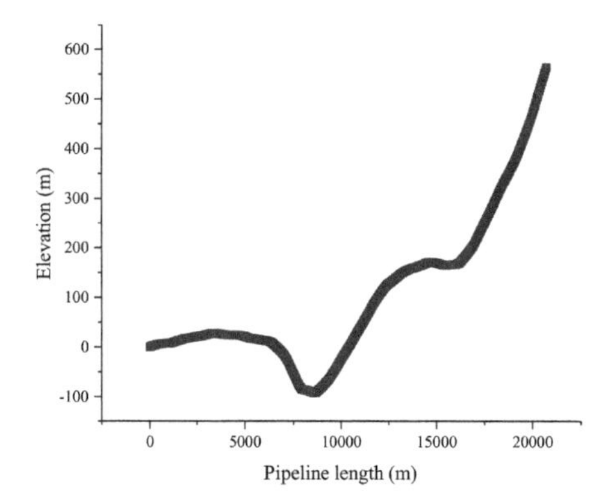

Figure 2. Flowline profiles.

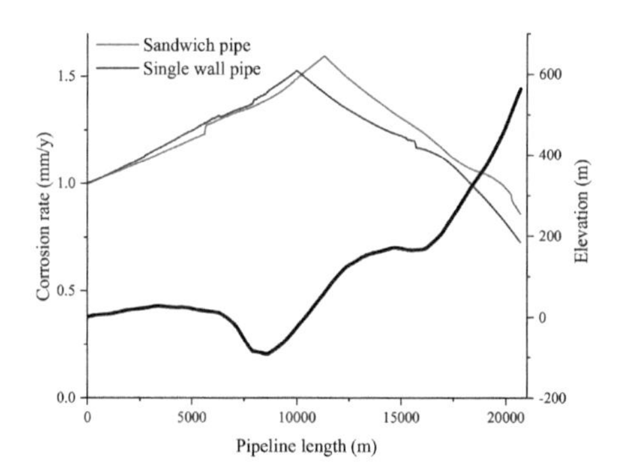

Figure 3. Corrosion rate comparison between sandwich pipe and single wall pipe.

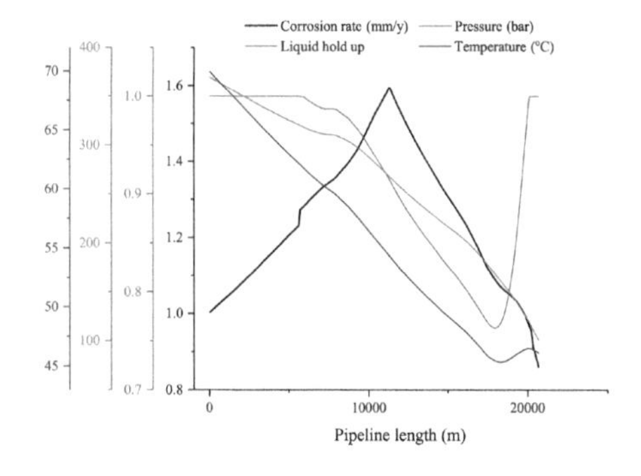

Figure 4. Distributions of corrosion rate, liquid hold up, pressure and temperature.

the corrosion rate of sandwich pipe related to many factors, including the pressure, the temperature, the liquid hold-up, so that the corrosion rate is possible to be reduced through changing these factors. Therefore, in the nextsection, parametric studies are conducted to analyze the detailed relationship among the above mentioned factors.

4 PARAMETRIC STUDY

4.1 *Pressure and temperature*

To see the effect of fluid temperature on CO_2 internal corrosion, the isothermal mode in OLGA is selected to obtain constant fluid temperature along the flowline. Besides, a short horizontal flowline interval with the length 100 m is used, so that the pressure drop along the flowline will be relatively small. The mean pressure of the inlet and outlet is approximately regarded as the pressure at the mid position of the flowline. The flow rate is 10000 sm^3/d, and the water cut is 0.2. Outlet pressure is chosen from 60bar to 100bar, and the temperature is chosen from 40°C to 80°C. With different combination of outlet pressure and temperature, the corresponding internal CO_2 corrosion rate along the flowline can be predicted by OLGA. Then, the corrosion rate, temperature and pressure at the middle position of the flowline are extracted, the relationship among these three parameters are presented by a contour map, as shown in Figure 5.

According to Figure 5, it is found that the increase of fluid pressure results in the increase of corrosion rate. The reason is that higher fluid pressure brings higher partial pressure of CO_2, resulting in lower fluid pH, which increases the corrosion reaction (Peng and Zeng 2015).

The effect of temperature on the corrosion rate is not monotonous. On the left side of the dotted line, the corrosion rate increases with the temperature, while on the right side, the corrosion rate decreases. This phenomenon is mainly due to the protective effect of corrosion product films and the scale formation are strengthened when the temperature increases (Nešić 2007, Peng and Zeng 2015). Therefore, the peak of corrosion rate occurs. For this case, the peak corrosion rate occurs when the temperature is about 65°C–70°C, which is the white dotted line. The trend presented by the simulation results are consistent with the experiment by Peng and Zeng (Peng and Zeng 2015).

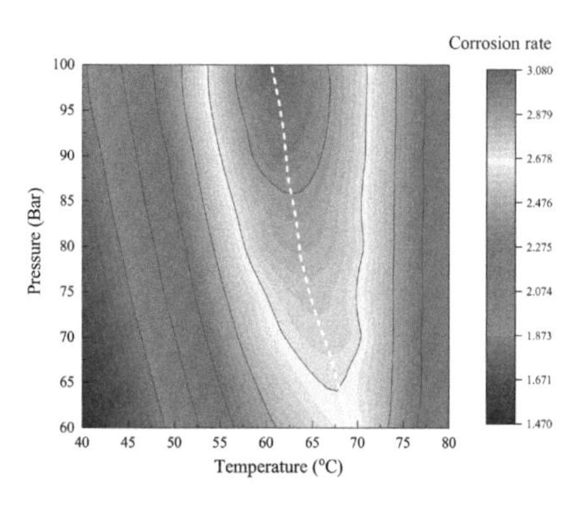

Figure 5. Internal corrosion rate under different temperature and pressure.

The effect of temperature and pressure on corrosion rate indicates that in practical situation, carefully monitored both temperature and pressure could help identify the position with high risk of CO_2 corrosion.

4.2 *Liquid flow rate and water cut*

The 100 m horizontal sandwich pipe is used again, as well as the isothermal calculation mode. The fluid temperature is chosen to be 70°C and the outlet pressure is 100bar. The corrosion rates under different liquid flow rate and water cut are simulated, which results are shown in Figure 6.

As can be seen from Figure 6, under the same flow rate, the higher water cut results in lower corrosion rate, due to the reduction of the volume of oil and CO_2. Practically, the water cut increases along the whole development life of an oil field. Therefore, the initial stage with relatively lower water cut may bring higher risk of internal corrosion. So, the corrosion monitoring should be conducted very carefully at the initial stage of development, and high efficient corrosion inhibitor could be considered to reduce the corrosion rate.

The higher liquid flow rate brings higher corrosion rate. Nešić summarized that high velocity of fluid flow enhances the transport of solid species (e.g. corrosion products, scales) towards and away from the metal surface, which may lead to an increase in the corrosion rate (Nešić 2007). As a result, if the liquid flow rate is high, high efficient corrosion inhibitor could be used, or if possible, water could be added into the fluid, which increases the fluid water cut and reduces the corrosion rate. However, this procedure needs to be further verified for its feasibility based on technical and economic issues.

4.3 *Flowline profile*

Due to the seabed topography, the subsea flowline is undulated. There are three basic profiles: horizontal, downhill-uphill, and uphill-downhill, as shown in Figure 7. A long flowline profile could be regarded as the combination of these ones. Toanalyze the effect of the flowline profile, the internal corrosion rate in these three basic profiles are simulated and compared. Horizontal distances between the inlet and outlet, L_h, of these three profiles are the same. And the vertical elevation L_v in Figures 7(b) and 7(c) are identical. It is assumed that L_h is 2000 m and L_v is 50 m. Since L_h is much larger than L_v, the total lengths of these three flowlines are relatively close, therefore, the pressure loss due to pipe wall friction are regarded as the same. For all the three cases, isothermal calculation mode in OLGA is selected, the temperature is set to be 55°C. The liquid flow rate is 9000 sm³/d, the water cut is 0.2, and the outlet pressure is 100 bar.

According to Figure 7(b), the relationship between the flowline interval length, horizontal distance and vertical elevation are obtained through equation (1) and (2):

$$L_1 = \sqrt{x^2 + L_v^2} \tag{1}$$

$$L_2 = \sqrt{(L_h - x)^2 + L_v^2} \tag{2}$$

The inflection points at $x = L_h/2$ in both Figure 7 (b) and (c) are selected as examples to compare the corrosion rates for the three cases. The results are presented in Figure 8.

From Figure 8, it is found that for horizontal profile, the corrosion rate distributes smoothly along the flowline. However, for the "downhill-uphill" and

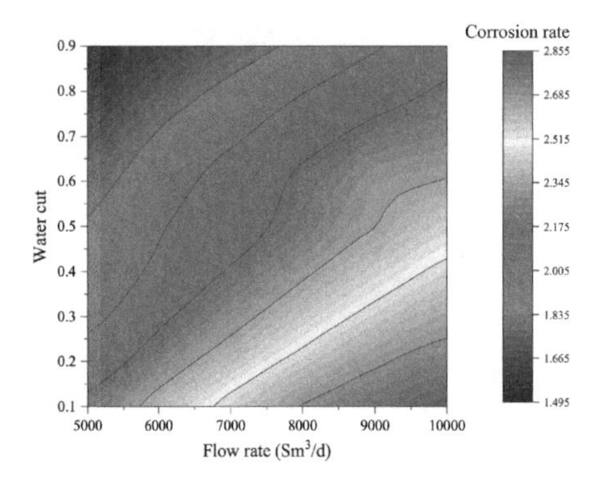

Figure 6. Internal corrosion rate under different flow rate and water cut.

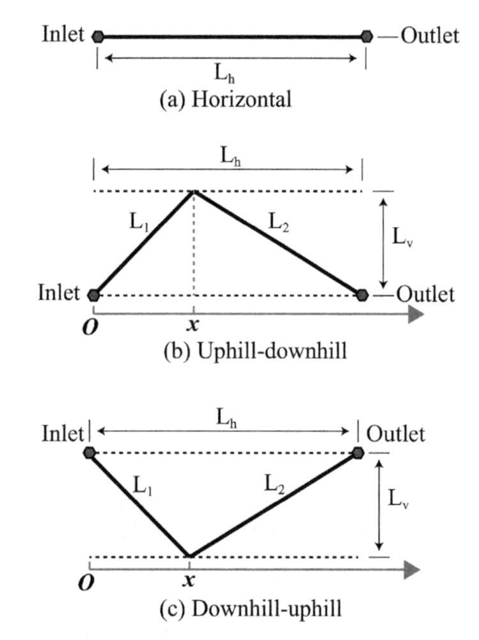

Figure 7. Basic profiles of subsea flowline.

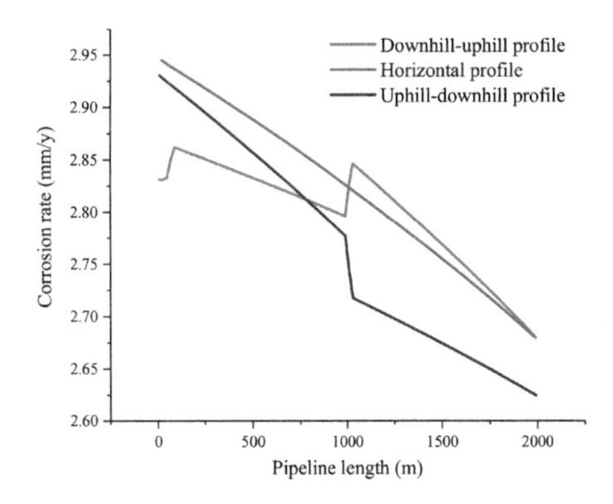

Figure 8. Corrosion rate distribution along the flowline for the three basic profiles.

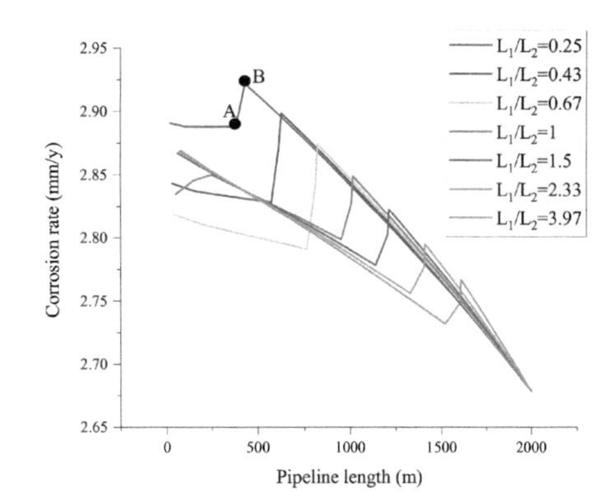

Figure 9. Corrosion rate distribution under different combination of pipe lengths.

Table 3. Different combinations of L_1 and L_2.

x (m)	L_1 (m)	L_2 (m)	L_1/L_2
400	403.11	1600.78	0.25
600	602.08	1400.89	0.43
800	801.56	1201.04	0.67
1000	1001.25	1001.25	1.00
1200	1201.04	801.56	1.5
1400	1400.89	602.08	2.33
1600	1600.78	403.11	3.97

"uphill-downhill" profiles, at the inflection point, sudden change of corrosion rate occurs. This is due to the multiphase flow characteristics changing at the inflection position, such as flow regime and liquid hold-up (Nešić 2007).

Besides, it is obvious that for the "downhill-uphill" profile, there is a local maximum corrosion rate around the inflection point, indicating higher risk of local corrosion. Therefore, in practical situation, the "canyon" part of a flowline should becarefully monitored. In addition, when determining the flow line route, we should pay attention to avoid letting the profile having too many undulated intervals, which may help reduce the flow line internal corrosion.

According to equation 1and2, different x bring different lengths of the two flowline intervals, L_1 and L_2. A series of x are selected and the corrosion simulation is conducted for each of them. The flowline interval lengths, L_1 and L_2 of each x is presented in Table 3.

For all these combinations, the local maximum corrosion rate occurs around the inflection position. However, different combinations result in different local maximum corrosion rates. As shown in Figure 9, the local maximum corrosion rate decreases with the increase of L_1/L_2.

Furthermore, each curve in Figure 9 has one local maximum corrosion rate, V_{cmax}, and one local minimum corrosion rate, V_{cmin}, for example, point B and

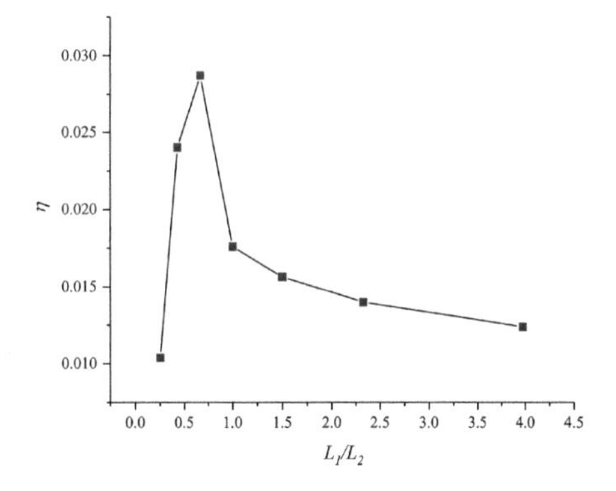

Figure 10. The relationship between L_1/L_2 and the corrosion rate sudden change amplitude η.

point A on the curve $L_1/L_2 = 0.25$. Then, the corrosion rate that sudden change amplitude η could be described by $(V_{cmax} - V_{cmin})/V_{cmin}$. High value of η means larger sudden change of the corrosion rate, indicating the possibility of localized corrosion. Different values of L_1/L_2 result in different values of η, as shown in Figure 10. With the increase of L_1/L_2, the value of η increases and, then, decreases. The largest value occurs when $L_1/L_2 = 0.67$. As a result, Figure 9 and Figure 10 reveal that the proper selection of the inflection position for the undulated flowline could reduce the internal corrosion rate.

5 CONCLUSIONS

In this paper, the CO_2 internal corrosion of sandwich pipe was analyzed. Software OLGA was used to simulate the internal corrosion rate. First of all, the corrosion rates for a long distance sandwich pipe and single wall pipe were compared. Although the corrosion rate of these two were similar, sandwich pipe is more economical. Then parametric studies were conducted to analyzed the effect of fluid temperature, pressure, flow rate, water cut, and flowline profile. Based on these results, several suggestions for dealing with the CO_2 internal corrosion were proposed:

1. Lower pressure results in lower corrosion rate, while the corrosion rate increases first and then decreas when the temperature increases. Therefore, careful monitoring of temperature and pressure is important, which could help identify the position with high risk of CO_2 corrosion.
2. Higher flow rate or lower water cut brings higher corrosion rate. High efficient corrosion inhibitor could be considered to reduce the corrosion rate under proper flow rate and water cut. In addition, water could be added into the fluid, which increases the fluid water cut and reduce the corrosion rate. But this method need to be further discussed for its feasibility.
3. Around the "canyon" part of the flowline, there will be the local maximum corrosion rate. The flowline route should avoid too many undulation intervals, which may help reduce the flow line internal corrosion.
4. For an undulation flowline, the inflection positions affect the corrosion rate due to the multiphase flow characteristic. As a result, proper selection of the inflection position of the undulated flowline could reduce the internal corrosion rate.
5. Some anti-corrosion materials could be used for the internal layer of sandwich pipe, to reduce the CO_2 corrosion rate. This is a relevant research topic considering that the inner layer of sandwich pipe is relatively thin, and, therefore, will not withstand the life cycle in conventional circumstances of fluids with CO_2 content.

ACKNOWLEDGEMENTS

The authors acknowledge the financial support from Petrogal Brasil (ANP R&D Program) and EMBRAPII COPPE Unit in Subsea Technology for the development of the research project Subsea Systems (SISTSUB). Special thanks to Schlumberger for providing the software OLGA to the Federal University of Rio de Janeiro.

REFERENCES

An, C., X. Castello, A.M. Oliveira, M. Duan, R.D.T. Filho, & S.F. Estefen (2012). Limit strength of new sandwich pipes with strain hardening cementitious composites (SHCC) core: Finite element modelling. In *International Conference on Ocean, Offshore and Arctic Engineering*, Rio de Janeiro, Brazil.

Beltrao, R., C.L. Sombra, A.C.V.M. Lage, J.R.F. Netto, & C.C.D. Henriques (2009). Challenges and new technologies for the development of the pre-salt cluster, Santos Basin, Brazil. In *Offshore Technology Conference*, Houston, Texas, U.S.A.

Costa Fraga, C.T., Q.A. Lara, A.C. Capeleiro Pinto, & C.C. Moreira Branco (2014). Brazilian Pre-Salt: An Impressive Journey from Plans and Challenges to Concrete Results. In *World Petroleum Congress*, Moscow, Russia.

de Abreu Campos, N., M.J. da Silva Faria, R.O. de Moraes Cruz, A.C.V. de Almeida, E.J. Rebeschini, H.P. Vaz, L.V. João, M.B. Rosa, & T.C. da Fonseca (2017). Lula Alto-Strategy and Execution of a Megaproject in Deep-Water Santos Basin Pre-Salt. In *Offshore Technology Conference Brasil*, Rio de Janeiro, Brazil.

DeWaard, C., U. Lotz, & A. Dugstad (1995). Influence of Liquid Flow Velocity on CO2 Corrosion: A Semi-Empirical Model. In *National assocaition of corrosion engineers anual conference*, Houston, Texas, USA.

Estefen, S.F., M.I. Lourenço, J. Feng, C.M. Paz, & D.B. de Lima Jr. (2016). Sandwich pipe for long distance pipelines: flow assurance and costs. In *International Conference on Ocean, Offshore and Arctic Engineering*.

Estefen, S.F., T.A. Netto, & I.P. Pasqualino (2005). Strength Analyses of Sandwich Pipes for Ultra Deepwaters. *Journal of Applied Mechanics 72*(4), 599–608.

Filho, J.M.F., A.C.C. Pinto, & A.S. Almeida (2009). Santos Basin's pre-salt reservoirs developmentThe way ahead. In *Offshore Technology Conference*, Houston, Texas, U.S.A.

Kahyarian, A., M. Singer, & S. Nesic (2016). Modeling of uniform CO2 corrosion of mild steel in gas transportation systems: A review. *Journal of Natural Gas Science and Engineering 29*, 530–549.

Kyriakides, S. & T. Netto (2000). On the dynamics of propagating buckles in pipelines. *International Journal of Solids and Structures 37*(46–47), 6843–6867.

Neto, E.A., P.S. Alonso, I.J.R. Filho,&F.G. Serpa (2009). Pre- Salt Cluster Long Term Supply Strategy. In *Offshore Technology Conference*, Houston, Texas, USA.

Nešić, S. (2007). Key issues related to modelling of internal corrosion of oil and gas pipelines - A review. *Corrosion Science 49*(12), 4308–4338.

Norsork (2005). CO2 Corrosion rate calculation model. *Majorstural, Norway: Norwegian Technological Standards Institute Oscarsgt 20.*

Nyborg, R. & A. Dugstad (2007). Top of line corrosion and water condensation rates in wet gas pipelines. In *NACE International corrosion conference and exhibition*, Nashville, Tennessee, USA. OLGA. *Schlumberger. OLGA user manual.*

Peng, S. & Z. Zeng (2015). An experimental study on the internal corrosion of a subsea multiphase pipeline. *Petroleum 1*(1), 75–81.

Su, J., D.R. Cerqueira, & S.F. Estefen (2005). Simulation of Transient Heat Transfer of Sandwich PipesWith Active Electrical Heating. *Journal of Offshore Mechanics and Arctic Engineering 127*(4), 366–370.

Progress in Maritime Technology and Engineering – Guedes Soares & Santos (Eds)
© 2018 Taylor & Francis Group, London, ISBN 978-1-138-58539-3

Ceramic coating solution for offshore structures

S. García, A. Trueba, L.M. Vega & E. Madariaga
Department of Sciences and Techniques of Navigation and Shipbuilding, University of Cantabria, Santander, Spain

ABSTRACT: This study highlights the need for offshore structures to consider the choice of material used in seawater applications to minimize microbial-associated and corrosion problems. Corrosion is a major problem in offshore environments due to extreme operating conditions and the presence of aggressive corrosive elements in a wide range of offshore structures. This study evaluated four different ceramic coatings were made of incorporating active ceramic particles against biofouling as copper, silver, zinc and titanium. These antifouling coatings were exposed for one year in shallow marine environments. Results were no significant differences in the biofouling attached to ceramic coatings. Instead, there were significantly higher biofouling attached to antifouling paints. Biofouling adhesion resistance was greatest when a coating thickness of 100 μm was used and when the substrate surface roughness was 0.04 μm. The results indicated up to more 30% total area covered by biofouling in paint coatings than ceramic coatings.

1 INTRODUCTION

The serious consequences of the corrosion process have become a major problem in offshore environments due to extreme operating conditions and the presence of aggressive corrosive elements. The corrosion resistance can represent the difference between trouble-free long-term operation and costly downtime. Steel structures situated above the seawater in the so-called atmospheric zone are in a high corrosivity category, with a corrosion rate in the range of 80–200 μm per year (De Baere et al. 2013). The conventional approach to prevent of external corrosion is to cover the exposed surfaces with a high-efficiency coating. However, the current polymeric coatings for offshore structures require periodical maintenance that includes a number of challenges associated with the physical environment in which the work takes place Thus, i.e., the application of coating systems for offshore steel towers in modern facilities can cost up to 15 to 25 €/m, depending on the work conditions and on the coating systems. Repairs of the corrosion protection can be from 5 to 10 times more expensive and total cost can easily rise to more than 1,000 €/m^2 (Momber 2011; Price & Figueira 2017).

Offshore constructions act as artificial reefs, supporting an undesirable accumulation of marine life by offering habitat for microorganism, algae, fish and invertebrates (Van Der Stap et al. 2016). Biofouling is defined as the undesirable phenomenon of adherence and accumulation of biotic deposits on a submerged artificial surface in contact with seawater (Trueba et al. 2015a). Corrosion and biofouling have mutual effects on each other: biofouling may induce corrosion, but corrosion may also induce biofouling. In that sense, both parameters need to be addressed in conjunction. This phenomenon applied in tidal turbines produces an important power loss. All current fouling control products are chemically active antifouling paints with biocides or non-stick fouling release coating. Today there is still debate regarding the most optimal coating systems for offshore structures (Singh 2014; Momber et al. 2015). The service life of antifouling paints is defined by the coating thickness and has a considerable environmental impact besides from continuously maintenance operations due to limited life cycle and durability (Yebra et al. 2010). While control of corrosion for shipping is an enormous market, dynamic performance coatings for ships are the subject of intense development by paint companies and other researchers and specific needs of this sector are intensely cost sensitive (Zargiel & Swain 2014). On the other hand, offshore coatings should be targeted towards static structures, where periodic dry docking is often impossible and in-situ maintenance is difficult and costly. Such structures require a long term solution to corrosion and biofouling and coating systems used for shipping are too short lived to be entirely suitable. Current antifouling paints for static offshore structures use similar formulations that are used in the shipping industry and rely either on biocides that gradually leach out of the coating (damaging surrounding ecosystems and only providing a limited time period of effective bio-activity) or on so called 'self-polishing' systems that require a certain water velocity to remove accumulated growth (unsuitable for static structures). Nowadays, coating solutions to combat corrosion offshore include marine paints supplemented by sacrificial anodes. Organic paint coatings have limited lifetimes (<10 years) before needing substantial maintenance and repair (Y. Zhang 2016). Any damage to the coating results in rapid corrosion of the underlying steel. Sacrificial anodes can provide added protection in immersed regions, but are costly to install and replace and provide no protection in splash and tidal zones where corrosion is most severe. In this sense, the offshore industry identifies research needs and opportunities for implementing new solutions to overcome the challenges of current painting systems in terms of durability associated to corrosion protection, mechanical resistance and fouling avoidance (Siewert 2005). A competitive approach used for similar and harsh environments are related to ceramic enamel coatings which have an excellent chemical and abrasion resistance due to their sintered vitreous structure.

Ceramic materials are attractive materials due to their characteristics like high chemical resistance, wear resistance, thermal resistance and corrosion resistance (Wahab et al. 2017). Bulk ceramic materials can be found in refractories, ceramic tiles, advanced ceramics like low temperature cofired ceramic but also ceramic coatings have this unique performance. Ceramic tiles are coated with ceramic glaze layers to provide an easy to clean, wear resistant and high aesthetic surface, result of this performance is an annual production on European level over 2000 Millions m^2 of ceramic coated tiles. Ceramic coatings are used for instance in the protection of airplane turbines, where ZrO_2 is applied by Thermal spray; also chemical reactors are protected with a ceramic enamel coating. The advantage of ceramic enamel coatings is the capability to tailor the properties of the vitreous matrix with other oxide and functional particles (Black, J.T., & Kohser 2017).

Ceramic coatings either enamels or glazes are made up of a mixture of molten glass and diverse additives. The molten glass can have diverse compositions, being made up of oxides like SiO_2, B_2O_3, Al_2O_3, Na_2O, K_2O, CeO_2, MgO, ZnO, CaO, ZrO_2, TiO2 and adhesion promoters for metal surface like, CoO_2, NiO, Fe_2O_3, MnO; the characteristic that a glass can be adjusted in composition allows modelling their final properties, like thermal expansion coefficient, melting point, chemical resistance and biofouling adhesion. Ceramic coatings can be modified by addition of raw materials like quartz, titanium oxide, zircon silicate or ceramic pigments to adjust to the metal type and final application product (Carter & Norton 2007). Until now the use of ceramic enamel to protect metallic structures has been limited by many factors. Conventional slurry application techniques (e.g. dip coating, wet spraying) are not suitable for large parts except for chemical reactors and panels. Even more importantly, a sintering thermal treatment is needed to consolidate the coating, that could be incompatible with some engineering materials (e.g. light alloys, quenched and tempered steels, etc.), as they would bring about unacceptable microstructural changes accompanied by a loss of mechanical strength apart from warpage and distortion of the coated items.

Thermal spray processes encompass a family of coating deposition techniques characterized by the use of a high-velocity and/or high-temperature gas stream to project softened/molten droplets of the coating material towards the substrate. Whilst the droplets may attain very high temperatures (hence, even refractory coating materials can be processed), the substrate remains relatively cold as it is rapidly scanned by the gas+droplets stream along typical raster patterns (Fauchais et al. 2014). A large variety of coating/substrate material combinations is therefore possible; in particular, ceramic enamels can be sprayed (Bao et al. 2013) onto relatively cold substrates, thus avoiding overheating, microstructural alterations and distortions. Moreover, thermal spraying techniques are applicable to large structures and, with due adjustments, they are portable for on-site work (Paul 2013). The challenge to obtain tight, corrosion-resistant ceramic enamel coatings by thermal spraying resides in the typical defectiveness which include voids and gaps between the flattened droplets (lamellae), microcracks within the lamellae (due to their rapid cooling after deposition), and entrained gases. These limits have, up to now, hindered the industrial uptake of thermal spray glass coatings which have been developed and validated at lab-scale for a variety of applications on metallic and ceramic substrates (Bolelli et al. 2008).

This study evaluated the antifouling (AF) action of a ceramic coated exposed to the seawater. The application of ceramic coated was developed by thermal spray. The aim was to minimise the biofilm adhesion on the surface and study the effect of new coated in composition and structures of the biofilms produced. The scientific relevance of this research in AF is very highlight because it involves a new environment friendly technology against biofouling to improve efficiency and productivity in offshore structures.

2 MATERIAL AND METHOD

2.1 *Area of study*

The geographic area in which the study was conducted and where was installed offshore Marine Laboratory of Biofouling Group Research was

Santander Bay (43°28′N and 3°48′W). Located in the Cantabrian Sea, it is the largest estuary on the northern Spanish coast, comprising an area of 22.42 km^2.

2.2 *Preparation of samples*

Four different ceramic coatings were deposited by thermal spraying technique over carbon steel (A569/A569M6, 3 mm thick by 250 mm × 250 mm) which will be visually examined and tested according to ASTM D790. Figure 1 shows how composition of each ceramic coating which were obtained by the optimal formulation parameters developed by University of Cantabria. In order to ensure that the coating not only serves an antifouling purpose, but also protects the substrate against corrosion, two approaches were explored in this research.

As per ISO 20340, steel structures for coastal and offshore areas are classified as C5-M (due to the high salinity) and should be coated where minimum requirements for protective paint systems. Therefore, Prior to coating application, the sample surface was blast-cleaned in order to get a final surface roughness of Sa2.5 or Sa3 (ISO 8503) and cleanliness (ISO 8501). The surface to be coated was clean, dry, free from oil/grease, and had the specified roughness and cleanliness until the first coat is applied. Dust and blast abrasives were removed from the surface after blast cleaning (not exceed rating 2 of ISO 8502–3). Then, paint coating was applied on sample surface with a total thickness of 300 μm.

The application of ceramic coatings were developed through atomistic deposition processes to get lower deposition rates and thinner coatings. Firstly, it was the deposition of dense metallic under-layers between the functional ceramic topcoat and the substrate. These serve a two-fold purpose: protecting the substrate from corrosion and enhancing the adhesion strength of the ceramic enamel top layer. They were deposited by the high velocity air-fuel spray process using compressed air instead of oxygen. HVAF torches indeed generated even higher particle velocities and lower particle temperatures, but they demonstrated as a viable means to deposit watertight metal coatings with low defectiveness. Furthermore, a different coating with a biocide free silicone coating (Silicone FR) were applied to carbon steel specimens as instructed by manufacturers. Samples were cleaned with FreeBact20 (AquaFix, Satsjbaden, Sweden) and sterile water and airdried before exposure. The samples were photographed prior to the experiment.

The surface topographies of the samples were denoted according to the standard specifications of the American Society of the International Association for Testing and Materials (ASTM). Their chemical compositions of coating are presented in Figure 2. Their surface roughness values were measured using a surface roughmeter (Mitutoyo, Surftest SJ-201 Series) in accordance with the guidelines established in the standard ASME/ANSI B46.1-2009.

2.3 *Biofouling assessment*

The biofouling assessment tested and analyze the behavior of different coatings against marine biofouling.

These samples assayed had 4 different coatings over carbon steel with same geometry manufactured. These samples were then submerged during 365 days in the shallow marine environments, at a depth of 0.5 m.

Biofouling analysis of samples in laboratory consisted of: i) Measurement of barnacle adhesion strength in shear as follow ASTM D5618-94. This test method covers the measurement of barnacle adhesion in shear to surfaces exposed in the marine environment, ii) analysis quantitative and qualitative of biofouling on sampled surfaces and analysis optical by microscope of biofouling as qualitative analysis.

The evaluating biofouling resistance and physical performance of marine coating system in related ASTM D6990-05. This practice establishes a practice for evaluating degree of biofouling

Base Material A-569													
Coating	Ceramic components												
	SiO_2	B_2O_3	Al_2O_3	Na_2O+K_2O	CaO+MgO	MnO	CoO	NiO	Fe_2O_3	TiO_2	ZrO2	ZnO	V_2O_5
Ceramic Coating (ZrO_2 base)	52	11	6	15	1	1	0.5	1	0.5		12		
Ceramic Coating (ZnO base)	50	10	5	13	5	0.5	1	1.5				14	
Ceramic Coating (TiO_2 base)	37	2	0.5	36						10			7
Paint Coating	Hempel® Hempasil X3 (Silicone FR)												

Figure 1. Ceramic composition adjustment.

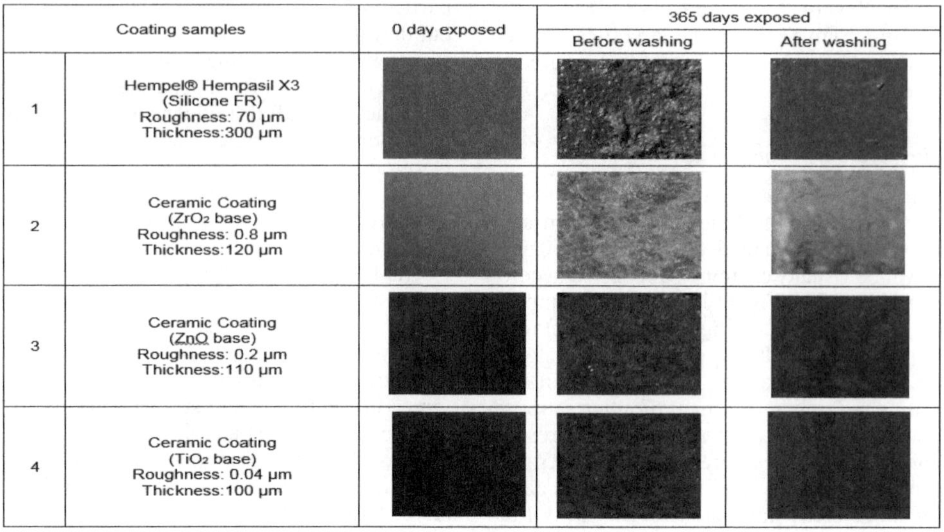

Coating samples		0 day exposed	365 days exposed	
			Before washing	After washing
1	Hempel® Hempasil X3 (Silicone FR) Roughness: 70 μm Thickness:300 μm			
2	Ceramic Coating (ZrO_2 base) Roughness: 0.8 μm Thickness:120 μm			
3	Ceramic Coating (ZnO base) Roughness: 0.2 μm Thickness:110 μm			
4	Ceramic Coating (TiO_2 base) Roughness: 0.04 μm Thickness:100 μm			

Figure 2. Samples of different coatings assayed.

settlement on and physical performance of marine coating systems when panels coated with such coating systems are subjected to immersion conditions in a marine environment.

2.4 *Experimental setup*

The coated samples were exposed between 01 October 2016 and 01 October 2017. Coated samples were tested under realistic conditions of exposure of the submerged zones in their offshore Marine Laboratory. They exposed in offshore conditions during 12 months and were checked monthly by visual inspection following the ASTM D 3623-78a.

Temperature of seawater were measured daily during the experiment period. The chemical parameters of seawater were measured daily a month during the experiment period.

3 RESULTS AND DISCUSION

The quantitative and qualitative analyses of marine microbial fouling communities is of great interest to design new AF coatings (Muthukrishnan et al. 2014). The antifouling performance of different coatings showed the typical images of the tested panels after immersion in natural seawater for 365 days. The cumulative release can be found in Figure 2. Surface chemistry is a significant factor in the formation, stability, and release of adhesion of fouling organisms to surfaces (Kim 2015). The sample coating No. 1 was covered by 100% of hard fouling organisms after 365 days and produced at 70% losses of AF coating on the surface. Furthermore, it was fouled by 10% of filamentous, 28% of barnacles, 42% of algae and 20% biofilm. The sample coating No. 2 was covered by 92% of biofouling organisms and produced at 24% losses of AF coating on the surface. Furthermore, it was fouled by 11% of filamentous, 3% of barnacles, 30% of algae and 50% biofilm. The sample coating No. 3 was covered by 82% of biofouling organisms and produced at 12% losses of AF coating on the surface. Besides, it was fouled by 20% of filamentous, 2% of barnacles, 30% of algae and 29% biofilm. The sample coating No. 4 was covered by 70% of biofouling organisms and produced at 4% losses of AF coating on the surface. Furthermore, it was fouled by 17% of filamentous, 24% of algae and 29% biofilm. Thus, sample coating No. 4 had the best antifouling release performance under static conditions.

The sample coating No. 1 was covered by some hard fouling organisms after 365 days and produced at 70% losses of AF coating on the surface (Figure 3). The effectiveness of non-stick coatings is based on a combination of surface free energy, surface structure and surface roughness (Bott 2011).

The sample coating No. 1 silicon-based produced the depletion and leaching of these silicon-based biofouling as the surface wears out, leading to changes in the surface chemical composition (eventually also topography) and lowering of the AF performance (Witucki & Pajk 2004). For this reason, coating 1 did not have long durability and high AF performance levels all through the coating lifecycle. On the other hand, according to Flemming (2009), ceramic coatings showed small damages on the coating. Ceramic coating 4 showed up to

Test Surface No.		Fouling rating %	Physical Condition (Defects) %	% Cover			
				filamentous bryozoans	barnacles	algae	Biofilm
1	Hempel® Hempasil X3 (Silicone FR)	100	70	10	28	42	20
2	Ceramic Coating (ZrO_2 base)	92	24	11	3	30	50
3	Ceramic Coating (ZnO base)	82	12	20	2	30	30
4	Ceramic Coating (TiO_2 base)	70	4	17	-	24	29

Figure 3. Report of antifouling performance.

20% of better effective biofouling-resistant than Ceramic coating 2, and 8% than Ceramic coating 3. As reported by Wagh (2016), these different AF characteristics of the glass in ceramic coating depended on the adjust of composition which allows modelling their final properties.

As previous research by García et al. (2016) and Trueba et al. (2015), the thickness and nature of the deposits forming the biofouling film depend on factors such as the surface roughness, the nutrient input, and the velocity of circulation of the seawater around the tube. It is not easy to associate the adhesion process of biofouling with any one of these factors. The initial surface roughness of sample coating No. 1 was 300 μm which reached the greater the final amounts of microbial adhesion and the faster the subsequent colonization of the surface than other sample coatings. Therefore, the initial surface roughness determined the level of bacterial adhesion.

The sample coating No. 4 with low roughness had less biofouling adhesion than other rougher sample coatings. The surface chemistry and physical properties of the substratum are both crucial to preventing the recruitment of biofouling organisms (Nurioglu et al. 2015). Surface chemistry was a significant factor in the formation, stability, and release of adhesion of biofouling organisms to surfaces. The change in wettability of a surface due to microtopographical roughness was likely to be a contributing factor to antifouling properties.

The physical-chemical properties of seawater were a temperature between 12–22°C, conductivity between 39–60 mS m^{-1}, between pH 8–8,2 and 8,15–8,65 oxygen dissolved for the experimentation, which they were appropriate for growing biological organisms (Dürr & Thomason 2009). According with Nuraini et al. (2017),the influence of sea water characteristics such as temperature, pH, conductivity and oxygen concentration on the biofouling of immersed solid surfaces played a significant role in the growing of the biofouling.

4 CONCLUSIONS

This study demonstrates that biofouling is extensive and formed by a diverse group of microorganisms in coatings with different compositions. Therefore, the addition of different compositions into glass of coating affects the number and species of microorganisms.

The roughness of the surface of coatings directly affects biofilm development on coating surface, thus maximizing the negative consequences of biofilm deposits. Marine field tests have demonstrated that the ceramic coatings have an excellent antifouling performance even under static conditions for more than 1 year. The functionality of ceramic coatings is based on antifouling efficiency relies on low adhesion strength and diverse AF additives to reject biofouling adhesion. The ceramic coatings with suitable compositions are recommended to the applications in marine anti-biofouling.

REFERENCES

Bao Y, Gawne DT, Gao J, Zhang T, Cuenca BD, Alberdi A. 2013. Thermal-spray deposition of enamel on aluminium alloys. Surf Coatings Technol. 232:150–158.

Black J.T., & Kohser RA. 2017. DeGarmo's materials and processes in manufacturing. John Wiley & Sons. New York, USA.

Bolelli G, Rauch J, Cannillo V, Killinger A, Lusvarghi L, Gadow R. 2008. Investigation of High-Velocity Suspension Flame Sprayed (HVSFS) glass coatings. Mater Lett. 62:2772–2775.

Bott TR. 2011. Industrial biofouling. Elsevier. Edgbaston, UK.
Carter B, Norton G. 2007. Ceramic Materials Science and Engineering. Springer. NY, USA.
De Baere K, Verstraelen H, Rigo P, Van Passel S, Lenaerts S, Potters G. 2013. Study on alternative approaches to corrosion protection of ballast tanks using an economic model. Mar Struct. 32:1–17.
Dürr S, Thomason JC. 2009. Biofouling. Newcastle-upon-Tyne, UK.
Fauchais PL, Heberlein JVR, Boulos MI. 2014. Thermal spray fundamentals: From powder to part. Springer. Minneapolis, USA.
Flemming H. 2009. Why Microorganisms live in biofilms and the problem of biofouling. Mar Ind biofouling. 4:3–12.
García S, Trueba A, Vega LM, Madariaga E. 2016. Impact of the surface roughness of AISI 316 L stainless steel on biofilm adhesion in a seawater-cooled tubular heat exchanger-condenser. Biofouling. 32:1185–1193.
Kim S-K. 2015. Handbook of Marine Biotechnology. Springer. Busan, Korea.
Momber A. 2011. Corrosion and corrosion protection of support structures for offshore wind energy devices (OWEA). Mater Corros. 62:391–404.
Momber AW, Plagemann P, Stenzel V. 2015. Performance and integrity of protective coating systems for offshore wind power structures after three years under offshore site conditions. Renew Energy. 74:606–617.
Muthukrishnan T, Abed RMM, Dobretsov S, Kidd B, Finnie AA. 2014. Long-term microfouling on commercial biocidal fouling control coatings. Biofouling. 30:1155–1164.
Nuraini L, Prifiharni S, Priyotomo G, Sundjono, Gunawan H. 2017. Evaluation of anticorrosion and antifouling paint performance after exposure under seawater Surabaya-Madura (Suramadu) bridge. In: AIP Conf Proc. Vol. 1823. Bandung, Indonesia.
Nurioglu AG, Esteves ACC, de With G. 2015. Non-toxic, non-biocide-release antifouling coatings based on molecular structure design for marine applications. J Mater Chem B. 3:6547–6570.
Paul S. 2013. Corrosion control for marine-and land-based infrastructure applications. Spray AH, Technology. Cambridge, UK.
Price S, Figueira R. 2017. Corrosion Protection Systems and Fatigue Corrosion in Offshore Wind Structures: Current Status and Future Perspectives. Coatings. 7:25.
Singh R. 2014. Corrosion Control for Offshore Structures. Elsevier. Oxford, UK.
Trueba A, García S, Otero FM, Vega LM, Madariaga E. 2015a. The effect of electromagnetic fields on biofouling in a heat exchange system using seawater. Biofouling. 31.
Trueba A, García S, Otero FM, Vega LM, Madariaga E. 2015b. Influence of flow velocity on biofilm growth in a tubular heat exchanger-condenser cooled by seawater. Biofouling. 31:527–534.
Van Der Stap T, Coolen JWP, Lindeboom HJ. 2016. Marine fouling assemblages on offshore gas platforms in the southern North Sea: Effects of depth and distance from shore on biodiversity. PLoS One. 11.
Wagh AS. 2016. Chemically Bonded Phosphate Ceramics: Twenty-First Century Materials with Diverse Applications: Second Edition. Elsevier. Oxford, UK.
Wahab JA, Ghazali MJ, Baharin AFS. 2017. Microstructure and Mechanical Properties of Plasma Sprayed Al2O3-13%TiO2 Ceramic Coating. MATEC Web Conf. 87:2027.
Witucki GL, Pajk T. 2004. The evolution of silicon-based technology in coatings. Pitture E Vernic. 1–10.
Yebra DM, Rasmussen SN, Weinell C, Pedersen LT. 2010. Marine Fouling and Corrosion Protection for Off-Shore Ocean Energy Setups. 3rd Int Conf Ocean Energy, 6 October, Bilbao.:1–6.
Zargiel KA, Swain GW. 2014. Static vs dynamic settlement and adhesion of diatoms to ship hull coatings. Biofouling. 30:115–129.
Zhang Y., 2016. Comparing the Robustness of Offshore Structures with Marine Deteriorations – A Fuzzy Approach. Adv Struct Eng. 18:1159–1171.

Maintenance

Progress in Maritime Technology and Engineering – Guedes Soares & Santos (Eds)
 ISBN 978-1-138-58539-3

Life cycle and cost performance analysis on ship structural maintenance strategy of a short route hybrid

H. Wang, E. Oguz & B. Jeong
Department of Naval Architecture, Ocean and Marine Engineering, University of Strathclyde, Glasgow, Scotland, UK

P. Zhou
Department of Naval Architecture, Ocean and Marine Engineering, University of Strathclyde, Glasgow, Scotland, UK
Zhejiang University, China

ABSTRACT: This paper presents the importance of coating maintenance and suggests an optimal strategy from economic and environmental points of view. Life cycle analysis is introduced to estimate the economy and environment impacts so particular decision can be made. A case study of a hybrid ferry is carried out where cash, energy and emission flows are tracked and evaluated. With different maintenance intervals, the consumptions of energy, materials and fuels are evaluated to estimate their cost benefits. Emissions normalization is also applied to determine environmental potentials to determine the relation of environmental impacts which are converted and compared in monetary terms. Annual-based hull inspection and re-coating is proved to reduce hull resistance and fuel consumptions which is a way to achieve cost-saving operations. The assessment has been proven to be able to make reliable decision, so it is suggested to facilitate life cycle assessment in the marine industry.

1 INTRODUCTION

Marine industry has been focusing on maritime emission control for decades and the Greenhouse Gases emission from international shipping has been reduced according to the third Greenhouse Gas Emission Study published by International Maritime Organization (IMO). The document not only indicated the emission reduction from 2009 to 2014 but also emphasised the methodologies of GHG emission estimations (IMO, 2015). Life cycle analysis (LCA) has already been practically applied in many industries and in a wide range of different products and recently LCA started to draw attention in the maritime field.

As a fact, there are still very limited numbers of research focusing on the application of LCA in marine application and most of them are especially for ship building and machinery operation. With consideration on the whole life span of a ship, LCA is a reasonable and suitable tool to evaluate the environmental impact, especially the global warming potential (GWP).

The research work done by Blanco-Davis has applied LCA to aid the shipyards to evaluate retrofitting performances of innovative ballast water treatment system and fouling release coating (Blanco-Davis et al, 2014; Blanco-Davis, Zhou, 2014). Alkaner and Zhou also investigated and compared the performance of fuel cell and diesel engines for marine applications with the help of LCA (Alkaner, Zhou, 2005). Research work done by Strazza's research team applied LCA to evaluate the environmental impact of paper stream on a cruise ship with implementation of different green practices (Strazza et al, 2015). Another LCA analysis carried out by Nicolae and his team determined the environmental impact related to commercial ships by optimization of raw material and energy consumption, and recycle processes (Nicolae et al, 2016). Ling-Chin and Roskilly have carried out two case ship studies comparing conventional and hybrid power system with a comprehensive consideration on construction, operation, maintenance and scrapping phase (Ling-Chin and Roskilly, 2016[a]; Ling-Chin and Roskilly, 2016[b]). With inspiration from these previous works, authors have carried out two case studies focusing on the propulsion system of a short-routed ferry and an offshore tug vessel. The studies have illustrated the lower cost and environmental impact with applications of battery packs for ferry and switching from

2 medium speed engines to 4 high speed engines for tug vessel (Wang et al, 2017; Oguz et al, 2017). These previous research works are striving to prove the availability of LCA tool in the field of shipping industry. This paper focuses on a more comprehensive LCA analysis to investigate the impacts of different alternative selections or decisions with consideration of life cycle cost and environment analysis for a whole ship life cycle.

Among ship life span, maintenance is one important phase, which usually is very much relevant and interesting to ship operators because the fuel consumption will be influenced by maintenance plan. As there are many research works carried out, the significance of hull coating on the operation fuel consumption is evidenced. Candries and his colleagues investigated three different coating and their impact on roughness and drag forces on ship hull (Candries et al, 2001). Dunnahoe indicated in his research that a more comprehensive dry-docking will help reduce the ship resistance. For example, with a 50% blasting and coating, the total resistance will be reduced by 20% (Dunnahoe, 2008). CFD model has been established by Demirel et al. to simulate different plate roughness due to different coating applied and experimental study has been carried out to determine the relationship between bio-fouling and ship resistance by Turan et al. (Demirel et al, 2014; Turan et al, 2016). The hull resistances were predicted for a tanker and a LNG tanker in their studies. From a long term of view, the hull roughness will impact the fuel consumption and is related to the bio-fouling on ship hull. It means with a regularly removal of bio-fouling the ship resistance could be kept low which will lead to a lower fuel cost. Hearin and his team tested the influence of mechanical grooming on coated panels which indicated that weekly grooming has a much lower fouling rate than a bi-weekly grooming (Hearin, 2015). Tribou and Swain investigated the effect of grooming on a copper ablative coating exposed statically for six years and their conclusions supports that more regular grooming can reduce more fouling on ship hull (Tribou and Swain, 2017). However, even though the dry-docking can greatly improve the energy efficiency, with the time going, the coating can be damaged or covered by bio-fouling which leads to the increasing of hull roughness. To avoid this situation, it is reasonable and practical to carried out regular re-coating to keep the hull roughness in an acceptable region, but the cost of re-coating will be increased which is due to hull washing, blasting and coating. This paper will evaluate the impact of re-coating interval on the ship life cycle financial and environmental performances and provide a guideline for shipyards and ship operators on their coating plans.

2 CASE STUDY GENERAL ASPECTS

2.1 *Introduction*

Since the environmental performance becomes one criteria of ship building, many shipyards ship operators and ship owners are keen to embrace new technologies and strategies to sustain their business. Maintenance plans have been seldom considered as a fact of complexity and long operation period. However, the impacts of maintenance plans cannot be neglected as they will eventually reflect the life cycle performances of the vessel. Inevitably, what values most to the shipyards may not be important to the ship operators and ship owners but as a fact of increasing and intensive competitions of ship-building bids, more cost efficient and environmental friendly the ship is, higher competitiveness a shipyard could be. Therefore, this paper focuses on the maintenance plan which considers from the construction phase to the end of life of a ship, recommending shipyards, ship operators and ship owner to assess the ship life cycle performances to mitigate the impact on the aspects of both cost and environmental.

2.2 *Case vessel description*

The case vessel considered in this paper is a short-routed ferry who regularly serves between islands in Scotland. The selection of this vessel is due to too many manoeuvring in shallow water which leads to more contacts than ocean going vessel. These contacts lead to more re-coating and hull maintenances. The specification of the vessel is listed in the following Table. To estimate the hull steel and coating area, equations and formulas are presented in the following sections.

2.2.1 *Steel weight estimation*

To estimate the steel weight in the ship hull structure, two methods are used: cubic number method and empirical equations.

Table 1. Case ship specification.

Name	MV Hallaig
Gross weight	499 tons
Length	43.5 m
Breadth	12.2 m
Depth	3 m
Draught	1.73 m
C_b	0.45
Power	360 kW*3
Superstructure decks	2
Builders	Ferguson Shipyard
Built year	2012

The first method uses a known base ship as a reference and applies block coefficient and length to depth ratio as corrections. The method can be described as following (Papanikolaou, 2014):

$$W_s = W_s' \times \frac{LBD}{L'B'D'} \times \frac{1 - \frac{1}{2} \times Cb}{1 - \frac{1}{2} \times Cb'} \times \frac{L/D}{L'/D'} \tag{1}$$

where W_s is the steel weight for case ship, ton;
W_s' is the steel weight for base ship, ton;
L and L′ are the lengths of case ship and base ship respectively, meter;
B and B′ are the breadth of case ship and base ship respectively, meter;
D and D′ are the depth of case ship and base ship respectively, meter;
C_b and C_b' are the block coefficient of case ship and base ship respectively.

The second method using the empirical equation developed by Garbatov's research team (Garbatov et al., 2017):

$$W_1 = 0.00072 \cdot Cb^{1/3} \cdot L^{2.5} \cdot T / D \cdot B \tag{2}$$

$$W_2 = 0.011 \cdot L \cdot B \cdot D \tag{3}$$

$$W_3 = 0.0198 \cdot L \cdot B \cdot D \tag{4}$$

$$W_4 = 0.0388 \cdot L \cdot B \cdot NJ \tag{5}$$

$$W_5 = 0.00275 \cdot L \cdot B \cdot D \tag{6}$$

$$W_s = W_1 + W_2 + W_3 + W_4 + W_5 \tag{7}$$

where,
W_s is the steel weight of case ship, ton;
W_1 is the weight of the main hull, ton;
W_2 is the weight of bulkheads in the main hull, ton;
W_3 is the weight of decks and platforms, ton;
W_4 is the weight of the superstructure, ton;
W_5 is the weight of the foundation and other, ton;
L is the length of the case ship, meter;
B is the breadth of the case ship, meter;
D is the depth of the case ship, meter;
T is the draft of the case ship, meter;
NJ is the deck number of the case ship superstructure;
C_b is the block coefficient of the case ship

Applying the ship's particulars listed in Table 1, the steel weight can be derived from both methods: applied with first method, the steel weight is about 126.38 ton; applied with second method, the weight of hull steel is approximate 126.22 ton. It is apparent that both methods give similar results, therefore, 126.38 is used as steel weight in this research.

2.2.2 *Coating area estimation*

Coating area in this paper is the wetted surface of the ship which will be merged in the water and attached by bio-fouling.

Figure 1 presents a hull wetted surface which is partially covered by bio-fouling and will be accumulated while staying in water.

The following Denny – Mumford formula (Molland et al., 2011) is applied to estimate the wetted surface:

$$S = 1.7L \times T + L \times B \times Cb \tag{8}$$

where
S is the wetted surface, m^2;
L is the length of the case ship, meter;
B is the breadth of the case ship, meter;
T is the draft of the case ship, meter;
C_b is the block coefficient of the case ship.

Since the steel weight and coating area are determined, the operation and maintenance principles will be discussed in the next section.

2.3 *Operation principle and maintenance plans*

The operation of the vessel is about 10 hours per day between two destinations. The operation can be divided into three parts: sailing, manoeuvring and in port. The daily operation hours are 6, 0.6 and 3.7 hours respectively.

Since this case study is based on a real case ship with its maintenance plans and operation profiles, the details are listed as following:

1. Partial coating: yearly;
2. Full coating: every five years.

The maintenance practice of partial coating is to annually remove bio-fouling accumulated on

Figure 1. Vessel hull with bio-fouling before cleaning.

ship external surface and re-paint the area which will help to reduce the roughness of ship hull to return to its initial condition so that the increasing of the energy efficiency of the vessel can be achieved. As a fact, the vessel is regulated to be dry-docked every five years which will carry out a full coating for the ship hull. Therefore, for every five years, the ship hull roughness is assumed to be returned to its initial condition which leads to the changes in fuel consumption. With the principle of applying different maintenance intervals, costs and energy consumptions due to maintenance will be varied and it is reasonable to determine an optimal maintenance plan to reach a minimum cost and environmental impact. The next section LCA model will be established to carry out life cycle analysis of these impacts on ship performances based on maintenance plans.

3 LCA MODELLING

The LCA model mainly comprises of four phases based on ship life span: construction, operation, maintenance and scrapping. The construction phase is defined as the ship building in shipyards, mainly including the hull construction and machinery installations; the operation phase is when the ship construction is completed and the ship is launched, in service and operated by ship operator; the maintenance of ship is carried out when the ship is in or off services by ship operator on ship or in shipyards, especially including hull and machinery maintenances; scrapping will be carried out when the ship is end of life in order to recycle or disposal the materials and machineries on board. Figure 2 presents an overall view of the four stages of a ship's life span.

Figure 2. Outline of the LCA process.

3.1 *Goal and scope of the study*

3.1.1 *Ship's maintenance plans*

The goal of this LCA modelling is to evaluate the performances of the case ship considering four life stages: construction, operation, maintenance and scrapping. The performances to be assessed include life cycle cost and environmental impacts which mainly focus on material purchases, energy consumption and emission release (CO2 equivalent). To evaluate the maintenance intervals on their impact on the LCA cost and environmental impact, several different intervals will be considered and under these conditions, the cost and environmental impacts will be derived using the LCA model established.

The reason behind the determination of the optimal ship performance is due to the relationships between maintenance plans with construction, operation and scrapping phase. If a long period coating maintenance is preferred, the hull roughness will be increased, and the fuel consumption will be increased which means the fuel cost in the operation phase will be higher. On the contrary, the dry-docking cost will be relatively reduced due to less frequent maintenance, considering coating materials investment and energy consumption.

3.1.2 *Boundary setting and data quality requirement*

In this study, four stages of ship life span: construction, operation and maintenance, and scrapping will be considered. After considering different ship maintenance plan with coating interval, the LCCA and LCA can be derived and compared to determine an optimal maintenance plan.

To carry out this study, some assumptions and boundaries are necessary, due to lack of data and simplification of the model:

1. After coating, the roughness of ship hull will be in the same condition as initially launched so that the fuel consumption will be the same as initial condition;
2. The other processes apply similar technologies as hull production processes which are provided by ship manufacturer (Ferguson shipyard);
3. The modelling uses GaBi 5 and its database, but the emission released due to engine running is estimated using emission factors;
4. The scrapping processes are referred to Ling-Chin and Roskilly's research (Ling-Chin and Roskilly, 2016a);
5. The manufacturing of steel plates and machineries from raw material are not considered;
6. The fuel consumption increment due to delayed coating maintenance is estimated by an empirical equation based on a half year fuel consumption data provided by the ship operator, CalMac;

7. Properties of coating and welding materials are based on reference and GaBi database;
8. Machinery maintenance is not considered in this study;
9. To keep the processes realistic, the transportation of materials and machinery are considered;
10. All the phases use the same electricity supply from wind farm which is commonly used in Scotland.

3.2 *Life cycle inventory analysis*

According to the goal and scope of the study, together with all the information from shipyard, ship operator and literature, the life cycle analysis for the case ship is carried out.

3.2.1 *Flow chart development*

To present a full LCA analysis, Figure 3 is introduced considering the following:

1. Hull constructions;
2. Engine and battery constructions;
3. Engine and battery operations;
4. Hull structure and coating maintenances;
5. Hull scrapping.
6. Machinery scrapping;

In Figure 3 the red coloured lines present the flow of fuel supply for the case ship, including heavy fuel oil and lubrication oil.

3.2.2 *Inventory results*

After establishment of the LCA model, the results for different phases are evaluated. In Table 2, the emission flows of significant emissions are presented. It is obvious that most of the emissions are from the operation phase. The application of less frequent maintenance will have an impact on the fuel consumption in operation phase which will lead to an increase in emission generation.

3.3 *Life cycle impact assessment*

The life cycle impact in this study was focused on global warming potential which has increasingly drawn attention from researchers. With the model and database in GaBi, three life cycle impact assessment results are derived in Figure 4–7, using CML, ReCiPe, TRACI and ILCD respectively (CML, 2016; RVIM, 2011; IERE, 2012, Wolf, 2012). It can be seen from these figures that there is no significant difference among CML, ReCiPe, TRACI and ILCD in GWP values (kg CO2 e). The equivalent CO_2 emission for the case ship is around 14 million

Table 2. Life cycle inventory analysis.

Inorganic emissions to air during all life phases (kg)					
Emission flows	Construction	Operation	Maintenance	Scrapping	Total
CO_2	1.07E+04	1.36E+07	1.71E+03	1.59E+03	1.36E+07
CO	13.1	3.10E+04	6.2	2.03	3.10E+04
NO_x	5.41	3.36E+05	2.45	1.55	3.36E+05
SO_2	5.91	6.37E+03	2.5	1.47	6.38E+03

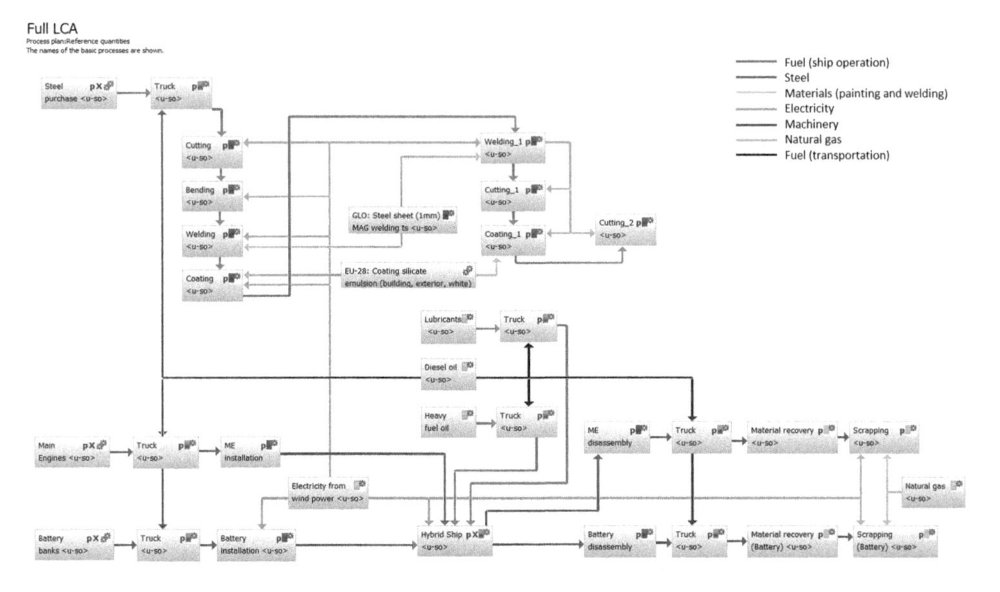

Figure 3. Flow chart of LCA model

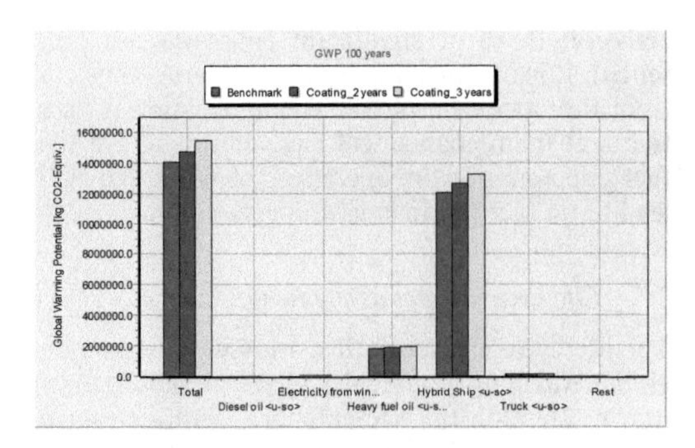

Figure 4. LCA results with application of CML 2001.

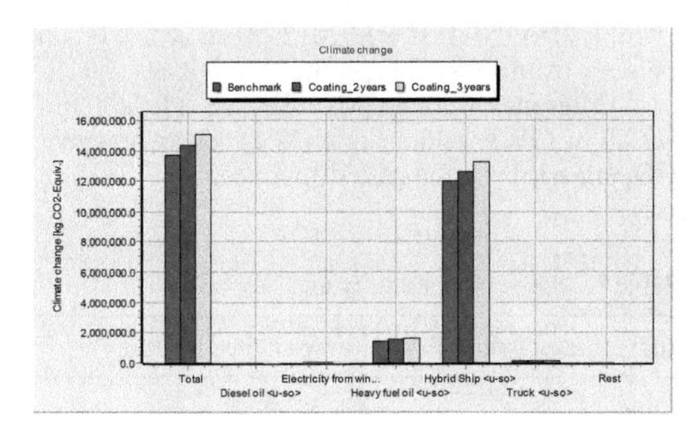

Figure 5. LCA results with application of ReCiPe.

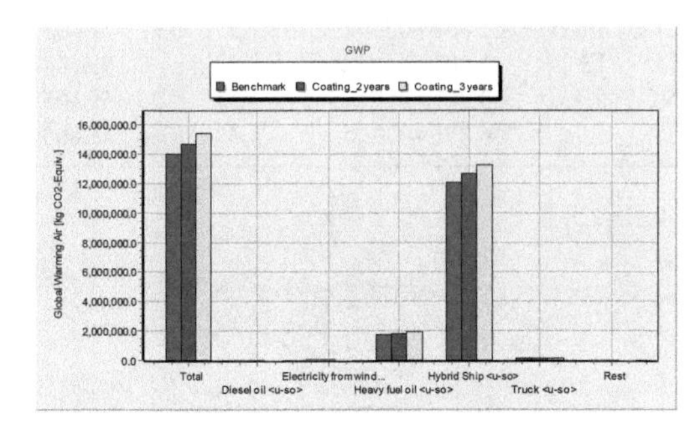

Figure 6. LCA results with application of TRACI.

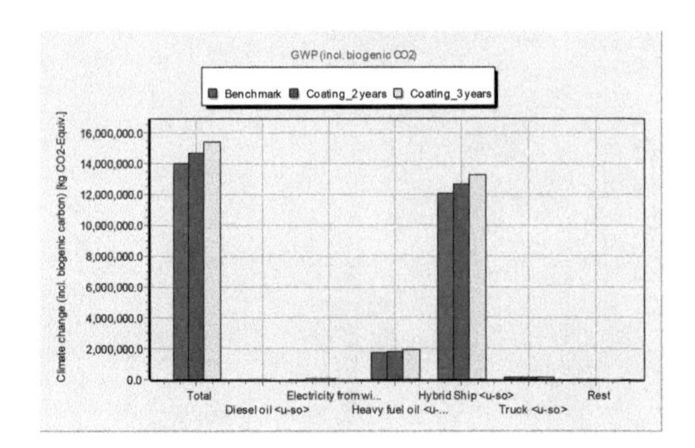

Figure 7. LCA results with application of ILCD.

ton. Furthermore, under different maintenance interval, these methods provide similar results and trends. Hence, for this LCA model, these methods have a good agreement with each other.

These figures also indicate that when the coating interval increased from 1 to 2 or from 1 to 3, the GWP value will be increased. It means more emission released compared to the case ship. However, with an increasing of steel renewal interval, the change between two results are minor because the steel renewal will not impact on the energy consumption of the vessel especially in the operation phase which occupies the more emission generation among all phases.

3.4 *Further results and identification of significant issues*

As the emission issue is not the only factor that affect the ship owners' decision, the cost of the vessel, including construction, operation, maintenance and scrapping may recommend them a different option. In this section, the emissions will be converted into cost or credits which can be compared between different scenarios together with life cycle cost.

3.4.1 *Conversion of environmental impact into costs*

The carbon credit policy in the UK is about \$29 per ton CO_2 emission (Maibach et al., 2008). Since the GWP for different scenarios has been determined, the difference value between cases can be applied for comparison.

For case 1 and case 2, the difference of GWPs (ΔGWP_1) is about 7E+5 kg/CO_2e. For case 1 and case 3, the difference of GWPs (ΔGWP_2) is about 1.4E+6 kg/CO_2e. As a consideration of carbon credit (\$29/ton CO_2), the emission credits increased for case 2 is \$2.03E+4 and for case 3 is \$3.06E+4.

3.4.2 *Optimal partial coating plan*

Considering coating maintenances, the fuel consumption increment due to late coating is estimated to be about 5% based on operator's data (provided by CalMac) and their experiences. In a life cycle, total investments of coating materials and activities are lower for late coating than that for early coating and the annual coating degraded area is advised to be 10% by CalMac. Due to increment of fuel consumption, the operation costs and emission released are significantly growing. In Figure 8, the increment of annual fuel consumption rate due to coating interval changes has been shown in order to indicate the significant relation between each other.

Figure 9 presents how the total life cycle cost increased with increasing coating interval. When

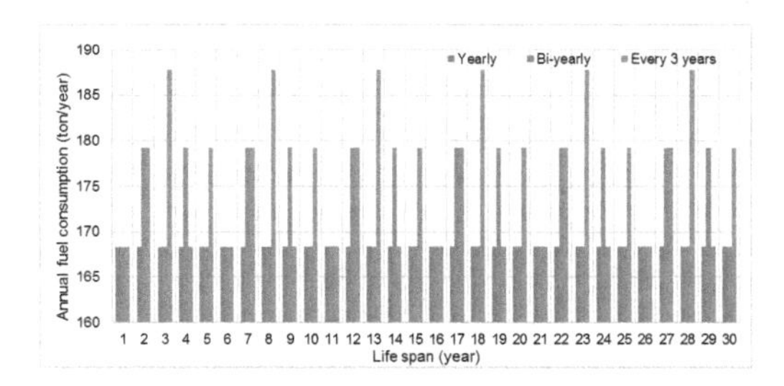

Figure 8. Specific fuel oil consumptions changes with coating interval.

Table 3. Cost increased due to the changes in partial coating interval.

Increased compared to yearly plan ($)	Partial coating interval (year)		
	1	2	3
Fuel cost	0	51989	98576
Maintenance cost	0	−3580	−4773
CO2 credit	0	12179	23092
Total cost	0	60588	116895

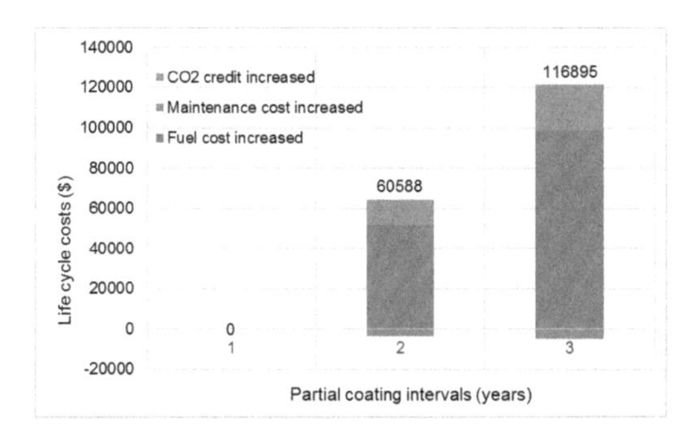

Figure 9. Costs increased under different coating intervals.

the coating interval increases from yearly to bi-yearly, the total cost increased by about $34000. Similarly, if we increase the coating frequency up to 3 years, the cost will be increased by around $75000 compared with yearly coating plan.

3.4.3 *Discussion on other stage of structural degradation*

Since this research is focusing on the operation and maintenance stages, the significance of regular coating is presented. It is because regular coating has the most environmental and economic impact on operation phase. However, the impacts of regular coating on other phases are important as well so this section will present the structural degradation if coating would be irregular in the ships life time.

For construction phase, as the operation and maintenance plan has been established or estimated, ship yard and ship owner should consider a proper initial coating, such as application of high performance coating which means high investment. This situation is not considered in this paper due to the practice of shipyard and ship owner and lack of relevant data and information.

The scraping phase will be influnced by coating activities. In this research, one assumption was made to simplify the impact estimation in scrapping phase: the hull conditions will be recovered to its initial condition after every entry to the drydocking which is not practical. It is reasonable to predict that more regular maintenances will not only make every single maintenance simpler in the drydocking shipyard but also recover the ship hull nearer to its initial condition.

4 CONCLUSIONS

This paper applies LCA methodology to evaluate the life cycle cost and environment impact of maintenance intervals on a short-route ferry and provides a guide for ship LCA analysis and proves the availability of LCA application in marine field especially for ships' life cycle assessment. This paper presents a comprehensive LCA analysis to investigate the impacts of different alternative selections or decisions with the consideration of life cycle cost and environment analysis for a whole ship life cycle. Investigation in this paper is to change the maintenance interval which related to the construction of ship to determine an optimal maintenance plan with a minimized impact in the aspect of financial and environmental. The impacts of re-coating interval are also evaluated on the ship life cycle financial and environmental performances to not only indicate an optimal partial coating interval but also to provide a guideline for shipyards and ship operators on their coating plans.

Considering four life stages of a vessel, including, construction, operation, maintenance and scrapping, the LCA model is established in GaBi software. The model covers activities for steel processing and machinery installations in the shipyard; operation of the engine and batteries on board; maintenance of ship hull (coating) and scrapping of hull materials and machineries. For coating interval, it is evidenced that a frequent coating leads a fewer cost in the case study. The results support the coating practices made by CalMac who carried out yearly partial coating to decrease the accumulations of bio-fouling and the hull roughness to reduce the fuel cost.

ACKNOWLEDGEMENTS

The authors wish to thank the Caledonian MacBrayne (CalMac Ferry Ltd) Ferries and Ferguson Marine for providing the data used in this paper. The authors also gratefully acknowledge that the research presented in this paper was partially generated as part of the HORIZON 2020 SHIPLYS (Ship life cycle software solutions) Project, Grant agreement number 690770.

REFERENCES

Alkaner, S. & Zhou, P. 2006. A comparative study on life cycle analysis of molten carbon fuel cells and diesel engines for marine application. Journal of power sources 158 (1), 188–199.

Blanco-Davis, E. & Zhou, P. 2014. LCA as a tool to aid in the selection of retrofitting alternatives. Ocean Eng 77:33 – 41.

Blanco-Davis, E. del Castillo, F. & Zhou, P. 2014. Fouling release coating application as an environmentally efficient retrofit: a case study of a ferry-type ship. Int. J. Life Cycle Assess (2014) 19:1705 – 1715, DOI 10.1007/s11367–014–0780–8.

Candries, M., Anderson, C. & Altlar, M. 2001. Foul release systems and drag: Observation on how the coating works. Journal of Protective Coatings, Linings, April 2001.

CML. (Institute of Environmental Sciences) 2016. CML-IA Characterisation Factors. https://www.universiteitleiden.nl/en/research/research-output/science/cml-ia-characterisation-factors Accessed on 27 Nov. 2017.

Demirel, Y., Khorasanchi, M., Turan, O., Incecik, A. & Schultz, M. 2014. A CFD model for the frictional resistance prediction of antifouling coatings. Ocean Engineering 89 (2014) 21–31.

Dunnahoe, T. 2008. International Marine Coatings Forum: Coatings and CO2. Materials Performance; Jun 2008; 47, 6; SciTech Premium Collection, pg. 92.

Garbatov, Y., Ventura, M., Geprgiev, P., Damyanliev, T. & Atanasova, I. 2017. Maritime Transportation and Harvesting of Sea Resources – Guedes Soares, Teixeira (Eds) © 2018 Taylor, Francis Group, London, ISBN 978–0-8153–7993–5.

Hearin, J., Hunsucker, K., Swain, G., Stephens, A., Gardner, H., Lieberman, K. & Harper, M. 2015. Analysis of long-term mechanical grooming on large-scale test panels coated with an antifouling and a fouling-release coating. Biofouling, the Journal of Bioadhesion and Biofilm Research, 2015, Vol. 31, No. 8, 625 – 638. ISSN: 0892–7014

IERE. (The Institute for Environmental Research and Education) 2012 TRACI Characterization Factors https://iere.org/programs/earthsure/TRACI-factors.htm Accessed on 27 Nov. 2017.

IMO, 2015. Third IMO Greenhouse Gas Study 2014. London: International Maritime Organization 2015.

Ling-Chin, J. & Roskilly, A. 2016a. Investigating the implications of a new-build hybrid power system for Roll-on/Roll-off cargo ships from a sustainability perspective – A life cycle assessment case study. Applied Energy 181 (2016) 416–434.

Ling-Chin, J. & Roskilly, A. 2016b. Investigating a conventional and retrofit power plant on-board a Roll-on/Roll-off cargo ship from a sustainability perspective – A life cycle assessment case study. Energy Conversion and Management 117 (2016) 305–318.

Maibach, M., Schreyer, C., Sutter, D., Essen, H.P., Boon, B.H., Smokers, R., Schroten, A., Doll, C., Pawlowska, B. & Bak, M. 2008. Handbook on estimation of external costs in the transport sector, Internalisation Measures and Policies for All External Cost of Transport (IMPACT) Version 1.1 Delft, CE, 2008 r.

Molland, A.F., Turnock, S.R. & Hudson, D.A. 2011. Ship Resistance and Propulsion: practical estimation of ship propulsive power. Cambridge University Press, NewYork. ISBN 978–0-521–76052–2

Nicolae, F., Popa, C. & Beizadea, H. 2016. Applications of life cycle assessment (LCA) in shipping industry, 14th International Multidisciplinary Scientific Geo-Conference SGEM 2014, Section name: Air Pollution and Climate Change.

Oguz, E., Wang, H., Jeong, B. & Zhou, P. 2017. Life cycle and cost assessment on engine selection for an offshore tug vessel. Maritime Transportation and Harvesting of Sea Resources – Guedes Soares, Teixeira (Eds) © 2018 Taylor, Francis Group, London, ISBN 978–0-8153–7993–5

Papanikolaou, A. 2014. Ship Design: Methodologies of Preliminary Design. National Technical University of Athens, Greece. ISBN 978–94–017–8750–5, DOI 10.1007/978–94–017–8751–2

RIVM. (The Dutch National Institute for Public Health and the Environment) 2011 Life Cycle Assessment (LCA) /LCIA: the ReCiPe model. http://www.rivm.nl/en/Topics/L/Life_Cycle_Assessment_LCA/ReCiPe Accessed on 27 Nov. 2017.

Strazza, C., Borghi, A., Gallo, M., Manariti, R. & Missanelli. E, 2015. Investigation of green practices for paper use reduction onboard a cruise ship — a life cycle approach. Int J Life Cycle Assess (2015) 20:982 – 993, DOI 10.1007/s11367–015–0900–0

Tribou, M. & Swain, G. 2017. The effects of grooming on a copper ablative coating: a six-year study. Biofouling, the Journal of Bioadhesion and Biofilm Research, 2017, Vol. 33, no. 6, 494–504. ISSN: 0892–7014

Turan, O., Demirel, Y., Day, S. & Tezdogan, T. 2016. Experimental determination of added hydrodynamic resistance caused by marine biofouling on ships. Transportation Research Procedia 14 (2016) 1649 – 1658.

Wang, H., Oguz, E., Jeong, B. & Zhou, P. 2017. Optimisation of operational modes of short-route hybrid ferry: A life cycle assessment case study. Maritime Transportation and Harvesting of Sea Resources – Guedes Soares, Teixeira (Eds) © 2018 Taylor, Francis Group, London, ISBN 978–0-8153–7993–5

Wolf, M.A., Pant, R., Chomkhamsri, K., Sala, S. & Pennington, D. 2012 The international reference life cycle data system (ILCD) handbook. Institute for Environment and Sustainability, Luxembourg. ISSN: 1831–9424. doi: 10.2788/85727

Progress in Maritime Technology and Engineering – Guedes Soares & Santos (Eds)
 ISBN 978-1-138-58539-3

An integrated operational system to reduce O&M cost of offshore wind farms

K. Wang
School of Energy and Power Engineering, Reliability Engineering Institute, Wuhan University of Technology, Wuhan, China
Key Laboratory of Marine Power Engineering and Technology, (MOT), Wuhan University of Technology, Wuhan, China
Faculty of Mechanical, Maritime and Materials Engineering, Delft University of Technology, Delft, The Netherlands

X. Jiang & R.R. Negenborn
Faculty of Mechanical, Maritime and Materials Engineering, Delft University of Technology, Delft, The Netherlands

X. Yan
School of Energy and Power Engineering, Reliability Engineering Institute, Wuhan University of Technology, Wuhan, China
Key Laboratory of Marine Power Engineering and Technology, (MOT), National Engineering Research Center for Water Transport Safety (WTSC), MOST, Wuhan University of Technology, Wuhan, China

Y. Yuan
School of Energy and Power Engineering, Reliability Engineering Institute, Wuhan University of Technology, Wuhan, China
Key Laboratory of Marine Power Engineering and Technology, (MOT), Wuhan University of Technology, Wuhan, China

ABSTRACT: Offshore wind is a relatively new industry and it is generally more expensive to generate electricity than many alternative renewable sources. Operation & Maintenance (O&M) makes up a significant part of the overall cost of running Offshore Wind Turbines (OWT). Since the O&M associated responsibility is shared among turbine manufacturers, wind farm operators and the offshore transmission owners, this has inevitably led to lack of information, duplication of effort and less efficiency. Big data analytics is one great technique that will drive future growth. In this paper, an integrated operational system of offshore wind farm is proposed deploying big data analytics. Firstly, the current state of the O&M of offshore wind farm and the big data analytics are introduced. Afterwards, a predictive maintenance model and a maintenance implementation model are proposed, and an integrated operational system is developed incorporating those two models in order to optimize maintenance planning and implementation. Finally, the possible contribution of such a system to a more effective O&M of offshore wind farm is discussed.

1 INTRODUCTION

The future development of Dutch wind energy will for a large part take place in the sea: the Dutch government plans to realize an offshore wind energy capacity of 3,450MW by 2023 (Gebraad, 2014; Broek, 2014).

The size of wind turbine has been enlarged continuously to increase the power output, see Figure 1 (Chemnews); Wind turbines are being placed further offshore, see Figure 2 (Rohrig, 2014).

The harsh offshore environment leads to more intense mechanical stress within the turbine (Ribrant, 2006), and the annual failure rate of wind turbines has arisen substantially as shown in Figure 3 (Echavarria et al., 2008).

The operational unavailability reaches 3% of the lifetime of WTs. Statistics show that the O&M cost makes up a significant part of the overall cost of running offshore wind farms(Nabati & Thoben, 2017), ranging from 10%–20% of the total cost at the initial stage to 35% at the end stage of a wind project (Tchakoua et al., 2014)

In general, the O&M efficiency of offshore wind farms can be improved from three aspects, namely wind turbine monitoring, supply chain

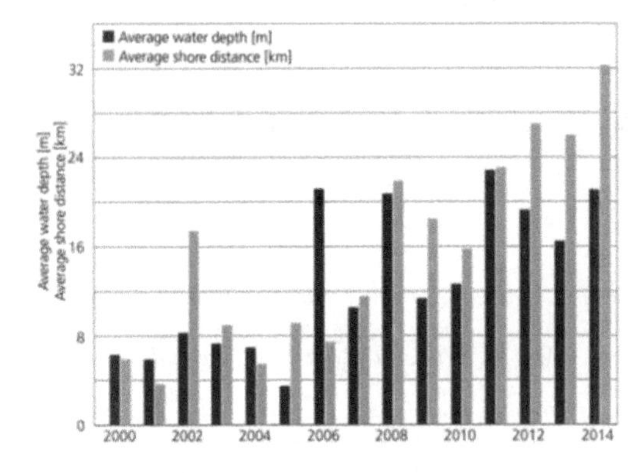

Figure 1. Wind turbine size and power output.

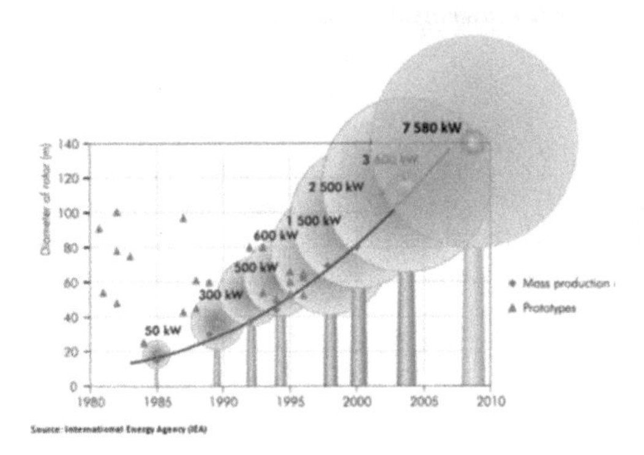

Figure 2. Wind turbines are placed further offshore.

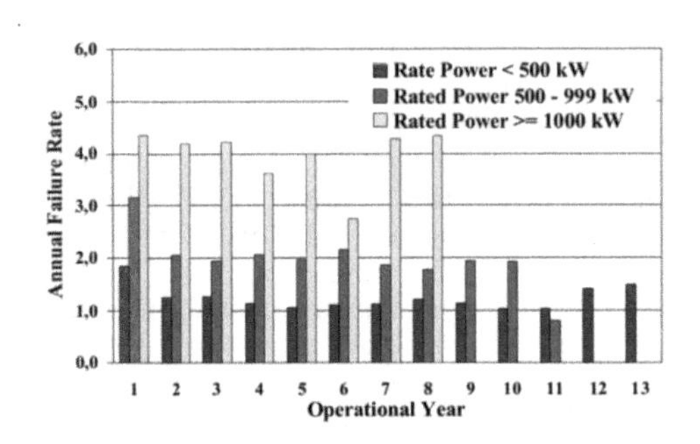

Figure 3. Annual failure rate of different rated power.

management and marine operations (Coronado & Fisher, 2015) To date, academic research on O&M of offshore wind tends to concentrate on the wind turbine monitoring. (Márquez et al., 2012; Kusiak & Li, 2011) However, although abundant data becomes available, the data itself is not identical to useful information. There are remaining challenges about how to extract valuable information from a large number of data. Comparatively, supply chain management and marine operations have received little attention. The supply chain management is more a business specific activity on a top level, it will not be covered in this paper. The marine operation contributes 12.5% to the total cost of O&M, it presents a potential for immediate efficiency gains and offers significant cost savings for minimal investment (Koopstra, 2015; Zheng H et al., 2016; Wang K et al., 2017a). In those regards, it is valuable to develop an integrated system to improve the effectiveness of O&M by accounting for both conditional monitoring of WTs and the marine operation.

This paper aims to propose such an integrated system to optimize O&M deploying big data analytics. This system would take the wind turbine monitoring as well as marine operation into account, promote information sharing and finally improve the overall O&M effectiveness of offshore wind. The remainder of this paper is organized as follows. The current state of O&M of offshore wind and the remaining issues/challenges are explained in Section 2. Big data analytics and its applicability for offshore wind are illustrated in Section 3. A predictive maintenance model and a maintenance implementation model are proposed and integrated into an operational system in Section 4. The implication of this integrated operational system is discussed in Section 5.

2 CURRENT STATE OF O&M OF OFFSHORE WIND

Maintenance strategies in the WT industry can be roughly classified into: 1) Corrective maintenance; 2) Preventive maintenance; 3) Predictive maintenance. Corrective maintenance is a kind of reactive repair after occurrence of a failure. Its main drawback lies in the consequent failure cost can be extremely high. Preventive maintenance is mainly a time-based periodic repair, and its weakness would be relatively high operational cost induced by unnecessary inspections and maintenances. Predictive maintenance is a kind of condition based preventive maintenance. It can improve the reliability, availability and maintainability of OWTs while simultaneously reducing the O&M cost (Byon et al., 2016). A comprehensive knowledge of the actual condition of WTs and thereafter estimation of the remaining useful lifetime (RUL) of critical failure components are essential in order to develop a feasible predictive maintenance strategy (Karyotakis, 2011).

Nowadays, various types of sensors, new measurement methods have been generated and applied to monitor the condition of WTs. A Supervisory Control and Data Acquisition (SCADA) system collects information extensively using sensors mounted on the WTs.

Normally a turbine contains 20–30 sensors, resulting in 60–100 different SCADA signals. With a sampling rate of 1 second with 8 byte values, around 1.8 GB raw data per turbine per month are produced. Upon analysis of those information, it is possible to identify critical failure components in terms of high probability of failure and/or severe failure consequence. Some studies have indicated that the electrical system, control system and rotor blades contribute more than 50% of number of failures, as shown in Figure 4 (Hahn et al., 2007).

Meanwhile, gearbox, drive train and generator cause the severest failure consequence in terms of annual downtime (Fischer et al., 2012), see Figure 5 (Crabtree et al., 2010).

From risk assessment viewpoint, it is reasonable to consider those six subsystems as critical failure components and the RUL estimation can be restricted to them in order to reduce O&M cost of WTs maintaining sufficient reliability level.

The failure modes of the critical components could generally be assigned to either mechanical or electrical associated failure. Mechanical failure is characterized with an increasing failure rate caused by aging factors, i.e., erosion, corrosion and fatigue, etc. An electrical failure rate is not necessarily increasing with age but could be relatively constant even decreasing, referring to a decreasing failure rate of electronic control system in Figure 6 (Faulstich et al., 2011).

Among above mentioned critical subsystems, rotor blade, gearbox and drive drain are mainly subjected to a mechanical failure. The evolution of a mechanical failure can be monitored using various techniques, as shown in Figure 7.

It can be noted that the sensitivity of various techniques to the failure prognosis is quite different. Vibration based technique can detect a potential mechanical failure a few months in advance but heat sensors can do the work days prior to the failure (Madsen, 2011).

Generator, electric system and control system are prone to mechanical failures as well as electrical failures. In general, electrical signals and the thermography based technique are used to detect failures occurring to those subsystems. However, different failure modes could lead to similar abnormality in signal, it is hard to tell one failure mode from another. And the abnormality can only be detected as early as days prior to the failure, thus in general, the condition monitoring of those subsystems are less effective from predictive maintenance strategy perspective.

In summary, offshore wind farms produce extremely large datasets, such as SCADA. However, not all data has been collected and stored properly, limited by the scalability of traditional databases. Furthermore, even with all data in place, there are remaining challenges faced—converting data into valuable information. Specifically, WTs are complex integrity made up of various sub-systems

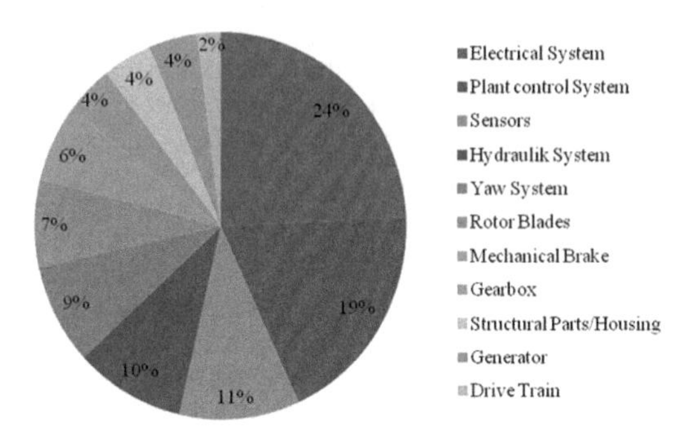

Figure 4. Failures percent of the main components.

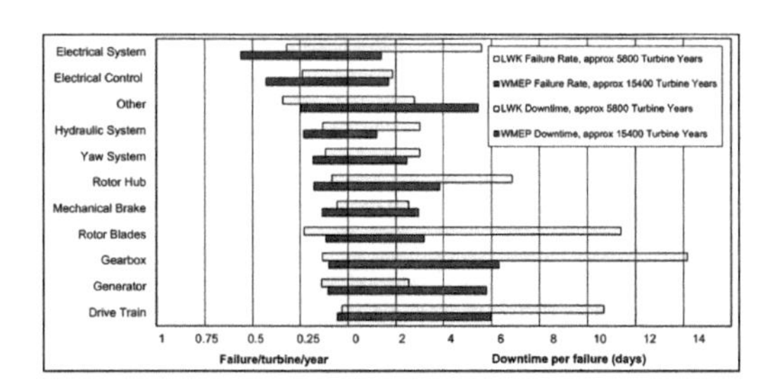

Figure 5. Failure rates and downtime from two large surveys of European WTs.

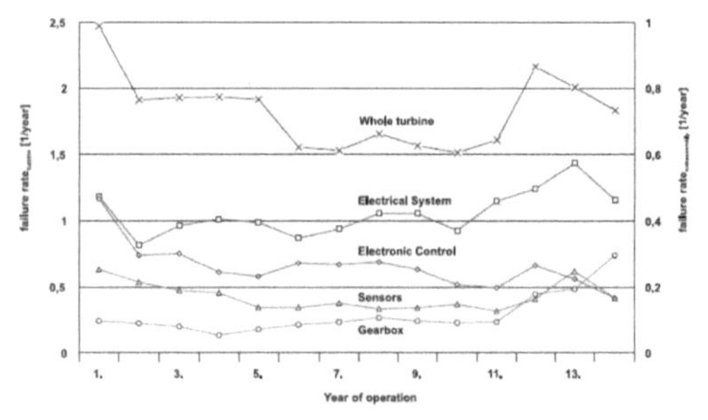

Figure 6. Failure rate with time of operation for onshore wind turbines in the WMEP study.

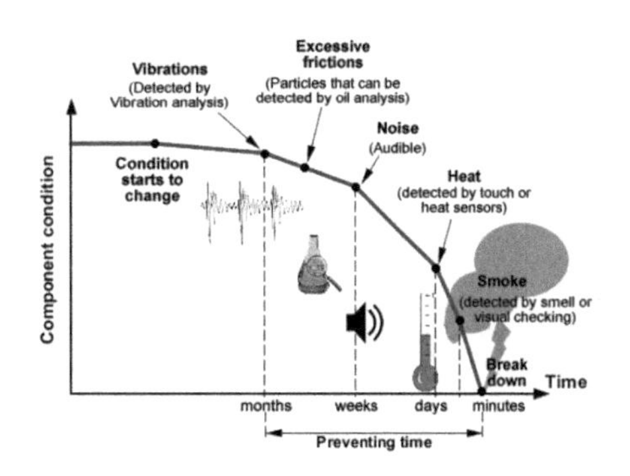

Figure 7. Typical development of a mechanical failure.

and components, one component can suffer from multiple failure modes, and different failure modes can interact with each other. The complication of failure mechanisms make it further more difficult to extract useful information based on traditional data analysis technique.

Last but not least, the maintenance of OWTs is normally implemented by fetching staff and equipment to and from the wind farm using marine vessels. This operation contributes 12.5% to the total cost of O&M and there is a potential to save cost by optimizing associated parameters accounting weather window, operational profile and available resources and deploying advanced data analytics.

3 LAYOUT OF TEXT BIG DATA ANALYTICS

Big data analytics is the one that can quickly access to valuable information from various types of data. The related technologies of big data processing generally include big data acquisition, preparation, storage, analysis and mining as well as display and visualization. The significance of big data lies in that the analysis and mining of large amounts of data are conducted through cloud computing and distributed parallel algorithm, etc., in order to obtain the underlying laws and values implied in the data, reveal complex relationships, assist in planning and improve real-time operations. In this respect, big data analytics is very suitable and capable to handle a large number of data, process complex information and achieve high computational efficiency (Wang et al., 2017b).

From O&M of offshore wind farm perspective, big data analytics could have following advantages:

1. to preprocess and visualize complex and diverse data for feature extraction and information mining;
2. to reveal the complex interrelationship between different failure modes so as to achieve a more accurate estimation of rul of critical failure components;
3. to support developing an integrated operational system accounting for both condition monitoring and marine operation by promoting information sharing and utilization.

4 AN INTEGRATED OPERATIONAL SYSTEM OF OWTS

Once various data about OWTs has been obtained, including condition of WTs, Wind farm, weather window, marine operational profile, available maintenance resource and its location, etc., an integrated operational system could be developed incorporating both maintenance planning and implementation.

A predictive maintenance model could be developed in order to pre-determine the demanded maintenance of the wind farm. As above mentioned, the big data analytics could make it possible to identify the correlation between different condition indicators (e.g. the speed and temperature of rotor, etc.) and the interaction of different failure modes effectively. Thereafter, the RUL of critical failure components can be predicted more accurately with indicated confidence/reliability level. Based on system reliability analysis (Jiang & Melchers, 2005), it is reasonable to develop a reliability centred maintenance planning. It can facilitate the planner/operator to decide when, where and which components need to be repaired; which maintenance strategies should be taken; and what technicians and resources (material, parts, equipment and vessel) are required to carry out the maintenance (Bajracharya G et al., 2009 & 2010). Santos, Teixeira & Guedes Soares (2015 & 2018) have deployed generalized stochastic Petri Nets (GSPN) method coupled with Monte Carlo to model the planning of O&M activities of an offshore wind turbine. The weather window, corrective maintenance based on replacements and age imperfect preventive maintenance were modeled and compared in terms of the wind turbine's performance and of the O&M costs.

In general, the implementation of the maintenance may not be able to conform to the maintenance plan constrained by actual weather and available technicians, equipment and other maintenance resources. Vessels are usually deployed to transport human and facilities to and from wind farm and the fuel consumption contributes mainly to the total cost of marine operation (Tchakoua et al., 2014; Wang et al., 2017a). The fuel consumption is closely related to the vessel type and size, the sailing route and speed, the weather window and demanded push-on force, etc. Based on big data analysis, it is possible to optimize the maintenance implementation in terms of minimal total cost of O&M, under given maintenance plan determined by the predictive maintenance model, including proper type and capacity of the vessel, proper number of technicians, proper amount of equipment and parts, the right component to be repaired at the right time in the right way, etc. On top of it, an optimal sailing route, speed as well as push-on power under given weather condition would be achieved. In this way, an integrated operational system can be proposed incorporating two models, as shown in Fig. 8.

The framework and the workflow of this system are illustrated in Fig. 9. The system is equipped with hardware, i.e., RAM and Processor etc., and software including Hadoop and Spark etc. The Hadoop consists of MapReduce and HDFS, it can meet the need for off-line batch processing of

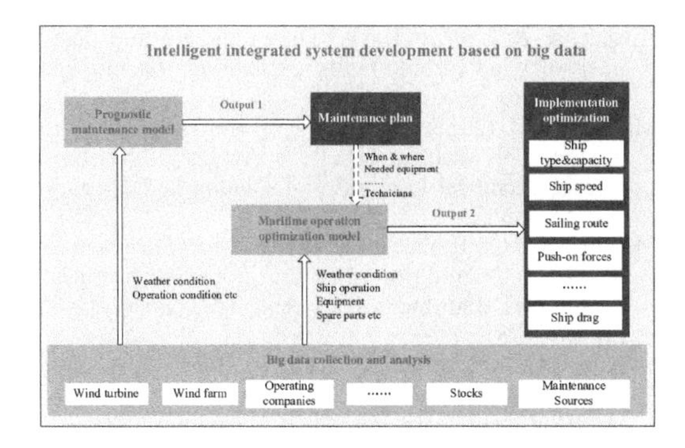

Figure 8. Models and strategies associated with O&M optimization.

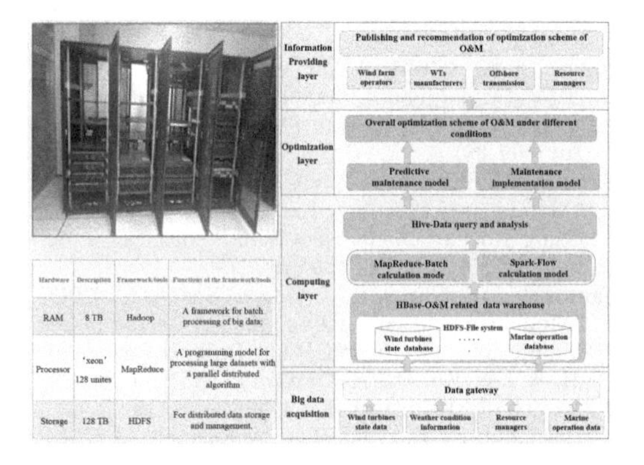

Hardware	Description	Framework/tools	Functions of the framework/tools
RAM	8 TB	Hadoop	A framework for batch processing of big data;
Processor	'xeon' 128 unites	MapReduce	A programming model for processing large datasets with a parallel distributed algorithm
Storage	128 TB	HDFS	For distributed data storage and management.

Figure 9. Framework and workflow of the integrated operational system.

big data; The Spark can achieve online real-time processing and support memory computing, which can meet the need for real-time optimization of O&M. Hence, the integrated system is capable to realize real-time parallel processing and effective data storage, and meet the requirement of system's adaptive scalability.

The proposed integrated operational system will work on four functional layers in a sequence, from the bottom to the top, namely data acquisition layer, computing layer, optimization layer and information providing layer, see Fig. 9. The data acquisition layer is responsible for the data collection, including wind turbine condition, weather condition, maintenance resources and marine operational profile. All data will be pre-processed and stored in the computing layer. The maintenance planning and implementation will be integrated and optimized in the core optimization layer. Once a mathematical modeling covering both maintenance planning and implementation is developed, it could be optimized deploying various numerical methods, such as Particle Swarm Optimization (Wang et al., 2017a) or GSPN(Santos et al., 2015 & 2018) among others. The finalized maintenance scheme will be present at the information providing layer in terms of visualized data, statistic result and technical report etc.

5 DISCUSSION

As a promising data analysis method, big data analytics has advantages for information extraction and accurate prediction. In this paper, an overall concept based on big data analytics is proposed in order to improve the effectiveness of O&M of OWTs. An integrated operational system is proposed to reduce the total O&M cost from the system-level point of view by incorporating a predictive maintenance model and a maintenance implementation. This big data based integrated system is expected to provide an optimal maintenance scheme with following advantages:

1. No data would be dropped in the data chain, and technical and operational conditions of turbines would be reviewed in detail with improved accuracy.
2. The complicated failure modes would be decoupled, and the root cause of certain failures could be identified.
3. Integrated and shared information can promote a more effective maintenance decision making with respect to both planning and implementation.
4. It is adaptable to a new or growing wind farm relatively easily and less costly.

Big data analytics will break through the limitation of traditional data analysis techniques. It is anticipated to play an significant role in improving effectiveness of O&M of OWTs. The collaboration among turbine manufacturers, wind farm operators and the offshore transmission owners is essential to decrease data blockage, promote information sharing and build up a big data platform in order to reduce areas of inefficiency and duplication of effort. Currently, the data sharing of OWTs among related stakeholders is at initial stage and rather limited. In this respective, a hybrid method combining both physics based failure model and data driven models can be deployed to optimize O&M of OWTs—maintaining acceptable reliability at reduced O&M cost.

REFERENCES

Bajracharya, G., Koltunowicz, T., Negenborn, R.R., Papp, Z., Djairam, D., Smit, J.J., De Schutter, B. (2009). "Optimization of maintenance of power system equipment based on a predictive health model". In *Proceedings of the 2009 IEEE Bucharest PowerTech (PT 2009)*, June, Bucharest, Romania.

Bajracharya, G., Koltunowicz, T., Negenborn, R.R., Djairam, D., De Schutter, B., Smit, J.J. (2010). "Optimization of transformer loading based on hot-spot temperature using a predictive health model". In *Proceedings of the 2010 International Conference on Condition Monitoring and Diagnosis (CMD 2010)*, September, Tokyo, Japan, pp. 914–917.

Broek, T. (2014). "Cost-sensitivity Analyses for Gearbox Condition Monitoring Systems Offshore, Wind Energy Research Group, Faculty of Aerospace Engineering—Delft University of Technology, Delft, the Netherlands.

Byon, E., Ntaimo, L., Ding, Y. (2010), "Optimal maintenance strategies for wind turbine systems under stochastic weather conditions.", *IEEE Trans. Reliab.*, 59, pp. 393–404.

Chemnews (2012), How wind turbines work, http://www.chemistryviews.org/details/ezine/1444481/How_Wind_Turbines_ Work.html.

Coronado, D. and Fisher, K. (2015) "Condition Monitoring of Wind Turbines: State of the Art", User Experience and Recommendations.

Crabtree, C.J., Feng, Y., Tavner, P.J. (2010), "Detecting Incipient Wind Turbine Gearbox Failure: A Signal Analysis Method for On-line Condition Monotoring". In *Proceedings of European Wind Energy Conference (EWEC 2010)*, 20–23 April,Warsaw, Poland, pp. 154–156.

Echavarria, E. et al. (2008), "Reliability of wind technology through time [J]". *Journal of Solar Energy Engineering*, 130(3), pp. 1047–1057.

Faulstich, S., Hahn, B., & Tavner, P. (2011), "Wind turbine downtime and its importance for offshore deployment", *Wind Energy* 14, pp. 327.

Fischer, K., Besnard, F., Bertling ,L. (2012), "Reliability-centered maintenance for wind turbines based on statistical analysis and practical experience", *IEEE Trans. Energy Convers.*, 27, 184–195.

Gebraad, P.M.O. (2014), "Data-driven wind plant control.", Delft University Of Technology, Doctoral thesis.doi:10.4233/uuid:5c37b2d7-c2da-4457-bff9-f6fd27fe8767.

Hahn, B., Durstewitz, M., Rohrig, K. (2007) "Reliability of Wind Turbines". In *Wind Energy*; Springer: Berlin/Heidelberg, Germany, pp. 329–332.

Jiang, X., Melchers, R.E. (2005), "Reliability analysis of maintained ships under correlated fatigue and corrosion", *International Journal of Maritime Engineering*, 147(A3), pp. 9–18

Karyotakis, A. (2011), "On the Optimisation of Operation and Maintenance Strategies for Offshore Wind Farms", Ph.D. Thesis, University College London (UCL), London, UK.

Koopstra, H. (2015), "An Integrated and Generic Approach for Effective Offshore Wind Farm Operations Maintenance", Master thesis, TU Delft, Delft, Netherlands.

Kusiak, A. & Li, W. (2011), "The prediction and diagnosis of wind turbine faults", *Renewable Energy*, 36, pp. 16–23.

Madsen, B.N. (2011), "Condition Monitoring of Wind Turbines by Electric Signature Analysis", Master's Thesis, Technical University of Denmark, Copenhagen, Denmark.

Márquez, F.P.G., Tobias, A.M., Pérez, J.M.P., & Papaelias, M. (2012), "Condition monitoring of wind turbines: Techniques and methods", *Renewable Energy*, 46, pp. 169–178.

Nabati, E.G., & Thoben, K.D. (2017). Big Data Analytics in the Maintenance of Off-Shore Wind Turbines: A Study on Data Characteristics. In Dynamics in Logistics (pp. 131–140). Springer International Publishing.

Ribrant, J. (2006), "Reliability Performance and Maintenance-A Survey of Failures in Wind Power Systems", Master's Thesis, School of Electrical Engineering, KTH Royal Institute of Technology, Stockholm, Sweden.

Rohrig, D.K. (2014), "Wind energy report Germany 2014", http://publica.fraunhofer.de/eprints/urn_nbn_de_0011-n-354656–16.pdf (2014), [Online; accessed 2017–09–09].

Santos, F.P., Teixeira, A.P. & Guedes Soares, C. (2018), "Maintenance Planning of an Offshore Wind Turbine Using Stochastic Petri Nets With Predicates", *J. Offshore Mech. Arct. Eng.*, 140, pp. 21904–1–9.

Santos, F., Teixeira, A.P., & Guedes Soares, C. (2015), "Modelling and Simulation of the Operation and Maintenance of Offshore Wind Turbines," *Proc. Inst. Mech. Eng. Part O J. Risk Reliab.*, 229(5), pp. 385–393.

Tchakoua, P., Wamkeue, R., Ouhrouche, M., Slaoui-Hasnaoui, F., Tameghe, T.A., & Ekembb, G. (2014), "Wind Turbine Condition Monitoring: State-of-the-Art Review, New Trends, and Future Challenges". *Energies*, 7, pp. 2596–2630.

Wang, K., Yan, X.P., Yuan Y.P., Jiang, X., Lodewijks, G. & Negenborn, R.R. (2017), "Study on Route Division for Ship Energy Efficiency Optimization Based on Big Environment Data", *4th International Conference on Transportation Information and Safety (ICTIS 2017)*. August 8th-10th, Banff, Alberta, Canada.

Wang, K., Jiang, X., Yan, X.P., Lodewijks, G., Yuan, YP., Negenborn, R.R. (2017), "PSO-based method for safe sailing route and efficient speeds decision-support for sea-going ships encountering accidents", *14th IEEE International Conference on Networking, Sensing and Control*, May, Calabria, Southern Italy.

Zheng, H., Negenborn, R.R., Lodewijks, G. (2016), "Closed-loop scheduling and control of waterborne AGVs for energy-efficient Inter Terminal Transport", *Transportation Research Part E Logistics & Transportation Review*,105, pp. 261–278.

Progress in Maritime Technology and Engineering – Guedes Soares & Santos (Eds)
© 2018 Taylor & Francis Group, London, ISBN 978-1-138-58539-3

Ships on condition data driven maintenance management

S. Lampreia & V. Lobo
Escola Naval e Centro de Investigação Naval (CINAV), Alfeite, Almada, Portugal

V. Vairinhos
Centro de Investigação Naval (CINAV), Alfeite, Almada, Portugal

J.G. Requeijo
Mechanical and Industrial Engineering Department, Faculty of Science and Technology, Universidade Nova of Lisbon, Lisbon, Caparica, Portugal

ABSTRACT: On condition maintenance management is gaining general acceptance both in ships as in other domains. This is a natural result given evolution of low cost sensors, statistical methodology, telecommunications, software and superabundance of observational data. In this paper we analyze the effects of digital revolution on the usual maintenance policies and anticipate its consequences on ships maintenance management. Specifically, we try to show that on condition maintenance is, intrinsically, a data driven maintenance policy and the natural solution that results from the convergence of those economic and technological realities. The development of statistical methodology capable of transforming, in real time, those data mountains in useful knowledge is, accomplished routinely using almost free software or, at least, easily accessible resources. The paper identifies and illustrates with real data examples some of the main consequences and issues associated to this new reality and its effects on main maintenance policies and management organizations.

1 INTRODUCTION

Data-Driven qualification applied to the description of a management or scientific activity means an opposition to the Model-Driven concept. Kobbaci et al (2008), Nabati et al (2017), Li, Honglei; Sorkhabi1 et al (2017). With Model-Driven decision-making process, it is assumed that scientific decisions or management activities are based on the previous formulation of mathematical models, for example distribution parametric models such as N(μ, σ) or exp(-λ) say, that are supposed to reproduce with reasonable fidelity the reality to be modeled; from machines functionality to nature behavior. This allows the use of logical reasoning, logical inference and statistical inference to predict systems or nature behavior. The data, in this context, is used to estimate the parameters and validate models. This is a classic setting that requires the elaboration of complex and costly mathematical models and their validation against available data. This is often not possible given the scarcity and cost of human skills needed to formulation and validation of such models and the fact that data is in general observational instead of experimental. On the other hand, when it is possible to obtain this kind of models, its operational performance is not always better than that of alternative non-parametric approaches, obtained by statistical analysis of the available data.

Superabundance of sensors, cheap computers and storage associated to the widespread availability of sophisticated data analysis methodologies and open and free programming environments such as R or Python, create unprecedented conditions for the development and use of data-driven maintenance management policies based in observational data.

Data collection is a process that cannot be separated from data analysis. To collect data and store it without analysis is not only useless as it can be harmful: there is no point in wasting valuable resources to collect data and not using it to ameliorate the reliability of systems and to avoid unnecessary risks and maintenance costs.

The paper has the following structure: in Part 2 we show that, current forms of Condition Based Maintenance (CBM) tend towards forms of maintenance based in data, exploring big data generated by CM sensors. Part 3 is dedicated to practical issues related with sensors and sensor networks. Parts 4 and 5 are used to present case studies (model-driven and data-driven) related with Condition Monitoring. Part 6 presents the conclusions and discussion.

2 CONDITION BASED MAINTENANCE

According with USA DEPARTMENT OF THE NAVY (1998), Condition-Based Maintenance (CBM) is "a methodology that stipulates the performance of maintenance only when there is objective evidence of need."

With more details:

Definition: *Condition-Based Maintenance (CBM) can be defined as a set of maintenance processes and capabilities derived from real-time assessment of weapon system condition obtained from embedded sensors and/or external test and measurements using portable equipment. The goal of CBM is to perform maintenance only upon evidence of need.*

According with American Bureau of Shipping, ABS (2016), Condition Based Maintenance (CBM) is "*A maintenance plan, conducted on a frequent or real-time basis, which is based on the use of Condition Monitoring to determine when part replacement or other corrective action is required. This process involves establishing a baseline and operating parameters, then frequently monitoring the machine and comparing any changes in operating conditions to the baseline. Repairs or replacement of parts are carried out before the machinery fails based upon the use of the tools prescribed for CM.*"

According with above definitions, this kind of maintenance is performed using the results of a systematic data gathering process named Condition Monitoring (CM) defined at ABS (2016) "*Condition monitoring comprises scheduled diagnostic technologies used to monitor machine condition to detect a potential failure. Practitioners in some countries refer to this term as an "on-condition task" or "predictive maintenance".*"

Data resulting from CM, obtained using several observation techniques (vibrations, thermic images and others) is used to decide 1) That some incipient failure process has been initiated 2) To predict, based in observations and accumulated experience that, if nothing is done, the failure will take place, with some probability, in some specific future time interval.

This means that CBM is a form of preventive maintenance, since it is meant to avoid functional failure (avoiding the risks associated to failure) and a predictive maintenance, since it is meant to predict, using monitoring data, the probability/possibility of system failure in a future time interval.

Another aspect to consider is the process used to decide (desirably well before the occurrence of failure) that some incipient failure process has been initiated. This means the CM data, combined with the statistical methodology used to detect this situation, are adequate and effective enough.

For example, how decide that the sequence of vibration measurements, observed in successive observation times, in a set of machine observation points can be interpreted as indicating that a failure process has been initiated? In other words, how to proceed to construct the diagnostic mentioned in the definition?

All this means that CBM is a methodology entirely dependent of sensor collected observational data and its interpretation using statistical methodologies able to decide the question: "the system/equipment has entered in an incipient failure process". According with ABS (2016) this is made using a Base Line Data resulting from observing this specific equipment (or others like it) in conditions that reflect a usual operating condition (same power, same ambiance, for example) and using this data to decide the issue. For that, Base Line Data is replaced with a convenient statistical synthesis. The "true state or condition", something very difficult to define and observe, is operationally replaced by that statistical synthesis of Baseline Data.

Having obtained this definition of a reference condition another statistical problem follows: how do we decide that current observational data resulting from Condition Monitoring (CM) is "the same" as the one represented by the Baseline data? Once more, this is the result of applying a statistical methodology to compare Baseline Data (its statistical synthesis) with current observation statistical synthesis resulting from CM, to formulate the diagnostic.

Obtained the diagnostic, the prediction part follows. The aim is to avoid the uncontrolled stopping of machine (catastrophic functional failure). For that, ideally, it would be necessary to predict, for a specific future interval of time, if (yes or no) that event would take place in that interval. Given the incertitude associated to this kind of situation, the usual procedure is to express that incertitude using probabilities, other possibilities being possible.

This analysis shows that CBM is only possible if based in a data collection process continuously interpreted using sophisticated statistical methodologies incorporated in software and the experience of qualified technicians.

3 SENSORS AND SENSOR NETWORKS

The analysis presented in Part2 shows that CBM is based in data collected by appropriate sensors installed preferably during the ship or system construction phase. The new installations must be equipped with the sensors needed to implement the chosen maintenance policies. This implies that the corresponding specifications must integrate the project from its early phases of conception and development, including the variables that must be observed by sensors to implement the chosen maintenance policies. Waiting for the ship or new system to be built to decide, then, what to observe and what sensors to install is the source of very disturbing troubles for the maintenance and operational use of the involved systems affecting its entire life cycle.

Another problem that cannot be overlooked is the fact that *à posteriori* installation of fixed sensors in a pre-existing installation can generate legal problems that affect the responsibilities assumed by the owner and manufacturer (insurances, for example), generating the interruption of guarantees with corresponding risks, costs and other penalties. This is, in generally overcome using portable collectors.

Maintenance sensor networks configuration and installation are, too often, totally determined by the builder, equipment manufactures and software suppliers, with little or no influence of the maintenance managers. The consequence is that, once the ship or specific system is built, it is very difficult to know which sensors exist, where they are installed what is that they measure, how the data is stored, its meaning and its format. This means that the maintenance computer system becomes a black box inaccessible to maintenance managers. To be sure, users cannot be allowed to modify data stored in system data bases, but it should be guaranteed, through appropriate specifications implemented by contract, the supply of that information and the exportation of data base content to ship-owner storage.

Nowadays, having access to the (BIG) data generated by the ship systems is a strategic issue both from an operational point of view as from a management point of view. The ship owner owns data. If, in a specific situation, it is not possible to disclose this data without compromising the manufacturer's responsibilities or eventual industrial secrets, it is preferable to choose suppliers that guarantee these functions even if with less sophisticated systems. Moreover, the usual free and unrestricted access and use to such data by systems manufacturers is a very important issue that must be regulated by contract.

The same happens with the acquisition of measurement equipment to carry out the CM tasks; it should be accorded preference to measurement equipment suppliers that guarantee the export of sensor data in relation to others that do not support.

This is crucial to guarantee that, in case of replacement of measurement equipment, there is integration of data, avoiding the interruption of data series.

The access to this kind of data is important to evaluate policies, definition of new maintenance policies, testing algorithms and, is, in sum, the condition allowing the possibility to develop, if necessary, new software to support empirically new projects. The minimum information that must be guaranteed is synthetized with the two following tables: Table1 – Sensors Identification and Table 2 – Sensor measurements data.

Table 1 – Identifies each sensor by an integer number and specifies its functional name, where it is installed and the variable it measures and its units. For example (Sensor Id, Sensor Function, Place, Units).

Table 2 – Stores the observed values measured by each one of the sensors and the observation time. For example (Sensor ID, Time, Observed Value).

Table 1. Parameters of phase 1.

Parameters	$\hat{\mu}$	$\hat{\sigma}$
Values	1,231	0,008

4 CASE STUDY 1: MODEL-DRIVEN CONTROL CHARTS BASED CM

4.1 *Methodology*

This is an example of Model-Driven (parametric) method, since inference rules are constructed assuming a population normal distribution for observations. CM observations are obtained by online sensors that stored data in a format useful for statistical treatment. The sequence is:

Phase I

- Test the variables independence, using the Estimated Autocorrelation Function (EACF) and the Estimated Partial Autocorrelation Function (EPACF).
- Because we found all the data is independent, the charts were built based on direct data measures;
- The process stability and data normality the vibration mean, and covariance are estimated.
- The control chart for individual observation and moving range for univariate data and modified T^2 chart for multivariate data are built and the vibration mean, variance and covariance are estimated;

Phase II

- To monitor the system vibration level, the modified CUSUM control chart for individual observation and modified T2 chart for multivariate data are built. In the modified control charts, the value (T_L) is used and not the mean, because we want to monitor not the pump parameters mean values but when it reaches the maximum values allowed. The methodology on phase II should be:
- – Define, based the equipment standard, the alert and maximum vibration level.
- – The Upper Control Limit (UCL) and the Alert Value (AL) are estimated to control the mean level of vibration.

- Establish rules for action on the system. The next are suggested:
  - Execute an intervention to detect any anomalous situation when 8 consecutive points above the AL are observed.
  - Proceed to a maintenance intervention when 4 consecutive points above UCL are observed.

4.2 *Case study*

The equipment under data driven maintenance study is a glycol-water eletropump installed on a ship. Vibration was measured in three points of electrical motor (Mot A vertical (V), Mot A horizontal(H) and Mot A axial(A)). For the univariate study only, the axial lecture is treated (point Mot A-A), for the multivariate study the three-axis lecture are studied (Points Mot A-A, Mot A-V and Mot A-H).

The vibration measures are expressed in speed units (mm/s) as global vibration level value (RMS).

The alert vibration value defined by the pump manufacturer is 1.8 mm/s, and the maximum, near the anomaly is 4.5 mm/s. In this study we use the 1.8 mm/s as the limit (T_L)

4.2.1 *Modified univariate control charts—CUSUM charts*

The modified CUSUM control charts would be applied on phase 2 of the univariate study. (Lampreia *et al*, 2013)

This charts are built based on cumulative sum (C) defined by:

$$C_t = \max\left(0,\ C_{t-1} + (Z_t - k)\right);\quad C_0 = 0 \tag{2}$$

where $Z_t = \left((\overline{X}_t - T_L)/\sigma_{\overline{X}}\right)$, $\sigma_{\overline{X}} = \sigma/\sqrt{n}$, $\Delta = \delta\sigma_{\overline{X}}$, $k = \delta/2$ e $T_L = (T_L)_{Standard} - \Delta_S$ and $\Delta_S = \delta_1\sigma$, where δ_1 is constant.

4.2.1.1 Phase 1

First, check data for independence using the ACF and PACF and the Statistica sofware. As can be seen from Figures 1 and 2, the obtained values are between the limits, meaning that the variable Mot A-A are independent, and so are the others, Mot A-V and Mot A-H.

Since data is independent, we can apply the individual and moving range control charts present on the statistica software to define the mean and variance, Figure 3 and Figure 4, where in the horizontal axis we can read the observation number and in the vertical the value of the observation and moving average for the respective charts.

For univariate study the estimated mean and variance are:

Now the modified CUSUM control chart can be applied on phase 2.

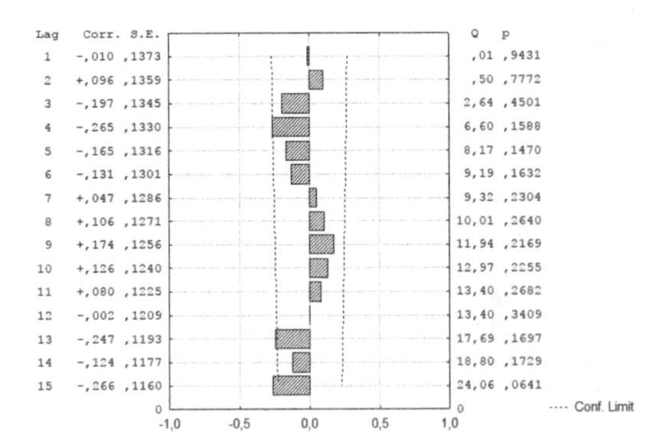

Figure 1. Autocorrelation function for point Mot A-A.

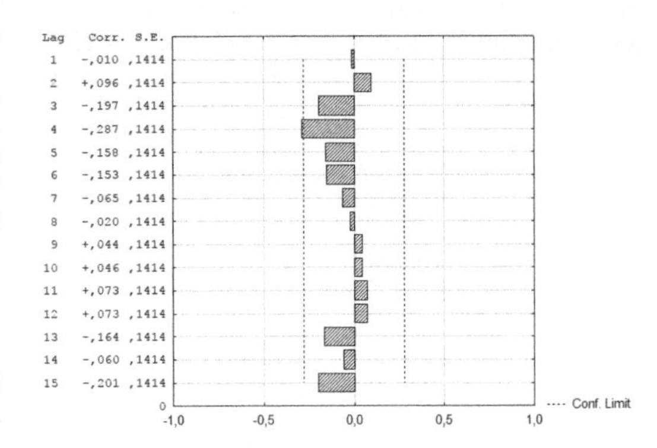

Figure 2. Partial autocorrelation function for point Mot A-A.

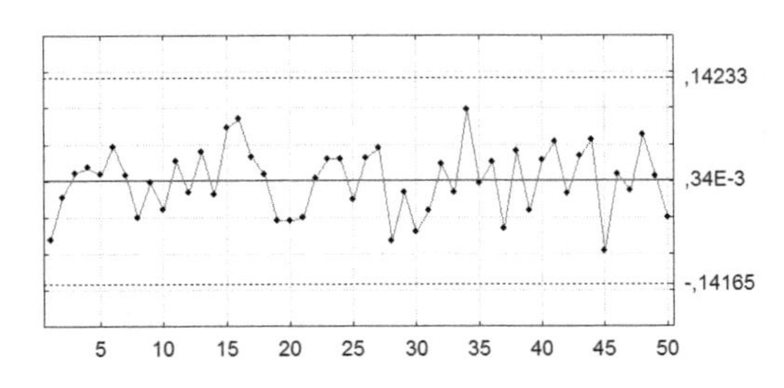

Figure 3. Individual observation control chart.

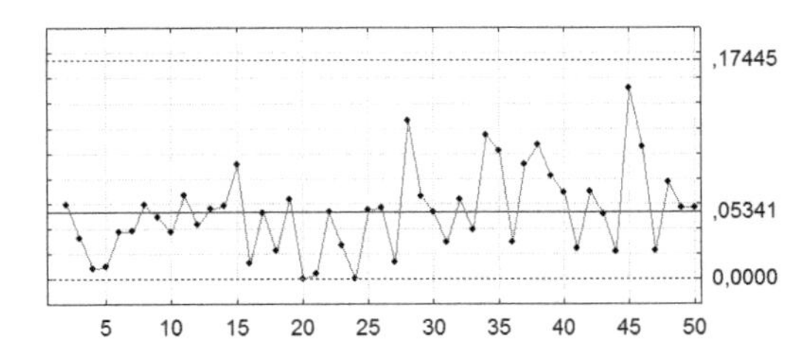

Figure 4. Moving range control chart.

4.2.1.2 Phase 2

In this phase we have to simulate data, in order to induce anomalies considering the T_L value, so 4 progressions of anomaly were considered. We will present two of them.

For the second progression using only the last four observations, accordingly to the specified methodology, we must proceed to a maintenance action. So, with the obtained data and combined other from other sensors, and with the correct analysis the intervention should be planned.

On the progression four, on the observation 29 we need to proceed to a maintenance action. This means that, using CM data, we may predict when an equipment intervention is needed.

4.2.2 *Multivariate methods—T^2 multivariate control charts*

With modified multivariate control charts T^2 we use, also, the value T_L for the chart setting (Lampreia *et al*, 2012).

4.2.2.1 Phase 1

Three of specified variables were used; as seen from phase 1, they are independent and T^2 control chart are set with the observed values.

As we can see from Figure 7, all the observations are under the maximum limit, since we defined the

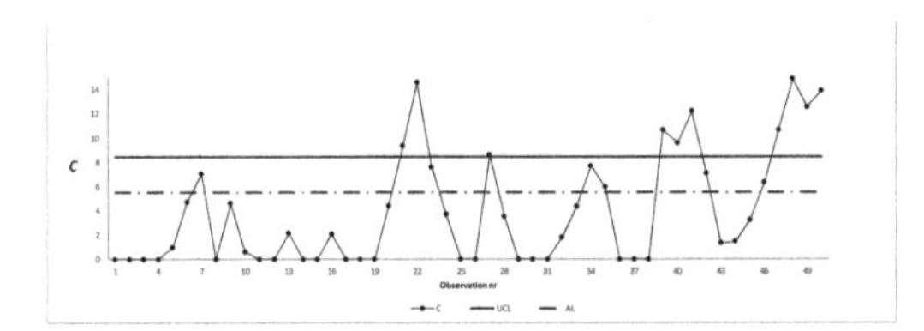

Figure 5. CUSUM control chart for progression 2.

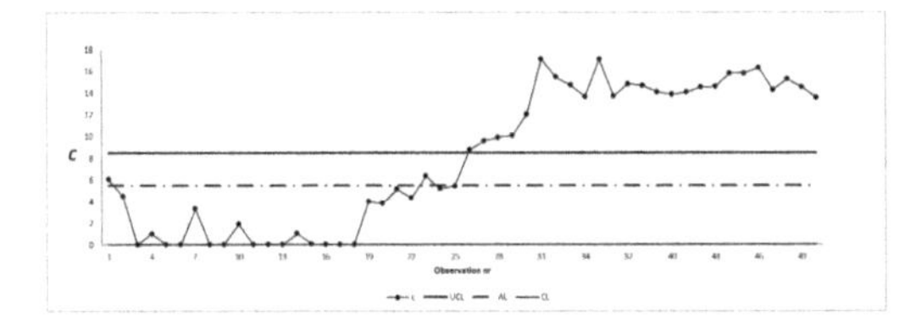

Figure 6. CUSUM control chart for progression 4.

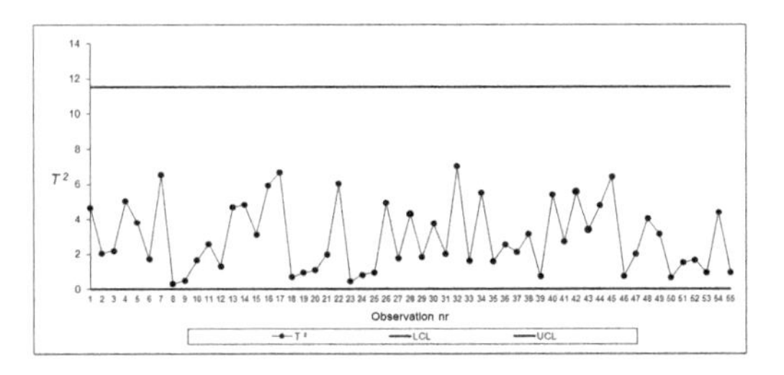

Figure 7. Phase 1 – T² control chart.

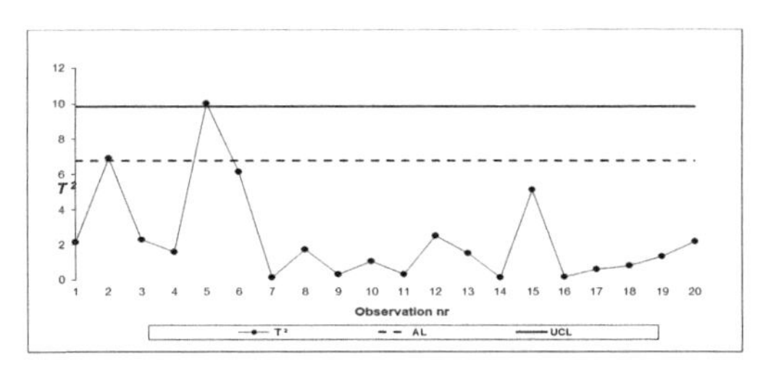

Figure 8. T² control charts for progression 2.

mean, variance and covariance matrix based on 200 observations as:

$$\bar{X} = \begin{bmatrix} 0,7979 \\ 1,250 \\ 1,297 \end{bmatrix}$$

$$S = \begin{bmatrix} 0,00189 & 0,00005 & -0,00050 \\ 0,00005 & 0,0018 & -0,00077 \\ -0,00050 & -0,00077 & 0,0027 \end{bmatrix}$$

$$S^{-1} = \begin{bmatrix} 560 & 35,11 & 114,68 \\ 35,11 & 648,482 & 195,8923 \\ 114,68 & 195,8923 & 455,504 \end{bmatrix}$$

4.2.2.2 Phase 2

Results for Phase 2 were unexpected: on the simultaneous study of three variables no observations were under the defined rules and so there is no need for intervention or inspection.

Despite the results, CM goes on and observations should be repeated to confirm the need of intervention showed on univariate study.

5 CASE STUDY 2: DATA DRIVEN BIPLOTS BASED CM

Here we use biplots (Gabriel, 1971; Galindo 1986) to illustrate its eventual use as an instrument o CM in complement or in addition of Water Fall graphs. This case is based in the same data as Case Study 1 but now using also observations at Points MOTOR B-A, MOTOR B-H and Motor B-V, using 6-dimension multivariate CM vibration data.

For each point (Motor A-A; Motor A-H: Motor A-V; Motor B-A; Motor B-H; Motor B-V) a vibration signal was observed and transformed, using FFT, in a vector with 50 amplitudes corresponding to the 50 retained Fourier Frequencies (FFs), labelled 1, 2,..50. This means that, for each one of those FFs, a vector of 6 amplitudes (one for each observation point) is available. Building a biplot for this 50 rows × 6 columns data set, Figure 9 was obtained.

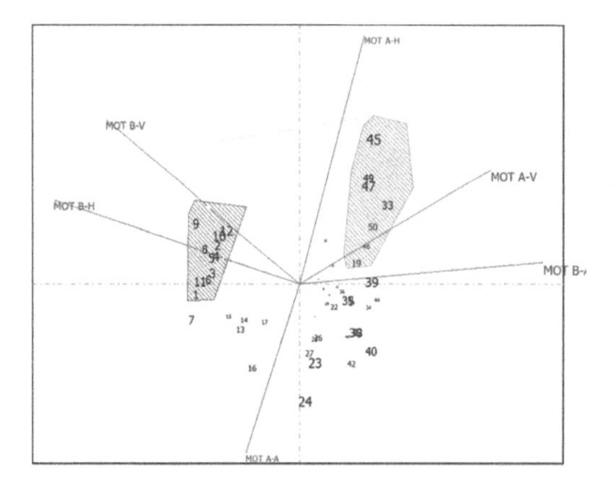

Figure 9. Biplot corresponding to FF's (Fourier frequencies) represented by points in the graph and Observation POINTS (spectrograms obtained at Observation points, represented by arrows).

For the construction and interpretation of these biplots see Gabriel (1971) Galindo (1986) and Vairinhos (2003). For the biplot at Figure 9, the labelled (1, 2,..50) points in the graph, represent Fourier Frequencies associated to the spectrograms resulting from the 6 observed vibration signals obtained at the 6 observation points mentioned above. These 6 observation points (or 6 observed variables) are represented by arrows (red arrows) in the graph. The angles among those arrows represented correlations among the spectrograms. The center of this graph corresponds to the mean of the 6 spectrograms.

6 CONCLUSIONS

CBM is controlled by the data streams associated to the corresponding CM.

The machine Condition (Reference Condition) to be controlled is operationally defined by a statistical convenient synthesis of the data set called Base Line Data. As examples of such methods two are presented: Modified Control Charts and Biplots.

Given the number, flexibility, price and features of available sensors, it is natural to expect that CBM will become the dominant future maintenance policy. This means a maintenance based in sophisticated forms of data analysis and statistical inference and prediction.

In this context, the importance of data collected by sensor networks, is the basic instrument available to the maintenance manager. This means that to the specifications covering these networks must be accorded a strategic importance when negotiating the ship construction contract.

As a contribution to a data driven maintenance, it has been suggested that biplots can be used as a CM data analysis methodology.

This is an ongoing project.

The use of the modified control charts on equipment condition monitoring may be useful but, we should use various statistical techniques, such as univariate and multivariate control charts, Biplots, and non-destructives analysis to confirm the results.

REFERENCES

ABS American Bureau of Shipping (2016) Guidance Notes On Equipment Condition Monitoring Techniques. Available at https://ww2.eagle.org/en/rules-and-resources/rules-and-guides.html. consulted at 218 JAN 18.

Department Of The Navy, Office of The Chief Of Naval Operations (1998) OPNAV INSTRUCTION 4790.16.

Gabriel K.R. (1971) The biplot graphic display of matrices with application to principal component analysis Biometrika (1971), 58, 3, p. 453 453.

Galindo, M.P. (1986) Una Alternativa de Representación Simultanea: HJ Biplot. Questio V.10 nº1 (marҫ 1986) pp-13–23 Available at http://diarium.usal.es/pgalindo/files/2012/07/0article-HJ–1986.pdf. consulted at 2016 Jan 18.

Kobbacy, Khairy Ahmed Helmy, Murthy, D.N. Prabhakar (Eds.) (2008) A Complex system maintenance handbook. Springer series in reliability engineering. Springer-Verlag.

Lampreia, Suzana, Requeijo, José, Dias, José, Vairinhos, Valter (2012). T^2 Charts Applied to Mechanical Equipment Condition Control, International Conference on Intelligent Engineering Systems 2012-INES2012. Caparica, june.

Lampreia, Suzana, Requeijo, José, Dias, José, Vairinhos, Valter (2013). Vibrations detection and analysis in equipments with MCUSUM charts and frequencies graphs, in Recent Advances in Integrity-Reliability-Failure, Editores: J.F. Silva Gomes e Shaker A. Meguid, Edições INEGI. International Conference on Integrity-Reliability-Failure-IRF2013, Funchal, Junho.

Li, Honglei; Garvan, Margaret R; Li, Jiaming; Echauz, Javier; Vachtsevanos, George J.; Brown, Douglas W.; Connolly, Richard J.; Zahiri Frank (2017) An Integrated Architecture for Corrosion Monitoring and Testing, Data mining, Modeling and Diagnostics/Prognostics. International Journal of Prognostics and Health Management, ISSN 2153–2648, 2017 005.

Nabati, E.G. & Thoben, K. (2016). Data Driven Decision Making in Planning the Maintenance Activities of Off-shore Wind Energy. The 5th International Confrence on Throught-life engineering Services (TESConf 2016), Elsevier procedia CIRP 59, pp 160–165.

Sorkhabil, Ali Ashasi; Fong Stanley; Prakash Guru, and Narasimha Sriram (2017) A Condition Based Maintenance Implementation for an Automated People Mover Gearbox International Journal of Prognostics and Health Management, ISSN 2153-2648, 2017 019.

Vairinhos,V. (2003) Basado en los Metodos Biplot. Desarrollo de Un Sistema para Minería de Datos Tesis de Doctorado, Departamento de Estadística de la Universidad de Salamanca.

Risk analysis

Progress in Maritime Technology and Engineering – Guedes Soares & Santos (Eds)
© 2018 Taylor & Francis Group, London, ISBN 978-1-138-58539-3

Risk analysis of ships & offshore wind turbines collision: Risk evaluation and case study

Qing Yu & Xuri Xin
School of Navigation, Wuhan University of Technology, Wuhan, China

Kezhong Liu
School of Navigation, Wuhan University of Technology, Hubei Key Laboratory of Inland Shipping Technology, National Engineering Research Center for Water Transport Safety, Wuhan, China

Jinfen Zhang
National Engineering Research Center for Water Transport Safety, Wuhan University of Technology, Wuhan, China

ABSTRACT: Wind energy is a renewable and green resource with the most profitable energy for electric generation. As a result, many Offshore Wind Farms (OWF) have been built in offshore sea areas. The impacts of collision risk between nearby ships and Offshore Wind Turbines (OWT) should be considered carefully. To address this issue, a risk analysis model is proposed based on Bayesian Belief Network (BBN). An initial qualitative analysis is conducted to identify the dimensions of the risk factors for the ship & OWT collision. Meanwhile, a case study is performed to investigate the collision risk of one OWF in China based on the multi-resource knowledge. The study concluded that the most influential factors for ship & OWT collision could be summarized, such as the interval distance between traffic flow and OWT, warming mark/buoyancy, and the human factor. The study provides with valuable evidence for OWF site selection and OWF safety management.

1 INTRODUCTION

With the increasing concerns on climate change, Carbon dioxide emission and environment pollution, wind energy could become a sustainable energy resource (Staid and Guikema, 2015). Hence, wind energy has been growing rapidly in China in recent years. It is known that the east coast possesses the most abundant wind resource in China. Offshore locations typically have higher and more steady wind speeds compared to the onshore wind. Denmark, UK, Netherlands, Germany Sweden are some countries that successfully operating large Offshore Wind Farms (OWF) (Ram, 2016). Chinese government and other stakeholder also have a growing interest in developing the OWF.

Those new projects and plans of building OWF will also have some impact to the nearby environment and the vessel traffic flow. Some factors should be fully considered before the site selection. Otherwise, it will result in significant damage and loss of productivity not only to the wind farm, but also for the navigational vessel. One of the challenges for renewable energy implementation is the disparity of the site selection, navigational impact and energy resources (Latinopoulos and Kechagia, 2015). The OWF located at east coasts could maximize the benefit from economic aspect. However, if the stakeholders consider the navigation safety issues, the OWF could become a significant threat at the world's busiest maritime traffic area. Thus, risk analysis and risk management are crucial for the long-term strategy. Some risk assessment methods, regulations and risk analysis tools from Europe or UK could serve as a starting point for the research (Rawson and Rogers, 2015), such as the MGN 543 from UK, the IWRAP risk analysis toolbox from IALA, etc. However, a risk management framework should contain the hazards present in specific location. Researches and stakeholders need to take a risk assessment before an OWF site selected.

Many risk analysis models were presented in the past researches or reports (Gudmestad, 2002, Mazaheri et al., 2016, Aneziris et al., 2016), which are proposed according to their research goals. In the initial stage, risk model for the OWF was originally developed from oil rigs risk analysis model, such as the COLLIDE model and the CRASH model. Then some improvements were taken for the model to make it more consistent with the environment around the OWF. However, most of the models are applied or described in a specific condition. They do not have a coherent risk analysis framework for ship & OWT collision that considering the human impact, rescue and security measurement in harmless. Such a framework would need to be design that including both OWF and Vessel Traffic Flow (VTF) impacts, it would need to explicitly include management and operations. Hence, such a framework could be developed to have substantial benefit for the wind farm stakeholders. This research is aimed to propose an assessment framework for evaluate the ships collision risk with offshore wind farm. In order to achieve the result, some questionnaires and statistics were performed to collect the database. A high standard database and professional judgments are collected from some exports and offshore windfarm managers. According to the feedback, a Bayesian Belief Network (BBN) model describes and evaluates the OWF framework. Then a case study of Chinese OWF is applied to examine the advantage of the model. Lastly, some preventions for the ship & Offshore Windfarm Turbines (OWT) collision were proposed relate to the research results.

2 HAZZARD INDENTIFICATION

Many risk analysis reports have been done to evaluate the OWF safety from different views, COLLIDE model from Safetec Nordic AS, CRASH/MARCS model from DNV, COLWT model from Germanischer Lloyd and COLLRISK mode from Anatec UK Ltd are some typical examples. The various models have been compared in previous studies. In the SAFESHIP (2005), a harmonization comparison was taken to discuss the model difference. Some results showed in the vicinity of the OWF, the VTF is light and the data on ship & OWT collision accident is limited (AB, 2008). Therefore, the statistical database is limited that cannot reflect enough information for the OWF risk analysis. In order to solve this problem, many collision possibility calculations for powered collision were proposed based on nearby VTF Gaussian distributed offset, which could estimate the number of collision candidates. The existing model mainly takes these aspects for ship & OWT collision:

1. VTF Complexity
The VTF complexity for assessing the effects of collision risk have been studied within different domain. Components such as vessel density, vessel number, vessel categories etc., which are closely influenced the number of collision candidates.

Meanwhile, there are some micro level factors that have been considered, such as ship length, distance between ship and OWT, load condition. The varying of these factors will affect the collision possibility of a vessel.

2. Accident causation mechanism
Database for ship & OWT collision is limited, therefore, some ship & manufactured structures

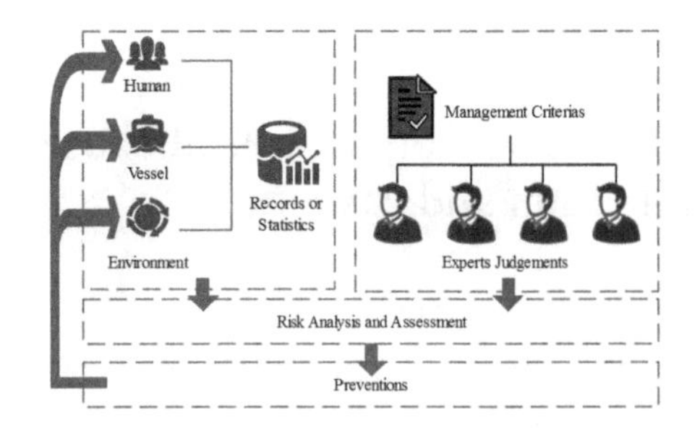

Figure 1. Accident causing mechanism.

collisions could be referenced. According to the accident data, in the past ten years, there are 26 ship & oil rigs collisions occurred, up to 75% of them were caused by human error.

Normally, the navigational risk for a ship is usually classified as four components, which are vessel, human, environment and management (Zhang et al., 2013) (See Fig. 1).

Most of the Collision Risk Factors (CRF) could be divided into these four categories as well. The data on CRF could be collected from two aspects, historical database or expert judgement. Both of them are apply to the BBN.

3. Environment condition
Environment condition includes many indicators. For this research, the main factors are the fog, the wind and the current. The fog will significantly influence the visual watch keeping, and reduce the position awareness. The wind and current would make the ship maneuvering difficult. It might be adjusted when ship is under working conditions, however, when the ship is drifting, the generate force of wind and flow will bring an adverse circumstance.

3 BAYESIAN BELIEF NETWORK

3.1 *BBN model*

BBN is a probabilistic evaluate model that present a target value under a conditional circumstance, the BBN is based on the Bayesian theory and the joint probability function is:

$$P(A|B)=\frac{P(B|A)\times P(A)}{P(B)} \tag{1}$$

For example, a prior probability $P(A)$ and $P(B)$ could be given by taking a statistical analysis or other methods, expert judgements or data interaction analysis could collect the conditional probability $P(B|A)$. By combining these parameters, then the posterior probability $P(A|B)$ could be calculated by referring to the prior probability.

3.2 *BBN processing*

In the BBN modeling, the interaction edges between each node are important as well as the selection of the relevant parameters (Zhang et al., 2016). The result of the BBN conduct after many iterative processes, which indicates the potential relation from the evidence nodes to the result nodes (Mazaheri et al., 2016).The data from the accident reports or the previous research, some parameters could be previously numbered to get the prior probability. This quantified data represents that according to the historical data or record or experts' beliefs, the condition probability distribution for a parameter before evidence is proved. However, during the investigation, some relevant evidences or information of an unknown event could be collected from some cases. These evidences are conditional and variable. After taking into account these evidences, the probability of an unknown event is the posterior probability. In this paper, all the nodes are selected from many parameters, this procedure has three stages:

Throughout all the parameter that may impact the collision risk by using the literature review and the expert questionnaire, the parameters were divided into different group based on their influence mechanism. Then a parameter list is given to the next stage.

Some experts were invited to investigate the parameters by taking Analytic Hierarchy Process (AHP) method, AHP could provide with a comprehensive and rational framework to quantify the importance of each parameters. After defining the importance of each parameter, all the parameters were ranked. Hence a criterion has been made to select the most important parameters. The criterions might be treated diversely when they are used for different propose. In this paper, the BBN model is trying to expresses the drifting collision risk. Therefore, the top important influence parameters are classified in different categories relate to these aspects.

Stating all the nodes in the BBN. The states of the nodes are established when more information is obtained. For instance, in the MGN 543, the distance between ship lanes and OWT is divided as shown in Table 1.

3.3 *BBN framework*

According to the historical reports, many ship & manufacturing structure collisions comply with an accident chain that combined with many sub risk factors under a certain step. Any break off within the accident chain could avoid the final accident occurrence (Wang and Sun, 2011). By taking some literature review, some CRFs within the accident chain have been selected, Then some experts were invited to rank and divide the CRFs into different categories based on their interaction and property. Meanwhile, the severity of each CRFs have been qualified and modified. As result, the risk model was composed of three portions, which are navigation difficulty modeling, position awareness modeling and rescue modeling. In each portion, the key nodes are followed with some sub-nodes. To deal with the database, the GeNIe software is used in this research for the modeling. The structure of the BBN is presented in Fig. 2.

3.3.1 *Navigation difficulty modeling*

To modify the structure of the BBN, some knowledge from different aspects are required. Navigation difficulty is a concept of evaluating the complexity of the marine environment for ship sailing. The PAWSA waterway risk model states the some categories, which include the vessel conditions, traffic conditions, navigational conditions and waterway conditions (Academy, 2009). Hence, some nodes were selected that based on the situation of the ship & OWT collision, and some factors that no effect on the model are eliminated.

Based on the interaction of each node, the links are shown in Table 2, where the VTF density,

Table 1. Recommended distance between wind farm boundaries and shipping routes.

Distance of turbine boundary from shipping route (90% of traffic, as per Distance C)	Factors for consideration	Tolerability
<0.5 nm (<926 m)	X-Band radar interference Vessels may generate multiple echoes on shore based radars	INTOLERABLE
0.5 nm – 3.5 nm (926 m – 6482 m)	Mariners' Ship Domain (vessel size and maneuverability) Distance to parallel boundary of a TSS S Band radar interference Effects on ARPA (or other automatic target tracking means) Compliance with COLREG	TOLERABLE IF ALARP Additional risk assessment and proposed mitigation measures required
>3.5 nm (>6482 m)	Minimum separation distance between turbines opposite sides of a route	BROADLY ACCEPTABLE

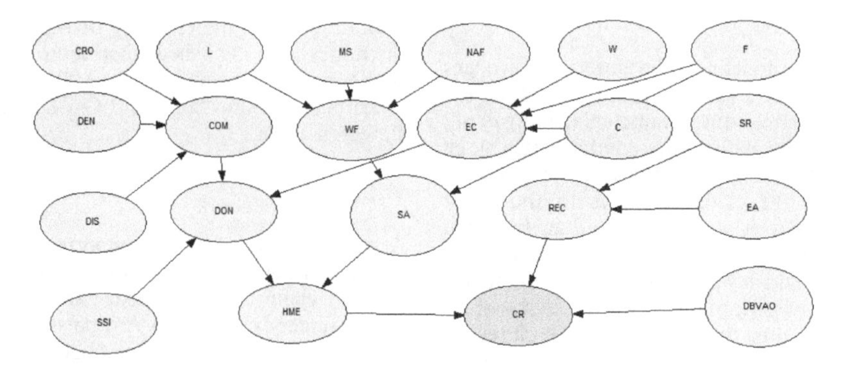

Figure 2. BBN model for ship & OWF Collision accidents.

Table 2. Nodes dependencies and resource in navigation difficulty modeling.

Nodes	Short form	Parent nodes	Source
VTF Complexity	Com	Den, Cro, Dis	S,E,L
Environment condition	EC	W, C, F	W,E
Difficulty of Navigation	DoN	Com, EC, SSI	S,E,L
Wind	W	N/A	W
Current	C	N/A	W
Fog	F	N/A	W
VTF density	Den	N/A	S
VTF Crossover	Cro	N/A	S,E,L
VTF distribution	Dis	N/A	S
Small ship impact	SSI	N/A	L,E

(AIS statistic = S, literature review = L, Experts judgement = E, Weather report = W).

crossover extent and distribution are the main indicators of the VTF complexity. The high density, critical crossover degree and wide distribution will increase the VTF complexity. The fog, wind and Current are the three important environment parameters. Serious current level, strong wind and dense fog will significantly reduce ship control capability and watching efficiency, and furthermore increase the navigation difficulty. Meanwhile, the OWF are normally built near the coast, and there are some working ship and fishing ship crossing the OWF and ship lane. This would influence other commercial ships in the ship lane. If the influence by small ships is critical, the navigation difficulty is high as well.

By setting the structure, the state of some nodes should be classified. Some information relate to the environment could be collected from past environment statistics report, the VTF information near the OWF can be obtained from AIS data statistics.

Table 3. Nodes dependencies and resource in position awareness modeling.

Nodes	Short form	Parent nodes	Source
Situation awareness	SA	F,NAF	E,L
Warning facility	WF	L,MS,NAF	E,I
Navigation aids facility	NAF	N/A	E,L
Lighting	L	N/A	I,S
Marks & signals	MS	N/A	I,S
Fog	F	N/A	W

(Expert judgement = E; Literature review = L; Spot Investigation = I; Historical statistics = S).

State criteria for each node have been set as:

- Wind: the proportion of days in a year that wind direction toward OWT and backward, which the wind scale should be larger than 3 knots.
- Current: the proportion of current direction toward OWT and backward, in which the current speed should be larger than 2 m/s.
- Fog: the proportion of fog days in a year that visual distance is below 2 nautical miles.

3.3.2 *Position awareness modeling*

Position awareness of the sailing ship is a category that reflects the ability for the ship's officer to realize the risk and keep the ship in the right course and positon. Some warning buoys, marks and lights could improve the position awareness that should be implement for OWF. The sufficient warning lights and marks with long visual range could help the sailing ship promote the ability of position awareness and observed the OWT easily.

Meanwhile, the visual distance is restricted during the fog day, and this impact can be mitigated by fitting the sound signals around OWF.

The structure of the position awareness model is presented in Table 3.

The criteria of each states is set based on it interaction from each nodes. For instance, the visual distance of lighting should at least more than 1 nautical mile. The installment should ensure all the ships could observe the warning from any direction near the OWF.

3.3.3 *Rescue modeling*

The risk candidate is the ship that out of control near the OWT due to mechanical failure or human error. The drifting ship have high threaten to the OWF safety when the wind and the current boost the risk. The emergency might be suspend if the ship could successfully save itself by taking self-rescue or by receiving external-rescue from others.

In the MARIN model (SAFESHIP, 2005), a self-repair function have been given to present the possibility of the repair failures varying with the duration after failure occurred. This function is based on the statistics of the incident reports, which is as follow:

$$F(t)=1 \quad \text{For } t<0.25$$
$$F(t)=1/(1.5\times(t-0.25)+1) \quad \text{For } t>0.25 \tag{2}$$

Fig. 4 shows that most of the ships success repair need at least 2 hours, even though the probability of failure is still very high. It is closely relates to the distance between ship lane and OWT and the ship drifting speed.

When the interval is close, drifting ships are hardly saving themselves by fixing the failure.

Meanwhile, the drifting ships could use their anchors to stop or ease drifting collision with OWTs, which has been discussed at many models (SAFESHIP, 2005, AB, 2008). But there some questions, the anchoring procedures are complicated. the anchoring machine should be acted within a short period; the ship cannot anchor if the ship speed higher than 2 knots; the drifting ship by towing its anchor could reduce its drifting speed to a certain extent, but the close distance between ship and OWTS make this way produce little effect.

The drifting ship could also be saved by receiving the external rescue. The success possibility depends on numerous issues, for instance, the rescue distance, rescue equipment and rescue training.

4 CASE STUDY OF ZGH OFFSHORE WINDFARM

In order to evaluate the BBN model, a case of ZGH offshore windfarm project is studied. The ZGH OWF project is building at Zhejiang province that is located at East Sea close to the Ningbo Zhoushan port, the distance between land and OWF is about 25 km. The ZGH OWF is separated as two areas. the total amount of power generation of the OWF will achieve 3.02 MKW. During the first stage, 54 offshore wind turbine will be distributed, and total power generation for the first stage is 0.216 MKW. There are many ship routes passing ZGH OWF, such as Western coastal route, Jinshan port entrance, Daishan Northern route and Donghai Bridge route etc. the nearest Western coastal route is the busiest for this area. The research data for this case is studied from two aspects: some quantifiable data such as VTF was come from historical AIS or statistics record. Other subjective

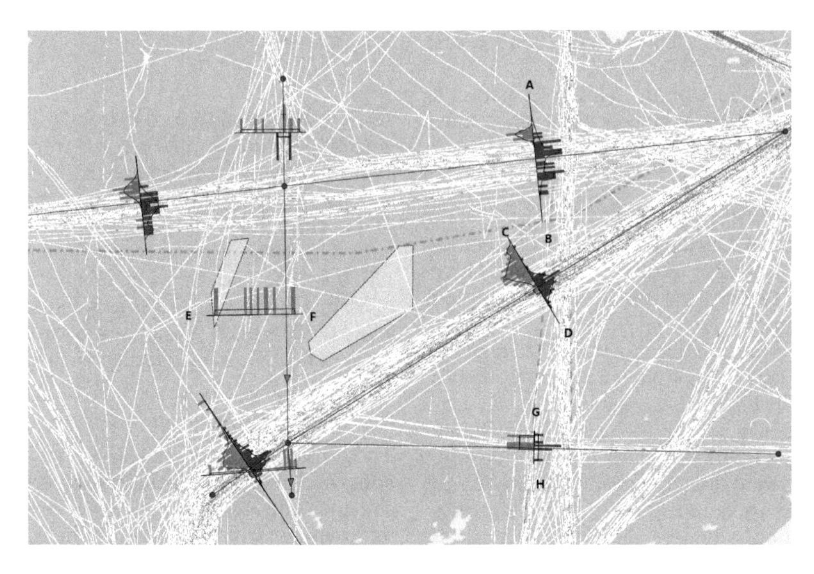

Figure 3. AIS Data record near ZGH OWF (30 days).

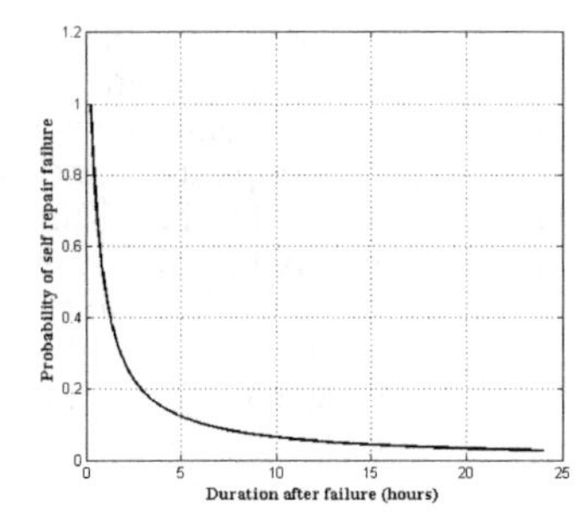

Figure 4. Probability of self-repair failure in MARIN model.

Table 4. Observation sections detail for each route.

Sections	Route	Position
A-B	Jinshan Port Entrance	From 30°32 N;122°01E to 30°29 N;122°01E
C-D	Western Coastal route	From 30°28 N;120°00E to 30°26 N;122°01E
E-F	Donghai Bridge route	From 30°26 N;121°49E to 30°26 N;121°53E
G-H	Daishan Northern route	From 30°23 N;122°01E to 30°22 N;122°01E

Table 5. Wind direction statistics for case area.

Month	Jan.	Feb.	Mar.	Apr.	May	Jun.	Jul.	Aug.	Sep.	Oct.	Nov.	Dec.
Wind direction	NNW	N	N	SSE	SSE	SSE	SSE	SSE	NNE	NNE	N	NNW
Possibility	19.2	21.6	15.3	15.8	15.9	17.5	25.4	21.1	16.9	17.2	15.6	20.9

data was obtained prove by experts' recommendation who are familiar with the ZGH OWF.

4.1 *Vessel traffic flow data*

To analysis the historical AIS data, an AIS risk toolbox software IWRAP is used to analysis the AIS data. As shown in Fig. 3, there are mainly 4 routes passing the ZGH OWF, and the Western coastal route cross in the middle of the OWF. The distance among the routes are from 0.25 nautical miles to 3 nautical miles. The closest is the Western coastal route and the farthest is the Daishan Northern route. On the basis of the interval criteria, most of the interval between VTF and OWT are at "Tolerable if ALARP" state.

To collect the VTF data for all the four routes, four observation sections are set for each routes, the detail information is shown in Table 4.

4.2 *Environment condition records*

From long period statistics of the actual weather records and hydrology records at case area. There are some validity statistical data could be analyzed for the BBN model.

There are 26.8 fog days during a year on average, the peak is 41 fog days and bottom is 13 fog days. Hence the fog possibility for a year is assumed as 7.3%.

The case area is located at East Sea of China, in which is under the influence of the monsoon obviously. The wind direction changes from NW to SE from winter to summer. The detailed statistics of wind direction is present in Table 5. The most influenced route by wind near the ZGH OWF is the Western Coastal Route, which is located at SE of the OWF. The SE wind usually appears from April to August and the average possibility is 19.1% according to the weather records from the Table 5.

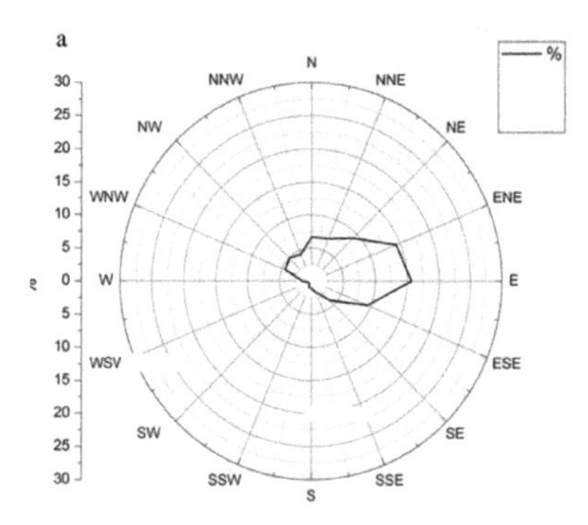

Figure 5. Rose diagram of current direction at case area.

Current is another feature that would affect the drifting ship collision. In this area, the peak month of the current force is August, when it is the active period of typhoon. Most of the current direction are NE and ESE, which takes about 50%. Direction of SE and W are rare. The impact current direction for ZGH OWF is the NW and the proportion is not high, the statistics of current direction is shown in Fig. 5.

4.3 *Expert judgments for subjective indicators*

The subjective data includes is consist from knowledge, understanding, beliefs and acceptance for a statement or a circumstance, which varies among different experts who have different background. This have been adopted in the earlier work in many maritime fields (Banda and Kujala, 2014). A multi-background judgment would result in high level of

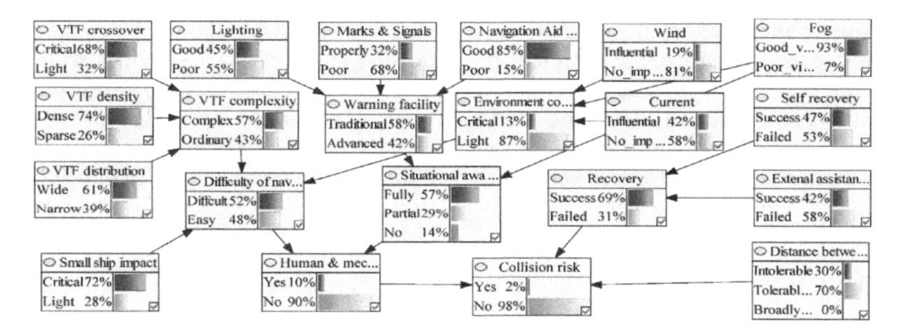

Figure 6. Case study evaluation by using BBN model.

relative and reliability (Banda and Kujala, 2014). In this way, some experts from different background were invited to evaluate the state for each parent node for the BBN model. All the evidence was considered as three levels that are high reliable, medium reliable and low reliable. For instance, recommendations from an OWF manager in OWF management aspect is supposed to be more reliable than a recommendation from a captain from a vessel in the same aspect.

By taking this concept, all the recommendations and judgements were collected. Every node was weighted based on the expert's background. Hence, the state values for each node can be easily input to the BBN model. The analysis of the case should, however, be decided by the further discussion according to the utility value and some prevention or recommendation could be proposed based on the results.

The evaluation result based on the BBN model for the case study is shows on the Fig. 6. The results show that under the expert judgements and present situation, the risk of ship and OWT collision is 1.7%, which is under a high risky circumstance.

5 RESULTS AND DISCUSSION

In general, it can be present by the BBN model that the collision risk for the ZGH OWF project is at a high-risk status at present due to the following reasons that should be taken into consideration.

1. The VTF nearby OWF is complicated and the passing ships are many. The node of VTF density shows the nearby VTF is dense. This is because of the Western Coastal Route crossing the middle of the windfarm, which is a recommended route for ships passing this area. Despite of ship numbers, there are some routes are crossover with each other around the OWF. The possibility for a ship encounter a crossover circumstance is rising. Meanwhile, the Traffic Separation Scheme is not fully implemented nearby the OWF. The distribution of the vessel flow is disperse and disordered.
2. The small ship nearby the OWF have some potential collision risk. There are some fishing areas that are close to the OWF and the fishing ship cross the OWF from one fishing area to another. Meanwhile, some engineering ships and maintenance ships passing though or even berthing inside the OWF for working operations. Therefore, the confined area within the OWF could make these small ships more difficult to maintain their course, while the collision risk of the small ships impact is higher than the commercial ship.
3. The warming light and marks are significant important issues that could reduce the collision risk for OWF. The model shows that the signals with good quality and quantity make the OWTs more easily be observed by nearby ships, especially under bad weather condition or at night. Some experts suggest that to improve the OWTs observation range, a radar beacon could be used for warming the passing ships.
4. According to the case study, in some area, the distance between ship route and OWTs is intolerable. As a result, a malfunctioning ship is hard to save herself by taking self-rescue, while the external rescue or collision prevention equipment can exert an outstanding influence to reduce the collision risk. Hence, a rescue station nearby OWF is advised if it is necessary.
5. Human error is an important issue. It is influenced by many aspects that relate to navigational environment, crew member training, risk awareness etc. The investigate and survey for the ship captain might present some useful information. Some captain point out that the OWTs impact is not critical, the reason is that the OWTs is easy to detect by use radar when it close. The ship & OWTs collisions are rare. But some arguments are put forward from some protesters. Human error are happened any time, but it is extraordinary danger when ship nearby the OWTs.

The BBN model of the ship & OWTs collision suggest that the collision risk analysis for OWF is an interrelated and complex system, that all the relate factors have direct or indirect impact the collision risk. The database for these factors are hardly to collect only from historical statistics or accident report, and the information only from subjective evaluation will make the model impractical. The case study shows a satisfied result that present the risk for the OWF by using multi-resource analysis. To combine the historical data and expert judgements, the BBN model could reflect the real situation, which are recommended for further risk assessment of OWF.

The small ship impact for OWTs collision is more critical than the passing commercial ships when measuring the uncertainty coefficient for these two nodes which do not similar with the record database. The reasons that given by experts could be summarized by two aspects: the operation measure of the small ship and AIS record omissions. Fishing ships operations near or inside the wind farm are unavoidable, and AIS has not been installed for some small fishing ships. Meanwhile, the construction ships for the OWF is considered as a collision candidate as well, which is worth to take prevention method when OWF is building.

Although the model analysis the collision risk based on the expert recommendation and accident causing mechanism, an examination between variables shows that the dynamic nature of ships & OWTs collision and further improvement band on feedback is not fully captured by the proposed model. This is due to the limitation of the background knowledge and recent database. It could be further improved by using the hybrid causal logic method on different views.

6 CONCLUSIONS

The study proposes a BBN model based on multi-resource knowledge of the OWF safety management that combine the view from the experts and accident causing mechanism in practice. A case study is performed to prove the reliability of the BBN model, and the results show that the BBN model could be satisfactorily applied to analyzing the collision risk for OWTs by processing different database from statistics and expert judgements. The description of the ships & OWTs collision causing mechanism and impact factors within the BBN frameworks provides with additional insight on wind farm planning and management, which could be referred for other OWF safety management aspects. While the proposed model could require further validation, the further researches could improve the model by combining different background knowledge.

Further improvement could be taken from many aspects. In regards to expanding the model contents by enhancing more elements or clarified the interaction between each elements, the collision mechanism for ship & OWTs collision should be fully studied. By taking more OWF record and expanding the investigation scope, the model could be more accurately and practically. A further study of the BBN structure and other modeling approach is necessary, which could improve the reliability of the model.

This model could be augmented with additional indicator and variables due to the different database from different aspects. A dynamic data with additional knowledge could better reflect the ships & OWTs collision and then implement for the OWF safety assessment.

REFERENCES

AB, S.S. (2008) Methodology for Assessing Risks to Ship Traffic from Offshore Wind Farms. Sweden, SSPA Sweden AB.

Academy, C.G. (2009) Ports and Waterways Safety Assessment Workshop Report. Kahului Harbor, Maui, Hawaii, 27–28 August 2009.

Aneziris, O.N., Papazoglou, I.A. & Psinias, A. (2016) Occupational risk for an onshore wind farm. Safety Science, 88, 188–198.

Banda, O.A.V. & Kujala, P. (2014) Bayesian network model of maritime safety management, Pergamon Press, Inc.

Gudmestad, O.T. (2002) Risk Assessment Tools for Use During Fabrication of Offshore Structures and in Marine Operations Projects. Journal of Offshore Mechanics & Arctic Engineering, 124, 153–161.

Latinopoulos, D. & Kechagia, K. (2015) A GIS-based multi-criteria evaluation for wind farm site selection. A regional scale application in Greece. Renewable Energy, 78, 550–560.

Mazaheri, A., Montewka, J. & Kujala, P. (2016) Towards an evidence-based probabilistic risk model for ship-grounding accidents. Safety Science, 86, 195–210.

Ram, B. (2016) Commentary on "Risk Analysis for U.S. Offshore Wind Farms: The Need for an Integrated Approach". Risk Analysis An Official Publication of the Society for Risk Analysis, 36, 641.

Rawson, A. & Rogers, E. (2015) Assessing the impacts to vessel traffic from offshore wind farms in the Thames Estuary.

SAFESHIP (2005) Reduction of Ship Collision Risks for Offshore Wind Farms. Collsision Frequencies. Germanischer Lloyd AG, Maritime Research Institute Netherlands MARIN, Technical University of Denmark.

Staid, A. & Guikema, S.D. (2015) Risk Analysis for U.S. Offshore Wind Farms: The Need for an Integrated Approach. Risk Analysis An Official Publication of the Society for Risk Analysis, 35, 587.

Wang, H. & Sun, S. (2011) Accident Causation Chain Analysis of Ship Collisions Based on Bayesian Networks. International Conference of Chinese Transportation Professionals.

Zhang, D., Yan, X.P., Yang, Z.L., Wall, A. & Wang, J. (2013) Incorporation of formal safety assessment and Bayesian network in navigational risk estimation of the Yangtze River. Reliability Engineering & System Safety, 118, 93–105.

Zhang, J., Teixeira, A.P., Guedes Soares, C., Yan, X. & Liu, K. (2016) Maritime Transportation Risk Assessment of Tianjin Port with Bayesian Belief Networks. Risk Anal, 36, 1171–87.

Progress in Maritime Technology and Engineering – Guedes Soares & Santos (Eds)
 ISBN 978-1-138-58539-3

Risk analysis of innovative maritime transport solutions using the extended Failure Mode and Effects Analysis (FMEA) methodology

E. Chalkia
Centre for Research and Technology Hellas, Hellenic Institute of Transport and Department of Transportation Planning and Engineering, Civil Engineering, National Technical University of Athens (NTUA), Athens, Greece

E. Sdoukopoulos
Centre for Research and Technology Hellas, Hellenic Institute of Transport, Athens, Greece
Department of Maritime Studies, University of Piraeus, Piraeus, Greece

E. Bekiaris
Centre for Research and Technology Hellas, Hellenic Institute of Transport, Athens, Greece

ABSTRACT: With 74% of EU's international trade by volume and 51% by value being carried by sea, maritime transport and ports play a predominant role in Europe. The increasingly dynamic business environment characterizing those sectors as well as the pressing need to efficiently and sustainably cope with increasing freight volumes, are driving the continuous investigation of innovative and promising solutions, the implementation of which can provide further efficiencies to the system as a whole and to the relevant stakeholders involved. Building upon the results of an expert consultation process, undertaken within the framework of the Mobility4EU project, this paper presents a validated set of such solutions analysing, based on the extended Failure Mode and Effects Analysis (FMEA) methodology, the main risks that could potentially hinder or delay their implementation, thus providing a set of appropriate strategies and measures that can be adopted for mitigating those risks.

1 INTRODUCTION

1.1 *Background*

With over 80% of world's cargo by volume and 70% by value being carried by sea, maritime transport is reasonably being acknowledged as the backbone of international trade heavily supporting the global economy (UNCTAD, 2017a). Over the past four decades, maritime transport volumes have been constantly increasing at an impressive rate, with the exception of 2009 when the impact of the global financial crisis became apparent also to the maritime transport industry among several other sectors.

The European continent proves to be attracting the majority of international freight flows (Figure 1), with 74% of the respective volumes being accommodated through an extended network of 329 seaports, among which 83 represent major hubs and are thus being acknowledged as the core part of the trans-European transport network (TEN-T) (European Parliament and EU Council, 2013). In addition, short sea shipping accounts for a considerable share of intra-European trade (37%), further stressing out the increased importance of maritime transport and ports in Europe, thus their substantial contribution to the European economy.

Both sectors are being characterized of a highly competitive and dynamic business environment, subject to increased technology penetration (e.g. new information and communication technologies, internet of things, automation and remote-controlling, augmented reality, etc.), facing disruptions as a result of global emerging market trends and developments (e.g. increasing vessel sizes, formation

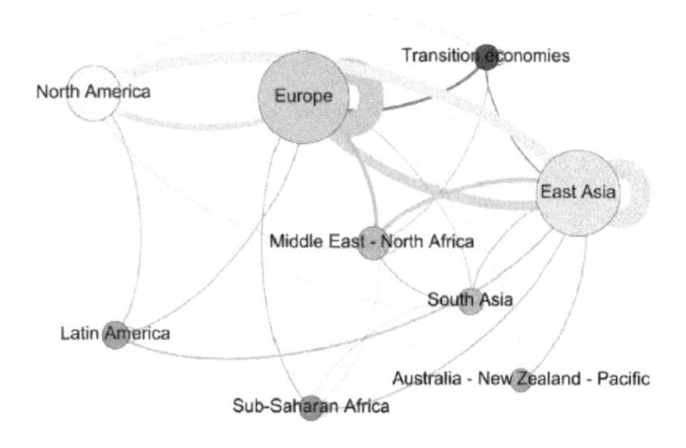

Figure 1. Trade flows across world's regions in 2015 (UNCTAD, 2017b).

of strategic shipping alliances, establishment of international terminal networks by global terminals operators, etc.) and being confronted with stricter environmental regulations coming into force (e.g. new emission limits in Sulphur Emission Control Areas, etc.), all of which prove to be significantly and rapidly changing business as it is (WATERBORNE TP, 2016). The intricate characteristics of this business environment along with the pressing need to efficiently accommodate ever increasing freight volumes, are continuously driving the investigation of new, innovative solutions (e.g. technology-, infrastructure-, policy-environment-oriented, etc.), the successful implementation and deployment of which may generate significant benefits to the system as a whole as well as to several stakeholders of the extended maritime transport and port communities.

The aforementioned considerations form the background of the EU-funded Mobility4EU project, which aims to deliver a vision for the European transport system in 2030 and an action plan, including a roadmap, for implementing that vision. The project follows a user-centeredness and cross-modality approach addressing, among other modes, waterborne transport systems covering both freight and passenger transport.

This vision and action plan of Mobility4EU is being based on the identification and assessment of societal challenges that are expected to influence future transport demand and supply, as well as on the compilation of a portfolio of emerging and promising cross-modal technical and organisational transport solutions likely to disrupt the current business environment. The entire process from studying new trends and developments, and assessing the potential of innovative transport solutions for developing the aforementioned vision and action plan, follows a structured participatory approach engaging, into the consultation process, a broad stakeholder community. For capitalizing upon these results and ensuring their sustainability, a European Transport Forum will be established at the end of the project, taking in this way a step further for efficiently complementing the delivered action plan.

Within the aforementioned context, this paper aims to provide a comprehensive overview of major risks, as identified and assessed by leading experts following a structured approach, that can potentially hinder or delay the development and uptake of promising and novel maritime transport solutions capable of delivering further efficiencies to the system, thus recommend a number of appropriate mitigation strategies and measures. To this end, the rest of the paper is structured as follows: a brief overview of the project's work plan for delivering the vision for the European transport system in 2030 is being provided first; the methodology and consecutive steps followed for defining and selecting the most promising solutions in the maritime transport sector and assessing implementation risks are being described in section 2; a description of the risks rated as most severe is being provided in section 3 along with possible mitigation measures that have been identified; section 4 concludes the paper proving a critical overview of all 58 risks identified (including both severe, moderate and insignificant ones) highlighting key points that need to be taken into consideration for moving towards a more efficient and sustainable European maritime transport system in 2030.

1.2 *Mobility4EU project overview*

Mobility4EU is a Coordination and Support Action funded under the Horizon 2020 programme of the European Commission from January 2016 to December 2018. As mentioned before, its overall objective is to deliver a vision for the European transport system in 2030 and an action plan, including a roadmap, for implementing that vision. To this end, recommendations for tangible measures in research, innovation and implementation targeting various stakeholder groups are being provided.

For meeting the project's overall objective, a number of societal challenges expected to influence future transport demand and supply were identified and assessed. More specifically, at the first phase of the project, 9 trends likely to shape the European transport system in 2030 were identified and were used as the starting point for devising the Mobility4EU context map (Mobility4EU, 2016). As a next step, with the support of European experts specializing in all fields of transport and covering both passenger and freight transport, a portfolio of 93 promising and innovative transport solutions addressing the identified user needs was formulated. The latter included solutions in concept or at research stage, but also incorporated recently implemented ones that need to be further supported for advancing their respective technologies or products and achieving wider implementation and deployment (Mobility4EU, 2018a).

Building upon the trends identified and the portfolio of innovative transport solutions formulated, a series of scenarios for the future of the transport system in Europe were developed utilizing the Multi-Actor Multi-Criteria Analysis (MAMCA) methodology (Macharis, 2007). In addition, the aforementioned results were also used as the basis for identifying and assessing implementation risks and barriers using the extended Failure Modes and Effects Analysis (FMEA) methodology, which was properly adjusted for meeting the needs of the Mobility4EU project. A large group of experts representing all key stakeholders, sourced from the

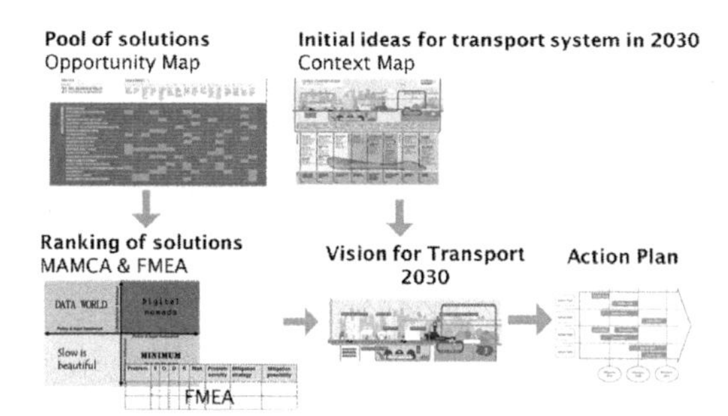

Figure 2. Steps followed in Mobility4EU for delivering the vision and action plan for the European transport system in 2030 (Mobility4EU, 2018a).

project's consortium and associated partners, were mobilized and supported the aforementioned process ensuring in that way the validity of the results, which are being presented for the maritime transport sector within the following sections.

The aforementioned consecutive steps comprising the project's work plan are clearly illustrated in the following figure (Figure 2).

2 RISK ASSESSMENT ANALYSIS FOR FUTURE TRANSPORT TRENDS AND INNOVATIVE SOLUTIONS

2.1 *The extended FMEA methodology*

The classical FMEA procedure is a tool that has been adapted in many different ways and for various purposes. It can contribute to improved designs for different products and processes, resulting in higher reliability, better quality, increased safety, enhanced customer satisfaction and reduced costs.

In the Mobility4EU project, the extended FMEA methodology, as developed in the ADVISORS project (Bekiaris and Stevens, 2005), was used for defining and assessing the risks of future transport trends and innovative solutions, since it proved to be fitting best the project's needs. It is based on the classical FMEA, which includes indicators of hazard consequence severity, occurrence probability, detectability and recoverability, but extends it by covering not only technical but also behavioural, legal and organizational—related risks. Risks are first identified and the level of risk is then assessed considering a number of characteristics for each risk type (i.e. technical, behavioural, legal and organisational). The significance of a risk overall depends on its consequences and the probability of its occurrence, but also on how easily it can be detected.

The overall process to be followed based on to the extended FMEA methodology is being presented in Figure 3.

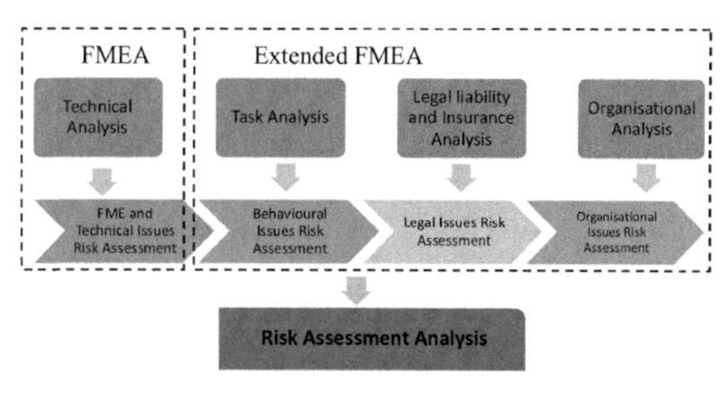

Figure 3. The extended FMEA (Bekiaris & Stevens, 2005).

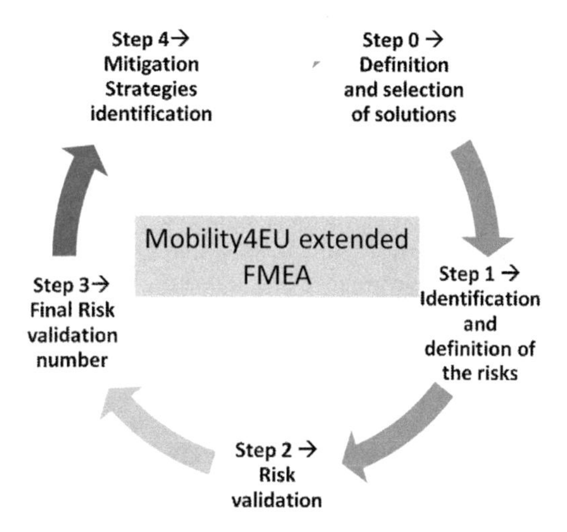

Figure 4. Steps of FMEA methodology, as implemented in Mobility4EU (Mobility4EU, 2018b).

As mentioned before, a number of risk types are being combined and incompatibilities or conflicts between different issues that may exist are being considered. Depending on which stakeholders are assessing and validating the risks, some risks may be unfavorable to all, whilst others may be inconvenient for specific stakeholders but benefit others. To this end, all stakeholder groups should be represented and get engaged in this process.

2.2 *Application of the extended FMEA methodology in Mobility4EU*

As mentioned above, a risk assessment consists of the identification and analysis of different risks (i.e. the identification of potential hazards and some estimation of their magnitude) and an evaluation of their tolerability in the relevant context. The steps that were followed, within the extended FMEA methodology for assessing possible implementation risks of the promising and innovative transport solutions that were selected in Mobility4EU are depicted in Figure 4 and are briefly described within the following sub-sections.

2.2.1 *Step 0 – Definition and selection of solutions*
For the application of the extended FMEA methodology, the 93 promising and innovative transport solutions that were identified at the first stage of the project had to be narrowed down in order to come-up with a feasible set of solutions to be assessed for each transport mode. To this end, experts were called to identify and rank the most critical solutions and the final sets were developed for each mode. For the maritime transport sector, the following 7 solutions were included in the final set:

1. *Alternative fuels*: With stricter emission limits on shipping being put forward (e.g. Emission Control Areas), the use of alternative fuels such as liquefied natural gas (LNG), bio-mass, methanol, etc. is being widely investigated since the relative reductions of air emissions are expected to be substantial. Such an operational environment highlights also the need for fuel-flexible vessels (i.e. engines and subsystems) that could effectively adapt to competitive market prices of certain fuels, thus meet current and planned (stricter) environmental regulations.
2. *Autonomous vessels for freight and passengers*: As vehicle automation progresses and considering the rather clear infrastructure of maritime transport compared to road, technical competences are being increasingly and rapidly transferred to vessels highlighting the latter as an optimal application field for automation. Systems' automation (e.g. navigation and route optimization), the availability of smart sensors and global networks for data transfer from ship to shore will promote remote-controlled and semi or fully autonomous vessels. The deployment of autonomous vessels is expected to significantly disrupt the shipping industry thus impose a significant impact on existing job profiles (e.g. remote control hubs operators) and the relevant skills needed.
3. *Blue modal shift—bringing transport to the waterways (in the urban environment)*: With road transport often reaching its maximum capacity and creating high levels of congestion, a modal shift towards waterborne transport in the urban environment (i.e. inland waterways or waterborne commuter solutions) can lead to the realization of significant economic and environmental benefits. For increasing the current relative low share of such modes, the relevant infrastructure needs to be modernized and be specialized in each context in terms of load (e.g. passengers and cars) and connectivity with land modes. Passenger, car or bike transfer on rivers in urban areas can for example significantly shorten urban routes compared to bridges. As a next step, and in line with solution 2, the platooning of vessels or ferries can be realized providing additional benefits and increasing the modes' attractiveness.
4. *Energy efficient and low emission ship*: Energy efficiency and increased environmental performance of ships can be reached through the deployment of various solutions including enhanced hydrodynamic performance, more efficient propulsion systems, reduced demand of on-board systems (e.g. lighting, working devices), employment of scrubbers, ballast water systems, etc.
5. *Hybrid and electrified ferries and vessels in ports*: Hybridization and electrification of ferries and of vessels in ports is already ongoing (e.g. 'Ampere' electric ferry in Norway, 'Copenhagen' and 'Berlin' hybrid electric Scandinavian ferries, etc.). However, further advancements are needed in order to enable longer electrified routes and higher loads. Ferries in particular operate on a fixed schedule with short docking times and would significantly benefit from wireless, inductive power transfer technologies.
6. *Multi-skilling and competence-based port labour training schemes*: Digitization at ports, a new working environment resulting from the introduction and development of global terminal operators as well as changing patterns of labour supply have led to the demand of new and/or combined skills, with training programs shifting from 'job analysis' to the identification of competences required for a given function.
7. *Smart connected vessels and ports*: Key ICT innovations in systems and software will affect almost all aspects of maritime transport processes. On-board increased communication between systems leading to vessels becoming 'system of systems' (i.e. smart connected vessels) together with critical infrastructure such as port and logistics sites will enable more efficient cargo handling processes, route planning etc. Technologies of augmented and virtual reality present also an increased potential application for managing for example vessel (bridge) operations, improving port and logistics infrastructure and operations, planning new terminals or assessing existing ones as well as for training purposes.

2.2.2 *Step 1 – Identification and definition of the risks*
For each of the aforementioned solutions, technical, legal, organizational and behavioural risks had to be identified and the following characteristics of the latter had to be defined:

- Risk mode → what is the possible risk.
- Risk effect → what is the effect if this risk occurs.
- Risk cause → what might trigger this risk to occur.
- Risk detection & recognition → how this risk is detected when it occurs.

2.2.3 *Step 2 – Risk validation*
For validating the identified risks, their severity, occurrence probability, detectability and recoverability were assessed using a 1–10 scale, where 10 represents an extremely severe, with high occurrence probability, improbable and non-recoverable risk. Within this process, different aspects were taken into consideration based on the risk type.

Risk severity (S)
- *Technical risks analysis*: Technical risks consist of technical (hardware and software) failures of the solutions of Step 0 or risks that are related to their technical maturity.
- *Behavioural risks analysis*: Behavioural risks are associated with the behavior of users and organizations that have a negative impact on the society and on the selected solution (e.g. human error issues).
- *Legal risks analysis*: Legal risks include significant legal issues that are likely to affect solution implementation and deployment (e.g. change to existing law required for solution implementation, significant legal cost for deployment, large potential liabilities, etc.).
- *Organizational risks analysis*: Lack of communication and reporting structures between actors can create a number of organizational risks that should be taken into consideration (e.g. accounting failures, frauds, internal control breaches, governance failures, etc.).

Risk occurrence probability (O)
Risk occurrence probability is the probability that all the risk causes related to the risk modes can occur. This is often a qualitative index especially when new technologies are concerned because of limited reliability data often available.

Risk detectability (D)
Risk detectability is the probability to detect the occurrence of a risk mode at an early stage. Detection of a developing risk is an important aspect in risk management, as early detection can facilitate the efficient application of mitigation strategies. With regard to technical, and to some extent behavioural risks, detection can be supported by sensors and data processing. For legal and organizational risks, surveys, monitoring and feedback are important tools.

Risk recoverability (R)
Risk recoverability is an efficacy index of the possible recovery action to be performed following risk management procedures. It estimates the ability of a solution to tolerate the risk.

2.2.4 *Step 3 – Final risk validation number*
Among the different risks identified, classified and validated, an overall relative indication of their significance is very useful, and to this end a risk number (RN) can be calculated within the extended FMEA, using the following formula:

$$Risk\ Number = S^{*}O^{*}\left[\frac{D+R}{2}\right]$$

This calculation is applied to each category of risks with the respective results ranging from 0–1000 depending on the validity of each risk, as indicated in the following table (Table 1).

Table 1. Risk evaluation based on the overall risk number (Mobility4EU, 2018b).

Overall risk number	Overall severity	Mitigation possibility
513–1000	Extremely severe	Very high
217–512	Severe	High
65–216	Moderate	Medium
9–64	Slight	Low
1–8	Insignificant	Improbable

Normally, organizations select a pre-defined range for the RN (i.e. the 513–1000 range is often selected since it includes the most severe risks but wider ranges can also be considered) and mitigation strategies are being implemented for the risks included in the selected range. Through this process, the use of available resources can be optimized and costs can be minimized.

Such a range (i.e. 513–1000) was also used in Mobility4EU, and the 12 most severe risks included for the 7 'critical' innovative maritime transport solutions reported in Step 0, are being described within the following section along with the proposed mitigation strategies/measures.

3 ASSESSMENT OF SEVERE RISKS FOR THE IMPLEMENTATION OF INNOVATIVE MARITIME TRANSPORT SOLUTIONS

For each of the 7 critical innovative maritime transport solutions identified in Step 0, the most severe implementation risks are being described below along with a set of proposed mitigation strategies/measures.

***Solution:* Alternative fuels**
Risk 1 (Organizational): The impact of the global financial crisis, investment uncertainty on both the supply and demand side for alternative fuels, and shipping economic cycles, have significantly hindered or have not enabled maritime transport stakeholders (e.g. ship-owners, port authorities, etc.) to have the required capital for investing on alternative fuel technologies. As a result, the share of alternatively fuelled vessels, port equipment, vehicles, etc.

in existing fleets is still quite small and their relative increase in the near future is expected to be very slow.

Mitigation strategy: This risk cannot be easily mitigated as investment decisions depend on a variety of different factors (e.g. economic environment, business dynamics, etc.). However, low interest rates may provide the necessary capital for investment.

Risk 2 (Organizational): The high investment uncertainty characterizing the alternative fuel market in shipping has led to inadequate supply or demand as well as to the absence of market leaders that could potentially drive their development and thus their faster and larger penetration in the marine fuel market.

Mitigation strategy: The introduction of more strict environmental regulations in shipping and in ports, as well as the formulation of the necessary legal framework for the bunkering and use of alternative fuels, will support the development of the required supply and demand and consequently the rise of the relevant market. The financial support provided by relevant programs (e.g. Connecting Europe Facility) for the development of the required infrastructure is also very important for mitigating this risk.

***Solution:* Hybrid and electrified ferries and vessels in ports**

Risk 3 (Organizational): The severe impact of the global financial crisis in shipping and the high investment uncertainty characterizing the sector have contributed towards ship-owners lacking the capital required for introducing hybrid and electrified vessels in their fleets. As a result, the relative share of such vessels is still very low and is expected to not increase considerably in the following years.

Mitigation strategy: This risk cannot be easily mitigated as investment decisions depend on a variety of different factors (e.g. economic environment, business dynamics, etc.). However, low interest rates may provide the necessary capital for investing on hybrid and electrified vessels.

***Solution:* Autonomous vessels for freight and passengers**

Risk 4 (Legal): For introducing autonomous vessels in the market, all relevant legal aspects related to their operation need to be carefully taken into consideration and tackled. This will require a long consultation process, where the engagement of all relevant stakeholders needs to be ensured, thus the formal approval processes by relevant bodies is often slow. At regional level, the relevant processes may be more complex since for example at European level national policies of Member States will need to be aligned. The delay in formulating the appropriate legal framework (i.e. at European level, national level, etc.) governing the operation of autonomous vessels at sea and in ports would lead to a longer, than expected, time horizon for autonomous vessels to enter into service.

Mitigation strategy: This risk can be mitigated by allocating targeted research funds to investigate the regulatory and legal frameworks/amendments (including liability regimes) required for the successful operation of autonomous vessels in the environment of early adopters (e.g. inland waterway transport) as well as that of followers. Industry pressures (especially of global players) and political support may place such an issue high in the political agenda of the EU and Member States and accelerate the required policy reform/introduction processes.

Risk 5 (Legal): Extended time periods are also required for specifying required revisions and additions in international shipping conventions so that the operation of autonomous vessels can be facilitated. Such a process will also require long consultations with all relevant stakeholders while the formal approval process by the responsible regulating authorities may also be slow.

Mitigation strategy: This risk can be mitigated by allocating targeted research funds to identify possible amendments required to international conventions for the safe and efficient operation of autonomous vessels in intercontinental shipping. The successful operation of such vessels in a smaller scale and other environments (e.g. first adopters) as well as the introduction of all necessary regulatory and legal requirements at EU and Member State level may facilitate and accelerate such a process, which may be further assisted by high industry demand and increased political support.

***Solution:* Energy efficient and low emission ships**

Risk 6 (Organizational): The severe impact of the global financial crisis, coupled with shipping economic cycles, has contributed towards a lack of capital of ship-owners to invest in measures for enhancing the energy and environmental performance of their vessels. Furthermore, the absence of more strict environmental regulations in shipping has resulted in a low demand of ship-owners for the implementation of such measures/technologies.

Mitigation strategy: This risk cannot be easily mitigated as investment decisions depend on a variety of different factors (e.g. economic environment, business dynamics, etc.). However, low interest rates may provide the necessary capital for investment while stricter environmental regulations in shipping may support a growth in demand for such measures/technologies.

Risk 7 (Organizational): The wider benefits that alternatively fuelled vessels may provide (i.e. energy, environmental, cost savings, etc.), the conformity of the latter with more strict shipping environmental regulations that may be enforced in the near future as well as the lower investment risk that such vessels present (on the long-term), may overrule the potential benefits to be achieved by a single or set (if appropriate) of energy and environmental efficiency improvement measures/technologies. As a result, the interest of ship-owners to invest on such measures can be low.

Mitigation strategy: This risk cannot be easily mitigated and depends on the development of the relevant market (i.e. for alternative fuels). However, the lower investment required for the implementation of such measures/technologies compared to alternatively fuelled vessels, vis-a-vis the environmental restrictions that are currently in force (more strict limits are planned to be enforced in the future) may withhold investments interests being transferred to alternatively-fuelled vessels.

***Solution:* Blue modal shift—bringing transport to the waterways (in the urban environment)**

Risk 8 (Organizational): Physical urban network limitations (e.g. inability to serve large urban areas), the need to be combined with other modes of transport for completing an urban trip, and the often low connectivity between the different modes (e.g. lack of intermodal interchanges or services—frequencies not well aligned, etc.) can lead to significant increases in travel time and consequently to a low demand for waterborne and inland waterway transport services in the urban environment decreasing in that way their modal share.

Mitigation strategy: This risk may be mitigated by modernizing current infrastructure (e.g. new vessels), ensuring reliable services (i.e. adherence to timetables) and improving connectivity with other transport modes time, location and fare-wise (i.e. integrated planning, intermodal interchanges, integrated fare systems, etc.).

Risk 9 (Organizational): High fares for passengers and additional handling costs for freight (although operational costs may be lower depending on capacity utilization) together with the combination of modes required for reaching final destinations, can lead to increases in travel costs, lowering as in the previous case the demand for waterborne and inland waterway transport services in the urban environment and consequently decreasing their modal share.

Mitigation strategy: This risk may be mitigated by ensuring high capacity utilization rates with regard to both passengers (i.e. appropriate service frequency) and freight (i.e. targeting low-value goods with higher lead times).

Risk 10 (Organizational): Services of low frequency, significant delays experienced and run-down vessels (i.e. not properly maintained) may lead towards low service quality, reliability and passenger comfort that would again result in low demand for waterborne and inland waterway transport modes in the urban environment and in decrease of their modal share.

Mitigation strategy: This risk can be mitigated by targeted infrastructure investments (e.g. new vessels, infrastructure in ports, ferry stations, etc.) and careful and integrated planning (e.g. service frequency, fare policy, efficient connections with other transport modes, etc.).

***Solution:* Smart connected vessels and automated ports**

Risk 11 (Legal): The implementation of smart connected vessels and port automation will require the collection of data via on-board/off-board monitors. Thus, it would bring in the scenery cloud computing, wireless communication technologies, Internet of Things and Big Data analytics; all these technologies integrated in smart connected vessels will trigger legal issues such as personal data sensitivity issues, cyber security, etc.

Mitigation strategy: The risk can be avoided by enforcing strict compliance with European Union and national legislations.

Risk 12 (Legal): Smart unmanned connected vessels may perform specific actions (via remote control). At the moment there is an absence of relevant legislative actions and problem of performance of legal obligations can thus be created.

Mitigation strategy: The risk can be only avoided by the necessary legislative actions being introduced by the European Union and Member States.

4 CONCLUSIONS

Within the context of the Mobility4EU project, a thorough risk analysis was conducted with regard to the implementation of 45 critical innovative transport solutions, out of total 93 that were identified, considering all modes of transport and covering both passengers and freight. The focus of this paper was confined on the maritime transport sector and for the 7 most critical solutions, as identified by relevant experts, the 12 most severe implementation risks were identified and presented. The majority included organizational risks (8), with several of them sharing a common ground that mainly refers to the heavy impact of the financial crisis on the shipping sector coupled with an investment uncertainty that the latter provides, which however differs based on the specific market that is being addressed (e.g. marine fuel market, shipbuilding, ship retrofitting, etc.). All remaining risks are legal ones highlighting the absence of appropriate regulatory frameworks that need to be established for tackling all relevant issues coupled with the introduction of new and innovative technologies that the respective solutions entail. No technical and behavioral risks were assessed by the relevant experts as severe.

Overall, a set of 58 risks (including severe, moderate and insignificant ones) were assessed. From these, 19 were categorized as organizational 19 as technical, 16 as legal and only 4 as behavioural (Figure 5).

It is clear from Figure 5 that organizational, technical and legal risks were equally represented in the assessment that was performed while behavioral ones prove to be lacking content. A possible explanation for this could be that the maritime transport sector is currently subject to increased penetration of new and innovative technologies which, as also mentioned before, for becoming operational more important barriers related to technical specifications, organizational issues and policies that need to be in place need to be overcome compared to the ones related with the behavior of the stakeholders involved. This does not imply however that such risks are not important, but before addressing them, experts need first to be sure that technical, organizational and legal challenges can be successfully overcome.

As mentioned before behavioral but also technical risks were not ranked as severe. With regard to the latter extensive research and pilot-testing may ensure that all technical inefficiencies can be detected as different development levels raising the technology readiness level of the solution under consideration. On the other hand, experts expressed a clear worry on the organizational part of the solutions, paying also increased attention on the relevant legal framework that should be formed or revised for efficiently tackling all relevant aspects. Changing business as it is integrates high risks which when coupled with an uncertain

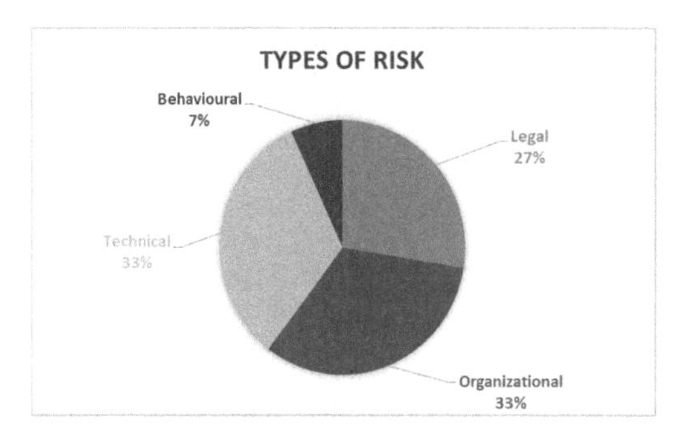

Figure 5. Classification of total implementation risks for innovative maritime transport solutions (Mobility4EU, 2018b).

economic and dynamic business environment as shipping is, further complicates and hardens investment decisions. This was clearly expressed by experts with regard to many of the solutions under consideration highlighting that the high costs for new vessels, infrastructure and equipment cannot be easily borne and different funding structures and initiatives may greatly contribute towards realizing a business shift towards for example more clear fuels and technologies.

ACKNOWLEDGEMENTS

This work has been realized in the context of the EU funded project Mobility4EU. Mobility4EU is a Coordination and Support Action funded by the European Commission under the Horizon 2020 Programme, with contract number 690732.

REFERENCES

Bekiaris, E. & Stevens, A. 2005. Common risk assessment methodology for advanced driver assistance systems. Transport Reviews, 25(3): 283–292.

European Parliament and the Council of the European Union. 2013. Regulation No 1315/2013 on Union guidelines for the development of the trans-European transport network and repealing Decision No 661/2010/EU. Official Journal of the European Union.

Macharis C. 2007. Multi-criteria Analysis as a Tool to Include Stakeholders in Project Evaluation: The MAMCA Method. Transport Project Evaluation: 115–131.

Mobility4EU project. 2016. Deliverable 2.1 – Societal Needs and Requirements for Future Transportation and Mobility as well as Opportunities and Challenges of Current Solutions. Technical Report. December 2016. http://www.mobility4eu.eu/wp-content/uploads/2017/01/M4EU_WP2_D21_v2_21Dec2016_final.pdf.

Mobility4EU project. 2018a. Deliverable 2.3 – Novel and Innovative Mobility Concepts and Solutions. Technical Report. January 2018. https://www.mobility4eu.eu/?wpdmdl=2069.

Mobility4EU project. 2018b. Deliverable 4.1 – Report on Risk Assessment and FMEA. Technical Report. January 2018. https://www.mobility4eu.eu/?wpdmdl=2070.

United Nations Conference on Trade and Development (UNCTAD). 2017a. Review of Maritime Transport 2016. http://unctad.org/en/pages/PublicationWebflyer.aspx?publicationid=1650.

United Nations Conference on Trade and Development (UNCTAD). 2017b. Key Statistics and Trends in International Trade 2016. http://unctad.org/en/PublicationsLibrary/ditctab2016d3_en.pdf.

WATERBORNE Technology Platform. 2016. Global Trends Driving Maritime Innovation. https://www.waterborne.eu/media/20004/global-trends-driving-maritime-innovation-brochure-august-2016.pdf.

Progress in Maritime Technology and Engineering – Guedes Soares & Santos (Eds)
 ISBN 978-1-138-58539-3

Sensitivity analysis of risk-based conceptual ship design

Y. Garbatov & F. Sisci

Centre for Marine Technology and Ocean Engineering (CENTEC), Instituto Superior Técnico, Universidade de Lisboa, Lisbon, Portugal

ABSTRACT: The presented work performs a sensitivity analysis of conceptual ship design of a multi-purpose vessel employing a risk-based framework for design. The risk-based framework includes conceptual design to define initial dimensions and characteristics of the ship based on the optimization of the required freight rate, stochastic load modelling, progressive collapse accounting for corrosion degradation during the service life, reliability and cost benefit analysis and finally risk management and sensitivity analysis. The impact of three type design modification factors related to ship design modification: length, block coefficient and service speed; structural design modification associated with the re-scantlings of the midship section leading to different stiffness of principal structural components and cost modification factors associated with the labour, steel, equipment and outfit, machinery costs, profit and overheads are analysed here. Several conclusions with respect to the most significant modification factors are derived. It is analysed the effect that ship design variables (L_{pp} and C_b) and building cost related variables (cost of a ton of steel, number of crew and more) have over the final estimation of the costs and is taken consideration of the sensitivity of these.

1 INTRODUCTION

Ship design and operation are predominantly governed by the ship owner's specification and applicable Regulations and Classification Rules.

The International Maritime Organization, IMO recognizes the importance of adopting the risk assessment procedures in their decision process by defining the Formal Safety Assessment, FSA (IMO, 2002, 2005, 2008, 2013) as a systematic methodology aimed at enhancing maritime safety, including the protection of life, health, maritime environment, cargo and ship integrity by using risk and cost-benefit assessments (see Figure 1).

Another development is the creation of the Port State Control, the PSC program with the objective of eliminating substandard ships from the waters. In addition, the maritime security is also an integral part of the IMO's responsibilities.

The FSA methodology (IMO, 2013), as presented in Figure 1 as stipulated by IMO, which is based on a Quantified Risk Analysis, QRA and provides widely application of QRA to the marine transportation. It is a structured methodology, aimed at enhancing the maritime safety, including the protection of life, health, the maritime environment and property.

Psaraftis (2012), Montewka et al. (2014) used FSA to create new rules and Papanikolaou et al. (2009) for design of ships in the damaged condition.

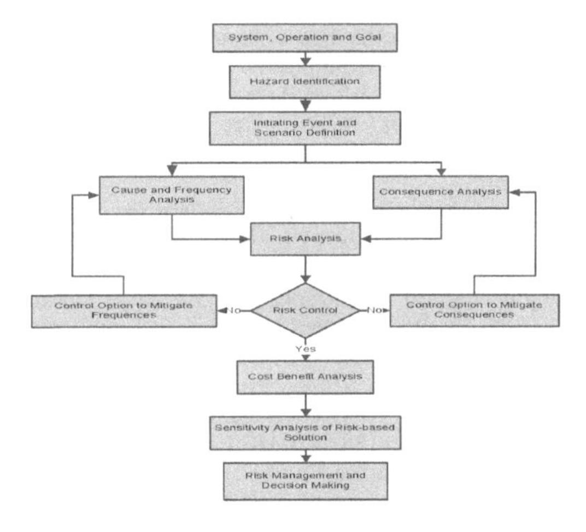

Figure 1. Formal safety assessment.

Recently this approach was used in sensitivity analysis on the optimum hull girder safety level of a Suezmax tanker by Guia et al. (2016).

The methodology employed here includes a synergistic decision models, ship hull reliability analysis algorithms, failure consequence assessment methods, and progressive collapse assessment methodology as has been given by the Common Structural Rules (IACS, 2015) and employed in the shipbuilding industry for the ship hull structural design and integrity assessment.

The ship's optimal safety level is assessed by performing a cost benefit analysis, CBA, where the objective is to establish an optimal safety level identified as a risk control option, RCO, which is represented by redesign of the initial ship design solution, including the main dimensions of the ship and midship section scantlings.

A quantified risk analysis, based on the formal safety assessment is used in the recently developed framework for conceptual ship design accounting for risk-based life cycle assessment in (Garbatov et al., 2017b).

The analysis focuses on sensitivity analysis of three categories design modification factors related to ship design, structural scantling and economical aspects of the capital investment cost, CAPEX and operational cost, OPEX with respect to the progressive ship hull structural collapse and the related probability of structural failure as well as the cost of collapse consequences, structural measures, human life, loss of cargo, accidental spills, where the last two are related to the environmental impact.

2 CONCEPTUAL SHIP DESIGN

The design is defined as a compromise decision support problem with multiple goal constraints.

Given the owner's requirements: cargo deadweight/containers, speed, range, regulations and data on similar ships to find the main dimensions of a ship.

The design solution has to satisfy the system constraints, where free-board has to be greater than the required classification free-board, metacentric height, natural period of roll and dimensional ratios (L/B, B/D and B/T) are within limits that reflect to the designer's experience-based insight (Damyanliev et al., 2017).

Satisfy bounds, where necessary, length of ship is between 95 and 130 m, breadth is between 17 and 25 m, depth is between 8 and 12 m, draft is between 4 and 8 m, velocity is between 10 and 20 knots, L/B is between 5 and 6.5, B/D is between 1.5 and 2.5.

Satisfy the design goal, the constraints are defined as the cargo deadweight/number of containers plus or minus deviation that equals to the owner's required cargo deadweight, design metacentric height plus or minus deviation, which is equal to the required meta-centric height, displacement minus a deviation equals to the required cargo deadweight, that is, the minimum theoretical displacement, shaft horsepower minus deviation is equal to zero, that is, the minimum theoretical power.

The mathematical model of Holtrop and Mennen (1982) is widely used that can provide an estimation of the hull resistance and engine power demand, which can be used to select a propeller engine set (Carlton, 1994). An alternative possible solution is the use of other methods as ITTC (1978) or BSRA (Patullo & Thomson, 1965) methods. The Holtrop's formulation is based on a statistical analysis of resistance data. A resistance service margin is included to provide the added power required to overcome in service the added resistance from hull fouling, waves and wind effects.

The propulsive power is calculated by estimating the resistance on the ship hull, R_{TOT}, calculating the required Effective Horsepower, EHP and Shaft Horsepower, SHP as:

$$R_{TOT} = R_F(1 + k_1) + R_{APP} + R_W + R_B + R_{TR} + R_A \quad (1)$$

$$EHP = R_{TOT}\, V/746 \quad (2)$$

$$SHP = EHP/0.65 \quad (3)$$

where R_{TOT} is the total resistance, N, R_F is the frictional resistance, N, $(1+k_1)$ is the form factor for the viscous resistance, R_{APP} is the appendages resistance, N, R_W is the wave resistance, N, R_B is the pressure resistance of the bulbous bow, R_{TR} is the transom resistance, N, R_A is the correlation resistance from a tested model to the real ship and V_s is the service speed, m/sec.

The total weight of the ship is split into the proper ship's weight, the lightweight, W_{light}, tons, and the consumables and cargo weight, the deadweight W_{dead}, tons. The lightweight is used as an input parameter to estimate the building costs and the deadweight is determinant in estimating the operational costs of the ship (Garbatov et al., 2017a).

The lightweight is the sum of the weights of the ship hull, W_{hull}, tons, which includes the main hull structure, superstructure and bulkheads of the ship, the outfit and hull engineering, W_{oh}, tons accounts the hull insulation, joiner bulkheads, pipes, deck fittings, cargo booms, anchors, rudder, galley equipment and hatch covers, and the machinery, W_m, tons is the sum of the weights of the entire propulsion system (Garbatov et al., 2017a).

The deadweight includes the weight of the cargo, W_{con}, tons (containers in this case), fuel weight, W_{fuel}, tons and weight of fresh water, lubricating oil, stores and crew and other weight related to the machinery being idle, W_{misc}, tons.

To estimate the number of containers, TEU that the ship can transport in one voyage, a regression analysis developed in (Chen, 1999) is used:

$$TEU_{bfloat} = S_b\,(0.0196\, L_{oa}\, B\, D - 148.6129) \quad (4)$$

$$TEU_{dfloat} = 0.050117\, L_{oa}\, B\, TN_d - 82.6702 \quad (5)$$

$$TEU = TEU_{bfloat} + TEU_{dfloat} \quad (6)$$

where TEU_{bfloat} is the number of containers below the deck, TEU_{dfloat} is the number of containers on the deck, S_b is a stowage factor for TEU below the deck, L_{oa} the length overall and TN_d is the number of rows of TEU above the deck.

The total cost of the ship is derived from the annual operating cost and capital cost, where the first is the sum of the salary of crew members, costs related to the stores and supplies, insurances, port expenses and annual fuel cost, and the second one accounts for all expenses of the building of the vessel (Garbatov et al., 2017a).

The required freight rate, RFR, €/ton has been calculated by dividing the discounted annual average cost of the investment, AAC, € by annual cargo capacity, ACC, ton/year:

$$RFR = AAC/ACC \quad (7)$$

The design problem is defined with multiple objectives and linear and nonlinear constraints and it is suitable for a solution by computer methods. A genetic algorithm with a termination criteria is employed (Deb et al., 2002, Wong et al., 2015) for a non-linear optimization problem in defining the best design solutions. The genetic algorithm of Deb et al. (2002) accommodates fast non-dominated sorting procedure, implementing an elitism for the multi-objective search, using an elitism preserving advanced approach allowing both continuous and discrete design variables. Pareto frontier (Komuro et al., 2006) is applied for a simultaneous minimization of the net sectional area and structural displacement. Employing the Pareto Frontier, an optimal solution accounting for the existing constraints may be chosen using a utility function to rank the different designs.

The Pareto optimal solution is defined as the solution for which any improvement in one objective will result in the worsening of at least in one other objective (Messac & Mullur, 2007). In this respect, recently a study of a stochastic structural optimization was presented in (Garbatov & Georgiev, 2017).

The Pareto solution for the ship length and breadth as a function of normalized RFR are shown in Figure 2 and Figure 3, where $RFR_{min} = 0$ and $RFR_{max} = 1$.

All Pareto design solutions comply with the design constraints and requirements and for the present analysis one design solution of a feeder multi-purpose/container vessel is considered, with main dimensions of $L_{pp} = 115.07$ m, B = 20.0 m, D = 10.4 m, T = 8.3 m, $C_b = 0.72$, Vs = 16 kn.

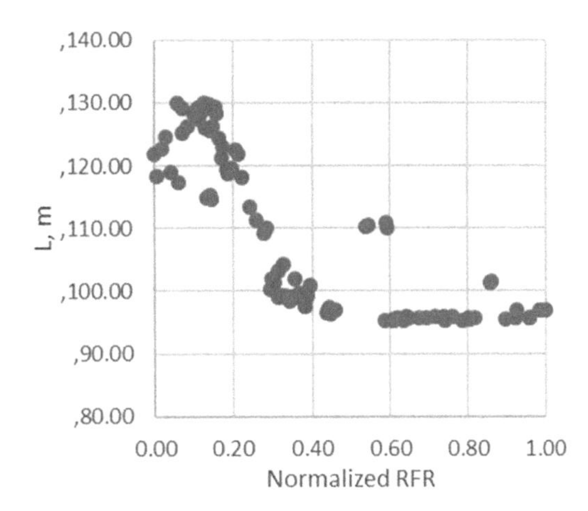

Figure 2. Pareto optimal design solutions of ship length.

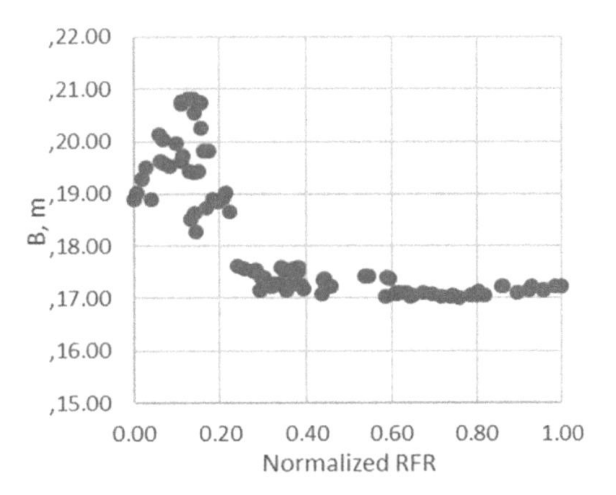

Figure 3. Pareto optimal design solutions of ship breadth.

3 STOCHASTIC LOAD MODELLING

The loads is defined for full, partial and full loads, where the long-term value of the still water and wave-induced bending moments are estimated based on IACS (2015). The primary total bending moment load on the ship hull can be decomposed into two components: the still water bending moment M_{SW} and wave induced bending moment M_W.

Statistical descriptors of the still water bending moment may be defined by using the regression equations as defined in (Guedes Soares & Moan, 1988):

The statistical descriptors of the still water bending moment in full, ballast and partial loads are following the Normal probability distribution, N_{FL} (160.8 MN.m, 54.4 MN.m), N_{BL} (295 MN.m, 72.6 MN.m) and N_{PL} (244.9 MN.m, 69.2 MN.m), where the first descriptor define the mean value

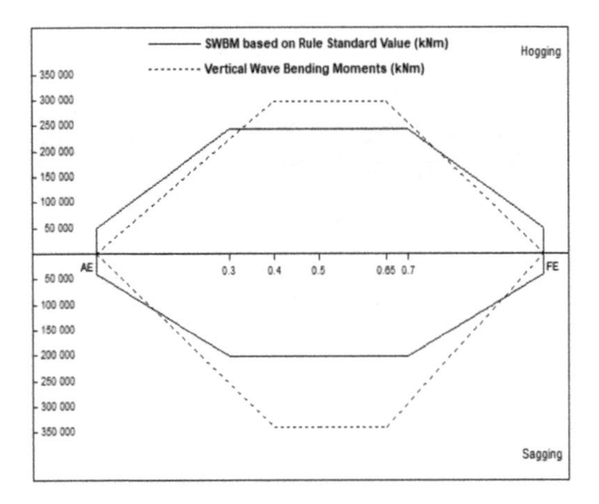

Figure 4. $M_{SW,CS}$ and $M_{W,CS}$ bending moments, (MARS2000, 2011).

and the second one the standard deviation. The still water bending moment is in a hogging condition for the full, ballast and partial loads.

The stochastic model for defining the vertical wave-induced bending moment, proposed in (Guedes Soares et al., 1996), is employed here. The mean value and standard deviation of the vertical wave-induced bending moment in the full, ballast and partial loading conditions are defined by the Gumbel distribution function as FG, FL(443.73 MN.m, 14.84 MN.m), FG, BL(341.19 MN.m, 13.15 MN.m), and FG,PL(373.32 MN.m, 13.04 MN.m) respectively.

4 PROGRESSIVE COLLAPSE

Assessing the ship hull structural risk of an ageing multipurpose ship requires the development of an ultimate limit state function with a reference to the progressive ship hull structural collapse of the primary ship hull structure, where the reference is made to the midship section.

The ship hull is considered to behave globally as a beam under transverse load subjected to still water and wave-induced effects.

The midship section scantling and the ultimate capacity is estimated using the progressive collapse method as stipulated by the Classification Society Rules and using the BV software (MARS2000, 2011).

The 5% confidence level value of the ultimate bending moment, MU5% = MUc is assumed as a characteristic one, which respect the value estimated by MARS2000 (2011) software and additionally it is assumed that COV equals to 0.08 and it is fitted to the Lognormal probability density function, fLN(MU).

5 CORROSION DEGRADATION

The non-linear time variant corrosion degradation model is used in the present study to estimate the structural degradation in time. The mean value, Mean value $[d^{cd}(t)]$ and standard deviation, St Dev $[d^{cd,1}(t)]$ of the corrosion depth as a function of time are defined as (Guedes Soares & Garbatov, 1999, Garbatov et al., 2007)

$$\text{Mean value } [d^{cd}(t)] = d_{\infty}[1-\exp(-(t-\tau_C)/\tau_t], \; t > \tau_C \quad (8)$$

$$\text{St Dev } [d^{cd}(t)] = a \,\text{Ln}(t-\tau_C - b) - c], \; t > \tau_C \quad (9)$$

where a, b and c are regression coefficients.

The analysed ship hull structural system is assumed to be subjected to general corrosion degradation, where the coating life, $\Delta C = 5$ years and transition life, $\Delta t = 7$ for all structural components and the long-term corrosion thickness of any individual structural component is defined based on the corrosion margins as defined by the Classification Society Rules and implemented in the BV software MARS2000 (2011).

6 RELIABILITY

The reliability of a ship hull structural system is defined as the likelihood of maintaining its ability to fulfil the design functions for some period. The objective is to estimate the reliability based on its ultimate strength when extreme loads act upon the ship hull structure subjected to corrosion degradation.

The probability of ship hull structural collapse here is estimating by using the FORM techniques (Hasofer & Lind, 1974). The limit state function is defined as:

$$g(\mathbf{X}|t) = x_U \, M_u - x_{SW} \, M_{SW} - x_W \, x_S \, M_W \quad (10)$$

where M_U is the ultimate capacity with a model uncertainty factor, x_U, which is assumed to be described by a Normal probability density function, $N_{xU}(1.05, 0.1)$. The model uncertainty factor, x_W accounts for the uncertainties in the linear response calculation of wave-induced bending moment, $N_{xW}(1, 0.1)$ and x_S is to account for the non-linear effects, $N_{xS}(1, 0.1)$. The model uncertainty factor in the steel water bending moment is accounted by xSW, NxSW(1, 0.1) (Silva et al., 2014, Garbatov & Guedes Soares, 2016).

The ultimate bending moment is estimated based on the BV software MARS2000 (2011) and employing the commercial software COMREL (2017), the beta reliability index is calculated.

The probability of failure P_f is obtained from the beta reliability index as:

$$P_f = \Phi(\beta) \quad (11)$$

where Φ is the standard normal probability distribution function.

The reliability index for the gross and net designs can be related assuming that the gross ship hull structural design respects the non-corroded ship hull structure up to the moment when the corrosion protection fails, and the net design respects the end of the service life when the structure is already corroded, and no maintenance actions took place. The service life of ship hull structural system is considered as $\tau_S = 25$ years.

The reliability index as a function time, $t \in [0, \tau_S]$ is defined as:

$$\beta(t) = \beta_{gross} - [\beta_{gross} - \beta_{net}][1-[\exp[-[(t-\tau_{C,ship})/\tau_{t,ship}]]]], \; t > \tau_C, \quad (12)$$

$$\beta(t) = \beta_{gross}, \; t < \tau_C \quad (13)$$

The importance of the contribution of each stochastic variable to the uncertainty of the limit state function $g(\mathbf{X})$ is assessed by the sensitivity factors, which can be defined as:

$$\alpha_i = -[\partial g(\mathbf{X}|t)/\partial x_i]/\sqrt{[\Sigma = -[\partial g(\mathbf{X}|t)/\partial x_i]^2]} \quad (14)$$

Figure 6 shows the sensitivities of the limit state function with respect to the changes in the stochastic variables. A positive sensitivity indicates that an increase in the stochastic variable reflects to an

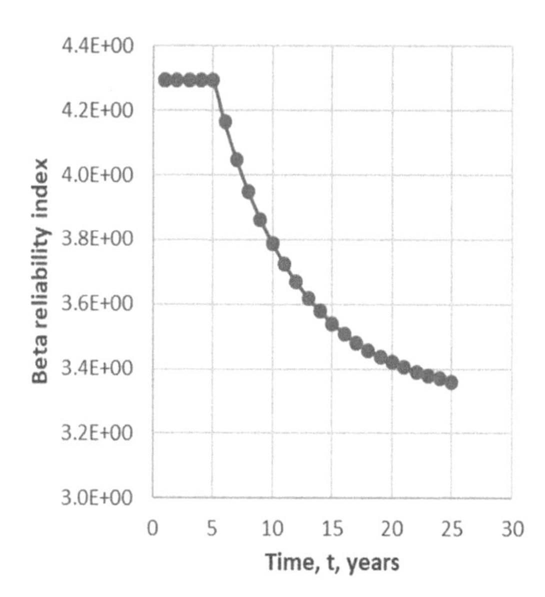

Figure 5. Beta reliability index as a function of time, DMFs = 1.

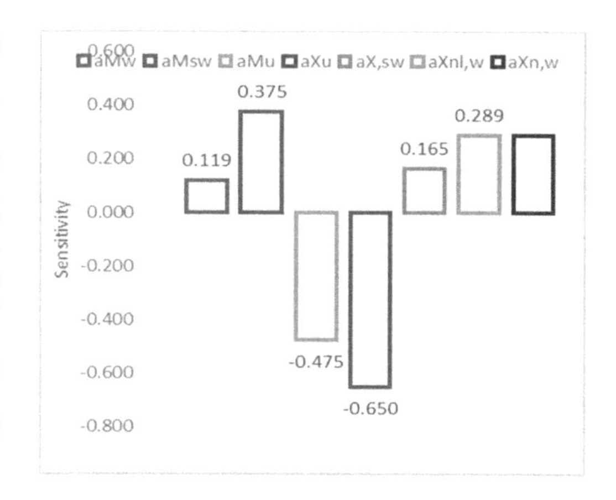

Figure 6. Sensitivities of stochastic variables, t = 0.

increase in the failure function and negatively contributes to the increase of reliability.

It can be seen from Figure 6 that the most important uncertainty on the analysed ship hull is the uncertainties related to the model used to estimate the ultimate bending moment followed by the ultimate bending moment value and still water bending moment etc.

7 COST BENEFIT ANALYSIS

The risk-benefit analysis is used to perform a risk management. This analysis compares the costs and risk to determine where the optimal risk value is on a cost basis. The optimal value occurs when the cost to control risk is equal to the cost of risk due to the ship hull structural collapse.

The ship's optimal safety level is assessed by performing a cost benefit analysis, where the objective is to establish an optimal safety level identified as a risk control option in changing the initial design.

The total expected cost is the sum of two distinct costs, one is the cost associated with the structural collapse of the ship and the other is the cost of implementing the risk control option. The first involves costs associated with the ship hull structural progressive collapse, environmental pollution and loss of human life, while the second involves the costs related to the constructional cost of the steel hull structure, where the amount of material and labour cost is a function of the weight of the structure. The methodology to obtain the optimum safety level, i.e. the optimum/target reliability index is employing

The cost benefit analysis of the modified midship section structure is performed based on the total expected cost, C_t and firstly, will be dealt with the structural design modification factor related to re-scantlings of the midship section leading to a different thickness of principal structural components, DMF_2:

$$C_t = C_{Tf} + C_{me} \quad (15)$$

where C_{Tf} is the total cost associated with the structural failure of the ship and C_{me} is the cost of the implemented structural safety measure as a function of DMF_2. Each of the costs is as a function of the reliability index, β, as this in return influences the cost of structural failure and the risk control option, estimate the safety target beta reliability level, β_t.

The cost associated with the ship hull structural failure is the cost related to the loss of the ship and cargo, environmental pollutions, clean-up related to oil spills and loss of human life.

The cost associated with the ship hull structural collapse is estimated over the service life of the ship, accounting for a discount rate of γ is defined as:

$$C_{Tf} = \Sigma P_f[C_n+C_c+C_d+C_v)]e^{-\gamma t} \quad (16)$$

where P_f is the probability of failure, C_n is the cost of the ship in the year t, C_c is the cost associated with the loss of cargo, C_d is the cost of accidental spill and C_v is the cost associated with the loss of human life.

The cost of the ship at any time $t \in [0, \tau_S]$, is a function of the initial cost of the ship at t = 0, and the scraping cost at $t = \tau_S$th year estimated as:

$$C_n(t) = C_n(0)-[C_n(0)-C_n(\tau_S)][1-[\exp[-[(t-\tau_{C,\,ship}) /\tau_{t,\,ship}]]]],\ t > \tau_C \quad (17)$$

$$C_n(t) = C_n(0),\ t < \tau_C \quad (18)$$

where $C_n(0)$ is the initial cost of the ship, $C_n(\tau_S)$ is the scrapping value of the ship and t is the year of operation, $t \in [0, \tau_S]$. The cost of ship as a function of time is shown in Figure 7.

In the present analysis, the cost of implementing a safety measure accounts for the redesign of the midship section structure, accounting for the cost of material and labour. Depending on the level of the modification, the cost of structural redesign, C_{me} may result in a positive or negative value respectively:

$$C_{me}(DMF_2) = \Delta W_{steel}(DMF_2)\ C_{steel} + C_{labor,steel}(DMF_2) \quad (19)$$

where $\Delta W_{steel}(DMF_2) = (DMF_2-1)W_{steel}$ is the weight of steel due to the design modifications, tons, DMF is the design modification factor, which is also associated with the beta reliability level, β, W_{steel} is the weight of the steel of the reference ship hull structural design, tons, C_{steel} is the cost of steel, €/ton and $C_{labour,steel}(DMF_2)$, €/ton is the labour cost of the constructing ΔW_{steel} (DMF_2), tons.

The cost associated with the loss of cargo, C_c, € is estimated by considering a part of the total amount of cargo of the ship in the case of ship hull structural failure.

In the case of ship hull structural failure, a part of the total amount of oil and fuel may be spilled. P_{spill} is the considered as a partial factor of spill, $P_{s,p}$ is the probability that the oil and fuel is reaching the shoreline (Sørgard et al., 1999). In the case of an accidental oil spill, $P_{spill} \cdot P_{s,p} \cdot W_{oil\ and\ fuel}$ is the weight of spill that needs to be cleaned up, which leads to a cost of:

$$C_d = P_{spill} \cdot P_{sl} \cdot CATS \cdot W_{oil\ and\ fuel} \quad (20)$$

where CATS · is the cost of one ton accidentally spilled oil and fuel that needs to be cleaned.

The cost of human life is accounted for by ICAF as used in a study performed in (Horte et al., 2007):

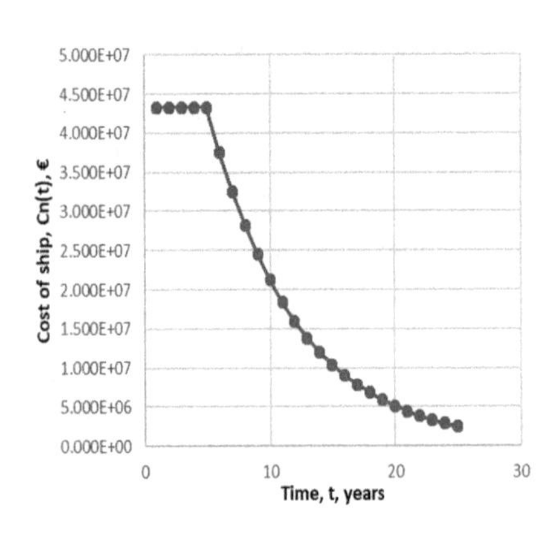

Figure 7. Cost of ship as a function of time, DMF_1.

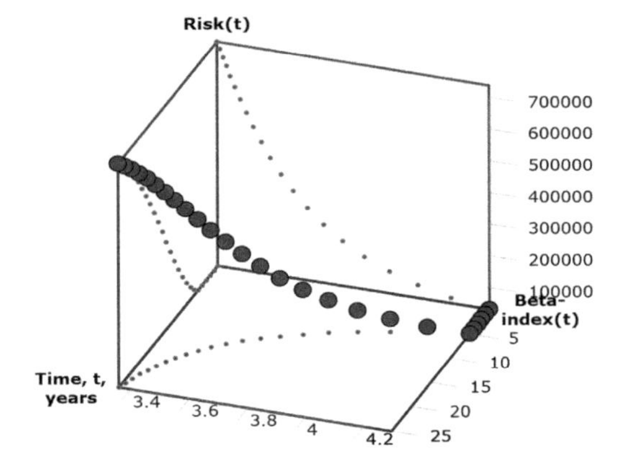

Figure 8. Risk-reliability relationship as a function of time.

$$C_v = n_{crew} \cdot P_{crew} \cdot ICAF \quad (21)$$

where n_{crew} is the number of crew members, P_{crew} is the probability of loss of the life of a crew member.

The risk-beta-reliability index relationship as a function of time is shown in Figure 8 where the risk is estimated as a product of the probability of failure, P_f and the consequential cost of failure represented by the total expected cost, C_t.

8 SENSITIVITY ANALYSIS OF RISK-BASED DESIGN

The Differential Sensitivity Analysis, DSA is used to evaluate a risk based conceptual design, enabling an instantaneous analysis of changes in the total cost due to changes in the input, defined here as design modification factors.

At the beginning, a simulation is conducted with all design modification factors at their original value. For each of the following simulations, one design modification factor is changed. The change of the output can be directly related to the design modification factor and reliability and risk level.

Three type design modification factors are analysed here DMF_1 to DMF_3.

DMF_1 includes the ship design modification factors related to L_{pp}, C_b and V_s. L_{pp} directly impact on the still and wave-induced loads and the L/B ratio, which is commonly referred to as an indicator of the ship propulsion and seakeeping and the L/D ratio as an indirect indicator of the stiffness of the ship hull. The reduction in the design speed results in a lowering in a fuel and oil consumption, which may reduce the OPEX up to 30%.

DMF_2 represents the structural design modification factor related to re-scantlings of the midship section leading to a different thickness of principal structural components.

DMF_3 is related to the cost modification factors associated with the labour, steel, equipment and outfit, machinery costs, profit and overheads.

All design modification factors vary in the range from −15% to 15%, are imposed to the ship design solutions and in the case of the length and the block coefficient of the ship is considered a redesign of the ship hull to satisfy the Classification Society Rules.

Figure 9 shows the sensitivity of the $DMF_{1,3}$ with respect to the RFR, which incorporate CAPEX and OPEX and representing the economic impact, demonstrating different gradients. The gradient of ΔRFR/ΔDMF can be read as a first qualitative estimation of the sensibility of the studied variables, which have higher positive values for the C_b and V_s and negative higher gradient in the case of the length between perpendiculars.

The cost of the steel, machinery, equipment, labour os construction, overhead and profit has a positive, but not significant effect on RFR.

It can be noticed that with increasing the ship length the required freight rate is reducing, which in accordance with the general acceptance that the bigger ships are more efficient in transporting cargo. It is to be also pointed out that the increasing the speed and block coefficient of the ship the required freight rate sharply increase. As for the other DMS, the contribute to the increase of the RFR but in a very reduced scale.

Figure 10 shows the sensitivity of DMF1,3 with respect to the normalized Beta reliably index, which incorporates the loads and resistance and related uncertainties, demonstrating only the most significant effects related to the length between perpendicular and block coefficient of the ship.

Increasing the length and block coefficient of the ship leads to an increase in the thickness of the deck structures to satisfy the Classification Society Rules,

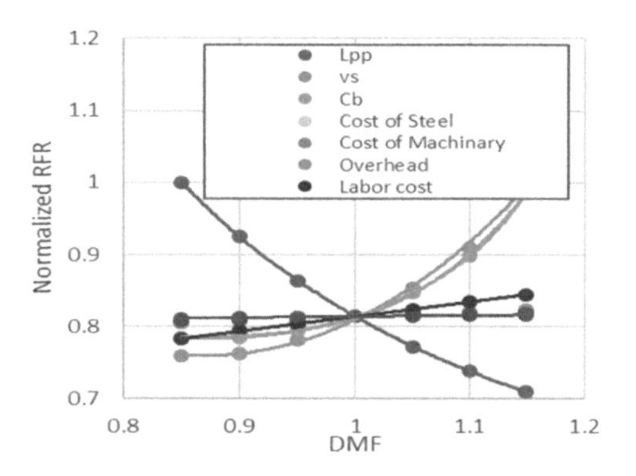

Figure 9. Normalized RFR as function of $DMF_{1,3}$.

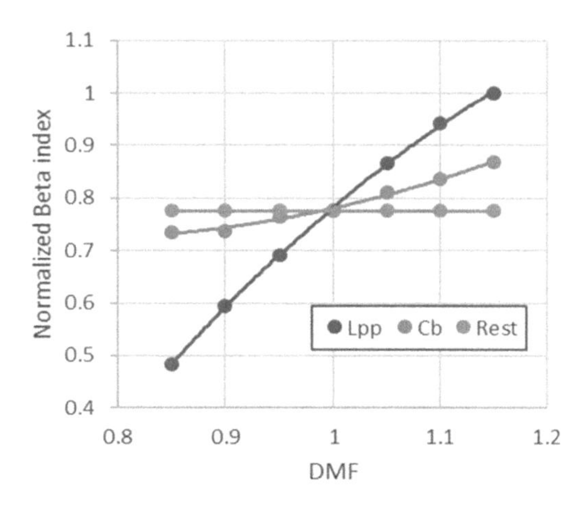

Figure 10. Normalized Beta reliability index as a function of DMFs.

which in turn increases the stiffness of the ship and the subjected load and results in an increase of the invested cost, CAPEX. Here a trade-off needs to be seen in identifying the optimum target reliability index. The weight of the steel, at the beginning of the service life, $t = 0$ years (gross), as a function of the most significant DMFs related to the ship design and scantling is shown in Figure 11.

The consequence cost of structural collapse, which is based on the probability of failure times the associated cost, Eqn (16), with respect to the same DMF is shown in Figure 12. It can be noticed that the most significant factor in reducing the consequence cost of structural failure is the thickness, followed by the length and the block coefficient of the ship.

Figure 12 also shows that a very high correlation between the cost of structural failure as a function of the thickness and length of the ship exists, estimated as 0.81 and practically no-correlation, about 0.01, between other relationships is observed.

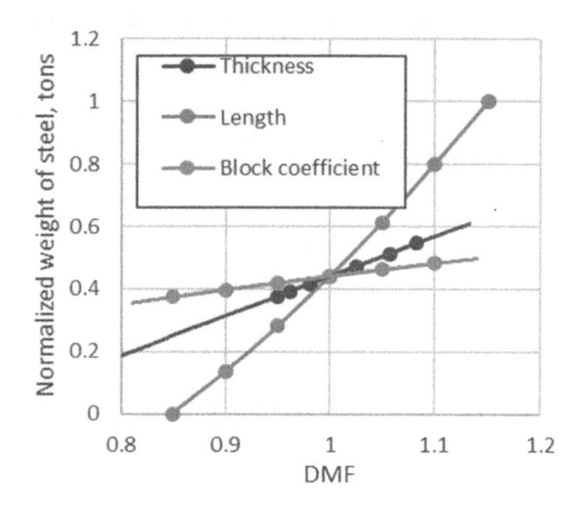

Figure 11. Weight of steel as a function of DMFs.

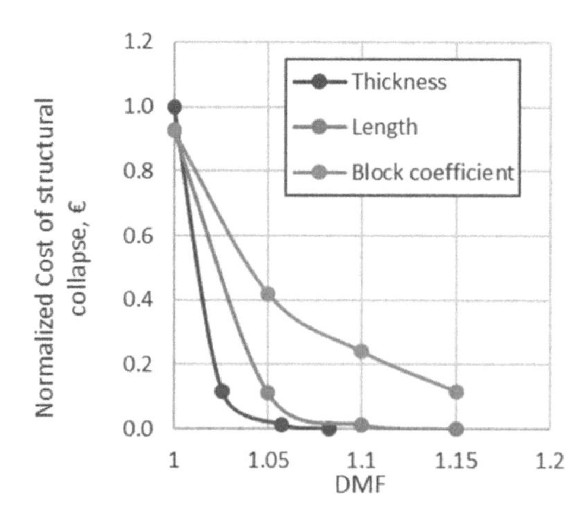

Figure 12. Cost of structural collapse as a function of DMFs.

9 RISK MANAGEMENT FOR DECISION MAKING

Coupling the risk control to the risk assessment, a risk management may be performed. The risk management is a process of making decisions for safety, regulatory changes, and choose different system structural configurations based on the output generated in the risk assessment.

The risk management requires an optimal allocation of the available capacity in supporting the objective and design functionality of the ship hull structural system. It also requires the definition of the acceptable risk level, and a comparative evaluation of alternative options for decision making. The goal of the management is to reduce the risk to an acceptable level.

The cost of ship hull structural collapse and design structural safety measure as a function of DMF2 are shown in Figure 13. It can be noticed that the cost of the control design structural safety measure equals to the cost of the ship hull structural collapse (consequence cost) at $DMF_2 \approx 1$, where the associated beta reliability index is the same and the crossing point may be assumed as an optimal risk value.

Three methods are normally used to select the target reliability level: (1) agreeing upon a reasonable level in the case of a novel structural system without prior history; (2) calibrating the beta reliability level implied in currently successfully used design codes; (3) choosing the target reliability level that minimizes a total consequence cost over the service life of the structural system in the case of design in which the failure results in economic losses and consequences.

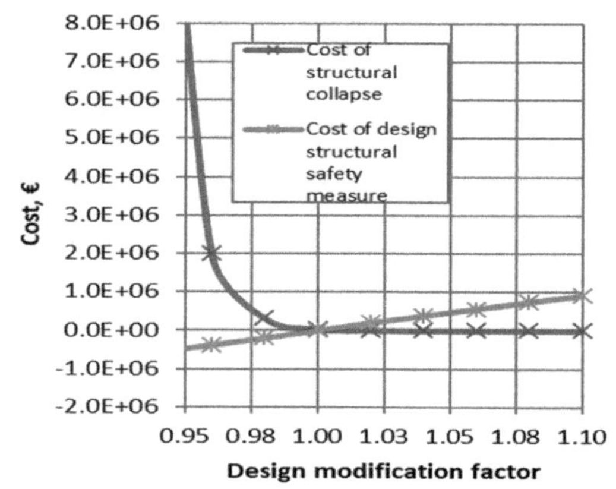

Figure 13. Cost of structural collapse and safety measure, DM_2.

The range of target beta reliability index, βt at the τ_Sth year of service life of the ship hull structural system may vary between 1.5 and 5.

The target beta reliability index is estimated by minimizing the total consequence cost, C_t defined as a function of the beta reliability index. The optimum/target reliability index is shown in Figure 14, where $\beta_t = 4.296$, corresponding to the minimum of the curve of the total consequence cost, $C_t(\beta)$.

A code calibration is a commonly used approach providing the means to design on previous experiences. It can be used to determine the implied reliability and risk levels in the code, then the target levels can be set in a consistent manner to be used in future designs (see Figure 15).

The partial safety factors, γ_R, γ_{SW}, γ_W are estimated based on the characteristic values of the ultimate, still water and wave-induced bending moments $M_U{}^c$, $M_{SW}{}^c$ and $M_W{}^c$, estimated at the 5% and 95% confidence level of the original probability density functions and the design values of all parameters involved in the limit state functions, $M_U{}^*$, $M_{SW}{}^*$, $M_W{}^*$, $x_U{}^*$, $x_{SW}{}^*$, $x_W{}^*$ and $x_S{}^*$ respecting the target reliability beta index level, β_t:

$$\gamma_R = M_U{}^C/(x_U{}^*M_U{}^*),\ \gamma_{SW} = (x_U{}^*M_{SW}{}^*)/M_{SW}{}^C,$$
$$\gamma_W = (x_S{}^*\ x_W{}^*\ M_W{}^*)/M_W{}^C \qquad (22)$$

The resulting partial safety factors can be used in the preliminary design, conditional on the imposed target reliability index, which represents an acceptable risk level and minimum cost by satisfying the following design criterion:

$$M_u/\gamma_R \geq \gamma_{SW}\, M_{SW} + \gamma_W\, M_W \qquad (23)$$

The estimated partial safety factors for the analysed ship hull structural system are presented in Figure 15, where for the target beta reliability index, $\beta_t = 4.296$, the partial safety factors for still water, wave-induced and ultimate bending moments are $\gamma_{SW} = 0.974$, $\gamma_W = 1.208$, $\gamma_R = 1.496$.

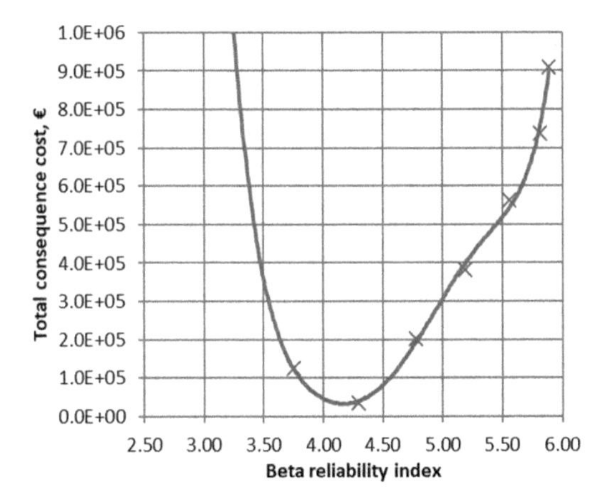

Figure 14. Total consequence cost, DMF_2.

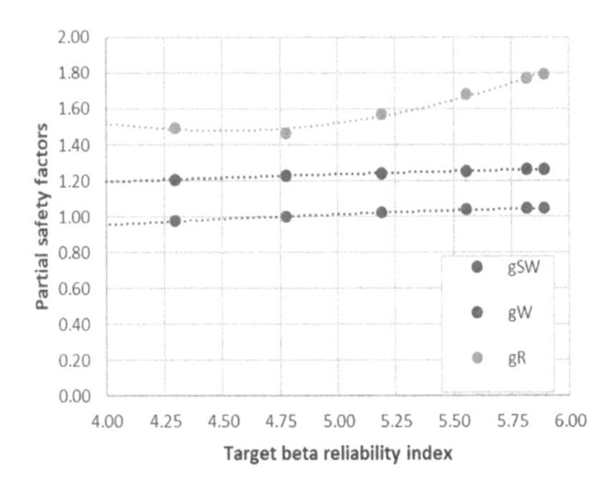

Figure 15. Partial safety factors.

10 CONCLUSIONS

A sensitivity analysis using a risk-based framework for the conceptual ship design of a multipurpose vessel was performed here evaluating the impact of three types of design modification factors related to ship design, structural scantling and cost descriptors. The sensitivity analysis demonstrated that the structural redesign factor and length and block coefficient of the ship have the most significative impact on the RFR, reliability and structural collapse consequence cost. Safety factors that can be used in the preliminary ship design, conditional on the imposed target reliability index, which represents an acceptable risk level and minimum cost were also developed. It has been also shown that the sensitivity analysis, which identifies the importance of the ship design parameters can be used to calibrate the target reliability level.

ACKNOWLEDGEMENTS

This paper reports a work developed in the project "Ship Lifecycle Software Solutions", (SHIPLYS), which was partially financed by the European Union through the Contract No 690770 – SHIPLYS-H2020-MG-2014-2015.

REFERENCES

Carlton, J. S. 1994. Marine Propellers and Propulsion.

Chen, Y. 1999. Formulation of a Multi-Disciplinary Design Optimization of Containerships. Faculty of the Virginia Polytechnic Institute and State University.
COMREL. 2017. Available: http://www.strurel.de/index.html.
Damyanliev, T.P., Georgiev, P. & Garbatov, Y. 2017. Conceptual ship design framework for designing new commercial ships. In: Guedes Soares, C. & Garbatov, Y. (eds.) Progress in the Analysis and Design of Marine Structures. London: Taylor & Francis Group, 183–191.
Deb, K., Pratap, A., Agrawal, S. & Meyarivan, T. 2002. A Fast and Elitist Multi-objective Genetic Algorithm: NSGA-II. IEEE Transactions on Evolutionary Computation, 6, 182–197.
Garbatov, Y. & Georgiev, P. 2017. Optimal design of stiffened plate subjected to combined stochastic loads. In: Guedes Soares, C. & Garbatov, Y. (eds.) Progress in the Analysis and Design of Marine Structures. London: Taylor & Francis Group, 243–252.
Garbatov, Y. & Guedes Soares, C., 2016, Reliability assessment of a container ship subjected to asymmetrical bending, Proceedings of the 13rd International Symposium on Practical design of ships and other floating structures (PRADS2016), Copenhagen, Denmark, Paper D155.
Garbatov, Y., Guedes Soares, C. & Wang, G. 2007. Nonlinear Time Dependent Corrosion Wastage of Deck Plates of Ballast and Cargo Tanks of Tankers. Journal of Offshore Mechanics and Arctic Engineering, 129, 48.
Garbatov, Y., Ventura, M., Georgiev, P., Damyanliev, T.P. & Atanasova, I. 2017a. Investment cost estimate accounting for shipbuilding constraints. In: Guedes Soares, C. & Teixeira, A. (eds.) Maritime Transportation and Harvesting of Sea Resources. London: Taylor & Francis Group, 913–921.
Garbatov, Y., Ventura, M., Guedes Soares, C., Georgiev, P., Koch, T. & Atanasova, I. 2017b. Framework for conceptual ship design accounting for risk-based life cycle assessment. In: Guedes Soares, C. & Teixeira, A. (eds.) Maritime Transportation and Harvesting of Sea Resources. London: Taylor & Francis Group, 921–931.
Guedes Soares, C., Dogliani, M., Ostergaard, C., Parmentier, G. & Pedersen, P.T. 1996. Reliability Based Ship Structural Design. Transactions of the Society of Naval Architects and Marine Engineers (SNAME), 104, 359–389.
Guedes Soares, C. & Garbatov, Y. 1999. Reliability of maintained, corrosion protected plates subjected to non-linear corrosion and compressive loads. Marine Structures, 12, 425–445.
Guedes Soares, C. & Moan, T. 1988. Statistical Analysis of Still-Water Load. Effects in Ship Structures. Transactions of the Society of Naval Architects and Marine Engineers (SNAME),, 96, 129–156.
Guia, J., Teixeira, A.P. & Guedes Soares, C. 2016. Sensitivity analysis on the optimum hull girder safety level of a Suezmax tanker. In: Guedes Soares, C. & Santos T.A. (eds.) Maritime Technology and Engineering 3. London, UK: Taylor & Francis Group, 823–830.
Hasofer, A.M. & Lind, N.C. 1974. An exact and invariant first-order reliability format. Journal of Engineering Mechanics Division, ASUE, 100, 111–121.
Holtrop, J. & Mennen, G.G.J. 1982. An approximate power prediction. International Shipbuilding Progress, 29, 166–170.
Horte, T., Wang, W. & White, N. 2007. Calibration of the hull girder ultimate capacity criterion for double hull tankers. Proceedings of the 10th International symposium on practical design of ships and other floating structures (PRADS), Houston, USA, 235–246.
IACS 2015. Common Structural Rules for Bulk Carriers and Oil Tankers. London: International Association of Classification Societies.
IMO 2002. Consolidated text of the guidelines for formal safety assessment (FSA) for use in the IMO rule-making process. MSC/Circ.1023/MEPC/Circ.392. 4 Albert Embankment, London SE1 7SR: International Maritime Organization Publishing.
IMO 2005. Amendments to the guidelines for formal safety assessment (FSA) for use in the IMO rule-making process. MSC/Circ.1180-MEPC/Circ.474. 4 Albert Embankment, London SE1 7SR: International Maritime Organization Publishing.
IMO 2008. Formal safety assessment on crude oil tankers. 4 Albert Embankment, London SE1 7SR.
IMO 2013. Revised guidelines for formal safety assessment (FSA) for use in the IMO rule-making process. MSC-MEPC.2/Circ.12. 4 Albert Embankment, London SE1 7SR: International Maritime Organization Publishing.
ITTC. 1978. ITTC Performance Prediction Method [Online]. Available: http://ittc.sname.org/2002_recomm_proc/7.5–02–03–01.4.pdf. [Accessed 13 10 2013].
Komuro, R., Ford, E.D. & Reynolds, J. 2006. The use of multi-criteria assessment in developing a process model. Ecological Modelling, 197, 320–330.
MARS2000 2011. Bureau Veritas, Rules for the Classification of Ships and IACS Common Structural Rules for Bulk Carriers and Tankers, Bureau Veritas.
Messac, A. & Mullur, A.A. 2007. Multi-objective optimization: Concepts and methods, Chapter 4. Optimization of structural and mechanical systems.
Montewka, J., Goerlandt, F. & Kujala, P. 2014. On a systematic perspective on risk for formal safety assessment (FSA). Reliability Engineering & System Safety, 127, 77–85.
Papanikolaou, A.D., Guedes Soares, C., Jasionowski, A., Jensen, J.J., McGeorge, D., Poylio, E. & Vassalos, D. 2009. Risk-Based Ship Design, Springer.
Patullo, R.N.M. & Thomson, G.R. 1965. The BSRA Trawler Series Beam-Draught and Length-Displacement Ratio Series resistance and propulsion tests: RINA.
Psaraftis, H.N. 2012. Formal Safety Assessment: an updated review. Journal of Marine Science and Technology, 17, 390–402.
Silva, J.E., Garbatov, Y. & Guedes Soares, C. 2014. Reliability assessment of a steel plate subjected to distributed and localized corrosion wastage. Engineering Structures, 59, 13–20.
Sørgard, E., Lehmann, M., Kristoffersen, M., Driver, W., Lyridis, D. & Anaxgorou, P. 1999. Data on consequences following ship accidents. Safety of shipping in coastal waters (SAFECO II). DNV.
Wong, J.Y.Q., Sharma, S. & Rangaiah, G.P. 2015. Design of Shell-and-Tube Heat Exchangers for Multiple Objectives using Elitist Non-dominated Sorting Genetic Algorithm with Termination Criteria.

Offshore and subsea technology

Progress in Maritime Technology and Engineering – Guedes Soares & Santos (Eds)
 ISBN 978-1-138-58539-3

Risk assessment of subsea oil and gas production systems at the concept selection phase

M. Abdelmalek
Centre for Marine Technology and Ocean Engineering (CENTEC), Instituto Superior Tecnico, Universidade de Lisboa, Lisbon, Portugal

C. Guedes Soares
Centre for Marine Technology and Ocean Engineering (CENTEC), Instituto Superior Técnico, Universidade de Lisboa, Lisbon, Portugal
Subsea Technology Laboratory, Department of Ocean Engineering, COPPE, Federal University of Rio de Janeiro, Rio de Janeiro, Brazil

ABSTRACT: This paper introduces a method of performing a semi-quantitative risk assessment of subsea production systems in the concept selection phase of the asset. The objective of the analysis is to support the decision of the involved parties for selecting the concept to be developed for producing and transferring the hydrocarbon. Accordingly, a review of the different failure modes, the underlying failure causes and their relevant reliability and Risk Influencing Factors (RIFs) are introduced, in order to give an extended explanation of failure of subsea equipment. Afterwards, the impact of the relevant RIFs is reflected on the different failure causes for the sake of producing more specific quantified failure rates of subsea systems. Quantified failure rates are then used as the start point of the Event Tree Analysis (ETA) in order to model the development of the possible accidental scenarios. Besides, the end terminals of the ETA are linked with different linguistic consequence groups, where each of these groups denotes a quantified magnitude of consequences to human, environment and assets. However, the above method of semi-quantitative risk assessment is implemented on a case study from offshore Brazil for the practical demonstration.

1 INTRODUCTION

The utilization of subsea system as a production method in deepwater oil and gas fields has increased significantly during the last three decades. According to DNV GL (2014), there are approximately 5000 X-mas trees (XT) installed worldwide. In addition, the utilization of subsea systems is deemed to be a safer alternative if compared with the other methods of producing oil and gas offshore. According to API (2015), the purpose of risk analysis in the concept selection phase is screening and prioritizing the different concepts in terms of risk level, as well as the selection between alternative scenarios and optimizing the configuration of the selected concept.

This paper presents a methodology of semi-quantitative risk assessment (semi-QRA) of subsea oil and gas production systems in the concept selection phase. The main steps of risk assessment are introduced by NORSOK Z-13 standard (NORSOK, 2010), which are system definition, hazard identification (HAZID), frequency and consequence analyses and finally, risk presentation. Rahimi & Rausand (2013), introduced a simplified method for HAZID and quantification of failure of subsea equipment reflecting the impact of the various RIFs on the produced failure rates. In addition, further approach of quantifying the impact of RIFs has been proposed by Aven et al (2006).

Different sources of historical failure data of subsea equipment are used in the analysis such as (Courban & Brouce, 2003; DNV, 2009; and Sintef, 2002). However, because probabilities of failure do not express all uncertainties about equipment failure (Aven, 2008), an additional qualitative assessment of the underlying RIFs is performed, based on Aven (2014) to highlight the strength of knowledge (SoK) where such factors are based.

Furthermore, Even Tree Analysis (ETA) is used to model the leakage escalation scenarios in terms of impact to human, environment and assets. A simplified semi-quantitative consequence groups of human, environment and asset are given according to Elsayed et al (2014) and Stendebakken et al.

(2014). Finally, a case study from offshore Brazil is used to demonstrate the proposed method of the semi-QRA in comparing between alternatives in the concept selection phase.

2 RISK ANALYSIS OF SUBSEA PRODUCTION SYSTEMS

2.1 *Risk analysis in the concept selection phase*

According to API (2015), in the concept selection phase, risk assessment is used to provide information about the risk level of the concepts under consideration. Afterwards, the decision makers are ranking the concepts based on a combination of its risk level and economical value, before selecting the optimum solution to be developed. Furthermore, it is also used to give information about the optimum configuration of the selected concept.

However, the analysis in this phase is not complex as it can be qualitative or semi-quantitative. In particular, it is requiring information about the failure rates of the main system or sub-systems such as, Christmas Trees (XT), manifolds, and templates. Detailed information and more complex risk analysis methods are used in the following phases of the asset's life cycles (e.g. detailed design and Front End Engineering Design (FEED) (API, 2015).

2.2 *Methodology of semi-QRA in the concept selection phase.*

Risk assessment is a systematic process that is used to establish and to provide judgment on risk picture of the undesired events that may occur in the future (Aven, 2008). According to NORSOK Z-13 (NORSOK, 2010), as well as several other standards (Kragh et al, 2010), the process of risk assessment consists of five steps, which are system definition, hazard identification, frequency analysis, consequence analysis and risk presentation.

The system definition stage identifies the objectives of risk assessment and the boundaries of the system under consideration, in addition to, the determination of the risk acceptance criteria. Subsequently, the hazard identification (HAZID) stage starts with the determination of the different Failure Modes (FMs) and their underlying Failure Causes (FCs). In the detailed risk assessment the Failure Mode, Effect and Criticality Analysis (FMECA) is widely used in the HAZID stage of the subsea systems (Bai and Bai, 2012). For more details regarding FMECA see (Rausand & Hoyland, 2004, p. 89). However, the FMECA is usually used in the detailed risk assessment as it is involving information on the failure of the component level (API, 2015). Therefore, FMs and the relevant FCs of the subsea equipment are determined based on the approach introduced by Rahimi & Rausand (2013).

Accordingly, failure rates of subsea equipment are calculated based on the following FMs (Bai & Bai, 2014a).

i. *Blockage FM (BLK),*
ii. *Structural failure FM (STU),*
iii. *Internal Leakage Process FM (ILP), and*
iv. *External Leakage Process FM (ELP).*

Subsequently, the frequency analysis stage starts with establishing the possible scenarios of failure (Spouge, 1999). In other words, it is modeling the development scenarios of failure of the analyzed system. Fault Tree Analysis (FTA) is usually used to model the development scenarios of failure, see (Rausand & Hoyland, 2004). In addition, quantitative values of failure of the underlying failure causes are used in order to produce the total frequency of the subsea system's failure. Such failure rates can be acquired from different historical data bases, for instance, OREDA Handbook (Sintef, 2002).

However, historical data are generic and represent the industrial average failure frequencies. Therefore, different approaches have been developed to produce plant specific failure rates. This is usually achieved by reflecting the impact of the different human, technical, and organizational (MTO) factors on the generic failure rates (Aven et al., 2006; and Rahimi & Rausand, 2013).

The produced frequencies of failure are used as the initiating events for the consequence analysis. The initiating events for the start points of the development of the different accidental scenarios of the subsea equipment. According to Spouge (1999), ETA, is used to model the consequence of failure of subsea equipment. Furthermore, and in the semi-QRA, the end terminals of the event trees are linked with linguistic consequence groups, where each of these groups is denoting simplified quantitative values of consequence to human, asset and environment, (Elsayed et al., 2014).

The last step of risk assessment is the presentation of the final results of risk of each concept. The provided risk picture is introduced as a combination of consequences and their associated probability of occurrence.

3 SUBSEA EQUIPMENT FAILURE QUANTIFICATION

3.1 *Failure of subsea equipment*

According to NORSOK Z-16 (Standard, 1998a), failure is defined as "the *termination of the ability of an item to perform its required function*". By adapting this definition to subsea systems, such systems fail when they lose the ability to contain or transfer the produced hydrocarbon under the normal operating conditions. Furthermore, failures of subsea equipment can occur due to the realization of one or more FMs, where each of these FMs might exist due to one FC or a combination of different FCs (Rahimi & Rausand, 2013).

Bai & Bai, (2014a) and Grusell & Fyrileiv, (2009) classified FCs into immediate FCs, and time-dependent or operational FCs. Immediate FCs occur due to accidental events; such as dropped objects. On the other hand, time-dependent FCs denotes failure mechanisms that develop over the time under the normal operational conditions, as for instance, internal corrosion and wax deposition. The relationship between failure rate and time is indicated by the bathtub curve (Aven, 1992; Rausand & Hoyland, 2004; Rinaldi et al., 2017)

Moreover, the impact of the associated RIFs of the various FCs needs to be considered in the analysis, since these RIFs represent the soft relationship between failure and the different MTO factors (Vinnem, 2014). This is also beneficial in terms of producing more specific failure rates for the system of interest. However, the relationship between FMs, the underlying FCs and the associated RIFs is illustrated in Figure 1.

3.2 *Subsea equipment failure modes, causes and the relevant RIFs*

3.2.1 *Failure modes*

Based on Bai & Bai, (2014a), subsea equipment fail due to several FMs. However, the focus in the current study will be on the STU, BLK, ELP and ILP failure modes. Since, these FMs are deemed to be the most critical in terms of consequences and repair costs and time.

Structural failure exists when the operating pressure exceeds the maximum allowable operating pressure (MAOP) of the subsea equipment (Bai & Bai, 2014b). Export pipelines are the most subsea equipment that is exposed to the risk of structural failure. For illustration, a minor landslide below the pipeline

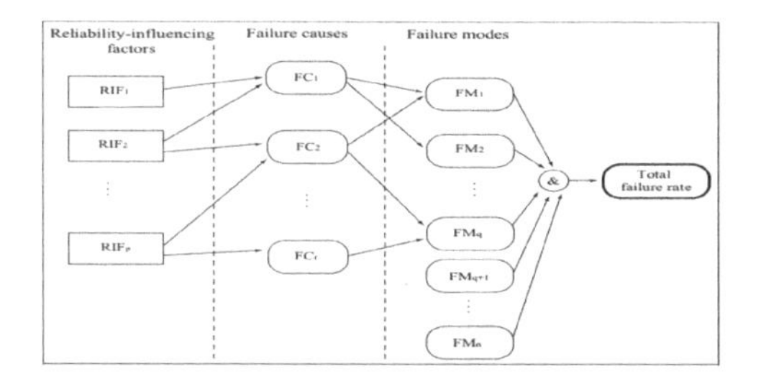

Figure 1. Factors contributing in the total failure rate of subsea systems (Rahimi and Rausand, 2013. p4).

will result in excessive bending force and over the time this can lead to collapse or local buckling to the bent section (Silva & Guedes Soares, 2016). Other causes also can contribute in the existence of the STU FM, such as impact loads, environmental load and installation errors (Bai & Bai, 2014b).

Blockage of subsea equipment is very challenging, especially for crude oil which contains high contents of impurities such as wax and asphalt (Bai & Bai, 2014a). The main consequence of this failure mode is the resulting economical loss by production reduction or stop, as well as, the costs of repair, retrieval or replacement of the defected component. In other scenarios, blockage might escalate and result in leakage if the backpressure exceeded the maximum structural capacity of the subsea equipment.

Moreover, process leak is the most critical failure mode of subsea equipment due to its harmful financial and environmental impact. Human impact can also be severe if a leak occurred and ignited in the vicinity of production facilities or commercial vessels. Hydrocarbon leaks can occur externally or internally. The external leaks represent the release of hydrocarbon to the environment, whereas internal leaks refer to leakage from component to another within the same equipment (Bai & Bai, 2014a).

3.2.2 *Failure causes*

Based on Grusell & Fyrileiv, (2009), "*failure causes are the underlying events which contribute in the occurrence of one or more failure modes*". Failure causes of subsea equipment can be classified into four major groups as indicated below:

i. *Operational failure causes (time dependent),*
ii. *Accidental failure causes,*
iii. *Environmental failure causes, and*
iv. *Design and material defect.*

The first group denotes FCs that developing over time under the normal operating state. Internal corrosion is very common FCs and it can lead to severe losses if not managed properly. While the accidental failure causes consist of all the sources of external impact to the submarine equipment e.g. dropped objects. Furthermore, the environmental causes represent all the natural phenomena that result in immediate or time dependent failure of subsea equipment. The last group denotes all other failure causes that are resulting from errors in design, manufacturing and installation activities. However, Table 1 mentions the various FCs of subsea equipment, according to Bai & Bai (2012) and DNV, (2010).

3.2.3 *Risk or Reliability Influencing Factors (RIFs)*

The development of failure causes are influenced by different MTO factors. These factors can act

Table 1. Failure causes groups vs. underlying failure causes.

Failure cause group	Operational failure causes	Accidental causes	Environmental causes	Material defects
Underlying failure causes	External corrosion, Internal corrosion, Erosion, Wax and hydrate formation, Human Error, Pressure variation, and Temperature variation	Dropped Object, Trawling impact, Dragged anchor, and Sinking vessel	Extreme weather, VIV, Subsea landslide, Marine growth, Free-span, and Loss of on-bottom stability	Manufacturing error, Installation error, and Inherent material defect

Table 2. Relevant RIFs of the different failure causes.

Blockage failure Mode (BLK)	
Failure Cause	RIFs
1. Wax Deposition (WD), 2. Asphaltene Formation (AF),	Oil temperature, Oil pressure and Oil characteristics
Structural Failure Mode (STU)	
Failure Cause	RIFs
1. Inherent Material Defects (IMD)	Manufacuring and inspection quality and Material properties.
2. Installation Error (INE)	Installation quality and Task supervision.
3. Loss of On-bottom Stability (LOS)	Environmental conditions, and Concrete layer quality
4. Free Spanning (FS)	Environmental condition, Seabed condition and Equipment technical condition
Internal and External Process leaks (ILP) and (ELP)	
Failure Cause	RIFs
1. Internal Corrosion (IC)	Oil temperature, Oil pressure, Corrosive medium and Material properties
2. External Corrosion (EC)	Corrosive environment, Material properties, External coating quality and Cathodic protection quality
3. Internal Erosion (IE)	Flow velocity, Sand contents and System geometry
4. External Impact (EI)	Intensity of commercial ship, Frequency of adjacent fishing activities, Supply vessels frequency and Frequency of lifting operation.

on increasing the development rate or the chance of occurrence of a particular failure cause. For example, high water content in the produced oil will increases the development rate of internal corrosion (Bai & Bai, 2012). Also, dense ship traffic nearby a transport pipeline will contribute to raising the probability of the external impact FC. On the other hand, these factors can also act on the reduction of failure probability if it is deemed to be less than the industrial average. For instance, an organizational factor such as high standard quality control will lead to reduction in the probability of having manufacturing error. However, Table 2 introduces a number of the relevant RIFs of the different FCs that are used in the risk assessment. The provided information in Table 2 is adapted from (Aven et al., 2006; Bai & Bai, 2012; Rahimi & Rausand, 2013).

3.3 *Failure rate quantification of subsea equipment*

Historical failure data are usually used in the early design phase risk assessment of the asset due to the lack of information about the system of interest (Rahimi & Rausand, 2013). Accordingly, different data sources are used in this phase for obtaining failure data of the different subsea equipment. In this study the OREDA handbook (Sintef, 2002) is used as the main source of failure rates of all subsea equipment other than export pipelines and risers. However, other failure data bases for submarine pipelines and risers are described later in this section.

Failure frequencies are given, in the OREDA handbook as failure per million hours. Furthermore, failure rates are classified by system, equipment and subunits, as well as FMs and FCs.

Furthermore, failures are also classified with respect to their nature or severity as indicated below.

- *"**Critical failure:** A failure that causes immediate and complete loss of an equipment/sub-item or components unit's capability of providing its output.*
- ***Degraded failure:** A failure that is not critical, but it prevents the equipment unit/sub-item or component from providing its output within specifications. Such a failure would usually, but not necessarily, be gradual or partial, and may develop into a critical failure in time.*
- ***Incipient failure:** A failure that does not immediately cause loss of an equipment unit/sub-item or components capability of providing its output, but which, if not attended to, could result in a critical or degraded failure in the near future."*

The transformation of failure frequencies from failure per million hours into failure per year is given by

$$f_{year} = \frac{f_{pmh}}{n \times 10^6} \times 8760 \quad (1)$$

where,

F_{year}: The annual frequency of failure per equipment;

F_{pmh}: The aggregated frequency of failure per million hours:

N: The total number of equipment observed.

On the other hand, information about failure rates and distribution of the underlying failure causes of subsea pipelines and risers is given in (Bai & Bai, 2012; DNV, 2010; Grusell & Fyrileiv, 2009).

Failure data of pipelines are classified by the type of flowlines (e.g. flexible and steel) and location of pipeline. The location of pipelines is categorized into four segments i) *Safety zone*, ii) *offshore midline*, iii) *landfall zone* and iv) *onshore zone*, see (DNV, 2010, p. 21) for more details. Accordingly, risers are considered to be in the safety zone.

3.4 *Reflection of RIFs impact on failure rates*

Aven et al., (2006), introduced a methodology for reflecting the impact of the various RIFs on failure rates. The method is called barriers and operational risk analysis (BORA). The BORA method consists of ten steps, starts with the identification of the different initiating events and end up with the calculation of the specific failure rates.

However, in the current study, only the adjustment process of the generic failure rates is used. This process consists of five steps which are, i) selection, ii) scoring and iii) weighting of RIFs, iv) SoK assessment of RIFs, and finally v) calculating the revised failure rates. The provided SoK assessment criteria are based on (Aven, 2014). Moreover, the SoK assessment step is not a part of the original BORA method, however, it is included in this study for highlighting the lack of knowledge about the different RIFs. Further information of the different steps is given below.

i. *Identification of RIFs:*
The first step is to select the relevant RIFs that influencing the development mechanism of the different FCs. The selected RIFs are obtained from the generic RIFs lists that introduced in (Aven et al., 2006; Rahimi & Rausand, 2013).

ii. *Scoring:*
Scoring of RIFs is based on six points scale A, B, C, D, E and F. Score A indicate the best value of the RIF which leads to lower failure rates, C indicates the industrial average with no effect on the generic failure rate, and B refers to conditions between A and C. Furthermore, scores D and E denotes negative and more negative value of RIFs, respectively, whereas, F denotes the worst impact which lead to duplication of failure rate, see (Aven et al., 2006) for further explanation.

iii. *Weighting of RIFs:*
Each failure cause is affected by different RIFs. Therefore, each RIF has to be assigned a value from 0 to 1 in order to reflect its importance to this particular failure cause. In addition, such weight is also used for normalization purposes. However, the sum of the different weights of RIFs of the particular failure cause has to be equal to 1. See the below example of weighting the different RIFs of the internal erosion FC.

iv. *Strength of knowledge assessment:*
The purpose of this step is to highlight the background knowledge that is used to identifying the impact of the various RIFs. The criteria for the SoK assessment is given in Table 4. However, the SoK can be Strong (S), Medium (M) or Poor (P). For further details see (Flage et al., 2014)

Table 3. Example of ranking the importance of internal corrosion FC RIFs.

RIF	Flow velocity	Sand content	System geometry
Score	C	D	B
Weight	0.3	0.4	0.3
SoK	M	S	S

Table 4. Strength of knowledge criteria (adapted from Aven, 2014).

Strength of knowledge scale		
Strong	Poor	Medium
All of the following conditions are met: a) The assumptions made in scoring and weighting RIFs are seen very reasonable b) Many reliable data are available c) There is broad agreement among the involved parties d) The phenomena involved are well understood.	One or more of the following conditions are met: a) The assumptions made in scoring and weighting RIFs present strong simplification b) Data are not available, or are unreliable c) There is lack of agreement among the involved parties d) The phenomena involved are poorly understood.	Conditions are between those characterizing high and low uncertainty

v. *Calculation of specific failure rates:*
The last step in this process is to calculate the revised failure frequencies of each cause based on the impact of the associated RIFs. The revised failure rates are calculated as given below:

$$f_{FC}(Rev)= f_{FC}(Gen) \,.\, k_i \tag{2}$$

$f_{FC}(Rev)$ denotes the revised failure rate of a particular failure causes, where:

$$k_i = \sum\nolimits_{i=1}^{n} w_i \,.\, Q_i \tag{3}$$

and w_i denotes the weight of RIF (i) for the FC, whereas, Q_i is quantitative value for measuring the status of the RIF (i), and n is the number of RIFs that influencing the FC. The sum of all weights of RIFs for a particular FC is equal to 1. On the other hand, the values of the different Q_i are calculated by the following method:

This step starts with identifying the upper and lower bounds of the generic failure rates. The lower and upper values can be assigned based on expert judgments, manufacturer values or 90% confidence interval as given in the OREDA handbook. Afterwards, denote the RIFs status s, and assign different numbers for each score. i.e. $S_A = 1, S_B = 2, \ldots$ and $S_F = 6$. Then, the different quantitative score values Q_i(s) can be calculated based on the following formulas:

$$Q_i(S) = \begin{cases} \dfrac{f_{low}}{f_{Avr}} \; if \; S = A \\ 1 \quad if \; S = C \\ \dfrac{f_{high}}{f_{Avr}} \; if \; S = F \end{cases} \tag{4a}$$

and

$$Q_i(B) = \frac{f_{low}}{f_{Avr}} + \frac{(S_B - S_A).\left(1 - \dfrac{f_{low}}{f_{Avr}}\right)}{S_C - S_A} \tag{4b}$$

$$Q_i(D) = 1 + \frac{(S_D - S_C).\left(\dfrac{f_{high}}{f_{Avr}} - 1\right)}{S_F - S_C} \tag{4c}$$

$$Q_i(E) = 1 + \frac{(S_E - S_C).\left(\dfrac{f_{high}}{f_{Avr}} - 1\right)}{S_F - S_C} \tag{4d}$$

This step is ending with the determination of the value of factor K and then calculating the revised failure frequency of the specific FC. The above process is repeated for all FCs of the different FMs in order to calculate the revised total failure rate of the equipment.

4 CASE STUDY

4.1 *Case definition*

The production field of interest is located 60 km off the coast of Brazil in water depth from 1700–1900 m. The expected production reserve is 400 MMbbl of API 27° oil with GOR 22.3 m3/bbl. The formation pressure is approximately 6000 psi at 100° C. The expected field production period is 18 years. Other than that, the hydrocarbon is to be produced through 19 subsea wells with maximum capacity of 8000 bbl/day for each well. In addition, there are 8 water injection wells are used to improve the extraction of the hydrocarbon from the reservoir.

There are two possible scenarios for developing the field of interest. The first is to use subsea wells with manifold connected to FPSO. While the second, is subsea to shore (S2S). In this concept, the subsea manifolds are connected with pipeline end manifold (PLEM) and then the produced oil to be transported to shore via an export flowline. The onshore facility is located 56 km off the subsea field.

The FPSO concept consists of 4 manifolds, where each of them connects 6 subsea trees. Afterwards, each manifold is connected with the FPSO's turret via 8" catenary riser. Alternatively, in the S2S scenario the production XTs are connected to 3 manifolds with 8 slots. Moreover, the subsea manifolds are connected to a PLEM and then connected to the onshore facility via 18" export pipeline. The export pipeline length is 56 km. A detailed inventory of equipment of both concepts is presented in Table 5.

4.2 *Hazard identification*

Hazard identification stage presents information about the possible scenarios of failure of subsea equipment. This includes the selection of FMs and its underlying FCs, in addition to, the relevant RIFs. In the light of the information provided in section 2.2, Table 6 presents the various FMs and the relevant FCs of subsea equipment. In addition, Table 7 introduces the abbreviations of the different FCs.

Table 5. Subsea equipment vs. equipment inventory of FPSO and S2S concepts.

	Quantity		Length/Unit (km)		Diameter (inch)	
Subsea Equipment	FPSO	S2S	FPSO	S2S	FPSO	S2S
Vertical XTs	19	19	NA	NA	11	11
Wellheads	19	19	NA	NA	11	11
Flowlines (XT – Manifold)	19	19	0.81	0.1	4	4
Flowlines (man – riser base (RB))	4	NA	2.9	NA	8	NA
Catenary risers (RB – Turret)	4	NA	1.9	NA	8	NA
Flowlines (manifold – PLEM)	NA	3	NA	2.53	NA	10.5
Flowlines (PLEM – Shore)	NA	1	NA	56	NA	18
6 slots manifolds	4	NA	NA	NA	NA	NA
8 slots manifolds	NA	3	NA	NA	NA	NA
PLEM	NA	1	NA	NA	NA	NA

Table 6. List of FMs and FCs of the different equipment of subsea system (Y = relevant and N = NOT).

	Failure modes and the associated failure causes											
	BLK		ELP				ILP		STU			
Equipment	WXD	ASD	IC	IE	EC	EI	IC	IE	INE	IMD	FS	LOS
Well stream flowlines	Y	Y	Y	Y	Y	Y	N	N	Y	Y	N	N
Export flowlines	Y	Y	Y	Y	Y	Y	N	N	Y	Y	Y	Y
Risers	Y	Y	Y	Y	Y	Y	N	N	Y	Y	N	N
PLEM	Y	Y	Y	Y	Y	Y	N	N	Y	Y	N	N
XTs & wellheads	Y	Y	Y	Y	Y	Y	Y	Y	Y	Y	N	N
Manifolds	Y	Y	Y	Y	Y	Y	Y	Y	Y	Y	N	N

Table 7. List of abbreviations of failure causes.

FC	Abbreviation	FC	Abbreviation	FC	Abbreviation
WXD	Wax deposition	EC	External corrosion	INE	Installation error
ASD	Asphaltene deposition	IC	Internal corrosion	IMD	Inherent Material defect
IC	Internal corrosion	EC	External corrosion	FS	Free-spanning
IE	Internal erosion	IC	Internal corrosion	LOS	Loss of on-bottom stability

4.3 *Frequency analysis*

The frequencies of the above FCs and FMs are calculated according to the information provided in section 3.2. However, the calculation of failure rates of subsea equipment and the distribution of frequencies of the various FCs of flowlines are based on the following assumptions.

1. ***OREDA handbook data:*** *Critical and 50% of the degraded failure rates of subsea equipment are used in computing failure rates. The reason of including 50% of the degraded rate is built on the supposition that half of the degraded failures are not detected and developed to become critical failures. Incipient failures are not considered as it is representing other FMs than what is included in this analysis.*
2. ***Distribution percentage of FCs of ILP and ELP FMs:*** *The distribution is based on the information provided in (DNV, 2010). The proposed distribution indicates that internal corrosion and external impacts are resulting in 38% and 33% of the total failure rate, respectively. The remaining percentage of failure rate is equally distributed on the rest of failure causes, which in this case are external corrosion and internal erosion. Similarly, the distribution is also applied on the lower and higher bounds of the various failure rates.*
3. ***Distribution percentage of FCs of STU FM:*** *In the current case, STU FM can occur due to INE, IMD, FS and LOS FCs. The contribution percentage of each FCs in the analysis is equally distributed on these FCs. The FS and LOS failure causes are only applicable on the export pipeline in the S2S scenario.*

Table 8. RIFs influencing the WXD failure cause.

RIF	Oil press.	Oil temp.	Oil characteristics
Score	C	C	D
Weight	0.2	0.4	0.4
SoK	S	S	S

Table 9. RIFs influencing the ASD failure cause.

RIF	Oil press.	Oil temp.	Oil characteristics
Score	C	C	D
Weight	0.4	0.2	0.4
SoK	S	S	S

Tables 8 and 9, illustrate the method of identifying FCs of the BLK FMs of well-steam flowlines, as well as selecting, scoring and analyzing the associated

RIFs. In addition, the application of the BORA method for revising the generic failure rates of the BLK FM is illustrated in Table 10.

The above method is repeated for all FMs of the different subsea equipment in order to obtain the revised annual failure rates. However, a summary of the results is presented later in the risk presentation section.

4.4 *Consequence analysis*

In the consequence analysis part, failure of subsea equipment is classified into two failure development groups. The first is loss of containment development scenario, while the latter is the development of the total failure risk. Loss of containment failure is using the total ELP failure frequencies as the initiating event of the ETA. In addition, each development concept is assessed against four possible leak sizes which are small, medium, large leaks and rupture of the equipment, see Table 11 for leak distributions. Subsequently,

Table 11. Leak sizes vs. leak probabilities (Grusell & Fyrileiv, 2009, p. 4).

Leak size	Hole size distribution
Small (<20 mm)	74%
Medium (20–80 mm)	16%
Large (>80 mm)	2%
Rupture	8%

Table 10. BLK failure mode well-stream flowlines.

	Annual failure rate								
Failure cause	Lower limit	Average	Upper limit	RIF	S	Q	W	K	Revised failure rate
WXD	4.0E-06	9.5E-06	4.8E-05	Oil pressure	C	1	0.2	1.3	1.24E-05
				Oil temperature	C	1	0.4		
				Oil characteristics	D	1.68	0.4		
ASD	2.2E-06	1.1E-05	5.5E-05	Oil pressure	C	1	0.4	1.27	1.4E-05
				Oil temperature	C	1	0.2		
				Oil characteristics	D	1.67	0.4		

the escalation of leaks is depending on the failure of detection and closure, as well as the ignition of the released oil. With regards to detection and closure, NORSOK U-001 (Norsok, 1998b) stated that the reservoir has to be controlled with two independent barriers against environmental releases. The XT as a complete system is accounted as one of these barriers, whereas, down-hole safety valve (DHSV) is the secondary mitigation measure. Moreover, a subsea isolation valve (SSIV) is installed between each manifold and the subsequent flowline i.e. export pipeline or riser.

The OLF-070 standard (OLF, 2004) provides information on probabilities of failure of the different high demand and low demand subsea safety instrumented systems (SIS) (i.e. leak detection and closure systems). Accordingly, Table 12 gives the different leak frequencies of the FPSO and S2S concepts as well as probabilities of detection and closure failure, and leak ignition. Hydrocarbon ignition probabilities are assigned based on (OGP, 2010, p16). However, for further information regarding classification of the different SIS and its associated failure probabilities see (Abrahamsen & Røed, 2011; OLF, 2004).

The above leak frequencies and the associated failure and ignition probabilities are used in the ETA of both concepts. For illustration, Figure 2 indicates the event tree that is used for modelling the development of medium leak of the FPSO concept. The end terminals of the event trees are linked to linguistic consequence groups (e.g. H1 = minor human impact and E3 = major environmental damage) where each of them has further definition given later in Table 14.

On the other hand, the development of the total failure risk is assessed against the economic consequences in terms of production delay and repair costs. In this case, the total failure frequency of each subsea development scenario is used as the start point of the ETA. However, the development steps of the total failure event tree are given in Table 13. In addition, probabilities of occurrence of these events are presented in the Table 14. However, such probabilities are assigned subjectively to reflect the analyst's degree of belief about the occurrence of the different events. Furthermore, Figure 3 indicates the ETA of the development of the total failure of the S2S concept.

Similar to the external leak event trees, the end terminals of the total failure event trees are linked to different linguistic economic consequence groups (e.g. P1 = minor economic effect and P5 = catastrophic economic effect). Definition of the different groups is given in Table 15 in the production loss consequence category

Table 12. Summary of the event tree frequencies.

Leak size	Leak frequency (λ/year)		Detection and closure failure probability	Ignition probabilities
	FPSO	S2S		
Small (<20 mm)	3.2E-02	1.0E-02	*5.0E-03*	0.09
Medium (20–80 mm)	6.9E-03	2.2E-03	*5.0E-03*	0.09
Large (>80 mm)	8.6E-04	2.8E-04	*5.0E-03*	0.09
Rupture	3.4E-03	1.1E-03	*5.0E-03*	0.09

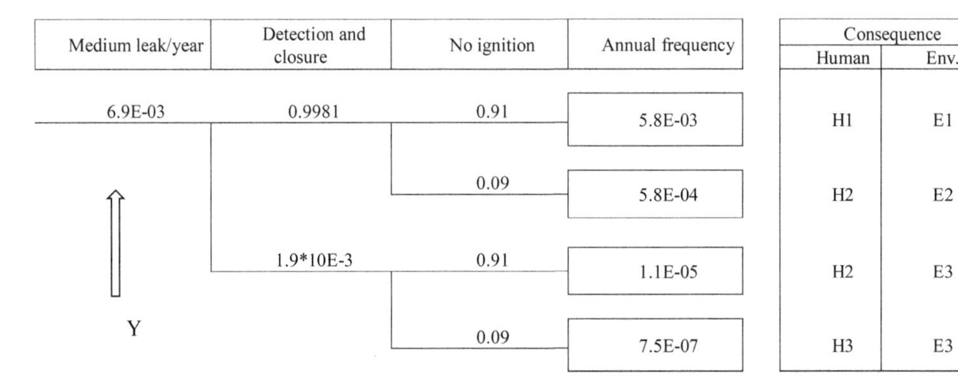

Figure 2. Event tree analysis of the medium leak of the FPSO scenario.

4.5 *Risk presentation*

This section presents the final results of the risk analysis of the two subsea development concepts. Table 16 indicates the generic and revised annual failure frequencies of the various equipment groups, as well as the entire system. In addition, the contribution values and percentages of the different FMs in the total failure rate of the FPSO and S2S concepts are provided in Figures 4 and 5, respectively. On the other hand, Table 17 introduces the annual frequencies of occurrence of the various consequences to human, environment and production loss. Furthermore, it is providing the associated expected losses of the two development concepts. The yearly expected loss to human, environment, and production are given as number of fatalities, restitution time in year and repair period in months, respectively. Expected losses are calculated by the ETA as indicated earlier in section 4.4.

Table 16 shows that, the annual frequency of failure of the FPSO system is 7.1E-02. The other alternative scenario has a 4.6E-02 annual failure rate which provides 35% reduction in frequency of failure per annum. XTs and wellheads are considered the highest contributor in risk of failure in the S2S scenario with 3.0E-02 failure per year, which is approximately 65% of the total failure frequency. On the other hand, in the FPSO scenario risers comes in the first place, in terms of failure rate, with 3.1E-02 failure per year. However, XTs and wellheads failure rate in the FPSO concept is quite similar to risers' failure frequency with 3.0E-02 failure per annum. Failure rate of the export pipeline in the S2S scenario is 1.1E-02 which accounts almost 25% of the total failure rate of the entire system.

On the other hand, by considering the effect of the different FMs on failure rates, it is observed that process leaks are representing 60% of the total failure rate of the FPSO scenario. However, this is dramatically decreased when looking at the S2S scenario, where leaks are denoting only 30% of the annual failure frequency of the system. Such increment in the FPSO scenario is drove by the higher frequencies of occurrence of the external impacts. This is due to the presence of supply vessels, lifting operations in addition to the large number of the infield flowlines, if compared to the S2S concept. Other than that, structural failures represents 44% of the annual failure frequency of the S2S scenario. The reason behind that is the presence of the export pipeline which is exposed to several

Table 13. Development scenario of the total failure rate.

Event 1	Production stop (the failure needs immediate intervention, since temporary solutions—e.g. pressure reduction—are not practical)
Event 2	Light intervention (ROV intervention for ELP, ILP & STU) or (Injection or pigging for BLK FM).
Event 3	Heavy intervention (retreival of the defected equipment for further repair) or (coil tubing for BLK FM)
Event 4	Replacement (Complete replacement of the defected equipment).

Table 14. Probabilities of failure events of subsea systems.

Event	Probability of occurrence
Production stop	0.65
Light intervention failure	0.3
Heavy intervention failure	0.3
Equipment replacement	0.3

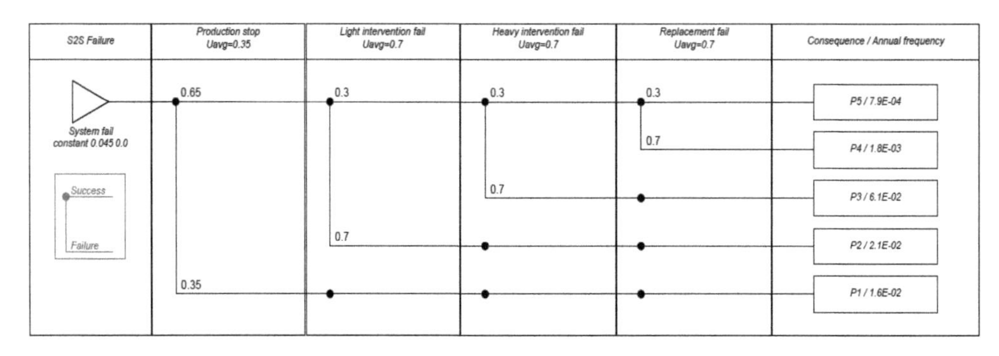

Figure 3. Event tree analysis of the total failure of the S2S concept.

Table 15. Definition of severity level of each consequence category to human, environment, and production (Adapted form, Elsayed et al., 2014; Stendebakken et al., 2014).

Consequence category	Consequence types		
	Human (H)	Production (P)	Environment (E)
C5. Catastrophic	Large number of fatalities Equivalent fatalities 100	Total production stop due to failure of critical equipment (export line) or multiple low redundant equipment (e.g. risers) —initiation of heavy repair vessel required—immediate heavy intervention. Complete replacement of blocked item required. Equivalent repair time 12 month *(including vessel's mobilization time)*	Uncontrolled pollution—long-term effect on recipients—long-term disruption of the ecosystem Equivalent restitution time 5 years
C4. Severe	Multiple fatalities Equivalent fatalities 10	Severe production stop due to failure of different low redundant equipment (e.g. manifold)—initiation of heavy repair vessel required—immediate heavy intervention. Retrieval of blocked item required. Equivalent repair time 6 month	Severe pollution—medium-term effect on recipients—medium-term disruption of the ecosystem Equivalent restitution time 3 years
C3. Major	Single fatality or multiple severe injuries Equivalent fatality 1	Significant production stop due to failure of low redundant equipment (e.g. manifold) or multiple high redundant equipment—initiation of light repair vessel required—immediate light intervention. Coil tubing required. *Equivalent repair time 3 month*	Major release—effects on recipients—short term disruption of the ecosystem. Equivalent restitution time 1 year
C2. Significant	Multiple or severe injuries Equivalent fatality 0.1	Partial production stop due to failure of high redundant equipment (e.g. XT)—initiation of repair vessel required—immediate light intervention. Pigging required. Equivalent repair time 1 month	Minor release—minimal acute environmental or public health impact—small, but detectable environmental consequences. Equivalent restitution time 0.1 year
C1. Minor	Single or minor injuries Equivalent fatality 0.01	No immediate production stop or initiation of repair vessel—light intervention required at the first chance of planned maintenance. chemical injection required Equivalent repair 0.25 month.	Negligible release—negligible pollution—no acute environmental or public health impact. Equivalent restitution time 0.01 year

Table 16. Annual failure rate of the equipment groups and total systems of the FPSO and S2S concepts.

	Annual failure rate per equipment group											
	XT and wellhead		Well-stream FL		Manifolds		Risers		Export pipeline		Failure rate per System	
Concept	Gen.	Rev.	Gen.	Rev.	Gen.	Rev.	Gen.	Rev.	Gen.	Rev.	Gen.	Rev.
FPSO	3.0E-2	3.0E-2	2.9E-3	3.5E-3	5.7E-3	6.5E-3	1.8E-2	3.1E-2	NA	NA	5.7E-2	7.1E-2
S2S	3.0E-2	*3.0E-2*	*2.0E-4*	*2.3E-4*	*4.3E-3*	*4.6E-3*	NA	NA	*7.4E-3*	*1.1E-2*	4.2E-2	4.6E-2

environmental conditions, in addition to, its length which has a direct proportion with failure rate.

Another element that needs to be considered is the impact of the different RIFs on the generic failure rates. It is clearly seen that RIFs have introduced significant value in adjusting the generic failure rates and make it more specific to the systems of interest. In the FPSO scenario, the adjustment process of RIFs has resulted in 20% rise in the annual failure frequency. Similarly, 9% increment in the annual failure frequency of the S2S has been achieved by accounting the impact of the various RIFs.

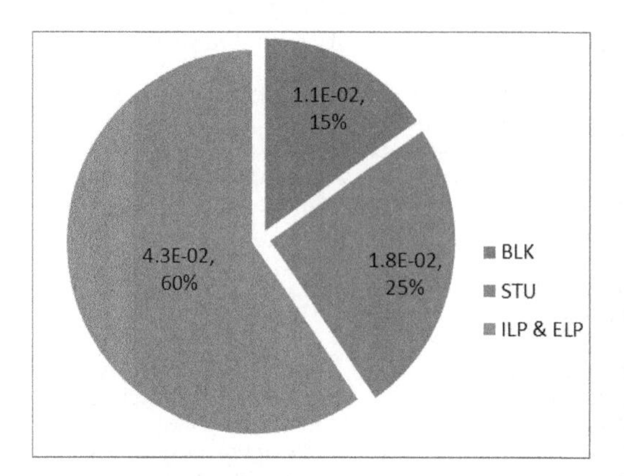

Figure 4. FMs contribution in the total annual failure rate of the FPSO development concept.

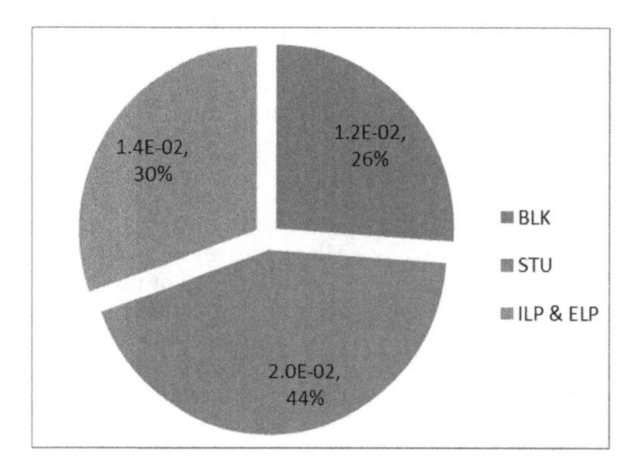

Figure 5. FMs contribution in the total annual failure rate of the S2S development concept.

In addition, this is also beneficial in terms of treating the lack of knowledge about the different operational conditions of the asset in such early phase. Moreover, it is providing the decision makers with information about the important elements that need improvements during design and intensive monitoring during the operation phase. Furthermore, the highlighting of the SoK of the various RIFs is also important to bring the focus on where to spend more efforts on gathering data and information in the subsequent risk assessments.

With regards to the consequence analysis results, Table 17 shows that the FPSO scenario pose more risk to human, environment and production than the S2S scenario. The expected loss of lives in the first concept is 1.5E-03, while in the latter it is 2.8E-04 fatalities per year. Correspondingly, the FPSO scenario introduces 1.85E-03 year (approx. 0.7 days) as an annual expected restitution time to the environmental damage. This is almost 25% higher than the annual expected recovery time of the S2S scenario. Lastly, the anticipated loss of production and repair time is due to asset damage is 9.8E-02 for the FPSO concept, while for the S2S concept it is 6.4E-02 month per year.

From all the above it is obvious that the S2S scenario provides lower failure rates and expected losses than the FPSO development concept. However, in the concept selection phase, other elements than risk results need to be addressed to formulate the decision of the involved parties about the selected concept (API, 2015). For example, investment analysis, logistic challenges and challenges in

Table 17. Expected loss to human, environment and production of the FPSO and S2S subsea development concepts.

		Severity class vs. Annual frequency of occurrence					Annual expected loss		
Subsea concept	Consequence groups	Minor	Significant	Severe	Major	Catastrophic	No. of Fatalities	Production Stop/Repair (Months)	Restitution time (years)
FPSO	H	3.5E-02	7.5E-03	3.8E-04	1.5E-07	5.8E-07			
	E	3.5E-02	7.7E-03	7.0E-04	7.6E-06	5.8E-07	1.5E-03	9.8E-02	1.85E-03
	P	2.5E-02	3.0E-02	9.7E-03	2.9E-03	1.3E-03			
S2S	H	1.3E-02	2.5E-04	9.9E-05	4.8E-08	1.9E-07			
	E	1.0E-02	2.2E-03	1.1E-03	2.4E-06	1.9E-07	2.8E-04	6.4E-02	1.4E-03
	P	1.6E-02	2.0E-02	6.3E-03	1.9E-03	8.1E-04			

construction and installation activities. However, under any circumstances the selected concept must not deviate from the minimum requirements of integrity and the safe operation (NORSOK, 2010).

On the other hand, by considering the effect of the different FMs on failure rates, it is observed that process leaks are representing 60% of the total failure rate of the FPSO scenario. However, this is dramatically decreased when looking at the S2S scenario, where leaks are denoting only 30% of the annual failure frequency of the system. Such increment in the FPSO scenario is drove by the higher frequencies of occurrence of the external impacts. This is due to the presence of supply vessels, lifting operations in addition to the large number of the infield flowlines, if compared to the S2S concept. Other than that, structural failures represents 44% of the annual failure frequency of the S2S scenario. The reason behind that is the presence of the export pipeline which is exposed to several environmental conditions, in addition to, its length which has a direct proportion with failure rate.

Another element that needs to be considered is the impact of the different RIFs on the generic failure rates. It is clearly seen that RIFs have introduced significant value in adjusting the generic failure rates and make it more specific to the systems of interest. In the FPSO scenario, the adjustment process of RIFs has resulted in 20% rise in the annual failure frequency. Similarly, 9% increment in the annual failure frequency of the S2S has been achieved by accounting the impact of the various RIFs.

In addition, this is also beneficial in terms of treating the lack of knowledge about the different operational conditions of the asset in such early phase. Moreover, it is providing the decision makers with information about the important elements that need improvements during design and intensive monitoring during the operation phase. Furthermore, the highlighting of the SoK of the various RIFs is also important to bring the focus on where to spend more efforts on gathering data and information in the subsequent risk assessments.

With regards to the consequence analysis results, Table 17 shows that the FPSO scenario pose more risk to human, environment and production than the S2S scenario. The expected loss of lives in the first concept is 1.5E-03, while in the latter it is 2.8E-04 fatalities per year. Correspondingly, the FPSO scenario introduces 1.85E-03 year (approx. 0.7 days) as an annual expected restitution time to the environmental damage. This is almost 25% higher than the annual expected recovery time of the S2S scenario. Lastly, the anticipated loss of production and repair time is due to asset damage is 9.8E-02 for the FPSO concept, while for the S2S concept it is 6.4E-02 month per year.

From all the above it is obvious that the S2S scenario provides lower failure rates and expected losses than the FPSO development concept. However, in the concept selection phase, other elements than risk results need to be addressed to formulate the decision of the involved parties about the selected concept (API, 2015). For example, investment analysis, logistic challenges and challenges in construction and installation activities. However, under any circumstances the selected concept must not deviate from the minimum requirements of integrity and the safe operation (NORSOK, 2010).

5 CONCLUSIONS

The paper introduces a semi-quantitative approach of risk analysis of subsea oil and gas production systems in the concept selection phase of the asset. In such early lifecycle phase, risk analysis involves information on failure rates of systems or subsystems level. Further detailed information is normally utilized in the FEED analysis and detailed design phase (API, 2015; NORSOK, 2010).

The identification process of the various FMs, FCs and RIFs of subsea equipment was carried out according to Rahimi & Rausand, (2013). In addition, in the failure rate quantification process, the BORA method (Aven et al., 2006) was used to reflect the impact of the different RIFs on the produced failure rates.

Moreover, in the consequence analysis, linguistic consequence groups were used in order to link consequence severities with the associated frequencies of occurrence. The provided information in the consequence groups definition were adapted from (Elsayed et al., 2014; Stendebakken et al., 2014).

Finally, the proposed method of risk analysis has been applied on a case study from offshore Brazil. The purpose of the analysis was on providing information on the risk level of two potential development concepts of subsea production field. The concepts are FPSO and S2S. The results showed that the S2S concept has lower risk of failure and expected losses than the FPSO scenario. However, it is also mentioned that other inputs than risk level are required in order to take the decision of the selection of the concept to be developed. It must be stressed that the case study aims to present an example of application and does not aim to reach definitive conclusions about the two generic concepts.

ACKNOWLEDGEMENTS

This work has been supported by EMBRAPII-COPPE Unit—Subsea Technology, within the project "Subsea Systems", which is conducted in cooperation with COPPE (UFRJ) and is financed by PETROGAL Brazil. The second author holds a visiting position at the Ocean Engineering Department, COPPE, Federal University of Rio de Janeiro, which is financed by the program "Ciências sem Fronteiras" of Conselho Nacional de Pesquisa of Brazil (CNPq).

REFERENCES

Abrahamsen, E.B., & Røed, W. (2011). A new approach for verification of safety integrity levels, Vol. *2*(20), 20–27.

API. (2015). General Overview of Subsea Production Systems, (January), 102.

Aven, T. (1992). *Reliability and Risk Analysis*. Elsevier Applied Science, Hampshire, England.

Aven, T. (2008). *Risk analysis. Assessing uncertainties beyond expected values and probabilities.* Chichester, UK: Willey.

Aven, T. (2014). *Risk, Surprises and Black Swans: Fundamental Ideas and Concepts in Risk Assessment and Risk Management.* New York: Rotledge.

Aven, T., Sklet, S., & Vinnem, J.E. (2006). Barrier and operational risk analysis of hydrocarbon releases (BORA-Release): Part I. Method description. *Journal of Hazardous Materials*, 681–691.

Bai, Yong. Bai, Q. (2012). *Subsea engineering handbook.* Houston, USA: Gulf Professional Publishing.

Bai, Yong. Bai, Q. (2014a). *Subsea pipeline design, analysis, and installation*. Houston, USA: Gulf Professional Publishing.

Bai, Yong. Bai, Q. (2014b). *Subsea pipeline integrity and risk management*. Houston, USA: Gulf Professional Publishing.

Courban, B., & Brouce, B. (2003). Pipeling and riser loss of containment (PARLOC) 2001. Offshore Research Focus 141.

DNV, G. (2014). Subsea Facilities-Technology Developments, Incidents and Future Trends. *Petroleum Safety Authority, Report*.

DNV. (2010). DNV-RP-F107: Risk Assessment of Pipeline Protection. *Materials Technology*, (October), 1–45.

Elsayed, T., Marghany, K., & Abdulkader, S. (2014). Risk assessment of liquefied natural gas carriers using fuzzy TOPSIS. *Ships and Offshore Structures*, *9*(4), 355–364. h.

Flage, R., Aven, T., Zio, E., & Baraldi, P. (2014). Concerns, Challenges, and Directions of Development for the Issue of Representing Uncertainty in Risk Assessment. *Risk Analysis*, *34*(7), 1196–1207.

Grusell, Christian. Fyrileiv, O. (2009). *Energy Report "Recommended Failure Rates for Pipelines."* Stavanger, Norway: DNV.

International Association of Oil and Gas Producers. (2010). Ignition probabilities. *Risk Assessment Data Directory*, (434), 30.

Kragh, E.; Faber, M.H., and Guedes Soares, C. Framework for Integrated Risk Assessment. Guedes Soares, C., (Ed.). Safety and Reliability of Industrial Products, Systems and Structures. London, U.K.: Taylor & Francis Group; 2010; pp. 7–19.

NORSOK. (1998a). Norsok standard Z-16 Regularity Management and Reliability Technology, Oslo, Norway.

NORSOK. (1998b). Norsok standard U-001 Subsea Production systems, Oslo, Norway.

NORSOK. (2010). Risk and emergency preparedness assessment—Norsok Standard Z-013, (Edition 3, October 2010), 1–107.

OLF 070. (2004). Application of IEC 61508 and IEC 61511 in the Norwegian Petroleum Industry. *The Norwegian Oil Industry Association*, *70*(October).

Rahimi, M., & Rausand, M. (2013). Prediction of failure rates for new subsea systems: a practical approach and an illustrative example. " Proceedings of the Institution of Mechanical Engineers, Part O. *Journal of Risk and Reliability*, *227.6*, 629–640.

Rausand, M., & Hoyland, A. (2004). *System reliability theory: models, statistical methods, and applications*. John Wiley & Sons.

Rinaldi, G., Thies, P.R., Walker, R., & Johanning, L. (2017). A Decision Support Model to Optimise the Operation and Maintenance of an Offshore Renewable Energy Farm. *Ocean Engineering*, 250–262.

Silva, L.M.R., & Guedes Soares, C. (2016). Study of the risk to export crude oil in pipeline systems, *Maritime Technology and Engineering 3*, Guedes Soares, C. & Santos T.A. (Eds.), Taylor & Francis Group, London, UK, pp. 1013–1018.

Sintef. (2002). Offshore Reliability Data Handbook. Trondheim, Norway: OREDA Participants.

Spouge, J. (1999). A guide to quantitative risk assessment for offshore installations. Aberdeen, UK: SD: CMPT.

Stendebakken, O.I., Vinnem, J.E., & Willmann, E. (2014). *A reliability study of a Deepwater Vertical Xmas Tree with attention to XT retrieval rate. Fakultet for ingeniørvitenskap og teknologi, Institutt for marin teknikk*. Trondheim, Norway.

Vinnem, J.E. (2014). *Offshore Risk Assessment Vol 1. "Principles, Modelling and Applications of QRA Studies."* Springer-Verlag London.

Progress in Maritime Technology and Engineering – Guedes Soares & Santos (Eds)
© 2018 Taylor & Francis Group, London, ISBN 978-1-138-58539-3

Availability assessment of a power plant working on the Allam cycle

U. Bhardwaj & A.P. Teixeira
Centre for Marine Technology and Ocean Engineering (CENTEC), Instituto Superior Técnico, Universidade de Lisboa, Lisbon, Portugal

C. Guedes Soares
Centre for Marine Technology and Ocean Engineering (CENTEC), Instituto Superior Técnico, Universidade de Lisboa, Lisbon, Portugal
Subsea Technology Laboratory, Department of Ocean Engineering, COPPE, Federal University of Rio de Janeiro, Rio de Janeiro, Brazil

ABSTRACT: The availability of a power production system utilizing the Allam cycle is modelled and analysed. First, a Failure Mode and Effects Analysis is conducted to identify the main failure modes of system's components. Generalized Stochastic Petri nets as dynamic model for supporting a Monte Carlo simulation are used to evaluate the availability of a test case of an Allam cycle system. The test case consists of a scenario of a power plant project using the Allam cycle, which can capture 100% CO_2 and therefore can be used in offshore oil and gas production units by reinjecting CO_2 in the wells to maintain reservoir pressure. The failure data for the components are taken from OREDA that contains failure rates of offshore equipment. Simulation results on the availability of the system and other production parameter are obtained considering corrective maintenance alone. Finally, a parametric study demonstrates the effect of variation of mean time to failure and mean time to repair on system performance. This study can help in optimizing operation and maintenance actions in complex production systems.

1 INTRODUCTION

The deep sea technology for natural resource production in the ocean is a specialized area, and is a field that requires high level of core technologies such as engineering and simulation. Subsea production system is extending its applicability in deeper water and this poses new challenges for the designs and operations of such systems. For example low temperature and high pressure are the two common characteristics of such environment that lead to high costs of maintenance and repair of such systems.

In many deep water oil fields, multiphase flow stream is transported from the wellhead to the topside where it is separated into oil gas water and other substituent. Sometimes the oil produced contains high content of carbon dioxide (CO_2) (Beltrao et al. 2009). The presence of CO_2 and water of the production fluid can cause internal corrosion in the subsea pipeline. Many coastal policies like Clean Air Act (CAA) emphasize on minimum release of CO_2 since ocean acidifications is primarily driven by increasing CO_2 in atmosphere (Abate, 2015), but the problem is rising in offshore industries. For example, offshore petroleum fields were responsible for up to 26% of the total CO_2-emissions of Norway in 2011 (Nguyen et al. 2016a). Also contemplating the greenhouse effect of CO_2 in atmosphere, greenhouse gases reduction is goal set forth in the Paris Agreement (IEA, 2014).

Recently, there have been significant mitigating initiatives. Nguyen et al. (2016b) have illustrated 15–20% reductions in CO_2-emissions for four actual cases located in the North and Norwegian Seas. However, there are other methods (Mathisen & Skagestad, 2017) developed to use of CO_2 for Enhanced Oil Recovery (EOR) and deep ocean storage of liquefied CO_2 (Goldthorpe, 2017 & Cumming et al. 2017). A more efficient solution can be provided by using Allam cycle, which is CO_2 and oxy-fuel power cycle that utilizes hydrocarbon fuels while inherently capturing approximately 100% of atmospheric emissions of CO_2 (Allam et al. 2017). The system drives a process called Carbon Capture and Storage (CCS), which would see the excess CO_2 from the fuel combustion funnelled into a pipeline or a tanker instead of being released into the air. It is believed that the process can be implemented in offshore oil and gas industry as the compressed CO_2 can be reinjected to subsea wells to maintain reservoir pressure.

The Allam cycle consists of mainly seven components whose details are mentioned in section 3. The challenge in maintenance planning and optimization of systems like the Allam cycle system is to include the realistic features and modelling the dynamics behaviour of systems' components (Zio et al. 2006, 2007 & 2009). Also, the system availability is the most relevant parameter for cost effective analysis of such system over a long period of time.

Availability is a key parameter that quantifies system performance and henceforth gaining popularity especially in the power plants, manufacturing system, offshore oil and gas industry etc.

Usually managers and engineers require a high availability and reliability to systems for maintaining production and service machinery operations. A relevant system availability analysis must reduce as much as possible the failure probability of any equipment in the system and add redundancy in order to minimize the impact of a single equipment failure on the availability.

Petri Nets (PNs) (Petri, 1962) have gained popularity for modelling manufacturing and production processes as well as it is finding significance in Operations Research. Hosseini et al. (1999) have utilized stochastic PNs to represent and analyse a hybrid maintenance model of systems with both deterioration and Poisson failures.

The modelling capability of PNs allows complex features to be easily incorporated into the model such as routine or on demand maintenance as well as dealing with different failure rates. This is why it can be used for real world systems.

In the field of reliability, more advanced transitions can be embedded like the Weibull distribution for representing component lifetimes (Rama & Andrews, 2013). To use this in a PN, the transitions need to be described by a distribution to generate a random transition time. Therefore, the PN must be simulated with a large sample size to assess the average system behaviour using Monte Carlo (MC) simulation.

This way Petri Net is further extended as Stochastic Petri Nets (SPN) (Reisig, 1985) by addition of timing and stochastic information. Stochastic Petri nets (Balbo 2001, Haas 2002, Marsan et al. 1995) facilitate the model to incorporate dynamic behaviour of complex systems. Several techniques based GSPN's (generalized SPN) (e.g. Malhotra & Trivedi 1995, Dutuit et al. 1997) coupled with MCS have been proposed for modelling and analysing systems in terms of reliability, availability, and especially the production availability of complex systems (e.g. Dutuit et al. 1997, Boiteau et al. 2006, Dutuit et al. 2008, Teixeira & Guedes Soares 2009). Using GSPN Santos et al. (2015) have estimated the effect of several design and operation and maintenance (O&M) parameters on the mean availability and production costs and revenues.

This study begins with the description of a system of Allam cycle, then a series of failure modes are determined by conventional FMEA method. A case study of a 300 MW pre-FEED design of commercial power plant project working on Allam cycle is chosen as case study. The failure and repair rates for corresponding components are taken from OREDA (offshore reliability data) considering their potential applicability in offshore sector.

Later, this paper presents a simulation method based on GSPN with predicates, coupled with MCS as a tool for assessing the availability of complex Allam cycle systems. The simulation model considers all the critical component models for the Allam cycle, their failure and repair rates are assumed to be exponentially distributed.

The aim of the present work is to demonstrate a simulation method based on GSPN model for availability assessment. This model can be updated further with more realistic values and can also be incorporated with more systems interdependencies.

2 THE ALLAM CYCLE

Originally the Allam cycle was presented in Kyoto at GHGT-11 and will be developed to commissioning stage in near future. The description presented below is based on the work of Allam et al. (2017).

Analyses of traditional power cycles have shown that the additional CO_2 removal systems can increase the cost of electricity by 50% to 70% when in order to capture 90% of the CO_2 generated from hydrocarbon fuel combustion (Bay & Bay, 2010). The Allam Cycle takes a new approach to reducing emissions by employing oxy-combustion and a high-pressure supercritical CO_2 working fluid in a highly recuperated cycle (Allam et al. 2013). Laumb et al., (2017) has demonstrated the Allam cycle working on solid fluid for high efficiency power production. The basic schematic arrangement of Allam cycle for natural gas fuel is shown in Figure 1. A pressurized gaseous fuel (14) is combusted in the presence of a hot oxidant flow containing a mixture of CO_2 and nominally pure oxygen (13, provided by a co-located Air Separation Unit (ASU)) and a hot CO_2 diluent recycle stream (9) at approximately 300 bars under lean combustion conditions. A pressurized gaseous fuel (14) enters in the turbine where it is combusted in the presence of a hot mixture of CO_2 and pure oxygen (13), coming from ASU and a hot CO_2 diluent recycle stream (9) at about 300 bar under lean combustion conditions. The exhaust flow is expanded in the turbine to approximately 30 bar (1).

From the turbine, the exhaust flow enters a recuperating heat exchanger which transfers heat from the hot exhaust flow to the high pressure CO_2 recycle stream which acts as diluent quench for the combustion products and as well as the oxidant flow providing oxygen to the combustor flame zone.

The exhaust flow from turbine then is cooled heat exchanger (2) and combustion derived water is separated (3) in water separator. The CO_2 stream is then recompressed (4), cooled (7), and pumped.

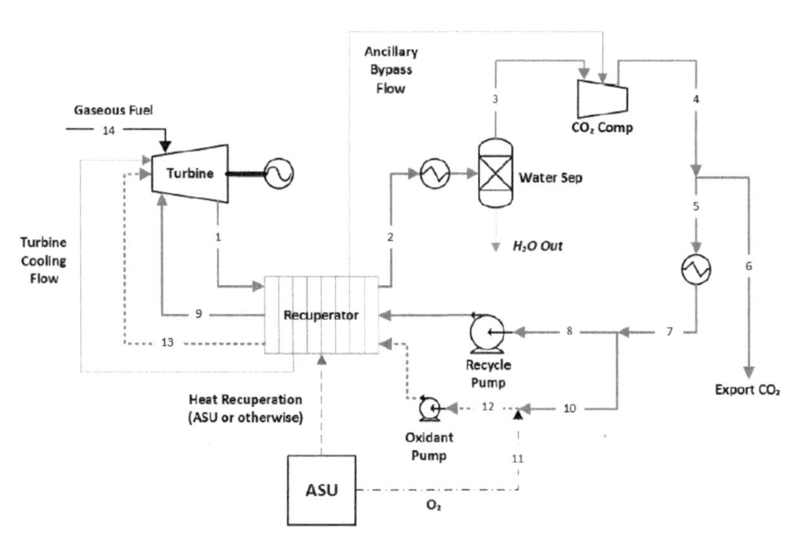

Figure 1. Schematic of the Allam cycle for gaseous fuel Test study of 300 MW natural gas plant in pre-FEED design.

It then re-enters the recuperative heat exchanger where some of its portion (10) is mixed with oxygen (11) to form an oxidant mix stream (12).This stream is fed separately to the heat exchanger and turbine.

2.1 *Test study of 300 MW natural gas plant in pre-FEED design*

A 50 MW demonstration-scale natural gas version of the plant, currently in construction by NET Power in order to understand operation of the Allam cycle and validate performance, control methodology, operational targets, and component durability. The NET Power is in the process of designing the 300 MWe commercial plant targeted to be completed in 2020 (Allam et al., 2017). The case study considered incorporates a production scenario of pre-FEED design of commercial power plant of 300 MW capacities.

2.2 *Major component with advanced specifications*

Allam cycles operate at high pressures throughout the cycle, resulting in a superheated CO_2 working fluid with a high density which may lead to smaller component sizes, smaller plant footprint, and therefore lower capital cost. Achieving the full benefits of the CO_2 cycle will depend on overcoming a number of engineering and materials science challenges that impact both the technical feasibility of the cycle as well as its economic viability.

Toshiba Corporation has developed a preliminary turbine design for a 500 MW system and high pressure combustor, which attained the required maximum test pressure of 300 bar in 2013. Also, Heatric (2018) has completed the design and manufacturing of the primary high pressure recuperative heat exchanger (printed circuit heat exchanger-PCHE).

The component specification and performance targets, at ISO conditions, for the first commercial plant are presented below:

- Turbine and combustor: input – 511 MW thermal energy, speed – 3600 rpm, diameter – 1.2 m and multistage (Turchi, C., 2014);
- Recuperator: type—Printed circuit heat exchanges with increased size (10,000 m^2) and capacity – 650 m^2/m^3 at 300 bar, plates size – 1.6 mm thick (Heatric 2018);
- ASU: production – 3550 MT/day of 99.5% O_2. Utilizes – 56 MW;
- Compressor: multi-stage, intercooled, centrifugal compressors, 310 bar, Utilizes – 77MW;
- Pumps: CO_2 and O_2 pumps as required for design duties.

2.3 *Electricity production and distribution*

At ISO conditions, it is expected that the plant will produce 303 MW of electricity. The thermal inputs from natural gas are 511 MW with LHV efficiency 59.30%. The turbine conditions for required production are as follows:

- Inflow: 923 kg/s;
- Inlet condition: 300 bar and 1158 C;
- Outlet condition: 30 bar and 727 C.

Most of the components are powered by electricity. The two main components ASU and Compressor utilize 56 and 77 MW of electricity. So at maximum capacity the plant will produce 170 MW of electricity.

The CO_2 production is predicted as 2400 MT/day at 150 bar pressure.

Table 1. FMEA of major component of the Allam cycle.

Component	Failure Modes	Effect	Causes
Turbine	Failure of rotor blades and disc	Equipment damage, Production stop	Centrifugal load Vibration induced Fatigue Erosion Thermal Creep Airborne Particles Foreign Object Corrosion
	Degradation of Stators	Reduced efficiency	Corrosion Erosion Fatigue Creep Buckling
	Casing fracture	Reduced safety	Fatigue
	Sticking of valves	Reduced efficiency	Sticking Leaking
	Vibration of Rotor	Loss of control	Inadequate flow Low pressure
Compressor	Zero flow and discharge pressure	Production stop	Wrong or no installation of the impeller key and the connecting shaft Closed suction valve and exhaust valve
	Inadequate flow and discharge pressure	Reduced efficiency	Abnormal circulation Low operating speed Inspiratory pressure Clogged inlet filter Molecular weight discrepancies Compressor reversal Circulating volume increasing on the suction side of the exhaust side Pressure gauge or flow meter failure Insufficient inspiratory pressure Clogged inlet filter
	Discharge pressure fluctuations	Loss of control, Vibration	Insufficient/small flow Flow regulating valve failure
	Surge	Vibration, misalignment	Impeller and runner fouling Ultra-high system pressure Insufficient inhalation flow Improper matching of import and export pipeline network Check valve failure or damage Anti-surge valve failure

	High gas temperature	Reduced safety	Insufficient cooling oil Decreased oil cooling capacity Loose or broken ceramic tubes Ceramic tube surface dirt Deviation of running point from the design
	Seal failure	Reduced safety, Reduced efficiency	O-ring failure Mechanical seal damage Low sealing pressure Sealing medium containing impurities Excessive damage to the seal ring clearance
	Abnormal vibration and noise	Misalignment	Critical speed Impeller friction and damage Coupling Rotor imbalance Fault Adverse gear Weak foundation Bending shaft Abnormal bearing Damaged alignment accuracy Misaligned
Separator	Loss of equipment casing	Reduced safety	Corrosion Stress-corrosion damage of equipment Pitting Erosion
	Malfunctioning valves	Reduced safety	Mechanical wear of the elements: fitting Fatigue
	Reduce of wall thickness of body, bottom and tube sheets	Reduced efficiency	Peeling Variation of cover thickness, breaks, cuts
	Leakage in piping	Reduced efficiency, Reduced safety	Surface wear Gas-abrasive Hydro-abrasive Cavitation
	Blockage of the fluid flow	Reduced efficiency	Paraffinn deposited
	Formation of foams	Reduced efficiency	Low pressure
Pump	Reduction in suction head	Reduced efficiency	Pump cavitation
	Reduction in pump head	Vibration	Pump cavitation
	Component corrosion	Reduced safety	Incorrect fluid—Excessive flow rate for fluid
	Shaft deflection	Misalignment	High radial thrust on pump Rotor
	Shaft unbalance	Vibration	Impeller wear
	Air leak through gasket/ stuffing box	Reduced safety	Damaged gasket

(*Continued*)

Table 1. (*Continued*)

Component	Failure Modes	Effect	Causes
	External Leakage	Reduced efficiency, Reduced safety	Seal failure Worn mechanical seal Scored shaft sleeve Stuffing box improperly packed
	Mechanical noise	Reduced efficiency, safety	Debris in the impeller Impeller out of balance Bent shaft Worn/damaged bearing Foundation not rigid Cavitation
	Positive suction head too low	Reduced efficiency, Vibration	Clogged suction pipe Valve on suction line only partially open
	Pump discharge head too high	Reduced efficiency, Vibration	Clogged discharge pipe Discharge line valve only partially open
	Suction line/impeller clogged	Production stop	Contaminants
	Worn/broken impeller	Reduced efficiency	Contaminants Wrong flow rate
	Thrust bearing failure	Production stop	Excessive axial load
Air Separator Unit (ASU)	Rapid Oxidation	Reduced efficiency, safety	Hydrocarbon enrichment (dry/hot boiling, distillation) Air compression Loss of pressure in HP column
	Pressure variation due to vaporizing liquids	Reduced efficiency	Polymer membrane and sieve failure
	Improper filtration	Reduced efficiency	Failure of sieve
	Improper purification	Reduced efficiency	Flow controller on Nitrogen product fault
	Embrittlement of interface pipes	Reduced safety	Rapid pressure drop
	Flooding of column	Production stop	Level controller fault
	ASU stops working	Production stop	Level controller malfunction or Low flow from high pressure column
	Abnormal pressure fluctuations	Reduced efficiency	Loss of feed
	Explosion	Production stop	Adsorption system fault Oxygen-enriched atmosphere

Recuperator	Fouling of flow plates	Poor cooling, Reduced efficiency	Corrosion from fluids (shell side) Galvanic and pitting Corrosion Dezincification
	Leakage through flow plates	Production stop	Corrosion from fluids Vibration of the tubes may cause the sheet to fail even if the tubes hold up
	Relief piping fails	Reduced safety	Mechanical failure Metal Erosion Steam or Water, Hammer Vibration Thermal Fatigue Freeze-Up Loss of Cooling Water Thermal Expansion
	Loss of structural integrity	Production stop	Steam or Water Hammer Vibration Thermal Fatigue Freeze-Up Thermal Expansion Loss of Cooling Water
	No heat transfer	No cooling	Various marine growths deposit—Scale, Mud, and Algae Funding

3 FMEA OF THE MAIN COMPONENTS OF AN ALLAM CYCLE

FMEA and FMECA are efficient tools for risk assessment to analyse and rank the risks associated with various products (or processes), failure modes (both existing and potential), prioritizing them for remedial action (Dailey 2004 & Barendsa 2012). Since FMEA method is based on finding and minimizing the failures, it has been broadly utilized in various kinds of industries (Rhee & Ishii 2003, Vandenbrande 1998). Hence, it is broadly applied on power generating system and proved to be an effective approach. It is obvious that there are various failures in all gas power plants projects like the present case study, which have potential risks.

Table 1 presents the main failure modes identified for the main components of the Allam cycle. Thereafter, all possible causes and effects are classified to the related failure modes. Moreover, this type of failures can be controlled by corrective actions and preventive maintenance. Though development of such actions is not on the scope of present study, still one case scenario with a maintenance policy is explained in a later section.

4 GENERALIZED ZED STOCHASTIC PETRI NETS MODEL

4.1 *Petri nets*

Petri nets are graphical and mathematical modelling tools first introduced by Carl Adam Petri in 1962 that can be applicable to many systems. The considerable development of a variety of Petri nets with enhanced modelling and simulation capabilities are nowadays applied to a wide range of scientific fields and engineering applications.

Many versions of the original Petri nets have been proposed for specific applications of production systems and for the fields of computer science, communications, and automation. GSPN coupled with MCS (Malhotra & Trivedi 1995, Dutuit et al. 1997, 2008, Boiteau et al. 2006, Santos et al. 2012) have been proposed with predicates namely guards and assignments. While "guards" are preconditions that enable or inhibit the transitions, "assignments" are post condition messages for updating variables used in the model when a transition fires.

Basically a Petri net has three elements: 1. places), which are used to model the conditions and 2. Transitions (nodes) events (e.g., fail) of a system; 3. directed arcs connected from places (input places) to transitions and vice versa (output places) establishing relations between network nodes. Usually, places, transitions, and arcs are graphically represented by circles, rectangles, and directed arrows, respectively (Murata, 1989).

Places are used to model the system's states (e.g., system functioning) and can also work as place of resources (e.g., number of maintenance crews available). Resources are graphically represented by small full marks named tokens, which are held inside places. The tokens in the network places characterize state of a system and its resources. Transitions symbolize the events (e.g., system failure), which manipulate the available resources and its firing causes the movement of tokens from place to place. Simple transitions are activated by a time delay. The relations between places and transitions are represented by arcs, connecting the places and transitions to form PN network. It is not possible to connect two transitions or places directly.

By such functions Petri nets represent the dynamic aspects of real systems, i.e., the dynamic relations between failure events, which were lacking in classical methods like static fault trees or event trees. More details on Petri nets can be found in Schneeweiss (2004) & Balbo (2001).

This paper adopts the Petri net with predicates module of the software GRIF (graphical interface for reliability forecasting) developed by TOTAL. This module has a Monte Carlo simulation engine MOCA-RP developed by Dutuit et al. (1997). The Allam cycle is modelled by GSPN and statistics related to the maintenance strategies are obtained by MCS. A simulation time of 20 years has been adopted to capture the steady-state behavior of the system without considering the end-of-life effects.

4.2 *Components failures*

The Allam cycle system consists of seven components in series. There is a possibility of failure of any component during the normal operation. System production stops if any of the component fails as they work in series. Each component fails independently of the state of the others and its time-to—failure follows an exponential distribution.

The basic system is considered with no redundancy. All components are considered to be repairable.

The basic system consists of seven main components. In this test case failures of the Turbine, compressor, pumps, recuperators and separator are taken into account. The components not included in the analysis (ASU etc.) are assumed to be working and do not fail. For simplification purposes, all units are assumed to be stochastically independent with constant failure and repair rates.

There are three different states for components: Working, under repair and Failed. The failed state of the component implies complete loss of its functionality. The failed component must be repaired by corrective maintenance.

In the present study for simplicity the failure and repair transitions are exponentially distributed. The values for the rates are shown in Table 2. Since the technology is new and advance components are utilized in the power plant, it is difficult to define actual failure rates for the components. For the sake of computation, highest failure rates of components, corresponding to the all failure modes are taken from OREDA (DNV 2002), as seen in Table 2.

4.3 *Corrective maintenance*

All corrective maintenances are performed by single maintenance team. Since all components are in series they are repaired when under a critical failure.

When several failures are waiting for repair at the same time, they are repaired along their level of priority 1, 2, 3 etc. as Table 3. When a repair begins, it is finished even if another one with more priority occurs. Moreover, a corrective maintenance team is available on site.

4.4 *Case study considering corrective maintenance*

Each of the six components, CMP#, of the system of Figure 1 is modelled like the *CMP1* in Figure 2. Since there are similarities among the CMP# models, only the GSPN model of one component is represented in Figure 2.

Each of the CMP# has three states (conditions): functioning (e.g. *CMP1_Work*), failed

Table 2. Failure and repair rates of the components.

Component	Unit ID	Failure rate (h^{-1})	Repair rate (h^{-1})
Turbine	CMP1	$2.884\ 10^{-3}$	0.0262
Compressor	CMP2	$4.75\ 10^{-4}$	0.067
Pumps	CMP3 & 4[a]	$1.41\ 10^{-4}$	0.0625
Recuperator	CMP5	$5.44\ 10^{-4}$	0.27
Separator	CMP6	$2.61\ 10^{-4}$	0.097

[a] There are two pumps in the Allam cycle.

Table 3. Repair priority levels of production components.

Component	Priority
Turbine	1
Compressor	2
Pumps	3
Recuperator	4
Separator	5

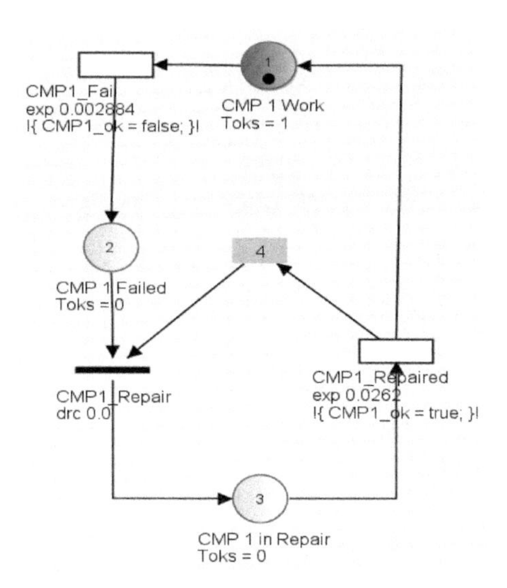

Figure 2. GSPN model for component 1 Table 2 (CMP 1).

(e.g. *CMP1_Failed*) and in repair (e.g. *CMP1_in Repair*). Next the following three operation events are modelled by as much transitions between states: component fails (e.g. *CMP1_Fail*), start repairing (e.g. *CMP1_Repair*) and completion of maintenance (e.g. *CMP1_Repaired*). The failure rates (e.g. event *CMP1_Fail*) and repair rates (e.g. event *CMP1_Repaired*) of the components are taken from Table 1. However, the starting of the repair of a failed component (e.g. event *CMP1_Repair*) is described by a Dirac function (*drc*) with a parameter of zero (hours). This means that the repair task begins immediately after the transition is enabled to be fired (e.g. *CMP1_Failed* with *token* = 1), i.e. the repair team starts repairing the component immediately (with no delays) after its failure.

Since it is assumed that all components (CMP#) are repaired by one team, each failed unit is repaired according to priority order as in Table 2 when multiple failures occur at same instant of time.

The initial condition of the system is considered to be "full production", i.e. all places that represent the functioning state of components have one token. At this state the outputs of system are 170 MW of electricity and 2400 MT/day of CO_2.

4.5 *Numerical results*

Once the Petri net model is built, a Monte Carlo simulation is run for analysing the behaviour of the system. For the calculation of average availabilities the MOCA-RP tool of GRIF software was used allowing to developed GSPN models and to perform its analysis. The simulation is for a period of 20 years, with steps of 200 hours for the iterations. The simulation approach can be as accurate as the analytical one, depending on the number of simulations executed. 1000 Monte Carlo simulations have been performed.

In the case study it was assumed that the components of the system can fail and are repaired by a corrective maintenance team but do not go through preventive maintenance.

The results show that the system availability is 88.5% (88.8% calculated analytically). The availabilities of individual components calculated by simulation and analytically are shown in Table 4. It can be observed a good agreement between the results calculated by these two methods. However, it can be seen that the values obtained analytically are a little higher than that obtained by software. This is due to the fact that the analytical method does not incorporate corrective maintenance policy i.e. one maintenance team for all components. In other words the Petri net model provides more flexibility and scope for complex systems. Figure 3 and 4 show the variation in production of electricity and CO_2 with time. The mean values of Electricity and CO_2 obtained are 150.39 MW and 2123.2 MT per day respectively.

4.6 *Parametric study*

A parametric analysis of the MCS results was performed to assess the influence of the variations

Table 4. Failure and repair rates of the components.

Component	Unit ID	Availability (by Petri Net)	Availability (analytically)
Turbine	CMP1	90.03	90,08
Compressor	CMP2	99.11	99,30
Pumps	CMP3 &4[a]	99.72	99,77
Recuperator	CMP5	99.72	99,80
Separator	CMP6	99.58	99,73

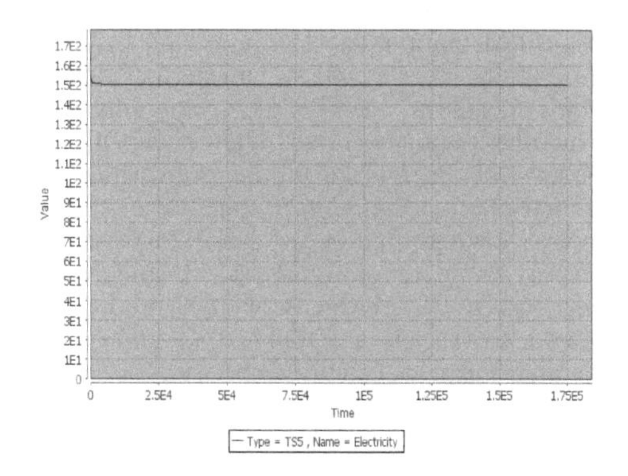

Figure 3. Electricity production with time.

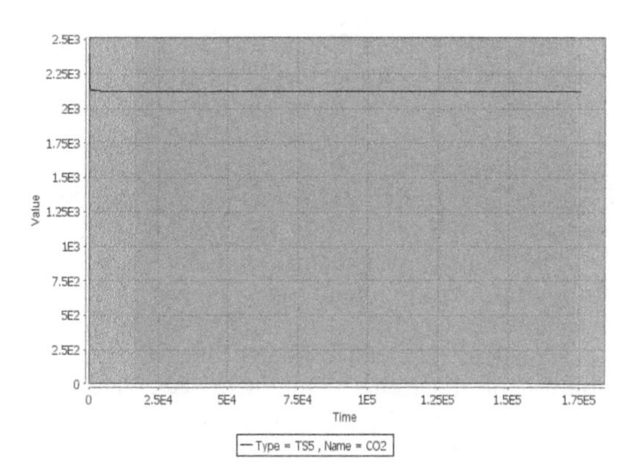

Figure 4. CO_2 production with time.

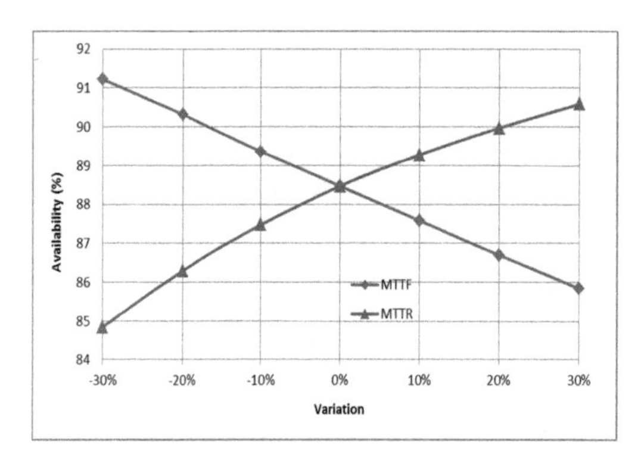

Figure 5. Effect of turbine's MTTF and MTTR on system availability.

of MTTF (=1/failure rate) and MTTR (=1/repair rate) on the availability and production of the system. Simulations of duration of 175,200 hours (i.e. 20 years) are considered.

It is easily understandable that all component of the Allam cycle are in series. Failure of a single component leads to production loss. The failure rate of the turbine is high and therefore it is the most critical component. Hence, the present parametric study assesses the system performance by varying MTTF and MTTR of turbine only, keeping all other parameters fixed. The effect of the turbine's MTTF and MTTR on the mean availability, electricity output and CO_2 production of the system are shown in Figures 5, 6 and 7, respectively.

It was found that varying MTTF and MTTR from -30% to 30% in steps of 10% produces an almost linear variation of only 6.08% and 6.5% in the system mean availability, respectively (Figure 5).

From Figure 5, 6 and 7 it is seen that the variation of the turbine's MTTR has the highest effect on the three performance measures. In absolute values, the ranges are around 5.75%, 9.83 MW and

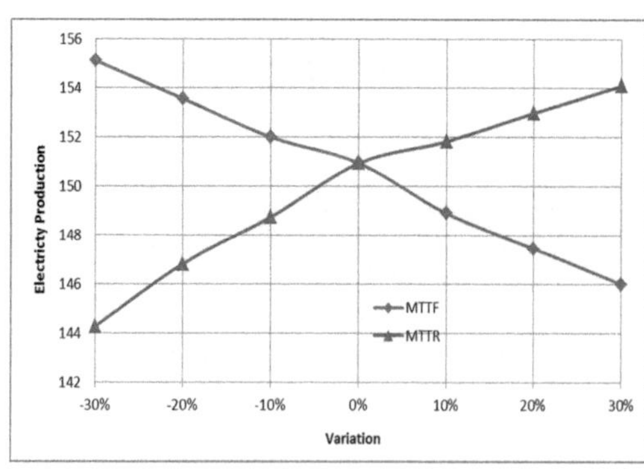

Figure 6. Effect of turbine's MTTF and MTTR on electricity production.

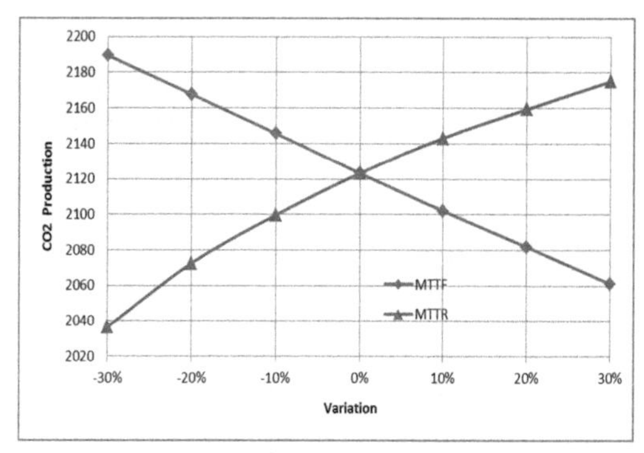

Figure 7. Effect on CO_2 production by varying parameters.

138.7 MT/day, respectively. It is interesting to see that the MTTR has almost the same effect with values 5.38, 9.09 MW and 128.3 MT/day, respectively. Finally, it can be seen that the impact of the variation of MTTF and MTTR can be expressed by a linear function.

A limitation of this work is the failure and repair rates adopted, which do not reflect the actual characteristics of the component as specific historical data are not available. Still this model can be updated with required preventive maintenance policy and more realistic input data to observe system performance and outputs.

5 CONCLUSIONS

It is believed that the use of the Allam cycle for gas stream processing can provide solution to various problems with CO_2 emission, in offshore installations.

First, a detailed FMEA is performed to understand main failure modes of the subsequent components of the Allam cycle. The aim of FMEA is to understand, prioritize the common failures and to implement corrective action and maintenance policy. Although, the detailed development of corrective actions is not in the scope of present work.

This paper focuses on one case study of 300 MW commercial power plant utilizing the Allam cycle. GSPN with predicates and MCS are used for modelling and simulation, as it is proved to be a useful modelling tool for complex systems like offshore installation. GSPN allow integrating failures and repairs and maintenance policies into the model.

Monte Carlo simulation is used to assess the key production indicators such as the average availability, electricity and CO_2 production. The results obtained are:

- System availability: 88.47%;
- Electricity production: 150.39 MW;
- CO_2 production: 2123.2 MT/day.

The availability obtained by Petri nets is compared with that obtained analytically. The similarity in their values validates the model developed.

Lastly, a parametric study is conducted to show the effect of the turbine's MTTF and MTTR on system performance. The present work can be extended in the future by using realistic failure and repair rates, maintenance policies, electricity and CO_2 production data to the model.

ACKNOWLEDGEMENTS

The contribution of the second and third authors has been financed by EMBRAPII-COPPE Unit—Subsea Technology, within the project "Subsea Systems", which is conducted in cooperation with COPPE (UFRJ) and is financed by PETROGAL Brasil. The third author holds a visiting position at the Ocean Engineering Department, COPPE, Federal University of Rio de Janeiro, which is financed by the program "Ciência sem Fronteiras" of Conselho Nacional de Pesquisa of Brazil (CNPq).

REFERENCES

Abate (2015), Climate Change Impacts on Ocean and Coastal Law: U.S. and International perspectives, Edt. Abate RS, *Oxford university press*. ISBN-13: 978-0199368747.

Allam R.J., Palmer M. & Brown, G.W. (2013), System and Method for High Efficiency Power Generation Using a Carbon Dioxide Circulating Working Fluid. *USA Patent 8, 596, 075 B2*.

Allam, R.J., Martin, S., Forrest, B., Fetvedt, J., Lu, X., Freed, D. & Manning, J. (2017), Demonstration of the Allam Cycle: An Update on the Development Status of a High Efficiency Supercritical Carbon Dioxide Power Process Employing Full Carbon Capture. *Energy Procedia*, vol. 114, pp. 5948–5966.

Archetti, F., Fagiuoli, E., & Sciomachen, A. (1987), Computation of the Makespan in a transfer line with station breakdowns using stochastic Petri Nets. *Computers and Operations Research,* vol. 14, pp. 409–414.

Balbo, G., (2001), Introduction to stochastic Petri nets, In: *Lectures on Formal Methods and Performance Analysis.* Springer, pp. 84–155.

Barendsa, D.M., Oldenhofa, M.T., Vredenbregta, M.J., Nautab M.J. (2012), Risk analysis of analytical validations by probabilistic modification of FMEA. *Journal of Pharmaceutical Biomedical Analysis*, vol. 64–65, pp. 82–86.

Bay, Y. & Bay, Q. (2010), *Subsea Engineering Handbook*, 1 ed., Elsevier.

Beltrao, R.L.C., Sombra, C.L., Lage, A.C.V.M., Netto, J.R.F., & Henriques C.C.D. (2009), Challenges and new technologies for the development of the pre-salt cluster, Santos Basin, Brazil. *Offshore Technology Conference 19880, Houston, Texas, U.S.A.*

Boiteau, M., Dutuit, Y., Rauzy, A. & Signoret, J.-P. (2006), The AltaRica data-flow language in use: modelling of production availability of a multi-state system. *Reliability Engineering and System Safety,* vol. 91, pp. 747–755.

Cumming L., Gupta N., Miller K., Lombardi C., Goldberg D., Brink U., Schrag D., Andreasen D. & Carter K. (2017), Mid-Atlantic U.S. Offshore Carbon Storage Resource Assessment. *Energy Procedia,* vol. 114, pp. 4629–4636.

Dailey, K.W. (2004), The FMEA pocket handbook, 1st ed. USA, DW Publishing Co.

DNV (2002), OREDA—Offshore Reliability Data Handbook, 4th ed. *DetNorsk Veritas*, Høvik, Norway.

Dutuit, Y., Châtelet, E., Signoret, J.-P. & Thomas, P. (1997), Dependability modelling and evaluation by using stochastic Petri nets: application to two test cases. *Reliability Engineering System Safety,* vol. 55, pp. 117–124.

Dutuit, Y., Innal, F., Rauzy, A. & Signoret, J.-P. (2008), Probabilistic assessments in relationship with safety integrity levels by using Fault Trees. *Reliability Engineering and System Safety,* vol. 93, pp. 1867–1876.

Goldthorpe, S., (2017), Potential for Very Deep Ocean Storage of CO_2 Without Ocean Acidification: A Discussion Paper, *Energy Procedia,* vol. 114, pp. 5417–5429.

Haas, P.J., (2002), Stochastic Petri Nets. Modelling, Stability, Simulation, *Springer-Verlag*.

Heatric (2018), Heatric-Resized Printed Circuit Heat Exchangers Avail at: https://www.heatric.com/diffusion_bonded heat exchangers.html.

Hosseini, M., Kerr, R., & Randall, R., (1999), Hybrid maintenance model with imperfect inspection for a system with deterioration and Poisson failure. *Journal of the Operational Research Society,* vol. 50, pp. 1229–1243.

IEA (2014), Energy Technology Perspectives: Harnessing Electricity's Potential. International Energy Agency, OECD/IEA, Paris. http://www.iea.org/publications/freepublications/publication/energy-technology-perspectives–2014.html

Laumb, J.D., Holmes, M.J., Stanislowski, J.J., Lu, X., Forrest B. & McGroddy, M. (2017), Supercritical CO_2

cycles for power production. *Energy Procedia,* vol. 114, pp. 573–580.
Malhotra, M. & Trivedi, K.S. (1995), Dependability modelling using Petri nets. *IEEE Transactions on Reliability,* vol. 44, pp. 428–440.
Marsan, M.A., Balbo, G., Conte, G., Donatelli, S. & Franceschinis, G. (1995), Modelling with Generalized Stochastic Petri Nets, *John Wiley & Sons.*
Mathisen A. & Skagestad, R. (2017), Utilization of CO_2 from emitters in Poland for CO_2-EOR. *Energy Procedia,* vol. 114, pp. 6721–6729.
Murata, T. (1989), Petri Nets: Properties, Analysis and Applications. *Proceedings of the IEEE,* vol. 77, pp. 541–580.
Nguyen, T., Tock, L., Breuhaus, P., Maréchal, F. & Elmegaard, B., (2016a), CO_2-mitigation options for the offshore oil and gas sector. *Applied Energy,* vol. 161, pp. 673–694.
Nguyen, T., Voldsund, M., Breuhaus, P. & Elmegaard, B. (2016b), CO_2-mitigation options for the offshore oil and gas sector. *Applied Energy,* vol. 161, pp. 673–694.
Petri, C.A. (1962), Communication with automation, *Bonn, Germany: Mathematical Institute of the University of Bonn,* Ph.D. thesis.
Rama, D. & Andrews, J. (2013), A reliability analysis of railway switches, *Proceedings of the Institution of Mechanical Engineers, Part F: Journal of Rail and Rapid Transit,* vol. 227, pp. 344–363.
Reisig, W. (1985), Petri Nets, An Introduction. EATCS, Monographson Theoretical Computer Science. *Springer Verlag.*
Rhee, S.J. & Ishii, K. (2003), Using cost based FMEA to enhance reliability and serviceability. *Advanced Engineering Informatics,* vol. 17, pp. 179–88.
Santos, F., Teixeira, A.P. & Guedes Soares, C. (2015), Modelling and Simulation of the Operation and Maintenance of Offshore Wind Turbines. *Proceedings of the Institution of Mechanical Engineers, Part O: Journal of Risk and Reliability,* vol. 229, pp. 385–393.
Santos, F.P., Teixeira, A.P. & Guedes Soares, C. (2012), Production Regularity Assessment Using Stochastic Petri Nets With Predicates, *Maritime Technology and Engineering,* C. Guedes Soares, Y. Garbatov, S. Sutulo, and T.A. Santos, eds., Taylor & Francis Group, London, pp. 441–450
SATODEV (2015), User Manual, GRIF, Petri Nets With Predicates, SATODEV, Merignac, France.
Schneeweiss, W.G. (2004), Petri Net picture book, LiLoLe-Verlag GmbH (Publ. Co. Ltd.), 2004, ISBN 3-934447-8-2.
Teixeira, A.P. & Soares, C.G. (2009), Modelling and analysis of the availability of production systems by stochastic Petri nets (in Portuguese). *Riscos industriais e emergentes* 1: 469–488. Lisboa: Edições Salamandra, lda.
Turchi, C. (2014), 10 MW Supercritical CO_2 Turbine Tes, NREL Nonproprietary Final Report, DE-EE0001589.
Vandenbrande W.W. (1998), How to use FMEA to reduce the size of your quality toolbox. *Quality Progress,* vol. 31, pp. 97–100.
Zio, E. (2009), Reliability Engineering: Old Problems and New Challenges. *Reliability Engineering & System Safety,* vol. 94, pp. 125–141.
Zio, E., Baraldi, P., & Patelli, E. (2006), Assessment of the Availability of an Offshore Installation by Monte Carlo Simulation. *International Journal of Pressure Vessels and Piping,* vol. 83, pp. 312–320.
Zio, E., Marella, M., & Podofillini, L. (2007), A Monte Carlo Simulation Approach to the Availability Assessment of Multi-State Systems with Operational Dependencies. *Reliability Engineering & System Safety,* vol. 92, pp. 871–882.

Progress in Maritime Technology and Engineering – Guedes Soares & Santos (Eds)
 ISBN 978-1-138-58539-3

Subsea water separation: A promising strategy for offshore field development

Y.X. Wang, C. Hong, J.K. Yang, S.F. Estefen & M.I. Lourenço
Ocean Engineering Department, COPPE, Federal University of Rio de Janeiro, Rio de Janeiro, Brazil

ABSTRACT: Subsea water separation dates back to Troll C pilot in the North Sea, in 2001, introduced to improve oil recovery for the brownfield. This paper presents a new conceptual design for a green field development considering subsea water separation. The deepwater target field locates 300 km from the coast in a water depth of more than 2000 m with limited field data provided. At this conceptual selection stage, several scenarios have been proposed for its economic development under low oil price. The concept presented in this paper combines FPSO application with subsea water removal and subsea boosting. Flow assurance issues, including hydraulics and thermal analysis, are discussed.

1 INTRODUCTION

Subsea water separation dates back to the Troll C pilot in the North Sea, in 2001 (Mikkelsen et al. 2005). It was initially introduced to improve oil recovery for brownfields as subsea water separation bears expectations such as lowering down back-pressure on wellheads and debottlenecking topside water processing capacity (Hendricks et al. 2016, Daigle et al. 2012, Alary et al. 2002). More recent installations of subsea three-phase separators in two brownfields, Tordis and Marlim, are both for the same reason (Gjerdseth et al. 2007, Orlowski et al. 2012). Up to now, the only commercial installation is still the Tordis separator, installed at a water depth of 210 m, in the year 2007, unlike subsea gas-liquid two-phase separation, which goes to deeper water, such as Pazflor, BC-10, and Perdido (Parshall et al. 2009, Gilyard et al. 2010).

All the commissioned subsea water separation projects are to solve the conventional and thorny problem encountered at late production life: excessive water production. Anikpo et al. (2015) used terminology Early SP and Late SP to illustrate whether subsea processing is considered at an early stage or it is proposed for mid or late life. In other words, Early SP is to have SP factored into the overall development life no matter SP installation is required in early, mid or late life, while Late SP considers modification of the field architecture in mid or late life only to accommodate SP installation to solve production problems. Anikpo et al. (2015) reported that, qualitatively, Early SP would bring about less total cost and better project scheduling.

The final development plan for a green field project usually goes through several stages. During concept selection, several scenarios shall be proposed and compared, seeking for economic and feasible solutions. This paper will present one of the subsea architecture concept design for a green deepwater field considering subsea water separation and subsea boosting, in an Early SP approach. The novel system not only considers SP in the field architecture but also shows its interaction with production sequence and production profile. Hydraulics and thermal analysis by software OLGA will also be presented. At last, a simple comparison with a conventional "FPSO + cluster manifold" system and some critical issues will be discussed.

2 BASIC INFORMATION OF THE TARGET OIL FIELD

The target field locates 300 km from the coast, in water depth between 2,100 m to 2,500 m. It covers an area of about 300 km^2. Only one well has been drilled in the area. Fluid samples have been analyzed, and crude composition is provided. However, core test or well test data is not available yet. According to estimated reserve and production life (27 years), the field is supposed to be developed with 66 production wells and 22 injection wells. Field boundary and water depth are illustrated in

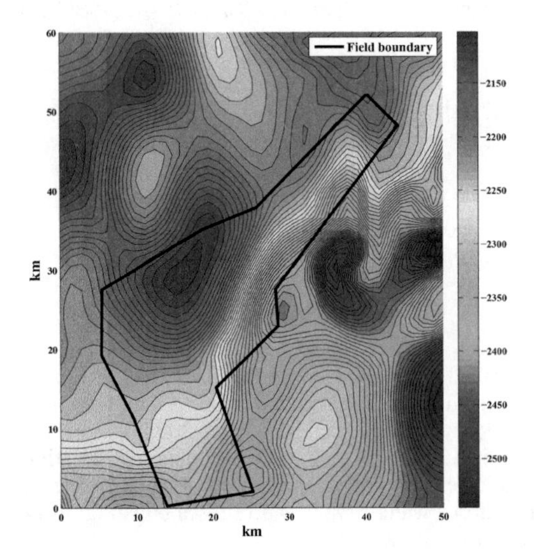

Figure 1. Field boundary and water depth contour.

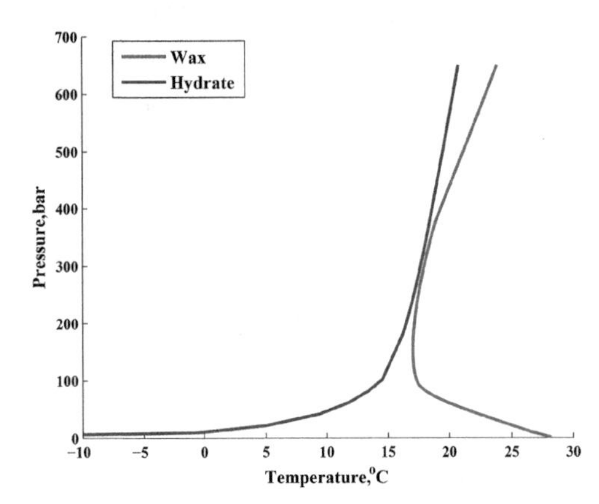

Figure 2. Hydrate and wax formation curve.

Figure 1. Hydrate and wax formation curves of the crude are shown in Figure 2.

3 CONCEPT DESIGN

A previous conceptual design was a conventional "FPSO + cluster manifold" system. The subsea layout of this system is shown in Figure 3 and Table 1. Each FPSO produces from three cluster manifolds, and each cluster manifold collects fluids from four or six production wells. Oil capacity and water capacity of each FPSO is supposed to be 150,000 bbl/day and 120,000 bbl/day, respectively. The production rate of each well is estimated to be 8,000 bbl/day. In the conventional design, due to limited water processing capacity of the FPSOs, each FPSO should produce from at most 18 production wells, or through three cluster manifolds. Thus, four FPSOs were planned to develop the field. However, with oil production decline, more and more water will be produced and treated on each FPSO in mid and late production life, which is expensive and unprofitable. If the bulk volume of water is separated subsea, each FPSO will be able to produce from more production wells. Therefore, less FPSOs will be needed. The conceptual subsea system which combines FPSO and subsea water separation is named hybrid system in this paper.

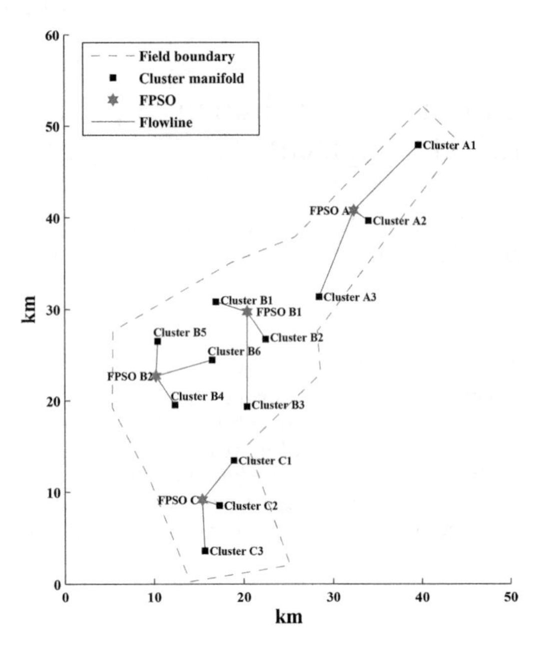

Figure 3. Subsea layout of the conventional system: 4 FPSOs combined with 12 clusters.

Table 1. Flowline length of the conventional FPSO system.

			Flowline length (m)	
FPSO	Manifold	Slots	Wellhead to manifold	Manifold to TDP
FPSO A	Cluster-A1	4	1000	9300
	Cluster-A2	4	1000	1100
	Cluster-A3	4	1000	9300
FPSO B1	Cluster-B1	6	1200	2800
	Cluster-B2	6	1200	2800
	Cluster-B3	6	1200	9500
FPSO B2	Cluster-B4	6	1200	2900
	Cluster-B5	6	1200	2900
	Cluster-B6	6	1200	5600
FPSO C	Cluster-C1	6	600	4700
	Cluster-C2	6	600	1100
	Cluster-C3	6	600	4700

3.1 *Subsea layout*

For the hybrid system, the amount and position of production wells and cluster manifolds are kept the same with the conventional system. Considering the large extension of the field area, the 12 production clusters are divided into two groups. Produced fluid from each group will be delivered to one corresponding FPSO. Group 1 includes 30 production wells, and Group 2 includes 36 production wells since each cluster manifold will connect four or six production wells as defined previously. Fluids from every two cluster manifolds will gather at and be separated by one subsea separator. The subsea separators are supposed to separate water from the hydrocarbon phases. After separation, oil and gas will be sent to the host FPSO and water will be reinjected subsea.

The six cluster manifolds in each group are divided into three sets based on their positions, and then the position of subsea separators and FPSOs are determined in such a way that the total length of subsea flowlines is minimized. The subsea layout for the hybrid system is shown in Figure 4.

3.2 *Production sequence*

This section takes Group 2 as an example to illustrate the production succession strategy. At the beginning of development, 3 cluster manifolds, producing from 18 production wells, will be connected to FPSO 2 since the oil processing capacity on the FPSO is assumed to be 150,000 bbl/day. At this production phase, little water production is expected, and oil production rate on the FPSO is 144,000 bbl/day. As output goes on, water will breakthrough in production wells. If all the production wells produce at a constant liquid rate, oil production declines as water cut increases. Therefore, when water cut rises to 30%, oil production rate on the FPSO will be 100,800 bbl/day. A new cluster will be on stream to deliver 48,000 bbl/day more oil to the FPSO. Therefore, at this production phase, the FPSO will receive oil from 4 cluster manifolds with oil production rate of 148,800 bbl/day, which is within the oil capacity of one FPSO. On the other hand, 73,200 bbl of water will be produced daily. This amount of water is supposed to be separated from the hydrocarbon phase in the subsea separators and be reinjected into the reservoir for pressure maintenance.

With further production, water cut in each well will increase continuously. Once the oil production rate on the FPSO declines to around 100,000 bbl/day, another cluster can be brought on stream to compensate oil production decline. Therefore, the other two untouched clusters will be brought on stream one by one in later production years until all the six clusters in the group are in production.

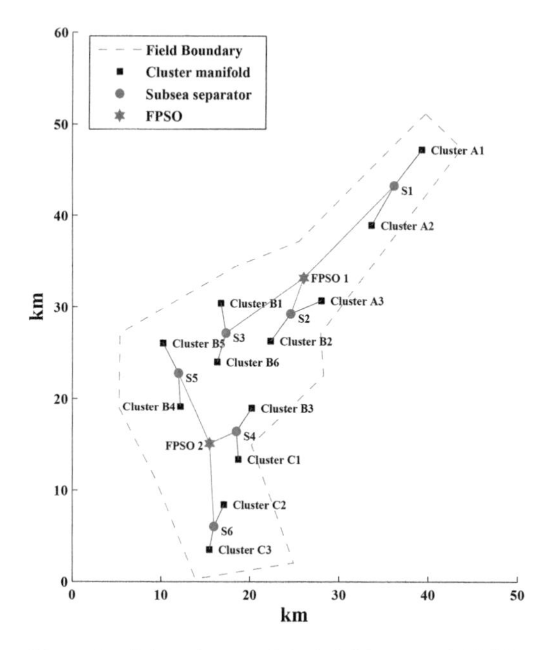

Figure 4. Subsea layout of the hybrid system: 2 FPSOs combined with 6 subsea separators and 12 clusters.

Table 2. Flowline length of the hybrid system.

		Flowline length (m)		
Separator	Manifold	Well to manifold	Manifold to separator	Separator to TDP
S1	Cluster-A1	1000	5010	14330
	Cluster-A2	1000	5010	
S2	Cluster-A3	1000	3750	4170
	Cluster-B2	1200	3750	
S3	Cluster-B1	1200	3300	10590
	Cluster-B6	1200	3300	
S4	Cluster-B3	1200	3060	3310
	Cluster-C1	600	3060	
S5	Cluster-B5	1200	3670	8470
	Cluster-B4	1200	3670	
S6	Cluster-C2	600	2620	9040
	Cluster-C3	600	2620	

3.3 *Production profile*

Due to the lack of field data, detailed reservoir engineering analysis and simulation were unavailable at this conceptual design stage. Therefore, several assumptions have been made for production profile prediction and drilling schedule.

1. It is assumed that only one FPSO should be commissioned every year. Thus, FPSO 1 is on stream in the first year and FPSO 2 in the second year.
2. Three clusters are commissioned in the same year when an FPSO is commissioned. That is, in the first year, 18 production wells should be drilled.
3. All the production wells follow the same decline rule as defined in conventional "FPSO + cluster manifold" system, regardless of its on-stream year set in the hybrid system. It is a quite rough assumption only for the conceptual design stage. If more detailed information of the reservoir is available, adjustments and improvements will be made.
4. Newly drilled production wells of the same cluster are commissioned at the same time. That is, each cluster is brought on stream as a unit and production wells are not added to the system one by one.
5. Clusters are brought on stream in sequence based on their distance to the corresponding host FPSO.

Table 3. Production sequence design of the hybrid system.

Group	Separator	Manifold	On-stream year
Group 1	S1	Cluster-A1	5
		Cluster-A2	7
	S2	Cluster-A3	1
		Cluster-B2	1
	S3	Cluster-B1	1
		Cluster-B6	4
Group 2	S4	Cluster-B3	2
		Cluster-C1	2
	S5	Cluster-B5	2
		Cluster-B4	5
	S6	Cluster-C2	8
		Cluster-C3	11

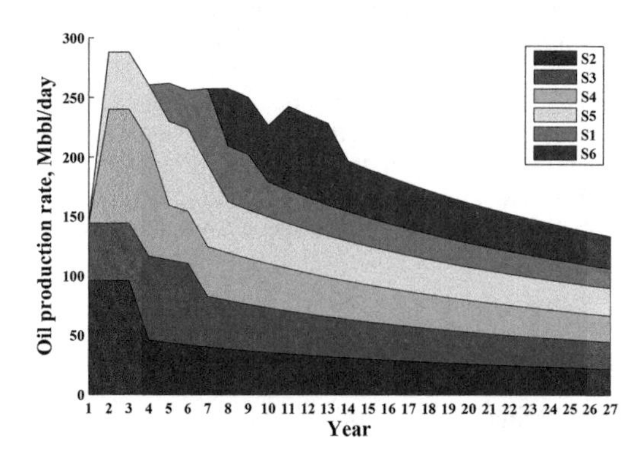

Figure 5. Production profile.

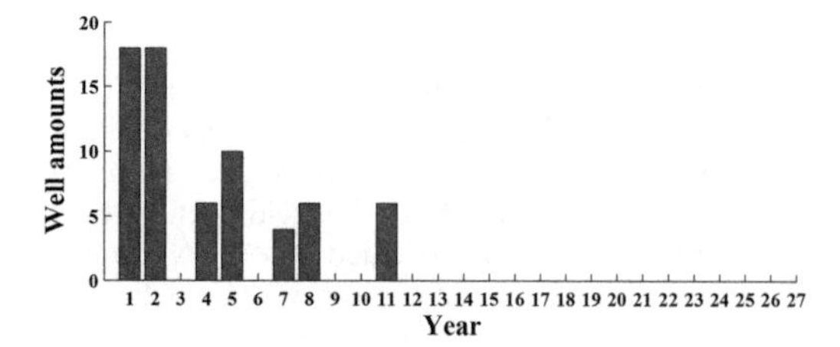

Figure 6. Drilling plan.

Based on the above assumptions, several production profiles were generated applying different decline rate. An optimistic result is chosen for further analysis following the largest estimated recovery factor. The corresponding production sequence is listed in Table 3. Predicted optimistic production profile and drilling plan are shown in Figure 5 and Figure 6, respectively.

4 FLOW ASSURANCE

The hybrid system consists of two sections: subsea separator to FPSO and production wells to a subsea separator, as illustrated in Figure 7. To simplify the simulation model, flow assurance analysis is carried out for each section separately under the same working condition set for subsea separators.

4.1 *Subsea separator to FPSO*

Ideally, all the six subsea separators are supposed to work under appropriate pressure and temperature conditions where only liquid phase exists (gas all dissolves in oil) and separation between oil and water happens. If not, gas should be separated from the liquid phase before the separation between oil and water. After separation, water will be reinjected subsea, and oil and gas will be transported to FPSOs. In flowlines connecting subsea separators and FPSOs, dissolved gas will gradually come out of the liquid phase as pressure drops.

An initial system pressure analysis by software OLGA showed that even though pressure drop between each production well and its corresponding subsea separator differs somewhat from one another due to different flowline lengths, the average value is around 30 bar. Also pressure analysis through flow in wells indicates that wellhead pressure will be around 280 bar at a production rate of 8,000 bbl/day. Therefore, the working pressure of the separators was estimated to be around 250 bar. At this pressure, liquid phase exists at a temperature below 68°C.

As shown in Table 2, the distance between each subsea separator and its host FPSO are different.

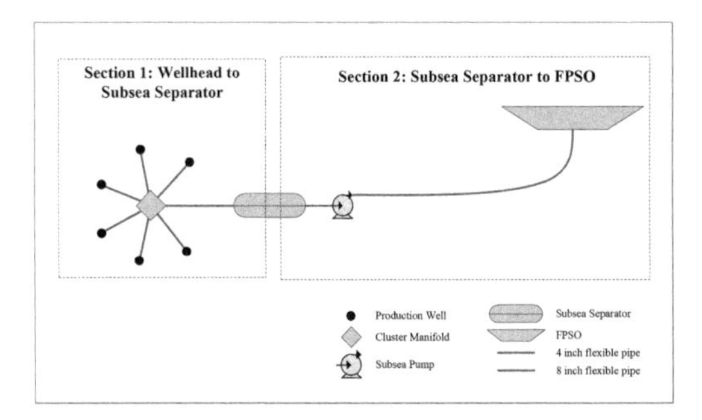

Figure 7. Sketch of the two sections for flow assurance analysis.

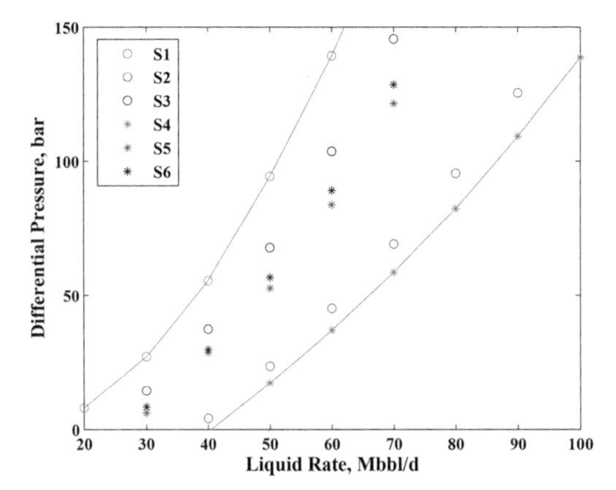

Figure 8. Required differential pressure of each subsea pump located downstream of the subsea separators.

Therefore, after separation, different pressure levels should be reached to boost the liquid to FPSOs in separate flowlines. This pressure difference will be provided by subsea pumps. Whats more, oil production rate will decline as production goes on, which will result in different pressure requirement at different production stages. Applying 8 inch flexible pipe between subsea separators and the FPSO, corresponding differential pressure at different flow rate is shown in Figure 8. The envelope shown in Figure 8 represents the variation of differential pressure along the whole production life, calculated by OLGA. For example, given that separator pressure is 250 bar, the maximum differential pressure of the subsea pump located downstream of subsea separator S3 is around 150 bar for boosting oil to the topside at the largest flow rate of 70,000 bbl/day. When the flow rate is below 50,000 bbl/day, which is the most possible along the whole production life, the necessary differential pressure is around 80 bar.

Thermal insulation keeps fluid flow along each flowline out of wax and hydrate formation region

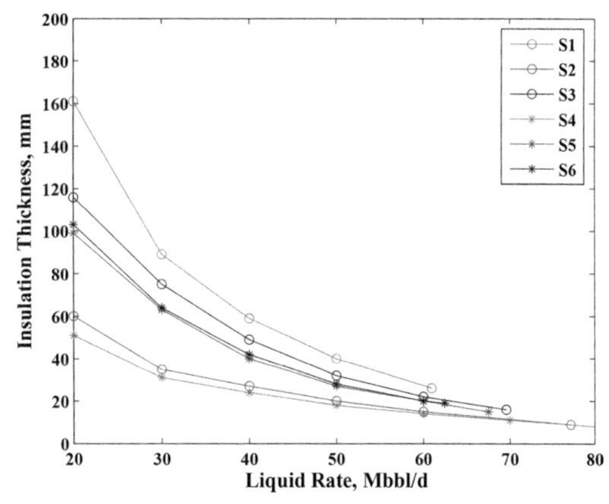

Figure 9. Minimum insulation thickness required at different flow rates.

(Figure 2). Minimum thermal insulation thickness for flowlines between each subsea separator and FPSO heavily depends on flowline length and minimum production rate. Lower production rate results in larger thermal loss along the flowlines and thus requiring thicker thermal insulation. As mentioned above, the ideal temperature at subsea separators should be less than 68°C. Therefore, temperature loss along the flowlines between each subsea separator and the FPSO should be less than 43°C given that fluid temperature reaching FPSO is around 25°C. Parametric studies were carried out setting fluid temperature at subsea separator equaling to 40°C, 45°C, 50°C, and 55°C. Minimum insulation thickness for separators working at 50°C is shown in Figure 9.

As an example, for subsea separator S1, the minimum insulation thickness at a flow rate of 60,000 bbl/day is only 30 mm. However, it increases to 160 mm at a flow rate of 20,000 bbl/day. Assuming that the thickness of insulation layer is limited to 30 mm, production at a flow rate lower than 60,000 bbl/day will drop into wax formation region and encounter a high risk of wax deposition and even blockage. Therefore, the minimum allowable flow rate in the flowline between S1 and FPSO with 30 mm external insulation is 60,000 bbl/day, which indicates that certain amount of water should be transported together with oil and gas up to the FPSO from flow assurance consideration. That is, when oil production rate is 20,000 bbl/day, around 40,000 bbl/day of water should be mixed with the oil.

After optimization, the insulation thickness for the flowlines between subsea separators (except S1) and FPSO were defined to be 50 mm. Since S1 locates farthest to its host FPSO with lowest flow rate, thicker insulation should be applied for this flowline. Figure 10 illustrates the yield components

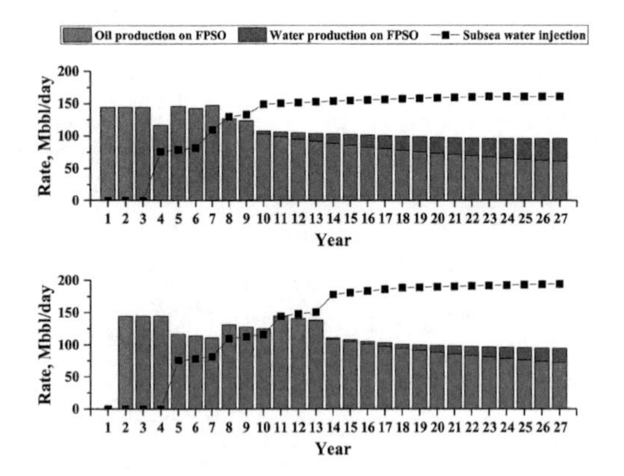

Figure 10. Yield components of each group: upper figure for Group 1 and lower figure for Group 2.

Table 4. Insulation design for flowlines between wellhead and subsea separators.

		Insulation thickness (mm)	
Separator	Manifold	Well to manifold	Manifold to separator
S1	Cluster-A1	40	20
	Cluster-A2	40	20
S2	Cluster-A3	30	10
	Cluster-B2	20	10
S3	Cluster-B1	20	5
	Cluster-B6	20	5
S4	Cluster-B3	15	5
	Cluster-C1	5	5
S5	Cluster-B5	20	5
	Cluster-B4	20	5
S6	Cluster-C2	10	0
	Cluster-C3	10	0

on each FPSO showing the amount of water that should be transported together with oil up to the host.

4.2 *Wellhead to subsea separator*

According to the analysis of flowlines between separators and FPSOs, it was recommended that fluid temperature reaching the subsea separators be 50°C. As can be seen in Figure 2, under condition 50°C and 250 bar, the crude stays in the hydrate and wax free region. Insulation for the flowlines between wellheads and subsea separators was to assure that fluid temperature at the subsea separators is around 50°C.

Major thermal properties of the insulation material used are: density 800 kg/m^3, thermal conductivity 0.18 W/(M·K), and heat capacity 800 J/(kg·K). OLGA was applied to calculate the flow process under different outside insulation thickness.

Six levels of thickness, 5 mm, 10 mm, 20 mm, 30 mm, 40 mm, 50 mm were simulated for both 4 inch and 8 inch pipes, and different thickness combination was tested. The best one had the minimum volume of insulation material. The insulation thickness design result is listed in Table 4.

5 COMPARISON BETWEEN THE CONVENTIONAL SYSTEM AND THE HYBRID SYSTEM

The design of the hybrid system aims at reducing the number of FPSOs used for developing the target field by separating the bulk volume of produced water on the seabed. Production on one FPSO for the conventional system and the hybrid system is shown in Figure 11. Oil production on FPSOs in the hybrid system remains at a relatively high level for a longer time. What is more, water production on the FPSO for the hybrid system will be substantially reduced and postponed.

Comparing the optimistic field production profile, as shown in Figure 12, peak production of the hybrid system reduces, but a slower decline is observed. In the optimistic estimation, the final recovery rate of the two systems is close. However, when well performance differs, no immutable

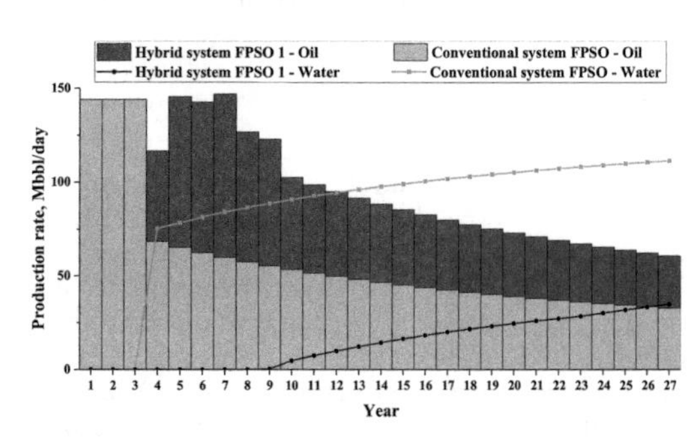

Figure 11. Oil and water production rate on an FPSO in the conventional system and the hybrid system.

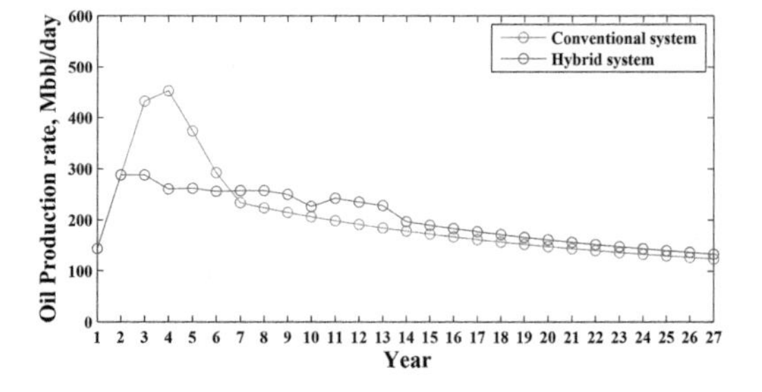

Figure 12. Field oil production rate.

conclusion on which system results in higher recovery factor can be drawn quickly unless precise reservoir description is informed. Qualitatively speaking, with earlier water breakthrough, a more rapid increment in water cut, and higher water cut at breakthrough, the hybrid system shows more potential. Nevertheless, cost reduction by reducing the number of FPSOs for development should be expected since leasing and operating FPSO is quite expensive.

6 FINAL REMARKS

The target field in this study is an offshore field with large extension, located far from the coast and in deep water. Several conceptual designs have been proposed for its economic development under low oil price. The hybrid system is a promising strategy as fewer and simpler topside facilities are required. However, several issues related to subsea technology shall be considered as point out below.

Capacity Requirement of Subsea Separators - Up to now, Tordis three phase subsea separator is still the largest one, which has a capacity of 50,000 bbl/day of oil and 100,000 bbl/day of water. The capacity requirement of the subsea separators proposed in the hybrid system is within these ranges. Therefore, similar vessels can be used or more compact separator design will further benefit installation and maintenance. As Tordis separator was installed at a water depth of 210 m, it is critical to validate the subsea technology down to water depth beyond 2000 m, which is the most important to turn the hybrid system design feasible.

Capacity Requirement of FPSOs - As most of the produced water shall be discarded on the seabed, a smaller water processing module on the FPSO is expected. For the field discussed in the paper, FPSOs with an oil capacity of 150,000 bbl/day and water capacity of 40,000 bbl/day are sufficient.

Water Reinjection - In water flood scheme, assuming that water injection begins immediately when production starts to maintain reservoir pressure at a high level and the ratio between injection and output equals to 1, the produced water separated at the seabed can serve as the water source for reinjection. A rough water injection design was also proposed. For both groups, the produced water can feed more than half of the water injection requirement while processed seawater injected from FPSO complements the rest. The injection capacity of an FPSO should be around 120,000 bbl/day.

Cost Estimation - Compared with the conventional system, the most significant improvement of the hybrid system is that four FPSOs with oil capacity of 150,000 bbl/day and water capacity of 120,000 bbl/day are replaced by two FPSOs with the same oil capacity but smaller water capacity (40,000 bbl/day) and six subsea separators. Cost estimation and reliability analysis are underway to verify the feasibility of the proposed hybrid system.

ACKNOWLEDGMENT

The authors acknowledge the financial support from Petrogal Brasil (ANP R&D Program) and EMBRAPII COPPE Unit in Subsea Technology for the development of the research project Subsea Systems (SISTSUB). Special thanks to Schlumberger for providing the software OLGA to the Federal University of Rio de Janeiro.

REFERENCES

Alary, V., J. Falcimaigne, et al. (2002). Subsea water separation: a cost-effective solution for ultra deep water production. In *17th World Petroleum Congress*. World Petroleum Congress.

Anikpo, A., F. Beltrami, et al. (2015). Subsea processing versus host selection: An imperative correlation. In *Offshore Technology Conference*. Offshore Technology Conference.

Daigle, T.P., S.N. Hantz, B. Phillips, R. Janjua, et al. (2012). Treating and releasing produced water at the ultra deepwater seabed. In *Offshore Technology Conference*. Offshore Technology Conference.

Gilyard, D.T., E.B. Brookbank, et al. (2010). The development of subsea boosting capabilities for deepwater perdido and bc 10 assets. In *SPE Annual Technical Conference and Exhibition*. Society of Petroleum Engineers.

Gjerdseth, A.C., A. Faanes, R. Ramberg, et al. (2007). The tordis ior project. In *Offshore technology conference*. Offshore Technology Conference.

Hendricks, R., L. McKenzie, O. Jahnsen, M. Storvik, Z. Hasan, et al. (2016). Subsea separation—an undervalued tool for increased oil recovery ior. In *SPE Asia Pacific Oil & Gas Conference and Exhibition*. Society of Petroleum Engineers.

Mikkelsen, J., T. Norheim, S. Sagatun, et al. (2005). The troll story. In *Offshore Technology Conference*. Offshore Technology Conference.

Orlowski, R., M.L.L. Euphemio, M.L. Euphemio, C.A. Andrade, F. Guedes, L.C. Tosta da Silva, R.G. Pestana, G. de Cerqueira, I. Lourenço, A. Pivari, et al. (2012). Marlim 3 phase subsea separation system-challenges and solutions for the subsea separation station to cope with process requirements. In *Offshore Technology Conference*. Offshore Technology Conference.

Parshall, J. et al. (2009). Pazflor project pushes technology frontier. *Journal of petroleum technology 61*(01), 40–44.

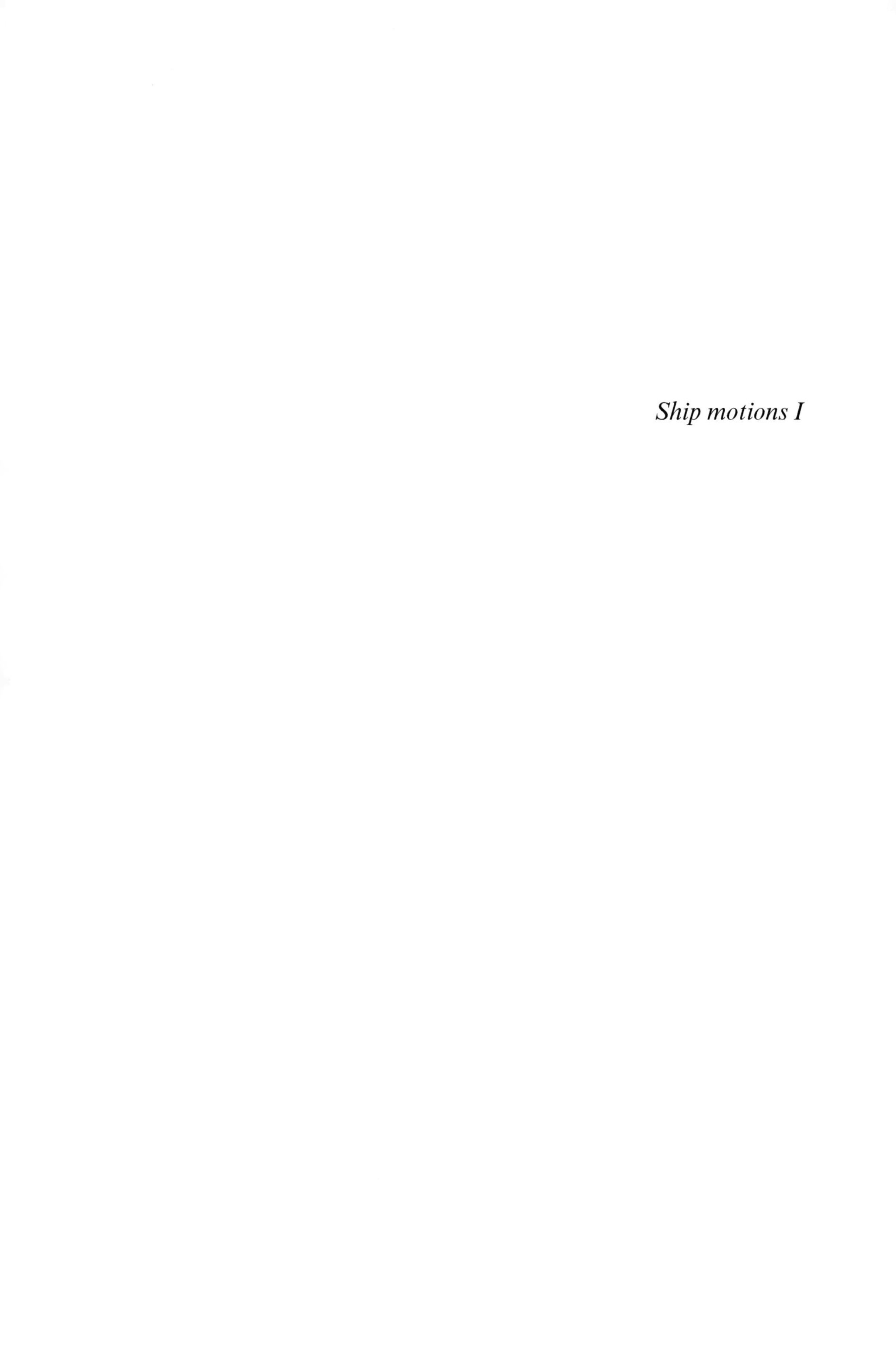

Ship motions I

Progress in Maritime Technology and Engineering – Guedes Soares & Santos (Eds)
© 2018 Taylor & Francis Group, London, ISBN 978-1-138-58539-3

Characterization of ship motions induced by wake waves

F.G.L. Pedro, L.V. Pinheiro & C.J.E.M. Fortes
Laboratório Nacional de Engenharia Civil (LNEC), Lisbon, Portugal

J.A. Santos
Instituto Superior de Engenharia de Lisboa, Instituto Politécnico de Lisboa, Lisbon, Portugal
Centre for Marine Technology and Ocean Engineering (CENTEC), Instituto Superior Técnico, Universidade de Lisboa, Lisbon, Portugal

M.A. Hinostroza
Centre for Marine Technology and Ocean Engineering (CENTEC), Instituto Superior Técnico, Universidade de Lisboa, Lisbon, Portugal

ABSTRACT: This paper analyses the hydrodynamic interaction between two ships, one stationary and the other navigating nearby, using physical and numerical modelling. The passing ship is a self-propelled scale model of the "Aurora" chemical tanker whereas the stationary ship is a scale model of the "Esso Osaka" oil tanker. The tests were carried out for several values of the "Aurora" advance velocity and for different water levels. The data obtained for the different test conditions was analyzed in what concerns both the characterization of the generated wake waves and induced movements of the "Esso Osaka". These motions were compared with results from the numerical package SWAMS, which simulates the movements of a ship (moored or free-floating) under different wave, wind, and current actions. The time series of the "Aurora" wake waves were used to define the incident waves in the SWAMS package. An analysis on the water depth influence on the ship movements is made, for both numerical and physical results.

1 INTRODUCTION

Froude (1877) identifies the existence of two wave systems in the wave field generated by ships sailing along a straight line: the transverse waves and the divergent waves. Equal phase lines of these two systems meet along two straight lines that are symmetrical with relation to the ship trajectory, the highest free-surface elevation values occurring along those lines.

The theoretical wave pattern generated by a point surface pressure disturbance that moves with a constant speed obtained by Lord Kelvin (1887) is similar to the ship-wake wave pattern. In that work, it is shown that the maximum of the free-surface elevation occurs along two straight lines symmetrical with relation to the point source trajectory making an angle of 38° 56' with each other (19° 28' with the point source trajectory).

To understand the physical properties of the waves generated by ships at critical and supercritical speeds, the Maritime and Coastguard Agency—MCA (1998) carried out experiments with different ship models in a towing tank. The models were towed at a range of constant speeds over several water depths, resulting in a wide range of wave heights and patterns which have been measured at depth limited Froude numbers from 0.8 to 2.6. The analysis of the decay rate of the maximum wave height and the depth influence on this decay, confirmed the Havelock's (1908) studies on the ship-wake waves potential decay with the distance from the disturbance.

This paper aims at comparing experimental results with numerical simulations using SWAMS package for moored ship behavior (Pinheiro et al., 2013), to simulate the behavior of ships subjected to ship-wake waves. For this use is made of scale model tests were a ship model sails with constant speed along a straight path at a constant distance from an otherwise motionless ship.

The scale-model tests were carried out at one of the wave tanks of the Portuguese Civil Engineering Laboratory (LNEC). The moving ship is a self-propelled scale model of the "Aurora" chemical ship whereas the otherwise motionless ship is a scale model of the "Esso Osaka" tanker.

After this introduction, the tests made with the moving "Aurora" ship model and the otherwise motionless "Esso Osaka" ship model are presented.

Then the main features of the numerical package SWAMS as well as its components are described. The results from the scale-model tests are used in the following chapter to assess the capabilities of the numerical models.

Port and Maritime Structures division of LNEC.

2 SCALE MODEL TESTS

2.1 *Equipment and measuring techniques*

The main objective of the tests was to characterize the wave field generated by the movement of the "Aurora" model ship as well as the motions induced by such wave field on the "Esso Osaka" model ship that would be otherwise motionless, near the "Aurora" trajectory.

The tests were carried out in a wave tank 23 m long and 22 m wide, Figure 1, of the Maritime Hydraulics Testing Hall of the Port and Maritime Structures division of LNEC.

The equipment used in the tests included, besides the "Aurora" and the "Esso Osaka" model ships, 7 resistive wave probes, 2 acoustic Doppler velocimeters (ADV), one gyroscope, one laser scan and one action camera.

The "Esso Osaka" model ship was deployed at x = 11.10 m, with its longitudinal axis perpendicular to the largest side of the tank. The 7 resistive wave probes and the 2 acoustic velocimeters were deployed as shown in Figure 1. The "Aurora" model ship made voyages along the smallest side of the tank with different velocities at an intended distance to the "Esso Osaka" of 0.59 m (this is the distance between the ship sides, 1.07 m between the ships longitudinal axes).

To measure the flow velocity associated to the ship-wake wave field, two acoustic Doppler velocimeters were used. They can measure the velocity components along each of the three axes (x, y and z) 6 cm below the undisturbed free surface level, with a sampling rate of 25 Hz. These velocimeters were deployed as shown in Figure 1, close to the wave gauges S2 and S4, Figure 2.

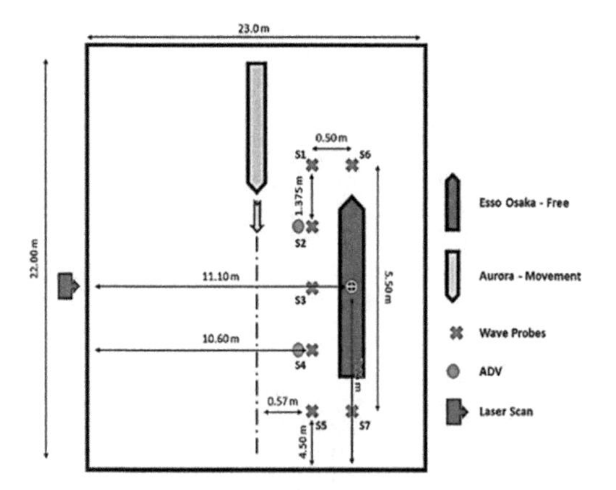

Figure 1. Physical model and equipment set-up.

The fiber-optic gyroscope installed at "Esso Osaka" can measure the ship's heading as well as the ship's accelerations along the surge, sway, heave, roll, pitch, and yaw movements. From these accelerations, the velocities along the same movements can be estimated. This equipment was deployed in the ship at a position as close as possible to the center of gravity of the ship, namely in what concerns the longitudinal and transversal axes. To manage the synchronization of the acquisition and storage of the gyroscope data, a set of LAB-VIEW routines was created (Hinostroza & Guedes Soares, 2016). The gyroscope data acquisition rate is adaptive, but a post-processing method enabled a 25 Hz data collection.

The laser scan was deployed outside the wave tank and its beam was perpendicular to the "Esso Osaka" side. The goal was not only to measure the distance between the "Esso Osaka" and the "Aurora" trajectory, but also to estimate the velocity

Figure 2. Top) "Esso Osaka" model ship and resistive wave gauges and velocimeters; Bottom) "Aurora" model ship.

of the "Aurora" when passing in front of the "Esso Osaka". The action camera was placed also outside the wave tank to film all the tests.

2.2 *Test conditions*

The "Esso Osaka" is a scale model at 1:100 of the hull of the homonym tanker. Its total length is 3.45 m, its beam is 0.54 m and its maximum draft is 0.23 m.

The "Aurora" is a scale model at 1:65.7 of the hull of the homonym chemical tanker. Its total length is 2.59 m, its beam is 0.43 m and its maximum draft is 0.11 m.

It must be pointed out that the goal of these experiments was not the simulation of the behavior of existing ships—hence the models used are not the reproduction at the same scale of real ships—but rather to characterize the ship wake wave field and the movements it induces in an otherwise motionless ship. Likewise, the mass distribution in the ship models does not aim to reproduce the inertia of real ships but rather the ship drafts leading to the selected Froude numbers and water depth/draft values.

The mass of the hull of the empty "Esso Osaka" (the hull alone) measured at a digital scale was 56.06 kg. The mass of the gyroscope installed at the "Esso Osaka", to measure the ship's linear and angular accelerations, is 5.44 kg. To simulate an intermediate loading condition, 8 cubic concrete blocks, with a total mass of 152.65 kg, were deployed inside the ship's hull. Then the total mass of the ship model (ship hull + concrete blocks + equipment) was 214.15 kg.

The "Aurora" is a self-propelled model. Its engine, rudder and gyroscope enable its navigation along a selected heading with a given advance speed. The propulsion of the "Aurora" is provided by a DC electric engine whose power is 22 W and enables a maximum speed of 2.0 ms^{-1} when the ships sails over a deep-water region with a draft of 0.11 m, (Perera *et al.* 2015). For the tested loading condition of the "Aurora" the mass of the ensemble ship + equipment + loading was 83.78 kg, which corresponds to a draft of 0.105 m and a displacement of 83.78×10^{-3} m^3.

Each test consisted of the following steps: a) Adjust the water level in the tank; b) Placing of blocks, weights and equipment on each ship; c) Connection and/or placement of equipment for measuring free surface elevation and flow velocity; d) Calibration of resistive probes; e) Voyage of the "Aurora" ship along the linear path with a pre-set engine power.

All journeys started with the "Aurora" at rest and its longitudinal axis 0.57 m away from the S1 to S5 wave gauges alignment and 4 m away from wave gauge S5. Having in mind the angle of the ship trajectory with the lines of maximum height for the ship generated waves (19° 28', meaning Fs below 0.7), to get meaningful measurements of ship generated waves in all wave gauges, the maximum distance between the ship and the wave-gauges alignment should not exceed 1.60 m. In the tests, three values for the fraction of the maximum engine power were considered (0.90, 0.80 and 0.70) and the same happened for the water depth values (0.820 m, 0.345 m and 0.255 m). For each test condition (fraction of maximum engine power and water depth) up to 10 repetitions were carried out. A total of 48 tests was carried out in this experiment.

The first water depth corresponds to a deep-water situation for both ships since it is larger than 4.2 times the draft of both ships. This means that the flow around the hull is not influenced by the bottom. The other two water depths correspond to an intermediate water depth situation. An example of a deep-water situation can be seen in Figure 1 with the free-surface elevation measured at wave gauge 3 during test T07 (engine power = 0.9 × maximum engine power, water depth = 0.82 m and distance between ships of 0.93 m) whereas an example of an intermediate water depth situation is presented in Figure 2 with the free-surface elevation measured at wave gauge 3 during test T15 (engine power = 0.8 × maximum engine power, water depth = 0.35 m and distance between ships of 0.34 m). In fact, the depression in this figure between instants t = 9 s and t = 12 s corresponds to the passage of the ship model in front of wave gauge S3 and is due to the pressure decrease associated to the flow between the keel and the tank bottom.

Ideally, the control devices of the "Aurora" ship model would imply almost identical repetitions. However, slight changes in the initial heading or in the ship model position make this almost impossible. Observed distances between the two ships (i.e. measured with the laser scan) ranged from 0.34 m up to 2.30 m. Such extreme values can be due not only to the "Aurora" deviation from the desired trajectory but also to the "Esso Osaka" drift caused by the ship-wake waves that reach it or to the wave tank oscillations, since the "Esso Osaka" was held in its position by springs with a very low stiffness.

"Aurora's" velocity is estimated from the quotient of its length to the time spent in front of the laser scan. But the real velocity of the "Aurora" model ship, when it passes in front of the "Esso Osaka" can hardly be considered constant. In fact, there are just a few tests whose model ship velocity obtained from the propagation of the ship-wake wave between consecutive wave probes matches the estimated velocity. Again, the deviation from the desired trajectory may be the culprit.

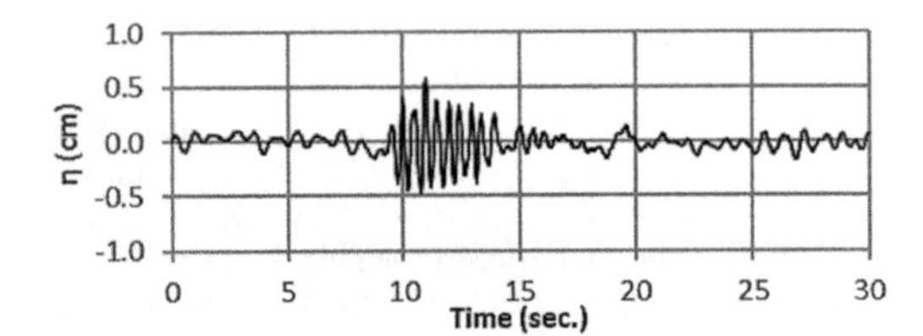

Figure 3. Time series of the free-surface elevation measured in probe S3 during test T07 (engine power = 0.9 × maximum engine power, water depth = 0.82 m and distance between ships of 0.93 m).

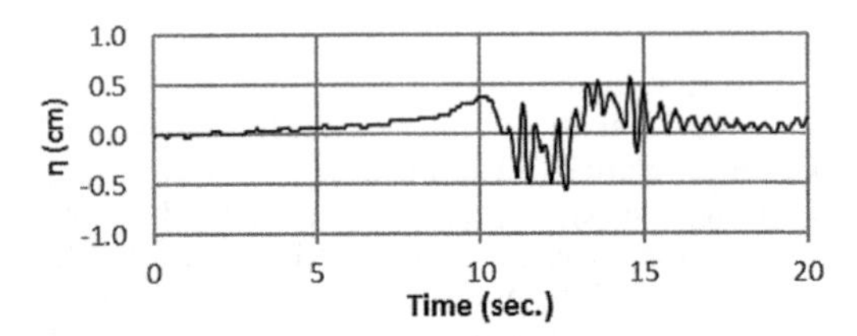

Figure 4. Time series of the free-surface elevation measured in probe S3 during test T15 (engine power = 0.8 × maximum engine power, water depth = 0.34 m and distance between ships of 0.34 m).

3 SHIP-WAVE INTERACTION

If the ship movements are small enough, the interaction between the free-floating ship and incident waves can be considered linear the. Consequently, one can decompose this interaction into two simpler problems, the so-called diffraction problem, where the ship is held motionless subjected to incident waves, and the so-called radiation problem where the ship moves in otherwise calm waters in such a way that the forces acting on the ship match the other forces on the ship (inertia and diffraction).

The motion equations along each generalized coordinate X_k can be written as

$$\sum_{j=1}^{6}\left[\left(M_{kj}+m_{kj}\right)X_j+\int_{-\infty}^{t}K_{kj}(t-\tau)X_j(\tau)d\tau + C_{kj}X_j\right]=F_k^D+F_k^{ext} \tag{1}$$

where M and C are respectively the mass and the hydrostatic restitution matrices of the ship, which depend only on the mass distribution in the ship and on the hull shape. m and K are respectively the added mass and the retardation function matrices. They are associated to the radiation problem, m_{kj} being the force along the generalized coordinate X_k due to unit acceleration of the ship along coordinate X_j ($X_i = 1$) and $K_{kj}(\tau)$ being the time series of the force along X_k due to an impulsive velocity along X_j at time $t = 0$ ($X_i(t) = \delta(0)$). F_k^D is the time series of the diffraction force along the X_k coordinate whereas F_k^{ext} represents the time series of the remaining external forces along the same coordinate.

The SWAMS numerical package, Pinheiro *et al.* (2013), is the result of coupling numerical models for sea-wave propagation with numerical models for moored ship behavior, thus enabling the identification of dangerous sea states and a better planning of port activities.

To simulate the behavior of moored ships it uses the BAS numerical model (Mynett *et al.* 1985) which assembles and solves equation (1) considering several sources for the external forces on the moored ship. It includes the constitutive relations for the mooring system elements (mooring lines and fenders) and the wind and current forces, as well as non-linear drift forces and damping not considered in the linear numerical models used to characterize the ship interaction with incident waves. The results of this numerical model are the time-series of the motions along each of the generalized coordinates and of the efforts in the mooring system elements. The key data for this model is the time series of the external forces. Should one have this time series for the ship-wake waves then it would be possible to get the motions of the moored (or of the free-floating) ship subjected to these waves.

3.1 *Frequency domain approach*

For a free-floating ship subjected to monochromatic waves of frequency ω, such that the free-surface elevation at one point can be given by $\eta(t)=\zeta_0\cos(\omega t)$, the diffraction forces can be written as $F_k^D(t)=\zeta_0 F_k^D(\omega)\cos(\omega t+\phi_k)$ and the motion of the ship along each generalized coordinate X_j will be also sinusoidal with frequency ω, $X_j=\zeta_j\cos(\omega t+\alpha_j)$. Then, the motion equations become

$$\begin{aligned}&\sum_{j=1}^{6}\{-\omega^2[(M_{kj}+a_{kj}(\omega)]\zeta_j\cos(\omega t+\alpha_j)\\&\quad+b_{kj}(\omega)\zeta_j\omega\sin(\omega t+\alpha_j)+C_{kj}\zeta_j\cos(\omega t+\alpha_j)\}\\&=\zeta_0F_k^D(\omega)\cos(\omega t+\phi_k(\omega))\end{aligned} \tag{2}$$

where $a_{kj}(\omega)$ and $b_{kj}(\omega)$ are the components of the radiation force, respectively, in phase with the motion acceleration and velocity, i.e. the so-called added mass coefficients

$$a_{kj}(\omega)=m_{kj}-\frac{1}{\omega}\int_0^\infty K_{kj}(\tau)\sin(\omega\tau)d\tau \tag{3}$$

and damping coefficients

$$b_{kj}(\omega) = \frac{1}{\omega}\int_0^{\infty} K_{kj}(\tau)\cos(\omega\tau)d\tau \quad (4)$$

This means that one can use the results from numerical models for ship-wave interaction in the frequency domain to get the added mass matrix, m, and the matrix of retardation functions, K.

The WAMIT model, Korsemeyer *et al.* (1988) is one of such models. It uses a panel method (i.e. the wetted hull surface is discretized with rectangular or triangular panels) to solve the integral equations for the strength of the dipoles distributed on each of those panels in the radiation and diffraction problems in the frequency domain. These are irrotational flows of ideal fluids, i.e. no viscosity is considered. From the solutions of those problems it is possible to compute the added mass and damping coefficients ($a_{kj}(\omega)$ and $b_{kj}(\omega)$) as well as the amplitude and phase shift of the diffraction forces ($F_k^D(\omega)$ and $\phi_k(\omega)$) for incident waves of unit amplitude ($\zeta_0 = 1$). The WAMIT model only computes diffraction forces due to long crested waves at some distance from the ship.

The number of frequencies considered does condition the quality of the estimates of the added mass and of the retardation functions matrices. The number of panels per wave length influences the solution of the radiation and diffraction problems in the frequency domain.

4 RESULTS AND DISCUSSION

The first set of comparisons between numerical and scale model results does not involve waves. It rather tries to fine tune the parameters corresponding to the masses distribution in the "Esso Osaka" ship model, especially the roll and pitch inertias and to investigate the need for including a viscous damping force in the time domain simulations. For this, use is made of the results of the "Esso Osaka" ship-model decay tests.

Such tests were carried out in still water conditions, for a water depth of 0.474 m and consisted on imposing an initial roll or pitch angle and measuring the time series of the resulting motions.

The first guess for the roll gyration radius, r_x, (distance from X-axis at which the mass of the ship may be assumed to be concentrated) is obtained assuming a uniform mass distribution on the ship's submerged hull, thus $r_x = 0.105\ m$. As expected, the resulting roll motion from BAS model does not agree with the measured roll motion in the decay test. Using for instance, an initial roll angle of −11.58°, Figure 5, the simulated roll decay period is smaller than the measured one, which implies that the roll gyration radius must be increased. By trial and error, one settles on a value of 0.1705 m for the roll gyration radius (i.e. an inertia value of 6.897×10^{-3} kg.m^2 around the X-axis).

It is worth pointing out that the wave generation by the roll motion is responsible for the decay observed in time series of the roll motion of Figure 5. Actually, the figure shows that the predicted roll decay fits quite well the measured one, hence there is no need to add a viscous roll damping factor (b_{4visc}),

$$F_4^{visc} = -b_{4visc}\dot{X}_4 \quad (5)$$

considered in the BAS model to account for the viscous torque associated to roll.

Using the same approach for the pitch decay test, one concludes that a gyration radius, $r_y = 0.7947\ m$ (i.e. an inertia value of 149.829×10^{-3} kg.m^2 around the Y-axis) provides a good fit between measured and numerical pitch decay natural period (see Figure 6). Again, there is no need to add a viscous pitch damping factor (b_{5visc}).

$$F_5^{visc} = -b_{5visc}\dot{X}_5\left|\dot{X}_5\right| \quad (6)$$

The numerical simulations with the BAS model were carried out with retardations functions obtained from the discrete Fourier transform of damping coefficients computed with the WAMIT model for frequencies between 0.20 Hz and 18.00 Hz, equally spaced of 0.20 Hz, using 3803 panels to discretize the submerged ship hull.

As a first approach, the ship-wake waves were considered long-crested and with the free-surface elevation measured at wave gauge S1. It was also assumed that the crests of those waves travelled

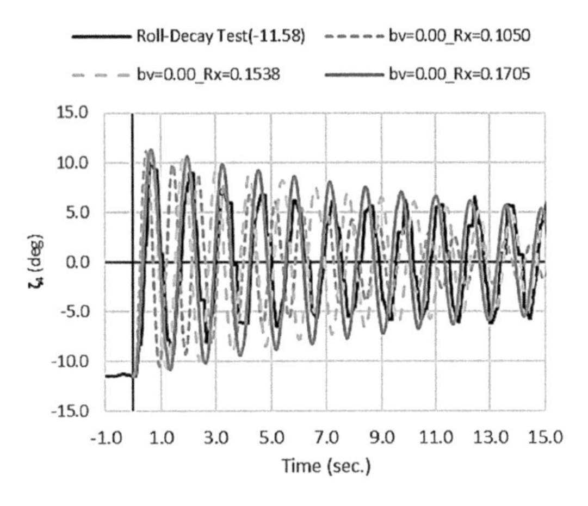

Figure 5. Time series of the "Esso Osaka" ship roll decay test. Initial roll angle of −11.58 degrees. Comparison between measurements and numerical model.

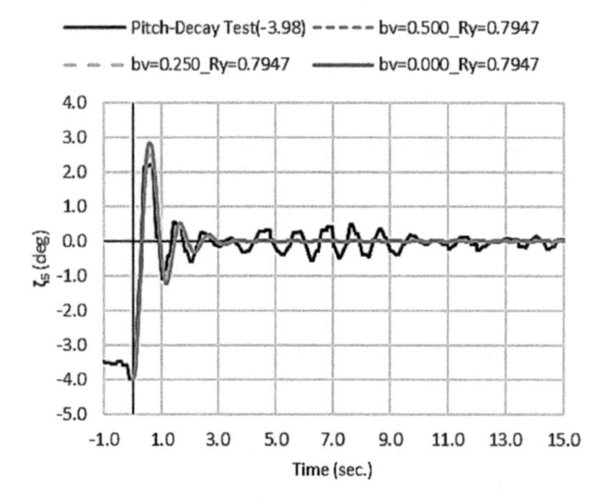

Figure 6. Time series of the "Esso Osaka" ship pitch decay test. Initial roll angle of –3.98 degrees. Comparison between measurements and numerical model.

with a 20° angle to the longitudinal axis of the "Esso Osaka" (i.e. 110° angle between the longitudinal axis of the "Esso Osaka" and the wave number vector).

Despite being a quite simple approach, it has the advantage of enabling the use of the response amplitude operators, RAOs, obtained directly from the frequency-domain WAMIT results

$$RAO_k(\omega) = \frac{\zeta_k(\omega)}{\zeta_0(\omega)} \tag{7}$$

$RAO_k(\omega)$ is a quotient between the amplitude of the response, ζ_k, and the amplitude of the excitation, ζ_0, for a given frequency, ω. They are computed in WAMIT for a long-crested wave that makes a given angle with the local x-axis of the ship (here the "Esso Osaka" ship model). This means that, as long as the angle between the incident wave and the ship does not change, the ratio between the amplitude of the response components—which can be obtained from the Fourier transform of the response time series—and the amplitude of the free-surface elevation components—which can be obtained from the Fourier transform of the free-surface elevation time series—should fall on the curve for the response amplitude operator of the free-floating ship for that incidence angle and water depth.

Due to the problems encountered in the scale model tests, with the repeatability of the "Aurora" voyages, Pedro *et al.* (2017), only a selected subset of the 48 tests carried out is studied in this paper. Table 1 presents the selected subset of 9 tests, that includes three repetitions of each test for a given water depth (0.82 m, 0.34 m and 0.255 m.). In all tests, the engine power, P_E, was set to 90% of the maximum engine power.

Figure 7 and Figure 8 present the RAOs for roll and pitch motions, respectively, for three water depths. The figures show that for low-frequency components, as the water depth decreases there is an increase in the response amplitude. The opposite happens for high-frequency components.

Table 1. Selected tests for analysis. $P_E = 90\%\ P_{Emax}$.

Test	Water depth (m)	Passing distance (m)	Advance speed (m/s)
T03	0.82	1.04	0.83
T08	0.82	0.93	0.86
T09	0.82	1.16	0.83
T12	0.34	1.00	0.83
T24	0.34	0.86	0.87
T27	0.34	0.96	0.87
T34	0.255	1.01	0.85
T35	0.255	0.85	0.88
T46	0.255	1.00	0.82

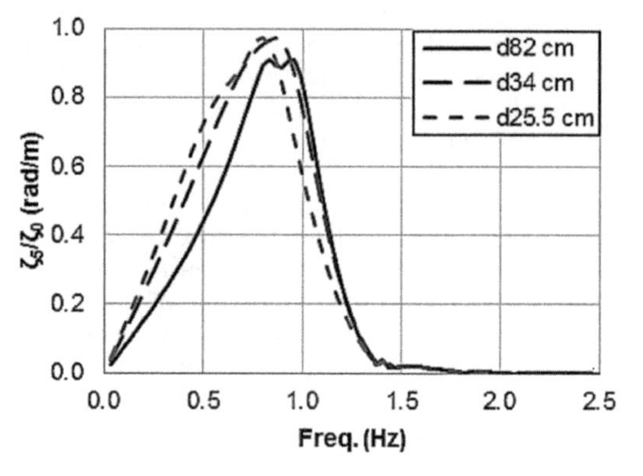

Figure 7. WAMIT's roll motion RAOs.

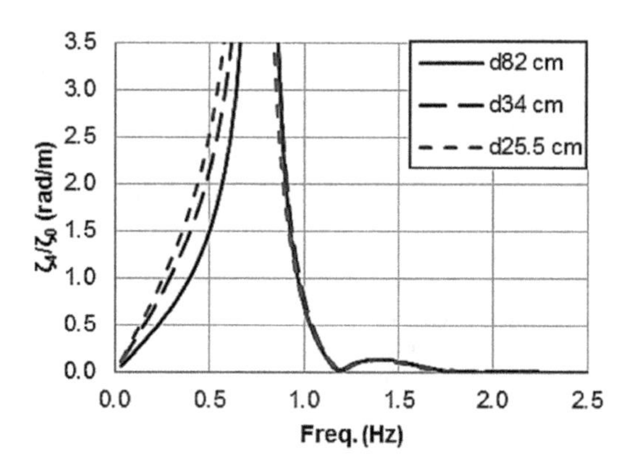

Figure 8. WAMIT's pitch motion RAOs.

Figure 9, Figure 10 and Figure 11 present the amplitude spectra and the time series of the measured free-surface elevation (gauge S1), roll and pitch movements, respectively, for test T08.

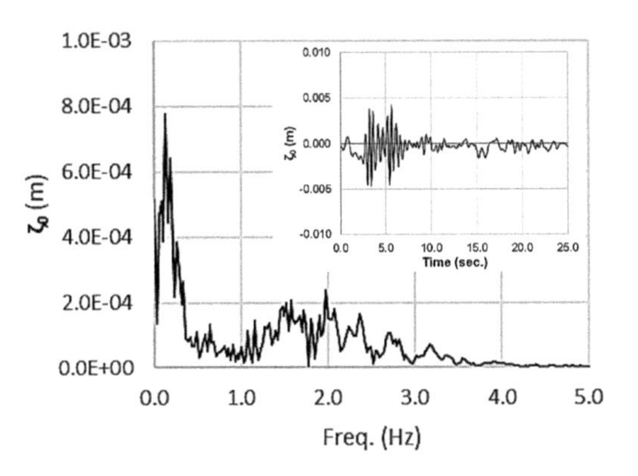

Figure 9. Amplitude spectrum and time series of free-surface elevation measured at wave gauge S1. Test T08.

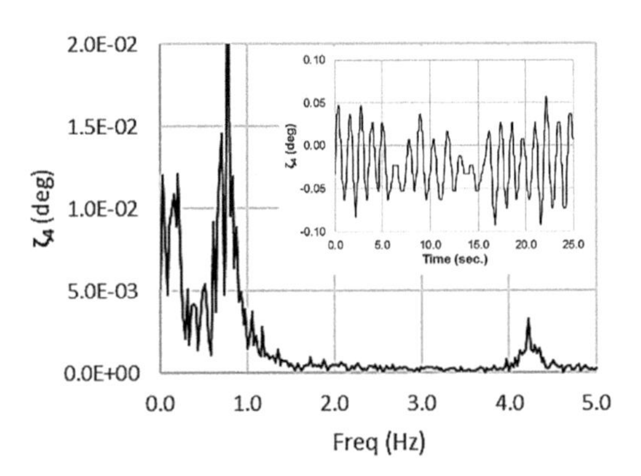

Figure 10. Amplitude spectrum and time series of roll motion. Test T08.

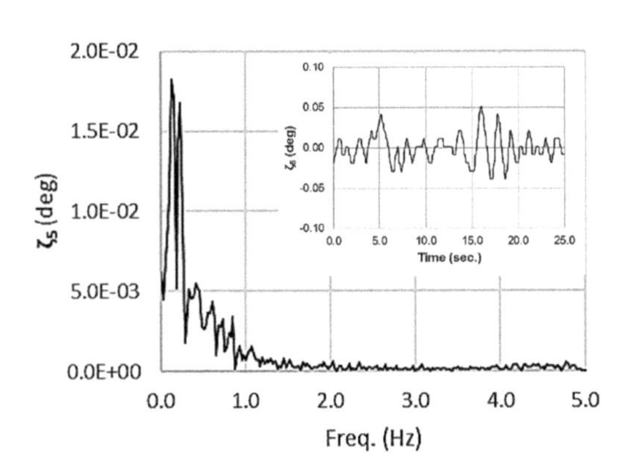

Figure 11. Amplitude spectrum and time series of pitch motion measured. Test T08.

Only the 10 largest values of the free-surface elevation amplitude spectrum were used to compute the RAOs, to avoid extremely small values of elevation amplifying RAOs, that may not correspond to actual physical phenomena.

Figure 12 to Figure 14 present the values of RAOs for the roll motion estimated with this

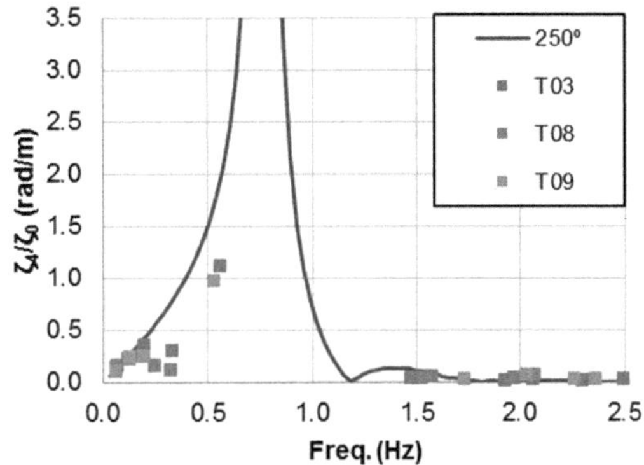

Figure 12. RAOs for roll motion, (h = 82 cm).

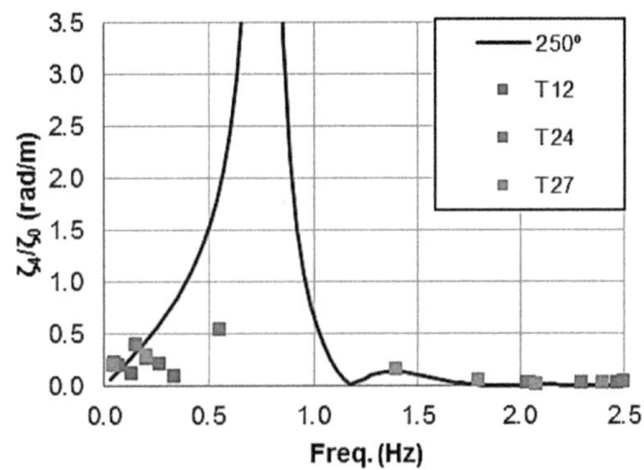

Figure 13. RAOs for roll motion, (h = 34 cm).

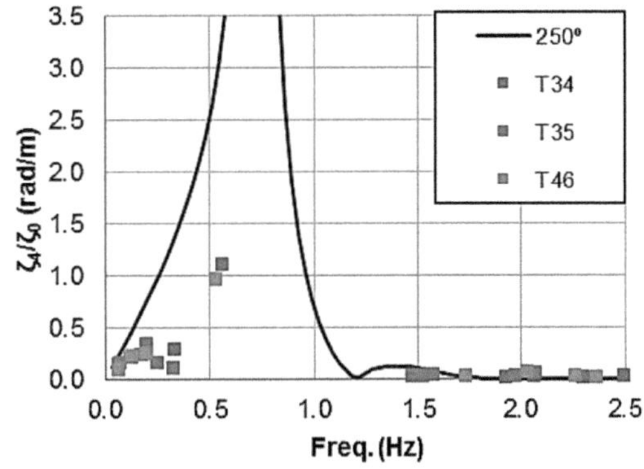

Figure 14. RAOs for roll motion, (h = 25.5 cm).

procedure, for the three water depths, whereas Figure 15 to Figure 17 present the corresponding RAOs for the pitch motion.

Roll motion is highly amplified around 0.75 Hz. However, the wake waves generated in the presented tests all had higher frequencies, for which the resulting RAO's are very small and, as can be seen in Figures 12 to 14, there is a very good agreement between physical tests and numerical ones. In addition to that, a significant part of the waves energy spectrum is in the low-frequency range, denoting the existence of, maybe, free-standing tank related long waves. The RAO's for these long waves were also well captured by the numerical model. The increase of RAO magnitude around 0.6 Hz is also clearly present in the physical tests measurements.

Regarding the pitch RAOs, there seems to be an underestimation of the experimental results given by the numerical model. The tests repetitions display a large dispersion of RAO values, evidencing the difficulty of reproducing the test characteristics. Mainly this is due to the difficulty of maintaining the ship on the desired path. Nevertheless, the biggest differences occur in the low-frequency range. In the frequency range of the wake waves (1.5 Hz to 2.5 Hz) the results are quite similar.

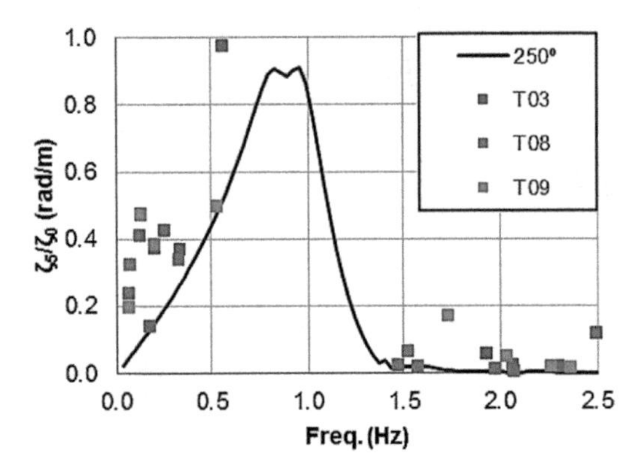

Figure 15. RAOs for pitch motion, (h = 82 cm).

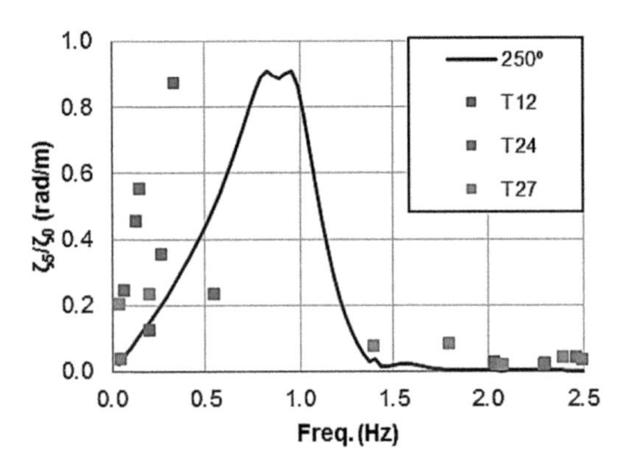

Figure 16. RAOs for pitch motion, (h = 34 cm).

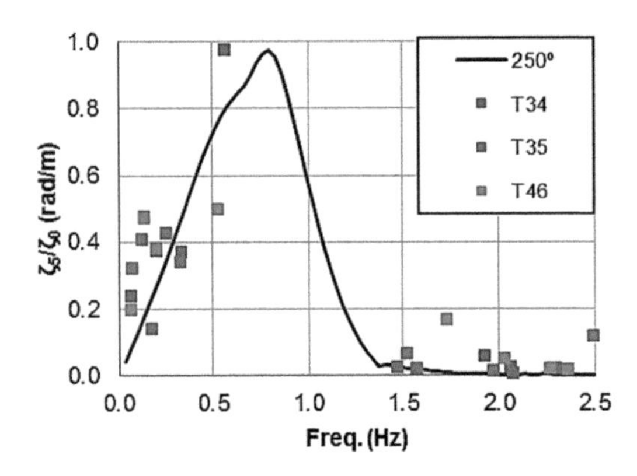

Figure 17. RAOs for pitch motion, (h = 25.5 cm).

5 CONCLUSIONS

The hydrodynamic interaction between two ships, one stationary and the other navigating, was analyzed using physical and numerical modelling.

The obtained data was analyzed in what concerns both the characterization of the generated wake waves and the induced movements of the still ship, using the response amplitude operators. All the analyzed tests in this paper had an advance velocity of about 0.85 m/s. Different water depths were simulated.

An analysis on the water depth influence on the ship movements showed that, for low-frequency components, as the water depth decreases there is an increase in the response amplitude, and for high-frequency components, as the water depth decreases the response amplitude also decreases, although, RAOs for high-frequencies are very small. This is due to the low values of the wave length to ship length ratio. As the otherwise motionless ship is a large crude carrier, small length waves, such as ship wake waves have very little impact. On the contrary, long waves can produce significant movements on that ship. Although, not intentionally, long waves were present in the physical tank, and therefore their influence on the ship was noticeable. The comparison of the RAOs on the low-frequency spectrum range showed that the RAO's were also well reproduced by the numerical model.

ACKNOWLEDGEMENTS

This work is a contribution to the M&MSHIPS project "Maneuvering & Moored SHIPS in ports"

(PTDC/EMSTRA/5628/2014) funded by the Portuguese Foundation for Science and Technology.

REFERENCES

Froude, W. 1877. Experiments upon the effect produced on the wave-making resistance of ships by length of parallel middle body. *Trans. Inst. of Naval Architects*, vol.18, pp.77–87.

Havelock, T.H. 1908. "The propagation of groups of waves in dispersive media, with application to waves on water produced by a travelling disturbance". *Proc. Royal Society A*, vol.81, n.°549, pp. 398–430.

Hinostroza, M.A. and Guedes Soares, C. 2016. "Parametric estimation of the directional wave spectrum from ship motions". *Int. J. Marit. Eng*, vol. 158, pp. A 121-A-130.

Korsemeyer F.T., Lee C.-H., Newman J.N. and Sclavounos P.D. 1988. The analysis of wave effects on tension-leg platforms, *7th International Conference on Offshore Mechanics and Arctic Engineering*, Houston, Texas, pp. 1–14.

Lord Kelvin (Sir William Thomson). 1887. "On ship waves". *Proc. Inst. of Mech. Engineers*, pp. 409–433.

Maritime Coastguard Agency (MCA) 1998. *Research Project 420 – Investigation of High-speed Craft on Routes near to Land or Enclosed Estuaries. Final Report.*

Mynett, A.E., Keunig, P.J. e Vis, F.C. 1985. The dynamic behaviour of moored ships inside a harbour configuration. *Int. Conf. on Numerical Modelling of Ports and Harbours*, Birmingham, England: 23–25 April 1985. Cranfield: BHRA, The Fluid Engineering Centre.

Pedro, F., Santos, J.A., Hinostroza, M., Pinheiro, L., Fortes, C.J.E.M. (2017). Experimental Characterization of Ship Motions Induced by Passing Ships. *International Short Course and Conference on Applied Coastal Research* (SCACR2017), IH Cantabria, Santander, Spain.

Perera, L.P., Ferrari, V., Santos, F.P., Hinostroza, M.A. and Guedes Soares, C. 2015. "Experimental evaluations on ship autonomous navigation and collision avoidance by intelligent guidance". *IEEE Journal of Oceanic Engineering*, 40(2), 374–387.

Pinheiro, L.V.; Fortes, C.J.E.M.; Santos, J.A.; Fernandes, J.L.M. Numerical Simulation of The Behaviour of a Moored Ship Inside an Open Coast Harbour. 2013. *V International Conference on Computational Methods in Marine Engineering MARINE 2013*. B. Brinkmann and P. Wriggers (Eds).

Progress in Maritime Technology and Engineering – Guedes Soares & Santos (Eds)
© 2018 Taylor & Francis Group, London, ISBN 978-1-138-58539-3

Motions and mooring loads of a tanker moored at open jetty in long crested irregular waves including second order effects

H.S. Abdelwahab & C. Guedes Soares
Centre for Marine Technology and Ocean Engineering (CENTEC), Instituto Superior Técnico, Universidade de Lisboa, Lisbon, Portugal

ABSTRACT: The motions of a moored vessel and the loads on the mooring lines and fenders are evaluated in frequency domain in an accurate and simple way. The linearized stiffness matrix for the mooring system including pre-tension and coupling is calculated using an existing numerical code based on the differential linearization. A Green's function panel method is used for the hydrodynamic calculations. The second order calculations are performed with an approximate method where the second order velocity potential is calculated without considering the forcing function on the free surface boundary condition. Results for the motions and mooring forces in long crested irregular waves are compared with experimental data available in the literature for a loaded 200 kDWT tanker, moored to an open jetty in water with a depth amounting to 1.2 times the draft of the vessel. The results show a good agreement with the experimental data and show that the second order low frequency loads should not be neglected when calculating the loads on the mooring lines and fenders in irregular waves.

1 INTRODUCTION

Each vessel will have mooring arrangements such that ropes and wires of recommended strength that can help her to be moored safely alongside a berth, floating platform, buoy or a jetty. Fundamentally a vessel has to be positioned alongside a jetty or a berth and the mooring system must resist forces and moments due to some, or possibly all, of the external exciting effects such as wind and current, passing ship and effects of waves, swells, and seiches (Schellin & Östergaard 1995). The mooring system must be designed to keep the resultant displacement of the ship within the required operational limits and to keep the loads on the mooring lines at an acceptable level. The forces on the moored ship have impacts on cargo handling operations, mooring and fender systems and safety on board. For berthed tankers, the cargo handling manifolds are the limiting factor for safe operation, it allows 3 m in surge and sway (PIANC 1995).

Large tankers and LNG carriers tend to be moored to cargo-handling jetties built close to deep water to keep away from the high costs of dredging navigation channels. As a result, mooring locations for these ships are more exposed to irregular wave action other than the other effects.

Floating moored vessels in irregular waves are subjected to large first order wave forces and moments, which are linearly proportional to the wave height and cause of the well-known first order motions with the same frequency of the exciting wave. In case of ships, moored by means of a linear system in random seas, it is likely that the second order drift force is always important, since it is then the only low-frequency excitation. These second order forces can be decomposed into three different terms. The first component is the steady drift force, which is a mean value that is frequency dependent, but time independent. The other components are due to the summation or subtraction of the single frequency components of the sea state. The sum of frequencies component is out of the scope of the current study as it may be important for cable or hull vibrations.

When the steady and low frequency forces are acting on a moored floating body, the steady drift force will result in an offset of the average position of the body and the difference frequency component will induce slowly varying motions. These forces are proportional to the square of the wave height and they could excite undesirable large amplitude horizontal motions, as they contain components that may fall within the range of the natural frequencies of the horizontal motion of the moored vessel. As the damping of low frequency horizontal motions of moored structures is generally low, this result in a large amplitude resonant response that may exceed the permissible limits of safe operation and cause a breakage of the mooring lines or structural damage.

The calculation of mean drift forces can be carried out using three methods, the so-called far-field solution, pressure integration solution and middle field method. The far-field method is based on the momentum assessment and has been applied by Maruo (1960) to represent the horizontal drift forces and by Newman (1967) to derive the horizontal drift moment. This momentum method is limited for horizontal drift forces. The pressure integration approach introduced by Pinkster (1980) is based on direct integration of the pressure over the wetted surface correctly up to the second order on the wave amplitude. All six degrees of freedom components of the drift forces can be computed, and the difference and sum frequency second order components can also be computed. A middle-field formulation was more recently proposed by Chen (2005) based on the integration of the momentum flux on a control surface located which surrounds the floating body, which allows all six components of the steady drift forces to be computed.

In the past, the only way to obtain accurate results, for the motions of a moored ship and mooring loads, was to perform expensive scale model tests. During the last decades, numerical methods became available, brought about by powerful computers and progress in the development of mathematical models. Van Oortmerssen (1976) was one of the first to develop a model (DIFFRAC), capable of analysing the behaviour of a moored tanker. An advanced model has been presented by Van der Molen (2006) based on a combination of a Boussinesq-type wave model with a time domain panel model to determine the wave forces on a moored ship in a complex geometry including the contributions of second-order waves and harbour oscillations.

The present study addresses the problem of first and second order to calculate motions and mooring loads for a moored vessel in the frequency domain. It was proposed by Chen (1994) and later by Pessoa et al. (2013) that the free surface forcing can be omitted from the boundary conditions in cases resulting in low frequency oscillations.

Today, the validity of the application of superposition to ship motion and sea loads is generally accepted in the field, so that the frequency domain method presented herein is useful since, it provides an efficient and fast solution.

The second order difference frequency responses to irregular wave excitations are calculated with an inverse Fourier transform based methodology suggested by Pessoa et al. (2016). This methodology is an extension to a scheme suggested by Duncan & Drake (1995) for calculating the sum frequency component of the second order wave elevation time series. With the extended methodology, the low frequency wave exciting loads spectra and the motions response spectra can be calculated. The main advantage of this approach is that the computational burden is significantly reduced. The downside is that the exciting forces need to be linearized, which is acceptable with the assumption of small amplitude of motions.

The tension to which a line is subjected is highly dependent on the position where the line is attached to the vessel and on the motion of the moored ship. This tension is also dependent on the stiffness of the line itself. So, it is very important to take in consideration all the physical phenomena possible to affect the results of motion. For frequency domain calculations, it is necessary to linearize the mooring systems. Pessoa (2013) introduced a formulation to calculate the stiffness matrix for arbitrary side-by-side mooring/fender systems that includes all the coupling terms and pretension effects. This formulation is applied in this study to calculate the 6×6 stiffness matrix for mooring system for a ship moored at open jetty.

In this paper, the hydrodynamic problem is carried out with WAMIT® V6.1S, which calculates linear and second order responses. The mooring forces are represented by a 6×6 stiffness matrix obtained with the implemented method by Pessoa (2013). The problem is linearized in order to be solved in the frequency domain. The boundary value problem is solved using Green's functions to derive the integral equation. The wetted surface needs to be represented by panels to find the numerical solution of the first order boundary value problem, but to obtain the complete second order solution it is required that the free surface close around the body is also represented by panels. For the current study it is sufficient to use Chen's approximation to solve the problem for only the low frequency second order quantities. The theoretical background, the implemented numerical methods and some computed results can be found in Lee (1995).

2 THEORETICAL BACKGROUND

The theory for the numerical calculation of linear and second order wave exciting forces on floating or fixed bodies has been thoroughly presented and discussed by many authors and is widely available in the literature. It can be found for instance in Pinkster (1980) or Lee (1995). In this section, only a brief explanation of the formulation of the first order and the second order hydrodynamic problem of the floating body is made. The methodology for calculating the linear and second order motions or mooring loads in irregular seas in the frequency domain will also be briefly discussed. Finally, the method for obtaining the 6×6 linearized stiffness matrix for the mooring system is discussed.

A boundary value problem is formulated for the analysis of the hydrodynamic forces induced by the waves on the body. The fluid is assumed to be inviscid, incompressible and the flow is irrotational, so that the flow can be described by a velocity potential within the domain enclosed by the boundaries defined for the problem. These include the wetted body surface, the free surface, vertical walls far away from the body and the sea bottom. To solve the problem, one must include conditions on these boundaries, and to verify the Laplace equation in the whole domain.

The solution can be obtained in three stages; first the linear boundary value problem is solved to obtain the first order velocity potentials, forces and motions. Then the first order results are used as input to solve the second order problem to get the second order motions and drift forces and moments. Finally, the linear response spectrum and slowly response spectrum for the motions and mooring loads can be calculated in long crested irregular wave. Also, the modelling of the mooring system is an additional challenge to solve the problem in accurate and simple way. A brief presentation of these stages is given in the following paragraphs.

2.1 *Linear hydrodynamic problem*

Incident waves are considered and within a linear approach all time dependent quantities change harmonically in time with the frequency of the incident waves (zero speed case). A fixed Cartesian coordinate system is defined with the vertical z-axis pointing upwards and zero incidence of the waves when they propagate along the direction of the positive x-axis. The problem is formulated in terms of a velocity potential which is linearly decomposed into independent components, namely incident, scattered, and radiation potential. The radiation potential can be further decomposed into components from all six degree of rigid body motions ξ_j.

The potentials must satisfy the Laplace equation in the fluid domain, the free surface boundary condition, and the body boundary conditions. Then the radiation and diffracted potentials can be obtained solving two independent boundary value problems. The solution to the problem will result in the velocity potentials from which the fluid pressures and resulting forces can be calculated and then the linear transfer function for the motion can be derived from

$$\sum_{j=1}^{6}\left\{-\omega^2\left[M_{kj}+A_{kj}^{(1)}\right]+i\omega\left[B_{kj}^{(1)}+B_v\right]+\left[C_{kj}+C_{moor}\right]\right\}\xi_j=F_k^{(1)} \tag{1}$$

where, A and B are the added mass and damping coefficients related to the exciting wave frequency ω, B_v is the linearized viscous damping (see section 2.4). C is the hydrostatic restoring matrix, C_{moor} is the additional restoring matrix from the mooring system as specified in section 2.3, M is the generalized mass matrix and $F_k^{(1)}$ are the first order amplitudes of the exciting force and moment in each mode of motion.

2.2 *Second order hydrodynamic problem*

The second order problem is due to interactions between pairs of incident harmonic waves with different frequencies ω_m, ω_n and the oscillating body. The aim is to calculate the second order velocity potential, which is decomposed into mean, sum and difference frequency components.

The current study considers only the mean and difference frequency components of the second order potential. Besides Laplace's equation, bottom and radiation conditions, the second order potential must comply with the free surface and body boundary conditions. Following Lee (1995) the conditions are:

$$\frac{\partial\varphi^{(2)}}{\partial z}-\frac{\omega^2}{g}\varphi^{(2)}=Q_F \quad \text{on } z=0 \tag{2}$$

$$\frac{\partial\varphi^{(2)}}{\partial n}=Q_B \quad \text{on } S_0 \tag{3}$$

where Q_F and Q_B are defined respectively as free surface forcing and body boundary forcing. The former functions are obtained from Taylor expansions about the mean positions of the boundaries and applied on these known positions. The present work applies an approximation for the second order boundary value problem solution, which consists of neglecting the free surface forcing Q_F in Equation 2. Basically, the calculation neglects the free surface integral. This method was suggested by Chen (1994) and it is acceptable in the current study as the solution is at low frequency oscillations and the sum frequency component will not be calculated. Once the velocity potential is known, the pressure can be calculated from Bernoulli's equation. The second order hydrodynamic forces are then calculated by direct integration of the pressure over the body's wetted surface. The second order forces may be separated into components related to quadratic products of first order quantities and components due to second order potentials and then the quadratic transfer function for the motion can be derived from

$$\begin{aligned}&\left\{-\left(\omega_m-\omega_n\right)^2\left[A_{kj}^{(2)}+M_{kj}\right]+i\left(\omega_m-\omega_n\right)\right.\\&\left.\left[B_{kj}^{(2)}+B_v\right]+\left[C_{kj}+C_{moor}\right]\right\}\xi_j^{(2)}=\left\{F_k^{(2)}\right\}\end{aligned} \quad (4)$$

where, A and B are the added mass and damping coefficients related to the difference of frequencies (ω_m–ω_n). $F_k^{(2)}$ includes the difference frequencies component of the second order force, both the quadratic and the second order potential contributions. Since these forces are evaluated in a reference system fixed in space at the equilibrium body position in calm waters, which is not an inertial reference system, several terms arise due to the rotation of the body. It requires proper expansions of the forces and the coordinate's transformations between this reference system and the instantaneous position of the body to derive them as described by Lee (1995).

2.3 *Mooring system modelling*

There are two linearization processes that must be taken into account to obtain the mooring system restoring matrix. The first is the mooring line stiffness properties linearization. As a quick engineering solution, since large extension of lines is not expected, it is possible to simply take the average stiffness k_i around a reasonable line stretching range. The same procedure can be applied to the fender reaction. In the second linearization process, it is then necessary to impose an infinitesimal displacement or rotation in each degree of freedom, calculate the resulting mooring line force and evaluate the effect of this force on the other degrees of freedom.

If F_k is the spring force or moment along direction k of the coordinate system (k = 1,..,6) and since what we want to account for in the equation of motion are variations of motions around an equilibrium position, it is necessary to subtract the initial loadings, then the stiffness coefficient associated the mode of motion j is given applying the Equation 5.

$$c_{kj}=\lim_{\xi_j\to 0}\frac{\partial F_k}{\partial \xi_j}-t_{Lo}n \quad (5)$$

where n is the normal direction of the resulting spring force applied in the mooring point on the moored ship and t_{L0} is the pre-tension assigned on the mooring line. By assuming a small value of motion in each degree of freedom, the 6 × 6 stiffness matrix C_{Li} for a mooring line can be obtained. Since the numerical model is linear, this procedure can be done separately for each mooring line and fender, then the result can be summed in an equivalent mooring system restoring matrix C_{moor} as shown in Equation 6;

$$C_{moor}=\sum_{i=1}^{N_L}C_{L_i} \quad (6)$$

where N_L is the total number of mooring lines and fenders. When the lines extended by δLi, the total tension each line i needs to endure is given by;

$$T_i=\delta L_i k_i+t_{Lo} \quad (7)$$

2.4 *Roll damping*

The results for roll, sway and yaw motions are significantly affected by viscous effects. Therefore, the potential flow calculations overestimate the roll amplitude in resonance and it is impractical. Due to coupling between sway and roll motions, it is expected that the results of the mean drift forces and moments will be affected by the overestimation of roll amplitude, as the mean drift forces mainly depends on the results of the first order quantities. So, it is very important to account for the roll damping effect. The amplitude of roll can be computed with reasonable accuracy by calculation of the linearized viscous roll damping B_v. This contribution can be calculated using different methods available in literature, and then it can be included in the external damping matrix of the system.

In regular wave problem, harmonic linearization can be used to linearize nonlinear variables such as the viscous roll damping. However, in irregular waves the system has multiple nonlinear variables, so that the stochastic linearization should be used for each variable to make sure that all the variables are linearized independently. This procedure is applied to obtain the linearized viscous damping and the linearized stiffness matrix for the mooring system.

2.5 *Motions and loads in irregular waves*

An irregular sea state can be defined within the linear theory as the summation of an infinite number of wave components with frequencies ω, amplitudes ζ and random phasing angle θ. In practice it is impossible to consider an infinite number of wave components, so the sea state is decomposed into N components. The linear motions responses in irregular waves are calculated based on linear spectral theory. The complex motion linear response spectra are calculated in the frequency domain as:

$$S_{\xi_j}^{(1)}(\omega_i)=S_\zeta(\omega_i)\cdot\left|\xi_j^{(1)}(\omega_i)\right|^2 \quad (8)$$

where S_ζ is the complex incident wave spectrum, and $\xi^{(1)}$ is the complex linear transfer function resulting from the solution of the linear hydrodynamic problem. Note that in this case we have used $\xi^{(1)}$ which represents the linear motion transfer function in Equation 8, but it could be any other linear quantity that may be described by a linear transfer function, such as load, pressure or mooring line tension.

As the N components that represent the sea state will be interacting with each other and the body, causing second order loads, a complex quadratic transfer function matrix with N x N elements is usually used to describe these interactions in the frequency domain. The method proposed by Pessoa et al. (2016) is applied to calculating the second order low frequency response spectrum based on the inverse of the Fourier transform. The goal of the method is to calculate the Fourier coefficients X_i whose Fourier discrete inverse transformation will result in the second order response time series. It calculates the second order spectrum by equating the contributions from each combination of wave frequencies in a geometric manner. The low frequency response matrix can be obtained as;

$$Y^{(2)}(\omega_m,\omega_n)=\{\zeta_w\}\cdot\{\zeta_w\}^T\cdot\xi_j^{(2)}(\omega_m,\omega_n) \tag{9}$$

where $\xi_j^{(2)}$ is the complex quadratic transfer function matrix obtained from the solution of the second order problem, $\{\zeta_W\}\cdot\{\zeta_W\}^T$ is complex bi-frequency wave amplitude matrix by assuming stochastic distributions related to S_ζ and T superscript stands for the transpose. As the harmonic frequencies ω_m and ω_n of the discrete wave spectrum are assumed to be uniformly distributed along the ω axis, the diagonal of $Y^{(2)}(\omega_m - \omega_n)$ will be related to a fixed oscillatory frequency equal to $(\omega_m - \omega_n)$. Then the Fourier coefficients X_i can be obtained by summing the elements of $Y^{(2)}$ whose pairs (m, n) yield the same result for $(\omega_m - \omega_n)$, which are basically the diagonals of $Y^{(2)}$ and it can be represented as;

$$X_i(\omega_m,\omega_n)=\sum_{j=1}^{L_{diagonal}} Y_j^{(2)}(\omega_m,\omega_n) \tag{10}$$

The quantity $\xi_j^{(2)}$ in Equation 9 could be any other second order quantity that can be described by a quadratic transfer function as motion, load, pressure or mooring line tension. The obtained complex linear response spectra and complex slowly varying response spectra can be used to calculate the time series for total motion/load by using the inverse Fourier transform algorithm (IFFT).

3 CASE STUDY

In the present study, the computational results are compared with model test measurements and previous numerical calculations for a loaded 200 kDWT tanker. This ship is chosen, because computational results as well as model test data are readily available from literature to compare with. The above described method is applied to the loaded tanker ship, with the specified loading condition. The principal particulars and body plan are shown in Table 1 and Figure 1, respectively.

The comparison is carried out in two parts. First, the obtained RAOs and mean drift forces are compared with the experimental results and numerical calculations obtained by Pinkster (1980) at 82.5 m water depth without mooring. The comparison herein is with regards to surge, sway and yaw modes of motion. In the second part, the response spectra for the same modes of motions and mooring loads are compared with the experimental results carried out by Van Oortmerssen (1976) on the same loaded 200 kDWT tanker moored to an open jetty in water with a depth amounting to 1.2 times the draft of the vessel. In this latter case, the mathematical representation of the calculated

Table 1. Main dimensions of the 200 kDWT tanker.

Main dimensions	Tanker
Length between perpendiculars (m)	310
Beam (m)	47.17
Depth (m)	29.7
Draft (m)	18.9
Displacement (m^3)	234,826
Longitudinal center of gravity (m)*	6.61
Block coefficient	0.85
Prismatic coefficient	0.855
Mid-ship section coefficient	0.995
Vertical center of gravity (m)	13.32
Metacenteric height (m)	5.78
Pitch radius of gyration (m)	77.5
Roll radius of gyration (m)	17.02

*Measured from the mid-ship section.

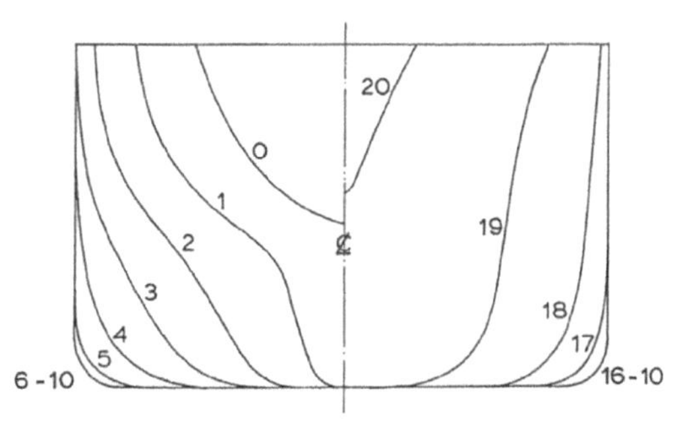

Figure 1. Body plan for the loaded 200 kDWT tanker ship (Van Oortmerssen 1976).

input wave spectrum used for comparison is the same spectrum, as measured at the data available in the literature (Van Oortmerssen 1976).

3.1 *Discretization and convergence analysis*

The hydrodynamic problem was solved with a panel method code (WAMIT). The hull-form is discretized up to the still water. The velocity potential was distributed over the body using high order panels based on B-spline basis functions, the characteristic panel length or (panel size) is used to define the discretization of the body. In most applications this provides a more accurate solution, with a smaller number of unknowns, compared to low-order discretization.

In linear hydrodynamic problem, it is sufficient to analyse a linear quantity, such as the diffraction loads, or a hydrodynamic coefficient, which converge quickly. For second order calculations, the steady drift load is calculated with the pressure integration method which requires a higher precision of the mesh. A convergence analysis has been carried out as shown in Figure 2, for the results sway mean drift forces in bow quartering wave 135°, based on the characteristic panel length. The analysis of the hull forms take into consideration that the discretization of the ship hull ship is

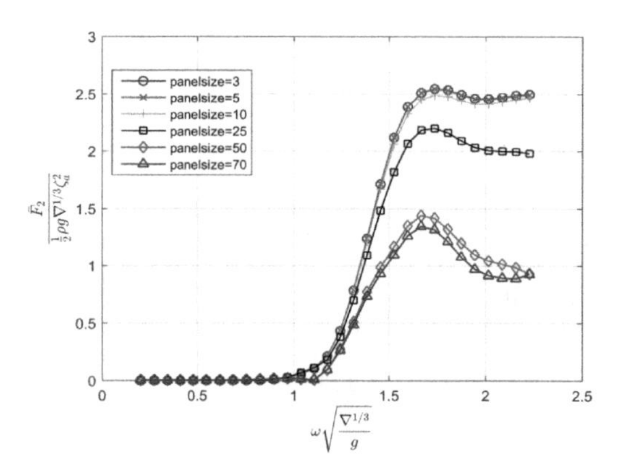

Figure 2. Mesh convergence analysis for sway mean drift force in bow quartering wave 135°.

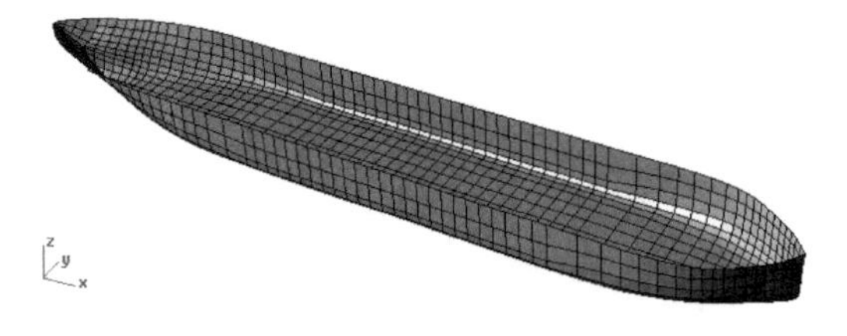

Figure 3. Mesh of the tanker hull with the characteristic length 5 m for the panels.

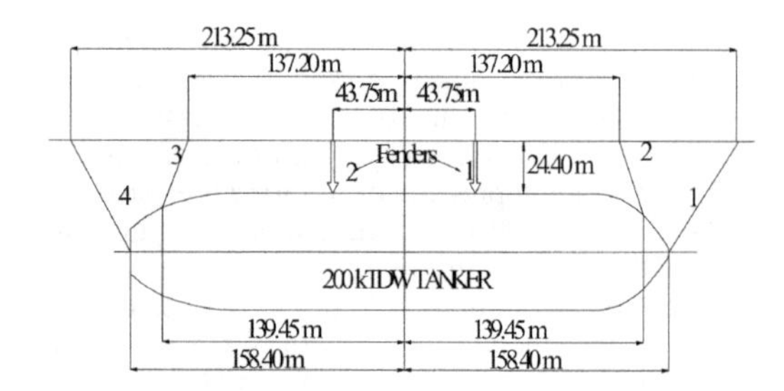

Figure 4. Mooring configurations (Van Oortmerssen 1976).

considered to be acceptable when the maximum panel length dimension should not be greater than 20% of the shortest studied wave length. Based on the convergence analysis, the characteristic panel length is 5 m which results the panel distribution shown in Figure 3.

3.2 *Mooring configurations*

The study was conducted for a loaded 200 kDWT tanker, moored to an open jetty in water with a depth amounting to 1.2 times the draft of the vessel. The mooring layout is shown in Figure 4. The vessel was moored by means of 4 lines, each representing two or three wires with nylon tails in reality. In each line a pretension of 196.13 kN was applied, ensuring their linear spring-like behaviour, at least within the small amplitude motions range.

In the experiment, the fenders were simulated by means of rocking arms, to which linear springs were connected. Vertical wheels have been used in order to minimize the friction between the model and fenders. The pair of stiff fenders had a linear elasticity amounting to15445.5 kN/m.

PIANC (1995) specifies the allowable range for surge or sway motion by 3 m which can be considered as small amplitudes. According to this allowable range of motion, the average stiffness can be obtained from the nonlinear load-elongation curves of the mooring lines which are presented by Van Oortmerssen (1976) in terms of elongation percentage. Using the length of each mooring line as specified in Figure 4, the linearized stiffness for each mooring line can be determined as 362 kN for line 4 and line 1 and 383.5 kN for line 2 and line 3.

4 COMPUTATIONAL AND EXPERIMENTAL RESULTS

4.1 *Regular waves results*

This section presents a comparison between experimental data and calculations in regular waves for the loaded 200 kDWT tanker floating in 82.5 m water depth without mooring. Results are presented as RAOs of surge, sway and yaw about the centre of gravity in bow quartering wave 135°, in non-dimensional value as follows: surge and sway per wave amplitude and yaw per wave slope. Also, the mean drift forces and moments for the same modes of motions are presented in non-dimensional form. The validation for the above modes of motion tanker ship in regular waves is carried out by comparing the computational results from the numerical method with experimental and numerical data of Pinkster (1980). The results are plotted against the non-dimensional wave frequency $\omega\sqrt{(\nabla^{1/3}/g)}$ (where g is acceleration due to gravity and ∇ is the underwater volume) for the same frequencies simulated in regular waves by Pinkster (1980). Surge, sway and yaw motions, Figures 5–7, are presented for tanker in bow quartering wave 135°.

The non-dimensional mean drift forces and moment are measured in bow quartering wave 135° as shown in Figures 8–10.

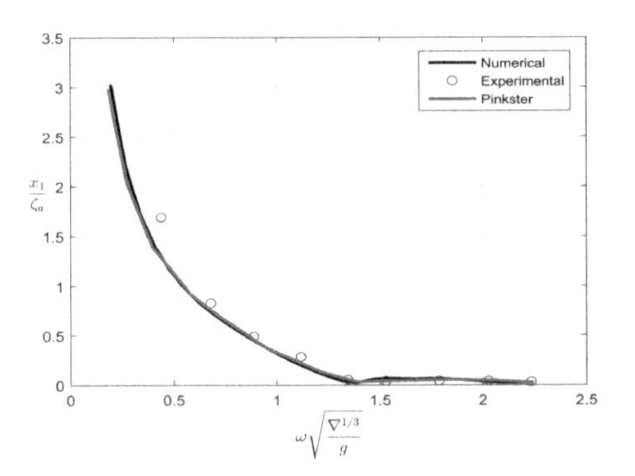

Figure 5. Surge motion RAO in bow quartering wave135°.

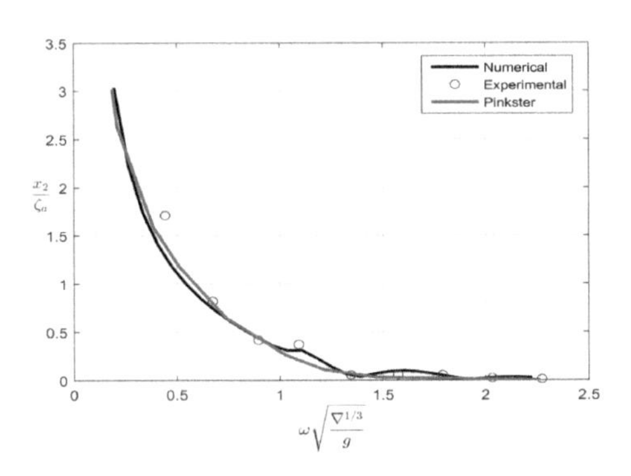

Figure 6. Sway motion RAO in bow quartering wave 135°.

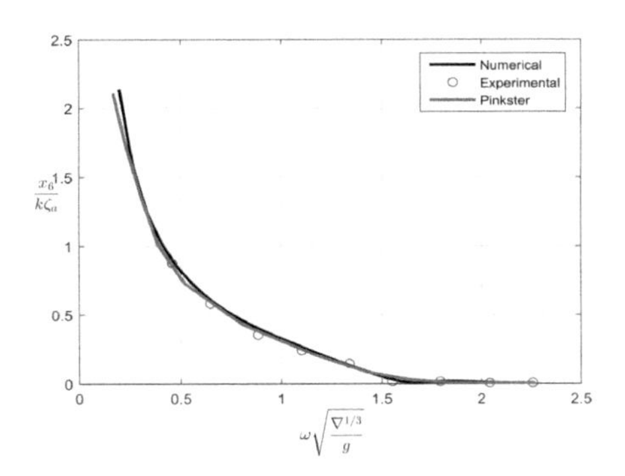

Figure 7. Yaw motion RAO in bow quartering wave 135°.

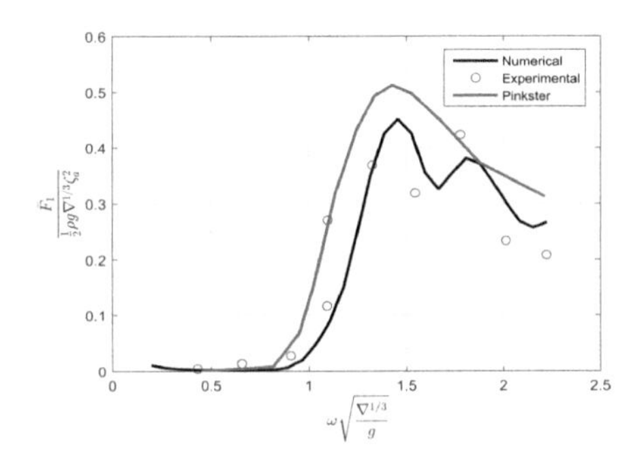

Figure 8. Surge mean drift force in bow quartering wave 135°.

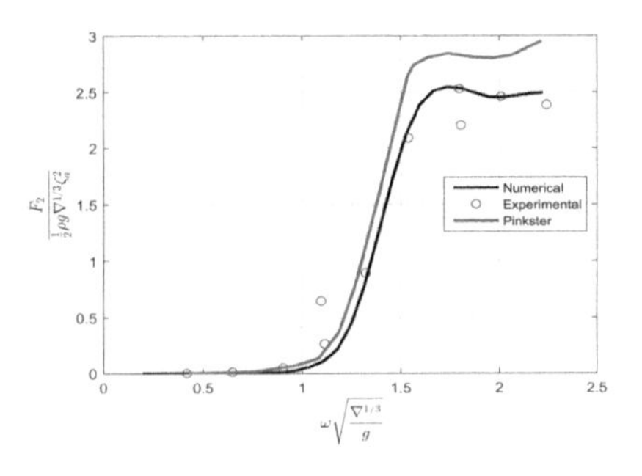

Figure 9. Sway mean drift force in bow quartering wave 135°.

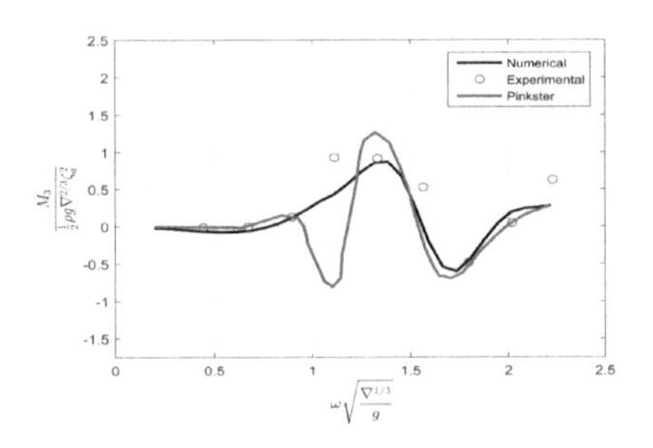

Figure 10. Yaw mean drift moment in bow quartering wave 135°.

4.2 *Irregular waves results*

This section presents a comparison between the numerical results for the variance spectrum for the motions and mooring loads and experimental and numerical results presented by Van Oortmerssen (1976) for the tanker with the mooring configurations presented in section 3.2. The simulation for the motions and mooring loads are carried out in long crested sea state with a significant wave height of 2.6 m, mean period 8.9 sec and with heading angle corresponding to bow quartering wave 135°.

The target irregular sea state was successfully reproduced. The outcome can be seen in Figure 11 that shows the numerical spectrum in a black line and the measured experimental spectrum in red. The details of the experimental program are described in Van Oortmerssen (1976).

The wave spectrum was simulated in the computations by means of 15 sine waves, ranging in frequency from 0.425 to 1.125 rad/sec. The measured wave spectrum, shown in Figure 11 was subdivided into 15 bands of constant width. Each band was represented by a sine component, having the centre frequency of that band and random wave height that satisfy the same amount of energy in the target irregular sea state. These wave components were summed with arbitrary phase angles, and with the aid of the computed transfer functions (linear and quadratic) for the motions and mooring loads, the variance spectra were determined in the 6 modes motions and for all mooring load.

The time history for the simulated wave, output motions and mooring loads can be calculated using inverse Fourier transformation of the variance spectrum. Due to space limitation the time history is presented only for the simulated wave as shown in Figure 12. The variance response spectrum is presented only for surge, sway and yaw motions as shown in Figure 13–15. The results for all mooring lines and fenders are shown in Figures 16–21. Besides spectra, Table 2 presents a comparison between the numerical and experimental statistical quantity RMS value of the variance spectrum.

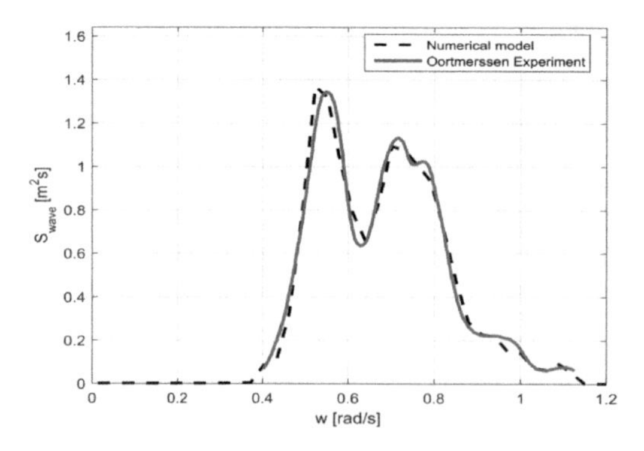

Figure 11. The simulated spectrum and the experimental wave spectrum with significant wave height 2.6 m and mean period 8.9 sec.

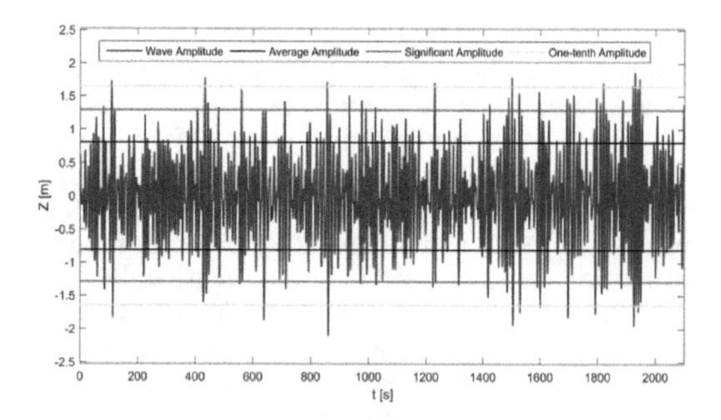

Figure 12. Time history (2100 sec) for the simulated wave with a significant wave height 2.6 m.

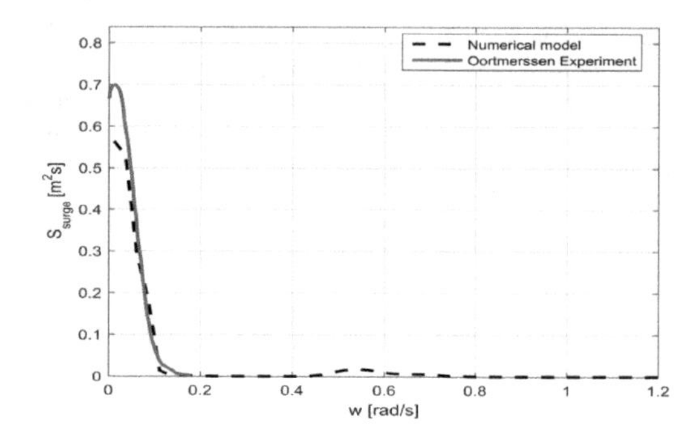

Figure 13. Computed and experimental response spectra for surge motion in long crested irregular wave 135°.

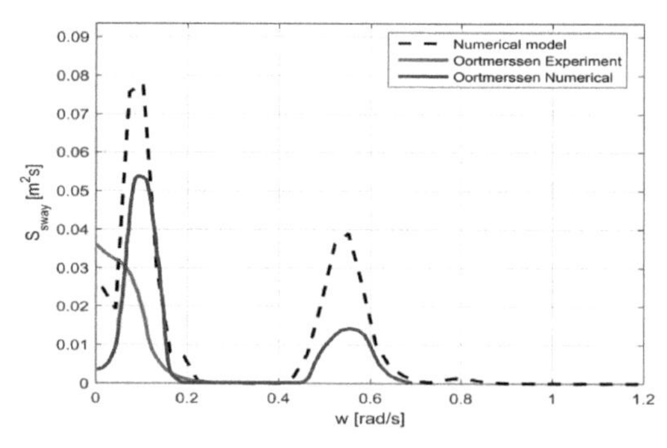

Figure 14. Computed and experimental response spectra for sway motion in long crested irregular wave 135°.

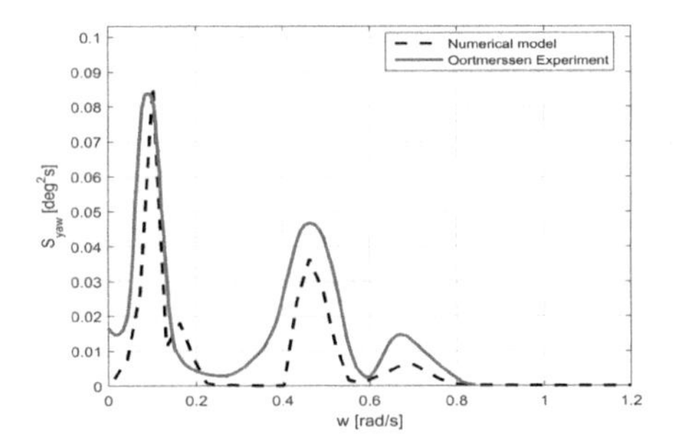

Figure 15. Computed and experimental response spectra for yaw motion in long crested irregular wave 135°.

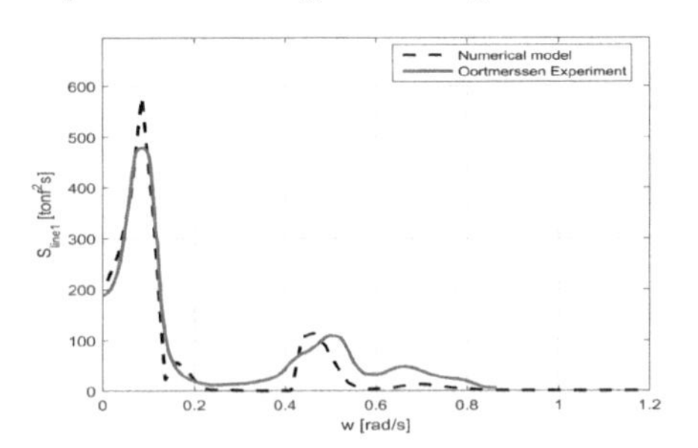

Figure 16. Computed and experimental response spectra for tension load on line 1 in long crested irregular wave 135°.

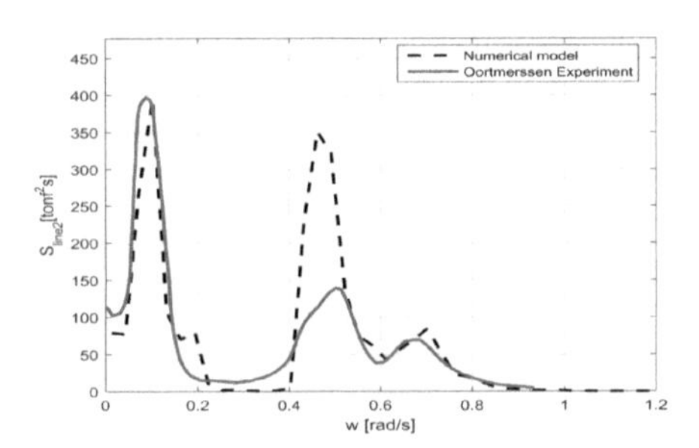

Figure 17. Computed and experimental response spectra for tension load on line 2 in long crested irregular wave 135°.

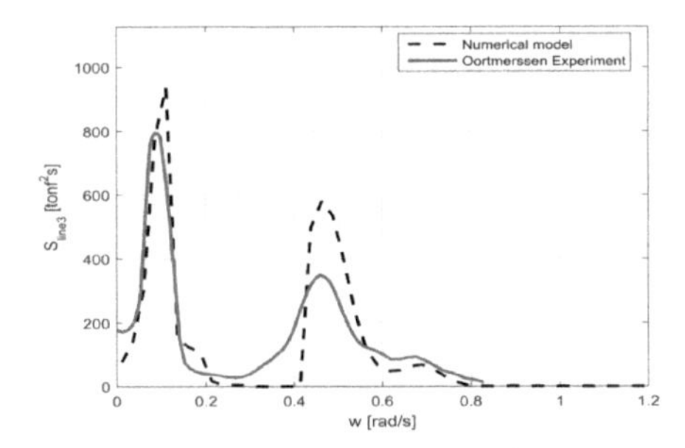

Figure 18. Computed and experimental response spectra for tension load on line 3 in long crested irregular wave 135°.

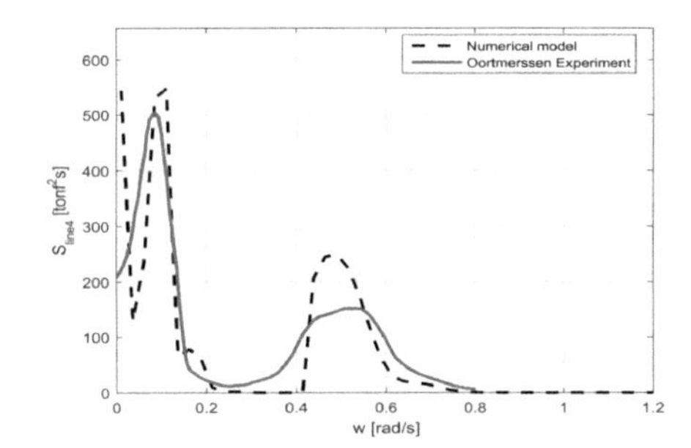

Figure 19. Computed and experimental response spectra for tension load on line 4 in long crested irregular wave 135°.

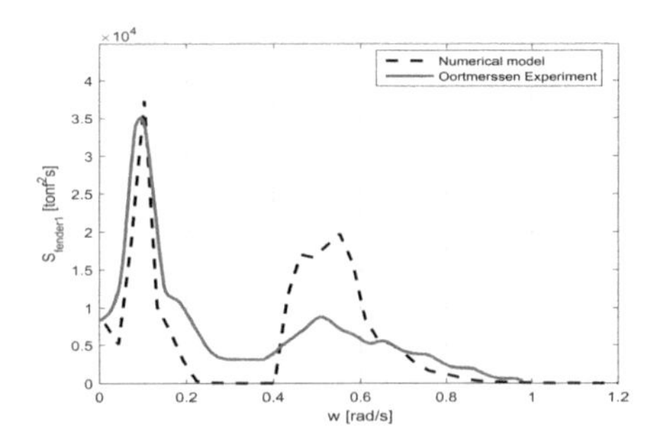

Figure 20. Computed and experimental response spectra for the load on fender 1 in long crested irregular wave 135°.

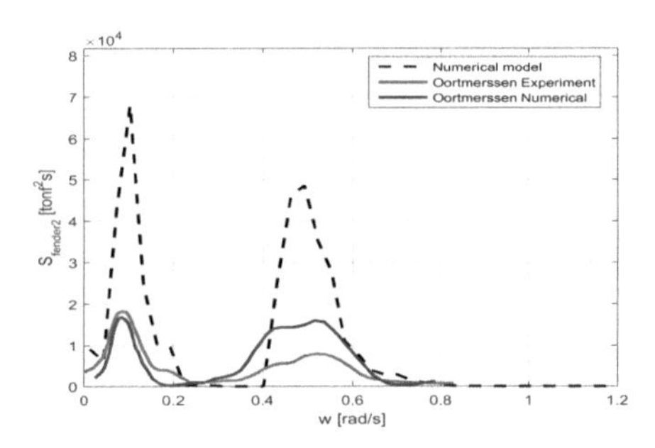

Figure 21. Computed and experimental response spectra for the load on fender 1 in long crested irregular wave 135°.

Table 2. Comparison between numerical and experimental RMS variance in bow quartering wave 135°.

Signal	Experimental	Numerical
Wave (m)	0.65	0.649
Surge (m)	0.22	0.201
Sway (m)	0.07	0.107
Yaw (degree)	0.13	0.096
Line 1(ton)	8.00	7.792
Line 2 (ton)	8.00	8.882
Line 3 (ton)	12.0	11.66
Line 4 (ton)	10.0	9.205
Fender 1 (ton)	71.0	80.42
Fender 2 (ton)	63.0	107.0

5 DISCUSSION OF RESULTS

The main goal of this work is to evaluate the motions of a moored vessel and the loads on the mooring lines and fenders in frequency domain in accurate and simple way in irregular waves. This is achieved via the presented method while taking into account the influence of second order forces and motions and the role it plays in representing the moored ship problem.

5.1 *Regular wave results*

For surge, sway and yaw motions in Figure 5–7, there is good agreement with the experimental and numerical results, for the simulated wave heading 135°. Surge, sway and yaw amplitudes decrease rapidly with increasing wave frequency. Results for sway were supposed to approach the component of the wave orbit radius at very low frequencies. However the calculations are carried out in shallow water and the incident waves are changed, so the wave exciting forces exerted on the ship differ from those in deep water; and the hydrodynamic coefficients of the ship are changed by the effect of the sea bottom.

The results in Figure 8–10 indicate that in general the mean second order forces and moments are reasonably well predicted. The consistency between the presented numerical results and experimental results by Pinkster (1980) is clear more than the agreement with numerical results. Results for the mean yaw moment show a difference between computations and experiments at the non-dimensional wave frequency corresponding to the natural roll frequency, this effect is due to the coupling between sway and roll motion. The presented results did not account the additional roll damping; however, it can be calculated with the presented numerical method.

Analysis of results in other heading angle such as beam waves show that, the incorporation of the viscous roll damping will result in a good prediction in the first order roll motions. So, the discrepancies, for the mean drift forces in beam wave at frequency corresponding to the natural frequency, can be avoided and better results obtained.

5.2 *Irregular wave results*

The target irregular sea state was successfully reproduced with the same amount of energy and the same significant wave amplitude 2.6 m as shown in Figures 11–12 and Table 2.

It must be noted that stochastic linearization has been applied in order to obtain the linearized stiffness matrix for the mooring system. The results for motions response spectra bow quartering waves, 135 degrees, are presented in Figures 13–15. The agreement with the experimental results is once again good, without incorporation of additional roll damping. The shapes of the computed spectra are similar to the experimental spectra, the exception is the over estimation of the peaks in sway. This may be due non-potential sway viscous damping, which is not accounted for in the present result. A way to overcome this could be to introduce some additional linearized viscous damping in this mode in a similar way as it can be done for the roll motion. Results for the sway response spectrum agree with numerical results from Van Oortmerssen (1976) other than the experimental result.

In general, results for surge sway and yaw response spectra show that the second order effects have a main contribution for the motions of the moored ship in irregular waves and it should not be neglected. This can be clearer from the results of the motion response spectra in head waves.

The analysis of results of the mooring load response spectra in Figures 16–21 show that, even with all the simplifications assumed in the numerical modelling system and with all the limitations of the implemented theory, the agreement, between the presented numerical results and the experimental results by Van Oortmerssen (1976), is good. Results show that the overestimation of the sway motion may affect the mooring loads at high frequency especially for the fenders and mooring lines 2–3 that work as breast lines. Although the motions are small enough to be on the linear behaviour of the response of the fenders, the simplification may induce overestimation in the predicted load of fenders.

The mooring loads are highly dependent on the position where the lines and fenders are attached to the vessel and on the motion of the moored ship. These loads are also dependent of the stiffness of the line and fender itself. However, we are uncertain of the details of the mooring configurations, Allowing for these uncertainties, our results agree acceptably well with Van Oortmerssen (1976) experimental results with respect to the shape and RMS variance.

6 CONCLUSIONS

From the above discussion it can be concluded that the main goal of the presented study has been accomplished, in accurate and simple way, based on the underlying assumptions of the numerical method.

As a general conclusion, one may say that the linear method applied is able to represent reasonably well the motion responses and mooring loads for a moored ship in irregular sea states. The hydrodynamic calculations are based on a higher order boundary element method, while the 6×6 degrees of freedom linearized stiffness matrix for the mooring system is calculated. The second order low frequency loads cannot be neglected when calculating the motions and loads on the mooring lines and fenders in irregular waves. It may be concluded that incorporation of the roll damping tends to affect both first and second order quantities. The numerical method predicts well the measured responses and loads, however, there is a tendency to overestimate the sway motion which leads to an overestimation of the tensions on the breast lines and fenders at high frequencies. The linear formulation of the hydrodynamic problem necessitates the simplification of the fenders, which sometimes is not enough.

ACKNOWLEDGEMENTS

This work is a contribution to the M&MSHIPS project "Maneuvering & Moored SHIPS in ports" (PTDC/EMSTRA/5628/2014) funded by the Portuguese Foundation for Science and Technology.

REFERENCES

Chen, X. B. (1994), "Approximation on the quadratic transfer function of low-frequency loads. The 7th International Conference on Behaviour of Offshore Structures, 1994 Massachusetts, USA. Pergamon, 289–302.

Chen, X. B. (2005), "Hydrodynamic analysis for offshore LNG terminals", The 2nd International Workshop on Applied Offshore Hydrodynamics, Rio de Janeiro, 2005.

Duncan, P. E. & Drake, K. R. 1995. A note on the simulation and analysis of irregular non-linear waves. *Applied Ocean Research,* Vol. 17, pp. 1–8.

Lee, C.-H. (1995), *WAMIT theory manual*, Massachusetts Institute of Technology, Department of Ocean Engineering.

Maruo, H. (1960), "The drift of a body floating on waves", *Journal of Ship Research,* Vol. 4, pp. 1–10.

Newman, J. N. (1967), "The drift force and moment on ship in waves". *Journal of Ship Research,* Vol. 11, pp. 51–60.

Pessoa, J. (2013), *Second order wave exciting loads and operability of side by side floating vessels in waves.* PhD, Technical University of Lisbon.

Pessoa, J., Fonseca, N. & Guedes Soares, C. (2013),. "Analysis of the first order and slowly varying motions of an axisymmetric floating body in bichromatic waves". *J. Offshore Mechanics and Arctic Engineering,* 135 (1) pp 135:0011601.1-11.

Pessoa, J., Fonseca, N. & Guedes Soares, C. (2016), "Side-by-side FLNG and shuttle tanker linear and second order low frequency wave induced dynamics". *Ocean Engineering,* Vol. 111, pp. 234–253.

PIANC (1995), *Criteria for Movements of Moored Ships in Harbours: A Practical Guide*, Report of working group PTC II-24.

Pinkster, J. A. (1980), *Low frequency second order wave exciting forces on floating structures*. PhD, Delft University of Technology.

Schellin, T. E. & Östergaard, C. (1995), "The vessel in port: Mooring problems". *Marine Structures,* Vol. 8, pp. 451–479.

Van der Molen, W. (2006), *Behaviour of moored ships in harbours.* PhD, Delft University of Technology.

Van Oortmerssen, G. (1976), *The motions of a moored ship in waves.* NSMB publication No. 510, Delft University of Technology.

Progress in Maritime Technology and Engineering – Guedes Soares & Santos (Eds)
 ISBN 978-1-138-58539-3

Numerical and experimental study of ship-generated waves

S.R.A. Rodrigues & C. Guedes Soares
Centre for Marine Technology and Ocean Engineering (CENTEC), Instituto Superior Técnico, Universidade de Lisboa, Lisbon, Portugal

J.A. Santos
Instituto Superior de Engenharia de Lisboa, Instituto Politécnico de Lisboa, Lisbon, Portugal

ABSTRACT: This paper presents the results of the numerical and experimental study on the propagation of ship-generated waves. In the numerical modified FUNWAVE code, which is able to reproduce most of the phenomena involved in wave propagation, the ship is represented by a moving pressure distribution function at the free surface. The experimental tests were carried out with a ship model at a 1:50 scale that moving at constant speed across a constant depth region. The values of wave heights obtained in the scale ship model 1:50 were used to compare with the predictions of wave heights provided by the numerical code. The comparison of the results obtained by the numerical code and the experimental tests shows very similar values for the maximum wave height.

1 INTRODUCTION

The design of various waterway features such as bank-erosion control structures and marine protection works, as well as the establishment of allowable ship speeds in navigable waterways, require knowledge of the characteristics of the waves generated by the ship traffic using the waterway.

Ship waves in river, lakes, estuaries, and coastal areas are of great importance to the environment, engineering applications as well as ships. In coastal shallow water, the waves generated by marine traffic, the interaction with structures and the reflection of land boundaries have become a crucial factor that affects the water environment and engineering applications. In comparison with the wind waves and ocean swells, ship waves inside a harbor can exhibit anomalous wave heights in determined areas. Ship waves are hardly dissipated in the harbor by interaction with the shoreline and may cause wave resonance. This accumulated wave energy may result in severe increase of wave height which causes damage to port facilities, degradation of harbor infrastructures and excessive disturbance to moored and moving ships.

The force associated with wave generation by a moving ship is an important part of ship's resistance, that is, the force along the longitudinal axis of the ship necessary for it to move with the desired speed setting. Several numerical models have been developed to determine this resistance to advance, which nowadays is often calculated with Computational Fluid Dynamics codes as shown for example by Ahmed & Guedes Soares, (2009) and Ciortan et al. (2012). In spite of their features, such models are not used to simulate the propagation of these waves away from the ship, because the growth of computing time with the simulated area size leads to excessive time requirements.

The numerical models developed to study the propagation of these waves are based on the shallow water equations, Stockstill & Berger (2001), or on improved versions of the Boussinesq equations, Nwogu & Demirbilek (2004), Dam et al. (2006) and Nascimento et al. (2008). In most of these models the ship is represented by a pressure distribution on the free surface at the position occupied by the ship. Any of the mentioned models reproduces in the horizontal plan the patterns of waves generated by ships in water areas both not limited and limited horizontally, in this case, reflections of those waves may occur.

There are some works that use scale models to characterize the free-surface elevation associated to the waves generated by ships, starting with the work of Sorensen (1967), which measured the wave field in a flat unbounded region to the work of Schipper (2007) where some results of the propagation of wave generated by a ship moving in a constant depth region are presented, through the work of Cornett et al. (2008), that studied the movements of ships docked inside a port induced by the nearby passage of another ship.

The objective of this paper is to present the results of the propagation of ship-generated waves given by the modified FUNWAVE code of Nascimento et al. (2008) and validate its wave-height predictions through the comparison with experimental data of the free-surface elevation obtained in scale-model tests.

The structure of this paper is as follows. After this introductory section, section 2 describes the main features of the modified FUNWAVE code and the pressure distribution function used in the numerical code. Section 3 presents the experimental scale model procedure to collect data of the free-surface elevation of the waves generated by a ship model at a 1:50 scale. Section 4 presents the results based on numerical simulations of the propagation of ship-generated waves as well as the comparison of the numerical and the experimental times series. Finally, the paper closes with some conclusions of the presented work.

2 NUMERICAL CODE FOR SHIP WAVES

FUNWAVE is a hydrodynamic numerical model based on the nonlinear Boussinesq equations derived by Wei & Kirby (1995). Their approach includes additional terms to the Boussinesq equations developed by Nwogu (1993). The original equations are extended up to intermediate water depth and it allows the simulation of wave propagation in situations of strong nonlinear interactions. This model can simulate most phenomena associated to wave propagation into shallow water. In order to include ship generated waves, a pressure gradient, which causes disturbances on the free surface, Figure 1, was added to the momentum equation. The continuity equation (Equation 1) and the momentum equation (Equation 2) for the modified model are:

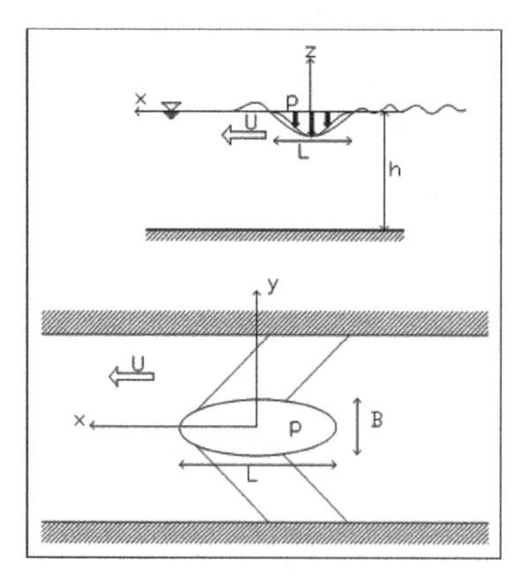

Figure 1. Representation of the moving pressure as the source function which generates the ship waves. (Adapted from Nascimento et al. (2008)).

$$\eta_t + \nabla\cdot\left\{(h+\eta)\left[\mathbf{u}_\alpha + \left(z_\alpha + \frac{1}{2}(h-\eta)\right)\nabla\left(\nabla\cdot(h\mathbf{u}_\alpha)\right) + \left(\frac{1}{2}z_\alpha^2 - \frac{1}{6}(h^2 - h\eta + \eta^2)\right)\nabla(\nabla\cdot\mathbf{u}_\alpha)\right]\right\} = 0 \tag{1}$$

$$\begin{aligned}
&\mathbf{u}_{\alpha t} + (\mathbf{u}_\alpha\cdot\nabla)\,\mathbf{u}_\alpha + g\,\nabla\eta + z_\alpha\left\{\frac{1}{2}z_\alpha\nabla(\nabla\cdot\mathbf{u}_{\alpha t}) + \nabla\left(\nabla\cdot(h\mathbf{u}_{\alpha t})\right)\right\} + \nabla\left\{\frac{1}{2}(z_\alpha^2 - \eta^2)(\mathbf{u}_\alpha\cdot\nabla)(\nabla\cdot\mathbf{u}_\alpha)\right.\\
&\left.+\frac{1}{2}\left[\nabla\cdot(h\mathbf{u}_\alpha) + \eta\nabla\cdot\mathbf{u}_\alpha\right]^2\right\}\\
&+\nabla\left\{(z_\alpha - \eta)(\mathbf{u}_\alpha\cdot\nabla)\left(\nabla\cdot(h\mathbf{u}_\alpha)\right) - \eta\left[\frac{1}{2}\eta\nabla\cdot\mathbf{u}_{\alpha t} + \nabla\cdot(h\mathbf{u}_{\alpha t})\right]\right\} = -\frac{\nabla P}{\rho}
\end{aligned} \tag{2}$$

where η is the surface elevation, h is the water depth, $\mathbf{u}_\alpha = (u,v)$ is the horizontal velocity vector at level $z = z_\alpha = -0.531h$, g is the gravitational acceleration, t index represents the partial time derivative, $\nabla = (\partial/\partial x, \partial/\partial y)$ is the horizontal gradient operator and P is the moving pressure source function.

Although it is not difficult to include the pressure gradient in the Boussinesq equations to act as the driver for ship waves, one must ensure that the pressure distribution implemented in the model is compatible with the nonlinear and dispersive properties of the Boussinesq equations, i.e. to the same order of approximation.

The Boussinesq equations can be deduced from the dimensionless form of the Navier-Stokes equations assuming the shallow water hypothesis $h/l << 1$. Such a procedure induces two dimensionless independent parameters in equations, ε and μ^2. The parameter ε represents the ratio between the wave amplitude and the local water depth, $\varepsilon = \eta/h$, that determines the magnitude of the nonlinear effects, whereas the magnitude of the dispersive effects are represented by the ratio between the local depth and the wave length, $\mu = h/l$.

Several assumptions can be made, resulting in different models of approximation which depend on the magnitude of these parameters. The relative importance of these effects is proportional to $\varepsilon/\mu^2 = \eta l^2/h^3$. For the Boussinesq equations there is a balance between dispersive and nonlinear effects, $\varepsilon = \mu^2$.

The first step in the analysis of the influence in the momentum equation is the expansion of the dependent variables in Equations 1 and 2 into power series of the forcing dominant parameters. Such procedure allows the determination of the magnitude of the perturbation source that is compatible with the characteristics of the equations. The dimensionless variables were defined using l_0 as the characteristic length, a_0 as the typical wave amplitude and h_0 as the characteristic depth and have the following relationships:

$$
\begin{aligned}
&x' = \frac{x}{l_0};\quad y' = \frac{y}{l_0};\quad z' = \frac{z}{h_0};\quad t' = \frac{\sqrt{gh_0}}{l_0}t \\
&u' = \frac{h_0}{a_0\sqrt{gh_0}}u;\quad v' = \frac{h_0}{a_0\sqrt{gh_0}}v \\
&\eta' = \frac{n}{a_0};\quad h' = \frac{h}{h_0};\quad P' = \frac{P}{\rho g h_0}
\end{aligned}
\tag{3}
$$

Substituting Equation 3 into the Equation 1 and 2, results the following continuity and momentum equations:

$$
\begin{aligned}
&\eta'_t + \nabla\cdot\Bigg\{(h'+\varepsilon\eta')\Bigg[\mathbf{u}'_\alpha + \mu^2\Bigg\{\Bigg[\frac{1}{2}z'^2_\alpha \\
&-\frac{1}{6}\left(h'^2 - h'\varepsilon\eta' + (\varepsilon\eta')^2\right)\Bigg]\nabla(\nabla\cdot\mathbf{u}'_\alpha) \\
&+\left[z'_\alpha + \frac{1}{2}(h'-\varepsilon\eta')\right]\nabla\left(\nabla\cdot(h'\mathbf{u}'_\alpha)\right)\Bigg\}\Bigg]\Bigg\} + O(\mu^4)
\end{aligned}
\tag{4}
$$

$$
\begin{aligned}
&\mathbf{u}'_{\alpha t} + \varepsilon(\mathbf{u}'_\alpha\cdot\nabla)\mathbf{u}'_\alpha + \nabla\eta' + \mu^2 V_1 + \varepsilon\mu^2 V_2 \\
&\quad + \nabla P' = O(\mu^4) \\
&V_1 = \frac{1}{2}z'^2_\alpha\nabla(\nabla\cdot\mathbf{u}'_{\alpha t}) + z'_\alpha\nabla\left(\nabla\cdot(h'\mathbf{u}'_{\alpha t})\right) \\
&\quad - \nabla\left[\frac{1}{2}(\varepsilon\eta')^2\nabla\cdot\mathbf{u}'_{\alpha t} + \varepsilon\eta'\nabla\cdot(h'\mathbf{u}'_{\alpha t})\right] \\
&V_2 = \nabla\left[(z'_\alpha - \varepsilon\eta')(\mathbf{u}'_\alpha\cdot\nabla)\left(\nabla(h'\mathbf{u}'_\alpha)\right)\right. \\
&\quad \left. + \frac{1}{2}\left(z'^2_\alpha - (\varepsilon\eta')^2\right)(u'_\alpha\cdot\nabla)(\nabla\cdot\mathbf{u}'_\alpha)\right] \\
&\quad + \frac{1}{2}\nabla\left[\left(\nabla\cdot(h'\mathbf{u}'_\alpha) + \varepsilon\eta'\nabla\cdot\mathbf{u}'_\alpha\right)^2\right]
\end{aligned}
\tag{5}
$$

In Equations 4 and 5 there is the need of a forcing term of order O(ε) for pulse generation, so $\eta', \mathbf{u}'_\alpha$ and P' terms are expanded in power series of O(ε). As the real forcing in the momentum equation is $\nabla P'$ that moves along the x axis, this also implies the influence of dissipative terms and it is appropriate that the pressure distribution magnitude relates with ε and μ smaller that the magnitude of μ^4. Expanding the $\eta', \mathbf{u}'_\alpha, P'$ and $\nabla P'$ terms in power series (the commas being omitted for convenience) one has:

$$
\begin{aligned}
&\eta = \eta_1 + \varepsilon\eta_2 + O(\varepsilon^2) \\
&\mathbf{u}_\alpha = (\mathbf{u}_\alpha)_1 + \varepsilon(\mathbf{u}_\alpha)_2 + O(\varepsilon^2) \\
&P = P_1 + \varepsilon P_2 + O(\varepsilon^2) \\
&\nabla P = \nabla P_1 + \varepsilon\mu\nabla P_2 + O(\varepsilon^2)
\end{aligned}
\tag{6}
$$

Substituting Equation 6 in Equation 5, one gets a finite sequence of linear problems. For the first order approximation one has:

$$
\begin{aligned}
&\mathbf{u}_{\alpha 1t} + \nabla\eta_1 + \nabla P_1 + \left\{\frac{1}{2}z^2_\alpha\nabla(\nabla\mathbf{u}_{\alpha 1t})\right. \\
&\quad \left. + z_\alpha\nabla\left(\nabla(h\mathbf{u}_{\alpha 1t})\right)\right\} = 0
\end{aligned}
\tag{7}
$$

In a first approximation ∇P_1 in the Equation 7 has to be equal to zero. For the second order approximation one has:

$$
\begin{aligned}
&\mathbf{u}_{\alpha 2t} + \nabla\eta_2 + \mu^2\left\{\frac{1}{2}z^2_\alpha\nabla(\nabla\mathbf{u}_{\alpha 2t}) + \nabla(h\mathbf{u}_{\alpha 2t})\right\} \\
&= -\left[(\mathbf{u}_{\alpha 1}\nabla)(\mathbf{u}_{\alpha 1})\right] - \mu^2\left[\eta_1\nabla(h\mathbf{u}_{\alpha 1t})\right] - \mu\nabla P_2
\end{aligned}
\tag{8}
$$

The terms on the right side represent the interaction of nonlinear terms the first order of η_1 and $\mathbf{u}_{\alpha 1}$ and the second order of P_2 that serve as forcing terms for the second order wave η_2 and $\mathbf{u}_{\alpha 2}$.

In case of ship-wave generation it is assumed that at time t_0 the velocity and the free surface elevation are equal to zero and the pressure P is an impulsive force that generates the waves, where

$$P = \varepsilon P_2 + O(\varepsilon^2) \tag{9}$$

The pressure P being dependent of P_2, due to the condition applied in equation 7, $\nabla P_1 = 0$. In Equation 9 it becomes evident that P is of order ε, demonstrating the need for the source to be a nonlinear disturbance, ensuring that ∇P is of $\varepsilon\mu$ order.

As boundary conditions, the numerical model includes absorbing sponge layers to simulate energy dissipation in the momentum equation to control shore reflection effects.

2.1 *Pressure distribution function*

Several pressure distributions are presented in the literature as nonlinear functions of O(ε) approximation at a rectangle of width B and length L. Such pressure distributions on a rectangle can represent the effects of a moving ship along a water body whose bottom has topographical changes, where such movement can cause initial disturbances at the free surface inside the domain rectangle, that propagate along time with a pattern similar to the ones of ship generated waves. Such patterns are stationary, from the point of view of an observer moving with the ship.

Li & Sclavounos (2002) proposed Equation 10 to study the solitary wave generated in front of a ship that sails over a deep-water region. In the equation *pa* designates the maximum value for the pressure distribution and it is assumed as constant value for the same order of ε, indicated in the non-linear source in Equation 9.

$$P(x,y) = pa\cos^2\left(\pi\frac{x}{L}\right)\cos^2\left(\pi\frac{y}{B}\right) \tag{10}$$

This equation is valid for $-L/2 \le x \le L/2$ and $-B/2 \le y \le B/2$, where L is the ship length and B is the ship beam. (x,y) represents the coordinates of a point on the floating ship in a body fixed reference and the origin of the coordinate reference (x_p, y_p) for an instant t is given by $y_p = y_0$ and $x_p = x_0+Vt$, where V is the speed of the ship, x_0 and y_0 represent the position of the reference origin to t = 0.

The *pa* value corresponding to the maximum of the pressure distribution was established using the procedure of Nascimento et al. (2008), which made an analogy between ship generated waves and the waves generated by a solid body that slides into the fluid. Since the wave generated on the first case are also due to the motion of a solid body in the fluid, such analogy is possible because the wave patterns are similar.

The classification of waves generated by the impact of a solid body is defined by theoretical solutions (Noda, 1970) and experimental results (Wiegel et al., 1970). This classification was based in the relationship between the body height and the Froude number for sliding. The same concept was applied by Nascimento et al. (2008) which observes the same relationship between the height of the body and the Froude number in the characteristics of the generated wave pattern.

The dimensional analysis of the ship wave problem results in:

$$\frac{H}{h} = f\left(\frac{B}{h},\frac{D}{h},\frac{L}{h},\frac{U}{\sqrt{gh}},\frac{L_c}{h},\frac{x}{h},\frac{\rho_w}{\rho_{SE}},T\sqrt{\frac{g}{h}},\frac{l}{h}\right) \tag{11}$$

where B is a ship beam, D is the ship draft, L is the ship length, U is the ship velocity, L_c is the channel width, x is the distance from the ship to shoreline, ρ_w is the density of the fluid, ρ_{SE} is the density of the ship, h is the wave height, T the wave period and l is the wave length. The dimensionless parameters for this problem are similar to ones of the waves generated by the impact of a block. This confirms the analogy between the two phenomena and so studying this problem may be a viable way to analyze the problem of generated ship waves.

Making a transformation in Equation 11 to include the block coefficient and the shoreline slope, quantities that are present in all relations that involve the ship characteristics and the local navigation, one has:

$$\frac{H}{h} = f\left(S,\frac{D}{h},\frac{h}{L},F_h,\frac{h}{L_c},I_s,T\sqrt{\frac{g}{h}}\right) \tag{12}$$

with $S = (BD)/A, F_h = U/\sqrt{gh}$ and $I_s = h/x$, where A is the cross sectional area of the channel, F_h is the depth Froude number and I_s the shoreline slope.

From Equation 12 it is expected that *D*/*h* has the same relation that the ratio between the block height and the depth of the channel, *e*/*h*. In this work, it will be assumed that the maximum pressure value equals the ratio between the draft and the channel depth:

$$pa = \frac{D}{h} \tag{13}$$

where $\frac{D}{h} = O(\varepsilon)$ and $\frac{h}{L} = O(\mu)$.

3 EXPERIMENTAL PROCEDURES

In experimental work, a systematic procedure is performed to collect data of the free-surface elevation of the waves generated by a ship model be used for verifying the numerical results.

The experiments were carried out at the National Laboratory of Civil Engineering (LNEC) in Lisbon, at a tank 17.45 m long and 16 m wide, with a water depth of 0.3 m.

The ship model used in the tests has a length of 1.6 m, a beam of 0.248 m and a draft of 0.0475 m that corresponds to the light ship condition of the "Gulf Stream" at a 1:50 scale. The model was pulled by a towing system connected to a motor in order to drive the desired speed for moving the ship model. This model was fixed at the fore and aft by strings to a taut line which passed over sheaves at two ends of the tank.

The wash wave profiles produced by ship model were measured by resistive gauges distributed along the sailing line and placed at different positions away from that line, as shown in the Figure 2.

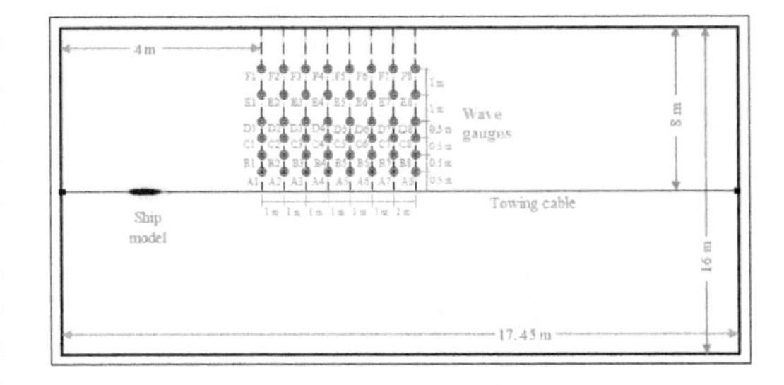

Figure 2. Location of the gauges for the experiments.

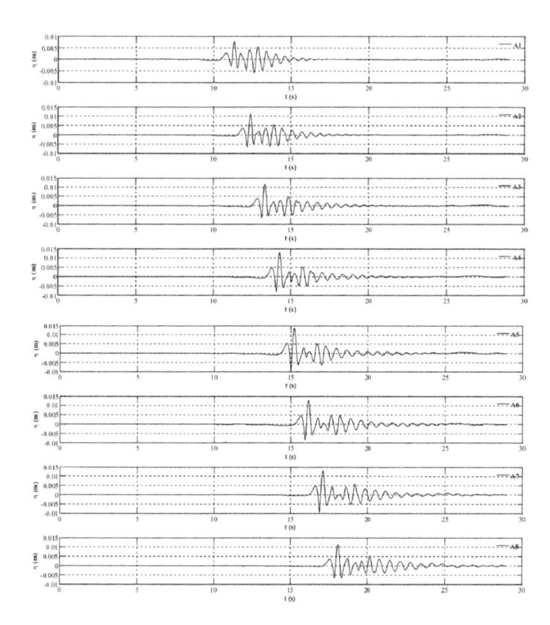

Figure 3. Free-surface elevation for the waves gauges A1 to A8 parallel to the sailing.

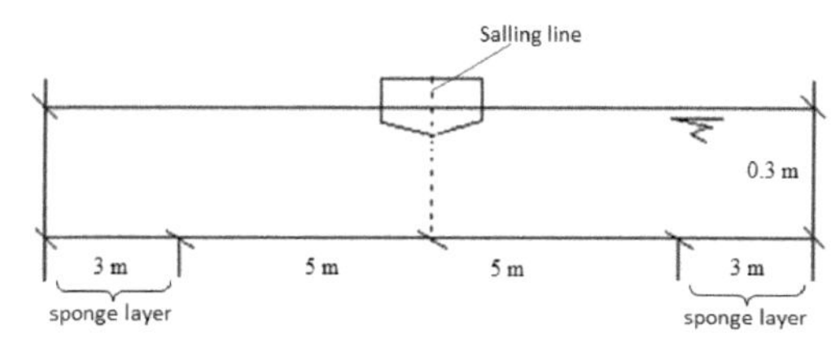

Figure 4. Channel cross-sectional profile used in the numerical simulation.

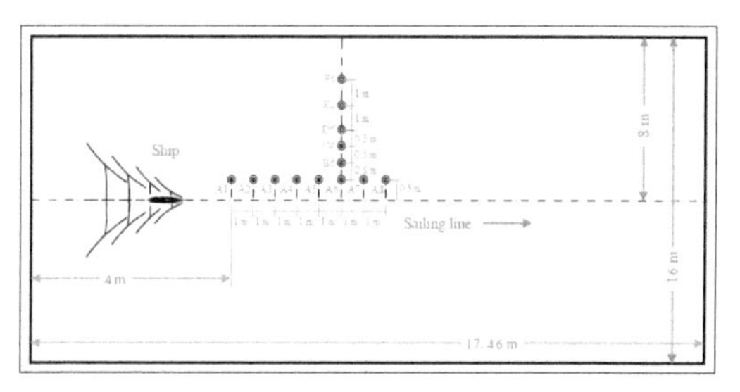

Figure 5. Wave gauges positions in the computational domain.

The resistive gauges were connected to a computer for data storage and analysis.

The ship model speed can be introduced through an interface developed in LabView that allows sending the speed signal to the motor.

Figure 3 show the development of the leading waves at 1 m intervals along the track of the ship model measured at 0.5 m from the sailing line (gauge A1 to gauge A8), for the Froude number 0.8.

4 RESULTS

4.1 *Numerical tests*

In order to evaluate the performance of the modified FUNWAVE, this section presents the results based on numerical simulations of the propagation of ship-generated waves. The ship is represented by a moving pressure distribution function at the free surface.

The numerically results are obtained for the pressure distribution proposed by Li & Sclavounos (2002) for a ship with a length of 1.6 m, a beam of 0.248 m and a draft of 0.0475 m that sails along the centerline of the channel with a constant velocity over a constant depth of 0.3 m. Parallel to the sailing line, on both sides of the channel, there are sponge layers 3 m wide to prevent the reflection back into the domain of the waves propagated up to them, Figure 4.

To analyse the propagation of ship waves over a constant depth bottom, eight equidistant points

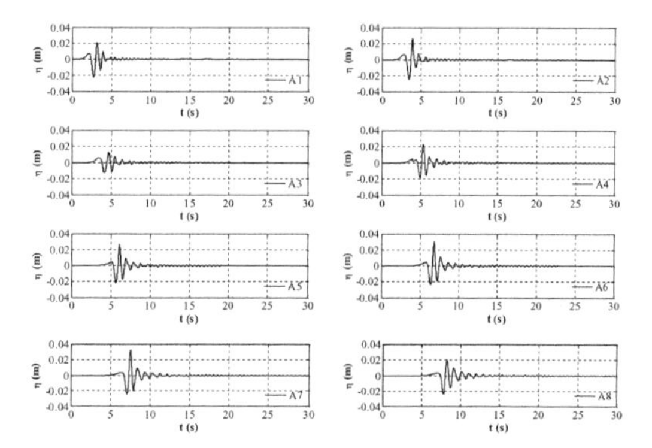

Figure 6. Time series of the free surface elevation for the gauges parallel to the sailing line, for $F_h = 0.8$.

are located at 0.5 m distance from the sailing line (gauges A1 to A8) and six points located at different distances from the sailing line (gauges A6, B6, C6, D6, E6 and F6 located at 0.5 m, 1 m, 1.5 m, 2 m, 3 m and 4 m, respectively), were considered as indicated in Figure 5.

The numerical simulations have been performed for different speeds of 1.37 m/s, 1.72 m/s and 1.89 m/s, which correspond to the depth Froude number, F_h, of 0.8, 1.0 and 1.1, respectively.

Figures 6 to 8 show the time series of the free surface elevation for the points parallel to the sailing line (gauges A1 to A8) for the three

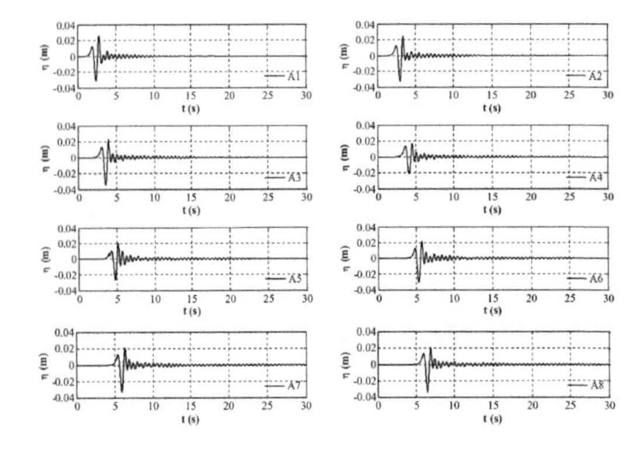

Figure 7. Time series of the free surface elevation for the gauges parallel to the sailing line, for $F_h = 1.0$.

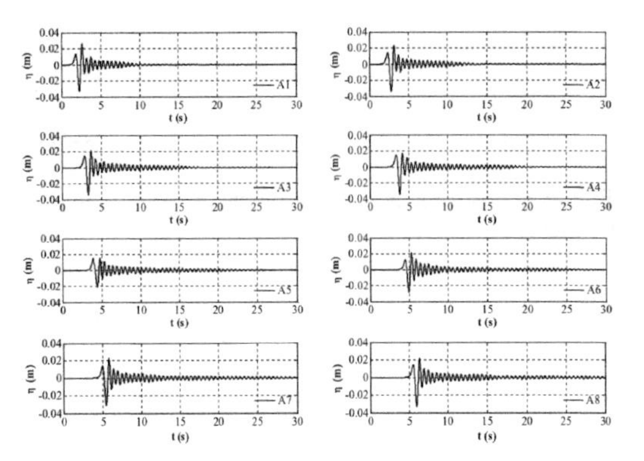

Figure 8. Time series of the free surface elevation for the gauges parallel to the sailing line, for $F_h = 1.1$.

depths Froude numbers of 0.8, 1.0 and 1.1. These figures show the typical wave generated by a sailing ship that has an initial peak at the bow, a deep trough that is referred to as drawdown and another peak at the stern followed by a train of waves.

Thus, according to the Figures 6 to 8, as the ship speed increases, so does the amplitude of the deep trough or drawdown increases. For example, for the gauge A1 the drawdown amplitude increases from roughly 0.023 m to 0.033 m when the depth Froude number increases from 0.8 to 1.1, being 0.031 m for the depth Froude number 1.0. The period of the drawdown also decreases with increasing ship speed, as expected. So, for the gauge A1 the period of the drawdown decreases from roughly 0.84 s to 0.68 s, when the depth Froude number increases from 0.8 to 1.1, being 0.72 s for the depth Froude number 1.0.

The figures also shows for the three depth Froude numbers a secondary wave train generated that appear after the ship passes. These secondary waves consisted mainly of the divergent waves and as the ship velocity increasing the height of the secondary waves increases.

Figures 9 to 11 show the time series of the free surface elevation for the waves located at different distances from the sailing line (gauges A6, B6, C6, D6, E6 and F6) for the three depths Froude numbers of 0.8, 1.0 and 1.1. Gauge A6, located at 0.5 m away from the sailing line, is the closest to that line, whereas gauge F6, located 4 m away the same line, is the farthest to the sailing line. The simulated results for the three depths Froude numbers show that the maximum wave height decreases as the distance from the sailing line increases. This height attenuation is due to diffraction—spreading the wave energy along the wave crest. Thus, for the depth Froude number

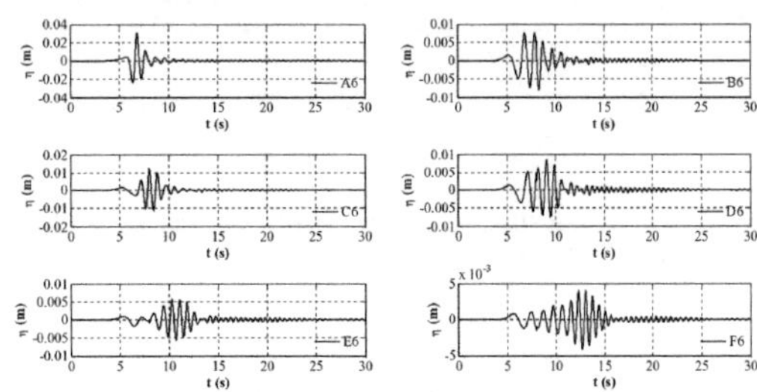

Figure 9. Time series of the free surface elevation for the gauges perpendicular to the sailing line, for $F_h = 0.8$.

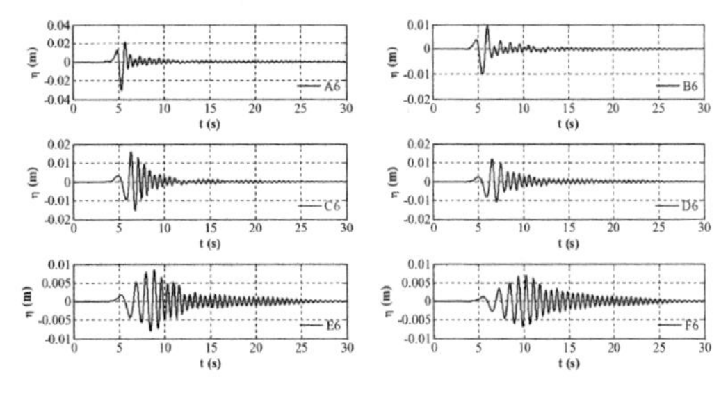

Figure 10. Time series of the free surface elevation for the gauges perpendicular to the sailing line, for $F_h = 1.0$.

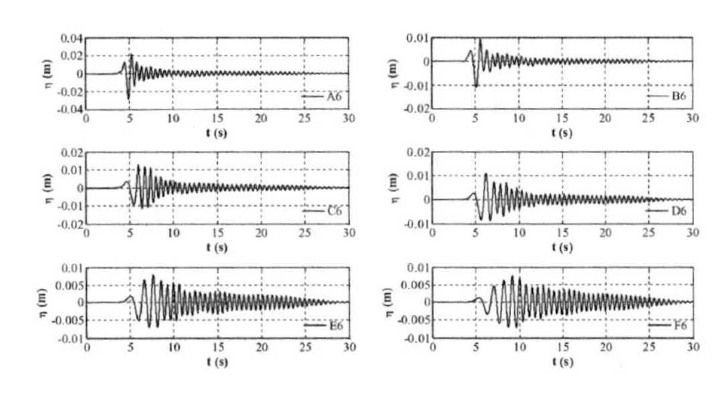

Figure 11. Time series of the free surface elevation for the gauges perpendicular to the sailing line, for $F_h = 1.1$.

of 0.8 the maximum wave height decrease from 0.054 m, at the gauge A6, to 0.008 m, at the gauge F6. For the depth Froude number 1.0 and 1.1 the maximum wave height decrease from 0.052 m to 0.014 m and from 0.050 m to 0.015 m, respectively.

The figures also show that the magnitude of the drawdown observed at gauge A6 near the sailing line decreases as the distance from the sailing line increases, as can be observed for the gauge F6 located farthest to that line. Thus, for the depth Froude number of 0.8 the drawdown amplitude decrease from 0.0236 m, at the gauge A6, to 0.0013 m, at the gauge F6. For the depth Froude number 1.0 and 1.1 the drawdown amplitude decrease from 0.0305 m to 0.0027 m and from 0.0283 m to 0.0033 m, respectively.

4.2 *Numerical comparison with experimental data*

This section presents some results of the numerical validation of the predictions of wave heights given by the modified FUNWAVE code trough the comparison with experimental data of the free-surface elevation, described in section 3, obtained in the scale ship model 1:50.

The Figure 13 compares the results obtained by the numerical code and the experimental tests of the maximum free-surface elevation, for the depth Froude number 0.8. The figure presents the evolution with the distance to the sailing line (y-coordinate) of the maximum free surface elevation along the tank. The numerical code and the experiments show very similar results for the maximum wave height. Although in the experimental tests, the ship model has some speed fluctuations along its trajectory, the maximum wave height values approach to the numerical results.

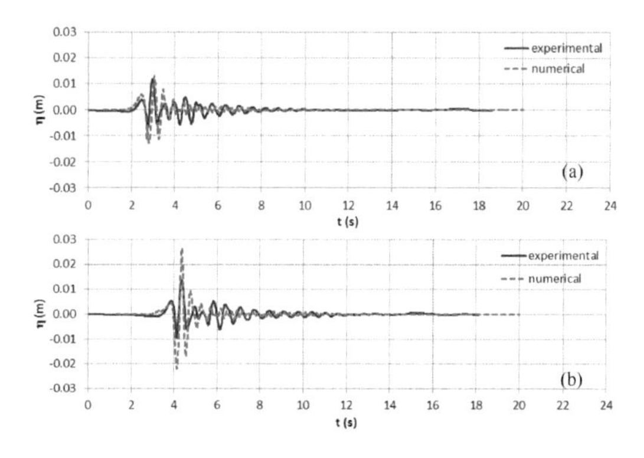

Figure 12. Time series comparison of the free-surface elevation for numerical (dashed line) and experimental data (solid line), for (a) the gauge A3 and (b) the gauge A5.

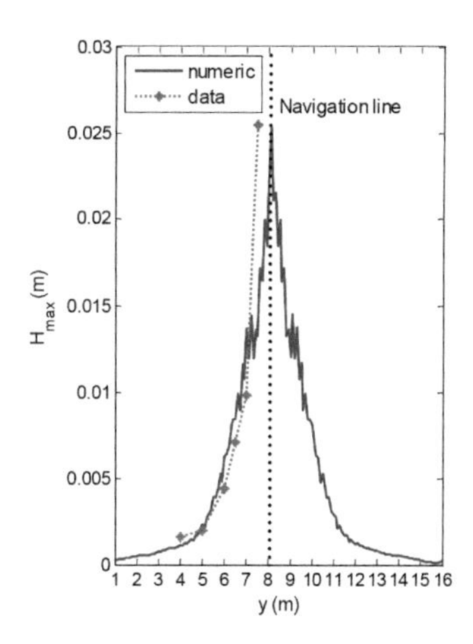

Figure 13. Comparison of the numerical results and experimental data of the maximum wave heights for $F_h = 0.8$.

5 CONCLUSIONS

This paper presents the results of the modified FUNWAVE code with the pressure distribution function at the free surface proposed by Li and Sclavounos (2002), to simulate the waves generated by a sailing ship.

Tests were made for a ship sailing along the centerline of the channel with a constant velocity over a constant depth.

The time series of the free surface elevation for the points parallel to the sailing line (gauges A1 to A8) for the three depth Froude numbers of 0.8, 1.0 and 1.1, show the typical wave generated by a sailing ship that has an initial peak at the bow, a deep trough that is referred to as drawdown and another peak at the stern followed by a train of waves. As the ship speed increases, so does the amplitude of the deep trough or drawdown increases. The period of the drawdown also decreases with increasing ship speed, as expected.

During the experimental tests there was some aspects that can be limitations of the experimental model, such as: the towing cable makes "belly" at middle, whereby the linear speed of the cable and consequently of the model may be smaller than if the cable did not do "belly"; the cable elasticity tends to stretch and recover periodically at least until the model reaches constant speed.

In order to validate the predictions of wave heights given by the modified FUNWAVE when the ship moves in limited depth, the time series of the free-surface elevation obtained in the scale model were compared with those provided by the numerical FUNWAVE code. The comparison of the numerical and the experimental times series shows some differences between them, which can be due to limitations in the experiments. However, comparison of the maximum wave heights between the results obtained by the numerical code and the experimental tests shows very similar results for the maximum wave height.

ACKNOWLEDGEMENTS

The first author acknowledges the support of Fundação para a Ciência e a Tecnologia, Portugal through the PhD grant SFRH/BD/75223/2010.

The first author also would like to thank to Maria Francisca do Nascimento from Universidade Estadual da Zona Oeste of Rio de Janeiro, Brazil, and to Professor Cláudio Neves from Universidade Federal of Rio de Janeiro for the availability of the modified FUNWAVE model to perform this work that is part of her PhD thesis work.

The experiments with a ship model were not possible without the collaboration of the Ports and Maritime Structures Unit of Nacional Laboratory of Civil Engineering (LNEC) in Lisbon.

This work contributes to the M&MShips project "Manoeuvring & Moored SHIPS in Ports" (PTDC/EMS-TRA/5628/2014) funded by the Portuguese Foundation for Science and Technology.

REFERENCES

Ahmed, Y. & Guedes Soares, C. 2009. Simulation of Free Surface Flow around a VLCC Hull using Viscous and Potential Flow Methods. *Ocean Engineering.* 36(9–10):691–696.

Ciortan, C., Wanderley, J. & Guedes Soares, C. 2012. Free surface flow around a ship model using an interface-capturing method. *Ocean Engineering.* 44:57–67.

Cornett, A., Tschirky, P., Knox, P., Rollings S. 2008. Moored Ship Motions due to Passing Vessels in a Narrow Inland Waterway. *Proceedings of the Coastal Engineering*, Hamburg, Germany, 722–734.

Dam, K.T., Tanimoto, K., Nguyen, B.T. & Akagawa, Y. 2006. Numerical study of propagation of ship waves on a sloping coast. *Ocean Engineering*, 33, pp. 350–364.

Li, Y. & Sclavounos, P.D. 2002. Three-dimensional nonlinear solitary waves in shallow water generated by advancing disturbance. *Journal of Fluid Mechanics*, v. 470, pp. 383–410.

Nascimento, M.F., Neves, C.F. & Maciel, G.F. 2008. Propagation of ship waves on a sloping bottom. *Proceedings of the 31st International Conference on Coastal Engineering*, Hamburg, Germany.

Noda, E. 1970. Water waves generated by landslides. *Journal of Waterway, Port, Coastal, Ocean Div., Am. Soc. Civ. Eng.*, v. 96(4), pp. 835–855.

Nwogu, O. 1993. Alternative form of Boussinesq equations for near-shore wave propagation. *Journal of Waterway, Port, Coastal, and Ocean Engineering*, 119(6), pp. 9–6, pp. 618–638.

Nwogu, G.O. & Demirbilek, Z. 2004. Numerical modeling of ship-induced currents in confined waterways. *Proceedings of the 29th International Conference on Coastal Engineering*, pp. 256–268, Lisbon.

Schipper, de M.A. 2007. On the generation of surfable ship waves in a circular pool: Part I; Physical background and wave pool design. MSc. Thesis, Delft University of Technology.

Sorensen, R.M. 1967. Investigation of ship-generated waves, *Journal of the Waterways and Harbor Division*, ASCE, pp. 85–99.

Stockstill, R.L. & Berger, R.C. 2001. Simulating barge drawdown and currents in channel and backwater areas, *Journal of Waterway, Port, Coastal and Ocean Engineering*, ASCE, pp. 290–298.

Wei, G. & Kirby, J.T. 1995. Time-dependent numerical code for extended Boussinesq equations. *Journal of Waterway, Port, Coastal and Ocean Engineering*, ASCE, pp. 251–261.

Wiegel, R.L., Noda, E.K., Kuba, E.M., Gee, D.M. & Tornberg, G.F. 1970. Water waves generated by landslide in reservoirs, *Journal of the Waterways and Harbors Division*, 96(2), pp. 307–333.

Ship motions II

Progress in Maritime Technology and Engineering – Guedes Soares & Santos (Eds)
© 2018 Taylor & Francis Group, London, ISBN 978-1-138-58539-3

Hydrodynamic study of the influence of bow and stern appendages in the performance of the vessel OPV 93

B. Verma, D. Fuentes, L. Leal & F. Zarate
COTECMAR, Cartagena, Colombia

ABSTRACT: Computational Fluid Dynamics (CFD) has become nowadays an important tool in the process of hydrodynamic design of modern ships. CFD is used to model any phenomena related to fluid flow in a control volume like a ship or any offshore structure in the sea. In the present study, the impact of bow and stern appendages on hydrodynamic performance for an Offshore Supply Vessel (OPV) is investigated through application of CFD. The study shows the process of modeling, validation and qualitative analysis of the OPV 93 design, by de-signing three types of bulbous bow and their integration with the hull and also describes the sizing and analysis of three different aft appendages (Interceptor, Flap and Wedge), taking into account the operational profile of the ship with the aim to reduce the ship drag, improve trim control and heave movement for high speed ships and to improve the propulsion efficiency.

1 INTRODUCTION

The Corporation of Science and Technology for the Development of the Maritime and Fluvial Naval Industry, Cotecmar has built three offshore patrol vessels (OPV) that are presently being operated by the Colombian navy. In order to comply with new mission requirements, capacities and growth margins, a new design of a second generation OPV type vessel was initiated. The preliminary design phase of this vessel is currently under development.

This preliminary phase of the project is aimed at optimizing the overall drag or resistance of the vessel, which influences the amount of installed power, engine weight and fuel consumption.

The CFD (Computational Fluid Dynamics) has become very important tool for carrying out the ship resistance analysis, as applied in (Ahmed et al. 2015), (Remolà et al. 2013), (Aksenov et al. 2015) to determine the ship resistance of advance considering the hull and the skeg, where the use of CFD tools has proven to be a suitable solution for doing analysis in the preliminary design phase.

Ship resistance analysis is of vital importance in giving greater viability to the development of a design project of a ship. These analyses are normally carried out in towing tank with the aim to optimize the ship resistance curve, to do seakeeping analysis and for hull form optimization. These studies can be done in advance using CFD tool, as presented in (Aksenov et al. 2015).

The selection and sizing of bow and stern appendages and their final integration with the optimized hull was made using the CFD tool based on similar analyzes as those presented in (John et al. 2012) y (Sharma and Sha 2005). In addition, the effectiveness of the appendices has been discussed taking as a criterion the reduction of resistance using respective appendage compared to the naked hull resistance (Karimi et al. 2013). Additionally, in the present study the appendices will be selected considering the operational profile of the vessel OPV 93C.

2 CFD SIMULATION TOOL

In this study, the CFD software used was a commercial viscous code based on a finite volume discretization. An unsteady Reynolds Average Navier-Stokes equation (RANS) approach was applied using the Star-CCM+ CFD software to resolve hydrodynamic problems.

2.1 *Problem description*

Initially the validation of the computational model will be carried out using the experimental results of a ship model tested in towing tank with the CFD model for the same ship, for which the resistance curve will be obtained and the percentage difference will be compared between the CFD results and experimental results up to the point of having maximum of 5% difference as one of the main objective of the quantitative analysis presented in this paper.

On the other hand, after the correct validation of the model, the hull of the OPV 93 vessel will be analyzed for the maximum design speed of 20 knots. The analysis was carried out in calm and deep water, and five different advance speeds were simulated to generate the resistance curve of the bare hull model. Subsequently, three alternative designs of bulbous bows and three stern appendages will be evaluated. This evaluation will be done simulating under the same configuration of the CFD model. Finally, the resistance curve for each model will be obtained, and based on the results, the effectiveness of the bulbous bow and aft appendage alternatives will be defined, selecting the appendices showing greater reduction in ship resistance.

2.2 *Applied methodology*

Figure 1 describes the algorithm applied for the validation of the computational model configuration, with CAD model as an input, then the meshing is performed, then evaluation through CFD and the verification of the validity of the model, making additional iterations if necessary.

2.3 *Geometrical model configuration*

The equations that govern the phenomenon to be analyzed are the Reynolds Averaged Navier Stokes (RANS) equations for an unstable, three-dimensional and incompressible flow.

To model the fluid flow, the software solver uses a finite volume method that discretizes the integral formulation of the Navier-Stokes equations. The RANS solver uses a predictor-corrector approach to link the equations of continuity and momentum.

The turbulence model selected in this study was a standard k–ε model. The use of the turbulence model formulation of the standard k–ε equation is reasonably robust, reliable near solid boundaries and recirculation regions such as ship boundary layers.

The height of the computational domain is 3.5 LWL (Length at waterline) and its width is taken as 2 LWL. Due to the symmetry of the problem, only the port side of the hull was modeled and meshed, significantly reducing the number of grid cells.

The velocity inlet limit of the computational domain is at a distance of one LWL ahead of the ship, while the pressure outlet limit is three LWL from the stern of the vessel. The computational domain dimensions satisfy the ITTC procedure, which recommends that the inlet velocity limit must be 1–2 LWL and the pressure outlet limit must be between 3–5 LWL away from the hull to avoid wave reflections (ITTC 2011).

The vessel under analysis corresponds to the hull and deck geometry for the vessel OPV 93, as shown in Figure 2 and whose main dimensions are shown in Table 1.

CAD Model
Surface & Volume Mesh
Boundary Conditions
CFD Solver
Results interpretation
Expt./CFD < 5%
NO
YES
Acceptance of CFD Model Setup

Figure 1. CFD model validation algorithm.

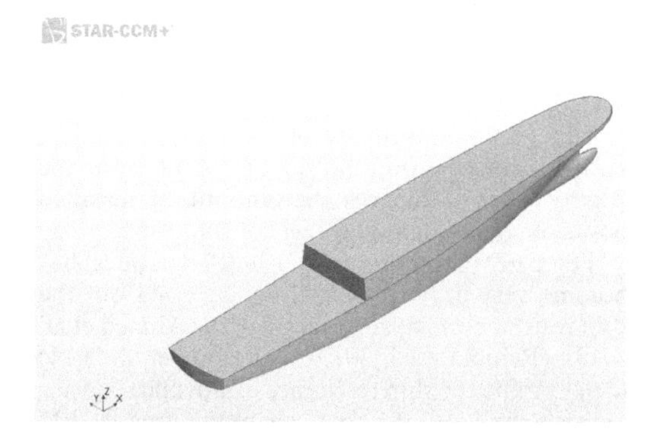

Figure 2. 3D hull model—OPV 93.

Table 1. Ship main particulars—OPV 93.

Parameter	Value
LWL	82.60 m
B (Breadth)	14.00 m
D (Depth)	7.00 m
T (Draft)	4.10 m
Displacement	2669 tons

2.4 *Model boundary conditions*

Given the geometric symmetry of the hull with respect to the ship centerline and making assumptions such as the sea condition, which for the model is calm water where there are no reflections or sum of waves that can alter the results of the calculations on the free surface, the declaration of symmetry of the domain is valid and adequate for this study. Therefore, the symmetry plane condition only reflects the results of half of the analyzed or simulated domain.

The boundary conditions applied at each boundary for analysis using Star CCM+ are summarized in Table 2 and shown in Figure 3.

2.5 *Mesh settings*

For the validation of the model, a mesh of approximately 1.8 million cells has been configured with 6 regions, each with at least 2 levels of refinement. On the other hand, for the OPV 93 vessel case study, a mesh of approximately 2.1 million cells was generated, adding the bow and stern appendages to the model, since the geometry of the model for validation does not include them.

Table 2. Boundary conditions.

Parts	Boundary type
Inlet	Velocity Inlet
Outlet	Pressure Outlet
Side	No-slip Wall
Hull	No-slip Wall
Symmetry	Symmetry plane
Top	No-slip Wall
Bottom	Stationary No-Slip Wall

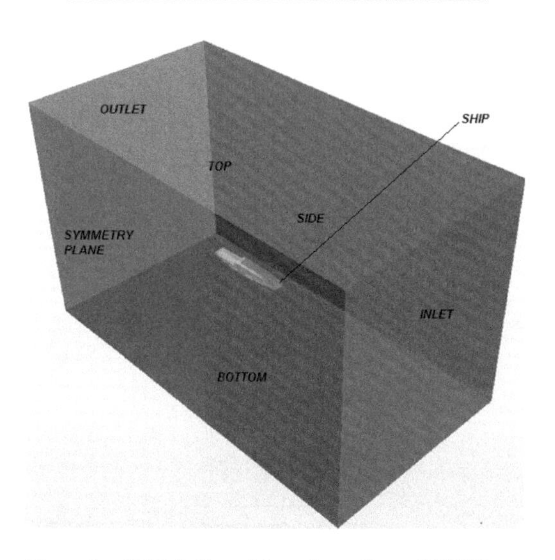

Figure 3. CFD hull model configuration—OPV 93.

3 LITERATURE REVIEW

3.1 *Bulbous bow design*

The main objective of the bulbous bow is to reduce bow wave height caused by local pressure disturbance formed in the bow of the vessel during its motion as indicated by Figure 4. As can be seen that the height of the bow wave is substantially reduced, which reduces the hull drag associated with the bow wave, thereby improving the fuel economy and increases ship range.

Bulbous bow are not efficient for all speed ranges. Their applicability and effectiveness is related to the range of Froude number between 0.238 up to 0.563 as proposed by Wigley (Wigley 1935–6).

Considering the fact that the above mentioned range of Fn is very wide, ship refinement parameter which can also affect the adoption of bulbous bow was considered in the present study, where it indicates that the bulbous bow is not recommended if the relation (Cb*B)/LPP > 0.135 (Wigley 1935–6), where Cb is the block coefficient, B is the beam and LPP is the length between perpendiculars.

The procedure and methodology for determining the dimensions of the bulbous bow is the same as explained in (Sharma et al. 2005) y (Kracht 1978) using 6 parameters (linear & nonlinear) shown in Figure 5.

3.1.1 *Types of bulb sections*

The bulb sections can be broadly classified into 3 main types: Δ-type, O-type and ∇-type as shown in Figure 6, as presented in (Kracht, 1978).

Not all the types of bulbs indicated in Figure 5 are suitable for the OPV 93 project, since each one has its function based on the hull form and ship service type.

As per the research of (Kracht, 1978), it has been found that the Δ-type & O-type bulb sections are more susceptible to suffer slamming because of its

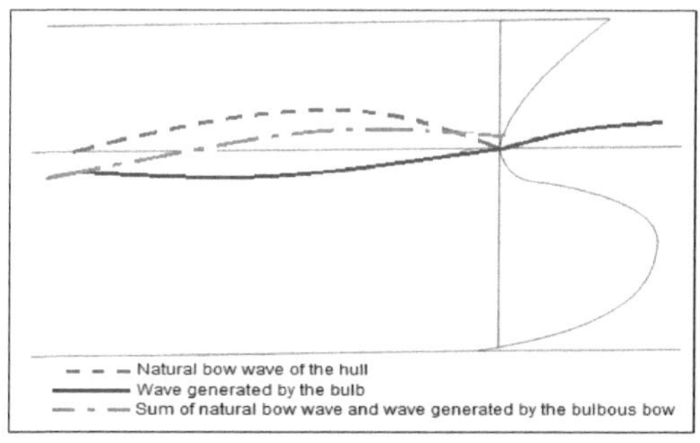

Figure 4. Bow wave attenuation by use of bulbous bow.

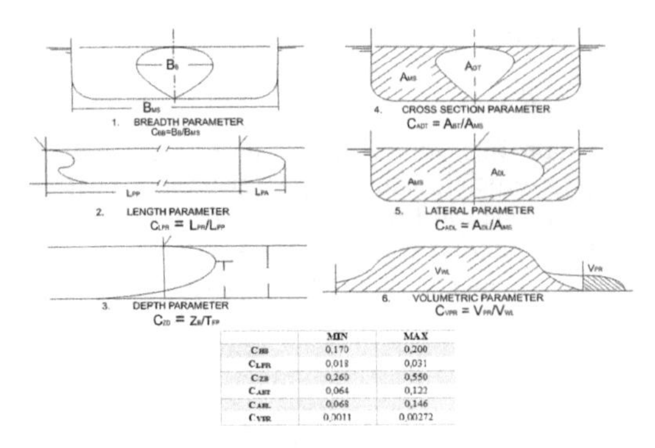

	MIN	MAX
C_{BB}	0,170	0,200
C_{LPR}	0,018	0,031
C_{ZB}	0,260	0,550
C_{ABT}	0,064	0,122
C_{ABL}	0,068	0,146
C_{VPR}	0,0011	0,00272

Figure 5. Parameters for bulbous bow design (Kracht 1978).

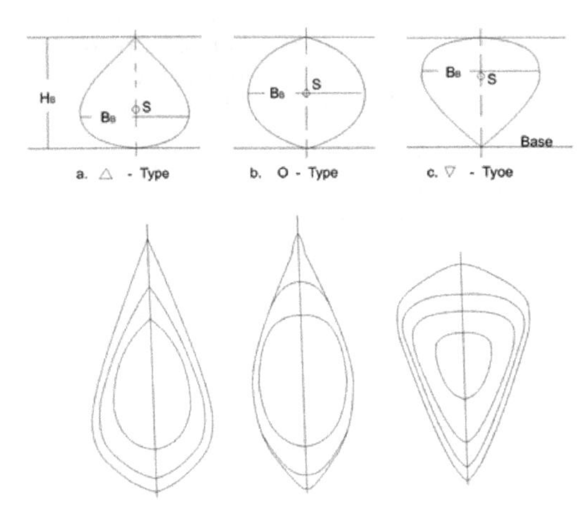

Figure 6. Bulbous bow types. (Kracht 1978).

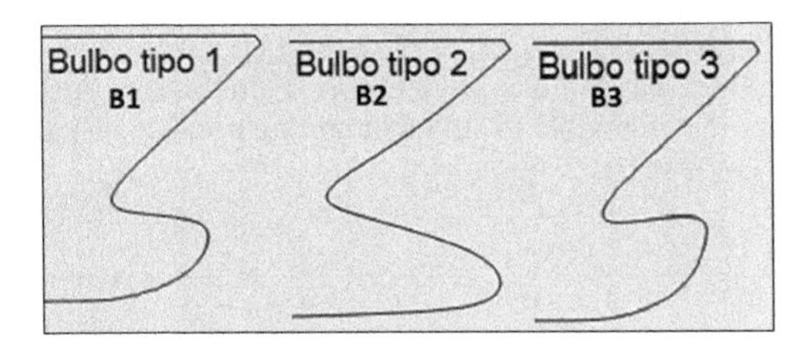

Figure 7. Bulb variants for evaluation.

lower part being flat. Looking for an OPV 93 design suitable for navigating in rough seas, Δ-type & O-type would be discarded and the ∇-type bulbs that maintain their fine entrance angle and higher center of gravity allowing the ship to improve its performance at maximum load and keep the slamming impacts low.

With this bulb type selection approach, we proceed to analyze three variants of ∇-type bulbs namely B1, B2 and B3, as indicated in Figure 7. These variations will be evaluated to identify how to improve the dissipation of the wave train generated by the hull.

3.2 *Stern appendages design*

The aft appendages are classified into three main types: Flap, Wedge and Interceptor. For the case of medium or heavy vessels, the main function is to smooth the stern waves and reduce the pressure drag resulting from changes in the flow field, thereby reducing the ship resistance and improve overall propulsive efficiency.

The design of these aft appendages uses different design parameters: the wedge and flap considers the chord length (CW or CF), entry angle (α_W & α_F) and the width; whereas for the interceptor the height (h) and the width are used. Figure 8 shows these appendices with their main dimensions, expect for the width, which is defined for all as 100% of the transom beam on the design waterline.

The selection of the aft appendages is made based on parametric studies that allow identifying the appropriate dimensions for the new design based on the values of similar vessels tested experimentally.

For the sizing of the flap and the wedge, trend graphs shown in Figure 9 and Figure 10 based on the experimental results of ship resistance developed by U.S. navy (Gabor K & Cheng W 1999) for models of naval ships that have used and implemented these type of appendages (flap and wedge) were used. The dimensions for the wedge and flap obtained from this study are presented in the Table 3.

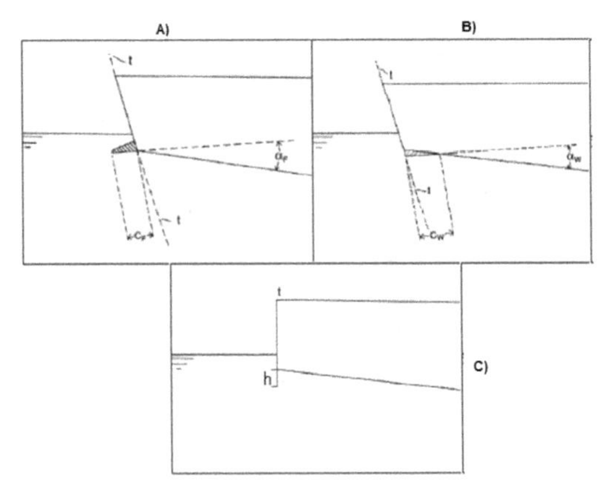

Figure 8. Stern appendages type and design (Jiménez R. 2009).

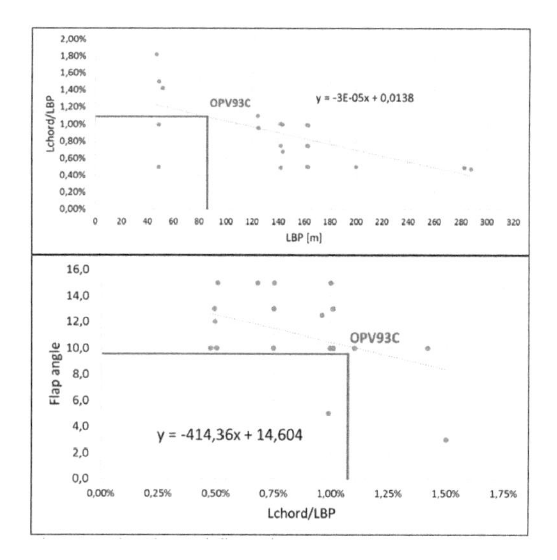

Figure 9. Flap size and dimensions.

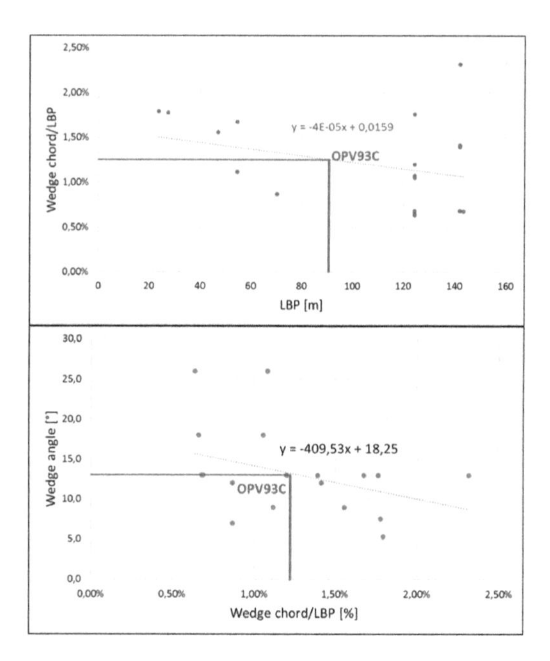

Figure 10. Wedge size and dimensions.

For the dimensioning of the interceptor, the experimental results developed by the Sharif University of Technology (Karimi et al. 2013) for ship models that have implemented this type of appendage (interceptor) were taken as a basis, making use of trend graphs as shown in Figure 11. The dimensions for interceptor obtained from this study are presented in the Table 3.

Table 3. Stern appendages sizing.

Stern appendage	Height (h) or Chord length (C_W/C_F)	Entry Angle (α)
Flap	1.12% LPP	10°
Wedge	1.25% LPP	13°
Interceptor	0.38% LPP	N/A

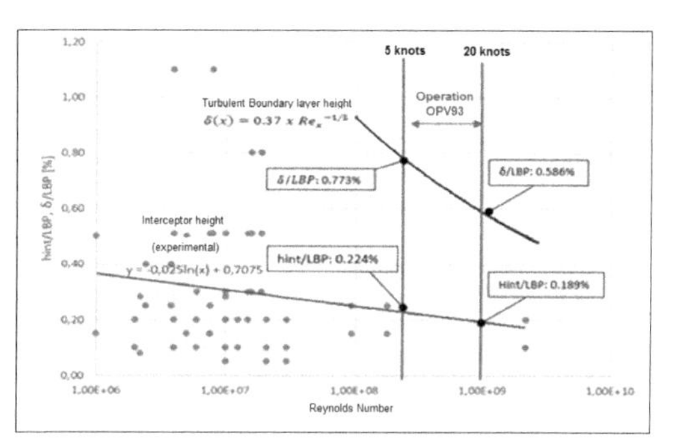

Figure 11. Interceptor size and dimensions.

4 ANALYSIS

4.1 *Model validation*

Reference vessel used for model validation is the first offshore patrol vessel constructed by COTECMAR, the OPV 80 or ARC 20 de Julio, currently in service. For this vessel the towing tank tests results of ship resistance are available for the hull with appendages (without propellers), with which the comparison was made with the results obtained from the CFD model configuration. The main dimensions of the OPV 80 model are presented in the Table 4.

4.1.1 *Model validation results*

For the purpose of model validation, contour plots of the free surface wake pattern obtained using CFD simulation of the OPV 80 model was compared with the theoretical kelvin wave pattern as presented in (Pritam & Premchand 2015), where in the whole pattern of divergent waves is roughly contained within two straight lines, which start from the pressure point and make angles of 19°28' on each side of the line of the motion, as shown in Figure 12.

As can be seen in Figure 12, the wave pattern generated by the CFD analysis obeys the behavior similar to the one presented in (Pritam & Premchand 2015), which shows that the computational model is adequately representing the physical phenomenon.

In all calculation variants convergence of iteration process was controlled both for residual values of all equations and coefficients of forces generated

Table 4. Ship main particulars—OPV 80.

Parameter	Value
LWL	74.40 m
B (Breadth)	13.00 m
D (Depth)	6.50 m
T (Draft)	3.90 m
Displacement	1800 tons

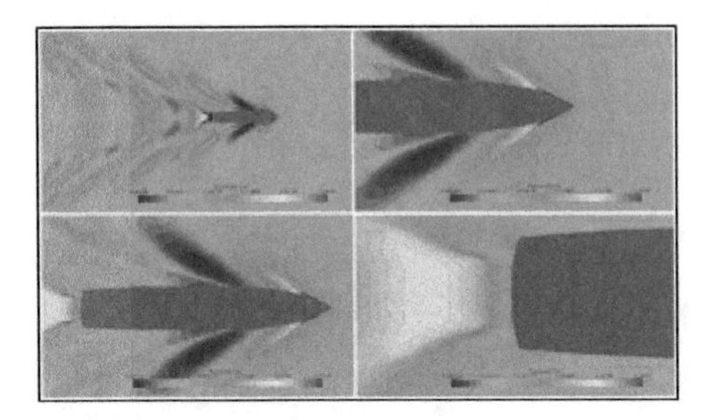

Figure 12. View of the free surface wave pattern of the vessel OPV 80.

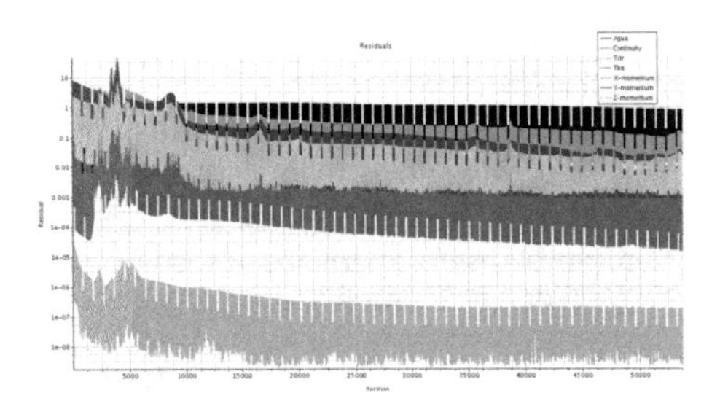

Figure 13. Typical diagram of iteration process convergence of residuals.

on the hull. An example convergence run (which was similar for all the tests) is presented in Figure 13. Iteration test was terminated when all the remainders (residuals) were below the value of 1e-04 and resistance coefficient value did not undergo changes any further.

In Figure 14 it is important to note that there are no excessive deformations of the streamlines surrounding the hull. The wave height generated in the bow is a key factor, because the main objective of the ship resistance optimization is to reduce the height of bow waves generated by the ship.

Graph 1 shows the comparison between the resistance curves obtained by experimental tests and by using CFD, and the percentage difference between the results is observed. This difference is always within ± 5% of the experimental value, therefore, based on the criterion defined in the validation algorithm; this CFD model closely validates the phenomenon under study. The results are presented in Table 5.

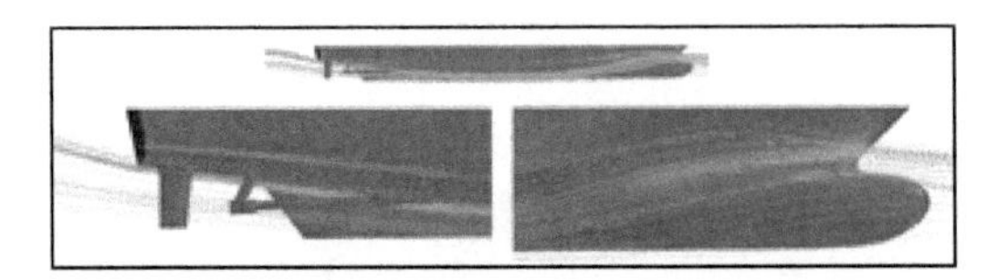

Figure 14. View of streamlines of vessel OPV 80.

Table 5. Ship resistance (Experimental vs. CFD).

Speed (knots)	Resistance OPV80 CFD (N)	Resistance OPV80 Experimental (N)	% difference
14	85.99	86.29	–0.34
16	122.33	121.78	0.45
18	167.40	160.99	3.83
20	205.15	214.88	–4.74

4.2 *Bulbous bow evaluation*

Model validation was followed with the study of three alternative designs of the bulbous bow, taking the bare hull resistance as a reference value.

4.2.1 *Bare hull*

The bare hull is the geometry of the appendage-free model and is the starting point for the analysis, since with this first simulation it will be possible to compare whether the use of a bow bulb is efficient and functional.

For the evaluation of bulbous bows, each of the three selected alternative designs is added to the bare hull. Streamlines around bare hull and three alternative designs are shown in Figure 16.

4.2.2 *Bulbous bow alternatives*

The three bulb alternatives were evaluated for five different speeds between 12 and 20 knots. Pressure field and streamlines around bare hull and three alternative designs are shown in Figure 16, the lowest bow wave can be observed for the bulb B1, and this means a decrease in wave-making resistance.

Graph 2 shows the diminution of total resistance for the each of the bulbous bow alternatives in comparison to bare hull. For high speeds, alternative B2 has the lowest total resistance, but for 12 knots, as being the economic speed of the OPV 93 defined by its operational profile, the alternative B1 offers the lowest total resistance, as shown in Table 6.

The geometry of the OPV 93C including B1 alternative is presented in Figure 17. The wave pattern generated by the hull with the B1 alternative is shown in Figure 18.

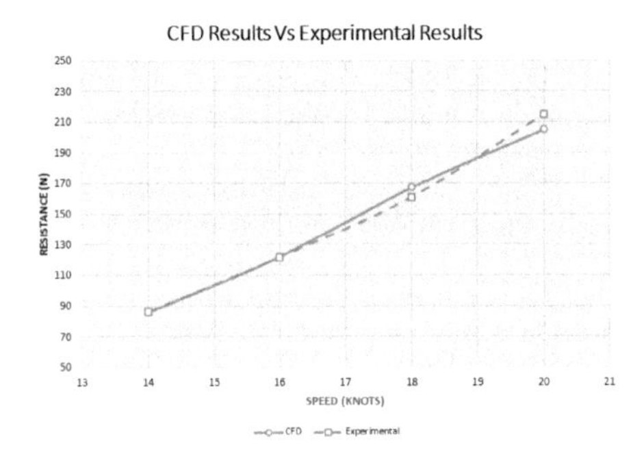

Graph 1. Ship resistance (Experimental vs. CFD).

Figure 15. OPV 93 bare hull sketch.

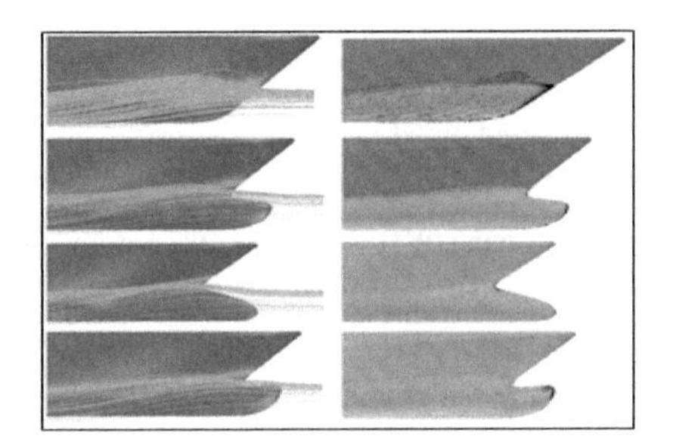

Figure 16. Streamlines & Pressure field—OPV 93.

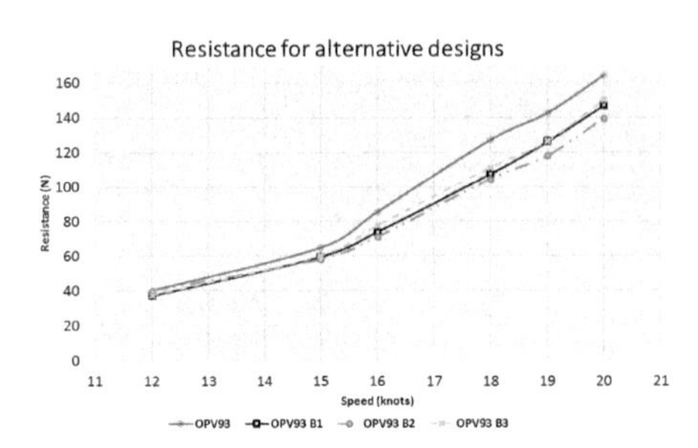

Graph 2. Ship resistance for 3 bulb alternatives.

Table 6. Ship resistance (Bare hull & 3 bulbous bow alternatives).

Speed (knots)	OPV 93 bare hull (N)	OPV 93 & bulb B1 (N)	OPV 93 & bulb B2 (N)	OPV 93 & bulb B3 (N)
12	40.00	37.05	39.97	37.27
15	64.73	59.50	57.99	60.10
16	85.32	73.82	70.87	77.52
18	126.91	107.39	104.45	111.08
20	164.03	146.67	139.46	150.05

4.3 *Evaluation of stern appendages*

After the selection of bulbous bow, the stern appendages are evaluated in conjunction with the selected bulb alternative B1. These appendages are added to the transom or behind it.

These devices create a vertical lift force at the transom and cause the flow to slow down under the hull at a location extending from its position

Figure 17. Sketch of OPV 93 with bulb B1.

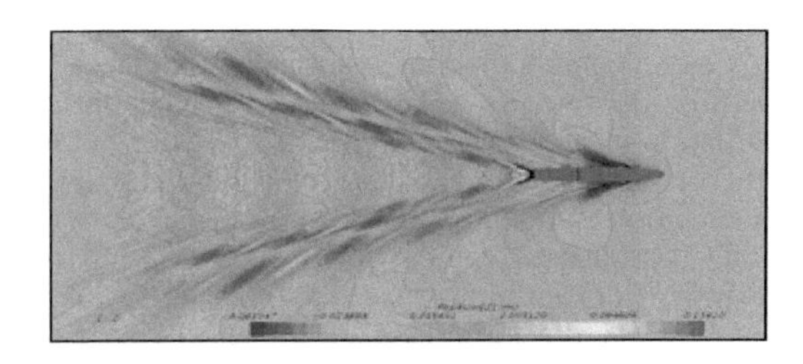

Figure 18. Kelvin wave pattern of OPV 93 with B1.

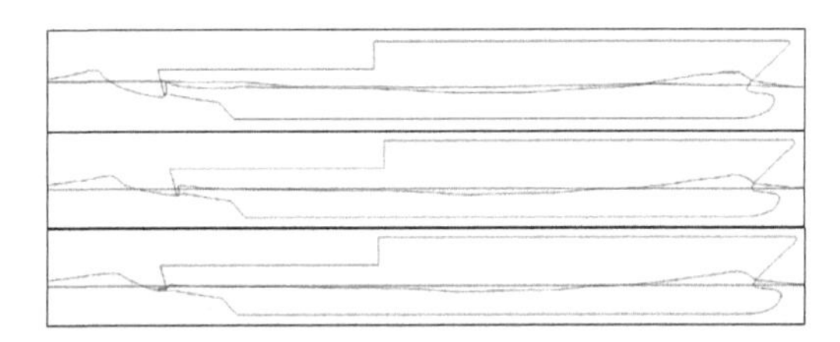

Figure 19. Wave profile generated by the three stern appendages: interceptor (top), wedge (middle) and flap (bottom).

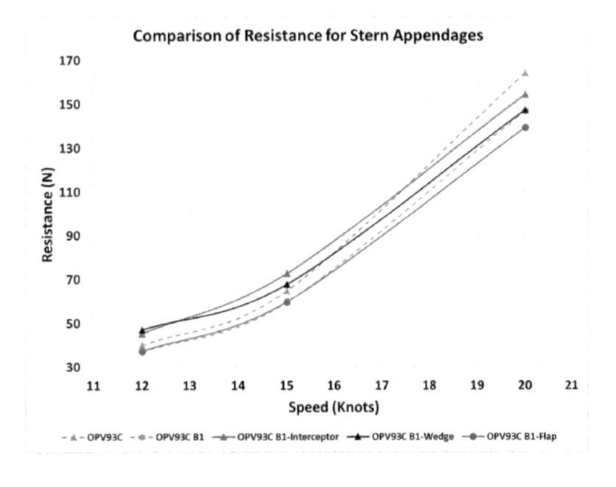

Graph 3. Ship resistance for stern appendages.

Table 7. Ship resistance (Bare hull & 3 bulbous bow alternatives).

Speed (knots)	OPV 93 bare hull (N)	OPV 93 B1 (N)	With Interceptor (N)	With Wedge (N)	With Flap (N)
12	40.00	37.05	45.28	47.06	37.32
15	64.73	59.50	72.60	67.65	59.76
20	164.03	146.67	154.28	147.20	139.15

to a point generally forward of the propellers. This decreased flow velocity will cause an increase in pressure under the hull, which in turn, causes reduced resistance due to the reduced after body suction force (reduced form drag) [John et al., 2011] as shown in Figure 19.

Graph 3 shows the total resistance curve generated by the three alternatives of stern appendage in comparison with bare hull and hull with bulb alternative B1 (OPV 93 B1). Implementation of interceptor and wedge causes an increase of the total resistance in comparison with OPV 93 B1. However flap addition reduces the total resistance for high speed ranges. This appendage is selected for the OPV 93 vessel which currently continues in its preliminary design phase. The results are presented in Table 7.

5 CONCLUSIONS

Model validation was made achieving a difference lower than 10% compared with towing tank results of OPV 80 hull. This validation was used as calibration of the CFD tool setup for the analysis of several appendages for the OPV 93 hull.

Bulbous bow implementation was probed as a method to reduce the resistance of a hull for this type of vessels. Results showed resistance reduction up to 15%.

The best bulb for the OPV 93 is the bulb B1 for the economic speed of 12 knots based on its operational profile. This selection should be revaluated if a different operational profile is required for this vessel.

Based on CFD evaluation, the best stern appendage was the flap; this appendage will reduce the resistance at high speeds.

6 FUTURE WORK

In the contractual phase of the project, towing tank tests for ship resistance calculation of the OPV 93 model with and without hull appendages will be performed. For future work it is recommended to use the towing tank results for CFD model validation and for the final selection of the aft appendages.

The present study was performed without activation of the rigid body motion. For future work the present CFD model will be improved by considering the rigid body motion using DFBI rotational and translational model activation to include the trim and sinkage effects. It is expected that this model will represent more accurately the hydrodynamic phenomenon of the vessel OPV 93.

Another piece of future work could be to include the effects of sea waves in the resistance calculations, since the ship rarely sail on calm water.

REFERENCES

Ahmed, Y, et al. 2015. Determining ship resistance using computational fluid dynamics (CFD). Journal of Transport System Engineering 2(1): 20–25.

Aksenov, A. et al. 2015. Ship hull form design and optimization based on CFD. Towards green marine technology and transport: 215–223. London: Taylor and Francis Group.

Gabor K & Cheng W 1999. Stern Wedges and Stern Flaps for Improved Powering—U.S. Navy Experience. SNAME Transactions 107: 67–99.

International Towing Tank Conference (ITTC) 2011. Practical guidelines for ship CFD applications. Proceedings of the 26th International Towing Tank Conference. Brazil.

John, S. 2012. Ship hull appendages: a case study. International Journal of Innovative Research and Development 1(10): 74–89.

Karimi, M. et al. 2013. An experimental study of interceptor's effectiveness on hydrodynamic performance of high-speed planning crafts. Polish Maritime Research 2(78) 20: 21–29.

Kracht, A. 1978. Design of bulbous bows. SNAME Transactions 86: 197–217.

Pritam K & M Premchand 2015. Numerical investigation of the influence of water depth on ship resistance. International Journal of computer Applications 116 (17): 11.

Rèmola, A. et al. 2014. A contribution to appendage drag extrapolation using computational tools. Developments in Maritime Transportation and Exploitation of Sea Resources: 67–72. A Coruña: Taylor and Francis Group.

Sharma, R. et al. 2005. Hydrodynamic design of integrated bulbous bow/sonar dome for naval ships. Defense science journal 55(1): 21–36.

Wigley, W. 1935–6. The theory of the bulbous bow and its practical application. Trans. N.E. Coast: 64–88. Britain: Institution of Engineers and Shipbuilders.

Jiménez R. 2009. Análisis experimental de una serie de flaps de popa en unidad de desplazamiento: 3–8. Universidad Austral de Chile.

Progress in Maritime Technology and Engineering – Guedes Soares & Santos (Eds)
© 2018 Taylor & Francis Group, London, ISBN 978-1-138-58539-3

Seakeeping optimization of a catamaran to operate as fast crew supplier at the Alentejo basin

F. Belga, M. Ventura & C. Guedes Soares
Centre for Marine Technology and Ocean Engineering (CENTEC), Instituto Superior Técnico (IST), Universidade de Lisboa, Lisbon, Portugal

ABSTRACT: This work optimizes the seakeeping performance of a displacement catamaran to operate as fast crew supplier for an offshore platform at the Alentejo basin, Portugal. The strip-theory code PDStrip was used to predict heave and pitch motions, exploiting an assumption of negligible interaction between demi-hulls in head seas. RMS vertical acceleration responses at the bow and the average MSI at the passenger area were selected as objective functions to minimize. Also, the dimensions and position of the passenger area on deck were designed in order to minimize MSI. Constraints in terms of stability and ship resistance were applied. Horizontal clearance ratios between 0.2 and 0.4 were studied with respect to resistance, stability and MSI. Hull variations were generated from a parent model using Lackenby's method, varying *LCB* and C_b within +–10%. An operability assessment of the optimized catamaran was carried out based on limiting seakeeping criteria imposed by the classification societies.

1 INTRODUCTION

Despite a continuous misuse of fossil fuels, which has been contributing to serious environmental problems, the ever increasing rate of energy consumption still makes us highly dependent. In Portugal there has been a lot of speculation and debate regarding the existence of economically viable reservoirs but there seems to be evidence of potential for hydrocarbon exploration (Martins 2015). In fact, several companies already hold concession rights over Portuguese basins to perform prospection studies (ENMC 2015).

Inspired by a concept for a deep offshore hydrocarbon field located 50 km off the coast (Carvalho 2016), at the Alentejo basin, the present research work addresses the design of a fast displacement catamaran to operate as crew supplier. Multi-hulls, in particular catamarans, take advantage of a high transverse stability, reduced roll and large deck areas to carry more cargo over narrower hulls without carrying ballast. As a result, their shallow drafts and small hydrodynamic resistance allow them to be designed for high speeds. Together with a high provision of nonsinkability and seaworthiness, it is assured an effective application of catamarans as high-speed crafts for the transport of passengers (Dubrovsky 2014). Based on a parent model kindly provided by DAMEN Shipyards, an optimization routine was developed with the intent of improving its seakeeping performance. In fact, as a high-speed craft, the requirement to operate well at high speeds, often in adverse weather conditions, is paramount. Furthermore, high accelerations are known to significantly decrease the operability level of such vessels as they often lead to structural damage and jeopardize safety and welfare on-board. Therefore, acceleration responses and seasickness of passengers upon a specified seaway were object of analysis. Extreme effects such as slamming and green water were neglected in this study, even though they might occur. Only heave and pitch motions in head seas have been considered since they promote the most critical situations for a catamaran, which is particularly stable transverse wise. The absolute vertical displacement ξ_z at a remote location (x, y, z) can then be calculated as in equation (1), assuming motions of small amplitude and that vertical ship responses do not vary with y.

$$\xi_z(x,\omega_e) = \Re\{[\xi_3^{\xi A}(\omega_e) - x\xi_5^{\xi A}(\omega_e)]e^{i\omega_e t}\} \quad (1)$$

where ξ_j^A is the complex amplitude of the harmonic heave ($j = 3$) and pitch ($j = 5$) motion. The frequency of encounter ω_e is the same as the frequency of the response and relates to the wave frequency ω_0 by $\omega_e = \omega_0 - k_0 U \cos\beta$, where $k_0 = \omega_0^2 / g$ is the wave number, g is the acceleration of gravity and β is defined as the ship heading relative to the waves, for which the convention used here assumes $\beta = 180°$ for head seas. t stands for the time variable.

Regarding the numerical tool for seakeeping analysis, despite a definite shift to 3-D codes observed nowadays, strip theories remain valuable as they allow reasonably accurate and relatively fast predictions of ship motions in most situations, which becomes particularly useful within optimization procedures where multiple calculations must be repeated, as was the case here. A previous work of comparison between three strip-theory based codes by Belga et al. (2018) suggested that, at high Froude numbers, the open-source code PDStrip (with transom terms) would be the most suited one to predict heave and pitch motions of fast displacement ships in head seas, motivating its use for the seakeeping optimization of the DAMEN catamaran. In this regard, an assumption about negligible interaction between demihulls was exploited, which is acceptable in head and following seas. In fact, PDStrip does not include such considerations and thus, using it to predict catamaran motions upon a larger range of wave directions would compromise the results. In any case, it is important to keep in mind the limitations of the work performed here.

RMS vertical acceleration responses at the bow and the average Motion Sickness Incidence (MSI) at the passenger area were selected as objective functions to minimize. Stability criteria from the High Speed Craft (HSC) code (IMO 2008) were applied, as well as a constraint on the maximum total ship resistance. Apart from PDStrip (Söding & Bertram 2009), a few Maxsurf modules were used (Bentley Systems, Inc. 2013, 1016), as well as MATLAB which served as main computational environment.

2 RELATED WORK

Up until the late 1980's, despite the significant volume of research on hydrodynamic optimization of ship designs that included seakeeping considerations, the conclusions did not bear substantial developments (Sarioz 1993, Kukner & Sarioz 1995), as the bulk of the work would mainly focus on ship resistance, e.g., Salvesen et al. (1985), Papanikolaou et al. (1991). With the widespread development of high-speed vessels and its use to transport passengers, light was shed upon the relevance of seakeeping behaviour as hydrodynamic parameter to optimize. As stated before, the speed provided by strip methods is especially valuable when the seakeeping tool is to be embedded within optimization routines. Thus, they have been recurrently used, for instance, to assess the sensitivity of certain seakeeping characteristics of a vessel (e.g., heave/pitch amplitudes, vertical accelerations, relative motions, slamming) upon the variation of the main particulars, which is often carried out with the intent of improving existing designs. Such studies commonly resort to systematic variations of the hull form parameters or main dimensions, as it is a quick and easy way to generate geometries from a parent model, e.g., Kukner & Sarioz (1995), Scamardella & Piscopo (2014), Piscopo & Scamardella (2015). Alternatively, parametric modelling and the use of form deformation techniques, often combined with Genetic Algorithms (GA) to improve efficiency when searching for the optimal solution, stand as the state of the art in terms of optimization problems (Maisonneuve et al. 2003, Kapsenberg 2005, Bagheri et al. 2014, Ang et al. 2015). Again, the development of more advanced and demanding 3-D numerical tools such as CFD solvers and their inclusion into routines to optimize both the seakeeping performance and ship resistance is, today, a common trend, e.g., Ang et al. (2015), Vernengo & Bruzzone (2016).

Welfare and comfort are, especially for passenger ships, two very important aspects. From this perspective, several authors have evaluated seakeeping criteria with respect to motion sickness and habitability on-board, some of which focused on the influence of changes in hull geometry on such parameters, e.g., Scamardella & Piscopo (2014), Piscopo & Scamardella (2015) who resorted to the MSI index.

Considerations in terms of resistance, stability, etc., can be integrated either in the form of constraints to the optimization problem, e.g., Kukner & Sarioz (1995), Dudson & Rambech (2003), or taken into account at a final stage where the optimum hull is further refined with respect to those characteristics, as Grigoropoulos (2004) is example.

3 OVERVIEW OF THE OPTIMIZATION PROBLEM

3.1 *Parent model and generation of hull variations*

The 30 metres parent catamaran (Figure 8) operates at a service speed of 25 knots (30 knots maximum) and was set to carry 12 passengers/crew, typical for such vessels, which classes them as cargo craft according to the HSC code (IMO 2008). To generate variations from the parent model, the method of Lackenby (Maxsurf Modeler) was used to vary the block coefficient (C_b) and the longitudinal centre of buoyancy (*LCB*) both by +–10% of the parent values (in a total of 225 models evaluated) – Figure 1 – while maintaining the displacement (Δ), waterline length (*LWL*) and maximum beam at the waterline (*BWL*).

Thus, a few differences in the overall dimensions might be observed. Given the lack of information in terms of weight distribution at this early design stage, values for *KG* and *LCG* were assumed (*KG* = 1.5 meters from the baseline and *LCG* = *LCB* for all generated models). In principle it should not significantly affect the seakeeping characteristics (Kukner & Sarioz 1995). Furthermore, it is important to note that not all generated models resulted into feasible solutions from the point of view of geometry. Not only excessively distorted models were discarded, as well as models not suited for numerical discretization by PDStrip which outputs warnings in such cases (Söding & Bertram 2009). The flowchart in Figure 5 illustrates the procedure that systematically generates hull variations and evaluates the solutions in terms of seakeeping, resistance and stability, as well as the numerical tools used for that purpose.

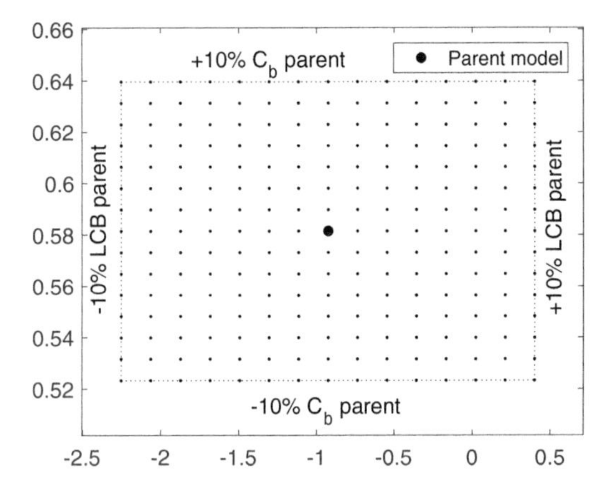

Figure 1. Combinations of parameters used to generate hull variations and the +−10% boundaries.

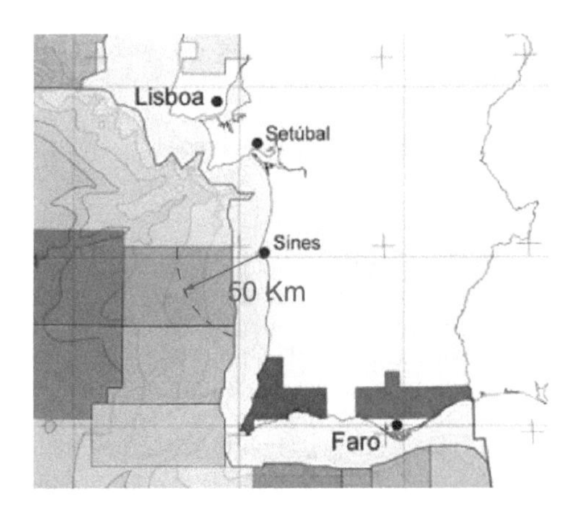

Figure 2. Location of the hydrocarbon field proposed by Carvalho (2016) (source: ENMC (2015), adapted).

3.2 *Seastates at the Alentejo basin*

The catamaran is designed to operate at the Alentejo basin, transporting crew between shore and an offshore platform located 50 km off the coast of Sines, as illustrated by Figure 2.

The wave regime along the Portuguese coast has been studied by Costa et al. (2001) and the measurements from Sines station were used to characterize the seastates at the Alentejo basin (Table 1). Although these measurements do not translate the sea environment at the site of interest (the station is located near the coast—water depths up to 200 metres), this was the information found available and thus, it was used nonetheless. The JONSWAP spectrum, extensively applied in the offshore industry, was used to represent the wave spectrum at the location. The formulation presented here (Guedes Soares 1998) depends on the significant wave height $H_{1/3}$ and peak period T_p.

$$S_\zeta(\omega_0) = \frac{\alpha}{\omega_0^5} e^{-1.25\left(\frac{\omega_p}{\omega_0}\right)^4} \gamma^{e^{-\frac{1}{2\sigma^2}\left(\frac{\omega_0}{\omega_p}-1\right)^2}} \tag{2}$$

where,

$$\omega_p = \frac{2\pi}{T_p} \tag{3}$$

$$\alpha = 5\pi^4(1-0.287\log\gamma)\frac{H_{1/3}^2}{T_p^4} \tag{4}$$

$$\begin{aligned}\gamma &= 5, \quad \text{if} \quad T_p \le 3.6\sqrt{H_{1/3}} \\ &= 1, \quad \text{if} \quad T_p > 5\sqrt{H_{1/3}} \\ &= e^{5.75-1.15\left(\frac{T_p}{\sqrt{H_{1/3}}}\right)}, \quad \text{otherwise}\end{aligned} \tag{5}$$

$$\begin{aligned}\sigma &= 0.07, \quad \text{if} \quad \omega_0 \le \omega_p \\ &= 0.09, \quad \text{if} \quad \omega_0 > \omega_p\end{aligned} \tag{6}$$

Table 1. Scatter diagram % for the Alentejo basin, measured at Sines station between 1988–2000 (Costa et al. 2001).

$H_{1/3}$ [m]	T_p[s] 3–5	5–7	7–9	9–11	11–13	13–15	15–17	>17	
0–1	0.480	1.210	2.620	6.760	6.790	3.520	1.350	0.110	22.840
1–2	1.020	2.590	5.600	14.450	14.510	7.530	2.880	0.230	48.810
2–3	0.390	0.990	2.140	5.530	5.550	2.880	1.100	0.090	18.670
3–4	0.140	0.360	0.780	2.020	2.030	1.050	0.400	0.030	6.810
4–5	0.040	0.110	0.240	0.610	0.610	0.320	0.120	0.010	2.060
5–6	0.020	0.040	0.090	0.240	0.240	0.120	0.050	0.000	0.800
	2.090	5.300	11.470	29.610	29.730	15.420	5.900	0.470	100%

In order to account with the relative velocity between the ship and the encountering head waves, the sea spectrum $S_\zeta(\omega_e) = S_\zeta(\omega_0)/(1-\frac{2\omega_0 U}{g})$ is defined. The basis for calculating responses is the transfer function of that response. Considering the absolute vertical displacement ξ_z at (x, y, z) from equation (1), the corresponding response spectrum S_z is given by:

$$S_x(x,\omega_e) = |\xi_z(x,\omega_e)|^2 \, S_\zeta(\omega_e) \tag{7}$$

3.3 *Objective functions*

Vertical accelerations were selected as parameter to minimize. RMS (Root Mean Squared) values, in particular, are considered good statistical measures as they provide useful and immediate (although inevitably less detailed) information, without the need to consider the whole frequency spectrum. For the case of the acceleration responses, RMS_{a_z} are calculated as:

$$RMS_{a_z}(x) = \sqrt{\int_{\omega_e} \omega_e^4 S_z(x,\omega_e)\, d\omega_e} \tag{8}$$

The minimization of maximum RMS_{a_z} value on deck, given the most probable seastate the vessel would have to face sailing at service speed was set as the main objective function of this optimization procedure, corresponding to a location at the forward extremity of the vessel. Since the length overall (LOA) of the generated models might change, RMS_{a_z} was instead evaluated at a fixed location near the bow (x = 28.5 metres), approximately at the same vertical of LWL, the same for all models. Motion sickness was also considered and the time dependent Motion Sickness Incidence (MSI) model, which assesses the percentage of passengers who vomit after a given time of exposure to a certain motion (McCauley et al. 1976), was used. This was useful as the actual voyage time of the catamaran is known (65 minutes to sail 50 km at 25 knots) – Figure 3. Here, the formulation described in Colwell (1989) is presented and depends on the average RMS acceleration, $|RMS_{a_z}| = 0.798 RMS_{a_z}$, and the average peak frequency, $|f_e| = |RMS_{a_z}/2\pi\sqrt{\int_{\omega_e} \omega_e^2 S_z(x,\omega_e)\, d\omega_e}$.

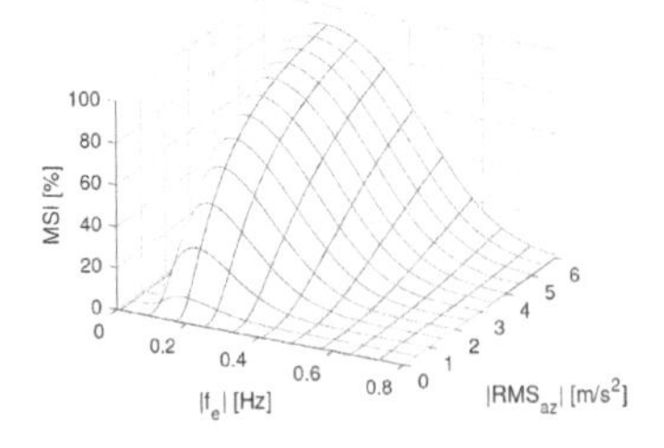

Figure 3. MSI model considering t = 65 min.

$$MSI = 100\Phi(z_a)\Phi(z_t) \tag{9}$$

where $\Phi(z)$ is the standard normal distribution function.

$$\Phi(z) = \frac{1}{\sqrt{2\pi}} e^{-\frac{z^2}{2}} \tag{10}$$

$$z_a = 2.128\log_{10}\left(\frac{|RMS_{a_z}|}{g}\right) - 9.277\log_{10}|f_e| - 5.809(\log_{10}|f_e|)^2 - 1.851 \tag{11}$$

$$z_t = 1.134 z_a + 1.989\log_{10} t - 2.904 \tag{12}$$

MSI was then computed at multiple locations on deck. For each generated model, the position of the area reserved to carry passengers was fixed as illustrated in Figure 4 and the minimum area needed to transport 12 passengers was estimated from a database of similar ships (48.24 m^2). Thus, for a constant passenger area with fixed width, its length changes depending on the horizontal clearances ratio S/LWL that is being considered (a range between 0.2 and 0.4 was studied) - S refers to the distance between the centrelines of the demi-hulls. The final position of the passenger area on deck was determined by the region that allowed minimizing the average MSI along its length. The minimization of the average MSI value at the optimized position of the passenger area was set as second objective function.

3.4 *Constraints*

The total ship resistance of the parent catamaran with the original horizontal clearance (76.114 kN) was set as maximum allowed value. To compute the ship resistance coefficient, $C_T = (1 + \beta k)C_f + \tau C_w$, an empirical method (Jamaluddin et al. 2013) was used to estimate the interference components—equations (13) and (14). $(1 + \beta k)\, C_f$ is the viscous resistance component and C_w the wave resistance calculated with slender-body theory (Maxsurf Resistance). β and τ are the respective interference factors. $(1 + \beta k)$ is the viscous form factor and C_f

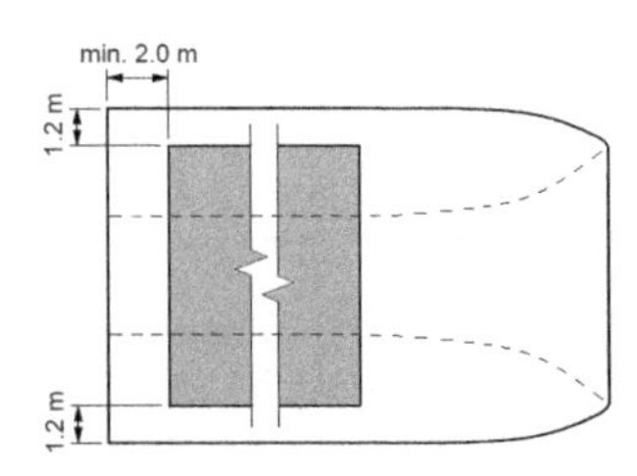

Figure 4. Sketch of the position of the passenger area (in grey) on deck with respect to the ship sides and the aft perpendicular.

the coefficient of friction resistance, obtained with the ITTC-1957 formula, $C_f = 0.075/(\log_{10}Re - 2)^2$, function of Reynolds number Re.

$$(1+\beta k) = 3.03(LWL/\nabla^{1/3})^{-0.40} + 0.016(S/LWL)^{-0.65} \tag{13}$$

$$\begin{aligned} \tau &= 0.068(S/LWL)^{-1.38}, \text{ at } F_n = 0.19 \\ &= 0.359(S/LWL)^{-0.87}, \text{ at } F_n = 0.28 \\ &= 0.574(S/LWL)^{-0.33}, \text{ at } F_n = 0.37 \\ &= 0.790(S/LWL)^{-0.14}, \text{ at } F_n = 0.47 \\ &= 0.504(S/LWL)^{-0.31}, \text{ at } F_n = 0.56 \\ &= 0.501(S/LWL)^{-0.18}, \text{ at } F_n = 0.65 \end{aligned} \tag{14}$$

The second constraint relates with stability requirements imposed by the HSC code (IMO 2008). Points 1.1 and 1.2 of *Annex 7 Stability of Multi-hull Craft, 1 Stability Criteria in the Intact Condition* were applied. Maxsurf Stability was used to compute the GZ curves. It is important to note that in this case, in order to study horizontal clearances different than the parent one ($S = LWL = 0.2872$) each generated hull variation would have to be modelled individually with different S values prior to the stability analysis, which was not feasible. For this reason only $S/LWL = 0.2872$ was considered.

3.5 *Synthesis model*

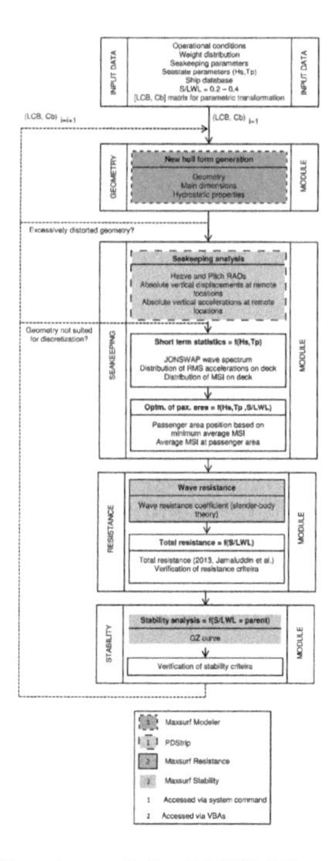

Figure 5. Flowchart of the MATLAB procedure that generates hull variations from the parent model and evaluates them in terms of seakeeping, resistance and stability.

4 OPTIMIZATION RESULTS

The optimization results are here shown resorting to colour plots in order to facilitate the visualisation of the gradient of the studied parameters as functions of LCB (measured with respect to $LPP/2$) and C_b, each pair representing a different generated model. Results are shown only for the case of the original horizontal clearance ratio S/LWL = 0.2872, Figure 6(a) for RMS vertical acceleration (in this case independent of S/LWL), Figure 6(b) for MSI and Figure 6(c) for total hull resistance R_T. Figure 6(d), which shows the percentage of models that satisfied the resistance criteria upon different horizontal clearance ratios, is presented as well, as it points out that for $S/LWL \leq$ 0.28 there are no feasible solutions. Furthermore, as all models fulfilled the HSC code requirements (IMO 2008), stability criteria did not affect the selection of the optimum solution and thus such results will not be shown. Also, it is interesting to compare Figures 6(a)–6(b)–6(c) with the square matrix shape of Figure 1, which reveals that about 15% of the generated hull forms have been neglected for the reasons explained before.

Figure 6(a) indicates that for the same C_b, an increase of LCB to an aft position will improve seakeeping performance, evaluated here in terms of RMS vertical accelerations. Regarding C_b, it seems that models with smaller values experience

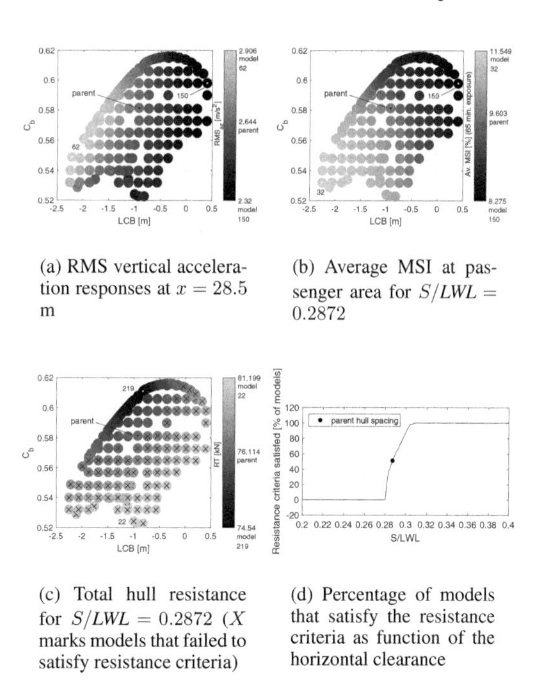

(a) RMS vertical acceleration responses at $x = 28.5$ m

(b) Average MSI at passenger area for $S/LWL = 0.2872$

(c) Total hull resistance for $S/LWL = 0.2872$ (X marks models that failed to satisfy resistance criteria)

(d) Percentage of models that satisfy the resistance criteria as function of the horizontal clearance

Figure 6. Optimization results considering the most probable seastate ($H_{1/3}$ = 1–2 m, T_p = 11–13 s, prob = 14.51%) at V_s = 25 knots as function of (LCB, C_b).

lower accelerations. However, the influence of block coefficient appears to be less significant, particularly for higher values of *LCB*. In any case, given that several other hull shape parameters are being varied as well, it is not possible to associate seakeeping improvements to single causes. In terms of motion sickness, similar results were observed in Figure 6(b) in the sense that a decrease of motion sickness is obtained by shifting *LCB* forward (for constant C_b values). However, it seems that decreasing C_b actually contributes for higher MSI values, an effect that, once again, becomes less significant as *LCB* increases. This has to do with the role of frequency on the MSI model. In fact, humans can apparently tolerate higher accelerations at higher frequencies without experiencing the same tendency towards motion sickness (O'Hanlon & McCauley 1973). Figure 3 indicates that above 0.16 Hz, where the frequency region experienced by all studied models is located, the lower $|f_e|$ is, the higher the experienced MSI for the same $|RMS_{az}|$ value. Furthermore, for a given frequency, MSI decreases with $|RMS_{az}|$. Since, as concluded before, decreasing Cb is not a very effective way to reduce $|RMS_{az}|$, the resulting reduction of $|f_e|$ leads to a higher incidence of motion sickness in that specific region of the model. In any case, model 150 seems to be the most suited in terms of seakeeping. From the point of view of ship resistance (Figure 6(c)), the optimum hull forms appear in regions of the figure that do not particularly favour the minimization of both RMS_{az} and MSI (Figures Figure 6(a) and 6(b)), suggesting incompatible trends (improvements by lowering *LCB* and increasing C_b). In fact, with $S/LWL = 0.2872$, resistance criteria does not allow the selection of the optimum seakeeping solution (model 150).

4.1 *Selection of the optimum hull*

Figure 7 shows all generated models that satisfied the imposed criteria. Again, stability was not restrictive. Even though the objective functions only concern the minimization of RMS_{a_z} and MSI, a third axis with R_T was included. This provides information about the horizontal clearance that is being used, which in reality generates complete new sets of possible solutions, each assigned to a different marker.

For simplicity, only 5 horizontal clearance ratios were displayed, ranging from the parent value (0.287) to 0.4. Again, the lower the horizontal clearance ratio, the lower the number of possible solutions generated, which is in agreement with Figure 6(d). It has been previously shown that model 150 is the most suited in terms of seakeeping (Figures 6(a) and 6(b)), appearing in the detail view of Figure 7(b) with the lowest RMS acceleration values (2.32 m/s²) and different *S/LWL* ratios.

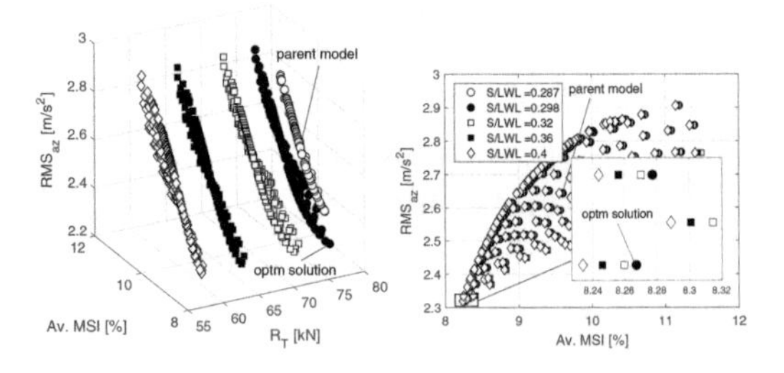

(a) 3-D representation of the family of possible solutions

(b) RMS vertical acceleration responses at $x = 28.5$ m as function of the average MSI at passenger area with detail view of the optimum seakeeping region

Figure 7. Family of possible solutions considering the most probable seastate ($H_{1/3} = 1$–2 m, $T_p = 11$–13 s, prob = 14.51%) at $V_s = 25$ knots.

Although the figure indicates that the horizontal clearance does not significantly affect MSI, it is a quite dominant parameter in terms of ship resistance as Figure 7(a) suggests. In fact, the wider the catamaran hulls are set apart, the lower the total resistance which, of course, carries additional construction costs. Since seakeeping was the main focus of this work, no efforts have been put into improving ship resistance and thus, model 150 with $S/LWL = 0.298$ was selected as optimum solution, as it was the minimum horizontal clearance ratio that allowed its selection.

4.2 *Overview of model 150 (S/LWL = 0.298)*

The main characteristics of both the parent model and the optimized model (model 150) are summarised in Table 2.

Figure 8 shows the rendered hull forms and Figure 9 compares the body lines of both models. Model 150 seems "bulgier" at the forward part, as a result of an increased *LCB*. Also, it seems that the parametric transformation of Lackenby affected the geometry of the bulbous bow significantly and it would have been interesting to use PDStrip to study alternative solutions for that region (e.g., a regular shaped bow without bulbous, an axe-bow or an inverted bow-type configuration typical in high-speed crafts).

Figure 10 confirms that by optimizing for RMS vertical response accelerations at the bow, improvements in terms of heave and pitch Response Amplitude Operators (RAOs) have been achieved. Moving *LCB* forward by about 10% (as seen before, C_b has a weaker impact on heave and pitch motions) seems to have had a positive impact. In any case, as pointed out by Blok & Beukelman (1984), the verified increase of waterplane area

Table 2. Main characteristics of the parent model and model 150.

	Parent model	Model 150
LOA [m]	30	28.8
LWL [m]	28.4	28.4
BWL, demi-hull [m]	2.4	2.4
S [m]	8.15	8.46
S/LWL [–]	0.287	0.298
BOA [m]	10.8	11
Deck area [m^2]	319.6	318.6
T (*DWL*) [m]	1.41	1.37
Δ [t]	112.6	112.5
KB [m]	0.87	0.88
KG [m]	1.5	1.5
LCB = *LCG* [m]	–0.93	0.4
C_b [–]	0.581	0.598
C_{wp} [–]	0.816	0.914
RMS_{az} at $x = 28.5$ m [m/s^2]	2.644	2.320
$\overline{MSI}$ at pax. area [%]	9.603	8.267
Pax. area [m^2], [% of deck area]	50.7,16	50, 16
Length of pax. area [m]	6.1	5.8
Free area aft [% of deck area]	10	11
Free area fwd. [% of deck area]	69	65
R_T [kN]	76.114	75.968

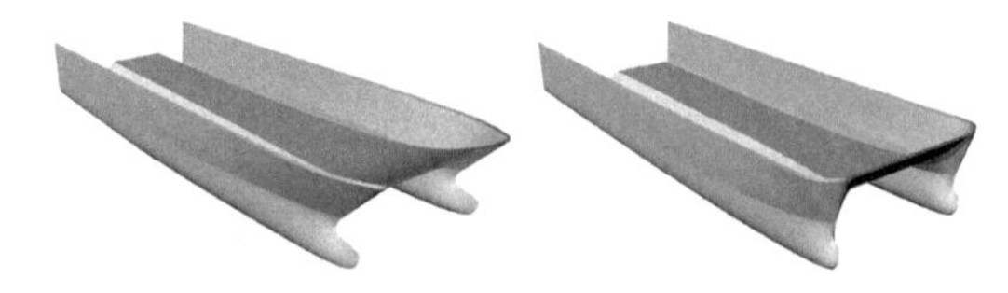

Figure 8. Models rendered in Rhino: parent (left) and model 150 (right).

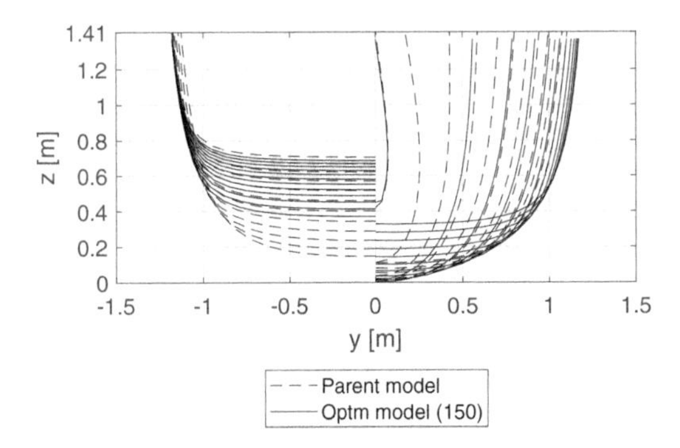

Figure 9. Hull form comparison between the parent model and model 150.

(C_{wp} raised 12%) and *BWL*/*DWL* ratio (refer to Table 2) probably influenced significantly these results as well. Thus, it is difficult to draw definite conclusions.

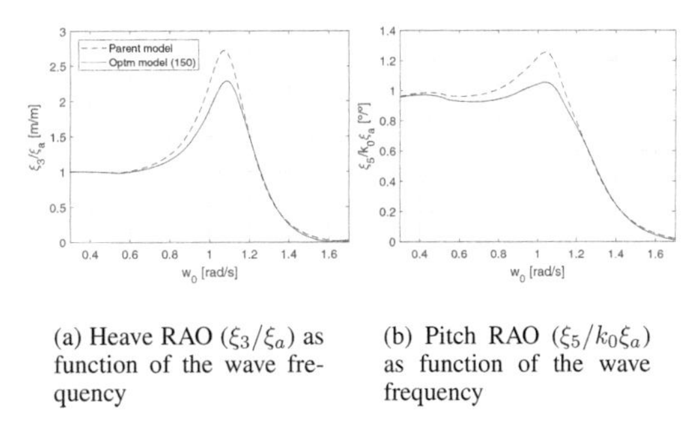

(a) Heave RAO (ξ_3/ξ_a) as function of the wave frequency

(b) Pitch RAO ($\xi_5/k_0\xi_a$) as function of the wave frequency

Figure 10. Comparison of heave and pitch RAOs at $V_s = 25$ knots.

Figure 10(a) indicates that at the resonance frequency (wavelengths of about 1.8–1.9*LWL*), model 150 experiences heaving amplitudes 16.5% smaller than the parent model. A decrease of the same order is observed for pitch at wavelengths of about 2*LWL* (Figure 10(b)). Regarding motion sickness, Table 2 indicates a shortening of the passenger area (due to a larger horizontal clearance ratio) and a consequent decrease of the average MSI by about 1.3%, as expected since the ability to capture a more advantageous region on deck increases. Note that the final passenger area surpasses, in both cases, the minimum of 48.24 m^2 estimated from the ship database. Since MSI has been computed at a discrete number of remote locations along the deck, dimensioning the passenger area for a well defined value was difficult. Finally, a small reduction of the total ship resistance was achieved as well with the optimization.

Figures 11 and 12 show some of the effects that different seastates have on seakeeping. Figures 11(a) and 12(a) indicate that the most severe period is $T_p = 5$–7 seconds, as it generates the variance peaks nearest to the resonance frequency of the vertical motions. The passenger area has been positioned in order to minimize the distribution of incidence of motion sickness along its length (by averaging the results) considering the most probable seastate ($H_{1/3} = 1$–2 m, $T_p = 11$–13 s). Figures 11(b) and 12(b) show the optimized location of the passenger area (shaded region), indicating that it begins 4.24 metres forward of the aft end of the ship (exceeding the minimum distance of 2 meters). For $H_{1/3} = 1$–2 metres, Figure 12(a) corresponds to the MSI values of Figure 12(b) averaged for the passenger area. It is interesting to note in Figure 11(b) that the optimized location of the passenger area does not coincide with the minimum region of RMS accerations, near the centre of gravity. Figure 12(b), on the other hand, confirms the method used to optimize the position of the passenger area. For $T_p = 11$–13 seconds, the passenger area indeed captures the minimum MSI values. Given this discrepancy with respect to RMS

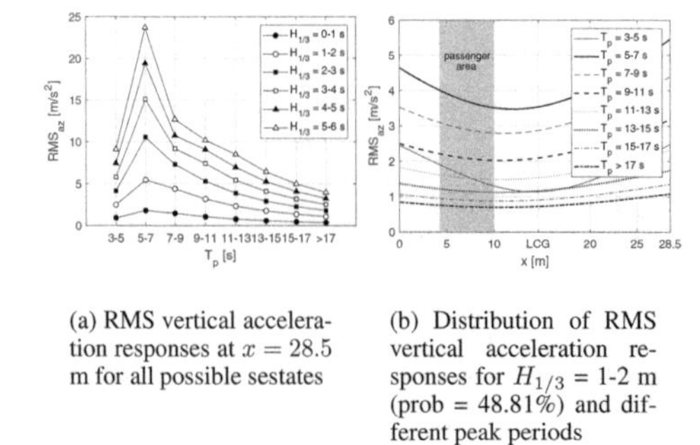

(a) RMS vertical acceleration responses at $x = 28.5$ m for all possible sestates

(b) Distribution of RMS vertical acceleration responses for $H_{1/3}$ = 1-2 m (prob = 48.81%) and different peak periods

Figure 11. Seastate effects on the RMS vertical acceleration responses of model 150 at $V_s = 25$ knots.

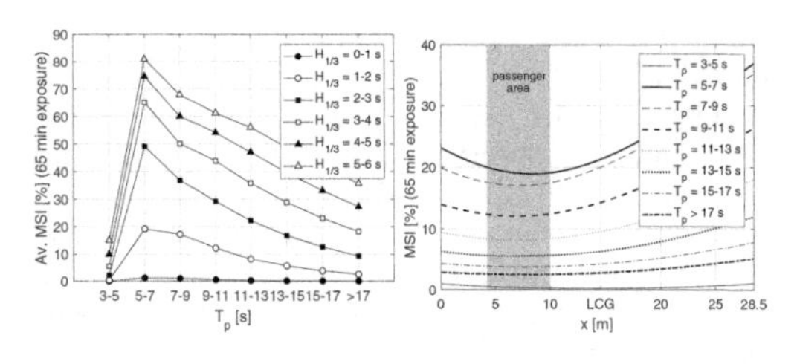

(a) Average MSI at passenger area for all possible sestates ($S/LWL = 0.298$)

(b) Distribution of MSI for $H_{1/3}$ = 1-2 m (prob = 48.81%) and different peak periods ($S/LWL = 0.298$)

Figure 12. Seastate effects on motion sickness of model 150 at $V_s = 25$ knots.

accelerations, this raises questions about whether using the incidence of motion sickness is truly the best approach to optimize comfort and welfare on-board. However, as discussed before, the MSI model considers the fact that humans are capable of withstanding severe accelerations as long as the frequency is high enough. Moving the passenger area to an aft position increases the average frequencies, thus generating lower motion sickness incidences, even though accelerations worsen as seen in Figure 11(b).

4.3 *Operability assessment of model 150 based on seakeeping criteria*

An operability assessment considers human comfort criteria, being particularly vital in the design of passenger vessels. In Guedes Soares (1998), the author suggested that, due to linearity assumptions, the wave spectrum could be represented as the product of the wave spectrum in terms of the unitary significant wave height, $S_{\zeta 1}(\omega_e)$, and the square of the significant wave height $H_{1/3}$.

$$S_\zeta(\omega_e, H_{1/3}, T_p) = H_{1/3}^2 S_{\zeta 1}(\omega_e, T_p) \tag{15}$$

Using the previous result, equation (7) for the vertical response of the vessel at a remote location (x, y, z) can be rewritten as follows:

$$S_z(x, \omega_e, H_{1/3}, T_p) = |\xi_z(x, \omega_e)|^2 H_{1/3}^2 S_{\zeta 1}(\omega_e, T_p) \tag{16}$$

Recalling equation (8), given a seakeeping criterion defined in terms of RMS accelerations, $RMS_{a_z,criteria}$, the limiting significant wave height as function of the period (peak period T_p in this case) is given by:

$$H_{1/3}^{\lim}(T_p) = \frac{RMS_{a_z,criteria}}{RMS_{a_z,1}} \tag{17}$$

where $RMS_{a_z,1}$ is the normalized root mean squared acceleration function of T_p.

The limiting criteria applied here are defined in the HSC code (IMO 2008) (*Chapter 4 Accommodation and Escape Measures, 4.3 Design acceleration*) and in the DNV GL rules for the classification of high speed crafts (DNV-GL 2012) (*Section 3 Structures, C3.3 Design Acceleration*). The first is associated with the maximum superimposed vertical accelerations at the centre of gravity of 1 g. The second relates with the longitudinal distribution of vertical accelerations. In both cases, the accelerations refer to average 1% highest accelerations rather than the RMS values, for which equation (18) from Hoffman & Karst (1975) has been applied using n = 100 (*erf* is the error function). For this operability assessment, 25 knots (service speed) and 27 knots (90% of the maximum speed of 30 knots) have been the considered vessel speeds, in conformity with the guidelines of the HSC code (IMO 2008). Figure 13 shows the obtained limiting seastates for both operational conditions.

$$az_{1/n} = \left[n\sqrt{2} \left[\frac{\sqrt{\ln n}}{n} + \sqrt{\pi} \left(\frac{1}{2} - erf \sqrt{2 \ln n} \right) \right] \right] RMS_{a_z} \tag{18}$$

Finally, the operability index is defined as "the percentage of time during which the ship is operational" (Fonseca & Guedes Soares 2002). For that purpose, a scatter diagram can be used to sum up the probabilities of occurrence of the seastates upon which the catamaran is suited to operate according to Figure 13(b). However, the scatter diagram in Table 1, which presents wave data as function of ranges of the parameters, does not present discretized enough data that allow a reliable computation of the operability index. In any case, boundaries can be estimated based on the available information. From a conservative perspective, the fast crew supplier will be able to

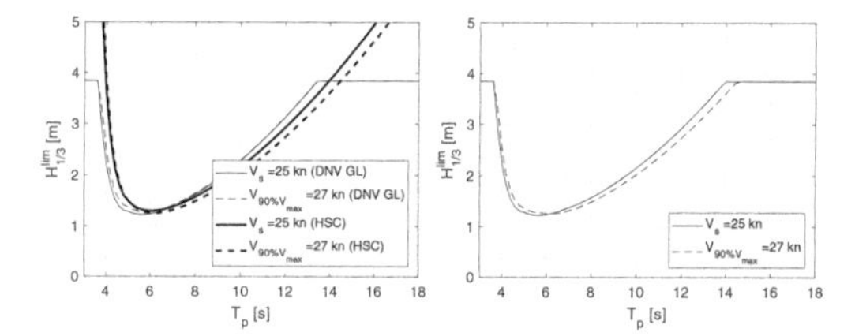

(a) Comparison of seakeeping criterion from DNV GL and HSC code for service speed and 90% of maximum speed

(b) Final limiting seastates for the operation of the fast crew supplier at service speed and 90% of maximum speed

Figure 13. Maximum allowed $H_{1/3}$ as function of the peak period based on criterion imposed by the classification societies.

operate at the Alentejo basin only 52% of the time. In a more optimistic scenario, the author expects an operability index of 91%. Furthermore, it is likely that the effective operability is much closer to the upper boundary. Also, a decrease in operability is expected at higher speeds, given the more strict limitations, particularly at higher T_p values.

5 CONCLUSIONS

The seakeeping optimization of a 30 metres parent catamaran to operate as crew supplier for an offshore platform at the Alentejo basin, Portugal, was the objective of this research work. The strip-theory based open-source code PDStrip has been used as tool for seakeeping analysis. The optimum model, generated with the method of Lackenby, was obtained by increasing C_b +3% and by shifting *LCB* forward by +10%. Due to the inherent properties of the method, the resulting geometry also benefited from an increased waterplane area (+12% of C_{wp}) and a lower design draft (–2.8%). Also, the transformation led to a bulbous bow geometry without much purpose, which seemed to suggest that a regular shaped bow, an axebow or an inverted bow-type configuration (typical in high-speed crafts) would be more appropriate. In fact, a proper description of the bow cross sections would probably allow an evaluation using PDStrip.

Results showed that increasing *LCB* improves the seakeeping performance, while the influence of C_b seemed less significant. Still, the large number of parameters varied simultaneously made it hard to draw definite conclusions.

It has been shown that the optimum model experienced lower RMS accelerations at the bow by about 12%. In terms of motion sickness, the improvements were more subtle. With the optimum model, less 1.3% people experienced MSI when compared to the parent model, which was also achieved by increasing the horizontal clearance by about 30 cm. An analysis of the RAOs of the optimum model allowed to conclude that at the resonance frequency, it experiences heaving amplitudes 16.5% smaller than the parent one. A decrease of the same order has been observed for pitch.

An upper bound in terms of ship resistance (equal to the parent model) was applied as constraint to the problem. Results showed that larger horizontal clearances allow decreasing the interference effect between demi-hulls, thus leading to smaller total resistance. In order to comply with the resistance criteria, $S/LWL = 0.298$ was selected, which corresponds to the already referred increase of 30 cm. A decrease in consumption of less 1.88 kW of effective power at 25 knots has been achieved. With respect to stability, the second applied constraint, only the parent horizontal clearance has been evaluated, with which all models satisfied the criteria imposed by the classification societies. Given the small difference to the parent value, no significant changes are expected with the selected *S/LWL* ratio, although larger horizontal clearances generally imply a decrease of the heeling angle at which the maximum GZ value occurs even if the ship becomes more stable transverse wise. The stability module of this work could certainly be improved by using a software that allow automated analyses with variable horizontal clearance ratios as input.

Finally, an operability assessment of the optimized catamaran based on limiting criteria from the classification societies allowed to estimate an operability index at the Alentejo basin of 91%, although a quite conservative scenario predicted that it would only be able to operate 52% of the time. A more detailed scatter diagram of the region would be required in order to compute these indexes more accurately.

Within the presented assumptions and limitations, the procedure applied here to systematically generate hull variations and evaluate them in terms of seakeeping, resistance and stability proved to be an effective method of optimization. It can be applied in a wide variety of cases as long as there is access to a parent model. The objective functions, constraints and numerical tools can be easily changed or added (e.g., a module for economical assessment) if necessary within the main MATLAB code. In any case, a discussion regarding efficiency is relevant in the sense that an alternative method to search solutions like GAs could have been applied. Also, results seemed to suggest that C_{wp} rather than C_b should be used to generate hull variations parametrically, as its influence in the seakeeping behaviour appears to be higher. Another note has to do with the limited level of automation generally allowed by commercial software like Maxsurf. For robustness, using simple executable files, preferably with open-source code, is more advisable. Finally, with respect to the seakeeping analysis, the fact that the catamaran was restricted to heave and pitch upon head seas (as a result of the assumption about negligible demi-hull interaction imposed by the limitations of PDStrip) clearly affects the credibility of the method. Evaluating the seakeeping behaviour upon a larger range of wave directions is fundamental for a trustworthy study of the behaviour of a ship in a realistic seaway. This would imply analysing rolling motions since vertical ship responses become increasingly dependent on y as the wave headings come closer to beam seas. Furthermore, as a high-speed catamaran, the probability of occurrence of slamming and green water is high (both neglected in this work) and thus, the optimization procedure could be improved with such considerations.

ACKNOWLEDGEMENTS

To DAMEN Shipyards Group for having kindly provided the parent catamaran model used in this optimization procedure.

REFERENCES

Ang, J.H., C. Goh, & Y. Li (2015). Hull form design optimisation for improved efficiency and hydrodynamic performance of 'ship-shaped' offshore vessels. In *International Conference on Computer Applications in Shipbuilding (ICCAS) 2015*, Bremen, Germany. Royal Institution of Naval Architects, RINA.

Bagheri, L., H. Ghassemi, & A. Dehghanian (2014). Optimizing the seakeeping performance of ship hull forms using genetic algorithm. *The International Journal on Marine Navigation and Safety of Sea Transportation (TransNav) 8*(1), 49–57.

Belga, F., S. Sutulo, & C. Guedes Soares (2018). Comparative study of various strip-theory seakeeping codes in predicting heave and pitch motions of fast displacement ships in head seas. In C. Guedes Soares and T.A. Santos (Eds.), *Progress in Maritime Technology and Engineering*, London, UK: Taylor & Francis.

Bentley Systems, Inc. (2013). *Maxsurf Motions and Resistance, Windows Version 20, User Manuals.*

Bentley Systems, Inc. (2016). *Maxsurf Modeler and Stability, Windows Version 21, User Manuals.*

Blok, J.J. & W. Beukelman (1984). The high-speed displacement ship systematic series hull forms—seakeeping characteristics. *Transactions of the Society of Naval Architects and Marine Engineers, SNAME 92*, 125–150.

Carvalho, A. (2016). Análise do modelo de desenvolvimento de campos de hidrocarbonetos aplicado à bacia do alentejo. Master's thesis, Universidade de Aveiro, Aveiro, Portugal.

Colwell, J.L. (1989). Human factors in the naval environment: a review of motion sickness and biodynamic problems. Technical Report DREA-TM-89-220, Defence Research Establishment Atlantic, Dartmouth, Nova Scotia, Canada.

Costa, M., R. Silva, & J. Vitorino (2001). Contribuição para o estudo do clima de agitação marítima na costa portuguesa. In *2as Jornadas Portuguesas de Engenharia Costeira e Portuária*, Number 20, Sines, Portugal.

DNV-GL (2012). *Rules for Classification and Construction, I Ship Technology, 3 Special Craft, 1 High Speed Craft* (2012 ed.). DNV GL.

Dubrovsky, V.A. (2014). Application and development of multihulls. *Journal of Ocean, Mechanical and Aerospace 6*, 1–7.

Dudson, E. & H.J. Rambech (2003). Optimisation of the catamaran hull to minimise motions and maximise operability. In *Proceedings of the 7th International Conference on Fast Sea Transportation, FAST2003*, Number P2003-7, Ischia, Italy.

ENMC (2015). Mapa de concessões em portugal continental. http://www.enmc.pt/. Entidade Nacional para o Mercado de Combustíveis (ENMC).

Fonseca, N. & C. Guedes Soares (2002). Sensitivity of the expected ships availability to different seakeeping criteria. In *Proceedings of 21st International Conference on Offshore Mechanics and Artic Engineering (OMAE'02)*, Oslo, Norway, pp. 595–603.

Grigoropoulos, G.J. (2004). Hull form optimization for hydrodynamic performance. *Marine Technology 41*(4), 167–182.

Guedes Soares, C. (1998). Stochastic modelling of waves and wave induced loads. In C. Guedes Soares (Ed.), *Risk and Reliability in Maritime Technology*, pp. 197–211. CRC Press/Balkema.

Hoffman, D. & O.J. Karst (1975). The theory of the Rayleigh distribution and some of its applications. *Journal of Ship Research 19*(3), 12–191.

IMO (2008). *International code of safety for High-Speed Craft (2000)* (MSIS 34 (2008) ed.). IMO, Maritime Safety Committee (MSC).

Jamaluddin, A., I.K.A.P. Utama, B. Widodo, & A.F. Molland (2013). Experimental and numerical study of the resistance component interactions of catamarans. *Journal of Engineering for the Maritime Environment 227*(1), 51–60.

Kapsenberg, G.K. (2005). Finding the hull form for given seakeeping characteristics. Technical report, MARIN, Wageningen, Netherlands.

Kukner, A. & K. Sarioz (1995). High speed hull form optimisation for seakeeping. *Advances in Engineering Software 22*(3), 179–189.

Maisonneuve, J.J., S. Harries, J. Marzi, H.C. Raven, U. Viviani, & H. Piippo (2003). Towards optimal design of ship hull shapes. In *Proceedings of the 8th International Marine Design Conference, IMDC 2003*, Athens, Greece, pp. 31–42.

Martins, J.M. (2015). Conferência: Pesquisa de petróleo em portugal. Entidade Nacional para o Mercado de Combustíveis (ENMC). Fundação Calouste Gulbenkian, Lisboa, Portugal.

McCauley, M.E., J.W. Royal, C.D. Wylie, J.F. O'Hanlon, & R.R. Mackie (1976). Motion sickness incidence: exploratory studies of habituation, pitch and roll, and the refinement of a mathematical model. Technical Report 1733-2, Office of Naval Research, Department of the Navy, Goleta, California, USA.

O'Hanlon, J.F. & M.E. McCauley (1973). Motion sickness incidence as a function of the frequency and acceleration of vertical sinusoidal motion. Technical Report 1733-1, Office of Naval Research, Department of the Navy, Goleta, California, USA.

Papanikolaou, A., G. Zaraphonitis, & M. Androulakakis (1991). Preliminary design of a high-speed SWATH passenger/car ferry. *Marine technology 28*(3), 129–141.

Piscopo, V. & A. Scamardella (2015). The overall motion sickness incidence applied to catamarans. *International Journal of Naval Architecture and Ocean Engineering (IJNAOE) 7*(4), 655–669.

Salvesen, N., C.H.V. Kerczek, C.A. Scragg, C.P. Cressy, & M.J. Melnhold (1985). Hydro-numeric design of SWATH ships. *Transactions of the Society of Naval Architects and Marine Engineers, SNAME 93*, 325–346.

Sarioz, K. (1993). *A hydrodynamic hull form design methodology for concept and preliminary design stage*. Ph.D. thesis, University of Newcastle, Newcastle Upon Tyne, United Kingdom.

Scamardella, A. & V. Piscopo (2014). Passenger ship seakeeping optimization by the overall motion sickness incidence. *Ocean Engineering 76*, 86–97.

Söding, H. & V. Bertram (2009). Program PDSTRIP: public domain strip method. https://sourceforge.net/projects/pdstrip/.

Vernengo, G. & D. Bruzzone (2016). Resistance and seakeeping numerical performance analyses of a semi-small waterplane area twin hull at medium to high speeds. *Journal of Marine Science and Application 15*(1), 1–7.

Progress in Maritime Technology and Engineering – Guedes Soares & Santos (Eds)
 ISBN 978-1-138-58539-3

Comparative study of various strip-theory seakeeping codes in predicting heave and pitch motions of fast displacement ships in head seas

F. Belga, S. Sutulo & C. Guedes Soares
Centre for Marine Technology and Ocean Engineering (CENTEC), Instituto Superior Técnico, Universidade de Lisboa, Lisbon, Portugal

ABSTRACT: The study focuses on a comparative investigation of three seakeeping codes based on the ordinary strip method: the open-source code PDstrip, the commercial code MaxSurf and an in-house code earlier developed at CENTEC. All programs were applied to two fast displacement ship forms upon head seas: a mono-hull and a catamaran, for which experimental data from model testing were available. In the latter case, an assumption about negligible hull-to-hull interaction, acceptable in head and following seas, was exploited. The peculiarity of these vessels is reaching Froude numbers up to 1.14, which increases possible uncertainties caused by the strip method assumptions. All codes are linear and formulated in the frequency domain. Heave and pitch RAOs have been compared, computed whenever possible both with and without transom terms activated. The observed differences between the three codes and from the experimental data motivated an analysis of the computed added masses and damping coefficients.

1 INTRODUCTION

Despite a definite shift to 3D seakeeping codes observed nowadays, see e.g., (Söding & Volker 2009), strip-theory based codes remain rather valuable as they allow reasonably accurate and relatively fast predictions of ship motions in most situations, which becomes particularly useful within optimization contexts where multiple calculations must be repeated or in any other cases when multiple computations are inevitable, see e.g., (Davis & Holloway 2003).

Another cause for continuing popularity of strip methods is their perfect compatibility with the problem of predicting ship global hull strength in a seaway on the level of beam theory: a relatively simple modification and extension of the code is required for computing the shear forces and bending moments for every transverse section of the hull, which can be viewed as rigid or flexible. In the later case, the study of sea-induced longitudinal vibrations become possible. An example of such extended algorithm can be found in (Fonseca & Guedes Soares 1998).

A wide number of strip-theory variants applied to the seakeeping of ships has been developed and proposed since the 50s. A rather comprehensive review of such variations can be found in the books (Beck et al. 1989), (Newman 1977), (Faltinsen 2005) and (Bertram 2012) and only the most important and relevant conclusions will be exposed below.

The first strip model was devised by Korvin-Kroukovsky and Jacobs in 1957, see (Korvin-Kroukovsky 1961), (Lugovsky 1999), as an alternative to the thin ship theory earlier developed, or around the same time, by Haskind, Peters & Stoker and Newman. Although initially viewed as promising and as a natural extension of the thin ship wave resistance theory, it suffered from heavy overprediction of the wave damping. The early strip model exploited, still on an intuitive level, slenderness of the ship hull, which made possible neglecting the longitudinal induced velocities and treating the ship hull as a set of transverse sections, each subjected to a simpler 2D hydrodynamic problem. While this model provided satisfactory results for zero or small speed of advance, typical for most merchant displacement ships, it was criticised for certain theoretical deficiencies. Remarkably, 5 independent (!) groups of ship hydrodynamicists from different countries developed, almost simultaneously (years 1968–70), an advanced variant of the strip method (Beck et al. 1989), (Lugovsky 1999). The most known, cited and comprehensive related publication (Salvesen, Tuck, & Faltinsen 1970) describes the most complete version of the method, developed for a ship with five degrees of freedom (5DOF), as the surge motion was ignored due to incompatibility with the strip model. It comprised the case of relatively fast ships with the transom becoming dry at certain speeds. The initials of the authors of this article inspired the popular designation of this model as STF. Although at zero ship speed this model becomes identical to that of Korvin-Kroukovsky, for many low-speed cases (Froude number Fn ≤ 0.2), the older model actually demonstrated better agreement with experimental data. However, its performance becomes definitely inferior for higher values of the Froude number when compared to the STF. As the STF method was also not free of certain imperfections, more sophisticated variants of the slender body theory were later developed (Newman 1977), (Bertram 2012). Yet, not only these methods are substantially more complicated for practical use but they failed to demonstrate definite superiority in terms of agreement with experiments as well. The latest progress focus mainly on the development of 3D models with increased practical applications, driven by the growth of the available computing power. Meanwhile, when a less accurate but faster method is required, the STF method is viewed as the most preferable and a winner in the natural selection process.

In general, it is recognized that the STF model is not suitable for high-speed crafts with Fn $\geq$ 0.6–0.7. In this case, a more precise, though more complicated 2D+t theory is recommended (Faltinsen 2005). However, (Sokolov & Sutulo 2005) indicates that at least at Fn ≤ 0.8, the plain STF method can provide quite reasonable predictions of longitudinal motions.

The STF model is relatively simple and easy to implement and thus, the large number of in-house computer codes written in various programming languages is hard to estimate. In theory, such codes must provide identical results. Yet, this does not always happen in practice, see e.g., (Guedes Soares 1990), (Guedes Soares 1991). The following reasons for the observed discrepancies can be imagined:

1. Human errors in coding combined with insufficient debugging and testing. Nowadays such flaws are not likely, especially when it goes about commercial codes.
2. Compiler and run-time library errors. Must be extremely unlikely but the second author of the present article had twice (!), in 1975 and 1999, definitely faced erroneous functioning of the complex division operation with 2 different Fortran compilers on two absolutely different types of hardware. Ironically, in the both cases the problem occurred inside seakeeping codes and was bypassed by coding definition of the malfunctioning operation.
3. Different length of the real variables and constants (typically 4 vs 8 bytes). However, hardly significant for STF algorithms.
4. Various codes may use different methods for estimating hydrodynamic characteristics of the hull sections. Again, all correct methods are supposed to produce identical results but in practice certain differences may be observed, see e.g., (Sutulo & Guedes Soares 2004).
5. Some uncertainties can be related to peculiarities of interpolation of the sectional characteristics between different frequencies and to careless treatment of very small encounter frequencies: the STF model has a singularity at the zero frequency and thus, a special asymptotic form must be developed for small frequencies.
6. Uncertainties related to the "transom effects". For relatively high-speed vessels the transom terms accounting for the dry transom should be accounted for, as they are responsible for the socalled hull lift damping (Faltinsen 2005). However, it was discovered that, in some cases, the transom terms worsen the agreement with the experimental data (Beck et al. 1989) and, as result, this option has been suppressed in some of the codes.

As the STF method keeps its practical significance, investigations focusing on uncertainties caused by the use of different implementations are of considerable value. An example of such studies is described in the present contribution. Three easily available codes were chosen for comparative computations: (1) PDstrip which is a public-domain code (Söding & Bertram 2009); (2) an in-house code developed by Fonseca (Fonseca & Guedes Soares 1998); and (3) the commercially available program MaxSurf Motions (Bentley Systems, Inc. 2013).

The code PDstrip is a Fortran 90 remake of the earlier Fortran 77 code STRIP. The main algorithm is the same in both cases, corresponding to the STF model and being characterised by careful treatment of the low end frequencies and by a sophisticated interpolation method between frequencies, performed separately for the modules and arguments of the complex variables of interest. The sectional complex added masses are estimated by the procedure HMASSE and the sectional excitation forces by the procedure WERREG (the Haskind–Newman relations are not used). Both procedures are based on a variant of the boundary integral equation (BIE) method described in (Bertram 2012). The code STRIP was using another BIE method earlier developed by Yeung (1973), see also (Sutulo & Guedes Soares 2004). Both are based on the theory developed in (Jaswon 1953).

The Fonseca linear code was written in Fortran 77 and is using the Frank boundary singularities method (Frank 1967). This method is not free from singular frequencies but in most cases this phenomenon did not emerge. The transom terms were suppressed in the used version of the program. Since the source code was not available to the authors, this deficiency could not be fixed.

The code MaxSurf Motions is claimed to be applicable for Fn ≤ 0.7 and it uses the multiparametric conformal mapping method (up to 15 parameters but 2-parameter Lewis forms are used by default) to evaluate the characteristics of the 2D sections, see e.g., (Ramos & Guedes Soares 1997). The heave and pitch motions are modelled with the strip method with somewhat simplified formulae for excitation forces. The program has a catamaran option but hull-to-hull interactions are neglected, i.e., the catamaran is assumed to be hydrodynamically equivalent to its single hull. The code allows empiric tuning of the model by introducing additional user-defined damping in heave and pitch, which is uniformly distributed between sections.

The latter option reflects the influence of a rather popular view that ignoring viscous damping in heave and pitch may result in overestimated responses. However, it has been long ago established with indisputable certainty that possible relative contribution of viscosity is negligible in the entire frequency range of interest and, of course, no such attempts were carried out in the present study.

An attempt to compare the three mentioned codes from the viewpoint of their applicability to the design optimization problem is undertaken in the present article. For this purpose, a number of comparative computations of ship responses in head regular waves have been carried out. Two ship forms were selected: the fast monohull "Model 5" (Blok & Beukelman 1984) and the fast displacement catamaran described in (Guedes Soares et al. 1999), for which experimental data from model testing were available. In the second case, any hydrodynamic interaction between the hulls was neglected, which is in accordance with the common practice and recommendations (Bertram 2012). Again, this presumes that the catamaran is hydrodynamically equivalent to its single hull. The conclusions drawn in this study helped in the selection of the most suitable program for the seakeeping optimization of a fast displacement catamaran to operate as crew supplier for an offshore platform at the Alentejo basin, in Portugal (Belga et al. 2018).

2 BRIEF OVERVIEW OF THE LINEAR SEAKEEPING MATHEMATICAL MODEL

The seakeeping inertial Cartesian frames, advancing with the ship speed U but not involved in the ship motions excited by waves can be somewhat different. In the present study the primary frame $O\xi\eta\zeta$ used in (Beck et al. 1989) was considered, i.e. the frame with the ζ-axis direct vertically upward, the η-axis to the port side and the ξ-axis pointing forward. The origin O is located at the waterplane area but on the same vertical as the ship centre of mass (gravity), C_G.

The linear mathematical model for ship motions in regular head seas can be described by the following set of complex linear algebraic equations for heave and pitch complex amplitudes ζ^* and θ^*:

$$
\begin{aligned}
&[-\omega_e^2(m+A_{33})+C_{33}+i\omega_e B_{33}]\zeta^* \\
&\quad+[\omega_e^2 A_{35}+C_{35}+i\omega_e B_{35}]\theta^* = F_\zeta^*, \\
&[-\omega_e^2 A_{53}+C_{53}+i\omega_e B_{53}]\zeta^* \\
&\quad+[\omega_e^2(I_{yy}+A_{55})+C_{55}+i\omega_e B_{55}]\theta^* = F_\theta^*,
\end{aligned}
\tag{1}
$$

where $\omega_e = \omega + kU$ is the encounter frequency in head seas, ω is the absolute wave frequency, $k = \omega^2/g$ is the wave number, g is the acceleration of gravity, m is the ship mass and I_{yy} its moment of inertia in pitch; A_{ij}, $i, j = 3, 5$ are the hydrodynamic inertial coefficients and B_{ij} are the damping coefficients; C_{ij} the restoring coefficients; F_ζ^* and F_θ^* are the heave force and pitch moment complex amplitudes, respectively (see Table 1).

Here, all the integrals are along the length of the ship L. a_{33} and b_{33} represent the sectional added mass and damping coefficients in ζ-direction. A_{33}^0 and B_{33}^0 refer to the speed independent components of A_{33} and B_{33}. ξ_{tr} is the ξ-coordinate of the aftermost cross section of the ship and a_{33}^{tr} and b_{33}^{tr} are the added mass and damping coefficients evaluated at that section. ρ is the fluid density, A_{wp} is the static waterplane area, M_{yy} is the first

Table 1. STF model: hydrodynamic coefficients.

	Main part	Transom terms
A_{33}	$\int_L a_{33} d\xi$	$-\frac{U}{\omega_e^2} b_{33}^{tr}$
A_{35}	$-\int_L \xi a_{33} d\xi - \frac{U}{\omega_e^2} B_{33}^0$	$+\frac{U}{\omega_e^2}\xi_{tr} b_{33}^{tr} - \frac{U^2}{\omega_e^2} a_{33}^{tr}$
A_{53}	$-\int_L \xi a_{33} d\xi + \frac{U}{\omega_e^2} B_{33}^0$	$+\frac{U}{\omega_e^2}\xi_{tr} b_{33}^{tr}$
A_{55}	$\int_L \xi^2 a_{33} d\xi + \frac{U^2}{\omega_e^2} A_{33}^0$	$-\frac{U}{\omega_e^2}\xi_{tr}^2 b_{33}^{tr} + \frac{U^2}{\omega_e^2}\xi_{tr} a_{33}^{tr}$
B_{33}	$\int_L b_{33} d\xi$	$+U a_{33}^{tr}$
B_{35}	$-\int_L \xi b_{33} d\xi + U\ A_{33}^0$	$-U\xi_{tr} a_{33}^{tr} - \frac{U^2}{\omega_e^2} b_{33}^{tr}$
B_{53}	$-\int_L \xi b_{33} d\xi - U\ A_{33}^0$	$-U\xi_{tr} a_{33}^{tr}$
B_{55}	$\int_L \xi^2 b_{33} d\xi + \frac{U^2}{\omega_e^2} B_{33}^0$	$+U\xi_{tr}^2 a_{33}^{tr} + \frac{U^2}{\omega_e^2}\xi_{tr} b_{33}^{tr}$
C_{33}	$\rho g A_{wp}$	–
C_{35}	$-\rho g M_{yy}$	–
C_{53}	$-\rho g M_{yy}$	–
C_{55}	$gmGML$	–
F_ζ^*	$\rho a_w \int_L (f_3^D + f_3^K) d\xi$	$+\rho a_w \frac{U}{i\omega_e} f^{D^{tr}}$
F_θ^*	$-\rho a_w \int_L \left[\xi(f_3^D + f_3^K) + \frac{U}{i\omega_e} f_3^D\right] d\xi$	$-\rho a_w \frac{U}{i\omega_e}\xi_{tr} f_3^{D^{tr}}$

area moment of the static waterplane and GM_L the longitudinal metacentric height. f_3^K and f_3^D represent, respectively, the sectional Froude–Krylov and diffraction forces in the vertical direction (ζ-axis) for unit amplitude incident waves. Regarding the ormer component, associated with the field of incident waves, the classical theory of linear gravity-waves defines, within deep water assumptions, the potential of a progressive incident wave with an arbitrary direction, which makes f_3^K easy to compute by evaluating it at the mean wetted cross section. Alternatively, it can be computed at each time step, an approach commonly followed by time domain formulations. As for f_3^D, which measures the perturbation of the field of incident waves due to the presence of the ship, instead of determining directly the 2-D diffraction potential, Green theorem is usually applied in order to solve it as function of the 2-D radiation potential (Salvesen et al. 1970). Finally, f_3^{Dtr} is the diffraction force evaluated at the aftermost section.

The set (1) is solved straightforward and the solution is typically represented in the form of frequency transfer functions (FTF):

$$\Phi_\zeta(\omega) = \frac{\zeta^*}{a_w}; \quad \Phi_\theta(\omega) = \frac{\theta^*}{ika_w}, \tag{2}$$

where a_w is the wave amplitude.

The FTF are complex and each of them can be represented in the form:

$$\Phi_{\zeta,\theta}(\omega) = |\Phi_{\zeta,\theta}(\omega)|\, e^{-i\varepsilon_{\zeta,\theta}(\omega)}, \tag{3}$$

where $|\Phi_{\zeta,\theta}(\omega)|$ is the amplitude FTF, traditionally called the Response Amplitude Operator (RAO) and

$$\varepsilon_{\zeta,\theta}(\omega) = -\arg \Phi_{\zeta,\theta}(\omega) \in [0, 2\pi], \tag{4}$$

is the response phase lag which is always positive for any physically meaningful causal system, although serious incorrect definitions of this parameter are rather common in the literature on seakeeping and vibrations.

3 OVERVIEW OF THE CODES

The main features of the three strip method seakeeping programs mentioned in Introduction are summarized in Tables 2–3.

It must be noted that various codes are using different frames. Of course, this must not affect the results but, as it is often important to know exactly which frame is used, the summary of the used frames is given in Table 4.

Table 2. Seakeeping codes: general data.

Code, Version	Designation	Main author	Reference
NA	Fonseca	Nuno Fonseca (CENTEC)	(Fonseca & Guedes Soares 1998)
PDSTRIP, rev. 33	PDstrip	Heinrich Söding	(Söding & Bertram 2009)
Maxsurf Motions, V20	MaxSurf	Bentley Systems	(Bentley Systems, Inc. 2013)

Table 3. Seakeeping codes: main characteristics.

Features	Fonseca	PDstrip	MaxSurf
Transom terms	–	+	+
Maximum number of strips	40	100	200
Minimum recommended number of strips	21	30–40	15–30
Offsets/parameters[1] per strip, max.	20	100	15
Offsets/parameters per strip, min.	8–10	10	3
Same number of offsets/ parameters for all strips	+	–	+
Equidistant offsets	+	+	–
Maximum number wavelengths	30	200	500
User defined wavelength range	+	+	–
Computational speed relative to Fonseca using its max. settings	100%	95%	30%

[1]In MaxSurf the sections are described by conformal mapping parameters rather than offset points.

Table 4. Reference systems used by the codes.

	Code	Origin (ξ, ζ)	ξ to:	η to:	ζ to:
Hull Geometry	Fonseca	fwdPP, WL	aft	port	up
	PDstrip	$L/2$, baseline	fwd	port	up
	MaxSurf	non applicable	fwd	stbd	up
Inputs (e.g. remote locations)	Fonseca	$L/2$, WL	fwd	port	up
	PDstrip	$L/2$, baseline	fwd	port	up
	MaxSurf	$L/2$, baseline	fwd	stbd	up
RAOs	Fonseca	C_G, baseline	fwd	port	up
	PDstrip	$L/2$, baseline	fwd	stbd	down
	MaxSurf	C_G, baseline	fwd	stbd	up

The Fonseca code is, in fact, a linear counterpart of a more complex time-domain semi-linear program described in (Fonseca & Guedes Soares

1998). Despite the substantial number of research papers successfully validating the results of the semi-linear code, mainly for moderately slow ships, e.g., (Fonseca & Guedes Soares 2002), (Fonseca & Guedes Soares 2004a), (Fonseca & Guedes Soares 2004b), (Fonseca & Guedes Soares 2005), the linear Fonseca code tested here lacks documentation and validation, as it was not planned to be used on its own in view of the availability of more advanced non-linear codes at the time.

The commercial code MaxSurf was presumably validated on a wide range of vessel types and some of these results can be found in the appendices of the manual (Bentley Systems, Inc. 2013) or in independent publications such as (Ghassemi et al. 2015). As could be expected from a commercial code, it possesses a rather developed graphic interface facilitating its use by less qualified specialists. At the same time, this code runs, due to unclear reasons, substantially slower than its counterparts. Although PDstrip is a linear frequency domain program, it can account for some nonlinear effects described in (Söding & Bertram 2009). However, these options were not activated in the present study. As stated by Palladino et al. (2006), the validation of PDstrip continues to be a work in progress. Although it has been used by many researchers for a wide variety of purposes including manoeuvring simulations (Schoop-Zipfel & Abdel-Maksoud 2011), analysis of wave-induced dynamic effects on sailing yachts (Bordogna 2013), computation of drift forces on offshore wind farm installation vessels (Augener & Krüger 2014) or even integrated into a global design tool (Salio et al. 2013), a thorough work of comparison with both experimental results and other seakeeping codes is still to be done (Palladino et al. 2006), (Gourlay et al. 2015).

In terms of the method, PDstrip is the only code which the formulation presents noteworthy variations from the previously presented STF method. Recalling Table 4, note that the inertial reference frame $O\xi\eta\zeta$ is, according to Siöding & Bertram (2009), positioned at amidships, rather than at the centre of gravity with respect to the motion results. The method computes hydrodynamic coefficients in a somewhat different way as compared to the formulae shown in Table 1. Namely, the terms related to the longitudinal derivative of the added masses and damping coefficients, which were handled directly, without their elimination through integration by parts and application of the Stokes–Tuck transformation that results in the original STF formulae. In addition, this code handles the surge motion using a simplified surge equation accounting for only Froude–Krylov and inertial forces. This equation is, however, coupled with the pitch motion through the modifications of the coefficients shown bellow. Yet, these contributions are, according to Söding, much smaller than errors of the strip-method.

$$\delta A_{55} = a_{11}\zeta_0^2, \tag{5}$$

where δA_{55} denotes the additional contribution to A_{55}, ζ_0 refers to the waterline level and a_{11} yields the generalized longitudinal force acting on the ship computed using the empirical formula (6), as function of the ship length L and the displacement m.

$$a_{11} = m / \left[\pi\sqrt{\rho L^3 / m} - 14 \right] \tag{6}$$

$$\delta F_\theta^{*K} \int_L \zeta p \frac{dA}{d\xi} d\xi + \zeta_{tr} p_{tr} A_{tr} \tag{7}$$

where δF_θ^{*K} is the additional Froude–Krylov contribution to the pitch moment F_θ^*, p refers to the pressure calculated at the centre of each cross section $(\xi, 0, \zeta)$, in head seas, A is the corresponding immersed area and the index tr denotes values evaluated at the transom stern. Therefore, the term $\zeta_{tr}\, p_{tr}\, A_{tr}$ vanishes if the transom corrections are to be neglected from the calculations. At the centre of a cross section, the pressure is given by:

$$p = -\rho g e^{-k(\zeta + T)} a_w e^{ik\xi}, \tag{8}$$

where T is the mean draft of the ship cross section.

Regarding the restoring coefficients, they are effectively computed has stated in Table 1, considering the mean position of the wetted surface. However, as a result of programming decisions from the developers, PDstrip computes the first area moment of the static waterplane with respect to the midship section, resulting in different coefficients for C_{35} and C_{53}, which apparently does not affect the motion results. Furthermore, there is an addition in C_{53} if transom terms are to be included in the calculations:

$$\delta C_{53} = \zeta_{tr} A_{tr}, \tag{9}$$

where ζ_{tr} is the ζ-coordinate of the centroid of the area of wetted transom A_{tr}.

The version of PDstrip that is being used here has been revised by the authors (rev. 33). The most significant modifications relate with the handling of transom terms. In PDstrip rev.32, the latest publicly available version, these terms were always kept for added masses, damping coefficients and exciting forces and moments, while the option to include or exclude them only affected equations (7) and (9). As this seemed to lack consistency, the source-code was improved in order to enable that option for the coefficients of Table 1 as well.

4 RESULTS OF COMPARATIVE COMPUTATIONS

The three codes briefly described above were applied to two ship forms showed in Figure 1: the fast monohull "Model 5" (Blok & Beukelman 1984) and a fast displacement river-going catamaran (Guedes Soares et al. 1999), for which experimental data from model testing were available. In the latter case, an assumption about negligible interaction between hulls, acceptable in head and following seas, was exploited, which presumes that the catamaran is hydrodynamically equivalent to its single hull. The main particulars of both models are shown in Table 5.

In order to obtain comparable results between the different codes, both hull forms were discretized using 40 strips, imposed by the maximum settings allowed by Fonseca which has the lowest capacity. For the same reason, 20 equally spaced offsets per strip were used, as none of the cross sections suffered from abrupt changes in curvature, which would require a higher density of points in those regions. Regarding wavelengths it was used the maximum number allowed by each code, in order to obtain the best definition of the frequency range possible.

In the case of the catamaran, four Froude numbers ranging between 0 and 0.6 were analysed, while for "Model 5" only the Froude number values 0.57 and 1.14 have been considered. The peculiarity of the studied vessels is the high Froude numbers at which they were tested, which increases possible uncertainties caused by the strip method assumptions. Comparisons have been performed

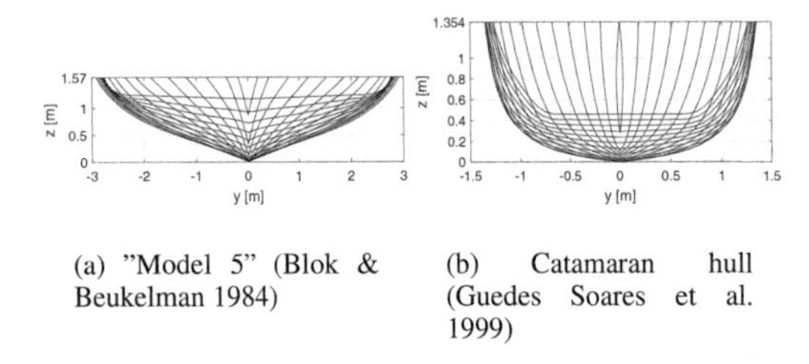

(a) "Model 5" (Blok & Beukelman 1984)

(b) Catamaran hull (Guedes Soares et al. 1999)

Figure 1. Underwater body lines of the full-scale vessels.

Table 5. Main dimensions of the full-scale vessels.

	LWL [m]	BWL [m]	T [m]
"Model 5" (Blok & Beukelman 1984)	50	5.83	1.57
Catamaran (Guedes Soares et al. 1999)	43	2.7	1.35

for the heave and pitch RAOs computed, whenever possible, both with and without transom terms activated. These are shown bellow, divided by the wave amplitude (a_w) for heave and by the wave slope (ka_w) for pitch, as function of the wave frequency ω and each plot corresponds to a different Froude number. Results for "Model 5" can be seen in Figure 2 for heave and Figure 3 for pitch, while for the catamaran case one should refer to Figures 6 and 7 for heave and pitch, respectively.

Figures 4 and 5 summarise the absolute difference (in the same units as the RAOs) between

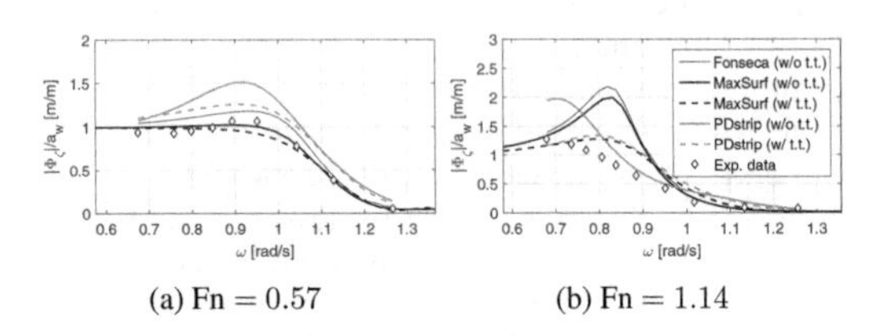

(a) Fn = 0.57 (b) Fn = 1.14

Figure 2. Heave RAOs as function of the wave frequency, "Model 5".

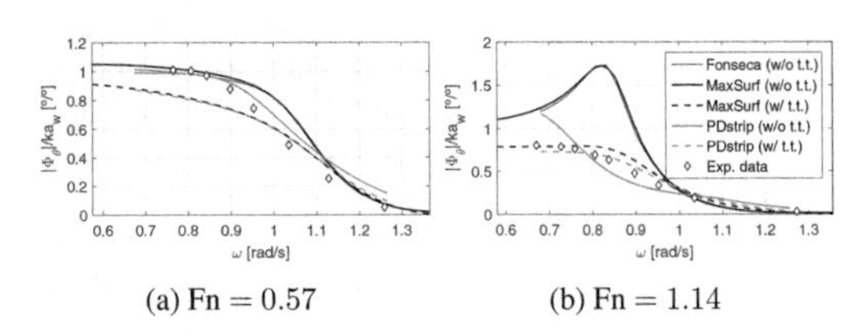

(a) Fn = 0.57 (b) Fn = 1.14

Figure 3. Pitch RAOs as function of the wave frequency, "Model 5".

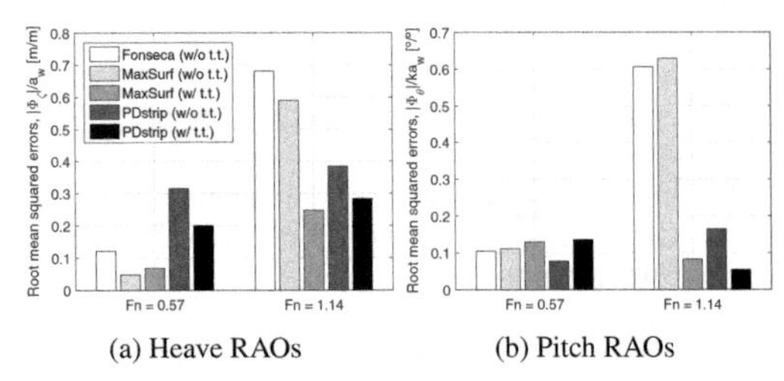

(a) Heave RAOs (b) Pitch RAOs

Figure 4. Root mean squared absolute differences between numerical and experimental results, "Model 5".

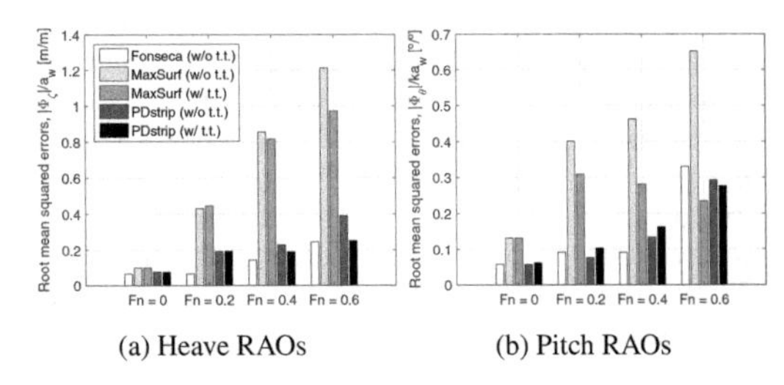

(a) Heave RAOs (b) Pitch RAOs

Figure 5. Root mean squared absolute differences between numerical and experimental results, Catamaran.

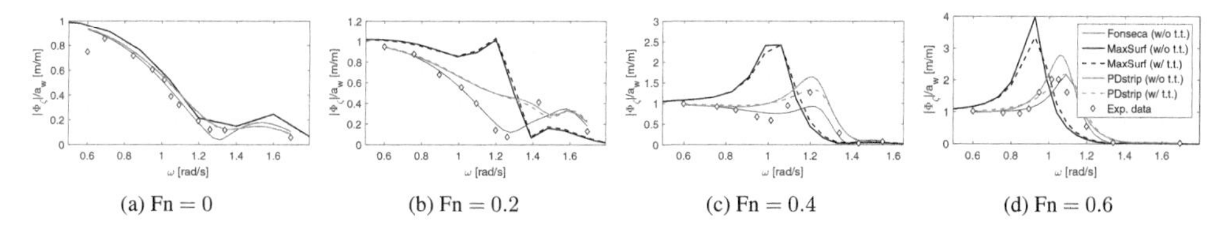

(a) Fn = 0 (b) Fn = 0.2 (c) Fn = 0.4 (d) Fn = 0.6

Figure 6. Heave RAOs as function of the wave frequency, Catamaran.

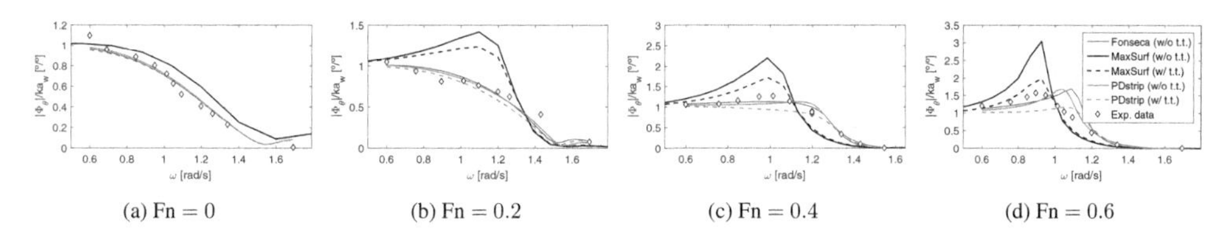

(a) Fn = 0 (b) Fn = 0.2 (c) Fn = 0.4 (d) Fn = 0.6

Figure 7. Pitch RAOs as function of the wave frequency, Catamaran.

the numerical results obtained with each code and the experimental data, averaged over all measured frequencies. In order to average these results, the square root of the average squared differences was used. Since the errors are squared before they are averaged, a relatively higher weight is given to larger errors. This is beneficial for this study where big deviations from the experimental results are particularly undesirable.

It can be observed that with increasing vessel speed, the accuracy in predicting heave and pitch motions with strip-theory decreases, a well-known limitation of such theories and a general principle that must be taken into account. Figures 4 and 5 illustrate this effect particularly well.

Given the nature of the coefficients of the coupled heave and pitch equations as shown in Table 1, transom terms have no effect at zero speed. Such terms are only meaningful as the ship starts moving. However, since PDstrip considers speed-independent transom corrections for the restoring coefficient C_{53} as shown in equation (9) and for the Froude–Krylov force as in (7), there is at Fn = 0 a small visible difference between its predictions both in Heave, Figure 6(a), and in Pitch, Figure 7(a). This seems to be adverse for pitching motion. With MaxSurf, this difference does not exist. In any case, at zero speed, Fonseca seems to make the most accurate predictions. PDstrip, without transom terms in particular, also performs reasonably well.

Following Figures 4 and 5, for Fn = 0.2, heave predictions seem to have been adversely affected by the inclusion of transom terms in the equations. At Fn = 0.4, the opposite happened. At Fn = 0.57, the inclusion of transom terms improve the heave predictions of PDstrip but that is not the case with MaxSurf. Regarding pitch RAOs, at Fn = 0.2, 0.4, the prediction of MaxSurf improve with the inclusion of transom terms while with PDstrip this decreases accuracy. For Fn = 0.57, the inclusion of transom terms generally makes pitch predictions worse. On the contrary, at the highest speeds (Fn = 0.6, 1.14), it is advantageous both for heave and pitch motions, to include such terms in the equations.

In a general sense, at higher Froude numbers, the inclusion of transom terms is beneficial, as they add damping to the system. It is interesting to note that the semi-linear version of Fonseca (Fonseca & Guedes Soares 1998) has been mainly used to predict motions and wave loads of slower large displacement vessels (e.g., containerships) and, in fact, neglects transom terms. The conclusions of (Marón et al. 2004) and (Fonseca & Guedes Soares 2004c), where that code was applied to a fast mono-hull, showed precisely that at higher Froude numbers its accuracy decreases significantly. This seems to indicate that if the transom terms are neglected the range of applications should focus on slower vessels. The same applies to Fonseca (linear version), which also excludes such terms.

Table 6 summarises the drawn conclusions, supported by Figures 4 and 5. A few comments on the capacity of the codes to accurately predict the resonance peaks and the respective frequencies have been added as well. In this regard, note that for the case of the catamaran, it is possible that they might not have been accurately captured by the experiments because of the frequency separation between experimental points (Guedes Soares et al. 1999).

It seems that at lower speeds, using a strip-theory code without transom terms is the best approach for optimal results. Table 6 suggests that Fonseca is, overall, the most suited for Fn = 0,

Table 6. Summary of code preferences.

Fn	Heave	Pitch
0	Fonseca[1]	Fonseca
0.2	Fonseca	PDstrip (w/o t. terms)
0.4	Fonseca[2]	Fonseca[3]
0.57	MaxSurf (w/o t. terms)	PDstrip (w/o t. terms)
0.6	Fonseca[4]	MaxSurf (w/t. terms)
1.14	MaxSurf (w/t. terms)[5]	PDstrip (w/t. terms)

[1]Differences between the codes are negligible
[2]Resonance peak more accurately predicted with PDstrip (w/t. terms)
[3]Resonance frequency more accurately predicted with MaxSurf
[4]Similar results with PDstrip (w/t. terms). Resonance frequency overestimated with both methods.
[5]Similar results with PDstrip (w/t. terms). Amplitudes at the middle frequency range poorly estimated with both methods.

0.2, 0.4. For Fn = 0.57, it appears that although not using transom terms proves to be more effective, MaxSurf and PDstrip work the best. For the highest speeds, i.e., Fn = 0.6, 1.14, it becomes clear that the results improve by including transom corrections in the equations. Overall, PDstrip seems to be the most consistent and reasonably accurate, even though MaxSurf produces somewhat similar results. However, from the perspective of embedding the code into optimization procedures, computational speed plays a crucial role and thus, PDstrip becomes the most natural choice (refer to Table 3). In addition, PDstrip allows: a substantial number of ship sections and offset points for geometry discretization, which improves accuracy during seakeeping computations; a large number of wavelengths to be used for motion results, being advantageous for a proper definition of the whole frequency range. Also, the method used to handle the 2D problem at each cross sections is allegedly superior (Söding & Bertram 2009), when compared to conformal mapping techniques.

The differences between the numerical predictions and the experimental data motivated an analysis of the global added masses and damping coefficients. As an example, these are shown below for Fn = 1.14 ("Model 5" case). Note that care was taken into displaying the same frequency range that was captured by the experiments, as in Figures 2 and 3.

It has been verified that the discrepancies between the codes increase with speed, as expected. The best results were obtained for the coefficients A_{33} and B_{33}, while for the remainders the results tend to differ more. Also, at higher speeds, the existence of irregular frequencies when using Fonseca becomes clear, as a result of the method used

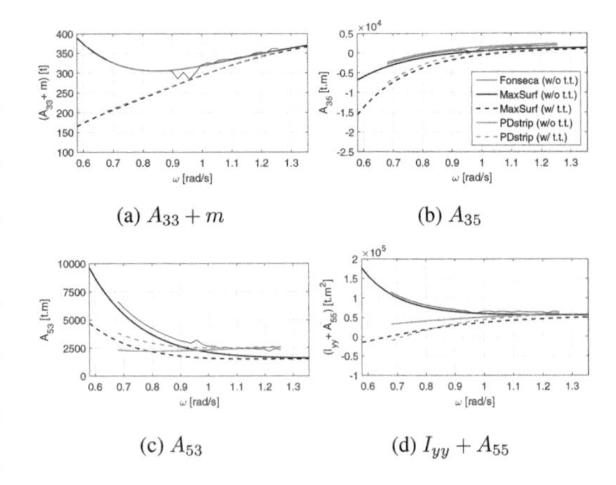

(a) $A_{33} + m$ (b) A_{35}

(c) A_{53} (d) $I_{yy} + A_{55}$

Figure 8. Global added masses for Fn = 1.14, "Model 5".

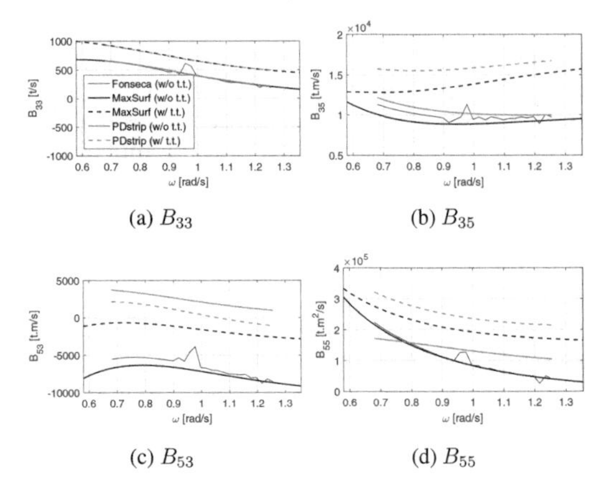

(a) B_{33} (b) B_{35}

(c) B_{53} (d) B_{55}

Figure 9. Global damping coefficients for Fn = 1.14, "Model 5".

to solve the 2D problem (Frank 1967). As hypothesized in Introduction, there is a number of possible reasons for the observed differences. In order to fully understand this problem, an in-depth study of the specifics of the methods applied by each code is required, which stands as suggestion for future work.

5 CONCLUSIONS

A comparative study of three strip-theory based codes is presented here, comprising the open-source code PDstrip, the commercial software MaxSurf Motions and an in-house code earlier developed at CENTEC referred to as Fonseca. Experimental data from model testing of two ship hull forms was used to verify the numerical predictions of the three codes in terms of heave and pitch RAOs upon head seas.

With increasing vessel speed, the accuracy in predicting heave and pitch motions with strip-theory decreases, a well-known limitation of such theories and a general principle that must be taken into account. At lower speeds, not using transom terms is preferable, Fonseca in particular, as the version of the code used here neglects them. In any case, both PDstrip and MaxSurf (without transom terms) also performed reasonably well in specific situations. At higher speeds (Fn = 0.6, 1.14), heave and pitch response peaks were generally overestimated but the inclusion of transom terms adds damping to the system, being advantageous for predicting heave and pitch motions. PDstrip and MaxSurf with transom terms seem the most suited codes in these cases.

Considering the high computational speed of PDstrip, as well as other useful features, it seems natural to select it as seakeeping tool for design procedures intended to optimize fast displacement vessels in head seas. Also, as an open-source Fortran code, it is possible to edit and further improve PDstrip if necessary, a useful characteristic bearing future work in mind. For an example of practical applications of the drawn conclusions refer to (Belga et al. 2018), where the hull form of a crew supply catamaran was optimized with respect to seakeeping.

ACKNOWLEDGEMENTS

The study was performed within the project PTDC/EMSTRA/5628/2014 "Manoeuvring and moored ships in ports–physical and numerical modelling", funded by the Portuguese Foundation for Science and Technology (FCT).

REFERENCES

Augener, P.H. & S. Krüger (2014). Computation of drift forces for dynamic positioning within the very early design stage of offshore wind farm installation vessels. In *Proceedings of the ASME 2014, 33rd International Conference on Ocean, Offshore and Arctic Engineering*, San Francisco, California, USA. American Society of Mechanical Engineers.

Beck, R.F., W.E. Cummins, J.F. Dalzell, & P. Mandel (1989). Motions in waves. In E.V. Lewis (Ed.), *Principles of Naval Architecture*, Volume 3, pp. 1–190. Jersey City, NJ: SNAME.

Belga, F., M. Ventura, & C. Guedes Soares (2018). Seakeeping optimization of a catamaran to operate as fast crew supplier at the Alentejo basin. In C. Guedes Soares and T.A. Santos (Eds.), *Progress in Maritime Technology and Engineering*, London, UK: Taylor & Francis.

Bentley Systems, Inc. (2013). *Maxsurf Motions, Windows Version 20, User Manuals.*

Bertram, V. (2012). *Practical Ship Hydrodynamics*. Oxford, UK: Butterworth–Heinemann.

Blok, J.J. & W. Beukelman (1984). The high-speed displacement ship systematic series hull forms – seakeeping characteristics. *Transactions of the Society of Naval Architects and Marine Engineers, SNAME 92*, 125–150.

Bordogna, G. (2013). The aero-hydrodynamic characteristics of yachts sailing upwind in waves. Master's thesis, TU Delft, Delft University of Technology.

Davis, M.R. & D.S. Holloway (2003). The influence of hull form on the motions of high speed vessels in head seas. *Ocean Engineering 30*, 2091–2115.

Faltinsen, O.M. (2005). *Hydrodynamics of High-Speed Marine Vehicles*. Cambridge, UK: Cambridge University Press.

Fonseca, N. & C. Guedes Soares (1998). Time-domain analysis of large-amplitude vertical ship motions and wave loads. *Journal of Ship Research 42*(2), 39–153.

Fonseca, N. & C. Guedes Soares (2002). Comparison of numerical and experimental results of nonlinear wave-induced vertical ship motions and loads. *Journal of Marine Science and Technology 6*(4), 193–204.

Fonseca, N. & C. Guedes Soares (2004a). Experimental investigation of the nonlinear effects on the statistics of vertical motions and loads of a containership in irregular waves. *Journal of Ship Research 48*(2), 148–167.

Fonseca, N. & C. Guedes Soares (2004b). Experimental investigation of the nonlinear effects on the vertical motions and loads of a containership in regular waves. *Journal of Ship Research 48*(2), 118–147.

Fonseca, N. & C. Guedes Soares (2004c). Validation of a time-domain strip method to calculate the motions and loads on a fast monohull. *Applied Ocean Research 26*(6), 256–273.

Fonseca, N. & C. Guedes Soares (2005). Comparison between experimental and numerical results of the nonlinear vertical ship motions and loads on a containership in regular waves. *International Shipbuilding Progress 52*(1), 57–89.

Frank, W. (1967). Oscillation of cylinders in or below the free surface of deep fluids. Technical Report 2375, Naval Ship Research and Development Centre, Washington DC, USA.

Ghassemi, H., S. Majdfar, & V. Gill (2015). Calculations of the heave and pitch RAO's for three different ship's hull forms. *Journal of Ocean, Mechanical and Aerospace – Science and Engineering 22*, 1–8.

Gourlay, T., A. von Graefe, V. Shigunov, & E. Lataire (2015). Comparison of aqwa, gl rankine, moses, octopus, pdstrip and wamit with model test results for cargo ship wave-induced motions in shallow water. In *Proceedings of the ASME 2015, 34th International Conference on Ocean, Offshore and Arctic Engineering*, St. John's, Newfoundland, Canada. American Society of Mechanical Engineers.

Guedes Soares, C. (1990). Comparison of measurements and calculations of wave induced vertical bending moments in ship models. *International Shipbuilding Progress 37*(412), 353–374.

Guedes Soares, C. (1991). Effect of transfer function uncertainty on short term ship responses. *Ocean Engineering 18*(4), 329–362.

Guedes Soares, C., N. Fonseca, P. Santos, & A. Maron (1999). Model tests of the motions of a catamaran hull in waves. In *Proceedings of the International Conference on Hydrodynamics of High-Speed Craft*, London, UK, pp. 1–10. Royal Institution of Naval Architects (RINA).

Jaswon, M. (1953). Integral equation method in potential theory. *Proceedings of Royal Society A275/360*, 23–32.

Korvin-Kroukovsky, B. (1961). *Theory of Seakeeping*. New York, NY: SNAME.

Lugovsky, V.V. (1999). *Ship Motions (in Russian)*. St. Petersburg: St. Petersburg State Marine Technical University Publishing Centre.

Marón, A., J. Ponce, N. Fonseca, & C. Guedes Soares (2004). Experimental investigation of a fast monohull in forced harmonic motions. *Applied Ocean Research 26*(6), 241–255.

Newman, J.N. (1977). *Marine Hydrodynamics*. Cambridge, MA: MIT Press.

Palladino, F., B. Bouscasse, C. Lugni, & V. Bertram (2006, October). Validation of ship motion functions of pdstrip for some standard test cases. In *9th Numerical Towing Tank Symposium*, Le Croisic, France.

Ramos, J. & C. Guedes Soares (1997). On the assessment of hydrodynamic coefficients of cylinders in heaving. *Ocean Engineering 24*(8), 743–763.

Salio, M.P., F. Taddei, P. Gualeni, A. Guagnano, & F. Perra (2013). Ship performance and sea state condition: an assessment methodology integrated in an early design stage tool. In *Proceedings of tthe 10th International Conference on Maritime Systems and Technologies, MAST 2013*, Gdansk, Poland.

Salvesen, N., E.O. Tuck, & O. Faltinsen (1970). Ship motions and sea loads. In *Transactions of the Society of Naval Architects and Marine Engineers, SNAME*, Volume 78, pp. 250–287.

Schoop-Zipfel, J. & M. Abdel-Maksoud (2011). A numerical model to determine ship manoeuvring motion in regular waves. In L. Eça, E. Oñate, J. García, T. Kvamsdal, and P. Bergan (Eds.), *4th International Conference on Computational Methods in Marine Engineering, MARINE 2011*, Lisbon, Portugal. Springer.

Söding, H. & V. Bertram (2009). Program PDSTRIP: public domain strip method. https://sourceforge.net/projects/pdstrip/.

Söding, H. & B. Volker (2009). A 3-d rankine source seakeeping method. *Ship Technology Research 56*(2), 50–58.

Sokolov, V. & S. Sutulo (2005, 27–29 June). A practical approach to a fast displacement ship's stabilization in head seas. In *Proceedings of International Conference on Fast Sea Transportation FAST-2005*, St. Petersburg, Russia.

Sutulo, S. & C. Guedes Soares (2004). A boundary integral equation method for computing inertial and damping characteristics of arbitrary contours in deep fluid. *Ship Technology Research/Schiffstechnik 51*, 69–93.

Yeung, R. (1973). *A Singularity Method for Free-Surface Flow Problems with an Oscillatory Body*. Ph. D. thesis, University of California, College of Engineering, Berkeley.

Ships in transit

Progress in Maritime Technology and Engineering – Guedes Soares & Santos (Eds)
 ISBN 978-1-138-58539-3

The transit state evaluation of a large floating dock by seakeeping criteria

E. Burlacu & L. Domnisoru
Faculty of Naval Architecture, University "Dunarea de Jos" of Galati, Galati, Romania

ABSTRACT: In this study the transit condition including river—costal navigation of a large floating dock, with 209.2 m length, is evaluated by short-term seakeeping criteria. The irregular waves are modelled by the one parameter ITTC power density spectra, with design limit significant wave height of 2 m for river and 4.942 m for costal conditions. The transit reference speed is 12 km/h, being analyzed also the 0 and 6 km/h speed cases. The transit of the floating dock is assessed for several ballasting cases, with draught 5.2, 6.2, 7.2 m and dock system vertical gravity centre position between 6 and 16 m. For the numerical analysis we have used an in-house seakeeping program, based on a linear strip theory. The seakeeping criteria are formulated in terms of motion and acceleration amplitudes statistical limit values. The numerical results of this study deliver the evaluation of the transit state of the floating dock by seakeeping safety criteria.

1 INTRODUCTION

At the design of the floating docks, besides the standard docking operation analyses, also the transit states between several locations have to be analyzed.

Usual the transit routes of the floating docks can include river and costal water-ways, where wave conditions may occur. Besides the structural evaluations (DNVGL, 2017), the assessment of navigation safety during the transit operations by seakeeping criteria have to be considered from the initial design stage (DNV, 2012; Solas, 2014).

In this study a large floating dock DOCKV is considered, designed by VARD Shipyard Tulcea (Burlacu, 2017), with the main characteristics in Table 1 and the offset lines in Figure 1.

The transit towing is done by a maritime 4000 HP tug (ANR, 2006). The resistance of the tug—floating dock system is analyzed by a theoretical model (Voitkunski, 1985; Obreja, 2005). Figure 2 presents the resistance diagrams of the tug and the floating dock in transit operation, in still water condition. From the tug-dock system resistance analysis a reference speed of 12 km/h results, in the analysis being included also the 0 and 6 km/h speed cases.

During the transit operation, the floating dock is in ballast condition. There are considered three ballast cases, with draught values between 5.2–7.2 m (Table 1). According to the dock ballasting scheme, the gravity centre of the floating dock is changing the vertical position *KG* between 6 and 16 m, resulting significant differences on transversal stability characteristics (Bidoaie & Ionas, 1998), presented in Table 2 and Figures 3. a,b,c, that can be considered in any loading case linear for heel angle $\varphi \le 6$ deg.

Table 1. Main characteristics of the floating dock (Burlacu, 2017).

Main characteristics		Case 1	Case 2	Case 3
Length overall	L_{OA} [m]		209.200	
Length of waterline	L_{WL} [m]	208.850	208.125	207.375
Breadth maximum	B_{max} [m]		61.090	
Height	H_{side} [m]		10.100	
Draught mean	T_m [m]	7.2	6.2	5.2
Volumetric displacement	∇ [m^3]	77587	66338	55162
Long.centre of gravity	LCG [m]	100.103	100.139	100.120
Vertical centre of gravity	KG[m](z_G)		6; 8; 10; 12; 14; 16	
Waterline area	A_{WL}[m^2]	11287	11211	11132
Number of elements	N_{EL}		280	
Average element length	d_x [m]		0.750	
Gravity acceleration	g [m/s^2]		9.81	
Water density	ρ [t/m^3]		1.000–1.025	

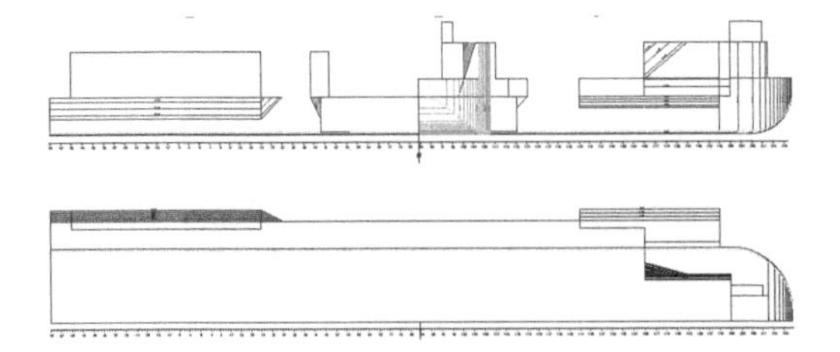

Figure 1. The body plane of the floating dock (Burlacu, 2017).

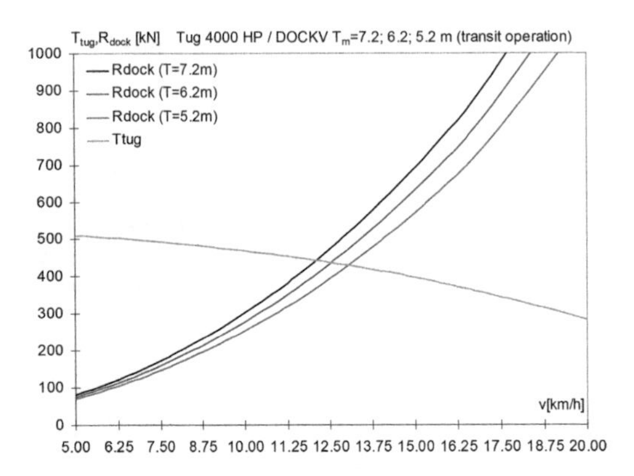

Figure 2. The prediction of resistance for tug-floating dock system, three loading dock conditions (T_m = 5.2, 6.2, 7.2 m).

Table 2. Transverse metacentre height of floating dock DOCKV.

	GM_{T0} [m]			$\varphi_{max\ GZ}$ [deg]		
	Case 1	Case 2	Case 3	Case 1	Case 2	Case 3
KG[m]	Fig.3.a	Fig.3.b	Fig.3.c	Fig.3.a	Fig.3.b	Fig.3.c
6	34.531	39.453	46.579	27.50	30.25	31.50
8	32.531	37.453	44.579	26.75	29.25	27.25
10	30.531	35.453	42.579	25.75	27.25	24.50
12	28.531	33.453	40.579	25.00	23.50	22.50
14	26.531	31.453	38.579	24.00	21.00	21.00
16	24.531	29.453	36.579	18.75	19.75	20.00

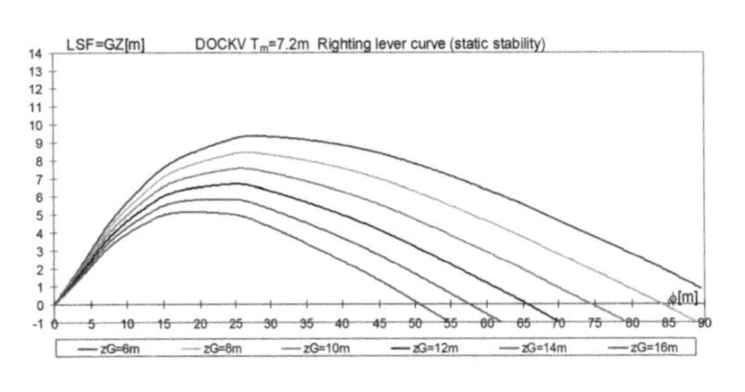

Figure 3a. GZ[m], DOCKV, T_m = 7.2 m, KG = 6–16 m.

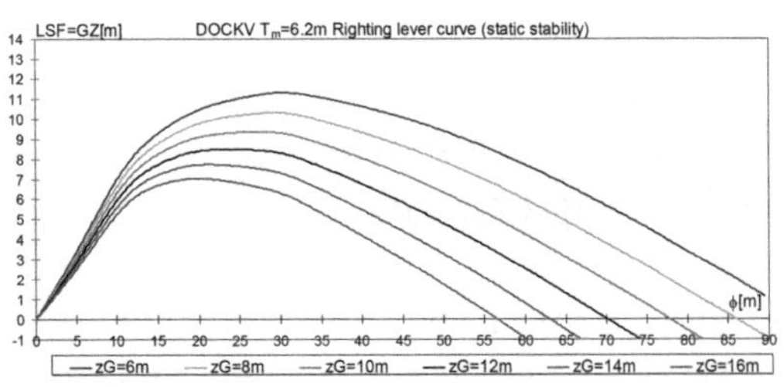

Figure 3b. GZ[m], DOCKV, T_m = 6.2 m, KG = 6–16 m.

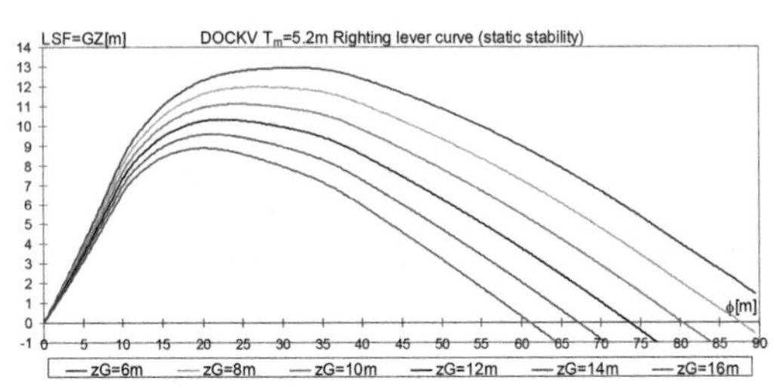

Figure 3c. GZ[m], DOCKV, T_m = 5.2 m, KG = 6–16 m.

The numerical seakeeping prediction of the large floating dock during transit on river-costal scenario is based on in-house program DYN (Domnisoru, 2001), making possible to analyse the influence of the loading case on the navigation safety limits.

2 THEORETICAL MODEL

For the numerical seakeeping analysis we have developed an in-house code DYN based on a linear hydrodynamic strip theory (Bhattacharyya, 1978; Söding, 1982; Voitkunski, 1985; Domnisoru, 2001), and experimental validated by a fishing vessel scaled model, on follow, head, beam and quartering sea conditions (Obreja et al., 2017).

The numerical analyses include two main steps: the computation of the response amplitude operators RAO, as the solution of the motion equations with regular wave excitation, and the computation of the response spectra and short-term statistical values, as the seakeeping solution in irregular waves with a specific wave spectrum.

In time domain the linearized motion equations system, with regular wave excitation, is:

$$\begin{aligned}&([M]+[A])\{\ddot{X}(t)\}+[B]\{\dot{X}(t)\}+[C]\{X(t)\}\\&=\{\bar{F}_w\}e^{-i\omega_e t}\omega_e=\omega-\omega^2/g\cdot v\cdot\cos\mu;\{X(t)\}=\{\bar{X}\}e^{-i\cdot\omega_e t}\end{aligned} \quad (1)$$

where, [A], [B], [C] are the hydrodynamic added mass, damping and restoring matrix; [M] is the ship's mass matrix; $\{F_w\}$ is the amplitude vector of wave generalized excitation forces; $\{X\}$ is the ship's motions amplitude vector; ω, ω_e are the wave and the encountering ship-wave circular frequencies; v is the ship speed; μ is the heading ship-wave angle.

The steady state linear ship's dynamic response in regular waves results from the Equation 1 solution:

$$\begin{aligned}
&[D(\omega_e)]\{\bar{X}\}e^{-i\omega_e t} = \{\bar{F}_w\}e^{-i\omega_e t} \rightarrow \{\bar{X}\} \\
&\quad = [D(\omega_e)]^{-1}\{\bar{F}_w\} \\
&[D(\omega_e)] = -\omega_e^2([M]+[A(\omega_e)]) \\
&\quad - i\omega_e[B(\omega_e)] + [C(\omega_e)] \\
&\{\bar{X}\} = \{X_1\} + i\{X_2\} \rightarrow x_j(t) \\
&\quad = X_{1j}\cos\omega_e t + X_{2j}\sin\omega_e t
\end{aligned} \tag{2}$$

and for a j motion component in the frequency domain the response amplitude operator RAO results:

$$x_j^a = \sqrt{X_{1j}^2 + X_{2j}^2} \rightarrow RAO_j = \frac{x_j^a}{a_w}\Big|_{\omega,\omega_e,\mu} for\ j = 1...6 \tag{3}$$

were a_w = 1 m is the regular wave with unit amplitude.

The short-term most probable statistical response in irregular waves with $S_w(\omega)$ spectrum (Price & Bishop, 1974; Betram, 2000; DNV, 2012), on j motion and acceleration component, using the RAO functions from Equation 3, results:

$$\begin{aligned}
&S_j = RAO_j^2 \cdot S_w^e\Big|_{\omega,\omega_e,\mu};\ S_w^e(\omega_e) = S_w(\omega)\cdot|d\omega_e/d\omega|^{-1} \\
&x_j^{mp} = RMS_j = \sqrt{m_{0j}};\ x_{acj}^{mp} = RMS_{acj} = \sqrt{m_{4j}} \\
&m_{0j} = \int_0^\infty S_j(\omega_e)d\omega_e;\ m_{4j} = \int_0^\infty \omega_e^4 S_j(\omega_e)d\omega_e
\end{aligned} \tag{4}$$

were RMS root mean square value is equal with the statistical most probable amplitude response.

For this study in Equation 4 the irregular waves are modelled by the one parameter ITTC power density spectra, Figure 4, (Price & Bishop, 1974; Betram, 2000; DNV, 2012):

$$S_w(\omega) = \frac{\alpha}{\omega^5}e^{-\frac{\beta}{\omega^4}};\ \alpha = 0{,}7795;\ \beta = \frac{3{,}11}{H_s^2} \tag{5}$$

were, H_s is the wave significant height, considering the design limits 2 m for river and 4.942 m for costal navigation conditions.

For the floating dock from this study (Fig. 1), the navigation safety on transit operation, river-costal, with different loading cases (Table 2), is evaluated in terms of limit sea state, $H_{s\ limit}$[m] or Beaufort level B_{limit} (Domnisoru, 2001; Obreja, 2005). The seakeeping limit criteria (Equation 6) are formulated in terms of maximum statistical most probable values RMS (Equation 4), for the main motions and accelerations components of the large floating dock: heave (ζ), pitch (θ) and roll (φ) (Table 3). The limit values (max) are imposed by the operational conditions from the design project. The RMS_z criteria for avoiding the deck wetness is formulated as statistical short-term extreme condition for a positive superposing of the dock motions and waves at side dock, aft or fore.

$$\begin{aligned}
&RMS_{z\max} = H_{side} - f_s - T_{fore,aft} \geq RMS_z \\
&RMS_z = RMS_\zeta + \frac{L}{2}\cdot RMS_\theta + \frac{B}{2}\cdot RMS_\varphi + \frac{1}{4}\cdot H_s \\
&RMS_{\theta\max} = 2^0 \geq RMS_\theta \\
&RMS_{\varphi\max} = 4^0 \geq RMS_\varphi \\
&RMS_{ac\zeta\max} = 0.1g \geq RMS_{ac\zeta} \\
&RMS_{ac\theta\max} = 0.1g/(L/2) \geq RMS_{ac\theta} \\
&RMS_{ac\varphi\max} = 0.1g/(B/2) \geq RMS_{ac\varphi}
\end{aligned} \tag{6}$$

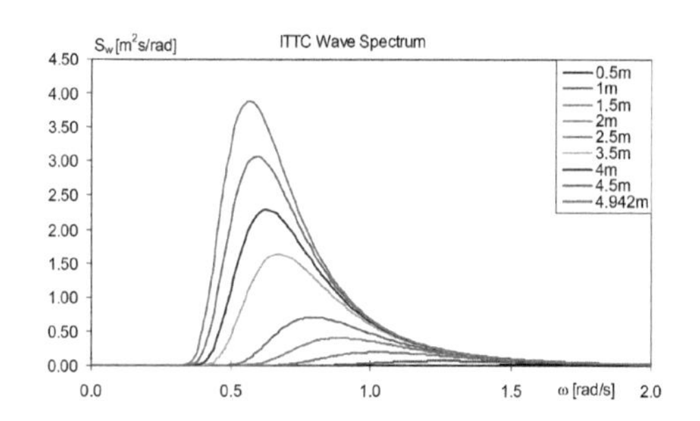

Figure 4. The ITTC wave spectrum, $H_s \leq 4.942$ m.

Table 3. Seakeeping limit criteria for the floating dock, on heave, pitch and roll motion and acceleration amplitudes.

Case	$RMS_{z\,max}$ [m]	$RMS_{\theta\,max}$ [rad]	$RMS_{\varphi\,max}$ [rad]	$RMS_{axz\,max}$ [m/s²]	$RMS_{ac\theta\,max}$ [rad/s²]	$RMS_{ac\varphi\,max}$ [rad/s²]
1	2.6					
2	3.6	0.03491	0.06981	0.981	0.00938	0.03212
3	4.6					

where, $f_s = 0.3$ m is the safety side height for floating docks with non-continuous wing tanks (DNVGL, 2017).

3 THE RESPONSE AMPLITUDE OPERATORS FOR THE LARGE FLOATING DOCK

For the large floating dock (Fig 1, Table 1), based on the theoretical model, Equations 1–3, with DYN code, the response amplitude operators are obtained.

The floating dock is in transit operation on a river-costal route, for three trial speeds v = 0, 6, 12 km/h. Three ballast conditions are considered, with six gravity centre vertical positions (Tables 1, 2). The dock-wave heading angle is in the range μ = 0–180 deg, with step $\delta\mu$ = 5 deg. For the response on heading angle range μ = 180–360 deg the centre line symmetry condition of the floating dock hull is used (Fig. 1). The heave, pitch and roll *RAO* functions are computed, for wave circular frequency range ω = 0–3 rad/s and step $\delta\omega$ = 0.001 rad/s. So, the best results for peak *RAO* values and later spectral moments (Equation 4) computation are obtained.

Figure 5.a presents the floating dock heave RAO_ζ (Equation 3) for v = 12 km/h, μ = 0,45,90,135,180 deg, loading case 1 (T_m = 7.2 m). Figure 5.b presents the heave RAO_ζ for v = 12 km/h, μ = 90 deg, for all three loading cases.

From the heave functions analysis (total 9 cases) the maximum RAO_ζ are in the case of beam sea (Fig. 5.a), being smaller on the other heading angle conditions. Due to almost prismatic hull shape (Fig. 1) the change of the loading case brings small differences on heave RAO_ζ functions (Fig. 5.b).

Figure 6.a presents the floating dock pitch RAO_θ (Equation 3) for v = 12 km/h, μ = 0,45,90,135,180 deg, loading case 1 (T_m = 7.2 m). Figure 6.b presents the pitch RAO_θ for v = 12 km/h, μ = 180 deg, on all three loading cases.

From the pitch functions analysis (total 9 cases) the maximum RAO_θ are in the case of head sea

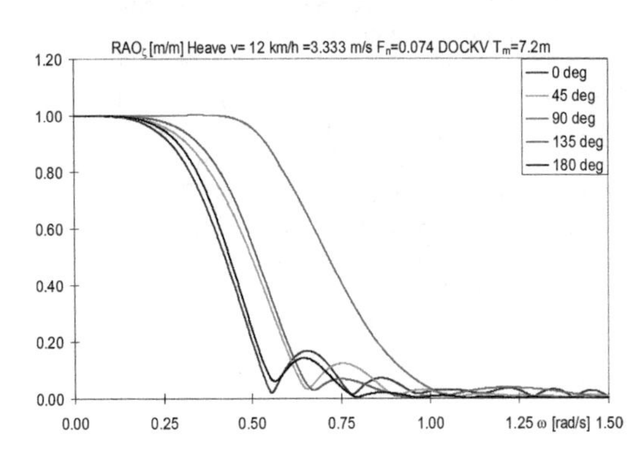

Figure 5a. Heave RAO_ζ [m/m], DV, T_m = 7.2 m, v = 12 km/h.

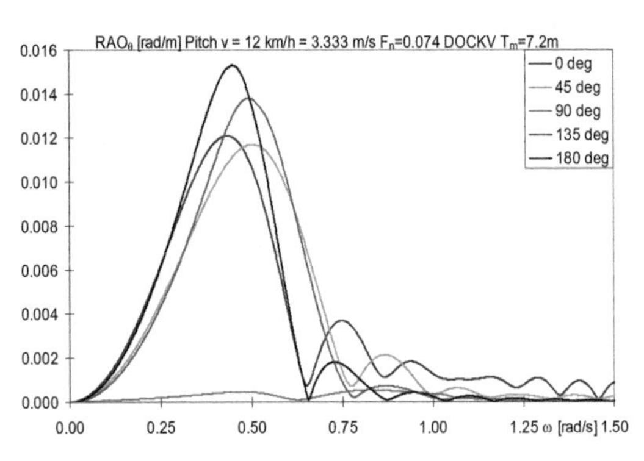

Figure 6a. Pitch RAO_θ [rad/m], DV, T_m = 7.2 m, v = 12 km/h.

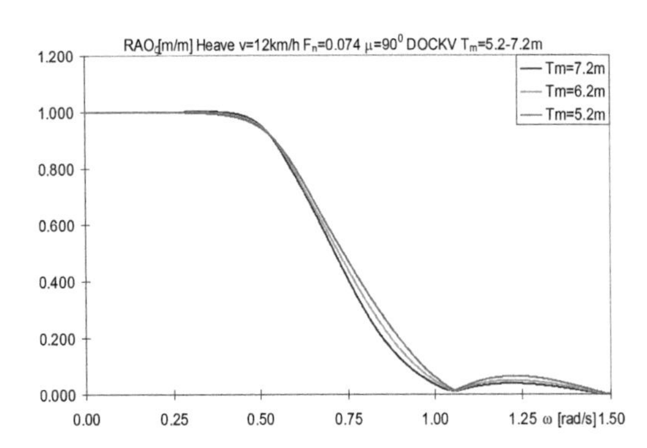

Figure 5b. Heave RAO_ζ [m/m], T_m = 5.2–7.2 m, v = 12 km/h, μ = 90°.

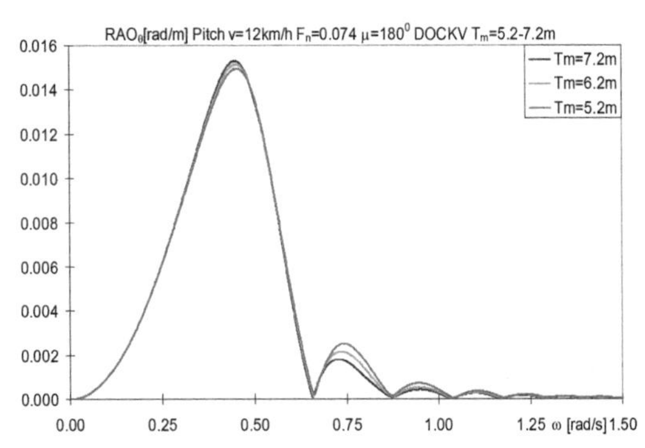

Figure 6b. Pitch RAO_θ [rad/m], T_m = 5.2–7.2 m, v = 12 km/h, μ = 180°.

(Fig. 6.a), having also comparable values on follow and quartering sea conditions and very reduced on beam sea case. Due to the almost prismatic floating dock hull shape (Fig. 1) small differences on pitch RAO_θ (Fig. 6.b) are obtained between the three loading cases.

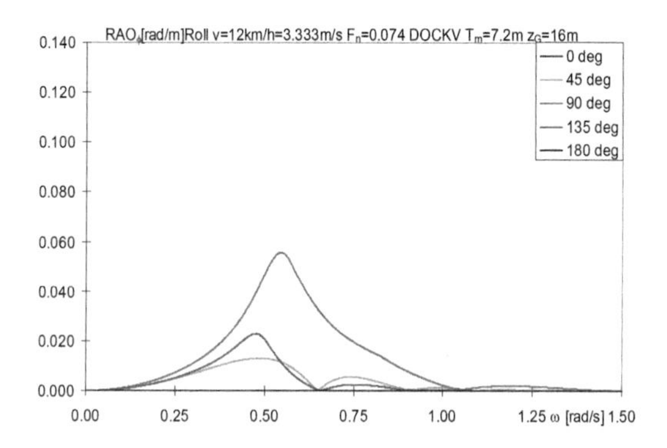

Figure 7a. Roll RAO_ϕ [rad/m], DV, T_m = 7.2 m, v = 12 km/h, KG = 16 m.

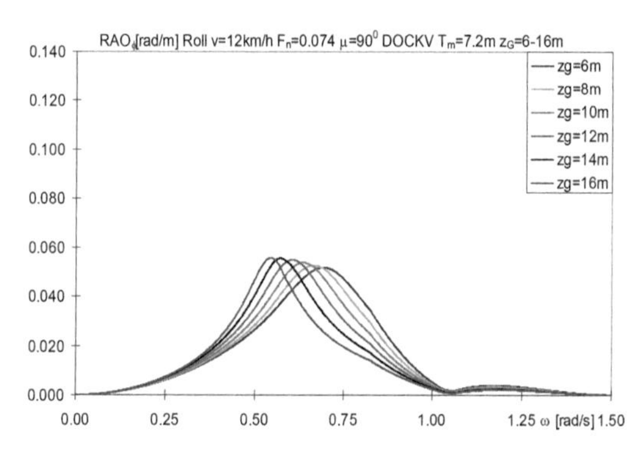

Figure 7b. RAO_ϕ [rad/m], T_m = 7.2 m, v = 12 km/h, KG = 6–16 m, μ = 90°.

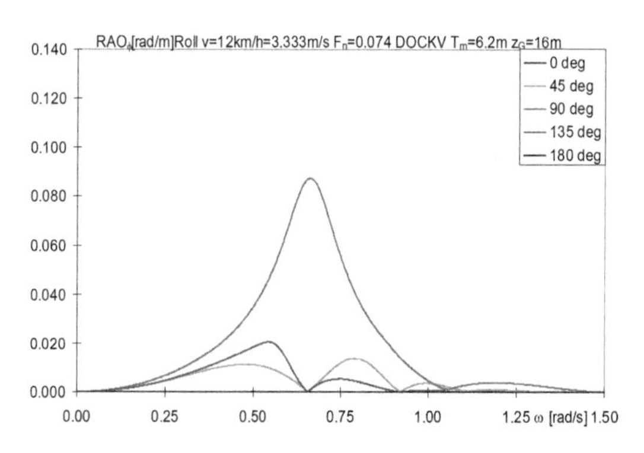

Figure 8a. Roll RAO_ϕ [rad/m], DV, T_m = 6.2 m, v = 12 km/h, KG = 16 m.

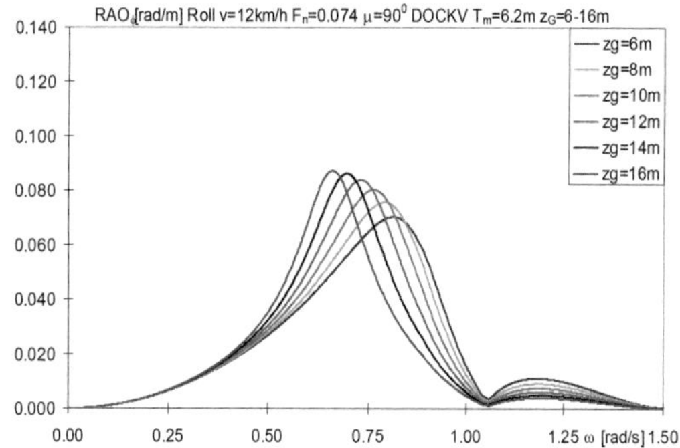

Figure 8b. RAO_ϕ [rad/m], T_m = 6.2 m, v = 12 km/h, KG = 6–16 m, μ = 90°.

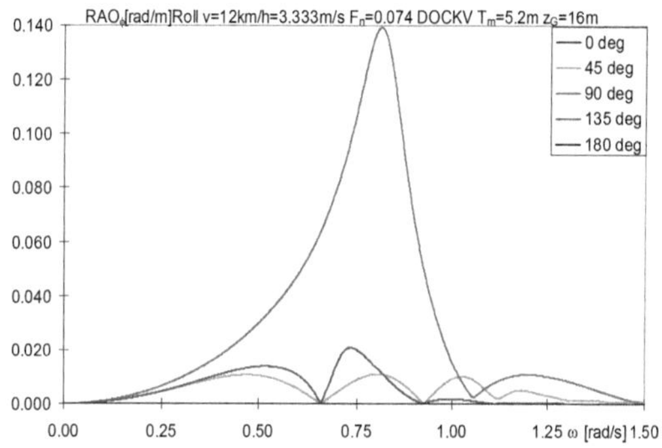

Figure 9a. Roll RAO_ϕ [rad/m], DV, T_m = 5.2 m, v = 12 km/h, KG = 16 m.

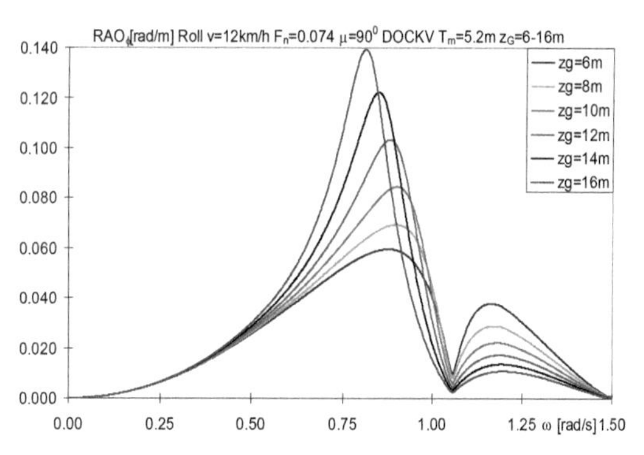

Figure 9b. RAO_ϕ [rad/m], T_m = 5.2 m, v = 12 km/h, KG = 6–16 m, μ = 90°.

Figures 7–9.a present the floating dock roll RAO_φ (Equation 3) for v = 12 km/h, μ = 0,45,90,135,180 deg, loading cases 1–3 (T_m = 7.2, 6.2, 5.2 m) and KG = 16 m. Figures 7–9.b presents the roll RAO_φ for v = 12 km/h, μ = 90 deg, on all three loading cases and KG = 6–16 m.

From the roll functions analysis (total 54 cases) the maximum RAO_φ are in the case of beam sea (Figs. 7–9.a), smaller in quartering condition and zero on head and follow sea cases. Due to the variation of the floating dock transverse metacentre height values (Table 2), corresponding to gravity centre vertical position range KG = 6–16 m, significant changes on roll RAO_φ (Figs. 7–9.b) are obtained between the three loading and six KG cases. The maximum roll dynamic response is predicted on loading case 3 (T_m = 5.2 m) and the minimum on loading case 1 (T_m = 7.2 m).

4 SHORT-TERM SEAKEEPING PREDICTION OF FLOATING DOCK TRANSIT OPERATION

For the evaluation of the large floating dock (Fig. 1) behaviour in irregular waves during the transit river-costal scenario, based on the numerical RAO functions from previous section and the input wave power density spectrum ITTC (Equation 5, Fig. 4) the heave, pitch and roll response spectra and the most probable response amplitudes (Equation 4) are obtained. Imposing the seakeeping criteria (Table 3, Equation 6), the floating dock seakeeping limits in terms of H_{slimit}[m] and Beaufort B_{limit} are obtained for all the loading and speed cases (Table 1).

Tables 4, 6, 8 present the maximum, on overall heading angles, RMS most probable roll motion and acceleration amplitudes, for the three loading cases. For each loading case and KG value results that the speed, in range 0–12 km/h, has a reduced hydrodynamic influence on the roll response. Considering the reference to the roll limits (Table 3), the roll response is maximum for loading case 3 (−29.26% to +47.83%), medium for loading case 2 (−22.77% to −2.32%) and minimum for loading case 1 (−58.04% to −31.53%).

Table 4. The maximum RMS for roll, DOCKV, T_m = 7.2 m.

v [km/h]	KG [m]	ϕ_{RMS} [rad]	%	ϕ_{acRMS} [rad/s^2]	%
adm	–	0.06981	–	0.03212	–
0 (F_n = 0)	6	0.039018	−44.11	0.013475	−58.04
	8	0.042231	−39.51	0.015616	−51.38
	10	0.044683	−36.00	0.017628	−45.11
	12	0.046299	−33.68	0.019383	−39.65
	14	0.047132	−32.49	0.020838	−35.12
	16	0.047321	−32.22	0.021977	−31.57
6 (Fn = 0.037)	6	0.039213	−43.83	0.013517	−57.91
	8	0.042431	−39.22	0.015670	−51.21
	10	0.044883	−35.71	0.017692	−44.91
	12	0.046489	−33.41	0.019453	−39.43
	14	0.047306	−32.24	0.020871	−35.01
	16	0.047475	−32.00	0.021916	−31.76
12(F_n = 0.074)	6	0.039412	−43.55	0.013561	−57.78
	8	0.042636	−38.93	0.015726	−51.04
	10	0.045086	−35.42	0.017758	−44.71
	12	0.046682	−33.13	0.019525	−39.21
	14	0.047482	−31.99	0.020945	−34.78
	16	0.047631	−31.77	0.021989	−31.53

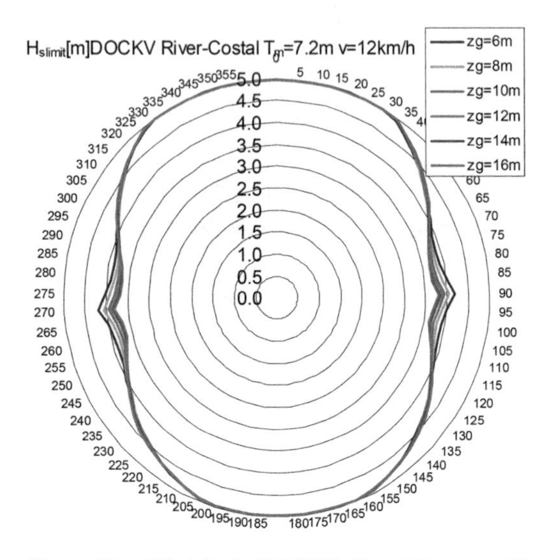

Figure 10. H_s[m] limit, DOCKV, T_m = 7.2 m, v = 12 km/h, KG = 6–16 m.

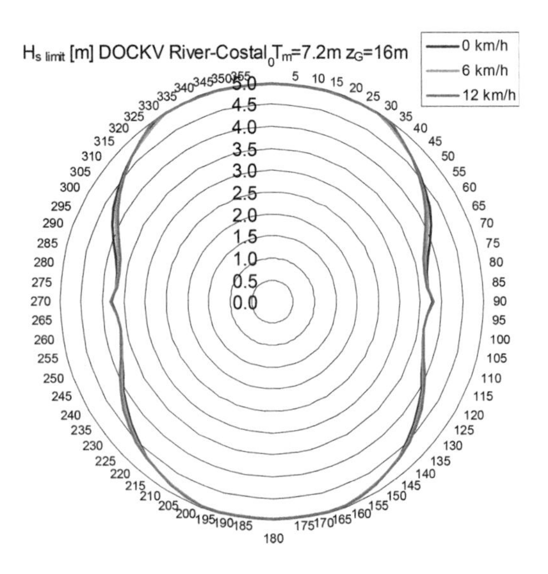

Figure 11a. H_s[m] limit, DOCKV, T_m = 7.2 m, v = 0,6,12 km/h, KG = 16 m.

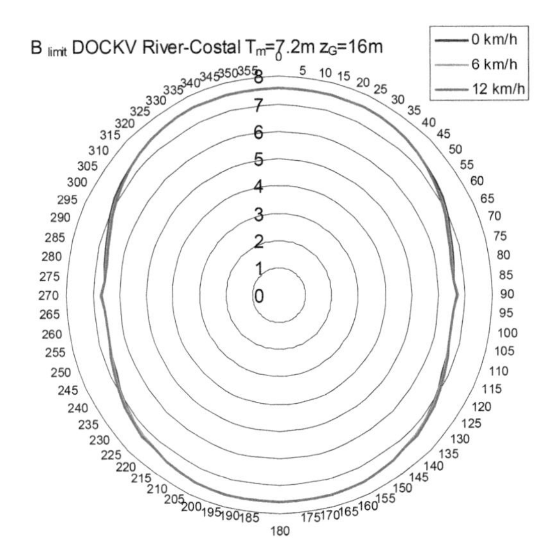

Figure 11b. B limit, DOCKV, T_m = 7.2 m, v = 0,6,12 km/h, KG = 16 m.

Table 5. H_s [m] and Beaufort limits, DOCKV, T_m = 7.2 m.

v [km/h]	KG [m]	$H_{s\ limit}$ [m]	B_{limit}	Seakeeping criteria
0 (F_n = 0)	6	3.872÷4.942	6.76÷7.55	heave/beam & quarter
	8	3.810÷4.942	6.71÷7.55	heave/beam & quarter
	10	3.750÷4.942	6.65÷7.55	heave/beam & quarter
	12	3.697÷4.942	6.61÷7.55	heave/beam & quarter
	14	3.650÷4.942	6.57÷7.55	heave/beam & quarter
	16	3.622÷4.942	6.54÷7.55	heave/beam & quarter
6 (Fn = 0.037)	6	3.869÷4.942	6.76÷7.55	heave/beam & quarter
	8	3.809÷4.942	6.71÷7.55	heave/beam & quarter
	10	3.743÷4.942	6.65÷7.55	heave/beam & quarter
	12	3.683÷4.942	6.59÷7.55	heave/beam & quarter
	14	3.642÷4.942	6.56÷7.55	heave/beam & quarter
	16	3.621÷4.942	6.54÷7.55	heave/beam & quarter
12 (F_n = 0.074)	6	3.865÷4.942	6.76÷7.55	heave/beam & quarter
	8	3.791÷4.942	6.69÷7.55	heave/beam & quarter
	10	3.723÷4.942	6.63÷7.55	heave/beam & quarter
	12	3.669÷4.942	6.58÷7.55	heave/beam & quarter
	14	3.636÷4.942	6.55÷7.55	heave/beam & quarter
	16	3.620÷4.942	6.54÷7.55	heave/beam & quarter
limits	–	3.620	6.54	heave/beam & quarter

Figures 10, 11.a,b and Table 5 present the seakeeping limits for the first ballasting case and the three trial speed values. In this case the pitch and roll motions and accelerations most probable values are satisfying the limit criteria. Although the heave accelerations criteria is satisfied, due to the small free board ($RMS_{z\ max}$ = 2.6 m), at ship's bow and stern, the heave motion criteria becomes a restriction, in the case of beam and quartering sea states, μ = 30–150 deg. On the seakeeping limits, the influence of the gravity centre vertical position KG is medium on beam sea, small on quartering sea and with no influence on head and follow seas, μ = 155–180 and 0–25 deg, when roll motion is very reduced or zero (Fig. 10).

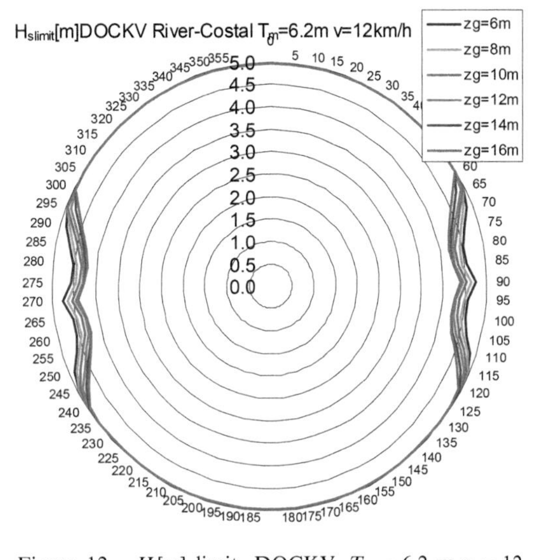

Figure 12. H_s[m] limit, DOCKV, T_m = 6.2 m,v = 12 km/h, KG = 6–16 m.

Table 6. The maximum RMS for roll, DOCKV, T_m = 6.2 m.

v [km/h]	KG [m]	ϕ_{RMS} [rad]	%	ϕ_{acRMS} [rad/s²]	%
adm	–	0.06981	–	0.03212	–
0 (F_n = 0)	6	0.053920	−22.77	0.028110	−12.48
	8	0.056936	−18.45	0.029900	−6.90
	10	0.059886	−14.22	0.030023	−6.52
	12	0.062410	−10.60	0.030829	−4.01
	14	0.064140	−8.13	0.031030	−3.38
	16	0.064711	−7.31	0.031267	−2.65
6 (Fn = 0.037)	6	0.054013	−22.63	0.028235	−12.09
	8	0.057062	−18.26	0.029926	−6.82
	10	0.060051	−13.98	0.030147	−6.13
	12	0.062617	−10.31	0.030916	−3.74
	14	0.064389	−7.77	0.031047	−3.33
	16	0.064996	−6.90	0.031290	−2.57
12 (F_n = 0.074)	6	0.054108	−22.50	0.028362	−11.69
	8	0.057190	−18.08	0.029995	−6.60
	10	0.060218	−13.74	0.030304	−5.64
	12	0.062826	−10.01	0.030915	−3.74
	14	0.064640	−7.41	0.031147	−3.02
	16	0.065286	−6.48	0.031370	−2.32

Figures 12, 13.a,b and Table 7 present the seakeeping limits for the second ballasting case and the three trial speed values. Similar to the first ballasting case, the pitch and roll motions and all the accelerations criteria are satisfied on all heading angles. The free board has intermediate size ($RMS_{z\,max}$ = 3.6 m) and the heave motion criteria becomes the only restriction, in the range μ = 60–120 deg, beam and quartering sea state cases. Compared to the first ballasting case, in second case the significant wave height limit is larger H_{slimit}[m] = 4.204 > 3.620 m (Tables 7,5), because free board is 1 m higher (Table 3), although the heave and roll motions are larger (Tables 6, 4). For the seakeeping limits, the influence of the gravity centre vertical position *KG* is medium on beam and quartering sea, having no influence on head, follow and partial quartering seas, μ = 125–180 and 0–55 deg. (Fig. 12).

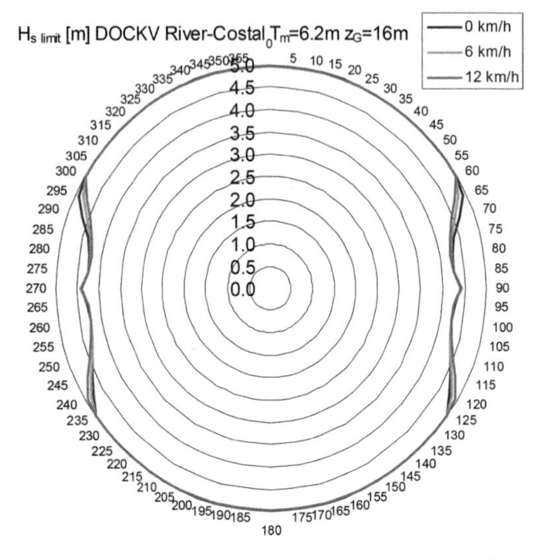

Figure 13a. H_s[m] limit, DOCKV, T_m = 6.2 m, v = 0,6,12 km/h, *KG* = 16 m.

Table 7. H_s [m] and Beaufort limits, DOCKV, T_m = 6.2 m.

v [km/h]	*KG* [m]	$H_{s\,limit}$ [m]	B_{limit}	Seakeeping criteria
0 (F_n = 0)	6	4.529÷4.942	7.27÷7.55	heave/beam-quarter
	8	4.435÷4.942	7.20÷7.55	heave/beam-quarter
	10	4.344÷4.942	7.14÷7.55	heave/beam-quarter
	12	4.267÷4.942	7.09÷7.55	heave/beam-quarter
	14	4.232÷4.942	7.06÷7.55	heave/beam-quarter
	16	4.219÷4.942	7.05÷7.55	heave/beam-quarter
6 (*Fn* = 0.037)	6	4.486÷4.942	7.24÷7.55	heave/beam-quarter
	8	4.398÷4.942	7.18÷7.55	heave/beam-quarter
	10	4.316÷4.942	7.12÷7.55	heave/beam-quarter
	12	4.253÷4.942	7.08÷7.55	heave/beam-quarter
	14	4.222÷4.942	7.06÷7.55	heave/beam-quarter
	16	4.215÷4.942	7.05÷7.55	heave/beam-quarter
12 (F_n = 0.074)	6	4.434÷4.942	7.20÷7.55	heave/beam-quarter
	8	4.354÷4.942	7.15÷7.55	heave/beam-quarter
	10	4.284÷4.942	7.10÷7.55	heave/beam-quarter
	12	4.235÷4.942	7.06÷7.55	heave/beam-quarter
	14	4.218÷4.942	7.05÷7.55	heave/beam-quarter
	16	4.204÷4.942	7.04÷7.55	heave/beam-quarter
limits	–	4.204	7.04	heave/beam-quarter

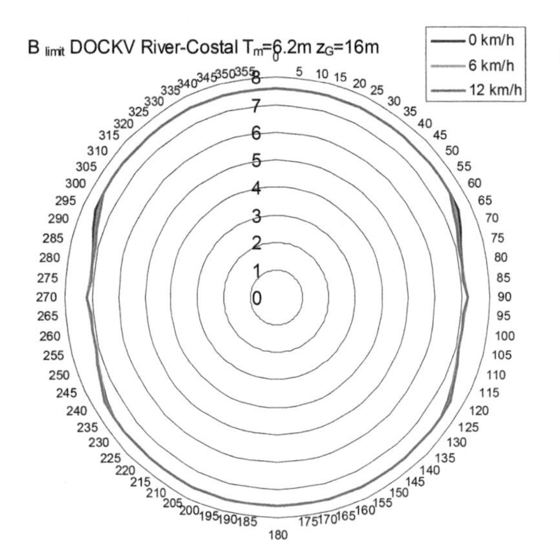

Figure 13b. *B* limit, DOCKV, T_m = 6.2 m, v = 0,6,12 km/h, *KG* = 16 m.

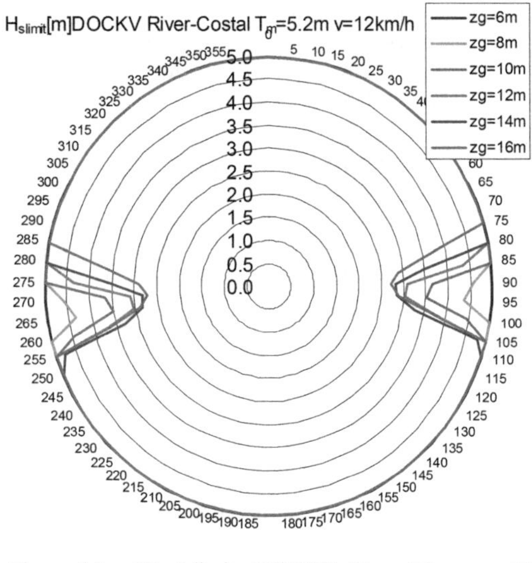

Figure 14. H_s[m] limit, DOCKV, T_m = 5.2 m, v = 12 km/h, *KG* = 6–16 m.

Table 8. The maximum *RMS* for roll, DOCKV, T_m = 5.2 m.

v [km/h]	KG [m]	ϕ_{RMS} [rad]	%	ϕ_{acRMS} [rad/s²]	%
adm	–	0.06981	–	0.03212	–
0 ($F_n = 0$)	6	0.049386	−29.26	0.031108	−3.14
	8	0.053044	−24.02	0.033109	3.09
	10	0.058344	−16.43	0.036480	13.59
	12	0.065212	−6.59	0.040557	26.28
	14	0.072999	4.56	0.044416	38.30
	16	0.081248	16.38	0.047316	47.33
6 (Fn = 0.037)	6	0.049402	−29.24	0.031129	−3.08
	8	0.053074	−23.98	0.033141	3.19
	10	0.058401	−16.35	0.036536	13.76
	12	0.065312	−6.45	0.040648	26.56
	14	0.073155	4.79	0.044544	38.69
	16	0.081472	16.70	0.047398	47.58
12 (F_n = 0.074)	6	0.049419	−29.21	0.031149	−3.01
	8	0.053110	−23.93	0.033798	5.23
	10	0.058597	−16.07	0.037525	16.84
	12	0.065367	−6.37	0.041504	29.23
	14	0.073312	5.01	0.044982	40.06
	16	0.081698	17.02	0.047479	47.83

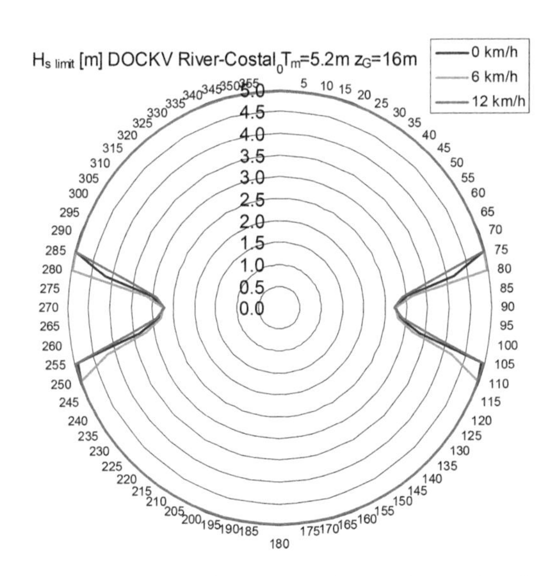

Figure 15a. H_s[m] limit, DOCKV, T_m = 5.2 m, v = 0,6,12 km/h, KG = 16 m.

Figures 14, 15.a,b and Table 9 present the seakeeping limits for the third ballasting case and the three trial speed values. In this case de free board size is the largest ($RMS_{z\ max}$ = 4.6 m), so that the heave motion and acceleration criteria are satisfied in all conditions. Also, the pitch motion and acceleration criteria are satisfied. Because the roll motion and acceleration have the maximum values (Tables 8, 6, 4), the only restrictions result from both roll criteria, in the range μ = 75–105 deg, beam sea state. For the seakeeping limits, the influence of the gravity centre vertical position KG is high on beam sea, having no influence on head, follow and quartering sea, μ = 110–180 and 0–70 deg. (Fig. 14), with H_{slimit} = 2.713 m.

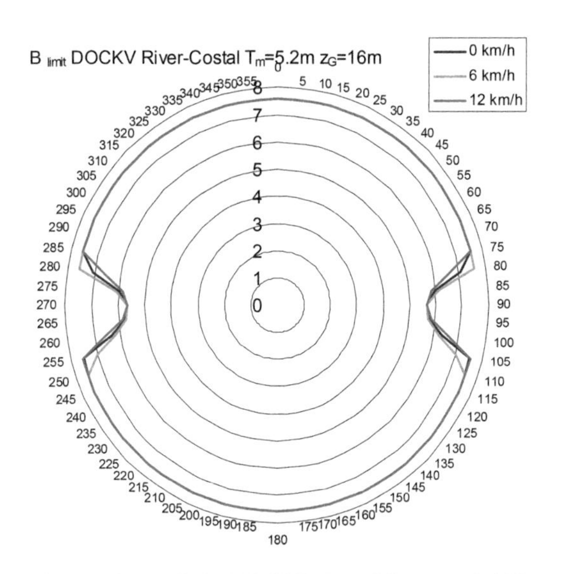

Figure 15b. B limit, DOCKV, T_m = 5.2 m, v = 0,6,12 km/h, KG = 16 m.

Table 9. H_s [m] and Beaufort limits, DOCKV, T_m = 5.2 m.

v [km/h]	KG [m]	$H_{s\ limit}$ m]	B_{limit}	Seakeeping criteria
0 ($F_n = 0$)	6	4.942	7.55	no restrictions
	8	4.528÷4.942	7.27÷7.55	roll acc./beam sea
	10	3.632÷4.942	6.55÷7.55	roll acc./beam sea
	12	3.069÷4.942	6.05÷7.55	roll acc./beam sea
	14	2.808÷4.942	5.75÷7.55	roll criteria/beam sea
	16	2.733÷4.942	5.65÷7.55	roll criteria/beam sea
6 (Fn = 0.037)	6	4.942	7.55	no restrictions
	8	4.516÷4.942	7.26÷7.55	roll acc./beam sea
	10	3.620÷4.942	6.54÷7.55	roll acc./beam sea
	12	3.057÷4.942	6.04÷7.55	roll acc./beam sea
	14	2.798÷4.942	5.74÷7.55	roll criteria/beam sea
	16	2.723÷4.942	5.64÷7.55	roll criteria/beam sea
12 (F_n = 0.074)	6	4.942	7.55	no restrictions
	8	4.320÷4.942	7.12÷7.55	roll acc./beam sea
	10	3.491÷4.942	6.42÷7.55	roll acc./beam sea
	12	3.028÷4.942	6.01÷7.55	roll acc./beam sea
	14	2.788÷4.942	5.72÷7.55	roll criteria/beam sea
	16	2.713÷4.942	5.63÷7.55	roll criteria/beam sea
limits	–	2.713	5.63	roll criteria/beam sea

Table 10. Seakeeping limits synthesize on floating dock transit.

Case	T_m [m]	$H_{s\,limit}$ [m]	B_{limit}	μ [deg]	Seakeeping criteria
1	7.2	3.620	6.54	30–150	heave/beam & quarter
2	6.2	4.204	7.04	60–120	heave/beam & quarter
3	5.2	2.713	5.63	75–105	roll mot & acc/beam

5 CONCLUSIONS

For the analysis of the transit operation of a large floating dock (Fig. 1, Table 1) a numerical model has been developed with 280 stations and the DYN in-house code, based on a linear hydrodynamic strip theory (section 2), is used for *RAO* functions computation (section 3).

For a river-costal transit scenario, on irregular waves with ITTC power density spectrum, imposing the seakeeping criteria (Equation 6, Table 3), in terms of admissible statistical most probable heave, pitch, roll motions and accelerations amplitudes, the short-term seakeeping limits prediction $H_{s\,limit}$ and B_{limit} for the large floating dock are obtained (section 4), with synthesize results in Table 10.

The short-term seakeeping transit analysis results are pointing out that the speed, in range 0–12 km/h, has a reduced influence (Tables 5, 7, 9 and Figs. 11, 13, 15.a,b). The influence of the gravity centre KG = 6–16 m is significant at beam sea, decreases on quartering sea and none on head and follow seas.

For cases 1 and 2 the heave motion criteria is the restriction, due to free board limits. Instead, for case 3 the roll motions and accelerations are maximum (Tables 8, 6, 4) and the roll restrictions are acting. The seakeeping limits occur always on beam sea and also quarter sea when free board is decreasing (Figs. 11, 13, 15.a,b). On river no transit restrictions occur (H_{slimit} > 2 m). On coastal the beam sea case must be avoided. If the roll criteria can be relaxed ($RMS_{\varphi} \geq 5°$, $RMS_{ac\varphi} \geq 0.15$ g/(B/2)) then case 3 has no more restrictions on costal route.

ACKNOWLEDGEMENTS

The authors wish to express their thanks to VARD Shipyard Tulcea, who granted us the large floating dock technical data. The research was supported by the Research Centre of the Naval Architecture Faculty at "Dunarea de Jos" University of Galati.

REFERENCES

ANR. 2006. *Album of ship types. Maritime Tug 4000 HP.* Constantza: Romanian Naval Authority.

Bertram, V. 2000. *Practical ship hydrodynamics*. Oxford: Butterworth Heinemann.

Bhattacharyya, R. 1978. *Dynamics of marine vehicles*. New York: John Wiley & Sons Publication.

Bidoaie, I. & Ionas, O. 1998. *Naval architecture complements*. Galati: Porto-Franco.

Burlacu, E. 2017. *Steady art of floating docks design. Technical data for large floating dock from VARD Shipyard Tulcea*. Galati: University "Dunarea de Jos" of Galati.

DNV. 2012. *Modelling and analysis of marine operations*, Recommended practice, DNV-RP-H103. Hovik: Det Norske Veritas, https://rules.dnvgl.com.

DNVGL. 2017. *Rules for classification. Floating docks.* Hovik: Det Norske Veritas, https://rules.dnvgl.com.

Domnisoru, L. 2001. *Ship dynamics. Oscillations and vibrations*. Bucharest: Technical Publishing House.

Obreja, D. 2005. *Ship theory. Concepts and methods for the navigation performances analysis*. Bucharest: Didactic and Pedagogic Publishing House.

Obreja, D., Nabergoj, R., Crudu, L. & Domnisoru, L. 2017. "Seakeeping performance of a Mediterranean fishing vessel", *Maritime Transportation and Harvesting of Sea Resources*, C. Guedes Soares & A.P. Teixeira (Eds.), Taylor & Francis Group, London, UK, pp. 483–491.

Price, W.G. & Bishop, R.E.D. 1974. *Probabilistic theory of ship dynamics*. London: Chapman and Hall.

Solas, 2014. *International convention for the safety of life at sea. Safety of navigation*. IMO, http://www.imo.org.

Söding, H. 1982. *Bewegungen und Belastungen der Schiffe im Seegang*. Hamburg: Institut für Schiffbau Hamburg.

Voitkunski, Y.I. 1985. *Ship theory handbook*. Sankt Petersburg: Sudostroenie.

Progress in Maritime Technology and Engineering – Guedes Soares & Santos (Eds)
 ISBN 978-1-138-58539-3

Comparison of dynamic and quasi-static towline model for evaluation of wave-induced towed ship motions

I. Ćatipović
Faculty of Mechanical Engineering and Naval Architecture, University of Zagreb, Zagreb, Croatia

ABSTRACT: Evaluation of wave-induced motions of a towed ship is necessary step in defining hull loads. This is especially important for a damaged ship when structural assessment with safety consideration needs to be performed for towing condition. Preferred way to evaluate the towed ship motions is solving coupled model that incorporates hydrodynamics of the towed ship as well as towline forces. In this study, two approaches for estimation of the towline response are compared. First approach takes into account geometric, axial and flexural stiffness of the towing line along with inertial and damping forces. Second approach only considers restoring forces due to axial and geometric stiffness. Therefore, dynamics and quasi-static model for evaluation of the towline response are compared. Obtained numerical results from both models are also compared with experimental data from available literature where model of the towed ship is observed on regular head waves.

1 INTRODUCTION

After an accident, the damaged ship should be removed from crash site. This is usually achieved by towing since the damaged ship loses ability of self propulsion. In order to ensure safety, it is necessary to evaluate loads on the hull of the damaged ship during towing operation. Time-domain simulation model for wave-induced motions of towed ship and towline tension is a solution of this problem. In such model, hydrodynamics of the towed ship is coupled with towline response.

Bhattacharyya et al. (2000) studied the dynamics of underwater towing of flexible cylindrical structures as a fluid-structure interaction problem commonly referred as "cylinders in axial flow". So in his study, inertial and damping forces on towline due to surrounding fluid are incorporated in governing equation. Further, he developed finite element method for the dynamics of the flexible towed cylinder. Special attention was on hydroelastic instabilities exhibited by the structure at certain critical tow speeds. Raman-Nair et al. (2009) developed motion equations for the coupled dynamics of a long flexible life raft and fast rescue craft in an irregular ocean waves. The flexible raft is modelled as spring-connected lumped masses, and it is assumed that the motion normal to the wave surface is small and can be neglected. Therefore, in his model, bodies move along the propagating wave profile. The wave forces are applied using Morison's equation for bodies in accelerated flow. Developed numerical model provides guidelines for predicting the tow loads and motions in severe sea states. Nakayama et al. (2012) presented a time domain simulation method for vertical motions (surge, heave and pitch motions) of a towed ship in head seas. Two-dimensional lumped mass method with gravity effect is used for expressing the dynamics of towline. So, towline is assumed to be a series of lumped masses and weightless rigid trusses that are connected by hinge connection. The torsion and elongation of the trusses are ignored. Regular head waves are employed as incident waves. Ćatipović et al. (2014a) developed a time domain numerical model for the wave-induced motions of a towed ship and towline tensions in regular head waves. All considerations are done in vertical plane so the ship is modelled as a rigid body with three degrees of freedom. Besides hydrodynamic loads due to waves, added mass and damping are also considered. Dynamics of the towline is described by finite elements method that incorporates inertial, damping and restoring forces acting in the towline. Desroches (1997) presented an alternative approach to the problem. He developed a new database of extreme towline tension for open ocean towing. In his research, the extreme towline tension consists of the mean towline tension and the peak dynamic tension with 0.001 chance of exceedance in a single day. The statistics of the nonlinear dynamic tensions are determined by a numerical simulation method based on the seakeeping motion of the tug and tow. The calculation of the statistics for the dynamic tension is computationally intensive. Therefore he developed the database of extreme tensions (for specified tug, tow, towing speed, sea state, wave angle, and type of towline) to quickly determine peak tensions.

The overview of published papers reveals that numerical models for towed ship are computationally demanding. This is even more emphasized by the fact that numerical evaluation should be done for certain number of sea states in combination with ship heading angles. On the other side, in the case of damaged ship, hull structural assessment (with safety considerations) during towing should be done in several days. Therefore, Ćatipović et al. (2014b) developed a simplified model to ensure fast evaluation of towline influence on towed ship motions. For approximation of towline forces quasistatic approach is used. Towed ship is considered in usual manner with hydrodynamic loading and reactions along with influence of towline.

To gain better insight how quasistatic approach is useful, its numerical results are compared with the results of dynamic model presented in Ćatipović et al. (2014a) where inertial and damping forces acting on towline are considered. Further, obtained numerical results from both approaches are also compared with experimental data from Nakyama et al. (2012) where towed ship model is observed in regular head waves.

2 MATHEMATICAL MODEL

2.1 *Towline dynamics model*

Elastic rod theory developed by Nordgen (1974) and Garret (1982) is a cornerstone for this mathematical model. Dynamic response of towline is defined in terms of centerline position. Large displacements are considered but with small deformation of towline material (which is treated as isotropic). Within restoring forces of towline, geometric stiffness is taken into account as well as axial and flexural stiffness. Inertial forces due to own mass are modelled. Morison's equation is used to approximate towline damping due to movement through air while towline water immersion is ignored.

The centerline of deformed towline is described by a space curve. In governing equations the space curve is defined by a position vector **r**. Further, a point at the curve is defined by an arc length of stress free towline s. Within dynamics analysis the position vector **r** is also function of the time t. In such terms, motion equation takes the following form (Ćatipović 2014a)

$$-\frac{\mathrm{d}^2}{\mathrm{d}s^2}\left(EI\frac{\mathrm{d}^2\mathbf{r}}{\mathrm{d}^2s^2}\right)+\frac{\mathrm{d}}{\mathrm{d}s}\left(\frac{T_E}{1+\varepsilon}\frac{\mathrm{d}\mathbf{r}}{\mathrm{d}s}\right)+m\mathbf{g}-\mathbf{q}_D=m\ddot{\mathbf{r}} \tag{1}$$

with

$$\frac{1}{(1+\varepsilon)^2}\frac{\mathrm{d}\mathbf{r}}{\mathrm{d}s}\cdot\frac{\mathrm{d}\mathbf{r}}{\mathrm{d}s}=1 \tag{2}$$

and with

$$\varepsilon=\frac{T_E}{AE} \tag{3}$$

where superposed dot denotes differentiation with respect to time, EI is the flexural stiffness of the towline, AE is the axial stiffness, T_E is the cross-section effective tension force, ε is the towline elongation, m is the towline distributed mass, **g** is the gravity acceleration vector. Distributed damping load $\mathbf{q}_D$ on the towline is derived using Morison's equation as follows.

$$\mathbf{q}_D=\frac{1}{2}C_D\rho_a D\left|\dot{\mathbf{r}}^n\right|\dot{\mathbf{r}}^n \tag{4}$$

with

$$\dot{\mathbf{r}}^n=\dot{\mathbf{r}}-\left(\dot{\mathbf{r}}\cdot\frac{\mathrm{d}\mathbf{r}}{\mathrm{d}s}\right)\cdot\frac{\mathrm{d}\mathbf{r}}{\mathrm{d}s} \tag{5}$$

where $|\;|$ denotes a vector length, D is the towline diameter, C_D is the drag coefficient and ρ_a is the air density. In Equation 5, $\dot{\mathbf{r}}^n$ is component of the towline velocity normal to the towline deformed position. Equation 2 is obtained from the condition that relates the deformed position and the elongation of the towline and should be solved in coupled manner with Equations 1 and 3 (Nordgen 1974 and Garrett 1982).

Based on Equations 1 to 5 nonlinear finite element for towline dynamics is formulated, by using Galerkin method (Ćatipović 2014a). Here unknown variables are position vector **r** and effective tension force T_E. As usual, Hermite polynomials are used for shape functions. Time integration is based on trapezoidal integration. Obtained equations for the time integration are highly nonlinear so for their solution Newton-Raphson iterative method is implemented.

2.2 *Towline quasi-static model*

This simplified model takes into account just geometric, axial and flexural stiffness of the towline. Also, small displacements are considered. Based on these propositions, governing differential equation is defined in form (Ćatipović 2014b)

$$EI\frac{\mathrm{d}^4y}{\mathrm{d}x^4}-T_E\frac{\mathrm{d}^2y}{\mathrm{d}x^2}=q \tag{6}$$

where y (x) is the towline deflection in vertical plane and q is the towline distributed weight due to own mass. In this equation, buoyancy effect on the towline is neglected. As in dynamic model, EI is the towline flexural stiffness and T_E is effective tension force. Similar equation is used for static response calculation of marine risers and it is usually called tensioned beam equation (API 1993). In this approach, q is assumed to be constant and it is distributed evenly over horizontal axis x. As a consequence of this simplification effective tension force is also constant. Same simplification and consequence is found in case when catenary is approximated by parabola.

Discretization of Equation 6 is conducted by Galerkin with shape function in the form (Kirk et al. 1979)

$$\phi(x)=\sin\left(\frac{\pi}{d}x\right) \tag{7}$$

where d denotes the towline span i.e. horizontal distance between suspension points. So the final form of discretized governing equation has the form

$$\frac{\pi^2}{2d}\left(\frac{\pi^2}{d^2}EI+T_E\right)a=\frac{2qd}{\pi} \tag{8}$$

where a is the unknown displacement coefficient or value of deflection in the middle of the towline.

During response evaluation towline stretched length $\tilde{s}$ should be defined. When towline deflection y (x) is known, stretched length is calculated in usual way (Kreyszig 1993)

$$\tilde{s}=\int_0^d\sqrt{1+\left(\frac{\mathrm{d}y}{\mathrm{d}x}\right)^2}\,\mathrm{d}x \tag{9}$$

Above expression is simplified by Taylor series expansion

$$\sqrt{1+\epsilon^2}\approx1+\frac{1}{2}\epsilon^2 \tag{10}$$

Finally, expression for stretched length is

$$\tilde{s}=\frac{\pi^2a^2}{4d}+d \tag{11}$$

On the other side, stretched length $\tilde{s}$ can be related to stress free length of towline by usual engineering expression

$$\tilde{s}=s\left(1+\frac{T_E}{AE}\right) \tag{12}$$

where AE is axial towline stiffness (same as for dynamics model).

Combination of Equations 8, 11 and 12 leads to expression that fully describes quasi-static model because brings into connection effective tension force T_E and towline span d

$$s\left(1+\frac{T_E}{AE}\right)=\frac{4q^2d^7}{\pi^4(EI\pi^2+T_Ed^2)^2}+d \tag{13}$$

Since the Equation 13 is nonlinear, Newton-Raphson method is applied as presented in Ćatipović (2014b).

2.3 *Time domain hydrodynamics of a towed ship*

Due to the nonlinear properties of the towline, the hydrodynamics of the towed ship is defined in the time domain (Cummins 1962)

$$([M^m]+[A^\infty])\{\ddot{\xi}(t)\}+\int_0^\infty[K(t-\tau)]\{\dot{\xi}(t)\}\mathrm{d}\tau+ \\ +[C^h]\{\xi(t)\}=\{F(t)\}+\{F^{TL}(t)\} \tag{14}$$

where $\{\xi(t)\}$ is the displacement vector of towed ship dependent on time, $[M^m]$ is the mass matrix due to own mass of the towed ship, $[C^h]$ is the hydrostatic stiffness matrix. Wave loads of the first order and ship resistance are incorporated in excitation force vector $\{F(t)\}$. Added resistance in waves is also incorporated in this vector. To achieve coupling, towline forces are in vector $\{F^{TL}(t)\}$. The matrix of impulse response function (or memory function $[K(t)]$ is calculated in form

$$[K(t)]=\frac{2}{\pi}\int_0^\infty[B(\omega)]\cos(\omega t)\mathrm{d}\omega \tag{15}$$

where $[B(\omega)]$ is radiational damping matrix calculated in frequency domain. Added mass for infinite frequency $[A^\infty]$ is defined by Oglivie (1964) as follows

$$[A^\infty]=A(\omega_{ac})+\frac{1}{\omega_{ac}}\int_0^\infty[K(t)]\sin(\omega_{ac}t)\mathrm{d}t \tag{16}$$

where $[A(\omega)]$ is added mass matrix determined in frequency domain while ω_{ac} is arbitrary chosen wave frequency.

It is important to note that in real conditions the towed ship is pulled at almost constant speed during towing. This problem is equivalent to the set up where the ship is restrained by towline in a fluid inflow of constant speed, see Figure 1. Then this constant speed should be considered when

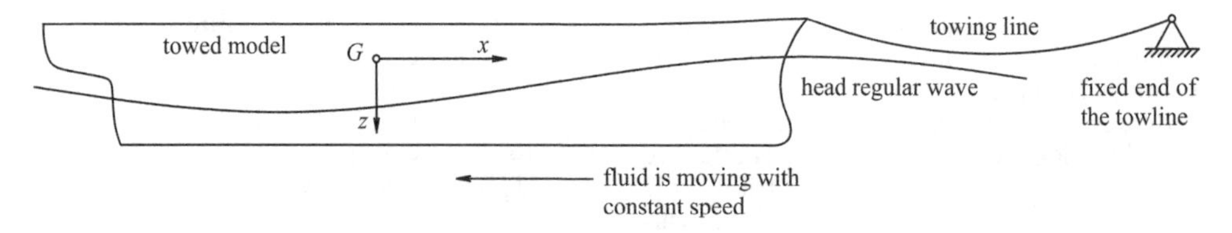

Figure 1. Calculation and experimental set up.

solving Equation 14. Therefore, constant speed is taken into account when calculating the added mass $[A(\omega)]$ and the radiational damping $[B(\omega)]$ in frequency domain as well as during evaluation of the added mass for infinite frequency $[A^{\perp}]$ and the impulse response function $[K(t)]$. The same principle is applied for evaluation of the first order wave forces $\{F(t)\}$.

In case of damaged ship, trim and heel due to hull damage and flooded sea water should also consider in Equation 14. Practically, trim and heel can be considered in usual manner when modelling wetted surface of the ship for calculating hydrodynamic loads and reactions. Further, flooded water can be modelled in simplified way as liquid cargo to include flooded water influence on damaged ship hydrodynamics.

3 CASE STUDY

Nakyama et al. (2012) carried out model tests using S-175 container ship model as a towed ship (Fig. 2). The principal dimensions of full scale ship and model are shown in Table 1 while body plan is shown in Figure 2. He measured wave-induced motions of the towed ship model in regular head waves (Fig. 1). The sway and yaw motions of the ship model were fixed by guide equipment. Therefore, surge, heave and pitch motions were measured. Also, towline tension was measured at towline suspension point on the ship. Diameter of wire rope used as towline was 2 mm with distributed mass 0.035 kg/m. The towline length was 1.9 m and the height of both suspension points was 0.255 m above still water surface. Suspension point on ship model was located on forward perpendicular. During measurements, towing speed was 0.471 m/s what corresponds to 7.0 knots in full scale ship. Wave height was set to 30 mm. The ratio between wave length and ship length was from 0.5 to 2.0. Properties of wire rope are presented in Table 2. For calculation procedure of wire rope properties see Appendix B.

Described model tests and measured data are used as reference point for comparison of towline dynamics and quasi-static model. Therefore, input values described by Nakayama et al. (2012) are used to carry out numerical calculations for both numerical models.

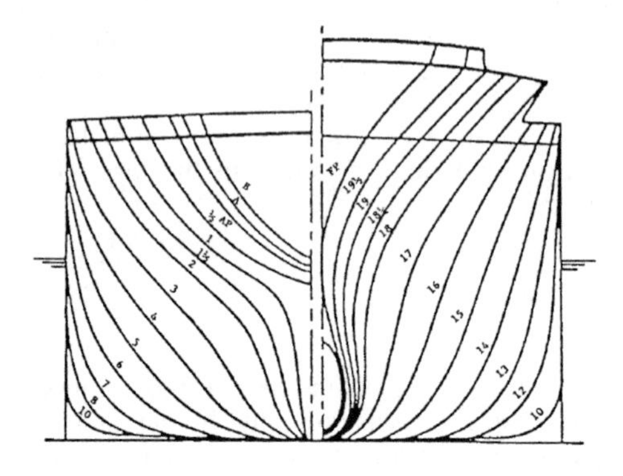

Figure 2. Body plan of S-175 container ship.

Table 1. Principal dimension of S-175 container ship.

	Full scale	Model
Length L (m)	175.00	3.00
Breadth B (m)	25.40	0.435
Draft of fore peak d_f (m)	7.00	0.120
Draft of midship d_m (m)	9.50	0.163
Draft of aft peak d_a (m)	12.02	0.206
Volume V (m^3)	24154.13	0.122
Block coefficient C_b	0.572	0.572
Position of LCG x_g (m)		–0.141
Radius of gyration k_{yy}/L(m)		0.239

Table 2. Properties of the wire rope.

Designation	Quantity	Unit
Wire rope diameter	3	mm
Distributed mass of wire rope	0.0346	kg/m
Young's modulus of wire rope material	193	GPa
Number of strands	49	
Strand diameter	0.335	mm
Sectional area	4.325	mm^2
Sectional moment of inertia (minimum value)	0.0304	mm^4

The calculation of excitation forces and hydrodynamic reaction (added mass and damping) in frequency domain is performed using HYDROSTAR (2010) developed by Bureau Veritas (France). HYDROSTAR is based on the potential flow theory and three-dimensional boundary element method and has the ability to include forward speed of a ship. Also it is capable to calculate added resistance in waves. Visualization of panel elements for S-175 container ship model as obtained by HYDROSTAR is shown in Figure 3.

The ship resistance is assessed by in-house free surface potential flow code with the addition of a viscous resistance approximated by the ITTC 1957 formulation (Appendix A). This code calculates the steady inviscid flow around a ship hull, the wave pattern and the wave resistance. It solves the exact, fully nonlinear potential flow problem by an iterative procedure. In case of S-175 container ship model, longitudinal symmetry of the problem is used to reduce the number of panels and calculation time. Therefore, only half of the fluid domain and ship hull is modelled. For this purpose, 1280 panels are used for ship hull and 1275 for free surface. Figure 4 shows hull and free surface for the nonlinear wave resistance calculation as generated by the automatic adaption. Presented free surface elevations are five times magnified.

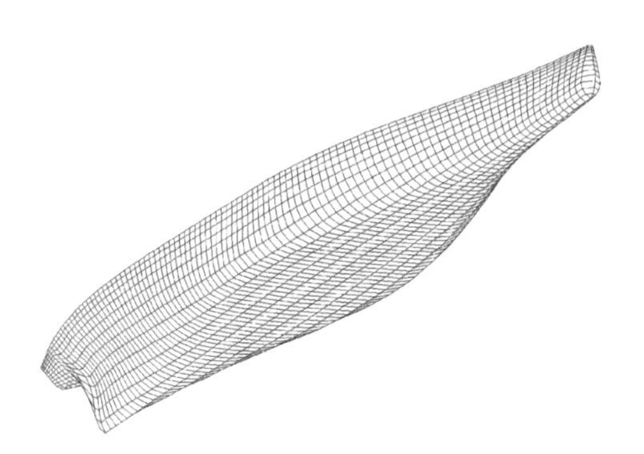

Figure 3. Panel elements of S-175 container ship model, HYDROSTAR.

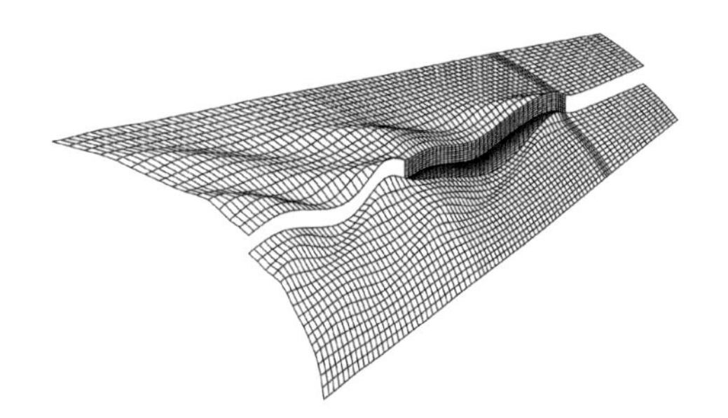

Figure 4. Hull and the free surface paneling for S-175 container ship model at $Fn = 0.35$.

Results of hydrodynamic and resistance calculations are used in coupled model for dynamics and quasi-static approach. Also added resistance in waves is considered. Within dynamics model, five finite elements are used for modeling of towline. Maximum tested time step size for time domain simulation was 0.1 s. For larger time steps numerical instability occurs and corrupts simulation results. The time step of 0.05 s has been used for all simulations, to be on the safe side. As usual, smaller time steps are acceptable but in such cases calculation time is significantly extended. For quasi-static model maximum tested time step size was 0.2 s. Same as for dynamics model, the time step of 0.05 s was used for all simulations.

For both models, the simulation time was 100 s per single wavelength. As expected, calculation time for quasi-static model was significantly less than for dynamics model, about two orders of magnitude.

4 RESULTS AND DISCUSSION

The calculation of wave-induced motions of the towed ship and the towline tension at suspension point on the ship is performed for regular head waves with range of wavelengths $\lambda/L = 0.5$–2.0.

The obtained numerical results from dynamics and quasi-static model are presented along with experimental results published by Nakayama et al. (2012.) Figures 5–7 show motions amplitudes of the towed ship. The surge and heave amplitudes,

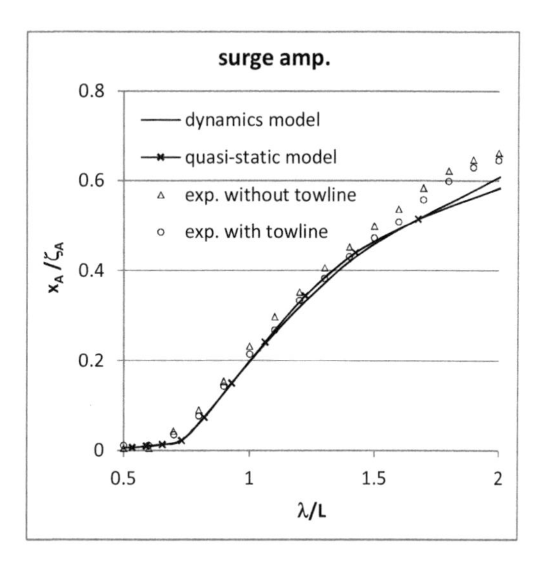

Figure 5. Comparison of surge amplitudes of the towed ship.

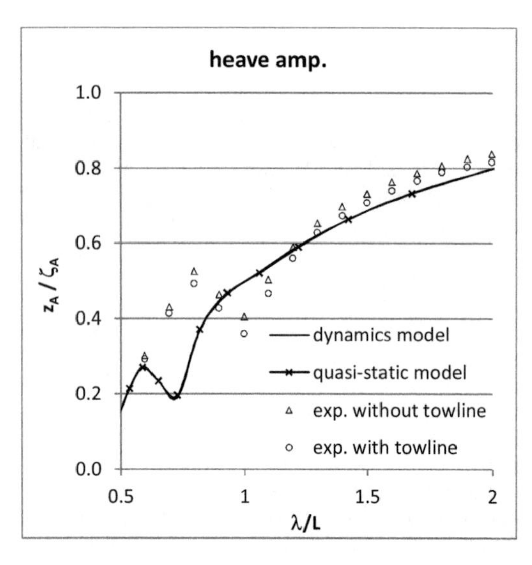

Figure 6. Comparison of heave amplitudes of the towed ship.

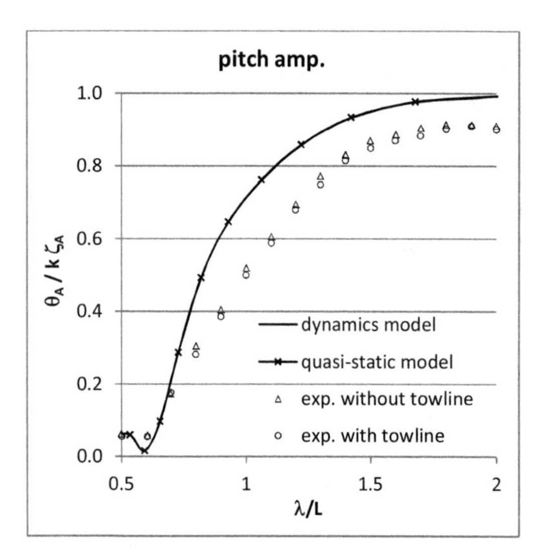

Figure 7. Comparison of pitch amplitudes of the towed ship.

respectively x_A and z_A, are made dimensionless using the wave amplitude ζ_A and the pitch amplitudes θ_A using wave slope $k\zeta_A$. Presented experimental data also contains measurements of ordinary seakeeping model tests without influence of the towline. As can be seen on these figures, motion amplitudes calculated by quasi-static model are almost equal to amplitudes obtained by dynamics model. Very small differences are found on Figure 5 where surge amplitudes are shown. This was expected since the surge is most affected by the towline. Therefore, the differences in approximation of the towline tension by dynamics and quasi-static model are most noticeable at the surge. Other motions amplitudes do not depend so much on towline influence. In more detail, the towline tension acts in surge direction. Also, the towline tension is perpendicular to heave direction. Since the vertical position of ship's center of gravity and the towline vertical position are very close, pitch moment caused by the towline cannot be high due to short moment arm, see Figure 1. Thus, the towline influence on the heave and the pitch is relatively small.

In this case study the towed ship is observed on regular head waves with wave amplitude set to 15 mm. That is less than 10% of the ship model draft (Table 1). So this is the case of low intensity wave loading. Such type of loading is favorable for presented comparison of dynamics and quasi-static model. For more complete comparison, a case study with some high intensity wave loading (like some extreme sea state) is needed. However, in such extreme loading good correlation of the towed ship motion approximated by these models is expected. The biggest influence on ship motion has restoring forces of the towline due to axial and geometric stiffness. The towline inertial forces are relatively small within coupled model because the towline mass is very small compared to the ship mass. Similarly, damping forces acting directly on the towline are minor to total damping forces of the ship. So, inertial and damping forces of the towline have small influence on the towed ship motion. In conclusion, as already described, both dynamic and quasi-static model take into account axial and geometric stiffness of the towline and hence the restoring forces that has the biggest influence on ship motion. This is also one more explanation of almost identical amplitudes of motion amplitudes (obtained by these models) presented in Figures 5–7.

As for the averaged towline tension the situation is similar, see Figure 8. Small differences are noticed for entire wavelength range. The cause of those differences is in simplified approach within quasi-static model, see Ćatipović (2014b). In this model coupling is achieved just trough the surge. So, the towline tension is included in excitation force just as horizontal component i.e. in the direction of x-coordinate, see Figure 1. In this way, the towline tension is equal to the horizontal force acting on the ship. In real conditions, this is not the case due to the towline slope at the suspension point. As a consequence, in quasi-static model the towline tension is equal to the sum of resistance (calculated by in-house code) and added resistance due to waves (calculated by HYDROSTAR), see Figure 8. In dynamic model, the towline slope at suspension point is taken into account and the towline tension is somewhat higher than overall resistance. This

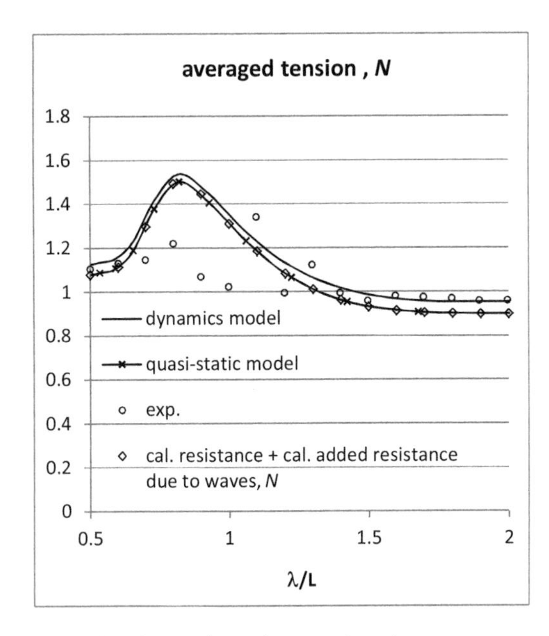

Figure 8. Comparison of averaged tension.

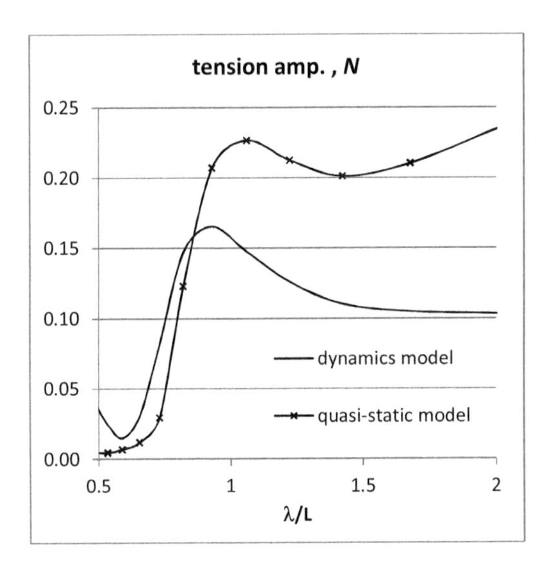

Figure 9. Comparison of tension amplitudes.

model has better representation of the real situation because the towline slope increases the towline tension (at suspension point). In conclusion, both models have good agreement with experimental data and can be used for evaluation of averaged towline tension. Also, better mutual agreement of presented models can be achieved by incorporating the towline slope in the quasi-static approach.

Figure 9 presents comparison of tension amplitudes. Results of quasi-static model are significantly different from those obtained by dynamics model. Since experimental data are not available, it is not easy to estimate which model better represents actual situation. However, advantage can be given to dynamics model because the tension amplitudes are affected by towline vibrations (besides towed ship motions). As known, vibrations can only be evaluated by considering inertial and damping forces which are only incorporated in dynamics model.

5 CONCLUSIONS

For purpose of damaged ship removal, dynamics and quasi-static approach for evaluation of towline response are compared within coupled model that contains hydrodynamics of towed ship. Published experimental data from model tests of towed S-175 container ship on regular head waves are used as a base for the comparison.

Motion amplitudes of towed ship model obtained by dynamics and quasi-static model are almost equal for whole range of tested wavelengths. Very small differences are found in surge amplitudes. This was expected since the surge of the towed ship is most affected by the towline. Thus differences of compared models are most noticeable at the surge. Also, both models are in good agreement with experimental results.

Comparison of the averaged towline tension reveals similar situation. Small differences are found in whole wavelength range. The cause of these differences is described simplification used in quasi-static model. However, both models are in good correlation with the experimental data. Significant differences are noticed in tensioned amplitudes. Experimental data for this item are not available, so it is not easy to estimate which model gives better results. Since the dynamics model is more complete and incorporates inertial and damping forces on the towline, it is assumed that this model has better representation of the towline response.

The quasi-static approach has advantage in short calculation time for time domain simulation. Compared to the dynamics model, the calculation time is less for two orders of magnitude for the same size of the time step. Case study showed that quasi-static approach is suitable for evaluation of towed ship motions and averaged towline tension. For evaluation of towline tension amplitudes, it is assumed that dynamics model is better since it is capable of modelling the towline vibrations.

This study presented that quasi-static model can be used as a fast tool for evaluation of wave-induced motions of the damaged ship in tow. Further, these wave-induced motions are necessary to define hull loads during tow. On the other hand, dynamics model is recommended way to investigate the towline response and to estimate peak values of the towline tension.

However, in further research validation of both models is necessary on irregular waves defined by some extreme sea state. To gain complete picture of towing problem, dynamics of tug vessel needs to be considered. Also, hydrodynamics of flooded water within the damaged ship should be more investigated.

ACKNOWLEDGEMENTS

This work has been supported by the Croatian Science Foundation within project 8658.

REFERENCES

American Petroleum Institute (API). 1993. Recommended Practice for Design, Selection, Operation and Maintenance of Marine Drilling Riser Systems; API RECOMMENDED PRACTICE 16Q (RP 16Q), First edition.

Bhattacharyya, S. K. & Vendhan C. P. 2000. The Finite Element Method for Hydroelastic Instability of Underwater Towed Cylindrical Structures. *Journal of Sound and Vibration* 237(1): 119–143.

Ćatipović, I., Degiuli, N., Werner, A., Čorić, V. & Radanović, J. 2014a. Numerical Model of Towing Line in Sea Transport. *33rd International Conference on Ocean, Offshore and Arctic Engineering, OMAE 2014, July 2014*, San Francisco.

Ćatipović, I., Degiuli, N., Werner, A., Čorić, V. & Radanović, J. 2014b. Approximation of Towline Influence on Towed Ship Motions. *Maritime Technology and Engineering,* C. Guedes Soares & T.A. Santos (Eds.), Taylor & Francis Group, London, UK, pp. 1095–1103.

Cummins, W.E. 1962. The Impulse Response Function and Ship Motions. *Schiffstechnik*.

Desroches, A.S.1997. Calculation of Extreme Towline Tension during Open Ocean Towing. M. Sc. Thesis, Massachusetts Institute of Technology.

Garrett, D.L., 1982. Dynamic Analysis of Slender Rods. *Journal of Energy Resources Technology*, Vol. 104: 302–307.

HYDROSTAR for Experts, v6.11 – User Manual. 2010. Bureau Veritas, Paris.

Kirk, C.L., Etok, E.U. & Cooper, M.T. 1979. Dynamic and static analysis of a marine riser. *Applied Ocean Research* 1 (3): 125–135. July, 1979.

Kiureghian, A.D., Hong, K.J. & Sackman, J.L., 1999. *Further Studies on Seismic Interaction in Interconnected Electrical Substation Equipment*. Pacific Earthquake Engineering Research Center, University of California, Berkeley.

Kreyszig, E. 1993. *Advanced Engineering Mathematics*, Seventh edition. New York: John Wiley & Sons, Inc.

Molland, A.F., Turnock, S.R. and Hudson, D.A. 2011. *Ship Resistance and Propulsion: Practical Estimation of Ship Propulsive Power*. New York: Cambridge University Press.

Nakayama, Y., Yasukawa, H., Hirata, N. & Hata, H. 2012. Time Domain Simulation of Wave-induced Motions of a Towed Ship in Head Seas. *Proceedings of the Twenty-second (2012) International Offshore and Polar Engineering Conference*, Rhodes, Greece, (June 2012): 901–907.

Nordgen, R.P. 1974. On Computation of the Motion of Elastic Rod. *Journal of Applied Mechanics*, Vol. 41, pp. 777–780.

Ogilvie, T. F. 1964. Recent Progress toward the Understanding and Prediction of Ship Motions. *Proceedings of 5th Symp. on Naval Hydrodynamics*, 2–128.

Raman-Nair, W., Power, J. & Simoes-Re, A. 2009. Numerical Model of Towing Dynamics of a Long Flexible Life Raft in Irregular Waves. *Marine Technology* 46(4): 213–218.

Raven, H.C. 1996. A Solution Method for the Nonlinear Ship Wave Resistance Problem. Ph.D. Thesis, Technische Universiteit Delft, The Netherlands.

Werner, A., Degiuli, N. & Sutlović, I. 2006. CFD as an Engineering Tool for Design and Analysis. *Strojarstvo* 48 (3-4): 115–121.

APPENDIX A—CALM WATER RESISTANCE CALCULATION

Calculation procedure for evaluation of calm water resistance is presented by Molland et al. (2011) as follows

$$R_T = R_V + R_W \tag{A1}$$

with

$$R_V = (1 + k)\, R_W \tag{A2}$$

where R_V is viscous resistance incorporating skin friction and viscous pressure resistance while R_W denotes wave resistance. In Equation A2, k is form factor which is dependent on hull form.

The form factor k can be calculated by different empirical formulas. Such empirical formulas can never represent anything else than the average of ships used in the regression analyses. Thus, using such empirical methods could give a distorded result that does not represent the actual relative resistance between different designs. One empirical formula that takes into account both the friction and the form effect in pressure is Holtrop's method for predicting the form factor

$$(1+k) = 0.93 + 0.487118(1+0.011C_{stern})\left(\frac{B}{L}\right)^{1.06806} \left(\frac{T}{L}\right)^{0.46106}\left(\frac{L_{WL}}{L_R}\right)^{0.121563}\left(\frac{L_{WL}^3}{\nabla}\right)^{0.36486}(1-C_P)^{-0.604247} \tag{A3}$$

Wave resistance can be calculated according to potential flow theory. The applied method is based on nonlinear ship wave calculation using the Raised Panel Iterative Dawson (RAPID) approach, developed by Raven (1996). An iterative procedure is used, consisting of a sequence of linear problems to evaluate steady nonlinear flow around ship hull. The Laplace problem in each iteration is solved using constant-strength source panels located at same distance above free surface. Combined free surface condition is treated using an essentially Dawson-like method while modeling velocity derivatives by difference scheme. Details of the procedure can be found in Werner et al. (2006).

The skin friction resistance coefficient is calculated according to the ITTC 1957 formula that incorporates some three-dimensional friction effects and is defined as

$$C_F = \frac{0.075}{(\log \mathrm{Re} - 2)^2} \tag{A4}$$

where Re is the Reynolds number. Then the skin friction resistance is calculated as

$$R_F = \frac{1}{2}\rho S v^2 C_F \tag{A5}$$

where ρ is fluid density, S is wetted surface and v is ship velocity.

APPENDIX B—PROPERTIES OF THE WIRE ROPE

In the case study, the towline is made of a wire rope. Thus, sectional properties of the wire rope need to be evaluated. This can be only done as a rough approximation especially in the case of sectional moment of inertia, as follows

$$A = \frac{n\pi d^2}{4} \tag{B1}$$

and

$$I_{\min} = \frac{n\pi d^4}{64} \tag{B2}$$

where A is sectional area and $I_{\min}$ is the minimum value of sectional moment of inertia of the wire rope. Number of strands in the wire rope is denoted by n and d is strand diameter. Equations B1 and B2 are taken from Kiureghian et al. (1999) where the approximation of sectional properties of flexible electrical conductors are studied. The minimum value of the moment of sectional moment of inertia is obtained by assuming that strands freely slide against one another within the wire rope. In that case the sectional moment of inertia $I_{\min}$ is simply the sum of the sectional moments of inertia of individual strands.

Wave-structure interaction

Progress in Maritime Technology and Engineering – Guedes Soares & Santos (Eds)
© 2018 Taylor & Francis Group, London, ISBN 978-1-138-58539-3

Comparisons of CFD, experimental and analytical simulations of a heaving box-type floating structure

H. Islam, S.C. Mohapatra & C. Guedes Soares
Centre for Marine Technology and Ocean Engineering (CENTEC), Instituto Superior Técnico, Universidade de Lisboa, Lisbon, Portugal

ABSTRACT: Wave radiation by a heaving box-type floating structure based on Computational Fluid Dynamics (CFD) simulations using OpenFOAM with a volume of fluid method is analysed. The numerical simulation results of vertical force obtained for a heaving box are compared against experimental data and linearized analytical solutions based on velocity potential flow theory. The comparison results include vertical force and response amplitude operator for the heaving box at different wave lengths. The comparison shows that the CFD results are in good agreement with the experimental ones and are very close to the analytical ones. Hence, it is indicated that CFD is well capable in simulating wave interaction behaviour of a heaving box-type floating structure.

1 INTRODUCTION

Over the last decade, there has been a significant interest on CFD modelling based on OpenFOAM in the application of a large number of engineering problems associated with fluid-structure interaction due to its level of accuracy and handing of complex problems arising in marine and ocean engineering. However, there is a slower progress in the literature of wave interaction with floating bodies, incorporating different motion characteristics for designing wave energy converters. Due to the recent advances in computational modelling, CFD simulations can give deeper insight into the fluid-floating body hydrodynamics, which can facilitate better design and optimized operating setup.

CFD is one of the popular approaches to model the hydrodynamic loads on the floating structures of different geometries. CFD methods which are based on solving Navier-Stokes equations, if properly solved, can capture most of the nonlinear hydrodynamic loads on an offshore structure and can lead to very reliable results. Therefore, recently, the scientific community is interested in comparing the results of motion characteristics of floating structures of various geometries obtained from CFD simulations against experimental data and analytical results to set benchmark.

Through literature, it is well-known that the box-type floating structures are frequently used as breakwaters for small harbours and marinas (see Black et al. 1971; Drimer et al. 1992; Sannasiraj et al. 1995; Williams et al. 2000; Bhattacharjee and Guedes Soares 2011; Huang et al. 2015). Recently, Mohapatra and Guedes Soares (2015) studied the wave forces on a two-dimensional rectangular floating structure based on linearized Boussinesq equations using eigenmode expansion method. On the other hand, the study of the radiation problem based on CFD modelling on box-type floating structures are of recent interest, which can provide the fundamental information about the hydrodynamic characteristics of added mass, radiation damping coefficients, and vertical wave forces. Under the excitation of waves, a rigid floating body will exhibit six degrees of freedom, namely three translations (heave, sway and surge) and three rotational motions (pitch, yaw and roll). Practically, vertical (heave) motion is of primary importance for the hydrodynamic analysis of a box-type floating structure for designing wave energy converters.

Recently, a few researchers have simulated box type floating structure based on CFD model using OpenFOAM solver. Jung et al. (2013) also studied numerically the interaction between a regular wave and a roll motion of a rectangular floating structure using volume of fluid method based on the finite volume method. Devolder et al. (2017) simulated the heaving floating point absorber wave energy converters using OpenFOAM based on two-phase Navier-Stokes fluid solver coupled with a motion solver. A similar problem was addressed by Lavrov and Guedes Soares (2016).

Connell and Cashman (2015) compared the heave response of a freely floating body between the mathematical analysis based on Navier-Stokes (N-S) time domain program, ANSYS Fluent and CFD model. Jaswar et al. (2015) compared the semi-submersible heave motion response prediction between diffraction, diffraction-viscous and diffraction-Morison methods based on diffraction potential flow theory. Bihs et al. (2017) simulated a horizontal cylinder in heave motion and the motion of a freely floating rectangular barge in waves using the CFD model FREE3D and compared the results with experimental data. Recently, Gadelho et al. (2017) simulated the floating fixed box based on CFD model and compared the results against analytical and experimental model results in different cases.

Recently, Rodriguez et al. (2016) investigated the numerical nonlinear heave response of a rectangular box concerning on the importance of the relative body dimensions. Further, Rodriguez and Spinneken (2016) performed a series of experiments to analyse the nonlinear loading and dynamic response of a heaving rectangular box in two-dimensions under regular and irregular wave conditions.

In the present paper, the wave radiation due to heaving motion of box-type floating structure in 3D based on CFD is studied and the obtained results are compared against experimental and linearized analytical results in order to analyse/measure the low/high fidelity of the model.

The governing equations for used CFD numerical model are based on the Navier-Stokes equation and continuity equation whilst, the linearized analytical model and its solutions are demonstrated briefly based on velocity potential flow theory. The floating structure is modelled as box-type structure over flat bottom with finite width and draft. The numerical solution is solved using OpenFOAM code with volume of fluid method whilst, the linearized analytical solution is based on eigenfunction expansion method.

In order to analyse the effect of waves on the floating structure, the results of the vertical wave force acting on the floating structure due to heave motion have been compared against the experimental model results (see Rodríguez & Spinneken 2016) and analytical solutions in 2D cases. Furthermore, the results of RAO obtained from CFD are compared with experimental and numerical results available in the literature. It is observed that the CFD results of vertical wave force are in good agreement with the linearized analytical results. Finally, the significant conclusions on the comparisons are discussed on the simulations of the hydrodynamic coefficients due to heave motion of the box-type floating structure.

2 CFD MODEL FORMULATION AND DESCRIPTION

The Computational Fluid Dynamics (CFD) software used for the presented study is the open source CFD toolkit OpenFOAM (Open Source Field Operation and Manipulation). The release version 2.4.0 of OpenFOAM was used together with waves2Foam utility in Linux environment. OpenFOAM has an extensive range of features to solve complex fluid flows and a wide variety of applications including offshore and coastal engineering problems. The solver has been elaborately discussed by Jasak (2009).

The governing equations for OpenFOAM are the Navier-Stokes equation and continuity equation for an incompressible laminar flow of a Newtonian fluid. In vector form, the Navier-Stokes equation is given by

$$\rho\left(\frac{\partial v}{\partial t} + v.\nabla v\right) = -\nabla p + \mu\nabla^2 v + \rho g, \tag{1}$$

where v is the velocity, p is the pressure, μ is the dynamic viscosity, g is acceleration due to gravity, and ∇^2 is the Laplace operator. Further, the continuity equation is of the form

$$\nabla v = 0. \tag{2}$$

The solver follows Cartesian coordinate system and the volume of fluid method to track the free surface elevation. This volume of fluid method determines the fraction of each fluid that exists in each cell. The equation for the volume fraction is obtained as

$$\frac{\partial \alpha}{\partial t} + \nabla(\alpha U) = 0, \tag{3}$$

where U is the velocity field, α is the volume fraction of water in the cell and varies from 0 to 1, full of air to full of water, respectively.

The governing equation is initially discretized using Finite Volume Method (FVM), and pressure velocity coupling is with PISO algorithm. Regarding the meshing, OpenFOAM can solve both structured and unstructured meshes and generally a hybrid mesh topology is adopted for optimum results. For the presented simulation cases, blockMesh utility was used to generate a structured outer domain with hexahedral mesh. The generated block mesh had higher resolution near the free surface in z-axis and successively lower resolution near the bottom line. In case of x-y plane, the even mesh distribution was applied. Next, two successive refinements were done near the free surface and around the geometry (heaving box) for better capturing the propagating waves, motion of the box and resistance encountered by it.

Finally, the box was integrated into the domain using the snappyHexMesh utility.

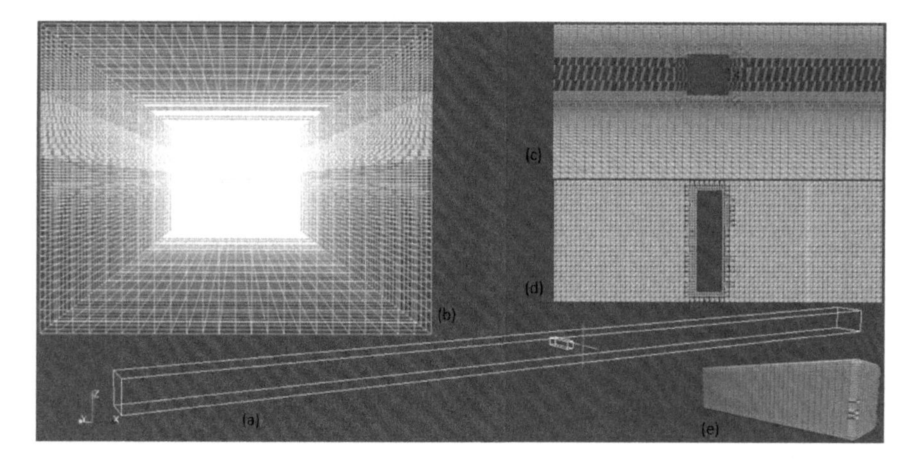

Figure 1. General mesh assembly for the heaving box case.

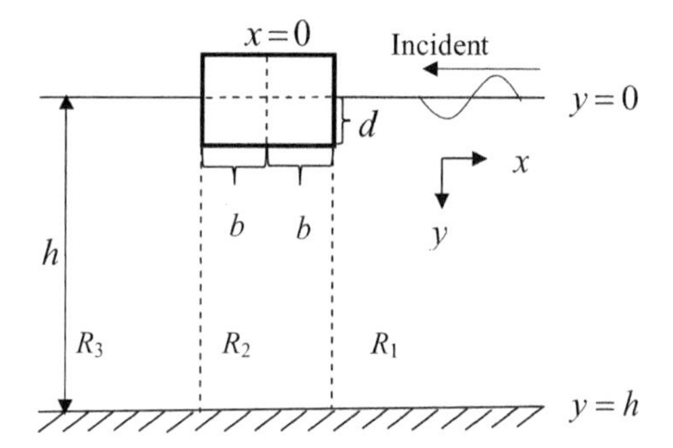

Figure 2. Schematic diagram of wave radiation due to heaving box-type floating structure.

Total mesh resolution of the domain was roughly 2.8 million. The general arrangement of the mesh is shown in Figure 1. In the figure, (a) shows total simulation domain, (b) mesh distribution in the domain in y-z plane, (c) mesh distribution near the heaving box in *z-x* plane, (d) mesh distribution near the heaving box in x-y plane, and (e) mesh distribution on the heaving box.

The numerical setup consists of a geometry that represents a 3D wave flume with 45.0 m length and 1.25 m water depth, and a rectangular floating box with 2m×0.5m×0.25m dimensions, like the one used in the experimental setup presented by Rodriguez & Spinneken (2016). The chosen Cartesian coordinates, *x*-axis is parallel to the flume length, y-axis is parallel to the flume width and *z*-axis is parallel to the flume height. The origin in *x*-axis and y-axis is located in the center of the domain and origin in *z*-axis is located in the initial free surface elevation. A schematic diagram of the setup is show in Figure 2.

In the next section, the mathematical formulation and its solution associated with wave interaction with heaving box-type floating structure will be demonstrated briefly to compare with CFD model results of vertical force acting on the structure in different cases.

3 ANALYTICAL MODEL FORMULATION AND LINEARIZED SOLUTION

A rectangular box-type structure of width $2b$, draft d and infinitely long in the *z*-direction is floating on the upper surface $-b < x < b$ is shown in Figure 2. Further, the origin O is assumed to be the middle point of the box. Hence, the whole domain is divided into three regions defined by: $(b < x < \infty, 0 < y < h)$, $(-b < x < b, d < y < h)$ and $(-\infty < x < -b, 0 < y < h)$, which are referred as R_1, R_2, and R_3 respectively.

The total potential $\phi(x, y, z)$ can be expressed as

$$\varphi = \varphi_I + \varphi_D + \varphi_R, \tag{4}$$

where φ_I is the incident wave potential, φ_D is the diffracted potential and φ_R is the radiation potential, respectively. Assuming that a progressive wave interacts with a floating rectangular structure making an oblique angle θ with *x*-axis and angular frequency ω, thus, the spatial incident velocity potential is of the form $\phi\ (x, y, z) = \varphi_1(x, y)e^{i\gamma z}$, $\gamma = k_0 \sin\theta$ is the *z*-component of the wave number k_0 associated with the incident waves.

Therefore, the incident wave potential as given by

$$\varphi_I = \frac{igI}{\omega}\frac{\cosh k_0(h-y)}{\cosh k_0 h}e^{-ipx}, \tag{5}$$

where $p = k_0\cos\theta$ with I is the incident wave amplitude and k_0 satisfy the gravity wave dispersion relation $\omega^2 = gk_0\tanh(k_0h)$.

The radiation potential ϕ_R can be expressed as (see Sannasiraj et al. 1995, Abul-Azm and Gesraha 2000).

$$\phi_R(x,y,z) = -i\omega I_R\varphi_R(x,y)e^{i\gamma z}, \tag{6}$$

where I_R is the amplitude of motion of the floating structure and φ_R is the spatial velocity potential which satisfies the reduced wave equation

$$\left(\frac{\partial^2}{\partial x^2}+\frac{\partial^2}{\partial y^2}-\gamma^2\right)\varphi_R = 0 \text{ in the fluid domain.} \tag{7}$$

The free surface boundary condition as

$$\frac{\partial\varphi_R}{\partial y}+\frac{\omega^2}{g}\varphi_R = 0 \text{ on } y=0 \text{ for } R_1 \text{ and } R_3. \tag{8}$$

The no-flow condition at the rigid uniform bottom boundary is

$$\frac{\partial\varphi_R}{\partial y} = 0 \text{ on } y = h \text{ in } R_1, R_2 \text{ and } R_3. \tag{9}$$

The non-homogeneous boundary condition due to the heave motion and the heave condition on the structural boundary are given by

$$\frac{\partial\varphi_R}{\partial y} = 1 \text{ on } -b \le x \le b, y = d \tag{10}$$

and

$$\frac{\partial\varphi_R}{\partial x} = 0, \text{ at } x = \pm b, 0 < y < d, \text{ respectively.} \tag{11}$$

The radiation condition as given by

$$\lim_{x\to\mp\infty}\left[\frac{\partial\varphi_R}{\partial x}\pm ip\varphi_R\right] = 0. \tag{12}$$

Using governing equation and along with relevant boundary conditions, the radiation potentials in R_1, R_2, and R_3 are denoted by ϕ_{R1}, ϕ_{R2}, and ϕ_{R3} respectively, one can obtain as

$$\varphi_{R1} = A_{10}e^{-ip_0(x-b)}\frac{\cosh k_0(h-y)}{\cosh k_0h} + \sum_{n=1}^{\infty}A_{1n}e^{p_n(x-b)}\frac{\cos k_n(h-y)}{\cos k_nh}, \tag{13}$$

$$\varphi_{R2} = \frac{-\cosh\alpha_0(h-y)}{\alpha_0\sinh\alpha_0(h-d)} + \sum_{n=0}^{\infty}\left[A_{2n}e^{-\alpha_n(x-b)}+B_{2n}e^{\alpha_n(x+b)}\right]\cos\beta_n(h-y), \tag{14}$$

$$\varphi_{R3} = A_{30}e^{-i\lambda_0(x+b)}\frac{\cosh\mu_0(h-y)}{\cosh\mu_0h} + \sum_{n=1}^{\infty}A_{3n}e^{\lambda_n(x+b)}\frac{\cos\mu_n(h-y)}{\cos\mu_nh}, \tag{15}$$

where the eigenfunctions in (13–15) are orthogonal over water depth associated with the eigenvalues p_n, α_n, β_n, and λ_n (as in Bhattacharjee and Guedes Soares 2011).

Using the conditions of continuity of velocity and pressure across the vertical interfaces at $x = \pm b$, a linear system of equations $4(N + 1)$ are obtained to solve for $4(N + 1)$ number of unknown coefficients A_{1n}, A_{2n}, A_{3n}, and B_{2n} for $n = 0, 1, 2,\ldots N$ using matlab code. Once the unknown coefficients are determined, then the full solution of radiation potentials will be obtained.

The vertical wave force due to the heave motion of the box can be calculated as

$$F_e = i\rho\omega\left[\int_{S_b}\varphi_I(x,y)n\,ds - \int_{S_b}\varphi_R(x,y)\frac{\partial\varphi_I}{\partial n}ds\right], \tag{16}$$

where Sb denotes the wetted surface of the floating structure in xy plane and n is the inward normal to the structure.

Now, the vertical force which are non-dimensionalized as

$$F = |F_e|/(2\rho gbI). \tag{17}$$

It may be noted that the present analytical solution converges faster (upto first 30 terms in the infinite series) than available in the literature (see Zheng et al., 2004).

In the next section, the numerical CFD model associated with heaving box under incoming waves will be validated with the experimental model results of non-dimensional vertical force in two different cases and then compared with WAMIT results.

4 COMPARISON OF NUMERICAL AGAINST EXPERIMENTAL MODEL RESULTS

To simulate the heaving box because of incoming waves, the library waves2Foam was used to generate and absorb free surface water waves (Jacobsen et al. 2012). The method applies the relaxation zone technique (active sponge layers) and supports a large number of wave theories, and the relaxation zones can be of arbitrary shapes. The main solver is waveDyMFoam, which is based on the native solver interDyMFoam and is described as a fully viscous solver for two incompressible and immiscible fluids.To non-dimentionalize the heave response RAO (Response Amplitude Operator), and the heaving force, following equations were used.

$$\mathrm{RAO}_{\mathrm{Heave}} = \frac{\zeta}{A_I}, \tag{18}$$

$$F_z' = \frac{F_z}{A_I}. \tag{19}$$

In case of experimental study, for some cases, complex response with multiple harmonics were found. Thus, results were filtered using Fourier Transformation (FT) and decomposed into first and second harmonics. For heaving motion, only first harmonics was considered. As for heaving force, two set of data for first and second harmonics were presented. However, in case of simulation, in most cases, a simple sinusoidal response was recorded, thus FT was not performed. Furthermore, in case of experimental results, 90% of the peak force was presented as heaving force results, which has been avoided in the CFD case.

One of the major difference between the experimental and CFD study is that, in the experimental study, for measuring the heave force, the box was held static. For measuring the heave motion, the box was set free to heave, but the resistance was not measured at that time. As a result, the measured heave resistance only accounts the diffracted wave forces, and not the radiated wave forces, whereas, in case of CFD, the forces were measured with the free heaving motion, thus accounting for both the diffracted and radiated forces.

The comparison among the CFD, numerical and experimental results for heave RAO and non-dimensional heaving force are shown in Figure 3 and 4, respectively. In the figures, the solid line (_______) represents potential flow prediction (WAMIT); dotted lines represent time domain simulation for $A_Ik = 0.05$ (_ _) and $A_Ik = 0.10$ (_. _); circular (o) and star (*) symbol represent experimental data for $A_Ik = 0.05$ and $A_Ik = 0.10$ respectively; and block (■) signs represent OpenFOAM prediction results.

As can be seen from the figures, the CFD results agree reasonably well with the experimental data. The overall trend is well captured, however relatively high deviations are observed for high kb values, or small wave length cases. The deviation here may be explained by insufficiency of mesh resolution

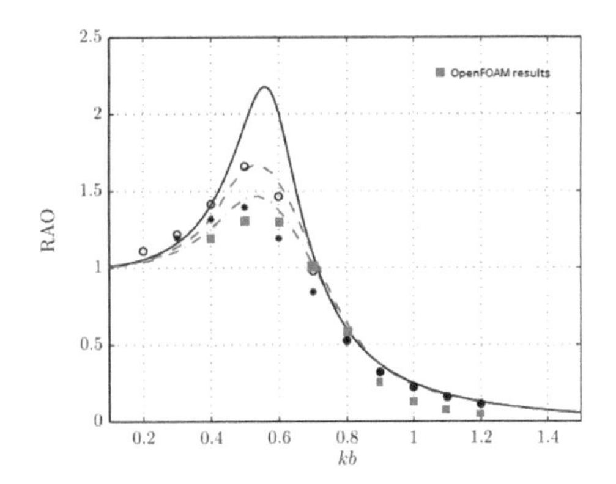

Figure 3. Comparison of RAO between the CFD, numerical and experimental model.

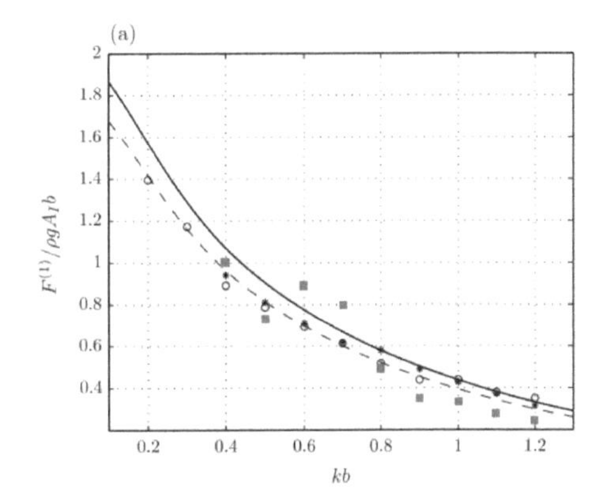

Figure 4. Comparison between the CFD and experimental model force time series on the box.

in z-axis. In case of small wave length cases, the heave amplitude is also small. If mesh resolution is insufficient in z-direction, the solver miss-predicts the heaving motion and the force in z direction.

The intention of simulating the heaving box is to evaluate its performance as a simplified Wave Energy Converter (WEC). Although, in actual sea, despite mooring, a WEC will have higher degrees of freedom, at initial stage, only heave motion and related force are considered. As can be seen from the z-force prediction, although the simulated forces follow the experimental trend, but there is deviation among the values. As mentioned before, in experiment, the heaving force was measured on a static box. Thus, radiated force were not considered. Furthermore, due to the combined effect of the radiated and diffracted waves, the incoming wave amplitude also loses some steepness, which lowers the encountered force. As a result, CFD predicted lower z-forces comparing to experimental cases. Near the resonance area, the radiated forces become comparatively larger, thus, higher encountered force is predicted by CFD. Although the present CFD results for heave force were able to capture some of the impact of radiated forces, the actual impact should be more pronounced. Thus, further investigation should be performed with better mesh resolution and distribution to improve the present results.

To further illustrate the results, hydrodynamic pressure distribution on free surface, and side cross-section of the domain are shown in Figure 5 and Figure 6.

The figures show pressure distribution in the domain during the encounter of an entire wave period. As described above, the destruction of incoming waves caused by the deflected and radiated waves from the heaving box can be observed in the figures. Because of wave destruction, the incoming waves slightly lose their amplitude and the box encounters reduced impact from the incoming waves.

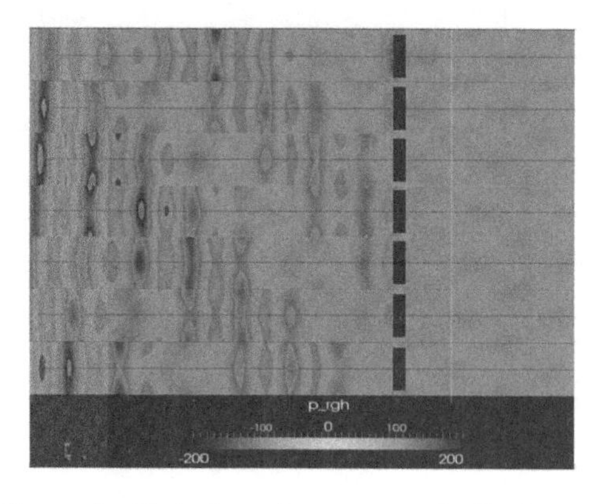

Figure 5. Hydrodynamic pressure distribution on the zero-water level of a heaving box, at kb = 0.8, wave height = 0.05.

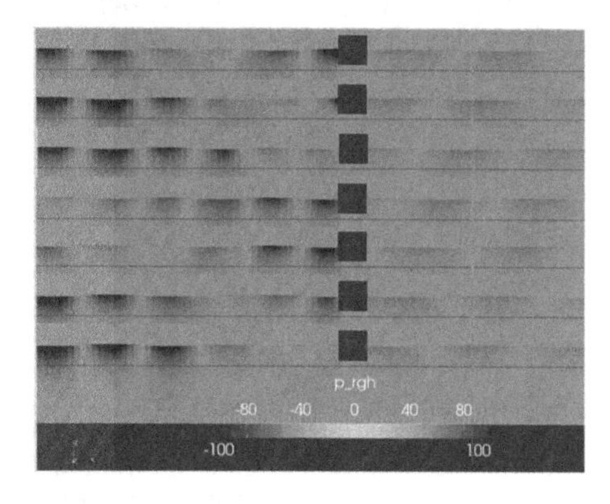

Figure 6. Hydrodynamic pressure distribution in the z-x plane of the domain of the heaving box at kb = 0.8, wave height = 0.05.

5 COMPARISION OF NUMERICAL AGAINST LINEARIZED ANALYTICAL RESULTS

In this section, a comparison between the CFD, experimental (Rodriguez and Spinneken 2016), and linearized analytical model (as obtained in Section 3) of vertical wave force acting on the floating box were made.

Figure 7 shows the comparison of CFD against linearized analytical results of non-dimensional vertical wave force acting on the floating box versus *kb*. It is observed that the CFD result is almost same with analytical model results for lower value and higher values of *kb* and deviation occurred for intermediate *kb*. On the other hand, from Figure 4 and 7, it is seen that the vertical wave force from WAMIT is similar in their pattern and very close with the linearized analytical results this may be due to the fact that both models are under potential flow theory assumption.

Figure 8 compares the results of vertical wave force acting on the floating structure between analytical and experimental model results versus *kb*. It is found that the non-dimensional vertical force is well agreed for intermediate values of *kb* and the

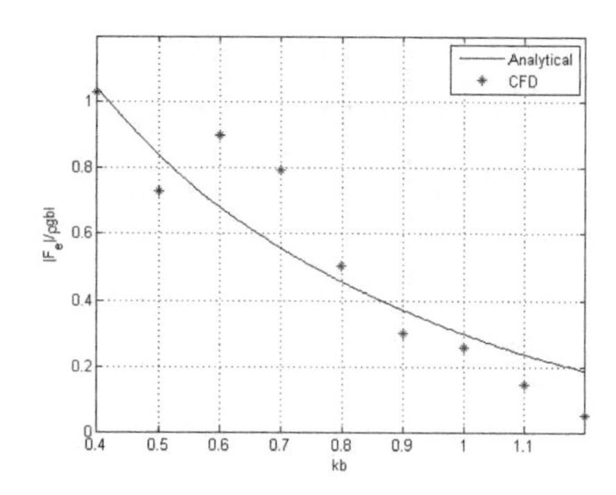

Figure 7. Comparison between the analytical and CFD model of vertical wave force on the box-type floating structure.

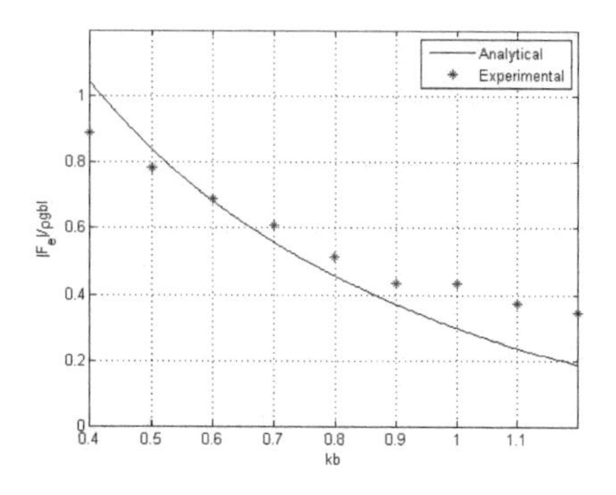

Figure 8. Comparison between the analytical and experimental model of vertical wave force on the box-type floating structure.

vertical force behavior pattern is similar in nature for both models. In both cases, the analytical results well capture the trend lines, however fails to capture the values properly. This is mostly because of the linear assumptions involved in the analytical method which are better modeled in CFD, and are properly evaluated in the experimental study. However, the results indicate that the analytical method is well reliable for concept validation and initial design development for WEC devices.

6 CONCLUSIONS

This paper studied the wave radiation due to a heaving box-type floating structure by comparing the vertical force results between the CFD, experimental and linearized analytical model solution results in different case. Further, the simulated results of RAO in CFD are also compared with experimental and numerical results available in the literature. The comparison and validation of the numerical wave flume to work with a heaving box-type floating structure based on experimental model data and the linearized velocity potential theory results are demonstrated to be successful. The vertical wave force on the heaving box-type floating structure agreed well with experimental model data and linearized analytical solution results.

The results obtained demonstrate the potential of CFD modelling, to investigate the interaction between water waves and wave energy converters in a 3D wave flume. As a future scope of the present study, introducing the multiple degrees of freedom of box in waves to perform the assessment performance of WECs.

ACKNOWLEDGEMENTS

This work was performed within the Project MIDWEST, Multi-fidelity Decision making tools for Wave Energy Systems, which is financed by the Portuguese Foundation for Science and Technology as part of the OCEANERA program.

REFERENCES

Abul-Azm, A.G. & Gesraha, M.R. 2000. Approximation to the hydrodynamics of floating pontoons under oblique waves. *Ocean Engineering* 27: 365–384.

Bihs, H., Kamata, A., Lu, Z.J. & Arntsen, I.A. 2017. Simulation of floating bodies using a combined immersed boundary with the level set method in REEF3D. *VII international Conference on Computational Methods in Marine Engineering MARINE* 2017.

Black, J.L., Mei, C.C. & Bray, M.C.G. 1971. Radiation and scattering of water waves by rigid bodies. *Journal of Fluid Mechanics* 46: 151–164.

Bhattacharjee, J. & Guedes Soares, C. 2011. Oblique Wave Interaction with a Floating Structure near a Wall with Stepped Bottom. *Ocean Engineering* 38(13):1528–1544.

Connell, K.O. & Cashman, A. 2015. Mathematical and CFD analysis of free floating heave-only body. *4th International Conference on Renewable Energy Research and Applications*, Palermo, Italy, Nov, 22–25.

Drimer, N., Agnon, Y. & Stiassnia, M. 1992. A simplified analytical model for a floating breakwater in water of finite depth. *Applied Ocean Research* 14(1): 33–41.

Devolder, B., Rauwoens, P. & Troch, P. 2017. Numerical simulation of an array of heaving floating point absorber wave energy converters using OpenFOAM. M. Visonneau, P. Queutey and D. Le Touzé (Eds). *VII International Conference on Computational Methods in Marine Engineering MARINE* 2017.

Gadelho, J.F.M., Mohapatra, S.C. & Guedes Soares, C. 2017. CFD analysis of a fixed floating box-type structure under regular waves. *In*: Guedes Soares, C. and Teixeira, Â.P. (Eds.), *Development in Maritime Transportation and Harvesting of Sea Resources*. London: Taylor & Francis Group, pp. 513–520.

Huang, Z., He, F. & Zhang, W. 2015. A floating box-type breakwater with slotted barriers. *Journal of Hydraulic Research* 52(5): 720–727.

Jacobsen, N.G., Fuhrman, D.R., & Fredsøe, J. 2012. A Wave Generation Toolbox for the Open-Source CFD Library: OpenFoam. *International Journal of Numerical Methods in Fluids* 70(9): 1073–1088.

Jasak, H. 2009. OpenFOAM: Open Source CFD in research and industry. *International Journal of Naval Architecture and Ocean Engineering* 1(2): 89–94.

Jaswar, K., Siow, C.L., Khairuddin, N.M., Abyn, H. & Guedes Soares, C. 2015. Comparison of floating structures motion prediction between diffraction, diffraction-viscous and diffraction-Morison methods. In: Guedes Soares & Santos (Eds.), *Maritime Technology and Engineering*. London: Taylor & Francis Group, pp. 1145–1152.

Jung, J.H., Yoon, H.S., Chun, H.H., Lee, I. & Park, H. 2013. Numerical simulation of wane interacting with a free rolling body. *International Journal of Naval Architecture and Ocean Engineering* 5: 333–347.

Lavrov, A. & Guedes Soares, C. 2016. Modelling the Heave Oscillations of Vertical Cylinders with Damping Plates. *International Journal of Maritime Engineering* 158(A3):A187 – A197.

Mohapatra, S.C. & Guedes Soares, C. 2015. Wave forces on a floating structure over flat bottom based on Boussinesq formulation. *In*: Guedes Soares, C. (ed.), *Renewable Energies Offshore*. London: Taylor & Francis Group, pp. 335–342.

Rodriguez, M., Spinneken, J. & Swan, C. 2016. Nonlinear loading of a two-dimensional heaving box. *Journal of Fluids and Structures* 60: 80–96.

Rodriguez, M. & Spinneken, J. 2016. A laboratory study on the loading and motion of a heaving box. *Journal of Fluids and Structures* 64: 107–126.

Sannasiraj, S.A., Sundar, V. & Sundaravadivelu, R. 1995. The hydrodynamic behavior of long floating structures in directional seas. *Applied Ocean Research* 17(4): 233–243.

Williams, A.N., Lee, H.S. & Huang, Z. 2000. Floating pontoon breakwater. *Ocean Engineering* 27: 221–240.

Zheng, Y.H., You, Y.G. & Shen, Y.M. 2004. On the radiation and diffraction of water waves by a rectangular buoy. *Ocean Engineering* 33:1063–1082.

Progress in Maritime Technology and Engineering – Guedes Soares & Santos (Eds)
© 2018 Taylor & Francis Group, London, ISBN 978-1-138-58539-3

TLP surge motion: A nonlinear dynamic analysis

S. Amat
Polytechnic University of Cartagena, Cartagena, Spain

M.J. Legaz
University of Cádiz, Cádiz, Spain

ABSTRACT: One of the most common structures used for oil exploitation in deep water are Tension Leg Platforms (TLPs). Compliant structures in the sea are subject to variation in frequency and structural response because of a nonlinear parameter in the equation of motion. Variation of frequency is important in fatigue life study of tethers. In this document, an alternative approach for the analysis and the numerical approximation of this equation of motion, using a variational framework, is presented. This method does not require a small parameter for finding surge motion of TLP. This method can be used to solve a highly nonlinear differential equation of surge motion. We prove that our procedure can never get stuck in local minima, and the error decreases until getting to the original solution independent of the perturbation parameter. A theoretical study of this method is presented. From a design engineer point of view is useful to have a sight of nonlinear vibration behavior of systems which are naturally nonlinear like TLP.

1 INTRODUCTION

Offshore platforms have many uses including oil exploration and production, navigation, ship loading and unloading, and to support bridges and causeways. Offshore oil production is one of the most visible of these applications and represents a significant challenge to the design engineer. These offshore structures must function safely for design lifetimes of twenty-five years or more and are subject to very harsh marine environments. Some important design considerations are peak loads created by hurricane wind and waves, fatigue loads generated by waves over the platform lifetime and the motion of the platform. The platforms are sometimes subjected to strong currents which create loads on the mooring system and can induce vortex shedding.

Offshore platforms are huge steel or concrete structures used for the exploration and extraction of oil and gas from the earth's crust. Offshore structures are designed for installation in the open sea, lakes, gulfs, etc., many kilometers from shorelines. These structures may be made of steel, reinforced concrete or a combination of both. The offshore oil and gas platforms are generally made of various grades of steel, from mild steel to high-strength steel, although some of the older structures were made of reinforced concrete.

Within the category of steel platforms, there are various types of structures, depending on their use and primarily on the water depth in which they will work.

Offshore platforms are very heavy and are among the tallest manmade structures on the earth. The oil and gas are separated at the platform and transported through pipelines or by tankers to shore.

A classification of offshore structures can be made in three groups fixed, floating and compliant structures. Fixed structures can be divided into three groups: jackets, jack up and gravity structures. Floating structures could be classificated as semisubmersible, drilling ship, grane barge. And compliant structures can be categorized as articulated columns, guyed towers and tension legs platforms.

Fixed structures are those which extend to the Seabed. Types of fixed structures are: steel jackets, concrete gravity structures, compliant towers.

Structures that float near the water surface, some of the recent developments are: tension leg platforms, semi submersibles, spars, ship shaped vessels (FPSO).

Tension Leg Platform (TLP) is larger versions of the Seastar platform. The long, flexible legs are attached to the seafloor, and run up to the platform itself. As with the Seastar platform, these legs allow for significant side to side movement, with little vertical movement.

Tension leg platform has excess buoyancy which keeps tethers in tension. Topside facilities, number of risers have to fixed at pre-design stage.

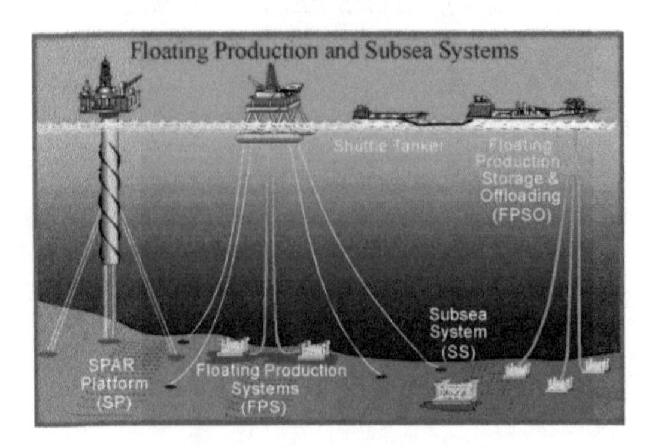

Figure 1. Floating offshore structures from www.wikimedia.org.

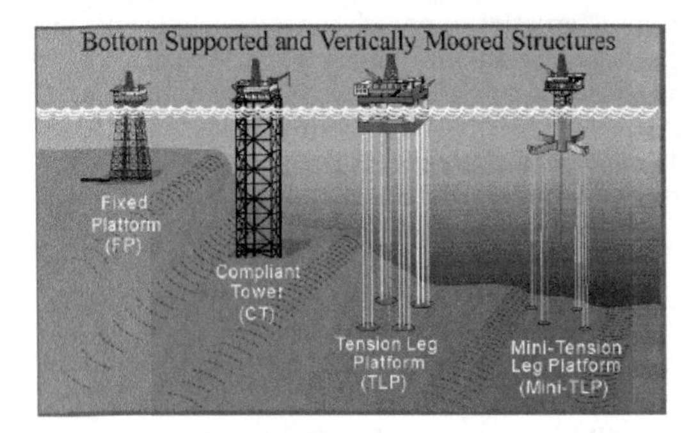

Figure 2. Fixed offshore structures from www.wikimedia.org.

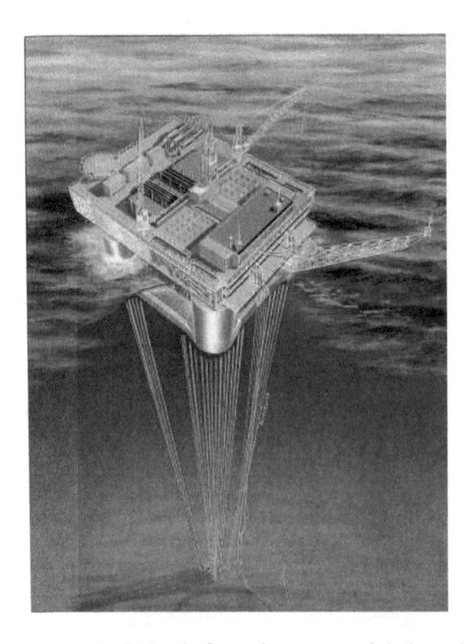

Figure 3. A TLP platform from www.globalsecurity.org.

It is used for deep water up to 1200 meters, it has no integral storage, it is sensitive to topside load/draught variations as tether tensions are affected.

2 TENSION LEG PLATFORMS (TLPS)

Tension leg platforms (TLPs) are well-known structures for oil exploitation in deep water and are becoming increasingly popular for oil drilling at very deep-water sites. In these structures, the maximum motion amplitude belongs to surge motion. A dynamic analysis of TLP model under wave is presented. Finding nonlinear equation of surge motion that contains geometrical nonlinear term, according to available methods to solve nonlinear equations the surge motion equation should be solved with a large parameter in time domain. When the surge motion amplitude of TLP is small, we have weak nonlinear equation of motion in surge direction. Many close form methods are available to solve this nonlinear equation. Rising the motion amplitude due to wave force, the traditional methods are not applicable to solve motion equation with a high nonlinear term.

In this paper, a variational method is presented to solve a nonlinear motion equation of TLP. In contrast to the traditional, this technique does not require a small parameter for finding surge motion of TLP. Therefore, the obtained results are valid not only for small parameter, but also for very large value of perturbation parameter. This effect is more important when the amplitude of vibration is large. This effect is also important in fatigue life study of tethers. The structural model used in this paper is simple. Many studies have been carried out by Ahmad (1996), Jain (1997) and Chandrasekaran and Jain (2002) to understand the structural behaviour of a TLP and to determine the effect of several parameters on the dynamic response and average life time of the structure.

3 EQUATION OF MOTION

A structural model of a TLP as a moored structure is shown in Figure 5.

Since the buoyancy of the TLP exceeds its weight, the vertical equilibrium of the platform requires taut moorings connecting the upper structure to the seabed.

The extra buoyancy over the platform weight ensures that the tendons are always kept in tension T_0 is initial pre-tension in each tether. By giving an arbitrary displacement, x, in the surge direction (see Figure 4), the increase in the initial pre-tension in each leg is given by the Figure 4.

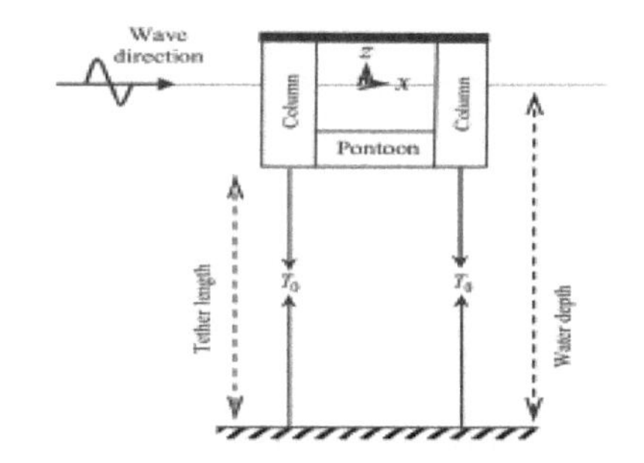

Figure 4. TLP by given arbitrary displacement.

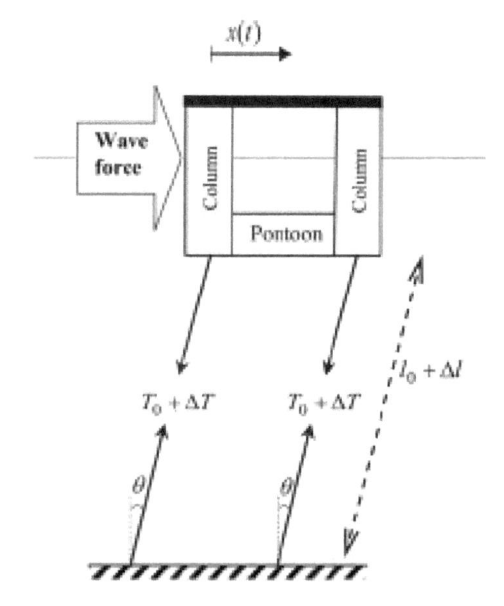

Figure 5. TLP as moored structure.

$$\Delta T(x) = A_t E\left(\frac{\sqrt{l_0^2+x^2}-l_0}{l_0}\right) \tag{1}$$

Considering $x \ll l_0$

Obtain: $$\Delta T(x) = \frac{0.5A_t E x^2}{l_0^2} \tag{2}$$

where: $$sin\theta = \frac{x}{\sqrt{l_0^2+x^2}} \tag{3}$$

Considering $x \ll l_0$

Obtain: $$sin\theta \approx \frac{x}{l_0} \tag{4}$$

where:

x displacement in the surge direction

θ angle between the initial and the displaced position of the tether

l_0 initial length of each tether

E Young's modulus of the tether,

$\Delta T(x)$ increase in the initial pre-tension due to the arbitrary displacement,

F_s tension of tendon

F_{wave} wave force

n number of tendon and

A_t cross-sectional area of tether.

The equation of motion in surge direction under wave takes the following form:

$$\sum F_x = M_{st}\ddot{x} \tag{5}$$

$$M_{st}\ddot{x} + c\dot{x} + F_s\sin(\theta) = F_{wave} \tag{6}$$

Substituting Equations (2) and (4) into Equation (6), one obtains

$$M_{st}\ddot{x} + c\dot{x} + \left(nT_0 + n\frac{A_t E x^2}{2l_0^2}\right)\left(\frac{x}{l_0}\right) = F_{wave} \tag{7}$$

Considering structural damping, c, to be equal to zero, one obtains

$$M_{st}\ddot{x} + \left(\frac{nT_0}{l_0}\right)x + \left(\frac{nA_t E}{2l_0^3}\right)x^3 = Fcos(\Omega t) \tag{8}$$

$$M_{st}\ddot{x} + k_1 x + k_3 x^3 = Fcos(\Omega t) \tag{9}$$

(i.e. $k_1 = nT_0/l_0$ and $k_3 = A_t E/(2l_0^3)$, where M_{st}, k_1 and k_3 are structural mass, linear and nonlinear stiffen parameter, respectively. F and Ω are the amplitude of intensity and frequency of wave force, respectively.

Defining

$$\frac{k_3}{M_{st}} = \varepsilon \tag{10}$$

$$\frac{k_1}{M_{st}} = \omega_n^2 \tag{11}$$

and

$$\frac{F}{M_{st}} = f \tag{12}$$

the equation of motion can be written as follows:

$$\ddot{x} + \omega_n^2 x + \varepsilon x^3 = fcos(\Omega t) \tag{13}$$

where ε and ω_n is the perturbation parameter and linear system frequency respectively.

A variational method is used to solve Equation (13), which is advantageous to traditional classic methods.

4 VARIATIONAL APPROACH

We start with the singular perturbation problem

$$\varepsilon x'(t) = f(x(t)) \quad in(0,T), \quad x(0) = x_0 \tag{14}$$

The structure of the solutions of these problems is well understood (Hairer and Wanner 1991). They are a superposition of smooth solution having an ε expansion plus some rapidly decay functions. Of course, we should assume some hypotheses on the function f is invertible and his logarithmic norm is negative in a ε-independent neighborhood of the solution.

To approximate these equations, we should use implicit methods. For explicit methods, the discretization parameters depend on the perturbation parameters and they have not practical interest. The error analysis and the existence and uniqueness of the implicit Runge–Kutta equations are also analyzed in Hairer and Wanner (1991). The results are supported in the Newton–Kantorovich theory. In particular, we have the difficult task to find good initial guesses to approximate the nonlinear Runge–Kutta equations by Newton-type methods.

The ideas which we would like to introduce for the treatment of singular perturbation problems are based on the analysis of a certain error functional of the form:

$$E(x) = \frac{1}{2}\int_0^T \left|\varepsilon x'(t) - f(x(t))\right|^2 dt \tag{15}$$

to be minimized among the absolutely continuous paths $x:(0,T) \to \mathbb{R}^N$ with $x(0) = x_0$.

Note that if $E(x)$ is finite for one such path x, then automatically, x' is square integrable.

This error functional is associated in a natural way with the original singular perturbation problem.

One main property of this functional is the absence of local minima different from the solution of the problem. Thus, the typical minimization schemes like (steepest) descent methods will work fine as they can never get stuck in local minima, and converge steadily to the solution of the problem, no matter what the initialization is. In our approach, based on optimality conditions, we only need to approximate linear singular perturbation problems.

In particular, the problem of having good initial guesses in the application of Newton-type methods, to approximate the associated nonlinear equations, is avoided. In fact, we will able to ensure global convergence independent of ε.

We should mention that we have already explored in some previous papers this point of view. Since the initial contributions (Amat and Pedregal 2009, 2013), we have also treated the reverse mechanism of using first discretization and then optimality (Amat et al. 2014). We also have addressed, from this viewpoint, other problems like DAEs (Amat, Legaz & Pedregal, 2013) or problems with retarded arguments (Amat Legaz & Pedregal, 2012).

5 COMPUTATIONAL IMPLEMENTATION OF THE VARIATIONAL APPROACH

In order to solve the perturbation equations (13), a new variational approach is used. A theoretical study is presented.

Since this optimization approach is really constructive, iterative numerical procedures are easily implementable.

1. Start with an initial approximation $x^{(0)}(t)$ compatible with the initial conditions (ex. $x^{(0)}(t) = x_0 + tf(x_0)$).
2. Assume the approximation $x^{(j)}(t)$ in [0,T], is known.
3. Compute its derivative $\varepsilon(x^{(j)})(t)$ in [0, T].
4. Compute the auxiliary function $y^{(j)}(t)$ as the numerical solution of the problem

$$\varepsilon y'(t) - \nabla f(x^{(j)}(t))y(t) = f(x^{(j)}(t)) - \varepsilon(x^{(j)})'(t) \quad in(0,T), \quad y(0) = 0 \tag{16}$$

by making use of a numerical scheme for algebraic equations with dense output (like collocation methods).

5. Change $x^{(j)}$ to $x^{(j+1)}$ by using the update formula

$$x^{(j+1)}(t) = x^{(j)}(t) + y^{(j)}(t) \tag{17}$$

6. Iterate (3), (4) and (5), until numerical convergence.

In practice, the stopping criterium used is

$$max\left\{\left\|y^{(j)}\right\|_\infty, \sqrt{2E(x^{(j)})}\right\} \le TOL \tag{18}$$

In particular, this numerical procedure can be implemented, in a very easy way, using a problem-solving environment like MATLAB (MathWorks, 2017).

6 CONCLUSIONS

The type and the importance of offshore structures have been highlighted.

The principal characteristic of a tension leg platforms have been reviewed.

The equation of surge motion has been analyzed.

A variational approach to solving the perturbation equations has been presented. This method never become blocked in a local minima and always converges to the solution sought regardless of the initial point. This is an important improvement over the methods that need a close stating point to the solution sought.

This method reaches the solution with independence of the value of parameter ε. This present a big Improvement respect to classical methods of resolution.

ACKNOWLEDGEMENTS

Research supported in part by Programa de Apoyo a la investigacion de la fundacion Seneca-Agencia de Ciencia y Tecnología de la Region de Murcia 19374/PI/14 and MTM2015-64382-P (MINECO/FEDER).

REFERENCES

Amat S., Legaz M.J. & Pedregal P., 2012. On a Newton-type method for differential algebraic equations. Journal of applied mathematics 2012: 1–15.

Amat S, Legaz MJ, Pedregal P., 2013. Linearizing stiff delay differential equations. Appl Math Inf Sci 2013, 7(1):229232.

Amat S, López DJ, Pedregal P., 2014. An optimization approach for the numerical approximation of differential equations. Optim 2014, 63(3):337–358.

Amat S, Pedregal P., 2009. A variational approach to implicit ODEs and differential inclusions. ESAIM-COCV 2009, 15(1):139–148.

Amat S., Pedregal P., 2013. On a variational approach for the analysis and numerical simulation of ODEs. Discret Contin Dyn Syst. 2013, 33(4):12751291.

Ahmad S. 1996. Stochastic TLP response under long crested random sea. Comput Struct. 1996, 61:975–993.

Chandrasekaran S, Jain AK. 2002. Dynamic behavior of square and triangular offshore tension leg platform under regular wave loads. Ocean Eng. 29:279–313.

Hairer E. and Wanner G., 1991. Solving ordinary differential equations II: stiff and differential algebraic problems. Berlin. Springer-Verlag.

https://upload.wikimedia.org/wikipedia/commons/4/4b/Snorre_A_TLP_illustration_%28NOMF_02764_009%29.jpg.

https://www.globalsecurity.org/military/systems/ship/images/offshore.jpg.

Jain AK. 1997. Nonlinear coupled response of offshore tension leg platforms to regular wave forces. Ocean Eng. 24:577–592.

The MathWorks, 2017. Inc. MATLAB and SIMULINK, Natick, MA.

Progress in Maritime Technology and Engineering – Guedes Soares & Santos (Eds)
© 2018 Taylor & Francis Group, London, ISBN 978-1-138-58539-3

Wave interaction with a rectangular long floating structure over flat bottom

Y. Guo, S.C. Mohapatra & C. Guedes Soares
Centre for Marine Technology and Ocean Engineering (CENTEC), Instituto Superior Técnico, Universidade de Lisboa, Lisbon, Portugal

ABSTRACT: Oblique wave diffraction by a long rectangular rigid floating structure over flat bottom is analysed based on linearized water wave theory. The analytical expressions for the velocity potentials are obtained using the separable variable method along with relevant boundary conditions. The solutions are determined by applying the eigenfunction expansion method with the matching technique. The accuracy of the numerical computation is demonstrated by analysing the convergence of the horizontal wave force. The correctness of the present method is validated by comparing with one available in the literature and the results are in very well-agreed. Several numerical results on horizontal and vertical wave forces acting on the floating structure and reflection and transmission coefficients are analysed in different cases. It is observed that with increase in oblique angle, both the horizontal and vertical forces decreases and the effect of draft and width on the forces are significant and the reflection and transmission coefficients satisfy the energy relation.

1 INTRODUCTION

Over the years, there have been significant studies on the different types of floating breakwater due to their significant importance in the field of ocean and coastal engineering. The floating breakwaters such as cylindrical, box-type, pontoon type, mat-type, and tethered float are typically used to protect small marinas or for shoreline erosion control. One of the most interesting floating breakwater types is the box type which is effective in moderate conditions (as in McCartney 1985). In spite of the large amount of literature on floating breakwaters by both of numerical and experimental, there is a still interest of analytical solutions associated with wave interaction with floating breakwaters to set the benchmarks.

Different methods based on theoretical analysis have been carried out to investigate the various aspects of wave-interaction with floating rigid structures of different geometry. Using variational formulation, the scattering of linear waves normally incident on a rectangular obstacle in a channel of finite water depth was investigated by Mei and Black (1969). The finite element formulations are used by Sannasiraj et al. (1998) to study the diffraction problem, radiation problem, and hydrodynamic forces of a pontoon-type floating breakwaters based on the Galerkin approximation. Andersen and Wuzhou (1985) presented the added mass and damping coefficients for finite water depth based on Green's theorem for arbitrary cross sections. Williams et al. (1997, 2000) solved the fluid motion by the boundary integral equation method using an appropriate Green's function and the reflection coefficient, the exciting forces for floating breakwaters were presented. With simplifying assumptions, the eigenfunction expansion method is used by Drimer et al. (1992) to study the performance of a box-type floating breakwater.

The classical eigenfunction expansion method along with the matching technique has been widely used in the study of wave interaction with floating structures due to its considerable accuracy and less use of the computer memory and as well as processing time. The solutions using the eigenfunction expansion method attends the convergence for the horizontal force in the context of the present work.

Using an eigenfunction expansion method, Abul-Azm and Gesraha (2000) investigated the hydrodynamic properties of long rigid floating pontoon interacting with linear oblique waves in finite water depth. Zheng et al. (2004, 2006) gave analytical expressions for the radiated potentials, wave forces, and hydrodynamic coefficients in beam and oblique waves of finite depth respectively. Gesraha (2004, 2006) investigated the reflection and transmission of incident waves interacting with a π-type floating breakwater in oblique waves without considering the flexibility of the structure. Bhattacharjee and Guedes Soares (2011) analysed the diffraction of oblique waves by a rectangular box-type floating structure near a wall with stepped bottom in both finite water and shallow water approximations. Malara et al. (2012) investigated the random wave field to study the free surface displacement based on eigenfunction expansion matching method. Mohapatra and Guedes Soares (2015) studied the wave forces acting on the floating structure over flat bottom based on Boussinesq-type equations using eigenmode expansion method. Recently, Gadelho et al. (2018) compared the wave forces acting on the floating box and free surface elevations before and after the structure between the analytical, numerical and experimental model results.

On the other hand, recently, there is little study on the rigid horizontal cylindrical floating breakwater based on numerical and experimental methods have been carried out (see Ji et al. 2015, 2016a, 2016b).

Although, more studies are being directed towards linear analysis, the analytical results are still in demand to set the benchmarks. An interesting aspect of wave interaction with floating structure is the study of wave forces and wave angle in order to design for floating breakwater. Therefore, the objective of the present work is to analyse the effect of wave forces and oblique wave angle on the floating rectangular structure analytically.

In order to achieve the objectives of the present study, a problem of oblique wave interaction with a fixed rigid infinitely long floating structure over flat bottom is formulated in finite water depth. The detail of mathematical formulation and solution procedure are described in a lucid manner. The velocity potentials associated with each region are determined using separable of variable method and its solutions are obtained by applying the eigenfunction expansion method along with the matching conditions at the edges of the floating structure.The accuracy of the numerical computation is demonstrated. Further, to verify the method, the obtained results are compared with one by use of the same method but under symmetric and anti-symmetric consideration and observed that are well agreed with each other. Further, several numerical results on horizontal and vertical forces are computed to analyse the effect of oblique angle, draft, and width of the floating structure. In addition, the numerical results on the reflection and transmission coefficients are presented and observed that they satisfy the energy balance relation. It is observed that with the increase of oblique angle, both the horizontal and vertical force decreases.

2 MATHEMATICAL FORMULATION

The mathematical modelling of the problem is three-dimensional Cartesian coordinate system (x, y, z), x–z is being the horizontal plane that coincides with the undisturbed free surface and the y-axis vertical downward positive direction. A rigid rectangular box infinitely long in z-direction floating on the water surface of width $2l$ and draft d, the water depth is h. Therefore, the fluid domain is divided into three regions such as: region 1 ($l < x < \infty$, $0 < y < h$), region 2 ($-l < x < l$, $d < y < h$) and region 3 ($-\infty < x < -l$, $0 < y < h$). A progressive wave with angular frequency ω is obliquely incident on the floating structure making an angle θ with the negative x-axis (see Figure 1). Further, it is assumed that the fluid is inviscid, incompressible and the motion is irrotational, then the velocity potentials exist $\Phi_j(x, y, z, t) = \mathrm{Re}\{\phi_j(x, y)e^{-i\omega t + i\nu z}\}$, where $j = 1, 2$ and 3 refer to region 1, region 2 and region 3, respectively, and $\nu = k_0 \sin\theta$. The velocity potentials ϕ_j in each region include two parts, the incident wave potential ϕ_I and the diffracted wave potential ϕ_{Dj}, it can be expressed as $\phi_j = \phi_I + \phi_{Dj}$. All these velocity potentials satisfy the reduced wave equation as

$$\frac{\partial^2\phi_j}{\partial x^2} + \frac{\partial^2\phi_j}{\partial y^2} - \nu^2\phi_j = 0 \text{ for } j = 1, 2, 3. \tag{1}$$

The linearized incident wave potential is given by

$$\phi_I = \frac{igI_0}{\omega}\frac{\cosh k_0(h-y)}{\cosh k_0 h}e^{-ik_0 x\cos\theta}, \tag{2}$$

where $i = \sqrt{-1}$, I_0 is the incident wave amplitude and h is the water depth. Further, the wave number k_0 satisfies the dispersion relation $\omega^2 = gk_0\tanh(k_0 h)$.

The diffracted potential satisfies the free surface and the bottom boundary conditions as

$$\frac{\partial\phi_{Dj}}{\partial y} + K\phi_{Dj} = 0, y = 0,\ x \in (-\infty, -l) \cup (l, \infty), \tag{3}$$

$$\frac{\partial\phi_{Dj}}{\partial y} = 0, y = h,\ x \in (-\infty, \infty), \tag{4}$$

where $K = \omega^2/g$ with g is the gravity acceleration.

The no flow condition on the structural boundary is given by

$$\frac{\partial\phi_{Dj}}{\partial x} = -\frac{\partial\phi_I}{\partial x},\ y \in (0, d),\ x = l \text{ and } x = -l, \tag{5}$$

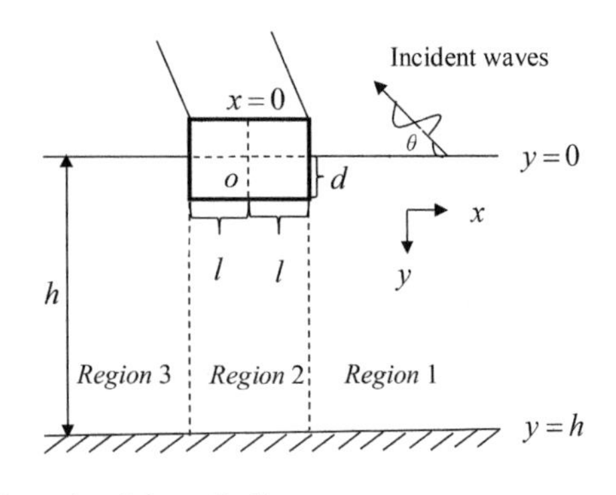

Figure 1. Schematic diagram.

$$\frac{\partial \phi_{Dj}}{\partial y} = -\frac{\partial \phi_I}{\partial y}, y = d,\ x \in (-l, l). \quad (6)$$

The continuity of pressure and velocity across the vertical interfaces at $x = \pm l$ are given by

$$\phi_{D1} = \phi_{D2}, x = l,\ y \in (d, h) \quad (7)$$

$$\phi_{D2} = \phi_{D3}, x = -l,\ y \in (d, h), \quad (8)$$

$$\frac{\partial \phi_{D1}}{\partial x} = \frac{\partial \phi_{D2}}{\partial x}, x = l,\ y \in (d, h), \quad (9)$$

$$\frac{\partial \phi_{D2}}{\partial x} = \frac{\partial \phi_{D3}}{\partial x}, x = -l, y \in (d, h), \quad (10)$$

where ϕ_{D1}, ϕ_{D2} and ϕ_{D3} refer to the spatial diffracted potential of region 1, region 2 and region 3 respectively.

Finally, the far field radiation condition is assumed to take of the form

$$(\phi \approx \{e^{-ik_0 \cos\theta(x-l)} + Ae^{ik_0 \cos\theta(x-l)}\} f_0(y),\ x \to \infty). \quad (11)$$

Next section will describe the solution procedure for the determination of velocity potentials in each region.

3 SOLUTION

Method of separation of variables has been used to obtain the analytical expressions of the diffracted potentials and associated eigenfunctions.

Using Eq. (1) and along with the relevant boundary conditions (3–4), the diffracted potential for region 1 is obtained as

$$\phi_{D1} = \sum_{n=0}^{\infty} A_{1n} \cos\{\lambda_n (h - y)\}\, e^{-\gamma_n (x-l)}, \quad (12)$$

where $\gamma_n^2 = v^2 + \lambda_n^2$.

Since the no flow conditions (6) are non-homogenous and the reduced wave equation and other boundary conditions are linear, the solution to the diffracted potential ϕ_{D2} in region 2 is a problem of partial differential equation with non-homogenous boundary conditions. Therefore, it consists of a general solution and a particular solution. With the same method as in region 1, the diffracted potential in region 2 can be obtained as

$$\phi_{D2} = -\phi_I + \sum_{n=0}^{\infty} \{A_{2n} e^{\alpha_n (x+l)} + A'_{2n} e^{-\alpha_n (x-l)}\} \cos p_n (h - y). \quad (13)$$

In case of region 3, proceeding in a similar manner as in case of region 1, diffracted potential in region 3 can be obtained as

$$\phi_{D3} = \sum_{n=0}^{\infty} A_{3n} \cos\{\lambda_n (h - y)\} e^{\gamma_n (x+l)}. \quad (14)$$

Considering velocity potential in every region include two parts one is for incident wave potential and other one is for the diffracted wave potential, the total velocity potential in every region can be expressed as

$$\phi_1 = \{(igI_0 / \omega) e^{-ik_0 \cos\theta(x-l)} + A_0 e^{ik_0 \cos\theta(x-l)}\} f_0(y) + \sum_{n=1}^{\infty} A_n e^{-\gamma_n (x-l)} f_n(y), \quad (15)$$

$$\phi_2 = \sum_{n=0}^{\infty} \{C_n \cosh(\alpha_n x) + D_n \sinh(\alpha_n x)\} \chi_n(y), \quad (16)$$

$$\phi_3 = \{(igI_0 / \omega) e^{-ik_0 \cos\theta(x+l)} + B_0 e^{-ik_0 \cos\theta(x+l)}\} f_0(y) + \sum_{n=1}^{\infty} B_n e^{\gamma_n (x+l)} f_n(y), \quad (17)$$

where

$$f_0(y) = \frac{\cosh k_0 (h - y)}{\cosh(k_0 h)},\ n = 0, \quad (18)$$

$$f_n(y) = \frac{\cos \lambda_n (h - y)}{\cos \lambda_n h},\ n = 1, 2, \ldots, \quad (19)$$

$$\chi_n(y) = \cos p_n (h - y),\ n = 0, 1, 2, \ldots, \quad (20)$$

$$\lambda_0 = -ik_0, \omega^2 = gk_0 \tanh(k_0 h) \text{ for } n = 0, \quad (21)$$

$$\omega^2 = -g\lambda_n \tan(\lambda_n h), n = 1, 2, \ldots \quad (22)$$

$$\gamma_n = \begin{cases} -ik_0 \cos\theta, & n = 0 \\ \sqrt{(k_0 \sin\theta)^2 + \lambda_n^2}, & n = 1, 2 \ldots \end{cases}, \quad (23)$$

$$p_n = \frac{n\pi}{h - d},\ n = 0, 1, 2, \ldots, \quad (24)$$

$$\alpha_n = \sqrt{(k_0 \sin\theta)^2 + p_n^2},\ n = 0, 1, 2, \ldots. \quad (25)$$

The eigenfunctions are orthogonal with respect to their own inner products. Therefore, the eigenfunctions of f_n and χ_n for $n = 0, 1, 2, \ldots$ are orthogonal as given by

$$\langle f_m, f_n \rangle = \int_0^h f_m(y) f_n(y) dy = \begin{cases} 0, & m \neq n, \\ F_n, & m = n, \end{cases} \quad (26)$$

$$\langle \chi_m, \chi_n \rangle = \int_d^h \chi_m(y) \chi_n(y) dy = \begin{cases} 0, & m \neq n, \\ X_n, & m = n, \end{cases} \quad (27)$$

where

$$F_n = \begin{cases} \dfrac{2k_0 h + \sinh(2k_0 h)}{4k_0 \cosh^2(k_0 h)}, & n = 0, \\ \dfrac{2\lambda_n h + \sin(2\lambda_n h)}{4\lambda_n \cos^2(\lambda_n h)}, & n = 1, 2, \ldots \end{cases} \quad (28)$$

$$X_n = \begin{cases} h-d & \text{for } n=0, \\ (h-d)/2 & \text{for } n=1,2,... \end{cases} \tag{29}$$

The continuity conditions for pressure as given by Eq. (7) at $x = l$ yields

$$(A_0 + \mathrm{i}gI_0/\omega)f_0(y) + \sum_{n=1}^{\infty} A_n f_n(y) = \sum_{n=0}^{\infty}\{C_n \cosh(l\alpha_n) + D_n \sinh(l\alpha_n)\}\chi_n(y). \tag{30}$$

Multiplying $\chi_m(y)$ in Eq. (30) and integrating from d to h along with the orthogonal relations, result in

$$\left(A_0 + \frac{\mathrm{i}gI_0}{\omega}\right)\int_d^h f_0(y)\chi_m(y)dy + \sum_{n=1}^{\infty} A_n \int_d^h f_n(y)\chi_m(y)dy = \{C_m \cosh(l\alpha_m) + D_m \sinh(l\alpha_m)\}X_m, \tag{31}$$

where $m = 0, 1, 2, \ldots$. The continuity conditions for horizontal velocity as given by Eq. (9) at $x = l$ yields

$$\mathrm{i}k_0 \cos\theta(A_0 - \mathrm{i}gI_0/\omega)f_0(y) - \sum_{n=1}^{\infty} A_n\gamma_n f_n(y) = \sum_{n=0}^{\infty}\alpha_n\{C_n \sinh(l\alpha_n) + D_n \cosh(l\alpha_n)\}\chi_n(y). \tag{32}$$

Multiplying both sides of Eq. (32) by the eigenfunction $f_m(y)$ and integrating from d to h, result in for $m = 0, 1, 2, \ldots$ as

$$\mathrm{i}k_0 \cos\theta\left(A_0 - \frac{\mathrm{i}gI_0}{\omega}\right)\int_d^h f_0(y)f_m(y)dy - \sum_{n=1}^{\infty} A_n\gamma_n \int_d^h f_n(y)f_m(y)dy = \sum_{n=0}^{\infty}\alpha_n\{C_n \sinh(l\alpha_n) + D_n \cosh(l\alpha_n)\}\int_d^h \chi_n(y)f_m(y)dy. \tag{33}$$

Application of the no flow condition (5) at $x = l$ yields

$$\mathrm{i}k_0 \cos\theta(A_0 - \mathrm{i}gI_0/\omega)f_0(y) - \sum_{n=1}^{\infty} A_n\gamma_n f_n(y) = 0. \tag{34}$$

Multiplying both sides of Eq. (34) by $f_m(y)$ and integrating from 0 to d, obtained as

$$\mathrm{i}k_0 \cos\theta\left(A_0 - \frac{\mathrm{i}gI_0}{\omega}\right)\int_0^d f_0(y)f_m(y)dy - \sum_{n=1}^{\infty} A_n\gamma_n \int_0^d f_n(y)f_m(y) = 0, \; m = 0,1,2,... \tag{35}$$

Adding the Eq. (35) to Eq. (33) for $m = 0$ and $m = 1, 2, \ldots$, yields

$$\mathrm{i}k_0 \cos\theta(A_0 - \mathrm{i}gI_0/\omega)F_0 = \sum_{n=0}^{\infty}\alpha_n\{C_n \sinh(l\alpha_n) + D_n \cosh(l\alpha_n)\}\int_d^h \chi_n(y)f_0(y)dy \tag{36}$$

$$-A_m\gamma_m F_m = \sum_{n=0}^{\infty}\alpha_n\{C_n \sinh(l\alpha_n) + D_n \cosh(l\alpha_n)\}\int_d^h \chi_n(y)f_m(y)dy \tag{37}$$

Similarly, using the continuity conditions of pressure and velocity in Eqs. (8) and (10) at $x = -l$ and the no flow condition (5) at $x = -l$ as well as the corresponding eigenfunctions and the orthogonal relationship, obtained the following equations for $m = 0,1,2,\ldots$ as

$$\left(\frac{\mathrm{i}gI_0}{\omega} + B_0\right)\int_d^h f_0(y)\chi_m(y)dy + \sum_{n=1}^{\infty} B_n \int_d^h f_n(y)\chi_m(y)dy = \{C_m \cosh(l\alpha_m) - D_m \sinh(l\alpha_m)\}X_m, \tag{38}$$

$$-\mathrm{i}k_0 \cos\theta(B_0 + \mathrm{i}gI_0/\omega)F_0 = \sum_{n=0}^{\infty}\alpha_n\{-C_n \sinh(l\alpha_n) + D_n \cosh(l\alpha_n)\} \times \int_d^h \chi_n(y)f_0(y)dy, \; m = 0, \tag{39}$$

$$B_m\gamma_m F_m = \sum_{n=0}^{\infty}\alpha_n\{-C_n \sinh(l\alpha_n) + D_n \cosh(l\alpha_n)\} \times \int_d^h \chi_n(y)f_m(y)dy, \; m = 1,2,..... \tag{40}$$

The infinite series must be truncated to finite number of term N in order to obtain the specific solutions of the unknown coefficients A_n, B_n, C_n and D_n. Then, a linear system of $4(N+1)$ algebraic equations can be obtained to solve those unknown coefficients for $n = 0, 1, 2, \ldots, N$. Once the unknown coefficients have been determined, the full velocity potentials in each region will be determined. Therefore, the wave forces, the reflection and transmission coefficients can be computed.

4 WAVE FORCES, REFLECTION AND TRANSMISSION COEFFICIENTS

Under the consideration of the wave forces acting on a unit length of the structure along the y-direction, the wave force includes two parts, the vertical force and the horizontal force, which can be obtained as

$$F_v = -\int_{-l}^{l} \mathrm{i}\rho\omega\phi_2 dx, \tag{41}$$

$$F_h = -\int_0^d i\rho\omega\phi_1 dy + \int_0^d i\rho\omega\phi_3 dy. \quad (42)$$

Substituting expressions (15)–(17) for the velocity potentials into Eqs. (41)–(42), the vertical and horizontal wave force can be expressed as

$$F_v = -i\rho\omega\sum_{n=0}^{\infty}\frac{2C_n}{\alpha_n}\sinh(l\alpha_n)\cos p_n(h-d), \quad (43)$$

$$F_h = i\rho\omega\left[\frac{B_0 - A_0}{k_0\cosh(k_0 h)}\{\sinh(k_0 h) - \sinh k_0(\mathrm{h-d})\}\right] + i\rho\omega\left[\sum_{n=1}^{\infty}\frac{B_n - A_n}{\lambda_n\cos(\lambda_n h)}\{\sin(\lambda_n h) - \sin\lambda_n(\mathrm{h-d})\}\right]. \quad (44)$$

In addition, the reflection coefficient K_r and transmission coefficient K_t can be computed by using the following formulae:

$$K_r = |i\omega A_0/(gI_0)|, \quad (45a)$$

$$K_t = |1 - i\omega B_0/(gI_0)|. \quad (45b)$$

Next section will present several numerical results on horizontal and vertical wave forces acting on the floating structure will be presented.

5 NUMERICAL RESULTS AND DISCUSSION

From Table 1, it is clear that the horizontal force converges up to three-decimal places for $N \geq 30$. Hence, the number of terms in the series solution is restricted to $N = 30$. Therefore, all numerical computations on horizontal force, vertical force, reflection coefficient and transmission coefficient are performed throughout the paper by considering $N = 30$ and the water density $\rho = 1025$ kgm^{-3} and acceleration due to gravity $g = 9.8$ ms^{-2} unless stated otherwise.

In order to check the correctness of the analytical method, the present results are compared with one relevant as in Abul-Azm and Gesraha (2000) for horizontal and vertical wave forces in Figure 2. It is observed that the present result is well-agreed with each other.

Figure 2 shows the comparison between the present and the results available in the literature for (a) horizontal and (b) vertical forces. The comparison is made for three different incident wave angles on exciting wave forces as a function of wave number kh. As for the case of $\theta = 0°$, the result of $\theta = 1°$ is used to replace as the wave is not beam. It is observed that the exciting wave forces obtained by the present method are well-agreed with Abul-Azm and Gesraha (2000).

Table 1. Convergence study for non-dimensional horizontal force $|F_h|/(2\rho g l I_0)$ for $l/h = 0.125$, $d/h = 0.875$ and $\theta = 60°$.

Non-dimensional frequency ($\omega\sqrt{d}/\sqrt{g}$)	N	Non-dimensional Horizontal force
1.483	5	2.8980
	10	2.8998
	15	2.9006
	20	2.9012
	25	2.9014
	30	2.9015
	31	2.9015
	32	2.9016
1.078	5	4.3134
	10	4.3197
	15	4.3225
	20	4.3245
	25	4.3250
	30	4.3255
	31	4.3255
	32	4.3256
0.847	5	4.7048
	10	4.7198
	15	4.7264
	20	4.7313
	25	4.7324
	30	4.7336
	31	4.7337
	32	4.7337

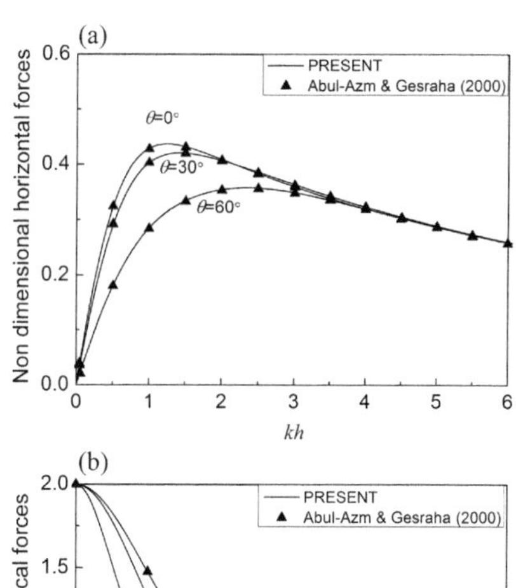

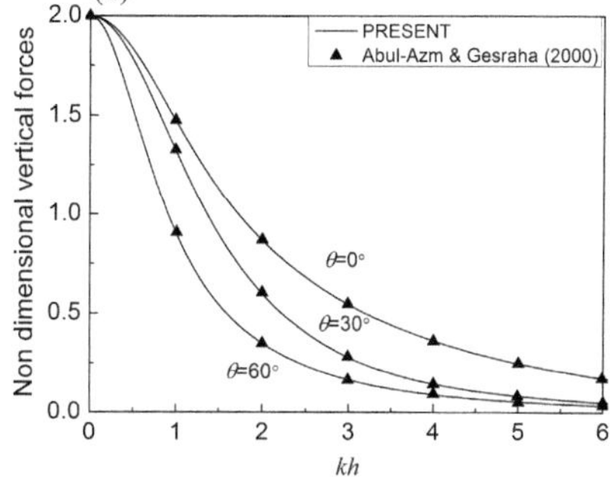

Figure 2. Comparison between the present (a) horizontal force and (b) vertical force and Abul-Azm and Gesraha (2000) for $l/h = 1.0$ and $d/h = 0.25$.

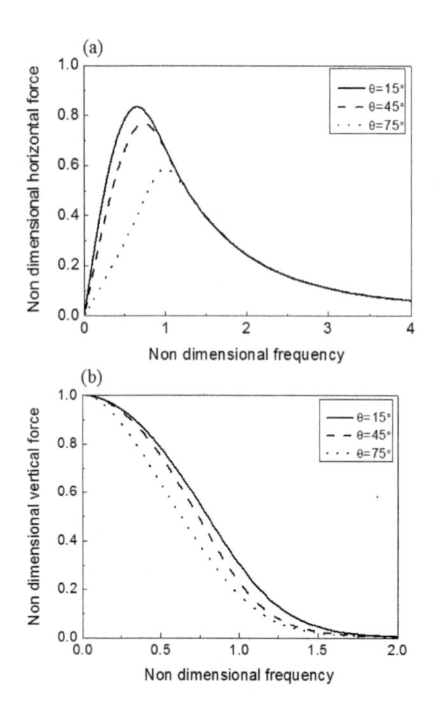

Figure 3. Variations of the (a) horizontal force and (b) vertical forces for different angle of incidence with $l/h = 0.5$ and $d/h = 0.5$.

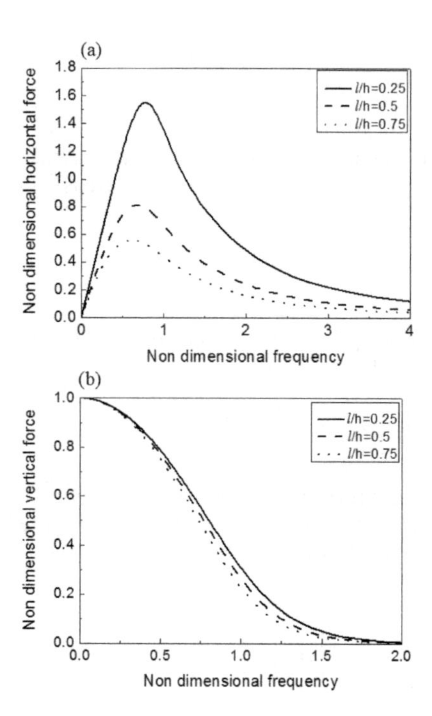

Figure 4. Effect of length on non-dimensional (a) horizontal force and (b) vertical force for $d/h = 0.5$ and $\theta = 30°$.

Figure 3 plots the non-dimensional (a) horizontal force $|F_h|/(2\rho glI_0)$ and (b) vertical force $|F_v|/(2\rho gl I_0)$ versus non-dimensional frequency $\omega\sqrt{d}/\sqrt{g}$ for different incident wave angles. Clearly, it is seen that the effect of incident wave angle on horizontal force is negligible as the frequency increases whilst, the variation and peak is occurred for smaller non-dimensional frequency. The vertical force keeps on decreasing as the incident angle increases. Further, the angle of incidence has great effect on the exciting wave forces for lower non-dimension frequencies (longer waves) and the effects on the exciting wave forces can be ignored for higher frequencies (shorter waves).

Figure 4 represents the variation of non-dimensional (a) horizontal force $|F_h|/(2\rho glI_0)$ and (b) vertical force $|F_v|/(2\rho glI_0)$ for different non-dimensional width of the structure with $d/h = 0.5$ and $\theta = 30°$ versus non-dimensional frequency $\omega\sqrt{d}/\sqrt{g}$. The horizontal force decreases with increase in width and peak occurred for a certain value of non-dimensional frequency. This may be due the fact that the interaction of incident and reflected waves in region 1. The variation of vertical forces for different width is negligible. It may be observed that without attending peak, as the non-dimensional frequency increases the vertical force keep on decreasing.

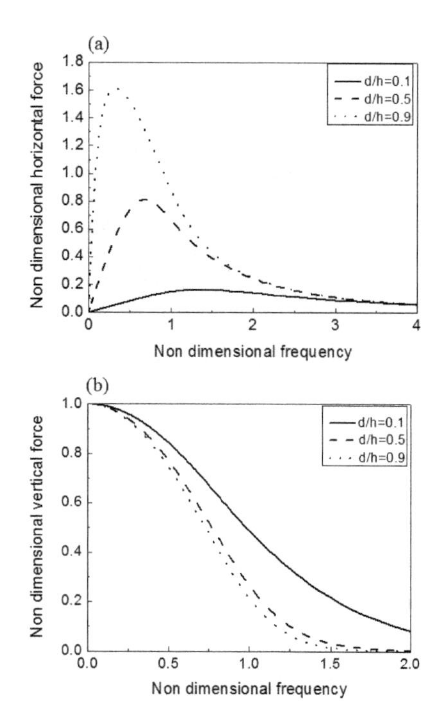

Figure 5. Effect of non-dimensional (a) horizontal and (b) vertical force for different non-dimensional draft with $l/h = 0.5$ and $\theta = 30°$.

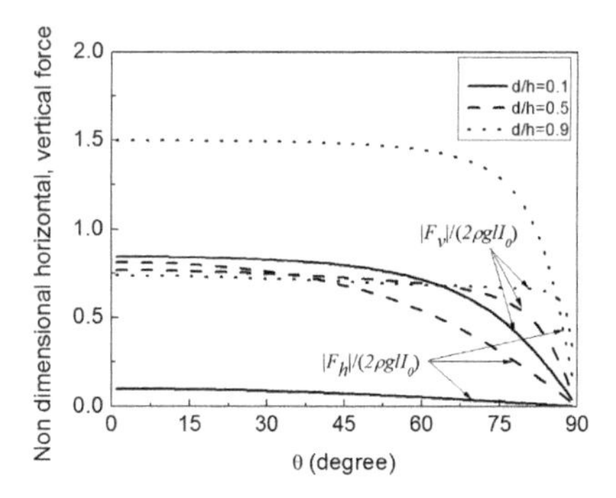

Figure 6. Comparsion of horizontal and vertical wave force on different draft for $\omega\sqrt{l}/\sqrt{g} = 0.527$ and $l/h = 0.5$.

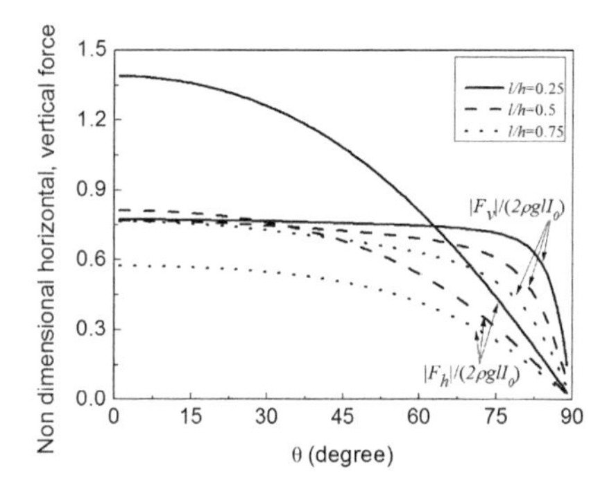

Figure 7. Comparison of horizontal force and vertical force on different width for $\omega\sqrt{d}/\sqrt{g} = 0.527$ and $d/h = 0.5$.

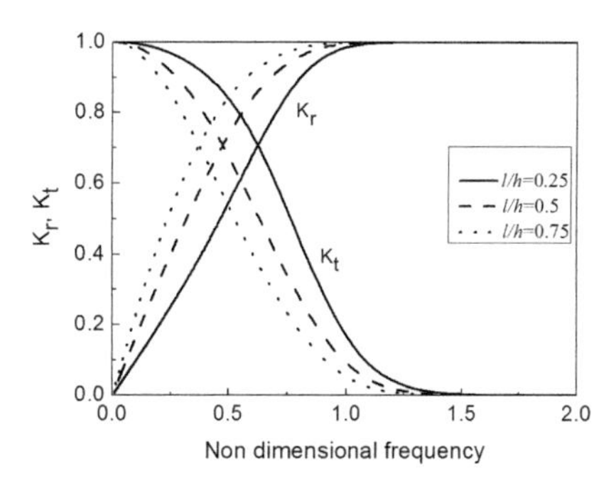

Figure 8. Variations of the K_r and K_t for different non-dimensional width with $d/h = 0.5$ and $\theta = 30°$.

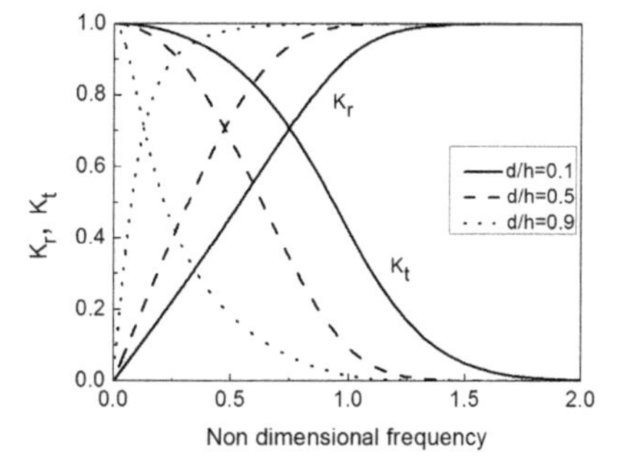

Figure 9. Variation of the K_r and K_t for different non-dimensional draft with $l/h = 0.5$ and $\theta = 30°$.

Figure 5 plots the variation of the non-dimensional (a) horizontal force $|F_h|/(2\rho glI_0)$ and (b) vertical force $|F_v|/(2\rho glI_0)$ for different draft of the structure versus non-dimensional frequency $\omega\sqrt{l}/\sqrt{g}$ with $l/h = 0.5$ and $\theta = 30°$. It is observed that the horizontal force increases with increase in the value of draft. However, without attending peak the pattern of vertical force is opposite in trend to that of horizontal force. This may be due to the deeper drafts causes more horizontal load on the structure and smaller load in vertical.

Figure 6 depicts the effect of incident angle on the non-dimensional horizontal force $|F_h|/(2\rho glI_0)$ and vertical force $|F_v|/(2\rho glI_0)$ for different values of the draft with $\omega\sqrt{l}/\sqrt{g} = 0.527$ and $l/h = 0.5$. It is found that the horizontal force increases with the increase in draft. However, the horizontal force becomes zero for $\theta = 90°$ this concludes that the incident wave angle is parallel to the floating structure which leads no effect.

Figure 7 shows the comparison of the non-dimensional horizontal force $|F_h|/(2\rho glI_0)$ and vertical force $|F_v|/(2\rho glI_0)$ versus θ for different width of the structure with $\omega\sqrt{d}/\sqrt{g} = 0.527$ and $d/h = 0.5$. The horizontal force increases with decrease in width of the structure. Further, the observation in horizontal force for incident wave angle is similar to that of Figure 6. However, the effect/variation in vertical force is negligible for smaller wave angle. From Figures (6–7), it may be concluded that the vertical force is higher than the horizontal force for higher wave angle.

In Figure 8, the variations of the reflection coefficient K_r and transmission coefficient K_t versus $\omega\sqrt{d}/\sqrt{g}$ for different width of the structure with $d/h = 0.5$ and $\theta = 30°$ are plotted. The K_r increases with increase in the width of the structure whilst, the pattern of K_t is opposite to that of K_r. This suggests that wider structure reflect more wave energy and wave transmission is less.

Figure 9 shows the variations of the reflection coefficient K_r and transmission coefficient K_t versus non-dimensional frequency $\omega\sqrt{l}/\sqrt{g}$ for different drafts of the structure with $l/h = 0.5$ and $\theta = 30°$. It is observed that with increase in the non-dimensional frequency the K_r increases and K_t decreases. It can be seen that structures with deeper drafts reflect more waves and transmit lesser waves. It is clear from the Figure (8–9) that the K_r and K_t satisfy the energy relation $K_r^2 + K_t^2 = 1$.

6 CONCLUSIONS

This paper dealt with wave diffraction by a long rectangular floating structure over flat bottom under linearized water wave theory in finite water depth. The analytical solutions of the BVP are obtained using the eigenfunction expansion method and along with the application of orthogonal relations. The accuracy of the numerical computation is studied. Further, in order check the correctness of the method, the obtained results are compared with one available in the literature. Numerical results on the horizontal and vertical wave forces, reflection and transmission coefficients are studied to analyze the effect of oblique wave angle, draft and width of the structure in different cases.

In numerical results, it is observed that the results of the present method are well-agreed with one available in literature. The exciting forces decrease with the increase of the oblique wave angle. In addition, the draft and the half width of the structure as well as the wave frequency have great influence on the exciting forces too. In case of reflection and transmission coefficients, the more waves gets reflected for larger width of the structure whilst, the observation is opposite in transmission coefficient to that of reflection coefficient. It is cleared from the numerical results of reflection and transmission coefficients that it satisfies the energy balance relation $K_r^2 + K_t^2 = 1$. The obtained results may be helpful to understand the accuracy of the method and design of floating breakwater. Furthermore, the present mathematical formulation can be modified/generalized to the problem of wave-interaction with a 3D (three-dimensional) floating structure of cylindrical geometry over the analytical and numerical methods in the coming future.

ACKNOWLEDGEMENTS

This work was performed within the Strategic Research Plan of the Centre for Marine Technology and Ocean Engineering which is financed by Portuguese Foundation for Science and Technology (Fundação para a Ciência e Tecnologia-FCT).

REFERENCES

Abul-Azm, A.G. & Gesraha, M.R. 2000. Approximation to the hydrodynamics of floating pontoons under oblique waves. *Ocean Engineering* 27: 365–384.

Andersen, P. & Wuzhou, He. 1985. On the Calculation of Two-Dimensional Added Mass and Damping Coefficients by Simple Green's Function Technique. *Ocean Engineering* 12 (5): 425–451.

Bhattacharjee, J. & Guedes Soares, C. 2011. Oblique wave interaction with a floating structure near a wall with stepped bottom. *Ocean Engineering* 38: 1528–1544.

Drimer, N., Agnon, Y. & Stiassnie, M. 1992. A simplified analytical model for a floating breakwater in water of finite depth, *Applied Ocean Research* 14: 33–41.

Gadelho, J.F.M.; Mohapatra, S.C., and Guedes Soares, C. 2018. CFD analysis of a fixed floating box-type structure under regular waves. *In*: Guedes Soares, C. and Ângelo P. Teixeira, (eds.), *Developments in Maritime Transportation and Harvesting of Sea Resources*. London: Taylor and Francis Group, pp. 513–520.

Gesraha, M.R. 2006. Analysis of π shaped floating breakwater in oblique waves: I. Impervious rigid wave boards. *Applied Ocean Research* 28: 327–338.

Gesraha, Mohamed R. 2004. An eigenfunction expansion solution for extremely flexible floating pontoons in oblique waves. *Applied Ocean Research* 26:171–192.

Ji, C.Y., Chen, X., Cui, J., Gaidai, O. & Incecik, A. 2016a. Experimental study on configuration optimization of floating breakwaters. *Ocean Engineering* 117: 302–310.

Ji, C.Y., Chen, X., Cui, J., Yuan, Z.M. & Incecik, A. 2015. Experimental study of a new type of floating breakwater. *Ocean Engineering* 105: 295–303.

Ji, C.Y., Guo, Y.C., Cui, J., Yuan, Z.M. & Ma, X.J. 2016b. 3D experimental study on a cylindrical floating breakwater system. *Ocean Engineering* 125: 38–50.

Malara G., Arena, F. & Spanos, P.D. 2012. On the interaction between random sea waves and a floating structure of rectangular cross section. In: Rizzuto & Guedes Soares (eds), *Sustainable Maritime Transportation and Exploitation of Sea Resources*. London: Taylor & Francis Group, pp. 189–196.

McCartney, B.L. 1985. Floating breakwater design. *Journal of Waterway, Port, Coastal, and Ocean Engineering* 111(2), 304–317.

Mei, C.C. & Black, J.L. 1969. Scattering of Surface Waves by Rectangular Obstacles in Water of Finite Depth. Journal of Fluid Mechanics 38 (3): 499–511.

Mohapatra, S.C. & Guedes Soares, C. 2015. Wave forces on a floating structure over flat bottom based on Boussinesq formulation. *In*: Guedes Soares, C. (ed.), *Renewable Energies Offshore*. London: Taylor & Francis Group, pp. 335–342.

Sannasiraj, S.A., Sundar, V. & Sundaravadivelu, R. 1998. Mooring Forces and Motion Responses of Pontoon-Type Floating Breakwaters. *Ocean Engineering* 25(1):27–48.

Williams, A.N. & Abul-Azm, A.G. 1997. Dual pontoon floating breakwater. *Ocean Engineering* 24(5): 465–478.

Williams, A.N., Lee, H.S. & Huang, Z. 2000. Floating pontoon breakwater. *Ocean Engineering* 27: 221–240.

Zheng, Y.H., Shen, Y.M., You, Y.G., Wu, B.J. & Jie, D.S. 2006. Wave radiation by a floating rectangular structure in oblique seas. *Ocean Engineering* 33:59–81.

Zheng, Y.H., You, Y.G. & Shen, Y.M. 2004. On the radiation and diffraction of water waves by a rectangular buoy. *Ocean Engineering* 33:1063–1082.

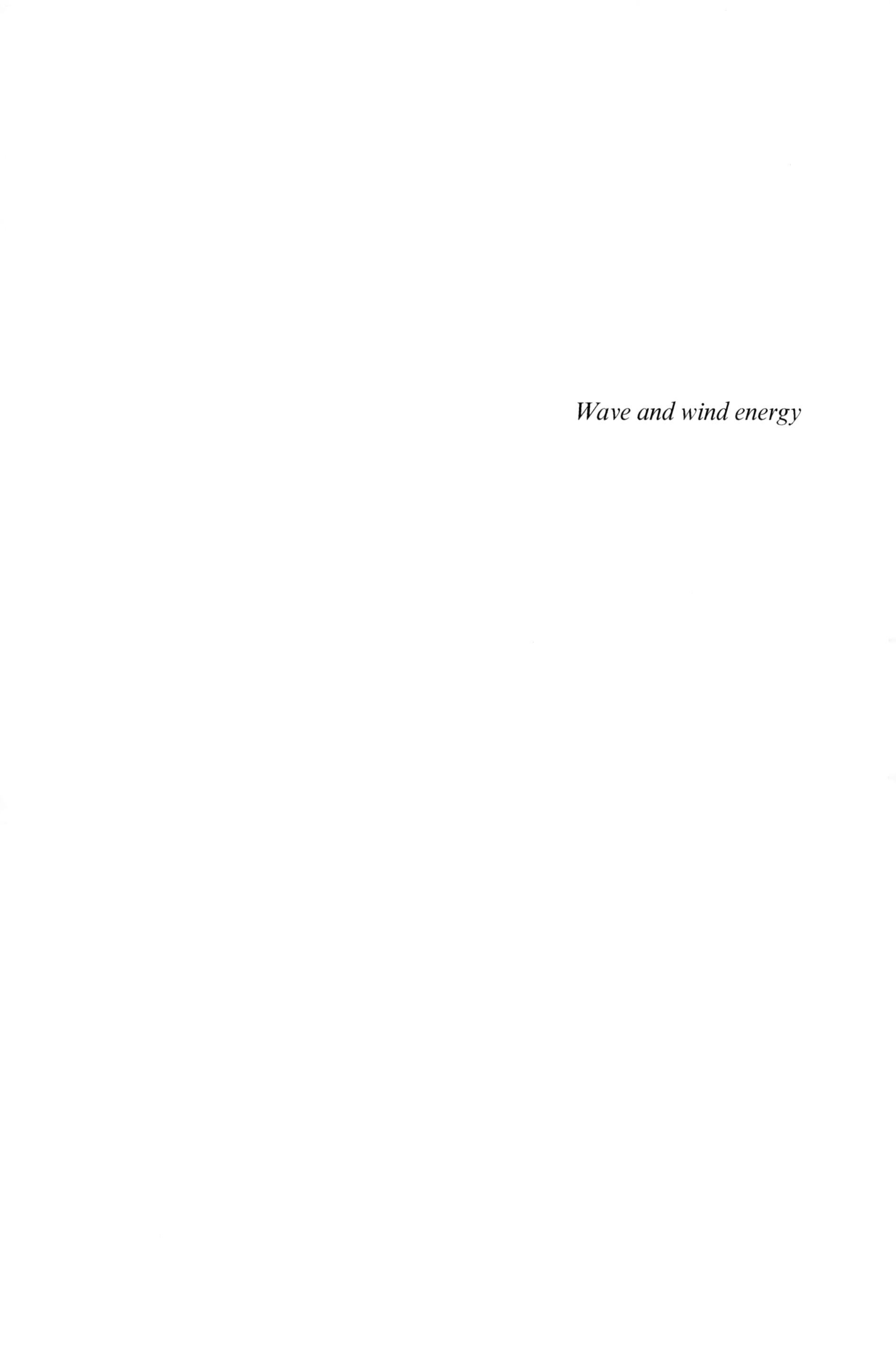

Wave and wind energy

Progress in Maritime Technology and Engineering – Guedes Soares & Santos (Eds)
© 2018 Taylor & Francis Group, London, ISBN 978-1-138-58539-3

Optimization of wave energy converters in the OPWEC project

F. Taveira-Pinto & P. Rosa-Santos
Interdisciplinary Centre of Marine and Environmental Research, Faculty of Engineering, University of Porto, Porto, Portugal

C.A. Rodríguez
Interdisciplinary Centre of Marine and Environmental Research, Faculty of Engineering, University of Porto, Porto, Portugal
Laboratory of Ocean Technology (Laboceano), Department of Naval Engineering and Ocean Engineering, Federal University of Rio de Janeiro, Rio de Janeiro, Brazil

M. López
Department of Construction and Manufacturing Engineering, University of Oviedo, Spain

V. Ramos
Universidade de Santiago de Compostela, EPS, Lugo, Spain,

S. Xu, K. Rezanejad, S. Wang & C. Guedes Soares
Centre for Marine Technology and Ocean Engineering (CENTEC), Instituto Superior Técnico, Universidade de Lisboa, Lisboa, Portugal,

ABSTRACT: The R&D project OPWEC—Optimizing Wave Energy Converters intends to develop, optimize and technically evaluate two different technologies to harness wave energy, using a composite modelling approach that combines physical and numerical models. The paper describes the main objectives of the project and the most significant features of one of the concepts – the CECO wave energy converter – with respect to its hydrodynamic response, as well as the implementation and validation of a time domain model. In addition, the next research steps are presented and discussed.

1 INTRODUCTION

The world ever-growing need for sustainable, non-polluting electricity led to the development of a wide variety of technologies to harvest wave energy, but only a few of them reached a demonstration testing phase at (near) full scale, in real sea conditions, after several years of R&D studies, in order to minimize investment risks. However, none is economically competitive yet (Day *et al.*, 2015).

The harsh marine environment and the space- and time-variability of the wave characteristics still pose several challenges to the development of wave energy technologies. However, the huge amount of untapped wave resource clearly justifies efforts and investments in the development of novel and existing technologies in order to reach a higher level of efficiency and reliability, to bring electricity production costs down to levels that could compete with the traditional sources (Taveira Pinto *et al.*, 2015).

This was the main motivation of the R&D project OPWEC—Optimizing Wave Energy Converters, focused on the development, optimization and preliminary economic viability assessment of two wave energy-harvesting technologies, with different working principles and application conditions. However, both technologies require the study of appropriate mooring systems and foundations for deep-water applications, where the wave resource is higher, and the development and the implementation of numerical modelling approaches to allow the testing and the optimization of the power-take-off systems in a realistic way.

The concepts being studied are the Floating Oscillating Water Column (FOWC) and the CECO wave energy converter. Oscillating Water Column (OWC) devices are amongst the concepts most successfully deployed worldwide, but the higher wave energy resource offshore justifies the study of FOWC deployed in deep waters. Some advances have been made in the study and improvement of the performance of bottom fixed OWC by changing its geometry and also the entrance with a step change in bottom surface (Rezanejad *et al.*,

2013, 2017). Furthermore the concept of multiple chambers was also studied as a way to improve the efficiency of the device (Rezanejad *et al.*, 2015, 2016). The experience and ideas obtained in this study have begun to be explored in floating devices (Rezanejad and Guedes Soares, 2015) and this project will allow pursuing the study of the floating solution of the OWC, which includes also the mooring solution (Zaroudi *et al.*, 2015). The initial steps in the development of this concept were concentrated on experimental work and modelling validation of moorings, which are described in an accompanying paper (Xu *et al.*, 2018).

The CECO concept is also very promising as it was designed to absorb, simultaneously, the kinetic and the potential wave energy and presents a simple wave-to-wire energy conversion principle.

The present paper describes the OPWEC project and selected results resulting from the development of CECO wave energy converter, as well as the next R&D steps.

2 OPWEC PROJECT

The OPWEC project intends to develop and technically assess two different technologies for wave energy harnessing, using either numerical or physical modelling: CECO and a FOWC.

The development and optimization of those two technologies has to take into account: (i) the hydrodynamic behavior of the Wave Energy Converter (WEC), (ii) the dynamics of its mooring system (if applied), (iii) the influence of the PTO, and (iv) the characteristics of the wave resource at the installation site. Due to the cross interference between those components, a wave-to-wire modelling approach is advisable.

Therefore, the work involves the application of a suite of numerical models, to analyze all components at once and estimating the efficiency and power generated from given local wave conditions. However, experimental data is needed for calibration and validation of the numerical models. Figure 1 presents the work plan of the OPWEC R&D project.

In the project, both 2D and 3D Boundary Equation Methods (BEM) based on linear potential flow theory are applied to study the interaction between waves and WECs, in normal operation, during which phenomena such as wave slamming or overtopping are not expected to occur. In fact, the BEM-based models, also known as panel models, might present some discrepancies with reality, especially for rough wave conditions, because of non-linear and viscous effects (Barbarit *et al.*, 2012). Nevertheless, due to the good compromise between the accuracy and computational cost,

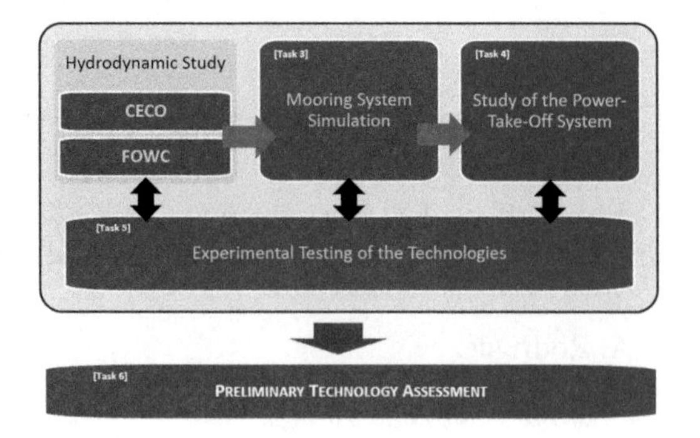

Figure 1. Work plan of the OPWEC R&D project.

BEM models are still the most recommended for investigating the behavior of large structures in waves (Payne *et al.*, 2008). In addition, panel models have already been used in the wave energy field to model different types of WECs, including point absorbers, such as single buoys (Pastor *et al.*, 2014), multibody floating devices (De Andres *et al.*, 2013) or the combined motion technologies (Rhinefrank *et al.*, 2011).

Generally, the application of BEM models to create a frequency domain model of the WEC is a faster but sufficiently accurate approach to obtain relevant insight in the system dynamics. Hence, the application of frequency domain models is also included in the project, since a wide range of conditions is to be tested. On the other hand, time domain methods are applied to account for non-linear effects associated to the PTO or the mooring system, to determine non-linear motion responses as well as to include any additional physical constraint. Integrated numerical approaches have to be used, as the WEC and mooring system dynamics are interdependent, *i.e.*, the mooring loads depend on WEC motions that, in turn, are conditioned by the mooring system. This logic is also valid for the modelling of the PTO system.

With this reasoning, the project was divided into 6 tasks. An initial hydrodynamic study of both technologies is carried out in Tasks 1 and 2. Task 3 deals with the numerical simulation of the mooring system and its effect on the WEC's response. In Task 4, the PTO system is considered and its influence in the response of the WECs in the time domain is assessed. Since the main interest is in the overall system dynamics, the PTO system will be modeled in a simplified way, to avoid describing the whole electrical system.

Firstly, the numerical models will assist in the definition of the PTO and mooring system characteristics and PTO control strategies to be reproduced in the physical model tests (*e.g.*, the air-flow control of turbines, type of generator and associated damping)

in Task 5, which, in turn, will provide experimental data for numerical model calibration and validation, still within Task 4. Specific mooring systems will be selected for both technologies (different mooring requirements), and then fine-tuned using the numerical models and later on reproduced in the experiments of Task 5.

In the composite modelling approach described, the numerical models of the two technologies are enhanced to properly reproduce their hydrodynamic response and to include the non-linear effects related to the mooring system and the PTO in the time domain. Then, the numerical models will be applied to find optimal operating conditions and define control strategies to maximize the power output, for specific conditions and to correctly assess the performance of the devices in Task 6.

The Preliminary Technology Assessment (Task 6) involves the detailed analysis of the mooring system and the PTO system characteristics and control strategies, in order to improve and optimize the performance of the two technologies for the conditions of the case studies. Finally, a preliminary economic viability assessment of both technologies will be carried out, which requires the estimation of the electricity production based on the WECs' power matrices, obtained from numerical and experimental results together with the available wave resource estimated at the selected sea sites. Estimations of construction, installation and decommissioning costs will also be included. Due to the lack of full-scale data, some scenarios in terms of O&M costs for project lifetime will be considered in the economic viability assessment.

Some of the most significant conclusions regarding the development of CECO with respect to its hydrodynamic response and implementation and validation of numerical models are presented next. In addition, the next research steps are also discussed.

3 CECO WAVE ENERGY CONVERTER

3.1 *Description of the concept*

CECO is a WEC designed to convert simultaneously the kinetic and the potential energy of ocean waves into electricity. It is composed of a floating part that moves under the action of waves in relation to a central element, which should be as fixed as possible, to explore the relative motions between the two parts, Figure 2.

The most innovative characteristic of this WEC is the oblique motion of its floating part, composed of two Lateral Mobile Modules (LMM), Figure 2. CECO may be installed near the coastline, attached to piles and other bottom-fixed nearshore structures, and offshore, in deep waters, fixed to an

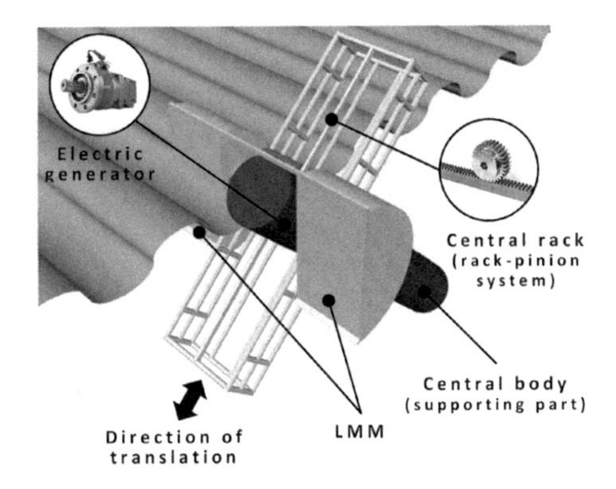

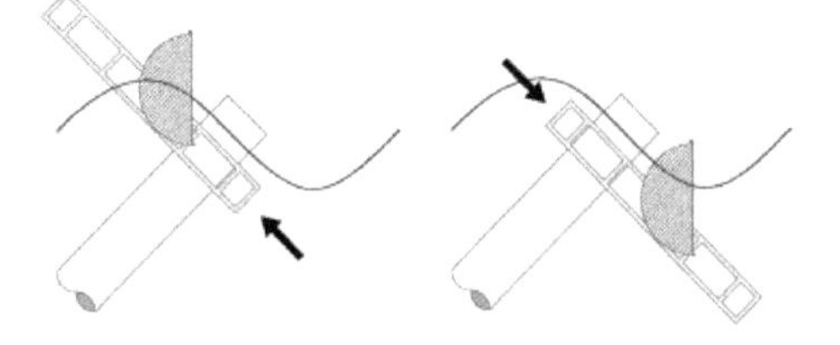

Figure 2. Three-dimensional sketch of the CECO concept (upper panel) and CECO motion under wave action (lower panel).

offshore structure. Until now, the R&D works carried out only involved the study of CECO attached to fixed supports (*e.g.*, Rosa-Santos *et al.*, 2015; López *et al.*, 2017a,b).

The additional strengths of CECO are its scalability, simple wave-to-wire energy conversion schema, production with standard components, flexibility of installation and high compatibility with other marine energy harnessing technologies and structures. This would result in synergies in terms of construction, O&M costs, but also in the sharing of the same physical space and infrastructures and in a more even combined electrical power output.

The inclination and direction of CECO's motions can be modified in according to the characteristics of the incoming waves, in order to ensure the best possible performance of this one degree of freedom device. This can be done by changing the inclination of the supporting elements and by aligning the device with the direction of incident waves.

The classification of CECO might not be straightforward. Its dimensions are quite small in comparison to typical wavelengths, but most point absorbers (*e.g.*, heaving-type axisymmetric) are able to naturally absorb wave energy from any direction, whereas CECO has to be properly orientated to waves to not jeopardize its performance. Moreover, point absorbers are usually designed to resonate for typical wave frequencies and present a narrow frequency response curve, with a large peak. The obtained results until now suggest that CECO frequency response curve might be flatter than that of typical point absorber and also more easily tuned, not only by the modification of the mass, geometry and volume of its floating part, but also by changing the inclination of the device. Therefore, it is possible to extend the good efficiencies of CECO to a relatively large range of sea conditions. This will be discussed latter in more detail.

Currently, CECO is at the third Technology Readiness Level and there is ongoing research to move the concept to the next level.

3.2 *Experimental proof of concept studies*

The experimental proof of concept studies of CECO were carried out in the Laboratory of the Hydraulics, Water Resources and Environment Division, of the Faculty of Engineering of the University of Porto (FEUP), Portugal. Two experimental campaigns have been carried out at the FEUP's wave basin (28.0 m long, 12.0 m wide and 1.2 m deep). The tested models were scaled-down reproductions of the idealized CECO prototype built on a geometric scale of 1:20.

The objective of the first study was to validate the concept and working principle of CECO, and to analyze the feasibility of some initial constructive solutions (Teixeira, 2012). In the second study, based on the knowledge acquired in the preceding study, several improvements were introduced in the physical model, mostly in the geometry of the LMM, the guiding system of the main rods, the cross-section of the central element, as well as in the simulation and monitoring of the power-take-off (PTO) system, which started to allow the direct measurement of the instantaneous absorbed power (Marinheiro, 2013).

The first study allowed concluding that the response of CECO occurs, essentially, in the frequency of incident waves and that its performance strongly depends on the incident wave's characteristics. The relative capture width reached at most 14% showing that CECO is a valid technology to harvest wave energy (Rosa-Santos *et al.*, 2015), though the calculated absorbed power did not take yet into account the losses related to the conversion of mechanical energy into electricity.

In the second phase, with an improved model, the electrical potential differences produced by the PTO were measured, hence considering the losses in the conversion of wave energy into electricity. The relative capture widths increased to values in the range between 10 and 30% (Marinheiro *et al.*, 2015). The tested CECO geometry and PTO system seemed to be more favorable for the shortest wave period tested (8 s), with the efficiency decreasing with wave period. The inclination of 45° was the most favorable one for the tested geometry and wave conditions.

It was also concluded that the inclination was an important factor controlling the behavior of CECO. For all tested conditions, results were better for 45° than for 30°, but opposite trends were observed in terms of evolution of the mean absorbed power with the wave period, for the two tested inclinations, suggesting a more complex relationship between the performance of CECO and its inclination, that was not explored at that time. However, it was stated that the 30° angle could become more advantageous for some sea wave conditions (not tested).

The PTO system used in the second study allowed varying the damping level, to simulate different conditions of wave energy extraction. In general, the amplitude of CECO's motions increased with the reduction of the damping. However, larger motions do not mean, necessarily, larger power outputs as the damping introduced in the system is related to the amount of wave energy converted. The results suggested that the damping in the system influences significantly CECO's response and that this parameter should be studied in more detail to determine the most suitable values for the design wave conditions.

The results for regular waves were slightly better than those with irregular waves. Possibly, due to the small dimensions of this WEC, the use of short crested waves, with a mean direction perpendicular to the alignment of its LMM, resulted only in a small reduction of the relative capture widths when compared to those with long crested waves (Marinheiro *et al.*, 2015). To accommodate changes in the incoming wave direction, the central element may be rotated to modify CECO's orientation. This adjustment may be done using a controlled mechanical system or a central element with a streamlined profile designed to orientate automatically CECO with the incoming sea.

Summing up, CECO performance depends significantly on the wave characteristics (wave period and wave height), the PTO damping and the inclination of the device. In the continuity of the experimental proof of concept studies, numerical simulations were used to better understand the CECO response under wave action and to optimize its performance.

4 NUMERICAL MODELLING OF CECO

4.1 *Introduction*

The most recent developments comprise the application of numerical modelling tools, based on the potential flow theory (Boundary Equation

Methods—BEM), to simulate the response of CECO in order to: i) analyse the advantage of constraining the oscillation of CECO to an inclined direction; ii) to assess the efficiency of CECO through the different energy conversion stages and the impact of the damping induced by the PTO system in the energy conversion performance and iii) to obtain the power matrix required to the assessment of the power production in the coast of Iberian Peninsula. From this, it will be possible to obtain some valuable insights on the next steeps aiming the optimization of CECO.

4.2 *Numerical modelling of CECO*

The response of CECO in the time domain was investigated by means of AQWA. Two approaches were applied to model the different CECO elements depending on their relative size to the wavelength. The large-volume elements (*i.e.*, the LMMs and the fixed central cylinder) were discretised in panels to carry out a frequency domain analysis with a 3D BEM (AQWA-Line), while the elements with a small cross-section (*i.e.*, the supporting frame and the longitudinal rods) were modelled as Morison's elements.

The BEM model is based on the potential theory and assumes an incompressible and inviscid fluid (irrotacional flow field). After solving the velocity potentials using a boundary integration approach, the frequency-dependent hydrodynamic coefficients—radiation damping and added mass—can be determined to be used in subsequent analyses.

The response of CECO in the time-domain was obtained by means of AQWA-Naut, for which the frequency-dependent coefficients were used to calculate the diffraction and radiation forces on the WEC (f_d and f_r, respectively). As the wetted surface of CECO mobile parts change significantly during the simulation, the non-linear *Froude-Krylov* and the hydrostatic forces beneath the incident waves (f_{FK} and f_{st}, respectively) were calculated and applied to the different parts of the structure in each time step. As for the forces acting on the slender elements (f_m), Morison's equation was used.

In sum, the dynamic equation of the CECO floating (mobile) part can be expressed as:

$$m\ddot{x} = f_{st} + f_{FK} + f_d + f_r + f_m + f_{PTO} \tag{1}$$

where m is the mass of the moving parts; x corresponds to the CECO's moving part displacement in its 1-DOF and f_{PTO} includes the forces transmitted to the PTO—which includes, not only the sliding and bearing friction, the backlash and the structural flexibility in both the guiding system and the rack-pinion mechanism, but also the power losses in the electric generator. The term f_{PTO} was modelled as,

$$f_{PTO} = f_0 + (C_1 + C_2)\dot{x} \tag{2}$$

where f_0 is a constant friction force, and C_1 and C_2 are two damping coefficients, related to the losses in the energy conversion machinery and in the effects of the electric generator, respectively. Accordingly, the total damping of the PTO is the sum of the latter two coefficients: $C_{PTO} = C_1 + C_2$.

The aforementioned numerical method was applied to simulate the response of a 1:20 scale CECO, corresponding to that one tested in the second proof of concept study (*c.f.*, section 3). To achieve this, a mesh with a spatial resolution of 0.04 ± 0.02 m was built, which, in total, accounted for a total of 2349 panels and 304 Morison elements. The time step of the simulations was set to 0.1 s (model scale).

As a performance indicator, the relative capture width (C_N) was used, which is defined as the ratio of the wave power captured by the PTO system (P_{abs}) to the wave energy flux per unit crest length (J) normalized with respect to the overall width of the device or beam (B),

$$C_N = \frac{P_{abs}}{J\,B} \tag{3}$$

where the average wave power captured by the PTO system of CECO for a certain sea state is given by

$$P_{abs} = \overline{f_{PTO}\dot{x}} \tag{4}$$

4.3 *Calibration and validation of the model*

Prior to any analysis, the numerical model was calibrated with the results from previous experimental tests. In particular, realistic values of the parameters f_0, C_1 and C_2 were determined. To achieve this, the time series of CECO displacements, $x(t)$, from the experimental tests were compared to their numerical counterpart in an interactive procedure to determine those parameters that minimize the error between the physical and the numerical approaches. The Normalized Root Mean Square Error (NRMSE) was used as a measure of the error.

First, the values of the parameters f_0 and C_1 were determined jointly. The experimental data that corresponds to a CECO setup without the electric generator and different regular wave conditions were used to achieve this. In the numerical model, the parameter C_2 was set equal to zero and different couples of values for f_0 and C_1 tested. The calibration results are presented in Figure 3. The values that minimized the error are: $f_0 = 24$ kN and $C_1 = 53$ kN.s.m^{-1} (NRMSE = 6.6%). Once fixed the values of f_0 and

C_1, the damping induced by the electric generator in the system was calibrated (parameter C_2). This time, physical model tests with regular wave conditions corresponding to 3 different configurations of the electric generator were used. Further details about how the damping induced by the electric generator was varied in the laboratory experiments can be found in Rosa-Santos *et al.* (2015) and López *et al.* (2017b). The values of the damping minimizing the error for each configuration resulted: C_2 = 90, 111 and 202 kN.s.m^{-1} (Figure 4).

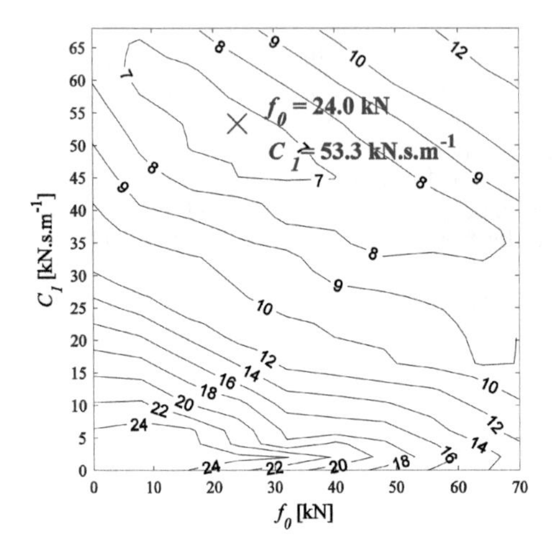

Figure 3. Average NRMSE (in%) obtained in the calibration of the parameters C_1 and f_0. All values are in prototype scale (based on López *et al.*, 2017a).

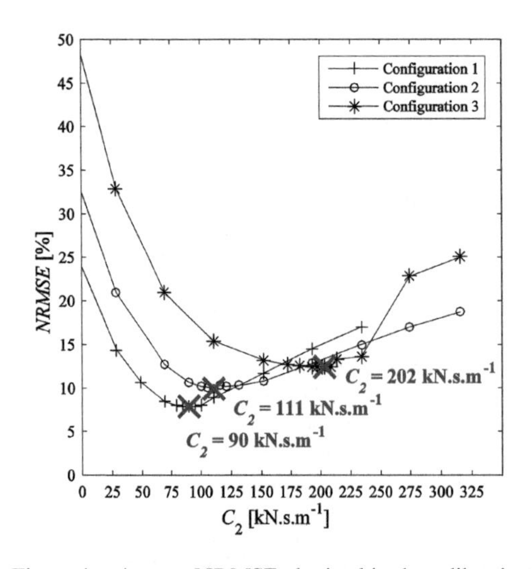

Figure 4. Average NRMSE obtained in the calibration of the parameter C_2 for three configurations of the generator. All values are in prototype scale (based on López *et al.*, 2017b).

4.4 *Inclination 45° versus 90°*

While most of the WECs based on the principles of the floating oscillating bodies are basically heaving buoys (e.g. Bosma *et al.*, 2015), the CECO concept is based on the motion of its floating part along an inclined direction of translation. Although this feature was initially considered an advantage of the proposed WEC, supporting evidence was required.

On these grounds, the numerical model described in the previous section was used to investigate the advantage of constraining the motion of the floating part to translations along an inclined axis. For this purpose, two configurations with different inclination angles were tested. One configuration corresponded to the CECO concept itself, with an inclination of β = 45°, while the second configuration, with an inclination of β = 90°, corresponded to a heaving buoy.

Each configuration was tested for forty different sea states corresponding to irregular wave conditions with $H_s \in [0.5, 4]$ m and $T_p \in [6,12]$ s. The friction and mechanical losses in the energy conversion machinery were simulated with the already calibrated values of f_0 and C_1 from Eq. 2. As for the effect of the electric generator, it was neglected in this analysis by setting the coefficient C_2 equal to

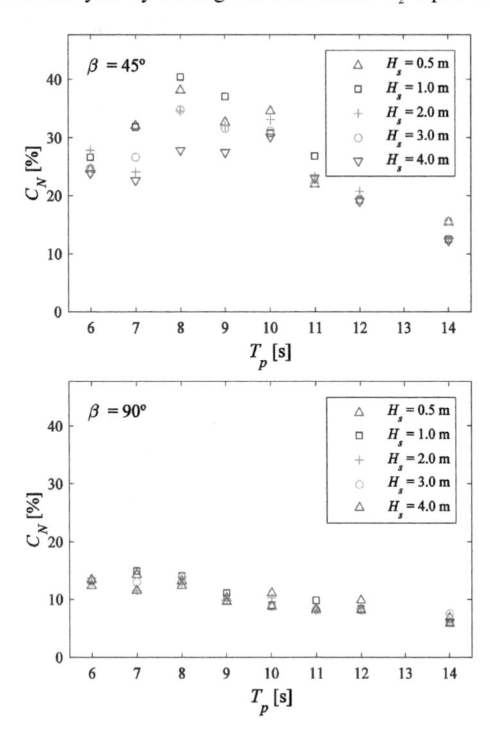

Figure 5. Relative capture width (in%) as a function of the significant wave height (H_S) and the wave peak period (T_p) for two different directions of translation of the floating part: β = 45° (upper panel) and β = 90° (lower panel).

zero. Figure 5 presents the values of the relative capture width obtained for each configuration (Eq. 3).

While CECO captures between 20 and 40% of incident wave energy for $\beta = 45°$, the C_N drops to values between 8 and 15% for $\beta = 90°$. When the inclination is set to $\beta = 45°$, the highest values of C_N are reached for those sea states with $T_p \in [8, 10]$ s. In addition, the value of C_N depends on H_s to a large extent. However, when the inclination is set to $\beta = 90°$, C_N increases as the value of T_p decreases, with little influence of H_s.

The results clearly indicate that an inclined direction of translation of the floating part outperforms in terms of wave power absorption a vertical one. Therefore, it can be stated that the CECO concept is able to capture a highest amount of wave power than a traditional heaving buoy configuration.

4.5 *Power matrix for one water level*

The results with irregular waves were also used to build the power matrix of CECO. If local wave conditions are known, the power matrices of a WEC result very useful to estimate the absorbed wave energy over a given period of time. It was considered that the CECO enters the survival mode for $H_s > 9$ m and the typical values of T_P in the nearshore of the western coast of the Iberian Peninsula. Bearing in mind this, the power matrix was limited to values of H_s between 0 and 9 m, and values of T_p between 4 and 18 s.

To construct the power matrices of absorbed wave power, 57 irregular wave conditions were simulated by combining different values of the spectral significant wave heights (H_{m0}) and peak wave periods (T_P), assuming a JONSWAP spectrum. Figure 6 presents, as an example, the power matrix

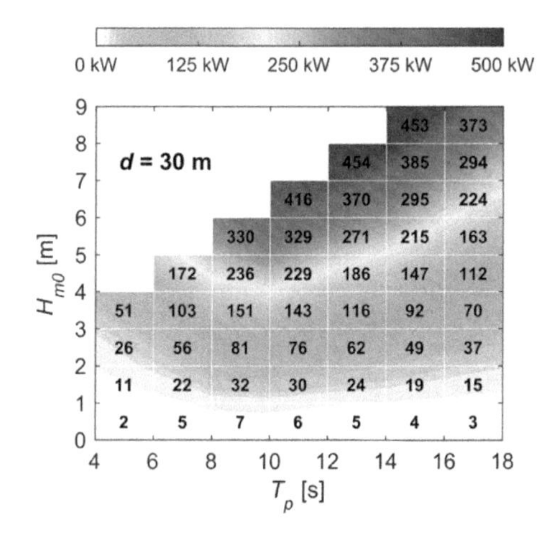

Figure 6. Wave power matrix of CECO for 30 m of water depth.

for the 30 m water depth. The results show that the performance of the device strongly depends on the wave parameters considered, with the absorbed wave power ranging between 2 and 454 kW. In general, the efficiency of the device absorbing wave power increases with the value of H_{m0} and for values of T_P around 10 s (results not presented here).

4.6 *Frequency domain model*

Time domain numerical models are the most direct way to perform numerical simulations because of their capability to take into account nonlinearities and deal with complex (non-analytical) user-defined effects. However, when a large number of simulations are required, as in the case of optimization studies or simulations in long-duration stochastic seas, this time domain (nonlinear) models may not be suitable due to the high required computational capacity.

In the particular case of CECO, the frequency domain model was not adopted in the first stages of development because of limitations in the linear module of the available numerical simulation toolkit. These limitations were related to the computation of hydrodynamic coefficients in Degrees of Freedom (DOFs) other than the six classical rigid-body modes and, the restraining of motions in some DOFs. Both limitations conflicted directly with CECO main feature, i.e., motion in single DOF along an inclined axis, so the linear model was abandoned at that time.

More recently, another hydrodynamic computational tool has become available, in which it is possible to compute hydrodynamics coefficients along generalized modes of motion. Under this conditions, it is now possible to focus on the development of a linear numerical model for the simulation of CECO motions and the estimation of absorbed power.

In comparison to the (time domain) nonlinear model, a few simplifying assumptions have been necessary to obtain a linear model. For instance, damping, restoring and Froude-Krylov excitation forces have been assumed linear. Losses and drag were also presumed linear and included as part of the linear damping. Under these hypotheses, the equation of motion of CECO took the form of a linear second order differential equation and, a frequency domain solution could be obtained.

A first outcome of the frequency domain model has been the possibility to obtain the CECO's numerical RAO for a broad range of wave frequencies, allowing the clear identification of the natural frequency of CECO. For the CECO nominal design characteristics (60% of submergence and 45° of inclination), the natural frequency obtained was around 8 s. As this value was quite different from

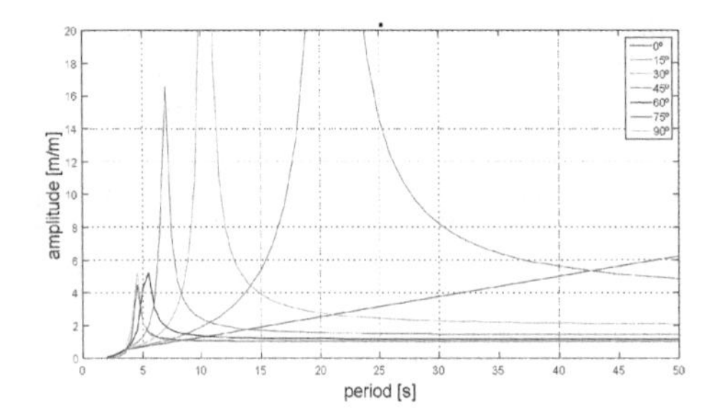

Figure 7. Effect of CECO's inclination on its natural period.

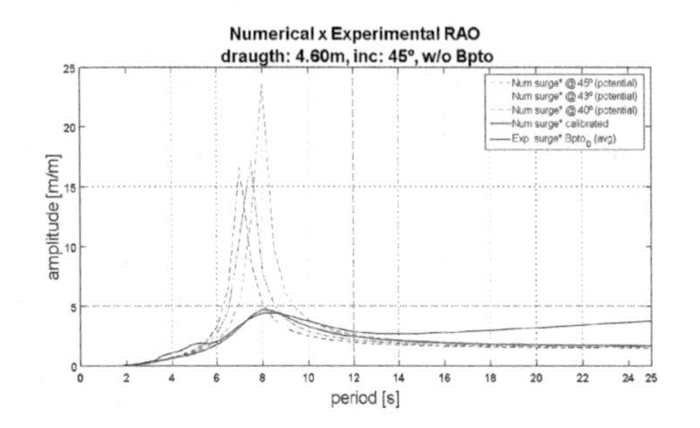

Figure 8. Calibration of CECO's RAO of motion.

the one expected for a pure heaving CECO (90° of inclination) at the same loading conditions, it was concluded that inclination was crucial for the definition of the CECO natural frequency. Several other numerical runs have been performed in order to confirm this hypothesis. Figure 7 shows some preliminary results considering only the potential damping.

Moreover, taking the experimental CECO's RAO of motion as reference, it has been possible to calibrate an external damping (representing losses and the PTO damping) for the linear numerical model. Figure 8 displays an example of the intermediary and final results in the calibration process of the CECO's numerical RAO. These preliminary results show a good agreement between the linear model and experiments, especially around the frequencies of interest.

The preliminary results of the linear numerical model are promissory for further investigation of the system dynamics to explore the possibilities of fine-tuning the device to reach an enhanced harvesting performance, taking into account the wave characteristics of the installation site.

5 PRELIMINARY ASSESSMENT OF CECO

Finally, the preliminary assessment of CECO along the coast of the Iberian Peninsula was carried out, by combining the previously obtained power matrix with the available wave resource. From the results it is possible to obtain valuable insights on the performance of CECO to drive the next research works.

The performance of CECO was analysed using the captured energy efficiency, CEE_{ff}, which is determined by combining the CECO power matrix and the wave resource matrix as follows,

$$CEE_{ff} = \frac{\sum_{i=1}^{H} P_{abs,i} h_i}{\sum_{i=1}^{H} J_i h_i B} \tag{5}$$

where $P_{abs,i}$ = wave power captured by CECO for the *ith* sea state (Eq. 3), h_i = number of hours of occurrence of the *ith* sea state, J_i = available wave power per meter of wave front for the *ith* sea state, B = capture width of CECO (14.6 m). For this purpose, the wave conditions for the near-shore of the Iberian Peninsula were computed for a period of ten years (2005–2014) using the spectral wave model SWAN. Figure 9 presents the mean annual distribution of CEE_{ff} in the area of study for the water depth

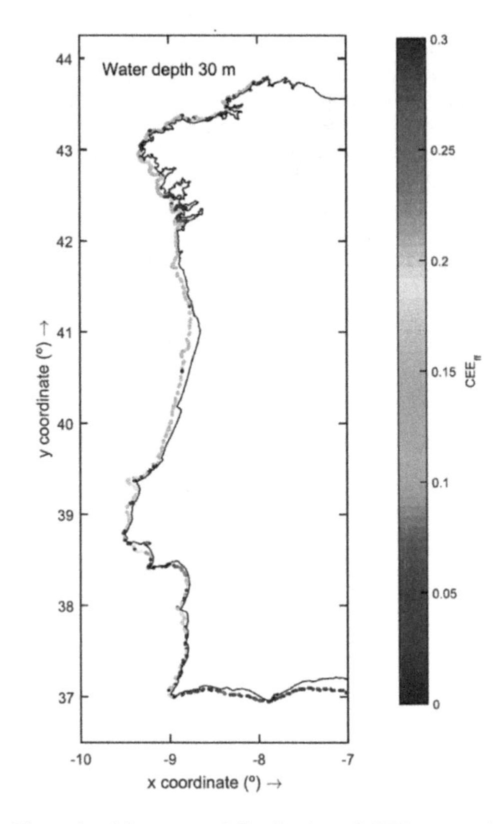

Figure 9. Mean annual distribution of CEE_{ff} across the area of study for the water depth of 30 m.

of 30 m. It can be seen that the CEE_{ff} changes significantly along the coast of the Iberian peninsula (from 0.35 to 0.10), being much higher in the southern locations, *i.e.*, CECO performs better in areas with a moderate wave resource. Ramos *et al.* (2017) also concluded that CECO performs better for milder wave climates, namely during the summer and spring seasons. These results may be explained by the range of peak periods (TP around 10 s) and wave heights at which CECO maximizes the wave power absorption, which correspond to moderate sea states. Hence, the current design of the CECO is not optimized for highly energetic wave conditions and has to be adapted, namely by tuning the geometry of its LMM.

6 CONCLUSIONS

The project OPWEC was presented and the most significant and recent conclusions regarding the development of CECO were described and discussed.

The numerical approach implemented in AQWA to simulate the response of CECO in the time domain was successfully validated with experimental results. Then, it was possible to demonstrate that the inclined direction of translation of CECO outperforms the vertical one in terms of wave power absorption. The preliminary assessment of CECO along the coast of the Iberian Peninsula allowed concluding that it is not yet optimized for the highly energetic wave conditions characteristic of northern locations, where the bulk of wave energy is spread in the range of 9 to 16 s of T_P.

The frequency domain model developed will be very important for the optimization of CECO's geometry and PTO damping. As the wave resource of the Iberian Peninsula presents significant seasonal variations, adaptive CECO setups should be explored too.

In terms of physical models tests, a new campaign should be performed in order to experimentally assess in detail the influence of the inclination angle. Some alternative geometries should be also explored, as well as new CECO configurations in order to reach deeper waters. Furthermore, the influence of scalability in CECO performance should be investigated by testing similar shapes with different dimensions.

ACKNOWLEDGEMENTS

This work was financially supported by the Project PTDC/MAR-TEC/6984/2014 – POCI-01–0145-FEDER-016882 – funded by FEDER funds through COMPETE2020 – Programa Operacional Competitividade e Internacionalização (POCI) and by national funds through FCT—Fundação para a Ciência e a Tecnologia, I.P.

REFERENCES

Babarit, A., Hals, J., Muliawan, M.J., Kurniawan, A., Moan., T., Krokstad, J., 2012. Numerical benchmarking study of a selection of wave energy converters. *Renewable Energy* 41(0):44–63.

Bosma, B., Sheng, W, Thiebaut, F., 2014. Performance Assessment of a Floating Power System for the Galway Bay Wave Energy Test Site. *International Conference on Ocean Energy* (ICOE). Halifax, Canada2014.

Day A, Babarit A, Fontaine A, He Y-P, Kraskowski M, Murai M, et al. (2015). Hydrodynamic modelling of marine renewable energy devices: A state of the art review. *Ocean Engineering*. 2015;108:46–69.

De Andres A, Guanche R, Armesto J, Del Jesus F, Vidal C, Losada I. 2013. Time domain model for a two-body heave converter: Model and applications. *Ocean Engineering*. 72:116–23.

Marinheiro, J., 2013. Optimization study of an innovative wave energy converter, M.Sc. thesis, Faculty of Engineering of the University of Porto, Portugal [in Portuguese].

Marinheiro, J.; Rosa-Santos, P.; Taveira-Pinto, F.; Ribeiro J.; 2015. Feasibility study of the CECO wave energy converter. *Maritime Technology and Engineering* (ed: C. Guedes Soares & T.A. Santos), CRC Press (Taylor and Francis Group), pp.1259–1267.

López, M., Taveira-Pinto, F., Rosa-Santos, P., 2017a. Numerical modelling of the CECO wave energy converter. *Renewable Energy*. Vol. 113. p. 202–210.

López, M., Taveira-Pinto, F., Rosa-Santos, P., 2017b. Influence of the power take-off characteristics on the performance of CECO wave energy converter. *Energy*. Vol. 120. p. 686–697.

Payne, G.S., Taylor, J.R., Bruce, T., Parkin, P. 2008. Assessment of boundary-element method for modelling a free-floating sloped wave energy device. *Part 1: Numerical modelling. Ocean Engineering*; 35(3):333–41.

Pastor J, Liu Y. 2014. Power absorption modeling and optimization of a point absorbing wave energy converter using numerical method. *Journal of Energy Resources Technology*. 136(2):021207.

Ramos, V., López, M., Taveira-Pinto, F., Rosa-Santos, P., 2017. Influence of the wave climate seasonality on the performance of a wave energy converter: A case study. *Energy*, 135:303–316.

Rezanejad, K. and Guedes Soares, C. 2015, Hydrodynamic performance assessment of a floating oscillating water column. Guedes Soares, C. & Santos T.A. (Eds.), *Maritime Technology and Engineering, London*, UK: Taylor & Francis Group; pp. 1287–1296.

Rezanejad, K.; Bhattacharjee, J., and Guedes Soares, C. 2013; Stepped sea bottom effects on the efficiency of nearshore oscillating water column device. *Ocean Engineering*. 70:25–38.

Rezanejad, K.; Bhattacharjee, J., and Guedes Soares, C. 2015; Analytical and numerical study of dual-chamber oscillating water columns on stepped bottom. *Renewable Energy*. 75:272–282.

Rezanejad, K.; Bhattacharjee, J., and Guedes Soares, C. 2016; Analytical and numerical study of nearshore multiple oscillating water columns. *Journal of Offshore Mechanics and Arctic Engineering*. 138:021901-1-021901-7.

Rezanejad, K.; Guedes Soares, C.; López, I., and Carballo, R. 2017; Experimental and numerical investigation of the hydrodynamic performance of an oscillating water column wave energy converter. *Renewable Energy*. 106:1–16.

Rhinefrank, K., Schacher, A., Prudell, J., Hammagren, E., Zhang Z., Stillinger, C., et al. 2011. Development of a Novel 1: 7 Scale Wave Energy Converter. *30th International Conference on Ocean, Offshore and Arctic Engineering*. Volume 5: Ocean Space Utilization – Ocean Renewable Energy ASME 2011, p. 935–944.

Rosa-Santos, P., Taveira-Pinto, F., Teixeira, L., Ribeiro J., 2015. CECO wave energy converter: Experimental proof of concept. *Journal of Renewable and Sustainable Energy*, 7, 061704, 14p, ISSN: 1941-7012.

Taveira-Pinto F., Iglesias G., Rosa-Santos P., Deng ZD., 2015. Preface to Special Topic: Marine Renewable Energy. *Journal of Renewable and Sustainable Energy* 2015; 7:061601.

Teixeira, L., 2012. Experimental study of a new wave energy converter. M.Sc. thesis, Faculty of Engineering, University of Porto, Portugal, 156p [in Portuguese].

Xu, S., Wang, S., Hallak, T., Rezanejad, K., Hinostroza, M., Guedes Soares, C., Rodríguez, C.A., Rosa-Santos, P., Taveira Pinto, F., 2018. Experimental study of two mooring systems for a floating point absorber wave energy converter. *Progress in Maritime Technology and Engineering,* C. Guedes Soares & T.A. Santos (Eds), London, UK: Taylor & Francis Group.

Zaroudi, H.G.; Rezanejad, K., and Guedes Soares, C. 2015; Assessment of mooring configurations on the performance of a floating oscillating water column energy convertor. Guedes Soares, C. (Ed.), *Renewable Energies Offshore*, London, UK: Taylor & Francis Group, pp. 921–928.

Progress in Maritime Technology and Engineering – Guedes Soares & Santos (Eds)
 ISBN 978-1-138-58539-3

Experimental study of two mooring systems for wave energy converters

S. Xu, S. Wang, T.S. Hallak, K. Rezanejad, M.A. Hinostroza & C. Guedes Soares
Centre for Marine Technology and Ocean Engineering (CENTEC), Instituto Superior Técnico, Universidade de Lisboa, Lisbon, Portugal

C.A. Rodríguez
Interdisciplinary Centre of Marine and Environmental Research, Faculty of Engineering, University of Porto, Porto, Portugal
Laboratory of Ocean Technology (Laboceano), Department of Naval Architecture and Ocean Engineering, Federal University of Rio de Janeiro, Rio de Janeiro, Brazil

P. Rosa-Santos & F. Taveira-Pinto
Interdisciplinary Centre of Marine and Environmental Research, Faculty of Engineering, University of Porto, Porto, Portugal

ABSTRACT: Model tests have been conducted to study the performance of alternative mooring systems aimed to be used at a later stage in a wave energy converter under development. The initial set of tests are intended to establish the properties of single lines of different configuration and thus a simple buoy has been chosen as the floater. Two mooring systems have been tested: a traditional catenary mooring system and a hybrid mooring system that behaves as a slack system. The performance of both mooring systems have been addressed by comparing the measured buoy motions in six degrees of freedom and mooring line tension, under regular and irregular head waves. In terms of mean drift motions, the obtained results demonstrated that the catenary mooring system may perform better for the operational conditions, while the slack system gave better results for the survival conditions. Regarding mooring line tensions in irregular seas, the slack system provides significantly lower spectral peaks.

1 INTRODUCTION

Wave energy extraction is getting more and more attention in recent years due to its high energy density as described in some review papers (Guedes Soares, 2012, Falcão 2015). According to Hayward & Osman (2011), more than two hundred different devices have been developed to absorb energy from waves.

The mooring system is of vital importance no matter to which floating Wave Energy Converter (WEC) is applied. The mooring system needs to be designed, among other things, to ensure the device's station-keeping and to contribute to its survivability in storm conditions. Furthermore, its influence on the device power absorption capacity should also be considered and it should be economical in terms of manufacture, installation and maintenance.

Plenty of studies can be found regarding WECs' mooring system design (Fitzgerald & Bergdahl (2008; Johanning et al. 2006; Casaubieilh et al. 2015). Catenary mooring system is the generally adopted. For instance, Harnois et al. (2015) used this type of mooring in a series of 1:5 scale model tests with a point absorber and Zanuttigh et al. (2013) applied it for the DEXA hinged attenuator. However, the catenary system may not be the preferred mooring system due to several drawbacks, including its large weight, large footprint as well as high costs. So, alternative mooring systems for WECs need to be considered and investigated.

The use of synthetic mooring ropes has allowed the development of alternative mooring systems (Weller et al. 2014, 2015). A hybrid mooring system for FLOW wave energy converter was proposed by Fonseca et al. (2009), consisting of four hybrid mooring lines arranged symmetrically. Each line was composed of double braid nylon rope and R4 stud link seabed chain.

Cerveira et al. (2013) studied the power performance of a WEC buoy under three mooring configurations: without mooring lines, moored with slack lines and moored with lines of moderate stiffness. Their results showed that the effects of mooring system on the annual captured wave energy are negligible. However, in another study it was found that the effect of pretension of catenary mooring line on energy production can be considerable, leading to 16% variation in absorbed energy (Angelelli et al., 2013).

Paredes et al. (2016) conducted regular and irregular wave model experiments to investigate the dynamics of a WEC point absorber under three alternative mooring systems: (i) catenary mooring lines, (ii) taut mooring lines with floaters and (iii) "S" shape mooring lines with floaters and weights. Their experimental results showed that the taut mooring with floaters was the best choice for energy harvesting, while the catenary mooring system worked better in extreme sea states by constraining more the motion responses compared to the other tested mooring systems.

Sergiienko et al. (2016) studied the effect of mooring configuration on the performance of a three-tether submerged point-absorbing WEC. Two configurations were studied: (i) a WEC located close to the sea surface with a low mass ratio (ratio of WEC mass to mass of water displaced by the WEC); and (ii) a heavy and deeply submerged buoy. It was found that, in the first configuration, the optimal mooring angle was sensitive to wave frequency. For the second case, the produced energy is highly dependent on the inclination angle.

After an extensive review of recent works, it was concluded that current research mainly focus on studying the influence of mooring parameters (such as pretension and mooring arrangement) on WEC energy performance. The study of impacts of different mooring systems on WECs is rare. Besides, the existing mooring systems for WECs still need to be improved.

The present study aims at providing background for the development of a mooring system for a Floating Oscillating Water Column (FOWC), which derived from studies on bottom fixed OWC by changing its geometry and the entrance with a step change in bottom surface (Rezanejad et al. 2013, 2017). Furthermore the concept of multiple chambers was also studied as a way to improve the efficiency of the device (Rezanejad et al. 2015, 2016). The experience obtained led to a floating device concept (Rezanejad & Guedes Soares, 2015) and initial studies of a mooring solution (Nava et al. 2013, Zaroudi, et al. 2015).

In this work, a mooring system consisting of a slack mooring line is considered. The mooring line is composed of polyester and chain. This type of mooring line is expected to combine the advantages of taut and catenary mooring lines, by reducing the mooring line weight, footprint and costs. Furthermore, it is expected to reduce the risk of damage of the mooring line compared to traditional taut mooring lines by the reduction of the line's mean tension level. Through a series of experimental model tests with a buoy in waves, the performance of the proposed mooring system has been compared with model tests with a traditional catenary mooring system.

2 METHODOLOGY

2.1 *Decay analysis*

Natural periods and linear damping coefficients were obtained from the decay curves. The time interval between two successive peaks in the decay time history has been defined as the natural period of a given motion. The linear damping coefficient was determined by log decrement ratio of peaks, i.e.:

$$\delta(\zeta) = \frac{1}{n}\ln\frac{x_i}{x_{i+n}} \tag{1}$$

$$\zeta(\delta) = \frac{\delta}{\sqrt{4\pi^2 + \delta^2}} \tag{2}$$

where, δ is the log decrement ratio, ζ is the linear damping coefficient, x is the motion amplitude, subscript i and n denote the index of motion peaks.

2.2 *Absorbed energy*

The instantaneous absorbed energy and time averaged absorbed energy can be calculated as (Falnes, 2002),

$$P(t) = B_{33}^{PTO}\left[\dot{z}(t)\right]^2 \tag{3}$$

$$\bar{P}(t) = \frac{1}{T}\int_0^T B_{33}^{PTO}\left[\dot{z}(t)\right]^2 dt \tag{4}$$

where z(t) is the heave motion response of the WEC's buoy, the dot notation represents the derivative with respect to time, B_{33}^{PTO} is the PTO (Power Take Off) linear damping coefficient. In this study, B_{33}^{PTO} will be assumed to be 1 N.s/m.

3 MODEL TESTS

3.1 *Point absorber*

The model of the cylinder buoy investigated in Paredes et al. (2016) was used to represent the floater of the present model tests campaign. The main dimensions of the buoy are given in Table 1, where m is the mass of the buoy, D is the diameter of buoy, H is the height of buoy and Zg is the vertical coordinate of the center of gravity relative to the buoy's bottom. Figure 1 illustrates the model.

3.2 *Mooring system*

Two mooring systems were investigated: a typical catenary mooring system and the proposed slack mooring system. The layout of both mooring systems were identical and consisted of three equally radially spaced mooring lines, i.e., 120° between two successive lines. The mooring system set-up is shown in Figure 2.

The coordinates of fairleads are listed in Table 2, where, the *z* values are referenced to the buoy bottom.

Table 1. Main dimensions of the buoy (Paredes et al., 2016).

m (kg)	*D* (m)	*H* (m)	Draft (m)	*Zg* (m)
45.84	0.515	0.400	0.220	0.140

Figure 1. Buoy model.

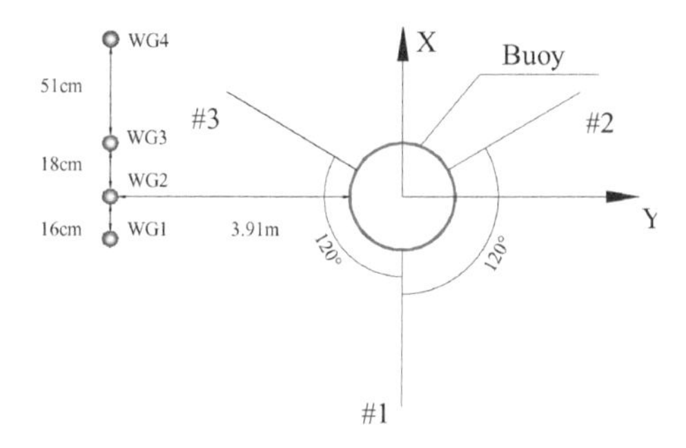

Figure 2. Top view of the mooring system arrangement.

Table 2. Coordinates of mooring lines fairleads.

Fairlead	X, Y, Z (cm)
F1	(–25.75, 0, 27.50)
F2	(12.88, 22.30, 27.50)
F3	(12.88, –22.30, 27.50)

The slack mooring lines were hybrid, i.e., each line was composed of two segments: a steel chain (lower) segment and a polyester rope (upper) segment. At the static equilibrium position of the buoy-mooring system, the chain segment laid completely on the seabed (i.e., the basin's bottom), while the polyester segment was slack. When a small offset is applied to the buoy (i.e., it moves away from its initial static position), the chain lifts partially, giving rise to a catenary-type restoring force. Indeed, the suspended (chain) weight becomes the major contributor to the mooring restoring force. If the buoy offset further increases, the chain may be lifted completely and, now, the polyester line will be stretched, inducing an additional (elastic) restoring force. The main parameters of the slack mooring line are shown in Table 3, where, *d* is the nominal diameter, *Wa* is the mass per unit length in air and *Ws* is the submerged mass per unit length.

The catenary mooring lines were composed of a single chain segment, each. The parameters of the chain lines are identical to those of the chain segments shown in Table 2, except for the length, which is equal to 4.5 m. The mooring radius of the slack and catenary mooring systems are 1.4 m and 4.28 m, respectively. The physical models of a slack and a catenary mooring lines are shown in Figures 3. Figure 4 shows the profiles of both types of mooring lines.

3.3 *Experimental set-up*

The model tests were carried out at the wave basin of the Laboratory of the Hydraulics, Water Resources and Environment Division of the Fac-

Table 3. Main parameters of the slack mooring line.

Segment	*L* (m)	*d* (mm)	*Wa* (kg/m)	*Ws* (kg/m)
Polyester	1	0.8	3.20E-04	8.16E-04
Chain	0.7	4	0.16	0.12

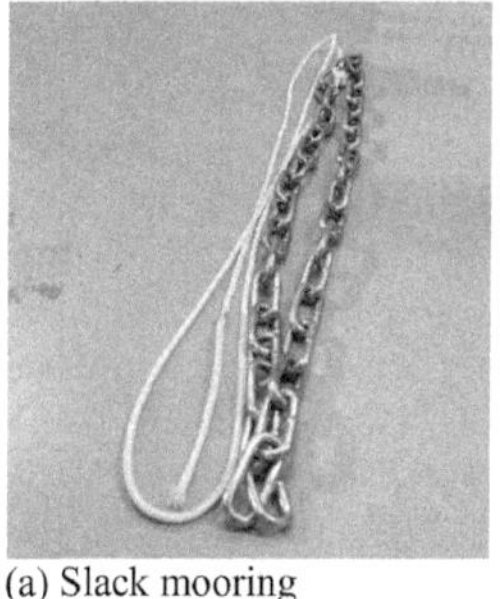

(a) Slack mooring

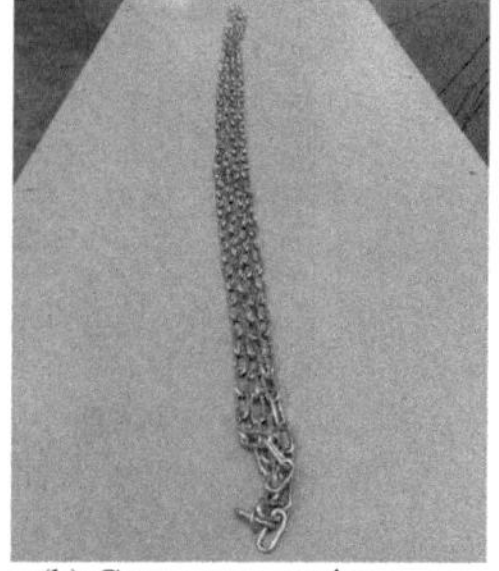

(b) Catenary mooring

Figure 3. Mooring line models.

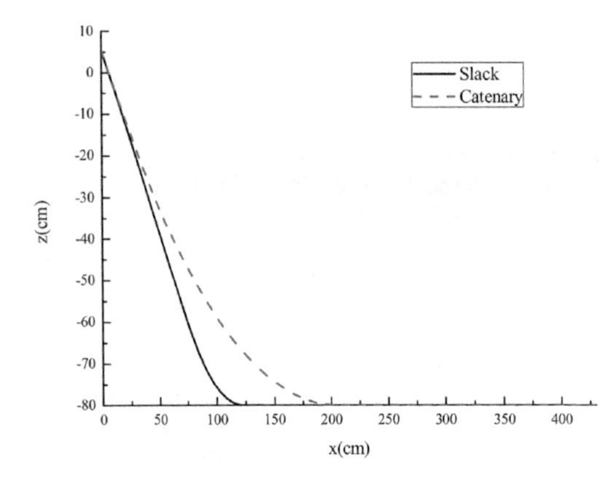

Figure 4. Profiles of the tested mooring lines.

Figure 5. Buoy during physical model tests.

ulty of Engineering of the University of Porto. The wave basin is 28 m long, 12 m wide and 1.2 m deep, and is equipped with a multidirectional wavemaker capable of generating regular and irregular waves in any direction. On the opposite side of the wavemaker, a wave dissipation beach is installed to absorb incoming wave. The water depth during the tests was set to 0.80 m.

The model scale was determined as 1:100, due to the limitations of wave maker for large waves at bigger scale.

The six-degree-of-freedom motions were measured using the Qualisys® system, an optical tracking system. Wave elevations in the basin were recorded by four Wave Gauges (WGs), schematized in Figure 2. A load cell was installed on line #1's fairlead and contacted to data acquisition system to measure #1 mooring tensions. This mooring line is expected to be the most loaded in head seas (wave propagation direction along positive X-axis in Figure 2). The sampling frequency for all measurements was set to 40 Hz. Figure 5 shows the tests layout with the model in the basin under irregular waves.

4 RESULTS AND DISCUSSIONS

4.1 *Decay tests*

Decay tests in calm-water were carried out to verify the natural periods of the buoy's rigid-body motions and to measure the corresponding damping coefficients. The decay model tests were implemented by pushing model to a position in the concerned DOF, then release the buoy to make it oscillate freely in still water. The motion responses were recorded by Qualisys® system, an optical tracking system.

Figures 6 to 8 show heave, pitch and surge time series of the decay tests of the buoy with the slack and catenary mooring systems. A summary of the tests results is given in Table 4. In terms of motion period, the effects of mooring system on heave and pitch are insignificant, since the mooring system restoring force are relatively small compared to buoy hydrostatic restoring force. However, in terms of damping, the catenary mooring seems to introduce additional damping to the buoy motions. In

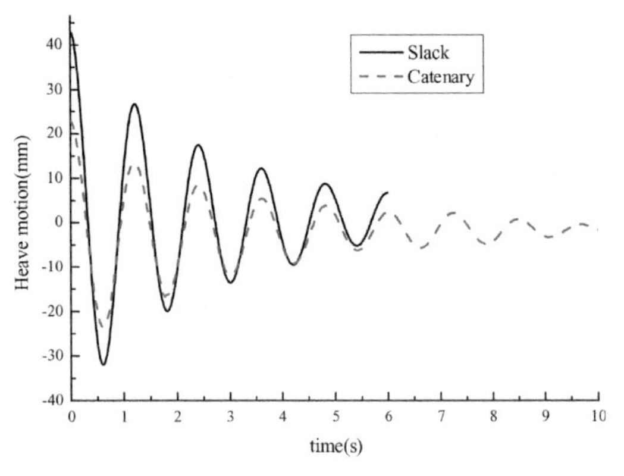

Figure 6. Heave decay tests.

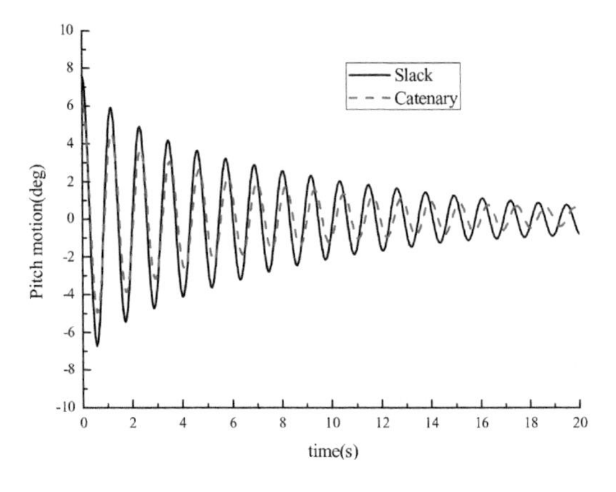

Figure 7. Pitch decay curves.

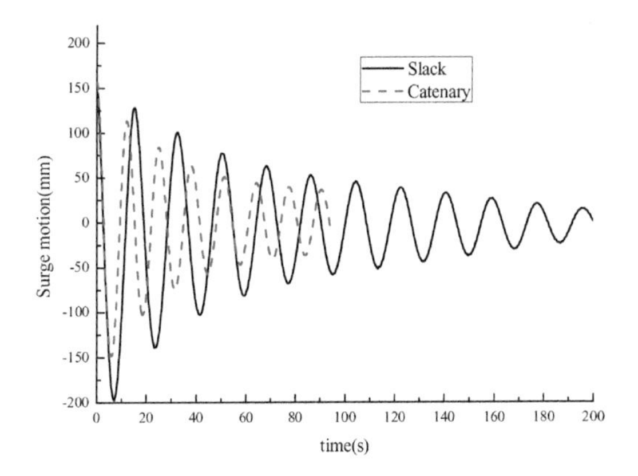

Figure 8. Surge decay curves.

Table 4. Summary results of decay tests.

	Values	Catenary	Slack
Heave	T(s)	1.21	1.20
	ζ(%)	6.8%	6.8%
Pitch	T(s)	1.17	1.15
	ζ(%)	2.8%	2.8%
Surge	T(s)	12.74	14.31
	ζ(%)	2.4%	2.7%

the case of surge, the period of motion is shorter for the catenary system, indicating that additional stiffness is present (associated to the catenary restoring force). The catenary system is also inducing extra-damping to the surge motion. It is important to notice that for short excursions (as the ones induced for the decay tests), the buoy under the slack system may display a behavior closer to that of a free floating system (without mooring lines).

4.2 *Regular wave tests*

Three regular wave conditions have been considered, representing two operational wave conditions and one survival condition. The corresponding wave parameters are given in Table 5.

As expected, the influence of the mooring system on heave and pitch motions was limited. So, only surge motions will be discussed here. Figure 9 (a, b and c) shows the time series of surge for the different wave conditions shown in Table 5.

For a better analysis of the surge motions, statistical results are shown in Table 6, where *Amp* is the motion amplitude (averaged maximum motion amplitude minus the mean motion amplitude).

In terms of mean drift, for the operational conditions, the buoy under the catenary system display smaller offsets than the slack system. This behavior

Table 5. Regular wave test cases.

Case	H (m)	T (s)
ROP1	0.04	0.90
ROP2	0.04	1.30
RSURV1	0.09	1.16

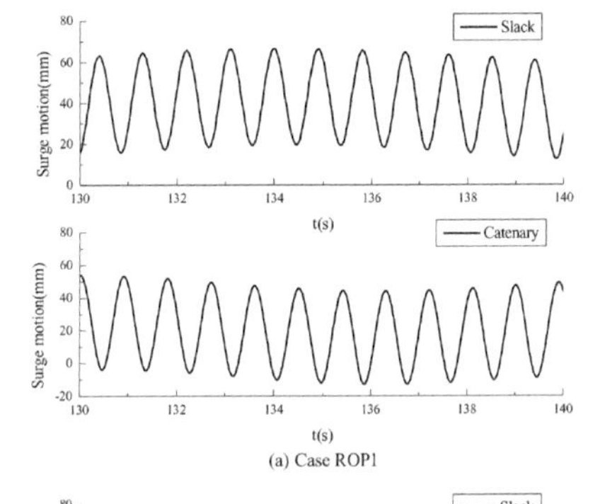

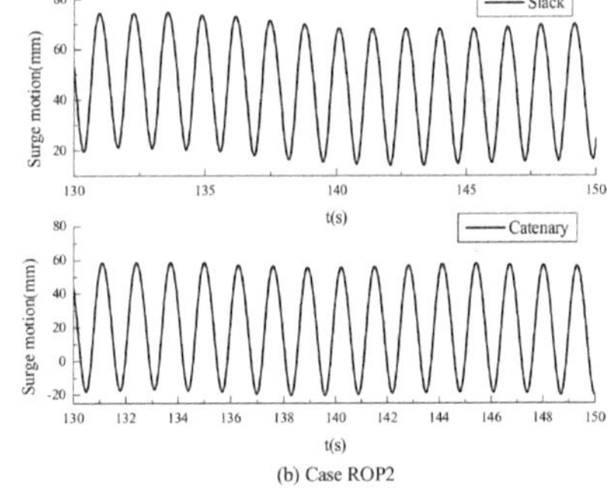

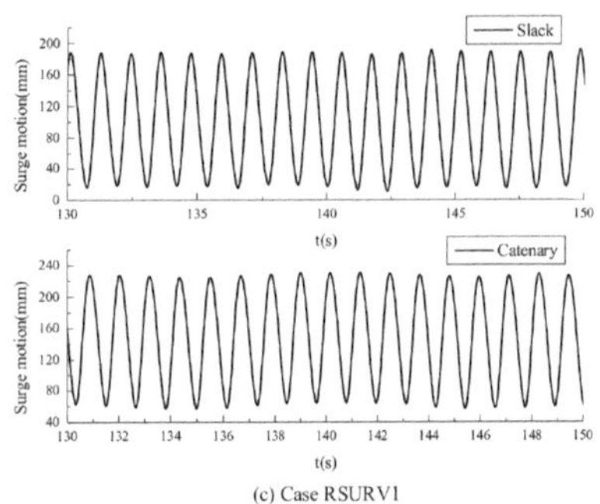

Figure 9. Surge motion responses in regular waves.

Table 6. Statistics of surge responses in regular waves.

	Catenary mooring		Slack mooring	
Case	Mean (mm)	Amp (mm)	Mean (mm)	Amp (mm)
ROP1	19.95	29.32	37.12	23.87
ROP2	20.76	37.94	44.13	27.15
RSURV1	147.00	82.43	100.47	86.06

was expected because the slack system offers less restoring than the catenary system for mild to moderate excursions. Notice that, at very small excursions, the slack mooring system is almost "slack", then, progressively (as the excursion increases), it achieves (catenary) restoring force due to its chain segment.

On the other hand, under survival conditions (i.e., when large excursions take place), the polyester segment of the slack mooring stretches and additional restoring force is introduced to slack mooring system. Depending on the amount of restoring provided by the polyester segment, the total restoring force of the slack system (under large excursions) can be larger than that of the catenary system. Thus, as observed in Figure 9c, inducing smaller mean drift.

In terms of drift amplitude, an interesting phenomenon was observed when comparing both mooring systems. When the buoy displayed larger mean drifts, the associated drift amplitudes were smaller. For instance, under operational conditions, when the mean drift was smaller for the catenary system, its drift amplitudes followed an opposite tendency, i.e., larger drift amplitudes than those of the slack system. For the survival conditions, when mean drift was smaller for the slack system, its drift amplitudes were larger than those of the catenary system.

The mooring tension amplitudes achieved under regular waves are shown in Figure 10 (a, b and c). It can be clearly seen that the slack mooring tension amplitudes are smaller than catenary mooring line in operational wave conditions. Indeed, for these conditions the suspended chain length of slack mooring line was less than the catenary. An opposite result was found for survival wave conditions. Slack mooring tensions increased dramatically due to the stretching of the polyester segment, giving rise to snap loads.

According to the results in Table 7, the tension amplitude of the catenary mooring line was about two times that of the slack mooring line for operational conditions, while an opposite result was observed for the survival condition.

4.3 *Irregular wave tests*

The operation and extreme wave conditions were from the wave statistics of region offshore Figueira da Foz, Portugal.

For the irregular wave tests a wave described by a JONSWAP spectrum with significant wave height of 0.045 m, peak period of 1.3 s and peak enhancement factor of 2.5 was adopted. The wave duration was 1200s. Figure 11 shows the measured time series of the wave elevation and its corresponding spectrum.

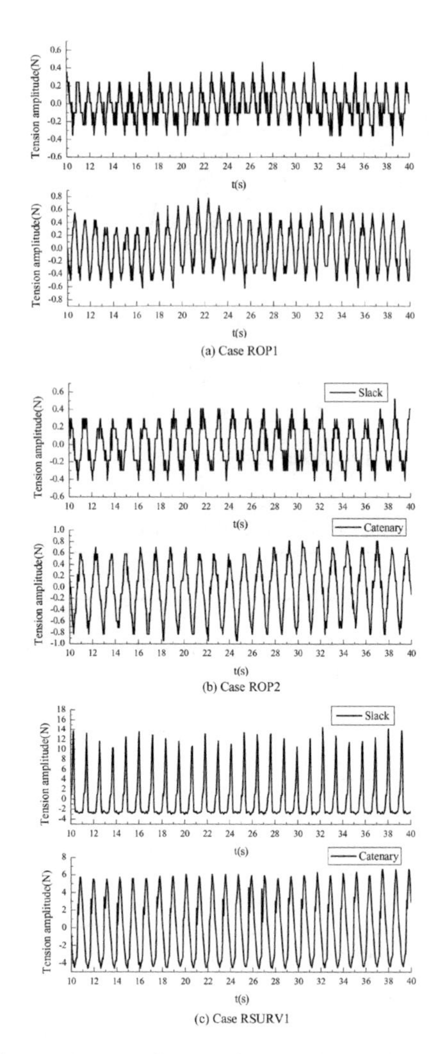

Figure 10. Mooring tension amplitudes in regular waves.

Table 7. Mooring tension amplitudes in regular waves.

Case	Tension Amp (N)	
	Catenary	Slack
ROP1	0.78	0.47
ROP2	0.93	0.53
RSURV1	7.13	14.41

The time series of the three main motion responses of the buoy, i.e. surge, heave and pitch, for both mooring systems are shown in Figure 12. Surge motion response is larger than that of the slack mooring system. On the other hand, heave and pitch responses show similar behaviors for

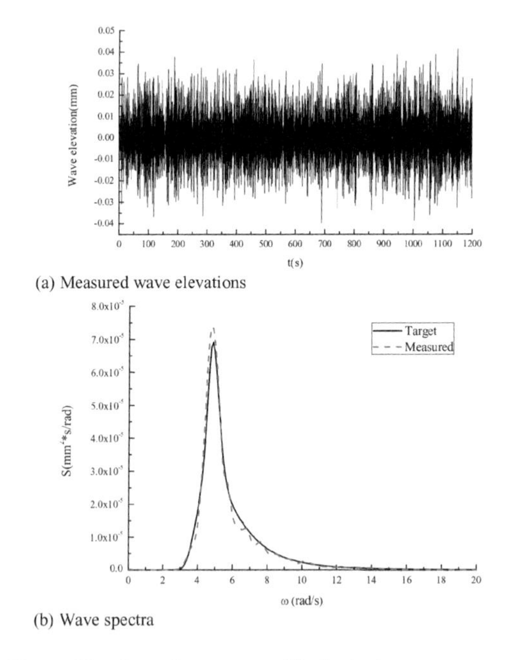

(a) Measured wave elevations

(b) Wave spectra

Figure 11. Irregular wave used in the tests.

both mooring systems. So, it may be concluded that the influence of the mooring system on these motions is limited.

For a better comparison, Table 8 shows the statistical results of the buoy motions in irregular waves. There, it becomes clearer that the buoy surge motion under the slack mooring system is greater than that under the catenary mooring. A possible explanation for these results is that the significant wave height of this sea condition is not large enough, so that the chain segment of the slack mooring line is not fully lifted up most of the time. In the other words, the restoring force of the polyester segment is not activated and the weight of suspended chain is still the major contribution to the stiffness of the slack mooring system. Apparently, the weight of suspended chain in the slack mooring line is lower than that of the catenary mooring line.

Furthermore, it is found that the extremes values of pitch response under the slack mooring are a bit greater than those of the catenary system. It seems that mooring line #1 was fully lifted up abruptly and, a great instantaneous mooring tension was induced together with a great pitch motion. For heave motion, no significant differences between the two mooring systems were observed.

Spectral analyses of the motion responses of the buoy in irregular waves for both mooring systems have also been performed. The results are shown in Figure 13. There, it is evident that surge motions for both mooring systems were domi-

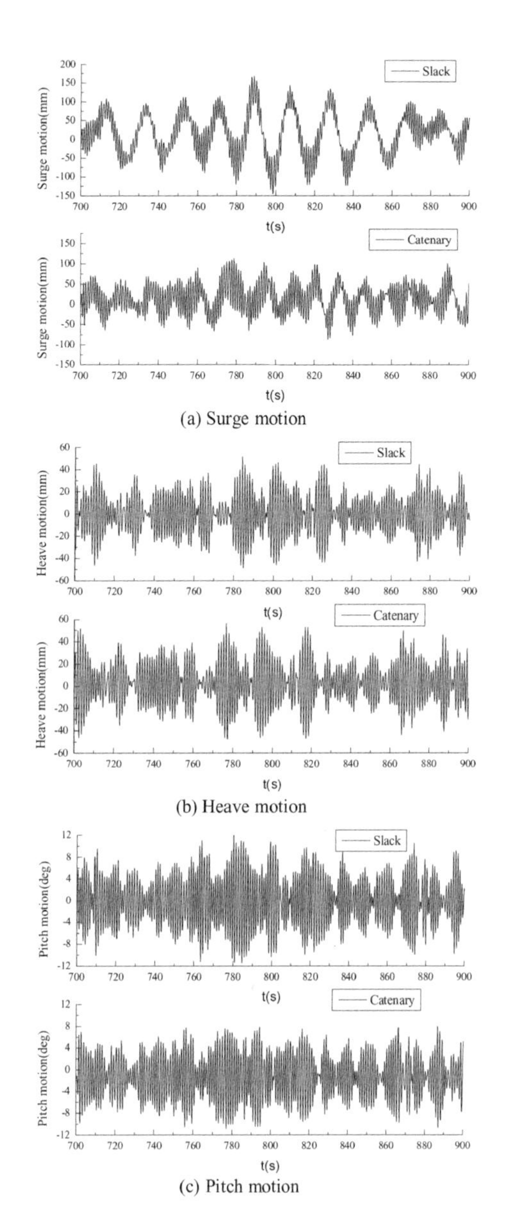

(a) Surge motion

(b) Heave motion

(c) Pitch motion

Figure 12. Buoy motion responses in irregular waves.

Table 8. Statistics of buoy responses in irregular waves.

Case	Value	Surge (mm)	Heave (mm)	Pitch (deg)
Catenary	Max	151.84	58.44	9.48
	Min	−112.72	−51.04	−12.00
	Mean	14.22	2.21	−1.37
	Std	39.24	19.06	4.08
Slack	Max	168.64	57.04	12.34
	Min	−144.65	−55.29	−12.56
	Mean	24.80	−0.77	−0.13
	Std	45.76	18.19	4.48

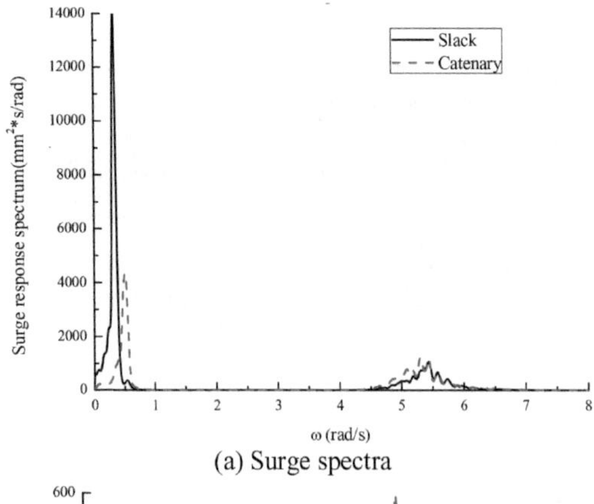

(a) Surge spectra

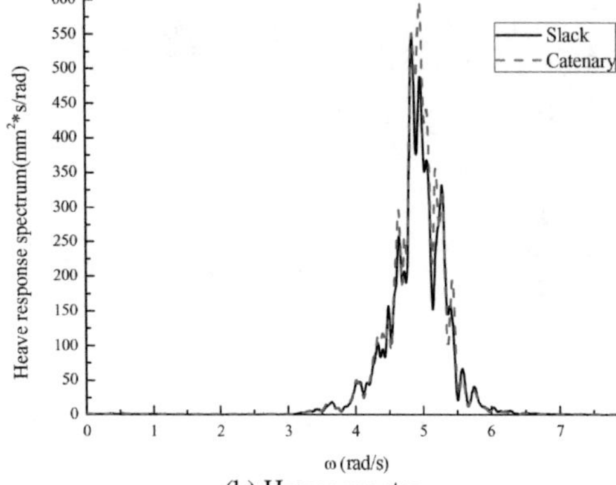

(b) Heave spectra

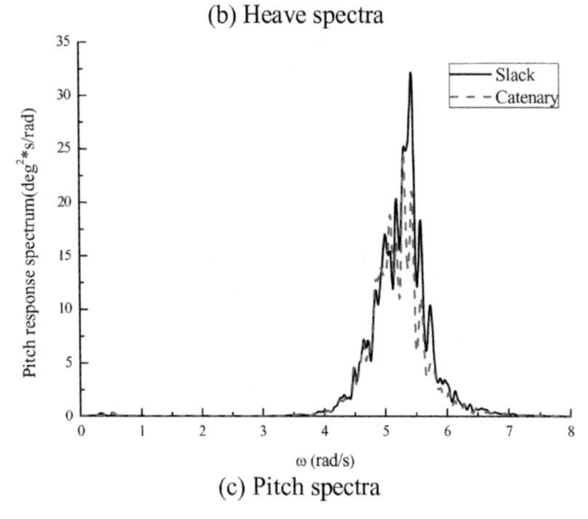

(c) Pitch spectra

Figure 13. Buoy motion response spectra.

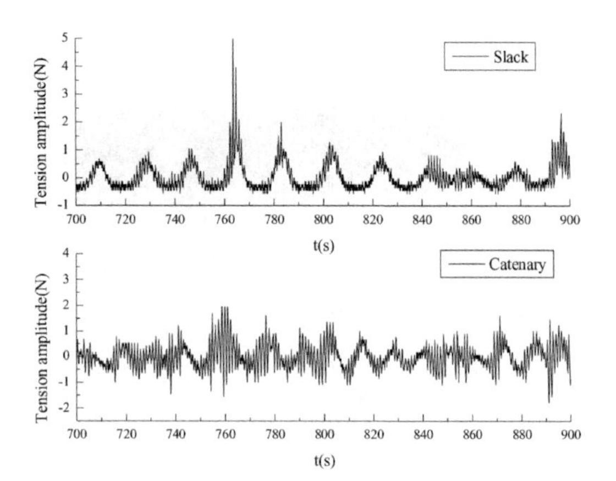

Figure 14. Mooring tension time series in irregular waves.

Table 9. Statistics of mooring tension in irregular waves.

Value	Catenary	Slack
Max(N)	3.69	5.39
Min(N)	−1.80	−0.58
Std(N)	0.52	0.37

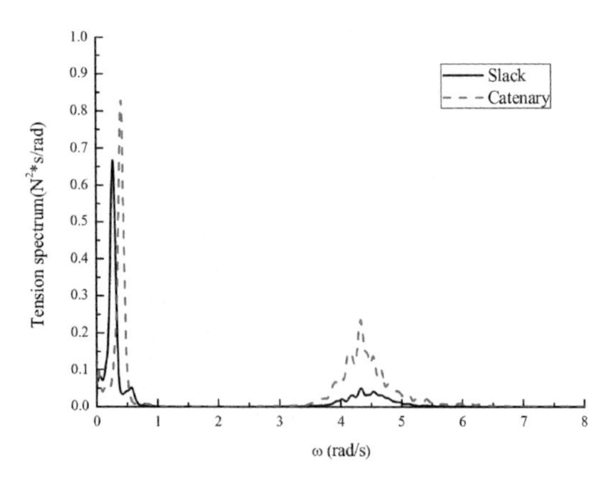

Figure 15. Mooring tension spectra.

nated by their low frequency component, whose corresponding peaks occurred around the natural surge period of each moored system. Furthermore, it is confirmed that the catenary mooring system is more efficient in restricting surge drift motions, at least for the tested wave condition. Heave and pitch motions are dominated by wave frequency components. Heave spectra are similar for both mooring systems, while pitch spectrum of the slack system displays a higher peak than that of the corresponding spectrum for the catenary system.

Figure 14 shows the time series of the mooring tension for both mooring systems. Several abrupt tension peaks are evident under the slack mooring system. The explanation for these results is the activation of the elastic restoring force associated to

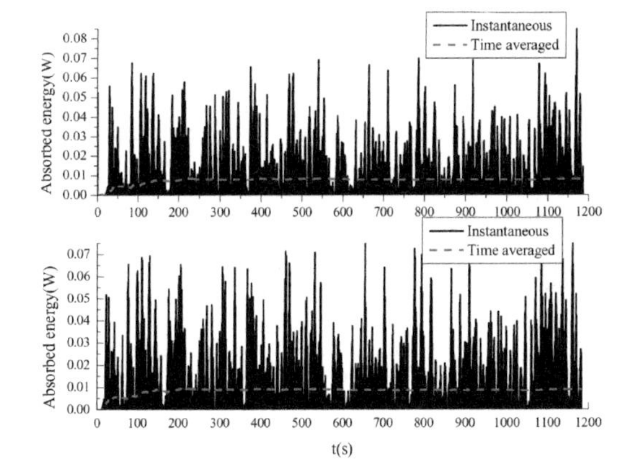

Figure 16. Estimated absorbed energy in irregular waves.

the polyester segment, which occurs when the chain segment is fully lifted up. Table 9 shows the statistical results of the mooring tension time record.

The extreme values of tension in the slack mooring system are greater than those in the catenary system. However, in terms of standard deviation, an opposite result is observed. The variation of the catenary mooring tension is more severe than that of the slack mooring tension. This result is confirmed by the spectra of mooring tensions shown in Figure 15, where the spectral peaks of the catenary system, indeed, are greater than those of slack system, for low frequency and wave frequency components.

Assuming that the buoy is working as a WEC, it is possible to compute the absorbed energy from the heave motion time series. The hypothetical results are shown in Figure 16, and are based on Eqs. 3 and 4. As evidenced by the experimental results shown above, the influence of the mooring system on heave motions is not significant, so it can be expected that the absorbed energy (that depends on heave motions) will not be greatly affected by the choice of the mooring system. The estimated absorbed energy for this WEC model under the tested sea condition was 0.008 W (the assumed B_{pto} was 1 N.s/m).

5 CONCLUSIONS

The performance of two mooring systems for typical point absorbers has been addressed experimentally. One mooring system is the typical system with steel catenary mooring lines, while the other is an alternative approach that uses hybrid lines, named as slack mooring system.

The performance of both mooring systems has been assessed through the buoy motions in six degrees of freedom and the mooring line tension in regular and irregular wave conditions.

In general, no significant differences were observed in the heave and pitch motions. However, surge responses and mooring line tensions were significantly affected both in regular an irregular waves.

For mild regular waves (operational conditions), mean drift motions are lower for the catenary system. However, in terms of the oscillatory drift amplitudes, the catenary system display greater values. For severe regular waves (survival conditions), the opposite tendency was observed when comparing both mooring systems. The main reason for this behavior is the two-stage restoring force displayed by the slack system. For small to moderate excursions, the stiffness of the slack mooring system is basically restricted to the catenary (lower) segment of the slack mooring line. However, when large excursions take place, the elastic stiffness associated to the polyester segment of the mooring line is "activated", resulting in a larger total stiffness when compared to the catenary system. Snap loads have been recorded in the lines of the slack mooring system under survival conditions.

A test run in irregular waves, corresponding to an operational condition have also been performed and, in terms of motions, similar conclusions to those in regular wave cases were obtained. However, in terms of mooring line tension, the spectral analyses have demonstrated that the peaks can be significantly lower for the slack mooring system, especially in the wave-frequency range.

Regarding wave energy absorption, as the experimental results indicate that heave motions are not significantly affected by the choice of the mooring system, the absorbed energy for a heaving point absorber may not be affected by the adopted mooring system, too.

ACKNOWLEDGEMENTS

This work was financially supported by the Project PTDC/MAR-TEC/6984/2014 – POCI-01-0145-FED ER-016882 – funded by FEDER funds through COMPETE2020 – Programa Operacional Competitividade e Internacionalização (POCI) and by national funds through FCT – Fundação para a Ciência e a Tecnologia, I.P.

COMPETE 2020

REFERENCES

Angelelli, E., Zanuttigh, B., Ferri, F., Kofoed, J.P. 2013. Experimental assessment of the mooring influence on the power output of floating Wave Activated Body WECs. *Proc; 10th EWTEC* (DK).

Casaubieilh, P., Thiebaut, F., Retzler, C., Shaw, M. and Sheng, W. "Performance improvements of mooring systems for wave energy converters.", in: Guedes Soares, C. (Ed.). *Renewable Energies Offshore*, London, UK: Taylor & Francis Group; pp. 897–903.

Cerveira, F., Fonseca, N., Pascoal, R. 2013. Mooring system influence on the efficiency of wave energy converters. *International Journal of Marine Energy*, 3, pp. 65–81.

Falcão, A.F., 2015, "Developments in oscillating water column wave energy converters and air turbines", in: Guedes Soares, C. (Ed.). *Renewable Energies Offshore*, London, UK: Taylor & Francis Group; pp. 3–11.

Falnes, J. 2002. *Wave-energy absorption by oscillating bodies, in: Ocean Waves and Oscillating Systems: Linear Interactions Including Wave-Energy Extraction.* Cambridge: Cambridge University Press, pp. 196–224.

Fonseca, N., Pascoal, R., Morais, T. and Dias, R. 2009. Design of a mooring system with synthetic ropes for the FLOW wave energy converter. In *ASME 2009 28th International Conference on Ocean, Offshore and Arctic Engineering* (pp. 1189–1198). American Society of Mechanical Engineers.

Fitzgerald, J, & Bergdahl, L. (2008), "Including moorings in the Assessment of a Generic Offshore Wave Energy Converter: a Frequency Domain approach", *Marine Structures*, 21, pp 23–46.

Guedes Soares, C.; Bhattacharjee, J.; Tello, M., and Pietra, L. 2012; Review and classification of Wave Energy Converters. Guedes Soares, C. Garbatov Y. Sutulo S. & Santos T.A., (Eds.). *Maritime Engineering and Technology.* London, UK: Taylor & Francis Group; pp. 585–594.

Harnois, V., Weller, S.D., Johanning, L., Thies, P.R., Le Boulluec, M., Le Roux, D., Soule, V., Ohana, J. 2015. Numerical model validation for mooring systems: Method and application for wave energy converters. *Renewable Energy*, 75, pp. 869–887.

Hayward, J. & Osman, P. 2011. *The Potential of Wave Energy*, Report, CSIRO.

Johanning I, Smith GH, Wolfram J., 2006. Mooring design approach for wave energy converters. *Journal of Engineering for the Maritime Environment* 220:159–74.

Nava, V.; Rajic, M., and Guedes Soares, C. 2013. Effects of the mooring line configuration on the dynamics of a point absorber. *32nd International Conference on Ocean, Offshore and Arctic Engineering (OMAE 2013);* Nantes, France. OMAE2013-11141.

Paredes, G.M., Palm, J., Eskilsson, C., Bergdahl, L., Taveira-Pinto, F. 2016. Experimental investigation of mooring configurations for wave energy converters. *International Journal of Marine Energy*, 15, pp. 56–67.

Rezanejad, K. and Guedes Soares, C. 2015, Hydrodynamic performance assessment of a floating oscillating water column. Guedes Soares, C. & Santos T.A. (Eds.), *Maritime Technology and Engineering*, London, UK: Taylor & Francis Group; pp. 1287–1296.

Rezanejad, K.; Bhattacharjee, J., and Guedes Soares, C. 2013; Stepped sea bottom effects on the efficiency of nearshore oscillating water column device. *Ocean Engineering*. 70:25–38.

Rezanejad, K.; Bhattacharjee, J., and Guedes Soares, C. 2015; Analytical and numerical study of dual-chamber oscillating water columns on stepped bottom. *Renewable Energy*. 75:272–282.

Rezanejad, K.; Bhattacharjee, J., and Guedes Soares, C. 2016; Analytical and numerical study of nearshore multiple oscillating water columns. *Journal of Offshore Mechanics and Arctic Engineering*. 138:021901-1 -021901-7.

Rezanejad, K.; Guedes Soares, C.; López, I., and Carballo, R. 2017; Experimental and numerical investigation of the hydrodynamic performance of an oscillating water column wave energy converter. *Renewable Energy*. 106:1–16.

Sergiienko, N.Y., Cazzolato, B.S., Ding, B., Arjomandi, M. 2016. An optimal arrangement of mooring lines for the three-tether submerged point-absorbing wave energy converter. *Renewable Energy*, 93, pp. 27–37.

Weller, S.D., Davies, P., Vickers, A.W., Johanning, L., 2014 "Synthetic rope responses in the context of load history: operational performance" *Ocean Eng*, 83, pp. 111–124.

Weller, S.D., Johanning, L. Davies, P., Banfield S.J., "Synthetic mooring ropes for marine renewable energy applications" *Renewable Energy*, 83, pp. 1268–1278.

Zanuttigh, B., Angelelli, E., Kofoed, J.P. 2013. "Effects of mooring systems on the performance of a wave activated body energy converter", *Renewable Energy*, 57, 422–431.

Zaroudi, H.G., Rezanejad, K. & Guedes Soares, C. (2015). Assessment of mooring configurations on the performance of a floating oscillating water column energy converter.", in: Guedes Soares, C. (Ed.). *Renewable Energies Offshore*, London, UK: Taylor & Francis Group; pp. 921–928.

Progress in Maritime Technology and Engineering – Guedes Soares & Santos (Eds)
 ISBN 978-1-138-58539-3

Experimental study on auto-parametrically excited heaving motion of a spar-buoy

T. Iseki
Tokyo University of Marine Science and Technology, Tokyo, Japan

ABSTRACT: Experimental studies were carried out to investigate the possibility of utilization of auto-parametrically excited oscillation based on the Mathieu-type instability. For the simplicity, the subject is limited to the heaving motion of a spar-buoy type point absorber. The device consists of an inner cylinder and 12 outer cylinders. Each outer cylinder is equipped a movable floating column controlled by a ball screw mechanism and the buoyancy of the outer cylinders can be dynamically changed to induce the Mathieu-type instability. Based on results of the numerical simulation and the model tests, the possibility of utilization of auto-parametrically excited oscillation for WEC was discussed. Furthermore, the effect of the proposed control system in irregular waves was investigated.

1 INTRODUCTION

Wave energy is widely considered as one of the most possible renewable and sustainable energy. Many types of energy-extracting technologies have been developed and some full-scale devices were built (Falnes 2002a, 2007, Falcão 2010). Based on the operating principle, the technologies can be divided into three categories, Oscillating water column, Oscillating bodies and Over-topping. The most advantageous point of the oscillating body type wave energy converters is the huge kinetic energy. To maximize the energy, the oscillation characteristics of the floating device, such as point absorbers, should be optimized to the most frequent waves at the local site. However, the frequency band of the response is relatively narrow and the natural frequency is much higher than the typical ocean wave frequencies. Therefore, several phase-control methods have been proposed (Falnes 2002b). One of the prominent measures to improve the efficiency is phase control by latching (Falnes & Budal 1978, Budal & Falnes 1980). The control by latching is applied to two-body devices and showing the practicality (Henriques et al. 2012).

On the other hand, it is known that the spar-buoy type platform shows extreme heave motions at resonance. Moreover, parametric resonance of the pitch motion has been confirmed. The phenomenon can be explained by the auto-parametrically excited oscillation based on the Mathieu-type instability. When the period of the heave motion is half of the pitch natural period, the equation of pitch motion can be expressed by Mathieu's equation. In this problem, it is essential to analyze the Mathieu-type instability (Rho et al. 2005) and preventive measures should be investigated (Koo et al. 2006).

By changing the view point, the author proposed a new concept to utilize the Mathieu-type instability for WEC (Iseki 2014, 2015). It is well known that the auto-parametrically excited oscillation based on the Mathieu's equation can be induced by the fluctuating restoring force or moment. For the simplicity, the subject was limited to the heaving motion of a spar-buoy type point absorber. The period of heaving motion of a buoy is determined by the ratio of the restoring force to its mass. A possible measure to change the ratio is alteration on the restoring force (buoyancy). For that purpose, the spar-buoy consists of an inner cylinder and 12 outer cylinders. The water plane area was changed by axially-sliding valves equipped to the outer cylinders in order to control the restoring force. Based on the numerical simulations, the possibility of utilization of auto-parametrically excited oscillation was shown in the effective range of the linear simulation. However, it was revealed that the axially-sliding valves were insufficient to induce an auto-parametrically excited oscillation in later experiments.

The author proposed a different type of dynamic control system in order to utilize the auto-parametrically excited oscillation based on the Mathieu-type instability (Iseki 2017). The new spar-buoy model was equipped the movable floating column systems to change the heaving restoring force and induce the Mathieu-type instability. Model tests were conducted to investigate the possibility of utilization of the Mathieu-type instability and the effect of the control system was confirmed.

In this paper, the detailed power balance is discussed and the effect of the proposed control system in irregular waves is investigated. This concludes that further effective control system will be required to develop a real system for WEC.

2 SPAR-BUOY MODEL

2.1 *Configuration*

In the previous report (Iseki 2014, 2015), a small spar-buoy model was made and axially-sliding valves were equipped in order to change the restoring force. Based on numerical simulations, it seemed to be possible to induce auto-parametrically excited oscillations. However, it was revealed that the axially-sliding valves were insufficient to induce an auto-parametrically excited oscillation in later experiments. The reason was that the sliding valves can reduce the restoring force but cannot increase that. Therefore, the spar-buoy model has been modified to realize Mathieu-type instability.

The model consists of an inner cylinder and 12 outer cylinders as well. The cylinders are made of transparent acrylic acid resin. The configuration of the model is shown in Figure 1 and the principal particulars are listed in Table 1. The inner cylinder is the main body of the spar-buoy and the outer cylinders are equipped with floating bodies inside to control the heaving restoring force. The floating bodies were made of styrene foam and connected to a control mechanism to change their drafts.

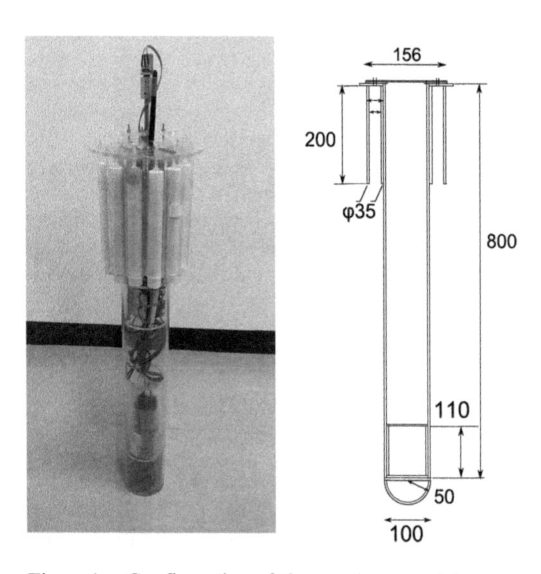

Figure 1. Configuration of the spar-buoy model.

Table 1. Principal particulars of the buoy model.

	Inner cylinder	Outer/ cylinder
Depth	0.8 m	0.2 m
Diameter	0.1 m	0.035 m
Draft	0.7 m	0.1 m
Total water plane area	1.74×10^{-2} m^2	
Total mass	6.18 kg	

2.2 *Controlling mechanism*

The spar-buoy model has a very simple mechanism in order to change its restoring force. Two circular plates are attached to the upper opening of the outer cylinders to make up a controlling mechanism. 12 columns of styrene form are put to the upper circular plate by screws and inserted in each outer cylinder. The relative vertical position of the upper circular plate is controlled by a ball screw mechanism and measured by a potentiometer equipped on the top of the mechanism. The driving force of the ball screw mechanism is supplied from a stepping motor controlled by an Arduino UNO R3. Nine axes accelerometer (Arduino 9 Axes Sensor Shield) is also installed to measure motions of the buoy. The configuration of the mechanism is shown in Figure 2 and 3.

2.3 *Realization of the parametric oscillation*

The floating columns controlled by the ball screw mechanism were equipped to the spar-buoy model to realize the parametrically excited oscillation. In this section, the method of realization is theoretically explained.

Expressing the vertical displacement by z, the equation of heaving motion can be written as follows:

$$(m+m')\ddot{z} + N_e\dot{z} + \rho g A_{WM} z + \rho g A_{WF}(z+z_C) = 0 \quad (1)$$

where m and m′ are the mass and the added mass of the buoy, N_e the equivalent linear damping coefficient, ρ the density of water, g the acceleration of gravity, z_c the relative position of the floating columns to the main body, and A_{WM} and A_{WF} are the water plane areas of the main body and the 12 floating columns.

Taking into account the relation, $A_W = A_{WF}+A_{WM}$, Equation 1 can be rewritten as follows:

$$(m+m')\ddot{z} + N_e\dot{z} + \rho g A_W z + \rho g A_{WF} z_C = 0 \quad (2)$$

Dividing the both side by m+m’, the equation becomes as follows:

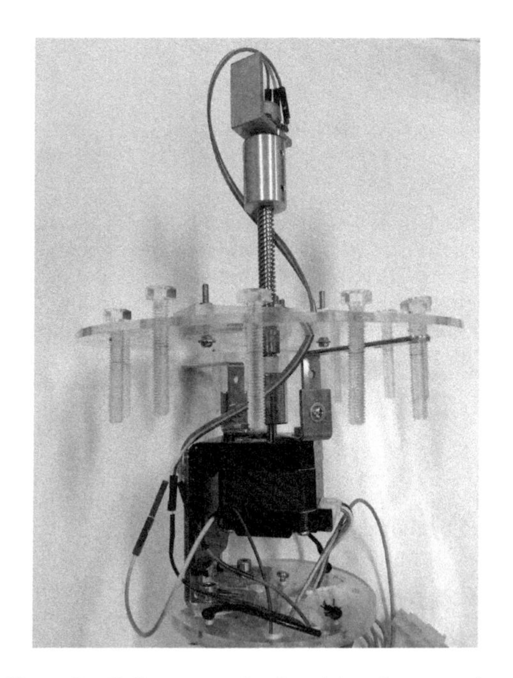

Figure 2. Ball screw mechanism driven by a stepping motor.

Figure 3. The upper circular plate is connected to the 12 floating columns and reciprocated by the ball screw mechanism.

$$\ddot{z}+2\alpha\dot{z}+\omega_n^2 z+\frac{\rho g A_{WF}}{m+m'}z_C=0 \tag{3}$$

where the damping factor α and the natural frequency ω_n are expressed as follows:

$$\omega_n=\sqrt{\frac{\rho g A_W}{m+m'}},\quad \alpha=\frac{N_e}{2(m+m')} \tag{4}$$

On the other hand, Mathieu's equation and the approximated solution are expressed as follows:

$$\ddot{z}+2\alpha\dot{z}+\left(\omega_{n0}^2+a\cos 2\omega_n t\right)z=0 \tag{5}$$

$$z=Ce^{\left(\frac{a}{4\omega_n}-\alpha\right)t}\sin\left(\omega_n t+\phi\right),\quad \phi=\frac{3\pi}{4} \tag{6}$$

where a is the amplitude of the fluctuating component of the restoring force and ϕ is the phase angle which is closely related to the component.

Therefore, the relative position z_c, which can realize the parametrically excited oscillation, is expressed as follows:

$$z_C=\frac{m+m'}{\rho g A_{WF}}az\cos 2\omega_n t \tag{7}$$

The actual timing of the restoring force, relative position z_c and the heaving motion during an auto-parametrically excited oscillation are illustrated in Figure 4. It should be noted that the relative position z_c expressed by Equation 7 includes vertical displacement of the main body z and the z_c becomes larger and larger with time.

In this report, as a practical realization of the parametric oscillation, the relative position z_c is defined as follows:

$$z_C=z_{C0}\sin(\omega_n t+\phi)\cos 2\omega_n t \tag{8}$$

where z_{c0} is the constant amplitude of the columns' movement generated by the ball screw mechanism. In this report, the amplitude was set to 2.8 mm.

Figure 5 shows the measured time histories of the heaving acceleration and the relative position of the columns. It should be noted that the heaving motion is indicated by the measured vertical

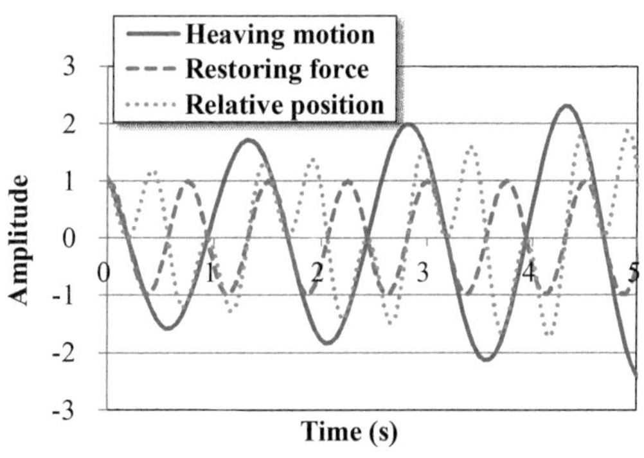

Figure 4. Time histories of the restoring force, relative position z_C and the heaving motion during an auto-parametrically excited oscillation.

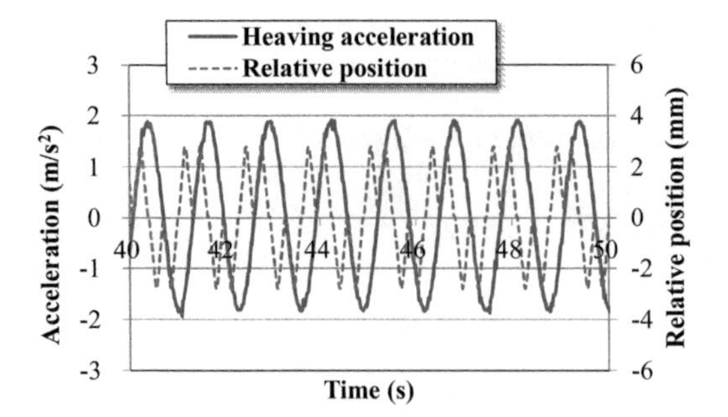

Figure 5. Time histories of the heaving acceleration and the relative position of the columns.

acceleration unlike Figure 4. Compared to Figure 4, the curve of the relative position seems to be linear sequence because of a ball screw mechanism and a stepping motor.

2.4 *Power estimation*

In this section, the power increased by the proposed control system is theoretically estimated. First of all, we consider a basic resonance condition in regular waves without any control. The equation of motion can be expressed as follows:

$$\ddot{z} + 2\nu\omega_n\dot{z} + \omega_n^2 z = \omega_n^2 F_{z0}\cos(\omega_n t + \phi) \tag{9}$$

where F_{z0} denotes the amplitude of the wave exciting force and ν denotes the damping ratio defined by the following relationship:

$$\alpha = \nu\omega_n \tag{10}$$

The solution at the resonance condition is expressed as follows:

$$z = z_0\sin(\omega_n t + \phi), \quad z_0 = \frac{F_{z0}}{2\nu} \tag{11}$$

Secondly, we consider the solution under the dynamic controlled condition. Denoting the increased heaving amplitude by z_ε, the equation of motion and the solution can be expressed as follows:

$$\ddot{z} + 2\nu\omega_n\dot{z} + \omega_n^2 z + \frac{A_{WF}}{A_W}\omega_n^2 z_C = \omega_n^2 F_{z0}\cos(\omega_n t + \phi) \tag{12}$$

$$z = (z_0 + z_\varepsilon)\sin(\omega_n t + \phi) \tag{13}$$

Thirdly, multiplying both side of Equation 12 by *dz*/*dt*, the following relationship can be obtained.

$$\frac{d}{dt}\left(\frac{1}{2}\dot{z}^2 + \frac{1}{2}\omega_n^2 z^2\right) = -2\nu\omega_n\dot{z}^2 - \frac{A_{WF}}{A_W}\omega_n^2 z_C\dot{z} + \omega_n^2 F_{z0}\cos(\omega_n t + \phi)\dot{z} \tag{14}$$

The left side of Equation 14 denotes the time derivative of the kinetic and potential energy of the buoy. Therefore, the right side of Equation 14 must become zero in the steady state. Taking the one cycle average of the right, the increased heaving amplitude z_ε can be obtained as follows:

$$z_\varepsilon = \frac{A_{WF}}{A_W}\frac{z_{C0}}{4\nu} \tag{15}$$

Finally, the increased power ΔW can be calculated by the following equation.

$$\Delta W = (m + m')\left\{\frac{1}{2}(z_0 + z_\varepsilon)^2\omega_n^3 - \frac{1}{2}z_0^2\omega_n^3\right\} \tag{16}$$

3 MOTION OF THE SPAR-BUOY

3.1 *Frequency response*

Frequency response function of heaving motion was calculated by in-house software (Iseki et al. 1993). The program is based on the three dimensional boundary integral method and the motion is assumed to be single degree of freedom. The Green function for infinite depth (Kim 1965) is introduced and the Froude-Krylov force and the diffraction force are considered as wave exciting forces.

The surface of the spar-buoy is divided into 612 triangular elements and the outer cylinders are approximated by a change of diameter of the inner cylinder that has the same water plane area as the sum of the inner and outer cylinders.

The added mass and the damping coefficients estimated by solving the radiation problem are shown in Figure 6. The estimated radiation damping coefficients are very small and it is considered that the actual damping is dominated by frictional resistance. Therefore, the actual damping coefficients evaluated by the previously conducted model experiments were used in the calculation of the response functions.

3.2 *Numerical simulation*

In order to investigate the possibility of utilization of auto-parametrically excited oscillation based on the Mathieu-type instability, simulations in the

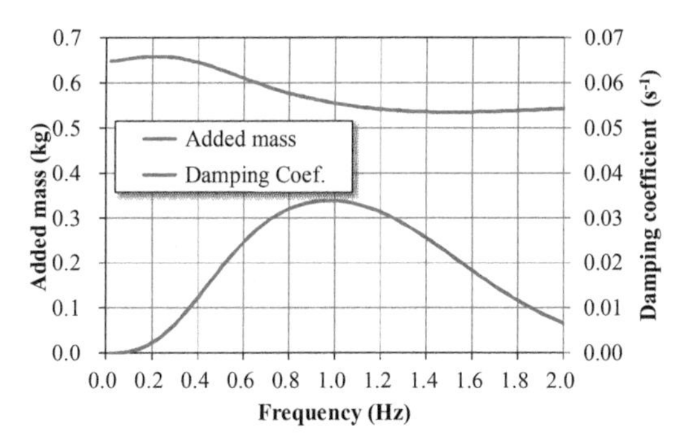

Figure 6. Added mass of the heaving motion under the four conditions.

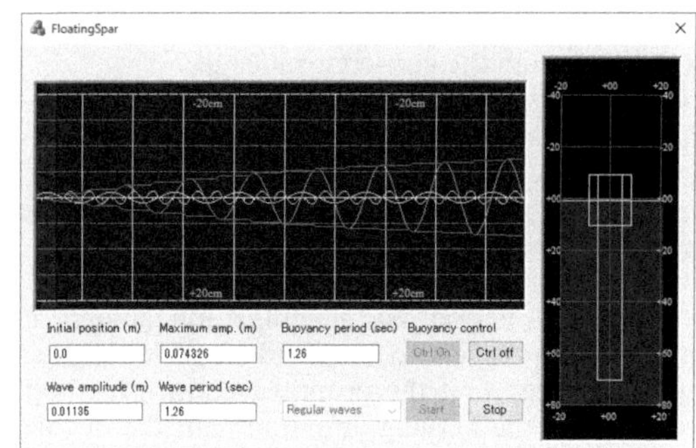

Figure 7. Screenshot of the simulation software.

Table 2. Basic characteristics of the spar-buoy model.

Natural frequency	0.793 Hz
Natural period	1.26 s
Damping coefficient	0.166 s^{-1}
Added mass at the natural period	0.578 kg

Table 3. Principal dimensions of ship maneuvering research basin of TUMSAT.

Length	54 m
Breadth	10 m
Depth	2 m
Wave maker	Flat plate type

time domain are conducted according to Equation 3. The damping coefficient, which was used in the simulation, was evaluated by the previously conducted model experiments. The added mass in the simulation program was based on the three dimensional boundary integral method described in the previous section. The actual value of the parameters are listed in Table 2. The wave exciting force consists of Froude-Krylov force and diffraction force. The time integration is carried out by the Runge-Kutta method.

Figure 7 shows the graphical user interface of the simulation program. In the right part of the screen, the heaving motion of the spar-buoy model is shown and time histories of heaving motion, wave elevation and the relative position z_c are indicated in the left part of the dialog box.

3.3 *Experimental results*

In order to prove the possibility of utilization of auto-parametrically excited oscillation, model experiments in regular waves were conducted in the Ship Maneuvering Research Basin of TUMSAT. The principal dimensions and the photo are shown in Table 3 and Figure 8.

Figure 9 shows the measured time histories of the heaving accelerations. The incident wave period and height were 1.26 s and 5.8 mm.

The bold solid line denotes the result with fixed floating columns ("Non-control"). On the other hand, the dotted line denotes the result with position controlled columns ("Controlled"). As shown

Figure 8. Ship maneuvering research basin of TUMSAT.

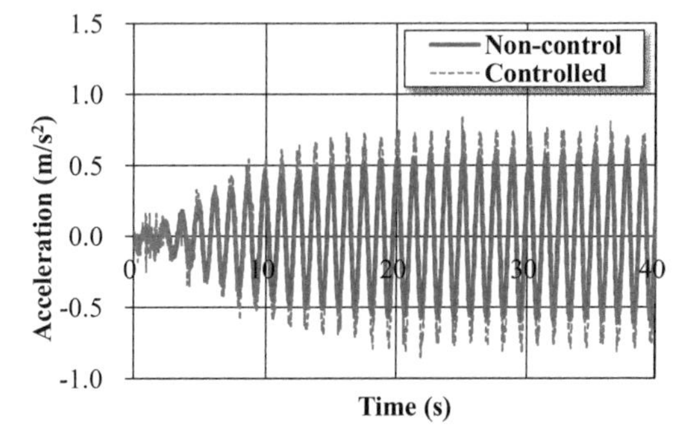

Figure 9. Measured time histories of heaving acceleration under the controlled and non-controlled conditions. The incident wave period and the height were 1.26 s and 5.8 mm.

in the figure, the controlled results have larger amplitude than the non-control result.

Figure 10 shows the comparison of the magnification factor between the simulations and the experiments. The magnification factor is defined by the ratio of the controlled amplitude to the non-control one. The vertical axis denotes the magnification factor and the horizontal axis denotes the incident wave period. The magnification factors are plotted with respect to the incident wave heights. It can be seen that the proposed control method can enlarge the heaving amplitude but the degree is affected by the incident wave height. The constant amplitude of the floating columns' movement can be considered as the reason why the magnification factors are affected. On the other hand, the magnification factors are not sensitive to the incident wave periods. Compared to the experimental results, it can be said that the rough trend match each other but experimental results are scattered around the simulated results. Therefore, the accuracy of the experiments should be improved more.

Figure 11 shows the comparison of the increased power based on the proposed control method. It can be seen that the experimental results are smaller than that of the simulations. It can be considered that the cause of the disagreement comes from the linear assumption of the damping coefficient. On the other hand, the consumed power of the controlling system was 1.87 W (1.20 W: Stepping motor, 0.57 W: Arduino). That was measured by the power supply voltage and the current under a non-load condition. Based on the results, it is revealed that around ten percent of the consumed power was recovered as the increased kinetic energy of the buoy. This concludes that the proposed control system can induce the Mathieu-type instability but is not suitable for wave energy converters yet. Further effective concept of the control system is required to develop a system for real WEC.

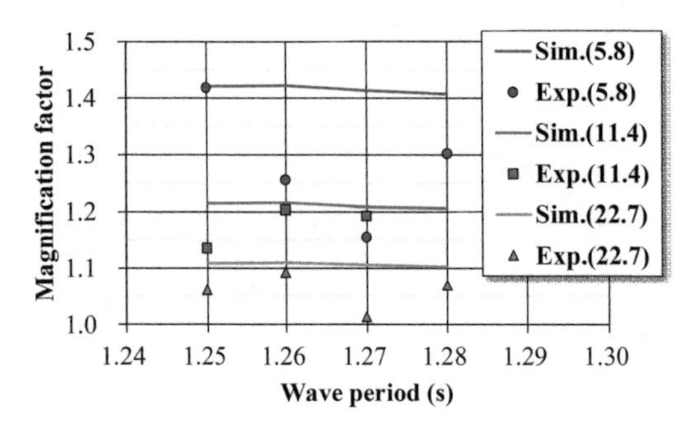

Figure 10. Comparison of the magnification factor between simulations and experiments with respect to the incident wave heights (5.8, 11.4 and 22.7 mm).

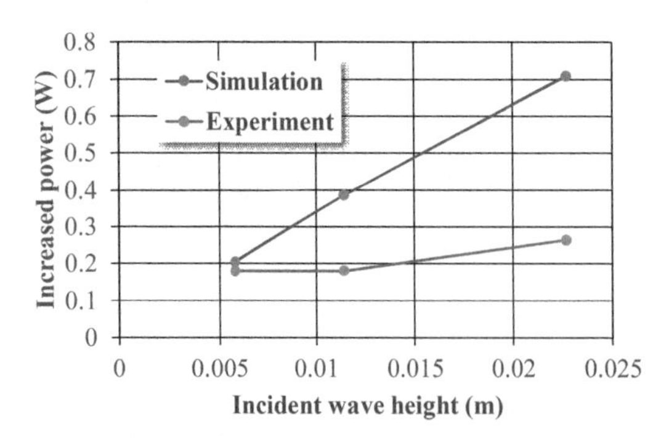

Figure 11. Increased power of the buoy compared between the experiments and the simulations.

3.4 *Simulations in irregular waves*

The irregular waves were generated by the assumption of linear superposition. In the simulation program, 500 component waves are calculated with random phase angles and amplitudes calculated from ISSC spectrum. The simulation was conducted in the model scale. The scale ratio, therefore, was assumed to be 1/20 temporally.

The spectrum of the numerically generated irregular waves and the ISSC spectrum are compared in Figure 12. The simulated time span is 409.6 s with 0.1 s sampling time. The mean wave period is 1.25 s and the significant wave height is 0.02 m.

Figure 13 shows the comparison of heave spectra under the "Non-control" and "Controlled" conditions. Looking at the graph, it is confirmed that the frequency responses at the natural frequency are very narrow. The spectral area of the controlled condition is 34.3% larger than that of the non-control condition. It concludes that the proposed control method is still effective even in irregular waves.

Figure 14 shows the comparisons of the simulated time histories with "Non-control" and

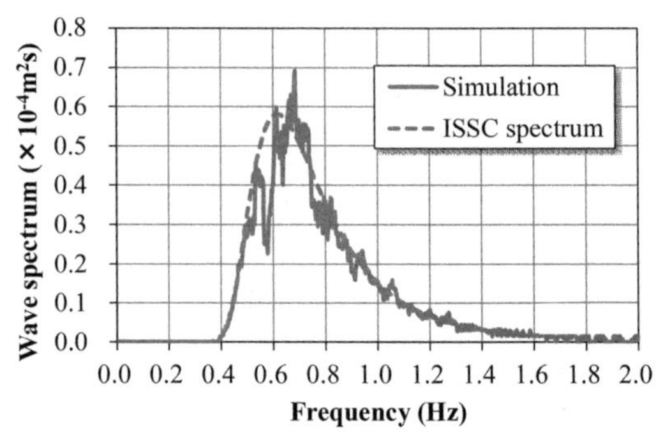

Figure 12. Comparison of the ISSC spectrum and the wave spectrum (T_{01} = 1.25 s, f_{01} = 0.8, $H_{1/3}$ = 0.02 m).

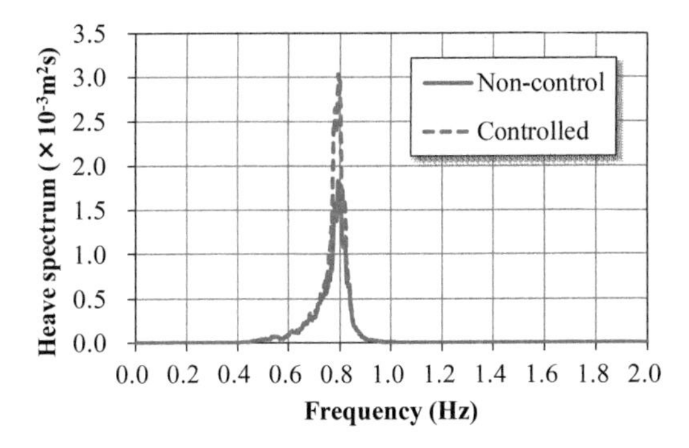

Figure 13. Comparison between the heave spectra under the "Non-control" condition and the "Controlled" condition.

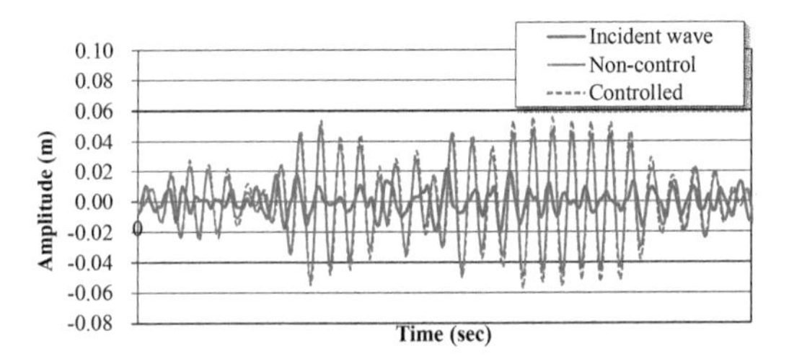

Figure 14. Comparison between the heave spectra under the "Non-control" condition and the "Controlled" condition.

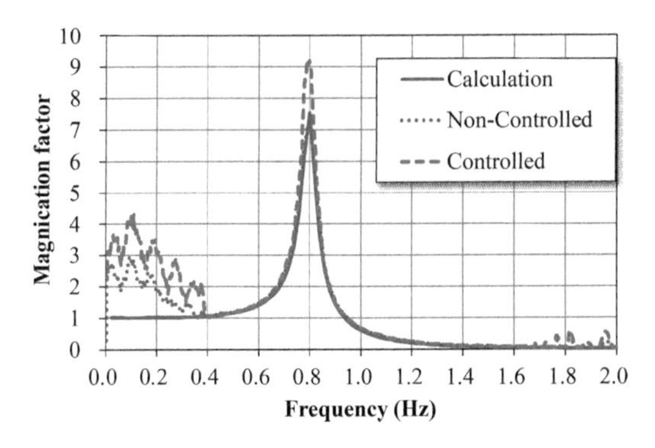

Figure 15. Comparison of the RAOs of the heaving motion estimated by the frequency response calculation and the simulations in time domain.

"Controlled" conditions. The graph is showing results of the first 40 seconds. It can be seen that the heaving oscillations have only a single frequency. Looking at Figure 13 again, the frequency response of the buoy is very narrow. Therefore, this concludes that the proposed control method is effective even in irregular waves because the heaving motion can be considered as a single frequency oscillation.

Figure 15 shows the comparison of RAOs of the heaving motion estimated by the frequency response calculation and the simulations in time domain. The RAOs are agree well and it can be concluded that the numerical simulation program is reliable.

4 CONCLUSIONS

Experimental studies were carried out to investigate the effectiveness of a dynamic control of oscillation characteristics. The device is a spar-buoy type point absorber and consists of an inner cylinder and 12 outer cylinders. Each outer cylinder is equipped a movable floating column controlled by a ball screw mechanism. The heaving restoring force can be changed dynamically by the floating column's movement. The numerical simulation and model tests were conducted in order to investigate the possibility for utilizing auto-parametrically excited oscillation based on the Mathieu-type instability. The results are summarized below:

1. It is possible to control the heaving restoring force by the ball screw mechanism driven by a stepping motor.
2. Based on the model experiments, the possibility of utilization of auto-parametrically excited oscillation was shown under several conditions. However, it was also revealed that the efficiency is far from practical applications at this stage.
3. Based on the simulations in irregular waves, it was confirmed that the proposed controlling method was effective even in irregular waves.
4. In order to induce the real auto-parametrically excited oscillation based on the Mathieu-type instability, the amplitude of the fluctuating restoring force must be magnified with the motion amplitude.

The dynamic control mechanism proposed in this report requires some amount of energy to drive itself. In order to induce a real auto-parametrically excited oscillation, the controlling energy have to be supplied from the oscillating response itself like the parametric rolling. Further effective concept is required to develop a system for real WEC.

ACKNOWLEDGEMENTS

This work was supported by JSPS KAKENHI Grant Number 17 K06960. The author expresses sincere gratitude to the above organization and thanks Mr. Yuki Uchibori, graduate student of Tokyo University of Marine Science and Technology.

REFERENCES

Budal, K. and Falnes, J. 1980, "Interacting point absorbers with controlled motion", in: B. Count (Ed.), Power form the Waves, Academic Press, London, 381–399.

Falcão, A.F.O. 2010, "Wave energy utilization: A review of the technologies", Renewable and Sustainable Energy Reviews 14, 899–918.

Falnes, J. and Budal, K. 1978, "Wave-power conversion by power absorbers, Norwegian Maritime Re-search 6 (4), 2–11.

Falnes, J. 2002a, "Ocean waves and oscillating systems", Cambridge University Press.

Falnes, J. 2002b, "Optimum control of oscillation of wave energy converters", International Journal of Offshore and Polar Engineering 12 (2), 147–155.

Falnes, J. 2007, "A review of wave-energy ex-traction", Marine Structures 20, 185–201.

Henriques, J.C.C., Falcão, A.F.O., Gomes R.P.F. and Gato, L.M.C. 2012, "Latching Control of an OWC Spar-buoy Wave Energy Converter in Regular Waves", Proc. OMAE2012-83631.

Iseki, T. 2014, "Optimization Method for Oscillation Characteristics of a Spar-buoy", Proc. OMAE2014-23223.

Iseki, T. 2015, "Dynamic Control of Oscillation Characteristics of a Spar-buoy", Maritime Technology and Engineering – Guedes Soares & Santos (Eds), 2, CRC Press, Taylor & Francis Group (London), 1243–1250.

Iseki, T. 2017, "Experimental Study on Dynamic Control of Oscillation Characteristics of a Spar-buoy", Proc. OMAE2017-61612.

Iseki, T., Ohtsu, K. and Minami, K. 1993, "A Study on Distributions of Significant Wave Height around a Ship in Irregular Waves – Considerations in Approach for small Boat –", J. Japan Institute of Navigation, 89, 63–70.

Kim, W. D. 1965, "On the Harmonic Oscillation of a Rigid Body and a Free Surface", J. F.M, 21, Part 3, 427–451. Larch, A.A. 1996b. Facilities ...

Koo, B., Kim, M. and Randall, R. 2006, "Mathieu instability of a spar platform with mooring and risers," Ocean Engineering, 31(2), 249–256.

Rho, J.B., Choi, H.S., Shin, H. S. and Park I. K. 2005, "A Study on Mathieu-type Instability of Conventional Spar Platform in Regular Waves", Int. J. Offshore Polar Eng., 15(2), pp. 104–108.

Waves

Progress in Maritime Technology and Engineering – Guedes Soares & Santos (Eds)
 ISBN 978-1-138-58539-3

Numerical analysis of waves attenuation by vegetation in enclosed waters

G.O. Mattosinho
Departamento de Engenharia Civil, Instituto Federal de Minas Gerais, Piumhi, Brazil

G.F. Maciel
Departamento de Engenharia Civil, Universidade Estadual Paulista (UNESP), Ilha Solteira, Brazil

A.S. Vieira
Universidade Federal de Mato Grosso do Sul (UFMS), Department de Sistemas de Informação, Coxim, Brazil

C.J.E.M. Fortes
Laboratório Nacional de Engenharia Civil (LNEC), Departamento de Hidráulica e Ambiente (DHE), Lisbon, Portugal

ABSTRACT: Numerical simulations using the SWAN-VEG model (Simulating **WA**ves **N**earshore, with vegetation module) were performed considering the case of Ilha Solteira (SP) reservoir, a navigable stretch of the Tietê—Paraná waterway. The objective was to analyse the attenuation of wind-generated wave energy by the presence of vegetation in the left margin of reservoir (place of higher simulated wave heights). In this analysis, we varied the vegetation parameter, V_f, different submergence ratios and the wind intensity for NE wind direction. The results pointed up a neglegible influence of mesh size but high wave attenuation coefficients due the presence of vegetation. Application of non-structural measures to the attenuation of waves in these areas is plausible and should be encouraged because in addition to producing high attenuation coefficients, this is in general a cheaper and environmentally friendly solution.

1 INTRODUCTION

The Tietê-Paraná Waterway is a navigation route that crosses the south, southeast and central-west regions of Brazil. It allows the transportation of cargo and passengers along the Paraná and Tietê rivers, being extremely important to helping to channel the agricultural production in the states of Mato Grosso, Mato Grosso do Sul, Goiás and for a portion of Rondônia, Tocantins and Minas Gerais. It owns twelve terminals, spread over 76 million hectares. The beginning of this system promoted the implementation of 23 industrial centers, 17 tourist centers and 12 distribution centers, thus directly creating about 4,000 jobs.

Figure 1 shows the general plan of the Tietê-Paraná Waterway and Figure 2 shows the representation of the stretch in the Ilha Solteira reservoir—SP.

In terms of market competitiveness, the waterways are one of the best solutions to reduce transport costs, since Brazil has an extensive hydrographic network, which is little used. However in those waterways, the waves generated by the winds, besides being difficult to navigate, are responsible also for the erosion of the margins and landslides.

Figure 1. General Plan of the Paraná River Waterway—Itaipú—São Simão.

So, to solve those problems (and therefore to increase cargo transportation and Tietê-Paraná occupancy rate), it is necessary to increase and deepen studies on: a) meteorological-hydrody-

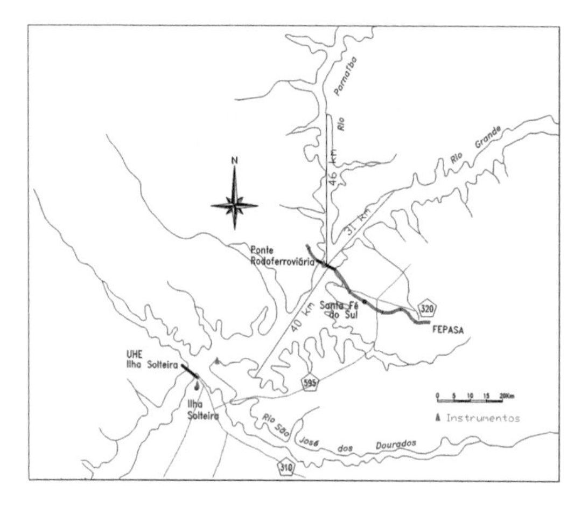

Figure 2. Representation of the stretch/route of the Tietê-Paraná waterway in the Ilha Solteira reservoir—SP.

namic coupling, with prediction of waves generated by winds in the reservoir of the Ilha Solteira dam, b) its effects on the safety of the navigation; c) the effect of waves on erosion processes at the of margins, transport of sediments, among others.

Moreover, it is necessary to find solutions for those problems. In general, rigid structures such as dikes and breakwaters (structural measures) are commonly used but they interfere drastically with the natural environment. Nowadays, alternative solutions (non-structural measures) that do not significantly alter the ecosystem in which they are inserted are gaining strength, in the face of the constant call for the maintenance and protection of the environment as a whole.

It is observed in the case of the reservoir of the Ilha Solteira Hydroelectric Power Plant, that nature created mechanisms to protect itself from the impacts generated by waves. One of those mechanisms is the vegetation along riverbanks and lakes (Figure 3). Natural vegetation may attenuate the waves and their effects, which is an interesting bioengineering option, without environmental and economic damages. Therefore, it is fundamental to analyse the influence of vegetation on the reduction of wave energy near the lake banks.

However, this dissipation mechanism (wave × vegetation interaction) is not yet fully understood.

Numerical models can be applied to study the effect of vegetation on the wave propagation and dissipation. These models can give a wind-wave characterization of all the study region and they can be used to long term studies. Particularly, the use of the spectral nonlinear model SWAN-VEG numerical model (OUDE, 2010 and SUZUKI et al., 2011) is an alternative, that solves the wave

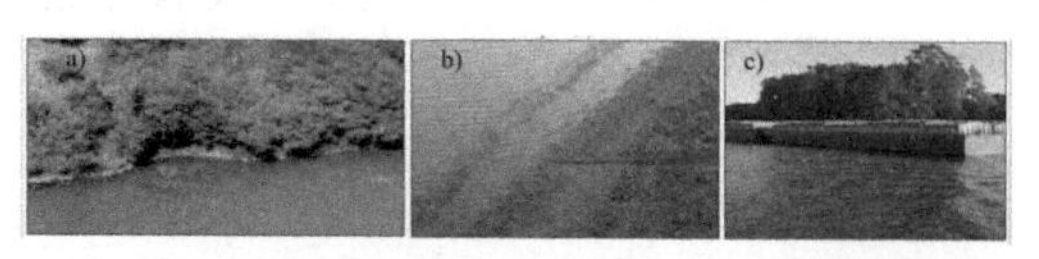

Figure 3. Erosive processes on the banks of the Ilha Solteira reservoir—SP: a) erosion of the margin by the waves; b) damping of waves by vegetation; c) generation of waves per vessel.

action equation and takes into account several processes on the wave propagation such as, for instance, wave generation, refraction, diffraction, shoaling, dissipation by wave breaking due to bottom influence, by whitecapping, by bottom friction or by vegetation, non-linear interactions between waves (triad and quadruple wave-wave interactions).

In the present work, this model is used to determine and quantify the dissipation of waves generated by winds at the Ilha Solteira reservoir, considering its margins without and with background vegetation. In this way, it is evaluated the importance of the vegetation on the attenuation of wind waves. Tests were performed considering several wind velocity conditions (for NE direction), and different values of the vegetation parameter factor, V_f (drag coefficient × diameter × density of plants) and the height of plants (different submergence ratios). Note that the wind characteristics are based upon measurements obtained under the ONDISA project, Morais et al. 2009.

Finally, it should be remembered that the numerical proposal, aiming at the determination of the wave attenuation coefficients by the vegetation, is an important element in determining the fragility matrix of the reservoir environment.

This paper begins with the model input conditions (section 2). Then, a discussion of the results obtained with the model is presented in section 3. Finally, the conclusions are drawn in section 4.

2 MODEL INPUT CONDITIONS

In the present study, the numerical model SWAN-VEG is applied considering different wind velocities (from 5.0 m/s to 20 m/s) from NE. For these case studies, different bathymetric meshes of the reservoir were elaborated with 250 m, 150 m and 100 m resolution.

In addition, vegetation meshes with different characteristics (according to the resolution of the bathymetric mesh) and located close to the banks of the reservoir were also considered, with the purpose of analysing the damping of waves generated by winds through the vegetated field.

The directional spectrum in the SWAN-VEG calculations was initially defined with a frequency discretization of 30 intervals of 0.05 to 3.0 Hz with a logarithmic distribution and a directional discretization of 2.5° covering the 360° (resulting in 144 steering ranges). All tests were performed with the SWAN-VEG adapted from SWAN version 41.01 (SWAN Team 2014), in stationary mode, without the presence of currents. The physical phenomena considered in the simulations were: refraction, diffraction, shoaling, and triad and quadruple wave-wave interactions. All relevant parameters were introduced in the SOPRO platform (FORTES et al., 2006).

Case studies were carried out with the different meshes. It was concluded that the attenuation was very similar for the three meshes tested, Mattosinho (2016). So, it was chosen the most refined mesh, namely 100 m.

Two analyses were made. The first one analyses the effects of the vegetation parameters. In this case, the vegetation factor characteristics V_f · [plant/m] (drag coefficient × diameter × density of plants) was varied from 0.0 (without vegetation) to 12.8 and the drag coefficients (C_D) were equal to 0.02, 1.00 and 1.60. The plant height (α_h) was considered equal to 0.5 m with 5 mm diameter. The coordinates and depths of the points used for is analysis are presented at Table 1.

Figures 4, 5 and 6 present the study area as well as the sketches of the points used to obtain the numerical results.

In relation to the Hs decay with submergence (phase 1), simulations were performed for the study points of Table 2, considering a wind intensity of 20 m/s (extreme velocity) in order to verify the decay of the significant wave height, Hs, according to the plant height for each point. The results will be shown in section 3.

Furthermore, in phase 2 (points PM1 – PM5) we analysed the variation of the significant wave height in relation to the vegetation parameter (0.0 to 12.8, see section 3) considering the characteristic wind speed of the region. The study points of phase 2 are shown in Table 3.

Table 1. Characteristics of the study points and simulated results of significant wave height [m] with $\alpha h = 0.5$ and wind of 5 m/s.

		Coordinates	
Pts	depth [m]	X	Y
PM1	0.57	479822	7751013
PM2	0.85	479872	7751063
PM3	1.40	479922	7751113
PM4	1.98	479972	7751163
PM5	2.98	480022	7751213

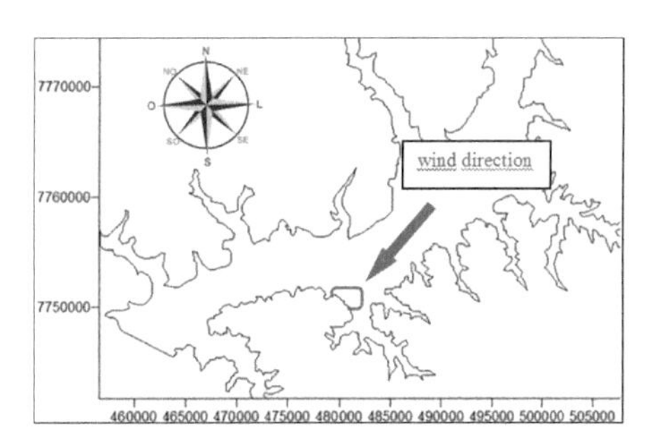

Figure 4. Georeferenced representation of the Ilha Solteira reservoir (study area, within which layouts of vegetated fields are arranged, around the reservoir).

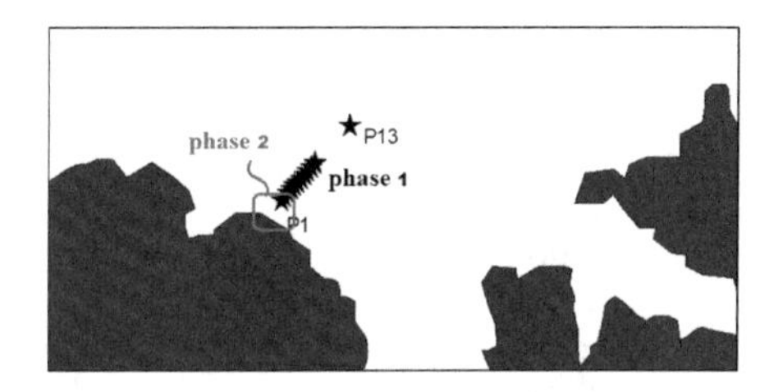

Figure 5. Sketch of the points to analyse the decay of Hs in relation to submergence for maximum wind velocity 20 m/s.

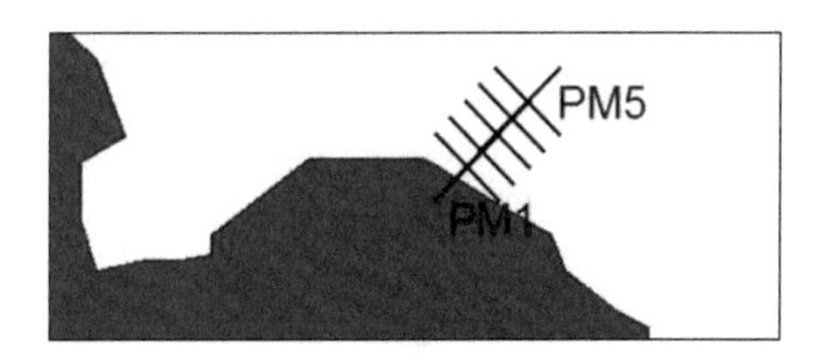

Figure 6. Sketch of the points to analyse the impact of vegetation parameter factor V_f.

Table 2. Coordinates and depth of the study points of phase 1.

	Points		
	X	Y	Depth [m]
P1	479992	7751213	5.37
P2	480042	7751263	5.93
P3	480092	7751313	6.83
P4	480142	7751363	7.84
P5	480192	7751413	8.83
P6	480242	7751463	10.06
P7	480292	7751513	10.98
P8	480342	7751563	11.91
P9	480392	7751613	12.94
P10	480442	7751663	13.96
P11	480492	7751713	15.18
P12	480503	7751726	15.44
P13	481014	7752238	25.54

Table 3. Coordinates and depth of the study points of phase 2.

	Points		
	X	Y	Depth [m]
PM1	479822	7751013	0.57
PM2	479872	7751063	0.85
PM3	479922	7751113	1.40
PM4	479972	7751163	1.98
PM5	480022	7751213	2.98

3 RESULTS AND DISCUSSIONS

Considering the most refined mesh (100 m resolution) and the input conditions of a wind intensity of 5 m/s (characteristic speed of the region), a plant height (αh) of 0.5 m with 5 mm diameter, and the vegetation parameters (V_f) varying from 0 to 12.8 with the drag coefficients (CD) were equal to 0.02, 1.00 and 1.60, it was possible to calculate the significant wave heights at point PM1 to PM5, see Table 4.

For the same parameters and input conditions, percentages of wave dampening were calculated. Figure 7 relates the damping of waves at the points of interest as a function of the vegetation parameter.

In Figure 7, it is shown that an increase of the vegetation parameter leads to an attenuation increase.

Rearranging the data, we obtain Figure 8, which shows the relation between vegetation parameter, V_f, with damping at the several points of interest.

Figure 8 shows that points PM4 and PM5 have a linear behavior. Point PM5 does not show any variation since it is a point outside the vegetation (reference point).

Moreover, it is also shown that for vegetation parameters close to zero the model shows a higher rate of growth in the attenuation. However, for V_f greater than 2 a linear trend becomes evident.

High damping rates are observed. Note that these percentages are related to the total damping of significant wave height. In this way, we have in fact the damping due to the bottom (margin slope) added to the effect of the vegetation. Returning to Table 4, we can infer the damping due to the bottom in the cases without vegetation.

In relation to the submergence analysis, simulations were performed for the study points of Table 2, considering wind intensity of 20 m/s (extreme velocity) in order to verify the decay of the significant wave height, Hs, according to the plant height for each point. The results are presented in Figure 9.

Table 4. Numerical results of significant wave height [m] with $\alpha h = 0.5$ and wind of 5 m/s.

	Without vegetation	With vegetation					
	N	400			1600		
Pts	C_D	0.02	1.00	1.60	0.02	1.00	1.60
PM1	0.21	0.20	0.09	0.08	0.18	0.05	0.04
PM2	0.24	0.23	0.15	0.14	0.22	0.11	0.10
PM3	0.26	0.26	0.23	0.22	0.26	0.21	0.21
PM4	0.27	0.27	0.27	0.27	0.27	0.27	0.27
PM5	0.28	0.28	0.28	0.28	0.28	0.28	0.28
	V_f	0.04	2.00	3.20	0.16	8.00	12.80

Hs: Significant heigh; V_f: Vegetation parameter; C_D: drag coefficients; αh: plant height; N: Plant density.
Note: submergence = plant height (αh)/local depth.

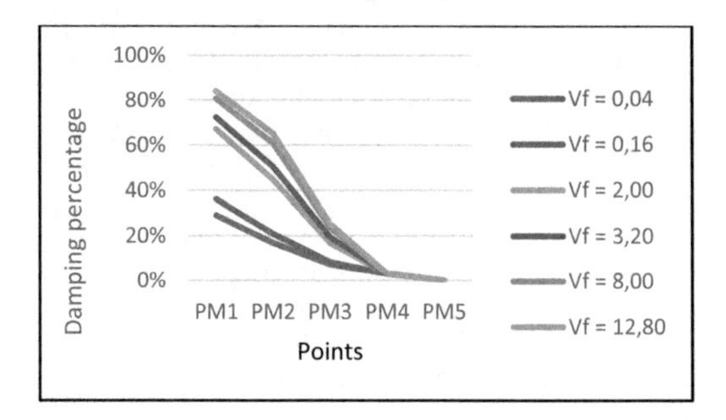

Figure 7. Damping of waves in the margin for wind velocity of 5 m/s.

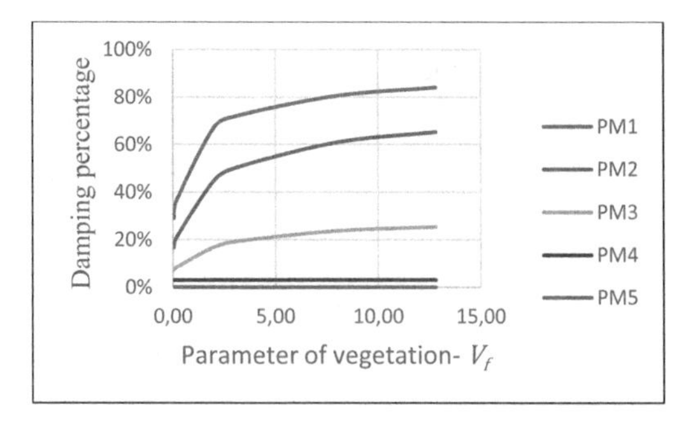

Figure 8. Increased damping as a function of V_f for wind velocity 5 m/s.

Figure 9 shows the decay of the significant wave height as the plant height was increased, which is in agreement with the physics of the problem, because the smaller the submergence is (ratio between plant height (αh) and local depth), the greater its damping effect. This behaviour was expected, since Oude (2010) and Suzuki et al. (2011) state that the emergent part of the plants do not affect the damping of the waves in the SWAN-VEG model.

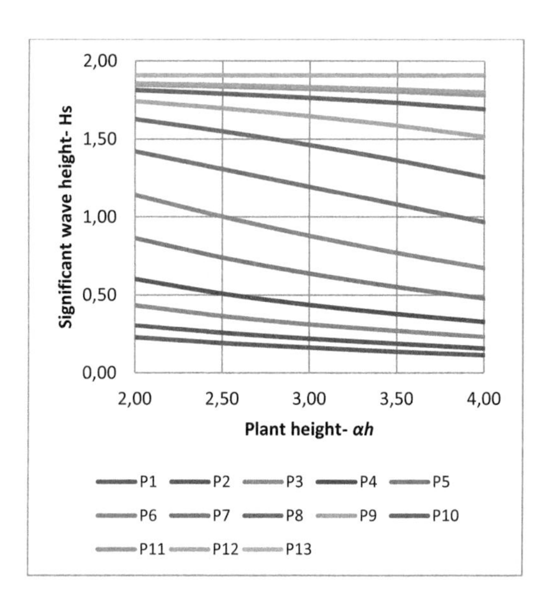

Figure 9. Decay of Hs in relation to submergence for maximum wind velocity 20 m/s.

4 CONCLUSIONS

The Ilha Solteira reservoir—SP is a navigable stretch of the Tietê—Paraná waterway, and so it has a significant role on the transportation of cargo and passengers along it. Safety conditions for that navigation and transportation are of major importance. Within this framework, it is quite essential to evaluate the waves generated by the winds in those waterways (which lead to difficulties to navigation and are responsible also for the erosion of the margins and landslides). Moreover, it is also important to analyse natural based solutions to mitigate those problems, such as vegetation.

SWAN-VEG is a wind-wave generation model that takes into account the presence of vegetation in the wave propagation and the present paper addresses this subject since it describes the application of that model, to the Ilha Solteira reservoir—SP.

In this work it was verified that wave attenuation due to vegetation is a highly dynamic and complex process whose quantification is important in the modelling of coastal hydrodynamics as well as in the definition of protection measures for the reservoirs banks, which are often exposed and subjected to erosive and overlapping processes due to waves.

As a cumulative result of the literature, the experience of the Brazilian group with the subject both at laboratory and in situ level (wave measurements in the lake inside the vegetation, at the Unesp farm (Lima, 2005) and Vasco (2005)), and in possession of the numerical simulations performed by Mattosinho (2016), the results from the SWAN-VEG model point to wave attenuation factors that depend heavily on the density of habitat plants and the submergence ratios.

Also, we observe that the adoption of non-structural measures, such as vegetated fields, are plausible, due to the sensible increase of wave damping as V_f increases. However, more in-depth studies, with field measurements and updating of bathymetry for mesh refinement, are of paramount importance in order to confront the numerical data.

From this work, one also can conclude that SWAN-VEG model is an important tool to test natural based solutions such as vegetation to mitigate navigation problems related with wind waves generated in large and huge waterways, as it is the Ilha Solteira reservoir.

REFERENCES

Antaq (agência nacional de transporte aquaviário). Situação atual da hidrovia tietê – paraná. 2012. Disponível em http://www.antaq.gov.br/portal/pdf/palestras/palestrajulho2012.pdf. Acesso em 24 set. 2015.

Fortes, C.J.; Pinheiro, L.; Santos, J.A.; Neves, M.G.; Capitão, R. Sopro – pacote integrado de modelos de avaliação dos efeitos das ondas em portos. Tecnologias da água, v. 1, p. 51–61, 2006.

Mattosinho, G.O. Dissipação de energia de ondas geradas por ventos em reservatórios de barragens, devido à presença de vegetação. Dissertação (mestrado em engenharia mecânica) – Faculdade de Engenharia, Universidade Estadual Paulista, Ilha Solteira, 2016. 85p.

Morais, V.S., Cunha, E.F., Maciel, G F., 2009. Medição, previsão e análise numérica dos mecanismos de geração de ondas a partir da cinética de ventos e dissipação de ondas na presença de fundos com vegetação, em lagos de barragens, proc. Xviii simpósio brasileiro de recursos hídricos. Campo grande - Mato Grosso do Sul - Brasil: Anais do XVIII Simpósio brasileiro de recursos hídricos.

Oude, R. Modelling wave attenuation by vegetation with swan-veg. 2010. Dissertation (master thesis: water engineering & management) – University of Twente, delft, 2010.

Suzuki, T.; Zijlema, M.; Burger, B.; Meijer, M.C.; Narayan, S. Wave dissipation by vegetation with layer schematization in swan. Coastal Engineering, Amsterdam, v. 59, p. 64–71, 2011.

Vasco, J.R.G. Modelo conceitual de dissipação da energia da onda que se propaga por fundos vegetados. 2005. 108 f. Dissertação (mestrado em engenharia civil) – Faculdade de Engenharia, Universidade Estadual Paulista, Ilha Solteira, 2005.

Vieira, A.S. Análises, aplicações e validações – numérico/experimentais do modelo swan em áreas restritas e ao largo. 251 f. Tese (doutorado em engenharia elétrica) – Faculdade de Engenharia, Universidade Estadual Paulista, Ilha Solteira, 2013.

Progress in Maritime Technology and Engineering – Guedes Soares & Santos (Eds)
 ISBN 978-1-138-58539-3

Peak period statistics associated with significant wave heights by conditional mean functions of the distributions

G. Muraleedharan, C. Lucas & C. Guedes Soares
Centre for Marine Technology and Ocean Engineering (CENTEC), Instituto Superior Técnico, Universidade de Lisboa, Lisbon, Portugal

ABSTRACT: This study uses the 21 years of daily maximum significant wave height and associated peak period distributions for winter and summer of a test site in the North Atlantic Ocean. Conditional mean functions of generalized *Pareto*, three-parameter *Weibull* and *Erlang* distributions are used to model the average conditional exceedances of peak periods associated with significant wave heights. The method of L-moments is used for model estimation. Mean maximum peak periods and most probable maximum peak periods are also computed from the data and estimated from the models. A general formula based on order-statistics is provided for estimation of average of one-third the highest wave periods and an expression is derived from generalized *Pareto* distribution.

1 INTRODUCTION

Albeit peak wave period is unstable in the estimation process (Rodríguez et al., 1999) it is an important wave parameter. Floating structures require bivariate distributions of significant wave heights (H_s) and characteristic (mean or peak) periods as design and operational criteria (Haver, 1985; Bitner-Gregersen, et al., 1998; Ferreira and Guedes Soares, 2002). Even though floating wave energy converters (Falcão, 2015) operate in shallow waters and their dynamic response depends on the significant wave height and associated peak periods of the prevailing sea state, their performance depends mainly on the wave period. The peak period is also important in evaluating the wave induced loads on marine structures such as floating platforms.

The IEC Standard states that the extreme sea state should "take account of the range of peak period appropriate to the 50-year significant wave height. Design calculations should be based on values of the peak period which result in the highest loads acting on an offshore wind turbine" (IEC, 2009). Hence estimates of peak period statistics associated with significant wave heights are highly essential in the design and operation of the ocean renewable energy devices.

Therefore a probability distribution of characteristic wave periods is essential to understand the efficiency of these devices. The marginal distributions of wave periods have narrow band frequency spectra (Rodríguez et al., 2004). Sea states often have two wave systems (Guedes Soares, 1984) and the wave period distributions will be bimodal and will not fit the distribution appropriate to single sea states. Longuet-Higgins (1975, 1983), obtained the wave period distribution as the marginal distribution of the joint distribution of individual wave heights (H) and periods (T). An inherent weakness of this model is the mean is zero and variance does not exist.

Estimation of certain wave period statistics of interest from these models are difficult due to their complexities. Hence applications of these models are constrained such as the fitting to observed wave period distributions.

Further, it is to be noted that when the data are generated from a physical process that can produce outliers, then a distribution that is a close fit to the current observed data will not ensure the same to the future data. So, it is preferable to use a robust approach based on a distribution that can yield reasonably accurate quantiles even when the true at site frequency distribution deviate from the fitted frequency distribution. Hence, in this study, the distributions of significant wave height (H_s) associated peak periods (Tp) of the test site ($0.25 \times 0.25°$) offshore Portugal in the North Atlantic Ocean are ascertained by modelling the average conditional exceedances of the parameters by Conditional Mean Functions (*CMF*) of the distributions: *Erlang*, three-parameter *Weibull* and generalized *Pareto* distributions. Since *CMF* determines the distributions of the parameters uniquely, it is sufficient to find the functional forms of *CMF* consistent with the data (Muraleedharan et al., 2009, 2015).

In this study, month wise grouped, everyday maximum significant wave heights and associated peak periods of the winter (December to February) and summer (June to August) extracted from 21 years (1958 to 1978) of the HIPOCAS (Guedes Soares, 2008) (Hindcast of the dynamic processes of the ocean and coastal areas of Europe) database are considered. The empirically computed average conditional exceedances of the peak periods *Tp* are simulated by the Conditional Mean Functions derived from three- parameter *Weibull,* generalized *Pareto* (*GP3*) and *Erlang* (shape parameter of gamma is assigned to the nearest integer) distributions. The method of L-moments described by Hosking and Wallis (1997) is used for model estimations. The L-moment estimation procedure for three parameter *Weibull* distribution provided by Muraleedharan and Guedes Soares (2014) is considered here.

L-moments or linear combination of probability weighted moments (*PWMs*) form the basis of an elegant mathematical theory and it facilitates the estimation process. L-moment methods are superior to *MLE*, and method of moments. L-moment ratios measure the shape of a distribution independent of its scale of measurement. L-moments are more robust to the presence of outliers in the data and are less subjected to bias in the estimation (Hosking and Wallis, 1997). Incorrect data values, outliers, trends and shifts in the mean of a sample can all be reflected in the L-moments of the sample. The mean ($\bar{T}_p$) and average of the one-third the highest peak periods ($\bar{T}_{p(1/3)}$) are deduced from the Conditional Mean Functions of the distributions. Most probable maximum peak periods ($T_{p(mpm)}$) and mean maximum peak periods ($\bar{T}_{p(max)}$) are also estimated from the parametric relations derived from three-parameter *Weibull* and generalized *Pareto* distributions.

A general formula based on order-statistics (Muraleedharan et al., 2015) is also provided for the estimation of significant wave height or average of one-third the highest wave periods and was able to derive a parametric relation from *GP*3 distribution to be used in the general formula. There exist significant functional relationships between the average peak of one-third the highest significant wave heights ($\bar{T}_{p(H1/3)}$) computed from the data and estimates of average of one-third the highest peak periods ($\bar{T}_{p(1/3)}$) estimates from the distributions. The ratio of significant wave period to mean wave period of 1.2 (Kitano et al., 2002) or 0.9–1.4 (Goda, 2010) are well interpreted by the ratio of $\bar{T}_{p(H1/3)}$ to $\bar{T}_p$.

2 METHODOLOGY

The empirical average conditional exceedances of peak periods t_p is computed as:

$$m\left(t_{p(r)}\right)=\frac{1}{N-r+1}\sum_{k=r}^{N}t_{p(k)} \tag{1}$$

where $t_{p(r)}$ is the rth order period t_p of a sample size N arranged in ascending order.

The conditional mean function, *CMF*, of the distribution function $F(t_p)$ is defined as:

$$m\left(t_p\right)=E\left(T_p \mid T_p>t_p\right)=\frac{1}{\bar{F}\left(t_p\right)}\int_{t_p}^{\infty}x\left[-\bar{F}'(x)\right]dx \tag{2}$$

where E is the usual expectation operator, and $\bar{F}'(x)$ is the derivative of $\bar{F}(x)$,

$$\bar{F}(x)=1-F(x) \tag{3}$$

Accordingly, $m(t_p)$ derived from *Erlang*, three-parameter *Weibull* and generalized *Pareto* distributions are given respectively as:

Erlang distribution function

$$F\left(t_p\right)=1-\sum_{i=0}^{\alpha-1}exp\left[-\left(\lambda t_p\right)\right]\frac{\left(\lambda t_p\right)^i}{i!} \tag{4}$$

where λ is the rate parameter ($\lambda > 0$) and α is the shape parameter.

$$m\left(t_p\right)=t_p+\lambda^{-1}\frac{\sum_{i=0}^{\alpha-1}\sum_{j=0}^{i}\frac{\left(\lambda t_p\right)^j}{j!}}{\sum_{i=0}^{\alpha-1}\frac{\left(\lambda t_p\right)^i}{i!}} \tag{5}$$

Three-parameter Weibull distribution function

$$F\left(t_p\right)=1-exp\left[-\left(\frac{t_p-\mu}{\lambda}\right)^{\alpha}\right], t_p\geq\mu; \alpha>0 \tag{6}$$

where μ is the location parameter ($-\infty<\mu<\infty$), λ is the scale parameter ($\lambda > 0$) and α is the shape parameter.

$$m\left(t_p\right)=t_p+\frac{\lambda}{\alpha}\frac{\Gamma\left(\frac{1}{\alpha},\left(\frac{t_{p-\mu}}{\lambda}\right)^{\alpha}\right)}{\bar{F}\left(t_p\right)} \tag{7}$$

$$\Gamma(s,x)=\int_x^{\infty}exp\left[-(t)\right]t^{s-1}dt \tag{8}$$

Generalized Pareto distribution

$$F\left(t_p\right)=1-\left[1-\frac{\alpha}{\lambda}\left(t_p-\mu\right)\right]^{\frac{1}{\alpha}},\ \alpha\neq0 \tag{9}$$

$$\mu\leq t_p<\infty\ if\ \alpha<0; \mu\leq t_p\leq\mu+\frac{\lambda}{\alpha}\ if\ \alpha>0$$

where μ is the location parameter ($-\infty<\mu<\infty$), λ the scale parameter ($\lambda > 0$) and α the shape parameter.

$$m\left(t_p\right)=t_p+\frac{\lambda}{(\alpha+1)}\left[1-\frac{\alpha}{\lambda}\left(t_p-\mu\right)\right], \alpha>0\ or\ \alpha<0 \tag{10}$$

The knowledge of $m(t_p)$ will enable to ascertain the distribution of Tp through the relationship:

$$F(t_p) = 1 - exp\left[-\int_0^{t_p}\left(\frac{m'(x)}{m(x)-x}dx\right)\right] \quad (11)$$

where $m'(x)$ is the derivative of $m(x)$.

Accordingly, the distribution function of T can be *Erlang*, three-parameter *Weibull* or generalized *Pareto* distributions.

3 MEAN WAVE PERIODS AND AVERAGE OF ON-THIRD THE HIGHEST WAVE PERIODS FROM CONDITIONAL MEAN FUNCTIONS

If $t_1, t_2, \ldots, t_N$ are N wave periods in a sample that are arranged in descending order of magnitude, $\bar{T}_F$ stands for the mean of the first FN values, where $0 \le F \le 1$. The mean value $\bar{T}_F$ of those periods that are larger than t is given by *CMF* of the distribution of *T*. Then $\bar{T}_{1/3}$ gives the average of the one-third the highest periods. Thus, t in the *CMF* is decided by the quantile function $Q(\cdot)$ of the distribution, i.e.,

$$P(T > t) = \bar{F}(t) = 1/3 \quad (12)$$

$$t = Q(1 - \bar{F}(t)) \quad (13)$$

where $\bar{T}_1$ gives the mean of the wave periods.

3.1 *General formula based on order-statistics to estimate average of one-third the highest wave periods* $(\bar{T}_{1/3})$

If $T_1, T_2, \ldots, T_n$ are n wave periods from a random sample arranged in ascending order of magnitude, denote by $T_{(r)}$ the rth largest among them so that $T_{(1)} < T_{(2)}, \ldots, T_{(n)}$, then the average of one-third the highest wave periods is given by:

$$\bar{T}_{1/3} = \frac{1}{(n/3+1)}\left[E\left(T_{(2n/3)} + T_{(2n/3+1)} + \ldots + T_n\right)\right] \quad (14)$$

2n/3 is chosen as the largest integer value in it if 2n/3 is not an integer.

The distribution of $T_{(r)}$ is specified by the density:

$$h(t_r) = \frac{n!}{(r-1)!(n-r)!}\left[F(t_r)\right]^{r-1}\left[1-F(t_r)\right]^{n-r} f(t_r) \quad (15)$$

$$r = 1, 2, \ldots, n$$

$F(\cdot)$ and $f(\cdot)$ are the distribution and density functions of T.

3.1.1 *Estimation of the average of one-third the highest wave periods from GP3 distribution using the general formula based on order-statistics*

If generalized *Pareto* distribution is chosen as the distribution function of T by conditional mean function of the distribution, then:

$$E(T_{(r)}) = \mu + \frac{\lambda}{\alpha}\left[1 - \frac{n!}{(n-r)!}\frac{\Gamma(n-r+1+\alpha)}{\Gamma(n+1+\alpha)}\right]; \alpha > 0 \quad (16)$$

If the factorial terms are replaced by gamma functions, then:

$$E(T_{(r)}) = \mu + \frac{\lambda}{\alpha}\left[1 - \frac{\Gamma(n+1)}{\Gamma(n-r+1)}\frac{\Gamma(n-r+1+\alpha)}{\Gamma(n+1+\alpha)}\right] \quad (17)$$

The product of the gamma functions can be easily evaluated by applying the laws of logarithm.

3.1.2 *Mean maximum wave period* $(\bar{T}_{p(max)})$

The density function of the distribution of the wave period $T_{(n)}$, the maximum wave period $T_{(max)}$ is given by:

$$g(t_{max}) = n\left[F(t_{max})\right]^{n-1} f(t_{max}) \quad (18)$$

Hence mean maximum peak period:

$$T_{p(max)} = E\left(T_{p(max)}\right) = \int t_{p(max)} g\left(t_{p(max)}\right) dt_{p(max)} \quad (19)$$

$\bar{T}_{p(max)}$ derived from generalized *Pareto* distribution ($\alpha > 0$) is given by:

$$\bar{T}_{p(max)} = \mu + \frac{\lambda}{\alpha}\left[1 - \Gamma(1+\alpha)\frac{\Gamma(n+1)}{\Gamma(n+1+\alpha)}\right] \quad (20)$$

$\bar{T}_{p(max)}$ derived from three-parameter *Weibull* distribution is given by:

$$\bar{T}_{p(max)} = \mu + \frac{\lambda}{\alpha}\Gamma\left(\frac{1}{\alpha}\right)\sum_{r=1}^{n}\binom{n}{r}(-1)^{r-1} r^{-\frac{1}{\alpha}} \quad (21)$$

3.1.3 *Most probable maximum wave period* $(T_{p(mpm)})$

Most probable maximum wave period is given by the mode of the distribution $g(\cdot)$ given by equation (18). If $F(\cdot)$ and $f(\cdot)$ are the distribution and density functions of generalized *Pareto*, and differentiating with respect to $t_{p(max)}$, the mode is obtained as:

$$T_{p(mpm)} = \mu + \frac{\lambda}{\alpha}\left[1 - \left(\frac{1-\alpha}{n-\alpha}\right)^{\alpha}\right] \quad (22)$$

For $\alpha > 1$ and for large n, Equation (22) holds well when α in the exponent is chosen as the largest integer value in it, if α is not an integer. $T_{p(mpm)}$ from *Weibull* distribution is given as the solution of the equation:

$$\left[\frac{(\alpha-1)-y\alpha}{(\alpha-1)-ny\alpha}\right]\exp(y)-1=0 \tag{23}$$

$$y=\left(\frac{t_{p(\max)}-\mu}{\lambda}\right)^{\mu} \tag{24}$$

Equation (23) can be solved by numerical methods like the *Newton-Raphson* technique. Alternatively it can also be solved by a standard numerical program executed in this work. If y_0 is an approximate solution, the:

$$T_{p(mpm)}=t_{p(max)}=\mu+\lambda(y_0)^{\frac{1}{\alpha}} \tag{25}$$

4 RESULTS AND DISCUSSIONS

This study considers the 21 years (1958–1978) month wise clustered daily maximum significant wave height (H_s) and associated peak periods (Tp) in winter (December to February) and summer (June to August) of a test site offshore Portugal in the North Atlantic Ocean and extracted from 44 years HIPOCAS database (Pilar et al., 2008). The plots of the average conditional exceedances of t_p computed from empirical data versus corresponding t_p are compared with the simulations by the conditional mean functions (*CMFs*) of the *Erlang*, three-parameter *Weibull* and generalized *Pareto* (*GP*3) distributions. *Erlang* distribution can be considered as a two parameter gamma distribution with a positive integer shape parameter. The nonlinearity of the distributions of the computed average conditional exceedances of peak periods Tp in winter and summer are well generated by the *Erlang* and three-parameter *Weibull CMFs* (Figures 1(a–c)

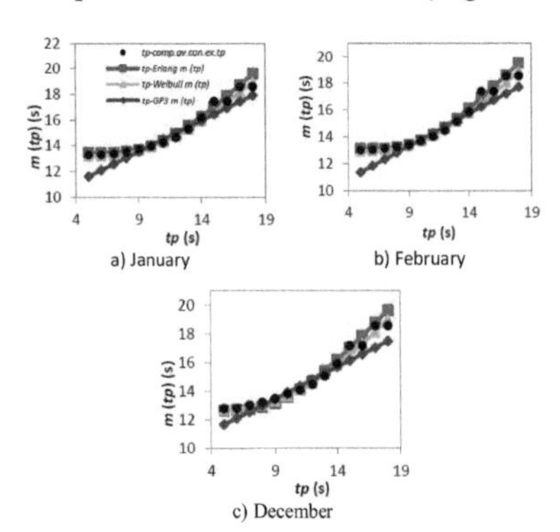

Figure 1. Simulation of computed average conditional exceedances of peak periods (t_p) by conditional mean function $m(t_p)$ of the distributions in winter season a) to c).

and Figures 2(d–f)). Since the *GP*3 *CMF* provides a linear fit, mainly the nonlinearity of the lower tail of the computed values are not apprehended by the fit, but, the peak period statistics estimated from the parametric relations derived from *GP*3 distribution are comparable with the computed values from the data. It can be seen that the functional forms of the *CMFs*, that determine the distribution uniquely, are consistent with the average conditional exceedances of t_p computed from the data. The model parameters are estimated by the method of L-moments and are shown in Table 1. The shape parameters of the *Weibull* distribution in winter are nearly 4.0 and hence, the peak periods are approximately normally distributed in this season. Whereas in summer, the peak period distributions can be assumed to follow roughly a *Rayleigh* curve (Rayleigh distribution is a special case of *Weibull* distribution for ($SH = 2$).

There exists significant linear functional relationship between average peak period of one third the highest wave heights ($\bar{T}_{p(H1/3)}$) and average of one-third the highest peak periods ($\bar{T}_{p(1/3)}$ and $\bar{T}_s$ (by Equation 17 from *GP*3)). The relationships are given in Equations 26 to 30. The coefficient of determination is > 90% in all cases and the relationships hold good in both winter and summer.

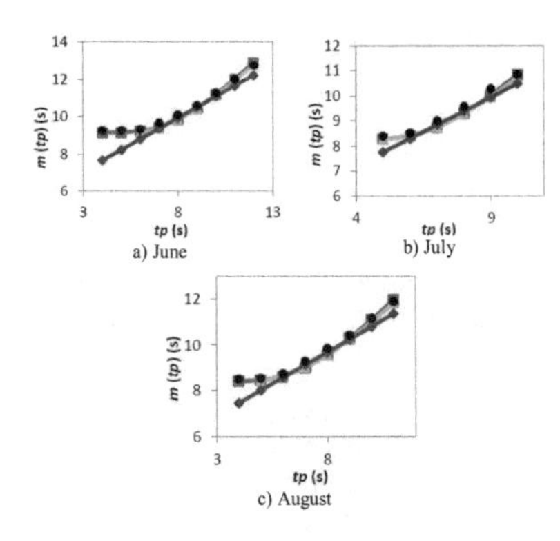

Figure 2. Simulation of computed average conditional exceedances of peak periods (t_p) by conditional mean function $m(t_p)$ of the distributions in summer season d) to f).

Table 1. Model parameter estimates (Location-LO, Scale-SC (Rate (RA) = 1/SC), and Shape-SH) by the method of L-moments.

	Weibull			GP3			Erlang	
Mon	LO	SC	SH	LO	SC	SH	RA	SH
Jan	3.4	10.8	3.9	8.3	10.1	1.1	1.7	22
Feb	3.6	10.3	3.8	8.0	9.7	1.1	1.7	22
Dec	2.5	11.2	3.9	7.3	12.0	1.2	1.4	18
Jun	5.3	4.3	2.5	6.5	4.5	0.8	3.3	30
Jul	4.6	4.1	2.8	6.0	4.2	0.8	3.3	33
Aug	4.3	4.6	2.6	5.7	4.8	0.8	2.5	24

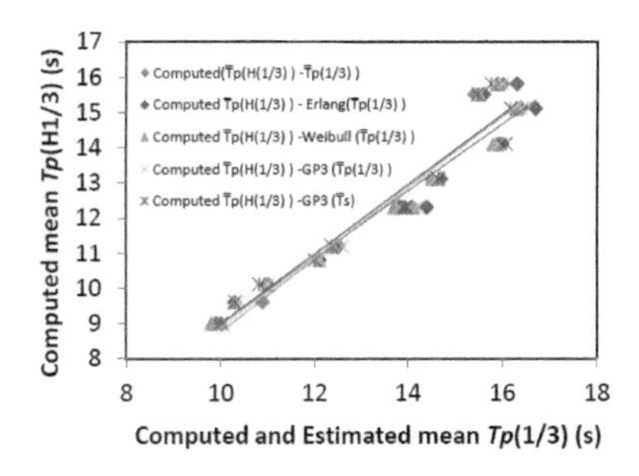

Figure 3. Computed mean $\overline{T}_{p(H1/3)}$ versus the computed and estimated mean $\overline{T}_{p(1/3)}$ and $\overline{T}_s$.

Table 2. Computed and estimated wave periods statistics; mean ($\overline{T}_p$), average of one-third the highest peak periods ($\overline{T}_s$), and average peak period of one-third the highest wave heights ($\overline{T}_{p(H1/3)}$) of a test site in the North Atlantic Ocean.

Statistics $T_p(s)$	Months					
Comp	Jan	Feb	Dec	Jun	Jul	Aug
$\overline{T}_p$	13.2	12.9	12.7	9.1	8.3	8.4
$\overline{T}_{p(1/3)}$	16.4	16.0	15.9	11.0	10.0	10.9
$\overline{T}_{p(H1/3)}$	15.1	15.8	14.1	10.1	9.0	9.6
$\overline{T}_{p(H1/3)} / \overline{T}_p$	1.1	1.2	1.1	1.1	1.1	1.1
Erlang						
$\overline{T}_p$	13.5	13.1	12.7	9.2	8.3	8.4
$\overline{T}_{p(1/3)}$	16.7	16.3	16.0	11.0	9.9	10.3
$\left(\frac{\overline{T}_{p(H1/3)}}{\overline{T}_p}\right)(\overline{T}_{p(H1/3)})$[1]	1.1	1.1	1.1	1.1	1.1	1.1
Weibull						
$\overline{T}_p$	12.6	12.4	12.1	9.1	8.2	8.3
$\overline{T}_{p(1/3)}$	16.3	15.9	15.8	11.0	9.8	10.3
$\left(\frac{\overline{T}_{p(H1/3)}}{\overline{T}_p}\right)(\overline{T}_{p(H1/3)})$[2]	1.2	1.2	1.2	1.1	1.1	1.1
GP3						
$\overline{T}_p$	13.2	12.9	12.7	9.1	8.3	8.4
$\overline{T}_{p(1/3)}$	16.4	16.0	15.9	11.0	10.0	10.4
$\overline{T}_s$	15.9	15.5	15.6	11.2	10.4	10.7
$\left(\frac{\overline{T}_{p(H1/3)}}{\overline{T}_p}\right)(\overline{T}_{p(H1/3)})$[3]	1.2	1.2	1.2	1.1	1.1	1.1
$\left(\frac{\overline{T}_{p(H1/3)}}{\overline{T}_p}\right)(\overline{T}_{p(H1/3)})$[4]	1.1	1.1	1.2	1.1	1.0	1.1

[1]From Equation 27; [2]from Equation 28; [3]from Equation 29; [4]from Equation 30.

These results are significant as $\overline{T}_{p(1/3)}$ and $\overline{T}_s$ can easily be computed from the marginal distributions of the peak periods whereas $\overline{T}_{p(H1/3)}$ requires joint distribution of significant wave heights and associated peak periods. Hence, from the knowledge of $\overline{T}_{p(1/3)}$ or $\overline{T}_s, \overline{T}_{p(H1/3)}$ can be readily estimated from the linear functional relationship between them. The functional relations are also shown in Figure 3. Computed $\overline{T}_{p(1/3)}$ and $\overline{T}_{p(H1/3)}$:

$$\overline{T}_{p(H1/3)} = 1.0299\overline{T}_{p(1/3)} - 1.5469\left(R^2 = 0.940\right) \qquad (26)$$

Erlang estimated $\overline{T}_{p(1/3)}$ and computed $\overline{T}_{p(H1/3)}$:

$$\overline{T}_{p(H1/3)} = 0.9466\overline{T}_{p(1/3)} - 0.4806\left(R^2 = 0.938\right) \qquad (27)$$

Weibull estimated $\overline{T}_{p(1/3)}$ and computed $\overline{T}_{p(H1/3)}$:

$$\overline{T}_{p(H1/3)} = 0.9947\overline{T}_{p(1/3)} - 0.9705\left(R^2 = 0.941\right) \qquad (28)$$

GP3 estimated $\overline{T}_{p(1/3)}$ and computed $\overline{T}_{p(H1/3)}$:

$$\overline{T}_{p(H1/3)} = 0.9975\overline{T}_{p(1/3)} - 1.0993\left(R^2 = 0.938\right) \qquad (29)$$

GP3 estimated $\overline{T}_s$ and computed $\overline{T}_{p(H1/3)}$:

$$\overline{T}_{p(H1/3)} = 0.9899\overline{T}_s - 0.9153\left(R^2 = 0.927\right) \qquad (30)$$

Certain peak period statistics of relevance such as the computed and estimated mean peak periods ($\overline{T}_p$), average of one-third the highest peak periods $[(\overline{T}_{p(1/3)}),(\overline{T}_s)]$ and average peak periods of one-third the highest significant wave heights ($\overline{T}_{p(1/3)}$) of winter and summer in the North Atlantic Ocean are provided in Table 2. The peak period statistics in winter are comparatively higher than the values in summer, i.e., the high waves with low frequencies prevail in winter. The peak period statistics estimates by all the three models are consistent with the computed values except the $\overline{T}_p$ *Weibull* estimates

Table 3. Computed (average) and estimated mean maximum peak periods ($\overline{T}_{p(max)}$) and most probable maximum peak periods $T_{p(mpm)}$ in winter and summer of a test site in the North Atlantic Ocean.

Comp	Jan	Feb	Dec	Jun	Jul	Aug
$\overline{T}_{p(max)}$		19.8 (average)			11.9 (average)	
$T_{p(mpm)}$		19.8 (average)			11.2 (average)	
Weibull						
$\overline{T}_{p(max)}$	20.1	20.7	19.8	13.0	13.5	12.6
$T_{p(mpm)}$	20.9	20.5	20.7	16.4	12.6	13.8
GP3						
$\overline{T}_{p(max)}$	17.5	17.0	17.3	12.1	11.2	11.7
$T_{p(mpm)}$	17.5	17.0	17.3	12.1	11.2	11.7

in winter. The ratio of significant wave period to mean wave period of 0.9 to 1.4 (Goda, 2010) is well interpreted by the ratio of $\bar{T}_{p(H1/3)}$ to $\bar{T}_p$.

The computed (average) and estimated peak period statistics: mean maximum peak periods ($\bar{T}_{p(max)}$) and most probable maximum peak periods ($T_{p(mpm)}$) in winter and summer are also provided in Table 3. The averages of the computed $\bar{T}_{p(max)}$ and $T_{p(mpm)}$ are both 19.8 s in winter whereas they have much low values in summer. The three-parameter *Weibull* estimates in winter are proximate to the computed values. But the *GP*3 $\bar{T}_{p(max)}$ and $T_{p(mpm)}$ estimates in summer are contiguous to the averages of the empirical values.

5 CONCLUSIONS

Daily maximum significant wave heights and associated peak periods extracted from month wise clubbed HIPOCAS database for winter (December to February) and summer (June to August) spread over a time span of 21 years (1958–1978) of a test site (0.25 × 0.25°) in the North Atlantic Ocean are considered in this study. The computed monthly average conditional exceedances of peak periods are precisely simulated by the conditional mean functions (*CMF*) of *Erlang*, three-parameter *Weibull* and generalized *Pareto* (*GP*3) distributions.

Mean peak periods ($\bar{T}_p$) and average of the one-third the highest peak periods ($\bar{T}_{p(1/3)}$) are deduced from the *CMF* of the models and are consistent with the computed values. The computed average peak period of one-third the highest wave heights ($\bar{T}_{p(H1/3)}$) have significant linear functional relationship ($R^2 > 90.0\%$) with $\bar{T}_{p(1/3)}$. The parametric relation derived from *GP*3 distribution based on order-statistics for the estimation of $\bar{T}_{p(1/3)}$ is simple for practical applications and also provide accurate estimates of $\bar{T}_{p(1/3)}$. The empirical ratio of significant wave period to mean wave period of 0.9–1.4 is well interpreted by $\bar{T}_{p(H1/3)}$ to $\bar{T}_p$. The averages of the mean maximum peak periods ($\bar{T}_{p(max)}$) and most probable maximum peak periods ($T_{p(mpm)}$) are also computed for the winter and summer seasons and are compared with estimated values. It will be reasonable to consider the parametric relations derived from three-parameter *Weibull* distribution for $\bar{T}_{p(max)}$ and $T_{p(mpm)}$ estimates in winter and that derived from *GP3* distribution for estimates in summer. The probabilistic approach adopted in this case study has universal application that contributes to clear the peak period climate of an oceanic region.

ACKNOWLEDGEMENTS

This work was performed within the Strategic Research Plan of the Centre for Marine Technology and Ocean Engineering (CENTEC), which is financed by Portuguese Foundation for Science and Technology (Fundação para a Ciência e Tecnologia-FCT).

REFERENCES

Bitner-Gregersen, E., Guedes Soares, C., Machado, U., Cavaco, P., 1998. Comparison of different approaches to joint environmental modelling. *Proceedings of the 17th international conference on offshore mechanics and Arctic Engineering* (OMAE98). ASME Paper OMAE98-1495.

Falcão, A.F., 2015, "Developments in oscillating water column wave energy converters and air turbines", in: Guedes Soares, C. (Ed.). *Renewable Energies Offshore*, London, UK: Taylor & Francis Group; pp. 3–11.

Ferreira, J.A., Guedes Soares, C., 2002. Modelling bivariate distributions of significant wave height and mean wave period. *Applied Ocean Research*, Vol. 24 (1), pp. 31–45.

Goda, Y., 2010. Random seas and design of maritime structures. *Advanced Series in Ocean Engineering*. Singapore: World Scientific, Singapore.

Guedes Soares, C. 2008. Hindcast of Dynamic Processes of the Ocean and Coastal Areas of Europe. *Coastal Engineering*. 55(11):825–826.

Guedes Soares, C., 1984. Representation of double-peaked sea wave spectra. *Ocean Engineering*, Vol. 11 (2), pp. 185–207.

Guedes Soares, C.; Bhattacharjee, J.; Tello, M., and Pietra, L. 2012; Review and classification of Wave Energy Converters. Guedes Soares, C. Garbatov Y. Sutulo S. & Santos T.A., (Eds.). *Maritime Engineering and Technology.* London, UK: Taylor & Francis Group; pp. 585–594.

Haver, S., 1985. Wave climate off northern Norway. *Applied Ocean Research*, Vol. 7 (2), pp. 85–92.

Hosking, J.R.M., Wallis, J.R., 1997. Regional frequency analysis: an approach based on L-moments. Cambridge University Press, Cambridge.

Kitano, T., Mase, H., Kioka, W. 2002. Theory of significant wave period based on spectral integrals. *Proceedings of the 4th international conference on ocean wave measurement and analysis*, 2001. ASCE, San Francisco.

Longuet-Higgins, M.S., 1975. On the joint distribution of wave periods and amplitudes of sea waves. *Journal of Geophysical Research*, Vol. 80, pp. 2688–2694.

Longuet-Higgins, M.S., 1983. On the joint distribution of wave periods and amplitudes in a random wave field. *Proceedings of the Royal Society of London*, Vol. 389 (A), pp. 241–258.

Muraleedharan, G., Guedes Soares, C., 2014. Characteristic and moment generating functions of generalised Pareto (GP3) and Weibull distributions. *Journal of Scientific Research and Reports*, Vol. 3 (14), pp. 1861–1874.

Muraleedharan, G., Lucas, C., Martins, D., Guedes Soares, C., Kurup, P.G., 2015. On the distribution of significant wave height and associated peak periods. *Coastal Engineering*, Vol. 103, pp. 42–51.

Muraleedharan, G., Mourani, S., Rao, A.D., Unnikrishnan, N.N., Kurup, P.G., 2009. Estimation of wave period statistics using numerical coastal wave model. *Natural Hazards* Vol. 49, pp. 165–186.

Pilar, P., Guedes Soares, C., Carretero, J.C., 2008. 44 year wave hind cast for the North East Atlantic European Coast. *Coastal Engineering*, Vol. 55 (11), pp. 861–871.

Rodríguez, G., Guedes Soares, C., Machado, U., 1999. Uncertainty of the Sea State Parameters resulting from the methods of spectral estimation. *Ocean Engineering*, Vol. 26 (10), pp. 991–1002.

Rodríguez., G., Guedes Soares, C., Pacheco, M., 2004. Wave period distribution in mixed sea states. *Journal of Offshore Mechanics and Arctic Engineering*, Vo. 126, pp. 105–112.

Progress in Maritime Technology and Engineering – Guedes Soares & Santos (Eds)
 ISBN 978-1-138-58539-3

Analysis of extreme storms in the Black Sea

L. Rusu
'Dunarea de Jos' University of Galati, Galati, Romania

M. Bernardino & C. Guedes Soares
Centre for Marine Technology and Ocean Engineering (CENTEC), Instituto Superior Técnico, Universidade de Lisboa, Lisbon, Portugal

ABSTRACT: The main objective of this work is to evaluate the extreme storms in the Black Sea using a Lagrangean approach. Results of hindcast simulations with a 3-hour temporal resolution are provided by the SWAN model implemented on the entire sea basin. The methodology is applied to 30 years of wave simulations (from 1987 to 2016) to identify the extreme storms, considering significant wave heights over a threshold of 5 m. From the 151 storms that were identified considering this threshold two, which were considered more relevant, are chosen for a further analysis. The statistical characteristics of these storms as maximum significant wave height, maximum area affected by the storm, storm life time, maximum area in the track, length of the track, are investigated as well as its evolution of along the track. An analysis of the areas more affected by the extreme events is also made.

1 INTRODUCTION

An improved understanding of the marine climatology is necessary for numerous marine and coastal activities, such as offshore platforms, the assessment of flooding risk, the design of marine structures and the evaluation of wave energy resources (Guedes Soares, 2008; Pilar et al., 2008; Rangel-Buitrago & Anfuso, 2013; Laugel et al., 2014). Extreme waves, storms and storminess have, in general, a major impact upon coastal populations and ecosystems. Also, they are one the phenomena that determines the short term evolution of coast by beach erosion or destruction of infrastructures (Trifonova et al., 2012).

The storm climate and related ocean roughness depend on the storm tracks, intensity, duration and frequency of occurrence (Dupuis et al., 2006). Also, one possible consequence of a change in climate over the 21st century or the past decades is an increase in wave heights over the ocean.

The purpose of this work is to contribute to the characterization of the wave climate in the Black Sea, in particular, regarding the occurrence of extreme storm events. Several studies have been made recently to evaluate storm conditions in the Black Sea. Valchev et al. (2012) studied the changes in occurrence and magnitude of storms in the western Black Sea during 60 years also investigating possible connections to the North Atlantic Oscillation (NAO) index. Recently, Zainescu et al. (2017) investigated the storm climate on the Danube delta coast, using wind and wave data and developed a storm severity index.

An ocean storm can be considered as an event in both space and time where the significant wave height rises above a predetermined critical level. The usual methodology is an Eulerian approach based on joint analysis of the variability of the field of significant wave height at given location.

The availability of databases with spatial information of the characteristics of the wave fields, covering periods of several decades, that can be obtained from large-scale simulations, allows an assessment of both the temporal and the spatial evolution of ocean storms, providing a Lagrangean description of storm characteristics.

This alternative process, to identify and track a storm along its trajectory, assessing the modifications of their characteristics over time was described in Bernardino et al. (2008) and Bernardino & Guedes Soares (2015, 2016) and had previously been applied to a closed area, the Barents Sea by Boukhanovsky et al. (2003).

This Lagrangean methodology will be applied, in this work, to a wave hindcast dataset, that resulted from SWAN model implementation over the Black Sea, (Rusu, 2015; Rusu et al., 2014a,b; Butunoiu & Rusu, 2012), forced by 10 m wind fields (U10) provided by the U.S. National Centers for Environmental Prediction, Climate Forecast System Reanalysis (NCEP-CFSR) (Saha et al., 2014).

2 DATA AND METHODOLOGY

Storms are considered as high intensity events of the significant wave height time series above a given level (Petruaskas & Aagaard, 1971; Angelides et al., 1981). The statistics of storm duration (see, e.g. Jardine et al., 1981; Graham, 1982; Mathiesen, 1994), and the corresponding maximum significant wave height in a storm (as the measure of storm intensity) are studied in Borgman (1973), Boukhanovsky et al. (1988), Arena & Pavone (2006), Bernardino et al. (2008) and Bernardino & Guedes Soares (2015, 2016).

Eulerian approach, the usual methodology is based on joint analysis of the variability of the field of significant wave height $h(x,y,t)$, at a given location (x_i,y_i). This approach results in a joint analysis of multivariate time series of intensities and durations of storms at each point. In theory, joint analysis of the durations of the storms, would identify the same storm in several places, but in practice since the actual characteristics of the storms will be changing over time, the process of identifying points that are affected by a same storm is very difficult.

With the availability of hindcast wave datasets the study of the spatio-temporal evolution of significant wave height fields is possible. Such data source allows obtaining fundamentally information concerning the storm variability over the ocean in an approach that is traditional for meteorological researches, where the spatio-temporal variability of cyclones is considered.

The Lagrangean approach considers the dynamics of storms as spatial structures in a field. With respect to the classical definition of storms in a wave time series, the storm area is considered in the space domain, for each time t, as the constrained set of grid points where

$$\Omega(t) = \{(x,y): h(x,y,t) > z\}, \tag{1}$$

with $h(x,y,t)$ also representing the significant wave height field and z the storm definition level.

The function

$$W(x,y,t) = \begin{cases} h(x,y,t) - z, (x,y) \in \Omega(t) \\ 0, [(x,y) \notin \Omega(t)] or [t \notin [0,D]] \end{cases} \tag{2}$$

for each time moment $0 \le t \le D$, where D is the total duration (or life time) of storm above the level z, may be interpreted as a spatial stochastic pulse. Pulse dynamics $W(x,y,t)$ depends of a set of parameters $\Xi(t)$ that characterize the spatial shape of the storm.

In theory, the Eulerian and Lagrangean approaches to storm dynamics represent the same physical situation in different ways, but for practical reasons, and for each concrete problem one approach could be more appropriate than the other.

In general, authors use the Eulerian approach, but in the present work, the described Lagrangean methodology is applied to the wave data set to perform a study of the storms in the Black Sea during the 30 year period 1987–2016. A threshold of 5 m for significant wave height and a minimum duration of 12 hours were imposed to perform the storm identification. Different restrictions regarding the minimum duration or the value of the threshold were used when analysing storms in other locations. Bernardino & Guedes Soares (2015, 2016), for instance, used a threshold of 7 m when analysing storms in the North Atlantic.

3 EXTREME WAVE CLIMATE

Based on the SWAN model results with a 3-hour temporal resolution for all period considered, various analyses were performed to characterize the storm conditions in the Black Sea.

Thus, considering a threshold of 5 m for the significant wave height to characterize the extreme events from the Black Sea, the geographical distribution of the storm conditions was determined as the percentage of the waves greater than this value. The corresponding results are illustrated in Figure 1 and they show that the western part of the basin is more affected by higher waves. At this annual scale, the extreme values have a frequency less than 0.5% in most of the Black Sea, only reaching 0.75% in the western part of the basin.

Looking for season distribution of the wave extremes, percentage of the waves greater 5 m in each season were computed following seasonal partition: December-January-February (DJF—winter), March-April-May (MAM—spring), June-July-August (JJA—summer) and September-October-November (SON—autumn). The maps computed for each season are given in Figure 2.

Winter is the season when most extreme significant wave height occurs (reaching almost 2% frequency in

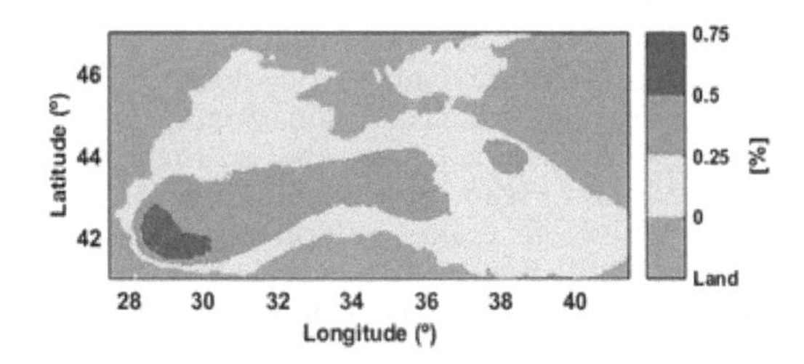

Figure 1. The geographical distribution of the storm conditions in percentage (significant wave heights greater than 5 m) as resulted from the 30-year wave simulations (1987–2016).

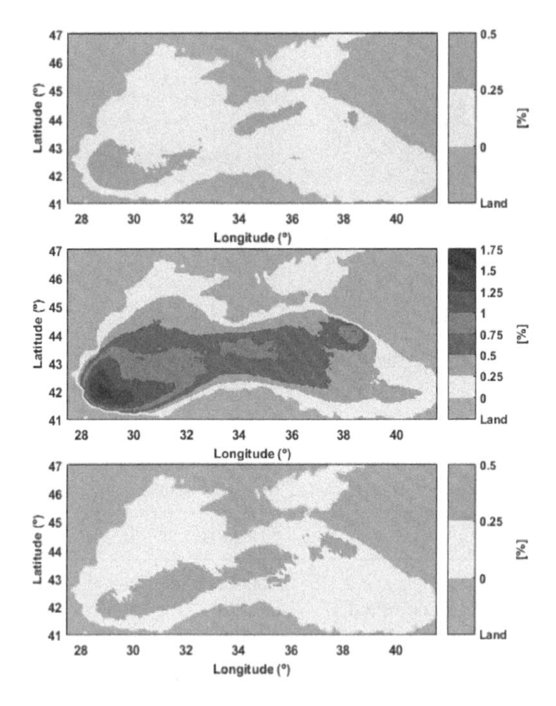

Figure 2. The geographical distribution of the storm conditions in percentage (significant wave heights greater than 5 m) for autumn (top), winter (center) and spring (bottom).

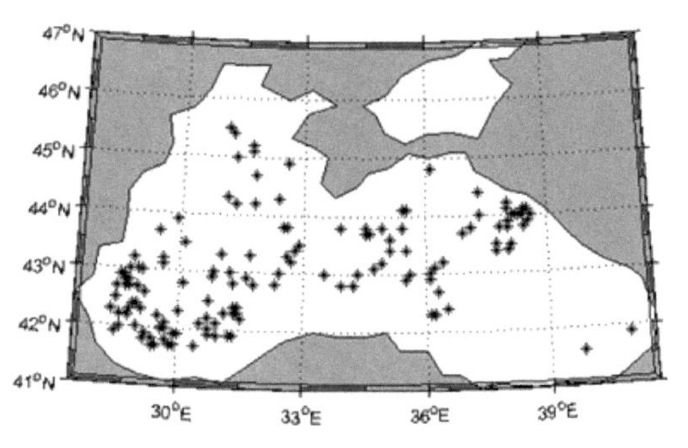

Figure 3. Initial location of the storm events.

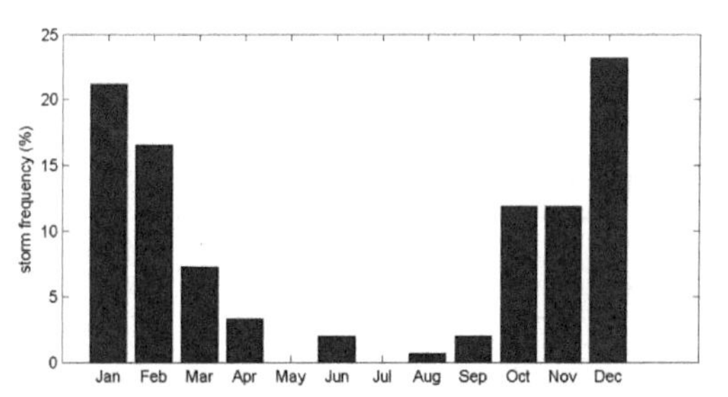

Figure 4. Seasonal distribution of extreme storms.

the west), followed by spring and autumn (less than 0.5%). During summer no extreme values above 5 m were identified. The geographical distribution is similar among seasons but nevertheless there is not a complete agreement between the two transition season, spring and autumn.

4 STORM ANALYSIS

The Lagrangean methodology for storm identification that was already described is applied to 0.32° × 0.32° gridded significant wave height obtained from model simulation, with a 3 hour temporal resolution, covering a 30 years period, from 1987 to 2016. Using a 5 m threshold for significant wave height and 12 hours minimum duration, 151 storms were identified in the Black Sea during the 30-year period.

Figure 3 shows the location of the beginning of each identified storm. It can be seen that storms do not start homogeneously through the Black Sea. The south western site extreme storm are more frequent and in the eastern almost non-existent.

The number of storms in each calendar month was counted and it can be seen in Figure 4. The frequency of storms has a seasonal distribution, with storms being more frequent between October and March. The stormiest month is December, closely followed by January. During the summer months, from May to September, the frequency of storms is very low.

For each storm, duration, maximal Hs in the track, area associated with maximal Hs, time of maximal Hs arising, maximal storm area in the track, time of maximal area arising, length of the track, mean direction of the track, arise length of the track, mean direction of track arising, decay length of the track and mean direction of track decay can be obtained.

The total number of storms identified each year was evaluated and can be observed in Figure 5. It can be seen that there is a high inter annual variability, with some years having only two storms but others reaching a total of nine.

The annual mean number of storms during the 30 years was five with a standard deviation of 2.19 (Table 1). No significant linear trend seems to be present.

Considering mean annual values of storm duration and maximum Hs in the storm, some statistics were taken over the period under study and are presented in Table 1.

From the 151 identified storms, two were selected for further analysis. Storm A, occurred during the end of January 2004 and lasted 36 hours and storm B occurred during the last days of January 2006, lasting 69 hours.

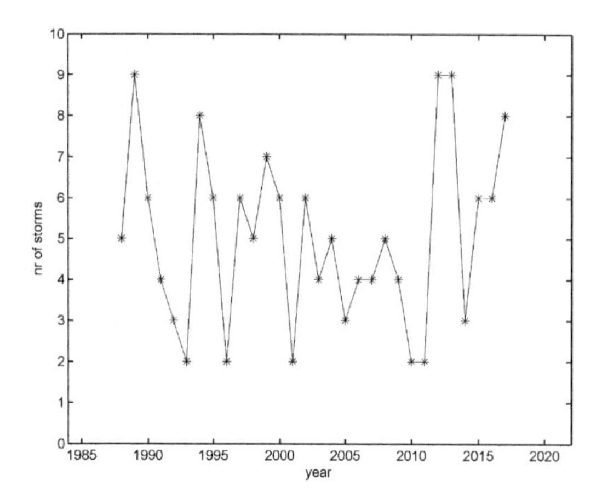

Figure 5. Total number of storms identified in each year.

Table 1. Annual statistics of the storm events.

Annual statistics	Mean	Max	Min	Standard deviation
Total nr of storms	5.03	9.00	2.00	2.19
Storm duration (h)	67.53	90.00	45.00	24.00
Max Hs (m)	6.94	9.00	5.83	0.62

The two storms have different characteristics although they were generated almost in the same area. The path of the storms can be observed in Figure 6. Storm A was also identified by Zainescu et al. (2017) as a severe storm using both observational and modelled data at Gloria platform that is located in the western coast. Storm A starts west of 29°E and travels north-eastwards until almost the end when it slightly backs down, vanishing after 36 hours.

Storm B is generated in the same area but lasts much longer and travels a more complex path. First it travels north-westwards, then turns back, travels east, and finally changes direction again and returns west. The two storms occurred in the same area of the Black Sea during the same season, by had different characteristics as it can be observed from Table 2.

Storm A, although shorter and travelling only 276 km, was more intense reaching a maximum Hs value in the track of 11.8 m and maximum affected area of 147.2×10^3 km^2. Storm B, lasts longer, travels almost twice the distance of storm A, but the maximum Hs in the track was only 7.6 m and it affected a smaller area, reaching only 89.3×10^3 km^2. Both storms took longer to reach the maximum Hs than to decay.

The evolution of the maximum Hs in storm A can be seem in Figure 7. The maximum value in the storm increases steadily during 21 hours and the decreases in approximately the same rate.

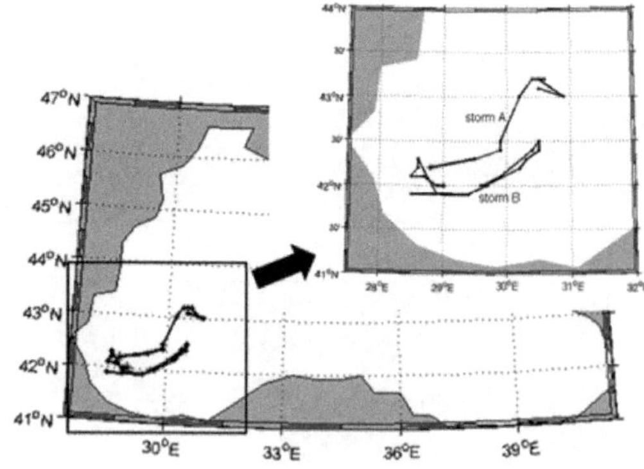

Figure 6. Trajectories of the two extreme storms.

Table 2. Characteristics of the two storms.

	Storm A	Storm B
Initial date	2004012203	2006012306
Storm duration (hours)	36	69
Maximal Hs in the track (m)	11.80	7.60
Storm area associated with maximal Hs	136.1	89.3
Time of maximal Hs arising (hours)	21	51
Maximal storm area in the track	147.2	89.3
Time of maximal area arising (hours)	27	51
Length of the track (km)	276	505
Mean direction of the track	233	70
Arise length of the track (km)	194	306
Mean direction of track arising	229	248
Decay length of the track (km)	82	199
Mean direction of track decay	342	69

Maximum Hs in storm B evolves in a different way. Although there are values above 5 m before the 24th of January, which fulfils the requirements for storm identification, the maximum Hs in the storm does not start to actually rise before the 24th. Then it starts to increase significantly reaching a maximum of 7.6 m on the 25th of January, decreasing after that date (Figure 8).

The evolution of the area affected by each storm can also be assessed. The area affected by storm A increases in time until a maximum value of 147.2×10^3 km^2 and then decreases (Figure 9). It can be noticed that the storm reaches the maximum Hs before it reaches the maximum affected area.

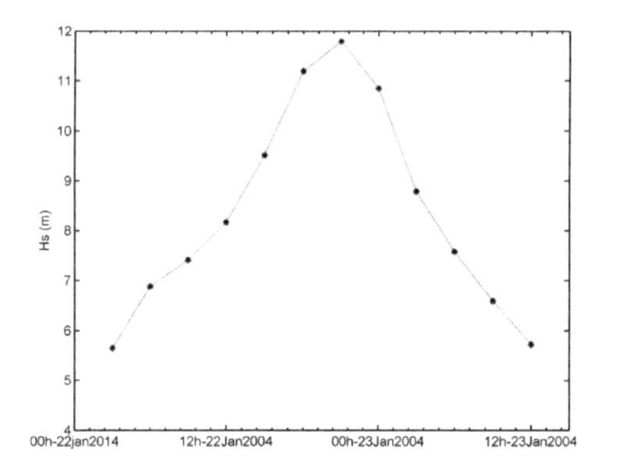

Figure 7. Evolution of the maximum significant wave height value in storm A.

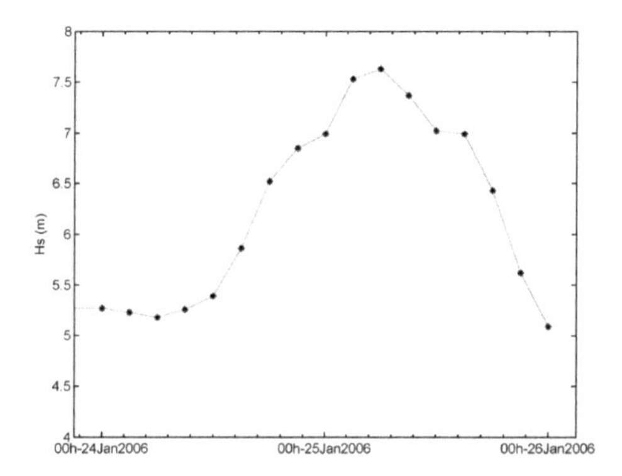

Figure 8. Evolution of the maximum significant wave height value in storm B.

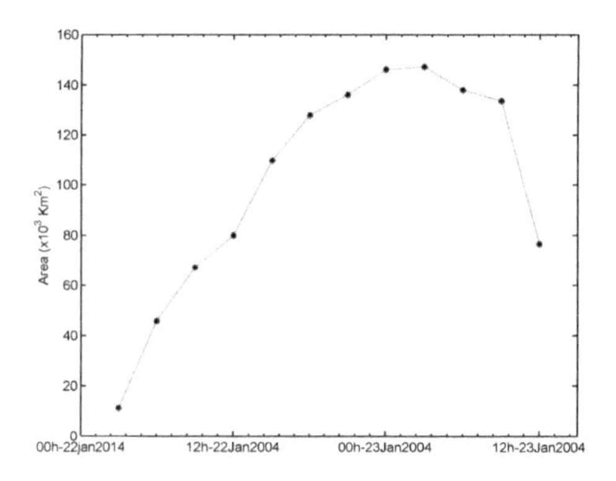

Figure 9. Evolution of the area affected by storm A.

Evolution of the area affected by storm B is similar; to the evolution of maximum Hs in the same storm as it maintains low values for more than a day and only starts to increase at as fast rate from the 24th onwards until it reaches it maximum approximately at the same time that reaches the maximum Hs value on the 25th of January (Figure 10). After the maximum is

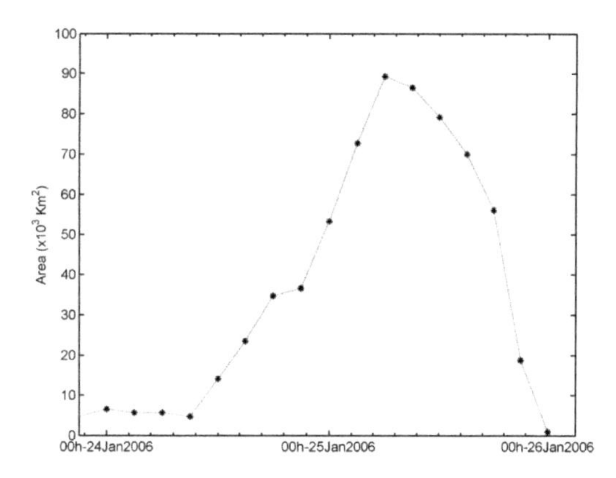

Figure 10. Evolution of the area affected by storm B.

reached, the storm starts to vanish and dissipated in 12 hours.

5 CONCLUSIONS

Using the results of the wave model simulations, performed with SWAN in the entire Black Sea basin for a 30-year period (1987–2016), an analysis of the geographical distribution of extreme storms has been performed. A threshold of 5 m between the extreme storm conditions and the normal sea states was considered for the significant wave height.

As noticed also in some previous studies, the highest significant wave height values are encountered in the western side of the Black Sea. Moreover, a strong seasonality can be also noticed. Thus, during spring and autumn the percentage of Hs greater than 5 m is about 0.5%, while in the summer time a lower percentage is characteristic. In the winter, the higher percentage of significant wave heights greater than 5 m, can be noticed (about 2%), with a clear localization in the western side of the basin.

Considering the same Hs value greater than 5 m for identifying the storms and also minimum storm duration of 12 hours, a Lagrangean methodology was used to perform an analysis of the storm dynamics. As a general trend, it was noticed that most of the storms initiate in the south-western side of the basin and further propagate to the central area. The period of the year when they have the highest frequency is between October and March.

Considering annual statistics, there is a high interannual variability in the number of storms and in some of the analysed characteristics. No apparent trend in annual number of extreme storms seems to be present. These results are in agreement with Valchev et al. (2012).

Finally, it can be also concluded that the Lagrangean methodology for storm identification proved to be an adequate tool to study extreme wave climate in the Black Sea. One of the storms identified and studied in some detail "Storm A", was classified by Valchev et al. (2012) as the forth severest storm in the period 1948–2010, ranked according to significant wave height maxima and also identified by Zainescu et al. (2017).

ACKNOWLEDGMENTS

This work was supported by a grant of Portuguese Foundation for Science and Technology, number SFRH/BPD/112391/2015 and other one of Romanian Ministry of Research and Innovation, CNCS—UEFISCDI, project number PN-III-P4-ID-PCE-2016-0028, within PNCDI III.

REFERENCES

Angelides D.C., Veneziano D. & Shyam Sunder S. 1981 Random sea and reliability of offshore foundations. *J. Eng. Mech. Div*. 107: 131–148.

Arena F. & Pavone D. 2006. Some statistical properties of random waves in a sea storm. Proceedings of the *25th International Conference on Offshore Mechanics and Arctic Engineering*, ASME Paper OMAE2005-92478.

Bernardino, M., Boukhanovsky, A.V. & Guedes Soares, C. 2008. Alternative approaches to storm statistics in the ocean. Proceedings of the *27th International Conference on Offshore Mechanics and Arctic Engineering*, ASME Paper OMAE2008-58053.

Bernardino, M. & Guedes Soares, C. 2015. A Lagrangian perspective of the 2013/2014 winter wave storms in the North Atlantic, Maritime Technology and Engineering, Guedes Soares, C. & Santos T.A. (Eds.), Taylor & Francis Group, London, UK, pp. 1381–1388.

Bernardino, M. and Guedes Soares, C. 2016. A climatological analysis of storms in the North Atlantic, Maritime Technology and Engineering 3, Guedes Soares, C. & Santos T.A. (Eds.), Taylor & Francis Group, London, UK, pp. 1021–1026.

Bidlot, J.-R., Janseen, P.A.E.M. & Abdalla, S. 2007. A revised formulation of ocean wave dissipation and its model impact. ECMWF Tech. Memo. 509, 27 pp.

Borgman, L. 1973. Probabilities for the highest wave in a hurricane. ASCE J. Waterways, Harbors and Coastal Engng., pp. 185–207.

Boukhanovsky A.V., Krogstad H.E., Lopatoukhin L.J., Rozhkov V.A., Athanassoulis G.A., Stephanakos C.N. 2003. Stochastic simulation of inhomogeneous metocean fields. Part II: Synoptic variability and rare events, Lecture Notes in Computer Science, Vol. 2658: 223–233.

Boukhanovsky A.V., Lopatoukhin L.J. & Ryabinin V.E. 1998. Evaluation of the highest wave in a storm. Reports of World Meteorological Organization, WMO/TD—Vol. 858. – 21 p.

Butunoiu, D. & Rusu, E. 2012. Sensitivity tests with two coastal wave models. *Journal of Environmental Protection and Ecology* 13(3): 1332–1349.

Campos, R. & Guedes Soares C. 2016. Comparison of HIPOCAS and ERA wind and wave reanalysis in the North Atlantic ocean. *Ocean Engineering* 112: 320–334.

Dupuis, H., Michel, D., & Sottolichio, A. 2006. Wave climate evolution in the Bay of Biscay over two decades. *Journal of Marine Systems* 63(3): 105–114.

Graham, C. 1982. The parameterization and prediction of wave height and wind speed persistence statistics for oil industry operational planning purposes. *Coastal Eng*. 6: 303–329.

Guedes Soares, C. 2008. Hindcast of Dynamic Processes of the Ocean and Coastal Areas of Europe. *Coastal Engineering* 55: 825–826.

Jardine, T.P. & Latham, F.R. 1981. An analysis of wave heights records for the NE Atlantic. *Quarterly Journal of Roy. Met. Soc.* 107: 415–426.

Laugel, A., Menendez, M., Benoit, M., Mattarolo, G. & Méndez, F. 2014. Wave climate projections along the French coastline: dynamical versus statistical downscaling methods. *Ocean Modelling* 84: 35–50.

Mathiesen, M., 1994. Estimation of wave height duration statistics. *Coastal Engineering* 23: 167–181.

Petruaskas, C. & Aagaard, P.M. 1971. Extrapolation of Historical Storm Data for Estimating Design Wave Heights. *J. Soc. Petroleum Engineering* 11: 23–37.

Pilar, P., Guedes Soares, C. and Carretero, J.C. 2008. 44-year wave hindcast for the North East Atlantic European coast. *Coastal Engineering* 55: 861–871.

Rangel-Buitrago, N. & Anfuso, G. 2013. Winter wave climate, storms and regional cycles: the SW Spanish Atlantic coast. *International journal of climatology* 33(9): 2142–2156.

Rusu, L. 2015. Assessment of the Wave Energy in the Black Sea Based on a 15-Year Hindcast with Data Assimilation. *Energies* 8(9): 10370–10388.

Rusu, L., Butunoiu, D. & Rusu, E. 2014a. Analysis of the extreme storm events in the Black Sea considering the results of a ten-year wave hindcast. *Journal of Environmental Protection and Ecology* 15(2): 445–454.

Rusu, L., Bernardino, M. & Guedes Soares, C. 2014b. Wind and wave modelling in the Black Sea. *Journal of Operational Oceanography* 7(1): 5–20.

Saha, S., Moorthi, S., Wu, X., et al. 2014. The NCEP climate forecast system version 2. *Journal of Climate* 27(6): 2185–2208.

Trifonova, E.V., Valchev, N.N., Andreeva, N.K. & Eftimova, P.T. 2012. Critical storm thresholds for morphological changes in the western Black Sea coastal zone. *Geomorphology* 143: 81–94.

Valchev, N.N., Trifonova, E.V. & Andreeva, N.K. 2012. Past and recent trends in the western Black Sea storminess. *Natural Hazards and Earth System Sciences* 12(4): 961–977.

Zăinescu, F.I., Tătui, F., Valchev, N.N. & Vespremeanu-Stroe, A. 2017. Storm climate on the Danube delta coast: evidence of recent storminess change and links with large-scale teleconnection patterns. *Natural Hazards*, 87(2): 599–621.

Progress in Maritime Technology and Engineering – Guedes Soares & Santos (Eds)
 ISBN 978-1-138-58539-3

Robust estimation and representation of climatic wave spectrum

G. Rodriguez
Applied Marine Physics and Remote Sensing Group, Institute of Environmental Studies and Natural Resources (iUNAT), Department of Physics, Universidad de Las Palmas, Las Palmas, Spain

G. Clarindo & C. Guedes Soares
Centre for Marine Technology and Ocean Engineering (CENTEC), Instituto Superior Técnico, Universidade de Lisboa, Lisbon, Portugal

ABSTRACT: The design and operation planning of marine facilities have to be developed taking into account dominant wave conditions and their variability over long periods of time. Climatic wave spectrum can be considered as a component of the long-term statistical characterization of wave conditions, which is paramount when dealing with the dynamic response of marine structures to wave loads. This study presents a methodology for robust estimation and representation of the climatic wave spectrum from a given set of measured wave spectra that is consistent with the statistical properties of the spectral density estimates at a given frequency. It provides information on location, spread, asymmetry, and extremes of the spectral density estimations by using robust and resistant methods for summarizing the data sets, and uses related graphical techniques for the efficient display of the information. The approach is exposed by analysing a large collection of measured wave spectra, used as a methodological example.

1 INTRODUCTION

The adequate knowledge of wave climate is crucial in any field concerning the behaviour of natural or man-made structures exposed to wave action, such as ocean engineering, naval architecture, coastal engineering, and coastal geomorphology. In particular, it is essential for long term dynamic response estimation of offshore and ship structures when exposed to wave loads and, consequently, for its reliable design (e.g. Haver & Nyhus, 1986; Hu, 1991; Teixeira & Guedes Soares, 2009; Vogel et al., 2016).

Wave climate can be defined simply as the long-term (commonly a decade or more) statistical characterization of the behaviour of waves caused by prevailing winds and storms for a particular location (Holthuijsen, 2007). In this context, sea surface elevation measured at a given point, $\eta(x,y,t)$, over a long period of time results in a wave record may be considered as a realization of a slowly varying nonstationary process. So, it may be split into many segments of short duration (usually from several tens of minutes to a few hours), commonly referred to as sea states, during which the process remains statistically stationary. Therefore, the stationarity hypothesis establishes two time scales in the study of wave conditions. Short-term statistics concerns the behaviour of waves during a single sea-state while long-term statistics deals with the characterization of wave conditions over time periods covering multiple sea-states and represents the wave climate when the period of time considered spans over years. Consequently, wave climate description requires the previous stochastic characterization of every sea state recorded during the time period considered (Guedes Soares, 1998).

The energy content, frequency composition, and shape of the spectra derived from sea states measured during several years shows a significant time variability, revealing the existence of patterns of various time scales, as well as the appearance of episodic events, related to the variability of the synoptic atmospheric conditions generating the different types of wave fields observed in a specific area. The wave climate statistical characterization is generally based on the representation of every particular sea state only by means of a few characteristic sea state parameters derived from the wave spectrum and not by the spectrum itself. The most common parameters used to characterise a sea state are the significant wave height, H_S, which accounts for the sea state severity, the mean period, T_m, or the peak period, T_p, and the mean wave direction, θ_m (Ferreira & Guedes Soares, 2000).

Over time, the probabilistic description of wave climate has been developed in many different ways, such as through the representation of the significant wave height empirical probability density function by some parametric model, or by extending this approach to the bivariate empirical probability distribution of the significant wave height and some characteristic wave period, commonly represented in the form of a scatter diagram, or sometimes even by conditioning this bivariate distribution to several directional sectors (e.g. Haver & Natvig, 1991, Ferreira & Guedes Soares, 2002). That is, the wave spectrum associated to a given group defined by a set of parameters may present a significant variability of spectral shapes, mainly reflected in a variable peakedness of the wind generated component and in the possible presence of swell contributions (e.g. Haver, 1985).

Accordingly, various authors have claimed the need to complement the traditional wave climate description by including information on the long-term behaviour of the wave spectrum. That is, introducing the long-term average or climatic wave spectrum (e.g. Scott, 1968; Haver and Nyhus, 1986; Buckley, 1988; Hu, 1991; Teng et al. 1994; Lucas et al. 2011; Vogel et al. 2016), which has been defined by the International Maritime Organization as a spectrum averaged over the ensemble of spectra, characterized by some probability of occurrence, and corresponding to dominant wave generation conditions over the whole area (Buckley, 1993).

The wave climate or, in particular, the assessment of the long term behaviour or response of marine structures should be performed by taking into consideration each one of the individual wave spectrum. However, it is clear that this approach represents a considerable computational effort and that the potential information provided would not be easily manageable and comprehensible. As a consequence, several approaches have been suggested to alleviate this inconvenient but all of them have a common main objective that consists in grouping multiple sea states, or wave spectra, in a relatively small set of representative wave conditions each one containing those wave spectra with similar energy content, frequency composition, and spectral shape.

Various approaches can be used to reduce the large variability observed in the collections of wave spectra considered for characterizing the wave climate, grouping the individual wave spectra into homogeneous sets, or classes, which only include those elements with a certain degree of similarity. The common approach to estimate the climatic wave spectrum from a given set of measured spectra has been to compute the average of the spectral density estimations associated to each frequency band, which is consistent with the name of long-term average wave spectrum, sometimes used as a synonym, and with the definition given by the International Maritime Organization as a spectrum averaged over the ensemble of spectra (Buckley, 1993). Additionally, its variability is commonly expressed in terms of the range (minimum and maximum values) and the standard deviation, or its variance (e.g. Scott, 1968; Buckley, 1988; Hu, 1991; Teng, 2001).

In the light of the foregoing, two successive and complementary steps can be distinguished in the procedure to estimate a climatic wave spectrum. In the first step, the complete collection of measured wave spectra are organized in homogeneous groups according to some specific criteria related to the particular goals of the study. In the second step, wave spectra associated to each group are used to derive the corresponding characteristic wave spectrum, as well as the associated statistical variability. With the previous considerations in mind, this work aims to clarify the general procedure of climatic wave spectrum estimation, providing a review of the existing methods to implement both steps, and to suggest a robust approach for estimating and representing the climatic wave spectrum from a given set of measured wave spectra, by using robust and resistant measures for summarizing large data sets, based on order statistics, and making use of related graphical techniques to present the information in a compact, comprehensive and informative way. A large set of measured wave spectra is used as a basis to expose the suggested methodological approach.

2 THEORETICAL BACKGROUND AND METHODOLOGY

2.1 *Brief review of methods to obtain homogeneous representative sets*

A classical approach to reach this goal, i.e. to reduce the variability among the sea states belonging to any representative set, consists in classifying the whole collection of measured wave spectra in ranges of significant wave height and estimating a representative wave spectrum for each one of these subsets (e.g. Buckley, 1988; Teng et al. 1994; Teng 2001). Other related and natural approaches are to decompose the full sea states collection in separate groups according to their severity, considering the sea states above and below a significant wave height threshold, or by identifying hydroclimatic seasons regarding sea state severity (e.g. Ewans and Kibblewhite, 1992). Such as previously commented, the most common approach is to categorize the sea states in terms of the significant wave height and some characteristic wave period (e.g. Haver and Nyhus, 1986) and, in this line of work, a natural extension is to increase of the number of representative parameters used to characterise each sea state, thus obtaining more homogeneous sets (e.g. Haver, 1985; Saulnier et al, 2011).

Data mining provides many possible approaches to estimate a few representative subsets of a large set of individuals, some of which have been recently used for the characterization of wave climate. In particular, several clustering and classification approaches have been applied to characterize multivariate wave climates (e.g. Abadie et al. 2006; Boukhanovsky, et al. 2007; Camus et al. 2011; Mortlock and Goodwin, 2015). An interesting classification approach is to organize the spectra by identifying the types of wave systems giving rise to every observed wave field as wind sea, swell, or combinations of these types of wind-generated waves (Boukhanovsky et al, 2007 and Lucas et al., 2011). It is interesting to note that many of these approaches are based on physical and statistical arguments whereas some of them are based only on statistical criteria.

2.2 *Climatic wave spectrum estimation*

The conventional approach to estimate the climatic wave spectrum from a given set of measured spectra is to compute the average of the spectral density estimations associated to each frequency band and express its variability in terms of the range and the standard deviation, or variance. To clarify the concepts and ideas underlying this approach it is interesting to identify the theoretical framework. Thus, to achieve this objective, consider that a large set of n measured wave spectra are available to characterise the wave climate in a given location. The spectral densities of each spectra have been evaluated at m specific frequencies equally spaced at Δf Hz intervals.

Then, let η_i be the i-th sea state in the sequence of n successive sea states and $S_{\eta_i}(f)$ the associated spectral density function. Each one of the n wave spectra represents a discrete function evaluated at m specific frequencies. Conceptually, any particular wave spectrum, $S_{\eta_i}(f)$ can be interpreted as a realization of a random process, denoted as, $\{S_\eta(f)\}$ and the complete collection of wave spectra, or realizations, can be named the ensemble.

$$\{S_{\eta_i}(f)\} \quad i = 1, 2, \ldots, n$$

Alternatively, the stochastic process $\{S_\eta(f)\}$ can be construed as a collection of random variables labelled by the parameter f. Then, considering that the sample space is the ensemble of realizations (wave spectrum), for any fixed frequency it is possible to define a random variable, $S_\eta(f_j)$, which represents the spectral density estimations at frequency f_j and is described by its probability density function (pdf). Consequently, the ensemble of wave spectra can be understood as a set of m random variables given by

$$\{S_\eta(f_j)\} \quad j = 1, 2, \ldots, m$$

A complete description of each random variable requires the identification of the pdf from which the data have been drawn. So, it should be necessary to find the theoretical probability distribution model representing the statistical behaviour of the spectral density estimations corresponding to any given value of the frequency. Although details concerning the data are very important, it is possible and common to summarise the data through the use of a few well-chosen parameters, easily computed from the data and understood, that typify them for the purpose of practical applications. Such a summary should include information about the range of the data set, its location, spread, and symmetry.

The most commonly used measures of location, which describe the central tendency of the data and represent the typical value for a dataset, and dispersion, assessing the variability or spread degree in the original data, are the sample mean and the standard deviation, or the variance. Accordingly, in practice, the most common approach is to estimate the average or mean spectrum, given by

$$\bar{S}(f_j) = \frac{1}{n}\sum_{i=1}^{n} S_i(f_j) \quad j = 1, 2, \ldots, m$$

and the corresponding variance spectrum,

$$Var\{S(f_j)\} = \frac{1}{n-1}\sum_{i=1}^{n}\left[S_i(f_j) - \bar{S}(f_j)\right]^2$$
$$j = 1, 2, \ldots, m$$

It is also common to estimate the maximum wave spectra by selecting the maximum value of the spectral density associated to each frequency. That is to say,

$$max\{S_i(f_j)\} \quad i = 1, 2, \ldots, n; \quad j = 1, 2, \ldots, m$$

2.3 *Limitations of conventional approach and robust alternative*

The climatic wave spectrum estimated through the above exposed conventional approach is given by the average value of spectral densities associated to each frequency and its statistical variability characterised by the corresponding variance. This classical approach to estimate the climatic wave spectrum is based on the normality assumption for each one of the random variables, $S_\eta(f_j)$ that constitute the ensemble of wave spectra. However, in general, the absence of prior information about the data does not allow one to assume that data are normally distributed. On the contrary, it is much more likely that a given data set is not normal. Therefore, simple computation of parameters that describe a set of data completely only if the data are normal will not generally be satisfactory, because the distribution of non-normal variables may lack symmetry and may have extreme values, or outliers, giving rise to extreme right or left tails. Non-normal variables, (particularly those with extreme right or left tails), may be better summarized and exposed by means of robust and resistant statistical measures, obtained from a small set of sample quantiles, and related graphical techniques.

An estimator is said to be robust, in general, if it presents a lack of susceptibility to the effects of incorrect assumptions regarding the underlying assumed probability model, usually the effects of non-normality. On the other hand, an estimator is said to be resistant when a change in a few data will not substantially change the value of the estimate. Particularly, a resistant estimator is not unduly influenced by a small number of outliers (Hoagling et al., 1983). Among the large variety of robust estimators proposed in the statistical literature those based on sample quantiles are of particular interest, due to its simple computation and direct interpretation, as well as no requirement of distributional assumptions. Moreover, these are closely linked with the construction of box plots, a valuable graphical tool to uncover the distributional characteristics of the data.

The q-th sample quantile, x_q, of a continuous random variable x, for any q ($0 < q < 1$) is defined by $P(X \leq x_q) = q$. In other words, it represents a value such that q% of the data lie below x_q. The quantile associated with $q = 0.5$ is called the median, or second quartile, denoted as Q_2; the quantiles associated with $q = 0.25$ and $q = 0.75$ are the lower and upper quartiles, denoted by Q_1 and Q_3, and those associated with $q = 0.1$ and $q = 0.9$ are the lower and upper deciles, denoted as D_1 and D_9, respectively.

Furthermore, skewed distributions affect the mean, making it less representative of the sample. Thus, robust alternatives have been proposed, of which the median is the most well-known. It is a good indicator of central tendency, or location, because it is fairly robust, remaining stable under a wide variety of circumstances. The range, represented by the extreme values of the variable, is the simplest measure of variability in a sample set but may give a distorted impression of the spread of the data if outliers are present, since only two observations are included in the estimate. Similarly, data dispersion is controlled by the existence of values considerably higher and lower than the mean value. Then, when the variable is not normally distributed another statistic called interquartile range, IQR, represents a robust indicator of dispersion, or scale, and reduces the problem by considering the variability within the middle 50% of the dataset. Therefore, it is defined in terms of the data values holding 25% and 75% of the values below them. That is, the interquartile range is defined as the difference between the upper and lower quartiles, $IQR = Q_3 - Q_1$.

Information provided by quartiles can be used to generate a clear and compact visual representation of the sample distribution known as boxplot, a graphical tool which strikingly summarize and display the distribution of a set of continuous data (Tukey, 1977). However, in its most common version, also known as box-and-whisker plot, the thin lines drawn from the edges of the box do not reach the extremes but up to some inner boundaries named as whiskers, which can be defined in various ways. Particularly, in this study, the whisker boundaries have been placed at D_1 and D_9, but can be changed by some other couple of percentiles (e.g. 5th and 95th or 1th and 99th). Note that the box emphasizes the part of the data set where the middle 50% of the data lie, whereas whiskers provides information on that part of the data set which lies in the usually large intervals between the quartiles and these boundaries, in this case the middle 80%. Further details on the construction of a boxplot can be found in many statistical reference books (e.g., Hoaglin, et al., 1983).

In line with the aforementioned, this study suggests the use robust and resistant measures of central tendency to estimate the climatic wave spectrum from a collection of measured wave spectra, as well as the use of boxplots for representing the climatic wave spectrum and its statistical variability in a compact, comprehensive, and informative way, providing a clear distributional overview in terms of robust and resistant measures of the central tendency, dispersion, skewness, extremes, and tail lengths of the spectral estimates at each frequency band.

3 APPLICATION TO MEASURED WAVE SPECTRA

Experimental data used to show the suggested methodology for estimating the climatic or representative wave spectrum from a large set of measured spectra consists of fifteen years (1996–2011) of hourly wave spectra from the 44005 buoy station, U.S. National Data Buoy Center, located in the North Atlantic Ocean (43°12′42″ N–69°7′42″W) in open waters about 200 meters depth. Wave spectra associated to mild sea states (significant wave height lower than 1 meter) and extreme wave conditions (Hs greater than 6 meters) have not been considered for further analysis.

In this sense, it should be noted that sea states with Hs higher than 1 m represent less than 0.1% of the total. Thus, the wave spectra set used in the study includes 59430 sea states covering the range of significant wave height from 1 to 6 meters.

Wave spectra associated to sea states in the range ($1 \leq Hs \leq 6$) are illustrated in Figure 1a., together with the mean, maximum and minimum spectra. Maximum (minimum) spectrum is obtained as the curve joining the maxima (minima) of the spectral density estimations corresponding to each frequency band. Naturally, maximum and minimum spectra are heavily affected by the presence of outliers and, consequently, their representativeness is very low, because the envelope curve is defined by just one spectral estimation for every frequency band. Nevertheless, only the maximum spectrum is used in practice (e.g., Buckley, 1988; Teng, et al., 1994). Regarding the mean spectrum, it is observed that the number of null, or very close to zero, spectral densities is significantly higher than that of very large values. As a consequence, the mean for each frequency band adopts very low values, so that it is hardly noticeable when plotted together with the measured spectra. This fact is due to the large variability among the multiple measured spectra included in the examined set. The same effect is also observed in Figure 1b, which depicts the full set of individual spectra together with the median spectrum, as well as that defined by the first, Q_1, and the third, Q_3, quartiles, the first, D_1, and ninth, D_9, deciles of the spectral densities for each frequency band. It can be observed that only Q_3 and D_9 curves are noticeable and that their representativeness is still inadequate. These spectra, curves joining the D_1, Q_1, Q_2, Q_3, and D_9 percentiles of the spectral estimations, are shown separately in Figure 1c, for clarity, while Figure 1d depicts the same information as Figure 1c, but using box-plots, thus providing a more intuitive idea of the variability associated to each frequency band and, hence, on the representativeness of the climatic (median) spectra.

A common and natural approach to reduce the variability in a given set of environmental data is to decompose the full data set in terms of the climatic season. In principle, any of the four resulting data sets should be more homogeneous than the whole set. Therefore, as a preliminary step to explore the usefulness of the suggested procedure to estimate a representative, or climatic, wave spectrum from a set of measured wave spectra, providing a clear distributional overview of the spectral estimates at each frequency band, the whole set of wave spectra has been split into four subsets, regarding climatic seasons. So, the individual spectra are distributed as follows: Spring (March, April, May), 14297, summer (June, July, August), 10903, autumn (September, October, November), 16848, and winter (December, January, February), 17382.

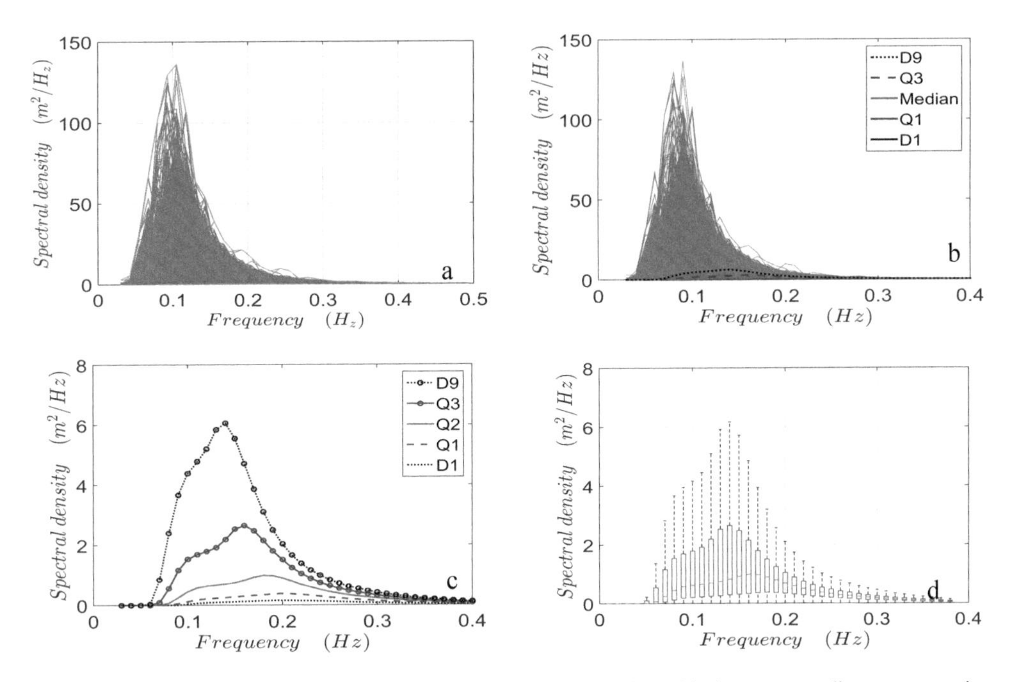

Figure 1. Individual wave spectra for all the sea states considered, together with the corresponding mean, maximum, and minimum spectra (a), individual wave spectra together with the D1, Q1, Q2, Q3, and D9 spectral curves (b), D1, Q1, Q2, Q3, and D9 spectral curves (c), and median wave spectra and its variability illustrated by using box-plots (d).

The results, shown in Figure 2, reveals that the variability of the spectral estimates is still very important, even after clustering spectra by climatic seasons. This fact is evidenced by the large values taken by the IQR and, especially by the distance between whisker boundaries ($D_9 - D_1$). It is interesting to note that, as expected, the variability of $S_\eta(f_j)$ reduces for very low and high frequency values and increases in the range of energy-containing frequencies. Another important feature to note is the large asymmetry of the $S_\eta(f_j)$ distribution in the more energetic frequency bands, clearly evidenced by the difference in the distances between the median and the first and third quartiles, respectively, and underlined by the differences in length of the segments $Q_1 - D_1$ and $D_9 - Q_3$.

These facts, also observed for the full data set (Figure 1d), strongly highlight the non-Gaussianity of $S_\eta(f_j)$, proves the inconsistency of using the mean value and the variance as indicators of its central tendency and dispersion, respectively, and brings to light the need for using robust and resistant estimators of these and other distributional properties.

Additionally, it is observed and is important to remark that both variability and asymmetry reduces considerably for frequencies significantly large. This finding is consistent with the theoretically and experimentally demonstrated existence of an equilibrium range in the high frequency tail of the wind generated waves spectrum, which is characterized by an asymptotic fall off with a power law of the frequency but with some uncertainty in the slope due to the intrinsic nature of the spectral estimations (Rodriguez and Guedes Soares, 1999).

In addition, it can be observed that median spectra, as well as the D_1, Q_1, Q_3, and D_9 spectral curves, adopts a bimodal structure, indicating the presence in any season of sea states with notably different spectral structure. A more detailed analysis has revealed the existence of young swell, partially and almost fully developed wind seas, as well as sea states resulting from the superposition of wind sea and swell wave fields. This is especially true for the period from March to November, a reduces during the winter months. However, the variability of the $S_\eta(f_j)$ increases substantially during winter due to the occurrence of episodic storms interspersed with moderate and mild wave conditions.

It clearly follows from the above comments that the simple separation according to climatic seasons does not provide, at least for this location, an adequate reduction in the variability among individual spectra for obtaining a satisfactory representative spectrum for each season.

As pointed out above, a classical and simple approach to reduce the variability among the sea states belonging to a given set consists in classifying the whole collection of measured wave spectra

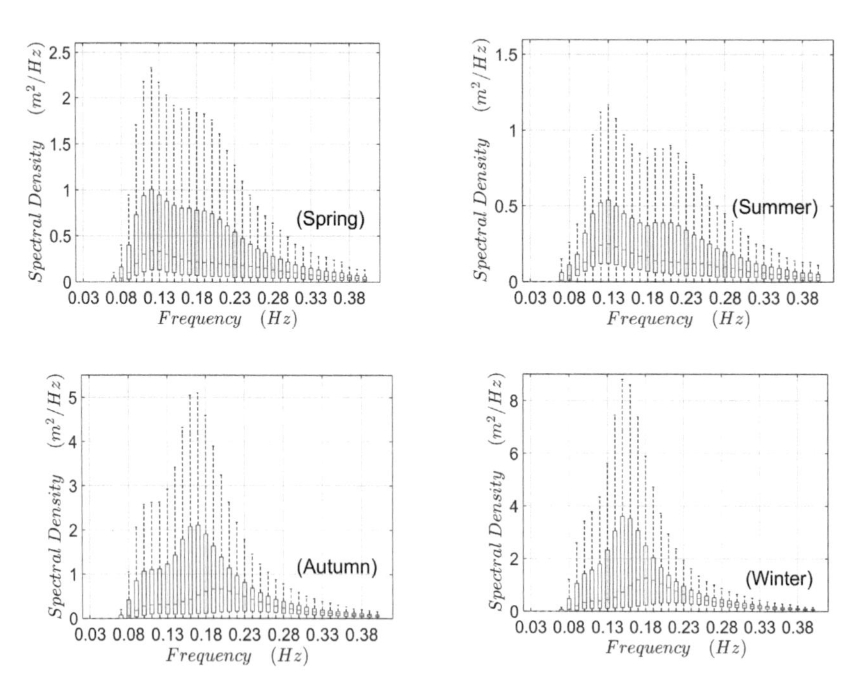

Figure 2. Representative (median) wave spectra and its variability, estimated in terms of D_1, Q_1, Q_3, and D_9 spectral curves, illustrated by using box-plots for individual spectra grouped according climatic seasons.

Table 1. Number of sea states in groups defined in function of Hm0.

H_{m0}(m)	Sea states
1.0–1.5	24259
1.5–2.0	14687
2.0–2.5	8760
2.5–3.0	5246
3.0–3.5	2869
3.5–4.0	1732
4.0–4.5	937
4.5–5.0	508
5.0–5.5	289
5.5–6.0	143
Total	59430

according to the corresponding sea state severity (i.e., in ranges of significant wave height) and estimating a representative wave spectrum for each one of these subsets (e.g. Buckley, 1988; Teng et al. 1994; Teng 2001).

Accordingly, the individual wave spectra have been grouped in terms of the associated H_{m0} with increments of 0.5 m. The number of sea states corresponding to every class is indicated in Table 1.

Individual wave spectra together with the spectral curves defined by D_1, Q_1, Q_2, Q_3, and D_9 are shown in left panels of Fig. 3 for each of the H_{m0} subranges, in increasing order of sea state severity, while the corresponding box-plots, illustrating

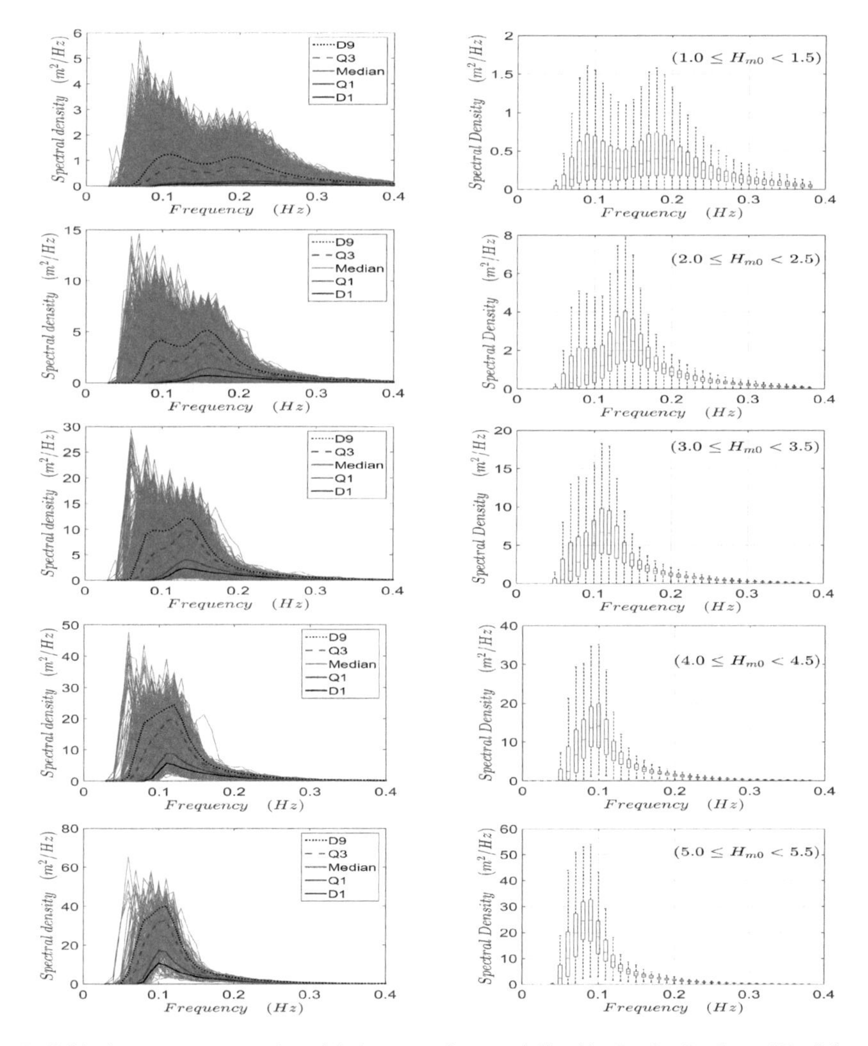

Figure 3. Individual wave spectra together with the spectral curves defined by D_1, Q_1, Q_2, Q_3, and D_9 (left panels) and median wave spectra and its variability illustrated by means of box-plots (right panels), for measured spectra grouped according to H_{m0}.

the variability associated to the spectral density estimations for each frequency band, are shown in the right panels. At first glance, results shown in this figure reveal a general improvement in the representativeness of the median spectrum in relation to that observed for the whole data set (Fig. 1) and for the seasonal subsets (Fig. 2). A more detailed examination unveils that the spreading of the $S_\eta(f_j)$ estimations reduces as H_{m0} increases, so that the distance between the outliers and the curves defined by the whisker-boundaries reduces considerably, improving the representativeness of the median spectrum.

This fact is clearer evidenced in the right panels of Fig. 3, which illustrate the statistical variability of the spectral estimations for each frequency band in form of box-plots. It becomes clear that the dispersion and the asymmetry of the $S_\eta(f_j)$ reduce as the wave height increases. This is true for high frequency bands and becomes also clearly observable in the range of more energetic frequencies, as the wave height increases

Spectral estimations corresponding to lower frequency bands undergo significant changes both in dispersion and asymmetry due to two factors. On the one hand, the above mentioned decrease in the number of swell wave fields as the wave height increases, also reflected in the associated decrease in the bimodal structure of the different representative spectral curves (such as demonstrated by Guedes Soares, 1984) and, on the other hand, the elongation of the equilibrium frequency range as the energy content of the wind sea wave fields increases. The latter can be clearly observed in Fig. 4, which depicts the median spectra for the different wave height ranges and reveals the enlargement of the high frequency range, the peak frequency shift to ward lower frequencies and the enhancement of the spectral peak, as expected.

Naturally, subsets formed by using a so simple approach, just considering the sea state severity, withhold a significant variability among sea states

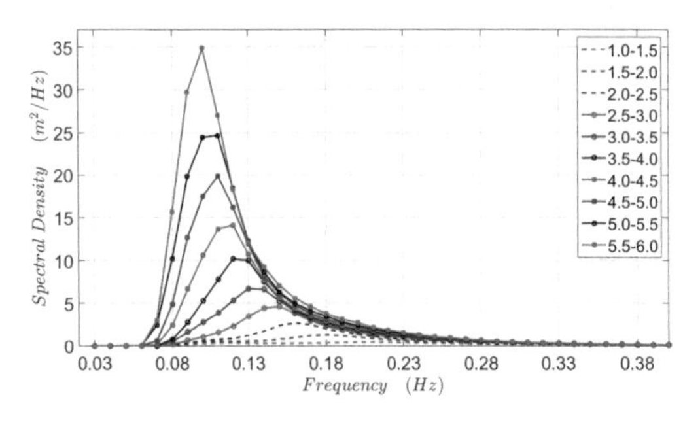

Figure 4. Climatic (median) wave spectra for various significant wave height ranges.

included in each group. However, the use of this simple and commonly used classification approach seems suitable enough to show the usefulness of the suggested methodology. The use of more sophisticated methods to cluster the sea states in more homogeneous groups should reduce the statistical variability associated to each frequency band and, therefore, an increase of the representativeness of the median spectrum.

The choice of a particular method to cluster the sea states will depend on the specific application for which the climatic spectrum is required. Nevertheless, keeping in mind that que $S_\eta(f_j)$ is in fact a random variable, it is always of great interest to provide adequate information on its statistical variability and representativeness, in addition to using robust and resistant estimators for the characteristic spectrum. Furthermore, results shown in Fig. 3 evidence that, depending on the particular objectives, it is possible to use the curves joining the D_1, Q_1, Q_2, Q_3, and D_9 percentiles, or any other quantile (e.g. 95%, 99% quantiles), of the spectral estimations as representative spectra.

4 CONCLUSIONS

This paper highlights that the climatic or representative wave spectrum estimation from a large collection of measured spectra is a two-stage procedure. The first phase, which can be based on physical and/or statistical arguments, consists in implementing methods for grouping multiple wave spectra into homogeneous representative sets. The degree of sophistication of the methodology used to form subsets with similar wave spectra may vary depending on the specific application demanding the climatic wave spectrum estimation.

The second phase of the procedure, which must be based on statistical principles, concerns the estimation of a statistically representative spectrum characterizing every subset. A procedure to estimate the climatic wave spectrum from a collection of measured wave spectra, based on robust and resistant statistical measures, as well as to summarize its statistical variability in a compact, comprehensive, and informative way, making use of boxplots, to provide a clear distributional overview of the spectral estimates at each frequency band has been suggested and implemented. Results demonstrate the robustness of the approach.

ACKNOWLEDGEMENTS

This work was performed within the Strategic Research Plan of the Centre for Marine Technology and Ocean Engineering (CENTEC), which is financed by Portuguese Foundation for Science and Technology (Fundação para a Ciência e Tecnologia-FCT), from where the second author was funded.

REFERENCES

Abadie, R. Butel, S. Mauriet, D. Morichon, H. Dupuis (2006). "Wave climate and longshore drift on the South Aquitaine coast"., *Cont. Shelf Res.*, 26, pp.1924–1939.

Boukhanovsky, A.V., Lopatoukhin, L.J. & Guedes Soares, C., (2007), "Spectral wave climate of the North Sea", *Applied Ocean Research,* 29 (3), pp. 146–154.

Buckley, W.N. (1988), "Extreme and climatic wave spectra for use in structural design of ships", *Naval Engineering Journal*, 100 (5), pp. 36–57.

Buckley W.H. (1993), "Design Wave Climates for the World Wide Operations of Ships. Part 1: Establishments of Design Wave Climate", Int. Maritime Organisation (IMO), Selected Publications, October 1993.

Camus, P., F. J. Mendez, R. Medina, A. S. Cofiño (2011), "Analysis of clustering and selection algorithms for the study of multivariate wave climate", *Coastal Eng.*, 58(6), pp. 453–462.

Chakrabarti, S.K. (1987), *Hydrodynamics of offshore structures*. Springer-Verlag.

Ewans, K. C.; Kibblewhite, A. C. (1992): "Spectral features of the New Zealand deep-water ocean wave climate", *New Zealand Journal of Marine and Freshwater Research* 26 (3-): pp. 323–338.

Ferreira, J. A. and Guedes Soares, C. (2000) Modelling Distributions of Significant Wave Height. *Coastal Engineering*. 40(4):361–374.

Ferreira, J. A. and Guedes Soares, C. (2002). Modelling Bivariate Distributions of Significant Wave Height and Mean Wave Period. *Applied Ocean Research.* 24(1):31–45.

Guedes Soares, C. (1984). "Representation of double peaked sea wave spectra", *Ocean Engineering*, 11(2), pp. 185–207.

Guedes Soares, C. (1998) Stochastic Modelling of Waves and Wave Induced Loads. Guedes Soares, C., (Editor). *Risk and Reliability in Marine Structures.* Balkema; pp. 197–212.

Haver, S., (1985) "Wave climate off northern Norway", *Applied Ocean Research*, 7(2): 85–92.

Haver, S., Natvig, B.J. (1991), "On Some Uncertainties in the Modelling of Ocean Waves and their Effects on TLP Response", *Int. J. Offshore and Polar Eng.*, 1(2): 2343–2352.

Haver S and Nyhus K A (1986) "A wave climate description for long term response calculations Proc. 5th Int" *Offshore Mechanics and Arctic Engineering Symposium*. ASME, Tokio.

Hoaglin, D.C., Mosteller, F., Tukey, J. W. (1983), Understanding of Robust and Exploratory Data Analysis, Wiley, New York.

Holthuijsen, L.H., (2007). "*Waves in Oceanic and Coastal Waters*", Cambridge University Press.

Hu, S.J., (1991) "Probabilistic wave spectrum and fatigue estimation", *Applied Ocean Research*, 13 (2): 93–99.

Lucas, C.; Boukhanovsky, A., and Guedes Soares, C. (2011) Modelling the Climatic Variability of Directional Wave Spectra. *Ocean Engineering.* 38(11–12):1283–1290.

Mortlock, T. R. & Goodwin, I. D. (2015), "Directional wave climate and power variability along the Southeast Australian shelf", *Continental Shelf Research,* 98: pp. 36–53.

Rodríguez, G., Guedes Soares, C. 1999. Uncertainty in the estimation of the slope of the high frequency tail of wave spectra, Applied Ocean Research, 21(4), 207–2013.

Saulnier, J.B., A Clément, FO Antonio, T Pontes, M Prevosto, P Ricci, (2011). "Wave groupiness and spectral bandwidth as relevant parameters for the performance assessment of wave energy converters", *Ocean Engineering*, 38(1): 130–147.

Scott, J.R., (1968), "Some average sea spectra", *Transactions Royal Institution of Naval Architects* 110 (2), 233–245.

Teixeira, A. P. and Guedes Soares, C. (2009) Reliability Analysis of a Tanker Subjected to Combined Sea States. *Probabilistic Engineering Mechanics.* 24(4):493–503.

Teng, C.C., Timpe, G., and Palao, I. (1994), "The Development of Design Waves and Wave Spectra for Use in Ocean Structure Design", *Transactions Society of Naval Architects and Marine Engineers*, 102: pp. 475–499.

Teng, C.C., (2001), Climatic and maximum wave spectra from long-term measurements. Proc. Int. Conf. Ocean Wave Measurements and Analysis, ASCE, San Francisco, USA.

Tukey, J.W. (1977), Exploratory data analysis, Addison-Wesley, Reading.

Wilcox, R.R. (2012) Introduction to robust estimation and hypothesis testing. Academic Press.

Progress in Maritime Technology and Engineering – Guedes Soares & Santos (Eds)
© 2018 Taylor & Francis Group, London, ISBN 978-1-138-58539-3

Author index